The Pancreas: Biology and Physiology

The Pancreas: Biology and Physiology

Edited by
Fred S. Gorelick
John A. Williams

Published in the United States of America by Michigan Publishing
Manufactured in the United States of America

ISBN 978-1-60785-716-7 (hardcover),
978-1-60785-719-8 (e-book)

Cover image: uncopyrighted micrograph by George Palade, contributed by James Jamieson

CONTENTS

CONTRIBUTORS

Catharina M. Alam, *Department of Biosciences, Cell Biology, Åbo Akademi University, Turku Finland*

Thomas Albers, *Departments of Chemistry and Physics, Augusta University. Augusta, GA, USA*

Joydeep Aoun, *The Epithelial Signaling and Transport Section, Molecular Physiology and Therapeutics Branch, National Institute of Dental and Craniofacial Research, National Institutes of Health, Bethesda, MD, USA*

Minoti V. Apte, *Pancreatic Research Group, South Western Sydney Clinical School, Faculty of Medicine, University of New South Wales, Ingham Institute for Applied Medical Research, Sydney, Australia*

Tanja Babic, *Department of Neural and Behavioral Sciences, Penn State College of Medicine, Hershey, PA, USA*

Sarah Baghestani, *Turku Center for Disease Modeling, University of Turku, Turku, Finland*

Melena D. Bellin, *Division of Pediatric Endocrinology, Department of Pediatrics, University of Minnesota, Minneapolis, MN, USA*

Mohamad Bouhamdan, *Department of Physiology, Wayne State University School of Medicine, Detroit, MI, USA*

Jason I. E. Bruce, *Faculty of Biology, Medicine, and Health, School of Medical Sciences, University of Manchester, Manchester, UK*

Rashmi Chandra, *Department of Medicine, Division of Gastroenterology, Duke University and VA Medical Centers, Durham, NC, USA*

Xuequn Chen, *Department of Physiology, Wayne State University School of Medicine, Detroit, MI, USA*

Darwin I. Conwell, *Section of Pancreatic Disorders, Division of Gastroenterology, Hepatology, and Nutrition, Ohio State University Wexner Medical Center, Columbus, OH, USA*

Michelle M. Cooley, *Graduate Program in Biochemical and Molecular Nutrition, University of Wisconsin-Madison, Madison, WI, USA*

David N. Criddle, *Department of Molecular Physiology & Cell Signalling, Institute of Systems, Molecular & Integrative Biology, University of Liverpool, UK*

Robert C. De Lisle, *Department of Anatomy and Cell Biology, University of Kansas School of Medicine, Kansas City, KS, USA*

Subhankar Dolai, *Departments of Medicine and Physiology, University of Toronto, Toronto, Canada*

Yousef El-Gohary, *Department of Surgery, St. Jude Children's Research Hospital, 262 Danny Thomas Pl, Memphis, TN, USA*

Farzad Esni, *Departments of Surgery and Developmental Biology, Children's Hospital of Pittsburgh, University of Pittsburgh Medical Center, Pittsburgh, PA, USA*

Melissa Fenech, *University of Western Ontario, London, Ontario, Canada*

Herbert Y. Gaisano, *Departments of Medicine and Physiology, University of Toronto, Toronto, Canada*

George Gittes, *Department of Surgery, Division of Pediatric General and Thoracic Surgery, Children's Hospital of Pittsburgh of UPMC, Pittsburgh, PA, USA*

Fred S. Gorelick, *Yale School of Medicine, Yale University, New Haven, CT and VA CT, West Haven, CT, USA*

Michael A. Gray, *Biosciences Institute, Faculty of Medical Sciences, Newcastle University, Newcastle upon Tyne, UK.*

Guy E. Groblewski, *Graduate Program in Biochemical and Molecular Nutrition, University of Wisconsin-Madison, Madison, WI, USA*

Phil A. Hart, *Section of Pancreatic Disorders, Division of Gastroenterology, Hepatology, and Nutrition, Ohio State University Wexner Medical Center, Columbus, OH, USA*

Peter Hegyi, *Institute for Translational Medicine, Medical School, University of Pécs, Pécs, Hungary*

Hiroshi Ishiguro, *Department of Human Nutrition, Nagoya University Graduate School of Medicine, Nagoya, Japan*

Elaina K. Jones, *Graduate Program in Biochemical and Molecular Nutrition, University of Wisconsin-Madison, Madison, WI, USA*

Bhanu P. Jena, *Department of Physiology, Wayne State University School of Medicine, Detroit, MI, USA*

Ikhyun Jun, *Department of Pharmacology, Brain Korea 21 PLUS Project for Medical Sciences, Severance Biomedical*

Science Institute, Yonsei University College of Medicine, Korea

Yonjung Kim, *Department of Pharmacology, Brain Korea 21 PLUS Project for Medical Sciences, Severance Biomedical Science Institute, Yonsei University College of Medicine, Seoul, Korea*

Min Goo Lee, *Department of Pharmacology, Brain Korea 21 PLUS Project for Medical Sciences, Severance Biomedical Science Institute, Yonsei University College of Medicine, Seoul, Korea*

Rodger A. Liddle, *Department of Medicine, Division of Gastroenterology, Duke University and VA Medical Centers, Durham, NC, USA*

Daniel S. Longnecker, *Department of Pathology, Geisel School of Medicine at Dartmouth, Lebanon, NH, USA*

Elizabeth Mann, *Division of Pediatric Endocrinology and Diabetes, University of Wisconsin, Madison, WI, USA*

Alpha R. Mekapogu, *Pancreatic Research Group, South Western Sydney Clinical School, Faculty of Medicine, University of New South Wales, Ingham Institute for Applied Medical Research, Sydney, Australia*

Shmuel Muallem, *The Epithelial Signaling and Transport Section, Molecular Physiology and Therapeutics Branch, National Institute of Dental and Craniofacial Research, National Institutes of Health, Bethesda, MD, USA*

Christopher L. Pin, *University of Western Ontario, London, Ontario, Canada*

Romano C. Pirola, *Pancreatic Research Group, South Western Sydney Clinical School, Faculty of Medicine, University of New South Wales, Ingham Institute for Applied Medical Research, Sydney, Australia*

Srinivasa P. Pothula, *Pancreatic Research Group, South Western Sydney Clinical School, Faculty of Medicine, University of New South Wales, Ingham Institute for Applied Medical Research, Sydney, Australia*

Maria Eugenia Sabbatini, *Department of Biological Sciences, Augusta University, Augusta, GA, USA*

Maria Dolors Sans, *Department of Molecular and Integrative Physiology, University of Michigan, Ann Arbor, MI, USA*

Mairobys Socorro, *Department of Surgery, Division of Pediatric General and Thoracic Surgery, Children's Hospital of Pittsburgh, Pittsburgh, PA, USA*

Fei Sun, *Department of Physiology, Wayne State University School of Medicine, Detroit, MI, USA*

Muna Sunni, *Division of Pediatric Endocrinology, Department of Pediatrics, University of Minnesota, Minneapolis, MN, USA*

Toshimasa Takahashi, *Departments of Medicine and Physiology, University of Toronto, Toronto, Canada*

Alexei V. Tepikin, *Department of Molecular Physiology & Cell Signalling, Institute of Systems, Molecular & Integrative Biology, University of Liverpool, UK*

Frank Thévenod, *Institute of Physiology, Pathophysiology & Toxicology, Centre for Biomedical Training and Research (ZBAF), Faculty of Health, University of Witten/Herdecke, Witten, Germany*

Diana M. Toivola, *Department of Biosciences, Cell Biology, Åbo Akademi University, Turku Finland*

Alberto Travagli, *Department of Neural and Behavioral Sciences, Penn State College of Medicine, Hershey, PA, USA*

Viktoria Venglovecz, *Department of Pharmacology and Pharmacotherapy, University of Szeged, Szeged, Hungary*

John A. Williams, *Department of Molecular and Integrative Physiology, University of Michigan, Ann Arbor, MI, USA*

Jeremy S. Wilson, *Pancreatic Research Group, South Western Sydney Clinical School, Faculty of Medicine, University of New South Wales, Ingham Institute for Applied Medical Research, Sydney, Australia*

Makoto Yamaguchi, *Department of Human Nutrition, Nagoya University Graduate School of Medicine, Nagoya, Japan*

Akiko Yamamoto, *Department of Human Nutrition, Nagoya University Graduate School of Medicine, Nagoya, Japan*

Xie Youming, *Karmanos Cancer Institute and Department of Oncology, Wayne State University School of Medicine, Detroit, MI, USA*

David I. Yule, *Department of Pharmacology and Physiology, University of Rochester, Rochester, NY, USA*

PREFACE

This book is designed to summarize the current state of knowledge on the structure and function of the exocrine pancreas. The focus is on the biology and physiology of the gland, although biochemistry, molecular biology, genetics, and pathophysiology are also considered. The work grew out of review entries from the Pancreapedia site (www.pancreapedia.org) and follows the publication in 2016 of the book *Pancreatitis*. Both books contain material published by the American Pancreatic Association under a Creative Commons Attribution-Noncommercial-NoDerivatives license and are available as open-access electronic files on the Pancreapedia site. To help fill the gaps, some new chapters were solicited and all authors of existing material were given the opportunity and encouraged to update. There are 26 chapters presented in four sections: "Pancreatic exocrine structure and function," "Acinar cells," Exocrine pancreas integrated responses," and "Pancreatic islet and stellate cell structure and function." Our goal is to make the information available to the Pancreas community worldwide at no cost while those who wish can purchase a modestly priced bound copy.

As with any compendium of this type, some readers may wonder why some chapters are present and other areas have been omitted. This is in part due to the editor's choices but is also due in part to the fact that some areas come to an end or morph into other areas such as those that relate to a disease. There are a few areas without much recent progress where we just could not identify expert authors willing to devote time to summarizing that area. On the positive side, since the Pancreapedia is a living and expanding document, anyone seeing an area not present can volunteer to write an entry for the Pancreapedia.

ACKNOWLEDGMENTS

We thank all the authors who contributed their expertise and writing to a novel open-access form of publishing, the Pancreapedia. This book could not have been realized without the advice and assistance of the Michigan Publishing arm of the University of Michigan Library, especially Charles Watkinson and Jason Colman. We thank the other Pancreapedia editors, Ashok Saluja, Markus Lerch, and Stephen Pandol, for advice on selecting authors. We especially thank the students both graduate and undergraduate from the University of Michigan School of Information who organized and formatted material, including Emily Rinck, Ben Krawatz, Melissa Wu, and Juliana Lam. We especially thank M. Dolors Sans who took over this work on top of her own research for the last year and a half. We also thank the American Pancreatic Association, the National Library of Medicine, and the University of Michigan for their invaluable support.

Pancreatic Exocrine Structure and Function

Chapter 1

Anatomy and histology of the pancreas

Daniel S. Longnecker

Department of Pathology, Geisel School of Medicine at Dartmouth, Lebanon, NH

Introduction

The mandate for this chapter is to review the anatomy and histology of the pancreas. The pancreas (meaning all flesh) lies in the upper abdomen behind the stomach. The pancreas is a part of the gastrointestinal system that makes and secretes digestive enzymes into the intestine and also an endocrine organ that makes and secretes hormones into the blood to control energy metabolism and storage throughout the body.

It is worthwhile to mention a few definitions for key terms as used in the context of the pancreas:

Exocrine pancreas, the portion of the pancreas that makes and secretes digestive enzymes into the duodenum. This includes acinar and duct cells with associated connective tissue, vessels, and nerves. The exocrine components comprise more than 95 percent of the pancreatic mass.

Endocrine pancreas, the portions of the pancreas (the islets) that make and secrete insulin, glucagon, somatostatin, and pancreatic polypeptide into the blood. Islets comprise 1–2 percent of pancreatic mass.

Since we are dealing with a three-dimensional solid structure, the aphorism that "a picture is worth a thousand words" seems to pertain.[1] Accordingly, this chapter will largely consist of images with extended legends. The images range from classic work of skilled medical artists to original drawings and photomicrographs from leaders in the study of pancreatic anatomy and structure. Text is interspersed as appropriate. Additional useful images are available online at other websites. We provide a list of some of these sites at the end with the references.

Gross Anatomy

Figures 1–13 depict the gross anatomy of the pancreas and its relationship to surrounding organs in adults. It is customary to refer to various portions of the pancreas as head, body, and tail. The head lies near the duodenum and the tail extends to the hilum of the spleen.

When the terms anterior, posterior, front, and back are used, they pertain to relationships in the human, standing erect. Superior and inferior are used in the same context so that they mean toward the head and toward the feet, respectively. These usages obviously do not pertain in quadruped animals where dorsal, ventral, cephalad, and caudad are more useful terms.

Use of the terms left and right can be problematic. For example, the spleen is located in the upper portion of the abdomen on the left side of the body. When the abdomen is pictured from the front, this places the spleen on the viewer's right-hand side. We will adopt the convention that right and left (unqualified) will be used in the first sense in the legends for gross anatomy (indicating the subject's right and left side).

When we are designating location within an image, we will use "image right" and "image left" to denote relationships within the image.

Artwork in **Figures 3, 7–8**, and **11–13** is by Jennifer Parsons Brumbaugh. These drawings were originally published in the AFIP Fascicle on pancreatic neoplasms and are used with permission of the publisher.[2] Chapter 1 of the Fascicle is recommended as a source for additional detail regarding pancreatic anatomy and histology, and for discussion of the genetic control of pancreatic development.

The tail of the pancreas and spleen are in the left upper quadrant of the abdomen, and the head of the pancreas is in the right upper quadrant just to the right of the midline. If you place your right hand over your upper abdomen with fingers extending to the left over the lower portion of your rib cage and the tip of your thumb extended up over the lower portion of the sternum, then your pancreas lies behind your hand in the back (retroperitoneal) portion of the abdomen. This may be visualized by reference to the small image in the upper image right corner in **Figure 4**.

e-mail: daniel.s.longnecker@dartmouth.edu

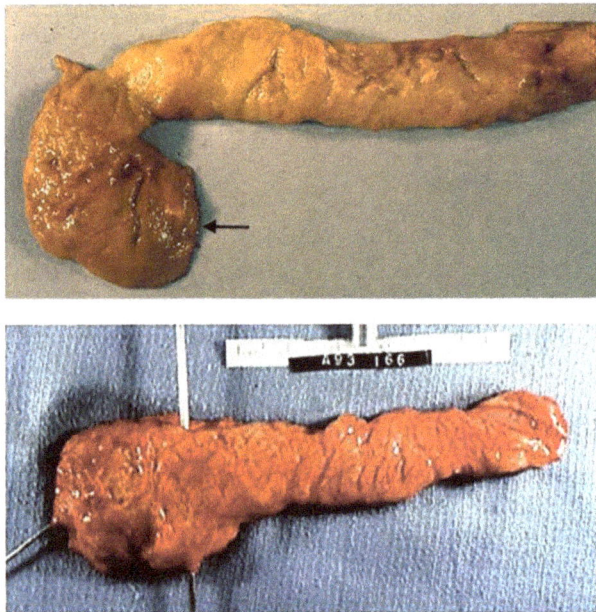

Figure 1. The gross anatomy of the human pancreas can vary. **Figures 1A** and **1B** are two normal human pancreases from autopsies of adults. Both pancreases have been dissected to remove fat and adjacent organs. The two photos illustrate that there is considerable individual variation in the shape of the pancreas.

(A) This pancreas has a conspicuous uncinate lobe that curves down and to the left (arrow). This is an unusual configuration since the uncinate process usually fuses more completely with the dorsal pancreas adding mass to the head of the pancreas as seen in **Figures 1B** and **2**.

(B) In this pancreas, the uncinate portion is fused to the remainder of the head. A probe (image left) has been put into the main pancreatic duct, and a second probe (vertical) is in the portal-superior mesenteric vein behind the pancreas. The diagonal groove in the tip of the tail (image right) marks the course of a branch of the splenic artery or vein.

The pancreas is about the size of the half of your hand that includes the index and third fingers excluding the thumb. The pancreas weighs about 100 grams and is 14–23 cm long [2].

- The head of the pancreas lies in the loop of the duodenum as it exits the stomach.
- The tail of the pancreas lies near the hilum of the spleen.
- The body of the pancreas lies posterior to the distal portion of the stomach between the tail and the neck and is unlabeled in this drawing.
- The portion of the pancreas that lies anterior to the aorta is somewhat thinner than the adjacent portions of the head and body of the pancreas. This region is sometimes designated as the neck of the pancreas and marks the junction of the head and body.

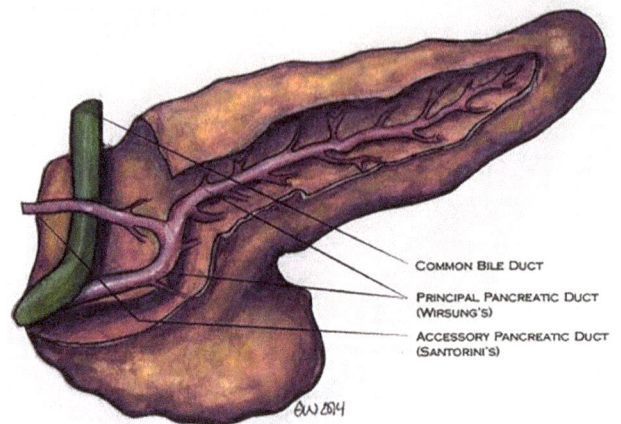

Figure 2. **Normal pancreas dissected to reveal the duct system.** The pancreas is viewed from the front and a portion of the parenchyma has been dissected away to reveal: (1) the main (principal) pancreatic duct (Wirsung's duct) with multiple branches; (2) the accessory duct (Santorini's duct); and (3) the distal common bile duct. Although the regions are not labeled, we see the head of the pancreas, image left, and tail of the pancreas, image right. This drawing depicts a configuration that is intermediate to those shown in **Figure 1** in regard to the degree of fusion of the uncinate process with the dorsal pancreas (drawing by Emily Weber for the Pancreapedia).

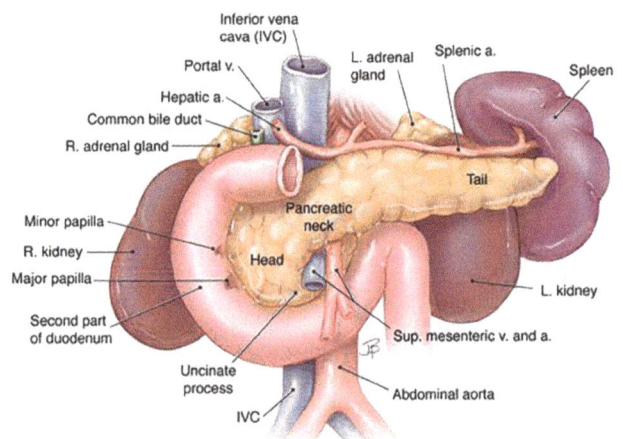

Figure 3. **Anatomic relationships of the pancreas with surrounding organs and structures** Several key relationships should be noted. Their recognition may be facilitated by also referring to **Figures 4** and **5**.

- The close proximity of the neck of the pancreas to major blood vessels posteriorly, including the superior mesenteric artery, superior mesenteric-portal vein, inferior vena cava, and aorta, limits the option for a wide surgical margin when pancreatectomy (surgical removal of the pancreas) is done.

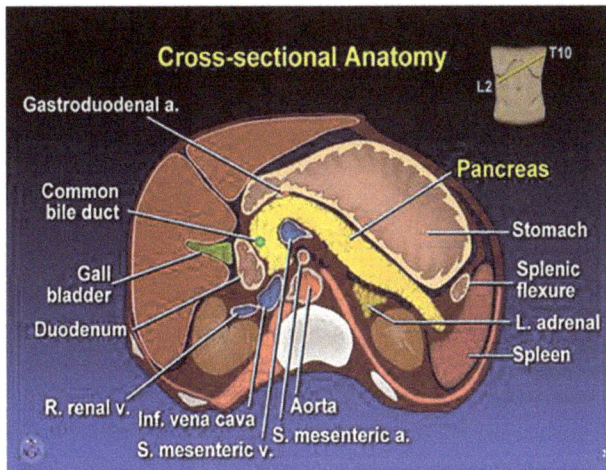

Figure 4. Cross-section of the upper abdomen at the level of the pancreas. Note that the plane of the transection is angled upward on the left as indicated in the drawing upper image right. The major organs except the liver (image left) and kidneys are labeled. Splenic flexure (image right) refers to the colon (used with permission. Copyright, American Gastroenterological Association).

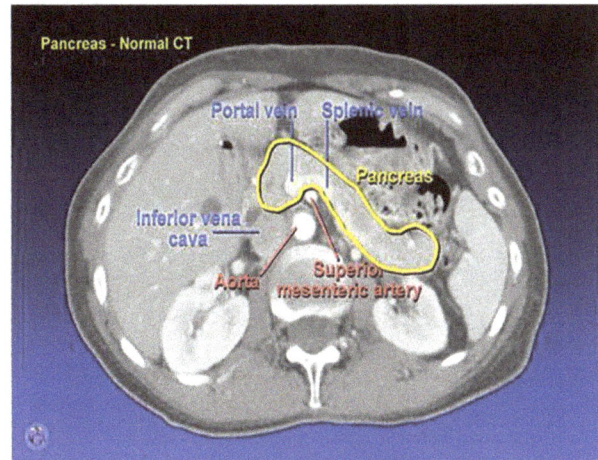

Figure 5. CT scan of the upper abdomen at the level of the pancreas. This annotated CT scan is oriented with the abdominal wall at the top and the spine and muscles of the back at the bottom viewing the cross-section from below. Thus, the spleen is at the extreme image right and the liver is image left inside the ribs that appear as white ovals in the abdominal wall. Kidneys lie lateral to the spinal column with the tail of the pancreas nearly touching the left kidney (used with permission. Copyright, American Gastroenterological Association).

- The common bile duct passes through the head of the pancreas to join the main duct of the pancreas near the duodenum as shown in **Figure 2**. The portion nearest to the liver lies in a groove on the dorsal aspect of the head (see **Figure 7B**).
- The minor papilla where the accessory pancreatic duct drains into the duodenum and the major papilla (ampulla of Vater) where the main pancreatic duct enters the duodenum are depicted, image left (image by Jennifer Parsons Brumbaugh; used with permission of the publisher[2]).

There is no anatomic landmark for the division between the body and tail of the pancreas, although the left border of the aorta is sometimes used to mark the junction.[2,3] Hellman defined the tail as the one-fourth of the pancreas from the tip of the tail to the head, whereas Wittingen defined the junction between the body and tail as the point where the gland sharply narrowed.[4,5] It would be difficult to define this point in the pancreases shown in **Figure 1.**

Embryology and Development

The pancreatobiliary anlagen appear at gestation week 5 in the human; fusion of the dorsal and ventral anlagen occurs during week 7.[6] Full development of acinar tissue extends into the postnatal period. In mice, pancreatic development begins at embryonic day 8.5 (e8.5) and is largely complete by day e14.5.[7]

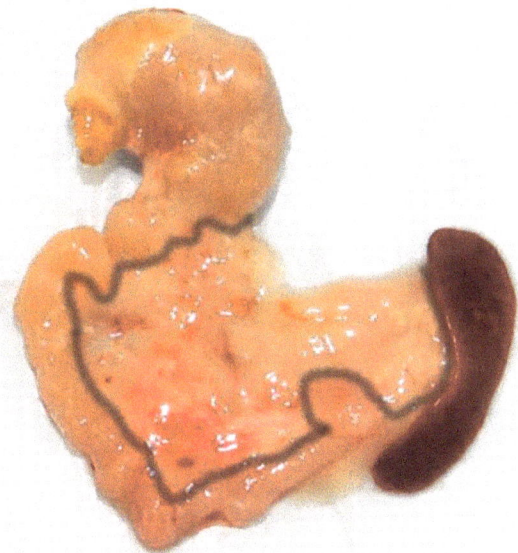

Figure 6. Mouse pancreas. The pancreas of an adult mouse is shown surrounded by the stomach (top), the duodenum and proximal jejunum (image left and bottom), and the spleen (image right). The duodenum wraps around the head of the pancreas (as demarcated by the line). Rodent pancreas is soft and diffuse compared with the human pancreas (photo provided by Catherine Carriere).

A

B

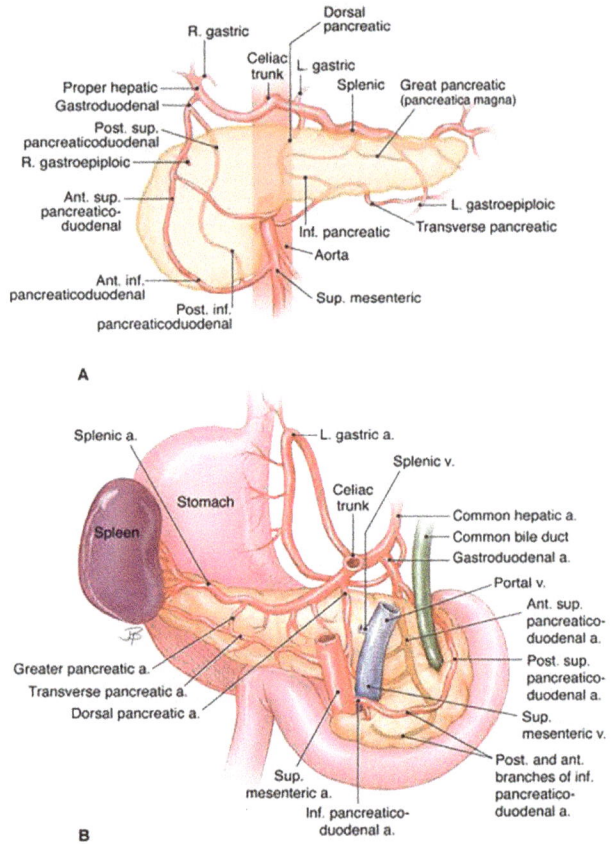

Figure 7. The arterial blood supply of the pancreas. The upper panel (**A**) is visualized from the front, and the lower panel (**B**) is seen from the back. The celiac trunk and the superior mesenteric artery both arise from the abdominal aorta. Both have multiple branches that supply several organs including the pancreas. The anastomosis of their branches around the pancreas provides collateral circulation that generally assures a secure arterial supply to the pancreas. Most of the arteries are accompanied by veins (not shown) that drain into the portal and splenic veins as they pass behind the pancreas as shown in **B**. The superior mesenteric vein becomes the portal vein when it joins the splenic vein (image by Jennifer Parsons Brumbaugh used with permission of the publisher[2]).

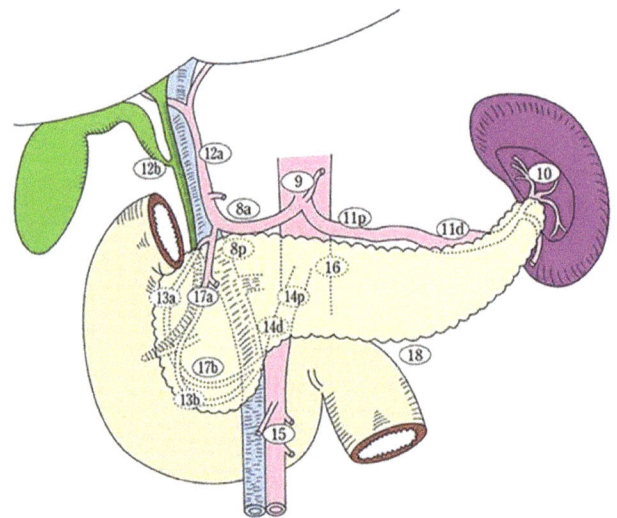

Eponymic names identify the anatomist, embryologist, or physician who is credited with first describing a structure. You may conclude that Wirsung, Santorini, and Vater were such scientists.

Histology and Ultrastructure

Figures 14–29 depict the histology of the exocrine pancreas at the light and electron microscopic levels. Most histologic images are from human tissue. Exceptions are usually noted in the legend. In Hematoxylin and Eosin (H&E)-stained sections, nucleic acids (DNA and RNA) stain blue; most proteins and carbohydrates stain pink

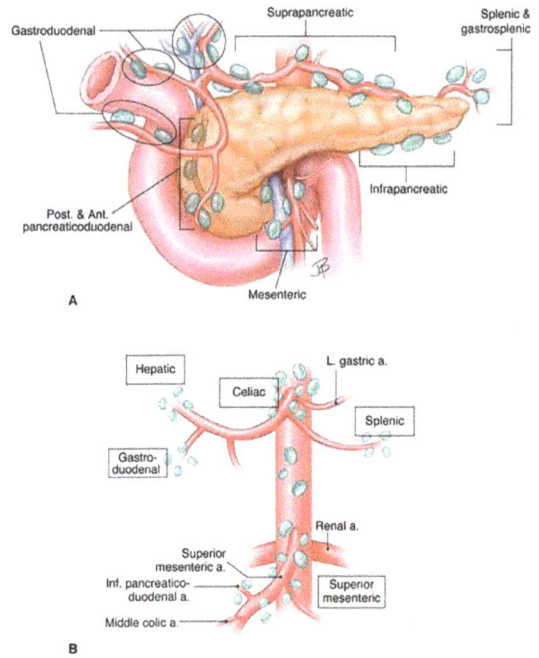

A

B

Figure 8. Lymph nodes draining the pancreas. This figure indicates the typical location of lymph nodes surrounding the pancreas. There is considerable individual variation in the location of lymph nodes and an image like this is idealized. Both, **A** and **B**, are anterior views. **B** includes some nodes that lie posterior to the pancreas (image by Jennifer Parsons Brumbaugh used with permission of the publisher[2]).

Figure 9. Pancreas-associated lymph nodes are assigned numerical codes (lymph node station numbers) that correspond to their anatomic location. This classification is used to denote the location of metastatic spread of pancreatic neoplasms or for other detailed studies. These station numbers are seldom used in Western publications and the image is provided primarily for reference (figure used with permission of the Japan Pancreas Association and the Kanehara publishers).

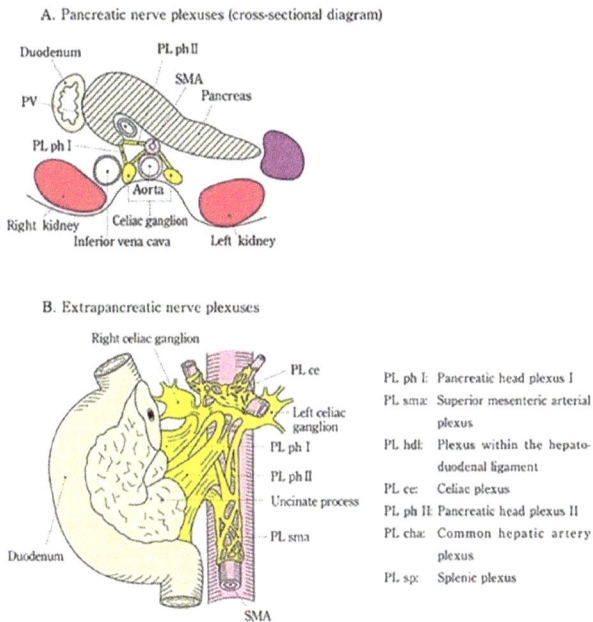

A. Pancreatic nerve plexuses (cross-sectional diagram)

B. Extrapancreatic nerve plexuses

PL ph I: Pancreatic head plexus I
PL sma: Superior mesenteric arterial plexus
PL hdl: Plexus within the hepato-duodenal ligament
PL ce: Celiac plexus
PL ph II: Pancreatic head plexus II
PL cha: Common hepatic artery plexus
PL sp: Splenic plexus

Figure 10. Nerves (yellow) serving the pancreas. The cross-sectional image (**A**) emphasizes the location of the celiac ganglia of the autonomic system lateral to the aorta while (**B**) emphasizes the rich nerve plexus that connects these ganglia to the pancreas. **SMA** (superior mesenteric artery). **PL** (plexus) (figure used with permission of the Japan Pancreas Association and the Kanehara publishers).

to red; fat is extracted by organic solvents used in tissue processing leaving unstained spaces. Sections for light microscopy are most often made from formalin-fixed par-affin-embedded tissue and the sections are usually 4 or 5 micrometers (μm) thick. Thinner (1-μm) sections of plastic-embedded tissues (prepared for electron microscopy) may also be used for light microscopy and a few such sections are also illustrated. For additional ultrastructural detail, the reader is referred to the chapter by Kern.[8]

Duct System

The components of the duct system are the main pancreatic duct (duct of Wirsung), interlobular ducts that drain into the main duct throughout the pancreas as depicted in **Figure 2**, and intralobular ducts (sometimes called intercalated duct-ules) that link acinar tubules to the interlobular ducts. The intralobular ducts and ductules are ordinarily seen only at the level of light and electron microscopy. Enzymes from acinar cells are released into a bicarbonate-rich solution that is secreted by the CACs and ductal cells and flows from the acini and acinar tubules to the intralobular ducts, then into the interlobular ducts and main duct, and finally into the duodenum at the major or minor papillae. This duct system is illustrated in **Figures 25–28**.

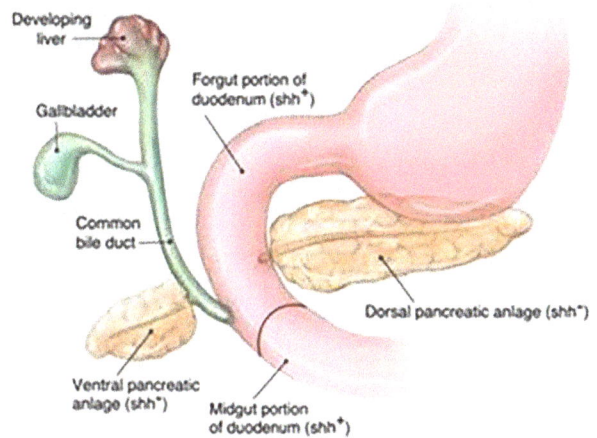

A

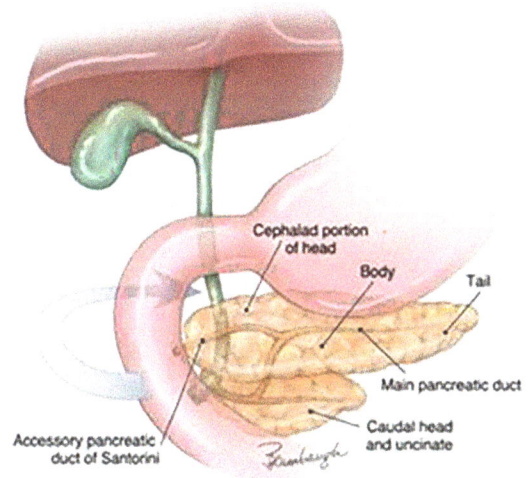

B

Figure 11. (**A**) **The figure reflects the embryonic development of the pancreas and biliary system in the human.** The pancreas develops from two outgrowths of the foregut distal to the stomach. The ventral diverticulum gives rise to the common bile duct, gallbladder, liver, and the ventral pancreatic anlage that becomes a portion of the head of the pancreas with its duct system including the uncinate portion of the pancreas. The dorsal pancreatic anlage gives rise to a portion of the head, the body, and tail of the pancreas including a major duct that is continuous through the three regions.

(**B**) **The figure depicts the rotation of the ventral anlage to fuse with the dorsal anlage and fusion of the duct systems such that the main pancreatic duct is formed from portions of the ducts of both dorsal and ventral anlagen.** The caudal portion of the head of the pancreas (uncinate) and the major papilla (ampulla of Vater) are derived from the ventral anlage. The minor papilla that drains the duct of Santorini is derived from the dorsal anlage (image by Jennifer Parsons Brumbaugh; used with permission of the publisher[2]).

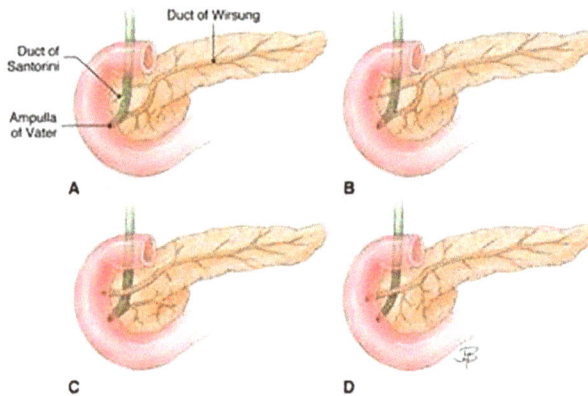

Figure 12. Anatomic variations in the pancreatic and common bile duct systems. The anatomic variations depicted provide additional examples of individual differences in pancreatic anatomy seen in adults. These will be most easily understood by comparing **Figures 11** and **12**. It becomes apparent that the duct of Santorini is derived from the dorsal anlage, whereas the duct of Wirsung (the main duct of the pancreas) is derived from the fusion of duct systems of both dorsal and ventral anlagen and drains into the duodenum at the ampulla of Vater as depicted in **A** and **B**. The connection of the duct of Santorini to the duodenum may regress as depicted in **A** or persist as in **B**, **C**, and **D**. The duct systems of the two anlagen may fail to fuse as depicted in **C** giving rise to "pancreas divisum." Rarely the duct systems may fuse but lose their connection to the ampulla as depicted in **D**. Pancreatic secretions then reach the duodenum through the duct of Santorini and the minor papilla (image by Jennifer Parsons Brumbaugh; used with permission of the publisher[2]).

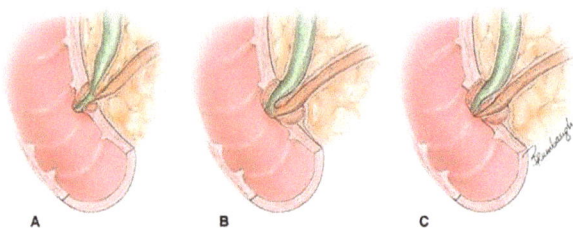

Figure 13. Anatomic variations in the union of the common bile duct and the main pancreatic duct at the major papilla (ampulla of Vater). "Common channel" refers to the fused portion of the bile and pancreatic ducts proximal to entry into the duodenum. The common channel may be long as depicted in **A** or short as in **B**. Less often, there is no common channel because the ducts open separately into the duodenum as depicted in **C**. The common channel has received much attention because stones in the biliary tract (gallstones) may lodge in the common channel causing obstruction of both pancreatic and biliary duct systems. Such an obstruction is frequently regarded as the cause of acute pancreatitis (image by Jennifer Parsons Brumbaugh; used with permission of the publisher[2]).

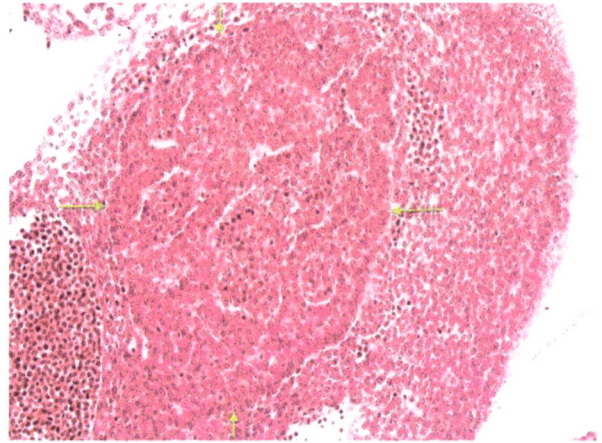

Figure 14. Fetal pancreas (H&E). This tissue section illustrates developing exocrine tissue in the center (arrows) surrounded by primitive mesenchymal and hematopoietic cells at an estimated gestational age of 5 weeks. The acinar tissue is composed of a network of interconnecting tubules (micrograph contributed by Dale E. Bockman).

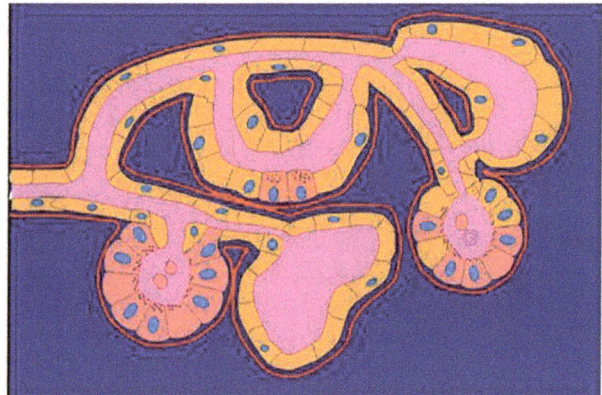

Figure 15. The exocrine pancreas is a complex tubular network. The point of this drawing is that pancreatic acini are not arranged in clusters like grapes at the ends of a branching duct system but rather as an anastomosing tubular network that at some termini form classic acini. Centroacinar cells (CACs) are typically located at the junction of an acinus or acinar tubule with a small ductule, but they may be interspersed within an acinar tubule. In this drawing, many acinar cells have been replaced by duct cells. This process, called acinar to ductal metaplasia (ADM), occurs in chronic pancreatitis[9] (also see **Figure 8** in the chapter by Bockman[3], image contributed by Dale E. Bockman).

The integrity of the duct system is of key importance in preventing entry of the exocrine enzymes into the interstitial space where they may be activated and cause tissue damage manifest as pancreatitis. The main and interlobular ducts have thick dense collagenous walls. The connective tissue component of the duct wall becomes progressively thinner as the ducts branch and become narrower. Intercellular tight junctions between duct cells, CACs, and acinar cells

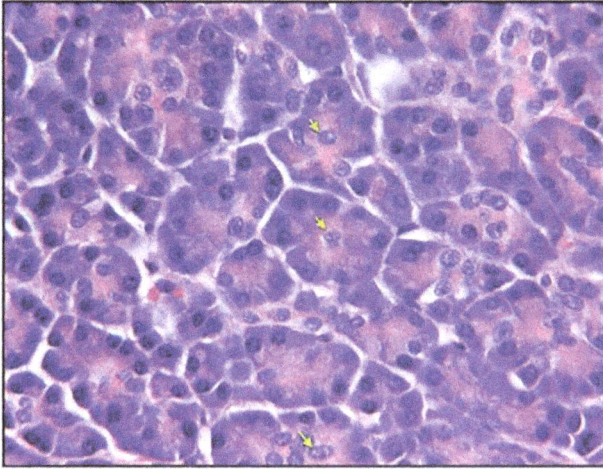

Figure 16. Acinar tissue, adult human pancreas (H&E). Acinar cells stain blue at their base because of the high content of RNA and the presence of nuclei. They are pink at their apex (lumenal aspect) where there is a high content of zymogen proteins (digestive enzymes). The nuclei of CACs are sometimes seen within an acinus (arrows).

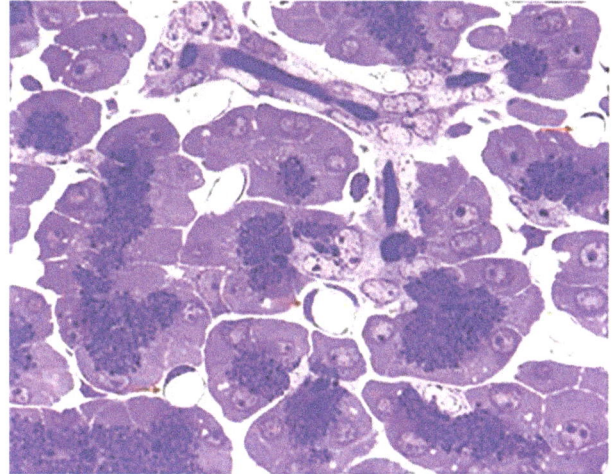

Figure 18. Pancreas with acinar and centroacinar cells with a small intralobular duct (toluidine blue stain, 1-μm-thick plastic-embedded tissue). The presence of numerous round empty capillaries (arrows) in the interstitial spaces indicates that the pancreas was perfused with fixative. A small branching intralobular duct is evident at the top of the field. Blue ZGs are conspicuous in the acinar cells (micrograph contributed by James Jamieson).

play a major role in preventing leakage of the duct system. These have not been well illustrated, although they can be seen in **Figures 21** and **22** as dark, thickened zones in the adjacent cell membranes near the acinar or duct lumen. The

chapter by Kern in *The Pancreas* provides excellent images and discussion of these tight junctions[8].

Interstitial Tissue

Interstitial tissue surrounds lobules of acinar tissue, ducts, and islets. The interstitium contains arteries, veins, capillaries, lymphatics, neural tissue, and stellate cells. Leukocytes may infiltrate the interstitium especially during pancreatitis or in reaction to neoplasms.

The pancreatic stellate cells (PSCs) are specialized connective tissue cells with characteristic structure (**Figure 29**). They secrete multiple components of the extracellular matrix and are activated by a multitude of factors including inflammatory mediators, alcohol and its metabolites, endotoxins, and cancer cell-derived factors.

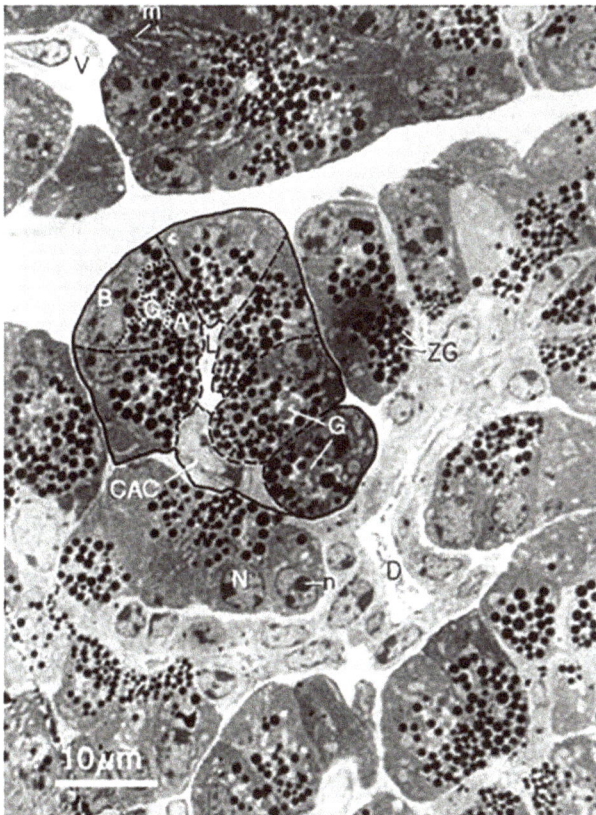

Figure 17. Pancreatic tissue with acinar, centroacinar, and ductal cells (EM thick section). The acinar cells are larger than CACs and are easily identified because of the darkly stained zymogen granules (ZGs). The basal portion (B) of the acinar cells lies next to the interstitial space that contains vessels (V), nerves, and connective tissue. Nuclei (N) with nucleoli (n) are in the basal portion of the acinar cells. The Golgi (G) lies at the junction of the basal and apical (A) portions of the cell. CACs have less rough endoplasmic reticulum and no secretory granules. Their cytoplasm is more lightly stained. A small ductule (D) extends from image right to below center. This is a 1-μm-thick section of plastic-embedded tissue prepared for electron microscopy that was stained with toluidine blue (micrograph contributed by James Jamieson).

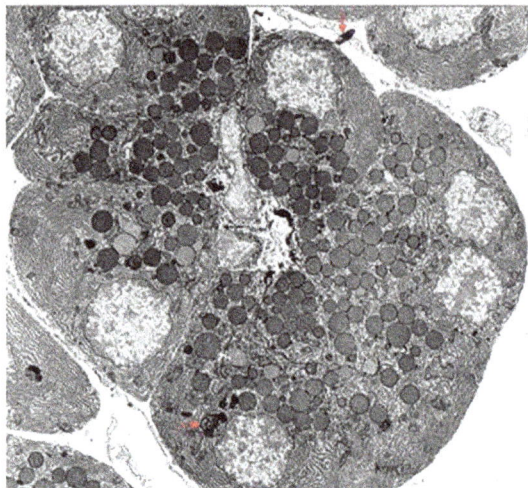

Figure 19. Acinar and centroacinar cells (low-power electron micrograph). ZGs, rough endoplasmic reticulum (RER), and nuclei are all identifiable in the acinar cells. In addition, several small dense inclusions of variable structure are present in the cytoplasm (lower red arrow). These are residual bodies derived from degradation of acinar cell organelles by lysosomal enzymes. The formation of such residual bodies is called autophagy, and large complex membrane-bound structures reflecting this process are called autophagic vacuoles. Such "cellular debris" is sometimes extruded into the interstitium as seen near the top of the field (upper red arrow). Residual bodies are also sometimes extruded into the acinar lumen providing a pathway for "garbage" disposal into the intestine. An acinar lumen is indicated by a small black arrow that lies between two CACs left of center. **Figures 21** and **22** show acinar lumens at higher magnification. ZGs vary in size from about 0.5 to 1.4 μm (micrograph contributed by James Jamieson).

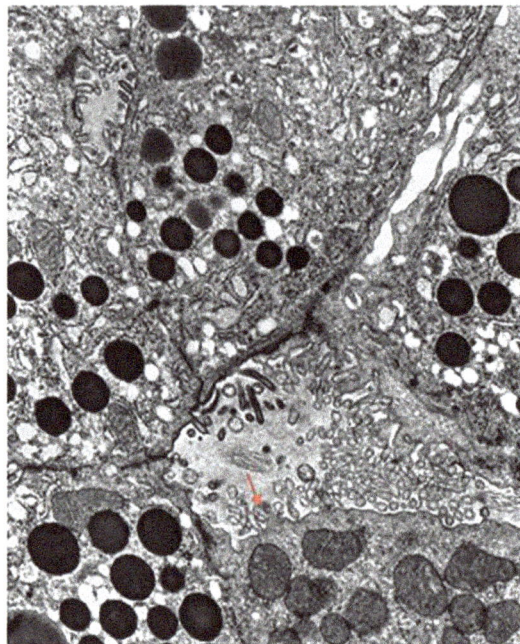

Figure 21. Apical portions of acinar cells abutting two acinar lumens (electron micrograph). A portion of a CAC forms part of the wall of the lower lumen (image right lower corner). The arrow in this lumen points to the CAC that has multiple mitochondria in the cytosol. Microvilli are evident protruding into the lumen from both CAC and acinar cells. A second smaller acinar lumen is near the image left upper corner. ZGs are heavily stained, so it is not possible to distinguish their membranes. RER is also evident in the acinar cells (micrograph contributed by James Jamieson).

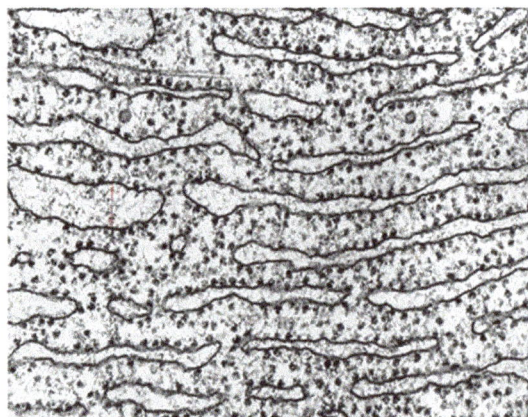

Figure 20. Rough endoplasmic reticulum (RER) shown by high-magnification electron micrograph. The ribosomes adhere to the cytosolic surface of the membrane whereas the cisternal (luminal) side is devoid of ribosomes. Arrows in the cisterna (image left; red arrows) point toward the interior side of the endoplasmic reticulum. A few ribosomes appear to be free in the cytosol (micrograph by George Palade, contributed by James Jamieson).

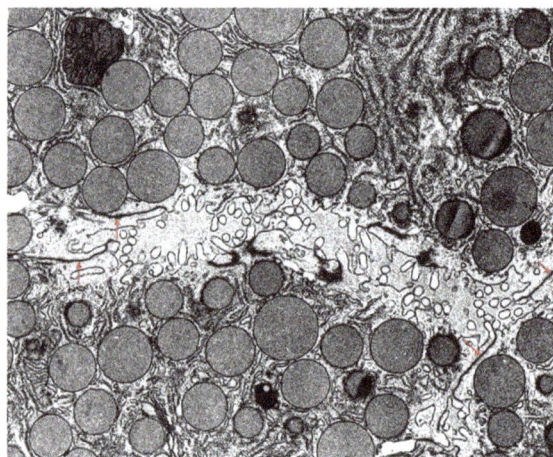

Figure 22. Apical domain of acinar cells is filled with zymogen granules (electron micrograph). The acinar cells abut a lumen near the center of the image. Microvilli protrude into the lumen. The section is lightly stained allowing visualization of the membrane of the ZGs. ZGs are typically about 1 μm in diameter. Tight junctions are present near the acinar lumen (arrows). A mitochondrion is evident upper image left and a smaller one is located lower image left (micrograph contributed by James Jamieson).

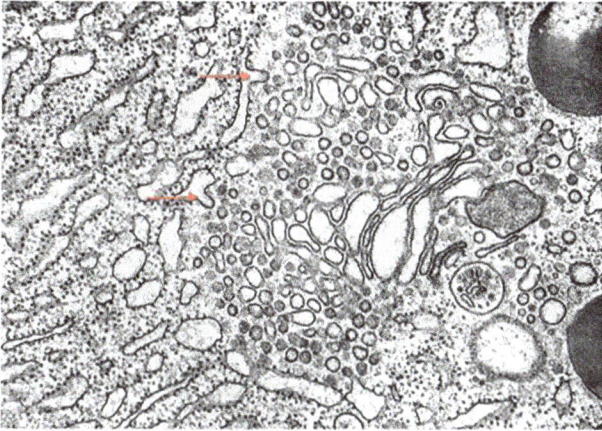

Figure 23. Key elements of the acinar cell protein synthetic pathway show a close physical relationship (transmission EM). Many of the vesicles seen in the middle of the field are likely involved in the transport of newly synthesized proteins from the RER (image left) to the Golgi (right of center). Arrows mark budding of vesicles from the RER and indicate the direction of protein transport by the vesicles to the Golgi and thence to the formation of ZGs (image right) (micrograph contributed by James Jamieson).

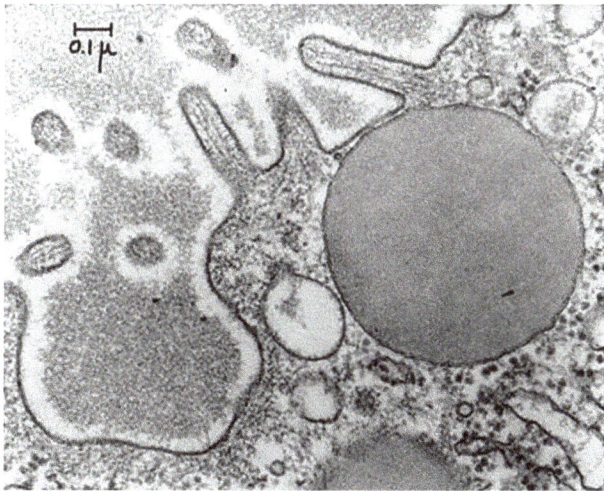

Figure 24. Steps of zymogen granule exocytosis at the apical membrane of the acinar cell are shown (transmission EM). Right of center there is a ZG with a hint of fusion of its membrane with the luminal cell membrane as an early step in secretion. To the left of this granule, there is a "cup" in the cell surface that apparently marks the site of excretion of a ZG after fusion of the membrane of zymogen granule with the luminal cell membrane. (The secretory process has been described in detail[5], micrograph by George Palade, contributed by James Jamieson).

Activated PSCs function as fibroblasts, thus contributing to fibrosis associated with chronic pancreatitis and pancreatic ductal adenocarcinoma as discussed in detail by Mekapogu et al.[10]

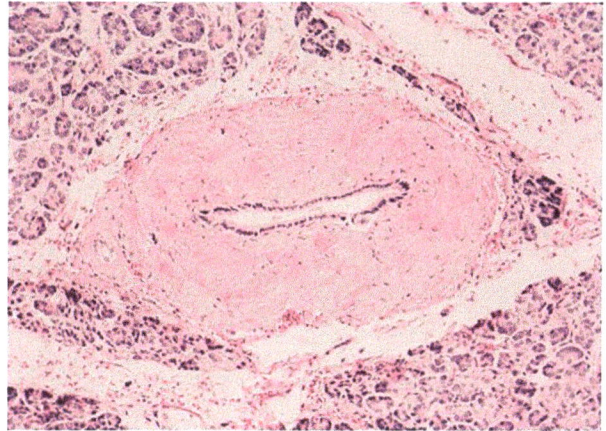

Figure 25. Main pancreatic duct, human (H&E). The lumen is lined by a single layer of cuboidal duct cells. The thickness of the collagenous duct wall is impressive and is probably accentuated because the lumen is empty and collapsed.

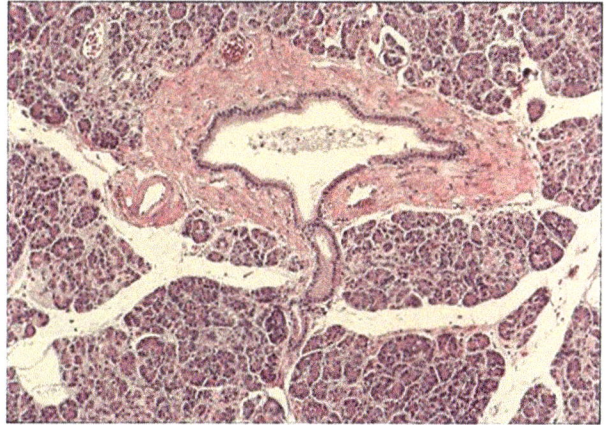

Figure 26. Interlobular duct, human (H&E). The lumen is lined by a single layer of duct cells. The collagenous wall is conspicuous but clearly thinner than that of the main duct. Near the center there is a smaller thin-walled intralobular duct joining the interlobular duct.

Endocrine Pancreas

Most islets (islets of Langerhans) that collectively comprise the endocrine pancreas are too small to be seen by gross examination, and thus they were not depicted in **Figures 1–13**. Islets vary greatly in size; ~70 percent are in the size range of 50–250 μm in diameter in humans with an average in the range of 100–150 μm.[4] Smaller islets are dispersed throughout the acinar lobules and most larger islets lie along the main and interlobular ducts of the pancreas. Most islets are spherical or ellipsoid, but they can be irregular in shape—sometimes reflecting the pressure of an adjacent structure, often a duct, or limitation by a tissue plane. Several reports provide support for the presence of a

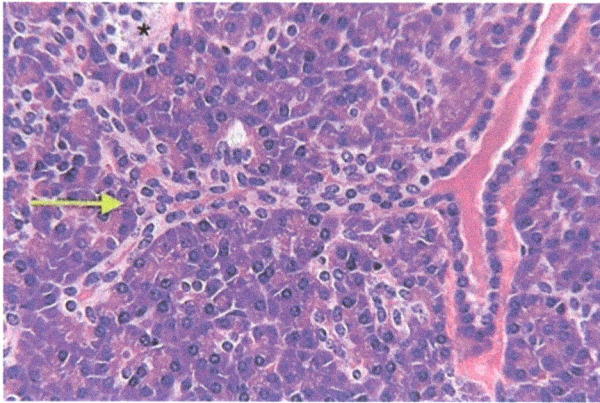

Figure 27. Intralobular ducts, human pancreas (H&E). An intralobular duct with a modest collagenous wall, image right, branches to give rise to an intralobular ductule that in turn branches, image left (arrow). The ductule is nearly devoid of collagen in its wall. The lumen of the small duct and ductule contains homogenous pink-staining protein-rich pancreatic juice. There is a small islet (small cells, pale cytoplasm) at the upper border, image left (asterisk).

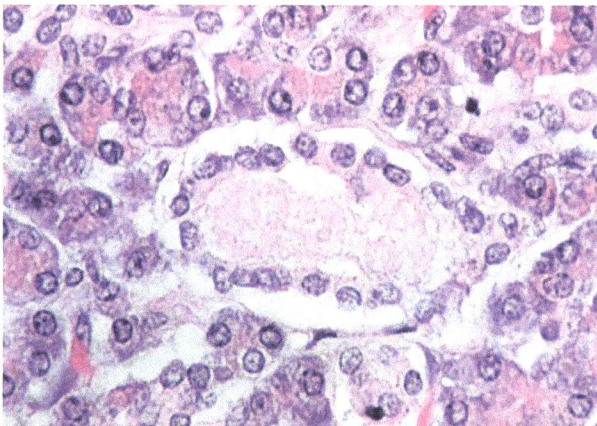

Figure 28. Acinar tissue with an intralobular ductule in cross-section, human pancreas (H&E). Note the single layer of cuboidal duct cells and the nearly complete absence of collagen in the wall of this ductule. Compare this with **Figures 19** and **27**, where intralobular ductules are shown in longitudinal section. The lumen of the ductule contains a pink granular proteinaceous precipitate from pancreatic juice. The clear spaces between the duct cells and the thin connective tissue wall of the ductule reflect artifactual separation of the cells from the basement membrane.

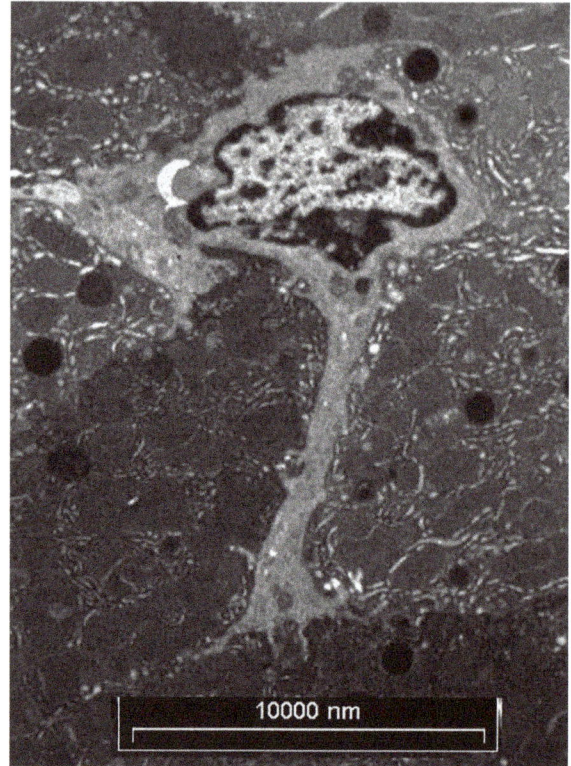

Figure 29. Pancreatic stellate cell in the interstitial space surrounded by acinar cells. An extension of the cell's cytoplasm extends downward between two acinar cells in the lower portion of the field. Dark inclusions in this extension are probably lipid droplets (micrograph contributed by the Pancreatic Research Group, UNSW, Australia; with special thanks to Dr. Murray Killingsworth).

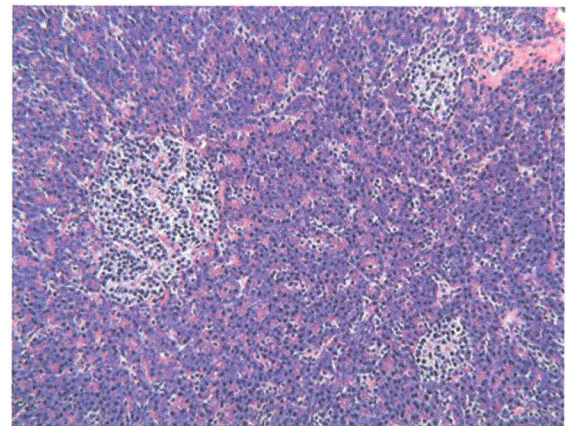

Figure 30. Human pancreas with three islets (H&E). This low-power histologic section illustrates three islets in the background of the more abundant acinar tissue with a small duct in the upper image right corner. The large islet, image left, may be two adjacent islets with a small islet conforming to the lower border of a large round islet. Two small oval islets are located image right at 2 and 4 o'clock. The islet cells are smaller and have paler cytoplasm than the surrounding acinar cells.

higher population density of islets in the tail of the pancreas than in the head and body, although others find no difference.[11–14] In adult humans, the number of islets is calculated to be 500,000–1 million, whereas there are far fewer in smaller animals.[15,16] Islets comprise 1–2 percent of the pancreas in adults of most mammalian species. In addition to the islets, isolated islet cells may be found dispersed in the acinar lobules or in association with ducts.

Photomicrographs of islets follow (**Figures 30–39**). Several of these have been immunostained using antibodies to specific islet peptide hormones to demonstrate various islet cell types including β-cells (insulin), α-cells (glucagon), δ-cells (somatostatin), and pancreatic polypeptide (PP) cells. The PP cells are commonly regarded as the fourth most prevalent endocrine cell type in the islets. Most PP cells are in the portion of the pancreas derived from the ventral pancreatic anlage, that is, the uncinate process that is reported to comprise about 10 percent of the pancreas.[17,18] In the portion of the pancreas derived from the dorsal pancreatic anlage, the majority of islet cells are β-cells (75–80%), followed by α-cells (about 15%), δ-cells (about 5%), and very few PP cells. In the uncinate process, there are few α-cells and many more PP cells. Stefan et al. present data from study of nondiabetic human pancreases showing that the PP cells comprise 54–94 percent of the volume of islets in the uncinate region, displacing most α-cells and some β-cells.[18] These investigators provide data indicating that PP cells are the second most prevalent endocrine cell type overall in the pancreas among their 13 nondiabetic subjects.

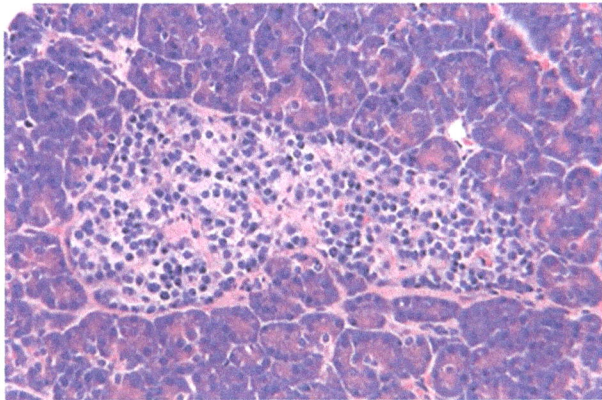

Figure 31. Human islet (H&E). This islet is elongate and nearly triangular in this cross-section. A thin fibrous septum lies along its lower border. Although most islets are oval or round in cross-section, islets vary greatly in shape as illustrated here.

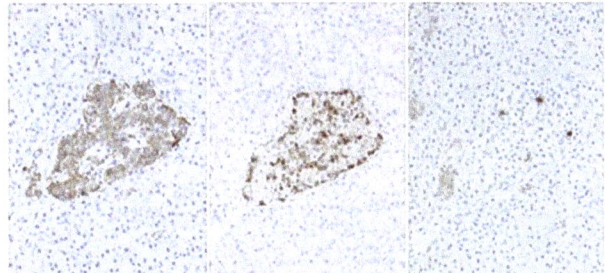

Figure 33. Islet cells store each hormone in distinct locations (immunoperoxidase). Serial sections of an islet have been immunostained using antibodies to insulin (image left), glucagon (center), and somatostatin (image right). The presence of the hormones is indicated by the brown stain. The predominance of insulin-secreting β-cells is obvious. In the center and image right photos, the location of α-cells and δ-cells is primarily at the border of groups of β-cells (photos provided by Arief A. Suriawinata).

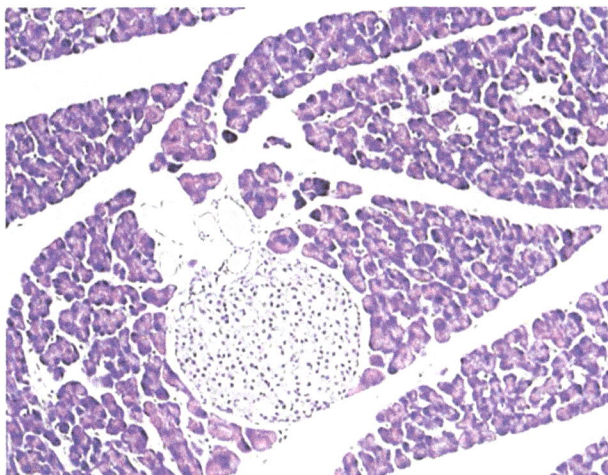

Figure 32. Exocrine pancreas with an islet, mouse pancreas (H&E). The top of a large islet abuts an intralobular duct that is slightly left of center (photomicrograph by Catherine Carriere).

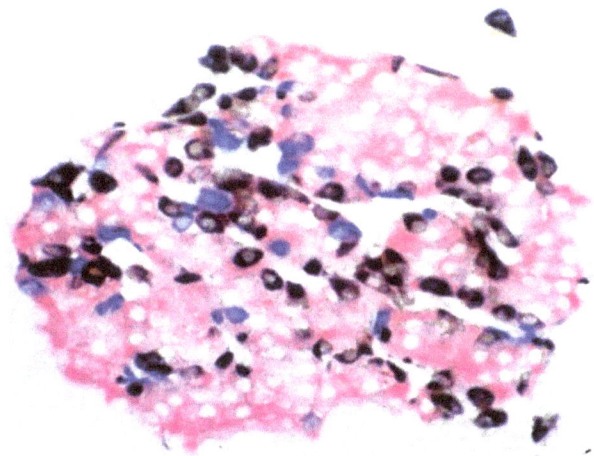

Figure 34. Triple immunolabeling of islet hormones shows the predominance of insulin-secreting cells and their distinct distributions. This islet was stained using antibodies to insulin, glucagon, and somatostatin to demonstrate beta cells (pink), alpha cells (brown), and delta cells (blue). The predominance of β-cells is obvious. α- and δ-cells are typically located at the periphery of clusters β-cells (image provided by Vincent A. Memoli and used with permission of the American Society of Clinical Pathology).

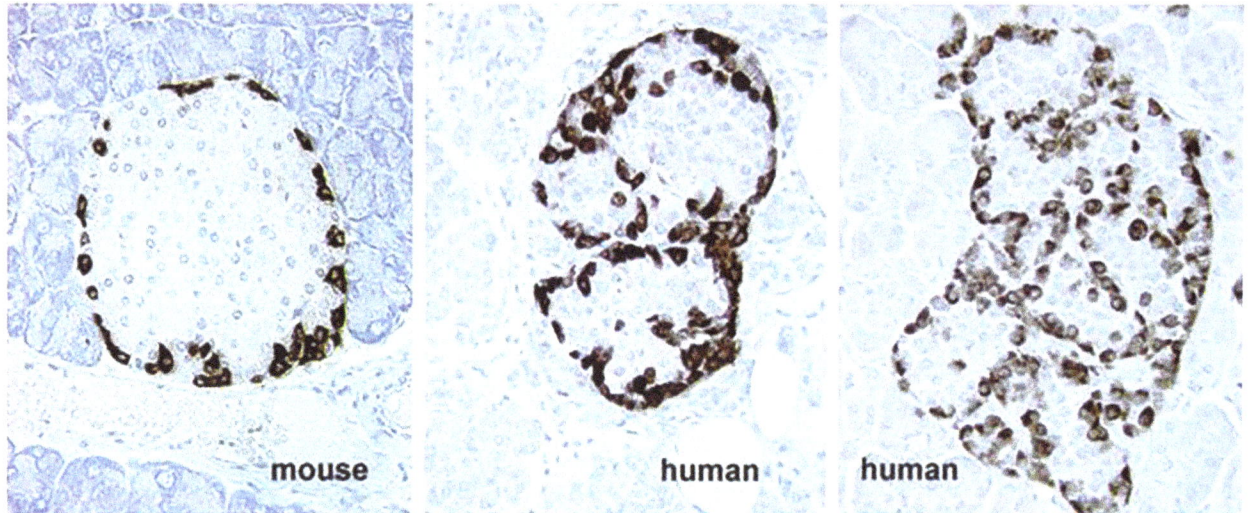

Figure 35. Mouse and human islets stained for glucagon (immunoperoxidase). These images show a minor species difference in the location of α-cells in mouse and human islets. In humans, α-cells appear within the islet, although they seem to be on the periphery of clusters of β-cells. Compare the staining in mouse islet to **Figure 36**, a mouse islet stained for insulin (micrographs provided by Susan Bonner-Weir).

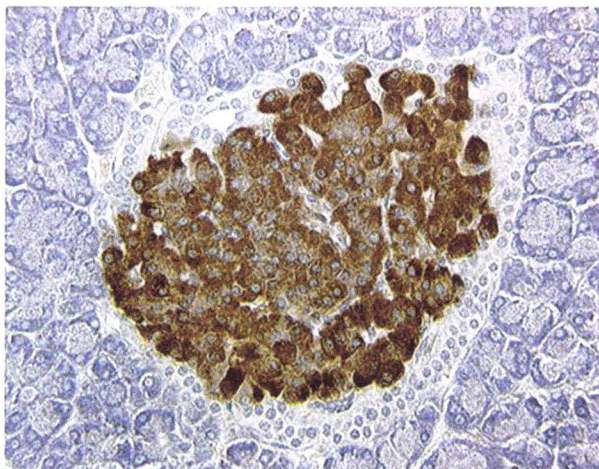

Figure 36. Mouse islet stained for insulin (immunoperoxidase). Note that unstained cells are located in the periphery of the islet corresponding to the location of glucagon staining in **Figure 35** (micrograph provided by Susan Bonner-Weir).

Online Resources

The following websites provide additional images of the pancreas. Some of the drawings are labeled in detail whereas others will challenge you to identify unlabeled structures. We recommend that you visit several of these after you review the text and images provided above. Due to the size of the files, it may take a minute or longer for some sites to open.

A drawing by Frank Netter similar to **Figure 3** is posted online at:

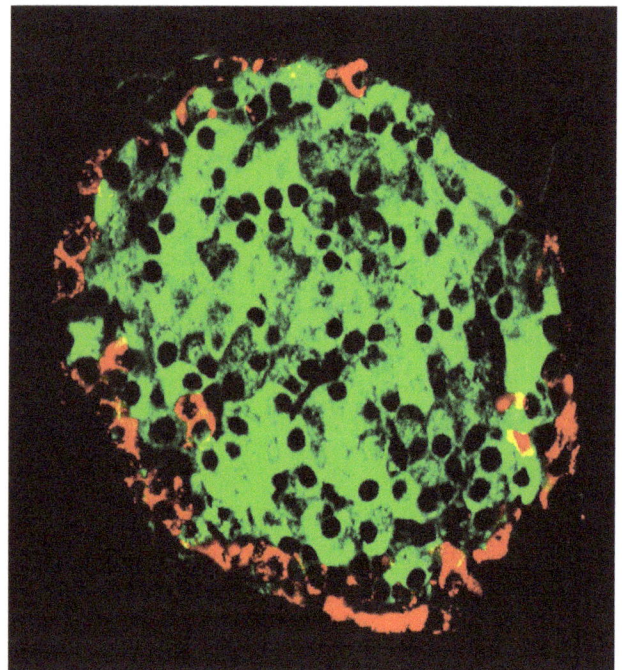

Figure 37. Mouse islet stained to demonstrate pancreatic polypeptide and insulin. Immunofluorescence using antibodies to insulin (green) and neuropeptide Y (NPY) that cross-reacts with PP (red) (micrograph contributed by Susan Bonner-Weir).

http://academicobgyn.files.wordpress.com/2010/11/pancreas-2.png&imgrefurl=http://academicobgyn.com/search/sep/feed/rss2/&h=514&w=782&sz=613&tbnid=i8zrTTDzdJwl7M:&tbnh=90&tbnw=137&zoom=1&usg=__BM8pPs5S5IKciuerloF-AgblkeE=&docid=qtye1NzwAozrdM&sa=X&ei=Rp17

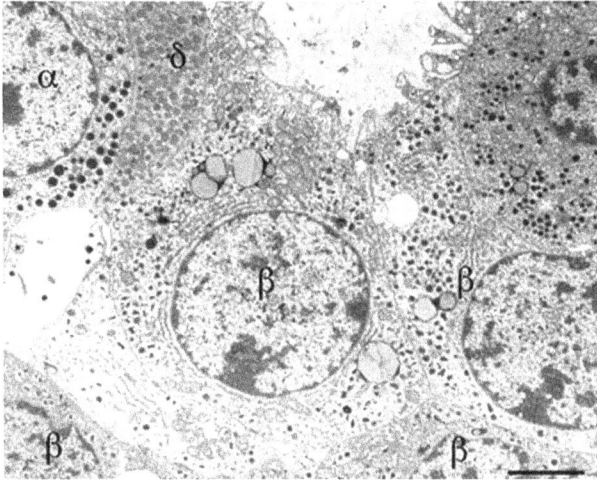

Figure 38. Human islet that was isolated during islet transplantation (electron micrograph). The α-, β-, and δ-cells are labeled. At the ultrastructural level, the cell types are distinguished primarily by differences in their granules. The α-cell granules are typically slightly larger than β-cell granules. δ-cell granules are typically less densely stained than the granules in α- and β-cells. At the image edge (12–1 and 9 o'clock) are interstitial (intercellular) spaces that were probably enlarged by perfusion during the islet isolation procedure. Multiple lipid vacuoles are in the cytoplasm of several islet cells—most notably in the central β-cell where lipid bodies lie at 4 and 11–12 o'clock around the nucleus. Scale bar = 4 μm (micrograph provided by Susan Bonner-Weir).

These two websites provide collections of drawings and photographs:

https://www.google.com/search?q=pancreas+anatom y&client=firefox-a&hs=kor&rls=org.mozilla:en-US:offi cial&tbm=isch&tbo=u&source=univ&sa=X&ei=ne57U q6gGtSr4AP7y4HQAw&ved=0CDEQsAQ&biw=1628& bih=953

https://www.google.com/search?q=Netter+drawing+of +pancreas&client=firefox-a&hs=7qR&rls=org.mozilla:en-US:official&tbm=isch&tbo=u&source=univ&sa=X&ei=G p97UoH8Hsu3sATlhoHYCQ&ved=0CC4QsAQ&biw=13 10&bih=916

Additional images of islets:

http://en.wikipedia.org/wiki/Islets_of_Langerhans

Acknowledgments

The author thanks the contributors of images listed in the text and legends: Susan Bonner-Weir for substantive input regarding the section on islets; Fred Gorelick for preliminary review and suggestions during the preparation of the manuscript; John Williams for all Pancreapedia matters; and Jane L. Weber for editorial suggestions.

References

Because this is an online publication, some references will be provided as website hyperlinks when this is appropriate. Several of the references are chapters in *The Pancreas: Biology, Pathobiology, and Disease.* Second Edition. Edited by V. L. W. Go et al. Raven Press Ltd. New York, 1993. These may be downloaded at http://journals.lww. com/pancreasjournal/Pages/the-pancreas_bio_pathobio_disease.aspx.

1. Barnard FR. Printer's Ink Dec 8 1921 :96–97.
2. Hruban RH, Pittman, MB, Klimstra, DS. AFIP atlas of tumor pathology, fourth series, fascicle tumors of the pancreas. Washington, DC: American Registry of Pathology in collaboration with the Armed Forces Institute of Pathology, 2007.
3. Bockman DE. Anatomy of the pancreas. In: Go VLW, ed. The pancreas: Biology, pathobiology, and disease. 2nd ed. New York: Raven Press, 1993:1–8.
4. Hellman B. Actual distribution of the number and volume of the islets of Langerhans in different size classes in non-diabetic humans of varying ages. Nature 1959; 184:1498–1499.
5. Valentijn K, Valentijn JA, Jamieson JD. Role of actin in regulated exocytosis and compensatory membrane retrieval: insights from an old acquaintance. Biochem Biophys Res Commun 1999; 266:652–661.
6. Lee PC, Lebenthal E. Prenatal and postnatal development of the human exocrine pancreas. In: Go VLW, ed. The pancreas: Biology, pathobiology, and disease. 2nd ed. New York: Raven Press, 1993:57–73.
7. Jorgensen MC, Ahnfelt-Ronne J, Hald J, Madsen OD, Serup P, Hecksher-Sorensen J. An illustrated review of early pancreas development in the mouse. Endocr Rev 2007; 28:685–705.
8. Kern HF. Fine structure of the human exocrine pancreas. In: Go VLW, ed. The pancreas: Biology, pathobiology, and disease. 2nd ed. New York: Raven Press, 1993:9–19.
9. Bockman DE. Morphology of the exocrine pancreas related to pancreatitis. Microsc Res Tech 1997; 37:509–519.
10. Mekapogu AR, Pothula, SP, Pirola, RC, Wilson, JS, Apte, MV. Pancreatic stellate cells in health and disease. Pancreapedia: Exocrine Pancreas Knowledge Base, 2020. DOI: 10.3998/panc.2020.08.
11. Hellman B. The frequency distribution of the number and volume of the islets Langerhans in man. I. Studies on non-diabetic adults. Acta Soc Med Ups 1959; 64:432–460.
12. Rahier J, Guiot Y, Goebbels RM, Sempoux C, Henquin JC. Pancreatic beta-cell mass in European subjects with type 2 diabetes. Diabetes Obes Metab 2008; 10:32–42.
13. Wittingen J, Frey CF. Islet concentration in the head, body, tail and uncinate process of the pancreas. Ann Surg 1974; 179:412–414.
14. Yoon KH, Ko SH, Cho JH, Lee JM, Ahn YB, Song KH, Yoo SJ, Kang MI, Cha BY, Lee KW, Son HY, Kang SK, et al. Selective beta-cell loss and alpha-cell expansion in patients with type 2 diabetes mellitus in Korea. J Clin Endocrinol Metab 2003; 88:2300–2308.

15. Korc M. Normal function of the endocrine pancreas. In: Go VLW, ed. The pancreas: Biology, pathobiology, and disease. New York: Raven Press, 1993:751–758.

16. Longnecker DS, Wilson GL. Pancreas. In: Haschek-Hock WM, Rousseaux CG, eds. Handbook of toxicologic pathology. San Diego: Academic Press, 1991:253–278.

17. Rahier J, Wallon J, Loozen S, Lefevre A, Gepts W, Haot J. The pancreatic polypeptide cells in the human pancreas: The effects of age and diabetes. J Clin Endocrinol Metab 1983; 56:441–444.

18. Stefan Y, Orci L, Malaisse-Lagae F, Perrelet A, Patel Y, Unger RH. Quantitation of endocrine cell content in the pancreas of nondiabetic and diabetic humans. Diabetes 1982; 31:694–700.

Chapter 2

Development of the pancreas

Christopher L. Pin and Melissa Fenech

University of Western Ontario, London, Ontario

Introduction

The adult pancreas is comprised of at least 10 different cells types including those with endocrine function (α, β, δ, γ, and ε), exocrine function (acinar and duct cells), vascular cells, neurons, and mesenchymal cells that are activated in response to injury. The endocrine and exocrine cells are derived from a common endodermal population of cells that also give rise to the liver and parts of the stomach and small intestine. The process by which these cells are specified to take on mature pancreatic phenotypes and the extrinsic and intrinsic factors that guide these processes will be discussed within this chapter. For a more in-depth discussion on the specific aspects of pancreatic development, there are a number of exceptional review articles that focus on transcriptional regulation, epigenetic regulation, cell lineage tracing, distinction from liver tissue, and signaling pathways involved.[1–7]

Morphological Development—from Endoderm to Definitive Pancreas

Specification of the mouse pancreas is initiated from the distal part of the endodermal foregut and the proximal part of the midgut beginning at embryonic day (E) 8.5 based on the initial expression of pancreas-specific genes.[8] By E9, the first morphological sign of pancreatic development is the condensation of mesenchyme over the endoderm that will give rise to the dorsal pancreatic bud, distal to where the stomach will form. In humans, this occurs just prior to 26 days of gestation.[9–11] At E9.5, a clustering or anlage of cells emerges from the dorsal aspect of the gut tube at the point where the notochord comes in contact with the gut (**Figure 1**). This relationship between the notochord and endoderm is of critical importance (described below). Shortly after, the paired dorsal aorta

move toward the midline and fuse, thereby displacing the notochord from the developing pancreas.[12] Less than one day later in mice, or six days later in humans, a second ventrally located anlage appears, just distal to the developing liver, extending from the common bile duct. The two buds continue to develop independently, affected by different surrounding tissues, until approximately E13 (days 37–42 in humans), at which point, the rotational movement of the developing gut tube and elongation of both buds cause alignment and fusion into a single organ. In the majority of cases, one of the pancreatic ducts becomes patent, leaving a single common pancreatic duct, the Duct of Wirsung, for the organ. In about 10 percent of people, the two buds do not fuse leading to pancreatic divisum and the presence of a second duct, termed the duct of Santorini or accessory duct.[13] Pancreatic divisum is the most common pancreatic developmental anomaly and, in some cases, is an increased risk for recurrent pancreatitis.[14,15]. By E15, clear acinar cell clusters and presumptive islets are apparent, and differentiation markers for all pancreatic cell types are expressed.

In general, pancreatic development can be separated into three time periods. The primary transition in mice encompasses E9–12 when specification of different pancreatic cell types takes place.[16] During the primary transition, the pancreatic buds initially appear as stratified epithelium with a centrally located lumen (**Figure 2**). Cells within the anlage maintain multipotency, with the ability to give rise to endocrine, acinar, and duct cells. The number of multipotent pancreatic progenitor cells (MPCs) is determined during the primary transition and appears to dictate the final organ size.[17] Beginning at E10.75, microlumen structures begin to form, with cells establishing an apical polarity eventually leading to duct formation.[18] Integrin-based interactions between outer cells of the developing buds

e-mail: cpin@uwo.ca

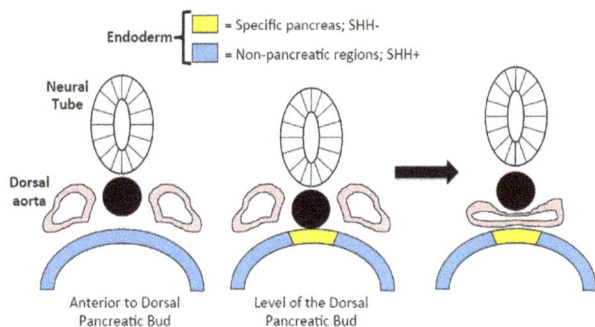

Figure 1. Specification of the dorsal pancreatic bud. Foregut endoderm anterior to the putative pancreatic bud expresses sonic hedgehog (SHH) that does not allow for pancreatic differentiation. Interactions with the notochord (filled black circle) lead to repression of SHH and progression to a pancreatic fate. This fate is maintained even after the paired dorsal aorta fuses at the midline displacing the notochord from direct interaction with the endoderm.

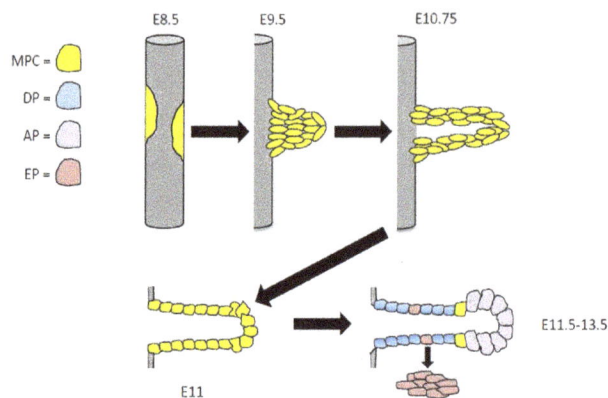

Figure 2. Early cell fate differentiation in pancreatic development. Multipotent pancreatic progenitor cells (MPCs) are first identified at embryonic day (E) 8.5. These cells maintain multipotency through to E10.5, at which point they begin to differentiate into acinar progenitor (AP), endocrine progenitor (EP), and duct progenitor (DP) cells. MPCs are maintained at the neck of pancreatic buds until E14.5.

and the adjacent basement membrane are required for initiation of branching.[19] Minimal expression of pancreatic hormones and digestive enzymes is observed, although endocrine cells expressing multiple hormones exist, present as clusters within the dorsal bud. Whether these initial endocrine cells are maintained in the adult is still debatable.

The secondary transition occurs between E12 and E15 and is characterized by significant expansion and branching of the pancreatic bud, delamination of individual presumptive islet cells from the pancreatic epithelium, and enhanced expression of both hormones and digestive enzymes. At this point, most of the cells have been specified to become either endocrine, acinar, or duct cells, with only a small proportion of cells maintaining multipotency.[20,21] The third stage of pancreatic development occurs after E15 and extends postnatally with continued maturation and expansion of the pancreas. While controversial, it appears that all cells within the pancreas after E15 are unipotent, determined to be either acinar, duct, or endocrine in nature. However, multipotency can be stimulated in the adult by injury.[21–23]. High-powered resolution of the developing pancreas reveals a relationship between cell location within the branching epithelium and the different cell types. Progenitor cells giving rise specifically to acinar cells are located predominantly at the tips of the branching duct network. Conversely, mature duct and endocrine cells are derived from the trunks of these branches (**Figure 2**). p120-catenin (p120ctn) mediates surface-tension cell sorting through heterogeneous expression. Low expression of p120ctn allows cells to migrate to tips, where they are influenced by their environment

to become acinar cells, while high p120ctn expressing remain in the trunk and become duct or endocrine cells.[24] A complete p120ctn deletion leads to defects branching and acinar differentiation.[25] MPCs exist at the transition position between the tips and the trunks.[20,26] Indeed, the location of these cells provided support for the hypothesis that centroacinar cells, which sit at the junction between acini and ducts, are pancreatic stem cells in the adult. However, studies examining centroacinar cells as stem cells have not supported such a role.[27] As mentioned above, by E15, MPCs are rare. Evidence suggests that presumptive endocrine cells undergo a mesenchymal-epithelial transition process and coalesce to form the primitive islets of Langerhans.[28,29] Further recruitment, proliferation, and maturation give rise to the mature islets postnatally. In mice, the islets consist of an inner core of β-cells, which secrete insulin, surrounded by a mantel of α (glucagon), δ (somatostatin), ε (ghrelin), and PP (pancreatic polypeptide) cells. In humans, these cells are positioned throughout the islets.

Differentiation of acinar cells is less overt. Acinar cells maintain attachment to the duct network and develop a classical epithelial phenotype based on their polarity and intercellular junctions. Expression of carboxypeptidase (CPA) is observed as early as E9,[8] but CPA expression can be observed at early time points in MPCs as well as in acinar progenitor cells.[20] More extensive expression of pancreatic enzymes begins around E12.5, and zymogen granules, the morphological hallmark of acinar cells, are first observed around E16.5. Acinar cells do exhibit polarity prior to birth based on the localization of tight and adherens junctions. However, the high degree of organization within acinar cells is first observed following birth.[30] At this point, mature acinar cell morphology has been obtained, while further proliferation and maturation at the molecular level continues to occur until after weaning, presumably due to alterations in dietary makeup.

The Three Main Pancreatic Cell types Are Derived from a Common Progenitor Pool

Elegant genetic lineage tracing experiments in mice have defined the progression of MPCs to individual unipotential populations as well as identified specific markers for this process. Mouse models were developed in which *cre recombinase* expression was driven in specific cell populations during development.[31,32] Initially, these experiments involved spatially expressing *cre recombinase* based solely on the promoter used and mating these "driver" mice to those containing a loxP-stop-loxP (LSL) site upstream of a reporter gene such as green fluorescent protein. Since this LSL site is downstream of a constitutively active and ubiquitous promoter, any cells derived from the initial expressing cells would be positive for the reporter gene. A number of groups used a system in which the cre recombinase could be manipulated temporally through the addition of a tamoxifen-inducible tag linked to *cre* (see **Figure 3** for scheme). From this work, several observations could be made. First, prior to E12, MPCs exist that give rise to all cell types.[20] Second, acinar progenitor cells are distinct from endocrine/duct progenitor cells indicating that duct cells have a closer developmental origin to endocrine cells.[20] Third, MPCs do not normally exist within the adult but can be stimulated to reappear by injury to the pancreas.[21] Fourth, acinar cells have the ability of dedifferentiation into duct-like cells and likely contribute to initial PanIN formation.[33] These studies were also instrumental in defining that MPCs are located at the junction between tip and trunk cells of the branching pancreatic duct and can be identified by a specific combination of transcription factors (TFs). MPCs identified during the primary transition express pancreatic transcription factor 1A (PTF1A), pancreatic-duodenal homeobox 1 (PDX1), Sry-related HMG box (SOX9), and NK homeobox 6.1 (NKX6.1). Hepatocyte nuclear factor 1β (HNF1β) defines epithelial cells in the trunk that give rise to endocrine and duct cells but not acinar cells,[20,34] although there is some controversy as Wright's group has suggested the HNF1β may actually be expressed in MPCs as well.[21] Whether this difference reflects the time points examined or differences in analysis is unclear. During the secondary transition, PTF1A and NKX6.1 become mutually exclusive with PTF1A cells specifying an acinar cell phenotype and NKX6.1 found in duct/endocrine precursors.[35]

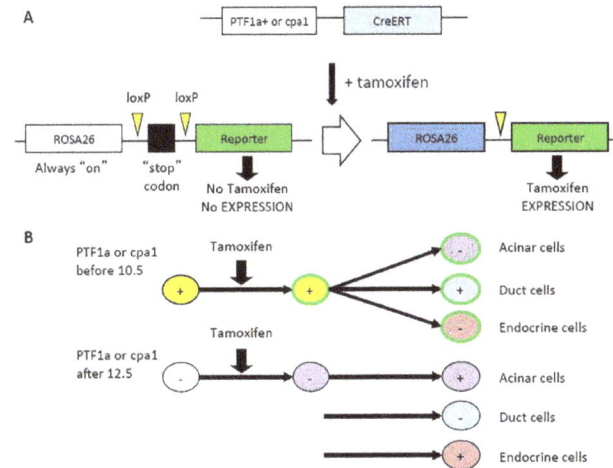

Figure 3. Lineage tracing during pancreatic development. (**A**) Scheme for permanently marking cells during development. Expression of cre recombinase is limited spatially by the promoter, in this case *Ptf1a* or *cpa1(carboxypeptidase 1)*, and temporally by the addition of a tamoxifen-inducible estrogen receptor (ERT) fused to cre. The constitutively active ROSA26 drives expression of the reporter gene, which is prevented by the presence of stop codons flanked by loxP sites (target for cre recombinase). Upon the introduction of tamoxifen, the stop codons are removed leading to reporter gene expression. (**B**) Using either the *Ptf1a* or *cpa1* promoter to drive creERT expression, introduction of tamoxifen early in development leads to reporter expression in all pancreatic cell types. Introduction of tamoxifen after E12.5 results in only acinar cells being labeled.

Developmental Relationships That Affect Pancreatic Development

Specification of the pancreas is dictated by interactions with a number of structures developing in close proximity to the developing endoderm. As mentioned, the ventral pancreatic duct shares a common cellular origin with the liver and, therefore, signals must be present that dictate one fate from the other. Interestingly, it appears that the dorsal and ventral aspects of the endoderm are governed by either significantly different signaling events or are able to process specific signaling pathways, such as fibroblast growth factor (FGF) signaling, in different ways.[36–39] This suggests that the default differentiation pattern differs for dorsal and ventral endoderm. Indeed, culturing ventral endoderm independent of other structures or signals results in pancreatic differentiation, based on the expression of both endocrine and exocrine cells[40,41] (**Figure 4**). This suggests that the default pattern of differentiation for the ventral endoderm is pancreas, and that surrounding tissues in the embryo repress pancreatic development to become other organs. For example, signals from the cardiac mesoderm and the septum transversum, the structures that eventually give rise to the diaphragm and parts of the ventral mesentery, promote liver differentiation at

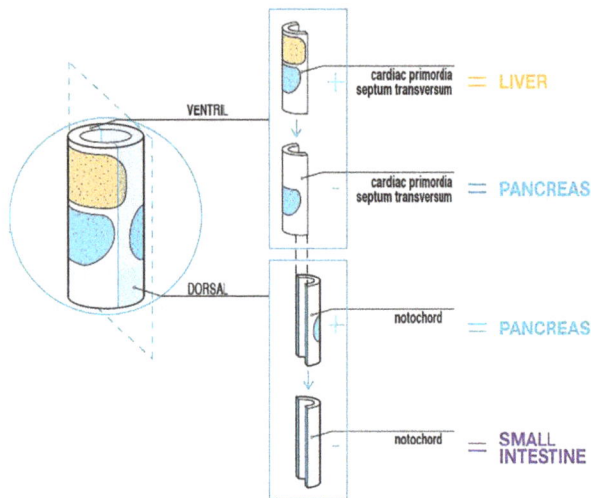

Figure 4. The dorsal and ventral pancreatic buds have different default patterns for development. The pancreas is derived from a ventral pancreatic bud (VPB) and a dorsal pancreatic bud (DPB) that first appear around E9.5 in mice. The ventral endoderm gives rise to liver tissue due to inhibitory signals from the septum transversum and cardiac primordia. In the absence of these signals, pancreatic development is the default pattern of differentiation. The dorsal endoderm requires inductive signals from the notochord to promote pancreas development. In the absence of these signals, the dorsal endoderm will differentiation into nonpancreatic structures, such as the small intestine (figure is modified from reference [43]).

the expense of the pancreas.[42] Conversely, culturing of the dorsal endoderm does not give rise to pancreas (**Figure 4**) suggesting that signals are required to promote differentiation and the default differentiation program is not pancreas.[12]

Mesenchyme

Coalescence of the mesenchyme at the level where the dorsal pancreas will form is the first morphological sign of pancreatic development. Removal of the mesoderm, or the fibroblasts within the mesoderm, prior to pancreatic specification results in pancreatic agenesis.[44-46] Mesoderm removal following specification results in a reduction of the total pancreatic size indicating an ongoing requirement for mesoderm signaling to attain complete organ development.[47] Interestingly, culturing of pancreatic mesenchyme with other sections of the dorsal endoderm can promote pancreatic differentiation, while mesenchyme from other regions of the anterior-posterior axis does not have this ability.[48] This suggests that the mesenchyme provides signals that promote pancreatic specification, yet limits differentiation, thereby allowing expansion of the organ. These signals come in a variety of sources.

Physical interactions between the mesoderm and developing pancreatic bud affect differentiation and maturation in multiple ways. Physical interaction is particularly important in specification of the exocrine pancreas. Glypicans and syndecans promote exocrine cell differentiation,[49] and Notch signaling promotes PTF1A's function while repressing Neurogenin (NGN)-3 expression.[50] PTF1A and NGN3 are key transcriptional regulators that promote exocrine and endocrine cell specification, respectively.[51,52] Laminin-1, a key extracellular component within the mesenchyme, promotes pancreatic duct formation,[53,54] and substrates that contain laminin-1, such as Matrigel, can promote duct formation from early pancreatic buds possibly through α6-containing integrin complexes.[55] Furthermore, direct interaction between the mesenchyme and endoderm represses islet formation. Only after extensive branching excludes the mesenchyme from the developing ductal network can progenitor cells undergo epithelial to mesenchymal transition (EMT) and coalesce to form definitive islets. Finally, direct interactions between the mesenchyme and developing pancreatic bud maintain a progenitor-like state. Inhibition of Epi-CAM (epithelial cell adhesion molecules) in pancreatic bud cultures induces premature endocrine cell maturation, suggesting Epi-CAM helps maintain a ductal endocrine progenitor cell type.[56] In addition to physical interactions, secreted factors from the surrounding mesenchyme have also been identified that either promote or repress pancreatic specification and differentiation. Signaling through the TGFβ superfamily affects many steps in pancreatic differentiation. TGFβ signaling from the mesenchyme promotes endocrine differentiation from MPCs,[57] and TGFβ treatment of pancreatic dorsal bud tissue leads to endocrine differentiation at the expense of exocrine tissue. Activin, another member of the TGFβ superfamily, also promotes endocrine differentiation while blocking branching and proliferation of the pancreatic bud and repressing exocrine differentiation. Inhibition of activin using follistatin promotes exocrine cell differentiation at the expense of endocrine tissue.[58] Disruption of TGFβ signaling affects the secondary transition[59] and once pancreatic fates have been determined, TGFβ signaling is required to maintain mature acinar cell fate as overexpression of dominant negative TGFβRII in the pancreas leads to acinar to duct cell metaplasia (ADM).[60] Additionally, a number of fibroblast growth factors (FGFs) are secreted by the mesenchyme, which affect pancreatic development. FGF4 represses endocrine differentiation and allows for expansion of the exocrine compartment,[61] while FGF7 and FGF10 repress exocrine differentiation, thereby allowing proliferation and expansion of the pancreatic bud.[62,63] *Fgf10*−/− mice exhibit pancreatic agenesis due to the inability to repress sonic hedgehog (SHH), a repressor of pancreatic differentiation (see below).[64]

Vasculature

At E8.5, de novo blood vessels coalesce from endothelial precursors immediately dorsal to the pre-pancreatic endoderm (reviewed in Risau et al.[65]). Paracrine signaling from these blood vessels is crucial in promoting early pancreatic organ specification, particularly in the dorsal pancreatic bud. The importance of blood flow was first identified in a genetically modified mouse model containing a targeted deletion of *Cdh2*, the gene encoding N-cadherin. Mice completely null for N-cadherin show dorsal pancreatic agenesis,[66] initially suggested to be due to impaired interactions between the mesenchyme and pancreatic endoderm. However, restoring N-cadherin specifically in the heart rescued both heart and pancreatic development indicating that maintaining blood flow to the pancreatic anlage is required for dorsal pancreatic formation.[67] Further studies showed that signals from the vasculature could be from both the circulating blood as well as the endothelial cells of the dorsal aorta and surrounding mesenchyme.[68] However, within the pancreas E-cadherin does not play a role in maintaining exocrine function postnatally. Four days after birth, pancreas-specific *Cadh1*-null mice failed to gain weight and were hypoglycemic. Four days after birth, *Cdh1*-null mice displayed disrupted exocrine tissue architecture and integrity and activation of the Wnt and Yap signaling pathways.[69]

Interactions between the vasculature and endoderm also limit pancreatic differentiation. Overexpression of VEGF-A with the *Pdx1* promoter leads to insulin production within the stomach. Conversely, overexpression of VEGF-A within the pancreatic anlage greatly enhances vasculature development and significantly reduces pancreas formation, specifically affecting exocrine pancreatic mass.[70] Ex vivo cultures in which the developing pancreas was treated with VEGF showed increased vasculature and greatly reduced size of the pancreas, stalling epithelial cells in a progenitor-like state.[71] Conversely, ablation of endothelial cells following pancreatic specification leads to more promiscuous differentiation of acinar cells, ultimately reducing the size of the pancreas as well. Therefore, the endothelium plays a dual role in dorsal pancreatic development. During the primary transition, the vasculature provides signals that specify pancreatic differentiation from the endoderm, yet maintain trunk progenitor populations in an undifferentiated state, preventing premature development of specification to an endocrine fate. During the secondary transition, the vasculature restricts exocrine differentiation.

Again, the signals derived from the vasculature are diverse. Lammert et al.[70] showed that direct interaction with the aortic endothelial cells induced pancreatic bud-like structures and PDX1 expression in the adjacent endoderm. Endothelial cells are also required for PTF1A expression in the dorsal foregut. Conversely, development of the ventral pancreatic bud does not appear to require direct endothelial interactions[41] but rather relies on paracrine factors for development. For example, sphingosine-1-phosphate, found in the blood, promotes recruitment to a pancreatic cell fate and induces proliferation of pancreatic mesenchymal and endothelial cells.[67]

Notochord and sonic hedgehog

A critical negative regulator of pancreatic development is SHH. This is evidenced by experiments in which SHH signaling is elevated either through a transgenic approach where the *Pdx1* promoter drives SHH expression or by ablating the patched receptor (Ptc). Ptc represses smoothen, a downstream mediator of SHH signaling. Both *Pdx1-Shh* and *Ptc*[−/−] mice exhibit pancreatic agenesis.[72] These experiments highlight the importance of repressing SHH at the initiation of pancreatic development. Forcing SHH expression, or the closely related Indian hedgehog (IHH), after pancreatic specification (E12.5) by using the *Pax4* promoter, still leads to a significant loss of exocrine and endocrine tissue that is replaced by mesenchyme and stroma.[73] Therefore, hedgehog suppression must be maintained to ensure proper pancreatic development. Indeed, activation of SHH signaling is observed during pancreatic injury correlating to acinar to duct cell metaplasia.[74] Surprisingly, in zebrafish endoderm, ectopic SHH signaling expands the Pdx1-positive pancreas domain, while inhibiting hedgehog activity results in pancreatic agenesis indicating species differences in signaling pathways.[75,76]

While the mesenchyme expresses factors that promote SHH repression, including activin and FGFs, the notochord appears to be the primary source of these signals at the initiation of dorsal pancreas induction. The foregut expresses SHH along its length except at the point where the notochord comes in close proximity to the endoderm (**Figure 4**). Ectopic notochord engraftment represses SHH expression in the lateral and ventral endoderm, and SHH-neutralizing antibodies can induce pancreas gene expression in dorsal endoderm explants.[77] However, blocking SHH in non-pancreatic endoderm does not lead to pancreatic development. This suggests that signals from the notochord are not instructive but, rather, permissive. In agreement with this permissive model, *Shh*[−/−] mice do not show expansion of the pancreas along the foregut, indicating that additional factors are required.

Ventral bud differentiation, cardiac mesenchyme, and septum transversum

The signals derived from the notochord and mesenchyme that permit pancreatic development are activin (TGFβ family) and FGFs. Surprisingly, these same factors repress ventral pancreatic differentiation, highlighting differences in the developmental processes of the two pancreatic buds. A number of other findings highlight differences between ventral and dorsal pancreatic bud development. *Hhex*[-/-] embryos fail to specify a ventral pancreas but the dorsal bud develops without issues.[78] Alternatively, loss of *Raldh2*, which is required for endogenous retinoic acid signaling, prevents development of the dorsal pancreas while the ventral portion of the organ remains unaffected.[79,80]

Since ventral pancreatic bud formation is the default pattern of ventral foregut differentiation, the factors that govern differentiation of this anlage initially repress pancreatic fate. Similar to the repression of SHH by FGFs and activin in the dorsal pancreatic bud, these factors repress pancreatic specification, thereby allowing differentiation of the liver. The ventral foregut endoderm lies in close juxtaposition to the cardiac mesoderm and septum transversum mesenchyme, and these tissues induce liver fate at the expense of the pancreas.[37] Bone morphogenetic proteins (BMPs) also promote epigenetic reprogramming of ventral endoderm to a liver fate. In the absence of these structures or signaling molecules, the endodermal cells activate pancreatic differentiation pathways.[42] Once liver development proceeds, the repressive signals are blocked allowing for pancreatic development from the more distal ventral endoderm to occur. In support of such a model, it has been suggested that *Hhex*[-/-] fail to develop a ventral pancreas due to lack of proliferation of the endoderm, which would displace the putative pancreatic bud from influences of the cardiac mesoderm. Supporting this hypothesis, *Hhex*[-/-] endoderm cultured in vitro is fully capable of expressing PDX1.[78]

While these various developmental structures have a significant impact on pancreatic development, their effects are governed by the permissive nature of the developing endoderm. In the next sections, the TFs and the signaling pathways that dictate pancreatic development will be discussed.

Transcription Factors Governing Pancreatic Development

In the previous sections, we discussed morphological events and extrinsic factors that regulate pancreatic development. In this section, we will discuss TFs that govern tissue and cell development. During pancreatic development, coordinated gene expression is critical for cell specification, expansion of the progenitor population, differentiation, and then final maturation and function. The TF network has been well characterized in the pancreas, especially regarding β-cell development, with many TFs exhibiting dual roles during development and adult function. In early development, specific TFs govern the establishment and maintenance of the MPCs, while in mature tissue, TFs affect cell-specific function. A complete description of

each TF is not possible. Therefore, this section will focus on identifying TFs linked to factors that define MPCs or are involved in exocrine tissue development as more extensive reviews on TFs that affect endocrine and specifically β-cell development can be found elsewhere.[81,82]

Transcription factors involved in specifying MPCs from endoderm

In the 1990s, two key TFs—PDX1 and PTF1a—were identified as critical for specification and differentiation of all cell types within the pancreas. Using genetically modified mice, both PDX1 and PTF1a were shown to be absolutely required for differentiation of both endocrine and exocrine pancreatic compartments indicating these factors are required for initial specification of MPCs from the endoderm. Since that time, several other TFs have been linked to specification of MPCs or shown to be required for expansion or maturation of the exocrine pancreas. Many of these proteins exhibit defined roles in development and mature cell function and have been implicated in various pathologies such as pancreatitis or pancreatic cancer. In this section, we have focused on their role in development.

Pancreatic and duodenal homeobox 1

Pancreatic and duodenal homeobox 1 (PDX1; alternatively known as insulin-promoting factor 1 (IPF1), or somatostatin-transactivating factor 1 (STF1)) is a homeobox containing TF. PDX1 is expressed in the developing mouse as early as E8.5 and maintained in all cells of the pancreas until E17.5, when it becomes restricted to islets.[83,84] Initial expression of PDX1 appears to delineate a region of the developing gut that later gives rise to the stomach, small intestine, and bile duct, suggesting that it is involved in specifying this region of the gut tube to a subset of tissues.[83] Single-cell RNA sequencing analysis of the dorsal and ventral pancreatic bud at E10.5 indicates PDX1 high- and low-expressing cells. PDX1-high cells from both ventral and dorsal pancreatic regions are pancreatic progenitors. PDX1-low cells from the dorsal pancreatic region are defined as premature endocrine linages, while those from the ventral pancreatic region are progenitors for hepatic or extrahepatic lineages.[85] Germline deletion of *Pdx1* in mice leads to an absence of all pancreatic tissues and neonatal lethality by postnatal day 5 due to hyperglycemia. Examination of the pancreas during development indicates that initial specification of the gut to become pancreas, as well as initial budding, still occurs in *Pdx1*[-/-] mice. However, subsequent expansion and branching of the tissue does not occur, indicating that expansion and complete differentiation of the pancreas requires PDX1 activity.[83,86,87] In the mature pancreas, PDX1 was initially described as limited to β- and δ-cells of the islets,[84] only being reexpressed in acinar cells upon pancreatic injury. However, recent studies have identified low levels of PDX1 in most mature acinar cells, and targeted deletion of *Pdx1* in post-developmental acinar cells shows that PDX1 is required for maintaining this compartment of the tissue.[88]

Heterozygous *Pdx1* (*Pdx1$^{+/-}$*) mice initially develop normally. However, over time *Pdx1$^{+/-}$* mice develop insulin insufficiency, indicating a requirement for high levels of PDX1 in insulin production. Transient loss of *PDX1* in adult tissue using a tet-inducible system supports the ongoing requirement for PDX1 in maintaining β-cell function, as loss of PDX1 leads to decreased insulin accumulation and apoptosis.[89] In humans, homozygous mutation of *PDX1* results in pancreatic agenesis [90], while heterozygous mutations of *PDX1* are found in patients with maturity-onset diabetes of the young (MODY). Genetic analysis identified the *PDX1* gene as MODY4,[91,92] and PDX1 can directly target and promote the expression of insulin and GLUT2 in β-cells.[89]

Pancreatic transcription factor 1A

Pancreatic transcription factor 1a (PTF1a; alternatively known as PTF1-p48 or p48) was first identified by its ability to bind combined E-box (CANNTG) and G-box (TGGGA) sequences found in promoters of digestive enzymes *trypsin*, *cpa1*, *amylase*, and *elastase*.[93,94] Early in pancreatic development, PTF1a is coexpressed with PDX1 in MPCs.[95] During the secondary transition, PTF1a becomes restricted to acinar progenitor cells.[21] PTF1a is a basic helix-loop-helix (bHLH) TF that forms an unconventional protein trimer with ubiquitously expressed E proteins, such as E12 and E47,[96] and recombinant binding protein J (RBPJ), a downstream mediator of NOTCH signaling. Combined, these factors make up the PTF1 complex, which specifies pancreatic acinar cells. However, while associated with RBPJ, these acinar cell progenitors do not undergo differentiation, thereby allowing for expansion of the progenitor population.[97] Upon commitment to acinar cell differentiation, PTF1a preferentially interacts with the RBPJ-like (RBPJL) protein and nuclear hormone receptor NR5A2 to drive pancreatic differentiation of acinar progenitor cells.[98,99]

Targeted deletion of *Ptf1a* in mice results in postnatal lethality immediately after birth,[52] likely due to neuronal functions, as PTF1a is also expressed in the neural tube and cerebellum.[100] *Ptf1a*-deficient mice have no pancreatic tissue at birth with PTF1a-deficient cells instead contributing to the common bile duct and duodenum.[101] Interestingly, the result of overexpressing PTF1a depends on developmental state. Forced PTF1a expression early in development in cells that would typically give rise to the duodenum and stomach leads to expansion of the pancreas, with increased differentiation of duct, islet, and acinar

cell compartments.[95] In ES cells, forced PTF1a expression requires MIST1 to activate digestive enzyme gene expression.[102]

Islet 1

Islet 1 (*Isl1*) is a member of the LIM/homeodomain family of TFs, expressed in a number of tissues during embryogenesis.[103] In the developing pancreas, ISL1 expression is initially observed within the mesenchyme and later in pancreatic islets. Targeted deletion of *Isl1* in mice leads to deficits in pancreatic development and provided initial evidence that the mesenchyme has a role in affecting pancreatic differentiation. Culturing putative mouse *Isl1$^{-/-}$* pancreatic endoderm with wild-type mesenchyme restored pancreatic differentiation confirming that ISL1 acts as noncell autonomous signal in early pancreas development.[104] While ISL1 is restricted to pancreatic islets in the adult, expressing *Isl1* in acinar cells using an adenoviral-mediated approach induces tubular complex formation through acinar to ductal metaplasia, in which exogenous expression facilitated this process.[105]

Hepatocyte nuclear factors

The hepatocyte nuclear factors (HNFs) are a homeodomain-containing superfamily of TFs that affect multiple stages of development. *HNF1b* encodes a protein that binds DNA as a homodimer or heterodimer along with HNF1a. During endoderm development, expression of HNF6 activates *HNF1b* expression. HNF1b is critical for endoderm development with HNF1b+ cells giving rise to liver, kidney, pancreas, bile ducts, urogenital tract, lung, thymus, and gut.[106] Mice with a germline deletion of the *Hnf1b* gene have a very similar phenotype to *Ptf1a$^{-/-}$* mice,[107] with no pancreatic development observed at birth. Temporal and spatial deletion of *HNF1b* throughout pancreatic development has demonstrated important roles in MPC expansion and acinar, ductal, and endocrine cell differentiation.[108] In humans, mutations in the *HNF1B* gene result in renal cysts, diabetes syndrome, and noninsulin-dependent diabetes mellitus. Some individuals with *HNF1B* mutations display pancreatic exocrine dysfunction, pancreatic hypoplasia, and, in some cases, complete absence of the pancreatic body and tail suggesting specific loss of dorsal pancreatic development.[109,110]

Hepatic nuclear factor 6 (HNF6; also known as ONECUT1 or OC-1) binds DNA to elicit transcription of genes that are important throughout endodermal development. In pancreatic development, *HNF6* expression begins prior to PDX1 in MPCs and is critical to several specific developmental functions depending on developmental time point.[111–113] After the onset of *Pdx1* expression, HNF6 is crucial for endocrine cell development. HNF6 induces HNF1B expression, which promotes a wave of proliferation and initiates *Neurogenin 3* (*Ngn3*) expression. Following this increase in *Ngn3* expression, HNF6 expression becomes limited to duct cells and is very marginally expressed in acinar cells.[113] *Hnf6⁻ʹ⁻* mice exhibit a hypoplastic pancreas with effects mostly limited to nonacinar cell compartments suggesting HNF6 is dispensable for acinar cell differentiation.[113] However, in a follow-up study, the pancreatic area of *OC1^{Δpanc}* at postnatal day (P) 2 decreased due to apoptosis. RNA-seq of *OC1^{Δpanc}* exocrine-enriched samples showed decreased expression of *Mist1*, *Gata4*, *Nr5a2*, and *Ptf1a*, all TFs promoting the acinar cell phenotype,[114] supporting a role for *OC1* in pancreatic development.

Sex-determining region Y-box (SRY-box) containing gene 9 (SOX9)

SOX9 is a member of the SRY-related, high mobility group box TFs. Early in development, between E9 and E12.5, SOX9 expression overlaps with PDX1 and marks MPCs. SOX9 is required for pancreatic cell fate and repression of intestinal differentiation through binding lineage-specific promoters and enhancers.[115] A FGF10/SOX9/FGFR2 feed-forward loop maintains pancreatic organ identity in MPCs.[116] While MPCs express SOX9, as cells begin to differentiate SOX9 expression decreases. By E14, SOX9 accumulation is restricted to a subpopulation of PDX1+ cells located within the epithelial cords.[117] Pancreas-specific knockout of *Sox9* using the *Pdx1-Cre* driver leads to stunted growth, elevated blood glucose levels, and death by postnatal day 4. Gross morphological assessment showed hypoplasia of both pancreatic buds beginning at E11.5, due to a reduction of PDX1+ cell proliferation and increase in cell death. Haplo-sufficient *Sox9* mice and individuals with campomelic dysplasia (CD), a semi-lethal skeletal malformation syndrome due to a heterozygous loss-of-function mutation in *SOX9*, have reduced islet cell mass, but the exocrine tissue remains unaffected.[118,119] In the adult, SOX9 accumulates in some duct and centroacinar cells,[118,120] and targeted deletion of *Sox9* in the adult pancreas of mice leads to a cystic phenotype and decreased expression of duct markers,[121,122] indicating SOX9's role in duct cell maintenance.

GATA transcription factors

GATA TFs are characterized by their ability to bind the DNA consensus site (A/T)GATA(A/G). GATA4 and GATA6 have ~85 percent homology, containing a highly conserved DNA-binding domain consisting of two zinc finger motifs.[123] Initial expression of GATA4 and GATA6 occurs early in development and is associated with the development of many organs of endodermal and mesodermal origin.[124] GATA4 and GATA6 are expressed between E9.5 and E15.5 in the developing pancreas. At E9.5, GATA4 and GATA6 are coexpressed throughout the pancreatic epithelium.[125] As development continues, GATA4 accumulation becomes restricted first to the tips of branching ducts and then acinar cells by E16.5 where expression is continually reduced until E18.5 when it is undetectable in exocrine cells. However, in the adult, GATA4 is detectable in subsets of α- and β-cells.[125] GATA6 expression overlaps GATA4 expression until E14.5–E15.5, when it then becomes restricted to the cells of the endocrine pancreas and duct epithelium.[125] Transgenic mouse models in which GATA6 targets are repressed result in pancreatic agenesis. This is not the case when GATA4 targets are repressed suggesting GATA6 is critical for pancreatic development, whereas GATA4 is dispensable.[125]

Transcription factors promoting acinar cell differentiation

Nuclear Receptor 5a2 (Nr5a2)

NR5A2, a member of the orphan nuclear receptor family, is critical for embryogenesis as germ line *Nr5a2* deletion results in early embryonic lethality.[126] NR5A2 maintains expression of embryonic stem cell identity genes such as *Oct4* and *Nanog*[127] as well as genes involved in self-renewal and growth.[128] Temporal and spatial expression of *Nr5a2* determines its function, suggesting that NR5A2 function may link early endoderm development and endoderm differentiation.[129] NR5A2 is expressed in the gastrointestinal endoderm and is required for the expansion of pancreatic epithelium. In pancreatic progenitor cells, NR5A2 promotes expression of genes involved in hepatic and pancreatic maturation.[130,131] In zebrafish, null *Nr5a2* mutations have no exocrine pancreas while development of the endocrine pancreas remains largely unaffected.[129] In the mature pancreas, NR5A2 regulates the expression of other protein complexes driving the assembly of digestive enzymes.

Hairy and enhancer of split-1 (Hes1)

Hes1 is one of seven *Hes* gene family members, which encode proteins that heterodimerize with other bHLH proteins, normally leading to target gene repression. HES1 is the key target of NOTCH signaling in pancreatic development. In the absence of NOTCH, RBPJ (also known as recombining binding protein Suppressor of Hairless) inhibits HES1 expression. When NOTCH signaling is activated, HES1 is expressed and represses expression of NOTCH ligands *Dll1* and *Jagged1* and the endocrine TF, *Ngn3*.[115] In pancreatic progenitor cells, HES1 expression inhibits the expression of *Ptf1a* and *Ngn3*. In *Hes1⁻ʹ⁻* mice, ectopic pancreatic tissue is found in the extrahepatic biliary system, posterior stomach, and proximal duodenum. A conversion of the biliary system into pancreatic tissue has also been noted in *Hes1*-deficient mice, suggesting HES1 restricts pancreatic differentiation from the endoderm.[132] Subsequent studies showed that HES1 and SOX17 are part of a complex regulatory network that controls organ cell fate commitment of pancreatobiliary progenitors in the ventral foregut, by restricting ectopic pancreatic development.[133]

MIST1

MIST1 is a member of the bHLH family of TFs expressed at high amounts in mature acinar cells. Activated shortly after PTF1a in a subset of pancreatic cells at E10.5, it is unclear whether MIST1 marks only potential acinar cells or MPCs. By E14.5, MIST1 clearly marks committed acinar cells and is excluded from duct cells through the rest of development and in the adult pancreas.[134] Deletion of *Mist1* leads to loss of acinar cell organization and incomplete maturation of acinar cells.[134] While still functional, *Mist1*−/− cells are more likely to undergo acinar-to-duct cell metaplasia, and these animals are more sensitive to pancreatic injury.[135]

Several targets of MIST1 transcriptional activity, including *Cx32*, *Rab3d*, and *Atp2c2*, have been identified.[30,136–139] However, it is believed that MIST1 does not initiate the transcription of these genes but rather functions to enhance expression of genes required for regulated exocytosis.[140] Therefore, MIST1 has been described as a scaling factor. Importantly, reactivation of MIST1 in adult *Mist1*−/− tissue restores expression and acinar cell organization, suggesting this is a primary role for MIST1 function.[141] The regulation of MIST1 is less understood. While it is clearly downstream of PTF1a and PDX1 in the pancreas, MIST1 is expressed in other serous exocrine tissues, including salivary glands and chief cells,[142] and no studies have shown direct targeting of MIST1 by PTF1a or PDX1. However, both NR5A2 and X-box binding protein 1 (XBP1) appear to directly regulate MIST1.[141,143]

X-box binding protein 1

X-box binding protein 1 (XBP1) is a bHLH TF and a key mediator of the unfolded protein response (UPR) pathway, activated under conditions in which protein folding capacity of the cells is adversely affected. In nonstressed cells, XBP1 shows limited transcriptional ability. Upon activation of the UPR, inositol-requiring enzyme 1 (IRE1) becomes active and deletes a 26-bp sequence from the *Xbp1* mRNA, producing spliced *Xbp1* transcript that encodes a TF.[144,145] sXbp1 targets genes involved in ER-associated degradation and increased protein folding as well as promoting apoptosis depending on the magnitude and length of the cell stress.

Likely due to the heavy protein folding capacity required in the pancreas, sXBP1 is present and functional even under physiological conditions.[146] Targeted deletion of *Xbp1* during development results in embryonic lethality.[147] Liver-specific rescue allows mice to survive for a short time postnatally to observe the phenotypic abnormalities that exist in other secretory organs. In the pancreas, *Xbp1* deletion leads to incomplete development of acinar cells and as a result undergo apoptosis during development.[148] Interestingly, XBP1 function is also required for maintenance of the acinar cell phenotype as induced deletion in adult cells leads to widespread apoptosis and regeneration specifically from acinar cells that escaped cre-mediated recombination.[149] It appears that XBP1 targets several important factors required for acinar cell differentiation including *Nr5a2* and *Mist1*.

Signaling Pathways Involved in Pancreatic Development

While many signaling pathways have been identified to regulate pancreatic development, we have focused on a few key pathways that have been identified in many studies.

β-catenin/Wnt

Wingless-related integration site (Wnt) signaling controls many embryonic developmental processes including body axis patterning, cell fate specification, and cell proliferation.[150–152] Wnt signaling encompasses three different signaling pathways including the canonical or β-catenin-mediated pathway and two noncanonical pathways that involve planar cell polarity and calcium. Signaling through the canonical Wnt pathway regulates expression of key mediators in pancreatic development,[153,154] while the noncanonical Wnt pathways regulate cell shape and intracellular calcium.[155] In pancreatic development, several WNT ligands are expressed including WNT2b, WNT4, WNT5a, and WNT7b.[156] Binding of WNT ligands to Frizzled (Fzd) receptors leads to Fzd interaction with Dishevelled (Dvl;[157]), at which point the WNT signaling pathway diverges into one of the three pathways mentioned above.[158,159] Activation of the canonical pathway leads to the dissociation of β-catenin from the APC/Axin/GSK3β complex, leading to β-catenin's translocation to the nucleus where it interacts with lymphoid enhancer factor (Lef-1)/T-cell factor (TCF) TFs to affect gene expression.[160–162]

The timing of β-catenin expression and activation of Wnt signaling is critical in pancreatic development. Typically, WNT signaling maintains a progenitor-like state, thereby allowing expansion of the tissue prior to differentiation.[154,155,163] Forced expression of WNT signaling prior to pancreatic specification prevents liver and pancreas development.[156,164]

Detectable amounts of nuclear β-catenin are observed as early as E12.5–E14.5 in the pancreas, prior to the secondary transition.[156] Activated β-catenin is critical during this time period as mice lacking β-catenin have smaller pancreata by E16.5.[153] Targeted deletion of β-catenin results in the loss of exocrine pancreatic tissue along with increased tubular structures, increased parenchymal fibrosis, and inflammatory infiltrate. The smaller pancreatic size in β-catenin null animals is due to a decrease in proliferation of PTF1A-expressing progenitor cells and most likely not due to a loss of adherence junctions. These studies indicate that β-catenin is initially required for expansion of the pancreatic epithelium prior to differentiation as well as for terminal differentiation of acinar cells. Although the exocrine pancreas was abolished, a loss of Wnt signaling did not disrupt the development of fully functional islets.[153]

Retinoic acid

Retinoic acid (RA) is generated through the metabolism of vitamin A by aldehyde dehydrogenases enzymes (ALDHs; [165,166]). RA is a ligand for the nuclear hormone receptor—retinoic acid receptor (RAR)—which exists as a DNA-bound repressor complex in the absence of RA. Heterodimers of RAR and retinoid X receptors (RXR) change conformation upon binding of RA to RAR, allowing recruitment of additional transcriptional regulators, thereby altering the expression of RAR-associated genes.[167]

In zebrafish, blocking RA signaling results in undetectable expression levels of endocrine (*insulin*) or exocrine (*trypsin*) markers and specifies initiation of pancreatic progenitors upstream of *Pdx1*.[168] In *Xenopus*, pharmacological and genetic approaches to block RA signaling diminished *Pdx1* expression only in the dorsal pancreas, suggesting that RA signaling might be required only for endocrine pancreatic development in *Xenopus*. However, RA inhibition at later developmental stages blocked expression of differentiated exocrine markers but retained Pdx1. Therefore, in *Xenopus*, RA signaling is not required for specification of ventral pancreas but is necessary for later exocrine cell differentiation.[169]

In mice, retinaldehyde dehydrogenase 2 (*Raldh2*) is expressed between E8.75 and E12.5 in the dorsolateral mesenchyme that is in contact with the dorsal pancreatic bud. RA is synthesized in dorsolateral mesodermal cells, and the developing tissues responsiveness to RA follows a general dorsal to ventral gradient. Targeted deletion of *Raldh2* in mice diminishes *Pdx1* and *Ptf1a* expression in the presumptive dorsal pancreatic bud, while ventral pancreatic development is largely unaffected. Absence of *Raldh2* during pancreatic development also decreases HB9 and ISL expression in the dorsal mesenchyme, supporting

a mechanism by which dorsal bud agenesis may be due to deficient mesodermal signaling.[79]

Hippo signaling

Hippo (or Salvador/Warts/Hippo (SWH)) signaling regulates organ size through controlling cell proliferation and apoptosis. Hippo signaling involves activation of mammalian Ste20-like kinases 1/2 (Mst1/2), also known as Hippo in *drosophila*, and Save family WW domain-containing protein 1 (SAV1), also known as Salvador in *drosophila*, complex. This leads to phosphorylation of large tumor suppressor 1/2 (LATS1/2). LATS1/2 maintains Yes-associated protein (YAP) and transcriptional coactivator with PDZ-binding motif (TAZ) in a cytoplasmic localization and inactive conformation. Therefore, upon LATS1/2 phosphorylation, dephosphorylated YAP and TAZ translocate to the nucleus where they interact with TAE domain family member (TEAD) 1–4 and alter gene expression that promotes proliferation and inhibits apoptosis.[170] Pancreata of *Mst1/2* double knockout mice have a decreased pancreatic mass due to a failure in exocrine cell maturation.[171] Additionally, overexpressing YAP in later pancreatic development expands the duct network while decreasing the number of acinar and endocrine cells.[171]

NOTCH signaling

The NOTCH pathway plays a role in dictating different developmental fates for closely associated cells. Since duct, exocrine, and endocrine cells arise from common epithelial compartments, NOTCH signaling has been suggested as a mechanism to ensure limited differentiation of endocrine cells. In neuronal development, NOTCH signaling triggers *lateral inhibition*, which prevents adjacent cells from having the same phenotype.[172] In pancreatic development, cells expressing NGN3 are destined to become endocrine cells, and NGN3-expressing cells have enhanced expression of NOTCH ligands, Delta (Dll), Jagged, and Serrated. NOTCH signaling is triggered by interaction of these ligands with NOTCH receptors on adjacent cells[173] resulting in receptor cleavage and release of the NOTCH intracellular domain (NICD). Translocation of NICD to the nucleus leads to modulation of transcriptional regulators *mindbomb* and RBPJ,[174] which enhances expression of HES1, a repressor of *Ngn3*. Therefore, lateral inhibition through increased HES1 has been suggested as a way of preventing widespread endocrine differentiation. In support of this theory, deletion of *Dll* or *Hes1* in the pancreas during development increased endocrine differentiation at the expense of acinar cells.[175,176] However, lateral inhibition cannot explain many of the effects of NOTCH signaling in pancreatic development, nor the final arrangement of exocrine versus duct versus endocrine cells. NOTCH signaling is time dependent, and forced early expression of NOTCH signaling also reduces the acinar cell population, implying a role for NOTCH in the expansion of this population of cells before exocrine differentiation.[177]

RBPJ activity can be modulated by NICD when it is present in the nucleus. RBPJ is a key component of the PTF1 complex, which includes PTF1a and the ubiquitous bHLH, transcription factor 3 (TCF3). During early pancreatic development, the PTF1 complex includes RBPJ, which allows for acinar cell expansion but limits differentiation.[178] When RBPJ-like (RBPJL) replaces RBPJ in the PTF1 complex, the role of this complex switches from one that maintains a specified acinar cell fate to a complex that drives and maintains differentiated acinar cells.[98] This RBPJ to RBPJL switch is consistent with targeted pancreatic RBPJL deletion, which decreases expression of genes involved in differentiated acinar cell processes such as digestive enzyme production, regulated exocytosis, or mitochondrial metabolism.[179]

NOTCH signaling has also been implicated in trunk cell (endocrine/duct progenitor cells) specification,[26] duct cell differentiation, and endocrine lineage commitment.[50,169,180] Centroacinar cells and a small percentage of duct cells seem to maintain HES1 expression[181] suggesting that these cells might make up a pancreatic progenitor/stem cell population.

Conclusion

The plethora of genetically modified mouse models that allow for targeted and inducible deletion or activation of genes has greatly improved our understanding of exocrine pancreas development and provided insight into disease processes in which the developmental processes are not activated properly (e.g., pancreatic agenesis or insufficiency) or reactivated (e.g., pancreatic cancer). Understanding how these various signaling pathways and transcriptional factors work in concert to cause acinar cell differentiation is still an active area of discussion and many of the findings still need to be translated to human development. In particular, research will be needed to understand the impact of environmental factors (such as diet) on exocrine pancreatic disorders, both during development and diseases such as pancreatitis. Likely, these studies will identify the importance of epigenetic regulators or events that mediate the environmental impact on pancreatic development.

References

1. Arnes L, Sussel L. Epigenetic modifications and long noncoding RNAs influence pancreas development and function. Trends Genet 2015; 31:290–299.
2. Carolan PJ, Melton DA. New findings in pancreatic and intestinal endocrine development to advance regenerative medicine. Curr Opin Endocrinol Diabetes Obes 2013; 20:1–7.
3. Gittes GK. Developmental biology of the pancreas: a comprehensive review. Dev Biol 2009; 326:4–35.
4. Kadison A, Kim J, Maldonado T, Crisera C, Prasadan K, Manna P, Preuett B, Hembree M, Longaker M, Gittes G. Retinoid signaling directs secondary lineage selection in pancreatic organogenesis. J Pediatr Surg 2001; 36:1150–1156.
5. McCracken KW, Wells JM. Molecular pathways controlling pancreas induction. Semin Cell Dev Biol 2012; 23:656–662.
6. Puri S, Hebrok M. Cellular plasticity within the pancreas-lessons learned from development. Dev Cell 2010; 18:342–356.
7. Serup P. Signaling pathways regulating murine pancreatic development. Semin Cell Dev Biol 2012; 23:663–672.
8. Gittes GK, Rutter WJ. Onset of cell-specific gene expression in the developing mouse pancreas. Proc Natl Acad Sci U S A 1992; 89:1128–1132.
9. Kallman F, Grobstein C. Fine structure of differentiating mouse pancreatic exocrine cells in transfilter culture. J Cell Biol 1964; 20:399–413.
10. Munger BL. Polarization optical properties of the pancreatic acinar cell of the mouse. J Biophys Biochem Cytol 1958; 4:177–186.
11. Pictet RL, Clark WR, Williams RH, Rutter WJ. An ultrastructural analysis of the developing embryonic pancreas. Dev Biol 1972; 29:436–467.
12. Kim SK, Hebrok M, Melton DA. Notochord to endoderm signaling is required for pancreas development. Development 1997; 124:4243–4252.
13. White JJ, Roberts ZN, Gest TR, Beale EG. Pancreas divisum: a common developmental variant that deserves attention in preclinical medical education. Clin Anat 2014; 27:1038–1045.
14. Gonoi W, Akai H, Hagiwara K, Akahane M, Hayashi N, Maeda E, Yoshikawa T, Tada M, Uno K, Ohtsu H, Koike K, Ohtomo K. Pancreas divisum as a predisposing factor for chronic and recurrent idiopathic pancreatitis: initial in vivo survey. Gut 2011; 60:1103–1108.
15. Moffatt DC, Cote GA, Avula H, Watkins JL, McHenry L, Sherman S, Lehman GA, Fogel EL. Risk factors for ERCP-related complications in patients with pancreas divisum: a retrospective study. Gastrointest Endosc 2011; 73:963–970.
16. Gu G, Dubauskaite J, Melton DA. Direct evidence for the pancreatic lineage: NGN3+ cells are islet progenitors and are distinct from duct progenitors. Development 2002; 129:2447–2457.
17. Stanger BZ, Tanaka AJ, Melton DA. Organ size is limited by the number of embryonic progenitor cells in the pancreas but not the liver. Nature 2007; 445:886–891.
18. Kesavan G, Sand FW, Greiner TU, Johansson JK, Kobberup S, Wu X, Brakebusch C, Semb H. Cdc42-mediated tubulogenesis controls cell specification. Cell 2009; 139:791–801.
19. Shih HP, Panlasigui D, Cirulli V, Sander M. ECM signaling regulates collective cellular dynamics to control pancreas branching morphogenesis. Cell Rep 2016; 14:169–179.

20. Zhou Q, Law AC, Rajagopal J, Anderson WJ, Gray PA, Melton DA. A multipotent progenitor domain guides pancreatic organogenesis. Dev Cell 2007; 13:103–114.

21. Pan FC, Bankaitis ED, Boyer D, Xu X, Van de Casteele M, Magnuson MA, Heimberg H, Wright CV. Spatiotemporal patterns of multipotentiality in Ptf1a-expressing cells during pancreas organogenesis and injury-induced facultative restoration. Development 2013; 140:751–764.

22. Houbracken I, de Waele E, Lardon J, Ling Z, Heimberg H, Rooman I, Bouwens L. Lineage tracing evidence for transdifferentiation of acinar to duct cells and plasticity of human pancreas. Gastroenterology 2011; 141:731–741.

23. Criscimanna A, Speicher JA, Houshmand G, Shiota C, Prasadan K, Ji B, Logsdon CD, Gittes GK, Esni F. Duct cells contribute to regeneration of endocrine and acinar cells following pancreatic damage in adult mice. Gastroenterology 2011; 141:1451–1462.

24. Nyeng P, Heilmann S, Löf-Öhlin ZM, Pettersson NF, Hermann FM, Reynolds AB, Semb H. p120ctn-mediated organ patterning precedes and determines pancreatic progenitor fate. Dev Cell 2019; 49:31–47.

25. Hendley AM, Provost E, Bailey JM, Wang YJ, Cleveland MH, Blake D, Bittman RW, Roeser JC, Maitra A, Reynolds AB, Leach SD. p120 catenin is required for normal tubulogenesis but not epithelial integrity in developing mouse pancreas. Dev Biol 2015; 399:41–53.

26. Afelik S, Qu X, Hasrouni E, Bukys MA, Deering T, Nieuwoudt S, Rogers W, Macdonald RJ, Jensen J. Notch-mediated patterning and cell fate allocation of pancreatic progenitor cells. Development 2012; 139:1744–1753.

27. Kopinke D, Brailsford M, Shea JE, Leavitt R, Scaife CL, Murtaugh LC. Lineage tracing reveals the dynamic contribution of Hes1+ cells to the developing and adult pancreas. Development 2011; 138:431–441.

28. Cole L, Anderson M, Antin PB, Limesand SW. One process for pancreatic beta-cell coalescence into islets involves an epithelial-mesenchymal transition. J Endocrinol 2009; 203:19–31.

29. Rukstalis JM, Habener JF. Snail2, a mediator of epithelial-mesenchymal transitions, expressed in progenitor cells of the developing endocrine pancreas. Gene Expr Patterns 2007; 7:471–479.

30. Johnson CL, Kowalik AS, Rajakumar N, Pin CL. Mist1 is necessary for the establishment of granule organization in serous exocrine cells of the gastrointestinal tract. Mech Dev 2004; 121:261–272.

31. Herrera PL, Nepote V, Delacour A. Pancreatic cell lineage analyses in mice. Endocrine 2002; 19:267–278.

32. Herrera PL. Defining the cell lineages of the islets of Langerhans using transgenic mice. Int J Dev Biol 2002; 46:97–103.

33. Habbe N, Shi G, Meguid RA, Fendrich V, Esni F, Chen H, Feldmann G, Stoffers DA, Konieczny SF, Leach SD, Maitra A. Spontaneous induction of murine pancreatic intraepithelial neoplasia (mPanIN) by acinar cell targeting of oncogenic Kras in adult mice. Proc Natl Acad Sci U S A 2008; 105:18913–18918.

34. Solar M, Cardalda C, Houbracken I, Martin M, Maestro MA, De Medts N, Xu X, Grau V, Heimberg H, Bouwens L, Ferrer J. Pancreatic exocrine duct cells give rise to insulin-producing beta cells during embryogenesis but not after birth. Dev Cell 2009; 17:849–860.

35. Schaffer AE, Freude KK, Nelson SB, Sander M. Nkx6 transcription factors and Ptf1a function as antagonistic lineage determinants in multipotent pancreatic progenitors. Dev Cell 2010; 18:1022–1029.

36. Berg T, Rountree CB, Lee L, Estrada J, Sala FG, Choe A, Veltmaat JM, De Langhe S, Lee R, Tsukamoto H, Crooks GM, Bellusci S, et al. Fibroblast growth factor 10 is critical for liver growth during embryogenesis and controls hepatoblast survival via beta-catenin activation. Hepatology 2007; 46:1187–1197.

37. Deutsch G, Jung J, Zheng M, Lora J, Zaret KS. A bipotential precursor population for pancreas and liver within the embryonic endoderm. Development 2001; 128:871–881.

38. Norgaard GA, Jensen JN, Jensen J. FGF10 signaling maintains the pancreatic progenitor cell state revealing a novel role of Notch in organ development. Dev Biol 2003; 264:323–338.

39. Zaret KS, Watts J, Xu J, Wandzioch E, Smale ST, Sekiya T. Pioneer factors, genetic competence, and inductive signaling: programming liver and pancreas progenitors from the endoderm. Cold Spring Harb Symp Quant Biol 2008; 73:119–126.

40. Li H, Arber S, Jessell TM, Edlund H. Selective agenesis of the dorsal pancreas in mice lacking homeobox gene Hlxb9. Nat Genet 1999; 23:67–70.

41. Yoshitomi H, Zaret KS. Endothelial cell interactions initiate dorsal pancreas development by selectively inducing the transcription factor Ptf1a. Development 2004; 131:807–817.

42. Rossi JM, Dunn NR, Hogan BL, Zaret KS. Distinct mesodermal signals, including BMPs from the septum transversum mesenchyme, are required in combination for hepatogenesis from the endoderm. Genes Dev 2001; 15:1998–2009.

43. Fernandez-Barrena MG, Pin CL Chromatin remodeling and epigenetic reprogramming in chronic disease and cancer in the liver and pancreas. In: Fernandez-Zapico ME, ed. Chromatin signaling and diseases. Boston: Academic Press, 2016:365–385.

44. Golosow N, Grobstein C. Epitheliomesenchymal interaction in pancreatic morphogenesis. Dev Biol 1962; 4:242–255.

45. Miralles F, Battelino T, Czernichow P, Scharfmann R. TGF-beta plays a key role in morphogenesis of the pancreatic islets of Langerhans by controlling the activity of the matrix metalloproteinase MMP-2. J Cell Biol 1998; 143:827–836.

46. Wessells NK, Cohen JH. Early pancreas organogenesis: morphogenesis, tissue interactions, and mass effects. Dev Biol 1967; 15:237–270.

47. Landsman L, Nijagal A, Whitchurch TJ, Vanderlaan RL, Zimmer WE, Mackenzie TC, Hebrok M. Pancreatic mesenchyme regulates epithelial organogenesis throughout development. PLoS Biol 2011; 9:e1001143.

48. Asayesh A, Sharpe J, Watson RP, Heckscher-Sorensen J, Hastie ND, Hill RE, Ahlgren U. Spleen versus pancreas: strict control of organ interrelationship revealed by analyses of Bapx1-/- mice. Genes Dev 2006; 20:2208–2213.

49. Zertal-Zidani S, Bounacer A, Scharfmann R. Regulation of pancreatic endocrine cell differentiation by sulphated proteoglycans. Diabetologia 2007; 50:585–595.

50. Ahnfelt-Ronne J, Jorgensen MC, Klinck R, Jensen JN, Fuchtbauer EM, Deering T, MacDonald RJ, Wright CV, Madsen OD, Serup P. Ptf1a-mediated control of Dll1 reveals an alternative to the lateral inhibition mechanism. Development 2012; 139:33–45.

51. Gradwohl G, Dierich A, LeMeur M, Guillemot F. neurogenin3 is required for the development of the four endocrine cell lineages of the pancreas. Proc Natl Acad Sci U S A 2000; 97:1607–1611.

52. Krapp A, Knofler M, Ledermann B, Burki K, Berney C, Zoerkler N, Hagenbuchle O, Wellauer PK. The bHLH protein PTF1-p48 is essential for the formation of the exocrine and the correct spatial organization of the endocrine pancreas. Genes Dev 1998; 12:3752–3763.

53. Crisera CA, Maldonado TS, Kadison AS, Li M, Alkasab SL, Longaker MT, Gittes GK. Transforming growth factor-beta 1 in the developing mouse pancreas: a potential regulator of exocrine differentiation. Differentiation 2000; 65:255–259.

54. Li Z, Manna P, Kobayashi H, Spilde T, Bhatia A, Preuett B, Prasadan K, Hembree M, Gittes GK. Multifaceted pancreatic mesenchymal control of epithelial lineage selection. Dev Biol 2004; 269:252–263.

55. Gittes GK, Galante PE, Hanahan D, Rutter WJ, Debase HT. Lineage-specific morphogenesis in the developing pancreas: role of mesenchymal factors. Development 1996; 122:439–447.

56. Cirulli V, Crisa L, Beattie GM, Mally MI, Lopez AD, Fannon A, Ptasznik A, Inverardi L, Ricordi C, Deerinck T, Ellisman M, Reisfeld RA, et al. KSA antigen Ep-CAM mediates cell-cell adhesion of pancreatic epithelial cells: morphoregulatory roles in pancreatic islet development. J Cell Biol 1998; 140:1519–1534.

57. Sanvito F, Herrera PL, Huarte J, Nichols A, Montesano R, Orci L, Vassalli JD. TGF-beta 1 influences the relative development of the exocrine and endocrine pancreas in vitro. Development 1994; 120:3451–3462.

58. Miralles F, Czernichow P, Scharfmann R. Follistatin regulates the relative proportions of endocrine versus exocrine tissue during pancreatic development. Development 1998; 125:1017–1024.

59. Tulachan SS, Tei E, Hembree M, Crisera C, Prasadan K, Koizumi M, Shah S, Guo P, Bottinger E, Gittes GK. TGF-beta isoform signaling regulates secondary transition and mesenchymal-induced endocrine development in the embryonic mouse pancreas. Dev Biol 2007; 305:508–521.

60. Bottinger EP, Jakubczak JL, Roberts IS, Mumy M, Hemmati P, Bagnall K, Merlino G, Wakefield LM. Expression of a dominant-negative mutant TGF-beta type II receptor in transgenic mice reveals essential roles for TGF-beta in regulation of growth and differentiation in the exocrine pancreas. EMBO J 1997; 16:2621–2633.

61. Dichmann DS, Miller CP, Jensen J, Scott Heller R, Serup P. Expression and misexpression of members of the FGF and TGFbeta families of growth factors in the developing mouse pancreas. Dev Dyn 2003; 226:663–674.

62. Elghazi L, Cras-Meneur C, Czernichow P, Scharfmann R. Role for FGFR2IIIb-mediated signals in controlling pancreatic endocrine progenitor cell proliferation. Proc Natl Acad Sci U S A 2002; 99:3884–3889.

63. Ye F, Duvillie B, Scharfmann R. Fibroblast growth factors 7 and 10 are expressed in the human embryonic pancreatic mesenchyme and promote the proliferation of embryonic pancreatic epithelial cells. Diabetologia 2005; 48:277–281.

64. Bhushan A, Itoh N, Kato S, Thiery JP, Czernichow P, Bellusci S, Scharfmann R. Fgf10 is essential for maintaining the proliferative capacity of epithelial progenitor cells during early pancreatic organogenesis. Development 2001; 128:5109–5117.

65. Risau W, Flamme I. Vasculogenesis. Vasculogenesis. Annu Rev Cell Dev Biol 1995; 11:73–91.

66. Esni F, Johansson BR, Radice GL, Semb H. Dorsal pancreas agenesis in N-cadherin- deficient mice. Dev Biol 2001; 238:202–212.

67. Edsbagge J, Johansson JK, Esni F, Luo Y, Radice GL, Semb H. Vascular function and sphingosine-1-phosphate regulate development of the dorsal pancreatic mesenchyme. Development 2005; 132:1085–1092.

68. Shah SR, Esni F, Jakub A, Paredes J, Lath N, Malek M, Potoka DA, Prasadan K, Mastroberardino PG, Shiota C, Guo P, Miller KA, et al. Embryonic mouse blood flow and oxygen correlate with early pancreatic differentiation. Dev Biol 2011; 349:342–349.

69. Serrill JD, Sander M, Shih HP. Pancreatic exocrine tissue architecture and integrity are maintained by E-cadherin during postnatal development. Sci Rep 2018; 8:13451.

70. Lammert E, Cleaver O, Melton D. Induction of pancreatic differentiation by signals from blood vessels. Science 2001; 294:564–567.

71. Magenheim J, Ilovich O, Lazarus A, Klochendler A, Ziv O, Werman R, Hija A, Cleaver O, Mishani E, Keshet E, Dor Y. Blood vessels restrain pancreas branching, differentiation and growth. Development 2011; 138:4743–4752.

72. Hebrok M, Kim SK, St Jacques B, McMahon AP, Melton DA. Regulation of pancreas development by hedgehog signaling. Development 2000; 127:4905–4913.

73. Kawahira H, Scheel DW, Smith SB, German MS, Hebrok M. Hedgehog signaling regulates expansion of pancreatic epithelial cells. Dev Biol 2005; 280:111–121.

74. Zhou X, Liu Z, Jang F, Xiang C, Li Y, He Y. Autocrine sonic hedgehog attenuates inflammation in cerulein-induced acute pancreatitis in mice via upregulation of IL-10. PLoS One 2012; 7:e44121.

75. Roy S, Qiao T, Wolff C, Ingham PW. Hedgehog signaling pathway is essential for pancreas specification in the zebrafish embryo. Curr Biol 2001; 11:1358–1363.

76. Jung IH, Jung DE, Park YN, Song SY, Park SW. Aberrant hedgehog ligands induce progressive pancreatic fibrosis by paracrine activation of myofibroblasts and ductular cells in transgenic zebrafish. PLoS One 2011; 6:e27941.

77. Hebrok M, Kim SK, Melton DA. Notochord repression of endodermal sonic hedgehog permits pancreas development. Genes Dev 1998; 12:1705–1713.

78. Bort R, Martinez-Barbera JP, Beddington RS, Zaret KS. Hex homeobox gene-dependent tissue positioning is required for

organogenesis of the ventral pancreas. Development 2004; 131:797–806.

79. Martin M, Gallego-Llamas J, Ribes V, Kedinger M, Niederreither K, Chambon P, Dolle P, Gradwohl G. Dorsal pancreas agenesis in retinoic acid-deficient Raldh2 mutant mice. Dev Biol 2005; 284:399–411.

80. Molotkov A, Molotkova N, Duester G. Retinoic acid generated by Raldh2 in mesoderm is required for mouse dorsal endodermal pancreas development. Dev Dyn 2005; 232:950–957.

81. Murtaugh LC, Leach SD. A case of mistaken identity? Nonductal origins of pancreatic "ductal" cancers. Cancer Cell 2007; 11:211–213.

82. Dassaye R, Naidoo S, Cerf ME. Transcription factor regulation of pancreatic organogenesis, differentiation and maturation. Islets 2016; 8:13–34.

83. Offield MF, Jetton TL, Labosky PA, Ray M, Stein RW, Magnuson MA, Hogan BL, Wright CV. PDX-1 is required for pancreatic outgrowth and differentiation of the rostral duodenum. Development 1996; 122:983–995.

84. Guz Y, Montminy MR, Stein R, Leonard J, Gamer LW, Wright CV, Teitelman G. Expression of murine STF-1, a putative insulin gene transcription factor, in beta cells of pancreas, duodenal epithelium and pancreatic exocrine and endocrine progenitors during ontogeny. Development 1995; 121:11–18.

85. Li LC, Qiu WL, Zhang YW, Xu ZR, Xiao YN, Hou C, Lamaoqiezhong, Yu P, Cheng X, Xu CR. Single-cell transcriptomic analyses reveal distinct dorsal/ventral pancreatic programs. EMBO Rep 2018; 19:e46148.

86. Ahlgren U, Jonsson J, Edlund H. The morphogenesis of the pancreatic mesenchyme is uncoupled from that of the pancreatic epithelium in IPF1/PDX1-deficient mice. Development 1996; 122:1409–1416.

87. Jonsson J, Carlsson L, Edlund T, Edlund H. Insulin-promoter-factor 1 is required for pancreas development in mice. Nature 1994; 371:606–609.

88. Roy N, Takeuchi KK, Ruggeri JM, Bailey P, Chang D, Li J, Leonhardt L, Puri S, Hoffman MT, Gao S, Halbrook CJ, Song Y, et al. PDX1 dynamically regulates pancreatic ductal adenocarcinoma initiation and maintenance. Genes Dev 2016; 30:2669–2683.

89. Lottmann H, Vanselow J, Hessabi B, Walther R. The Tet-On system in transgenic mice: inhibition of the mouse pdx-1 gene activity by antisense RNA expression in pancreatic beta-cells. J Mol Med (Berl) 2001; 79:321–328.

90. Stoffers DA, Zinkin NT, Stanojevic V, Clarke WL, Habener JF. Pancreatic agenesis attributable to a single nucleotide deletion in the human IPF1 gene coding sequence. Nat Genet 1997; 15:106–110.

91. Sachdeva MM, Claiborn KC, Khoo C, Yang J, Groff DN, Mirmira RG, Stoffers DA. Pdx1 (MODY4) regulates pancreatic beta cell susceptibility to ER stress. Proc Natl Acad Sci U S A 2009; 106:19090–19095.

92. Stoffers DA, Ferrer J, Clarke WL, Habener JF. Early-onset type-II diabetes mellitus (MODY4) linked to IPF1. Nat Genet 1997; 17:138–139.

93. Cockell M, Stevenson BJ, Strubin M, Hagenbuchle O, Wellauer PK. Identification of a cell-specific DNA-binding activity that interacts with a transcriptional activator of genes expressed in the acinar pancreas. Mol Cell Biol 1989; 9:2464–2476.

94. Petrucco S, Wellauer PK, Hagenbuchle O. The DNA-binding activity of transcription factor PTF1 parallels the synthesis of pancreas-specific mRNAs during mouse development. Mol Cell Biol 1990; 10:254–264.

95. Afelik S, Chen Y, Pieler T. Combined ectopic expression of Pdx1 and Ptf1a/p48 results in the stable conversion of posterior endoderm into endocrine and exocrine pancreatic tissue. Genes Dev 2006; 20:1441–1446.

96. Krapp A, Knofler M, Frutiger S, Hughes GJ, Hagenbuchle O, Wellauer PK. The p48 DNA-binding subunit of transcription factor PTF1 is a new exocrine pancreas-specific basic helix-loop-helix protein. Embo J 1996; 15:4317–4329.

97. Esni F, Ghosh B, Biankin AV, Lin JW, Albert MA, Yu X, MacDonald RJ, Civin CI, Real FX, Pack MA, Ball DW, Leach SD. Notch inhibits Ptf1 function and acinar cell differentiation in developing mouse and zebrafish pancreas. Development 2004; 131:4213–4224.

98. Masui T, Long Q, Beres TM, Magnuson MA, MacDonald RJ. Early pancreatic development requires the vertebrate Suppressor of Hairless (RBPJ) in the PTF1 bHLH complex. Genes Dev 2007; 21:2629–2643.

99. Beres TM, Masui T, Swift GH, Shi L, Henke RM, MacDonald RJ. PTF1 is an organ-specific and Notch-independent basic helix-loop-helix complex containing the mammalian Suppressor of Hairless (RBP-J) or its paralogue, RBP-L. Mol Cell Biol 2006; 26:117–130.

100. Sellick GS, Barker KT, Stolte-Dijkstra I, Fleischmann C, Coleman RJ, Garrett C, Gloyn AL, Edghill EL, Hattersley AT, Wellauer PK, Goodwin G, Houlston RS. Mutations in PTF1A cause pancreatic and cerebellar agenesis. Nat Genet 2004; 36:1301–1305.

101. Kawaguchi Y, Cooper B, Gannon M, Ray M, MacDonald RJ, Wright CV. The role of the transcriptional regulator Ptf1a in converting intestinal to pancreatic progenitors. Nat Genet 2002; 32:128–134.

102. Rovira M, Delaspre F, Massumi M, Serra SA, Valverde MA, Lloreta J, Dufresne M, Payre B, Konieczny SF, Savatier P, Real FX, Skoudy A. Murine embryonic stem cell-derived pancreatic acinar cells recapitulate features of early pancreatic differentiation. Gastroenterology 2008; 135:1301–1310.

103. Zhuang S, Zhang Q, Zhuang T, Evans SM, Liang X, Sun Y. Expression of Isl1 during mouse development. Gene Expr Patterns 2013; 13:407–412.

104. Ahlgren U, Pfaff SL, Jessell TM, Edlund T, Edlund H. Independent requirement for ISL1 in formation of pancreatic mesenchyme and islet cells. Nature 1997; 385:257–260.

105. Miyazaki S, Tashiro F, Fujikura J, Yamato E, Miyazaki J. Acinar-to-ductal metaplasia induced by adenovirus-mediated pancreatic expression of Isl1. PLoS One 2012; 7:e47536.

106. Poll AV, Pierreux CE, Lokmane L, Haumaitre C, Achouri Y, Jacquemin P, Rousseau GG, Cereghini S, Lemaigre FP. A vHNF1/TCF2-HNF6 cascade regulates the transcription factor network that controls generation of pancreatic precursor cells. Diabetes 2006; 55:61–69.

107. Haumaitre C, Barbacci E, Jenny M, Ott MO, Gradwohl G, Cereghini S. Lack of TCF2/vHNF1 in mice leads to pancreas agenesis. Proc Natl Acad Sci U S A 2005; 102:1490–1495.

108. De Vas MG, Kopp JL, Heliot C, Sander M, Cereghini S, Haumaitre C. Hnf1b controls pancreas morphogenesis and the generation of Ngn3+ endocrine progenitors. Development 2015; 142:871–882.

109. Bellanne-Chantelot C, Chauveau D, Gautier JF, Dubois-Laforgue D, Clauin S, Beaufils S, Wilhelm JM, Boitard C, Noel LH, Velho G, Timsit J. Clinical spectrum associated with hepatocyte nuclear factor-1beta mutations. Ann Intern Med 2004; 140:510–517.

110. Iwasaki N, Tsurumi M, Asai K, Shimuzu W, Watanabe A, Ogata M, Takizawa M, Ide R, Yamamoto T, Saito K. Pancreatic developmental defect evaluated by celiac artery angiography in a patient with MODY5. Hum Genome Var 2016; 3:16022.

111. Vanhorenbeeck V, Jenny M, Cornut JF, Gradwohl G, Lemaigre FP, Rousseau GG, Jacquemin P. Role of the Onecut transcription factors in pancreas morphogenesis and in pancreatic and enteric endocrine differentiation. Dev Biol 2007; 305:685–694.

112. Jacquemin P, Lemaigre FP, Rousseau GG. The Onecut transcription factor HNF-6 (OC-1) is required for timely specification of the pancreas and acts upstream of Pdx-1 in the specification cascade. Dev Biol 2003; 258:105–116.

113. Zhang H, Ables ET, Pope CF, Washington MK, Hipkens S, Means AL, Path G, Seufert J, Costa RH, Leiter AB, Magnuson MA, Gannon M. Multiple, temporal-specific roles for HNF6 in pancreatic endocrine and ductal differentiation. Mech Dev 2009; 126:958–973.

114. Kropp PA, Zhu X, Gannon M. Regulation of the pancreatic exocrine differentiation program and morphogenesis by Onecut 1/Hnf6. Cell Mol Gastroenterol Hepatol 2019; 7:841–856.

115. Shih HP, Seymour PA, Patel NA, Xie R, Wang A, Liu PP, Yeo GW, Magnuson MA, Sander M. A gene regulatory network cooperatively controlled by Pdx1 and Sox9 governs lineage allocation of foregut progenitor cells. Cell Rep 2015; 13:326–336.

116. Seymour PA, Shih HP, Patel NA, Freude KK, Xie R, Lim CJ, Sander M. A Sox9/Fgf feed-forward loop maintains pancreatic organ identity. Development 2012; 139:3363–3372.

117. Seymour PA, Freude KK, Tran MN, Mayes EE, Jensen J, Kist R, Scherer G, Sander M. SOX9 is required for maintenance of the pancreatic progenitor cell pool. Proc Natl Acad Sci U S A 2007; 104:1865–1870.

118. Seymour PA, Freude KK, Dubois CL, Shih HP, Patel NA, Sander M. A dosage-dependent requirement for Sox9 in pancreatic endocrine cell formation. Dev Biol 2008; 323:19–30.

119. Piper K, Ball SG, Keeling JW, Mansoor S, Wilson DI, Hanley NA. Novel SOX9 expression during human pancreas development correlates to abnormalities in Campomelic dysplasia. Mech Dev 2002; 116:223–226.

120. Kopp JL, Dubois CL, Schaffer AE, Hao E, Shih HP, Seymour PA, Ma J, Sander M. Sox9+ ductal cells are multipotent progenitors throughout development but do not produce new endocrine cells in the normal or injured adult pancreas. Development 2011; 138:653–665.

121. Delous M, Yin C, Shin D, Ninov N, Debrito Carten J, Pan L, Ma TP, Farber SA, Moens CB, Stainier DY. Sox9b is a key regulator of pancreaticobiliary ductal system development. PLoS Genet 2012; 8:e1002754.

122. Manfroid I, Ghaye A, Naye F, Detry N, Palm S, Pan L, Ma TP, Huang W, Rovira M, Martial JA, Parsons MJ, Moens CB, et al. Zebrafish sox9b is crucial for hepatopancreatic duct development and pancreatic endocrine cell regeneration. Dev Biol 2012; 366:268–278.

123. Molkentin JD. The zinc finger-containing transcription factors GATA-4, -5, and -6. Ubiquitously expressed regulators of tissue-specific gene expression. J Biol Chem 2000; 275:38949–38952.

124. Fujikura J, Yamato E, Yonemura S, Hosoda K, Masui S, Nakao K, Miyazaki Ji J, Niwa H. Differentiation of embryonic stem cells is induced by GATA factors. Genes Dev 2002; 16:784–789.

125. Decker K, Goldman DC, Grasch CL, Sussel L. Gata6 is an important regulator of mouse pancreas development. Dev Biol 2006; 298:415–429.

126. Labelle-Dumais C, Jacob-Wagner M, Pare JF, Belanger L, Dufort D. Nuclear receptor NR5A2 is required for proper primitive streak morphogenesis. Dev Dyn 2006; 235:3359–3369.

127. Gu P, Goodwin B, Chung AC, Xu X, Wheeler DA, Price RR, Galardi C, Peng L, Latour AM, Koller BH, Gossen J, Kliewer SA, et al. Orphan nuclear receptor LRH-1 is required to maintain Oct4 expression at the epiblast stage of embryonic development. Mol Cell Biol 2005; 25:3492–3505.

128. Benod C, Vinogradova MV, Jouravel N, Kim GE, Fletterick RJ, Sablin EP. Nuclear receptor liver receptor homologue 1 (LRH-1) regulates pancreatic cancer cell growth and proliferation. Proc Natl Acad Sci U S A 2011; 108:16927–16931.

129. Nissim S, Weeks O, Talbot JC, Hedgepeth JW, Wucherpfennig J, Schatzman-Bone S, Swinburne I, Cortes M, Alexa K, Megason S, North TE, Amacher SL, et al. Iterative use of nuclear receptor Nr5a2 regulates multiple stages of liver and pancreas development. Dev Biol 2016; 418:108–123.

130. Rausa FM, Galarneau L, Belanger L, Costa RH. The nuclear receptor fetoprotein transcription factor is coexpressed with its target gene HNF-3beta in the developing murine liver, intestine and pancreas. Mech Dev 1999; 89:185–188.

131. Galarneau L, Pare JF, Allard D, Hamel D, Levesque L, Tugwood JD, Green S, Belanger L. The alpha1-fetoprotein locus is activated by a nuclear receptor of the Drosophila FTZ-F1 family. Mol Cell Biol 1996; 16:3853–3865.

132. Sumazaki R, Shiojiri N, Isoyama S, Masu M, Keino-Masu K, Osawa M, Nakauchi H, Kageyama R, Matsui A. Conversion of biliary system to pancreatic tissue in Hes1-deficient mice. Nat Genet 2004; 36:83–87.

133. Spence JR, Lange AW, Lin SC, Kaestner KH, Lowy AM, Kim I, Whitsett JA, Wells JM. Sox17 regulates organ lineage segregation of ventral foregut progenitor cells. Dev Cell 2009; 17:62–74.

134. Pin CL, Rukstalis JM, Johnson C, Konieczny SF. The bHLH transcription factor Mist1 is required to maintain exocrine pancreas cell organization and acinar cell identity. J Cell Biol 2001; 155:519–530.

135. Kowalik AS, Johnson CL, Chadi SA, Weston JY, Fazio EN, Pin CL. Mice lacking the transcription factor Mist1 exhibit an altered stress response and increased sensitivity to caerulein-induced pancreatitis. Am J Physiol Gastrointest Liver Physiol 2007; 292:G1123–G1132.

136. Tian X, Jin RU, Bredemeyer AJ, Oates EJ, Blazewska KM, McKenna CE, Mills JC. RAB26 and RAB3D are direct transcriptional targets of MIST1 that regulate exocrine granule maturation. Mol Cell Biol 2010; 30:1269–1284.

137. Ramsey VG, Doherty JM, Chen CC, Stappenbeck TS, Konieczny SF, Mills JC. The maturation of mucus-secreting gastric epithelial progenitors into digestive-enzyme secreting zymogenic cells requires Mist1. Development 2007; 134:211–222.

138. Rukstalis JM, Kowalik A, Zhu L, Lidington D, Pin CL, Konieczny SF. Exocrine specific expression of Connexin32 is dependent on the basic helix-loop-helix transcription factor Mist1. J Cell Sci 2003; 116:3315–3325.

139. Fenech MA, Sullivan CM, Ferreira LT, Mehmood R, MacDonald WA, Stathopulos PB, Pin CL. Atp2c2 is transcribed from a unique transcriptional start site in mouse pancreatic acinar cells. J Cell Physiol 2016; 231:2768–2778.

140. Mills JC, Taghert PH. Scaling factors: transcription factors regulating subcellular domains. Bioessays 2011; 34:10–16.

141. Direnzo D, Hess DA, Damsz B, Hallett JE, Marshall B, Goswami C, Liu Y, Deering T, Macdonald RJ, Konieczny SF. Induced Mist1 expression promotes remodeling of mouse pancreatic acinar cells. Gastroenterology 2012; 143:469–480.

142. Pin CL, Bonvissuto AC, Konieczny SF. Mist1 expression is a common link among serous exocrine cells exhibiting regulated exocytosis. Anat Rec 2000; 259:157–167.

143. Acosta-Alvear D, Zhou Y, Blais A, Tsikitis M, Lents NH, Arias C, Lennon CJ, Kluger Y, Dynlacht BD. XBP1 controls diverse cell type- and condition-specific transcriptional regulatory networks. Mol Cell 2007; 27:53–66.

144. Yoshida H, Matsui T, Yamamoto A, Okada T, Mori K. XBP1 mRNA is induced by ATF6 and spliced by IRE1 in response to ER stress to produce a highly active transcription factor. Cell 2001; 107:881–891.

145. Calfon M, Zeng H, Urano F, Till JH, Hubbard SR, Harding HP, Clark SG, Ron D. IRE1 couples endoplasmic reticulum load to secretory capacity by processing the XBP-1 mRNA. Nature 2002; 415:92–96.

146. Kubisch CH, Sans MD, Arumugam T, Ernst SA, Williams JA, Logsdon CD. Early activation of endoplasmic reticulum stress is associated with arginine-induced acute pancreatitis. Am J Physiol Gastrointest Liver Physiol 2006; 291:G238–G245.

147. Huh WJ, Esen E, Geahlen JH, Bredemeyer AJ, Lee AH, Shi G, Konieczny SF, Glimcher LH, Mills JC. XBP1 controls maturation of gastric zymogenic cells by induction of MIST1 and expansion of the rough endoplasmic reticulum. Gastroenterology 2010; 139:2038–2049.

148. Lee AH, Chu GC, Iwakoshi NN, Glimcher LH. XBP-1 is required for biogenesis of cellular secretory machinery of exocrine glands. EMBO J 2005; 24:4368–4380.

149. Hess DA, Humphrey SE, Ishibashi J, Damsz B, Lee AH, Glimcher LH, Konieczny SF. Extensive pancreas regeneration following acinar-specific disruption of Xbp1 in Mice. Gastroenterology 2011; 141:1463–1472.

150. Parr BA, McMahon AP. Dorsalizing signal Wnt-7a required for normal polarity of D-V and A-P axes of mouse limb. Nature 1995; 374:350–353.

151. Stark K, Vainio S, Vassileva G, McMahon AP. Epithelial transformation of metanephric mesenchyme in the developing kidney regulated by Wnt-4. Nature 1994; 372:679–683.

152. McMahon AP, Bradley A. The W*nt*-1 (*int*-1) proto-oncogene is required for development of a large region of the mouse brain. Cell 1990; 62:1073–1085.

153. Wells JM, Esni F, Boivin GP, Aronow BJ, Stuart W, Combs C, Sklenka A, Leach SD, Lowy AM. Wnt/beta-catenin signaling is required for development of the exocrine pancreas. BMC Dev Biol 2007; 7:4.

154. Murtaugh LC, Law AC, Dor Y, Melton DA. Beta-catenin is essential for pancreatic acinar but not islet development. Development 2005; 132:4663–4674.

155. Dessimoz J, Bonnard C, Huelsken J, Grapin-Botton A. Pancreas-specific deletion of beta-catenin reveals Wnt-dependent and Wnt-independent functions during development. Curr Biol 2005; 15:1677–1683.

156. Heller RS, Dichmann DS, Jensen J, Miller C, Wong G, Madsen OD, Serup P. Expression patterns of Wnts, Frizzleds, sFRPs, and misexpression in transgenic mice suggesting a role for Wnts in pancreas and foregut pattern formation. Dev Dyn 2002; 225:260–270.

157. Wong HC, Bourdelas A, Krauss A, Lee HJ, Shao Y, Wu D, Mlodzik M, Shi DL, Zheng J. Direct binding of the PDZ domain of Disheveled to a conserved internal sequence in the C-terminal region of Frizzled. Mol Cell 2003; 12:1251–1260.

158. Jung H, Kim BG, Han WH, Lee JH, Cho JY, Park WS, Maurice MM, Han JK, Lee MJ, Finley D, Jho EH. Deubiquitination of Disheveled by Usp14 is required for Wnt signaling. Oncogenesis 2013; 2:e64.

159. Yanfeng WA, Berhane H, Mola M, Singh J, Jenny A, Mlodzik M. Functional dissection of phosphorylation of Disheveled in Drosophila. Dev Biol 2011; 360:132–142.

160. Brunner E, Peter O, Schweizer L, Basler K. pangolin encodes a Lef-1 homologue that acts downstream of Armadillo to transduce the Wingless signal in Drosophila. Nature 1997; 385:829–833.

161. Riese J, Yu X, Munnerlyn A, Eresh S, Hsu SC, Grosschedl R, Bienz M. LEF-1, a nuclear factor coordinating signaling inputs from wingless and decapentaplegic. Cell 1997; 88:777–787.

162. Behrens J, von Kries JP, Kuhl M, Bruhn L, Wedlich D, Grosschedl R, Birchmeier W. Functional interaction of beta-catenin with the transcription factor LEF-1. Nature 1996; 382:638–642.

163. Dor Y, Brown J, Martinez OI, Melton DA. Adult pancreatic beta-cells are formed by self-duplication rather than stem-cell differentiation. Nature 2004; 429:41–46.

164. Heiser PW, Lau J, Taketo MM, Herrera PL, Hebrok M. Stabilization of beta-catenin impacts pancreas growth. Development 2006; 133:2023–2032.

165. Arregi I, Climent M, Iliev D, Strasser J, Gouignard N, Johansson JK, Singh T, Mazur M, Semb H, Artner I,

Minichiello L, Pera EM. Retinol dehydrogenase-10 regulates pancreas organogenesis and endocrine cell differentiation via paracrine retinoic acid signaling. Endocrinology 2016; 157:4615–4631.

166. Sandell LL, Sanderson BW, Moiseyev G, Johnson T, Mushegian A, Young K, Rey JP, Ma JX, Staehling-Hampton K, Trainor PA. RDH10 is essential for synthesis of embryonic retinoic acid and is required for limb, craniofacial, and organ development. Genes Dev 2007; 21:1113–1124.

167. Germain P, Iyer J, Zechel C, Gronemeyer H. Co-regulator recruitment and the mechanism of retinoic acid receptor synergy. Nature 2002; 415:187–192.

168. Stafford D, Prince VE. Retinoic acid signaling is required for a critical early step in zebrafish pancreatic development. Curr Biol 2002; 12:1215–1220.

169. Stafford D, Hornbruch A, Mueller PR, Prince VE. A conserved role for retinoid signaling in vertebrate pancreas development. Dev Genes Evol 2004; 214:432–441.

170. Cebola I, Rodriguez-Segui SA, Cho CH, Bessa J, Rovira M, Luengo M, Chhatriwala M, Berry A, Ponsa-Cobas J, Maestro MA, Jennings RE, Pasquali L, et al. TEAD and YAP regulate the enhancer network of human embryonic pancreatic progenitors. Nat Cell Biol 2015; 17:615–626.

171. Gao T, Zhou D, Yang C, Singh T, Penzo-Mendez A, Maddipati R, Tzatsos A, Bardeesy N, Avruch J, Stanger BZ. Hippo signaling regulates differentiation and maintenance in the exocrine pancreas. Gastroenterology 2013; 144:1543–1553, 1553.

172. Axelrod JD. Delivering the lateral inhibition punchline: it's all about the timing. Sci Signal 2010; 3:pe38.

173. Jensen J, Heller RS, Funder-Nielsen T, Pedersen EE, Lindsell C, Weinmaster G, Madsen OD, Serup P. Independent development of pancreatic alpha- and beta-cells from neurogenin3-expressing precursors: a role for the notch pathway

in repression of premature differentiation. Diabetes 2000; 49:163–176.

174. Leach SD. Epithelial differentiation in pancreatic development and neoplasia: new niches for nestin and Notch. J Clin Gastroenterol 2005; 39:S78–82.

175. Apelqvist A, Li H, Sommer L, Beatus P, Anderson DJ, Honjo T, Hrabe de Angelis M, Lendahl U, Edlund H. Notch signalling controls pancreatic cell differentiation. Nature 1999; 400:877–881.

176. Jensen J, Pedersen EE, Galante P, Hald J, Heller RS, Ishibashi M, Kageyama R, Guillemot F, Serup P, Madsen OD. Control of endodermal endocrine development by Hes-1. Nat Genet 2000; 24:36–44.

177. Hald J, Hjorth JP, German MS, Madsen OD, Serup P, Jensen J. Activated Notch1 prevents differentiation of pancreatic acinar cells and attenuate endocrine development. Dev Biol 2003; 260:426–437.

178. Cleveland MH, Sawyer JM, Afelik S, Jensen J, Leach SD. Exocrine ontogenies: on the development of pancreatic acinar, ductal and centroacinar cells. Semin Cell Dev Biol 2012; 23:711–719.

179. Fujikura J, Hosoda K, Kawaguchi Y, Noguchi M, Iwakura H, Odori S, Mori E, Tomita T, Hirata M, Ebihara K, Masuzaki H, Fukuda A, et al. Rbp-j regulates expansion of pancreatic epithelial cells and their differentiation into exocrine cells during mouse development. Dev Dyn 2007; 236:2779–2791.

180. Shih HP, Kopp JL, Sandhu M, Dubois CL, Seymour PA, Grapin-Botton A, Sander M. A Notch-dependent molecular circuitry initiates pancreatic endocrine and ductal cell differentiation. Development 2012; 139:2488–2499.

181. Kopinke D, Brailsford M, Pan FC, Magnuson MA, Wright CV, Murtaugh LC. Ongoing Notch signaling maintains phenotypic fidelity in the adult exocrine pancreas. Dev Biol 2012; 362:57–64.

Chapter 3

Bioenergetics of the exocrine pancreas: Physiology to pathophysiology

David N. Criddle and Alexei V. Tepikin

Department of Molecular Physiology & Cell Signalling, Institute of Systems, Molecular & Integrative Biology, University of Liverpool, L69 3BX UK

Introduction

The major cell types defining exocrine functions of pancreas are pancreatic acinar and pancreatic ductal cells. Acinar cells secrete digestive enzymes, precursors of digestive enzymes (zymogens), and NaCl-rich fluid, while ductal cells secrete large volumes of bicarbonate-containing fluid and form a conduit for exocrine pancreatic secretion (reviewed in [1-3]). Both exocytotic secretion of proteins and fluid secretion are energy-consuming processes. Furthermore, pancreatic acinar cells have high levels of protein synthesis[4] supporting substantial protein export by secretion. The bioenergetics of this organ are designed to sustain these energy demands and adjust ATP generation when cells undergo shifts between resting and stimulated conditions. Disruption/insufficiency of ATP generation is an important feature of pancreatic damage in pathological conditions. In this chapter, we will consider both physiological adaptation of bioenergetics in exocrine pancreas and its disruption in acute pancreatitis.

Pancreatic acinar cells constitute more than 80 percent of the volume of the exocrine pancreas[5] and are thus likely to make major contributions to bioenergetics parameters measured in experiments utilizing undissociated pancreas. Signaling mechanisms and bioenergetics of this cell type have been extensively investigated and a substantial part of this chapter therefore will be focused on the bioenergetics of pancreatic acinar cells. There has also been recently significant progress in characterizing bioenergetics changes in ductal cells, particularly in conjunction with the pathophysiology of acute pancreatitis; we will discuss this progress in section "Alteration of Bioenergetics in Pancreatic Pathophysiology."

Mechanisms of ATP Generation

Information on the relative importance of different substrates (carbohydrates, amino acids, and fatty acids) for the bioenergetics of exocrine pancreas is somewhat limited.

Pancreatic acinar cells express a range of amino acid transporters,[6] allowing accumulation of amino acids against a significant concentration gradient; this is beneficial for prominent protein synthesis in this cell type and could also support utilization of amino acids as substrates for ATP production. Notably, in *in vivo* and *in vitro* experimental models of acute pancreatitis expression of several amino acid transporters was significantly downregulated.[6] A study by S. Araya and colleagues highlighted a high rate of glutamine transport in pancreatic acinar cells and compared CO_2 generation in response to addition of amino acids (glutamine and leucine), glucose, and palmitoyl.[7] The study concluded that glutamine is the preferred energy source for pancreatic acinar cells and that these cells operate an efficient conversion of glutamine to glutamate followed by glutamate utilization (via α-ketoglutarate) in the Krebs cycle.[7] Information on the contribution of other amino acids and fatty acids, however, is sparse. Glucose can support the bioenergetics of acinar cells, and glucose-containing extracellular solution has been frequently utilized in studies of glycolysis and oxidative phosphorylation in this cell type. The relative contribution of oxidative metabolism and glycolysis to the net ATP generation in exocrine pancreas and pancreatic acinar cells is still a subject for debate. Mitochondria occupy a significant proportion of cell volume in pancreatic acinar cells (approximately 8%[5]). An early study by Bauduin and colleagues concluded that mitochondria are by far the major source of ATP production in acinar cells while glycolysis makes only minor contribution.[8] This paper, however, reported that high concentration of oligomycin (inhibitor of the mitochondrial F_1F_0-ATP synthase: abbreviated to ATP synthase) reduced nucleoside triphosphate (presumably mainly ATP) content to approximately 41 percent of the control value, suggesting a significant contribution of glycolysis to ATP generation. However, the same authors reported that antimycin (inhibitor of the mitochondrial electron transport chain) reduced nucleoside triphosphate content to 16 percent and anaerobiosis to 8 percent.[8] H. Kosowski and colleagues from

e-mails: criddle@liverpool.ac.uk, A.Tepikin@liverpool.ac.uk

W. Halangk laboratory reported a decrease of ATP content to approximately 21 percent of the control value following 30 minuses of anoxia.[9] Experiments with anoxia and antimycin could be interpreted as evidence for a rather modest contribution of glycolysis to ATP generation in the acinar cells. However, these treatments are expected to trigger the reverse mode of ATP synthase, which can convert this enzyme into a powerful ATP consumer (i.e., in this mode the enzyme functions as an ATPase consuming ATP and transporting protons into the matrix; reviewed in [10–13]). The discrepancy between the ATP levels retained in the presence of oligomycin and after the inhibition of mitochondrial electron transport chain by anoxia or antimycin could be explained by a reverse mode of ATP synthase. ATP consumption by this reverse mode could be potentially prevented by the F_1F_0-ATP synthase inhibitory factor 1 (IF1).[13] However, it should be noted that normal pancreas contains relatively low levels of IF1 (e.g., in comparison with heart or pancreatic cancer tissue[14]) and therefore the effect of the reverse mode could be significant. Consequently, experiments with oligomycin or combination of oligomycin with an inhibitor of the electron transport chain should provide a better estimation of contribution made by glycolysis to ATP generation of the acinar cells. The early estimation of 41 percent[8] is consistent with more recent findings indicating a significant (at least 30–40 percent) contribution of glycolysis to ATP generation in acinar cells.[14,15] Furthermore, the balance of glycolytic and oxidative ATP production depends on external factors. In particular, secretagogues that produce moderate Ca^{2+} signals upregulate the rate of Krebs cycle and mitochondrial ATP generation,[15–17] while application of insulin strongly potentiates glycolytic ATP production, which can sustain bioenergetics of acinar cells with damaged mitochondria.[18,19]

Stimulus-Metabolism Coupling

In the late 1970s, an important and surprising observation was made in experiments on isolated pancreatic acini[20,21]; upon stimulation with secretagogues (caerulein, cholecystokinin, or carbachol), cellular ATP levels were stable or even slightly increased.[20,21] It would be expected that the increased energy demands imposed by the stimulation of secretion should reduce ATP content. The findings were therefore paradoxical and suggested existence of an effective mechanism(s), capable of compensating and even overcompensating for the increased ATP usage in the stimulated cells, that is, indicative of an efficient stimulus-metabolism coupling in this cell type. Early evidence for such stimulus-dependent upregulation of bioenergetics came from the study by M. Korc and colleagues demonstrating strong and rapid upregulation of glucose uptake in isolated pancreatic mouse acini stimulated with secretagogues[20] (see

Figure 1A). Caerulein, CCK, and carbachol were tested and all produced clear and substantial increases in glucose uptake (2.5–3 times in comparison with control).[20] Crucially, the increase in the glucose uptake could be mimicked by the Ca^{2+} ionophore A23187. Furthermore, incubation of cells in the solution containing the Ca^{2+} chelator EGTA suppressed the agonist-induced glucose uptake.[20] The authors therefore concluded that intracellular Ca^{2+} may "mediate hormone-stimulated glucose transport."[20] The authors' suggested role of cellular Ca^{2+} in the regulation of bioenergetics of the acinar cells was confirmed by the following 40 years of research. Important findings that characterized the role played by Ca^{2+} signaling in regulating bioenergetics were made at the end of the previous century in Denton's laboratory. These experiments revealed the ability of Ca^{2+} to upregulate the rate of Krebs cycle; the effect of Ca^{2+} was mediated by rate-limiting dehydrogenases of the Krebs cycle (reviewed in [22,23]). At the single cell level, the relationships between cytosolic Ca^{2+}, mitochondrial Ca^{2+}, and NAD(P)H (an important reducing equivalent produced in Krebs cycle that supplies electrons for the respiratory chain) were first demonstrated in hepatocytes.[24] This study utilized cellular autofluorescence to monitor the NAD(P)H responses and revealed fast upregulation of the NAD(P)H production by Ca^{2+} transients.[24] In isolated pancreatic acinar cells, relationships between Ca^{2+} signaling (in cytosol and mitochondria) and NAD(P)H were first characterized by Voronina and colleagues.[16] This study revealed fast and robust NAD(P)H transients associated with Ca^{2+} responses (**Figure 1B** and **C**), consistent with NAD(P)H responses recorded from perfused pancreas.[25]

Subsequently, rises of NAD(P)H in response to CCK-induced Ca^{2+} transients were shown to occur in isolated human pancreatic acinar cells, confirming a similar stimulus-metabolism coupling to that previously elucidated in murine cells.[26] Physiological concentrations of important secretagogues (e.g., CCK and ACh) induce oscillatory Ca^{2+} responses in acinar cells (reviewed in [27–29]). NAD(P)H responses were sufficiently fast to form oscillations (reflecting oscillations in the rate of the Krebs cycle) closely associated with the oscillations of cytosolic Ca^{2+} [16]). Upon stimulation with Ca^{2+}-releasing secretagogues, pancreatic acinar cells generate local and global Ca^{2+} transients[30,31] (reviewed in [27,28]). All global Ca^{2+} transients were able to trigger NAD(P)H transients (i.e., increased rate of the Krebs cycle). The majority of local Ca^{2+} responses were also able to trigger resolvable NAD(P)H responses except for the very short local Ca^{2+} transients (less than 2.9 seconds) that are probably quenched by cytosolic calcium buffers before reaching the mitochondria.[16] To upregulate activity of dehydrogenases of the Krebs cycle and accelerate its rate, Ca^{2+} must enter mitochondrial matrix. The phenomenon and properties of mitochondrial Ca^{2+} entry were

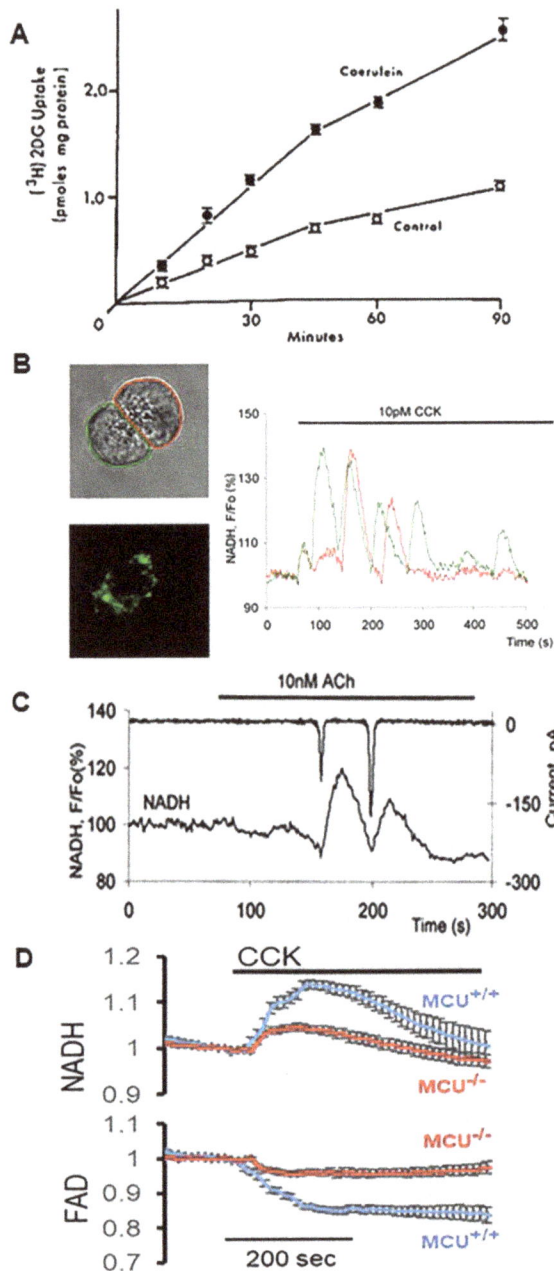

Figure 1. Stimulus-metabolism coupling in pancreatic acinar cells. (A) Changes in glucose uptake by pancreatic acini upon application of caerulein (adapted from Korc et al.[20]). **(B)** NADH oscillations in pancreatic acinar cells. Left panels show transmitted image of two pancreatic acinar cells and regions of interest for recording of fluorescence (corresponding to traces on the right panel). Lower left panel shows the distribution of NADH fluorescence (adapted from Voronina et al.[15]). **(C)** Correlation between cytosolic Ca^{2+} responses (revealed by negative deflections on the trace of Ca^{2+}-dependent Cl^- current) and changes of NADH fluorescence (adapted from Voronina et al.[15]). **(D)** Effect of mitochondrial Ca^{2+} uniporter (MCU) knockout on the NADH and FAD responses to CCK stimulation in pancreatic acinar cells (adapted from Chvanov et al.[17]).

extensively characterized in the second part of the previous century; however, the molecular mechanism was only discovered in 2011.

Mitochondrial Ca^{2+} entry is primarily mediated by the mitochondrial Ca^{2+} uniporter complex[32,33] (reviewed in [34]). In mammals, this complex is composed of the mitochondrial Ca^{2+} uniporter (MCU, pore-forming protein localized in the inner mitochondrial membrane), its Ca^{2+}-dependent regulators MICU1/MICU2,[35–38] and other associated and regulatory proteins (reviewed in reference [34]). An important function of MICU1/MICU2 is to provide a threshold Ca^{2+} concentration that limits Ca^{2+} entry into mitochondria at low cytosolic Ca^{2+} concentration. The voltage difference between mitochondria and cytosol is substantial (approximately 140 mV); it is sufficient to drive Ca^{2+} entry into mitochondria even at low resting Ca^{2+} levels. This would form a futile Ca^{2+} cycle necessitating continuous energy-dependent Ca^{2+} extrusion from mitochondria.[35–37,39–41] At higher Ca^{2+} concentrations, MICU1 and MICU2 bind Ca^{2+} and undergo conformational change allowing Ca^{2+} entry through the MCU channel.[35–37,41] Threshold Ca^{2+} concentrations necessary for such "opening" of the MCU complex are usually significantly higher than the resting cytosolic Ca^{2+} and usually also higher than Ca^{2+} concentrations in the areas distant from Ca^{2+}-releasing channels in the cellular organelles or Ca^{2+} influx channels in the plasma membrane. Therefore, Ca^{2+} entry into mitochondrial matrix in many systems depends on the mitochondrial localization and specifically on the proximity of the organelles and their MCU complex to the Ca^{2+}-releasing channels (reviewed in [42]). In pancreatic acinar cells, mitochondria are found in three major groups—subplasmalemmal, perigranular, and perinuclear.[43–46] These mitochondria are conveniently localized to respond to Ca^{2+} release from IP_3 receptors (IP_3Rs),[44] ryanodine receptors (RyR),[47] and store-operated Ca^{2+} (SOC) influx channels.[43,48] The mitochondrial Ca^{2+} uniporter is responsible for the bulk of Ca^{2+} entry into the mitochondrial matrix; however, other mitochondrial Ca^{2+} influx mechanisms have been reported (e.g., [49–51]). A recent study, utilizing *MCU* knockout (*MCU* KO) mice, demonstrated that the MCU complex is responsible for the major component of the mitochondrial Ca^{2+} entry in CCK-stimulated pancreatic acinar cells; however, very small mitochondrial Ca^{2+} signals were observed in CCK-stimulated acinar cells from *MCU* KO mice suggesting an alternative (and currently unknown) Ca^{2+} entry mechanism in pancreatic mitochondria.[17] These observations were consistent with the results of measurements of mitochondrial NAD(P)H and FAD responses. During the Krebs cycle, FAD is converted to $FADH_2$; FAD is fluorescent and the decrease of its content/fluorescence in mitochondria suggests its reduction and could be interpreted (with caution) as an acceleration of Krebs cycle.[52] NAD(P)H and FAD responses associated with Ca^{2+} signals have been measured in a number of cell types (e.g., [24,53]); in these studies, transient increases in NAD(P)H fluorescence were usually accompanied by decreases in FAD fluorescence. The amplitudes of both NAD(P)H and FAD responses to CCK stimulation drastically decreased in the acinar cells from *MCU* KO animals in comparison with wild-type (WT) mice[17] (**Figure 1D**). This confirms the major role of Ca^{2+} signals and MCU-mediated mitochondrial Ca^{2+} responses in stimulus-metabolism coupling in pancreatic acinar cells. Notably, small NAD(P)H and FAD responses were retained in the cells from *MCU* KO mice; this is consistent with the retention of small mitochondrial Ca^{2+} signals in cells of these animals.[17] Another important parameter reflecting cellular bioenergetics is oxygen consumption. In line with the observed upregulation of glucose uptake[20] and NAD(P)H increase,[16,17] oxygen consumption significantly increases in perfused pancreas[25] and isolated acinar cells stimulated with Ca^{2+}-releasing secretagogues.[17,25,54,55] Importantly, this effect was diminished in acinar cells from *MCU* KO mice,[17] again highlighting the role of mitochondrial Ca^{2+} in the stimulus-metabolism coupling of this cell type. The notion of the efficient stimulus-metabolism coupling in pancreatic acinar cells is consistent with the outcome of experiments involving ATP and creatine phosphate measurements in perfused rat pancreas, which indicated that "ATP and PCr remained largely unchanged" in spite of maximal stimulation of the organ with 100 pM CCK producing robust protein and fluid secretion.[56] In this study, a higher concentration of CCK (1nM) produced a small decrease in ATP levels, which may be associated with tissue-level effects of supramaximal doses of this agonist and will be discussed in the role of bioenergetics in pancreatitis (see below). While studies of the relationships between Ca^{2+}-releasing secretagogues and bioenergetics, conducted on isolated acinar cells and perfused pancreata, reported stable or minimally changed ATP levels, despite upregulation of energy-consuming processes,[15,20,21] results from experimental animal models, involving low "physiological" as well as supramaximal doses of secretagogues, are somewhat contradictory. In the study by R. Luthen and colleagues, all tested concentrations of caerulein (7 hourly intraperitoneal injections of caerulein each corresponding to 0.1, 1, 5, and 50 µg/kg body weight) produced significant and similar decreases in ATP concentration in mouse pancreata (by approximately 33% one hour after the first injection and 45% one hour after the last injection).[57] It is conceivable that in intact tissue the rate of ATP production is limited by tissue oxygen levels and nutrient supply (e.g., due to changes of pancreatic microcirculation, reviewed in [58]), which are usually not restricted in single cell preparations. It should be also noted that other studies involving animal models and stimulation with secretagogues reported

stable pancreatic ATP levels.[56,59] This apparent contradiction emphasizes the importance of continuing investigation of stimulus-metabolism coupling at cellular and tissue levels as well as in animal models. The pathophysiology of acute pancreatitis is frequently associated with disruption of bioenergetics and loss of ATP in pancreatic tissue. In experimental models of acute pancreatitis, this is particularly clear for bile-induced and fatty acid-induced acute pancreatitis; the relevant studies will be discussed below.

Metabolism—Signaling Feedback

Several pathophysiologically relevant substances (including fatty acids, reactive oxygen species, asparaginase, bile acids, and nonoxidative alcohol metabolites) can deplete cellular ATP (see references [15,18,19,60–64] and the next section). While Ca^{2+} signaling can influence metabolism and mediate efficient stimulus-metabolism coupling in pancreatic acinar cells, there is also a powerful effect of metabolism on signaling. Interesting early manifestation of such feedback mechanism came from study by A. Hofer and colleagues, which demonstrated inhibition of Ca^{2+} leak from the ER lumen by ATP depletion.[65] Further studies identified the prominent influence of ATP depletion on ACh-induced Ca^{2+} oscillations in pancreatic acinar cells[66]; a modest decrease in cytosolic ATP concentration (revealed by measuring luciferase bioluminescence in live cells) resulted in abrupt termination of Ca^{2+} oscillations ([66] and **Figure 2A**). The molecular phenomenon underlying this cellular response was identified in studies from the Yule laboratory that characterized ATP sensitivity of IP_3 receptors responsible for the Ca^{2+} oscillations in the acinar cells. All three types of IP_3Rs are expressed in pancreatic acinar cells[67]; however, the expression of $InsP_3R_1$ is relatively low[67] and it is unlikely to play significant role in physiologically relevant Ca^{2+} responses. $InsP_3R_2$ and $InsP_3R_3$ are the major receptor subtypes in acinar cells.[67] These are ATP-sensitive, and Ca^{2+} fluxes via these channels are inhibited by ATP reduction in mM or sub-mM range.[68,69] Further study utilizing acinar cells from $InsP_3R_2$ knockout mice revealed particular importance of this Ca^{2+}-releasing channel in determining ATP sensitivity of Ca^{2+} responses[68] (**Figure 2B**). Another vital component of calcium signaling, Ca^{2+} influx via the plasma membrane, is also ATP-sensitive. In pancreatic acinar cells, Ca^{2+} influx is primarily mediated by the store-operated Ca^{2+} entry (SOCE) based on the interaction of the ER Ca^{2+} sensor STIM1 and the plasma membrane channel-forming protein Orai.[48,70–73] A role for TRPC channels in SOCE has also been reported.[74]

ATP depletion resulted in the efficient inhibition of the SOCE in pancreatic acinar cells.[66] Inhibition of mitochondria and/or ATP depletion have been shown to suppress

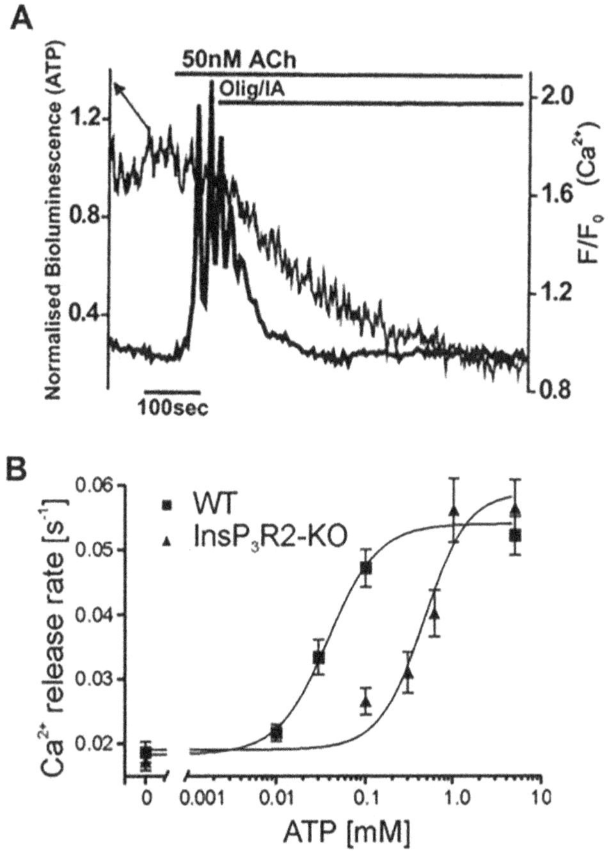

Figure 2. **Metabolism-signaling feedback.** (**A**) Inhibition of acetylcholine-induced cytosolic Ca^{2+} oscillations in a pancreatic acinar cell by ATP depletion (induced by application of oligomycin and iodoacetate) (adapted from Barrow et al.[66]). (**B**) ATP sensitivity of Ca^{2+} release from internal stores of $InsP_3$-stimulated permeabilized pancreatic acinar cells. The figure compares ATP sensitivity of cells isolated from wild-type (WT) mice and from mice with knocked out $InsP_3R2$. (The figure was adapted with modifications from Park HS et al.[68])

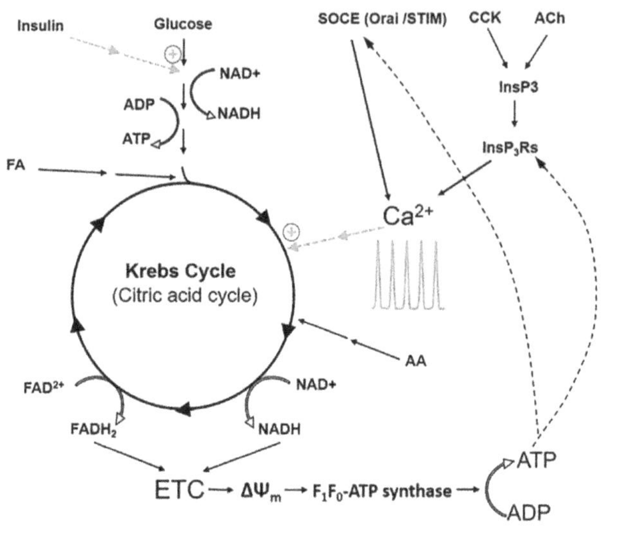

Figure 3. **Simplified diagram of stimulus-metabolism coupling and metabolism-signaling feedback in pancreatic acinar cells.** The schematic illustrates the mechanisms of physiological adaptation of bioenergetics to modulate stimulation by secretagogues. In the diagram $\Delta\Psi_m$ indicates mitochondrial membrane potential (major component of the proton-motive force), ETC—electron transport chain, AAs—amino acids, and FAs—fatty acids. Dashed green lines illustrate the stimulating effect of physiological Ca^{2+} signals (Ca^{2+} oscillations induced by ACh or CCK) on the Krebs cycle and of insulin on glycolysis. Dashed black lines highlight the role of ATP in the regulation of $InsP_3$ receptors ($InsP_3Rs$) and store-operated Ca^{2+} entry (SOCE). Both IP_3Rs and the SOCE pathway are ATP-sensitive mechanisms. In resting unstimulated cells and those stimulated by physiological concentrations of secretagogues, ATP levels are sufficient to sustain both $InsP_3R$-mediated Ca^{2+} release from the endoplasmic reticulum and SOCE via the STIM/Orai mechanism. However, ATP depletion (e.g., triggered by the inducers of acute pancreatitis) can suppress the InsP3R-mediated Ca^{2+} release and SOCE. This is also accompanied by the inhibition of Ca^{2+} leak from the ER, Ca^{2+} uptake into the ER by sarcoplasmic/endoplasmic reticulum Ca^{2+} pumps, and Ca^{2+} extrusion by Ca^{2+} pumps of the plasma membrane (not shown). For the sake of clarity/simplicity we have omitted some Ca^{2+}-mobilizing second messengers and corresponding intracellular Ca^{2+}-releasing channels. (For detailed reviews of Ca^{2+} signaling in this cell type, see references [27] and [28].) We have also simplified the entry mechanisms of AA and FA derivatives into the Krebs cycle. Feedback regulation of the rate of the Krebs cycle by ATP, ADP, and NAD(P)H was also omitted on this diagram.

SOCE entry in a number of other cell types[75–79] and could be therefore considered as a general phenomenon not limited to the pancreatic acinar cell. The specific molecular mechanism responsible for such inhibition is still debated. Stimulus-metabolism coupling and metabolism-signaling feedback are schematically illustrated in **Figure 3**. As expected, ATP depletion also suppressed Ca^{2+} uptake into the ER lumen via sarcoplasmic/endoplasmic reticulum Ca^{2+} ATPase (SERCA) and Ca^{2+} extrusion via plasma membrane Ca^{2+} ATPase (PMCA).[18,19,66] In pancreatic acinar cells, inhibition of SOCE by ATP depletion occurred in parallel with the inhibition of Ca^{2+} extrusion[66] suggesting an important and hitherto unidentified link between the two processes. The inhibition of Ca^{2+} influx by ATP depletion is, however, incomplete and prolonged depletion leads to Ca^{2+} overload and damage/death of pancreatic acinar cells.[18,19] Furthermore, some inducers of acute

pancreatitis (e.g., bile acids and fatty acids) seem to be capable of generating an alternative Ca^{2+} influx pathway which is not inhibited by ATP depletion.[18,19,66] A number of novel Ca^{2+} influx channels have been recently identified in pancreatic acinar cells[80–82] and the effects of mitochondrial inhibition and ATP depletion on these channels are currently unknown. Identifying the relationships between the recently identified Ca^{2+} entry pathways and cellular bioenergetics could be the subject of productive future investigation.

The pronounced effect of cellular metabolism on Ca^{2+} signaling limits energy expenditure associated with actual signaling responses and downstream reactions of this critical for the acinar cells signaling cascade. It also offers some protection against Ca^{2+} overload and associated Ca^{2+} toxicity. This protection is however incomplete and can be overwhelmed by the inducers of acute pancreatitis.

Alteration of Bioenergetics in Pancreatic Pathophysiology

Ca^{2+}-dependent mitochondrial dysfunction is a core feature of acute pancreatitis. In contrast to the spatially restricted, oscillatory Ca^{2+} signaling events that occur in response to physiological levels of secretagogues in isolated pancreatic acinar cells, supramaximal stimulation causes damaging global, sustained calcium elevations.[83–85] The ability to cause overload of cytosolic Ca^{2+} and thereby disrupt Ca^{2+} homeostasis in acinar cells is a principal pathological feature common to known precipitants of acute pancreatitis, including bile acids, nonoxidative ethanol metabolites, and fatty acids[74,86] (**Figure 4A**), and is an important trigger for mitochondrial damage that compromises cellular ATP production.[87] The importance of altered energy levels within the pancreas relevant to its pathology has been recognized for many years, although in vivo studies have produced somewhat varied results. In the early 1990s, substantial falls of cellular ATP in the pancreas were detected in experimental AP models using noninvasive NMR spectrometry of ^{31}P.[59] Interestingly, there was heterogeneity in responses dependent on the model employed. Thus, whereas significant falls of ATP were detected in AP induced by fatty acid (oleic acid) infusion or regional ischemia, levels were unchanged in caerulein hyperstimulation and partial ductal obstruction models. In contrast, a contemporaneous study in rat pancreas showed that ATP levels were more than halved from controls in the caerulein AP model.[88] A subsequent investigation by W. Halangk and colleagues supported these findings, demonstrating a reduction of phosphorylating respiration in acinar cells and fall of ATP to 57 percent in caerulein AP after 24 hours (using a Clark-type electrode).[89] The authors concluded that caerulein-induced acute pancreatitis is accompanied by "a drastic and long-lasting reduction of the capacity for mitochondrial ATP production." It should be noted that this phenomenon developed relatively slowly, that is, at early time points (up to 5 hours) the cellular ATP content reported was not significantly diminished from controls and therefore slower than other manifestations of experimental acute pancreatitis like vacuolization[90] and intracellular

trypsinogen activation.[91] In contrast, a recent study found that high levels of L-lysine caused mitochondrial damage which occurred before activation of trypsinogen and NFkB, potentially via impairment of electron transport chain coupling or a disbalance between hydrolysis and synthesis of ATP.[92] To date, the balance of evidence from *in vivo* models has demonstrated loss of pancreatic ATP as acute pancreatitis develops, indicating that early mitochondrial damage is a critically important factor in disease progression; variability between studies may reflect methodological differences, etiology, and/or severity of the experimental model and further comparative studies are warranted. The *in vivo* findings are therefore consistent with ATP measurements on isolated acinar cells showing rapid declines of ATP concentration in cells treated with fatty acid (POA) or bile acid (TLC-S).[15,87] In addition, the bile acid chenodeoxycholate has been shown to decrease intracellular ATP levels in pancreatic ductal cells, with inhibition of both oxidative and glycolytic metabolism linked to impaired bicarbonate secretion.[93,94] With respect to in vivo models, the bioenergetics of acinar and ductal cells in pancreatic tissue will be influenced by changes in local blood flow. This is particularly relevant to the pathophysiology of acute pancreatitis, which is associated with impairment of microvascular perfusion[95–98] (reviewed in [58]). Inducers of acute pancreatitis have been shown to increase the permeability of pancreatic capillaries[96–98] and reduce microvascular blood flow[99] and pancreatic oxygen tension.[99,100] Notably, different models of experimental acute pancreatitis are associated with distinct changes in the microcirculation, ranging from minimal responses and even modest increases of capillary perfusion in mild acute pancreatitis[101] to drastic decreases of capillary perfusion associated with more severe models.[99,101] The intrinsic abilities of the exocrine cells to adjust metabolism and signaling in response to secretagogues and the inducers of acute pancreatitis (discussed in the previous section) will be therefore "superimposed" on changes in oxygen supply and nutrient availability determined by the microcirculation.

The mechanisms underlying ATP depletion induced by acute pancreatitis precipitants have been addressed in detail using isolated pancreatic acinar cells. Confocal microscopy investigations demonstrated that Ca^{2+} overload of acinar cells caused rapid depolarization of the mitochondrial membrane potential, accompanied by decreased NAD(P)H and increased FAD^+ levels, with a resultant cellular ATP fall leading to necrosis. When applied at lower concentration (200 μM) the bile acid TLCS caused predominantly oscillatory calcium rises, accompanied by NADH elevations, in accord with stimulus-metabolism coupling (as described above). However, sustained Ca^{2+} elevations at higher concentration (500 μM) led to rises of mitochondrial Ca^{2+} with dramatic decrease of NAD(P)H[62,102] (**Figure 4B**). Similarly, the nonoxidative ethanol metabolite

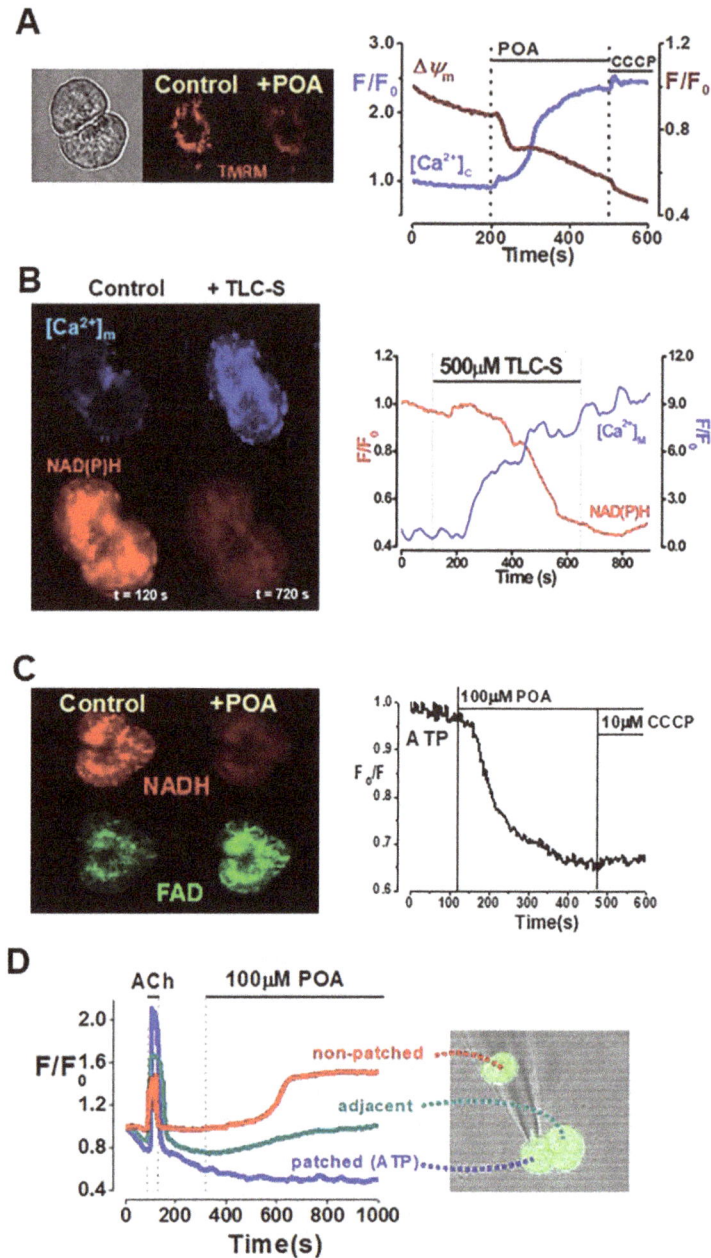

Figure 4. **Pathological Ca^{2+}-dependent mitochondrial bioenergetic dysfunction in pancreatic acinar cells.** (**A**) Correlation between palmitoleic acid (POA)-induced sustained cytosolic Ca^{2+} response (Fluo4) and mitochondrial depolarization (TMRM) (adapted from Criddle et al.[87]). (**B**) Correlation between bile acid (TLCS: taurolithocholic acid sulphate)-induced sustained mitochondrial Ca^{2+} rises (Rhod2) and decrease of NAD(P)H (autofluorescence) in a doublet of pancreatic acinar cells (left panel) and corresponding traces (right panel) (adapted from Booth et al.[62]). (**C**) Bioenergetics changes in response to POA in pancreatic acinar cells, shown as concurrent decrease of mitochondrial NADH and increase of FAD (autofluorescence: left panel) and depletion of ATP (Magnesium Green). (**D**) Provision of ATP via a patch-pipette prevents toxic POA-induced sustained Ca^{2+} elevation in pancreatic acinar cells, highlighting the vital importance of ATP-dependent Ca^{2+} pumps to Ca^{2+} clearance and prevention of cell damage (**C, D**) (adapted from Criddle et al.[87]).

palmitoleic acid ethyl ester (POAEE) and POA, produced by hydrolysis of the fatty acid ethyl ester,[103,104] induced sustained elevations of cytosolic Ca^{2+} that resulted in mito-chondrial dysfunction.[87] Simultaneous measurements of

cytosolic Ca^{2+} and $\Delta\psi_m$ revealed the temporal relationship of these events; as cytosolic Ca^{2+} progressively rose there was a mirrored mitochondrial depolarization and rapid decrease of NAD(P)H[105] (**Figure 4A**). The consequence of mitochondrial impairment was a profound depletion of intracellular ATP (measured using Magnesium Green); this effect was maximal and not further increased by the protonophore CCCP[87] (**Figure 4C**). Interestingly, a Ca^{2+}-independent action of POA was also detected, revealed by pretreatment with the Ca^{2+} chelator BAPTA, which prevented the loss of $\Delta\psi_m$, although NAD(P)H decrease still persisted. The rundown of mitochondrial ATP production has dramatic effects on Ca^{2+} homeostasis in the pancreatic acinar cell, which appears particularly vulnerable to sustained cytosolic Ca^{2+} increases; unlike excitable cells such as cardiomyocytes, the contribution of the Na^+/Ca^{2+} exchanger to Ca^{2+} homeostasis in the pancreatic acinar cell appears weak or absent.[106] The major routes for Ca^{2+} clearance, the SERCA and PMCA pumps, are fueled by cellular ATP and would be progressively impaired as energy levels fall during acute pancreatitis development.[107,108] This is likely to result in a vicious cycle in which Ca^{2+}-dependent mitochondrial dysfunction perpetuates sustained cytosolic Ca^{2+} elevation, resulting in augmented necrotic cell death.[109] The importance of maintaining ATP levels for acinar cell function was demonstrated by the provision of intracellular ATP via a patch pipette[87]; this prevented the damaging Ca^{2+} elevations induced by POA, confirming the crucial roles of the SERCA/PMCA for Ca^{2+} clearance in pancreatic acinar cells[87] (**Figure 4D**). In accord, ATP provision prevented the sustained Ca^{2+} rises and consequent acinar cell necrosis induced by bile acid,[62] while supplemental ATP was able to mitigate the detrimental effects of ethanol and its nonoxidative metabolites on CFTR in ductal cells that promote acute pancreatitis.[110,111]

How does ATP depletion occur?

The precise mechanisms of ATP depletion are still unclear, although a crucial role of MPTP opening in mediating mitochondrial dysfunction has been identified in the pancreas,[112–114] in common with other tissues such as heart, liver, and brain.[115–117] The formation of this megapore causes a profound permeabilization of the inner mitochondrial membrane that allows movement of solutes <1.5 kDal in and out of the organelle. Although the exact composition of the pore is contentious and currently unresolved,[118–122] prevention of MPTP opening is achieved via inhibition of the modulatory protein cyclophilin D (CypD); this protein is therefore considered an important target for acute pancreatitis therapy.[113,123–126] Our group has shown that knockout and pharmacological inhibition of CypD ameliorated damage in isolated murine and human acinar cells induced by acute pancreatitis precipitants and was protective in multiple in vivo experimental models reflecting the principal etiologies (including biliary: TLCS-AP[127] and alcoholic: FAEE-AP[104]).[113] Furthermore, CypD-sensitive

MPTP formation mediated mitochondrial dysfunction in pancreatic ductal cells, indicating a common mechanism operating in multiple cell types implicated in the pathogenesis of acute pancreatitis.[3,126] The principal trigger for MPTP opening is the sustained elevation of mitochondrial Ca^{2+} [128]; however, reactive oxygen species (ROS) have also been implicated in triggering MPTP channel formation.[129,130] In human and murine pancreatic acinar cells, bile acid-induced effects are complex since sustained elevations of cytosolic and mitochondrial Ca^{2+} lead to sustained ROS increases; both Ca^{2+} and ROS formation are therefore likely to be important to the outcome of bioenergetics changes that underlie mitochondrial dysfunction in acute pancreatitis.[62] Physiological and pathophysiological changes of acinar cell bioenergetics are schematically illustrated in **Figure 5**.

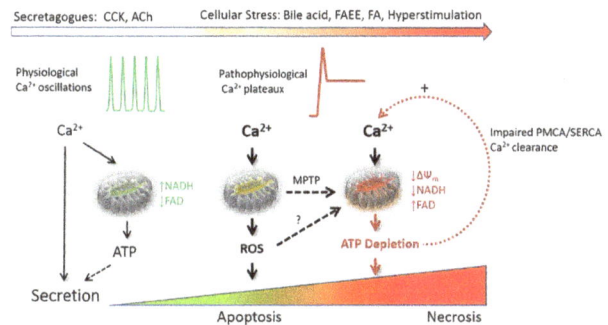

Figure 5. Diagram illustrating the effects of physiological and pathophysiological Ca^{2+} signals on mitochondrial bioenergetics that determine pancreatic acinar cell responses and fate. Physiological stimulation of pancreatic acinar cells with secretagogues cholecystokinin (CCK) and acetylcholine (ACh) elicits oscillatory Ca^{2+} signals that stimulate mitochondrial ATP production (stimulus-metabolism coupling) that fuels secretion. However, when subjected to stress, cellular bioenergetics are perturbed leading to instigation of cell death patterns. Precipitants of acute pancreatitis (AP), including bile acids, nonoxidative ethanol metabolites (fatty acid ethyl esters: FAEEs), fatty acids (FAs), and secretagogue hyperstimulation, cause sustained cytosolic Ca^{2+} elevations raise mitochondrial matrix Ca^{2+}. Ca^{2+}-dependent ROS generation in the mitochondria preferentially promotes apoptotic cell death, which may constitute a local protective mechanism whereby necrosis is avoided. However, as the level and/or duration of stress increases, a shift from apoptotic to necrotic cell death ensues. Ca^{2+}-dependent formation of the mitochondrial permeability transition pore (MPTP) is a primary event that results in mitochondrial depolarization, altered NADH/FAD levels, ATP fall, and necrotic cell death development. Elevated ROS levels also drive necrosis in a manner that is independent of the MPTP. Since clearance of cytosolic Ca^{2+} from the acinar cell is acutely dependent on ATP-requiring pumps (plasmalemmal Ca^{2+}-ATPase (PMCA) and sarco-endoplasmic reticulum Ca^{2+}-ATPase (SERCA)), the depletion of ATP evokes a vicious cycle that perpetuates Ca^{2+} overload and mitochondrial dysfunction. The figure is adapted from Criddle DN.[131]

Role of ROS in bioenergetic alterations

Secretagogues and bile acids increase ROS generation in pancreatic acinar cells[62,132] (reviewed in [131]). In pancreatic acinar cells, augmented ROS production promotes apoptosis,[64] which may be an endogenous protective mechanism that enables the cell to deal with cellular stress by avoiding necrosis.[62,64,133,134] Necrosis is considered a largely uncontrolled form of cell death and in acute pancreatitis is associated with development of systemic inflammation[135]; accordingly, inhibition of caspases causes a more severe necrotizing pancreatitis.[133,136] Our recent work has demonstrated that oxidative stress induced by exogenous H_2O_2 or menadione, which generates intracellular ROS via redox cycling,[64] caused bioenergetics changes that were independent of CypD-sensitive MPTP opening.[137] Increasing levels of oxidants elicited an apoptosis to necrosis shift that was unaffected by knockout of CypD or pharmacological inhibition with cyclosporin A. This lack of sensitivity to CypD is consistent with evidence in primary hepatocytes showing that CypD deletion affected MPTP formation induced by Ca^{2+} but not by thiol oxidants[138] and a model in which there is a regulated (Ca^{2+}-activated and cyclosporin-sensitive) and unregulated (cyclosporin-insensitive) MPTP.[139,140] Therefore, counteracting excessive ROS in the pancreas with antioxidants might be desirable. However, despite many clinical trials for acute pancreatitis, results have generally been disappointing or inconclusive.[141] A more recent approach to combat oxidative stress is the targeting of antioxidants to mitochondria in order to produce more localized, specific actions.[142] For example, MitoQ, which accumulates in mitochondria due to its positively charged triphosphonium ion (TPP^+),[143] prevented ROS-mediated damage in diverse cell types,[144–148] although efficacy in clinical investigations so far has been disappointing.[149,150] In isolated acinar cells, MitoQ reduced H_2O_2-mediated ROS elevations yet afforded no protection against loss of mitochondrial membrane potential induced by CCK or bile acid.[151] Furthermore, MitoQ treatment produced mixed effects in *in vivo* acute pancreatitis models. When applied as a treatment after disease induction, it was unprotective in TLCS-AP but caused partial amelioration of histopathology scores in CER-AP, albeit without concomitant reduction of biochemical parameters (pancreatic trypsin and serum amylase). Interestingly, lung myeloperoxidase and interleukin-6 were concurrently increased by MitoQ in CER-AP, which may involve biphasic effects on ROS in isolated polymorphonuclear leukocytes, inhibiting acute increases but causing elevation at later time points. In pancreatic acinar cells, MitoQ exerted complex actions on mitochondrial bioenergetics that may be in part mediated by the targeting TPP^+ moiety.[63] Thus, MitoQ and the general antioxidant n-acetyl cysteine caused sustained elevations of basal respiration and inhibition of spare respiratory

capacity, an important indicator of the cell's ability to respond to stress,[152,153] actions attributable to an antioxidant action. However, mitochondrial ATP turnover capacity and cellular ATP concentrations were also markedly reduced by both MitoQ and DecylTPP, a nonantioxidant control, suggesting nonspecific actions of the positively charged mitochondrial-targeting moiety.[154–156] Such events were associated with a compensatory elevation of glycolysis and induction of acinar cell death. The results of this study also indicated that ROS exert a significant negative feedback control of basal respiration, potentially by effects on uncoupling proteins, thereby determining basal metabolic function.[157–159] Increasing evidence points to important physiological signaling roles of ROS,[160–162]; therefore, an important consideration when addressing mitochondrial dysfunction by quenching free radicals is the potential consequences of disrupting normal physiological organellar function.

Inflammatory response: bioenergetics changes in blood cells

While most work elucidating bioenergetics changes related to acute pancreatitis has been carried out in pancreatic cells and tissues, changes in blood cells linked to the generation and propagation of a systemic inflammatory response are likely to be of significance.[163,164] The bioenergetics changes in circulating leukocytes are only partially understood, not least because investigations encounter significant technical difficulties during isolation, including potential damage to cells that might induce artifacts. Previously, a modest (1.5-fold) increase in basal respiration was found in total peripheral blood mononuclear cells isolated from acute pancreatitis patients, although ATP production was unaltered.[165] Our recent work has shown subset-specific bioenergetics alterations occur in leukocytes from acute pancreatitis patients that imply functional alterations linked to clinical disease progression.[166] Separation of blood cell subtypes from acute pancreatitis patients was feasible for detailed bioenergetics evaluations using Seahorse XF flux analysis. The study found no difference in basal respiration in blood leukocyte subtypes between acute pancreatitis patients and healthy controls. However, when the mitochondria were challenged with a "stress test" (by manipulation of the respiratory chain with rotenone and antimycin (complexes I and III), oligomycin (ATP synthase), and FCCP (protonophore)[137,167,168]), it was revealed that patient lymphocytes possessed decreased maximal respiration and spare respiratory capacity, indicating functional impairment. Furthermore, a diminished oxidative burst was evident in neutrophils from acute pancreatitis patients, compared to healthy controls, that may indicate compromised activity in this cell type, whereas this was enhanced in both monocytes and lymphocytes; further work is necessary to determine the bases for such altered bioenergetics in circulating blood cells and whether bioenergetics analysis might be predictive of acute pancreatitis severity and thereby assist prognosis.

Therapeutic aspects: maintenance of ATP provision

How may detrimental bioenergetics changes linked to development of acute pancreatitis be prevented? Recent attention has focused on inhibition of the Ca^{2+} overload machinery that causes mitochondrial dysfunction. For example, preclinical studies in murine pancreatic acinar cells have shown that inhibition of Orai1, expressed in pancreatic acinar cells,[48] prevented sustained Ca^{2+} increases induced by fatty acid ethyl esters, coupled with maintenance of mitochondrial membrane potential and a reduction of necrosis.[71] Prevention of the toxic effects of acute pancreatitis precipitants by Orai1 inhibitors GSK-7975A (GlaxoSmithKline) and CM4620 (CalciMedica) was further demonstrated in human pancreatic acinar cells, and the potential for acute pancreatitis treatment indicated from their efficacy in multiple in vivo experimental models (CER-AP, TLCS-AP, and FAEE-AP).[169] More recently, important protective actions of the Orai1 inhibitor CM4620 on immune cells were shown that contribute to the beneficial outcomes in acute pancreatitis,[72] and this agent is currently being evaluated in phase II clinical trials.[170] Interestingly, a recent study has also identified the involvement of TRPM2 channels in Ca^{2+} entry in pancreatic acinar cells and inhibition of these redox-sensitive channels may provide a novel approach to treat biliary acute pancreatitis.[82] As mentioned above, prevention of the downstream effects of Ca^{2+} overload induced by acute pancreatitis precipitants on mitochondria is considered a promising therapeutic strategy, with new MPTP inhibitors in development and evaluation in preclinical models.[113,124,126,171,172] Despite a lack of clarity about the composition of the MPTP, most effort has been directed toward inhibition of the modulatory protein CypD, a prime target in multiple pathologies.[115] Genetic deletion of CypD or pharmacological inhibition with cyclosporin A or its analogue NIM811 treatment effectively reduced histological damage and biochemical alterations in multiple acute pancreatitis models.[113,114,126] Furthermore, since ATP supplementation ameliorated Ca^{2+} clearance from the cytosol and prevented damage in acinar and ductal cells, there is potential to prevent necrotic damage in acute pancreatitis by boosting cellular energy provision.[87,110,173] However, development of an effective means to achieve this in patients is challenging; ATP does not effectively cross cell membranes and is subject to degradation by enzymatic action. Nevertheless, studies using liposomes as a vehicle to deliver ATP have been shown to

exert protective actions against damage in heart, liver, and brain.[174–176] Notably a recent study showed that enhancing cellular energy production with galactose ameliorated asparaginase- and FAEE-induced acute pancreatitis,[61] indicating potential therapeutic value of manipulating energy levels in acute pancreatitis patients. Currently, a multicenter randomized double-blind clinical (GOULASH) trial is being conducted to assess high- versus low-energy administration in the early phase of acute pancreatitis.[177] Previously, enteral nutrition has been shown to significantly decrease mortality and frequency of multiorgan failure in severe acute pancreatitis,[178,179] and results of the GOULASH trial are awaited with interest.

References

1. Pallagi P, Hegyi P, Rakonczay Z, Jr. The physiology and pathophysiology of pancreatic ductal secretion: the background for clinicians. Pancreas 2015; 44:1211–1233.
2. Chandra R, Liddle RA. Regulation of pancreatic secretion. Pancreapedia: Exocrine Pancreas Knowledge Base, 2020. DOI: 10.3998/panc.2020.14.
3. Hegyi P, Petersen OH. The exocrine pancreas: the acinar-ductal tango in physiology and pathophysiology. Rev Physiol Biochem Pharmacol 2013; 165:1–30.
4. van Dijk DPJ, Horstman AMH, Smeets JSJ, den Dulk M, Grabsch HI, Dejong CHC, Rensen SS, Olde Damink SWM, van Loon LJC. Tumour-specific and organ-specific protein synthesis rates in patients with pancreatic cancer. J Cachexia Sarcopenia Muscle 2019; 10:549–556.
5. Bolender RP. Stereological analysis of the guinea pig pancreas. I. Analytical model and quantitative description of nonstimulated pancreatic exocrine cells. J Cell Biol 1974; 61:269–287.
6. Rooman I, Lutz C, Pinho AV, Huggel K, Reding T, Lahoutte T, Verrey F, Graf R, Camargo SM. Amino acid transporters expression in acinar cells is changed during acute pancreatitis. Pancreatology 2013; 13:475–485.
7. Araya S, Kuster E, Gluch D, Mariotta L, Lutz C, Reding TV, Graf R, Verrey F, Camargo SMR. Exocrine pancreas glutamate secretion help to sustain enterocyte nutritional needs under protein restriction. Am J Physiol Gastrointest Liver Physiol 2018; 314:G517–G536.
8. Bauduin H, Colin M, Dumont JE. Energy sources for protein synthesis and enzymatic secretion in rat pancreas in vitro. Biochim Biophys Acta 1969; 174:722–733.
9. Kosowski H, Schild L, Kunz D, Halangk W. Energy metabolism in rat pancreatic acinar cells during anoxia and reoxygenation. Biochim Biophys Acta 1998; 1367:118–126.
10. Vinogradov AD. Steady-state and pre-steady-state kinetics of the mitochondrial F_1F_o ATPase: is ATP synthase a reversible molecular machine? J Exp Biol 2000; 203:41–49.
11. Grover GJ, Marone PA, Koetzner L, Seto-Young D. Energetic signalling in the control of mitochondrial F1F0 ATP synthase activity in health and disease. Int J Biochem Cell Biol 2008; 40:2698–2701.

12. Lehninger AL, Wadkins CL. Oxidative phosphorylation. Annu Rev Biochem 1962; 31:47–78.

13. Campanella M, Casswell E, Chong S, Farah Z, Wieckowski MR, Abramov AY, Tinker A, Duchen MR. Regulation of mitochondrial structure and function by the F1Fo-ATPase inhibitor protein, IF1. Cell Metab 2008; 8:13–25.

14. Tanton H, Voronina S, Evans A, Armstrong J, Sutton R, Criddle DN, Haynes L, Schmid MC, Campbell F, Costello E, Tepikin AV. F1F0-ATP synthase inhibitory factor 1 in the normal pancreas and in pancreatic ductal adenocarcinoma: effects on bioenergetics, invasion and proliferation. Front Physiol 2018; 9:833.

15. Voronina SG, Barrow SL, Simpson AW, Gerasimenko OV, da Silva Xavier G, Rutter GA, Petersen OH, Tepikin AV. Dynamic changes in cytosolic and mitochondrial ATP levels in pancreatic acinar cells. Gastroenterology 2010; 138:1976–1987.

16. Voronina S, Sukhomlin T, Johnson PR, Erdemli G, Petersen OH, Tepikin A. Correlation of NADH and Ca^{2+} signals in mouse pancreatic acinar cells. J Physiol 2002; 539:41–52.

17. Chvanov M, Voronina S, Zhang X, Telnova S, Chard R, Ouyang Y, Armstrong J, Tanton H, Awais M, Latawiec D, Sutton R, Criddle DN, et al. Knockout of the mitochondrial calcium uniporter strongly suppresses stimulus-metabolism coupling in pancreatic acinar cells but does not reduce severity of experimental acute pancreatitis. Cells 2020; 9, 1407: 1–16.

18. Mankad P, James A, Siriwardena AK, Elliott AC, Bruce JI. Insulin protects pancreatic acinar cells from cytosolic calcium overload and inhibition of plasma membrane calcium pump. J Biol Chem 2012; 287:1823–1836.

19. Samad A, James A, Wong J, Mankad P, Whitehouse J, Patel W, Alves-Simoes M, Siriwardena AK, Bruce JI. Insulin protects pancreatic acinar cells from palmitoleic acid-induced cellular injury. J Biol Chem 2014; 289:23582–23595.

20. Korc M, Williams JA, Goldfine ID. Stimulation of the glucose transport system in isolated mouse pancreatic acini by cholecystokinin and analogues. J Biol Chem 1979; 254:7624–7629.

21. Williams JA, Korc M, Dormer RL. Action of secretagogues on a new preparation of functionally intact, isolated pancreatic acini. Am J Physiol 1978; 235:517–524.

22. Denton RM. Regulation of mitochondrial dehydrogenases by calcium ions. Biochim Biophys Acta 2009; 1787:1309–1316.

23. McCormack JG, Halestrap AP, Denton RM. Role of calcium ions in regulation of mammalian intramitochondrial metabolism. Physiol Rev 1990; 70:391–425.

24. Hajnoczky G, Robb-Gaspers LD, Seitz MB, Thomas AP. Decoding of cytosolic calcium oscillations in the mitochondria. Cell 1995; 82:415–424.

25. Kanno T, Saito A, Ikei N. Dose-dependent effect of acetylcholine stimulating respiratory chain and secretion of isolated perfused rat pancreas. Biomed Res 1983; 4:175–186.

26. Murphy JA, Criddle DN, Sherwood M, Chvanov M, Mukherjee R, McLaughlin E, Booth D, Gerasimenko JV, Raraty MG, Ghaneh P, Neoptolemos JP, Gerasimenko OV, et al. Direct activation of cytosolic Ca2+ signaling and enzyme secretion by cholecystokinin in human pancreatic acinar cells. Gastroenterology 2008; 135:632–641.

27. Yule DI. Ca^{2+} signalling in pancreatic acinar cells. Pancreapedia: Exocrine Pancreas Knowledge Base, 2020. DOI: 10.3998/panc.2020.12.

28. Petersen OH, Tepikin AV. Polarized calcium signaling in exocrine gland cells. Annu Rev Physiol 2008; 70:273–299.

29. Williams JA. Cholecystokinin (CCK) regulation of pancreatic acinar cells: physiological actions and signal transduction mechanisms. Compr Physiol 2019; 9:535–564.

30. Thorn P, Lawrie AM, Smith PM, Gallacher DV, Petersen OH. Local and global cytosolic Ca^{2+} oscillations in exocrine cells evoked by agonists and inositol trisphosphate. Cell 1993; 74:661–668.

31. Kasai H, Li YX, Miyashita Y. Subcellular distribution of Ca^{2+} release channels underlying Ca^{2+} waves and oscillations in exocrine pancreas. Cell 1993; 74:669–677.

32. Baughman JM, Perocchi F, Girgis HS, Plovanich M, Belcher-Timme CA, Sancak Y, Bao XR, Strittmatter L, Goldberger O, Bogorad RL, Koteliansky V, Mootha VK. Integrative genomics identifies MCU as an essential component of the mitochondrial calcium uniporter. Nature 2011; 476:341–345.

33. De Stefani D, Raffaello A, Teardo E, Szabo I, Rizzuto R. A forty-kilodalton protein of the inner membrane is the mitochondrial calcium uniporter. Nature 2011; 476:336–340.

34. Mammucari C, Raffaello A, Vecellio Reane D, Gherardi G, De Mario A, Rizzuto R. Mitochondrial calcium uptake in organ physiology: from molecular mechanism to animal models. Pflugers Arch 2018; 470:1165–1179.

35. Csordas G, Golenar T, Seifert EL, Kamer KJ, Sancak Y, Perocchi F, Moffat C, Weaver D, Perez SF, Bogorad R, Koteliansky V, Adijanto J, et al. MICU1 controls both the threshold and cooperative activation of the mitochondrial Ca^{2+} uniporter. Cell Metab 2013; 17:976–987.

36. Kamer KJ, Mootha VK. MICU1 and MICU2 play nonredundant roles in the regulation of the mitochondrial calcium uniporter. EMBO Rep 2014; 15:299–307.

37. Patron M, Checchetto V, Raffaello A, Teardo E, Vecellio Reane D, Mantoan M, Granatiero V, Szabo I, De Stefani D, Rizzuto R. MICU1 and MICU2 finely tune the mitochondrial Ca^{2+} uniporter by exerting opposite effects on MCU activity. Mol Cell 2014; 53:726–737.

38. Perocchi F, Gohil VM, Girgis HS, Bao XR, McCombs JE, Palmer AE, Mootha VK. MICU1 encodes a mitochondrial EF hand protein required for Ca^{2+} uptake. Nature 2010; 467:291–296.

39. Bhosale G, Sharpe JA, Sundier SY, Duchen MR. Calcium signaling as a mediator of cell energy demand and a trigger to cell death. Ann N Y Acad Sci 2015; 1350:107–116.

40. Bhosale G, Sharpe JA, Koh A, Kouli A, Szabadkai G, Duchen MR. Pathological consequences of MICU1 mutations on mitochondrial calcium signalling and bioenergetics. Biochim Biophys Acta Mol Cell Res 2017; 1864:1009–1017.

41. Mallilankaraman K, Doonan P, Cardenas C, Chandramoorthy HC, Muller M, Miller R, Hoffman NE, Gandhirajan RK, Molgo J, Birnbaum MJ, Rothberg BS, Mak DO, et al. MICU1 is an essential gatekeeper for MCU-mediated mitochondrial Ca^{2+} uptake that regulates cell survival. Cell 2012; 151:630–644.

42. Tepikin AV. Mitochondrial junctions with cellular organelles: Ca^{2+} signalling perspective. Pflugers Arch 2018; 470:1181–1192.

43. Park MK, Ashby MC, Erdemli G, Petersen OH, Tepikin AV. Perinuclear, perigranular and sub-plasmalemmal mitochondria have distinct functions in the regulation of cellular calcium transport. EMBO J 2001; 20:1863–1874.

44. Tinel H, Cancela JM, Mogami H, Gerasimenko JV, Gerasimenko OV, Tepikin AV, Petersen OH. Active mitochondria surrounding the pancreatic acinar granule region prevent spreading of inositol trisphosphate-evoked local cytosolic Ca^{2+} signals. EMBO J 1999; 18:4999–5008.

45. Dolman NJ, Gerasimenko JV, Gerasimenko OV, Voronina SG, Petersen OH, Tepikin AV. Stable Golgi-mitochondria complexes and formation of Golgi Ca^{2+} gradients in pancreatic acinar cells. J Biol Chem 2005; 280:15794–15799.

46. Johnson PR, Dolman NJ, Pope M, Vaillant C, Petersen OH, Tepikin AV, Erdemli G. Non-uniform distribution of mitochondria in pancreatic acinar cells. Cell Tissue Res 2003; 313:37–45.

47. Straub SV, Giovannucci DR, Yule DI. Calcium wave propagation in pancreatic acinar cells: functional interaction of inositol 1,4,5-trisphosphate receptors, ryanodine receptors, and mitochondria. J Gen Physiol 2000; 116:547–560.

48. Lur G, Haynes LP, Prior IA, Gerasimenko OV, Feske S, Petersen OH, Burgoyne RD, Tepikin AV. Ribosome-free terminals of rough ER allow formation of STIM1 puncta and segregation of STIM1 from IP_3 receptors. Curr Biol 2009; 19:1648–1653.

49. Bondarenko AI, Jean-Quartier C, Parichatikanond W, Alam MR, Waldeck-Weiermair M, Malli R, Graier WF. Mitochondrial Ca^{2+} uniporter (MCU)-dependent and MCU-independent Ca^{2+} channels coexist in the inner mitochondrial membrane. Pflugers Arch 2014; 466:1411–1420.

50. Hamilton J, Brustovetsky T, Rysted JE, Lin Z, Usachev YM, Brustovetsky N. Deletion of mitochondrial calcium uniporter incompletely inhibits calcium uptake and induction of the permeability transition pore in brain mitochondria. J Biol Chem 2018; 293:15652–15663.

51. Samanta K, Mirams GR, Parekh AB. Sequential forward and reverse transport of the Na^+ Ca^{2+} exchanger generates Ca^{2+} oscillations within mitochondria. Nat Commun 2018; 9:156.

52. Duchen MR, Surin A, Jacobson J. Imaging mitochondrial function in intact cells. Methods Enzymol 2003; 361:353–389.

53. Duchen MR. Ca^{2+}-dependent changes in the mitochondrial energetics in single dissociated mouse sensory neurons. Biochem J 1992; 283:41–50.

54. Manko BO, Manko VV. Mechanisms of respiration intensification of rat pancreatic acini upon carbachol-induced Ca^{2+} release. Acta Physiol (Oxf) 2013; 208:387–399.

55. Morton JC, Armstrong J, Cash N, Ouyang Y, Tepikin A, Sutton R, Criddle DN. Pathophysiological modulation of pancreatic acinar cell bioenergetics by cholecystokinin. Pancreas 2016; 45:1494–1551: P1491–1437.

56. Matsumoto T, Kanno T, Seo Y, Murakami M, Watari H. Dose effects of cholecystokinin and acetylcholine on phosphorus compounds and secretory responses in isolated perfused pancreas of rat. Jpn J Physiol 1991; 41:483–492.

57. Lüthen RE, Niederau C, Ferrell LD, Grendell JH. Energy metabolism in mouse pancreas in response to different dosages of a CCK analogue. Pancreas 1995; 11:141–146.

58. Cuthbertson CM, Christophi C. Disturbances of the microcirculation in acute pancreatitis. Br J Surg 2006; 93:518–530.

59. Nordback IH, Clemens JA, Chacko VP, Olson JL, Cameron JL. Changes in high-energy phosphate metabolism and cell morphology in four models of acute experimental pancreatitis. Ann Surg 1991; 213:341–349.

60. Peng S, Gerasimenko JV, Tsugorka T, Gryshchenko O, Samarasinghe S, Petersen OH, Gerasimenko OV. Calcium and adenosine triphosphate control of cellular pathology: asparaginase-induced pancreatitis elicited via protease-activated receptor 2. Philos Trans R Soc Lond B Biol Sci 2016; 371: 20150423: 1–12.

61. Peng S, Gerasimenko JV, Tsugorka TM, Gryshchenko O, Samarasinghe S, Petersen OH, Gerasimenko OV. Galactose protects against cell damage in mouse models of acute pancreatitis. J Clin Invest 2018; 128:3769–3778.

62. Booth DM, Murphy JA, Mukherjee R, Awais M, Neoptolemos JP, Gerasimenko OV, Tepikin AV, Petersen OH, Sutton R, Criddle DN. Reactive oxygen species induced by bile acid induce apoptosis and protect against necrosis in pancreatic acinar cells. Gastroenterology 2011; 140:2116–2125.

63. Armstrong JA, Cash NJ, Morton JC, Tepikin AV, Sutton R, Criddle DN. Mitochondrial targeting of antioxidants alters pancreatic acinar cell bioenergetics and determines cell fate. Int J Mol Sci 2019; 20, 1700: 1–16.

64. Criddle DN, Gillies S, Baumgartner-Wilson HK, Jaffar M, Chinje EC, Passmore S, Chvanov M, Barrow S, Gerasimenko OV, Tepikin AV, Sutton R, Petersen OH. Menadione-induced reactive oxygen species generation via redox cycling promotes apoptosis of murine pancreatic acinar cells. J Biol Chem 2006; 281:40485–40492.

65. Hofer AM, Curci S, Machen TE, Schulz I. ATP regulates calcium leak from agonist-sensitive internal calcium stores. FASEB J 1996; 10:302–308.

66. Barrow SL, Voronina SG, da Silva Xavier G, Chvanov MA, Longbottom RE, Gerasimenko OV, Petersen OH, Rutter GA, Tepikin AV. ATP depletion inhibits Ca^{2+} release, influx and extrusion in pancreatic acinar cells but not pathological Ca^{2+} responses induced by bile. Pflugers Arch 2008; 455:1025–1039.

67. Futatsugi A, Nakamura T, Yamada MK, Ebisui E, Nakamura K, Uchida K, Kitaguchi T, Takahashi-Iwanaga H, Noda T, Aruga J, Mikoshiba K. IP3 receptor types 2 and 3 mediate exocrine secretion underlying energy metabolism. Science 2005; 309:2232–2234.

68. Park HS, Betzenhauser MJ, Won JH, Chen J, Yule DI. The type 2 inositol (1,4,5)-trisphosphate ($InsP_3$) receptor determines the sensitivity of $InsP_3$-induced Ca^{2+} release to ATP in pancreatic acinar cells. J Biol Chem 2008; 283:26081–26088.

69. Betzenhauser MJ, Wagner LE, 2nd, Iwai M, Michikawa T, Mikoshiba K, Yule DI. ATP modulation of Ca^{2+} release by type-2 and type-3 inositol (1, 4, 5)-triphosphate receptors. Differing ATP sensitivities and molecular determinants of action. J Biol Chem 2008; 283:21579–21587.

70. Voronina S, Collier D, Chvanov M, Middlehurst B, Beckett AJ, Prior IA, Criddle DN, Begg M, Mikoshiba K, Sutton

R, Tepikin AV. The role of Ca^{2+} influx in endocytic vacuole formation in pancreatic acinar cells. Biochem J 2015; 465:405–412.

71. Gerasimenko JV, Gryshchenko O, Ferdek PE, Stapleton E, Hebert TO, Bychkova S, Peng S, Begg M, Gerasimenko OV, Petersen OH. Ca^{2+} release-activated Ca^{2+} channel blockade as a potential tool in antipancreatitis therapy. Proc Natl Acad Sci U S A 2013; 110:13186–13191.

72. Waldron RT, Chen Y, Pham H, Go A, Su HY, Hu C, Wen L, Husain SZ, Sugar CA, Roos J, Ramos S, Lugea A, et al. The Orai Ca^{2+} channel inhibitor CM4620 targets both parenchymal and immune cells to reduce inflammation in experimental acute pancreatitis. J Physiol 2019; 597:3085–3105.

73. Ahuja M, Schwartz DM, Tandon M, Son A, Zeng M, Swaim W, Eckhaus M, Hoffman V, Cui Y, Xiao B, Worley PF, Muallem S. Orai1-mediated antimicrobial secretion from pancreatic acini shapes the gut microbiome and regulates gut innate immunity. Cell Metab 2017; 25:635–646.

74. Kim MS, Hong JH, Li Q, Shin DM, Abramowitz J, Birnbaumer L, Muallem S. Deletion of TRPC3 in mice reduces store-operated Ca^{2+} influx and the severity of acute pancreatitis. Gastroenterology 2009; 137:1509–1517.

75. Gamberucci A, Innocenti B, Fulceri R, Banhegyi G, Giunti R, Pozzan T, Benedetti A. Modulation of Ca^{2+} influx dependent on store depletion by intracellular adenine-guanine nucleotide levels. J Biol Chem 1994; 269:23597–23602.

76. Marriott I, Mason MJ. ATP depletion inhibits capacitative Ca^{2+} entry in rat thymic lymphocytes. Am J Physiol Cell Physiol 1995; 269:C766–C774.

77. Hoth M, Fanger CM, Lewis RS. Mitochondrial regulation of store-operated calcium signaling in T lymphocytes. J Cell Biol 1997; 137:633–648.

78. Glitsch MD, Bakowski D, Parekh AB. Store-operated Ca^{2+} entry depends on mitochondrial Ca^{2+} uptake. EMBO J 2002; 21:6744–6754.

79. Samanta K, Douglas S, Parekh AB. Mitochondrial calcium uniporter MCU supports cytoplasmic Ca^{2+} oscillations, store-operated Ca^{2+} entry and Ca^{2+}-dependent gene expression in response to receptor stimulation. PLoS One 2014; 9:e101188.

80. Romac JM, Shahid RA, Swain SM, Vigna SR, Liddle RA. Piezo1 is a mechanically activated ion channel and mediates pressure induced pancreatitis. Nat Commun 2018; 9:1715.

81. Swain SM, Romac JM, Shahid RA, Pandol SJ, Liedtke W, Vigna SR, Liddle RA. TRPV4 channel opening mediates pressure-induced pancreatitis initiated by Piezo1 activation. J Clin Invest 2020; 130:2527–2541.

82. Fanczal J, Pallagi P, Gorog M, Diszhazi G, Almassy J, Madacsy T, Varga A, Csernay-Biro P, Katona X, Toth E, Molnar R, Rakonczay Z, Jr., et al. TRPM2-mediated extracellular Ca^{2+} entry promotes acinar cell necrosis in biliary acute pancreatitis. J Physiol 2020; 598:1253–1270.

83. Petersen OH, Sutton R, Criddle DN. Failure of calcium microdomain generation and pathological consequences. Cell Calcium 2006; 40:593–600.

84. Raraty M, Ward J, Erdemli G, Vaillant C, Neoptolemos JP, Sutton R, Petersen OH. Calcium-dependent enzyme activation and vacuole formation in the apical granular region

of pancreatic acinar cells. Proc Natl Acad Sci U S A 2000; 97:13126–13131.

85. Ward JB, Petersen OH, Jenkins SA, Sutton R. Is an elevated concentration of acinar cytosolic free ionised calcium the trigger for acute pancreatitis? Lancet 1995; 346:1016–1019.

86. Criddle DN, Raraty MG, Neoptolemos JP, Tepikin AV, Petersen OH, Sutton R. Ethanol toxicity in pancreatic acinar cells: mediation by nonoxidative fatty acid metabolites. Proc Natl Acad Sci U S A 2004; 101:10738–10743.

87. Criddle DN, Murphy J, Fistetto G, Barrow S, Tepikin AV, Neoptolemos JP, Sutton R, Petersen OH. Fatty acid ethyl esters cause pancreatic calcium toxicity via inositol trisphosphate receptors and loss of ATP synthesis. Gastroenterology 2006; 130:781–793.

88. Hirano T, Manabe T, Tobe T. Protective effects of gabexate mesilate (FOY) against impaired pancreatic energy metabolism in rat acute pancreatitis induced by caerulein. Life Sci 1991; 49:PL179–PL184.

89. Halangk W, Matthias R, Schild L, Meyer F, Schulz HU, Lippert H. Effect of supramaximal cerulein stimulation on mitochondrial energy metabolism in rat pancreas. Pancreas 1998; 16:88–95.

90. Chvanov M, De Faveri F, Moore D, Sherwood MW, Awais M, Voronina S, Sutton R, Criddle DN, Haynes L, Tepikin AV. Intracellular rupture, exocytosis and actin interaction of endocytic vacuoles in pancreatic acinar cells: initiating events in acute pancreatitis. J Physiol 2018; 596:2547–2564.

91. Malla SR, Krueger B, Wartmann T, Sendler M, Mahajan UM, Weiss FU, Thiel FG, De Boni C, Gorelick FS, Halangk W, Aghdassi AA, Reinheckel T, et al. Early trypsin activation develops independently of autophagy in caerulein-induced pancreatitis in mice. Cell Mol Life Sci 2020; 77:1811–1825.

92. Biczo G, Hegyi P, Dosa S, Shalbuyeva N, Berczi S, Sinervirta R, Hracsko Z, Siska A, Kukor Z, Jarmay K, Venglovecz V, Varga IS, et al. The crucial role of early mitochondrial injury in L-lysine-induced acute pancreatitis. Antioxid Redox Signal 2011; 15:2669–2681.

93. Katona M, Hegyi P, Kui B, Balla Z, Rakonczay Z, Jr., Razga Z, Tiszlavicz L, Maleth J, Venglovecz V. A novel, protective role of ursodeoxycholate in bile-induced pancreatic ductal injury. Am J Physiol Gastrointest Liver Physiol 2016; 310:G193–G204.

94. Maleth J, Venglovecz V, Razga Z, Tiszlavicz L, Rakonczay Z, Hegyi P. Non-conjugated chenodeoxycholate induces severe mitochondrial damage and inhibits bicarbonate transport in pancreatic duct cells. Gut 2011; 60:136–138.

95. Weidenbach H, Lerch MM, Gress TM, Pfaff D, Turi S, Adler G. Vasoactive mediators and the progression from oedematous to necrotising experimental acute pancreatitis. Gut 1995; 37:434–440.

96. Sanfey H, Cameron JL. Increased capillary permeability: an early lesion in acute pancreatitis. Surgery 1984; 96:485–491.

97. McEntee G, Leahy A, Cottell D, Dervan P, McGeeney K, Fitzpatrick JM. Three-dimensional morphological study of the pancreatic microvasculature in caerulein-induced experimental pancreatitis. Br J Surg 1989; 76:853–855.

98. Plusczyk T, Westermann S, Rathgeb D, Feifel G. Acute pancreatitis in rats: effects of sodium taurocholate, CCK-8, and

Sec on pancreatic microcirculation. Am J Physiol Gastrointest Liver Physiol 1997; 272:G310–G320.

99. Tomkotter L, Erbes J, Trepte C, Hinsch A, Dupree A, Bockhorn M, Mann O, Izbicki JR, Bachmann K. The effects of pancreatic microcirculatory disturbances on histopathologic tissue damage and the outcome in severe acute pancreatitis. Pancreas 2016; 45:248–253.

100. Kinnala PJ, Kuttila KT, Gronroos JM, Havia TV, Nevalainen TJ, Niinikoski JH. Pancreatic tissue perfusion in experimental acute pancreatitis. Eur J Surg 2001; 167:689–694.

101. Knoefel WT, Kollias N, Warshaw AL, Waldner H, Nishioka NS, Rattner DW. Pancreatic microcirculatory changes in experimental pancreatitis of graded severity in the rat. Surgery 1994; 116:904–913.

102. Voronina S, Longbottom R, Sutton R, Petersen OH, Tepikin A. Bile acids induce calcium signals in mouse pancreatic acinar cells: implications for bile-induced pancreatic pathology. J Physiol 2002; 540:49–55.

103. Diczfalusy MA, Bjorkhem I, Einarsson C, Hillebrant CG, Alexson SE. Characterization of enzymes involved in formation of ethyl esters of long-chain fatty acids in humans. J Lipid Res. 2001; 42:1025–1032.

104. Huang W, Booth DM, Cane MC, Chvanov M, Javed MA, Elliott VL, Armstrong JA, Dingsdale H, Cash N, Li Y, Greenhalf W, Mukherjee R, et al. Fatty acid ethyl ester synthase inhibition ameliorates ethanol-induced Ca^{2+}-dependent mitochondrial dysfunction and acute pancreatitis. Gut 2014; 63:1313–1324.

105. Criddle DN. The role of fat and alcohol in acute pancreatitis: a dangerous liaison. Pancreatology 2015; 15:S6–S12.

106. Muallem S, Beeker T, Pandol SJ. Role of Na^+/Ca^{2+} exchange and the plasma membrane Ca^{2+} pump in hormone-mediated Ca^{2+} efflux from pancreatic acini. J Membr Biol 1988; 102:153–162.

107. Brini M, Carafoli E. Calcium pumps in health and disease. Physiol Rev 2009; 89:1341–1378.

108. Baggaley EM, Elliott AC, Bruce JI. Oxidant-induced inhibition of the plasma membrane Ca^{2+}-ATPase in pancreatic acinar cells: role of the mitochondria. Am J Physiol Cell Physiol 2008; 295:C1247–C1260.

109. Criddle DN, Gerasimenko JV, Baumgartner HK, Jaffar M, Voronina S, Sutton R, Petersen OH, Gerasimenko OV. Calcium signalling and pancreatic cell death: apoptosis or necrosis? Cell Death Differ 2007; 14:1285–1294.

110. Judak L, Hegyi P, Rakonczay Z, Jr., Maleth J, Gray MA, Venglovecz V. Ethanol and its non-oxidative metabolites profoundly inhibit CFTR function in pancreatic epithelial cells which is prevented by ATP supplementation. Pflugers Arch 2014; 466:549–562.

111. Maleth J, Balazs A, Pallagi P, Balla Z, Kui B, Katona M, Judak L, Nemeth I, Kemeny LV, Rakonczay Z, Jr., Venglovecz V, Foldesi I, et al. Alcohol disrupts levels and function of the cystic fibrosis transmembrane conductance regulator to promote development of pancreatitis. Gastroenterology 2015; 148:427–439.

112. Shalbueva N, Mareninova OA, Gerloff A, Yuan J, Waldron RT, Pandol SJ, Gukovskaya AS. Effects of oxidative alcohol metabolism on the mitochondrial permeability transition pore and necrosis in a mouse model of alcoholic pancreatitis. Gastroenterology 2012; 144:437–446.

113. Mukherjee R, Mareninova OA, Odinokova IV, Huang W, Murphy J, Chvanov M, Javed MA, Wen L, Booth DM, Cane MC, Awais M, Gavillet B, et al. Mechanism of mitochondrial permeability transition pore induction and damage in the pancreas: inhibition prevents acute pancreatitis by protecting production of ATP. Gut 2016; 65:1333–1346.

114. Biczo G, Vegh ET, Shalbueva N, Mareninova OA, Elperin J, Lotshaw E, Gretler S, Lugea A, Malla SR, Dawson D, Ruchala P, Whitelegge J, et al. Mitochondrial dysfunction, through impaired autophagy, leads to endoplasmic reticulum stress, deregulated lipid metabolism, and pancreatitis in animal models. Gastroenterology 2018; 154:689–703.

115. Briston T, Selwood DL, Szabadkai G, Duchen MR. Mitochondrial permeability transition: a molecular lesion with multiple drug targets. Trends Pharmacol Sci 2019; 40:50–70.

116. Karch J, Molkentin JD. Identity of the elusive mitochondrial permeability transition pore: what it might be, what it was, and what it still could be. Curr Op Physiol 2018; 3:57–62.

117. Šileikytė J, Forte M. The mitochondrial permeability transition in mitochondrial disorders. Oxid Med Cell Longev 2019; 2019:1–11.

118. Bernardi P. Why F-ATP synthase remains a strong candidate as the mitochondrial permeability transition pore. Front Physiol 2018; 9:1543.

119. He J, Ford HC, Carroll J, Ding S, Fearnley IM, Walker JE. Persistence of the mitochondrial permeability transition in the absence of subunit c of human ATP synthase. Proc Natl Acad Sci U S A 2017; 114:3409–3414.

120. He J, Carroll J, Ding S, Fearnley IM, Walker JE. Permeability transition in human mitochondria persists in the absence of peripheral stalk subunits of ATP synthase. Proc Natl Acad Sci U S A 2017; 114:9086–9091.

121. Bround MJ, Bers DM, Molkentin JD. A 20/20 view of ANT function in mitochondrial biology and necrotic cell death. J Mol Cell Cardiol 2020; 144:A3–A13.

122. Karch J, Bround MJ, Khalil H, Sargent MA, Latchman N, Terada N, Peixoto PM, Molkentin JD. Inhibition of mitochondrial permeability transition by deletion of the ANT family and CypD. Sci Adv 2019; 5:eaaw4597.

123. Criddle DN. Keeping mitochondria happy—benefits of a pore choice in acute pancreatitis. J Physiol 2019; 597:5741–5742.

124. Javed MA, Wen L, Awais M, Latawiec D, Huang W, Chvanov M, Schaller S, Bordet T, Michaud M, Pruss R, Tepikin A, Criddle D, et al. TRO40303 ameliorates alcohol-induced pancreatitis through reduction of fatty acid ethyl ester-induced mitochondrial injury and necrotic cell death. Pancreas 2018; 47:18–24.

125. Schild L, Matthias R, Stanarius A, Wolf G, Augustin W, Halangk W. Induction of permeability transition in pancreatic mitochondria by cerulein in rats. Mol Cell Biochem 1999; 195:191–197.

126. Toth E, Maleth J, Zavogyan N, Fanczal J, Grassalkovich A, Erdos R, Pallagi P, Horvath G, Tretter L, Balint ER, Rakonczay Z, Jr., Venglovecz V, et al. Novel mitochondrial transition pore inhibitor N-methyl-4-isoleucine cyclosporin

is a new therapeutic option in acute pancreatitis. J Physiol 2019; 597:5879–5898.

127. Laukkarinen JM, Van Acker GJ, Weiss ER, Steer ML, Perides G. A mouse model of acute biliary pancreatitis induced by retrograde pancreatic duct infusion of Na-taurocholate. Gut 2007; 56:1590–1598.

128. Giorgio V, Guo L, Bassot C, Petronilli V, Bernardi P. Calcium and regulation of the mitochondrial permeability transition. Cell Calcium 2018; 70:56–63.

129. Elrod JW, Molkentin JD. Physiologic functions of cyclophilin D and the mitochondrial permeability transition pore. Circulation Journal 2013; 77:1111–1122.

130. Bernardi P. The mitochondrial permeability transition pore: a mystery solved? Front Physiol 2013; 4:95.

131. Criddle DN. Reactive oxygen species, Ca^{2+} stores and acute pancreatitis; a step closer to therapy? Cell Calcium 2016; 60:180–189.

132. Chvanov M, Huang W, Jin T, Wen L, Armstrong J, Elliot V, Alston B, Burdyga A, Criddle DN, Sutton R, Tepikin AV. Novel lipophilic probe for detecting near-membrane reactive oxygen species responses and its application for studies of pancreatic acinar cells: effects of pyocyanin and L-ornithine. Antioxid Redox Signal 2015; 22:451–464.

133. Bhatia M, Wallig MA, Hofbauer B, Lee HS, Frossard JL, Steer ML, Saluja AK. Induction of apoptosis in pancreatic acinar cells reduces the severity of acute pancreatitis. Biochem Biophys Res Commun 1998; 246:476–483.

134. Kaiser AM, Saluja AK, Sengupta A, Saluja M, Steer ML. Relationship between severity, necrosis, and apoptosis in five models of experimental acute pancreatitis. Am J Physiol Cell Physiol 1995; 269:C1295–C1304.

135. Kloppel G, Maillet B. Pathology of acute and chronic pancreatitis. Pancreas 1993; 8:659–670.

136. Mareninova OA, Sung KF, Hong P, Lugea A, Pandol SJ, Gukovsky I, Gukovskaya AS. Cell death in pancreatitis: caspases protect from necrotizing pancreatitis. J Biol Chem 2006; 281:3370–3381.

137. Armstrong JA, Cash NJ, Ouyang Y, Morton JC, Chvanov M, Latawiec D, Awais M, Tepikin AV, Sutton R, Criddle DN. Oxidative stress alters mitochondrial bioenergetics and modifies pancreatic cell death independently of cyclophilin D, resulting in an apoptosis-to-necrosis shift. J Biol Chem 2018; 293:8032–8047.

138. Basso E, Fante L, Fowlkes J, Petronilli V, Forte MA, Bernardi P. Properties of the permeability transition pore in mitochondria devoid of Cyclophilin D. J Biol Chem 2005; 280:18558–18561.

139. He L, Lemasters JJ. Regulated and unregulated mitochondrial permeability transition pores: a new paradigm of pore structure and function? FEBS letters 2002; 512:1–7.

140. Armstrong JS, Yang H, Duan W, Whiteman M. Cytochrome bc_1 regulates the mitochondrial permeability transition by two distinct pathways. J Biol Chem 2004; 279:50420–50428.

141. Armstrong JA, Cash N, Soares PM, Souza MH, Sutton R, Criddle DN. Oxidative stress in acute pancreatitis: lost in translation? Free Radic Res 2013; 47:917–933.

142. Murphy MP, Hartley RC. Mitochondria as a therapeutic target for common pathologies. Nat Rev Drug Discov 2018; 17: 865–886.

143. Asin-Cayuela J, Manas AR, James AM, Smith RA, Murphy MP. Fine-tuning the hydrophobicity of a mitochondria-targeted antioxidant. FEBS Lett 2004; 571:9–16.

144. Chouchani ET, Methner C, Nadtochiy SM, Logan A, Pell VR, Ding S, James AM, Cocheme HM, Reinhold J, Lilley KS, Partridge L, Fearnley IM, et al. Cardioprotection by S-nitrosation of a cysteine switch on mitochondrial complex I. Nat Med 2013; 19:753–759.

145. Dare AJ, Bolton EA, Pettigrew GJ, Bradley JA, Saeb-Parsy K, Murphy MP. Protection against renal ischemia-reperfusion injury in vivo by the mitochondria targeted antioxidant MitoQ. Redox Biol 2015; 5:163–168.

146. Dashdorj A, Jyothi KR, Lim S, Jo A, Nguyen MN, Ha J, Yoon KS, Kim HJ, Park JH, Murphy MP, Kim SS. Mitochondria-targeted antioxidant MitoQ ameliorates experimental mouse colitis by suppressing NLRP3 inflammasome-mediated inflammatory cytokines. BMC Med 2013; 11:178.

147. Chacko BK, Reily C, Srivastava A, Johnson MS, Ye Y, Ulasova E, Agarwal A, Zinn KR, Murphy MP, Kalyanaraman B, Darley-Usmar V. Prevention of diabetic nephropathy in Ins2$^{+/-AkitaJ}$ mice by the mitochondria-targeted therapy MitoQ. Biochem J 2010; 432:9–19.

148. Lowes DA, Webster NR, Murphy MP, Galley HF. Antioxidants that protect mitochondria reduce interleukin-6 and oxidative stress, improve mitochondrial function, and reduce biochemical markers of organ dysfunction in a rat model of acute sepsis. Br J Anaesth 2013; 110:472–480.

149. Snow BJ, Rolfe FL, Lockhart MM, Frampton CM, O'Sullivan JD, Fung V, Smith RA, Murphy MP, Taylor KM, Protect Study G. A double-blind, placebo-controlled study to assess the mitochondria-targeted antioxidant MitoQ as a disease-modifying therapy in Parkinson's disease. Mov Disord 2010; 25:1670–1674.

150. Pham T, MacRae CL, Broome SC, D'souza RF, Narang R, Wang HW, Mori TA, Hickey AJR, Mitchell CJ, Merry TL. MitoQ and CoQ10 supplementation mildly suppresses skeletal muscle mitochondrial hydrogen peroxide levels without impacting mitochondrial function in middle-aged men. Eur J Appl Physiol 2020; 120:1657–1669.

151. Huang W, Cash N, Wen L, Szatmary P, Mukherjee R, Armstrong J, Chvanov M, Tepikin AV, Murphy MP, Sutton R, Criddle DN. Effects of the mitochondria-targeted antioxidant mitoquinone in murine acute pancreatitis. Mediators Inflamm 2015; 2015:901780.

152. Sansbury BE, Jones SP, Riggs DW, Darley-Usmar VM, Hill BG. Bioenergetic function in cardiovascular cells: the importance of the reserve capacity and its biological regulation. Chem Biol Interact 2011; 191:288–295.

153. Yadava N, Nicholls DG. Spare respiratory capacity rather than oxidative stress regulates glutamate excitotoxicity after partial respiratory inhibition of mitochondrial complex I with rotenone. J Neurosci 2007; 27:7310–7317.

154. Reily C, Mitchell T, Chacko BK, Benavides G, Murphy MP, Darley-Usmar V. Mitochondrially targeted compounds and their impact on cellular bioenergetics. Redox Biol 2013; 1:86–93.

155. Pokrzywinski KL, Biel TG, Kryndushkin D, Rao VA. Therapeutic targeting of the mitochondria initiates excessive superoxide production and mitochondrial depolarization causing decreased mtDNA integrity. PLoS One 2016; 11:e0168283.

156. Gottwald EM, Duss M, Bugarski M, Haenni D, Schuh CD, Landau EM, Hall AM. The targeted anti-oxidant MitoQ causes mitochondrial swelling and depolarization in kidney tissue. Physiol Rep 2018; 6:e13667.

157. Echtay KS, Murphy MP, Smith RA, Talbot DA, Brand MD. Superoxide activates mitochondrial uncoupling protein 2 from the matrix side. Studies using targeted antioxidants. J Biol Chem 2002; 277:47129–47135.

158. Brand MD, Affourtit C, Esteves TC, Green K, Lambert AJ, Miwa S, Pakay JL, Parker N. Mitochondrial superoxide: production, biological effects, and activation of uncoupling proteins. Free Radic Biol Med 2004; 37:755–767.

159. Arsenijevic D, Onuma H, Pecqueur C, Raimbault S, Manning BS, Miroux B, Couplan E, Alves-Guerra MC, Goubern M, Surwit R, Bouillaud F, Richard D, et al. Disruption of the uncoupling protein-2 gene in mice reveals a role in immunity and reactive oxygen species production. Nat Genet 2000; 26:435–439.

160. Hamanaka RB, Chandel NS. Mitochondrial reactive oxygen species regulate cellular signaling and dictate biological outcomes. Trends Biochem Sci 2010; 35:505–513.

161. Finkel T. Signal transduction by reactive oxygen species. J Cell Biol 2011; 194:7–15.

162. Droge W. Free radicals in the physiological control of cell function. Physiol Rev 2002; 82:47–95.

163. Mayerle J, Sendler M, Hegyi E, Beyer G, Lerch MM, Sahin-Toth M. Genetics, cell biology, and pathophysiology of pancreatitis. Gastroenterology 2019; 156:1951–1968 e1951.

164. Sendler M, Weiss FU, Golchert J, Homuth G, van den Brandt C, Mahajan UM, Partecke LI, Doring P, Gukovsky I, Gukovskaya AS, Wagh PR, Lerch MM, et al. Cathepsin B-mediated activation of trypsinogen in endocytosing macrophages increases severity of pancreatitis in mice. Gastroenterology 2018; 154:704–718.

165. Chakraborty M, Hickey AJ, Petrov MS, Macdonald JR, Thompson N, Newby L, Sim D, Windsor JA, Phillips AR. Mitochondrial dysfunction in peripheral blood mononuclear cells in early experimental and clinical acute pancreatitis. Pancreatology 2016; 16:739–747.

166. Morton JC, Armstrong JA, Sud A, Tepikin AV, Sutton R, Criddle DN. Altered bioenergetics of blood cell sub-populations in acute pancreatitis patients. J Clin Med 2019; 8:2201.

167. Armstrong JA, Sutton R, Criddle DN. Pancreatic acinar cell preparation for oxygen consumption and lactate production analysis. Bio-protocol 2020; 10:e3627.

168. Brand MD, Nicholls DG. Assessing mitochondrial dysfunction in cells. Biochem J 2011; 435:297–312.

169. Wen L, Voronina S, Javed MA, Awais M, Szatmary P, Latawiec D, Chvanov M, Collier D, Huang W, Barrett J,

Begg M, Stauderman K, et al. Inhibitors of ORAI1 prevent cytosolic calcium-associated injury of human pancreatic acinar cells and acute pancreatitis in 3 mouse models. Gastroenterology 2015; 149:481–492.

170. Stauderman KA. CRAC channels as targets for drug discovery and development. Cell Calcium 2018; 74:147–159.

171. Antonucci S, Di Sante M, Sileikyte J, Deveraux J, Bauer T, Bround MJ, Menabò R, Paillard M, Alanova P, Carraro M, Ovize M, Molkentin JD, et al. A novel class of cardioprotective small-molecule PTP inhibitors. Pharmacol Res 2020; 151:104548.

172. Valasani KR, Vangavaragu JR, Day VW, Yan SS. Structure based design, synthesis, pharmacophore modeling, virtual screening, and molecular docking studies for identification of novel cyclophilin D inhibitors. J Chem Inf Model 2014; 54:902–912.

173. Maleth J, Hegyi P, Rakonczay Z, Jr., Venglovecz V. Breakdown of bioenergetics evoked by mitochondrial damage in acute pancreatitis: mechanisms and consequences. Pancreatology 2015; 15:S18–22.

174. Verma DD, Hartner WC, Levchenko TS, Bernstein EA, Torchilin VP. ATP-loaded liposomes effectively protect the myocardium in rabbits with an acute experimental myocardial infarction. Pharm Res 2005; 22:2115–2120.

175. Konno H, Matin AF, Maruo Y, Nakamura S, Baba S. Liposomal ATP protects the liver from injury during shock. Eur Surg Res 1996; 28:140–145.

176. Laham A, Claperon N, Durussel JJ, Fattal E, Delattre J, Puisieux F, Couvreur P, Rossignol P. Intracarotidal administration of liposomally-entrapped ATP: improved efficiency against experimental brain ischemia. Pharmacol Res Commun 1988; 20:699–705.

177. Marta K, Szabo AN, Pecsi D, Varju P, Bajor J, Godi S, Sarlos P, Miko A, Szemes K, Papp M, Tornai T, Vincze A, et al. High versus low energy administration in the early phase of acute pancreatitis (GOULASH trial): protocol of a multicentre randomised double-blind clinical trial. BMJ Open 2017; 7:e015874.

178. Parniczky A, Kui B, Szentesi A, Balazs A, Szucs A, Mosztbacher D, Czimmer J, Sarlos P, Bajor J, Godi S, Vincze A, Illes A, et al. Prospective, multicentre, nationwide clinical data from 600 cases of acute pancreatitis. PLoS One 2016; 11:e0165309.

179. Marta K, Farkas N, Szabo I, Illes A, Vincze A, Par G, Sarlos P, Bajor J, Szucs A, Czimmer J, Mosztbacher D, Parniczky A, et al. Meta-analysis of early nutrition: the benefits of enteral feeding compared to a nil per os diet not only in severe, but also in mild and moderate acute pancreatitis. Int J Mol Sci 2016; 17:1691.

Chapter 4

Cyclic nucleotides as mediators of acinar and ductal function

Thomas Albers[1] and Maria Eugenia Sabbatini[2]

Departments of [1]Chemistry and Physics and [2]Biological Sciences, Augusta University

Cyclic Nucleotides and Their Biosynthesis

Cyclic nucleotides, like other nucleotides, are composed of three functional groups: a ribose sugar, a nitrogenous base, and a single phosphate group. There are two types of nitrogenous bases: purines (adenine and guanine) and pyrimidines (cytosine, uracil, and thymine). A cyclic nucleotide, unlike other nucleotides, has a cyclic bond arrangement between the ribose sugar and the phosphate group. There are two main groups of cyclic nucleotides: the canonical or well-established and the noncanonical or unknown-function cyclic nucleotides. The two well-established cyclic nucleotides are adenosine-3',5'-cyclic monophosphate (cyclic AMP) and guanine-3',5'-cyclic monophosphate (cyclic GMP). Both cyclic AMP and cyclic GMP are second messengers. The noncanonical cyclic nucleotides include the purines inosine-3',5'-cyclic monophosphate (cyclic IMP), xanthosine-3',5'-cyclic monophosphate (cyclic XMP), and the pyrimidines cytidine-3',5'-cyclic monophosphate (cyclic cCMP), uridine-3',5'-cyclic monophosphate (cyclic UMP), and thymidine-3',5'-cyclic monophosphate (cTMP).[1] An overview of the noncanonical cyclic nucleotides is provided later.

A cyclase enzyme (lyase) catalyzes the formation of the cyclic nucleotide from its nucleotide triphosphate precursor (**Figure 1**). Cyclic nucleotides form when the phosphate group of the molecule of nucleotide triphosphate (ATP or GTP) is attacked by the 3' hydroxyl group of the ribose, forming a cyclic 3',5'-phosphate ester with release of pyrophosphate. This cyclic conformation allows cyclic nucleotides to bind to proteins to which other nucleotides cannot. The reaction is an intramolecular nucleophilic reaction and requires Mg^{2+} as a cofactor, whose function is to decrease the overall negative charge on the ATP by complexing with two of its negatively charged oxygen. If its negative charge is not reduced, the nucleotide triphosphate cannot be approached by a nucleophile, which is, in this reaction, the 3' hydroxyl group of the ribose.[2] Soluble AC prefers Ca^{2+} to Mg^{2+} as the cofactor to coordinate ATP binding and catalysis.[3]

Canonical Cyclic Nucleotide Signaling in the Exocrine Pancreas

Cyclic nucleotide signaling can be initiated by two general mechanisms. One mechanism is the binding of an extracellular ligand to a transmembrane G-protein-coupled receptor (GPCR). The receptor protein has seven transmembrane α-helices connected by alternating cytosolic and extracellular loops. The N-terminus is in the extracellular space, while the C-terminus is in the cytosol. The ligand-binding site is in the extracellular domain and the cytosolic domain has a heterotrimeric G protein-binding site.[4] After a ligand binds to the GPCR, it activates a heterotrimeric G-protein, which is composed of three subunits: a guanine nucleotide-binding α-subunit and a $\beta\gamma$-heterodimer.[5] Depending on which family the G protein is, it goes on to activate ($G_{\alpha s}$ protein subunit) or inhibit ($G_{\alpha i}$ protein subunit) the membrane-bound cyclase.

The second signaling mechanism involves the binding of a small signaling molecule to a soluble cyclase. The signal source can be either extracellular, such as nitric oxide (NO),[6] or intracellular, such as bicarbonate.[7] The signaling by an extracellular ligand is limited by its ability to cross the plasma membrane. In the cytosol, the signal molecule binds to the heme-binding domain of the soluble cyclase. The cyclase, in turn, increases the intracellular levels of cyclic nucleotides.[8,9]

In the exocrine pancreas, adenylyl cyclases can be activated by either extracellular or intracellular signals. The extracellular signals can be a neurotransmitter, such as vasoactive intestinal polypeptide (VIP), a hormone, such as secretin,[10] or a gas, such as NO.[11] Intracellular signals include HCO_3-.[12] The increase in the cyclic nucleotide levels modifies the activity of downstream effectors such as kinases,[13,14] guanine-nucleotide-exchange factor (GEF),[15]

---

e-mail: msabbatini@augusta.edu

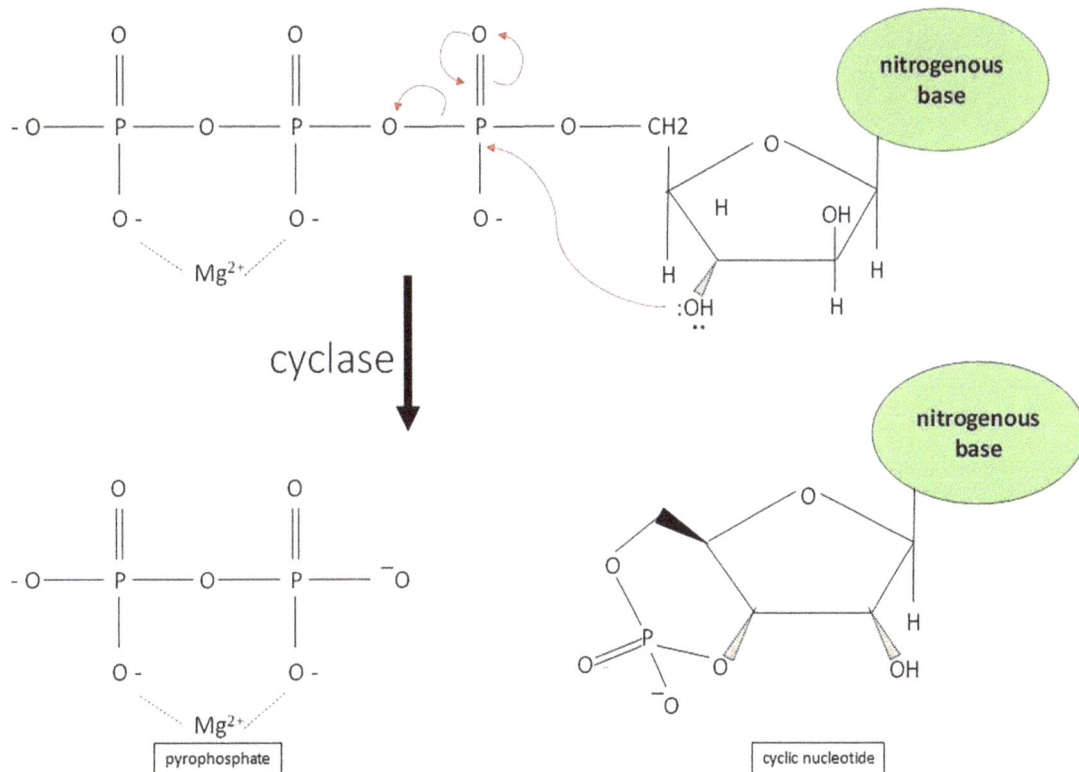

Figure 1. Two aspartic acid residues in the active site of the cyclase (AC or GC) promote the binding of ATP. Two Mg^{2+} ions are required to decrease the overall negative charge on the ATP by complexing with two of its negatively charged oxygen. Mg^{2+} is involved in the deprotonation of the 3-hydroxyl group in the ribose ring of ATP. Soluble AC uses Ca^{2+} rather than Mg^{2+} as a coenzyme. This step is necessary for the nucleophilic catalysis on the 5' α-phosphate by the newly formed oxyanion. The end products of this catalytic reaction are a cyclic nucleotide (cyclic AMP or cyclic GMP) and a pyrophosphate group.

RNA-binding protein,[16] ion channels,[17] and phosphodiesterases,[18] which are discussed later in this chapter.

Adenylyl Cyclase/Cyclic AMP Signaling

Cyclic AMP is formed from cytosolic ATP by the enzyme adenylyl cyclase. There are 10 isoforms of adenylyl cyclases; nine are anchored in the plasma membrane, with their catalytic portion protruding into the cytosol, and one is soluble.[19]

Transmembrane AC

The nine transmembrane AC isoforms are each coded by a different gene (**Figure 2**). The human *ADCY1* gene is located on chromosome 7 at p12.3, human *ADCY2* gene on chromosome 5 at p15.3, human *ADCY3* gene on chromosome 2 at p23.3, human *ADCY4* gene on chromosome 14 at q12, human *ADCY5* gene on chromosome 3 at q21.1, human *ADCY6* gene on chromosome 12 at q12-q13,

human *ADCY7* gene on chromosome 16 at q12.1, human *ADCY8* gene on chromosome 8 at q24, and human *ADCY9* gene on chromosome 16 at p13.3.[20] The transmembrane AC isoforms share a high-sequence homology in the primary structure of their catalytic site and the same three-dimensional structure. The AC structure is a pseudodimer that can be divided in two main regions, transmembrane and cytoplasmic regions, and further divided into five different domains: (1) the NH_2 terminus, (2) the first transmembrane cluster (TM1), (3) the first cytoplasmic loop composed of C1a and C1b, (4) the second transmembrane cluster (TM2) with extracellular N-glycosylation sites, and (5) the second cytoplasmic loop composed of C2a and C2b. The transmembrane regions (TM1 and TM2), whose function is to keep the enzyme anchored in the membrane, are composed of 12 membrane-spanning helices (6 membrane-spanning helices each with short interhelical loops). The cytoplasmic regions C1 and C2 are approximately 40 kDa each and can be further subdivided into C1a, C1b, C2a, and C2b. Both C1a and C2a are highly conserved catalytic ATP-binding regions,[21] which dimerize to form a

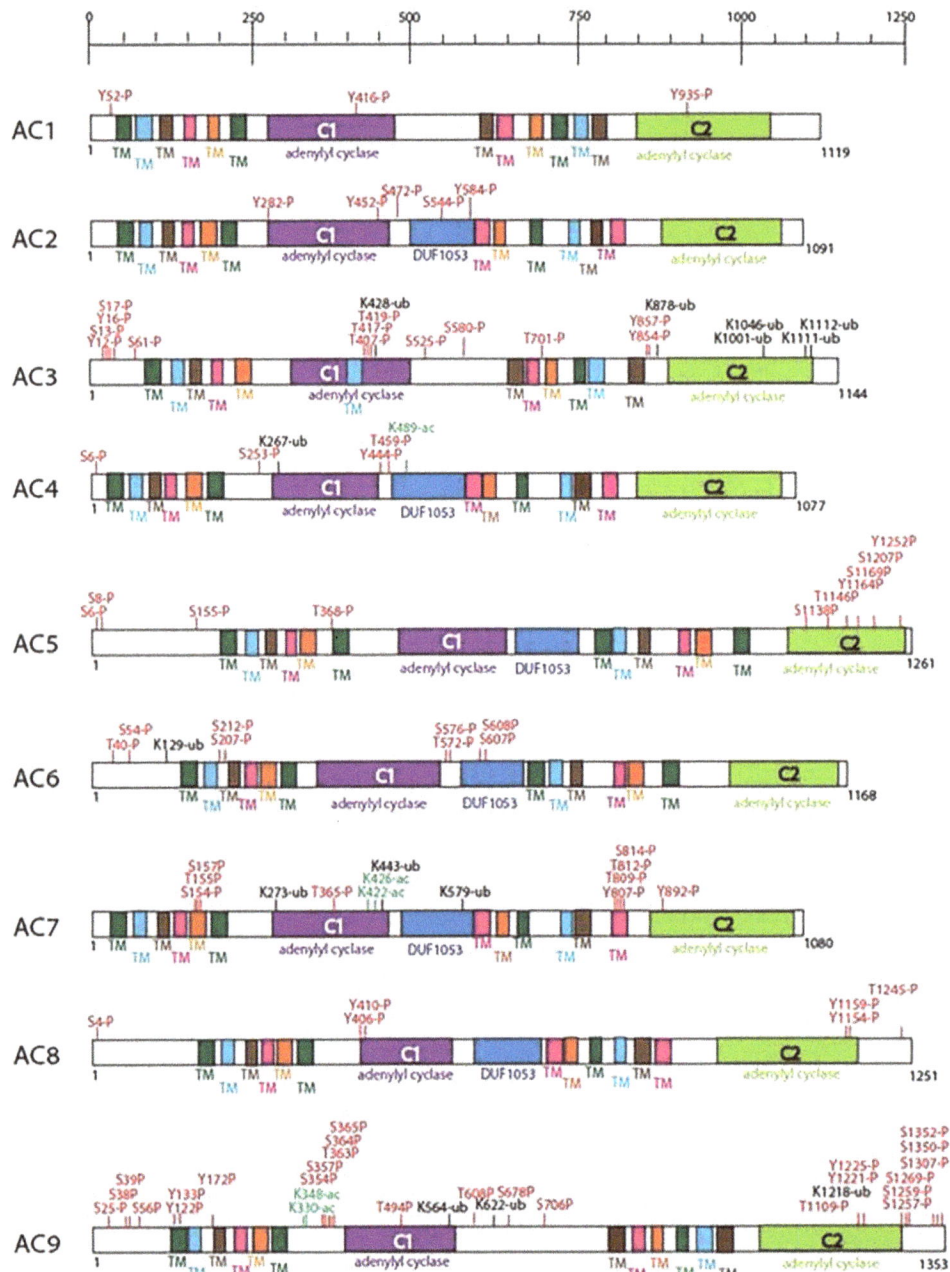

Figure 2. Schematic representation of the domain structure of the nine human transmembrane AC isoforms. The number of amino acid residues is reported on the side of each structure. Modification sites and domains are represented with different color. The transmembrane AC isoforms share a common structure composed of two cytosolic domains (C1 and C2) and six transmembrane segments organized in two tandem repeats. Both C1 and C2 domains contribute to ATP binding and formation of the catalytic core. **Abbreviations:** TM (transmembrane segments); DUF (domain of unknown function); ac (acetylation); P (phosphorylation site); ub (ubiquitination); S (serine); K (lysine); T (threonine); Y (tyrosine) (data obtained from PhosphoSitePlus).

pseudosymmetric enzyme, which forms the catalytic site. ATP binds at one of two pseudosymmetric binding sites of the C1-C2 interface. Two amino acid residues, Asn1025 and Arg1029, of AC2 are conserved among the C2 domains and critical for the catalytic activity of AC; mutation of either residue causes in a 30–100-fold reduction in the AC activity.[22] A second C1 domain subsite includes a P-loop that accommodates the nucleotide phosphates and two conserved acid residues that bind to ATP through interaction with two Mg^{2+}; one Mg^{2+} contributes to catalysis, whereas the second one interacts with nucleotide β- and γ-phosphates from substrate binding and possibly also for leaving-group stabilization. Both C2a and C2b are less conserved than the C1 domain.[21,23] The C1b domain is the largest domain, contains several regulatory sites, and has a variable structure across the isoforms. However, the C2b domain is essentially nonexistent in many isoforms and has not yet been associated with a function.[24] The overall domain structure of each human transmembrane AC isoform is shown in **Figure 2** and a detailed comparison of the cytoplasmic domains (C1 and C2), transmembrane segments, acetylation, phosphorylation, and ubiquitination sites of each isoform is indicated. **Figure 3** shows the three-dimensional model of AC and its relation to heterotrimeric G protein α subunit.

Without stimulation, the enzyme AC is constitutively inactive. There are at least two heterotrimeric G-proteins responsible for the regulation of transmembrane AC activity: G_s and G_i. When a secretagogue (e.g., secretin, vasoactive intestinal polypeptide) binds to its GPCR, it causes a change in the conformation of the receptor that stimulates the $G_{s\alpha}$ subunit to exchange GDP against GTP, which causes GTP-$G_{s\alpha}$ to detach from the $G_{\beta\gamma}$ subunits and bind to the two cytoplasmic regions of transmembrane AC.[25] With GTP-$G_{s\alpha}$ bound, AC becomes active and converts ATP to cyclic AMP in a process involving the release of water and a pyrophosphate. $G_{s\alpha}$ has shown to play an important role

in the exocrine pancreas because $G_{s\alpha}$-deficient mice show morphological changes in the exocrine pancreas, as well as malnutrition and dehydration.[26] Certain isoforms of transmembrane ACs are also positively (AC2, AC4, AC5, AC6, AC7) or negatively (AC1, AC3, AC8) regulated by the $G_{\beta\gamma}$ subunits, which also bind to the two cytoplasmic regions of transmembrane AC.[25]

When a GPCR is coupled to the heterotrimeric protein G_i, GTP-$G_{i\alpha}$ binds to adenylyl cyclase and, unlike GTP-$G_{s\alpha}$, GTP-$G_{i\alpha}$ inhibits the activity of the enzyme, causing lower levels of cyclic AMP in the cells. In the pancreas, somatostatin binds to its SS2 receptor and causes activation of $G_{i\alpha}$ subunit and inhibition of adenylyl cyclase.[28,29] Once the concentration of the ligand is below activation levels, the G_α subunit, which has an intrinsic GTPase activity, hydrolyzes GTP to GDP, reassociates with $G_{\beta\gamma}$ and becomes inactive. The cycle of GTP hydrolysis and inactivation occur within seconds after the G protein has been activated. Upon inactivation, G proteins are ready to be reactivated by another extracellular signal.

Transmembrane ACs are classified into four groups based on their regulatory properties (**Table 1 and Figure 4**):

– Group I, which consists of Ca^{2+}-stimulated isoforms: AC1, AC3, AC8.
– Group II, which consists of $G_{\beta\gamma}$-stimulated isoforms: AC2, AC4, AC7.

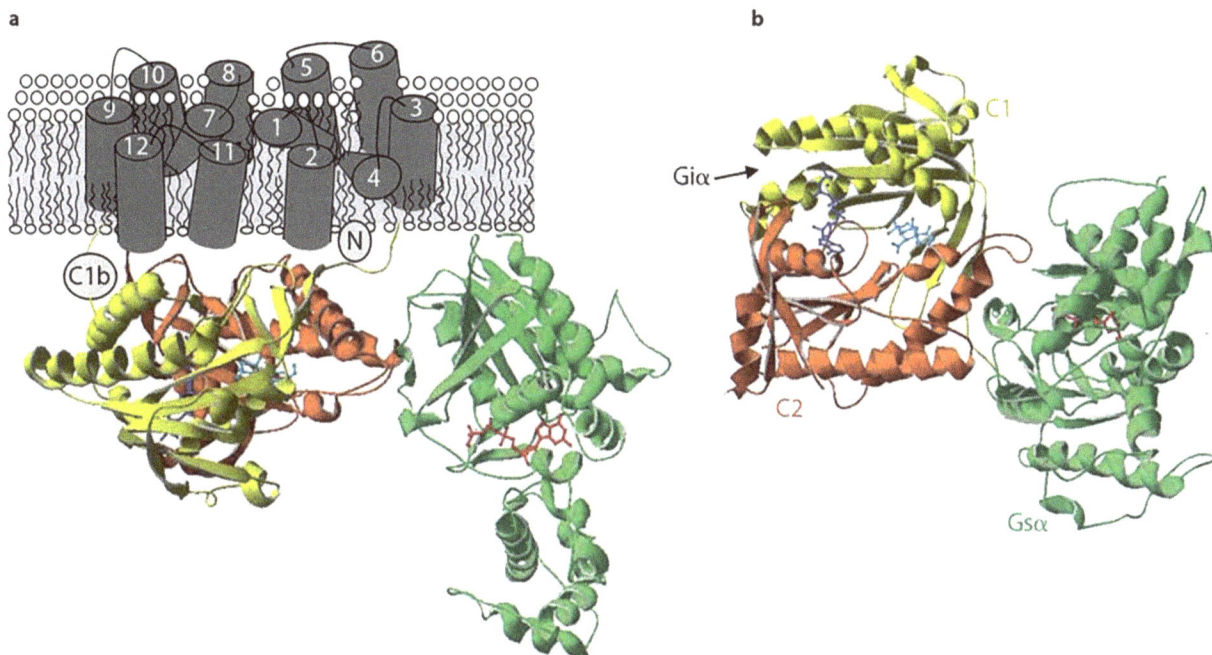

Figure 3. Crystal structure of adenylyl cyclase. (A) The figure shows the catalytic domains of mammalian AC C1 (yellow) and C2 (rust) with Gαs (green). The location of forskolin (cyan) and P-site inhibitor (dark blue) is also shown. **(B)** An alternate view from cytoplasmic side, showing forskolin and catalytic site. The interaction site of Giα with C1 domain is indicated by an arrow. (This figure was reproduced with permission from reference.[27])

– Group III, which consists of $G_{\alpha i}$/Ca^{2+}-inhibited isoforms: AC5, AC6.
– Group IV, which solely consists of Ca^{2+}-, $G_{\beta\gamma}$-insensitive isoform: AC9.

The expression profile of the transmembrane AC isoforms in intact mouse pancreas, isolated pancreatic acini, and duct fragment has been established using RT-PCR. Five different transmembrane AC isoforms were identified in pancreatic exocrine cells: AC3, AC4, AC6, and AC9 mRNAs were expressed in isolated pancreatic acini and sealed duct fragments, whereas AC7 mRNAs were only expressed in duct fragments.[30] Using real-time quantitative PCR analysis, the relative expression of each isoform in

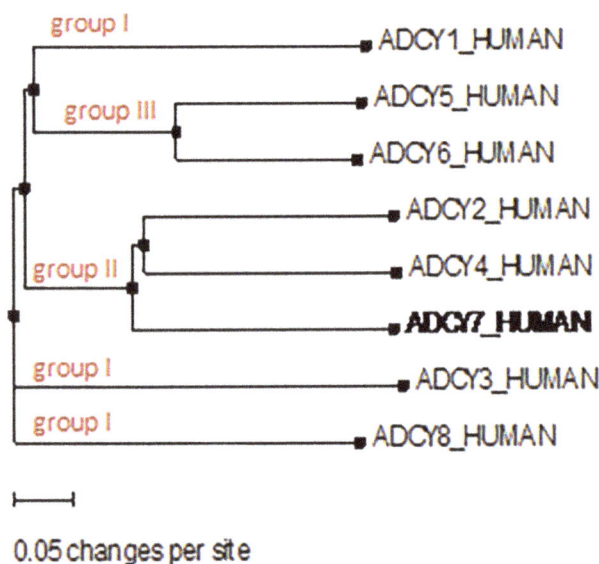

group I
group III
group II
group I
group I

ADCY1_HUMAN
ADCY5_HUMAN
ADCY6_HUMAN
ADCY2_HUMAN
ADCY4_HUMAN
ADCY7_HUMAN
ADCY3_HUMAN
ADCY8_HUMAN

0.05 changes per site

Figure 4. Phylogenetic tree of the transmembrane AC sequence. A dendrogram shows the degree of relatedness of sequence of transmembrane AC isoforms built by ClustalW. The length of each horizontal line indicates estimated evolutionary distance. Branches separate an individual subfamily. Groups defined by differences in regulatory properties are indicated in red. The order of evolution is as followed (from the most conserved to the less conserved): ADCY3 > ADCY1 > ADCY7 > ADCY4. ADCY5 and ADCY6, as well as ADCY2 and ADCY4, are coevolutionary isoforms.

Table 1. Regulatory properties of AC isoforms.

AC isoform	MW (kDa)[a] (mouse)	basal pI	Gαs	Gαi	Gβγ	FSK	Calcium	Protein Kinases
Group I:							Calcium-stimulated	
AC1	123.37	8.77	(+)	(-)	(-)	(+)	(+, CaM) or (-, CaM kinase	(+, PKCα)
AC3	129.08	6.15	(+)	(-)	(-)	(+)	IV)	(+, PKCα)
AC8	140.1	6.53	(+)	(-)	(-)	(+)	(+, CaM) or (-, CaM kinase II) (+, CaM)	(=)
Group II:							Calcium-insensitive	
AC2	123.27	8.4	(+)	(=)	(+)	(+)		(+, PKCα)
AC4	120.38	7.31	(+)	(=)	(+)	(+)		(+, PKC) or (-, PKCα)
AC7	122.71	8.49	(+)	(=)	(+)	(+)		(+, PKCδ)
Group III:							Calcium-inhibited	
AC5	139.12	6.9	(+)	(-)	(+,	(+)	(-, < 1 μM)	(-, PKA [b]) (+, PKCα/ζ)
AC6	130.61	8.56	(+)	(-)	β1γ2) (+, β1γ2)	(+)	(-, < 1 μM)	(-, PKA [b], PKCδ, ε)
Group IV:								
AC9	150.95	7.07	(+)	(-)	(=)	(=) or +,weak	(+, CaM kinase II) (-, calcineurin)	(-, novel PKC)
AC10	186/ 48 [c]	6.99	(=)	(=)	(=)	(=)	Calcium-stimulated	(=)

(+) AC activity is stimulated; (-) AC activity is inhibited; (=) AC activity is not modified. Data taken from [31-34]. (a) The molecular weight (MW) and basal isoelectric point (pI) data was obtained from PhosphoSitePlus (Cell Signaling Technology, Inc.). (b) AC6 is directly phosphorylated by PKA at Ser674 and thereby inhibited.[35] (c) 186 kDa (the full-length form) and 48 kDa (the truncated form).

pancreatic acini and ducts compared with the intact pancreas was assessed: isolated pancreatic acini were shown to have higher transcript levels of AC6 compared with intact pancreas, whereas isolated duct fragments were shown to have higher transcript levels of AC4, AC6, and AC7 compared with intact pancreas.

Similar transcript levels of AC3 and AC9 were observed in pancreas, acini, and ducts.[30] In conclusion, several adenylyl cyclase isoforms are expressed in pancreatic exocrine cells, with AC6 being highly expressed in both pancreatic acinar and duct cells.

Soluble AC

Soluble AC10 is unique in some many ways. The human *ADCY10* gene is located on chromosome 1 at q24. It is not anchored in the plasma membrane. As indicated in **Figure 5**, the catalytic domains C1 and C2 are located at the N-terminus and connected by a ~68-residue linker that forms a death domain-like subdomain with the ~33-residue N-terminus of the protein. The C-terminal from this C1-C2 tandem of the full-length mammalian soluble AC comprises a ~1,100-residue C-terminal region without a transmembrane region.[36] Its catalytic domain sequence is more closely related to some bacterial ACs than mammalian ACs.[34] Unlike transmembrane AC, soluble AC has no transmembrane domain. For that reason, its location is in the cytosol, but it can be associated with certain cellular organelles, such as the nucleus, mitochondria, and microtubules.[37] Unlike transmembrane ACs that are regulated by G proteins, forskolin, and calmodulin, among others, soluble AC is stimulated by HCO_3-.[7] The HCO_3- ion induces a conformational change of the active site of soluble AC similar to that observed in transmembrane ACs upon stimulation with $G\alpha s$.[3] Using RT-PCR and Western blotting, soluble AC has also been identified in acinar cells. By immunohistochemistry using a soluble AC antibody, AC10 has been localized just below the apical region of the cell in the non-stimulated condition and, after treatment with the CCK analog caerulein, a punctuate intracellular pattern was seen.[12]

In pancreatic acini, the activation of soluble AC with HCO_3- enhances secretagogue-stimulated cyclic AMP levels and inhibits secretagogue-stimulated zymogen activation and cell vacuolization.[12]

Intracellular targets of cyclic AMP

All the protein targets described below have a cyclic nucleotide-binding domain (CNBD) that has been conserved across a wide range of proteins, including the bacterial transcription factor catabolite activator protein.[38,39]

Protein kinase A

Cyclic AMP stimulates protein kinase A (PKA), which phosphorylates a number of cellular proteins by transferring a phosphate from ATP to a serine or a threonine located in sequence of residue of target protein. PKA contains two regulatory subunits, which possess the cyclic nucleotide-binding domain (CNBD), and two catalytic subunits, which are responsible for the Ser/Thr phosphorylation. Upon binding of cyclic AMP to the two regulatory subunits, the two catalytic subunits are detached from the regulatory subunits and become active.[13] The steps implicated in the activation of PKA by cyclic AMP are described in **Figure 6**.

The presence of PKA in pancreas was first reported in acinar cells from guinea pig.[40] PKA catalyzes the phosphorylation of regulatory proteins associated with the pancreatic exocytotic process.[41,42] However, PKA does not appear to directly participate in pancreatic amylase secretion because the inhibitor of PKA, H-89, does not modify either basal or cyclic AMP-dependent secretagogues-stimulated amylase secretion from mouse pancreatic acini.[43] Unlike in mouse acinar cells, in sealed mouse ducts PKA plays an essential role in the regulation of fluid secretion.[30]

One of the important targets of PKA is the transcription factor cyclic AMP response element binding protein (CREB). Similar to other cell types, in pancreatic acini CREB phosphorylation at Ser133 increases upon PKA activation.[43] The phosphorylation of CREB promotes the formation of a transcriptional complex on the cyclic AMP (cyclic AMP) response element (CRE) of certain promoters. The complex contains three proteins: (1) CREB, (2) the CREB-binding protein (CBP), and (3) CREB-regulated

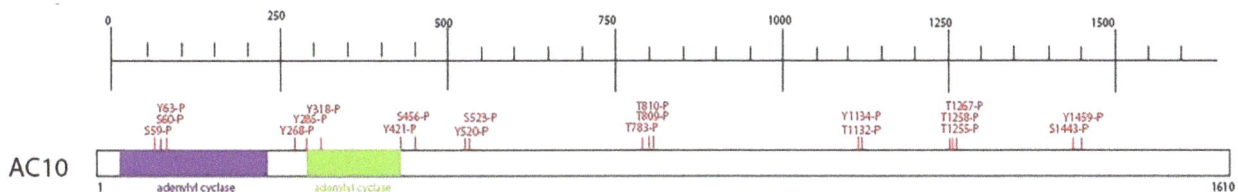

Figure 5. Schematic representation of the domain structure of the soluble AC isoform, AC10. The number of amino acid residues is reported on the side of structure. Modification sites and domains are represented with different color. **Abbreviations:** P (phosphorylation site); S (serine); K (lysine); T (threonine); Y (tyrosine) (data obtained from PhosphoSitePlus).

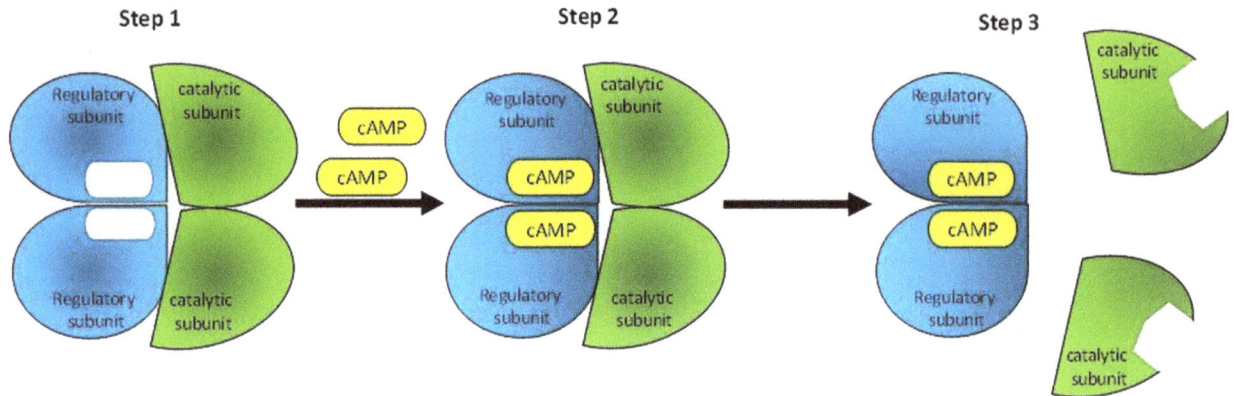

Figure 6. Schematic representation of the activation of protein kinase A (PKA) by cyclic AMP. PKA is composed of four subunits: two regulatory and two catalytic. In the absence of cyclic AMP, the regulatory subunit inhibits the catalytic subunit. Upon binding of cyclic AMP to regulatory subunits, the regulatory subunits change conformation. The catalytic subunits become detached and able to phosphorylate target proteins in the cell.

transcription coactivator 2. Its role is to stimulate the gene expression of certain proteins implicated in the regulation of metabolism, signaling, proliferation, differentiation, survival, and oncogenesis.

Other important targets of PKA are the cystic fibrosis transmembrane conductance regulator (CFTR), 1,-4,-5-inositol trisphosphate receptor (IP3R), A-kinase anchoring proteins (AKAPs), ERK 1/2, and some isoforms of phosphodiesterase.[44]

In pancreatic duct cells, PKA phosphorylates CFTR located in the apical membrane, on its regulatory domain, which then enables channel gating (opening and closing) and Cl– secretion.[45] Cyclic AMP evokes Cl– currents through CFTR, which mediates fluid transport across the luminal surfaces of pancreatic epithelial cells.[46] In pancreatic acinar cells, PKA phosphorylates only one of the three IP3R isoforms, IP_3R-3.[47,48] The phosphorylation of IP_3R-3 by PKA causes IP_3-induced Ca^{2+} release, which is decreased in terms of the magnitude and kinetics of Ca^{2+} release.[48,49] Another important target of PKA are the A-kinase anchor proteins (AKAPs), which are a family of structurally related proteins consisting of more than 50 members.[50] AKAP-150 has been shown to play a relevant role in the regulation of Na^+/K^+ ATPase pump activity in the homologous parotid gland.[51,52] Cyclic AMP also phosphorylates and thereby increases the activity of phosphodiesterases PDE3, PDE4, and PDE5 through PKA-induced phosphorylation.[18,53] Both PDE3 and PDE4 are cyclic AMP-specific PDEs, whereas PDE5 is a cyclic GMP-specific PDE.[18]

Exchange protein directly activated by cyclic AMP (Epac)

Cyclic AMP stimulates Epac.[15] There are two isoforms of Epac: Epac1 and Epac2.[54] Both isoforms are homologous proteins with Epac2 having an N-terminal extension. They share common domain structures within a N-terminal regulatory region and a C-terminal catalytic domain (**Figure 7**).[55–57] The N-terminal regulatory region possesses one (Epac1) or two (Epac2) cyclic nucleotide-binding domains (CNBDs) and a DEP (Dishevelled, Egl-10, and Pleckstrin) domain responsible for its localization to the plasma membrane. The C-terminal region contains CDC25-homology domain, a Ras exchange motif (REM) domain required for stabilizing GEF activity, and the GEF domain, which exerts GEF activity toward the small G proteins Rap1 and Rap2.[58] Epac1 is found in both pancreatic acini and ducts[30,43,59] and participates in cyclic AMP-stimulated amylase secretion.[43,59] Epac actives Rap1,[15] which is a small G-protein localized on zymogen granules as shown by both mass spectrometry and immunocytochemistry[60] and implicated in pancreatic amylase secretion.[43] In addition to its

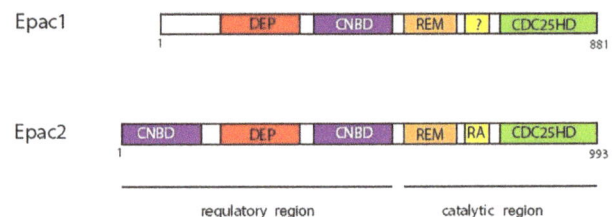

Figure 7. Domain structure of Epac1 and Epac2. Both Epac1 (881 amino acids) and Epac2 (993 amino acids) contain two regions: regulatory and catalytic regions. The **regulatory region** has the cyclic nucleotide-binding domain (CNB) and the Desheveled-Egl-10-Pleckstrin,[64] which is responsible for the membrane localization. The **catalytic region** has the CDC25-homology domain, which is responsible for the guanine-nucleotide exchange activity; the Ras exchange motif (REM), which stabilizes the catalytic helix of CDC25-HD; and the Ras-association (RA) domain, which is a protein interaction motif.

role in pancreatic amylase secretion, Epac regulates exocytosis in pancreatic beta cells. Incretin-induced insulin secretion is mediated by Epac2, the primary isoform of Epac in pancreatic beta cells.[61–63]

Cyclic nucleotide-gated channels

Cyclic nucleotide-gated (CNG) channels are nonselective tetrameric cation channels that mediate Ca^{2+} and sodium influx in response to direct binding of intracellular cyclic nucleotides.[65,66] The mammalian CNG channel genes fall into two different gene families. One subfamily consists of four members, CNGA1, CNGA2, CNGA3, and CNGA4, which represent the principal subunits that, except for CNGA4, form functional channels.[17] The core structural unit consists of six transmembrane segments, designated S1–S6, and a cyclic nucleotide-binding domain (CNBD) near the C-terminal region. A pore region of ~20–30 amino acids is located between S5 and S6. The S4 segment in CNG channels resembles the voltage-sensor motif found in the S4 segment of voltage-gated K^+, Na^+, and Ca^{2+} channels. Both N-terminal and C-terminal regions are located in the cytoplasmic side and a glycosylated segment connecting S5 to the pore region is extracellular.[17] The functional role of CNGs is well studied in retinal rod photoreceptors,[67] sperm,[68] central nervous system,[69] and cardiac excitability.[70] Studies of CNG channels in exocrine tissues have not been reported.

CNG channels belong to a heterogeneous gene superfamily of pore-loop cation channels that share a common transmembrane topology and pore structure. Other members of this superfamily are the hyperpolarization-activated cyclic nucleotide-gated channel (HCN),[71] the ether-a-gogo (EAG), and human eag-related gene[72] family of voltage-activated K^+ channels.[73] HCN channels are principally operated by voltage and permeable to both Na^+ and K^+. Opening of HCN channels causes hyperpolarization of the membrane. Unlike CNG, in which cyclic nucleotides are strictly required to open the channel in HCN, cyclic nucleotides facilitate the opening by shifting the voltage dependence of activation to more positive values.[71] Cyclic AMP has shown to modulate HCN channel activity through a PKA-dependent mechanism.[74,75]

The basolateral voltage-activated K^+ channels, which belong to the HCN channel subfamily, are necessary for the regulation of Cl– secretion from pancreatic acini. In the rat pancreatic acinar cells, the presence of K^+ channels in the basolateral membrane causes a membrane hyperpolarization, which provides the driving force for Cl– efflux. In addition, the efflux of K^+ balances the K^+ uptake by the Na^+, K^+ ATPase pump, and other cotransporters.[76] The functional and pharmacological properties of these channels are conferred once KCNE1 coassembles with KCNQ1.[77] Both KCNE1 and KCNQ1 genes are expressed in rodent pancreas.[78,79] Cyclic AMP[76,80] and carbachol[81] increase the

amplitude of the slowly activating voltage-dependent K^+ channel current (I_{Ks}) in rat pancreatic acinar cells.

RNA-binding protein

The Ca^{2+}-regulated heat-stable protein of 24 kDa (CRHSP-24, also known as CARHSP1) is a serine phosphoprotein originally identified as a physiological substrate for the Ca^{2+}-calmodulin-regulated protein phosphatase calcineurin (PP2B).[16] In pancreatic acini, cyclic AMP partially dephosphorylated CRHSP-24 on at least two sites[82] through the activation of a phosphatase inhibited by calyculin A and okadaic acid, namely a PP2A or PP4.[82]

Regulation of adenylyl cyclase/cyclic AMP signaling

The cytosolic levels of cyclic AMP are modulated by regulating GPCR activity, G protein activity, adenylyl cyclase activity, and cyclic AMP degradation.

Receptor regulation

GPCRs can be regulated in several ways. One way is through phosphorylation of specific amino acids in their cytosolic domain. When these amino acids are phosphorylated, the receptor becomes desensitized. G-protein-coupled receptor kinases (GRKs) are proteins that specifically phosphorylate GPCRs. Two GRKs have been found in the pancreas: GRK5[83] and GRK6.[84] PKA can also phosphorylate GPCRs. In mouse pancreatic acini, VPAC receptors appear to be regulated by PKA phosphorylation based on the inhibition of PKA activity using a PKA inhibitor (H-89) causing up to twofold increase in VIP-stimulated cyclic AMP formation.[30]

G-protein regulation

G-protein activity can be affected by various toxins, with the two best studied being *cholera toxin* and *pertussis toxin*. *Cholera toxin* in complex with NAD^+ and GTP-bound ADP-ribosylation factor 6 (ARF6-GTP) catalyzes the ADP-ribosylation of the α-subunit of G_s protein and prevents it from hydrolyzing its bound GTP, thereby locking the G_s protein in the active state, which causes the continuous activation of transmembrane AC.[85] In guinea pig pancreatic acini, *cholera toxin* increases cyclic AMP levels and amylase secretion.[86] In mouse pancreatic acini, *cholera toxin* increases amylase secretion.[87,88] Its effect is potentiated by cholecystokinin and is less marked than in guinea pig pancreatic acini.[87] Unlike *cholera toxin*, *pertussis toxin* modifies the α-subunit of G_i protein and locks the G_i protein in the inactive state. The toxin catalyzes the ADP-ribosylation of a cysteine residue at position-4 from the C-terminal of the α-subunit of G_i protein, inhibiting the interaction of this protein with the receptor and attenuating the intracellular transduction.[89–91] In rabbit pancreatic acini, *pertussis toxin* enhances CCK-induced cyclic AMP levels without affecting cholecystokinin (CCK)-induced Ca^{2+} mobilization or amylase secretion.[92] In rat pancreatic acinar cells, although the pretreatment with either *pertussis toxin* or *cholera toxin* does not modify CCK-stimulated intracellular Ca^{2+} levels or phosphoinositide hydrolysis,[27] *pertussis toxin* increases the basal levels of cyclic AMP and amylase secretion.[93] Regulators of G-protein signaling (RGS) molecules, which catalyze the GTP hydrolysis of heterotrimeric G-proteins, also play a critical role in regulating G-protein activity. RGS1, RGS2, RGS4, RGS16, and GAIP have been found in isolated pancreatic acinar cells using RT-PCR.[94] Although their function in the regulation of G_s activity in exocrine pancreas is still unknown, in olfactory neurons RGS2 decreases G_s-stimulated cyclic AMP levels.[95]

Adenylyl cyclase regulation

Adenylyl cyclase activity can be regulated by distinct intracellular signals. As previously indicated in **Table 1**, transmembrane ACs are classified into four groups: group I consists of Ca^{2+}-stimulated (AC1, AC3, AC8); group II consists of $G_{\beta\gamma}$-stimulated (AC2, AC4, AC7); group III consists of $G_{i\alpha}/Ca^{2+}/PKC/PKA$-inhibited (AC5, AC6); and group IV consists of Ca^{2+}-inhibited (AC9), which is forskolin-insensitive.[27,33,96] Recently, AC9 activity has also been shown to be inhibited by $G\alpha i/o$ proteins and PKC.[97] The Ca^{2+}-binding protein involved in the stimulatory effect of Ca^{2+} on the group I is calmodulin, which forms an active Ca^{2+}-calmodulin complex. Calmodulin is present in pancreatic acini and activated by CCK.[98] The Ca^{2+}-calmodulin complex binds to the calmodulin-binding site present in the group I isoform and increases its activity dramatically. AC9 is also stimulated by calmodulin.[99] The Ca^{2+}-binding protein involved in the inhibitory effect of Ca^{2+} on AC9 is calcineurin, which is a serine/threonine protein phosphatase activated by CCK[16,100] and involved in amylase secretion from rat pancreatic acini[101] as well as caerulein-induced intracellular pancreatic zymogen activation.[102]

Forskolin is a diterpene extracted from the root of the plant *Coleus forskohlii* that directly activates all transmembrane AC isoforms, except AC9,[20,103] by interacting with the two cytoplasmic domains (C1 and C2) that form the catalytic domain.[104] The lack of effect of forskolin on AC9 may be accounted for by the residues Tyr1082 and Ala1112.[105]

Unlike transmembrane AC, AC10 is not activated by either G protein or forskolin. Its activation is dependent on the HCO_3- levels,[106] though it can also be activated by divalent cations, such as Ca^{2+}, Mg^{2+}, and Mn^{2+}.[107] A combination of Ca^{2+} and HCO_3- activates soluble adenylyl cyclase

synergistically.[107] AC10 is also activated by changes in intracellular pH.[108]

The most common posttranslational modification of an AC isoform is the phosphorylation of a serine, threonine, or tyrosine residue (**Figures 3** and **4**). Phosphorylation of AC1 and AC3 by Ca^{2+}/calmodulin kinases inhibits the cyclase activity by blocking the binding site. Phosphorylation of ACs by either PKA or PKC causes an inhibition of the enzyme activity. Ubiquitination and acetylation are other modifications found in the human AC isoforms, though their consequences in AC activity are still unknown.

Regulation of cyclic AMP degradation

Cyclic AMP degradation is carried out by the enzyme phosphodiesterase,[44] which is an exonuclease capable of hydrolyzing a phosphate ester and pyrophosphate bonds and, thereby, converting cyclic AMP into 5'AMP.[18] Eleven PDE isoforms exist and each has unique biochemical properties. PDE1, which hydrolyzes both cyclic AMP and cyclic GMP, has been found in pancreatic acini using immunocytochemistry,[109] whereas PDE7B, a cyclic AMP-specific PDE, has been found in whole pancreas using Northern blotting.[110] PDE4, which is highly expressed in most immune and inflammatory cells and a cyclic AMP-specific PDE, is involved in the development of acute pancreatitis because the selective inhibitor rolipram attenuates the severity of acute pancreatitis in rats.[111]

Role for adenylyl cyclase/cyclase AMP pathway in pancreatic exocrine cells

The exocrine pancreas is primarily composed of pancreatic acini and ducts. Pancreatic acini synthesize and release digestive enzymes into the duodenum, whereas pancreatic ducts release a HCO_3-rich fluid to neutralize the acidic chyme released from the stomach. In this section, the roles for adenylyl cyclase/cyclic AMP pathway are described.

Pancreatic duct HCO₃-rich fluid

Secretagogues, such as secretin and vasoactive intestinal polypeptide (VIP), increase cyclic AMP and stimulate HCO_3-rich fluid secretion from pancreatic duct cells.[112–114] An increase in the levels of cyclic AMP, through PKA phosphorylation, activates CFTR to recirculate chloride back into the glandular lumen and, thereby, depolarizes both luminal and basolateral membranes. Depolarization of the basolateral membrane increases the driving force of an electrogenic sodium-HCO_3- cotransporter on the basolateral membrane leading to the entry of HCO_3-, which is then secreted at the apical membrane via the $Cl-/HCO_3-$ exchanger.[45] The AC6/cyclic AMP/PKA pathway has an important role in the physiological function of pancreatic ducts because the VIP-stimulated expansion of the lumen observed *in vitro* in pancreatic ducts from WT mice upon VIP stimulation was absent in duct fragments from AC6-deficient mice. *In vivo* collection of pancreatic fluid also showed a decrease in fluid secretion from AC6-deficient mice.[30] The secretory effect is highly dependent on PKA activation because in isolated pancreatic ducts from AC6-deficient mice PKA activation was abolished in response to VIP, secretin, and forskolin.[30,96]

Several ion channels are affected by cyclic AMP/PKA pathway. PKA phosphorylates CFTR located in the apical membrane of the pancreatic duct cells.[45] Elevation of intracellular cyclic AMP by stimulation with forskolin significantly inhibits the Na^+/H^+ exchanger (NHE) and this, like the stimulation of the apical anion exchanger, may occur through a direct physical interaction with CFTR.[45]. The basolateral Cl^-/HCO_3^- exchanger (AE) does not seem to be directly activated by forskolin.[115] For more details, see Chapter 19.

Pancreatic acini enzymatic-rich fluid

Early work showed that a number of compounds that increase cyclic AMP levels stimulate amylase secretion from pancreatic acini.[29,93,116–121] Phosphodiesterase inhibitors, such as 3-isobutyl-1-methylxantine, increase pancreatic amylase secretion.[122] *Pertussis toxin* catalyzes the ADP-ribosylation of a cysteine residue at postion-4 from the carboxyl-terminal domain of the α-subunit of Gi protein, inhibiting the interaction of this protein with the receptor and impairing intracellular transduction. Treatment with *pertussis toxin* causes an increase in cyclic AMP levels and amylase secretion from rat pancreatic acini,[92] where multiple *pertussis toxin*-sensitive G proteins have been found (e.g., G_{i1}, G_{i2}, G_{i3}, and G_o).[123] Forskolin interacts with the two cytosolic domains C1 and C2 of transmembrane ACs, except AC9.[27] Forskolin slightly stimulates amylase secretion in rat[124,125] and potentiates the response to Ca^{2+}-dependent secretatogues.[126] Recently, pancreatic acini from AC6-deficient mice showed a reduction in stimulated amylase secretion and PKA activity.[30] Because this inhibition was only partial, it is likely that other AC isoforms expressed in pancreatic acini are also responsible for the secretory role of pancreatic acini. The result showing the deletion of AC6 does not affect the response to the Epac1 analog 8-pCPT-2'-O-Me-cyclic AMP on amylase secretion supports this hypothesis.[30]

Glucagon-like peptide-1 (GLP-1) is a glucoincretin hormone secreted by intestinal L cells in response to nutrient ingestion that can act through its receptor (GLP-1R).[127] GLP-1R is expressed in isolated mouse pancreatic acini and mediates GLP-1-induced amylase secretion from isolated mouse pancreatic acini through cyclic AMP/PKA pathway activation.[128]

Both VIP and secretin, through the phosphorylation of p21-activated kinase (PAK) 4, but not PAK2, cause an activation of the Na^+/K^+ ATPase in isolated rat pancreatic acini and, as a consequence, an increase in pancreatic acinar fluid secretion; Epac mediates VIP-induced PAK4 activation, whereas PKA mediates secretin-induced PAK4 activation.[129]

Differentiation, transdifferentiation, and proliferation

Cyclic AMP plays an important role in differentiation, transdifferentiation, and proliferation of pancreatic cells. Isolated adult islets of Langerhans were able to transdifferentiate to duct epithelial-like cyst structures in presence of elevated cyclic AMP and a solid extracellular matrix (e.g., matrigel and collagen I).[130] The presence of intracellular cyclic AMP elevating factor, such as *cholera toxin*, was also required for the proliferation and maintenance of pancreatic epithelial duct cells.[131] However, transforming growth factor-β (TGF-β), which is an important regulator of growth and differentiation in the pancreas, can activate PKA without affecting cyclic AMP levels in pancreatic acini.[132] TGF-β-mediated growth inhibition and TGF-β-induced p21 and SnoN expression are mediated by PKA because both effects were blocked the PKA inhibitors H89 and PKI peptide.[132] A physical interaction between a Smad3/Smad4 complex and the regulatory subunits of PKA has been shown in pancreatic acini.[132]

Development of pancreatitis

Acute pancreatitis is an acute inflammatory disease of the pancreas. The disease appears to be initiated when a pathologic factor like alcohol or bile injuries the acinar cell and it responds by releasing inflammatory mediators and by activating digestive enzymes, especially proteinases, and restricting their secretion. These events initiate a cascade that leads to pancreatic inflammation and local and systemic tissue injury.[133] The participation of cyclic AMP in the development of pancreatitis has been studied by the Gorelick lab. An early study showed that cyclic AMP-dependent secretagogues sensitize the pancreatic acinar cells to zymogen activation induced by caerulein, a CCK analog.[134] The same research group in a subsequent work showed that cyclic AMP, by enhancing the release of pancreatic enzymes from the acinar cell, can overcome the acinar cell injury induced by high concentrations of carbachol, a cholinergic agonist.[135] Recently, the inhibition of soluble AC by KH7 was shown to enhance the activation evoked by caerulein of two important digestive enzymes chymotrypsinogen and trypsinogen, as well as caerulein-stimulated amylase secretion from rat pancreatic acini.[12] Together these studies suggest a complex role for cyclic AMP in acute pancreatitis in which it may enhance some pancreatitis responses while simultaneously lessening the effects of others.

Role for adenylyl cyclase/cyclase AMP pathway in pancreatic cancer

Of all the different cancers, pancreatic cancer is one of the major unsolved health problems. In 2020, the U.S. projected estimates report that the number of deaths from pancreatic cancer will be over 47,050.[136] This is a 2.8 percent increase in estimated deaths from 2019.[137] Unfortunately, the usual cancer treatment options do not have much of a positive effect. For that, it is of interest to develop new clinical strategies to increase the survival of patients with pancreatic cancer.

In 1996, receptors for VIP and β-adrenergic agonists (i.e., adrenaline and isoproterenol), which are functionally coupled to AC, were found in five human pancreatic adenocarcinoma cell lines BxPC-3, Hs 766T, Capan-2, Panc-1, and Capan-1 (Hs 766T and BxPC-3 were the most responsive, followed by Capan-2 and Capan-1, and finally Panc-1; MIA PaCa-2 cells did not respond to any of the agonists tested).[138] Because high concentration of secretin was necessary to stimulate AC, the process of neoplastic transformation has either downregulated the expression of secretin receptors or led to a defect in the receptor itself.[138]

Later, the $G_{\alpha s}$/AC/cyclic AMP pathway was shown to participate in the development of pancreatic pre-neoplastic lesions and the regulation of multiple pancreatic cancer cellular processes, including migration and invasion.[139–141] Support for the AC/cyclic AMP pathway in developing pancreatic pre-neoplastic lesions has come from a study showing the presence of a mutation in the oncogene *GNAS*, which results in constitutive activation of $G_{\alpha s}$[142] in two precursor lesions of pancreatic adenocarcinoma: the pancreatic intraepithelial neoplasia (PanIN) and the macroscopic intraductal papillary mucinous neoplasm (IPMN).[143] Two different groups have shown that the coexpression of *GNAS*^{R201C} and *KRas*^{G12D} promotes murine pancreatic tumorigenesis, mimicking the pre-neoplastic lesions PanIN and IPMN.[144,145] However, GNAS^{R201C} does not seem to cooperate with the oncogene KRas^{G12D} as other mutants do (e.g., TP53, SMAD4, and CDKN2A). GNAS^{R201C} functions as an "onco-modulator" that affects the phenotype of pre-neoplastic lesions arising in the context of mutant KRAS^{G12D} and leads to invasive carcinomas with a predominantly well-differentiated morphology.[145] The early activation of hippo pathway by Gαs/cyclic AMP/PKA is likely to be the downstream pathway responsible for the differentiation in *GNAS*^{R201C} and *KRas*^{G12D} double-positive cells because the coexpression of these mutants causes phosphorylation of the transcriptional coactivator YAP1, which is sequestered and inactivated in the cytoplasm, by Hippo kinase cascade.[145]

Support for the AC/cyclic AMP pathway in inhibiting migration and invasion has come from a number of studies using forskolin as a stimulator of AC.[139,140,146] From all the transmembrane isoforms present, AC3 was more highly expressed in the pancreatic tumor tissue, compared to the adjacent nontumor tissues, and in two pancreatic cancer cell lines, HPAC and PANC-1, compared to the normal pancreatic ductal cell line (hPDEC). AC3 was also the isoform responsible for the inhibitory effect of forskolin on cell migration and invasion in pancreatic cancer cell lines HPAC and PANC-1.[146] The mechanism by which forskolin/AC3/cyclic AMP pathway inhibits cell migration and invasion involves the quick formation of AC3/CAP1 complex and sequestration of G-actin, which leads to an inhibition of filopodia formation and cell motility.[146]

The stimulation of the AC/PKA pathway by forskolin and VIP increases expression and release of apomucin MUC5AC from pancreatic cancer cells SW1990.[147] Because the coexpression of mutants *GNAS*^{R201C} and *KRas*^{G12D} upregulates apomucins MUC1 and MUC5AC, but not MUC2, which resembles the expression of apomucins in human IPMNs,[145] *GNAS*^{R201C} mutation, by $G_{\alpha s}$/cyclic AMP/PKA pathway, might induce the expression of MUC5AC in IPMNs.

Guanylyl Cyclase/Cyclic GMP Signaling

Cyclic GMP is made from GTP through a catalytic reaction mediated by guanylyl cyclase (GC). Like AC, GC can be transmembrane or soluble. Unlike transmembrane ACs, all transmembrane GCs share a basic topology, which consists of an extracellular ligand-binding domain, a single transmembrane region, and an intracellular domain that contains a juxtamembranous protein kinase-homology domain (KHD), a coiled-coil amphipathic α-helical or hinge region, and the catalytic GC domain at its C-terminal end. The function of the KHD is still unknown. Although it binds ATP and contains several residues conserved in the catalytic domain of protein kinases, kinase activity has not been found. In fact, it regulates the GC activity at the C-terminal end. The coiled-coil hinge region is involved in the process of dimerization, which is essential for the activation of GC domain.[148] There are at least seven transmembrane guanylyl cyclases: GC-A, GC-B, GC-C, GC-D, GC-E, GC-F, and GC-G (**Figures 8 and 9**). Only GC-A, GC-B, and GC-C have shown to regulate the function of exocrine pancreas.

Transmembrane GC

There are at least two groups of ligands for transmembrane GC: natriuretic peptides and guanylin and uroguanylin peptides.

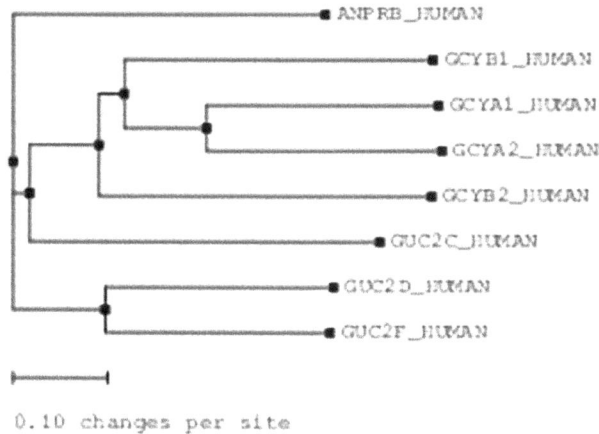

0.10 changes per site

Figure 8. Phylogenetic tree of the transmembrane and soluble GC sequence. A dendrogram shows the degree of relatedness of sequence of GC isoforms built by ClustalW. The length of each horizontal line indicates estimated evolutionary distance. Branches separate an individual subfamily. The order of evolution as followed (from the most conserved to the less conserved): ANP RB (also known as NPR-B or GC-B) > GUC2C (GC-C) > GUC2D (GC-D, retinal) > GCYB2 GC (GC soluble subunit β2) > GCYB1 > GCYA1 (GC soluble subunit α1). GUC2D and GUC2F, as well as GCYA1 and GCYA2, are coevolutionary isoforms.

Natriuretic peptides

There are three members of natriuretic peptide family: atrial natriuretic peptide (ANP), B-type natriuretic peptide (BNP), and C-type natriuretic peptide (CNP) (**Figure 10A**). The actions of natriuretic peptides are mediated by the activation of three transmembrane receptor subtypes: natriuretic peptide receptor type A (NPR-A, also known as GC-A), type B (NPR-B, also known as GC-B) (**Figure 10B**), and type C (NPR-C). The last one (NPR-C) is not a transmembrane GC; it was initially considered to be a clearance receptor whose only function was to regulate the plasma concentration of natriuretic peptides.[149] Later, NPR-C receptors have been found to be coupled to the inhibition of AC through α subunit of an inhibitory guanine nucleotide regulatory protein $(G\alpha_{i1}/G\alpha_{i2})$ and/or the activation of phospholipase C (PLC) through the βγ subunits of the same G_i protein[150-153] in a number of tissues, including pancreatic acini.[154] Because NPR-C receptors are not coupled to cGMP generation, NPR-C receptors are not further discussed in this chapter.

The NPR-A and NPR-B receptors, whose relative molecular mass is 130–180 kDa, have a similar structure that contains four domains: an extracellular ligand-binding domain, a single transmembrane domain, an intracellular tyrosine-like domain, an amphipathic region, and a GC catalytic domain. Upon ligand binding, the NPR-A and NPR-B receptors change their conformation, which results in GC activation and cyclic GMP generation.[155,156]

Guanylin and uroguanylin

Guanylin and uroguanylin are peptides secreted from the intestine, which influence electrolyte and fluid transport in the intestine and kidney, respectively.[157,158] Their effects are mediated by the NPR-C receptor, which is predominately expressed in the intestine.

Soluble GC

Soluble GC is a histidine-ligated hemoprotein that consists of two homologous subunits, α and β. The well-known isoform is the α1β1 protein; α2β2 subunits have also been identified.[159,160] Each soluble GC subunit consists of four domains, an N-terminal heme-nitric oxide oxygen (H-NOX) domain (also called a SONO domain), a central Per-ARNT-Sim[114] domain, a coiled-coil domain, and a C-terminal catalytic cyclase domain. The β1 subunit contains a N-terminal heme-binding domain, a Per/Arnt/Sim[114] domain, a coiled-coil domain, and a C-terminal catalytic domain,[161] as described in **Figure 11**.

Intracellular targets of cyclic GMP

Intracellular targets of cyclic GMP, like intracellular targets of cyclic AMP, have a cyclic nucleotide-binding domain (CNBD) in their structure.

Cyclic nucleotide-gated channels

Cyclic nucleotide-gated channels (CNGs) have been described above (see Intracellular Targets of Cyclic AMP). The physiological significance of cyclic GMP as activating agent of CNG has been described in photoreceptor cells and olfactory sensory neurons, where CNGs play an important role in sensory transduction.[66] There are no reports of CNG function in an exocrine tissue.

Cyclic GMP-dependent protein kinase

The increase in the levels of cyclic GMP activates cyclic GMP-dependent serine/threonine protein kinase (PKG). Two genes *prkg1* and *prkg2* code for the two isoforms PKGI and PKGII.[14] The human *prkg1* gene is located on chromosome 10 at p11.2 –q11.2 and has 15 exons. The N-terminus of PKGI is encoded by two alternative exons that produce the isoforms PKGIα and PKGIβ. The human *prkg2* gene is located on chromosome 4 at q13.1q21.1 and has 19 exons. Its transcript yields a protein with an apparent mass of 87.4 kDa.[14] Like PKA, PKG is composed of two functional domains: a regulatory domain and a catalytic domain. The regulatory domain is subdivided into the N-terminal domain and the cyclic nucleotide-binding domain (CNBD) containing the high and low cyclic GMP affinity binding

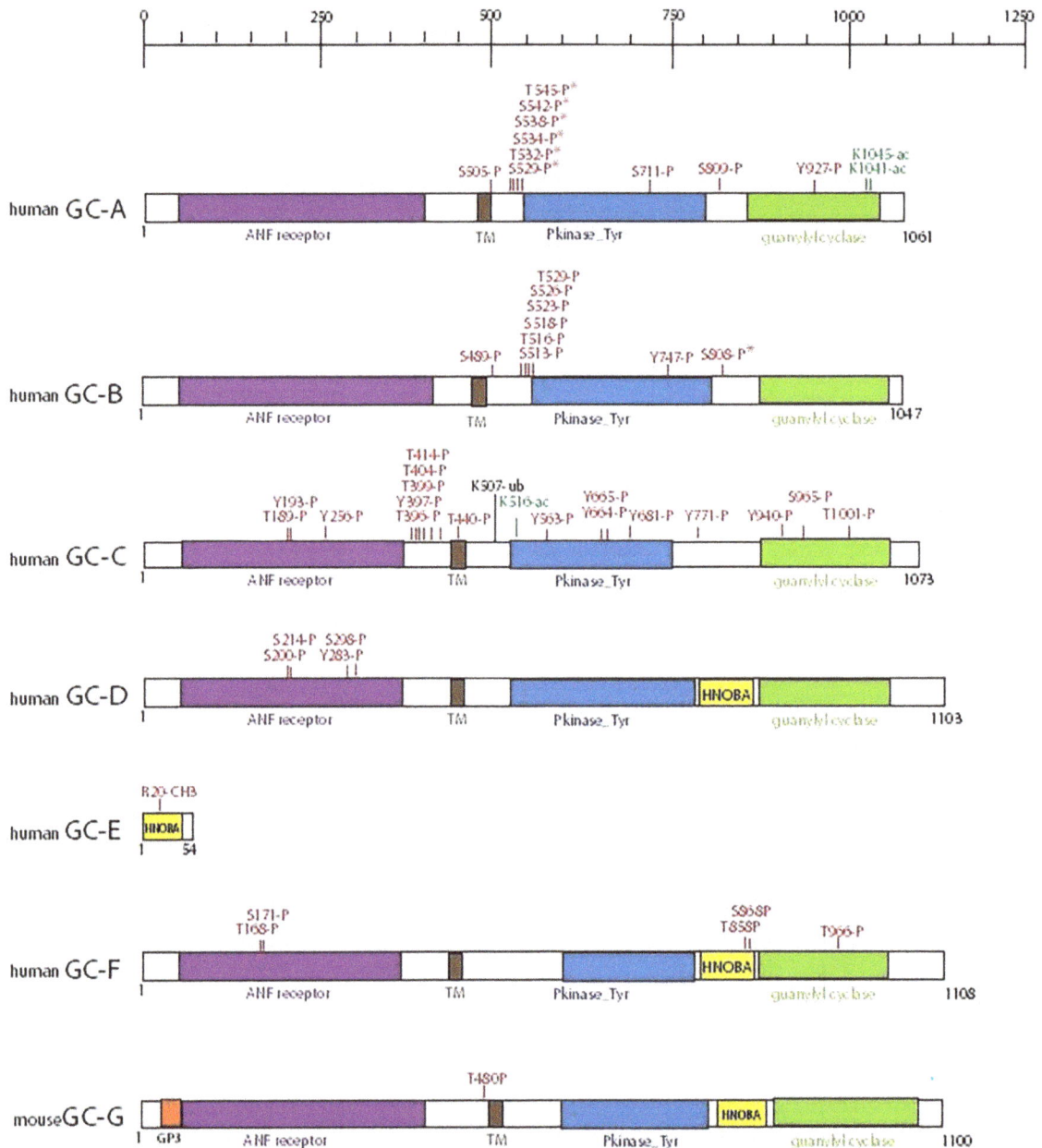

Figure 9. Schematic representation of the domain structure of the seven GC isoforms. The number of amino acid residues is reported on the side of each structure. Modification sites and domains are represented with different color. The transmembrane GC isoforms share a common structure consisting of an extracellular ligand-binding domain, a short transmembrane (TM) region, and an intracellular domain that contains the catalytic region (GC) at its C-terminal end. **Abbreviations:** TM (transmembrane segments); ac (acetylation); P (phosphorylation site); ub (ubiquitination); CH3- (methylation); S (serine); K (lysine); R (arginine); T (threonine); Y (tyrosine). * (sites implicated in the activity of the enzyme) (data obtained from PhosphoSitePlus).

pockets. The catalytic domain contains the Mg^{2+}-ATP- and peptide-binding pockets. Upon binding of cyclic GMP to the two regulatory subunits, the two catalytic subunits are released from the regulatory subunits and become active.[14] The substrates of this kinase are P240, P132, and phospholamban, though none of them is a specific PKG substrate.[166,167] The intracellular levels of cyclic GMP are regulated by PDE enzymes, which hydrolyze cyclic GMP

into 5'GMP.[167] In pancreatic acinar cells from guinea pig, the presence of PKG activity has been reported.[40]

Role for guanylyl cyclase/cyclic GMP pathway in pancreatic exocrine cells

The role of GC/cyclic GMP in the regulatory function of pancreatic exocrine cells is still controversial. One of the

A

B

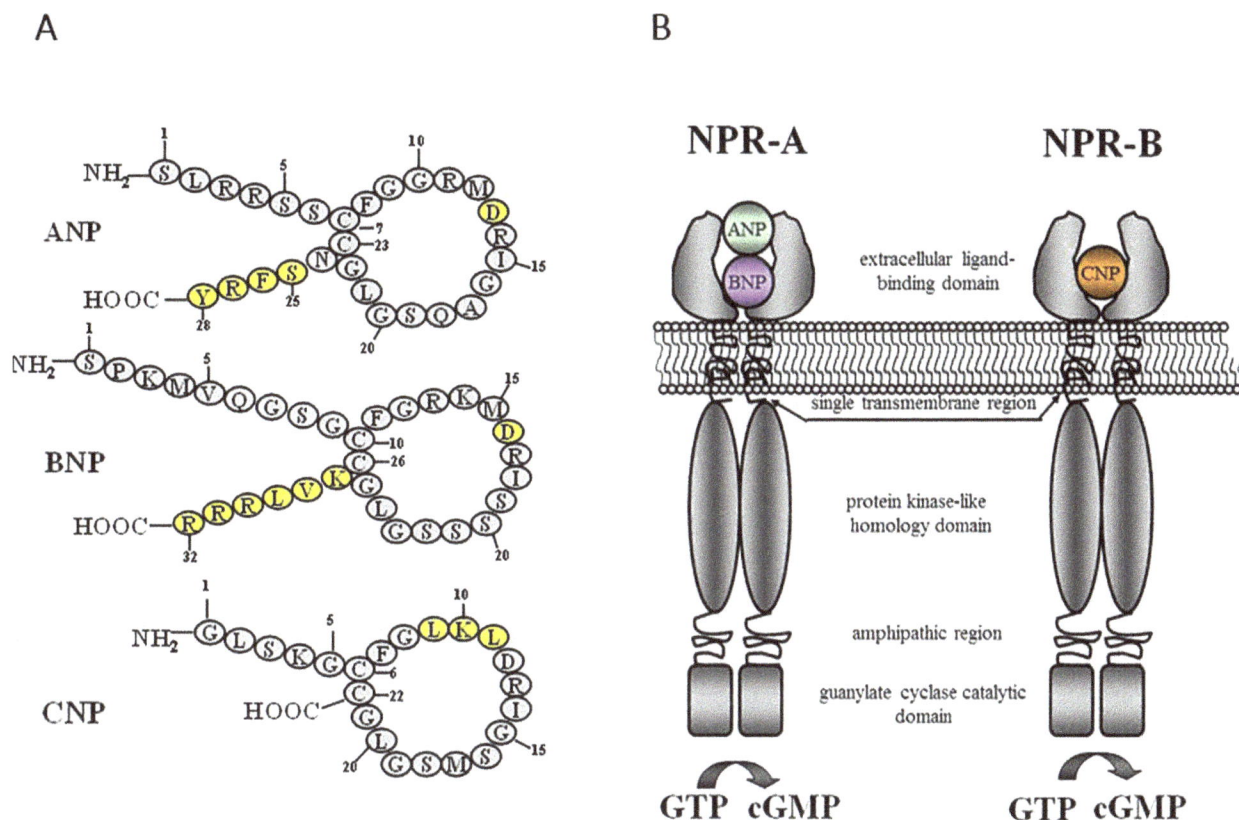

Figure 10. (**A**) **Schematic representation of the amino acid sequence and the structure of biological active natriuretic peptides ANP, BNP, and CNP.** The members of the natriuretic peptide family share a common structure, which consists of a 17-amino-acid-bonded loop bridge by intracellular disulfide bond required for the natriuretic and diuretic activity. Note the amino acids with a yellow color are also important for their activity. (**B**) **Schematic representation of natriuretic peptides receptors NPR-A and NPR-B.** The structure of NPR-A and NPR-B receptors possesses three domains: the extracellular ligand-binding domain, the intracellular protein kinase-like homology domain, and the GC catalytic domain.

first papers published on isolated pancreatic lobules from guinea pig and rabbit showed that carbamylcholine (carbachol), pancreozymin (now known as CCK), and caerulein all increased the levels of cyclic GMP without modifying the levels of cyclic AMP. The authors concluded that cyclic GMP is the second messenger involved in the process of stimulus-secretion coupling in the acinar cells of exocrine pancreas.[168] Later, Ca^{2+} was shown to be an important mediator of the stimulus-secretion coupling process.[169] Moreover, increased intracellular levels of cyclic GMP has a little or no effect on the stimulus-secretion coupling in pancreatic acinar cells.[11,169] However, cyclic GMP has been involved in the Ca^{2+} entry across the cell membrane to replenish the intracellular Ca^{2+} stores.[170,171]

The function of NO, the ligand for soluble GC, has been studied in the exocrine pancreas. NO can increase endogenous cyclic GMP and rat pancreatic secretory activity.[11] NO triggers an increase in intracellular Ca^{2+} levels via cyclic GMP and inositol trisphosphate in pancreatic acinar cells.[172] NO is localized in intrapancreatic ganglionic cells

and efferent nerve fibers[173] and implicated in the control of mesenteric circulation.[174] NO inhibits pancreatic exocrine secretion in dogs[175] and rats.[176] NO production regulates cyclic GMP formation and Ca^{2+} influx in rat and guinea pig isolated pancreatic acini.[177] Blocking NO production by chemical inhibitors of NO synthase, NG-monomethyl-L-arginine or NG-nitro-L-arginine, abolished cyclic GMP formation induced by the cholinergic agonist carbachol in a dose-dependent manner.[177] NO has shown to have a protective role in acute pancreatitis.[178–180]

The functions of two ligands for transmembrane GCs, natriuretic peptides, and guanylin have also been studied in exocrine pancreas.

Natriuretic peptides

All of the three receptors of natriuretic peptides are expressed in pancreatic acini,[154] and both ANP and CNP increase intracellular levels of cyclic GMP in isolated pancreatic acini.[154,181] However, the action of ANP and CNP on

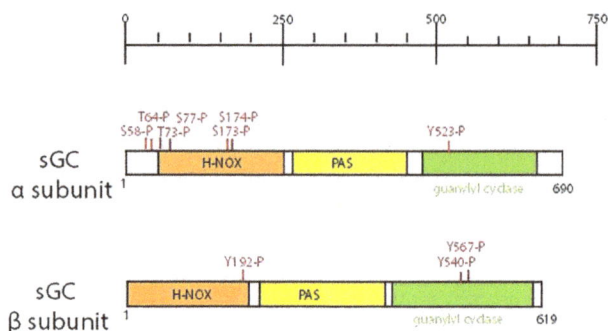

Figure 11. Schematic representation of the domain structure of soluble GC (α subunit and β subunit). The number of amino acid residues is reported on the side of each structure. Modification sites and domains are represented with different color. Each soluble GC subunit consists of four domains: a N-terminal heme-nitric oxide oxygen (H-NOX) domain (also called a SONO domain); a central Per-ARNT-Sim[114] domain; a coiled-coil domain; and a C-terminal catalytic cyclase domain. **Abbreviations:** P (phosphorylation site); S (serine); T (threonine); Y (tyrosine) (data obtained from PhosphoSitePlus).The PAS domain mediates protein–protein interactions and have often been found to bind heme, a flavin, or a nucleotide.[162] The coiled-coil domain appears to be unique to soluble GC.[163] The functions of PAS and coiled-coil domains are still unknown. The catalytic domain is localized at the C-terminal 467–690 and 414–619 residues of the α1 and β1 subunits, respectively.[164] The catalytic domains must form a heterodimer for cyclic GMP synthesis and in the full-length protein.[163] The C-terminal regions of the α1 and β1 subunits are highly homologous to the particulate GC and AC catalytic domains.[163] Soluble GC binds nitric oxide (NO), which is its primary activator,[165] and can also be activated by carbon monoxide, but not oxygen.[163] NO is a gaseous second messenger molecule synthesized from L-arginine and oxygen by the enzyme NO synthase. NO binds to the heme cofactor of soluble GC. The binding of NO to soluble GC leads to an increase in cyclic GMP.

pancreatic secretion is not mediated by an increase in cyclic GMP. Indeed, ANP and CNP increase pancreatic fluid and protein output through the NPR-C receptor activation/Ca^{2+} release.[154,182,183]

Guanylin and uroguanylin

In rat pancreatic acini, guanylin increases cyclic GMP levels, elicits a small amount of amylase secretion, and a small Ca^{2+} transient.[11] Guanylin is localized specifically to the centroacinar cells and proximal duct cells and released luminally into the pancreatic ducts based on its presence in the pancreatic juice.[184] Functional studies in two different human pancreatic duct cell lines revealed that guanylin is an intrinsic pancreatic regulator of Cl^- current activation in pancreatic duct cells via cyclic GMP. Using whole-cell patch-clamp, forskolin increased Cl^- conductance mediated by cyclic AMP, while guanylin increased Cl^- conductance mediated by cyclic GMP, but not cyclic AMP.[185]

The existence of both membrane and soluble GCs in pancreatic acini suggests that there are two distinct sources of cyclic GMP located in different compartments, which could have different effects in pancreatic acini.

Role for guanylyl gyclase/cyclic GMP pathway in pancreatic cancer

Guanylin, uroguanylin, and GC-C are expressed at mRNA and protein levels in pancreatic cancer specimens and cancer cell lines, and uroguanylin inhibits pancreatic cancer cell proliferation in a concentration-dependent manner.[186]

The transmembrane cell surface receptor GC-C has been identified in 60–70 percent of pancreatic cancer. An anti-GC-C antibody–drug conjugate TAK-264 (formally known as MLN0264) in patients with advanced or metastatic pancreatic cancer (NCT02202785) showed low efficacy and, for that, no further clinical investigation was undertaken.[187]

PKG1 is expressed in pancreatic cancer cells and its inhibitor, DT3, causes cytotoxicity through necrosis and inhibits proliferation and migration of pancreatic cancer cells.[188]

Noncanonical Cyclic Nucleotides

Cyclic IMP, cyclic XMP, cyclic CMP, cyclic UMP, and cyclic TMP are cyclic nucleotides whose function is well characterized (**Figure 12**). Using HPLC-MS/MS spectrometry, both cyclic CMP and cyclic UMP have been found in numerous cultured cell types and in human urine. Cyclic CMP and cyclic UMP concentrations are regulated by the cell proliferation status because growth arrest of cells resulted in preferential decrease of cellular cyclic CMP and cyclic UMP concentrations over cyclic AMP and cyclic GMP concentrations. Previous findings suggest that cyclic CMP and cyclic UMP could play a role as second messengers because cyclic CMP and cyclic UMP-hydrolyzing PDEs were found in mammalian tissues. Recently, soluble AC has shown to be responsible for the production of cyclic CMP and cyclic UMP in HEK293 and B103 cells because the soluble AC inhibitor KH7 decreased HCO_3-stimulated cyclic nucleotide levels in concentration-dependent manner. Forskolin, which is a stimulator of all transmembrane ACs except AC9, does not affect the levels of cyclic CMP and cyclic UMP. The authors conclude that soluble AC may likely have a distinct role in the regulation of cyclic nucleotide levels compared to soluble GC, membrane GC, and membrane AC.[1] In RFL6 lung fibroblasts endogenously expressing soluble GC, NO-stimulated cyclic UMP formation was similar to cyclic GMP formation.[189] In contrast to soluble GC, transmembrane GC does not induce cyclic UMP formation.[190] Recently, cyclic CMP was found in several mouse tissues including pancreas as assessed by HPLC-MS/MS and HPLC-MS/TOF.[191]

cyclic purine nucleotides

cyclic AMP   cyclic GMP   cyclic IMP   cyclic XMP

cyclic pyrimidine nucleotides

cyclic CMP   cyclic UMP   cyclic TMP

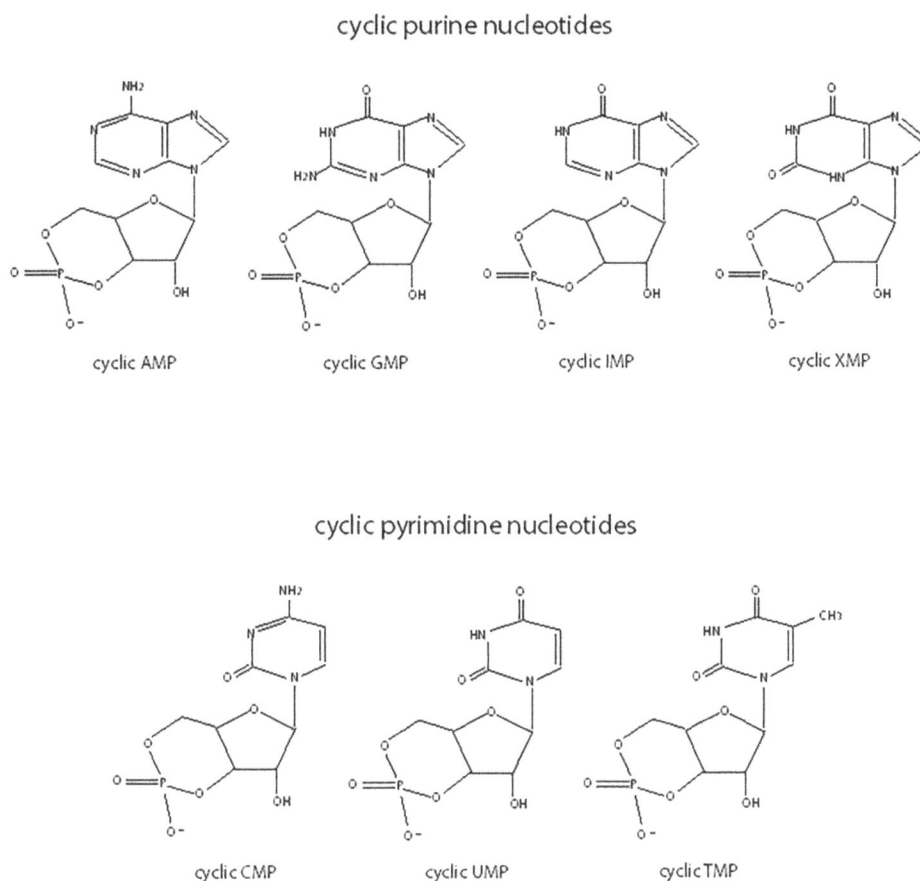

Figure 12. Schematic representation of cyclic purine (cyclic AMP, cyclic GMP, cyclic IMP, and cyclic XMP) and pyrimidine (cyclic CMP, cyclic UMP, and cyclic TMP) nucleotides.

Unlike cyclic CMP and cyclic UMP, cyclic TMP, cyclic IMP, and cyclic XMP levels are very low to be detectable in cultured cell lines.[190] Cyclic IMP levels increase in a hypoxic environment probably as a result of ATP deamination, which becomes ITP, and by soluble GC activity, ITP becomes cyclic IMP.[1]

Noncanonical cyclic nucleotides have been studied so far in cardiovascular system, central nervous system, and reproductive system. A description of their roles in the regulation of these system can be found in reference [157]. To the best of our knowledge, at the present there is no data available for the role of noncanonical cyclic nucleotides in the digestive system.

References

1. Seifert R, Schneider EH, Bahre H. From canonical to noncanonical cyclic nucleotides as second messengers: pharmacological implications. Pharmacol Ther 2015; 148:154–184.

2. Bruice PY. The chemistry of the nucleic acids. In: Zalesky J, ed. Organic chemistry. 7th ed. New York City, NY: Pearson Education, Inc., 2015:1155–1181.

3. Steegborn C, Litvin TN, Levin LR, Buck J, Wu H. Bicarbonate activation of adenylyl cyclase via promotion of catalytic active site closure and metal recruitment. Nat Struct Mol Biol 2005; 12:32–37.

4. Pierce KL, Premont RT, Lefkowitz RJ. Seven-transmembrane receptors. Nat Rev Mol Cell Biol 2002; 3:639–650.

5. Linder ME, Gilman AG. G proteins. Sci Am 1992; 267:56–61, 64–55.

6. Murad F. Shattuck Lecture. Nitric oxide and cyclic GMP in cell signaling and drug development. N Engl J Med 2006; 355:2003–2011.

7. Zippin JH, Levin LR, Buck J. CO_2/HCO_3-responsive soluble adenylyl cyclase as a putative metabolic sensor. Trends Endocrinol Metab 2001; 12:366–370.

8. Koesling D, Friebe A. Soluble guanylyl cyclase: structure and regulation. Rev Physiol Biochem Pharmacol 1999; 135:41–65.

9. Steegborn C. Structure, mechanism, and regulation of soluble adenylyl cyclases—similarities and differences to transmembrane adenylyl cyclases. Biochim Biophys Acta 2014; 1842:2535–2547.

10. Ulrich CD, 2nd, Holtmann M, Miller LJ. Secretin and vasoactive intestinal peptide receptors: members of a unique

family of G protein-coupled receptors. Gastroenterology 1998; 114:382–397.

11. Yoshida H, Tsunoda Y, Owyang C. Effect of uncoupling NO/cGMP pathways on carbachol- and CCK-stimulated Ca^{2+} entry and amylase secretion from the rat pancreas. Pflugers Arch 1997; 434:25–37.

12. Kolodecik TR, Shugrue CA, Thrower EC, Levin LR, Buck J, Gorelick FS. Activation of soluble adenylyl cyclase protects against secretagogue stimulated zymogen activation in rat pancreatic acinar cells. PLoS One 2012; 7:e41320.

13. Taylor SS, Zhang P, Steichen JM, Keshwani MM, Kornev AP. PKA: lessons learned after twenty years. Biochim Biophys Acta 2013; 1834:1271–1278.

14. Hofmann F, Bernhard D, Lukowski R, Weinmeister P. cGMP regulated protein kinases (cGK). Handb Exp Pharmacol 2009: 137–162.

15. de Rooij J, Zwartkruis FJ, Verheijen MH, Cool RH, Nijman SM, Wittinghofer A, Bos JL. Epac is a Rap1 guanine-nucleotide-exchange factor directly activated by cyclic AMP. Nature 1998; 396:474–477.

16. Groblewski GE, Yoshida M, Bragado MJ, Ernst SA, Leykam J, Williams JA. Purification and characterization of a novel physiological substrate for calcineurin in mammalian cells. J Biol Chem 1998; 273:22738–22744.

17. Kaupp UB, Seifert R. Cyclic nucleotide-gated ion channels. Physiol Rev 2002; 82:769–824.

18. Conti M, Beavo J. Biochemistry and physiology of cyclic nucleotide phosphodiesterases: essential components in cyclic nucleotide signaling. Annu Rev Biochem 2007; 76:481–511.

19. Tang WJ, Gilman AG. Adenylyl cyclases. Cell 1992; 70:869–872.

20. Premont RT, Matsuoka I, Mattei MG, Pouille Y, Defer N, Hanoune J. Identification and characterization of a widely expressed form of adenylyl cyclase. J Biol Chem 1996; 271:13900–13907.

21. Cooper DM, Mons N, Karpen JW. Adenylyl cyclases and the interaction between calcium and cAMP signalling. Nature 1995; 374:421–424.

22. Yan SZ, Huang ZH, Shaw RS, Tang WJ. The conserved asparagine and arginine are essential for catalysis of mammalian adenylyl cyclase. J Biol Chem 1997; 272:12342–12349.

23. Krupinski J, Coussen F, Bakalyar HA, Tang WJ, Feinstein PG, Orth K, Slaughter C, Reed RR, Gilman AG. Adenylyl cyclase amino acid sequence: possible channel- or transporter-like structure. Science 1989; 244:1558–1564.

24. Zhang G, Liu Y, Ruoho AE, Hurley JH. Structure of the adenylyl cyclase catalytic core. Nature 1997; 386:247–253.

25. Feinstein PG, Schrader KA, Bakalyar HA, Tang WJ, Krupinski J, Gilman AG, Reed RR. Molecular cloning and characterization of a Ca^{2+}/calmodulin-insensitive adenylyl cyclase from rat brain. Proc Natl Acad Sci U S A 1991; 88:10173–10177.

26. Xie T, Chen M, Weinstein LS. Pancreas-specific $G_s\alpha$ deficiency has divergent effects on pancreatic α- and β-cell proliferation. J Endocrinol 2010; 206:261–269.

27. Sadana R, Dessauer CWT. Physiological roles for G protein-regulated adenylyl cyclase isoforms: insights from knockout and overexpression studies. Neurosignals 2009; 17:5–22.

28. Singh P, Asada I, Owlia A, Collins TJ, Thompson JC. Somatostatin inhibits VIP-stimulated amylase release from perifused guinea pig pancreatic acini. Am J Physiol Gastrointest Liver Physiol 1988; 254:G217–G223.

29. Matsushita K, Okabayashi Y, Hasegawa H, Koide M, Kido Y, Okutani T, Sugimoto Y, Kasuga M. In vitro inhibitory effect of somatostatin on secretin action in exocrine pancreas of rats. Gastroenterology 1993; 104:1146–1152.

30. Sabbatini ME, D'Alecy L, Lentz SI, Tang T, Williams JA. Adenylyl cyclase 6 mediates the action of cyclic AMP-dependent secretagogues in mouse pancreatic exocrine cells via protein kinase A pathway activation. J Physiol 2013; 591:3693–3707.

31. Defer N, Best-Belpomme M, Hanoune J. Tissue specificity and physiological relevance of various isoforms of adenylyl cyclase. Am J Physiol Renal Physiol 2000; 279:F400–416.

32. Nelson EJ, Hellevuo K, Yoshimura M, Tabakoff B. Ethanol-induced phosphorylation and potentiation of the activity of type 7 adenylyl cyclase. Involvement of protein kinase C delta. J Biol Chem 2003; 278:4552–4560.

33. Willoughby D, Cooper DM. Organization and Ca^{2+} regulation of adenylyl cyclases in cAMP microdomains. Physiol Rev 2007; 87:965–1010.

34. Buck J, Sinclair ML, Schapal L, Cann MJ, Levin LR. Cytosolic adenylyl cyclase defines a unique signaling molecule in mammals. Proc Natl Acad Sci U S A 1999; 96:79–84.

35. Chen Y, Harry A, Li J, Smit MJ, Bai X, Magnusson R, Pieroni JP, Weng G, Iyengar R. Adenylyl cyclase 6 is selectively regulated by protein kinase A phosphorylation in a region involved in $G\alpha_s$ stimulation. Proc Natl Acad Sci U S A 1997; 94:14100–14104.

36. Kleinboelting S, Diaz A, Moniot S, van den Heuvel J, Weyand M, Levin LR, Buck J, Steegborn C. Crystal structures of human soluble adenylyl cyclase reveal mechanisms of catalysis and of its activation through bicarbonate. Proc Natl Acad Sci U S A 2014; 111:3727–3732.

37. Zippin JH, Chen Y, Nahirney P, Kamenetsky M, Wuttke MS, Fischman DA, Levin LR, Buck J. Compartmentalization of bicarbonate-sensitive adenylyl cyclase in distinct signaling microdomains. FASEB J 2003; 17:82–84.

38. Nagathihalli NS, Castellanos JA, Shi C, Beesetty Y, Reyzer ML, Caprioli R, Chen X, Walsh AJ, Skala MC, Moses HL, Merchant NB. Signal transducer and activator of transcription 3, mediated remodeling of the tumor microenvironment results in enhanced tumor drug delivery in a mouse model of pancreatic cancer. Gastroenterology 2015; 149:1932–1943.

39. Weber IT, Gilliland GL, Harman JG, Peterkofsky A. Crystal structure of a cyclic AMP-independent mutant of catabolite gene activator protein. J Biol Chem 1987; 262:5630–5636.

40. Jensen RT, Gardner JD. Cyclic nucleotide-dependent protein kinase activity in acinar cells from guinea pig pancreas. Gastroenterology 1978; 75:806–816.

41. Burnham DB, Williams JA. Activation of protein kinase activity in pancreatic acini by calcium and cAMP. Am J Physiol Gastrointest Liver Physiol 1984; 246:G500–G508.

42. Burnham DB, Sung CK, Munowitz P, Williams JA. Regulation of protein phosphorylation in pancreatic acini by

cyclic AMP-mediated secretagogues: interaction with carbamylcholine. Biochim Biophys Acta 1988; 969:33–39.

43. Sabbatini ME, Chen X, Ernst SA, Williams JA. Rap1 activation plays a regulatory role in pancreatic amylase secretion. J Biol Chem 2008; 283:23884–23894.

44. Grunwald B, Vandooren J, Gerg M, Ahomaa K, Hunger A, Berchtold S, Akbareian S, Schaten S, Knolle P, Edwards DR, Opdenakker G, Kruger A. Systemic ablation of MMP-9 triggers invasive growth and metastasis of pancreatic cancer via deregulation of IL6 expression in the bone marrow. Mol Cancer Res 2016; 14:1147–1158.

45. Argent BE, Gray MA, Steward MC, Case RM. Cell physiology of pancreatic ducts. In: Johnson LR, ed. Physiology of the gastrointestinal tract. Volume 2. 5th ed. Amsterdam, USA: Elsevier Inc., 2012:1399–1423.

46. Chanson M, Scerri I, Suter S. Defective regulation of gap junctional coupling in cystic fibrosis pancreatic duct cells. J Clin Invest 1999; 103:1677–1684.

47. LeBeau AP, Yule DI, Groblewski GE, Sneyd J. Agonist-dependent phosphorylation of the inositol 1,4,5-trisphosphate receptor: a possible mechanism for agonist-specific calcium oscillations in pancreatic acinar cells. J Gen Physiol 1999; 113:851–872.

48. Straub SV, Giovannucci DR, Bruce JI, Yule DI. A role for phosphorylation of inositol 1,4,5-trisphosphate receptors in defining calcium signals induced by peptide agonists in pancreatic acinar cells. J Biol Chem 2002; 277:31949–31956.

49. Giovannucci DR, Groblewski GE, Sneyd J, Yule DI. Targeted phosphorylation of inositol 1,4,5-trisphosphate receptors selectively inhibits localized Ca^{2+} release and shapes oscillatory Ca^{2+} signals. J Biol Chem 2000; 275:33704–33711.

50. Calejo AI, Tasken K. Targeting protein-protein interactions in complexes organized by A kinase anchoring proteins. Front Pharmacol 2015; 6:192.

51. Kurihara K, Nakanishi N. Regulation of Na,K-ATPase by cAMP-dependent protein kinase anchored on membrane via A-kinase anchoring protein subtype, AKAP-150, in rat parotid gland. Ann N Y Acad Sci 2003; 986:636–638.

52. Saino T, Watson EL. Inhibition of serine/threonine phosphatase enhances arachidonic acid-induced $[Ca^{2+}]_i$ via protein kinase A. Am J Physiol Cell Physiol 2009; 296:C88–C96.

53. Zaccolo M, Movsesian MA. cAMP and cGMP signaling cross-talk: role of phosphodiesterases and implications for cardiac pathophysiology. Circ Res 2007; 100:1569–1578.

54. Ueno H, Shibasaki T, Iwanaga T, Takahashi K, Yokoyama Y, Liu LM, Yokoi N, Ozaki N, Matsukura S, Yano H, Seino S. Characterization of the gene EPAC2: structure, chromosomal localization, tissue expression, and identification of the liver-specific isoform. Genomics 2001; 78:91–98.

55. Bos JL. Epac proteins: multi-purpose cAMP targets. Trends Biochem Sci 2006; 31:680–686.

56. Gloerich M, Bos JL. Epac: defining a new mechanism for cAMP action. Annu Rev Pharmacol Toxicol 2010; 50:355–375.

57. Holz GG, Kang G, Harbeck M, Roe MW, Chepurny OG. Cell physiology of cAMP sensor Epac. J Physiol 2006; 577:5–15.

58. Chen H, Wild C, Zhou X, Ye N, Cheng X, Zhou J. Recent advances in the discovery of small molecules targeting exchange proteins directly activated by cAMP (EPAC). J Med Chem 2014; 57:3651–3665.

59. Chaudhuri A, Husain SZ, Kolodecik TR, Grant WM, Gorelick FS. Cyclic AMP-dependent protein kinase and Epac mediate cyclic AMP responses in pancreatic acini. Am J Physiol Gastrointest Liver Physiol 2007; 292:G1403–G1410.

60. Chen X, Walker AK, Strahler JR, Simon ES, Tomanicek-Volk SL, Nelson BB, Hurley MC, Ernst SA, Williams JA, Andrews PC. Organellar proteomics: analysis of pancreatic zymogen granule membranes. Mol Cell Proteomics 2006; 5:306–312.

61. Kang G, Joseph JW, Chepurny OG, Monaco M, Wheeler MB, Bos JL, Schwede F, Genieser HG, Holz GG. Epac-selective cAMP analog 8-pCPT-2'-O-Me-cAMP as a stimulus for Ca^{2+}-induced Ca^{2+} release and exocytosis in pancreatic β-cells. J Biol Chem 2003; 278:8279–8285.

62. Ozaki N, Shibasaki T, Kashima Y, Miki T, Takahashi K, Ueno H, Sunaga Y, Yano H, Matsuura Y, Iwanaga T, Takai Y, Seino S. cAMP-GEFII is a direct target of cAMP in regulated exocytosis. Nat Cell Biol 2000; 2:805–811.

63. Seino S, Shibasaki T. PKA-dependent and PKA-independent pathways for cAMP-regulated exocytosis. Physiol Rev 2005; 85:1303–1342.

64. Ju HQ, Ying H, Tian T, Ling J, Fu J, Lu Y, Wu M, Yang L, Achreja A, Chen G, Zhuang Z, Wang H, et al. Mutant Kras- and p16-regulated NOX4 activation overcomes metabolic checkpoints in development of pancreatic ductal adenocarcinoma. Nat Commun 2017; 8:14437.

65. Benarroch EE. HCN channels: function and clinical implications. Neurology 2013; 80:304–310.

66. Biel M, Michalakis S. Cyclic nucleotide-gated channels. Handb Exp Pharmacol 2009; 191:111–136.

67. Ma H, Butler MR, Thapa A, Belcher J, Yang F, Baehr W, Biel M, Michalakis S, Ding XQ. cGMP/protein kinase G signaling suppresses inositol 1,4,5-trisphosphate receptor phosphorylation and promotes endoplasmic reticulum stress in photoreceptors of cyclic nucleotide-gated channel-deficient mice. J Biol Chem 2015; 290:20880–20892.

68. Weyand I, Godde M, Frings S, Weiner J, Muller F, Altenhofen W, Hatt H, Kaupp UB. Cloning and functional expression of a cyclic-nucleotide-gated channel from mammalian sperm. Nature 1994; 368:859–863.

69. DiFrancesco JC, DiFrancesco D. Dysfunctional HCN ion channels in neurological diseases. Front Cell Neurosci 2015; 6:174.

70. Herrmann S, Schnorr S, Ludwig A. HCN channels-modulators of cardiac and neuronal excitability. Int J Mol Sci 2015; 16:1429–1447.

71. Biel M, Wahl-Schott C, Michalakis S, Zong X. Hyperpolarization-activated cation channels: from genes to function. Physiol Rev 2009; 89:847–885.

72. Zhang X, Andren PE, Chergui K, Svenningsson P. Neurokinin B/NK3 receptors exert feedback inhibition on L-DOPA actions in the 6-OHDA lesion rat model of Parkinson's disease. Neuropharmacology 2008; 54:1143–1152.

73. Ficker E, Jarolimek W, Brown AM. Molecular determinants of inactivation and dofetilide block in ether a-go-go (EAG)

channels and EAG-related K$^+$ channels. Mol Pharmacol 2001; 60:1343–1348.

74. Chang F, Cohen IS, DiFrancesco D, Rosen MR, Tromba C. Effects of protein kinase inhibitors on canine Purkinje fibre pacemaker depolarization and the pacemaker current i_f. J Physiol 1991; 440:367–384.

75. Boulton S, Akimoto M, VanSchouwen B, Moleschi K, Selvaratnam R, Giri R, Melacini G. Tapping the translation potential of cAMP signalling: molecular basis for selectivity in cAMP agonism and antagonism as revealed by NMR. Biochem Soc Trans 2014; 42:302–307.

76. Kim SJ, Kim JK, Pavenstadt H, Greger R, Hug MJ, Bleich M. Regulation of slowly activating potassium current (I_{Ks}) by secretin in rat pancreatic acinar cells. J Physiol 2001; 535:349–358.

77. Warth R, Garcia Alzamora M, Kim JK, Zdebik A, Nitschke R, Bleich M, Gerlach U, Barhanin J, Kim SJ. The role of KCNQ1/KCNE1 K$^+$ channels in intestine and pancreas: lessons from the KCNE1 knockout mouse. Pflugers Arch 2002; 443:822–828.

78. Takumi T, Ohkubo H, Nakanishi S. Cloning of a membrane protein that induces a slow voltage-gated potassium current. Science 1988; 242:1042–1045.

79. Yang WP, Levesque PC, Little WA, Conder ML, Shalaby FY, Blanar MA. KvLQT1, a voltage-gated potassium channel responsible for human cardiac arrhythmias. Proc Natl Acad Sci U S A 1997; 94:4017–4021.

80. Lee E, Gerlach U, Uhm DY, Kim J. Inhibitory effect of somatostatin on secretin-induced augmentation of the slowly activating K$^+$ current (I_{Ks}) in the rat pancreatic acinar cell. Pflugers Arch 2002; 443:405–410.

81. Kim SJ, Greger R. Voltage-dependent, slowly activating K$^+$ current (I_{Ks}) and its augmentation by carbachol in rat pancreatic acini. Pflugers Arch 1999; 438:604–611.

82. Schafer C, Steffen H, Krzykowski KJ, Goke B, Groblewski GE. CRHSP-24 phosphorylation is regulated by multiple signaling pathways in pancreatic acinar cells. Am J Physiol Gastrointest Liver Physiol 2003; 285:G726–G734.

83. Kunapuli P, Benovic JL. Cloning and expression of GRK5: a member of the G protein-coupled receptor kinase family. Proc Natl Acad Sci U S A 1993; 90:5588–5592.

84. Benovic JL, Gomez J. Molecular cloning and expression of GRK6. A new member of the G protein-coupled receptor kinase family. J Biol Chem 1993; 268:19521–19527.

85. Jobling MG, Gotow LF, Yang Z, Holmes RK. A mutational analysis of residues in cholera toxin A1 necessary for interaction with its substrate, the stimulatory G protein Gsalpha. Toxins (Basel) 2015; 7:919–935.

86. Gardner JD, Rottman AJ. Action of cholera toxin on dispersed acini from guinea pig pancreas. Biochim Biophys Acta 1979; 585:250–265.

87. Singh M. Role of cyclic adenosine monophosphate in amylase release from dissociated rat pancreatic acini. J Physiol 1982; 331:547–555.

88. De Lisle RC, Howell GW. Evidence of heterotrimeric G-protein involvement in regulated exocytosis from permeabilized pancreatic acini. Pancreas 1995; 10:374–381.

89. Adamson PB, Hull SS, Jr., Vanoli E, De Ferrari GM, Wisler P, Foreman RD, Watanabe AM, Schwartz PJ. Pertussis toxin-induced ADP ribosylation of inhibitor G proteins alters vagal control of heart rate in vivo. Am J Physiol Heart Circ Physiol 1993; 265:H734–H740.

90. Fields TA, Casey PJ. Signalling functions and biochemical properties of pertussis toxin-resistant G-proteins. Biochem J 1997; 321:561–571.

91. Kost CK, Jr., Herzer WA, Li PJ, Jackson EK. Pertussis toxin-sensitive G-proteins and regulation of blood pressure in the spontaneously hypertensive rat. Clin Exp Pharmacol Physiol 1999; 26:449–455.

92. Willems PH, Tilly RH, de Pont JJ. Pertussis toxin stimulates cholecystokinin-induced cyclic AMP formation but is without effect on secretagogue-induced calcium mobilization in exocrine pancreas. Biochim Biophys Acta 1987; 928:179–185.

93. Stryjek-Kaminska D, Piiper A, Zeuzem S. EGF inhibits secretagogue-induced cAMP production and amylase secretion by G$_i$ proteins in pancreatic acini. Am J Physiol Gastrointest Liver Physiol 1995; 269:G676–G682.

94. Luo X, Ahn W, Muallem S, Zeng W. Analyses of RGS protein control of agonist-evoked Ca^{2+} signaling. Methods Enzymol 2004; 389:119–130.

95. Sinnarajah S, Dessauer CW, Srikumar D, Chen J, Yuen J, Yilma S, Dennis JC, Morrison EE, Vodyanoy V, Kehrl JH. RGS2 regulates signal transduction in olfactory neurons by attenuating activation of adenylyl cyclase III. Nature 2001; 409:1051–1055.

96. Sabbatini ME, Gorelick F, Glaser S. Adenylyl cyclases in the digestive system. Cell Signal 2014; 26:1173–1181.

97. Cumbay MG, Watts VJ. Novel regulatory properties of human type 9 adenylate cyclase. J Pharmacol Exp Ther 2004; 310:108–115.

98. Duan RD, Guo YJ, Williams JA. Conversion to Ca^{2+}-independent form of Ca^{2+}/calmodulin protein kinase II in rat pancreatic acini. Biochem Biophys Res Commun 1994; 199:368–373.

99. Cumbay MG, Watts VJ. Gα$_q$ potentiation of adenylate cyclase type 9 activity through a Ca^{2+}/calmodulin-dependent pathway. Biochem Pharmacol 2005; 69:1247–1256.

100. Gurda GT, Guo L, Lee SH, Molkentin JD, Williams JA. Cholecystokinin activates pancreatic calcineurin-NFAT signaling in vitro and in vivo. Mol Biol Cell 2008; 19:198–206.

101. Groblewski GE, Wagner AC, Williams JA. Cyclosporin A inhibits Ca^{2+}/calmodulin-dependent protein phosphatase and secretion in pancreatic acinar cells. J Biol Chem 1994; 269:15111–15117.

102. Husain SZ, Grant WM, Gorelick FS, Nathanson MH, Shah AU. Caerulein-induced intracellular pancreatic zymogen activation is dependent on calcineurin. Am J Physiol Gastrointest Liver Physiol 2007; 292:G1594–G1599.

103. Seamon KB, Daly JW. Forskolin: its biological and chemical properties. Adv Cyclic Nucleotide Protein Phosphorylation Res 1986; 20:1–150.

104. Sunahara RK, Taussig R. Isoforms of mammalian adenylyl cyclase: multiplicities of signaling. Mol Interv 2002; 2:168–184.

105. Yan SZ, Huang ZH, Andrews RK, Tang WJ. Conversion of forskolin-insensitive to forskolin-sensitive (mouse-type IX) adenylyl cyclase. Mol Pharmacol 1998; 53:182–187.

106. Chen Y, Cann MJ, Litvin TN, Iourgenko V, Sinclair ML, Levin LR, Buck J. Soluble adenylyl cyclase as an evolutionarily conserved bicarbonate sensor. Science 2000; 289:625–628.

107. Litvin TN, Kamenetsky M, Zarifyan A, Buck J, Levin LR. Kinetic properties of "soluble" adenylyl cyclase. Synergism between calcium and bicarbonate. J Biol Chem 2003; 278:15922–15926.

108. Nomura M, Beltran C, Darszon A, Vacquier VD. A soluble adenylyl cyclase from sea urchin spermatozoa. Gene 2005; 353:231–238.

109. Morley DJ, Hawley DM, Ulbright TM, Butler LG, Culp JS, Hodes ME. Distribution of phosphodiesterase I in normal human tissues. J Histochem Cytochem 1987; 35:75–82.

110. Hetman JM, Soderling SH, Glavas NA, Beavo JA. Cloning and characterization of PDE7B, a cAMP-specific phosphodiesterase. Proc Natl Acad Sci U S A 2000; 97:472–476.

111. Mersin H, Irkin F, Berberoglu U, Gulben K, Ozdemir H, Onguru O. The selective inhibition of type IV phosphodiesterase attenuates the severity of the acute pancreatitis in rats. Dig Dis Sci 2009; 54:2577–2582.

112. Ishiguro H, Steward MC, Lindsay AR, Case RM. Accumulation of intracellular HCO_3^- by Na^+-HCO_3^- cotransport in interlobular ducts from guinea-pig pancreas. J Physiol 1996; 495:169–178.

113. Ishiguro H, Naruse S, Steward MC, Kitagawa M, Ko SB, Hayakawa T, Case RM. Fluid secretion in interlobular ducts isolated from guinea-pig pancreas. J Physiol 1998; 511:407–422.

114. Pascua P, Garcia M, Fernandez-Salazar MP, Hernandez-Lorenzo MP, Calvo JJ, Colledge WH, Case RM, Steward MC, San Roman JI. Ducts isolated from the pancreas of CFTR-null mice secrete fluid. Pflugers Arch 2009; 459:203–214.

115. Lee MG, Wigley WC, Zeng W, Noel LE, Marino CR, Thomas PJ, Muallem S. Regulation of Cl^-/HCO_3^- exchange by cystic fibrosis transmembrane conductance regulator expressed in NIH 3T3 and HEK 293 cells. J Biol Chem 1999; 274:3414–3421.

116. Ohnishi H, Mine T, Kojima I. Inhibition by somatostatin of amylase secretion induced by calcium and cyclic AMP in rat pancreatic acini. Biochem J 1994; 304:531–536.

117. Akiyama T, Hirohata Y, Okabayashi Y, Imoto I, Otsuki M. Supramaximal CCK and CCh concentrations abolish VIP potentiation by inhibiting adenylyl cyclase activity. Am J Physiol Gastrointest Liver Physiol 1998; 275:G1202–G1208.

118. Burnham DB, McChesney DJ, Thurston KC, Williams JA. Interaction of cholecystokinin and vasoactive intestinal polypeptide on function of mouse pancreatic acini in vitro. J Physiol 1984; 349:475–482.

119. O'Sullivan AJ, Jamieson JD. Protein kinase A modulates Ca^{2+}- and protein kinase C-dependent amylase release in permeabilized rat pancreatic acini. Biochem J 1992; 287:403–406.

120. Gardner JD, Sutliff VE, Walker MD, Jensen RT. Effects of inhibitors of cyclic nucleotide phosphodiesterase on actions of cholecystokinin, bombesin, and carbachol on pancreatic acini. Am J Physiol Gastrointest Liver Physiol 1983; 245:G676–G680.

121. Collen MJ, Sutliff VE, Pan GZ, Gardner JD. Postreceptor modulation of action of VIP and secretin on pancreatic enzyme secretion by secretagogues that mobilize cellular calcium. Am J Physiol Gastrointest Liver Physiol 1982; 242:G423–G428.

122. Gardner JD, Korman LY, Walker MD, Sutliff VE. Effects of inhibitors of cyclic nucleotide phosphodiesterase on the actions of vasoactive intestinal peptide and secretin on pancreatic acini. Am J Physiol Gastrointest Liver Physiol 1982; 242:G547–G551.

123. Schnefel S, Profrock A, Hinsch KD, Schulz I. Cholecystokinin activates G_i1-, G_i2-, G_i3- and several G_s-proteins in rat pancreatic acinar cells. Biochem J 1990; 269:483–488.

124. Dehaye JP, Gillard M, Poloczek P, Stievenart M, Winand J, Christophe J. Effects of forskolin on adenylate cyclase activity and amylase secretion in the rat exocrine pancreas. J Cyclic Nucleotide Protein Phosphor Res 1985; 10:269–280.

125. Kimura T, Imamura K, Eckhardt L, Schulz I. Ca^{2+}-, phorbol ester-, and cAMP-stimulated enzyme secretion from permeabilized rat pancreatic acini. Am J Physiol Gastrointest Liver Physiol 1986; 250:G698–G708.

126. Heisler S. Forskolin potentiates calcium-dependent amylase secretion from rat pancreatic acinar cells. Can J Physiol Pharmacol 1983; 61:1168–1176.

127. Drucker DJ. Glucagon-like peptides. Diabetes 1998; 47:159–169.

128. Hou Y, Ernst SA, Heidenreich K, Williams JA. Glucagon-like peptide-1 receptor is present in pancreatic acinar cells and regulates amylase secretion through cAMP. Am J Physiol Gastrointest Liver Physiol 2016; 310:G26–G33.

129. Ramos-Alvarez I, Lee L, Jensen RT. Cyclic AMP-dependent protein kinase A and EPAC mediate VIP and secretin stimulation of PAK4 and activation of Na^+,K^+-ATPase in pancreatic acinar cells. Am J Physiol Gastrointest Liver Physiol 2019; 316:G263–G277.

130. Wang R, Li J, Rosenberg L. Factors mediating the transdifferentiation of islets of Langerhans to duct epithelial-like structures. J Endocrinol 2001; 171:309–318.

131. Yamamoto T, Yamato E, Taniguchi H, Shimoda M, Tashiro F, Hosoi M, Sato T, Fujii S, Miyazaki JI. Stimulation of cAMP signalling allows isolation of clonal pancreatic precursor cells from adult mouse pancreas. Diabetologia 2006; 49:2359–2367.

132. Yang H, Lee CJ, Zhang L, Sans MD, Simeone DM. Regulation of transforming growth factor β-induced responses by protein kinase A in pancreatic acinar cells. Am J Physiol Gastrointest Liver Physiol 2008; 295:G170–G178.

133. Yadav D, Whitcomb DC. The role of alcohol and smoking in pancreatitis. Nat Rev Gastroenterol Hepatol 2010; 7:131–145.

134. Lu Z, Kolodecik TR, Karne S, Nyce M, Gorelick F. Effect of ligands that increase cAMP on caerulein-induced zymogen activation in pancreatic acini. Am J Physiol Gastrointest Liver Physiol 2003; 285:G822–G828.

135. Chaudhuri A, Kolodecik TR, Gorelick FS. Effects of increased intracellular cAMP on carbachol-stimulated zymogen activation, secretion, and injury in the pancreatic acinar cell. Am J Physiol Gastrointest Liver Physiol 2005; 288:G235–G243.

136. Siegel RL, Miller KD, Jemal A. Cancer statistics, 2020. CA Cancer J Clin 2020; 70:7–30.

137. Siegel RL, Miller KD, Jemal A. Cancer statistics, 2019. CA Cancer J Clin 2019; 69:7–34.

138. al-Nakkash L, Simmons NL, Lingard JM, Argent BE. Adenylate cyclase activity in human pancreatic adenocarcinoma cell lines. Int J Pancreatol 1996; 19:39–47.

139. Burdyga A, Conant A, Haynes L, Zhang J, Jalink K, Sutton R, Neoptolemos J, Costello E, Tepikin A. cAMP inhibits migration, ruffling and paxillin accumulation in focal adhesions of pancreatic ductal adenocarcinoma cells: effects of PKA and EPAC. Biochim Biophys Acta 2013; 1833:2664–2672.

140. Zimmerman NP, Roy I, Hauser AD, Wilson JM, Williams CL, Dwinell MB. Cyclic AMP regulates the migration and invasion potential of human pancreatic cancer cells. Mol Carcinog 2015; 54:203–215.

141. Zhang D, Ma QY, Hu HT, Zhang M. beta2-adrenergic antagonists suppress pancreatic cancer cell invasion by inhibiting CREB, NFkappaB and AP-1. Cancer Biol Ther 2010; 10:19–29.

142. Kozasa T, Itoh H, Tsukamoto T, Kaziro Y. Isolation and characterization of the human G_s α gene. Proc Natl Acad Sci U S A 1988; 85:2081–2085.

143. Wu J, Matthaei H, Maitra A, Dal Molin M, Wood LD, Eshleman JR, Goggins M, Canto MI, Schulick RD, Edil BH, Wolfgang CL, Klein AP, et al. Recurrent GNAS mutations define an unexpected pathway for pancreatic cyst development. Sci Transl Med 2011; 3:92ra66.

144. Taki K, Ohmuraya M, Tanji E, Komatsu H, Hashimoto D, Semba K, Araki K, Kawaguchi Y, Baba H, Furukawa T. GNAS^{R201H} and Kras^{G12D} cooperate to promote murine pancreatic tumorigenesis recapitulating human intraductal papillary mucinous neoplasm. Oncogene 2016; 35:2407–2412.

145. Ideno N, Yamaguchi H, Ghosh B, Gupta S, Okumura T, Steffen DJ, Fisher CG, Wood LD, Singhi AD, Nakamura M, Gutkind JS, Maitra A. GNAS^{R201C} induces pancreatic cystic neoplasms in mice that express activated KRAS by inhibiting YAP1 signaling. Gastroenterology 2018; 155:1593–1607.

146. Quinn SN, Graves SH, Dains-McGahee C, Friedman EM, Hassan H, Witkowski P, Sabbatini ME. Adenylyl cyclase 3/adenylyl cyclase-associated protein 1 (CAP1) complex mediates the anti-migratory effect of forskolin in pancreatic cancer cells. Mol Carcinog 2017; 56:1344–1360.

147. Ho JJ, Crawley S, Pan PL, Farrelly ER, Kim YS. Secretion of MUC5AC mucin from pancreatic cancer cells in response to forskolin and VIP. Biochem Biophys Res Commun 2002; 294:680–686.

148. Kuhn M. Structure, regulation, and function of mammalian membrane guanylyl cyclase receptors, with a focus on guanylyl cyclase-A. Circ Res 2003; 93:700–709.

149. Maack T, Suzuki M, Almeida FA, Nussenzveig D, Scarborough RM, McEnroe GA, Lewicki JA. Physiological role of silent receptors of atrial natriuretic factor. Science 1987; 238:675–678.

150. Murthy KS, Teng BQ, Zhou H, Jin JG, Grider JR, Makhlouf GM. G_{i-1}/G_{i-2}-dependent signaling by single-transmembrane natriuretic peptide clearance receptor. Am J Physiol Gastrointest Liver Physiol 2000; 278:G974–G980.

151. Palaparti A, Li Y, Anand-Srivastava MB. Inhibition of atrial natriuretic peptide (ANP) C receptor expression by antisense oligodeoxynucleotides in A10 vascular smooth-muscle cells is associated with attenuation of ANP-C-receptor-mediated inhibition of adenylyl cyclase. Biochem J 2000; 346:313–320.

152. Pagano M, Anand-Srivastava MB. Cytoplasmic domain of natriuretic peptide receptor C constitutes G_i activator sequences that inhibit adenylyl cyclase activity. J Biol Chem 2001; 276:22064–22070.

153. Ventimiglia MS, Najenson AC, Rodríguez MR, Davio CA, Vatta MS, Bianciotti LG. Natriuretic peptides and their receptors. Pancreapedia: Exocrine Pancreas Knowledge Base, 2011. DOI: 10.3998/panc.2011.36.

154. Sabbatini ME, Rodriguez M, di Carlo MB, Davio CA, Vatta MS, Bianciotti LG. C-type natriuretic peptide enhances amylase release through NPR-C receptors in the exocrine pancreas. Am J Physiol Gastrointest Liver Physiol 2007; 293:G987–G994.

155. Sabbatini ME. Natriuretic peptides as regulatory mediators of secretory activity in the digestive system. Regul Pept 2009; 154:5–15.

156. Pandey KN. Natriuretic peptides and their receptors. Peptides 2005; 26:899–900.

157. Forte LR, Jr. Uroguanylin and guanylin peptides: pharmacology and experimental therapeutics. Pharmacol Ther 2004; 104:137–162.

158. Sindic A, Schlatter E. Renal electrolyte effects of guanylin and uroguanylin. Curr Opin Nephrol Hypertens 2007; 16:10–15.

159. Harteneck C, Wedel B, Koesling D, Malkewitz J, Bohme E, Schultz G. Molecular cloning and expression of a new alpha-subunit of soluble guanylyl cyclase. Interchangeability of the alpha-subunits of the enzyme. FEBS Lett 1991; 292:217–222.

160. Yuen PS, Potter LR, Garbers DL. A new form of guanylyl cyclase is preferentially expressed in rat kidney. Biochemistry 1990; 29:10872–10878.

161. Cary SP, Winger JA, Derbyshire ER, Marletta MA. Nitric oxide signaling: no longer simply on or off. Trends Biochem Sci 2006; 31:231–239.

162. Moglich A, Ayers RA, Moffat K. Structure and signaling mechanism of Per-ARNT-Sim domains. Structure 2009; 17:1282–1294.

163. Derbyshire ER, Marletta MA. Structure and regulation of soluble guanylate cyclase. Annu Rev Biochem 2012; 81:533–559.

164. Winger JA, Marletta MA. Expression and characterization of the catalytic domains of soluble guanylate cyclase: interaction with the heme domain. Biochemistry 2005; 44:4083–4090.

165. Lucas KA, Pitari GM, Kazerounian S, Ruiz-Stewart I, Park J, Schulz S, Chepenik KP, Waldman SA. Guanylyl cyclases and signaling by cyclic GMP. Pharmacol Rev 2000; 52:375–414.

166. Li H, Liu JP, Robinson PJ. Multiple substrates for cGMP-dependent protein kinase from bovine aortic smooth muscle: purification of P132. J Vasc Res 1996; 33:99–110.

167. Francis SH, Busch JL, Corbin JD, Sibley D. cGMP-dependent protein kinases and cGMP phosphodiesterases in nitric oxide and cGMP action. Pharmacol Rev 2010; 62:525–563.

168. Haymovits A, Scheele GA. Cellular cyclic nucleotides and enzyme secretion in the pancreatic acinar cell. Proc Natl Acad Sci U S A 1976; 73:156–160.

169. Williams JA, Yule DI. Stimulus-secretion coupling in pancreatic acinar cells. In: Johnson LR, ed. Physiology of the gastrointestinal tract. Volume 2. 5th ed. Amsterdam, USA: Elsevier Inc., 2012:1361–1398.

170. Pandol SJ, Schoeffield-Payne MS. Cyclic GMP mediates the agonist-stimulated increase in plasma membrane calcium entry in the pancreatic acinar cell. J Biol Chem 1990; 265:12846–12853.

171. Pandol SJ, Gukovskaya A, Bahnson TD, Dionne VE. Cellular mechanisms mediating agonist-stimulated calcium influx in the pancreatic acinar cell. Ann N Y Acad Sci 1994; 713:41–48.

172. Moustafa A, Sakamoto KQ, Habara Y. Nitric oxide stimulates IP_3 production via a cGMP/PKG-dependent pathway in rat pancreatic acinar cells. Jpn J Vet Res 2011; 59:5–14.

173. Love JA, Szebeni K. Histochemistry and electrophysiology of cultured adult rabbit pancreatic neurons. Pancreas 1999; 18:65–74.

174. Matrougui K, Maclouf J, Levy BI, Henrion D. Impaired nitric oxide- and prostaglandin-mediated responses to flow in resistance arteries of hypertensive rats. Hypertension 1997; 30:942–947.

175. Konturek SJ, Bilski J, Konturek PK, Cieszkowski M, Pawlik W. Role of endogenous nitric oxide in the control of canine pancreatic secretion and blood flow. Gastroenterology 1993; 104:896–902.

176. Wrenn RW, Currie MG, Herman LE. Nitric oxide participates in the regulation of pancreatic acinar cell secretion. Life Sci 1994; 55:511–518.

177. Gukovskaya A, Pandol SJ. Nitric oxide production regulates cGMP formation and calcium influx in pancreatic acinar cells. Am J Physiol Gastrointest Liver Physiol 1994; 266:G350–G356.

178. DiMagno MJ, Williams JA, Hao Y, Ernst SA, Owyang C. Endothelial nitric oxide synthase is protective in the initiation of caerulein-induced acute pancreatitis in mice. Am J Physiol Gastrointest Liver Physiol 2004; 287:G80–G87.

179. Jaworek J, Jachimczak B, Tomaszewska R, Konturek PC, Pawlik WW, Sendur R, Hahn EG, Stachura J, Konturek SJ. Protective action of lipopolysaccharides in rat caerulein-induced pancreatitis: role of nitric oxide. Digestion 2000; 62:1–13.

180. Jaworek J, Jachimczak B, Bonior J, Kot M, Tomaszewska R, Karczewska E, Stachura J, Pawlik W, Konturek SJ. Protective role of endogenous nitric oxide (NO) in lipopolysaccharide-induced pancreatic damage (a new experimental model of acute pancreatitis). J Physiol Pharmacol 2000; 51:85–102.

181. Heisler S, Kopelman H, Chabot JG, Morel G. Atrial natriuretic factor and exocrine pancreas: effects on the secretory process. Pancreas 1987; 2:243–251.

182. Sabbatini ME, Vatta MS, Vescina C, Gonzales S, Fernandez B, Bianciotti LG. NPR-C receptors are involved in C-type natriuretic peptide response on bile secretion. Regul Pept 2003; 116:13–20.

183. Sabbatini ME, Rodriguez MR, Dabas P, Vatta MS, Bianciotti LG. C-type natriuretic peptide stimulates pancreatic exocrine secretion in the rat: role of vagal afferent and efferent pathways. Eur J Pharmacol 2007; 577:192–202.

184. Kulaksiz H, Cetin Y. Uroguanylin and guanylate cyclase C in the human pancreas: expression and mutuality of ligand/receptor localization as indicators of intercellular paracrine signaling pathways. J Endocrinol 2001; 170:267–275.

185. Kulaksiz H, Schmid A, Honscheid M, Eissele R, Klempnauer J, Cetin Y. Guanylin in the human pancreas: a novel luminocrine regulatory pathway of electrolyte secretion via cGMP and CFTR in the ductal system. Histochem Cell Biol 2001; 115:131–145.

186. Kloeters O, Friess H, Giese N, Buechler MW, Cetin Y, Kulaksiz H. Uroguanylin inhibits proliferation of pancreatic cancer cells. Scand J Gastroenterol 2008; 43:447–455.

187. Almhanna K, Wright D, Mercade TM, Van Laethem JL, Gracian AC, Guillen-Ponce C, Faris J, Lopez CM, Hubner RA, Bendell J, Bols A, Feliu J, et al. A phase II study of antibody-drug conjugate, TAK-264 (MLN0264) in previously treated patients with advanced or metastatic pancreatic adenocarcinoma expressing guanylyl cyclase C. Invest New Drugs 2017; 35:634–641.

188. Karakhanova S, Golovastova M, Philippov PP, Werner J, Bazhin AV. Interlude of cGMP and cGMP/protein kinase G type 1 in pancreatic adenocarcinoma cells. Pancreas 2014; 43:784–794.

189. Bahre H, Kaever V. Measurement of 2',3'-cyclic nucleotides by liquid chromatography-tandem mass spectrometry in cells. J Chromatogr B Analyt Technol Biomed Life Sci 2014; 964:208–211.

190. Beste KY, Seifert R. cCMP, cUMP, cTMP, cIMP and cXMP as possible second messengers: development of a hypothesis based on studies with soluble guanylyl cyclase $\alpha_1\beta_1$. Biol Chem 2013; 394:261–270.

191. Schneider EH, Seifert R. Report on the third symposium "cCMP and cUMP as new second messengers." Naunyn Schmiedebergs Arch Pharmacol 2015; 388:1–3.

Chapter 5

Intermediate filament keratins in the exocrine pancreas

Catharina M. Alam,[1] Sarah Baghestani,[1] and Diana M. Toivola[1,2]

[1]Department of Biosciences, Cell biology, Åbo Akademi University, Tykistökatu 6A, [2]Turku Center for Disease Modeling, University of Turku, Kiinamyllynkatu 10, FIN-20520 Turku, Finland

Intermediate filament keratins are cytoskeletal components in epithelial tissues, including the exocrine pancreas. Keratin (K) intermediate filaments are highly conserved proteins that are expressed from early developmental stages up to the differentiated epithelial cells in adult individuals. A multitude of specific cellular functions have been identified for keratins expressed in simple epithelia, such as the pancreas, liver, lung, and intestine. These functions vary, depending on the cell and tissue type, as well as the developmental stage and changes in the cellular environment. Keratins are composed of dynamic, posttranslationally regulated cytoplasmic filaments built up of obligate heteropolymers of type I and type II keratins. In simple epithelia, the main keratins are type II K8 and K7, and type I K18, K19, and K20.

The exocrine pancreas comprises more than 85% of the pancreatic mass and consists of acinar and ductal cells. The acinar cells express mainly K8 and K18, which assemble into both cytoplasmic and apicolateral filaments, as well as minor levels of K19 and K20, which are confined to the apicolateral regions under basal conditions. Pancreatic duct cells express mainly K19 (type I) and K7 (type II). Pancreatic keratins respond quickly to cell stress by keratin phosphorylation and filament breakdown followed by keratin upregulation, de novo filament formation, and remodeling during the recovery phase in experimental exocrine pancreatic injury models. However, despite these dynamic stress responses, mutations or genetic deletion of K8 or K18 in humans or mouse models, only have modest effects on exocrine pancreatic health and stress tolerance. This is different from other simple epithelial tissues—most notably the liver—where K8, K18, and K19 mutations or deletions cause clear pathological outcomes. In contrast, overexpression of K8/K18 leads to pathological outcomes in the exocrine pancreas but not in the liver. These seemingly antagonistic outcomes in two cell types that have similar keratin expression patterns underscore the versatile and tissue-specific function of keratins. The biological reasons underlying the different susceptibilities of the exocrine acinar cells to keratin deficiencies, compared to other simple epithelial cells, are not fully understood but may, in part, be due to the increased levels of regeneration protein-II observed in the pancreas in several K8/K18-deficient mouse models.

Cytoskeletal Intermediate Filaments

The cytoskeleton is an organized network of proteins present in all cells. In eukaryotic cells, this network consists of three main filament systems: microtubules, microfilaments, and intermediate filaments (IFs). Among the main cytoskeletal filament groups, IFs are, as their name suggests, intermediate in size, measuring 10–12 nm in diameter; microfilaments are the smallest (6 nm) and microtubules the largest (25 nm).[1] IFs comprise a large and diverse group of proteins that are ubiquitously expressed. They are divided into six different types, where type I and type II comprise different keratins, type III includes the muscle IFs desmin and vimentin expressed in mesenchymal cells and type IV neurofilaments, including nestin and α-internexin expressed, for example, in nerve cells, and synemin, expressed widely in muscle cells. Type V comprises the lamins, which are nuclear IFs found in all nucleated cells, including all epithelial cells.[2] Finally, group VI comprises the highly specialized IFs phakinin and filensin, expressed only in the eye lens. The different types of IFs are typically expressed in a highly cell- and tissue-specific and/or in a developmentally specific manner.[3,4]

Keratins, as well as all other IF proteins, have a basic structure of a central coil-coil α-helical rod domain, flanked by an N-terminal head domain and a C-terminal tail domain of variable length (**Figure 1**). Two IF molecules dimerize to form tetramers, which are the building blocks of these mechanically strong, yet flexible filaments. The IF assembly does not require ATP and IF filaments are nonpolar, in contrast to microfilaments and microtubules.[4–6] Keratins, as well as other IFs, are dynamically regulated through various posttranslational modifications (PTMs), including

e-mails: Catharina.alam@gmail.com, diana.toivola@abo.fi

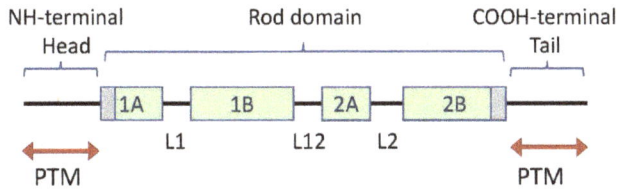

Figure 1. Keratin protein structure. All keratins are composed of a central helical rod domain, flanked by a head domain at the amino terminal (N-terminal) and a tail domain at the carboxy terminal (C-terminal) of the protein. The rod domain is segmented into four parts; 1A, 1B, 2A, and 2B, interlinked by short linker regions (L1, L12, L2). The peripheral sites of the rod domain (depicted in grey) contain highly conserved regions in keratins. The head and tail consist of nonhelical segments, which contain most of the sites for posttranslational modifications (PTMs), including multiple phosphorylation sites.[8,9]

phosphorylation, O-linked glycosylation, ubiquitination, sumoylation, acetylation, and transamidation.[7,8]

The keratin family includes 54 different functional keratin genes and proteins that are divided into two generic groups: type I, acidic keratins, and type II, neutral or basic keratins. Type I and type II keratins form obligate heteropolymers consisting of at least one of each type. Keratins have a molecular weight in the range of 44–66 kD and constitute the main cytosolic IFs in all epithelial cells such as pancreatic acinar, duct, and endocrine islet cells. The different types of keratins are further expressed in a cell- and tissue-specific manner[6,10] (**Figure 2**). The human type I and type II keratin genes are clustered on the human chromosomes 17q21.2 and 12q13, respectively, with the exception of type I K18, which is located adjacent to type II K8 on chromosome 12q13.13.[11] K8 and K18 are thought to be the

closest descendants to an "ancestral keratin pair," and the adjacent location of these two keratins may reflect an early divergence within the keratin gene family.[12,13] K8/K18 is also the earliest keratin pair expressed in embryogenesis.[11]

Simple layered epithelial cells found in the intestine, lung, liver, and pancreas predominantly express the type II keratin K8 (and to a lesser extent K7), type I keratins K18, K19, K20, and in a few epithelial cells, K23. The specific type I and type II expression pairs differ between different organs, as discussed below. Overall, K8/K18 is the predominant pair in the liver and pancreas, while K8/K19 heteropolymers predominate in intestinal epithelial cells. However, in hepatocytes, K8/K18 is the only keratin pair, whereas in the pancreatic acinar cells and in intestinal cells, K8 pairs with K18, K19, and K20.[9,14]

Of all simple epithelial organs, liver is the most affected by keratin abnormalities.[15] This disease susceptibility is probably due to that hepatocytes express only K8 and K18, with the exception of ductal cholangiocytes in the liver, which also express K7 and K19.[10,16] Several liver disease-associated mutations of K8, K18, and K19 have been identified in humans.[5,9,17] Mouse models expressing these mutations often phenocopy these humans disease,[5,15] and are therefore important models for studying keratinopathies. Since K8 is the main type II keratin in most simple epithelia, the K8 null mouse as well as mice overexpressing wild-type K8 or disease-related K8 or K18 mutations are valuable for studying the roles of keratins in simple epithelial organs. The liver and colon are the organs most affected by these keratin deficiencies and they are also the most studied in this context.[5] The exocrine pancreas is less affected by K8 deletion; the K8 null exocrine pancreas appears, in fact, to be modestly protected from

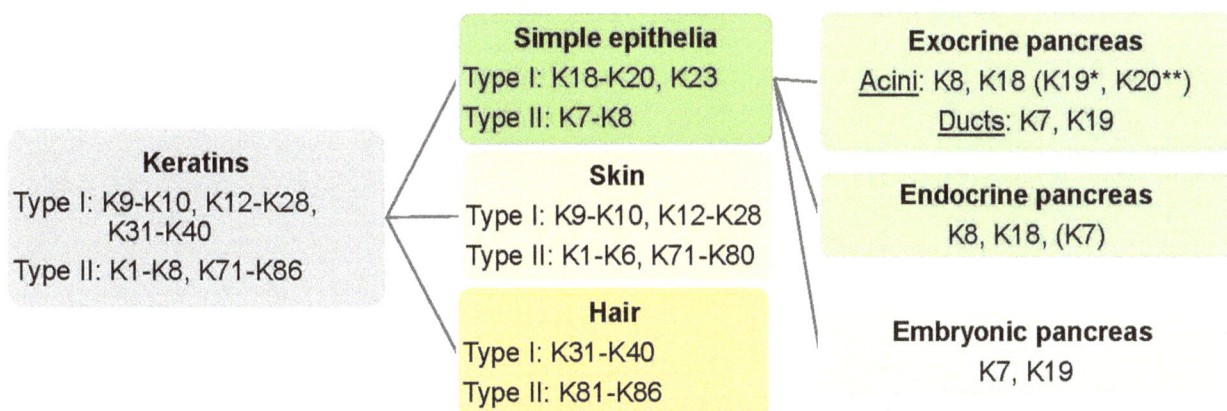

Figure 2. Keratin expression in epithelial tissues and the pancreas. Intermediate filament type I and type II keratin (gray box) proteins are divided in to skin keratins (light yellow box), hair keratins (yellow box), and simple epithelial keratins (green box). The main keratins in adult human exocrine pancreas acinar cells are K8, K18, with the addition of K19* in the centroacinar cells. K19 and K7 are expressed in mammalian exocrine ducts (rat ducts also express K20). Mouse acinar cells also express mainly K8 and K18 but in addition lower levels of K19* and K20** near the apicolateral membranes. The endocrine pancreatic cells (in mouse) express K8 and K18 and lower levels of K7. K7 and K19 are found in the embryonic pancreas.

experimental pancreatitis-induced injury.[18] Experimental K8 and K18 overexpression, conversely, conveys age-associated atrophic changes to the exocrine pancreas, yet does not result in any known pathological anomalies in the liver.[19] In this chapter, the expression and regulation of keratins in the exocrine pancreas during basal conditions and pancreatic injury as well as keratin-dependent regulation of cell stress-response proteins will be considered. Further, the implications of keratin defects on different types of pancreatic injury and the changes in keratin expression in pancreatitis and pancreatic cancer will be discussed.

Intermediate Filament Keratin Function, Regulation, and Disease Association

IFs, including keratins, were first recognized as structural components of cells, serving mainly as static mechanical support for the cells. Further investigations into the structure and function of keratins nevertheless revealed that they are, in fact, highly dynamic structures that assemble and disassemble very quickly, for example, by means of post-translational regulation in response to various stimuli.[20] Phosphorylation is the most frequently occurring—and also the most studied—PTM in keratins. The PTM sites are mainly located on the head and tail domains of keratins and these domains typically contain multiple phosphorylation sites (**Figure 1**). Keratin PTMs enable rapid and dynamic regulation of the keratin filament solubility, as well as network assembly and organization.[8] Keratin PTMs, moreover, regulate the association of keratins with multiple essential cellular proteins, such as the cytolinker protein plectin, the 14-3-3 adapter protein, as well as several protein kinases and phosphatases.[8,21,22] Keratins attach to the desmosomes and hemidesmosomes through cytokinker proteins, thus forming a cytosolic network that extends from the cell membrane to the nucleus, providing a flexible, yet mechanically stable cellular reinforcement.[23] Keratins are, as a consequence of their interaction with multiple cellular proteins, involved in diverse physiological processes, in addition to providing dynamically adaptable mechanistic reinforcement in the cell.[24] The multitude of different proteins that interact with keratins, many of which are essential cell-signaling mediators, imply a central regulatory role for epithelial keratins in intracellular organization, cell signaling, maintaining cell polarity, regulating translation, and targeting proteins and organelles in the cell.[5,20,25–31]

Keratins also serve a vital role in the protection from both mechanical and nonmechanical cell stress. A strong cell-specific upregulation of keratin protein and/or mRNA takes place after different types of injury. This upregulation, which often occurs in the regenerative phase after injury,

can be observed, for example, in response to skin damage, liver injury, and chemically induced pancreatic injury and may include de novo expression of stress-induced keratins,[32] as discussed below. The upregulation of keratins in response to cell stress provides mechanical reinforcement but may also regulate cellular responses on a more intricate level through keratin interactions with cellular pathways that determine cell survival, apoptosis, or regeneration.[33]

Given the multiple functions for keratin IFs, it is hardly surprising that keratin deficiencies or mutations play a role in several human diseases. Mutations in skin keratins cause several different diseases including epidermolysis bullosa simplex, and mutations in simple epithelial keratins increase the susceptibility to various liver diseases.[17] Research work based on animal models further links keratin abnormalities with skin and nail disorders, as well as dysfunctions in hair and liver.[17] K18 null mice develop age-dependent pathological hepatocyte anomalies, resembling chronic cirrhosis, while K8 ablation causes liver insufficiencies[34–36] as well as a microbiota-dependent colitis.[37,38] The K8 null colitis is accompanied by deficiencies in ion transport and energy metabolism, a Notch1-associated cell fate shifts toward a more secretory cell type, increased inflammasome signaling, as well as increased susceptibility to colorectal cancer.[39–42] In the murine thymus, K8 deletion causes increased apoptosis.[43] In the K8 null mouse endocrine pancreas, we have observed mislocalization of the β-cell glucose transporter 2 (GLUT2), defects in mitochondrial morphology and function, as well as decreased insulin levels associated with abnormal insulin vesicle morphology.[25,30] Moreover, mice with reduced K8 expression (K8 heterozygote null mice) show increased susceptibility to experimental type I diabetes.[44] In many cases, however, keratin mutations do not directly cause a disease, but they constitute a risk factors and render the affected individual more vulnerable to certain conditions.[5,9] These indirect disease associations are not always easy to verify, as the keratin mutation may be one of many factors contributing to a disease but may nevertheless be an important susceptibility factor for diseases/dysfunctions of epithelial cells and tissues.

Keratin Expression in the Exocrine Pancreas in Humans and Murine Models

The exocrine pancreas consists of acinar cells that secrete digestive enzymes and a network of ducts that transport these enzymes from the pancreas to the small intestine.[45] The acinar cells are pyramidal-shaped simple epithelial cells arranged in acini, whereas the duct cells are simple squamous or cuboidal-type epithelial cells.[45] Keratins make up the IFs of the pancreas and comprise 0.3% of the total pancreatic proteins.[46] The pattern of keratin expression differs between different species and developmental stages. The subcellular localization of the different keratins in the pancreas is moreover highly orchestrated and molded by the cell-specific conditions.

Under basal circumstances, adult mouse acinar cells express K8 and K18 with minor levels of K19 and K20 (**Figure 2**). K8 and K18 form cytoplasmic filaments throughout the acinar cell, including prominent keratin bundles (known as apicolateral filaments) running along the apical and lateral domains, in parallel to the F-actin layer closest to the cell membrane.[18] In contrast, K19 and K20 are only observed apicolaterally under basal circumstances (**Figure 3**).[18,46,47] K19 and K7 are the main keratins of exocrine pancreatic duct cells in mice,[18] and K19, K7, as well as K20, in rats.[48,50]

Outside the scope of this chapter, the mouse endocrine pancreas, consisting of islets of Langerhans, expresses mainly K8 and K18, but also some K7, which like K8, forms heteropolymers with K18.[25] Additionally, K20 expression has been reported in neonatal rat endocrine pancreas.[48,49]

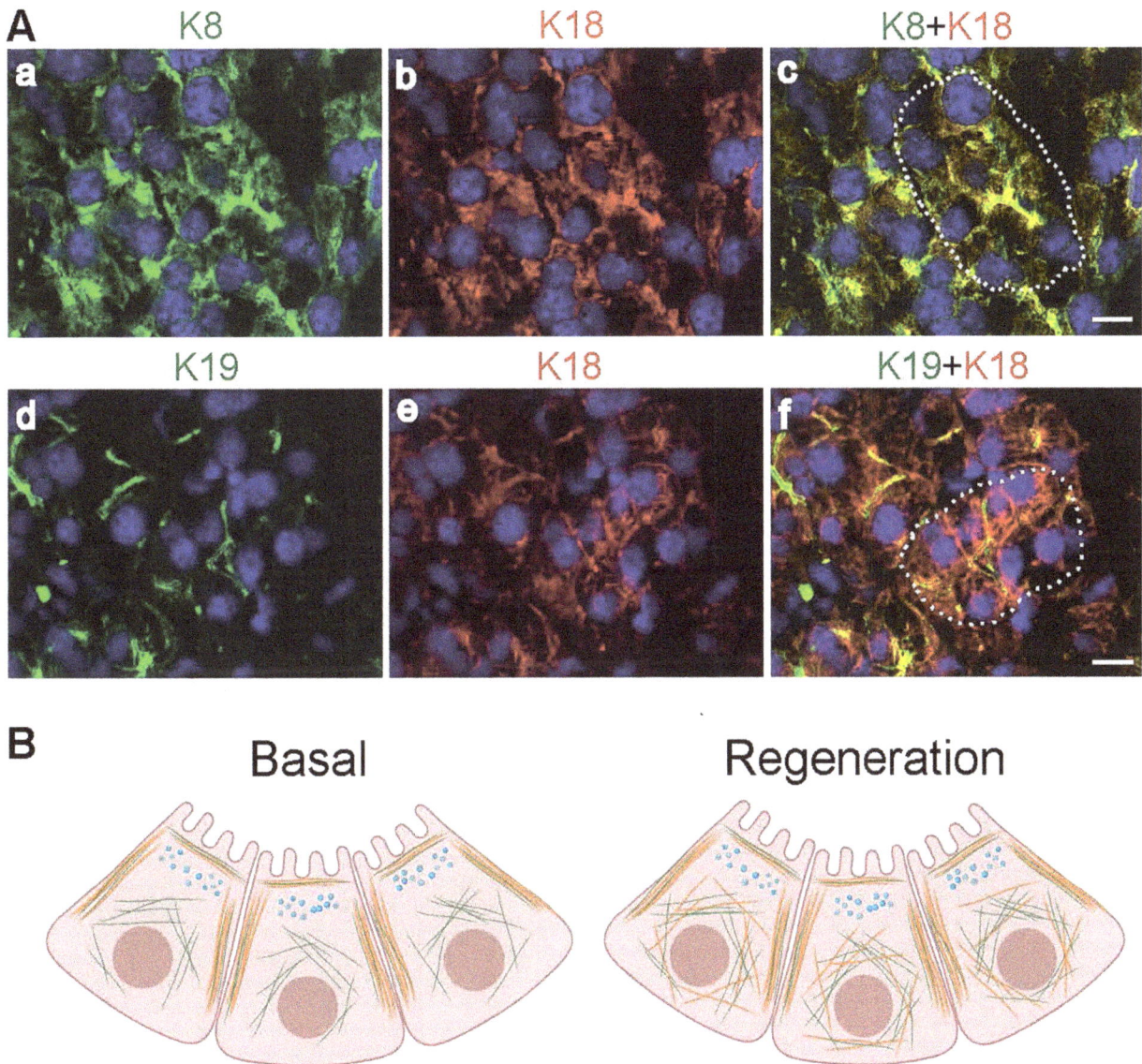

Figure 3. Keratin expression in mouse exocrine pancreatic acinar cells. (A) K8 (**a**, green) and K18 (**b**, red) form cytoplasmic heterodimeric filaments (merged image of K8/K18 is shown in **c**) in acinar cells whereas K19 (**d**, green), here costained with K18 (**e**, red; merged image of K19/K18 in **f**), is observed apicolaterally. Nuclei are shown in blue and an acinus in **c** and **f** with basally located nuclei are outlined with a dotted line. Scale bar = 20 μm. (**B**) The schematic illustration (created with BioRender.com) shows acinar cell cytoplasmic and apicolateral K8/K18 (green) and apicolateral K19 and K20 (orange) filament localization under basal conditions (left) and the apicolateral as well as de novo K19/K20 cytoplasmic filament localization during regeneration after acinar cell injury (right).

In humans, K7 and K19 are expressed in all epithelial cells of pancreas during fetal development.[50] Postnatally, differentiated human acinar cells only express K8 and K18, while K7 and K19 expression is retained in the pancreatic duct cells[50] and K19 in centroacinar cells.[18] K20 does not appear to be expressed under basal conditions in human pancreatic duct cells to any significant degree.[48,49]

The physiological importance of these subtle interspecies pancreatic cell differences in keratin expression is not known. However, given the ubiquitous expression of K7 and K19 in early fetal development, retention of K19 in rodent acinar cells may be indicative of a lower level of cellular differentiation. Furthermore, K20 expression in the neonatal rat endocrine pancreatic cells is associated with cell proliferation[50] and increased expression of mouse K19 and K20, with acinar cell regeneration. It may hence be speculated that the presence of K7, K19, and K20 expression in rodent pancreas reflects a lower level of cellular differentiation and perhaps higher degree of plasticity and regenerative capacity. This hypothesis is in line with the observations that pancreatic regeneration after injury, obesity, and during pregnancy is significantly higher in rodent than in human pancreatic cells.[51]

The Effects of Experimental Keratin Mutations, Deletion, or Overexpression in the Exocrine Pancreas under Basal Conditions

The exocrine pancreas has a strikingly high tolerance to keratin absence or mutations, when compared to similar keratin deficiencies in the liver, which also predominantly expresses K8 and K18. The subject matter deserves consideration since these findings challenge a simplistic view of keratins as static stress protectors in all cells and bid for a more scrutinizing analysis of the mechanisms underlying the stress protein functions of keratins. Several transgenic mouse models that either lack or overexpress wild-type keratins, or that overexpress specific human keratin mutations, have been used to explore the role of keratins in the exocrine pancreas (summarized in **Table 1**).

K8 and K18 null mice

The only type II keratin expressed in acinar cells is K8. Since keratins are obligate heteropolymeric proteins and require both a type I and type II keratin to form stable filaments, it is expected that acinar cells in mice lacking K8 are entirely void of keratins. Indeed, the absence of both K8

Table 1. Basal exocrine pancreatic phenotypes of keratin transgenic mice. H, human; m, mouse; pK, phospho-keratin; ↑ - ↑↑↑/+ - +++, level of increase from low to higher. (Note that the table depicts published analysis and that K20 and phospho-keratin or Reg-II levels have not been determined in all studies.) Moreover, wild-type mice strongly elevate Reg-II levels after CDD and caerulein pancreatitis treatment.[52]

Genotype:	Wild-type	K8 null	K18 null	hK18 over-expression	K18 glycosy-lation Deficient (S30/31/49 A)	Low & modest overexpression		High overexpression	
Phenotype:						hK18	mK8	hK8	mK8/hK18
Keratin filament network and subcellular location	K8/K18 cytoplasmic, apicolateral filaments K19/K20 apicolateral filaments	Absent	Absent cytoplasmic filaments Intact apicolateral filaments	Disrupted cytoplasmic filaments Intact apicolateral filaments	Normal	Normal but ↑K18	Normal but ↑↑K8/K18 ↑K19, K20 cytoplasmic filaments ↑pK18	Normal but ↑↑↑K8/K18 ND ND	Normal but ↑↑↑K8/K18 ↑K19/K20 cytoplasmic filaments ↑↑pK8/pK18
Phenotype	Normal	Normal histology Moderately decreased acinar cell viability Normal stimulated secretion	Normal histology Moderately decreased acinar cell viability Normal stimulated secretion	Normal histology Moderately decreased acinar cell viability Normal stimulated secretion	Normal histology	Normal histology	Normal histology	Age-dependent pancreatitis, dysplasia, fibrosis, ductal metaplasia, inflammation, acinar dedifferentiation Mislocalised zymogen granules	Age-dependent vacuolization, atrophy Rounded acini Numerous, small and mislocalised zymogen granules
Reg-II level	Normal	↑↑	↑	↑	ND	ND	ND	ND	ND
References	18, 46, 52, 53	18, 52	18, 52	47, 52	32	19, 47	19	54	19

and K18 has been demonstrated by immunofluorescence labeling and immune-electron microscopic analysis in K8 null mice. In contrast, K18 null mice still express K8, since K8 forms heteropolymers with K19 in the absence of K18. However, under basal circumstances, K8/K19 keratin filaments are located solely to the apicolateral region of the acinar cells in K18 null mice, as K19 cannot compensate for cytoplasmic K18 filament formation under basal conditions. However, the histology of the K18 null pancreas is normal, apart from some polynuclear areas that bear a resemblance to the histological observations in the livers of K8 null and K18 null mice.[18,35] The secretory response upon stimulation by cholecystokinin octapeptide (CCK-8) is normal in K8 and K18 null mice, but the acini are moderately less viable than in wild-type mice.[18] The modest phenotype changes in K18 null mouse pancreas may be explained, at least in part, by the existence of the K8/K19 heteropolymers. However, in K8 null mice, the resistance to injury probably comes down to other compensatory mechanisms.[18] One such suggested mechanism is the upregulation of regulatory protein II (Reg-II), which occurs in the K8 null pancreas. This will be discussed in greater detail later in this chapter.

Transgenic mice overexpressing wild-type keratins

In contrast to keratin deletion, keratin overexpression is well tolerated in the liver, while in the pancreas it is associated with various pathological changes.[19,54] In a study using transgenic mice overexpressing human K8, Casanova and colleagues reported severe age-dependent progressive abnormalities in the exocrine pancreas, demonstrated by a 30% loss of pancreatic mass, dysplasia, fibrosis, ductal metaplasia, inflammation, and dedifferentiation of acinar cells into duct cells.[54] Moreover, Toivola et al. (2008) showed that the extent of pancreatic damage correlates with the level of keratin overexpression.[19] This latter study used mice with different levels of keratin upregulation: human K18 overexpressing mice where keratin levels were only mildly increased, mouse K8 overexpressing mice with moderately upregulated K8, K18, K19, and K20, and mice that overexpressed both mouse K8 and human K18 (K8/K18 overexpressors), leading to a substantial keratin upregulation.[19] Though the effects of keratin overexpression were less severe in the study by Toivola and colleagues, similar acinar cell anomalies were observed by Casanova et al., such as age-dependent atrophy and fatty vacuole formation—particularly in the mice with the highest levels of keratin overexpression. Hence, it is likely that the aggravated injury in the earlier study may have been due to a higher level of keratin upregulation.[19]

Genetic overexpression of both K8 and K18 in pancreatic acinar cells changes the distribution and phosphorylation

level of keratins and causes alterations in the quantity, size, and distribution of the amylase-containing zymogen granules in the acinar cells. The zymogen granules in K8/K18 overexpressing cells, as well as in human K8 overexpressing mice, were smaller in size but more numerous than in wild-type mice.[19] Moreover, the granules were not retained in their characteristic apical location but instead diffusely localized in the cytoplasm in K8/K18 overexpressors.[19,54] In addition to these phenotypic differences, K8/K18 overexpressors also displayed increased keratin phosphorylation at K8 S79 and K18 S33, and thick K8/K18 bundles that localized not only around the apical lumen but also in perinuclear and cytoplasmic locations of the acinar cells.[19] Hence, overexpression of keratins in acinar cells appears to interfere with the intracellular organization of keratin and alter the exocrine function of the acinar cells. Interestingly, mice that overexpress a K18 S33A mutation (which inhibits serine 33 phosphorylation) display keratin filaments that are retracted from the basal and nuclear region and instead concentrated in the apical region of acinar cells. However, these mice do not display abnormal pancreatic histology or disease.[55]

These studies of the exocrine pancreas, using keratin overexpressor mice, further highlight the interrelationship between type I and type II keratins. Since K18 overexpression caused a minimal increase in keratin levels compared to K8 overexpression, it is evident that the level of type II keratins has a more profound effect on overall keratin regulation than type I keratins in the exocrine pancreas. Moreover, as the exocrine pancreas contains more than one type I keratin (K18, K19, K20), a deficiency or upregulation of one of these may be compensated, to a certain extent, by a down or upregulation of another type I keratin.

Transgenic mice overexpressing human diseases-related keratin mutations

The contribution of specific keratin mutations for increased susceptibility to injury or disease has been analyzed with the help of transgenic mouse models, expressing human keratin mutations. Many of these experimental models, including human K18 R90C mice (which have a mutation equivalent to K14 R125C, found in epidermolysis bullosa simplex patients) as well as human K8 G62C mice (expressing a common human keratin variant mutation in liver disease patients), display early-onset liver inflammation, necrosis, and increased susceptibility to hepatotoxicity.[15,36,56] The exocrine pancreas in these mice, on the contrary, appears far less affected by these mutations. Although the viability of the acini is somewhat compromised in K18 R90C mice displaying disrupted keratin filaments, these mice do not display increased pancreatic injury under basal circumstances.[18] K18 R90C mice lack intact cytoplasmic keratin filaments (displaying only K8/

K18 dots in the cytoplasm), but they express apicolateral filaments in acinar cells, since K8 polymerizes with K19 and K20 that are located apicolaterally under basal conditions. This is in contrast with K18 R90C mouse hepatocytes, in which apicolateral keratin filaments are scarcer. This difference between hepatocyte and pancreatic keratins has been suggested as one reason underlying the lesser disease susceptibility of the pancreas compared with the liver in the K18 R90C transgenic mouse model.[18]

Keratins in Mouse and Cell Models for Pancreatic Injury

Keratin networks are dynamic and respond swiftly to changes in the cellular environment. Several studies have investigated the roles of keratins in the exocrine pancreas by subjecting wild-type and K8- or K18-deficient mice to experimental pancreatitis models, including the caerulein model and the choline-deficient diet model[57] (summarized in **Table 2**). Typically, in these conditions, keratins are first rapidly broken down upon acinar cell injury but reform quickly and become highly upregulated during recovery. During the recovery process, keratins are also extensively phosphorylated, as is common in keratin stress responses. This dynamic process has been described after caerulein-induced exocrine pancreatic injury, which causes a rapid disassembly of the acinar cell keratin network within 1 hour of induction of injury, followed by a reassembly of keratin cytoplasmic filaments in the early recovery phase 7–24 hours after induction of injury. The cytoplasmic keratin network reforms during the recovery phase, and temporarily, strong *de novo* cytoplasmic filaments containing K19 and K20 become very prominent (**Figure 3B**), in addition to the K8/K18 cytoplasmic filaments.[18,32] This keratin upregulation is transcriptionally regulated since keratin mRNA levels also increase 48 hours after caerulein induction.[46]

Ultimately, in the late recovery phase (within 5 days), retrieval of the normal, baseline filament network structure occurs, and K19 and K20 filaments repossess their exclusively apicolateral localization. A similar keratin upregulation on protein and mRNA level, accompanied by *de novo*

Table 2. Experimental pancreatic injury phenotypes in wild-type and keratin transgenic mice. H, human; m, mouse; pK, phospho-keratin; ↑ - ↑↑↑/+ - +++, level of increase from low to higher; CDDcholine-deficient diet pancreatitis model; CV, coxsackievirus; ND, not determined; STZ, streptozotocin; * endocrine pancreas phenotype.

Mouse genotype: Disease model:	Wild-type	K8 null	K18 null	hK18 R90C over-expressor	K18 glycosylation deficiency (S30/31/49A)
Caerulein Susceptibility: Keratin phenotype:	+++ Degradation ↑pK8 ↑K8/K18/K19/K20 cytoplasmic filaments	++ Absent	++ Degradation ↑pK8 ↑ K8/K19 cytoplasmic filaments	+++ Rearrangements ND	ND
CDD Susceptibility: Keratin phenotype:	+++ ↑pK8 ↑ K8/K18/K19/K20 cytoplasmic filaments	+++ Absent	+++ ND	+++	ND
CV B4-V (high virulence)	No lethality	↑ lethality	No lethality	ND	ND
CV B4-P (low virulence)	Susceptible	Less susceptible, quicker recovery	Similar to Wild-type	ND	ND
STZ acute	Low lethality *Prone	ND *Moderately resistant	ND, ↑ Oedema *Islet cell necrosis	Low lethality	↑ Lethality ↑ Oedema ↑ Apoptosis
STZ chronic	Mild exocrine damage *Moderate damage	↑ exocrine damage: oedema, hyperplasia, atrophy *Moderately resistant *K8 htz null mice more sensitive	ND	ND	ND
Reference	18, 25, 32, 46, 47, 52, 53	18, 25, 44, 53	18, 32, 53	32, 47	32

cytoplasmic K19 and K20 filament formation, is seen 1–2 days after discontinuation of choline-deficient diet feeding, with keratin levels returning to baseline levels 7 days into the recovery phase.[46] A similar transitory, recovery phase apical to cytoplasmic localization shift also appears in caerulein-treated K18 null mice, in which the keratin network under basal circumstances is exclusively apicolateral in acinar cells. In this model as well, the filaments revert back to their apicolateral localization later in the recovery phase.[18] It is still unknown what regulates the induction of this remarkable transient K19 and K20 cytoplasmic filament formation, but it is suggested that it plays an important role in the recovery of acinar cells after injury. The existence of K19 filaments in K18 null mouse acinar cells has been postulated as a possible reason for the high tolerance to experimentally induced pancreatic injury in these mice.[18] If this is the case, it is possible that K18 mutations may be more detrimental for human acinar cells, which lack K19 filaments. Interestingly, the keratin upregulation appears to be specific to regeneration after pancreatitis, since generalized stress models, such as heat or water immersion, do not alter keratin pancreatic expression levels.[46] However, TGFβRII dominant-negative mutant mice, which develop a severe chronic pancreatitis phenotype similar to human K8 over-expressing mice, show highly increased K8 and K18 levels in the pancreas.[54] Keratin gene transcription in the pancreas has not been extensively studied, but the K19 gene in pancreatic duct cells is regulated by PDX1, GKLF/KL4, and SP1, indicating an association between K19 expression and the developmental stage in the pancreas.[58,59]

Interaction with the keratin-binding protein, epiplakin, contributes to the dynamics of pancreatic keratin remodeling after caerulein injury, as epiplakin-deficient mice display both a quicker keratin network breakdown after injury and impaired filament rearrangement, as demonstrated by the accumulation of keratin aggregates at the most severe phase of the cell injury. These defects might be caused by excessive hyperphosphorylation in the absence of epiplakin,[60,61] emphasizing the role of dynamic keratin regulation in stress responses. In addition to the stress-induced keratin filament remodeling, keratins could also be involved in caerulein-induced inflammatory responses in the acinar cells, since caerulein-induced keratin upregulation is associated with nuclear factor-κB (NF-κB) activation.[46]

The extensive rearrangement of keratin filaments in response to pancreatic injury in caerulein- and choline-deficient diet-induced pancreatitis suggests a protective role for keratins in these injury models. Yet intriguingly, apart from minor differences observed in certain disease parameters, K8 null (which lack acinar cell keratins entirely), K18 null (which express only K8/K19 acinar cell filaments), and K18 R90C mice are notoverall more sensitive to choline-deficient diet or caerulein-induced

pancreatitis, compared with wild-type mice.[18] In fact, even with more prominent vacuolization, the histological damage in terms of inflammation and edema, is slightly lower in K8 null and K18 null mice after caerulein, compared to wild-type mice.[18] Moreover, no differences in pancreas histology, serum amylase, or lipase levels were observed in K18 R90C mice compared with wild-type K18 over-expressing mice.[47,52] These seemingly near dispensable effects of keratin deficiencies in the exocrine pancreas are puzzling. However, some potential underlying reasons for the relative stress tolerance of the exocrine pancreas in keratin-deficient mice can be extrapolated from existing animal studies. These include differences in the cellular localization of keratins in the acinar cells and hepatocytes, upregulation of cytoprotective regulatory proteins (discussed in the next section), and differences in the function of keratins in different organs.[47,52] Further, the importance of keratins in pancreatic stress may also depend on the type and duration of the injury. For example, in a study where K8 null and K18 null mice were challenged with coxsackie B virus-induced pancreatitis, the vulnerability of keratin-deficient mice depended on the virulence of the coxsackie virus strain. When subjected to acute pancreatitis caused by the highly virulent coxsackie virus strain B4-V (CVB4-V), keratin null mice suffered significantly higher mortality compared with wild-type mice (40% mortality in K8 null mice and 0% in K8 wild-type mice). In contrast, K8 null recovered quicker than their wild-type counterparts from infection with a less virulent, B4-P coxsackie virus strain (CVB4-P). It has been proposed that this difference may come down to the effect of Reg-II-stimulated tissue regeneration, since Reg-II upregulation occurs after infection with CVB4-P virus, as well as in the caerulein- and choline-deficient diet pancreatitis models, but not CVB4-V in wild-type mice.[52,53] Interestingly, Reg-II is upregulated in the K8 null pancreas under basal conditions,[52] and this upregulation may assist in the injury response to CVB4-P-induced pancreatitis.

In addition to CVB4-V coxsackie virus-induced pancreatitis, keratin-deficient animals also appear to be more sensitive to exocrine pancreatic injury induced by streptozotocin (STZ). STZ is a commonly used toxin for inducing type I diabetes in experimental animals. This toxin is taken up by β-cells through the glucose-transporter 2 (GLUT2) and causes acute β-cell destruction if administered at high doses and partial β-cell depletion and inflammation if administered at multiple low doses.[25,62] Interestingly, K8 null mice develop widespread exocrine pancreatic edema, atrophy, vacuolization, and inflammation in response to the chronic stress induced by low-dose, STZ treatment (40 mg/kg/day, for five days), while wild-type animals display only modest exocrine damage after this treatment.[25] Interestingly, K8 null β-cells are less sensitive to acute high-dose STZ (200 mg/kg/day), likely due to mislocalization of GLUT2, which is needed for the uptake of STZ into β-cells. Yet, transgenic mice expressing a human K18 glycosylation-preventing (K18-Gly(-)) mutation suffer severe exocrine pancreatic injury after the same high-dose STZ treatment.[32] The STZ treatment in this study induced multi organ failure in the glycosylation-deficient K18 mice, hence the toxic effects on the pancreas may have been exacerbated by the keratin mutation-induced liver deficiency in these animals.[32] Taken together, these results demonstrate that the function of keratins, and their importance in protecting from cellular injury in the pancreas, may be crucially dependent on the disease mechanisms of a particular experimental model, the duration of the injury, and the type of keratin anomaly. Interestingly, in addition to keratin IFs, the type V intermediate filaments, nuclear lamins, may also contribute to pancreatic stress protection, since pancreas-specific conditional lamin A null mice show a spontaneous phenotype similar to chronic pancreatitis, and patients with mutations in lamin A have a higher risk of developing pancreatitis.[63,64]

Dynamic stress-induced regulation of keratins has also been described in acinar cancer cell cultures *in vitro*, under various experimental stress conditions. For example, K23 mRNA was highly induced by the histone deacetylase inhibitors sodium butyrate and trichostatin during differentiation in the pancreatic cancer cell line AsPC-1,[65] demonstrating the propensity for keratin *de novo* expression during stress conditions. In pancreatic carcinoma, PANC-1 cells, which express K8 and K18, caerulein treatment induced K8 S431 phosphorylation and a reorganization of the keratin filaments into a tight perinuclear network through activation of ERK and downregulation of PP2A and alpha 4, resulting in enhanced cell migration.[66] Interestingly, similar K8 phosphorylation and perinuclear reorganization have also been observed in PANC-1 and A549 cells (human carcinomic alveolar cells) exposed to metastasis-enhancing bioactive lipid sphingosyl-phosphorylcholine, 12-Otetradecanoylphorbol-13-acetate (TPA), leukotriene B4 (LTB4), or shearing force.[66,67]

Regenerating Protein II as a Cytoprotective Factor in Keratin-Deficient Exocrine Pancreas Models

The regenerating (Reg) gene was first isolated from islet β-cells in the pancreas.[68] The protein encoded by this gene (now termed Reg-I) was found to have a significant stimulating effect on β-cells in the endocrine pancreas and was able to ameliorate diabetes in 90% of pancreatectomized rats and in nonobese diabetic (NOD) mice.[69] It was later discovered that the Reg genes constitute a multigene family consisting of three types of genes (Reg-I, II, and III) that differ in expression pattern and functional characteristics, as reviewed in references.[52,70]

In the endocrine pancreas, Reg-I seems to play an important role in islet regeneration, while in the exocrine pancreas Reg-II is the predominant Reg-protein. However, Reg-II is substantially upregulated in the exocrine pancreas during recovery from caerulein- or choline-deficient diet-induced pancreatitis,[52] which indicates that it has an analogous function to that of Reg-I in the endocrine pancreas. Interestingly, in the caerulein model, Reg-II levels increase early on, while the K18 and K19 levels are at their highest later in the recovery phase, but in the choline-deficient model, Reg-II and K18 peak at around the same time.[52]

The unexpectedly high resistance to pancreatic damage in keratin-deficient or mutant mice discussed above has encouraged analysis of compensatory factors in these mice that might explain why keratin deletion causes less damage in the exocrine pancreas compared to the liver, despite a similar keratin expression pattern. In a microarray analysis of K8 null mouse pancreas, several genes were indeed found up- or downregulated compared to wild-type control mice. Among the significantly upregulated genes were several members of the Reggene family, but particularly the Reg-II gene. Upregulation of Reg-II was also observed in other mice with keratin deficiencies, such as K18 null, K18 R90C transgenic mice, and phosphorylation-deficient K18 S52A transgenic mice. However, this upregulation was not seen in K19 null mice, which express normal, cytoplasmic, and apicolateral keratin filaments.[52] It has thus been suggested that the resistance to pancreatitis in K8 null and K18 null mice, likely comes down to a compensatory protective effect of the Reg-II upregulation. This compensatory mechanism could explain the greater resistance of K8 null mice to the moderate pancreatic injury caused by infection with CVB4-P, which is accompanied by anti-apoptotic and cell regenerative responses, as well as to the acute injury by CVB4-V infections, which causes an upregulation of genes that favor apoptosis, metaplasia, and fibrosis.[18,52,71]

The benefit of acinar cell Reg-II overexpression for recovery from experimental pancreatitis and diabetes has been questioned in a study by Li and colleagues, since transgenic acinar cell-specific overexpression of Reg-II neither protected the mice from streptozotocin-induced diabetes nor from caerulein-induced pancreatitis.[72] In this study, Reg-II overexpression was moderate and roughly at a similar level to the spontaneous Reg-II upregulation after caerulein treatment. It is possible that the advantage of transgenic overexpression of Reg-II over an endogenic Reg-II upregulation in response to caerulein is not significant, if the experimentally induced overexpression levels do not exceed the endogenic stress-induced upregulation. Reg-II is nevertheless markedly upregulated in acinar cells upon injury and is hence evidently a stress response protein, albeit the precise function of Reg-II in pancreatic injury remains unclear. Apart from Reg-II upregulation, other factors that may contribute to the exceptional disease resistance in the exocrine pancreas of keratin-deficient mice under basal conditions could include the unique intracellular localization of keratins in these cells and the shift between apicolateral and cytoplasmic filaments in response to cell injury, which is characteristic for the exocrine pancreas.

Keratins in Human Pancreatitis and Pancreatic Cancer

Keratins and other IFs among the cytoskeletal proteins are interesting to the medical field since they are associated with over 100 human diseases.[9,17,73] Some keratin mutations are known to be directly causative of disease (e.g., several skin diseases), while mutations in simple epithelial keratins have been shown to predispose to various liver diseases. Moreover, since keratin expression and posttranslational modifications are frequently altered upon cell stress, keratins are also used as diagnostic biomarkers for diseases. One example of such histopathologies is the liver disease-associated formation of keratin aggregates, known as Mallory Denk bodies in hepatocytes.[9,74]

The prevalence of K8 mutations in human pancreatitis has been analyzed in a few studies, but no clear associations have been found. Two K8 variant mutations, Y54H and G62C, that predispose to cryptogenic liver disease[75] have been studied in this context. K8 G62C was first found to predispose to chronic pancreatitis.[76] A later study found no significant correlation between the frequency of either K8 G62C or K8 Y54H and acute or chronic pancreatitis or pancreatic cancer in a cohort of more than 2,400 patients.[77] Similarly, K8 G62C mutations were found not to be associated with familial, sporadic, or alcoholic pancreatitis.[78] However, a recent study has identified an association between the KRT8 gene and pancreatic cancer in a genome-wide association study in a Japanese population.[79]

Changes in keratin expression levels during pancreatitis and pancreatic cancer have been analyzed in a few studies. K20 is a keratin that under normal circumstances has an expression pattern restricted to a few cell types and can therefore be used as a marker for certain cancers. As mentioned earlier, K20 expression has been reported in normal pancreatic duct cells in rats,[48,50] but the expression in humans is very low in the pancreas under basal conditions.[49] However, K20 expression in human pancreatic duct cells is markedly induced in metastatic pancreatic cancer cells,[49] and detection of K20 in pancreatic tumors or in blood or bone marrow samples from patients with pancreatic duct cell carcinomas correlates with a worse prognosis.[80,81] Furthermore, K19, which in humans postnatally is restricted to duct cells, can be observed in precancerous acinar-like cells before the appearance of metaplastic changes in cell morphology. Hence, K19 expression in neoplastic acinar cells is an early sign of metaplasia, which precedes the fibrotic changes related to metaplasia.[82]

Conclusions and Vision

During the last decades, keratin IFs have emerged as important cellular regulators and stress proteins of epithelial cells in many different organs. As the complexity of keratin-related functions has been unraveled, it has become evident that the function of keratins depends on the type of keratins expressed in the cells and their subcellular localization, as well as on the specific cell type and the cell-specific functions and regulatory pathways.

The keratin expression in the liver and the exocrine pancreas is similar, yet the susceptibilities to keratin deficiency-related injury are evidently different. The published studies relating to the role of keratins in pancreas and liver cells have provided some clues that help to explain the differences in stress tolerance.

The dramatic upregulation of the injury-response protein Reg-II in the exocrine pancreas of K8 null and other keratin-deficient models is probably one factor that protects K8-deficient mice from exocrine pancreatic injury.[52] Indeed, this type of protection appears to take place also in skin cells, where keratin-mutant keratinocytes are similarly prepared for injury through upregulation of basal-level JUN-kinase activation and profuse activation of osmotic shock-induced stress pathways.[83] Keratin deficiency may thus induce a chronic injury response that puts cells in an "alert state" facilitating a swift cellular response to stress. The activation or inhibition of the same cell-signaling molecule may, however, have different consequences in different organs or under different circumstances, due to the complexity of the cell signaling pathways and downstream effects. It has, for instance, been shown that activation of transcription factor NF-κB has a protective, anti-apoptotic effect in liver injury and it has been suggested that the sensitivity of K8 null liver cells to apoptosis is linked to a defective activation of this transcription factor.[84] Activation of NF-κB nevertheless appears to have a negative effect on pancreatic cell survival in caerulein-induced pancreatitis, since it enhances the inflammatory response.[85] It is interesting to speculate whether the resistance of K8-deficient mice to caerulein-induced pancreatitis may be associated with interference of the NF-κB activation in pancreatitis,[46] but a link between K8 deficiency and impeded NF-κB activation has not yet been reported for the exocrine pancreas.

Keratins in the exocrine pancreas may at first appear redundant, given that keratin deficiency is remarkably well tolerated under basal conditions, as well as in some pancreatic injury models. The dynamic remodeling of keratins in response to exocrine pancreatic stress and the association of keratins with both Reg-II regulation and inflammatory responses, such as NF-κB activation, however, contradict the notion that keratins lack a role in exocrine pancreatic injury responses. Rather than presenting an exceptional model organ in which keratins do not matter for stress tolerance, perhaps the exocrine pancreas really demonstrates the complexity of keratin-associated cell biology and the remarkable adaptability of the cells to counterbalance inherent weaknesses.

Acknowledgments

The authors wish to acknowledge research funding during this time from the Novo Nordisk foundation, the Finnish Diabetes foundation (DMT), the Academy of Finland (DMT), the Sigrid Juselius foundation (DMT), the Medical Support foundation Liv och Hälsa (DMT), the Swedish Cultural foundation in Finland (CMA), and the Jalmari and Rauha Ahokas Foundation (CMA).

References

1. Herrmann H, Aebi U. Intermediate filaments: structure and assembly. Cold Spring Harb Perspect Biol 2016; 8; 1–22
2. Brady GF, Kwan R, Bragazzi Cunha J, Elenbaas JS, Omary MB. Lamins and lamin-associated proteins in gastrointestinal health and disease. Gastroenterology 2018; 154:1602–1619.
3. Eriksson JE, Dechat T, Grin B, Helfand B, Mendez M, Pallari HM, Goldman RD. Introducing intermediate filaments: from discovery to disease. J Clin Invest 2009; 119:1763–1771.
4. Herrmann H, Strelkov SV, Burkhard P, Aebi U. Intermediate filaments: primary determinants of cell architecture and plasticity. J Clin Invest 2009; 119:1772–1783.
5. Omary MB, Ku NO, Strnad P, Hanada S. Toward unraveling the complexity of simple epithelial keratins in human disease. J Clin Invest 2009; 119:1794–1805.
6. Schweizer J, Bowden PE, Coulombe PA, Langbein L, Lane EB, Magin TM, Maltais L, Omary MB, Parry DA, Rogers MA, Wright MW. New consensus nomenclature for mammalian keratins. J Cell Biol 2006; 174:169–174.
7. Snider NT, Omary MB. Assays for posttranslational modifications of intermediate filament proteins. Methods Enzymol 2016; 568:113–138.
8. Omary MB, Ku NO, Tao GZ, Toivola DM, Liao J. "Heads and tails" of intermediate filament phosphorylation: multiple sites and functional insights. Trends Biochem Sci 2006; 31:383–394.
9. Toivola DM, Boor P, Alam C, Strnad P. Keratins in health and disease. Curr Opin Cell Biol 2015; 32:73–81.
10. Ku NO, Toivola DM, Zhou Q, Tao G, Zhong B, Omary MB. Studying simple epithelial keratins in cells and tissues. In: Omary MB, Coulombe, PA, eds. Intermediate filament cytoskeleton. Volume 78 (Methods in Cell Biology). San Diego: Elsevier Academic Press, 2004:489–517.
11. Hesse M, Zimek A, Weber K, Magin TM. Comprehensive analysis of keratin gene clusters in humans and rodents. Eur J Cell Biol 2004; 83:19–26.
12. Neznanov NS, Oshima RG. *cis* regulation of the keratin 18 gene in transgenic mice. Mol Cell Biol 1993; 13:1815–1823.
13. Waseem A, Gough AC, Spurr NK, Lane EB. Localization of the gene for human simple epithelial keratin 18 to chromosome 12 using polymerase chain reaction. Genomics 1990; 7:188–194.
14. Zhou Q, Toivola DM, Feng N, Greenberg HB, Franke WW, Omary MB. Keratin 20 helps maintain intermediate filament organization in intestinal epithelia. Mol Biol Cell 2003; 14:2959–2971.
15. Ku NO, Strnad P, Zhong BH, Tao GZ, Omary MB. Keratins let liver live: mutations predispose to liver disease and crosslinking generates Mallory-Denk bodies. Hepatology 2007; 46:1639–1649.
16. Strnad P, Stumptner C, Zatloukal K, Denk H. Intermediate filament cytoskeleton of the liver in health and disease. Histochemistry and Cell Biology 2008; 129:735–749.
17. Omary MB. "IF-pathies": a broad spectrum of intermediate filament-associated diseases. J Clin Invest 2009; 119:1756–1762.
18. Toivola DM, Baribault H, Magin T, Michie SA, Omary MB. Simple epithelial keratins are dispensable for cytoprotection in two pancreatitis models. Am J Physiol Gastrointest Liver Physiol 2000; 279:G1343–G1354.
19. Toivola DM, Nakamichi I, Strnad P, Michie SA, Ghori N, Harada M, Zeh K, Oshima RG, Baribault H, Omary MB. Keratin overexpression levels correlate with the extent of spontaneous pancreatic injury. Am J Pathol 2008; 172:882–892.
20. Snider NT, Omary MB. Post-translational modifications of intermediate filament proteins: mechanisms and functions. Nat Rev Mol Cell Biol 2014; 15:163–177.
21. Bouameur JE, Favre B, Fontao L, Lingasamy P, Begre N, Borradori L. Interaction of plectin with keratins 5 and 14: dependence on several plectin domains and keratin quaternary structure. J Invest Dermatol 2014; 134:2776–2783.
22. Liao J, Omary MB. 14-3-3 proteins associate with phosphorylated simple epithelial keratins during cell cycle progression and act as a solubility cofactor. J Cell Biol 1996; 133:345–357.
23. Uitto J, Richard G, McGrath JA. Diseases of epidermal keratins and their linker proteins. Exp Cell Res 2007; 313:1995–2009.
24. Windoffer R, Beil M, Magin TM, Leube RE. Cytoskeleton in motion: the dynamics of keratin intermediate filaments in epithelia. Journal of Cell Biology 2011; 194:669–678.
25. Catharina M Alam 1, Jonas S G Silvander, Ebot N Daniel, Guo-Zhong Tao, Sofie M Kvarnström, Parvez Alam, M Bishr Omary, Arno Hänninen, Diana M Toivola Keratin 8 modulates β-cell stress responses and normoglycaemia J Cell Sci. 2013; 126:5635–5644.
26. Etienne-Manneville S. Cytoplasmic intermediate filaments in cell biology. Annu Rev Cell Dev Biol 2018; 34:1–28.
27. Kim S, Coulombe PA. Emerging role for the cytoskeleton as an organizer and regulator of translation. Nat Rev Mol Cell Biol 2010; 11:75–81.
28. Loschke F, Seltmann K, Bouameur JE, Magin TM. Regulation of keratin network organization. Curr Opin Cell Biol 2015; 32:56–64.
29. Oriolo AS, Wald FA, Ramsauer VP, Salas PJ. Intermediate filaments: a role in epithelial polarity. Exp Cell Res 2007; 313:2255–2264.

30. Silvander JSG, Kvarnstrom SM, Kumari-Ilieva A, Shrestha A, Alam CM, Toivola DM. Keratins regulate beta-cell mitochondrial morphology, motility, and homeostasis. FASEB J 2017; 31:4578–4587.

31. Toivola DM, Tao GZ, Habtezion A, Liao J, Omary MB. Cellular integrity plus: organelle-related and protein-targeting functions of intermediate filaments. Trends Cell Biol 2005; 15:608–617.

32. Ku NO, Toivola DM, Strnad P, Omary MB. Cytoskeletal keratin glycosylation protects epithelial tissue from injury. Nat Cell Biol 2010; 12:876–885.

33. Pan X, Hobbs RP, Coulombe PA. The expanding significance of keratin intermediate filaments in normal and diseased epithelia. Curr Opin Cell Biol 2013; 25:47–56.

34. Roux A, Gilbert S, Loranger A, Marceau N. Impact of keratin intermediate filaments on insulin-mediated glucose metabolism regulation in the liver and disease association. FSEB J 2016; 30:491–502.

35. Toivola DM, Nieminen MI, Hesse M, He T, Baribault H, Magin TM, Omary MB, Eriksson JE. Disturbances in hepatic cell-cycle regulation in mice with assembly-deficient keratins 8/18. Hepatology 2001; 34:1174–1183.

36. Toivola DM, Omary MB, Ku NO, Peltola O, Baribault H, Eriksson JE. Protein phosphatase inhibition in normal and keratin 8/18 assembly-incompetent mouse strains supports a functional role of keratin intermediate filaments in preserving hepatocyte integrity. Hepatology 1998; 28:116–128.

37. Habtezion A, Toivola DM, Asghar MN, Kronmal GS, Brooks JD, Butcher EC, Omary MB. Absence of keratin 8 confers a paradoxical microflora-dependent resistance to apoptosis in the colon. Proc Natl Acad Sci U S A 2011; 108:1445–1450.

38. Habtezion A, Toivola DM, Butcher EC, Omary MB. Keratin-8-deficient mice develop chronic spontaneous Th2 colitis amenable to antibiotic treatment. J Cell Sci 2005; 118:1971–1980.

39. Asghar MN, Priyamvada S, Nystrom JH, Anbazhagan AN, Dudeja PK, Toivola DM. Keratin 8 knockdown leads to loss of the chloride transporter DRA in the colon. Am J Physiol Gastrointest Liver Physiol 2016; 310:G1147–G1154.

40. Helenius TO, Misiorek JO, Nystrom JH, Fortelius LE, Habtezion A, Liao J, Asghar MN, Zhang H, Azhar S, Omary MB, Toivola DM. Keratin 8 absence down-regulates colonocyte HMGCS2 and modulates colonic ketogenesis and energy metabolism. Mol Biol Cell 2015; 26:2298–2310.

41. Lahdeniemi IAK, Misiorek JO, Antila CJM, Landor SK, Stenvall CA, Fortelius LE, Bergstrom LK, Sahlgren C, Toivola DM. Keratins regulate colonic epithelial cell differentiation through the Notch1 signalling pathway. Cell Death Differ 2017; 24:984–996.

42. Misiorek JO, Lahdeniemi IAK, Nystrom JH, Paramonov VM, Gullmets JA, Saarento H, Rivero-Muller A, Husoy T, Taimen P, Toivola DM. Keratin 8-deletion induced colitis predisposes to murine colorectal cancer enforced by the inflammasome and IL-22 pathway. Carcinogenesis 2016; 37:777–786.

43. Odaka C, Loranger A, Takizawa K, Ouellet M, Tremblay MJ, Murata S, Inoko A, Inagaki M, Marceau N. Keratin 8 is required for the maintenance of architectural structure in thymus epithelium. PLoS One 2013; 8:e75101.

44. Alam CM, Silvander JSG, Helenius TO, Toivola DM. Decreased levels of keratin 8 sensitize mice to streptozotocin-induced diabetes. Acta Physiol (Oxf) 2018; 224:e13085.

45. Pandol SJ. The exocrine pancreas. The exocrine pancreas. In: Granger DN, Granger JP, eds. Colloquium Series on Integrated Systems Physiology: From Molecule to Function to Disease. Volume 3. San Rafael, CA: Morgan & Claypool Life Sciences; 2011:1–64.

46. Zhong B, Zhou Q, Toivola DM, Tao GZ, Resurreccion EZ, Omary MB. Organ-specific stress induces mouse pancreatic keratin overexpression in association with NF-κB activation. J Cell Sci 2004; 117:1709–1719.

47. Toivola DM, Ku NO, Ghori N, Lowe AW, Michie SA, Omary MB. Effects of keratin filament disruption on exocrine pancreas-stimulated secretion and susceptibility to injury. Exp Cell Res 2000; 255:156–170.

48. Bouwens L, Braet F, Heimberg H. Identification of rat pancreatic duct cells by their expression of cytokeratins 7, 19, and 20 in vivo and after isolation and culture. J Histochem Cytochem 1995; 43:245–253.

49. Wildi S, Kleeff J, Maruyama H, Maurer CA, Friess H, Buchler MW, Lander AD, Korc M. Characterization of cytokeratin 20 expression in pancreatic and colorectal cancer. Clin Cancer Res 1999; 5:2840–2847.

50. Bouwens L. Cytokeratins and cell differentiation in the pancreas. J Pathol 1998; 184:234–239.

51. Carlotti F, Zaldumbide A, Ellenbroek JH, Spijker HS, Hoeben RC, de Koning EJ. β-cell generation: can rodent studies be translated to humans? J Transplant 2011; 2011:892453. doi:10.1155/2011/892453

52. Zhong B, Strnad P, Toivola DM, Tao GZ, Ji X, Greenberg HB, Omary MB. Reg-II is an exocrine pancreas injury-response product that is up-regulated by keratin absence or mutation. Mol Biol Cell 2007; 18:4969–4978.

53. Toivola DM, Ostrowski SE, Baribault H, Magin TM, Ramsingh AI, Omary MB. Keratins provide virus-dependent protection or predisposition to injury in coxsackievirus-induced pancreatitis. Cell Health and Cytoskeleton 2009; 1:51–65.

54. Casanova ML, Bravo A, Ramirez A, Morreale de Escobar G, Were F, Merlino G, Vidal M, Jorcano JL. Exocrine pancreatic disorders in transsgenic mice expressing human keratin 8. J Clin Invest 1999; 103:1587–1595.

55. Ku NO, Michie S, Resurreccion EZ, Broome RL, Omary MB. Keratin binding to 14–3-3 proteins modulates keratin filaments and hepatocyte mitotic progression. Proc Natl Acad Sci U S A 2002; 99:4373–4378.

56. Baribault H, Penner J, Iozzo RV, Wilson-Heiner M. Colorectal hyperplasia and inflammation in keratin 8-deficient FVB/N mice. Genes Dev 1994; 8:2964–2973.

57. Yi H, Yoon HN, Kim S, Ku NO. The role of keratins in the digestive system: lessons from transgenic mouse models. Histochem Cell Biol 2018; 150:351–359.

58. Brembeck FH, Rustgi AK. The tissue-dependent keratin 19 gene transcription is regulated by GKLF/KLF4 and Sp1. J Biol Chem 2000; 275:28230–28239.

59. Deramaudt TB, Sachdeva MM, Wescott MP, Chen Y, Stoffers DA, Rustgi AK. The PDX1 homeodomain transcription factor negatively regulates the pancreatic ductal cell-specific keratin 19 promoter. J Biol Chem 2006; 281:38385–38395.

60. Spazierer D, Raberger J, Gross K, Fuchs P, Wiche G. Stress-induced recruitment of epiplakin to keratin networks increases their resistance to hyperphosphorylation-induced disruption. J Cell Sci 2008; 121:825–833.

61. Wogenstein KL, Szabo S, Lunova M, Wiche G, Haybaeck J, Strnad P, Boor P, Wagner M, Fuchs P. Epiplakin deficiency aggravates murine caerulein-induced acute pancreatitis and favors the formation of acinar keratin granules. PLoS One 2014; 9:e108323.

62. Szkudelski T. The mechanism of alloxan and streptozotocin action in B cells of the rat pancreas. Physiol Res 2001; 50:537–546.

63. Elenbaas JS, Bragazzi Cunha J, Azuero-Dajud R, Nelson B, Oral EA, Williams JA, Stewart CL, Omary MB. Lamin A/C maintains exocrine pancreas homeostasis by regulating stability of RB and activity of E2F. Gastroenterology 2018; 154:1625–1629.

64. Haque WA, Vuitch F, Garg A. Post-mortem findings in familial partial lipodystrophy, Dunnigan variety. Diabet Med 2002; 19:1022–1025.

65. Zhang JS, Wang L, Huang H, Nelson M, Smith DI. Keratin 23 (K23), a novel acidic keratin, is highly induced by histone deacetylase inhibitors during differentiation of pancreatic cancer cells. Genes Chromosomes Cancer 2001; 30:123–135.

66. Park MK, Lee CH. Effects of cerulein on keratin 8 phosphorylation and perinuclear reorganization in pancreatic cancer cells: involvement of downregulation of protein phosphatase 2A and α4. Environ Toxicol 2016; 31:2090–2098.

67. Karantza V. Keratins in health and cancer: more than mere epithelial cell markers. Oncogene 2011; 30:127–138.

68. Terazono K, Yamamoto H, Takasawa S, Shiga K, Yonemura Y, Tochino Y, Okamoto H. A novel gene activated in regenerating islets. J Biol Chem 1988; 263:2111–2114.

69. Gross DJ, Weiss L, Reibstein I, van den Brand J, Okamoto H, Clark A, Slavin S. Amelioration of diabetes in nonobese diabetic mice with advanced disease by linomide-induced immunoregulation combined with Reg protein treatment. Endocrinology 1998; 139:2369–2374.

70. Abe M, Nata K, Akiyama T, Shervani NJ, Kobayashi S, Tomioka-Kumagai T, Ito S, Takasawa S, Okamoto H. Identification of a novel Reg family gene, Reg IIIdelta, and mapping of all three types of Reg family gene in a 75 kilobase mouse genomic region. Gene 2000; 246:111–122.

71. Ostrowski SE, Reilly AA, Collins DN, Ramsingh AI. Progression or resolution of coxsackievirus B4-induced pancreatitis: a genomic analysis. J Virol 2004; 78:8229–8237.

72. Li B, Wang X, Liu JL. Pancreatic acinar-specific overexpression of Reg2 gene offered no protection against either experimental diabetes or pancreatitis in mice. Am J Physiol Gastrointest Liver Physiol 2010; 299:G413–G421.

73. Szeverenyi I, Cassidy AJ, Chung CW, Lee BT, Common JE, Ogg SC, Chen H, Sim SY, Goh WL, Ng KW, Simpson JA, Chee LL, et al. The Human Intermediate Filament Database: comprehensive information on a gene family involved in many human diseases. Hum Mutat 2008; 29:351–360.

74. Strnad P, Paschke S, Jang KH, Ku NO. Keratins: markers and modulators of liver disease. Curr Opin Gastroenterol 2012; 28:209–216.

75. Ku NO, Gish R, Wright TL, Omary MB. Keratin 8 mutations in patients with cryptogenic liver disease. N Engl J Med 2001; 344:1580–1587.

76. Cavestro GM, Frulloni L, Nouvenne A, Neri TM, Calore B, Ferri B, Bovo P, Okolicsanyi L, Di Mario F, Cavallini G. Association of keratin 8 gene mutation with chronic pancreatitis. Dig Liver Dis 2003; 35:416–420.

77. Treiber M, Schulz HU, Landt O, Drenth JP, Castellani C, Real FX, Akar N, Ammann RW, Bargetzi M, Bhatia E, Demaine AG, Battagia C, et al. Keratin 8 sequence variants in patients with pancreatitis and pancreatic cancer. J Mol Med (Berl) 2006; 84:1015–1022.

78. Schneider A, Lamb J, Barmada MM, Cuneo A, Money ME, Whitcomb DC. Keratin 8 mutations are not associated with familial, sporadic and alcoholic pancreatitis in a population from the United States. Pancreatology 2006; 6:103–108.

79. Lin Y, Nakatochi M, Hosono Y, Ito H, Kamatani Y, Inoko A, Sakamoto H, Kinoshita F, Kobayashi Y, Ishii H, Ozaka M, Sasaki T, et al. Genome-wide association meta-analysis identifies GP2 gene risk variants for pancreatic cancer. Nat Commun 2020; 11:3175.

80. Schmitz-Winnenthal FH, Volk C, Helmke B, Berger S, Hinz U, Koch M, Weitz J, Kleeff J, Friess H, Zoller M, Buchler MW, Z'Graggen K. Expression of cytokeratin-20 in pancreatic cancer: an indicator of poor outcome after R0 resection. Surgery 2006; 139:104–108.

81. Soeth E, Grigoleit U, Moellmann B, Roder C, Schniewind B, Kremer B, Kalthoff H, Vogel I. Detection of tumor cell dissemination in pancreatic ductal carcinoma patients by CK 20 RT-PCR indicates poor survival. J Cancer Res Clin Oncol 2005; 131:669–676.

82. Grippo PJ, Sandgren EP. Acinar-to-ductal metaplasia accompanies c-myc-induced exocrine pancreatic cancer progression in transgenic rodents. Int J Cancer 2012; 131:1243–1248.

83. D'Alessandro M, Russell D, Morley SM, Davies AM, Lane EB. Keratin mutations of epidermolysis bullosa simplex alter the kinetics of stress response to osmotic shock. J Cell Sci 2002; 115:4341–4351.

84. Lee J, Jang KH, Kim H, Lim Y, Kim S, Yoon HN, Chung IK, Roth J, Ku NO. Predisposition to apoptosis in keratin 8-null liver is related to inactivation of NF-kappaB and SAPKs but not decreased c-Flip. Biol Open 2013; 2:695–702.

85. Chen X, Ji B, Han B, Ernst SA, Simeone D, Logsdon CD. NF-κB activation in pancreas induces pancreatic and systemic inflammatory response. Gastroenterology 2002; 122:448–457.

Acinar Cells

Chapter 6

Pancreatic acinar cell protein synthesis, intracellular transport, and export

Michelle M. Cooley,[1] Elaina K. Jones,[1] Fred S. Gorelick,[2] and Guy E. Groblewski[1]

[1]Graduate Program in Biochemical and Molecular Nutrition, University of Wisconsin-Madison, Madison, WI
[2]Yale School of Medicine, Yale University, New Haven, CT and VA CT, West Haven CT

Introduction

A key function of the exocrine pancreas is the production of digestive enzymes. The pancreatic acinar cell synthesizes, stores, and secretes the proenzymes and enzymes needed to digest dietary proteins, carbohydrates, and lipids. Meeting these functional requirements necessitates that the acinar cell has very high rates of protein synthesis and export. Nascent proteins undergo folding, select modifications, concentration, segregation from other classes of proteins, and vectorial movement before reaching their final destination in secretory granules. These are concentrated in the apical pole of the acinar cell. Eating stimulates neural and hormonal pathways that mediate acinar cell zymogen granule exocytosis into the pancreatic duct. The acinar cell has been a model system for foundational studies of protein synthesis and export. After electron microscopy was developed, for example, cell biologists first visualized organelles and established their function by studying acinar cells (**Figure 1**). Here, we focus on acinar cell protein synthesis, trafficking, and processing in the pancreatic acinar cell necessary for its central role in producing digestive enzymes. We present results primarily obtained using rodent acinar cells, though the limited data from human acinar cells suggest the functions are likely the same as in rodents.

Protein Synthesis

The ribosome

The ribosome is the central element of the protein synthetic machinery. The eukaryotic ribosome is composed of two subunits: a large 60S unit containing 28S, 5S, and 5.8S rRNA and approximately 49 proteins and a small subunit at 40S, which includes 18S rRNA and about 33 proteins. The two subunits form a groove wherein new protein synthesis directed by messenger RNA takes place, aided by associated protein complexes to initiate, elongate, and terminate

protein synthesis. Secretory proteins, including pancreatic digestive enzymes, contain a distinct n-terminal signal sequence which contains a 15- to 50-amino acid peptide that includes a hydrophobic core. Also known as a leader sequence, the signal sequence allows nascent proteins to cross the endoplasmic reticulum (ER) membrane by traversing the translocon, a large multiprotein channel.[1] The signal sequence binds to the signal recognition particle (SRP) complex and brings the ribosome, attached nascent protein, and mRNA to the ER membrane SRP receptor.[2] The SRP is then released and protein synthesis resumes with entry of the signal sequence into the translocon and its subsequent proteolytic cleavage from the nascent protein.[3] In addition to soluble export proteins, the translocon also mediates the insertion of transmembrane proteins into the ER membrane and has a role in protein degradation.

The endoplasmic reticulum

The synthesis of new secretory proteins occurs on and in the ER. This organelle has two morphologically distinct compartments: the rough ER (RER) and smooth ER (SER), which have specialized functions. The RER has ribosomes on its cytoplasmic surface, which are easily seen by electron microscopy, whereas the SER lacks ribosomes (**Figure 1**). Between the RER and SER are regions termed transitional elements. Vesicular transport of nascent proteins from the ER occurs at sites in the transitional ER known as ER exit sites (**Figure 2**). The intracisternal space is formed by RER, transitional elements, and SER; this is the site in which nascent proteins begin to fold and undergo export. To accommodate high rates of secretory protein synthesis, a dense RER occupies much of the basal region of the acinar cell and extends apically. Regularly placed groupings of electron dense ribosomes (~30 nm) mark the sites of active protein synthesis on the cytosolic face of the ER membrane.

---

e-mail: groby@nutrisci.wisc.edu

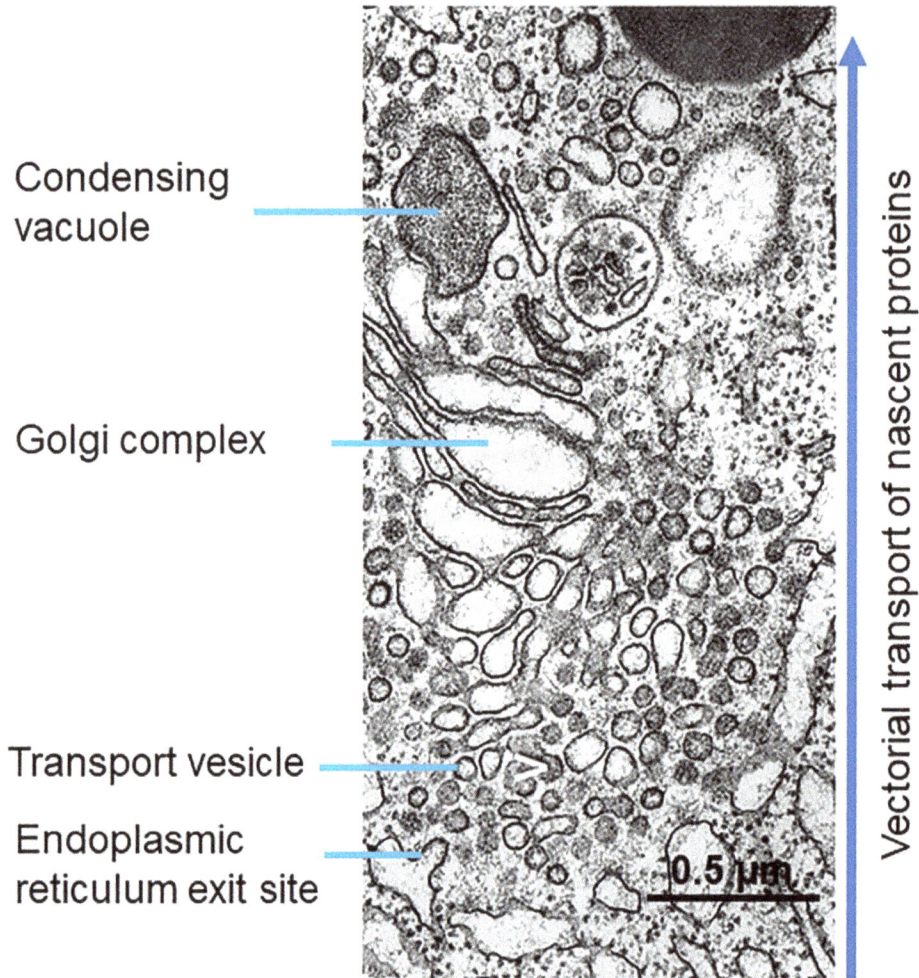

Figure 1. Electron micrograph of guinea pig pancreatic acinar cell showing compartments involved in nascent protein synthesis and storage. The vectorial features of this pathway from endoplasmic reticulum to secretory granule are evident. Nascent proteins are synthesized in about 5 minutes in the endoplasmic reticulum; then they move by transport vesicles to the Golgi complex, where they exit at about 20 minutes in condensing vacuoles (aka immature secretory granules), and mature into secretory granules (image from reference [4]).

The acinar cell ER shows regional functional specialization that also facilitates cell signaling. For example, the initial acinar cell cytoplasmic calcium signal that mediates protein secretion depends on an ordered, spatial release from distinct ER stores. In response to neurohumoral stimulation, the first calcium signal arises at the apex of the cell from select ER domains and is mediated by the inositol trisphosphate (IP3) receptor. Similarly, the propagation of the calcium wave throughout the acinar cell requires calcium release from ER ryanodine receptors which are distributed toward the base of the cell.

Distinct direct contacts between cell organelles have now been described in many tissues and include sites between the ER and mitochondria and ER and plasma membrane. These are known to be involved in lipid transfer and calcium signaling.[5] In the acinar cell, apically distributed IP3 receptors (ER) that release calcium into the cytosol are found to closely approximate regulators of calcium entry into the cell (Orai1, STIM, and others) that are activated when depletion of ER calcium is sensed.[6] These connections are thought to help coordinate both physiologic responses, such as protein secretion, and pathologic cell responses.

Modifications and Processing of Nascent Proteins

Nascent secretory proteins are subject to a variety of modifications; many of these occur within specific cellular organelles and are involved in forming the three-dimensional structure and covalent modifications necessary for proper protein function. Protein modifications often require the activity of accessory resident proteins that are concentrated in specific organelles.

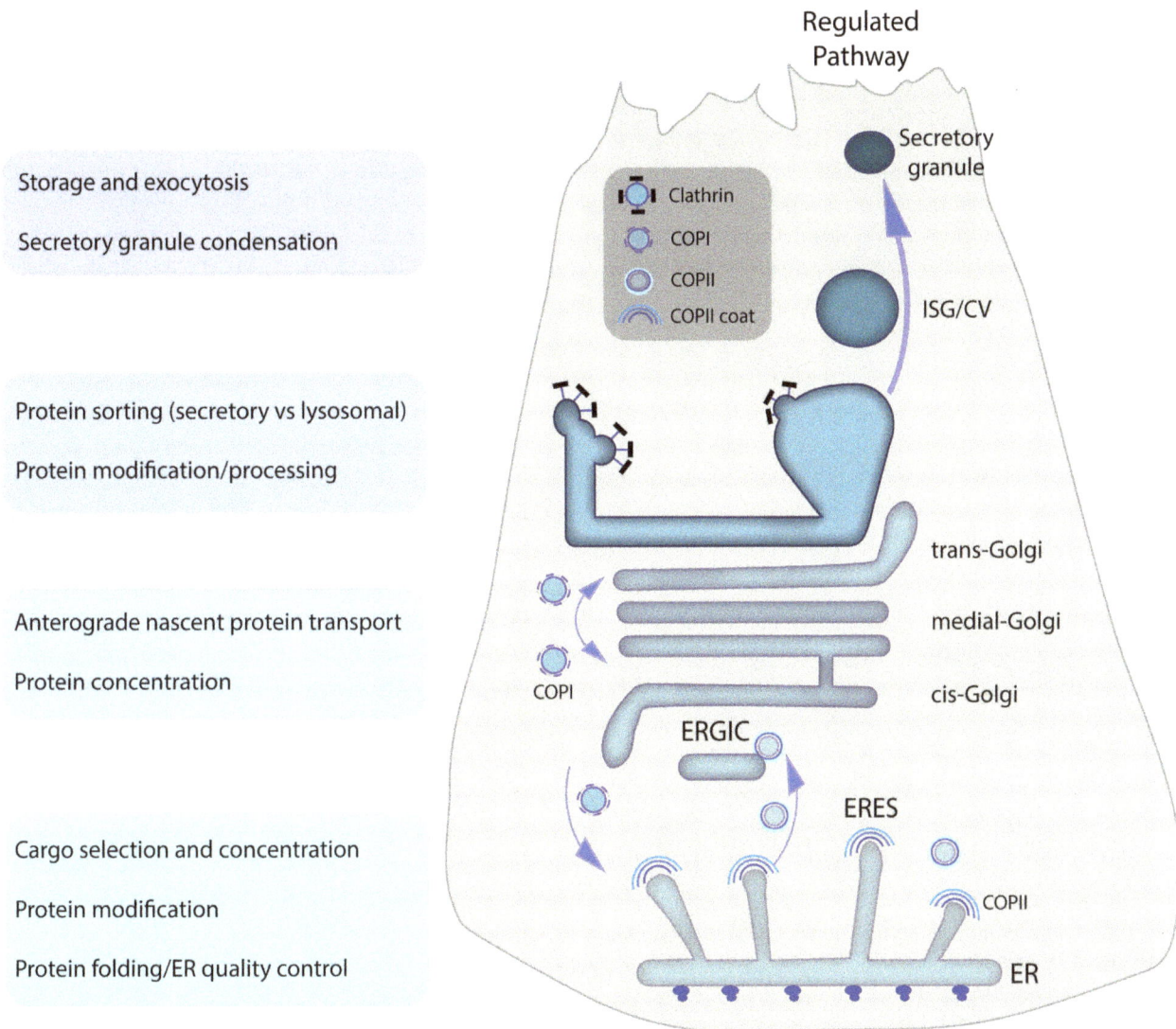

Figure 2. Secretory trafficking pathways in pancreatic acinar cells. Schematic depicting the classic secretory trafficking pathway in acini. Proteins are synthesized into the endoplasmic reticulum (**ER**) lumen, undergo folding and co/posttranslational modifications, and are selected and/or concentrated into COPII-coated vesicles at specialized ER exit sites (ERES). These vesicles interact with the ER-Golgi intermediate compartment (**ERGIC**), from which ER-resident proteins are sorted back to the ER via COPI-coated vesicles while secretory proteins are concentrated and trafficked through the Golgi stacks. Additional protein modifications occur in the Golgi. Secretory proteins are then packaged into immature secretory granules/condensing vacuoles (**ISGs/CVs**) through clathrin-mediated removal of membrane and cargo from the trans-Golgi; ISGs mature and condense by further removal of membrane/cargo. The resulting mature secretory granules are stored until signals for exocytosis are received by the cell.

Cleavage of the signal peptide

One of the first modifications of secretory proteins within the ER is the proteolytic removal of the signal peptide. The cleavage is mediated by a specific ER protease (signal peptidase) and occurs within the translocon after a large portion of the nascent protein has entered the ER cisterna.[3,7] Removal of the signal peptide traps the nascent protein in the secretory pathway unless the protein misfolds.

Protein modifications and folding

Secretory proteins undergo posttranslational modifications within the ER that directs their folding into a tertiary structure that shields their hydrophobic residues in the interior of the molecule and forms their functional domains. Several major protein modifications, some of which are covalent, occur within the ER—including disulfide bond formation, glycosylation, and acetylation. Accessory ER resident proteins also direct the folding of nascent proteins.

Disulfide bonds

Disulfide bonds form by oxidative linkage of sulfhydryl groups between two Cys residues. These bonds are necessary to form higher-order structures both within and between polypeptides and can occur either co- or posttranslationally. The ER redox environment is uniquely suited to promote disulfide bond formation; indeed, disulfide bonds are highly abundant among secretory proteins and rarely found in cytosolic proteins. It follows that changes in ER redox status affect protein folding and stability. For example, proteomic assessment of acinar ER proteins following ethanol feeding shows a shift toward increased oxidation, which suggests the likelihood of aberrant disulfide bonds—leading to protein misfolding and ER stress.[8,9] That said, disulfide bonds can occur spontaneously under normal conditions, and so ER-resident protein disulfide isomerases (PDIs) act to rearrange disulfide bonds and ensure their formation between the correct Cys residues. As such, PDIs are also considered chaperones (see below). One PDI family member specifically expressed in pancreatic acinar cells, termed PDIp, interacts with several digestive enzymes and proteolytic zymogens and prevents their aggregation.[10–12]

Glycosylation

Glycosylation describes the addition of carbohydrate chains called glycans to nascent proteins. Protein glycosylation in the ER is mediated by resident glycosyltransferases and glycosidases that generate a milieu of glycan moieties that can influence protein folding and function (reviewed in reference [13]). One such modification is the addition of N-acetylglucosamine (GlcNAc) to Asn residues, termed N-glycosylation. N-glycosylation of newly translated peptides is regulated by the oligosaccharyltransferase (OST) complex by recognition of an Asn-X-(Thr/Ser) consensus sequence and occurs cotranslationally via interactions between the OST and the translocon. Though N-glycosylation is understood to be critical for nascent protein folding, little is known about this particular modification in acinar cells. Studies in the early 1990s demonstrated the presence of N-glycosylation on human pancreatic elastase 1 (Ela1).[14] It was later shown that N-glycosylation of bile salt-dependent lipase (BSDL) was essential for its expression, association with molecular chaperones, and secretion.[15,16] More work has been done investigating the significance of O-glycosylation of acinar cell proteins, which occurs in the Golgi, including lysosome-associated membrane proteins and some digestive enzymes.

Acetylation

Studies over the last decade have identified a critical role for Nε-lysine acetylation and ER proteostasis—that is, the dynamic regulation of the biogenesis, folding, trafficking, and degradation of proteins within the ER. Posttranslational acetylation of proteins in the ER is regulated by the ER acetyl-CoA transporter, AT-1 (*SLC33A1*), and ER-resident acetyltransferases, ATase1 and ATase2 (*NAT8B* and *NAT8*, respectively), which utilize the luminal acetyl-CoA as acetylation substrate. In contrast to N-glycosylation, protein acetylation occurs only after successful protein folding and does not appear to involve a consensus sequence—though only outward-facing lysines are acetylated, suggesting a dependence on protein tertiary structure formation. Interestingly, the acetyltransferases interact with the OST complex to acetylate the targeted residues, suggesting a relationship between N-glycosylation and acetylation. The current view is that this acetylation identifies properly folded (i.e., glycosylated) proteins and promotes their entry into the secretory pathway, whereas improperly folded proteins cannot be acetylated appropriately and are thus subject to degradation. It has been recently shown that AT-1 expression increases at the onset of pancreatitis but falls as the disease progresses.[17] Furthermore, studies involving mice expressing mutant AT-1[S113R/+], a human mutation associated with hereditary spastic paraplegias,[18] or pancreatic acinar-specific AT-1 deletion, show elevated ER stress, inflammation, and fibrosis consistent with a chronic pancreatitis-like phenotype, which unexpectedly includes enhanced trypsin activation.[17] Prior analyses of the so-called "ER acetylome" in neuronal cells identified acetylated proteins that are of interest in pancreatic acinar physiology, including BiP (GRP78), LAMP2, and cathepsin D.[19] Given the recently established role of AT-1 in pancreatic acinar homeostasis, investigating the acetylation state of acinar cell proteins will provide further insight into the functional significance of this modification on pancreatic outcomes.

Molecular chaperones

Although very small peptides can spontaneously fold unassisted, the folding of larger and more complex proteins in the ER is hindered by the available space and necessity of precise luminal conditions. ER-resident chaperones assist newly translated polypeptides by recognizing and interacting with key protein regions and shielding them from outside interference while the necessary modifications are made. Two main chaperone systems regulate this process in the ER: heat shock chaperones (e.g., BiP, aka GRP78, and GRP94), which interact with exposed hydrophobic regions; and lectins (e.g., calreticulin/calnexin), which interact with N-linked glycans. Evidence shows that these chaperones often cycle between association, dissociation, and reassociation with their targets until folding is complete. It follows that molecular chaperones are critical for pancreatic acinar cell function. In mice, GRP78[+/-] pancreas exhibits alterations in ER morphology, reduction in the expression of calreticulin and calnexin, and experiences increased disease severity in response to caerulein-induced pancreatitis.[20] As previously mentioned, PDIs are also considered chaperones as they support protein folding by facilitating disulfide bond formation.

The unfolded protein response

To sustain high rates of ER protein synthesis and secretory trafficking efficiency, pancreatic acini rely on a robust ER stress response system to manage the accumulation of proteins in the ER lumen. Known as the unfolded protein response (UPR), this system is regulated by three ER transmembrane proteins that "sense" the load of unfolded proteins in the ER lumen: inositol-requiring enzyme 1 (IRE1), activating transcription factor 6 (ATF6), and protein kinase R-like ER kinase.[21] These sensors mediate signal transduction pathways that initiate cellular response mechanisms to alleviate the burden of ER stress (reviewed in references [21, 22]).

The adaptive UPR

During the initial stages of ER stress, or when ER stress is mild, IRE1 dimerizes and auto-transphosphorylates, activating its RNase activity which processes the transcription factor X-box binding protein 1 (XBP1) into its spliced product (XBP1s).[23,24] Effects of IRE1 on other types of mRNA have also been reported. Additionally, ATF6 translocates to the Golgi using the COPII complex-coated vesicles and is cleaved by site 1 and 2 proteases to generate its active cytosolic form cATF6.[25,26] Both XBP1s and cATF6 regulate the expression of genes encoding ER chaperones, ER biogenesis, and ER-associated degradation (ERAD); furthermore, cATF6 controls XBP1 expression.[23,24,27–30] These actions serve to increase the ER folding capacity and secretory output to clear accumulated proteins and are collectively referred to as the adaptive UPR.

The role of XBP1s in this protective mechanism has been extensively studied in the context of exocrine pancreas physiology and pancreatitis. XBP1s works in concert with the transcription factor Mist1 to drive terminal differentiation of pancreatic acini and other secretory cell types; XBP1 null mice are not viable.[29,31–36] XBP1 heterozygotes (XBP1[+/-]) exhibit increased pancreatic pathology, including oxidative stress that affects disulfide bond formation, in response to ethanol-mediated ER stress, which establishes a protective role for XBP1s during pancreatitis.[8,36] Moreover, XBP1s regulate the expression of AT-1, which is part of the ER protein acetylation machinery.[37] Although less is known about the role of ATF6 in acinar cells, a recent publication suggests a putative role for ATF6 and the apoptotic protein p53 in the development of chronic pancreatitis.[38]

The pathological UPR

In the event the adaptive UPR cannot ameliorate existing ER stress, subsequent pathways with more profound consequences are activated. Similar to IRE1, PERK autophosphorylates in response to ER stress and activates its kinase activity, which in turn phosphorylates the translation initiation factor 2α (eIF2α), inhibiting global cap-dependent translation. This provides the ER a reprieve from translational protein input. Should these actions fail to alleviate ER stress, PERK activates the apoptotic protein CCAAT/enhancer binding protein homologous protein (CHOP) by increasing expression of activating transcription factor 4 (ATF4). This duality in PERK function is illustrated by studies showing the necessity of both PERK and ATF4 in normal pancreatic development and function, whereas loss of CHOP results in normal exocrine function and protects against the acceleration of pancreatitis.[39–41] Sustained IRE1 signaling may also play a role in the pathological UPR by activating jun N-terminal kinase (JNK), which contributes to inflammation and apoptosis, though this pathway has not been fully studied in pancreatic acinar cells.

Regulated IRE1-dependent decay

Under conditions of sustained ER stress, the endoribonuclease activity of IRE1 may become broader, affecting the translation of mRNAs other than XBP1. Reports indicate a potential role for regulated IRE1-dependent decay (RIDD) in regulating insulin mRNA in pancreatic β cells under ER stress due to hyperglycemia.[42,43] The physiological role of RIDD in exocrine pancreatic function and response to stress has not yet been explored.

Removal and degradation of accumulated proteins

ER-associated degradation (ERAD) is a cellular pathway that removes terminally misfolded proteins from the ER lumen or membrane and regulates their degradation. ERAD may occur by proteasomal (ERAD-I) or autophagic (ERAD-II) mechanisms. It is believed that ERAD-I mediates the disposal of monomeric proteins. The best characterized branch of ERAD-I is regulated by the adaptor protein suppressor/enhancer of Lin-12-like (Sel1L) and the dislocon channel HMG-coa reductase degradation protein (Hrd1) complex, which retrotranslocates misfolded ER proteins to the cytosol where they are subsequently ubiquitinylated and targeted to the proteasome for degradation.[44] Interestingly, inducible Sel1L knockout mice exhibit classic exocrine insufficiency as well as persistent ER stress/UPR and, curiously, significantly smaller zymogen granules.[45] This study suggests a key role for Sel1L and ERAD-I in pancreatic acinar ER homeostasis.

ER homeostatic mechanisms utilizing autophagic/lysosomal pathways will be collectively referred to here as ERAD-II, including ER-phagy, reticulophagy, and/or

ER-quality control (ERQC) autophagy; whether these classifications represent similar or distinct processes is the subject of debate, which will not be dissected here. In contrast to ERAD-I, ERAD-II is proposed to degrade large protein aggregates and ER membrane rather than individual proteins (reviewed in reference [46]). In mammals, autophagy-mediated ER degradation utilizes ER membrane-associated receptor molecules. To date, six receptors have been identified: RTN3L, FAM134B, SEC62, CCPG1, ATL3, and TEX264. Although the various stimuli and protein targets for these receptors have not been completely characterized, they all interact with the autophagy protein LC3 through LC3-interacting regions (LIRs), which mediates the selective removal of ER through the autophagy pathway. Cell-cycle progression gene 1 (CCPG1) was first identified as an ER-phagy cargo receptor in pancreatic acini that is induced by UPR signaling, and loss of CCPG1 leads to ER disordering and loss of polarity in acini.[47-49]

Initiation of ERAD-II may be regulated by acetylation. The autophagy regulatory protein ATG9 is the only membrane-associated autophagy protein that resides in the ER. Studies of AT-1 function identified acetylation sites on ATG9 on the luminal side that restricts its activity, and loss of AT-1 function spurs increased autophagy.[37,50] These findings suggest that ATG9 may act as an ER lumen acetylation sensor wherein loss of acetyl-CoA availability, corresponding to accumulation of proteins, induces reticulophagy.

The extent to which ERAD-I and ERAD-II regulate the degradation of pancreatic acinar cell proteins under normal and disease states is unclear. Similarly, whether defects in either ERAD mechanism contribute to aberrant intracellular trypsin activation and/or pancreatitis pathology remains to be investigated.

Early Acinar Cell Secretory Protein Trafficking

Pancreatic acinar cell secretory proteins are translated into the highly expanded rough ER (**Figures 1** and **2**). Following folding and posttranslational modifications discussed above, soluble and membrane-associated secretory proteins adjacent to specialized ERES are packaged into coat protein complex II (COPII)-coated vesicles and directed to the ER-Golgi intermediate compartment (ERGIC). In the ERGIC, resident ER proteins are sorted back to the ER in COPI-coated vesicles, while secretory proteins are concentrated and delivered to the Golgi cisternae (*cis-*, *medial-*, and *trans-*Golgi), sometimes termed the Golgi stacks and/or Golgi ribbons (reviewed in [51-53]). Proteins may undergo further posttranslational modifications during their sequential movement through the Golgi. Ultimately, secretory proteins are concentrated into condensing vacuoles within the *trans-*Golgi compartment where they bud off as precursors of secretory or zymogen granules (ZGs). It should be noted that condensing vacuoles (CVs) and immature secretory granules (ISGs) are often used interchangeably in acinar cell literature (**Figures 2** and **3**).

ER to Golgi trafficking

Anterograde transport of ER membrane and protein to the ERGIC begins with the formation of COPII vesicles at specialized ERES subdomains.[52,54] Formation of the COPII coat is directed by the cytosolic proteins Sar1, Sec23, Sec24, Sec13, and Sec31.[53,55] In brief, the GTPase Sar1 is GTP-loaded and activated by the guanine exchange factor Sec12 near the ERES, prompting the insertion of Sar1 into the ER membrane using an amphipathic helix. Sar1 membrane insertion facilitates the membrane deformation ultimately required for vesicle formation and budding. Membrane-associated Sar1 also binds to the heterodimer Sec23-Sec24. Sec 24 interacts with transmembrane ER-associated receptors to facilitate soluble cargo loading. Sequential formation of these complexes at the ERES promotes the addition of an outer layer composed of a heterotetramer of two Sec13 and two Sec31 subunits which self-assemble into cage-like structures morphologically similar but biophysically distinct to clathrin coats.[56] In addition to the five core COPII regulatory proteins, a growing number of proteins that transiently interact during COPII coat formation and trafficking have been identified but will not be detailed here.[53,55] Vesicle scission occurs through an unclear process, and the COPII vesicles—typically 60–80 nm in diameter—are directed to the ERGIC where they tether and undergo fusion via soluble n-ethylmaleimide sensitive receptor (SNARE) protein interactions.[57]

Although the acinar cell is known for its massive protein secretory capacity and many studies have been directed at understanding the mechanisms of COPII vesicle formation in other cells, few have addressed this pathway in pancreatic acinar cells. In 1999, Martínez-Menárguez et al. used immuno-electron microscopy on pancreas thin sections to demonstrate that the membrane-associated SNARE protein rBet1 is concentrated in COPII-coated vesicles; however, the soluble secretory proteins amylase and chymotrypsin were not, suggesting that soluble cargo loading is not receptor-mediated but occurs by nonselective transport.[58] A later study in yeast followed the loading of soluble secretory proteins and COPII membrane proteins demonstrating receptor-mediated concentrative sorting of soluble proteins into COPII vesicles.[59] Whether additional acinar secretory proteins other than amylase and chymotrypsin may be selectively loaded in COPII vesicles remains unknown. A more recent study demonstrated that pancreas-specific deficiency of the Sec23B isoform, but not Sec23A, results in embryonic lethality due to acinar cell degeneration but, interestingly, did not disrupt the morphology of islets.[60,61] Clearly, further investigation of COPII formation, loading,

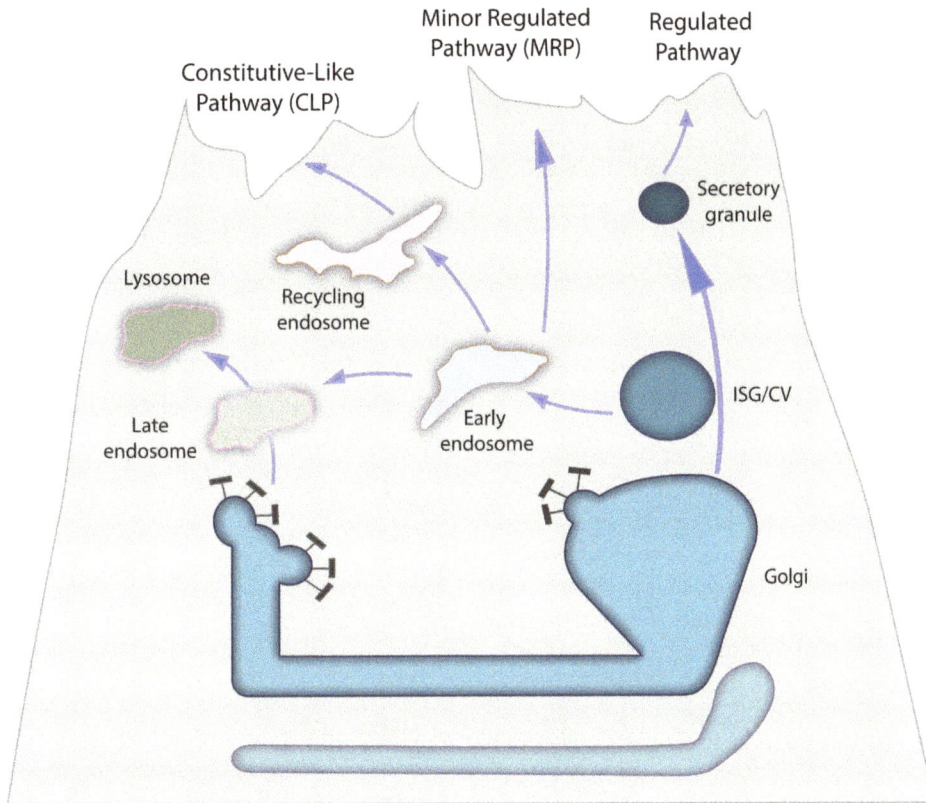

Figure 3. Additional secretory pathways identified in acinar cells. The classic regulated secretory pathway is represented by the formation of mature secretory granules from ISGs/CVs. Parallel secretory pathways are formed by the trafficking of membrane and secretory proteins through endosomal compartment intermediates prior to secretion, either directly from the early endosome in the minor-regulated pathway (**MRP**) or through the recycling endosome as in the constitutive-like pathway (**CLP**).

and trafficking in normal acinar cells and the importance of this pathway during pathologic ER stress is warranted.

ER-Golgi intermediate compartment

The ERGIC, also known as the vesicular tubular cluster due to its morphology, acts as sorting compartment between the ER and Golgi apparatus (**Figures 1 and 2**). The most reliable marker of this compartment is ERGIC-53, a mannose-specific membrane lectin that functions as a cargo receptor for the transport of glycoproteins from the ER to the ERGIC.[62] The ERGIC is formed and maintained by the continuous fusion of COPII vesicles arriving from the ER. Resident ER proteins are retrieved from the ERGIC back to the ER by their concentration and sorting into COPI-coated vesicles. In contrast to COPII coat formation, COPI coat proteins are composed of seven COP adapters that are assembled in response to activation of an ADP ribosylation factor (Arf1, Arf2, or Arf3) by GTP loading with guanine exchange factors. Activation triggers Arf myristoylation and membrane tethering identical to Arf function in clathrin coat formation (reviewed in reference [63]). Cargo selection

by COPI is directed by sorting signals present on cytoplasmic domains including di-lysine, KKxx, and KxKxx motifs present on many ER proteins. Likewise, the KDEL receptor, which functions to both retain ER resident proteins and recycle them from the ERGIC and *cis*-Golgi, interacts with KDEL-containing proteins, including ER chaperones, thereby concentrating them in COPI vesicles.[57]

In acinar cells, Martínez-Menárguez et al. showed that although amylase and chymotrypsin were not selectively sorted during COPII vesicle formation in the ER, these enzymes were concentrated in the ERGIC by their apparent exclusion from COPI vesicles recycling back to the ER.[58] This and a later study support the concept that the ERGIC functions as an initial concentrating compartment for acinar secretory proteins that are nonselectively transported from ER.[57,64] Interestingly, although COPI buds are most numerous in the ERGIC, they are also found in cisternae at all levels of the Golgi stacks as well as a minor presence on ISGs, indicating that either COPI recycling takes place at all levels of the early secretory pathway or that COPI vesicles have additional functions in the Golgi (see below).[58] Of note, an earlier study by Sesso et al. provided a three-dimensional reconstruction of the rough ER-Golgi interface in serial thin sections of rat pancreas. This was depicted as a series budding small vesicles that fused to form tubulovesicular elements that appeared to be interposed between the ER and *cis*-Golgi, although the nature of these tubulovesicles is uncertain, it seems plausible they represented what is now termed the ERGIC.[65]

Golgi complex

A number of theories have been postulated regarding Golgi formation, structural organization, and mechanisms of protein and membrane trafficking between the Golgi compartments.[66–69] Golgi architecture in mammalian cells consists of a morphologically heterogeneous set of membrane-limited compartments with a characteristic stack-like appearance (**Figure 1**). Membrane and proteins generally flow or mature in an anterograde direction from *cis*- to *trans*-cisternae. The *cis*-cisternae are oriented toward the ER and are associated with small vesicle formation, whereas the *trans*-Golgi is oriented toward the apical plasma membrane and is associated with production of secretory granules. These regions are most commonly described as zones of the *cis*-, *medial*-, and *trans*-Golgi, with movement between compartments by vesicle fission and fusion and so-called noncompact zones that are interconnected by lateral tubules between cisternae.[70,71] These interconnected regions appear as higher-order structures by light microscopy giving rise to the term "Golgi ribbon." There are numerous small, coated vesicles oriented both close to and budding off from the rims of the Golgi cisternae, the majority identified as COPI-coated by immuno-electron microscopy. COPI coats are most abundant in the ERGIC and *cis*-Golgi, which decrease in number moving toward the *trans*-Golgi.[70] Recent studies have described these as COP1b vesicles that mediate anterograde trafficking within the Golgi cisternae whereas the COPIa vesicles mediate retrograde recycling of ER proteins from the ERGIC and *cis*-Golgi, although the molecular determinants dictating COPIa versus COPIb coats are uncertain.[57,68]

The most distal *trans*-Golgi cisternae give rise to and are continuous with the *trans*-Golgi network (TGN), a series of branching tubules with many budding profiles and forming vesicles. The TGN is biochemically distinct from the rest of the Golgi complex, having a higher concentration of certain proteins like TGN-38 compared to the *cis*- and *medial*-Golgi. The TGN is highly dynamic in accordance with the rates of protein synthesis and secretory protein entry into the compartment. Multiple organelles have been shown to arise from the TGN including early, late, and recycling endosomes, as well as vesicles specifically destined for apical and basolateral plasma membrane or back to the ER.[70] Unique to the TGN is large number of clathrin coats and associated clathrin-coated vesicles (**Figure 3**).

Clathrin coat formation at Golgi membrane is mediated by assembly of clathrin adaptor protein complexes AP1, AP3, and AP4; these multisubunit complexes associate with cytoplasmic domains of specific Golgi transmembrane proteins to nucleate clathrin coat assembly.[72,73] In addition to heteromeric AP complexes, additional monomeric clathrin adaptors are present at the *trans*-Golgi including Golgi-localized γ-ear-containing Arf-binding proteins (GGAs) and enthoprotin/epsinR56.[74]

Protein Modification in the Golgi Complex

The Golgi is a major site of protein modification including terminal glycosylation, proteolytic processing, and sulfation, The Golgi fulfills its multiple functions using several classes of processing enzymes. These are primarily membrane proteins and may be regulated by their distinct localizations within the Golgi complex. For example, galactosyltransferase is restricted to the two or three *trans*-most Golgi cisternae.[75]

O-glycosylation

Complex *O*-linked oligosaccharides are attached an oxygen atom of serine or threonine by glycotransferases within the Golgi (as opposed to *N*-linked glycosylation in the ER). Lysosomal enzymes, including lysosome-associated membrane proteins 1 and 2 (LAMP1,2), are glycoproteins that are modified in this manner and are essential for the function and integrity of lysosomes.[76] Furthermore, a study perturbing *O*-glycosylation in pancreas showed that a number of pancreatic digestive enzymes (e.g., bile salt-activated lipase, pancreatic triacylglycerol lipase, pancreatic alpha-amylase) are *O*-glycosylated, and that loss of *O*-glycosylation results in exocrine and endocrine insufficiency.[77]

Mannose-6-P modification

In the *cis*-Golgi, mannose-6-phosphate (M6P) residues are added to proteins to direct sorting into the endolysosomal pathway.[75] These residues interact with the M6P receptors (MPRs) which are localized to the Golgi region of polarized cells, coated vesicles, endosomes, and lysosomes as identified by Brown and Farquhar in 1984.[78] Although an early study found MPRs to be restricted mainly to the *cis*-Golgi stacks, later work in acinar cells demonstrated AP1-mediated clathrin-coated vesicle retrieval of MPRs from immature secretory granules (see below).[78,79] Two distinct MPRs, a 46-kDa MRP46 (cation-dependent MPR) and a 300-kDa MPR300 (cation-independent MRP), have been identified. Although the majority of soluble acid hydrolases are modified with M6P residues allowing their recognition by MPRs, other soluble enzymes and nonenzymatic proteins are transported to lysosomes in a M6P-independent manner mediated by alternative receptors such as lysosomal integral membrane protein (LIMP-2) or sortilin.[80] Sorting of cargo receptors and lysosomal transmembrane proteins requires sorting signals present in cytosolic domains which interact with components of clathrin coats or an adaptor protein complex. Additionally, phosphorylation and lipid modifications can further regulate signal recognition and trafficking.[80]

Trafficking from the Golgi Complex Outward

Condensing vacuoles and formation of zymogen granules

The *trans*-Golgi is the site of condensing vacuole (or immature secretory granule, ISG) formation in pancreatic acinar cells. Early radiolabeling of acinar secretory proteins *in vivo* combined with electron microscopy radiographic analysis suggested the condensing vacuoles received secretory proteins directly transported on vesicular carriers from the so-called "ER transitional zone" that likely represented the ERES or ERGIC (**Figure 1**).[81] Later, higher-resolution studies by Jameson and Palade in pancreatic slices from guinea pig stimulated with secretagogue revealed the presence of label in the Golgi stacks[82]—although in his 1974 Nobel acceptance speech Palade still depicted the ER transitional compartment as a direct route to the condensing vacuoles.[4] Though many studies have examined condensing vacuole formation, a comprehensive understanding of how secretory cargo are concentrated into condensing vacuoles, how enlarged vacuoles form, and the molecular mechanisms of condensing vacuole fission from the TGN remains uncertain but is likely to be driven at least in part by the acidification of the compartment.

The details of secretory granule formation in specialized secretory cells have received considerable attention over the last 30 years. Evidence derived from *in vitro* fusion assays in subcellular fractions of neuroendocrine cells supports that newly formed ISGs undergo homotypic fusion in a process mediated by syntaxin 6 and synaptotagmin IV.[83,84] Most studies in acinar cells posit that ISGs formed in the TGN are the direct precursors of mature secretory granules. However, Hammel et al.[85] using a detailed morphometric analysis, proposed that small Golgi-derived vacuoles fuse to form ISGs.[85] They further propose that small granules undergo homotypic fusion to form larger granules, although the molecular and microscopic details of this theory are lacking. Ultimately, the immature granules further mature and become smaller by clathrin-coated vesicle-mediated removal of membrane and small amounts of digestive enzymes to the endosomal system resulting in concentration of the digestive enzyme content into a mature, electron-dense ZG.[79] AP1-mediated clathrin coat formation directed by MPRs also removes some, but not all, lysosomal enzymes from ISGs.[79]

Few studies have investigated the secretagogue regulation of post-Golgi ZG formation. Kostenko et al. demonstrated a role for the tyrosine kinase c-src in mediating Golgi morphology and secretory granule formation.[86] Overexpression of c-src caused Golgi expansion whereas pharmacological inhibition reduced granule formation in cultured AR42J acinar cells and isolated acinar cells. These results provide the first known signaling pathway for acute Golgi-mediated ZG formation in response to secretory stimulation.

Anterograde endosomal trafficking through the minor secretory compartment

The great majority of digestive enzyme secretion is mediated by ZG exocytosis at the apical membrane (**Figures 2 and 3**). However, there are two additional and unique parallel secretory pathways identified in acinar cells from the pancreas and parotid glands known as the constitutive-like (CLP) and minor regulated (MRP) pathways as shown in **Figure 3**.[87–92] Though these pathways only provide a small contribution to total protein secretion, they are likely important to acinar cell function. The CLP and MRP were identified by their rapid discharge (~2 hours) of newly synthesized secretory proteins in pulse-chase studies, whereas ZG proteins were secreted by ~10 hours under basal conditions.[88] Secretion from the CLP and MRP is acutely inhibited by brefeldin A (BFA), an inhibitor of guanine nucleotide exchange factors for class 1 ADP-ribosylation factors that function in vesicle formation from *trans*-Golgi and endosomal compartments.[87] The CLP and MRP were proposed to originate from vesicle fission at the TGN and ISGs and traffic through an endosomal intermediate, subsequently identified as the early endosome prior to secretion at the apical membrane.[93–96] The MRP traffics from early or sorting endosomes directly to the apical membrane upon low-level secretagogue stimulation, whereas the CLP may subsequently enter an endosomal recycling compartment prior to exocytosis. More recent studies have identified that the endosomal, TGN peripheral membrane protein TPD52 and its associated proteins Rab5 and EEA1 play an important role in CLP trafficking in acinar cells.[96] It was also demonstrated that ZGs containing the SNARE protein vesicle-associated membrane protein 8 (VAMP8) require an intact endosomal pathway expressing D52, Rab5, and EEA1 in order to mature and/or undergo exocytosis.[97]

Concentration of Nascent Proteins through the Secretory Pathway

Concentration of acinar cell nascent secretory proteins occurs throughout the pathway but is not uniform.

Differences of over an order of magnitude have been observed for amylase, chymotrypsin, and procarboxypeptidase A, and occur at multiple compartments in the secretory pathway.[64] The net effect of the enrichment mechanism is to concentrate soluble proteins such as amylase, trypsinogen, and chymotrypsinogen 10- to 20-fold between the ER lumen and the Golgi complex and even further enrichment as they move toward the ZG and may be as high as a hundred-fold over the ER lumen. Though the mechanisms of such concentration remain unclear and likely vary among compartments, the decreasing pH gradient within the Golgi complex (going from pH 7.0 to 6.0) may contribute to this effect. Selection could also occur through interactions between secretory protein moieties and receptors; for example, the putative Golgi receptor muclin may interact with sulfated, *O*-glycosylated zymogens as a means to concentrate them in budding compartments.[98]

Secretory Granule Exocytosis

Regulated secretion arises from a storage pool that excludes newly synthesized secretory proteins that accumulate during the intestinal interdigestive phase. It is able to release 15–30 percent of the gland's secretory protein content by classical regulated exocytosis and is maximally stimulated following ingestion of a meal when release of massive amounts of digestive enzymes and zymogens (considering zymogens as a category of digestive enzyme) are required at rates greater than can be attained by protein synthesis alone. ZG exocytosis does not appear to contribute to resting secretion.

The purpose of these distinct mechanisms of secretion is unknown, but they likely serve to both ensure that some digestive enzymes will be present in the small intestine at all times and provide a secretory response that is proportional to luminal nutrients. A minor regulated pathway could serve to increase enzyme secretion in response to smaller quantities of food than presented by a full meal. Finally, a novel role for the constitutive and minor regulated compartments is that they might deliver the t-SNARES necessary for ZG fusion with the apical membrane.[99]

The release of ZG content into the lumen of the acinus requires fusion of the vesicular membrane with the apical plasma membrane. Four key steps are likely involved in this process: approximation of secretory granules in the apical region of the acinar cell, near the plasma membrane; tethering of secretory granules to the plasma membrane; docking and priming that involves SNARE proteins; and the final, calcium-dependent fusion event.[97]

The initial step, movement of the secretory granule from its site of formation in the *trans*-Golgi to the apical region of the cell, likely requires active involvement of contractile elements, particularly actin and associated motor proteins, in movement of the ZG to its apical plasma membrane target. It should be noted that in the resting interphase between rounds of exocytosis, an apical actin terminal web presumably negatively regulates resting secretion in that an actin mesh is always found between ZGs and the apical plasma membrane.[100–102] However, other actin roles have been proposed.[103] For example, an additional actin pool, regulated by the actin-polymerizing formin mDia1, mediates the final movement of ZGs to the apical membrane.[104] After reaching this most apical domain, the membrane of the ZG must recognize and become tethered to the apical plasma membrane prior to fusion. This implies that the actin meshwork beneath the apical membrane must be dissociated for close membrane apposition to occur, though additional steps, including overcoming fusion barriers, are needed before fusion can occur.

The role of microtubules in pancreatic acinar cell secretion is less clear. Microtubules are long, dynamic cytoskeletal structures composed of heterodimer polymers of α- and β-tubulin that undergo cycles of regulated polymerization and depolymerization.[105,106] During polymerization, β-subunits of one tubulin heterodimer contact the α-subunits of the next dimer resulting in one end of the microfilament having the α-subunits exposed and the other end with β-subunits exposed; these ends are designated the minus (−) and plus (+) ends, respectively. Microtubule motor proteins, such as kinesins and dynein, associate with select intracellular cargo and utilize ATP to facilitate transport along the microtubules.[107] Studies in the mid-1970s found that treatment of rodent pancreas both *in vivo* and *ex vivo* with the microtubule-destabilizing compounds vinblastine and colchicine significantly inhibited, but did not fully prevent, secretagogue-stimulated amylase secretion.[108–111] Later work demonstrated that the minus ends of microtubules are anchored along the apical membrane and extend radially to the plus ends anchored in the basal cytoplasm.[112] Both kinesin and dynein have been identified in acini, though their localization (on ZGs, Golgi, ER) and effects on secretion are debated.[113–115] Interestingly, some studies show kinesin associating with ZGs in apical regions in response to secretory stimulation, an unexpected finding given that kinesins travel along microtubules in a minus-to-plus direction (i.e., anterograde in acini). Marlowe et al. speculate that organelles could contain both kinesin and dynein for bidirectional movement, and that kinesin could be involved in post-exocytosis membrane retrieval.[112] Additional high-resolution studies are needed to characterize the population and polarization of microtubules at the acinar apex to understand how kinesin shapes the secretory response.

The SNARE hypothesis for membrane recognition and fusion, which appears to be a generalized mechanism for all cells examined, is particularly relevant for the pancreatic acinar cell where specific interactions between the ZG membrane and the apical plasma membrane ensure that exocytosis of digestive enzymes and proenzymes occurs into the acinar lumen.[116]

Two populations of ZGs have been identified in pancreatic acinar cells, those enriched in the SNARE protein VAMP2 and those enriched in Endobrevin/VAMP8.[97] According to the SNARE hypothesis for exocytosis, VAMPs on the granule membrane interact with a syntaxin isoform and a SNAP isoform on the plasma membrane. Acinar cell apical plasma membrane contains syntaxins 2 and 4 and, interestingly, both apical plasma membrane and ZGs express SNAP23 and SNAP29.[117,118] Co-immunoprecipitation analysis revealed that VAMP2 ZGs interact with plasma membrane syntaxin 2 and SNAP23, whereas VAMP8 ZGs interact with apical membrane syntaxin 4 and SNAP23. ZG-ZG compound exocytosis that occurs during secretion was shown to involve VAMP8/syntaxin 3 and SNAP23 all of which are present on ZGs. A role for SNAP29 in acinar cell function has not been described.

To determine the roles of these VAMPs in the acinar cell, VAMP8 knockout mice together with adenoviral expression of tetanus toxin to selectively cleave VAMP2 were used to delineate the roles of VAMP2 versus VAMP8 ZG exocytosis during secretagogue stimulated secretion.[97] Results supported that VAMP2 and VAMP8 are the primary ZG SNAREs mediating stimulated but not basal secretion. Moreover, measuring acinar cell secretion over time in a perifusion apparatus revealed that VAMP2 ZG mediated an early immediate phase of secretion that peaks at 2 minutes and begins to decline followed by VAMP8 ZG-mediated second prolonged phase of secretion that peaks at 5 minutes and decays over 20 minutes. A subsequent study identified that the VAMP8, but not the VAMP2-mediated pathway, was primarily inhibited during high cholecystokinin (CCK)-induced acute acinar pancreatitis and that knockout of VAMP8 prevented most of the high-dose CCK-mediated secretory inhibition and fully blocked the accumulation of active trypsin in acinar cells.[119]

ZGs have been shown to undergo exocytosis at the basolateral plasma membrane during acute pancreatitis.[120] Evidence suggests that VAMP8 normally inhibits basolateral exocytosis; however, PKC-mediated phosphorylation of the SNARE accessory protein Munc18c allows VAMP8 to mediate basolateral exocytosis in a SNARE complex involving VAMP8/syntaxin 4/SNAP23.[121] Presumably, release of digestive enzymes to the extracellular space enhances tissue damage thereby exacerbating disease progress.

A unique mechanism that has been observed in acinar cells is that of sequential compound exocytosis, whereby secretory granules first fuse with apical plasma membrane (primary granule fusion) and this is followed by sequential fusion of other (secondary and tertiary) granules onto these primary granules.[118,122] Though the physiological meaning of compound exocytosis is not currently understood, it may facilitate more rapid and efficient release of ZG content since the acinar lumen is a small fraction of the total surface area limiting the number of granules that can fuse with the apical membrane at any one time. What triggers sequential granule fusion following primary granule fusion at the apical plasma membrane is unclear but may involve changes in pH, changes in phospholipid content, acquisition of different SNARE proteins, and so on, and needs further research.

References

1. Gemmer M, Förster F. A clearer picture of the ER translocon complex. J Cell Sci 2020; 133:1–11.
2. Akopian D, Shen K, Zhang X, Shan S. Signal recognition particle: an essential protein-targeting machine. Annu Rev Biochem 2013; 82:693–721.
3. Voss M, Schröder B, Fluhrer R. Mechanism, specificity, and physiology of signal peptide peptidase (SPP) and SPP-like proteases. Biochim Biophys Acta - Biomembranes 2013; 1828:2828–2839.
4. Palade G. Intracellular aspects of the process of protein synthesis. Science 1975; 189:347–358.
5. Reinisch KM, De Camilli P. SMP-domain proteins at membrane contact sites: structure and function. Biochim Biophys Acta 2016; 1861:924–927.
6. Son A, Park S, Shin DM, Muallem S. Orai1 and STIM1 in ER/PM junctions: roles in pancreatic cell function and dysfunction. Am J Physiol Cell Physiol 2016; 310:C414–C422.
7. Nyathi Y, Wilkinson BM, Pool MR. Co-translational targeting and translocation of proteins to the endoplasmic reticulum. Biochim Biophys Acta (BBA) - Mol Cell Res 2013; 1833:2392–2402.
8. Waldron RT, Su H-Y, Piplani H, Capri J, Cohn W, Whitelegge JP, Faull KF, Sakkiah S, Abrol R, Yang W, Zhou B, Freeman MR, et al. Ethanol induced disordering of pancreatic acinar cell endoplasmic reticulum: an ER stress/defective unfolded protein response model. Cell Mol Gastroenterol Hepatol 2018; 5:479–497.
9. Grey MJ. Proteomic study defines how alcohol alters ER structure and redox proteome to trigger ER stress and acinar cell pathology in pancreatitis. Cell Mol Gastroenterol Hepatol 2018; 5:640–641.
10. Desilva MG, Notkins AL, Lan MS. Molecular characterization of a pancreas-specific protein disulfide isomerase, PDIp. DNA Cell Biol 1997; 16:269–274.
11. Fu X-M, Dai X, Ding J, Zhu BT. Pancreas-specific protein disulfide isomerase has a cell type-specific expression in various mouse tissues and is absent in human pancreatic adenocarcinoma cells: implications for its functions. J Mol Histol 2009; 40:189–199.
12. Fujimoto T, Nakamura O, Saito M, Tsuru A, Matsumoto M, Kohno K, Inaba K, Kadokura H. Identification of the physiological substrates of PDIp, a pancreas-specific protein-disulfide isomerase family member. J Biol Chem 2018; 293:18421–18433.

13. Reily C, Stewart TJ, Renfrow MB, Novak J. Glycosylation in health and disease. Nat Rev Nephrol 2019; 15:346–366.

14. Wendorf P, Linder D, Sziegoleit A, Geyer R. Carbohydrate structure of human pancreatic elastase 1. Biochem J 1991; 278:505–514.

15. Abouakil N, Mas E, Bruneau N, Benajiba A, Lombardo D. Bile salt-dependent lipase biosynthesis in rat pancreatic AR 4–2 J cells. Essential requirement of N-linked oligosaccharide for secretion and expression of a fully active enzyme. J Biol Chem 1993; 268:25755–25763.

16. Bruneau N, Lombardo D. Chaperone function of a Grp 94-related protein for folding and transport of the pancreatic bile salt-dependent lipase. J Biol Chem 1995; 270:13524–13533.

17. Cooley MM, Thomas DDH, Deans K, Peng Y, Lugea A, Pandol SJ, Puglielli L, Groblewski GE. Deficient endoplasmic reticulum acetyl-CoA import in pancreatic acinar cells leads to chronic pancreatitis. Cell Mol Gastroenterol Hepatol 2020; 11:725–738

18. Peng Y, Li M, Clarkson BD, Pehar M, Lao PJ, Hillmer AT, Barnhart TE, Christian BT, Mitchell HA, Bendlin BB, Sandor M, Puglielli L. Deficient import of acetyl-CoA into the ER lumen causes neurodegeneration and propensity to infections, inflammation, and cancer. J Neurosci 2014; 34:6772–6789.

19. Pehar M, Lehnus M, Karst A, Puglielli L. Proteomic assessment shows that many endoplasmic reticulum (ER)-resident proteins are targeted by N^ε-lysine acetylation in the lumen of the organelle and predicts broad biological impact. J Biol Chem 2012; 287:22436–22440.

20. Ye R, Mareninova OA, Barron E, Wang M, Hinton DR, Pandol SJ, Lee AS. Grp78 heterozygosity regulates chaperone balance in exocrine pancreas with differential response to cerulein-induced acute pancreatitis. Am J Pathol 2010; 177:2827–2836.

21. Kubisch CH, Logsdon CD. Endoplasmic reticulum stress and the pancreatic acinar cell. Expert Rev Gastroenterol Hepatol 2008; 2:249–260.

22. Waldron RT, Pandol S, Lugea A, Groblewski G. Endoplasmic reticulum stress and the unfolded protein response in exocrine pancreas physiology and pancreatitis. Pancreapedia: The Exocrine Pancreas Knowledge Base, 2015. DOI: 10.3998/panc.2015.41.

23. Yoshida H, Matsui T, Yamamoto A, Okada T, Mori K. XBP1 mRNA is induced by ATF6 and spliced by IRE1 in response to ER stress to produce a highly active transcription factor. Cell 2001; 107:881–891.

24. Calfon M, Zeng H, Urano F, Till JH, Hubbard SR, Harding HP, Clark SG, Ron D. IRE1 couples endoplasmic reticulum load to secretory capacity by processing the XBP-1 mRNA. Nature 2002; 415:92–96.

25. Haze K, Yoshida H, Yanagi H, Yura T, Mori K. Mammalian transcription factor ATF6 is synthesized as a transmembrane protein and activated by proteolysis in response to endoplasmic reticulum stress. Mol Biol Cell 1999; 10:3787–3799.

26. Schindler AJ, Schekman R. In vitro reconstitution of ER-stress induced ATF6 transport in COPII vesicles. Proc Natl Acad Sci U S A 2009; 106:17775–17780.

27. Adachi Y, Yamamoto K, Okada T, Yoshida H, Harada A, Mori K. ATF6 is a transcription factor specializing in the regulation of quality control proteins in the endoplasmic reticulum. Cell Struct Funct 2008; 33:75–89.

28. Bommiasamy H, Back SH, Fagone P, Lee K, Meshinchi S, Vink E, Sriburi R, Frank M, Jackowski S, Kaufman RJ, Brewer JW. ATF6alpha induces XBP1-independent expansion of the endoplasmic reticulum. J Cell Sci 2009; 122:1626–1636.

29. Shaffer AL, Shapiro-Shelef M, Iwakoshi NN, Lee A-H, Qian S-B, Zhao H, Yu X, Yang L, Tan BK, Rosenwald A, Hurt EM, Petroulakis E, et al. XBP1, downstream of Blimp-1, expands the secretory apparatus and other organelles, and increases protein synthesis in plasma cell differentiation. Immunity 2004; 21:81–93.

30. Lee A-H, Iwakoshi NN, Glimcher LH. XBP-1 regulates a subset of endoplasmic reticulum resident chaperone genes in the unfolded protein response. Mol Cell Biol 2003; 23:7448–7459.

31. Lee A-H, Chu GC, Iwakoshi NN, Glimcher LH. XBP-1 is required for biogenesis of cellular secretory machinery of exocrine glands. EMBO J 2005; 24:4368–4380.

32. Kowalik AS, Johnson CL, Chadi SA, Weston JY, Fazio EN, Pin CL. Mice lacking the transcription factor Mist1 exhibit an altered stress response and increased sensitivity to caerulein-induced pancreatitis. Am J Physiol Gastrointest Liver Physiol 2007; 292:G1123–G1132.

33. Huh WJ, Esen E, Geahlen JH, Bredemeyer AJ, Lee A-H, Shi G, Konieczny SF, Glimcher LH, Mills JC. XBP1 controls maturation of gastric zymogenic cells by induction of MIST1 and expansion of the rough endoplasmic reticulum. Gastroenterology 2010; 139:2038–2049.

34. Tian X, Jin RU, Bredemeyer AJ, Oates EJ, Błażewska KM, McKenna CE, Mills JC. RAB26 and RAB3D are direct transcriptional targets of MIST1 that regulate exocrine granule maturation. Mol Cell Biol 2010; 30:1269–1284.

35. Reimold AM, Etkin A, Clauss I, Perkins A, Friend DS, Zhang J, Horton HF, Scott A, Orkin SH, Byrne MC, Grusby MJ, Glimcher LH. An essential role in liver development for transcription factor XBP-1. Genes Dev 2000; 14:152–157.

36. Lugea A, Tischler D, Nguyen J, Gong J, Gukovsky I, French SW, Gorelick FS, Pandol SJ. Adaptive unfolded protein response attenuates alcohol-induced pancreatic damage. Gastroenterology 2011; 140:987–997.

37. Pehar M, Jonas MC, Hare TM, Puglielli L. SLC33A1/AT-1 protein regulates the induction of autophagy downstream of IRE1/XBP1 pathway. J Biol Chem 2012; 287:29921–29930.

38. Zhou L, Tan J-h, Cao R-c, Xu J, Chen X-m, Qi Z-c, Zhou S-y, Li S-b, Mo Q-x, Li Z-w, Zhang G-w. ATF6 regulates the development of chronic pancreatitis by inducing p53-mediated apoptosis. Cell Death Dis 2019; 10:1–12.

39. Iida K, Li Y, McGrath BC, Frank A, Cavener DR. PERK eIF2α kinase is required to regulate the viability of the exocrine pancreas in mice. BMC Cell Biol 2007; 8:38.

40. Gao Y, Sartori DJ, Li C, Yu Q-C, Kushner JA, Simon MC, Diehl JA. PERK is required in the adult pancreas and is essential for maintenance of glucose homeostasis. Mol Cell Biol 2012; 32:5129–5139.

41. Suyama K, Ohmuraya M, Hirota M, Ozaki N, Ida S, Endo M, Araki K, Gotoh T, Baba H, Yamamura K-I. C/EBP homologous protein is crucial for the acceleration of

experimental pancreatitis. Biochem Biophys Res Commun 2008; 367:176–182.

42. Maurel M, Chevet E, Tavernier J, Gerlo S. Getting RIDD of RNA: IRE1 in cell fate regulation. Trends Biochem Sci 2014; 39:245–254.

43. Lipson KL, Ghosh R, Urano F. The role of IRE1α in the degradation of insulin mRNA in pancreatic beta-cells. PloS One 2008; 3:1–7.

44. Bhattacharya A, Qi L. ER-associated degradation in health and disease—from substrate to organism. J Cell Sci 2019; 132:1–10.

45. Sun S, Shi G, Han X, Francisco AB, Ji Y, Mendonça N, Liu X, Locasale JW, Simpson KW, Duhamel GE, Kersten S, Yates JR, et al. Sel1L is indispensable for mammalian endoplasmic reticulum-associated degradation, endoplasmic reticulum homeostasis, and survival. Proc Natl Acad Sci U S A 2014; 111:E582–E591.

46. Wilkinson S. Emerging principles of selective ER autophagy. J Mol Biol 2020; 432:185–205.

47. Smith MD, Harley ME, Kemp AJ, Wills J, Lee M, Arends M, von Kriegsheim A, Behrends C, Wilkinson S. CCPG1 is a non-canonical autophagy cargo receptor essential for ER-phagy and pancreatic ER proteostasis. Dev Cell 2018; 44:217–232.

48. Smith MD, Wilkinson S. CCPG1, an unconventional cargo receptor for ER-phagy, maintains pancreatic acinar cell health. Mol Cell Oncol 2018; 5:e1441631.

49. Lahiri V, Klionsky DJ. CCPG1 is a noncanonical autophagy cargo receptor essential for reticulophagy and pancreatic ER proteostasis. Autophagy 2018; 14:1107–1109.

50. Farrugia MA, Puglielli L. Nε-lysine acetylation in the endoplasmic reticulum—a novel cellular mechanism that regulates proteostasis and autophagy. J Cell Sci 2018; 131:1–10.

51. Barlowe CK, Miller EA. Secretory protein biogenesis and traffic in the early secretory pathway. Genetics 2013; 193:383–410.

52. Kurokawa K, Nakano A. The ER exit sites are specialized ER zones for the transport of cargo proteins from the ER to the Golgi apparatus. J Biochem 2019; 165:109–114.

53. Budnik A, Stephens DJ. ER exit sites-localization and control of COPII vesicle formation. FEBS Lett 2009; 583:3796–3803.

54. Bannykh SI, Rowe T, Balch WE. The organization of endoplasmic reticulum export complexes. J Cell Biol 1996; 135:19–35.

55. Jensen D, Schekman R. COPII-mediated vesicle formation at a glance. J Cell Sci 2011; 124:1–4.

56. Matsuoka K, Schekman R, Orci L, Heuser JE. Surface structure of the COPII-coated vesicle. Proc Natl Acad Sci U S A 2001; 98:13705–13709.

57. Appenzeller-Herzog C, Hauri H-P. The ER-Golgi intermediate compartment (ERGIC): in search of its identity and function. J Cell Sci 2006; 119:2173–2183.

58. Martínez-Menárguez JA, Geuze HJ, Slot JW, Klumperman J. Vesicular tubular clusters between the ER and Golgi mediate concentration of soluble secretory proteins by exclusion from COPI-coated vesicles. Cell 1999; 98:81–90.

59. Malkus P, Jiang F, Schekman R. Concentrative sorting of secretory cargo proteins into COPII-coated vesicles. J Cell Biol 2002; 159:915–921.

60. Khoriaty R, Everett L, Chase J, Zhu G, Hoenerhoff M, McKnight B, Vasievich MP, Zhang B, Tomberg K, Williams J, Maillard I, Ginsburg D. Pancreatic SEC23B deficiency is sufficient to explain the perinatal lethality of germline SEC23B deficiency in mice. Sci Rep 2016; 6:27802.

61. Khoriaty R, Vogel N, Hoenerhoff MJ, Sans MD, Zhu G, Everett L, Nelson B, Durairaj H, McKnight B, Zhang B, Ernst SA, Ginsburg D, et al. SEC23B is required for pancreatic acinar cell function in adult mice. Mol Biol Cell 2017; 28:2146–2154.

62. Hauri HP, Kappeler F, Andersson H, Appenzeller C. ERGIC-53 and traffic in the secretory pathway. J Cell Sci 2000; 113 (Pt 4):587–596.

63. Arakel EC, Schwappach B. Formation of COPI-coated vesicles at a glance. J Cell Sci 2018; 131:1–9.

64. Oprins A, Rabouille C, Posthuma G, Klumperman J, Geuze HJ, Slot JW. The ER to Golgi interface is the major concentration site of secretory proteins in the exocrine pancreatic cell. Traffic 2001; 2:831–838.

65. Sesso A, de Faria FP, Iwamura ES, Corrêa H. A three-dimensional reconstruction study of the rough ER-Golgi interface in serial thin sections of the pancreatic acinar cell of the rat. J Cell Sci 1994; 107:517–528.

66. Glick BS, Luini A. Models for Golgi traffic: a critical assessment. Cold Spring Harb Perspect Biol 2011; 1–15.

67. Papanikou E, Glick BS. Golgi compartmentation and identity. Curr Opin Cell Biol 2014; 29:74–81.

68. Pantazopoulou A, Glick BS. A kinetic view of membrane traffic pathways can transcend the classical view of Golgi compartments. Front Cell Dev Biol 2019; 7:153–165.

69. Saraste J, Prydz K. A new look at the functional organization of the Golgi ribbon. Front Cell Dev Biol 2019; 7:1–21.

70. Klumperman J. Architecture of the mammalian Golgi. Cold Spring Harb Perspect Biol 2011; 3:1–19

71. Rambourg A, Clermont Y, Hermo L. Formation of secretion granules in the Golgi apparatus of pancreatic acinar cells of the rat. Am J Anat 1988; 183:187–199.

72. McMahon HT, Mills IG. COP and clathrin-coated vesicle budding: different pathways, common approaches. Curr Opin Cell Biol 2004; 16:379–391.

73. Barois N, Bakke O. The adaptor protein AP-4 as a component of the clathrin coat machinery: a morphological study. Biochem J 2005; 385:503–510.

74. Bonifacino JS. The GGA proteins: adaptors on the move. Nat Rev Mol Cell Biol 2004; 5:23–32.

75. Slot JW, Geuze HJ. Immunoelectron microscopic exploration of the Golgi complex. J Histochem Cytochem 1983; 31:1049–1056.

76. Tokhtaeva E, Mareninova OA, Gukovskaya AS, Vagin O. Analysis of N- and O-glycosylation of lysosomal glycoproteins. Methods Mol Biol 2017; 1594:35–42.

77. Wolters-Eisfeld G, Mercanoglu B, Hofmann BT, Wolpers T, Schnabel C, Harder S, Steffen P, Bachmann K, Steglich B, Schrader J, Gagliani N, Schlüter H, et al. Loss of complex O-glycosylation impairs exocrine pancreatic function and induces MODY8-like diabetes in mice. Exp Mol Med 2018; 50:1–13.

78. Brown WJ, Farquhar MG. The mannose-6-phosphate receptor for lysosomal enzymes is concentrated in cis Golgi cisternae. Cell 1984; 36:295–307.

79. Klumperman J, Kuliawat R, Griffith JM, Geuze HJ, Arvan P. Mannose 6–phosphate receptors are sorted from immature secretory granules via adaptor protein AP-1, clathrin, and syntaxin 6–positive vesicles. J Cell Biol 1998; 141:359–371.

80. Braulke T, Bonifacino JS. Sorting of lysosomal proteins. Biochim Biophys Acta 2009; 1793:605–614.

81. Jamieson JD, Palade GE. Intracellular transport of secretory proteins in the pancreatic exocrine cell. I. Role of the peripheral elements of the Golgi complex. J Cell Biol 1967; 34:577–596.

82. Jamieson JD, Palade GE. Synthesis, intracellular transport, and discharge of secretory proteins in stimulated pancreatic exocrine cells. J Cell Biol 1971; 50:135–158.

83. Wendler F, Page L, Urbé S, Tooze SA. Homotypic fusion of immature secretory granules during maturation requires syntaxin 6. Mol Biol Cell 2001; 12:1699–1709.

84. Ahras M, Otto GP, Tooze SA. Synaptotagmin IV is necessary for the maturation of secretory granules in PC12 cells. J Cell Biol 2006; 173:241–251.

85. Hammel I, Lagunoff D, Galli SJ. Regulation of secretory granule size by the precise generation and fusion of unit granules. J Cell Mol Med 2010; 14:1904–1916.

86. Kostenko S, Heu CC, Yaron JR, Singh G, de Oliveira C, Muller WJ, Singh VP. c-Src regulates cargo transit via the Golgi in pancreatic acinar cells. Sci Rep 2018; 8:11903.

87. von Zastrow M, Castle JD. Protein sorting among two distinct export pathways occurs from the content of maturing exocrine storage granules. J Cell Biol 1987; 105:2675–2684.

88. Arvan P, Castle JD. Phasic release of newly synthesized secretory proteins in the unstimulated rat exocrine pancreas. J Cell Biol 1987; 104:243–252.

89. Castle JD. Sorting and secretory pathways in exocrine cells. Am J Respir Cell Mol Biol 1990; 2:119–126.

90. Castle JD, Castle AM. Two regulated secretory pathways for newly synthesized parotid salivary proteins are distinguished by doses of secretagogues. J Cell Sci 1996; 109:2591–2599.

91. Hendricks LC, McClanahan SL, Palade GE, Farquhar MG. Brefeldin A affects early events but does not affect late events along the exocytic pathway in pancreatic acinar cells. Proc Natl Acad Sci U S A 1992; 89:7242–7246.

92. Huang AY, Castle AM, Hinton BT, Castle JD. Resting (basal) secretion of proteins is provided by the minor regulated and constitutive-like pathways and not granule exocytosis in parotid acinar cells. J Biol Chem 2001; 276:22296–22306.

93. Thomas DDH, Kaspar KM, Taft WB, Weng N, Rodenkirch LA, Groblewski GE. Identification of annexin VI as a Ca²⁺-sensitive CRHSP-28-binding protein in pancreatic acinar cells. J Biol Chem 2002; 277:35496–35502.

94. Thomas DDH, Weng N, Groblewski GE. Secretagogue-induced translocation of CRHSP-28 within an early apical endosomal compartment in acinar cells. Am J Physiol Gastrointest Liver Physiol 2004; 287:G253–G263.

95. Thomas DDH, Martin CL, Weng N, Byrne JA, Groblewski GE. Tumor protein D52 expression and Ca²⁺-dependent phosphorylation modulates lysosomal membrane protein trafficking to the plasma membrane. Am J Physiol Cell Physiol 2010; 298:C725–C739.

96. Messenger SW, Thomas DDH, Falkowski MA, Byrne JA, Gorelick FS, Groblewski GE. Tumor protein D52 controls trafficking of an apical endolysosomal secretory pathway in pancreatic acinar cells. Am J Physiol Gastrointest Liver Physiol 2013; 305:G439–G452.

97. Messenger SW, Falkowski MA, Thomas DDH, Jones EK, Hong W, Gaisano HY, Giasano HY, Boulis NM, Groblewski GE. Vesicle associated membrane protein 8 (VAMP8)-mediated zymogen granule exocytosis is dependent on endosomal trafficking via the constitutive-like secretory pathway. J Biol Chem 2014; 289:28040–28053.

98. Boulatnikov I, De Lisle RC. Binding of the Golgi sorting receptor muclin to pancreatic zymogens through sulfated O-linked oligosaccharides. J Biol Chem 2004; 279:40918–40926.

99. Castle AM, Huang AY, Castle JD. The minor regulated pathway, a rapid component of salivary secretion, may provide docking/fusion sites for granule exocytosis at the apical surface of acinar cells. J Cell Sci 2002; 115:2963–2973.

100. O'Konski MS, Pandol SJ. Effects of caerulein on the apical cytoskeleton of the pancreatic acinar cell. J Clin Invest 1990; 86:1649–1657.

101. Muallem S, Kwiatkowska K, Xu X, Yin HL. Actin filament disassembly is a sufficient final trigger for exocytosis in non-excitable cells. J Cell Biol 1995; 128:589–598.

102. Valentijn KM, Gumkowski FD, Jamieson JD. The sub-apical actin cytoskeleton regulates secretion and membrane retrieval in pancreatic acinar cells. J Cell Sci 1999; 112:81–96.

103. Larina O, Bhat P, Pickett JA, Launikonis BS, Shah A, Kruger WA, Edwardson JM, Thorn P. Dynamic regulation of the large exocytotic fusion pore in pancreatic acinar cells. Mol Biol Cell 2007; 18:3502–3511.

104. Geron E, Schejter ED, Shilo B-Z. Directing exocrine secretory vesicles to the apical membrane by actin cables generated by the formin mDia1. Proc Natl Acad Sci U S A 2013; 110:10652–10657.

105. Brouhard GJ, Rice LM. Microtubule dynamics: an interplay of biochemistry and mechanics. Nat Rev Mol Cell Biol 2018; 19:451–463.

106. Jordan MA, Wilson L. Microtubules as a target for anticancer drugs. Nat Rev Cancer 2004; 4:253–265.

107. Caviston JP, Holzbaur ELF. Microtubule motors at the intersection of trafficking and transport. Trends Cell Biol 2006; 16:530–537.

108. Seybold J, Bieger W, Kern HF. Studies on intracellular transport of secretory proteins in the rat exocrine pancreas. II. Inhibition of antimicrotubular agents. Virchows Arch A Pathol Anat Histol 1975; 368:309–327.

109. Williams JA, Lee M. Microtubules and pancreatic amylase release by mouse pancreas in vitro. J Cell Biol 1976; 71:795–806.

110. Stock C, Launay JF, Grenier JF, Bauduin H. Pancreatic acinar cell changes induced by caerulein, vinblastine, deuterium oxide, and cytochalasin B in vitro. Lab Invest 1978; 38:157–164.

111. Nevalainen TJ. Inhibition of pancreatic exocrine secretion by vinblastine. Res Exp Med 1975; 165:163–168.

112. Marlowe KJ, Farshori P, Torgerson RR, Anderson KL, Miller LJ, McNiven MA. Changes in kinesin distribution and phosphorylation occur during regulated secretion in pancreatic acinar cells. Eur J Cell Biol 1998; 75:140–152.

113. Kraemer J, Schmitz F, Drenckhahn D. Cytoplasmic dynein and dynactin as likely candidates for microtubule-dependent apical targeting of pancreatic zymogen granules. Eur J Cell Biol 1999; 78:265–277.

114. Ueda N, Ohnishi H, Kanamaru C, Suzuki J, Tsuchida T, Mashima H, Yasuda H, Fujita T. Kinesin is involved in regulation of rat pancreatic amylase secretion. Gastroenterology 2000; 119:1123–1131.

115. Schnekenburger J, Weber I-A, Hahn D, Buchwalow I, Krüger B, Albrecht E, Domschke W, Lerch MM. The role of kinesin, dynein and microtubules in pancreatic secretion. Cell Mol Life Sci 2009; 66:2525–2537.

116. Rothman JE. The principle of membrane fusion in the cell (Nobel lecture). Angew Chem Int Ed Engl 2014; 53:12676–12694.

117. Falkowski MA, Thomas DDH, Messenger SW, Martin TF, Groblewski GE. Expression, localization, and functional role for synaptotagmins in pancreatic acinar cells. Am J Physiol Gastrointest Liver Physiol 2011; 301:G306–G316.

118. Behrendorff N, Dolai S, Hong W, Gaisano HY, Thorn P. Vesicle-associated membrane protein 8 (VAMP8) is a SNARE (soluble N-ethylmaleimide-sensitive factor attachment protein receptor) selectively required for sequential granule-to-granule fusion. J Biol Chem 2011; 286:29627–29634.

119. Messenger SW, Jones EK, Holthaus CL, Thomas DDH, Cooley MM, Byrne JA, Mareninova OA, Gukovskaya AS, Groblewski GE. Acute acinar pancreatitis blocks vesicle-associated membrane protein 8 (VAMP8)-dependent secretion, resulting in intracellular trypsin accumulation. J Biol Chem 2017; 292:7828–7839.

120. Gaisano HY, Gorelick FS. New insights into the mechanisms of pancreatitis. Gastroenterology 2009; 136:2040–2044.

121. Lam PPL, Cosen Binker LI, Lugea A, Pandol SJ, Gaisano HY. Alcohol redirects CCK-mediated apical exocytosis to the acinar basolateral membrane in alcoholic pancreatitis. Traffic 2007; 8:605–617.

122. Wäsle B, Turvey M, Larina O, Thorn P, Skepper J, Morton AJ, Edwardson JM. Syncollin is required for efficient zymogen granule exocytosis. Biochem J 2005; 385:721–727.

Chapter 7

Pancreatic digestive enzyme synthesis and its translational control

Maria Dolors Sans

Department of Molecular and Integrative Physiology, University of Michigan Medical School, Ann Arbor, MI-48109

Protein synthesis plays a central role in provision of pancreatic digestive enzymes and the maintenance of the pancreas. Both the mRNA profile and relative content of newly synthesized proteins are dominated by digestive enzymes. In the mature pancreas, about 90 percent of protein synthesis has been estimated to be devoted to a mixture of about 20 digestive enzymes.[1] Whether the acinar cell can regulate digestive enzyme synthesis independent of the synthesis of cellular structural proteins is unclear. Pancreatic protein synthesis is regulated to match digestive enzyme synthesis to dietary need. Although the identity of individual synthesized proteins and long-term regulation is primarily determined by transcriptional regulation, the short-term extent, or rate of protein synthesis, is regulated at the translational level,[2] because it needs to be immediate, flexible, and reversible.[3]

In general, the gastrointestinal tract, including the exocrine pancreas, atrophies in the absence of food, and protein synthesis that occurs in response to food intake is required to maintain normal function. Individual dietary components, like protein and amino acids,[4,5] regulate pancreatic protein synthesis in mice and rats.[6,7] In humans, feeding increases both the rate of secretion and synthesis of digestive enzymes, although the rate of zymogen turnover remains fairly constant during feeding and fasting.[8] Pancreatic protein synthesis is also regulated by hormones such as cholecystokinin (CCK) and insulin, as well as by neural stimulation, all of which are influenced by food intake (**Figure 1**).

This review highlights how dietary elements and hormones affect intracellular effectors, such as mTORC1 and intracellular calcium concentration, to regulate certain steps of the pancreatic protein translational machinery, at the initiation and elongation levels. When stimulatory conditions are present, pancreatic protein synthesis is mainly stimulated through the mTORC1 pathway, but when there is an overstimulation of the pancreas, or a dietary imbalance, there is an increase on intracellular calcium concentrations, and cell stress, that trigger a signal which inhibits protein synthesis. These mechanisms coincide with stress in the endoplasmic reticulum (ER stress), resulting in unfolded/misfolded protein products retained in the ER and the stimulation of several cell survival and cell death mechanisms.

Protein Synthesis Associated Mechanisms and Their Regulation

The acinar cell of the exocrine pancreas has the greatest rate of protein synthesis of any mammalian organ, and it has long been used as a cell model to study the protein synthesis mechanisms in mammalian cells.[9] Cooley et al. nicely reviewed the protein synthesis pathways to secretion in pancreatic acinar cells, and we refer the reader to their publication on the Pancreapedia,[10] and in Chapter 6 of this book, for more detailed information.

Briefly, in eukaryotic cells, the ribosomes—composed of two subunits: large (60S) and small (40S)—are the central elements of the protein synthesis machinery and translate the sequence of an mRNA into the amino acid sequence of a protein. To initiate the process, the two ribosomal subunits are recruited together, with the mRNA and tRNA, by different proteins and eukaryotic initiation factors (eIFs) on the cytosolic site of the ER membrane and start the addition of the first amino acid to the peptide chain. This complex continues the addition of amino acids to form a polypeptide, by a process called elongation, assisted by eukaryotic elongation factors,[8] and terminate the process by the help of termination factors (eTFs). The newly formed polypeptide chains will be inserted into the ER lumen through channels formed by proteins called translocons and are subject to several modifications. These include chaperoning and folding mechanisms that will yield mature proteins that are ready to be directed to their final destination.[11,12]

Pancreatic acinar cells devote most of their protein synthesis capacity to synthesize, and secrete, digestive enzymes.[1] These mechanisms are directed to match the need for digesting the amount and quality of the different dietary elements. Meal-to-meal regulation needs to be

e-mail: mdsansg@umich.edu

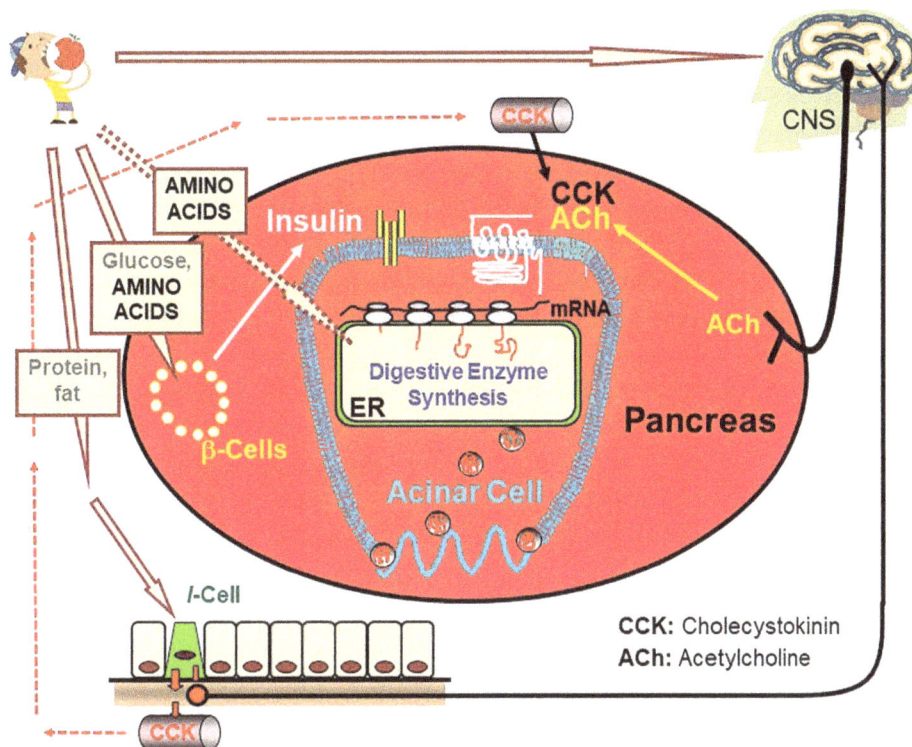

Figure 1. Overview of the main physiological stimulators of pancreatic digestive enzyme synthesis. Meal feeding stimulates the central nervous system (CNS), which, in turn, stimulates digestive enzyme synthesis via the vagus and acetylcholine release in the pancreas. Meal feeding stimulates the release of CCK from the "I" cells in the duodenum, which will directly stimulate the exocrine pancreas and vagal afferents, and insulin from the β-cells in the islets of Langerhans, which can also contribute to the regulation of pancreatic enzyme synthesis. Finally, some dietary nutrients, such as amino acids, will directly stimulate pancreatic digestive enzyme synthesis (figure modified from reference [4]).

immediate, reversible, and flexible, and this is achieved by posttranscriptional processes directed at the regulation of the mRNA translation into protein.[2]

The ER is an exceptionally prominent organelle in the acinar cell and is a major site of protein synthesis and transport, protein folding, lipid and steroid synthesis, carbohydrate metabolism, and calcium storage.[11]

mRNA translation

Translation of mRNA into protein can be divided into three phases: initiation, elongation, and termination.[13,14] During initiation (**Figure 2**), methionyl-tRNA initiator tRNA and several initiation factors associate with the 40S ribosomal subunit to form the 43S preinitiation complex. This complex binds to mRNA and migrates to the correct AUG initiation codon followed by the attachment of the 60S ribosomal subunit. Key regulated initiation factors in this process are the guanine nucleotide exchange factor eIF2B which activates eIF2 and eIF4E which recognizes the 5'(m7G)-cap of the mRNA. eIF4E is present in cells largely bound to its binding protein (4E-BP1) and is released when 4E-BP1

is phosphorylated on multiple sites. eIF4E interacts with the scaffolding protein eIF4G and eIF4A to form the eIF4F complex. The binding of eIF4F to an m7G cap commits the translational apparatus to the translation of the mRNA, and the ribosome will seek the start codon to start translation. The activation of S6 kinase and the phosphorylation of the ribosomal protein S6 enhance the translation of a specific set of mRNAs. In the elongation process, amino acids from amino acyl-tRNAs are added to the growing peptide in the order dictated by the mRNA bound to the ribosome. The key regulatory molecule is elongation factor 2 (eEF2) which catalyzes the translocation of the peptidyl tRNA from the A-site to the P-site on the ribosome (**Figure 3**).[15] In the termination phase, the completed protein is released from the ribosome[14] and translocated into the ER lumen to continue its maturation steps, involving many chaperones and folding processes.

Posttranslational steps and associated processes

As mentioned above, a protein destined for secretion must undergo proper folding and modifications, with the

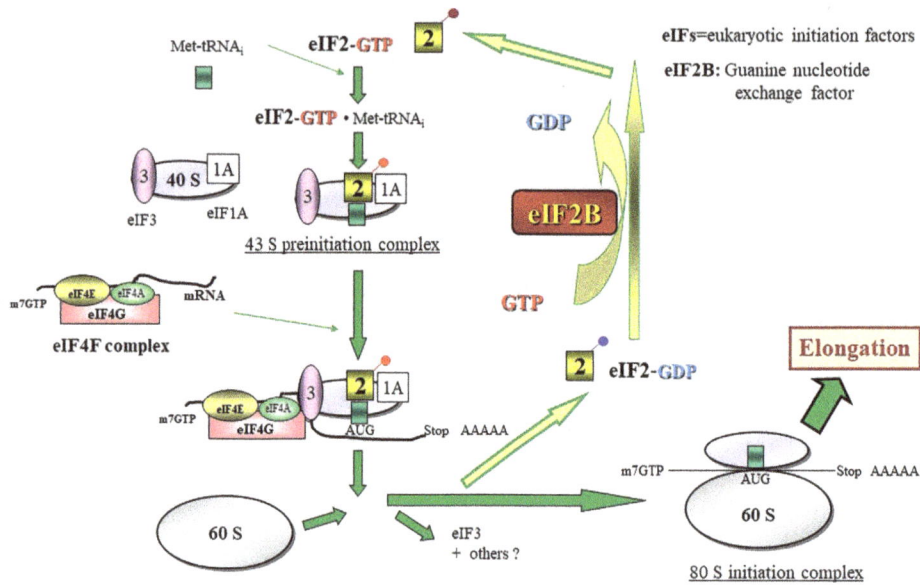

Figure 2. Initiation of protein synthesis in mammalian cells. During initiation, the initiator Met-tRNAi binds to a free 40S ribosomal subunit in a reaction requiring GTP bound to eIF2 to form the 43S preinitiation complex. In the next step, the eukaryotic initiation factor 4E (eIF4E) recognizes the capped end of the mRNA and interacts with the scaffolding protein eIF4G to form the eIF4F complex. After this, the 43S preinitiation complex will join the eIF4F complex and the ribosome will seek the start codon to start the process. A GTPase-activating protein promotes the hydrolysis of the GTP bound to eIF2 and releases eIF2-GDP. Regeneration of active eIF2-GTP is mediated by the guanine nucleotide-exchange factor eIF2B. The dissociation of the initiation factors allows the addition of the 60S subunit of the ribosome to form the 80S initiation complex, which is competent to enter elongation.

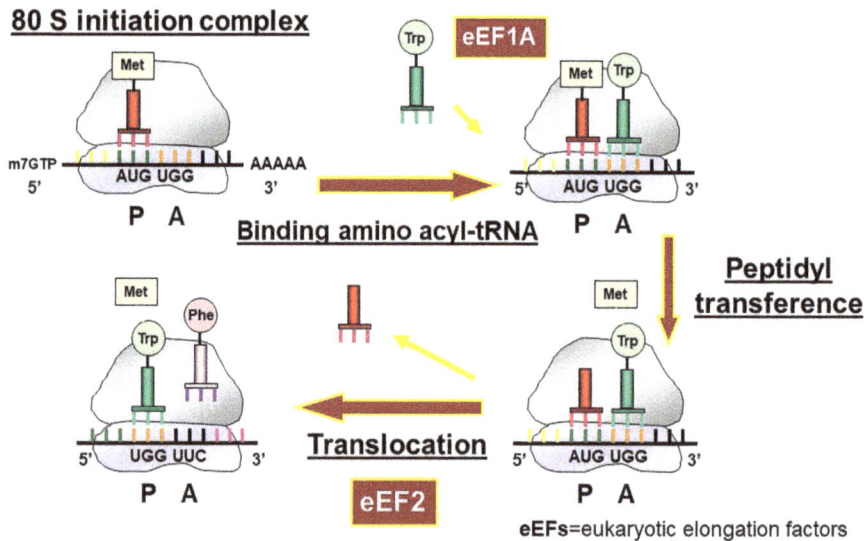

Figure 3. Protein elongation steps. The first step of elongation is the binding of an aminoacyl-tRNA to a vacant ribosomal A-site base pairing with the mRNA on the A-site. This process requires elongation factor 1A. The second step is peptidyl transference; the carboxyl end of the polypeptide chain uncouples from the tRNA molecule in the P-site and joins the amino acid linked to the tRNA molecule in the A-site by a peptide bond. The third step, translocation, is catalyzed by elongation factor 2 (eEF2) and involves the translocation of the peptidyl-tRNA in the A-site to the P-site as the ribosome moves exactly three nucleotides along the mRNA molecule. The elongation process will repeat this cycle of amino acid addition to elongate protein polypeptides.[14,16]

aid of chaperones and folding enzymes such as protein disulfide isomerase (PDI) and the immunoglobulin heavy-chain binding protein (BiP). These interactions occur after nascent protein synthesis and translocation into the ER lumen.[11] Despite all these specific and sophisticated controlled mechanisms, a fraction of proteins does not achieve native/mature and functional form and are either misfolded or aggregated.[17] At this point, these proteins can (1) remain in the ER or (2) enter the ER-associated degradation (ERAD) pathway mediated by the proteasome. These steps ensure that misfolded proteins do not inadvertently enter the secretory pathway.[18] Retention of misfolded proteins in the ER can induce ER stress[19,20] and a coordinated adaptive program called the Unfolded Protein Response (UPR).[21,22] The UPR activates specific mechanisms directed to retain balance and proper function of the ER and the cell by (1) inhibiting protein synthesis,[23] (2) upregulating protein folding by enhancing translation of ER chaperone and folding enzymes, and (3) activating degradation pathways associated with the ER—the ERAD.[24] If the balance is not restored, it can lead to cell death or apoptosis;[25] thus, achieving normal function is critical for cell's survival.

Three ER-resident transmembrane proteins have been identified as proximal sensors of ER stress: the kinase and endoribonuclease inositol-requiring element 1 (IRE1),[26] the PKR-like ER kinase (PERK)[27] that phosphorylates the α subunit of eukaryotic initiation factor 2 (eIF2α) on its Ser51 residue and inhibits translation initiation,[27] and the basic leucine-zipper-activating transcription factor 6 (ATF6).[28] Phosphorylation of eIF2α by PERK inhibits general protein translation but allows preferential translation of mRNAs encoding several short upstream open reading frames like the mRNA for the activating transcription factor 4 (ATF4).[29] IRE1α dimerization followed by autophosphorylation triggers its mRNase activity, which processes the mRNA encoding unspliced X box-binding protein 1 (XBP1u) to produce an active transcription factor, spliced (XBP1s), that controls the transcription of genes encoding proteins involved in protein folding.[30] The activation of the UPR may lead either to cell survival, by triggering the synthesis of ER chaperone proteins such as BiP and protein disulfide isomerase (PDI), along with a decrease in general protein translation, or to cell demise. The latter, occurs through the activation of programmed cell death signals.[21,30]

There are several connections to activation of ER stress response pathways and pathological human conditions.[31] A malfunction of the ER stress response caused by aging, genetic mutations, or environmental factors can result in various diseases such as diabetes, inflammation, cancer, pancreatitis (shown below), nonalcoholic fatty liver disease, and neurodegenerative disorders including Alzheimer's disease, Parkinson's disease, and bipolar disorder.[32–34] How ER stress response pathways play a role

in these pathologies is an active area of research, and various components of the stress response pathways are being investigated as potential therapeutic targets.[31,35]

Main intracellular regulators

Intracellular calcium concentration

Calcium is a widespread signaling molecule that can affect different processes, including localization, function, and association of proteins, either with other proteins, organelles, or nucleic acids. Ca^{2+}, in addition to its role as an intracellular mediator of cell-surface humoral interactions, may function prominently in the regulation of posttranscriptional protein synthesis in a variety of eukaryotic cell types.[11,36] Early studies hypothesized that the rate of protein synthesis could be modulated in intact cells by varying the concentration and subcellular distribution of intracellular calcium, and it was thought that it was controlled by free cytosolic calcium rather than the sequestered cation. However, Brostrom and Brostrom (1990)[37] proposed that maintenance of optimal rates of protein synthesis depends on the amount of calcium sequestered in the endoplasmic reticulum rather than free cytosolic calcium, and several other studies in different cell types confirmed this hypothesis.[38–40]

Maintenance of free versus bound Ca^{2+} balance in all cellular compartments is critical for many cellular functions. This balance can be achieved, in part, thanks to calcium-binding proteins located in the cytoplasm or in specific organelles that retain or release Ca^{2+} as needed. Some of these proteins (e.g., calnexin, calreticulin BiP, PDI) are located in the ER and, in addition to binding to calcium, function as chaperones and help with protein folding.[41] Integration of Ca^{2+} signaling in the lumen of the cellular reticular network including the ER, the mitochondria, the nucleus, and the cytoplasm provides integrated mechanisms for responding to cellular stresses by activation of appropriate coping responses.

In response to signal transduction-generated inositol trisphosphate, calcium is released to the cytoplasm as part of the signal transduction cascade.[11,42] Sequestered calcium can also be released experimentally by treating cells with calcium ionophores (A23187) or inhibitors of the microsomal calcium-dependent ATPase such as thapsigargin (Tg). All of these treatments disrupt protein folding in the ER and inhibit translation initiation, through an ER stress mechanism that involves phosphorylation of eIF2α by PKR and the inhibition of eIF2B activity.[21,37,43,44] It has also been hypothesized that perturbation of the translocon, rather than suppression of protein processing, initiates the signal emanating from the ER culminating in eIF2α phosphorylation and translational repression. Therefore, sequestered Ca^{2+} from the ER can be seen as moderating the rate of mRNA translation in many cell types, including pancreatic acinar cells.

The importance of maintaining Ca^{2+} homeostasis and appropriate adaptation to ER stress is underlined by the accumulating evidence that constant disturbing of Ca^{2+} homeostasis and chronic ER stress could lead to neurodegenerative disorders,[45] diabetes,[46,47] cardiac hypertrophy,[48] or cancer.[49]

The mTORC1 pathway

The mTORC1 pathway is essential for cells to maintain homeostasis, providing tight control between the synthesis and degradation of cellular components, and it is the allosteric target of the drug rapamycin that has clinical uses in organ transplantation, cardiology, and oncology.[50] mTORC1 connects environmental cues (nutrient and growth factor availability as well as stress) to metabolic processes in order to preserve cellular homeostasis. Under nutrient-rich conditions, mTORC1 promotes cell growth by stimulating biosynthetic pathways, including synthesis of proteins, lipids, and nucleotides, and by inhibiting cellular catabolism through repression of the autophagic pathway.[50] mTOR is an evolutionary conserved serine/threonine protein kinase that belongs to the PI3K-related kinase (PI3K) superfamily. This atypical kinase is the basis of two structurally and functionally different complexes termed mTOR complex 1 (mTORC1) and mTOR complex 2 (mTORC2).[51] In addition to the mTOR catalytic subunit, mTORC1 consists of regulatory-associated protein of mammalian target of rapamycin (Raptor) (a scaffold protein that is required for the correct subcellular localization of mTORC1),[52,53] mammalian lethal with Sec13 protein 8 (mLST8; also known as GßL) (which associates with the catalytic domain of mTOR and stabilizes the kinase activation loop),[52–54] and the two inhibitory subunits: proline-rich Akt substrate of 40 kDa (PRAS40)[55] and DEP domain containing mTOR-interacting protein (DEPTOR).[56] Both, DEPTOR and PRAS40 are inhibitory proteins; phosphorylation blocks this inhibition. PRAS40 represents an essential component for insulin activation of mTORC1. Raptor is essential to mTORC1 function and its genetic deletion leads to loss of activity.[1] mTORC2 contains, among other components, mTOR, mLST8, DEPTOR, the regulatory subunits mSin1, Rictor (rapamycin-insensitive companion of mTOR),[57] and regulates cell survival and proliferation primarily by phosphorylating several members of the AGC (PKA/PKG/PKC) family of protein kinases.[58,59]

Several proteins have been identified as substrates for phosphorylation by mTORC1. The first ones were proteins implicated in the control of mRNA translation, and this process remains one whose control by mTORC1 is best understood.[60] mTORC1 functions to activate several steps in mRNA translation (phosphorylates S6 kinases, the inhibitory eIF4E-binding proteins (4E-BPs), and the eIF4G). Its signaling requires amino acids (the precursors for protein synthesis) to activate these translation steps. Leucine is the most effective single amino acid that can stimulate mTORC1 signaling.[61] Finally, it has also been described that mTORC1 positively regulates translation elongation[60] and the protein degradation process through the proteasome, to increase the intracellular pool of amino acids, which will influence the rate of new protein synthesis[61] (**Figure 4**).

Deregulation of mTORC1 signaling increases the risk for metabolic diseases, including type 2 diabetes,[62] cancer, and epilepsy.[58] In the exocrine pancreas, stimulation of the mTORC1 pathway has been shown to be activated by several hormones and nutrients and leads the stimulation of acinar cell protein synthesis/digestive enzymes[63] (**Figure 4**). mTORC1 activation and signaling in pancreas is usually demonstrated by phosphorylation of downstream mediators, ribosomal protein S6 and 4E-BP1.

Experimental Analysis of Pancreatic Protein Synthesis

Pancreatic protein synthesis has been traditionally measured by administering isotopically labeled amino acids with subsequent measurement of the incorporation of label into newly synthesized protein. For animal studies, this is usually a radioactive isotope, while human studies have most often used stable isotopes.[8] Several animal studies have measured pancreatic protein synthesis after diet-induced stimulation,[2,64,65] hormonal stimulation (insulin, CCK),[5,66–74] or after acute pancreatitis.[75–80] Among these studies, there are some discrepancies on the outcome of the results that can be due to the different ways of administering the labeled amino acid as a tracer and because some were lacking an accurate assessment of the specific radioactivity of the precursor amino acid at the site of protein synthesis.

The primary assumption was that the experimental treatment does not alter the relationship between the labeling of the sampled pool and that of the aminoacyl-tRNA, the direct precursor of protein synthesis, on *in vivo* experiments, specially. For those studies, the readily accessible compartment pools, such as the intracellular free amino acids or plasma pools, to estimate precursor labeling were used. However, experimental conditions have the potential to alter precursor enrichment either by affecting the amount of labeled amino acid entering the cell or by affecting the contribution of unlabeled amino acids derived from protein degradation to the charging of aminoacyl-tRNA. In these cases, the problem of accurately determining precursor enrichment can be minimized by the *flooding dose* technique, originally described by Garlick et al.[1] and validated by Sweiry et al. in rat pancreas.[81]

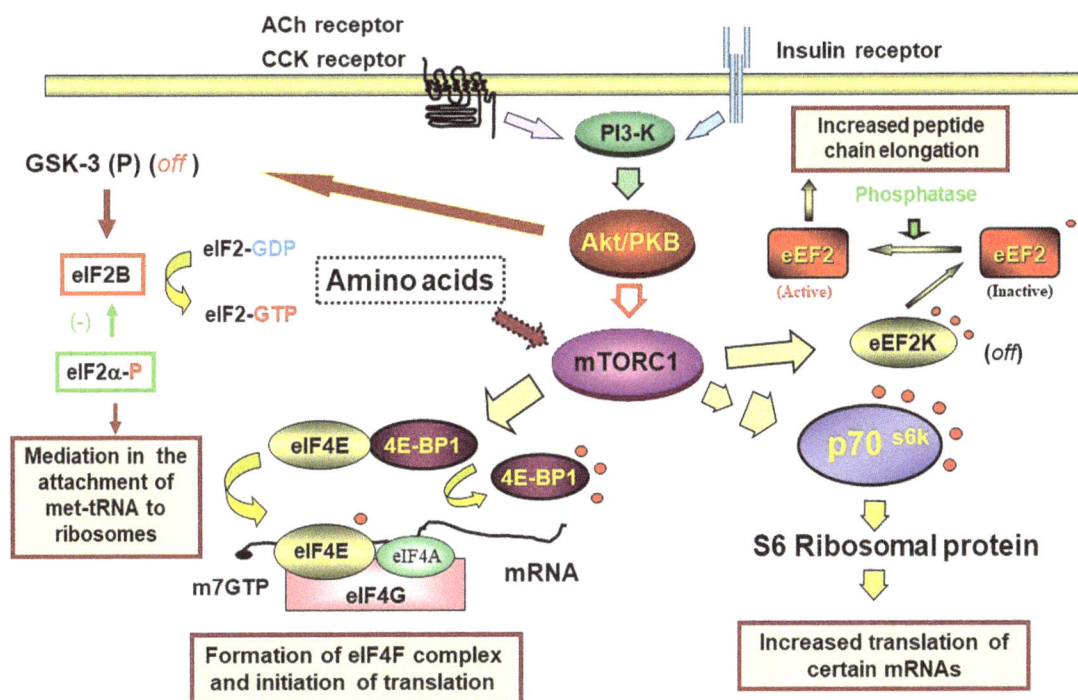

Figure 4. Scheme of the major mechanisms that regulate translation initiation in pancreatic acinar cells. G protein-coupled receptors, including CCK, stimulate translation initiation in pancreatic acinar cells through the phosphatidylinositol 3-kinase (PI3K) pathway. PI3K is activated in response to the active receptor and is believed to stimulate mTORC1 through phosphorylation of Akt/PKB, which, in turn, phosphorylates mTORC1. The round red knobs denote regulatory phosphate groups. mTORC1 is responsible (at least in part) for phosphorylating the eIF4E binding protein 4E-BP1 that allows the release of the mRNA cap-binding protein eIF4E, which is required for the formation of the eIF4F complex, which also includes eIF4G (a scaffolding protein) and eIF4A (a RNA helicase) and is necessary for the global increase in translation. mTOR also phosphorylates directly or through another kinase (multiple arrows) S6K1 (p70[S6k]), which is responsible for phosphorylating ribosomal protein S6 (S6) and thereby increasing the translation of mRNAs with polypyrimidine tracts. These mechanisms have been shown to be activated by meal feeding. eIF2 is required for Met-tRNA binding to the small ribosomal subunit. Phosphorylation of eIF2B (a GDP/GTP exchange factor) by GSK-3 results in inhibition of its GDP/GTP exchange activity. GSK-3 is itself inactivated (grey arrow) when it is phosphorylated by Akt/PKB. Thus, the inactivation (phosphorylation) of GSK-3 causes dephosphorylation of eIF2B, leading to its activation, which enhances translation initiation. The phosphorylation of the α-subunit of eIF2 on Ser51 by several stress-activated kinases can inhibit eIF2B activity. In the present study, meal feeding did not significantly alter eIF2B activity. In the figure, protein kinases are shown in blue, translation initiation factors in yellow, and scaffolding or structural proteins in purple (modified from reference [2]).

With the *flooding dose* technique, the labeled amino acid is injected, not as a tracer but contained in a large (i.e., much larger than the endogenous free amino acid pool) bolus of unlabeled amino acid, making the specific activities in all free amino acid compartments more alike than if the labeled compound is given as a tracer dose.[82] Thus the labeling of aminoacyl-tRNA is less likely to be affected by experimental manipulations. In addition, the large amount of amino acid injected ensures that the specific activity in the free pools remains almost constant for a period of time after injection.[83] Although various amino acids have been used, L-[3H]-phenylalanine was chosen as a radioactive tracer for experiments in the pancreas, because phenylalanine transport across the basolateral membrane of

the pancreatic acinar cells is not rate-limiting for protein synthesis, and because [3H] can be easily quantitated by liquid scintillation counting (LSC).[81] The *flooding dose* technique is advantageous not only because it is reliable but also because it can be used in unrestrained and unanesthetized animals[83] and has been well validated in muscle,[84] liver,[85] and pancreas.[2,5,81,86]

For *in vitro* studies, using isolated acini or pancreatic lobules, protein synthesis is measured by the incorporation of radioactively labeled amino acids, such as [35S]-Methionine[66,72] or [3H]-Leucine,[67] where the incorporation of these amino acids is linear during the experimental time because their intracellular pool is large and constant, due to the *flooding* of the extracellular space with an excess of amino acid, and therefore, there is no change on their specific activities.[82]

Stimulation of Pancreatic Protein Synthesis

Diet

Long-term regulation

Exocrine pancreatic adaptation to dietary changes has been observed in a variety of species.[87,88] The content and secretion of the digestive enzymes, proteases, amylase, and lipases change in proportion to the dietary content of their respective substrates, protein, carbohydrate, and fat. Changes in content of specific enzymes take place over 5–7 days. Similar changes occur in the synthesis of specific digestive enzymes measured by the incorporation of radioactive amino acids.[89–91] Changes in individual digestive enzymes occur at both the transcriptional and translational levels but the major effect results from changes in specific mRNA levels. Various hormones mediate many of these effects and in most cases their release is increased by the nutrients whose digestion they regulate.

Proteases: A high-protein diet (60–80% casein) increases the content of multiple proteases and the mRNA levels of trypsinogen, chymotrypsinogen, and proelastase in rodents.[87,92] There are, however, differential effects on different isoforms of enzymes such as trypsinogen, and this increase is not mimicked by feeding the individual amino acids.[93] This is consistent with a potential effect of dietary protein (but not amino acids) on CCK release and increasing its plasma levels. Feeding soybean trypsin inhibitor (SBTI), an indirect stimulant of endogenous CCK release, or infusing CCK, increased trypsinogen I and chymotrypsinogen B mRNA in rats.[94] Other studies using isolated pancreatic lobules showed that synthesis of proteases following infusion of the CCK analog caerulein *in vivo* was greatly increased, compared to a small increase in translatable mRNA, suggesting a posttranscriptional regulation of their synthesis.[95,96]

Lipases: The content and synthesis of pancreatic triglyceride lipase increases in response to a high-fat diet (around 40–70% of calories as triglycerides),[87] and this is accompanied by an increase in its mRNA.[97,98] Secretin has been proposed as the mediator of the effect of dietary lipids,[99] because fatty acids can stimulate secretin release, and infusion of secretin to rats, in vivo, induced an increase in the relative synthesis of pancreatic lipase.[87]

Amylase: The amount of carbohydrate in the diet has been shown to have significant effects on both pancreatic amylase content and amylase mRNA.[87,88] The effects of carbohydrate are believed to be primarily mediated by insulin. In early studies with diabetic animals, amylase content, synthesis, and mRNA levels fall dramatically, while lipase increases moderately.[88,100] Insulin administration restores the amylase synthesis, content, and mRNA levels in diabetic rats. However, insulin administration to normal rats either decreases or does not change pancreatic amylase levels. A more direct role for glucose on amylase synthesis, in addition to its indirect effects through insulin release, has been suggested.[101] In fact, a dietary response sequence in the promoter of the amylase Amy2.2 gene has been identified that mediates dietary adaptation and the effect of insulin.[102] The role of insulin in the regulation of pancreatic protein and digestive enzymes synthesis will be discussed in more detail later, in this review.

Short-term regulation

Meal stimulation: Only a few studies have evaluated the immediate regulation of the pancreatic translational synthetic machinery after food intake. Some early studies showed that fasting reduces total protein synthesis in the pancreas, and refeeding stimulates it.[1,65,103] In humans, feeding increases both, the rate of secretion and synthesis of digestive enzymes, although the rate of turnover of zymogens remains fairly constant during feeding and fasting.[8]

In mice, acute food intake stimulates pancreatic protein synthesis (**Figure 5A**) and translational effectors, without increasing digestive enzyme mRNA levels (**Figure 5B**).[2] These results indicate that acute feeding exerts a direct regulation of the mRNA translational machinery and a cellular reserve of untranslated mRNA, without changes in transcription.

The stimulation of total pancreatic protein synthesis in mice and rats is associated with the stimulation of the protein kinase B (PKB/Akt)/mTORC1 pathway and the phosphorylation of 4E-BP1 and ribosomal protein S6, downstream of mTORC1, as well as the formation of the eIF4F complex (**Figure 6**).[2]

Meal/food intake stimulation of pancreatic digestive enzymes triggers a series of stimulatory mechanism through dietary elements, hormones, and neural interactions (**Figure 1**) that are not completely understood, but this review summarizes what it is known about the most important ones.

Protein and amino acid stimulation: Protein and amino acids[104] are essential components of both human and animal nutrition and are involved in the maintenance of general health and well-being.[105] Amino acids are the building blocks of protein, cell structures, and tissues and can also act as regulators of protein metabolism and many physiological processes.[106] A long-term deficiency of the essential AAs can lead to *protein-energy malnutrition*, manifested in humans as *marasmus*, or *kwashiorkor*.[107] The therapeutic effects of amino acid when administered to patients, specially to infants, for growth as well as to maintain metabolism, have also been shown.[108] AA supplementation specially benefits patients with chronic pancreatitis

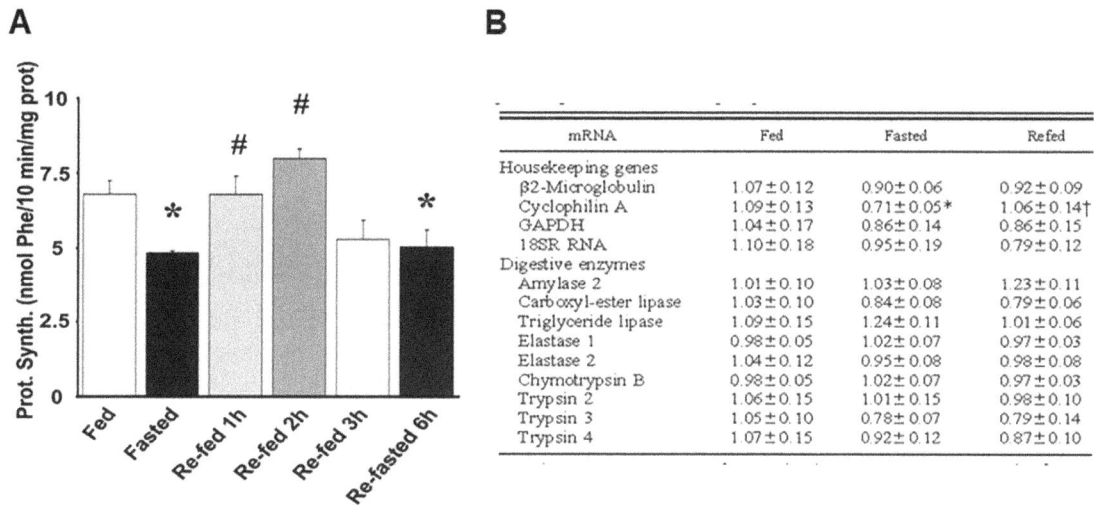

Figure 5. Acute feeding in mice stimulates pancreatic protein synthesis without an increase in the mRNA for the digestive enzymes. (**A**) Pancreatic protein synthesis analyzed by the flooding dose technique, measuring Phe incorporation into pancreatic protein. *Groups: Fed* refers to animals fed ad libitum, with pancreas removed at 9–10 AM. *Fasted* refers to animals fasted for 18 hours, and *refed* refers to animals fasted for 18 hours and refed for 1, 2, 3, or 6 hours. Values are expressed as nanomoles of incorporated phenylalanine per milligram of protein per 10 minutes and are the means ± SE. *P < 0.05 versus fed group; #P < 0.05 versus fasted group. (**B**) Relative mRNA quantities of digestive enzymes and housekeeping genes in pancreas of animals fed ad libitum, fasted for 18 hours and refed for 2 hours. Values are the means ± SE. *P < 0.05 versus fed group, †P < 0.05 versus fasted group (from reference [2]).

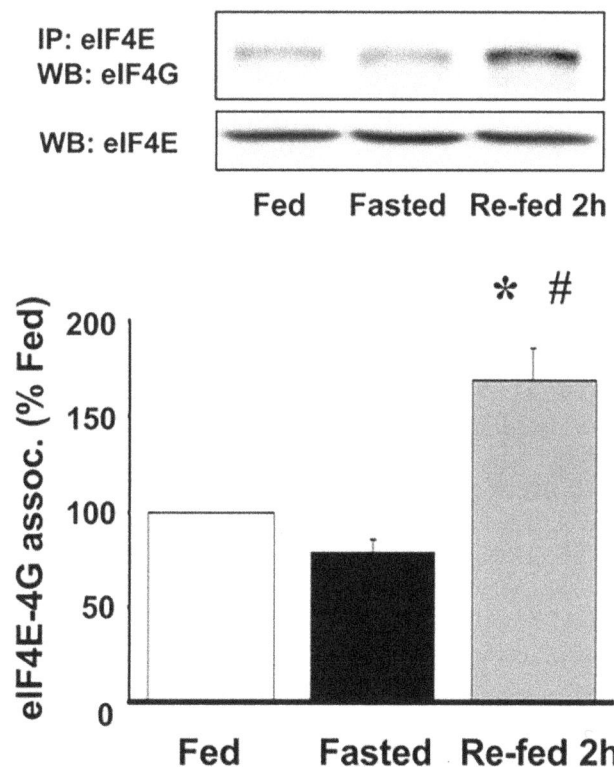

Figure 6. Effects of fasting and feeding on the eIF4F complex formation. eIF4F complex formation is measured as eIF4E and eIF4G coimmunoprecipitation. Data for coimmunoprecipitated eIF4E and eIF4G are expressed as percentage of fed group levels. Representative blots for eIF4G associated with eIF4E and total eIF4E are shown in the *insets*. *P < 0.05 versus fed group, # P < 0.05 versus fasted group (from reference [2]).

and pancreatic cancer because they suffer from malnutrition.[109] The branched-chain amino acids (BCAAs), leucine, isoleucine, and valine, in particular, are considered biological regulators,[110] in part, because they stimulate insulin release, play an important role in energy homeostasis,[111] and are involved in the stimulation of cell proliferation in certain types of cancer.[112] On the other hand, it has also been shown that the deficiency of BCAAs in the diet improves metabolic health in mice[113] and humans.[114]

The exocrine pancreas synthesizes and secretes between 6 and 20 grams of digestive enzymes per day in humans[115,116]; it is very vulnerable to protein deficiency because it requires optimal nutrition for enzyme synthesis.[116] The effects caused by dietary protein deficiency in the pancreas were shown by Crozier et al.,[117] in a study done in mice that mimicked the *kwashiorkor* syndrome. Mice were fed a protein-free diet for several days, resulting in pancreatic atrophy and the involvement of the mTORC1 pathway.

A high-protein diet[6,118] and dietary amino acids[7] have been shown to stimulate pancreatic protein synthesis at the translation initiation level in mice and rats and, at the same time, also stimulate pancreatic growth, because of their long-term treatment. Short-term effects of protein and AAs seem mainly to target the regulation of the protein synthetic machinery of pancreatic acinar cells through mTORC1 activation. BCAAs, especially leucine, can stimulate mTORC1 and the pancreatic protein synthesis machinery of rats, independently of CCK and insulin.[5] Other studies also have shown the role of leucine in the exocrine pancreas of rats,[7] pigs,[119] and ruminants.[120]

A short-term study in mice demonstrated that diets lacking protein or amino acids inhibited pancreatic protein synthesis (measured by the *flooding dose* technique) (**Figure 7**) and the amount of ribosomes engaged in translation on a given moment, by polysomal fractionation (**Figure 8**).[4] Polysomal profiling analysis performed on pancreatic homogenates showed a significant increase in the polysomal fraction (i.e., more actively translated mRNAs into protein) in the control refed group (**Figure 8B**), compared to the control fasted group (**Figure 8A**). This was reflected in the analysis of the area under the curve of the polysomal fractions (**Figure 8A** and **B**). The groups lacking protein or leucine in the diet showed a very similar polysomal profile with a large reduction of the polysomal fraction, compared to the refed and fasted control groups (**Figure 8C** and **D**). The analysis of the area under the curve of these groups confirmed this reduction. At the same time, the analysis of the subpolysomal fraction indicated that refeeding with a 20 percent protein diet reduced the amount of free ribosomal subunits 40S and 60S, consistent with the increase in the 80S ribosomal unit and the polysomal fraction (**Figure 8B**). The lack of protein or leucine in the diet increased the subpolysomal peaks (area under the curve

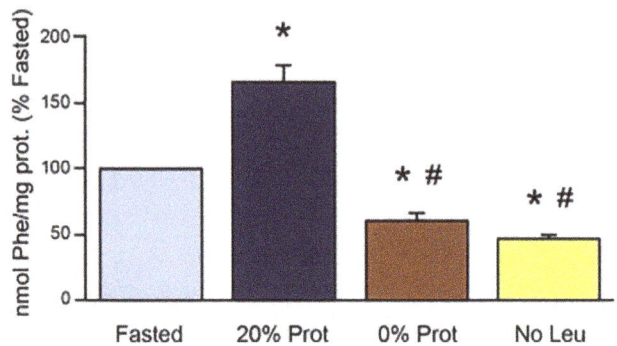

Figure 7. Effect of the different feeding treatments on total pancreatic protein synthesis. Mice were fasted for 16 hours and refed with three experimental diets: control diet (20% protein), protein-deficient diet (0% protein), and leucine-deficient diet (no leucine). #$P < 0.05$ vs. control fasted group. #$P < 0.05$ vs. control refed group (from reference [4]).

for both groups), indicating that there were more ribosomal subunits disengaged for translation (**Figure 8C** and **D**). These results confirm that total pancreatic protein synthesis was inhibited in the diets lacking protein or leucine, with a strong reduction of the mRNAs attached to the ribosomes for their translation into protein.

This inhibition was accomplished through different intracellular mechanisms: the protein-free diet inhibited pancreatic protein synthesis through a partial inhibition of the mTORC1 pathway, and of the eIF2B activity; the leucine-free diet inhibition was caused by the inhibition of the eIF2B activity and activation of the stress signals leading to the general control nonderepressible 2 (GCN2) kinase activation and eIF2α phosphorylation, induced by the amino acid imbalance in the system and in the pancreas.[4] This study shows that dietary amino acids are important regulators of postprandial digestive enzyme synthesis, and their deficiency could induce pancreatic insufficiency and malnutrition.

Amino acids regulate mRNA translation in the pancreas through multiple mechanisms. One mechanism involves activation of mTOR that controls several downstream effectors, including RNA polymerase I, S6K1, 4E-BP1, and the elongation factor 2 kinase (eEF2K) (inactive when phosphorylated).[4] Amino acids, then, stimulate dephosphorylation (activation) of eEF2 that may also prevent activation of GCN2 by enhancing the removal of deacylated tRNA from the P-site on the ribosome (**Figure 3**)—a potential activator of GCN2 or EIF2AK4 (eIF2α kinase 4). GCN2 is induced during amino acid deprivation by a mechanism that involves uncharged tRNA binding to a regulatory region homologous with HisRS (histidyl-tRNA synthetase) enzymes.[121,122] Phosphorylation of eIF2 impedes recycling of eIF2 to its active GTP-bound form (**Figure 2**), and the reduction in the levels of eIF2–GTP reduces global translation, allowing cells to conserve resources and to initiate

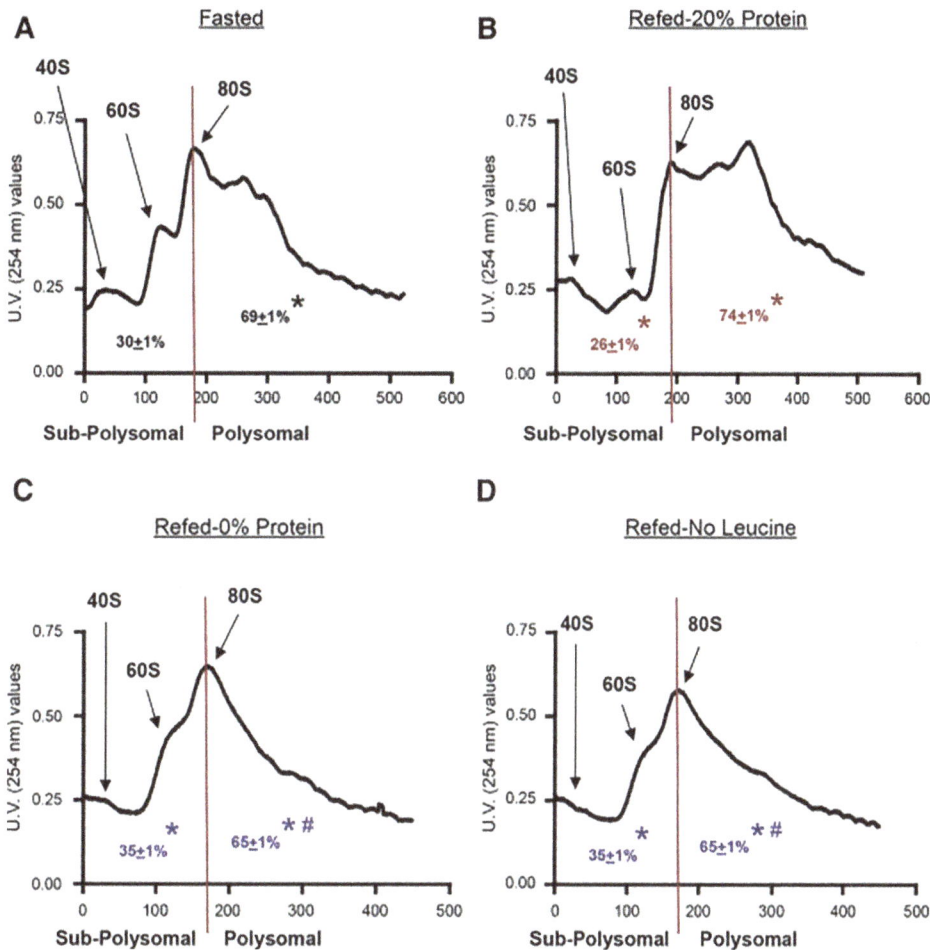

Figure 8. **Effect of the four different experimental treatments on pancreatic polysomal profiles**. Each profile shows the peaks for the 40S and 60S ribosomal subunits, as well as for the whole ribosome (80S) and the different polysomes, in the polysomal fraction for representative fractionations. The graphs also show the calculations of the area under the curve for the subpolysomal and the polysomal fractions. The experimental groups were as follows: (**A**) fasted; (**B**) refed (20% protein); (**C**) refed protein-deficient diet (0% protein); and (**D**) refed leucine-deficient diet (no leucine). *$P < 0.05$ vs. control group. #$P < 0.05$ vs. control refed group (from reference[4]).

a reconfiguration of gene expression to effectively manage stress conditions. A study using GCN2-deficient mice indicated an important effect of leucine deficiency in the translational machinery of the liver.[123] Preventing activation of GCN2, amino acids preserve eIF2B activity, which promotes translation of all mRNAs, that is, global protein synthesis is enhanced.

BCAAs, specifically leucine, have been shown to stimulate pancreatic protein synthesis in rats,[5] dairy goats,[124] and calves.[125] Our study demonstrated that BCAAs stimulate the activation of mTORC1, analyzed by the phosphorylation of 4EBP1 and S6K (**Figure 9A** and **B**) and the formation of the eIF4F complex, independently of the hormones cholecystokinin (CCK) (**Figure 9C**) and insulin.[5]

Thus, dietary amino acids, as well as a mixed meal can stimulate the synthesis of digestive enzymes through the translational machinery by activating the mTORC1 pathway (**Figures 5–9**).[2,4,5,7]

Cholecystokinin (CCK)

Regulation of translation initiation

As mentioned above, feeding stimulates pancreatic digestive enzyme synthesis at the translational level, and, in addition to dietary components, this process is also mediated by hormones such as CCK and insulin and neurotransmitters such as acetylcholine (**Figure 1**). There are no conclusive published studies about the role of postprandial CCK in the stimulation of pancreatic digestive enzyme synthesis, but unpublished results from our laboratory, using CCK-deficient mice[126] *in vivo*, demonstrate that CCK is, in

112 *Maria Dolors Sans*

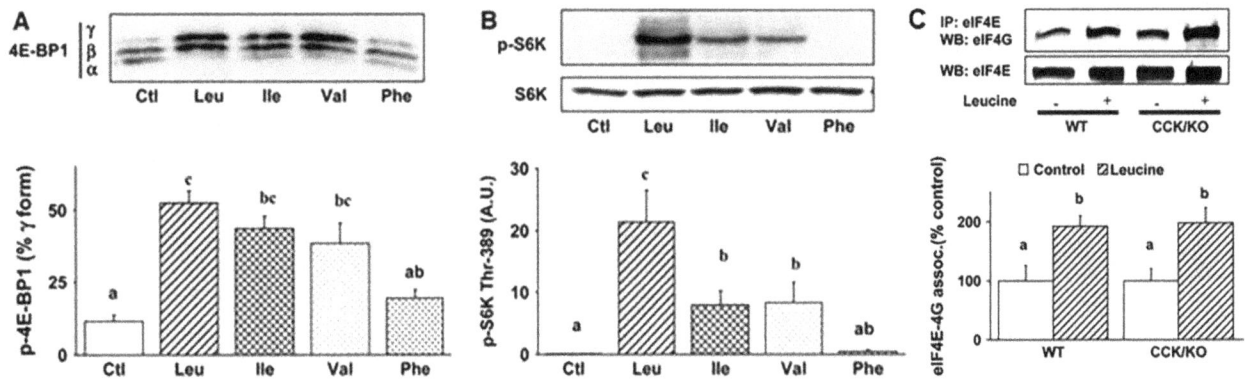

Figure 9. Effect of different amino acids on the activation of mTORC1 *in vivo* (A and B) and (C) effects of leucine on eIF4F complex formation in CCK/wild-type and CCK/KO mice. (A) 4E-BP1 phosphorylation and (B) S6K phosphorylation on Thr-389, in rats 30 minutes after oral administration. (A) 4E-BP1 phosphorylation is expressed as the percentage of the protein present in the γ form. *Inset* shows representative immunoblot. The most highly phosphorylated γ-form exhibits the slowest electrophoretic mobility and does not bind eIF4E. (B) S6K phosphorylation is expressed as arbitrary units (AUs). *Inset* shows a representative immunoblot for phosphorylated (p-S6K) and total S6K. Means not sharing a letter differ, $P < 0.05$. (C) eIF4F complex formation is expressed as a percentage of controls. *Inset*: representative immunoblot for eIF4G and eIF4E. Means not sharing a letter differ, $P < 0.05$ (from reference [5]).

part, necessary for the stimulation of pancreatic digestive enzymes synthesis after feeding a complete and balanced diet as well as for the stimulation of the mTORC1 pathway. On the other hand, there are multiple studies showing the involvement of CCK in the stimulation of pancreatic acinar cell protein synthesis and translational mechanisms, *in vivo*[127] and *in vitro*.[66,67,72,128]

In vivo: Bragado et al.[127] demonstrated that CCK, either administered by intraperitoneal injection to rats, or by stimulating endogenous CCK release by intragastric administration of a trypsin inhibitor, induced a time- and dose-dependent phosphorylation of pancreatic eIF4E and its binding protein 4EBP1 (or PHAS-I) as well as the formation of the eIF4F complex (**Figures 2 and 4**). These events occurred over a range of CCK doses from 0.2 to 5 μg/kg. The acute effects of endogenous CCK, released after gavaging a synthetic trypsin inhibitor, camostat (100 mg/kg), were similar. Thus, both exogenous and endogenous CCK activate translational initiation factors *in vivo*. The activation of translational machinery necessary for initiation of protein synthesis likely contributes to the normal postprandial synthesis of pancreatic digestive enzymes.

In vitro: CCK has a biphasic effect on enzyme secretion being stimulatory at physiological concentrations and inhibitory at higher concentrations, and follows the same biphasic pattern stimulating protein synthesis[66,72,73,128] (**Figure 10**). Thus, CCK increases the rate of translation initiation[66,72,73,128] and elongation processes[16] at physiological concentrations. That this acute stimulation of protein synthesis is at the translational level is shown by the fact that it occurs without a change in mRNA levels and

in the presence of actinomycin D.[69,70] In these studies, the stimulation of protein synthesis occurred within 30 minutes and showed additivity between insulin and CCK. Increased synthesis of both digestive enzymes and structural proteins were observed, although differences between individual proteins suggested nonparallel translational effects.[70] Additionally, CCK or its analogue caerulein activates several regulators of translation, like the S6 kinase (S6K),[86,128] the phosphorylation of eIF4E,[73,127] activate the formation of the eIF4F complex by stimulating the release of eIF4E from its binding protein 4E-BP1, and increase the association of eIF4E with eIF4G.[73,127] The activation of S6K, the formation of the eIF4F complex, and the activation of the elongation processes and eEF2 appear to be regulated through a rapamycin-sensitive pathway and to be downstream of PI3K[16,66,128] (**Figure 4**).

Bragado et al.[128] also showed that the expression and activity of p70s6k-p85s6k (S6K) was biphasic after CCK stimulation. Carbachol and bombesin, but not vasoactive intestinal peptide, also activated S6K. These study concludes that the S6K stimulation by CCK is not mediated by PKC or mobilization of intracellular calcium but by PI3K. At the same time, it is shown that S6K is not involved in the secretion of digestive enzymes induced by CCK[128] (**Figure 4**).

Other stimulants, besides CCK, stimulate total protein synthesis in pancreatic acinar cells at their physiological concentrations. Carbachol, insulin, and bombesin, in multiple species and preparations, all stimulate the synthesis of total protein, trypsinogen, chymotrypsinogen, lipase, and amylase by isolated acini.[67–69]

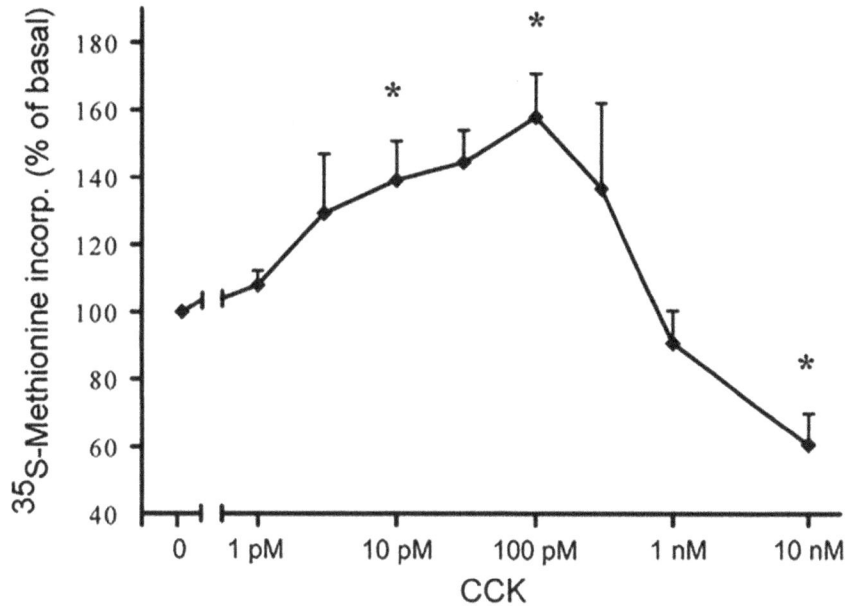

Figure 10. Effect of CCK octapeptide on L-[³⁵S]methionine incorporation into acinar protein. Rat pancreatic acini were incubated for 45 minutes with varying concentrations of CCK octapeptide before the addition of 2 μCi/ml of the labeled amino acid for 15 minutes. Incorporation was measured into TCA precipitable protein. *$P < 0.05$ vs. control (basal group) (from reference [72]).

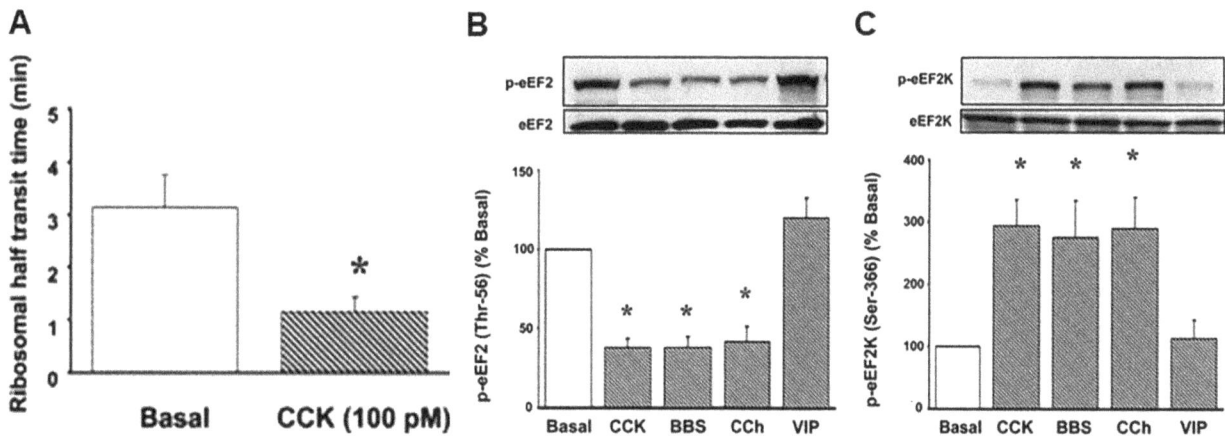

Figure 11. Effects of several secretagogues on key elongation steps. (**A**) Average of the ribosomal half-transit time values ± SE, obtained after calculating the separation (in time) between the post-mitochondrial supernatant (PMS) and post-ribosomal Supernatant (PRS) for each group. *$p < 0.05$ vs. basal. (**B**) and (**C**) Effect of stimulatory doses of CCK, bombesin (BBS), carbachol (CCh), and the vasoactive intestinal peptide (VIP) on eEF2 (**B**) and eEF2K (**C**) phosphorylation. Acini were incubated for 60 minutes with CCK at 100 pM, BBS at 10 nM, CCh at 10 lM, and VIP at 100 nM. Results are expressed as a percentage of basal levels. *$p < 0.05$ vs. basal. Blots are representative of phosphorylated and total eEF2 and eEF2K levels (from reference[16]).

Regulation of translation elongation

Stimulatory doses of cholecystokinin (CCK), bombesin, and carbachol increased elongation rates (measured as ribosomal half-transit time) in pancreatic acini *in vitro* (**Figure 11A**). At the same time, these secretagogues reduced elongation factor 2 (eEF2) phosphorylation (the main factor known to regulate elongation), and increased the phosphorylation of the eEF2 kinase (**Figure 11B** and **C**). The mTOR inhibitor rapamycin reversed the dephosphorylation of eEF2 induced by CCK, as did treatment with the p38 MAPK inhibitor SB202190, the MEK inhibitor PD98059, and the phosphatase inhibitor calyculin A. Neither rapamycin, SB202190, PD98059 nor calyculin A had an effect on CCK-mediated eEF2 kinase phosphorylation. Therefore, translation elongation in pancreatic acinar cells is likely regulated by eEF2 through the mTOR, p38, and MEK pathways and modulated through PP2A.[16]

Insulin

The pancreas is composed of two separate organ systems—the exocrine pancreas, responsible for synthesizing the enzymes that will enter the intestine and participate in digestion, and the endocrine pancreas, composed of islets of Langerhans, responsible for synthesizing insulin and other hormones that enter the blood and regulate glucose levels and metabolism. The physical location of both systems in the same organ and the common vascular supply, including a portal blood system by which venous blood from islets can perfuse the exocrine tissue, seems likely to have a purpose. In fact, the exocrine pancreas and the pancreatic islets have a multitude of complex anatomical and functional interrelations that can affect one another.[129,130] Among these, pancreatitis can induce diabetes, by damaging the endocrine cells of the islets,[131,132] and diabetes is correlated with exocrine pancreas insufficiency.[133,134] This indicates a role for insulin in the regulation of digestive enzyme synthesis and/or secretion.

For more detailed information on insulin secretion and action on pancreatic exocrine function, the reader is invited to check the reviews by Mann et al. on the Pancreapedia[135] and in Chapter 25 of this book; and the one from Sans et al. also on the Pancreapedia[136] and in Chapter 17 of this book.

Briefly, insulin is known to stimulate protein synthesis by translational effects in many tissues.[137] Early studies described the presence of insulin receptors in pancreatic acinar cells[138] and showed that insulin was involved in the incorporation of radioactive amino acids into total protein or specific enzymes of normal and diabetic rodents *in vivo*.[139–141] Most of the studies showed a positive effect of insulin, but due to technical differences between studies, the results are difficult to interpret. When protein synthesis in mice was quantitated by autoradiography, there was more incorporation into peri-insular acinar cells than into tele-insular acinar cells,[141] and this effect was lost after inducing diabetes with streptozotocin—implying that insulin had an effect on the peri-insular acinar cells. One study, with a different methodological approach, increased insulin secretion by treatment with sulfonylurea (a drug that stimulates insulin release) or by glucose infusion. Both treatments increased protein synthesis by 25–40 percent. When Zucker fatty rats were studied as a model of insulin resistance, overall pancreatic protein synthesis was reduced by nearly 50 percent.[142] There was considerable difference in synthesis between different digestive enzymes separated by 2-D-gel electrophoresis. A morphological study of prolonged diabetes in rats showed gross abnormality in the

acinar cell secretory pathway 28 days after STZ treatment that was partially reversed by insulin administration.[143] However, shorter studies have not shown significant structural alterations 1 week after STZ other than the appearance of cytoplasmic lipid droplets.[68]

More recent studies have been able to overcome some of the methodological issues by measuring the effects of insulin over short times to prevent changes in mRNA and under conditions where the precursor pool is large and constant.[82,144] *In vitro* studies have used isolated pancreatic acini under dilute labeling conditions to keep the precursor pool constant. Insulin stimulates the incorporation of multiple different amino acids (leucine, methionine, phenylalanine) into protein, in isolated acini from diabetic rats.[67–70] Similarly, insulin increased methionine incorporation into total protein and immunoprecipitated amylase in rat pancreas-derived AR42J cells.[145] The effect of insulin on both cell types occurred over concentrations from 30 pM to 100 nM and was mediated by the insulin receptor.[68,70] Insulin was also shown to have nonparallel effects on different digestive enzymes and structural proteins such as myosin and LDH.[70]

Although the mechanism of insulin signaling is well studied in a number of target tissues, especially liver, fat, and muscle, only a few studies have been carried out in the exocrine pancreas. The main signaling pathway regulating protein synthesis is the mTORC1 pathway and this pathway has been documented in pancreatic acinar cells primarily in mediating the actions of CCK and acetylcholine to stimulate digestive enzyme synthesis.[63,146] Insulin has been shown to be involved in protein synthesis in several studies[66,68,69] and has nonparallel effects on translation of specific proteins.[70] It also activates S6K and S6 phosphorylation downstream of TORC1 in rat and mouse acini.[147,148] Akt is upstream of mTORC1 and insulin activates the phosphorylation of Akt on S473 and T308 in rat acini.[149] Bragado et al.[66] showed that insulin stimulates protein synthesis in pancreatic acinar cells *in vitro*, through the PI3K/Akt/mTPRC1 stimulated mechanisms (**Figure 4**). However, almost all of these studies were dependent on first inducing diabetes with streptozotocin to reduce endogenous levels of insulin, and diabetes effects could confound some of the results.

One recent study from our laboratory (to be published shortly) done in genetically modified mice *in vivo* shows, for the first time, that insulin signaling through its receptor in pancreatic acinar cells stimulates the synthesis of digestive enzymes after a balanced meal feeding.[150] We developed a mouse model with a conditional insulin receptor (IR) KO in the pancreatic acinar cells: the pancreatic acinar cell insulin receptor knock out (PACIRKO) mice. It was obtained by crossing $IR^{lox/lox}$ mice with tamoxifen-regulated Ela-Cre mice. The lack of insulin signaling to pancreatic acinar cells induced a 40 percent reduction in pancreas weight of PACIRKO mice as well as total pancreatic protein and digestive enzymes content. The Akt/mTORC1 pathway activity was significantly reduced in both PACIRKO and diabetic mice, compared to normal control mice. Total pancreatic fluid secretion was not changed, but protein concentration and content of several digestive enzymes was reduced in the PACIRKO mice, compared to diabetic models. The specific deletion of the insulin receptor gene induced a decrease of pancreatic digestive enzymes content, an increase in the size of pancreatic acinar cell lumen, as well as apoptosis and activation of regeneration mechanisms in the pancreas with time. This study demonstrates that insulin directly regulates the synthesis of digestive enzymes in a postprandial situation as well as the short- and long-term exocrine pancreatic homeostasis.

This effect of insulin stimulating the synthesis of pancreatic digestive enzymes has clinical implications related to type 1 diabetes, where insulin is reduced. A significant number of patients with diabetes suffer exocrine pancreatic abnormalities especially pancreatic exocrine insufficiency due to a reduction of the digestive enzymes, which occurs in about 50 percent of patients with type 1 diabetes.[151] The decrease in pancreatic function correlates with the duration of both, type 1 and 2 diabetes.[104,152] In other studies with type 1 diabetes patients, atrophy of exocrine tissue and pancreatic fibrosis has been reported.[104,153–155] This pancreatic insufficiency during diabetes can affect physical development and metabolism in children and adolescent diabetic patients.[156,157]

Exocrine pancreatic abnormality has also been seen in animal models of diabetes.[158,159] Guinea pigs with spontaneous diabetes show fatty degeneration of pancreatic acinar cells and decreased digestive enzymes and bicarbonate secretion.[160] Streptozotocin-treated diabetic rats show altered digestive enzyme gene expression and reduced amylase secretion,[161–163] without a decrease in the amount of fluid secreted.[164] In isolated pancreatic acini, *in vitro*, there is a reduction of amylase secretion after STZ-induced diabetes[164–167] due to reduced amylase pancreatic content.[162,163] More clinically relevant, mouse models of type 1 diabetes like NOD mice or the Ins2^{Akita} mouse model with insulin misfolding also show alterations in the exocrine pancreas.[168,169]

From these studies, it can be concluded that in both animals and humans insulin (besides modulating the action of other hormones (CCK) or stimulators (ACh)[164,166]) is necessary for the synthesis of digestive enzymes, and that these functions are inhibited in T1D.

Intracellular Calcium Levels Affecting Pancreatic Protein Synthesis

CCK and other G protein-stimulating pancreatic secretagogues inhibit pancreatic acinar protein synthesis at concentrations that hyperstimulate their receptors.[72,73] All of them have a common intracellular mechanism, increasing intracellular calcium to high concentrations.[42,170] High intracellular calcium levels have been shown to inhibit protein synthesis in other cell types,[22] by ER stress and activation of the UPR.[171] On the other hand, it has also been shown that removing calcium from the media, or depleting calcium from inside the acinar cells, also inhibits pancreatic protein synthesis,[72,172] indicating a need for calcium in the regulation of protein synthesis in pancreatic acinar cells.

Calcium signaling is involved in the stimulation of pancreatic protein synthesis

Changes in physiological calcium levels inside pancreatic acinar cells act as a signaling mechanism that affects the activity of calcium-regulated enzymes.[42] Calcineurin, also known as protein phosphatase 2B (PP2B), is a serine/threonine protein phosphatase that is highly regulated by Ca^{2+}-calmodulin.[173] This ubiquitously expressed phosphatase controls Ca^{2+}-dependent processes in all human tissues; it has been found in the highest concentrations in the brain, but it has also been detected in many other mammalian tissues,[174,175] including the pancreas.[175,176] It is believed to be relatively inactive in cells under basal conditions of low intracellular calcium but becomes active after stimulation with calcium-mobilizing agonists.[173] Calcineurin is best known for driving the adaptive immune response by dephosphorylating the nuclear factor of the activated T-cells (NFAT) family of transcription factors. Therefore, calcineurin inhibitors, FK506 (tacrolimus), and cyclosporin A serve as immunosuppressants. FK506 and cyclosporin A (CsA) binding to their intracellular target proteins (12-kDa FK506-binding protein and cyclophilin A) inhibit calcineurin phosphatase activity.[177] The use of FK506 and CsA has been used to implicate calcineurin in a number of cellular processes, including pancreatic endocrine[178] and exocrine secretion.[176,179] Moreover, study of the side effects of CsA and FK506 in organ transplant therapy has implicated calcineurin in the protein synthesis mechanisms of some tissues, including kidney and liver.[180] However, in contrast to the apparent multitude of cellular substrates for the type 1 and 2A serine/threonine phosphatases, relatively few cellular targets of calcineurin have been described.[173,181]

High concentrations of CsA were found to inhibit amylase secretion,[176,182] but it is not clear whether FK506 inhibits pancreatic exocrine secretion[179,182] or not (unpublished observations). Both immunosuppressants were found to block pancreatic growth in response to chronic elevation of CCK induced by feeding trypsin inhibitor to mice,[183] indicating a role for calcineurin in pancreatic growth (see the review on pancreatic growth by John A. Williams in Chapter 22 of this book and in Pancreapedia).[184] Since, protein synthesis is an obligatory requirement for growth of all cells,[185] calcineurin could also be involved in the regulation of pancreatic protein synthesis.

Using FK506 and CsA, we determined that calcineurin is involved in the CCK stimulation of pancreatic acinar cell protein synthesis and translational machinery,[73] demonstrating the involvement of intracellular calcium in the stimulation of mRNA translation into protein. FK506 also inhibited protein synthesis stimulated by bombesin and carbachol (**Figure 12A**). FK506 did not significantly affect the activity of the initiation factor-2B or the phosphorylation of the initiation factor-2α, ribosomal protein S6, or the mRNA cap-binding protein eukaryotic initiation factor (eIF)4E. Instead, blockade of calcineurin with FK506 reduced the phosphorylation of the eIF4E-binding protein, decreased the formation of the eIF4F complex (**Figure 12B**), and increased the phosphorylation of eukaryotic elongation factor 2 (**Figure 12C**). From these results, it can be concluded that calcineurin activity and intracellular calcium are required for pancreatic protein synthesis, and this action may be related to effects on the formation of the mRNA cap-binding complex (**Figure 1**) and the elongation processes (**Figure 2**).[73]

High intracellular calcium levels inhibit pancreatic protein synthesis

In vitro

The inhibition of acinar protein synthesis and polysome size by CCK and the inhibitors of the microsomal Ca^{2+} ATPase, thapsigargin (Tg), and 2,5-di(tertbutyl)-hydroquinone (Bhq) occurs as a result of depletion of Ca^{2+} from the endoplasmic reticulum lumen, increasing intracellular cytoplasmic calcium.[172] The combination of administering the intracellular Ca^{2+} chelator, BAPTA, with Tg and Bhq depleted the pools of $[Ca^{2+}]$ without changing cytosolic $[Ca^{2+}]$ but with the same inhibitory effect on protein

synthesis and polysome formation, suggesting that depletion of intracellular Ca^{2+} stores, without changes in cytosolic $[Ca^{2+}]$, decreases protein synthesis at translation initiation.[172] It has also been demonstrated that removal of Ca^{2+} from the medium enhanced the inhibitory action of CCK on both protein synthesis and eIF2B activity as well as further increased eIF2α phosphorylation[72] (**Figure 13**).

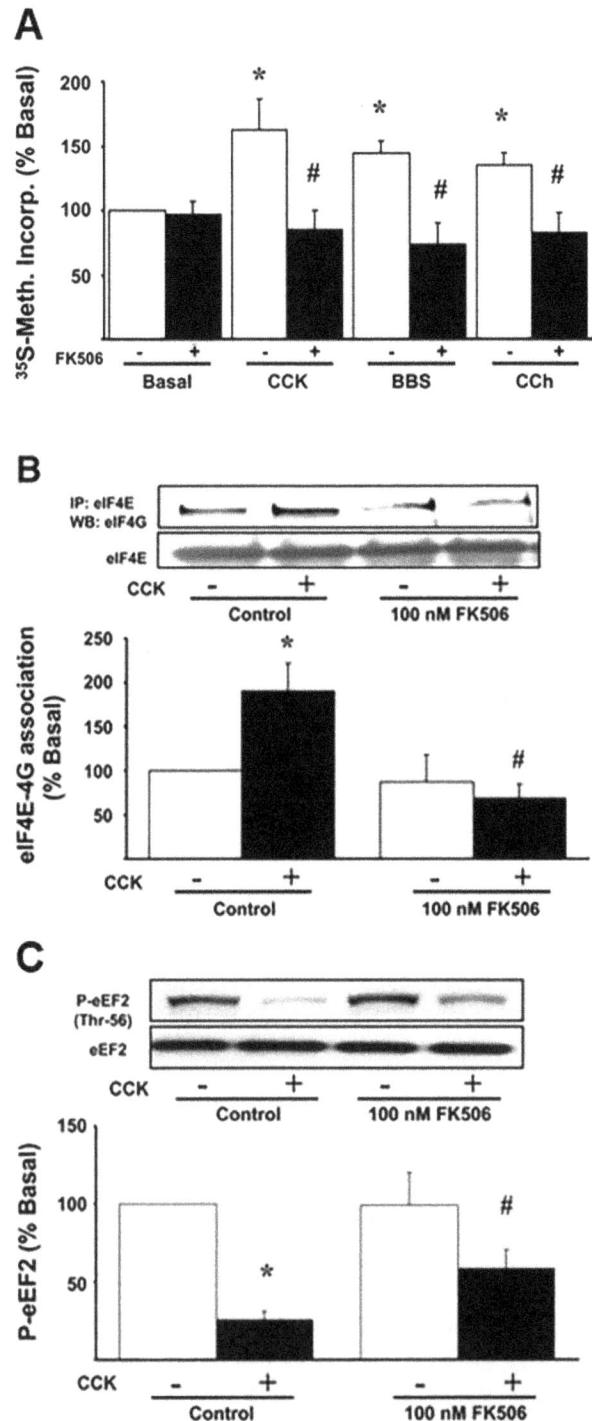

Figure 12. (**A**) **Effect of FK506 on pancreatic protein synthesis stimulated with different agonists.** The incorporation of [^{35}S]methionine was analyzed in basal, CCK-stimulated (100 pM), bombesin (BBS)-stimulated (10 nM), and carbachol (CCh)-stimulated (30 μM) acini, with or without 100-nM FK506. In each experiment, ^{35}S incorporation was normalized to control. *$P < 0.05$ vs. basal group; #$P < 0.05$ vs. its corresponding control stimulated group (CCK 100 pM; BBS 10 nM; CCh 30 μM). (**B**) **Effects of FK506 on the formation of eIF4F and (C) on eEF2 phosphorylation.** (**B**) Formation of the eIF4F complex, measured as coimmunoprecipitation of eIF4E and eIF4G. (**C**) **Phosphorylation of eEF2 (Thr-56) in response to CCK in the absence or presence of 100 nM FK506**. *Insets*: Representative Western blots for eIF4G and total eIF4E (**B**) and for phosphorylated and total eEF2 (**C**). *$P < 0.05$ vs. basal group; #$P < 0.05$ vs. control CCK-stimulated group (from reference [73]).

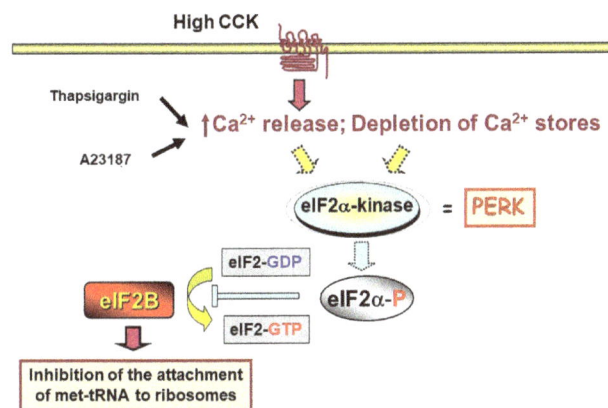

Figure 13. Mechanism by which high concentrations of CCK and induction of ER stress inhibit initiation of translation. eIF2B activity is inhibited at high doses of CCK by release of intracellular Ca^{2+} and depletion of the intracellular stores, most likely from the ER lumen. This activates a kinase (such as PERK) that phosphorylates eIF2α. The ionophore A23187 and thapsigargin (the inhibitor of the microsomal Ca^{2+}-ATPase) also inhibit eIF2B activity through depletion of Ca^{2+} stores and phosphorylation of eIF2α.[101,186]

In the same study, with rat acini, carbachol (CCh) and the ionophores A-23187 and Tg also inhibited eIF2B and protein synthesis, whereas bombesin and the CCK analog JMV-180, that do not cause depletion of intracellular calcium stores at high concentrations, were without effect (**Figure 14**).[72]

Previous studies have shown that eIF2B can be negatively regulated by glycogen synthase kinase-3 (GSK-3). However, GSK-3 activity, as assessed by its phosphorylation state, was inhibited at high concentrations of CCK, an effect that should have stimulated, rather than repressed, eIF2B activity. Therefore, CCK inhibits eIF2B activity only through eIF2alpha phosphorylation (**Figure 13**).

In vivo

Inhibition of pancreatic protein synthesis *in vivo* normally occurs during the development of caerulein-induced acute pancreatitis (AP),[56] but it has been shown that in some studies, total pancreatic protein synthesis is inhibited, whereas in other studies, protein synthesis is not affected. On the other hand, in a study done using minced rabbit pancreas samples, a decrease in protein synthesis was observed in response to stimulatory doses of CCK, and this was accompanied by a decrease in the number of polysomes.[71] Most of the differences seen on the effects of CCK *in vivo, ex vivo,* or *in vitro* could be explained by methodological differences on the way of administering and analyzing the incorporation of AAs into protein.

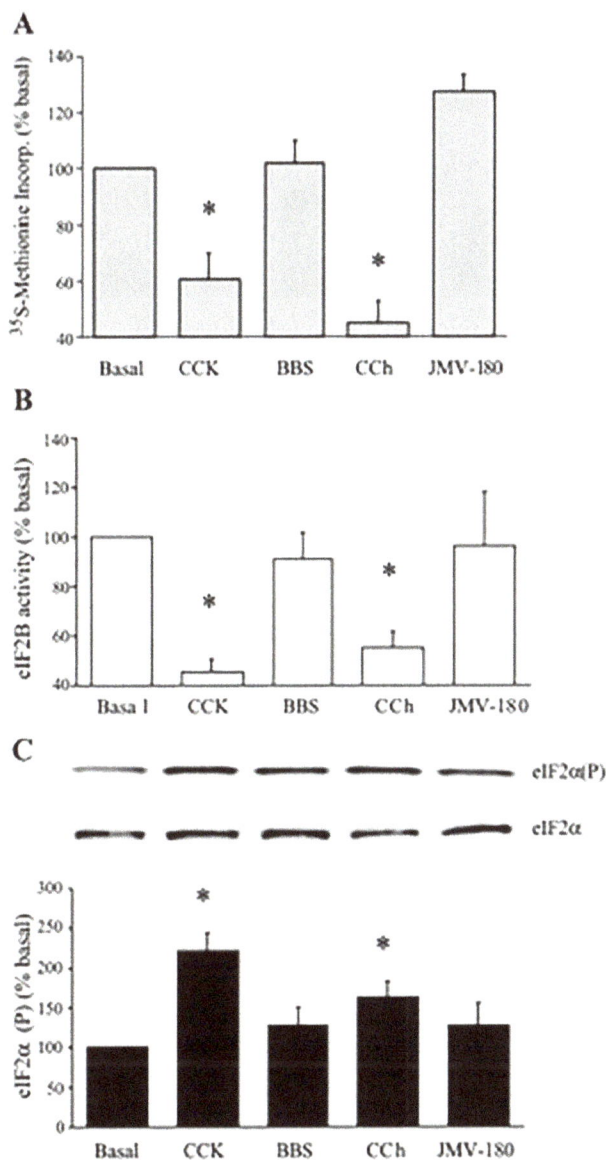

Figure 14. Effect of high doses of CCK octapeptide, carbachol (CCh), bombesin (BBS), and the CCK analog (JMV-180) on isolated rat acini. Effects on (**A**) protein synthesis (**B**), eIF2B activity, and (**C**) eIF2α phosphorylation. Acini were incubated for 60 minutes with CCK at 10 nM, CCh at 1 mM, and BBS and JMV-180 at 1 μM and L-[^{35}S]methionine incorporation determined as well as eIF2B activity, and eIF2α phosphorylation. Results are expressed as a percentage of basal levels. *$P < 0.05$ vs. control (basal group). The blots in (**C**) are representative for eIF2α phosphorylation levels (from reference [72]).

In a caerulein-induced AP study,[86] pancreatic protein synthesis was already reduced 10 minutes after the initial caerulein administration and was further inhibited after three and five hourly injections (**Figure 15A**). Caerulein inhibited the two major regulatory points of translation initiation: the activity of the guanine nucleotide exchange

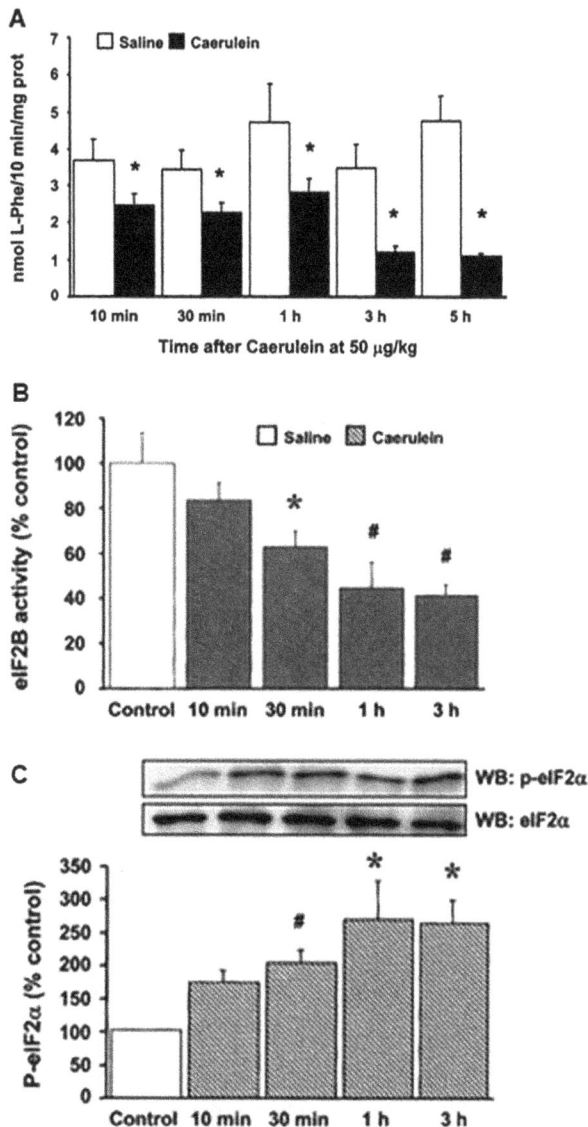

Figure 15. Effect of caerulein-induced acute pancreatitis on pancreatic protein synthesis and on the first step of translation initiation (Figure 1), in mice. (A) L-[³H]phenylalanine incorporation into pancreatic protein in C57BL/6 mice. Values are expressed as nanomoles of incorporated phenylalanine per 10 minute per milligram of protein and are means and SE. *P < 0.05 vs. control (saline groups), (B) eukaryotic initiation factor (eIF)2B activity, and (C) eIF2α phosphorylation. The pancreas was removed 10 minutes, 30 minutes, and 1 hour after a single-dose injection of caerulein at 50 μg/kg and 1 hour after 3 hourly injections. The results are expressed as a percentage of pooled control levels. In (C) the *insets* show representative blots for eIF2α phosphorylation levels and total eIF2α. For (B) and (C), data shown are means and SE. *P < 0.05, #P < 0.01 vs. pooled control group (from reference [86]).

factor eIF2B (with an increase of eIF2α phosphorylation) (**Figure 15B and C**) and the formation of the eIF4F complex due, in part, to degradation of eIF4G (**Figure 16**). This inhibition was not accounted for by changes upstream (caerulein activated Akt) or downstream of mTORC1. Caerulein also decreased the phosphorylation of eEF2, implying that elongation was not inhibited during AP. Thus, the inhibition of pancreatic protein synthesis in this model of AP results from the inhibition of translation initiation due to increased eIF2α phosphorylation, reduction of eIF2B activity, and the inhibition of eIF4F complex formation.[86] This inhibitory effect appears to be calcium-dependent and suggests that pancreatic acinar cells adapt to this short-term stress[187] by inhibiting the synthesis of pancreatic digestive enzymes.[71,72] Although not yet fully studied, an ER resident kinase such as PERK[27] (**Figure 13**) is likely to mediate eIF2α phosphorylation and it has been shown to be important for the translational control and cell survival of pancreatic acinar cells.[23,188,189]. The inhibition of protein synthesis associated with high concentrations of CCK could therefore be an adaptive or protective mechanism in response to stress localized in the ER[73,86]

Several early studies showed that pharmacological inhibition of protein synthesis during AP helped prevent acinar cell damage.[190,191] Later studies, on the contrary, have shown that a complete inhibition of pancreatic protein synthesis with cycloheximide during AP reduced pancreatic edema[192] but worsened the development of the disease and inhibited apoptotic pathways[193,194] that have been proposed to be more beneficial than the necrotic ones in the severity of AP.[195-199] Thus, a complete inhibition of pancreatic protein synthesis does not seem a likely event to happen during AP; rather, a selective inhibition of the synthesis of specific proteins would most likely account for the reduction seen in total protein synthesis during AP. Since synthesizing digestive enzymes in the exocrine pancreas is the main purpose of the gland under physiological situations, inhibition of this synthesis could likely happened during adverse situations such as AP, when secretion is also blocked, in an attempt to prevent or reduce more acinar cell damage.

Additionally, despite the fact that numerous studies have shown the possible interaction between lysosomal hydrolases and digestive enzyme zymogens in acinar cells in the development of AP,[200-203] whether lysosomal enzymes and pancreatic zymogens can be regulated at the translational level during AP has not been addressed.

Pancreatitis and the ER stress response

The events that regulate the severity of AP are, for the most part, unknown, and the exact mechanisms by which diverse etiological factors induce an attack are still unclear. The use of animal models of AP has permitted to advance in our understanding of the early cellular events that underlie the development of acute pancreatitis. However, there is

A

B

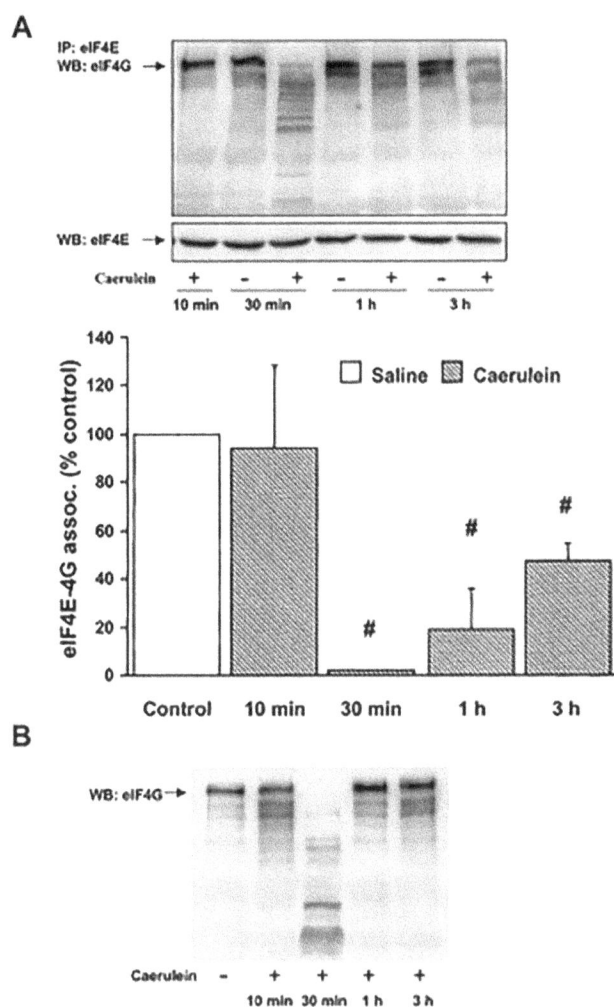

Figure 16. Effects of caerulein-induced acute pancreatitis on eIF4F complex formation, measured as eIF4E and eIF4G coimmunoprecipitation. (**A**) Data for coimmunoprecipitated eIF4G are expressed as percentage of pooled control levels and SE. Representative blots for eIF4G associated with eIF4E and total eIF4E are shown in the *insets*. #*P* < 0.01 vs. pooled control group. (**B**) Representative Western blotting showing the degradation of total pancreatic eIF4G 30 minutes after caerulein injection (from reference [86]).

still no understanding of a common triggering mechanism for the disease. These models share several common features including secretory blockade,[77,78,80,204] intracellular trypsin activation,[80,205,206] high levels of digestive enzymes in blood, cytoplasmic vacuolization,[78,205] and activation of NFκB with later induction of an inflammatory response.[207] Alterations in intracellular Ca^{2+} signaling, either by disturbances of calcium influx or by disturbances of calcium coming from the ER,[187,208–210] lead to the activation of multiple intracellular mechanisms that can involve the activation of ER stress, and it has been shown by us ([86]) and others[75–79] that total protein synthesis is inhibited in AP.

ER stress signals are already active in normal pancreas in conditions that are important for the development of the gland,[188] but these signals could likely be involved in the early stages of the development of the disease, to try to compensate any original imbalance. When intracellular calcium concentrations increase to supramaximal levels,[72] the compensatory mechanisms are overworked, and ER stress signals lead to the activation of programmed cell death mechanisms,[211] similar to the ones seen during acute pancreatitis.[212]

Kubisch et al. showed, for the first time, that all major ER stress sensing and signaling mechanisms are present in pancreatic acini and are activated early in the arginine model of experimental acute pancreatitis.[212] Arginine treatment caused an early activation of ER stress, as indicated by phosphorylation of PERK and its downstream target eIF2alpha, ATF6 translocation into the nucleus, and upregulation of BiP. XBP-1 splicing and CHOP expression were observed within 8 hours. After 24 hours, increased activation of the ER stress-related proapoptotic molecule caspase-12 was observed along with an increase in caspase-3 activity and TUNEL staining in exocrine acini. These results indicate that ER stress is an important early acinar cell event that could contribute to the development of acute pancreatitis and, with time, stimulate apoptosis in the arginine model.

Several studies directed at finding different physiologic and metabolic imbalances in the exocrine pancreas, such as oxidative stress, mitochondrial dysfunction and autophagy, deregulation of ATP generation and ER Acetyl-CoA (AT) availability, changes in lipid metabolism, and inflammatory and cell death responses,[213,214] have demonstrated that these mechanisms often first manifest by abnormal cytosolic Ca^{2+} signaling. After abnormal calcium cytosolic levels, there is a multifaceted set of organelles and cellular interactions that trigger ER stress in pancreatic acinar cells, leading the way to different degrees of cell damage[215] and, ultimately, to acute[216,217] and chronic[218] pancreatitis.

Other studies have demonstrated that mutations in some digestive enzymes, that can be hereditary, cause problems with their proper folding and trigger the ER stress response in the pancreas. This protein misfolding in the ER may also contribute to parenchymal damage by causing acinar cell death, and this may led to chronic pancreatitis.[219,220]

A proteomics study on the RER of pancreatic acinar cells, from rat pancreases of control and two experimental models of AP (arginine- and caerulein-induced), revealed an increase of many ER proteins during AP, compared to the control group, but showed differences in the amount of several others between the two models of AP, possibly due to the different degree of cell damage caused by the two models or to the different stages of the progression of cell damage.[221] For instance, there was a clear reduction of some chaperone proteins, like BiP, in the arginine model, possibly due to the activation of apoptotic processes, the latest steps of the UPR mechanisms, when chaperoning for unfolded proteins is no longer needed.

Along these lines, Lugea and coworkers suggest that the ER stress and UPR responses alone are not involved in the development of acute pancreatitis.[222] They argue that to be considered as triggering factors of the development of AP and cause cell damage, these mechanisms need to be combined with other environmental or genetic stressors. These authors study the effects of more clinically relevant causes of pancreatic injury in experimental models. They show that alcohol abuse triggers a UPR response that attenuates the ER stress caused by alcohol to the pancreas and suggest that early UPR mechanisms are beneficial for the well-being and survival of pancreatic acinar cells, because they are directed to compensate the original ER stress.[222,223] Lugea at al. demonstrate this hypothesis in another study[224] where the UPR was inhibited and cell death mechanisms were activated after applying, to AR42J cells, two well-known risk factors for clinical AP and chronic pancreatitis: cigarette smoke and alcohol.[225,226]

In summary, several cellular insults that cause damage to pancreatic acinar cells start with depleting Ca^{2+} from ER stores. This Ca^{2+} depletion triggers the initial ER stress response that inhibits digestive enzymes synthesis. The ER stress response tries to compensate for these original imbalances, through the UPR, and when the cell situation can no longer be counteracted, the UPR stimulates cell death mechanisms through apoptosis. Some of these early ER stress mechanisms could likely be involved in the triggering events of AP, but it is also very likely that several insults need to be present to develop disease. A balancing regulation in response to ER stress is essential for cell survival and may act as a protective mechanism in the pancreas, as it does in other tissues. Because of that, it has also been proposed that the ER protein folding machinery and the UPR responses could be potential therapeutic targets to prevent and treat pancreatic diseases.[227]

Acknowledgments

I want to express my gratitude to Prof. J. A. Williams for all his help, mentoring, scientific guidance, and collegiality during all these years working side by side. The research path is a hard road, and not always fairly compensated, but worth the ride, when you have such a mentor. Thank you, John, it wouldn't have been possible without you!

References

1. Webster PD, 3rd, Black O, Jr., Mainz DL, Singh M. Pancreatic acinar cell metabolism and function. Gastroenterology 1977; 73:1434–1449.

2. Sans MD, Lee SH, D'Alecy LG, Williams JA. Feeding activates protein synthesis in mouse pancreas at the translational level without increase in mRNA. Am J Physiol Gastrointest Liver Physiol 2004; 287:G667–G675.

3. Meyuhas O. Synthesis of the translational apparatus is regulated at the translational level. Eur J Biochem 2000; 267:6321–6330.

4. Sans MD, Crozier SJ, Vogel NL, D'Alecy LG, Williams JA. Dietary protein and amino acid deficiency inhibit pancreatic digestive enzyme mRNA translation by multiple mechanisms. Cell Mol Gastroenterol Hepatol 2021; 11:99–115.

5. Sans MD, Tashiro M, Vogel NL, Kimball SR, D'Alecy LG, Williams JA. Leucine activates pancreatic translational machinery in rats and mice through mTOR independently of CCK and insulin. J Nutr 2006; 136:1792–1799.

6. Hashi M, Yoshizawa F, Onozuka E, Ogata M, Hara H. Adaptive changes in translation initiation activities for rat pancreatic protein synthesis with feeding of a high-protein diet. J Nutr Biochem 2005; 16:507–512.

7. Hashimoto N, Hara H. Dietary amino acids promote pancreatic protease synthesis at the translation stage in rats. J Nutr 2003; 133:3052–3057.

8. O'Keefe S J, Lee RB, Li J, Zhou W, Stoll B, Dang Q. Trypsin and splanchnic protein turnover during feeding and fasting in human subjects. Am J Physiol Gastrointest Liver Physiol 2006; 290:G213–G221.

9. Jamieson JD, Palade GE. Synthesis, intracellular transport, and discharge of secretory proteins in stimulated pancreatic exocrine cells. J Cell Biol 1971; 50:135–158.

10. Cooley MM, Jones EK, Gorelick FS, Groblewski GE. Pancreatic acinar cell protein synthesis, intracellular transport, and export. Pancreapedia: Exocrine Pancreas Knowledge Base, 2020. DOI: 10.3998/panc.2020.15.

11. Schwarz DS, Blower MD. The endoplasmic reticulum: structure, function and response to cellular signaling. Cell Mol Life Sci 2016; 73:79–94.

12. Merrick WC, Pavitt GD. Protein synthesis initiation in eukaryotic cells. Cold Spring Harb Perspect Biol 2018; 10:a033092.

13. Hershey JWB, Merrick WC. Pathway and mechanism of initiation of protein synthesis. In: Translational control of gene expression. Volume 39. New York: Cold Spring Harbor Laboratory Press, 2000:33–88.

14. Merrick WC. The protein biosynthesis elongation cycle. In: Translational control of gene expression monograph. Volume 39. New York: Cold Spring Harb Lab Press, 2000:89–125.

15. Dever TE, Dinman JD, Green R. Translation elongation and recoding in eukaryotes. Cold Spring Harb Perspect Biol 2018; 10:a032649.

16. Sans MD, Xie Q, Williams JA. Regulation of translation elongation and phosphorylation of eEF2 in rat pancreatic acini. Biochem Biophys Res Commun 2004; 319:144–151.

17. Hartl FU, Hayer-Hartl M. Converging concepts of protein folding in vitro and in vivo. Nat Struct Mol Biol 2009; 16:574–581.

18. Ruggiano A, Foresti O, Carvalho P. Quality control: ER-associated degradation: protein quality control and beyond. J Cell Biol 2014; 204:869–879.

19. Harding HP, Calfon M, Urano F, Novoa I, Ron D. Transcriptional and translational control in the mammalian unfolded protein response. Annu Rev Cell Dev Biol 2002; 18:575–599.

20. Ron D, Harding HP. Protein-folding homeostasis in the endoplasmic reticulum and nutritional regulation. Cold Spring Harb Perspect Biol 2012; 4: 1–13, a013177

21. Walter P, Ron D. The unfolded protein response: from stress pathway to homeostatic regulation. Science 2011; 334:1081–1086.

22. Wek RC, Cavener DR. Translational control and the unfolded protein response. Antioxid Redox Signal 2007; 9:2357–2371.

23. Iida K, Li Y, McGrath BC, Frank A, Cavener DR. PERK eIF2α kinase is required to regulate the viability of the exocrine pancreas in mice. BMC Cell Biol 2007; 8:38.

24. Kim R, Emi M, Tanabe K, Murakami S. Role of the unfolded protein response in cell death. Apoptosis 2006; 11:5–13.

25. Tabas I, Ron D. Integrating the mechanisms of apoptosis induced by endoplasmic reticulum stress. Nat Cell Biol 2011; 13:184–190.

26. Wang XZ, Harding HP, Zhang Y, Jolicoeur EM, Kuroda M, Ron D. Cloning of mammalian Ire1 reveals diversity in the ER stress responses. Embo J 1998; 17:5708–5717.

27. Harding HP, Zhang Y, Ron D. Protein translation and folding are coupled by an endoplasmic-reticulum-resident kinase. Nature 1999; 397:271–274.

28. Yoshida H, Okada T, Haze K, Yanagi H, Yura T, Negishi M, Mori K. ATF6 activated by proteolysis binds in the presence of NF-Y (CBF) directly to the cis-acting element responsible for the mammalian unfolded protein response. Mol Cell Biol 2000; 20:6755–6767.

29. Harding HP, Novoa I, Zhang Y, Zeng H, Wek R, Schapira M, Ron D. Regulated translation initiation controls stress-induced gene expression in mammalian cells. Mol Cell 2000; 6:1099–1108.

30. Hetz C. The unfolded protein response: controlling cell fate decisions under ER stress and beyond. Nat Rev Mol Cell Biol 2012; 13:89–102.

31. Oakes SA, Papa FR. The role of endoplasmic reticulum stress in human pathology. Annu Rev Pathol 2015; 10:173–194.

32. Yoshida H. ER stress and diseases. FEBS J 2007; 274:630–658.

33. Lebeaupin C, Vallee D, Hazari Y, Hetz C, Chevet E, Bailly-Maitre B. Endoplasmic reticulum stress signalling and the pathogenesis of non-alcoholic fatty liver disease. J Hepatol 2018; 69:927–947.

34. Tahmasebi S, Khoutorsky A, Mathews MB, Sonenberg N. Translation deregulation in human disease. Nat Rev Mol Cell Biol 2018; 19:791–807.

35. Ryno LM, Wiseman RL, Kelly JW. Targeting unfolded protein response signaling pathways to ameliorate protein misfolding diseases. Curr Opin Chem Biol 2013; 17:346–352.

36. Mockett BG, Guevremont D, Wutte M, Hulme SR, Williams JM, Abraham WC. Calcium/calmodulin-dependent protein kinase II mediates group I metabotropic glutamate receptor-dependent protein synthesis and long-term depression in rat hippocampus. J Neurosci 2011; 31:7380–7391.

37. Brostrom CO, Brostrom MA. Calcium-dependent regulation of protein synthesis in intact mammalian cells. Annu Rev Physiol 1990; 52:577–590.

38. Miyamoto S, Patel P, Hershey JW. Changes in ribosomal binding activity of eIF3 correlate with increased translation rates during activation of T lymphocytes. J Biol Chem 2005; 280:28251–28264.

39. Chin KV, Cade C, Brostrom CO, Galuska EM, Brostrom MA. Calcium-dependent regulation of protein synthesis at translational initiation in eukaryotic cells. J Biol Chem 1987; 262:16509–16514.

40. Kimball SR, Jefferson LS. Regulation of protein synthesis by modulation of intracellular calcium in rat liver. Am J Physiol Endocrinol Metab 1992; 263:E958–E964.

41. Vassilakos A, Michalak M, Lehrman MA, Williams DB. Oligosaccharide binding characteristics of the molecular chaperones calnexin and calreticulin. Biochemistry 1998; 37:3480–3490.

42. Yule DI. Ca^{2+} signaling in pancreatic acinar cells. Pancreapedia: Exocrine Pancreas Knowledge Base, 2020. DOI: 10.3998/panc.2020.12.

43. Prostko CR, Brostrom MA, Brostrom CO. Reversible phosphorylation of eukaryotic initiation factor 2 alpha in response to endoplasmic reticular signaling. Mol Cell Biochem 1993; 127–128:255–265.

44. Srivastava SP, Davies MV, Kaufman RJ. Calcium depletion from the endoplasmic reticulum activates the double-stranded RNA-dependent protein kinase (PKR) to inhibit protein synthesis. J Biol Chem 1995; 270:16619–16624.

45. Tadic V, Prell T, Lautenschlaeger J, Grosskreutz J. The ER mitochondria calcium cycle and ER stress response as therapeutic targets in amyotrophic lateral sclerosis. Front Cell Neurosci 2014; 8:147.

46. Guerrero-Hernandez A, Verkhratsky A. Calcium signalling in diabetes. Cell Calcium 2014; 56:297–301.

47. Zhang IX, Raghavan M, Satin LS. The endoplasmic reticulum and calcium homeostasis in pancreatic beta cells. Endocrinology 2020; 161:bqz028.

48. Collins HE, Zhu-Mauldin X, Marchase RB, Chatham JC. STIM1/Orai1-mediated SOCE: current perspectives and potential roles in cardiac function and pathology. Am J Physiol Heart Circ Physiol 2013; 305:H446–H458.

49. Wang WA, Groenendyk J, Michalak M. Endoplasmic reticulum stress associated responses in cancer. Biochim Biophys Acta 2014; 1843:2143–2149.

50. Rabanal-Ruiz Y, Otten EG, Korolchuk VI. mTORC1 as the main gateway to autophagy. Essays Biochem 2017; 61:565–584.

51. Guertin DA, Sabatini DM. Defining the role of mTOR in cancer. Cancer Cell 2007; 12:9–22.

52. Hara K, Maruki Y, Long X, Yoshino K, Oshiro N, Hidayat S, Tokunaga C, Avruch J, Yonezawa K. Raptor, a binding partner of target of rapamycin (TOR), mediates TOR action. Cell 2002; 110:177–189.

53. Kim DH, Sarbassov DD, Ali SM, King JE, Latek RR, Erdjument-Bromage H, Tempst P, Sabatini DM. mTOR interacts with raptor to form a nutrient-sensitive complex that signals to the cell growth machinery. Cell 2002; 110:163–175.

54. Yang H, Rudge DG, Koos JD, Vaidialingam B, Yang HJ, Pavletich NP. mTOR kinase structure, mechanism and regulation. Nature 2013; 497:217–223.

55. Vander Haar E, Lee SI, Bandhakavi S, Griffin TJ, Kim DH. Insulin signalling to mTOR mediated by the Akt/PKB substrate PRAS40. Nat Cell Biol 2007; 9:316–323.

56. Peterson TR, Laplante M, Thoreen CC, Sancak Y, Kang SA, Kuehl WM, Gray NS, Sabatini DM. DEPTOR is an mTOR inhibitor frequently overexpressed in multiple myeloma cells and required for their survival. Cell 2009; 137:873–886.

57. Laplante M, Sabatini DM. mTOR signaling in growth control and disease. Cell 2012; 149:274–293.

58. Condon KJ, Sabatini DM. Nutrient regulation of mTORC1 at a glance. J Cell Sci 2019; 132:1–6

59. Fu W, Hall MN. Regulation of mTORC2 signaling. Genes 2020; 11:1–19

60. Proud CG. mTORC1 signalling and mRNA translation. Biochem Soc Trans 2009; 37:227–231.

61. Zhang Y, Nicholatos J, Dreier JR, Ricoult SJ, Widenmaier SB, Hotamisligil GS, Kwiatkowski DJ, Manning BD. Coordinated regulation of protein synthesis and degradation by mTORC1. Nature 2014; 513:440–443.

62. Blandino-Rosano M, Barbaresso R, Jimenez-Palomares M, Bozadjieva N, Werneck-de-Castro JP, Hatanaka M, Mirmira RG, Sonenberg N, Liu M, Ruegg MA, Hall MN, Bernal-Mizrachi E. Loss of mTORC1 signalling impairs beta-cell homeostasis and insulin processing. Nat Commun 2017; 8:16014.

63. Sans MD, Williams JA. The mTOR signaling pathway and regulation of pancreatic function. Pancreapedia: Exocrine Pancreas Knowledge Base, 2017. DOI: 10.3998/panc.2017.08.

64. Kern HF, Rausch U, Scheele GA. Regulation of gene expression in pancreatic adaptation to nutritional substrates or hormones. Gut 1987; 28:89–94.

65. Lahaie RG. Dietary regulation of protein synthesis in the exocrine pancreas. J Pediatr Gastroenterol Nutr 1984; 3:S43–50.

66. Bragado MJ, Groblewski GE, Williams JA. Regulation of protein synthesis by cholecystokinin in rat pancreatic acini involves PHAS-I and the p70 S6 kinase pathway. Gastroenterology 1998; 115:733–742.

67. Korc M, Bailey AC, Williams JA. Regulation of protein synthesis in normal and diabetic rat pancreas by cholecystokinin. Am J Physiol Gastrointest Liver Physiol 1981; 241:G116–G121.

68. Korc M, Iwamoto Y, Sankaran H, Williams JA, Goldfine ID. Insulin action in pancreatic acini from streptozotocin-treated rats. I. Stimulation of protein synthesis. Am J Physiol Gastrointest Liver Physiol 1981; 240:G56–G62.

69. Lahaie RG. Translational control of protein synthesis in isolated pancreatic acini: role of CCK8, carbachol, and insulin. Pancreas 1986; 1:403–410.

70. Okabayashi Y, Moessner J, Logsdon CD, Goldfine ID, Williams JA. Insulin and other stimulants have nonparallel translational effects on protein synthesis. Diabetes 1987; 36:1054–1060.

71. Perkins PS, Pandol SJ. Cholecystokinin-induced changes in polysome structure regulate protein synthesis in pancreas. Biochim Biophys Acta 1992; 1136:265–271.

72. Sans MD, Kimball SR, Williams JA. Effect of CCK and intracellular calcium to regulate eIF2B and protein synthesis in rat pancreatic acinar cells. Am J Physiol Gastrointest Liver Physiol 2002; 282:G267–G276.

73. Sans MD, Williams JA. Calcineurin is required for translational control of protein synthesis in rat pancreatic acini. Am J Physiol Cell Physiol 2004; 287:C310–C319.

74. Schick J, Kern H, Scheele G. Hormonal stimulation in the exocrine pancreas results in coordinate and anticoordinate regulation of protein synthesis. J Cell Biol 1984; 99:1569–1574.

75. Gilliland L, Steer ML. Effects of ethionine on digestive enzyme synthesis and discharge by mouse pancreas. Am J Physiol Gastrointest Liver Physiol 1980; 239:G418–G426.

76. Kern HF, Adler G, Scheele GA. Structural and biochemical characterization of maximal and supramaximal hormonal stimulation of rat exocrine pancreas. Scand J Gastroenterol 1985; 112:20–29.

77. Lampel M, Kern HF. Acute interstitial pancreatitis in the rat induced by excessive doses of a pancreatic secretagogue. Virchows Arch A Pathol Anat Histol 1977; 373:97–117.

78. Machovich R, Papp M, Fodor I. Protein synthesis in acute pancreatitis. Biochem Med 1970; 3:376–383.

79. Ogden JM, Modlin IM, Gorelick FS, Marks IN. Effect of buprenorphine on pancreatic enzyme synthesis and secretion in normal rats and rats with acute edematous pancreatitis. Dig Dis Sci 1994; 39:2407–2415.

80. Saluja AK, Saito I, Saluja M, Houlihan MJ, Powers RE, Meldolesi J, Steer M. In vivo rat pancreatic acinar cell function during supramaximal stimulation with caerulein. Am J Physiol Gastrointest Liver Physiol 1985; 249:G702–G710.

81. Sweiry JH, Emery PW, Munoz M, Doolabh K, Mann GE. Influx and incorporation into protein of L-phenylalanine in the perfused rat pancreas: effects of amino acid deprivation and carbachol. Biochim Biophys Acta 1991; 1070:135–142.

82. Sans MD. Measurement of pancreatic protein synthesis. Pancreapedia: Exocrine Pancreas Knowledge Base, 2010. DOI: 10.3998/panc.2010.16.

83. Davis TA, Fiorotto ML, Nguyen HV, Burrin DG. Aminoacyl-tRNA and tissue free amino acid pools are equilibrated after a flooding dose of phenylalanine. Am J Physiol Endocrinol Metab 1999; 277:E103–E109.

84. Otsuki M, Okabayashi Y, Ohki A, Suehiro I, Baba S. Dual effects of hydrocortisone on exocrine rat pancreas. Gastroenterology 1987; 93:1398–1403.

85. Lundholm K, Ternell M, Zachrisson H, Moldawer L, Lindstrom L. Measurement of hepatic protein synthesis in unrestrained mice-evaluation of the "flooding technique." Acta Physiol Scand 1991; 141:207–219.

86. Sans MD, DiMagno MJ, D'Alecy LG, Williams JA. Caerulein-induced acute pancreatitis inhibits protein synthesis through effects on eIF2B and eIF4F. Am J Physiol Gastrointest Liver Physiol 2003; 285:G517–G528.

87. Brannon PM. Adaptation of the exocrine pancreas to diet. Annu Rev Nutr 1990; 10:85–105.

88. Scheele GA. Regulation of pancreatic gene expression in response to hormone and nutritional substrates. In: Go VLW, ed. The pancreas: biology, pathobiology, and disease. 2nd ed. New York: Raven Press, Ltd., 1993:103–120.

89. Dagorn JC, Lahaie RG. Dietary regulation of pancreatic protein synthesis. I. Rapid and specific modulation of enzyme synthesis by changes in dietary composition. Biochim Biophys Acta 1981; 654:111–118.

90. Stockmann F, Soling HD. Regulation of biosynthesis of trypsinogen and chymotrypsinogen by nutritional and hormonal factors in the rat. Eur J Clin Invest 1981; 11:121–132.

91. Wicker C, Puigserver A. Effects of inverse changes in dietary lipid and carbohydrate on the synthesis of some pancreatic secretory proteins. Eur J Biochem 1987; 162:25–30.

92. Giorgi D, Renaud W, Bernard JP, Dagorn JC. Regulation of proteolytic enzyme activities and mRNA concentrations in rat pancreas by food content. Biochem Biophys Res Commun 1985; 127:937–942.

93. Hara H, Hashimoto N, Akatsuka N, Kasai T. Induction of pancreatic trypsin by dietary amino acids in rats: four trypsinogen isozymes and cholecystokinin messenger RNA. J Nutr Biochem 2000; 11:52–59.

94. Rosewicz S, Dunbar Lewis L, Wang X-Y, Liddle RA, Logsdon CD. Pancreatic digestive enzyme gene expression: effects of CCK and soybean trypsin inhibitor. Am J Physiol Gastrointest Liver Physiol 1989; 256:G733–G738.

95. Wicker C, Puigserver A, Rausch U, Scheele G, Kern H. Multiple-level caerulein control of the gene expression of secretory proteins in the rat pancreas. Eur J Biochem 1985; 151:461–466.

96. Steinhilber W, Poensgen J, Rausch U, Kern HF, Scheele GA. Translational control of anionic trypsinogen and amylase synthesis in rat pancreas in response to caerulein stimulation. Proc Natl Acad Sci U S A 1988; 85:6597–6601.

97. Wicker C, Puigserver A. Changes in mRNA levels of rat pancreatic lipase in the early days of consumption of a high-lipid diet. Eur J Biochem 1989; 180:563–567.

98. Ricketts J, Brannon PM. Amount and type of dietary fat regulate pancreatic lipase gene expression in rats. J Nutr 1994; 124:1166–1171.

99. Rausch U, Rudiger K, Vasiloudes P, Kern H, Scheele G. Lipase synthesis in the rat pancreas is regulated by secretin. Pancreas 1986; 1:522–528.

100. Williams JA, Goldfine ID. The insulin-acinar relationship. In: Go VLW, ed. The exocrine pancreas: biology, pathobiology, and diseases. New York: Raven Press, 1986:347–360.

101. Sans MD, Williams JA. Regulation of pancreatic protein synthesis and growth. In: Beger HG, Warshaw AL, Hruban RH, Büchler MW, Lerch MM, Neoptolemos JP, Shimosegawa T, Whitcomb DC, eds. The pancreas: an integrated textbook of basic science, medicine, and surgery. 3rd ed. Hoboken, NJ: John Wiley & Sons Ltd., 2018:95–105.

102. Schmid RM, Meisler MH. Dietary regulation of pancreatic amylase in transgenic mice mediated by a 126-base pair DNA fragment. Am J Physiol Gastrointest Liver Physiol 1992; 262:G971–G976.

103. Case RM. Synthesis, intracellular transport and discharge of exportable proteins in the pancreatic acinar cell and other cells. Biol Rev Camb Philos Soc 1978; 53:211–354.

104. Laass MW, Henker J, Thamm K, Neumeister V, Kuhlisch E. Exocrine pancreatic insufficiency and its consequences on physical development and metabolism in children and adolescents with type 1 diabetes mellitus. Eur J Pediatr 2004; 163:681–682.

105. Karau A, Grayson I. Amino acids in human and animal nutrition. Adv Biochem Eng Biotechnol 2014; 143:189–228.

106. Santos CS, Nascimento FEL. Isolated branched-chain amino acid intake and muscle protein synthesis in humans: a biochemical review. Einstein (Sao Paulo) 2019; 17:eRB4898.

107. Williams CD, Oxon BM, Lond H. Kwashiorkor: a nutritional disease of children associated with a maize diet. 1935. Bull World Health Organ 2003; 81:912–913.

108. Millward DJ. Identifying recommended dietary allowances for protein and amino acids: a critique of the 2007 WHO/FAO/UNU report. Br J Nutr 2012; 108:S3–21.

109. Schrader H, Menge BA, Belyaev O, Uhl W, Schmidt WE, Meier JJ. Amino acid malnutrition in patients with chronic pancreatitis and pancreatic carcinoma. Pancreas 2009; 38:416–421.

110. Wagenmakers AJ, Coakley JH, Edwards RH. Metabolism of branched-chain amino acids and ammonia during exercise: clues from McArdle's disease. Int J Sports Med 1990; 11 Suppl 2:S101–113.

111. Zhang S, Zeng X, Ren M, Mao X, Qiao S. Novel metabolic and physiological functions of branched chain amino acids: a review. J Anim Sci Biotechnol 2017; 8:10.

112. Xiao F, Wang C, Yin H, Yu J, Chen S, Fang J, Guo F. Leucine deprivation inhibits proliferation and induces apoptosis of human breast cancer cells via fatty acid synthase. Oncotarget 2016; 7:63679–63689.

113. Wei S, Zhao J, Wang S, Huang M, Wang Y, Chen Y. Intermittent administration of a leucine-deprived diet is able to intervene in type 2 diabetes in db/db mice. Heliyon 2018; 4:e00830.

114. Fontana L, Cummings NE, Arriola Apelo SI, Neuman JC, Kasza I, Schmidt BA, Cava E, Spelta F, Tosti V, Syed FA, Baar EL, Veronese N, et al. Decreased consumption of branched-chain amino acids improves metabolic health. Cell Rep 2016; 16:520–530.

115. El-Hodhod MA, Nassar MF, Hetta OA, Gomaa SM. Pancreatic size in protein energy malnutrition: a predictor of nutritional recovery. Eur J Clin Nutr 2005; 59:467–473.

116. Descos L, Duclieu J, Minaire Y. Exocrine pancreatic insufficiency and primitive malnutrition. Digestion 1977; 15:90–95.

117. Crozier SJ, D'Alecy LG, Ernst SA, Ginsburg LE, Williams JA. Molecular mechanisms of pancreatic dysfunction induced by protein malnutrition. Gastroenterology 2009; 137:1093–1101.

118. Crozier SJ, Sans MD, Wang JY, Lentz SI, Ernst SA, Williams JA. CCK-independent mTORC1 activation during dietary protein-induced exocrine pancreas growth. Am J Physiol Gastrointest Liver Physiol 2010; 299:G1154–G1163.

119. Morales A, Buenabad L, Castillo G, Vazquez L, Espinoza S, Htoo JK, Cervantes M. Dietary levels of protein and free amino acids affect pancreatic proteases activities, amino acids transporters expression and serum amino acid concentrations in starter pigs. J Anim Physiol Anim Nutr 2017; 101:723–732.

120. Cao YC, Yang XJ, Guo L, Zheng C, Wang DD, Cai CJ, Liu SM, Yao JH. Effects of dietary leucine and phenylalanine on pancreas development, enzyme activity, and relative gene

expression in milk-fed Holstein dairy calves. J Dairy Sci 2018; 101:4235–4244.

121. Wek S, Zhu S, Wek R. The histidyl-tRNA synthetase-related sequence in the eIF-2 alpha protein kinase GCN2 interacts with tRNA and is required for activation in response to starvation for different amino acids. Mol Cell Biol 1995; 15:4497–4506.

122. Wek RC, Jiang HY, Anthony TG. Coping with stress: eIF2 kinases and translational control. Biochem Soc Trans 2006; 34:7–11.

123. Anthony TG, McDaniel BJ, Byerley RL, McGrath BC, Cavener DR, McNurlan MA, Wek RC. Preservation of liver protein synthesis during dietary leucine deprivation occurs at the expense of skeletal muscle mass in mice deleted for eIF2 kinase GCN2. J Biol Chem 2004; 279:36553–36561.

124. Cao Y, Liu K, Liu S, Guo L, Yao J, Cai C. Leucine regulates the exocrine function in pancreatic tissue of dairy goats in vitro. Biomed Res Int 2019; 2019:7521715.

125. Guo L, Yao JH, Zheng C, Tian HB, Liu YL, Liu SM, Cai CJ, Xu XR, Cao YC. Leucine regulates α-amylase and trypsin synthesis in dairy calf pancreatic tissue in vitro via the mammalian target of rapamycin signalling pathway. Animal 2019; 13:1899–1906.

126. Lacourse KA, Swanberg LJ, Gillespie PJ, Rehfeld JF, Saunders TL, Samuelson LC. Pancreatic function in CCK-deficient mice: adaptation to dietary protein does not require CCK. Am J Physiol Gastrointest Liver Physiol 1999; 276:G1302–G1309.

127. Bragado MJ, Tashiro M, Williams JA. Regulation of the initiation of pancreatic digestive enzyme protein synthesis by cholecystokinin in rat pancreas in vivo. Gastroenterology 2000; 119:1731–1739.

128. Bragado MJ, Groblewski GE, Williams JA. p70s6k is activated by CCK in rat pancreatic acini. Am J Physiol Cell Physiol 1997; 273:C101–C109.

129. Williams JA, Goldfine ID. The insulin-pancreatic acinar axis. Diabetes 1985; 34:980–986.

130. Bertelli E, Regoli M, Orazioli D, Bendayan M. Association between islets of Langerhans and pancreatic ductal system in adult rat. Where endocrine and exocrine meet together? Diabetologia 2001; 44:575–584.

131. Gomez-Cerezo J, Garces MC, Codoceo R, Soto A, Arnalich F, Barbado J, Vazquez JJ. Postprandial glucose-dependent insulinotropic polypeptide and insulin responses in patients with chronic pancreatitis with and without secondary diabetes. Regul Pept 1996; 67:201–205.

132. Ewald N, Bretzel RG. Diabetes mellitus secondary to pancreatic diseases (Type 3c)—are we neglecting an important disease? Eur J Intern Med 2013; 24:203–206.

133. Hardt PD, Hauenschild A, Nalop J, Marzeion AM, Jaeger C, Teichmann J, Bretzel RG, Hollenhorst M, Kloer HU. High prevalence of exocrine pancreatic insufficiency in diabetes mellitus. A multicenter study screening fecal elastase 1 concentrations in 1,021 diabetic patients. Pancreatology 2003; 3:395–402.

134. Meisterfeld R, Ehehalt F, Saeger HD, Solimena M. Pancreatic disorders and diabetes mellitus. Exp Clin Endocrinol Diabetes 2008; 116 Suppl 1:S7–S12.

135. Mann E, Sunni M, Bellin MD. Secretion of insulin in response to diet and hormones. Pancreapedia: Exocrine Pancreas Knowledge Base, 2020. DOI: 10.3998/panc.2020.16.

136. Sans MD, Bruce JI, Williams JA. Regulation of pancreatic exocrine function by islet hormones. Pancreapedia: Exocrine Pancreas Knowledge Base, 2020. DOI: 10.3998/panc.2020.01.

137. Jefferson LS. Lilly Lecture 1979: role of insulin in the regulation of protein synthesis. Diabetes 1980; 29:487–496.

138. Korc M, Sankaran H, Wong KY, Williams JA, Goldfine ID. Insulin receptors in isolated mouse pancreatic acini. Biochem Biophys Res Commun 1978; 84:293–299.

139. Couture Y, Dunnigan J, Morisset. Stimulation of pancreatic amylase secretion and protein synthesis by insulin. Scand J Gastroenterol 1972; 7:257–263.

140. Duan RD, Wicker C, Erlanson-Albertsson C. Effect of insulin administration on contents, secretion, and synthesis of pancreatic lipase and colipase in rats. Pancreas 1991; 6:595–602.

141. Kramer MF, Poort C. The effect of insulin on amino acid incorporation into exocrine pancreatic cells of the rat. Horm Metab Res 1975; 7:389–393.

142. Trimble ER, Rausch U, Kern HF. Changes in individual rates of pancreatic enzyme and isoenzyme biosynthesis in the obese Zucker rat. Biochem J 1987; 248:771–777.

143. Yagihashi S. Exocrine pancreas of streptozotocin-diabetes rats treated with insulin. Tohoku J Exp Med 1976; 120:31–42.

144. Sankaran H, Iwamoto Y, Korc M, Williams JA, Goldfine ID. Insulin action in pancreatic acini from streptozotocin-treated rats. II. Binding of ^{125}I-insulin to receptors. Am J Physiol Gastrointest Liver Physiol 1981; 240:G63–G68.

145. Mossner J, Logsdon CD, Williams JA, Goldfine ID. Insulin, via its own receptor, regulates growth and amylase synthesis in pancreatic acinar AR42J cells. Diabetes 1985; 34:891–897.

146. Williams JA. Cholecystokinin (CCK) regulation of pancreatic acinar cells: Physiological actions and signal transduction mechanisms. Compr Physiol 2019; 9:535–564.

147. Sung CK, Williams JA. Insulin and ribosomal protein S6 kinase in rat pancreatic acini. Diabetes 1989; 38:544–549.

148. Burnham DB, Williams JA. Effects of carbachol, cholecystokinin, and insulin on protein phosphorylation in isolated pancreatic acini. J Biol Chem 1982; 257:10523–10528.

149. Berna MJ, Tapia JA, Sancho V, Thill M, Pace A, Hoffmann KM, Gonzalez-Fernandez L, Jensen RT. Gastrointestinal growth factors and hormones have divergent effects on Akt activation. Cell Signal 2009; 21:622–638.

150. Sans MD, Amin RK, Vogel NL, D'Alecy LG, Kahn RC, Williams JA. Specific deletion of insulin receptors on pancreatic acinar cells defines the insulin-acinar axis: implications for pancreatic insufficiency in diabetes. Gastroenterology 2011; 140:A-233.

151. Hardt PD, Ewald N. Exocrine pancreatic insufficiency in diabetes mellitus: a complication of diabetic neuropathy or a different type of diabetes? Exp Diabetes Res 2011; 2011:761950.

152. Foulis AK, Frier BM. Pancreatic endocrine-exocrine function in diabetes: an old alliance disturbed. Diabet Med 1984; 1:263–266.

153. Foulis AK, Stewart JA. The pancreas in recent-onset type 1 (insulin-dependent) diabetes mellitus: insulin content of islets, insulitis and associated changes in the exocrine acinar tissue. Diabetologia 1984; 26:456–461.

154. Gepts W. Pathologic anatomy of the pancreas in juvenile diabetes mellitus. Diabetes 1965; 14:619–633.

155. Mohapatra S, Majumder S, Smyrk TC, Zhang L, Matveyenko A, Kudva YC, Chari ST. Diabetes mellitus is associated with an exocrine pancreatopathy: conclusions from a review of literature. Pancreas 2016; 45:1104–1110.

156. Frier BM, Saunders JH, Wormsley KG, Bouchier IA. Exocrine pancreatic function in juvenile-onset diabetes mellitus. Gut 1976; 17:685–691.

157. Vesterhus M, Raeder H, Johansson S, Molven A, Njolstad PR. Pancreatic exocrine dysfunction in maturity-onset diabetes of the young type 3. Diabetes Care 2008; 31:306–310.

158. Kang JH, Na KJ, Mo IP, Chang D, Yang MP. Juvenile diabetes mellitus accompanied by exocrine pancreatic insufficiency in a dog. J Vet Med Sci 2008; 70:1337–1340.

159. Barreto SG, Carati CJ, Toouli J, Saccone GT. The islet-acinar axis of the pancreas: more than just insulin. Am J Physiol Gastrointest Liver Physiol 2010; 299:G10–G22.

160. Balk MW, Lang CM, White WJ, Munger BL. Exocrine pancreatic dysfunction in guinea pigs with diabetes mellitus. Lab Invest 1975; 32:28–32.

161. Snook JT. Effect of diet, adrenalectomy, diabetes, and actinomycin D on exocrine pancreas. Am J Physiol 1968; 215:1329–1333.

162. Soling HD, Unger KO. The role of insulin in the regulation of α-amylase synthesis in the rat pancreas. Eur J Clin Invest 1972; 2:199–212.

163. Korc M, Owerbach D, Quinto C, Rutter WJ. Pancreatic islet-acinar cell interaction: amylase messenger RNA levels are determined by insulin. Science 1981; 213:351–353.

164. Sofrankova A, Dockray GJ. Cholecystokinin- and secretin-induced pancreatic secretion in normal and diabetic rats. Am J Physiol Gastrointest Liver Physiol 1983; 244:G370–G374.

165. Han J, Liu YQ. Suppressed glucose metabolism in acinar cells might contribute to the development of exocrine pancreatic insufficiency in streptozotocin-induced diabetic mice. Metabolism 2010; 59:1257–1267.

166. Otsuki M, Williams JA. Effect of diabetes mellitus on the regulation of enzyme secretion by isolated rat pancreatic acini. J Clin Invest 1982; 70:148–156.

167. Williams JA, Sankaran H, Korc M, Goldfine ID. Receptors for cholecystokinin and insulin in isolated pancreatic acini: hormonal control of secretion and metabolism. Fed Proc 1981; 40:2497–2502.

168. Meagher C, Tang Q, Fife BT, Bour-Jordan H, Wu J, Pardoux C, Bi M, Melli K, Bluestone JA. Spontaneous development of a pancreatic exocrine disease in CD28-deficient NOD mice. J Immunol 2008; 180:7793–7803.

169. Wang J, Takeuchi T, Tanaka S, Kubo SK, Kayo T, Lu D, Takata K, Koizumi A, Izumi T. A mutation in the insulin 2 gene induces diabetes with severe pancreatic beta-cell dysfunction in the Mody mouse. J Clin Invest 1999; 103:27–37.

170. Yule DI. Cell calcium. Special issue: Ca²⁺ signaling and secretory function. Preface. Cell Calcium 2014; 55:281.

171. Sehgal P, Szalai P, Olesen C, Praetorius HA, Nissen P, Christensen SB, Engedal N, Moller JV. Inhibition of the sarco/endoplasmic reticulum (ER) Ca²⁺-ATPase by thapsigargin analogs induces cell death via ER Ca²⁺ depletion and the unfolded protein response. J Biol Chem 2017; 292:19656–19673.

172. Perkins PS, Park JH, Pandol SJ. The role of calcium in the regulation of protein synthesis in the exocrine pancreas. Pancreas 1997; 14:133–141.

173. Klee CB, Ren H, Wang X. Regulation of the calmodulin-stimulated protein phosphatase, calcineurin. J Biol Chem 1998; 273:13367–13370.

174. Rusnak F, Mertz P. Calcineurin: form and function. Physiol Rev 2000; 80:1483–1521.

175. Burnham DB. Characterization of Ca²⁺-activated protein phosphatase activity in exocrine pancreas. Biochem J 1985; 231:335–341.

176. Groblewski GE, Wagner AC, Williams JA. Cyclosporin A inhibits Ca²⁺/calmodulin-dependent protein phosphatase and secretion in pancreatic acinar cells. J Biol Chem 1994; 269:15111–15117.

177. Clipstone NA, Crabtree GR. Identification of calcineurin as a key signalling enzyme in T-lymphocyte activation. Nature 1992; 357:695–697.

178. Donelan MJ, Morfini G, Julyan R, Sommers S, Hays L, Kajio H, Briaud I, Easom RA, Molkentin JD, Brady ST, Rhodes CJ. Ca²⁺-dependent dephosphorylation of kinesin heavy chain on β-granules in pancreatic β-cells. Implications for regulated β-granule transport and insulin exocytosis. J Biol Chem 2002; 277:24232–24242.

179. Doi R, Inoue K, Chowdhury P, Kaji H, Rayford PL. Structural and functional changes of exocrine pancreas induced by FK506 in rats. Gastroenterology 1993; 104:1153–1164.

180. Buss WC, Stepanek J. Tissue specificity of translation inhibition in Sprague-Dawley rats following in vivo cyclosporin A. Int J Immunopharmacol 1993; 15:775–782.

181. Creamer TP. Calcineurin. Cell Commun Signal 2020; 18:137.

182. Waschulewski IH, Hall DV, Kern HF, Edwardson JM. Effects of the immunosuppressants cyclosporin A and FK 506 on exocytosis in the rat exocrine pancreas in vitro. Br J Pharmacol 1993; 108:892–900.

183. Tashiro M, Samuelson LC, Liddle RA, Williams JA. Calcineurin mediates pancreatic growth in protease inhibitor-treated mice. Am J Physiol Gastrointest Liver Physiol 2004; 286:G784–G790.

184. Williams JA. Regulation of normal and adaptive pancreatic growth. Pancreapedia: Exocrine Pancreas Knowledge Base, 2020. DOI: 10.3998/panc.2020.05.

185. Pardee AB. G1 events and regulation of cell proliferation. Science 1989; 246:603–608.

186. Sans MD, Williams JA. Translational control of protein synthesis in pancreatic acinar cells. Int J Gastrointest Cancer 2002; 31:107–115.

187. Sutton R, Petersen OH, Pandol SJ. Pancreatitis and calcium signalling: report of an international workshop. Pancreas 2008; 36:e1–14.

188. Harding HP, Zeng H, Zhang Y, Jungries R, Chung P, Plesken H, Sabatini DD, Ron D. Diabetes mellitus and exocrine

pancreatic dysfunction in perk⁻/⁻ mice reveals a role for translational control in secretory cell survival. Mol Cell 2001; 7:1153–1163.

189. Harding HP, Zhang Y, Bertolotti A, Zeng H, Ron D. Perk is essential for translational regulation and cell survival during the unfolded protein response. Mol Cell 2000; 5:897–904.

190. Gronroos JM, Aho HJ, Hietaranta AJ, Nevalainen TJ. Early acinar cell changes in caerulein-induced interstitial acute pancreatitis in the rat. Exp Pathol 1991; 41:21–30.

191. Korbova L, Kohout J, Malis F, Balas V, Cizkova J, Marek J, Cihak A. Inhibitory effect of various cytostatics and cycloheximide on acute experimental pancreatitis in rats. Gut 1977; 18:913–918.

192. Abe R, Shimosegawa T, Kimura K, Takasu A, Koizumi M, Toyota T. Lipopolysaccharide-induced desensitization to pancreatic edema formation in rat cerulein pancreatitis. Pancreas 1998; 16:539–544.

193. Kaiser AM, Saluja AK, Lu L, Yamanaka K, Yamaguchi Y, Steer ML. Effects of cycloheximide on pancreatic endonuclease activity, apoptosis, and severity of acute pancreatitis. Am J Physiol Cell Physiol 1996; 271:C982–C993.

194. Resau JH, Marzella L, Jones RT, Trump BF. What's new in in vitro studies of exocrine pancreatic cell injury? Pathol Res Pract 1985; 179:576–588.

195. Bhatia M. Apoptosis versus necrosis in acute pancreatitis. Am J Physiol Gastrointest Liver Physiol 2004; 286:G189–G196.

196. Bhatia M. Apoptosis of pancreatic acinar cells in acute pancreatitis: is it good or bad? J Cell Mol Med 2004; 8:402–409.

197. Frossard JL, Rubbia-Brandt L, Wallig MA, Benathan M, Ott T, Morel P, Hadengue A, Suter S, Willecke K, Chanson M. Severe acute pancreatitis and reduced acinar cell apoptosis in the exocrine pancreas of mice deficient for the Cx32 gene. Gastroenterology 2003; 124:481–493.

198. Bhatia M, Wallig MA, Hofbauer B, Lee HS, Frossard JL, Steer ML, Saluja AK. Induction of apoptosis in pancreatic acinar cells reduces the severity of acute pancreatitis. Biochem Biophys Res Commun 1998; 246:476–483.

199. Mareninova OA, Sung KF, Hong P, Lugea A, Pandol SJ, Gukovsky I, Gukovskaya AS. Cell death in pancreatitis: caspases protect from necrotizing pancreatitis. J Biol Chem 2006; 281:3370–3381.

200. Guillaumes S, Blanco I, Villanueva A, Sans MD, Clave P, Chabas A, Farre A, Lluis F. Chloroquine stabilizes pancreatic lysosomes and improves survival of mice with diet-induced acute pancreatitis. Pancreas 1997; 14:262–266.

201. van Acker GJ, Perides G, Steer ML. Co-localization hypothesis: a mechanism for the intrapancreatic activation of digestive enzymes during the early phases of acute pancreatitis. World J Gastroenterol 2006; 12:1985–1990.

202. Van Acker GJ, Weiss E, Steer ML, Perides G. Cause-effect relationships between zymogen activation and other early events in secretagogue-induced acute pancreatitis. Am J Physiol Gastrointest Liver Physiol 2007; 292:G1738–G1746.

203. Kukor Z, Mayerle J, Kruger B, Toth M, Steed PM, Halangk W, Lerch MM, Sahin-Toth M. Presence of cathepsin B in the human pancreatic secretory pathway and its role in trypsinogen activation during hereditary pancreatitis. J Biol Chem 2002; 277:21389–21396.

204. Steer ML. Pathogenesis of acute pancreatitis. Digestion 1997; 58 Suppl 1:46–49.

205. Steer ML, Saluja AK. Experimental acute pancreatitis: studies of the early events that lead to cell injury. In: Go VLW, DiMagno EP, Gardner JD, Lebenthal E, Reber HA, Scheele GA, eds. The pancreas. Biology, pathobiology and disease. 2nd ed. New York: Raven Press, 1993:489–500.

206. Hofbauer B, Saluja AK, Lerch MM, Bhagat L, Bhatia M, Lee HS, Frossard JL, Adler G, Steer ML. Intra-acinar cell activation of trypsinogen during caerulein-induced pancreatitis in rats. Am J Physiol Gastrointest Liver Physiol 1998; 275:G352–G362.

207. Chen X, Ji B, Han B, Ernst SA, Simeone D, Logsdon CD. NF-κB activation in pancreas induces pancreatic and systemic inflammatory response. Gastroenterology 2002; 122:448–457.

208. Fanczal J, Pallagi P, Gorog M, Diszhazi G, Almassy J, Madacsy T, Varga A, Csernay-Biro P, Katona X, Toth E, Molnar R, Rakonczay Z, Jr., et al. TRPM2-mediated extracellular Ca²⁺ entry promotes acinar cell necrosis in biliary acute pancreatitis. J Physiol 2020; 598:1253–1270.

209. Bruce JIE. Metabolic regulation of the PMCA: role in cell death and survival. Cell Calcium 2018; 69:28–36.

210. Bruce JI, Elliott AC. Oxidant-impaired intracellular Ca²⁺ signaling in pancreatic acinar cells: role of the plasma membrane Ca²⁺-ATPase. Am J Physiol Cell Physiol 2007; 293:C938–C950.

211. Pandol SJ, Gorelick FS, Lugea A. Environmental and genetic stressors and the unfolded protein response in exocrine pancreatic function—a hypothesis. Front Physiol 2011; 2:8.

212. Kubisch CH, Sans MD, Arumugam T, Ernst SA, Williams JA, Logsdon CD. Early activation of endoplasmic reticulum stress is associated with arginine-induced acute pancreatitis. Am J Physiol Gastrointest Liver Physiol 2006; 291:G238–G245.

213. Srinivasan MP, Bhopale KK, Caracheo AA, Kaphalia L, Loganathan G, Balamurugan AN, Rastellini C, Kaphalia BS. Differential cytotoxicity, ER/oxidative stress, dysregulated AMPKalpha signaling, and mitochondrial stress by ethanol and its metabolites in human pancreatic acinar cells. Alcohol Clin Exp Res 2021; 45:961–978.

214. Biczo G, Vegh ET, Shalbueva N, Mareninova OA, Elperin J, Lotshaw E, Gretler S, Lugea A, Malla SR, Dawson D, Ruchala P, Whitelegge J, et al. Mitochondrial dysfunction, through impaired autophagy, leads to endoplasmic reticulum stress, deregulated lipid metabolism, and pancreatitis in animal models. Gastroenterology 2018; 154:689–703.

215. Ito M, Nakagawa H, Okada T, Miyazaki S, Matsuo S. ER-stress caused by accumulated intracistanal granules activates autophagy through a different signal pathway from unfolded protein response in exocrine pancreas cells of rats exposed to fluoride. Arch Toxicol 2009; 83:151–159.

216. Gukovskaya AS, Gorelick FS, Groblewski GE, Mareninova OA, Lugea A, Antonucci L, Waldron RT, Habtezion A, Karin M, Pandol SJ, Gukovsky I. Recent insights into the pathogenic mechanism of pancreatitis: role of acinar cell organelle disorders. Pancreas 2019; 48:459–470.

217. Habtezion A, Gukovskaya AS, Pandol SJ. Acute pancreatitis: a multifaceted set of organelle and cellular interactions. Gastroenterology 2019; 156:1941–1950.

218. Cooley MM, Thomas DDH, Deans K, Peng Y, Lugea A, Pandol SJ, Puglielli L, Groblewski GE. Deficient endoplasmic reticulum acetyl-CoA import in pancreatic acinar cells leads to chronic pancreatitis. Cell Mol Gastroenterol Hepatol 2021; 11:725–738.

219. Szmola R, Sahin-Toth M. Pancreatitis-associated chymotrypsinogen C (CTRC) mutant elicits endoplasmic reticulum stress in pancreatic acinar cells. Gut 2010; 59:365–372.

220. Hegyi E, Sahin-Toth M. Human CPA1 mutation causes digestive enzyme misfolding and chronic pancreatitis in mice. Gut 2019; 68:301–312.

221. Chen X, Sans MD, Strahler JR, Karnovsky A, Ernst SA, Michailidis G, Andrews PC, Williams JA. Quantitative organellar proteomics analysis of rough endoplasmic reticulum from normal and acute pancreatitis rat pancreas. J Proteome Res 2010; 9:885–896.

222. Lugea A, Waldron RT, Pandol SJ. Pancreatic adaptive responses in alcohol abuse: role of the unfolded protein response. Pancreatology 2015; 15:S1–S5.

223. Pandol SJ, Gorelick FS, Gerloff A, Lugea A. Alcohol abuse, endoplasmic reticulum stress and pancreatitis. Dig Dis 2010; 28:776–782.

224. Lugea A, Gerloff A, Su HY, Xu Z, Go A, Hu C, French SW, Wilson JS, Apte MV, Waldron RT, Pandol SJ. The combination of alcohol and cigarette smoke induces endoplasmic reticulum stress and cell death in pancreatic acinar cells. Gastroenterology 2017; 153:1674–1686.

225. Lankisch PG, Breuer N, Bruns A, Weber-Dany B, Lowenfels AB, Maisonneuve P. Natural history of acute pancreatitis: a long-term population-based study. Am J Gastroenterol 2009; 104:2797–2805.

226. Cote GA, Yadav D, Slivka A, Hawes RH, Anderson MA, Burton FR, Brand RE, Banks PA, Lewis MD, Disario JA, Gardner TB, Gelrud A, et al. Alcohol and smoking as risk factors in an epidemiology study of patients with chronic pancreatitis. Clin Gastroenterol Hepatol 2011; 9:266–273.

227. Lukas J, Pospech J, Oppermann C, Hund C, Iwanov K, Pantoom S, Petters J, Frech M, Seemann S, Thiel FG, Modenbach JM, Bolsmann R, et al. Role of endoplasmic reticulum stress and protein misfolding in disorders of the liver and pancreas. Adv Med Sci 2019; 64:315–323.

Chapter 8

Calcium signaling in pancreatic acinar cells

David I. Yule

Department of Pharmacology and Physiology, University of Rochester. Rochester, NY 14642

Introduction

Intracellular Ca^{2+} signaling and the primary function of the exocrine pancreas to secrete digestive enzymes are intimately linked. A secretagogue-induced rise in the level of intracellular Ca^{2+} signaling is the pivotal intracellular event triggering both the molecular machinery that results in exocytosis of zymogen granules and also the activation of ion channels that leads to the secretion of the primary fluid from acinar cells. Given the need to ensure vectoral secretion of a potentially hazardous cargo into the lumen of the gland, the Ca^{2+} signals and activation of effectors are elaborately orchestrated in time and space. The importance of the physiological Ca^{2+} signal is reinforced by increasing evidence that disordered Ca^{2+} signaling is causally associated with pancreatic pathology.[1] The aim of this chapter is twofold—first to provide the reader with an overview of the spatiotemporal features of intracellular Ca^{2+} signals in acinar cells as measured by contemporary high-resolution fluorescence measurements. Second, a review will be presented of the current understanding of the molecular mechanisms underpinning Ca^{2+} release, Ca^{2+} influx, and clearance—processes that are fundamental for the specific properties of intracellular Ca^{2+} signals in acinar cells. The reader is referred to various "Method" submissions in *The Pancreapedia* regarding current imaging techniques used to monitor intracellular Ca^{2+} changes with high spatial and temporal resolution in acinar cells.

Signal transduction leading to secretagogue-induced cytosolic calcium signals in acinar cells

The primary pancreatic secretagogues, cholecystokinin (CCK), acetylcholine (ACh), and bombesin, share a common general mechanism to raise $[Ca^{2+}]_i$. It is well established that each agonist binds a specific, cognate, seven-transmembrane domain receptor in the plasma membrane (PM) that couples to the heterotrimeric G protein, $G\alpha q$. Activation of $G\alpha q$ results in the increased activity of phosphoinositide (PI)-specific phospholipase C which

cleaves phosphatidylinositol, 4,5-bisphosphate (PIP$_2$) in the PM and subsequently generates the second messengers inositol 1,4,5-trisphosphate (IP$_3$) and diacylglycerol. Historically, studies in pancreatic acinar cells have played important roles in establishing the signal transduction pathway utilized by so-called "calcium-mobilizing" agents. A seminal paper by Streb and colleagues[2] was the first to demonstrate that IP$_3$ could induce Ca^{2+} release in studies using permeabilized pancreatic acini.[2] Notably, release could only be evoked from rough ER vesicles but not mitochondria or vesicles prepared from plasma membrane.[3] These data confirmed earlier studies that secretagogues stimulation resulted in substantial Ca^{2+} loss from ER but not zymogen granules or mitochondria.[4] Finally, the causative "loop" was then closed by the demonstration that stimulation with pancreatic secretagogues resulted in the rapid formation of IP$_3$.[5] Notably, this pathway has subsequently been shown to be ubiquitous and accounts for the generation of Ca^{2+} signaling events important for a multitude of cellular events in all tissue systems.

The Characteristics of Secretagogue—Induced Cytosolic Calcium Signals in Acinar Cells

Temporal properties at high concentrations of secretagogues

In common with many nonelectrically excitable cell types, the spatiotemporal patterns of Ca^{2+} signals elicited in pancreatic acinar cells are dependent on the concentration of agonist. Regarding the temporal characteristics of signals in isolated pancreatic acinar cells, experiments monitoring the global intracellular Ca^{2+} concentration ($[Ca^{2+}]_i$) using fluorometric techniques in single rodent cells reported that maximum concentrations of CCK, Ach, or bombesin evoked a similar Ca^{2+} signal. Exposure to either agonist results in a rapid 5–12-fold increase in $[Ca^{2+}]_i$ from a basal value of $[Ca^{2+}]$ ~75–150 nM to reach a peak within seconds of approximately 1 μM (**Figure 1A**). This peak then declined over 2–5 minutes to reach a new plateau level

e-mail: David_Yule@urmc.rochester.edu

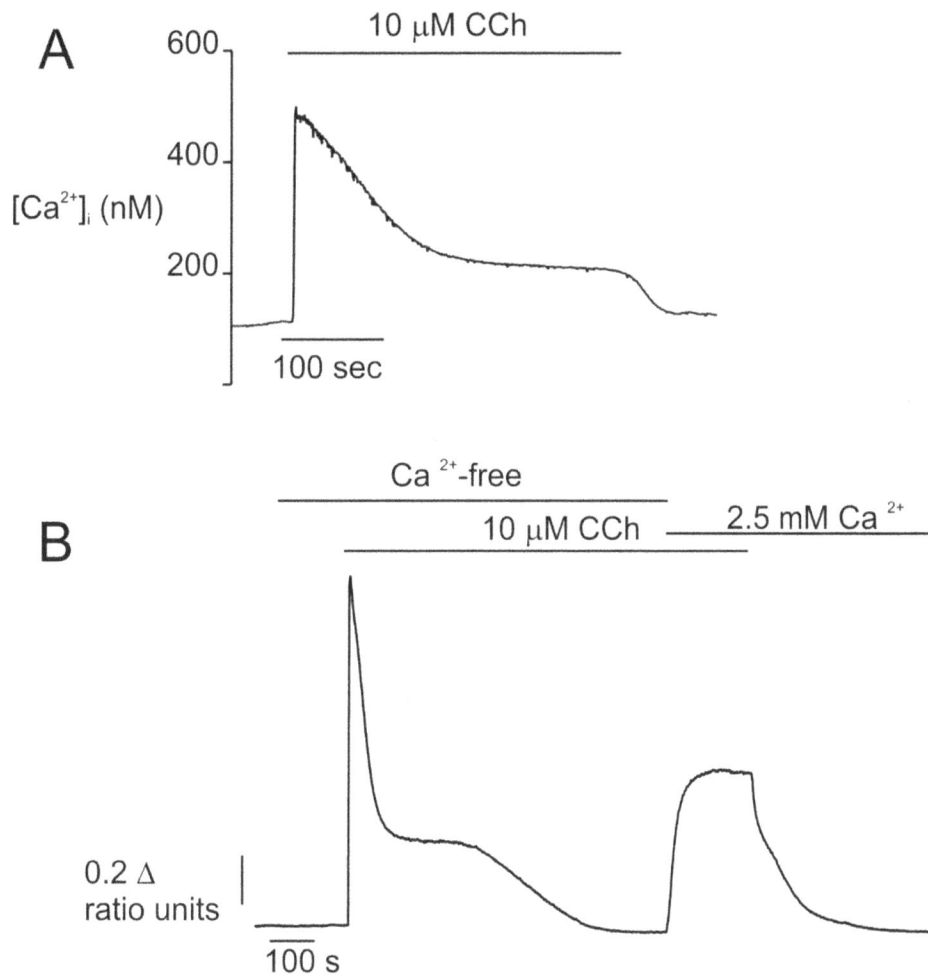

Figure 1. $[Ca^{2+}]_i$ changes evoked by maximal concentrations of secretagogues in single cells. Stimulation with maximal concentrations of secretagogues results in a characteristic "peak and plateau"-type Ca^{2+} signal. (**A**) In a single fura-2-loaded rat pancreatic acinar cell, stimulation with 10 μM CCh resulted in a sharp increase in $[Ca^{2+}]_i$, which subsequently declines to a new plateau level which was maintained throughout the period of secretagogue application. (**B**) In the absence of extracellular Ca^{2+}, the initial peak can be initiated but is only transient indicating that the initial phase of the response is a result of Ca^{2+} release from intracellular stores. Readmission of extracellular Ca^{2+} restores the "plateau" phase of the response indicating that Ca^{2+} influx is required to maintain this portion of the response.

around 100 nM above basal, which is maintained as long as the agonist is present.[6–9] The initial peak was shown to be the result of Ca^{2+} release from intracellular stores, since the early transient response was essentially unaffected by removal of extracellular Ca^{2+} while the later plateau phase was absent (**Figure 1B**).[8–11] Removal of extracellular Ca^{2+} during the plateau phase resulted in the rapid attenuation of the signal (**Figure 2B**), thus indicating an absolute dependence on extracellular Ca^{2+} for this maintained phase and implying the presence of a mechanism for Ca^{2+} influx from the extracellular milieu. A study monitoring ACh-induced Ca^{2+} signals in human pancreatic acinar cells confirmed this general paradigm; the initial phase of the signal occurs

independently of extracellular Ca^{2+}, while the sustained plateau requires the presence of Ca^{2+} in the extracellular milieu.[12]

Temporal properties at physiological concentrations of secretagogues

In contrast to the situation with maximal concentrations of agonists, similar experimental techniques demonstrated that lower, physiological concentrations of agonist resulted in the initiation of Ca^{2+} oscillations.[7–9,13,14] These oscillations are characterized by repetitive, regular cycles of elevated and subsequently decreasing Ca^{2+} levels. In single acinar cells or small acini, physiological concentrations of secretagogues (1–50 pM CCK; 50–300 nM ACh) induce after a latency of 30 seconds to 2 minutes fairly regular Ca^{2+} oscillations at a frequency of between 1 and 6 cycles per minute. The maximal global $[Ca^{2+}]_i$ reached during the release phase is generally between 200 nM and 1 μM. At least in mouse pancreatic acinar cells, the global temporal profile stimulated by ACh differs from that stimulated by CCK and bombesin.[15–18] Oscillations stimulated by peptide secretagogues tend to be characterized by slow, relatively long-lived transients originating and returning to basal levels between Ca^{2+} spikes while ACh-induced oscillations are characterized by faster short-lasting transients originating from an elevated plateau (**Figure 2A** and **B**).

Mechanisms underlying oscillating Ca^{2+} signals

Consistent with a prominent role for receptor-activated, Gαq-stimulated, PIP_2 hydrolysis as the underlying mechanism, Ca^{2+} oscillations in pancreatic acinar cells can be mimicked by agents, which activate heterotrimeric G proteins such as GTPγS, sodium fluoride, or mastoparan and by introduction of IP_3 into the cell, for example, via dialysis from a whole-cell patch clamp pipette.[15,16,20] Oscillations are primarily the result of cycles of intracellular Ca^{2+} release and ATP-dependent reuptake because the oscillations can be initiated in the absence of extracellular Ca^{2+} and are inhibited by agents, which deplete ATP or inhibit the Ca^{2+} pump on the endoplasmic reticulum.[9,13,21] Although Ca^{2+} release during each cycle only minimally depletes the intracellular Ca^{2+} store[22] and reuptake is efficient,[23] the maintenance of oscillatory behavior is dependent on extracellular Ca^{2+}. These observations again indicate an absolute requirement for Ca^{2+} influx to sustain the Ca^{2+} signal presumably primarily by maintaining the level of Ca^{2+} in the intracellular store.[9,15]

Oscillating levels of IP_3 are not necessary *per se* for oscillatory behavior, since nonmetabolizable IP_3 is capable

of initiating Ca^{2+} oscillations in mouse pancreatic acinar cells.[20] These data indicate that the mechanism underlying Ca^{2+} oscillations is most likely the result of an inherent property of the Ca^{2+}-release mechanism. Nevertheless, these data do not preclude the possibility that upon agonist stimulation the $[IP_3]$ itself fluctuates and contributes to the kinetics of Ca^{2+} release. Indeed mathematical predictions based on the experimental behavior of Ca^{2+} oscillations when Ca^{2+} or IP_3 is artificially raised during agonist exposure predict that the $[IP_3]$ is itself oscillating during CCK stimulation.[24] One mechanism whereby oscillating IP_3 could occur is through periodic activation cycles of RGS proteins. RGS proteins stimulate the GTPase activity of Gα subunits, thereby terminating the stimulus for activation of effectors such as PLC β. In rat pancreatic acinar cells, infusion of RGS proteins via the patch pipette results in dampening of Ca^{2+} signals.[25–28] Interestingly, the common catalytic core of RGS proteins, the so-called RGS box, is much less effective than infusion of full-length RGS proteins.[26] In addition, specific RGS proteins appear to affect the Ca^{2+} signals generated by different secretagogues differentially.[27] These observations may indicate that individual RGS proteins are associated in a signaling complex with specific secretagogues receptors and other signaling proteins. This interaction may impact the kinetics of IP_3 production and contribute to the agonist-specific characteristics of secretagogue-stimulated Ca^{2+} signals in pancreatic acinar cells.

Spatial properties of Ca^{2+} signals

Measurements of global Ca^{2+} signals have provided a wealth of data regarding the general temporal properties of Ca^{2+} signals in pancreatic acinar cells. However, by the nature of the measurements, the experiments report the $[Ca^{2+}]_i$ as a mean value integrated from throughout the cell. Modern imaging techniques, however, allow the monitoring of $[Ca^{2+}]_i$ at the subcellular level and in multiple cells

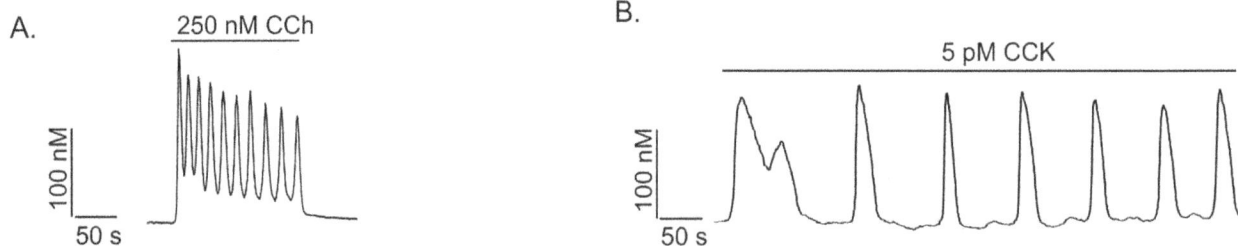

Figure 2. Physiological concentrations of agonists evoke Ca^{2+} oscillations. In single mouse pancreatic acinar cells, low concentrations of secretagogue (50–400 nM ACh or 1–50 pM CCK) evoke repetitive Ca^{2+} transients termed Ca^{2+} oscillations. In small clusters of mouse acinar cells, muscarinic receptor and CCK stimulation resulted in distinct global temporal patterns of Ca^{2+} signal. (**A**) CCh stimulation results in sinusoidal oscillations superimposed on an elevated baseline, while as shown in (**B**), CCK stimulation results in much less frequent, broader transients which originate and return to basal $[Ca^{2+}]_i$ levels between transients (from reference [19]).

of a coupled acinus. Probes, both chemical and genetically encoded sensors, are available with a choice of spectral characteristics and affinities for Ca^{2+} and with the ability to be targeted to various cellular compartments.[29] Importantly, transgenic animals have been generated expressing a variety of these probes.[30,31] Highlighting the promise of these techniques to enable the study of Ca^{2+} signals in living animals and thus in their native environment, an initial study using intravital two-photon microscopy in YC3.60 expressing animals reported Ca^{2+} signals following the injection of the muscarinic agonist bethanechol with characteristics similar to that seen in isolated acini.[32] The combination of the flexibility of the probes plus the imaginative use of digital imaging techniques, such as confocal, multiphoton, and total internal reflection fluorescence (TIRF) microscopy, has revealed that the Ca^{2+} signal displays remarkable spatial intricacy, which appears to be fundamental for the appropriate activation of downstream effectors. An early study by Kasai and Augustine utilizing digital imaging of small acinar clusters demonstrated a profound spatial heterogeneity in the Ca^{2+} signal following stimulation with a maximal concentration of ACh (**Figure 3A**). Stimulation resulted in the initiation of Ca^{2+} release in the apical region of the cell immediately below the luminal plasma membrane and the subsequent spread of the signal as a wave toward the basal aspects of the cell.[33] This Ca^{2+} wave has generally been reported to travel across the cell at a speed of between 5 and 45 µm/s, consistent with the velocity observed in other cell types.[33–38]

These data were somewhat counterintuitive since contemporary studies had demonstrated that secretagogue receptors were expressed on the basolateral face of the cell[40] and it was known that the ER, the presumed site of Ca^{2+} release, was present throughout the cell. Nathanson and colleagues, using line-scanning confocal microscopy, later confirmed that a similar pattern of apical to basal Ca^{2+} wave was initiated by high concentrations of CCK.[34]

At physiological concentrations of agonists using similar techniques, Ca^{2+} signals are also shown to initiate in the apical region of acinar cells.[39,41] This initial site of Ca^{2+} release has been termed the "trigger zone"[39] (**Figure 3B**). Ca^{2+} release invariably occurs at this specialized site even under conditions where stimulation of agonist is restricted to the basal region by focal application of agonist. This has been most elegantly demonstrated by focal flash photolysis of caged-carbachol contained in a whole-cell patch-clamp pipette isolated in the base of the cell.[42] It should be noted, however, that the apical portion of the acinus, immediately proximal to the tight junctions, may express a relatively high number of secretagogue receptors since this area has been reported to be most sensitive to focal agonist stimulation.[43,44]

At threshold concentrations of ACh, repetitive short-lasting Ca^{2+} transients are initiated, which strikingly are

contained to the apical third of the cell and do not propagate to the basal region.[39,41] These spikes, although short-lived, have been shown using low-affinity Ca^{2+} indicators to be of large amplitude in the order of 1–4 µM Ca^{2+}.[45]

At intermediate concentrations of ACh, apically initiated global Ca^{2+} transients dominate, often superimposed on a slight global elevation of $[Ca^{2+}]_i$. The frequency of these transients corresponds to the frequency of oscillations noted in microfluorimetry studies.

Low concentrations of CCK predominately result in apically initiated global Ca^{2+} signals, which are of longer duration.[15,16] In studies of whole-cell patch-clamped acinar cells, where the Ca^{2+} buffering of the cell is set by dialysis from the patch pipette, the broad CCK-induced transients have been reported to be preceded by short-lasting apically localized transients.[46] Studies indicate that the precise site of initiation of each transient by specific agonists is very similar but possibly not identical.[44] The site of initiation of each Ca^{2+} transient in the trigger zone is nevertheless tightly coupled functionally to both the exocytosis of zymogen granules and the activation of Cl^- channels required for the process of fluid secretion from the pancreatic acinar cells.[45,47] Notably, the entire spectrum of signals evoked by secretagogues, from localized apical release to propagation of global Ca^{2+} waves, can be mimicked by global, uniform application of IP_3 either through the patch pipette or via flash photolysis from a caged precursor (**Figure 3C**).[17,36,41,48] It is envisioned that future work utilizing intravital microscopy in animals expressing genetically encoded Ca^{2+} indicators will further our understanding of the spatial and temporal properties of Ca^{2+} signals evoked by physiological stimulation.

The impact of cell-cell communication on Ca^{2+} signaling

Individual cells in the pancreatic acinus are extensively coupled by the expression of gap junctional proteins.[49] These channels effectively allow the passage of small molecules up to a molecular mass of 1–2 kDa between cells and furthermore provide electrical coupling of large numbers of cells in the acinus. While stimulation of small acini with maximal concentrations of agonist results in the Ca^{2+}-dependent closure of these junctions,[50,51] at physiological levels of CCK and bombesin where global Ca^{2+} transients predominate the junctions remain open and Ca^{2+} signals appear to spread as waves between individual cells.[51,52] In contrast, brief, apically confined transients initiated by threshold concentrations of ACh have been reported not to propagate between coupled cells.[51] Each cycle of CCK-induced intercellular Ca^{2+} signaling has been reported to be initiated by a "pacemaker cell."[51] This pacemaker presumably represents the individual cell within the acinus most sensitive to agonist. The propagation of a Ca^{2+} wave between adjacent cells obviously requires relatively

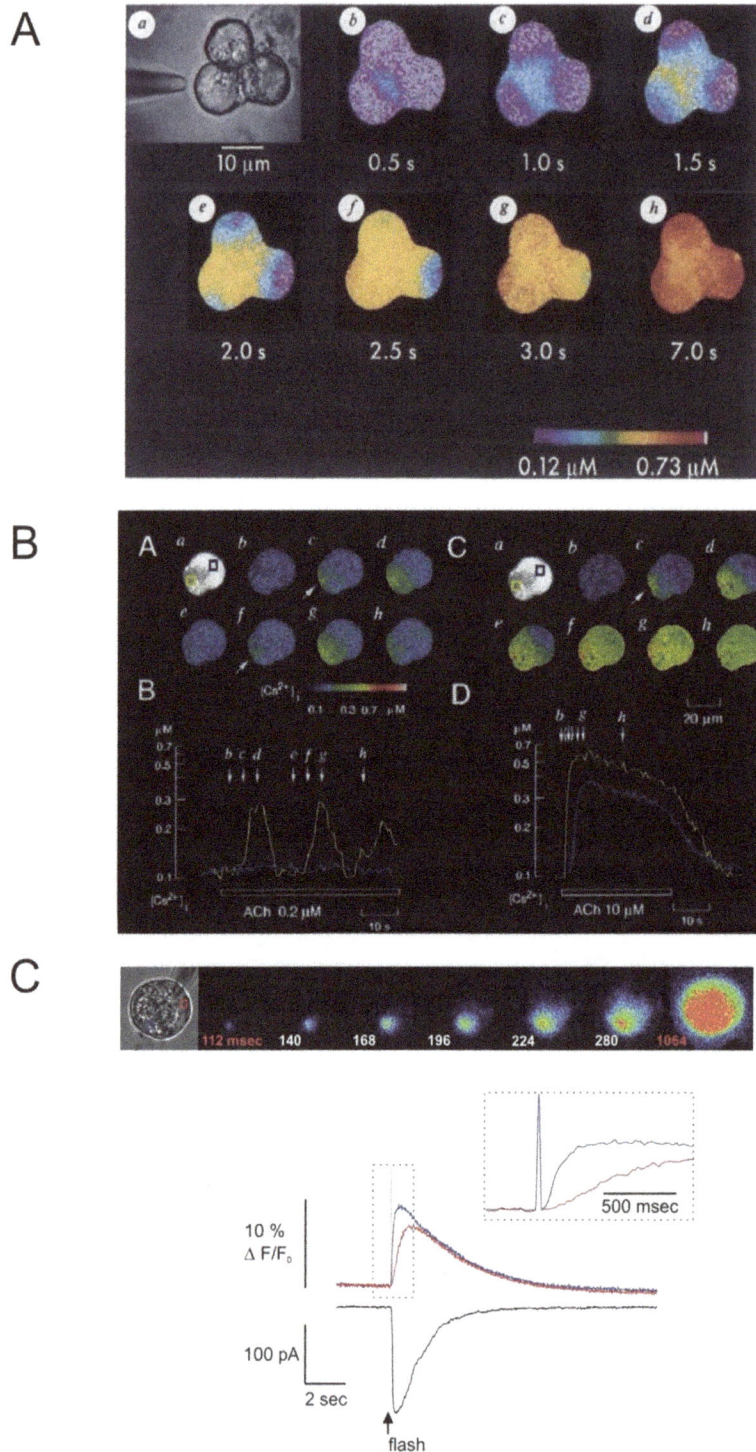

Figure 3. Spatial characteristics of Ca^{2+} signals in pancreatic acini. Digital imaging of Ca^{2+} indicators reveals spatial homogeneity in agonist-stimulated Ca^{2+} signals. (**A**) Stimulation of a triplet of mouse pancreatic acinar cells (shown in **a**) with a maximal concentration of ACh results in the initiation of the Ca^{2+} signal in the extreme apical portion of the acinar cells (shown in **b**). The signal subsequently spread toward the basal aspects of each cell (from reference [33]). The pseudocolor scale indicates the levels of [Ca^{2+}]$_i$. In (**B**), a single pancreatic acinar cell is stimulated with a threshold concentration of ACh. Ca^{2+} signals are again initiated in the apical portion of the cell but remain in the apical third of the cell without spreading to the basal aspects of the cell (images (**B**) **Ab-Ah**). The kinetic recorded from an apical region of interest (yellow trace in (**B**)) and from the basal region (blue trace) demonstrates that [Ca^{2+}]$_i$ elevations are only observed in the apical pole of the acinar cell under these conditions (from reference [39]). (**C**) Shows the changes in Ca^{2+} following photolytic liberation of 1,4,5 InsP$_3$ from a caged precursor induced into a single mouse acinar cell via a whole-cell patch-clamp pipette. Following global elevation of 1,4,5-InsP$_3$, Ca^{2+} changes initially occur at the apical pole of the acinar cell and spread to the basal pole in a similar fashion to secretagogue stimulation. The kinetic shows the Ca^{2+} changes in the apical (blue trace) vs. basal pole (red trace) of the cell together with the activation of a chloride conductance as measured by whole-cell patch clamp (from reference [17]).

long-range messengers. It appears that IP$_3$ and small amounts of Ca^{2+} are capable of diffusing between coupled cells to act in concert in this manner providing a signal to synchronize the intercellular Ca^{2+} wave. The primary evidence for this contention is that a Ca^{2+} signal can be observed in neighboring cells when 1,4,5-IP$_3$ is injected into an unstimulated individual cell. In addition, while Ca^{2+} injected into a resting cell fails to measurably increase [Ca^{2+}]$_i$ in adjacent cells, microinjection of Ca^{2+} into cells previously stimulated with threshold concentrations of CCK leads to a measurable increase in [Ca^{2+}]$_i$ in neighboring cells.[51] These data are consistent with Ca^{2+} acting to facilitate further Ca^{2+} release from intracellular stores as will be described in detail in the remainder of this chapter. The physiological function of propagating Ca^{2+} waves in pancreatic acinar cells is not at present firmly established. A reasonable proposal, however, is that gap-junctional communication represents a mechanism to increase the responsiveness of an acinus to threshold concentrations of agonist. In this scenario, the acinus is rendered as sensitive to secretagogues stimulation as the pacemaker cell. In support of this idea, isolated single cells are much less sensitive to secretagogues stimulation than isolated acini and in addition experimental maneuvers which increase gap-junctional permeability lead to increased secretagogue-induced amylase secretion.[52]

Molecular Mechanisms Underlying Ca^{2+} Signaling in Pancreatic Acinar Cells

Intracellular Ca^{2+} release

IP$_3$-induced Ca^{2+} release

The invariable apical initiation of Ca^{2+} release dictates the view that this trigger zone must represent a specialized region of ER, highly sensitive to IP$_3$. Studies have shown that this exquisite sensitivity to IP$_3$ is a result of the abundant expression of IP$_3$ receptors (IP$_3$Rs) in the extreme apical region of pancreatic acini.[53–55] IP$_3$Rs were first isolated and cloned from cerebellum and have subsequently been shown to represent a family of three proteins named the type-1 IP$_3$R (IP$_3$R1), type-2 IP$_3$R (IP$_3$R2), and type-3 IP$_3$R (IP$_3$R3), which are all related to the ryanodine receptor Ca^{2+} release channel.[56–58] Initially, it was reported that IP$_3$R3 was expressed in the apical pole.[54] Later studies, however, showed that all three subtypes had essentially identical expression; all IP$_3$Rs were excluded from areas containing zymogen granules and were apparent immediately below the apical and lateral plasma membrane[53,55] (**Figure 4A–D**). This localization is essentially identical to the "terminal web" of actin-based cytoskeleton in this region (**Figure 4E**). The localization to this region may be dependent on lipid rafts because cholesterol depletion

results in the redistribution of IP$_3$R and disruption of the apical to basal Ca^{2+} wave.[59] By this technique, no other significant localization of IP$_3$R was noted except for moderate expression on perinuclear structures.[53,55] The later distribution is consistent with a recent report of IP$_3$-induced Ca^{2+} release from isolated nuclei prepared from mouse pancreatic acinar cells.[60] Studies where IP$_3$ was released from a caged precursor in various localized regions of mouse acinar cells have also functionally confirmed that the apical region of the cell is more sensitive to IP$_3$ than the basal area of the cell.[48] These data were later confirmed by a study imaging permeabilized pancreatic acini.[61] Quantitative Western analysis and PCR have indicated that there is approximately equal expression of IP$_3$R3 and IP$_3$R2 in pancreatic acinar cells making up ~ 90 percent of the total complement of IP$_3$R.[62, 63] The pivotal importance of IP$_3$R2/3 for Ca^{2+} signaling and subsequent exocrine secretion has been confirmed by studies of transgenic animals in which both subtypes have been knocked out.[64] Individual knockdown of IP$_3$R2 or IP$_3$R3 resulted in a modest reduction in the magnitude of Ca^{2+} release[64,65] but no obvious global phenotype. However, the compound IP$_3$R2/3 null animal dies soon after weaning. The lack of viability results from a failure to ingest and subsequently digest food as a direct consequence of a general failure of exocrine gland secretion. In both salivary and pancreatic acinar cells from IP$_3$R2/3 null animals, secretagogue-stimulated Ca^{2+} signals and secretion of fluid and protein are essentially absent.[64] These data also reveal that the residual IP$_3$R-1 in the IP$_3$R-2/3 null animal cannot compensate for the loss of the other subtypes in pancreatic acinar cells.

IP$_3$R structure and regulation

The functional IP$_3$R is formed cotranslationally by the tetrameric association of four individual receptor subunits.[66,67] In pancreatic acinar cells, there is evidence that the channel can form a heterotetramer since multiple types of IP$_3$R can be detected in immunoprecipitates of specific individual receptor types.[68–70] Indeed, a recent study has shown by sequential immunodepletion of individual subtypes from pancreatic lysates that homotetrameric IP$_3$Rs likely constitute a minority of the functional channels.[69] Each subunit has a binding site for InsP$_3$ toward the n-terminus, which is formed by a cluster of positively charged amino acids thought to coordinate the negatively charged phosphate groups of IP$_3$.[71,72] The Kd for binding of IP$_3$ to pancreatic membranes has been reported to be between 1 and 7 nM, a figure similar to that reported for other peripheral tissues such as liver.[17,73] The c-terminus of each subunit is postulated to span intracellular membranes six times and a single cation selective pore is formed from this region of the protein in the tetrameric receptor. When purified or expressed in a heterologous system and then reconstituted

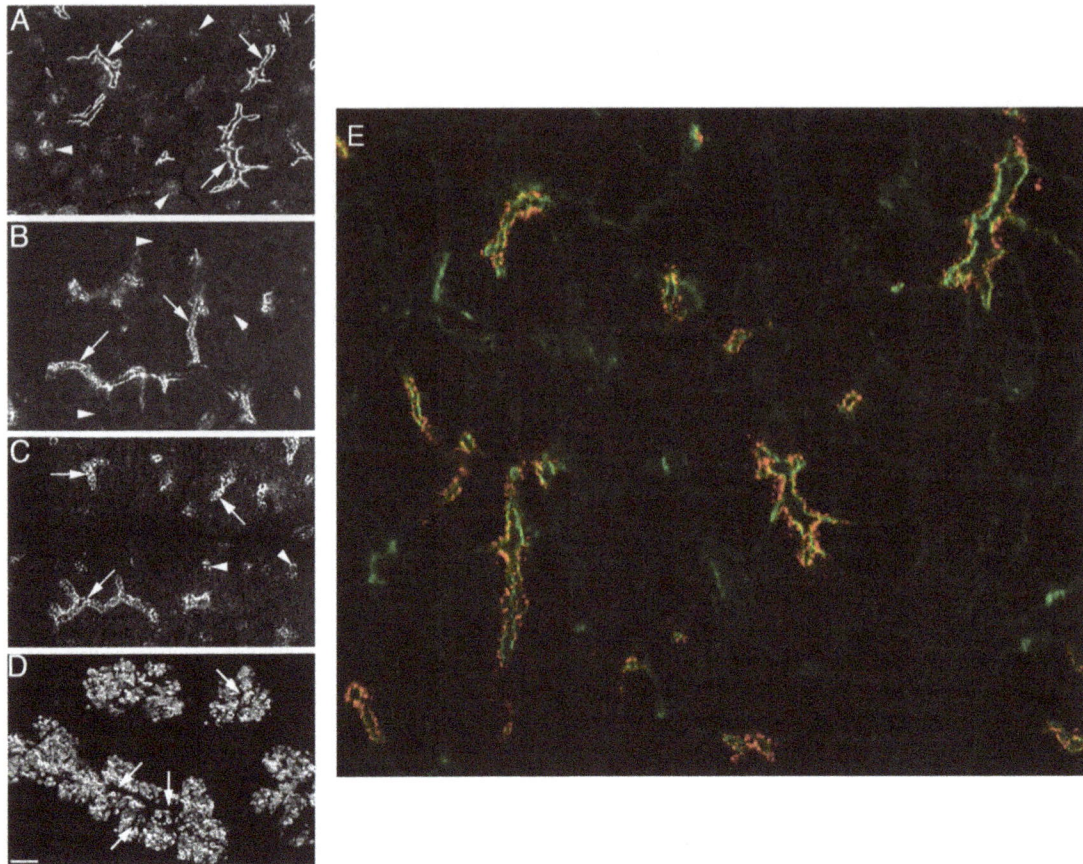

Figure 4. Localization of IP$_3$R in pancreatic slices. The localization of IP$_3$R1, (**A**) IP$_3$R2, (**B**) and IP$_3$R3 (**C**) was determined with specific antibodies to individual InsP$_3$R types and visualization by confocal microscopy. IP$_3$R of all types predominately localized to the extreme apical pole of acinar cells, immediately below the luminal plasma membrane (arrows in **A–D**; compare localization of zymogen granules visualized by staining for amylase in panel **D**). IP$_3$R1 and IP$_3$R3 also localized to perinuclear structures (arrowheads in **A** and **C**) (from reference [53]). (**E**) IP$_3$R3 (stained in red) are colocalized with the terminal web of the actin cytoskeleton (stained in green).

in planar lipid bilayers, the protein can be demonstrated to function as an InsP$_3$-gated cation channel with many of the characteristics of the release channel.[74] The fact that the outer nuclear membrane is continuous with the ER has been exploited in patch-clamp experiments to study IP$_3$R channel activity in isolated nuclei from Xenopus laevis oocytes, COS, and DT40–3KO cells.[75–77] These experiments have provided insight into the activity and regulation of the channel in a native membrane; however, to date, no information is available regarding the activity of IP$_3$R in native pancreatic membranes.

While the IP$_3$ binding pocket and channel pore are highly conserved between IP$_3$R family members, the intervening sequence between the binding region and pore is more divergent and consists of the so-called "regulatory and coupling" or "modulatory" domain. This region consisting of ~1,600 amino acids is thought to be important in modulating the Ca^{2+} release properties of the IP$_3$R. Indeed, Ca^{2+} release through the IP$_3$R is markedly influenced

by many factors, most importantly by Ca^{2+} itself.[78,79] The majority of studies have indicated that all forms of the IP$_3$R are biphasically regulated by Ca^{2+} in the range of 0.5–1 μM increases the steady-state open probability of the channel while at higher concentrations the activity decreases.[78,79] This property of the IP$_3$R is thought to be fundamentally important in the generation of the different spatial and temporal patterns of Ca^{2+} signals observed in cells.[80] In pancreatic acinar cells, IP$_3$-induced Ca^{2+} release has been shown to be inhibited when Ca^{2+} is elevated and enhanced when Ca^{2+} is buffered with chelators.[81,82] The [Ca^{2+}]$_i$, together with the range of action of Ca^{2+}, can be manipulated by dialyzing cells with buffers exhibiting differing on-rates for Ca^{2+} binding. In pancreatic acinar cells, restriction of the range of action of Ca^{2+} using the slow on-rate buffer EGTA resulted in spatially restricted IP$_3$-induced spikes and the attenuation of global waves consistent with EGTA inhibiting the positive effect of Ca^{2+} to facilitate Ca^{2+} release between spatially separated release sites. In contrast, the fast on-rate buffer BAPTA resulted in larger monotonic Ca^{2+} release and this was interpreted to reflect the loss of local Ca^{2+} inhibition of IP_3R.[83]

IP_3R activity is also influenced through interaction with numerous factors such as proteins, adenine nucleotides, and phosphorylation in particular by cyclic nucleotide-dependent kinases (for review see references[66,84]). The IP_3R1 represents one of the major substrates for phosphorylation by protein kinase A in brain and thus represents a potentially important locus for cross-talk between the cAMP and Ca^{2+} signaling systems.[85] In pancreatic acinar cells, PKA activation results in phosphorylation of IP_3R3.[18,19] Functionally, phosphorylation of IP_3R in pancreatic acinar cells correlates with IP_3-induced Ca^{2+} release which is decreased in terms of the magnitude and kinetics of Ca^{2+} release.[19,86]

Physiologically relevant concentrations of CCK, but not Ach, also result in PKA-dependent phosphorylation of IP_3R.[19] This observation is consistent with earlier reports that CCK stimulation leads to an increase in cAMP and PKA activation. CCK-induced phosphorylation of $InsP_3R$ may contribute to the specific characteristics of CCK-induced Ca^{2+} signals as maneuvers, which interfere with PKA activation and disrupt the pattern of CCK-induced but not Ach-mediated Ca^{2+} signaling. Conversely, raising cAMP converts ACh-induced Ca^{2+} signaling characteristics into signals, which resemble CCK stimulation.[86,87] CCK-stimulated signaling is not affected by raising cAMP or by stimulation with VIP presumably because PKA activation and phosphorylation of IP_3R has already occurred.[13,86] CCK and bombesin stimulation results in Ca^{2+} signals with similar characteristics and this may be related to the fact that bombesin stimulation also results in phosphorylation of IP_3R in mouse pancreatic acinar cells.[19,88] Paradoxically, although there are numerous examples of similar attenuated Ca^{2+} signaling following PKA phosphorylation in other cell types,[89] PKA phosphorylation of individual IP_3R subtypes studied in isolation has only been shown to increase Ca^{2+} release.[89] These data raise the possibility that the physiologically relevant phosphorylation event actually occurs on a tightly associated binding partner to inhibit Ca^{2+} release and not the IP_3R-3 directly. In vitro, IP_3R can also be phosphorylated by protein kinase C, Ca^{2+}/calmodulin-dependent kinase II, and tyrosine kinases of the src family.[66,84] While no direct evidence has been reported regarding phosphorylation of IP_3R by these pathways in pancreatic acinar cells, activation of PKC has been shown to inhibit Ca^{2+} release in permeabilized pancreatic acinar cells and to attenuate Ca^{2+} oscillations stimulated by secretagogues or by direct G-protein activation, without an effect on PI hydrolysis.[90,91] Thus, the possibility exists that the IP_3R is a substrate for other kinases in pancreatic acinar cells.

Cellular levels of ATP also modulate IP_3-induced Ca^{2+} release in permeabilized mouse pancreatic acinar cells.[92] ATP has characteristic distinguishing effects on individual

IP_3R subtypes[89] and despite an approximately equal complement of IP_3R2 and IP_3R3 in acinar cells, the modulation has essentially identical properties to IP_3R2 in wild-type animals and is indistinguishable from IP_3R3 in IP_3R2 null animals. These data indicate that the specific properties of the IP_3R2 appear to dominate the overall profile of IP_3-induced release in pancreatic acinar cells[92] and are consistent with the effects of ATP occurring on heterotetrameric IP_3R in which IP_3R2 is present.[69] This hypothesis was directly tested in a heterologous expression system where concatenated IP_3R constructs consisting of tandem cDNAs encoding defined heterotetramers were expressed in IP_3R null cells. The characteristics of Ca^{2+} release from heterotetrameric IP_3R consisting of at least two subunits of IP_3R2 were indistinguishable from homotetrameric IP_3R2.[69,93]

Introduction of G-protein βγ subunits into pancreatic acinar cells has also been shown to induce Ca^{2+} release. An initial report attributed this to the activation of PLC β2 by βγ and the subsequent production of IP_3.[94] A later report however demonstrated that the Gβγ-induced Ca^{2+} release was independent of $InsP_3$ production and the result of a direct interaction of Gβγ with IP_3R.[95] This association was confirmed by co-immunoprecipitation and shown to increase the open probability of IP_3R in a manner independent of IP_3. This interaction of IP_3R and βγ may be important for the action of Gαi-linked agonists in pancreatic acinar cells.

Ryanodine receptor-induced Ca^{2+} release

Evidence exists for Ca^{2+} release initiated through activation of ryanodine receptors (RyRs) in pancreatic acinar cells. This family of channels, best studied as the Ca^{2+} release channel in skeletal and cardiac muscle, is classified as belonging to the same gene superfamily as IP_3R. Indeed, RyRs are modulated by similar regulators and share some sequence homology with the IP_3R especially in the putative Ca^{2+}-conducting pore region of the c-terminus. However, while IP_3Rs have an absolute requirement for IP_3 with Ca^{2+} as an important coagonist for gating, RyRs only require Ca^{2+} to open through a process termed calcium-induced calcium release (CICR). The functional expression of RyRs in pancreatic acinar cells has been indicated by a number of studies; for example, microinjection of Ca^{2+} in the presence of the IP_3R antagonist heparin results in Ca^{2+} release in mouse pancreatic acinar cells.[39] In addition, treatment of pancreatic acinar cells with high concentrations of ryanodine known to block RyR dampens secretagogue-induced Ca^{2+} signals,[34,36] and one report has shown that low concentrations of ryanodine, which permanently opens the RyRs in a subconductance state, results in Ca^{2+} release.[96] In muscle cells, caffeine activates RyRs and results in emptying of sarcoplasmic reticulum Ca^{2+} stores. In contrast, the majority of reports from pancreatic acinar cells indicate that caffeine does not elicit Ca^{2+} release and actually inhibits secretagogue-induced Ca^{2+} signaling through an action to inhibit phospholipase C and IP_3R-mediated Ca^{2+} release.[96,97] This later effect of caffeine, together with presumably much lower numbers of RyRs in pancreatic acinar cells, probably explains the absence of caffeine-induced Ca^{2+} release.

The physical presence of RyR has, however, been difficult to demonstrate in pancreatic acinar cells with conflicting positive and negative reports of expression. For example, in one study RyR could be detected using western analysis in rodent salivary gland acinar cells but not in pancreatic acinar cells.[21] In contrast, an initial study reported the expression using PCR analysis of RyR type 2 (RyR2) but not RyR type 1 or 3 (RyR1 and RyR3, respectively) in mRNA extracted from rat pancreatic acini.[98] A later report, however, using single-cell PCR and western analysis demonstrated that all three types of RyR were expressed in pancreatic acini.[99] A study performed in human acini has reported mRNA for RyR1 but not RyR2/3. Notably, [^{3}H]-ryanodine binding to membrane preparations provides evidence that the RyR protein is indeed expressed in human pancreatic acini.[100] In all probability, the lack of consensus in these studies relates to the relatively low expression of RyR in pancreas when compared to muscle cells. While IP_3R have a well-defined localization in pancreatic acinar cells, several studies have reported that ryanodine receptors have a more diffuse distribution. Immunohistochemistry and studies with fluorescently labeled ryanodine have indicated that RyR are distributed throughout acinar cells with perhaps the greatest concentration in the basal aspect of the cell.[36,98,101] As a result of this localization to areas of the cell with low levels of IP_3R, it has been suggested that activation of RyR plays an important role in the propagation of Ca^{2+} signals from the initial release of Ca^{2+} in the trigger zone to the basal aspects of the cell. In support of this contention, high concentrations of ryanodine in some studies have been shown to slow or spatially limit the spread of Ca^{2+} waves in mouse pancreatic acinar cells (**Figure 5A** and **B**)[34,36] and also to selectively attenuate the peak $[Ca^{2+}]$ in the basal aspects of cells.[101]

Ca^{2+} release stimulated by cADPR

In a number of cell types, there is evidence that in addition to activation by Ca^{2+} the RyR activity are modulated by cyclic adenosine diphosphoribose (cADPr). cADPr was first suggested to be a Ca^{2+}-releasing second messenger based on experiments performed in sea urchin eggs where it was shown to release Ca^{2+} and function as a messenger during fertilization.[102] Subsequently, it has been shown to release Ca^{2+} and satisfy some of the criteria for a second messenger in mammalian systems including lymphocytes, pancreatic β-cells, and cardiac myocytes.[103–105] Ca^{2+} release following cADPr exposure is inhibited by blocking

concentrations of ryanodine and appears to reduce the threshold for CICR through RyR.[106] Opinion is however divided as to whether it functions through a direct effect on the RyR or indirectly through interaction with an accessory protein such as FKBP 12.6 or calmodulin.[107,108]

In pancreatic acinar cells, the intracellular application of cADPr results in Ca^{2+} release. This observation has been reported in both rat and mouse acini with either whole-cell pipette dialysis of cADPr, liberation of cADPr from a caged precursor by two-photon flash photolysis, and by direct application to permeabilized acini.[61,109–112] In addition, cADPr has been reported to release Ca^{2+} from a rat pancreatic microsomal preparation.[113] The spatial localization of cADPr-induced Ca^{2+} release remains to be conclusively resolved. In an initial report, cADPR introduced by dialysis from a patch-clamp pipette into mouse pancreatic acinar cells resulted in Ca^{2+} release from the apical pole of the cell.[109] This Ca^{2+} release was blocked by both ryanodine and interestingly the IP_3R antagonist heparin. These data were interpreted as indicating that Ca^{2+} release mediated by cADPr was dependent on both IP_3R and RyR, presumably as highly localized Ca^{2+} release initially through RyR-sensitized neighboring IP_3R to basal levels of IP_3.[109] Consistent with this view, it was also shown that low concentrations of ryanodine also resulted in Ca^{2+} release from the apical pole in a manner apparently dependent on functional IP_3R.[96] In contrast, in a study more consistent with the localization of the majority of RyR, selective local uncaging of cADPr in different regions of an acinar cell by two-photon photolysis reported that the basal aspect of the cell was more sensitive to cADPr.[110] These data are consistent with a report that in permeabilized acini the apical pole exhibited a higher affinity for IP_3 but cADPr released Ca^{2+} exclusively from the basal aspects of the cell.[61]

Important questions to address, which are required to establish cADPr as a bona fide Ca^{2+}-releasing second messenger in pancreatic acinar cells, are to demonstrate that the molecule is produced following secretagogues stimulation and that it is necessary for Ca^{2+} release. To these ends, it has been reported that CCK and ACh but not bombesin stimulate the activity of a cytosolic ADP-ribosyl cyclase activity resulting in the production of cADPr.[114] One such enzyme which possesses this activity is CD38. Ca^{2+} signaling events in acinar cells prepared from CD38 null mice are dampened and appear reminiscent of RyR blockade.[115] In addition, 8-NH_2-cADPr, a structural analog of cADPr which antagonizes the effect of cADPr, has been reported to block CCK and bombesin but not ACh or IP_3-induced Ca^{2+} signals in mouse pancreatic acinar cells.[116,117] This later result, while not internally consistent with the ability of ACh to induce the formation of cADPr, has been suggested to indicate that CCK stimulation preferentially couples to the generation of cADPr and thus may account for

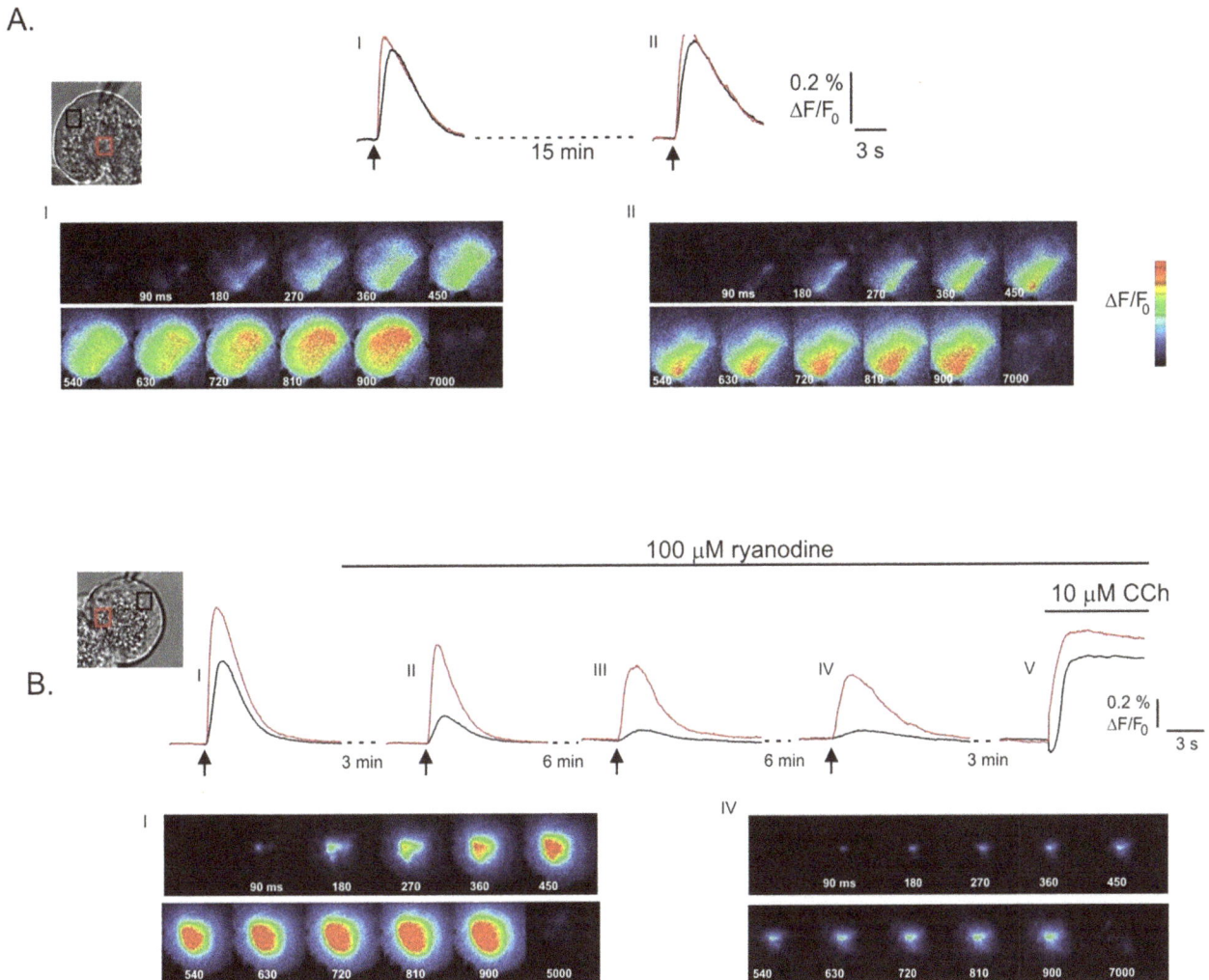

Figure 5. Contribution of RyR to global Ca^{2+} signals in pancreatic acinar cells. Global Ca^{2+} signals were initiated by photolysis of caged-IP$_3$. In (**A**) the images and kinetic traces show that global Ca^{2+} signals can be initiated multiple times following exposure to 1,4,5-IP$_3$. However, as shown in (**B**), exposure to a high concentration of ryanodine, known to inhibit RyR, leads to a slowing in the progression of the Ca^{2+} wave and to the restriction of the signal to the apical pole of the cell following stimulation with 1,4,5-IP$_3$ (compare (**B**) **I**; absence of ryanodine and (**B**) **IV**; exposure to ryanodine for 15 minutes) (from reference [36]).

some of the distinct characteristics of CCK-induced Ca^{2+} signals reported.

Nicotinic acid dinucleotide phosphate-induced Ca^{2+} release

An additional putative messenger that potently induces Ca^{2+} release in pancreatic acinar cells is nicotinic acid adenine dinucleotide phosphate (NAADP). Once again, the activity of this agent was first reported to play a role in invertebrate fertilization. Although it has been reported that NAADP-induced Ca^{2+} release from nuclei isolated from mouse pancreatic acini is dependent on RyR,[60] the majority of evidence suggests that the receptor is likely to represent a novel Ca^{2+} release channel since while the activity of IP$_3$R and RyR exhibits a bell-shaped dependence on Ca^{2+} the putative NAADP receptor does not support CICR.[118] While this property of NAADP is not well suited to play a role in the propagation of Ca^{2+} waves, it has been suggested that NAADP is required to initiate Ca^{2+} signals and this initial Ca^{2+} increase subsequently recruits IP$_3$R and RyR. This idea is supported by the observation that NAADP introduced via the patch pipette into mouse pancreatic acinar cells results in Ca^{2+} release in the apical pole but that this release is absolutely dependent on both InsP$_3$R and RyR.[111] These data suggest that Ca^{2+} release through

NAADP is quantitatively very small but ideally localized to sensitize IP_3R and RyR. While IP_3R and RyR reside in the ER, it has been suggested that NAADP primarily acts on a distinct store probably an acidic lysosome-related organelle.[119] This idea is primarily based on the observation that NAADP-induced Ca^{2+} signaling but not cADPr- or IP_3R-induced Ca^{2+} release is inhibited by experimental maneuvers which either inhibit vacuolar-type H^+ ATPase or result in osmotic disruption of lysosomes.[119]

The concentration-response relationship for NAADP-induced Ca^{2+} release also displays unique properties. NAADP-induced Ca^{2+} release is biphasic; nanomolar levels induce Ca^{2+} release while micromolar concentrations fail to release Ca^{2+} but render the mechanism refractory to subsequent stimulation.[120] Provocatively, exposure of mouse pancreatic acinar cells to inactivating concentrations of NAADP also renders cells refractory to stimulation with threshold concentrations of CCK but not ACh or bombesin, and disruption of the NAADP releasable store selectively disrupts CCK-induced Ca^{2+} signals.[117,120] In addition, NED-19 (an inhibitor of NAADP action) attenuates CCK-, but not ACh-induced Ca^{2+} responses.[121] These data suggest that CCK and NAADP under these conditions may utilize a common mechanism to induce Ca^{2+} release.

In support of this contention, physiologically relevant concentrations of CCK can also be shown to result in the production of NAADP while ACh stimulation does not result in measurable accumulation.[122] The molecular mechanism responsible for coupling CCK receptor occupation to NAADP production is at present not completely understood. Evidence suggests, however, that the enzyme CD38 may also be a candidate for the hormone-responsive NAADP synthase because CCK-stimulated NAADP formation is absent in CD38 null transgenic animals.[115] This activity of CD38 occurs at acidic pH and this requirement appears to be fulfilled by the expression of some of the enzymes in an endosomal compartment.[115]

The two-pore channel (TPC) family of proteins has recently been proposed as a candidate for the cognate receptor for NAADP.[118,123] The TPC gene family is related to the superfamily of voltage-gated ion channels and consists of three distinct gene products which are represented in both plants and animals.[123] The primary supporting evidence for the idea that TPC represents a NAADP receptor is that the expression of TPCs appears to confer Ca^{2+} release activity in the presence of NAADP.[118,123,124] Additionally, the expression of TPCs also leads to specific binding activity of radiolabeled NAADP, and TPCs are primarily expressed in an endolysosomal compartment, consistent with the prominent site of NAADP-induced Ca^{2+} release.[118,123] Since this specific binding is much lower affinity than the functional effects of NAADP and specific binding can be observed to co-immunoprecipated proteins,[125,126] it is questionable whether the effects of NAADP are mediated through

an accessory protein. In support of a role for TPC2 as a target of NAADP, Ca^{2+} release was reduced in acini prepared from TPC-2 knockout animals.[121] Of note, NAADP-induced Ca^{2+} release was also reduced in RYR3 knockout animals and in permeabilized cells incubated with α-RYR1 antibodies. The latter observation is consistent with reports from T lymphocytes that suggest RYR are the target of NAADP.[127] In total, these data led the authors to suggest that both RYRs and TPC channels might function in pancreatic acinar cells to provide the trigger for Ca^{2+} release following CCK stimulation.

Cellular mechanisms underlying Ca^{2+} influx across the plasma membrane

Ca^{2+} influx from the extracellular space is essential to sustain Ca^{2+} signaling in pancreatic acinar cells and in turn the long-term maintenance of secretion.[11] This influx is not blocked by antagonists of voltage-gated Ca^{2+} channels but is attenuated by lanthanides.[10,11] Ca^{2+} influx is also sensitive to changes in extracellular pH, being enhanced by alkaline and inhibited in acidic conditions.[10,11]

Store-operated Ca^{2+} entry and its molecular components

Functionally, Ca^{2+} influx can be initiated by substantial depletion of intracellular Ca^{2+} pools, the so-called "store-operated" or "capacitative" Ca^{2+} entry (SOCE) pathway.[128] This pathway can be readily demonstrated by inhibition of ER Ca^{2+} pumps with the plant sesquiterpene lactone thapsigargin, which results in Ca^{2+} influx independent of receptor activation and PI hydrolysis.[11,128–130] Recently, significant progress has been made regarding the molecular candidates responsible for the SOCE pathway. Using siRNA screens, the first breakthrough was the identification of stromal interaction molecule 1 (STIM-1) as the ER Ca^{2+} sensor responsible for coupling intracellular Ca^{2+} store depletion to the opening of the Ca^{2+} permeable conductance in the plasma membrane.[131–133] STIM-1 is an integral ER membrane protein, which harbors an EF hand-type Ca^{2+}-binding domain localized in the ER lumen. Following depletion of ER stores, Ca^{2+} dissociates from the EF hand and this results in the aggregation of STIM-1 molecules to areas of the ER close to the plasma membrane where it can physically gate the ion channel responsible for mediating Ca^{2+} influx.[134,135] Strong candidates for the actual pore-forming components of the pathway have also been identified. Using siRNA screens and complimented by linkage analysis of patients with severe combined immunodeficiency, studies have identified members of the Orai family of proteins as the channel constituents of the archetypal SOCE current, the so-called, calcium-release activated current (I_{CRAC}). The channel has been extensively characterized in lymphocytes as a highly selective Ca^{2+} channel.[136–138] In particular, homomultimers of Orai1 form channels with the biophysical characteristics of I_{CRAC}.[139] Both STIM-1 and Orai-1 are expressed in mouse pancreatic acinar cells and thus represent good candidates for mediating SOCE.[137,140,141] Lower levels of secretagogue stimulation result in only modest ER Ca^{2+} depletion, which may not be sufficient to activate STIM-1. While not directly demonstrated in pancreatic acinar cells, minor store depletion may be sensed by STIM-2, a family member which because of increased affinity for ER [Ca^{2+}] can activate Orai channels following physiological stimulation.[142] It is, however, clear that marked global depletion of ER Ca^{2+} occurs upon stimulation with maximal concentrations of secretagogues and upon exposure to pancreatic toxins, such as bile acids and fatty acid esters of ethanol.[1,22,143] Thus, the SOCE pathway may be important during strong stimulation and during pathological Ca^{2+} depletion of ER stores. Indeed, it has been suggested that activation of this pathway is responsible for the inappropriate intracellular activation of trypsin, which occurs in models of acute pancreatitis.[144–147] Consistent with this idea, inhibition of Ca^{2+} influx with pharmacology that targets SOCE has been shown to reduce the severity of pancreatitis in several experimental animal disease models[148–150] as well as mitigating the detrimental cellular hallmarks of pancreatitis *ex vivo*.[151] Further, maneuvers which inhibit Ca^{2+} release, including blockade of IP_3R with caffeine which prevents depletion of the ER and thus activation of SOCE, also are protective in experimental pancreatitis.[152]

Contribution of TRP channels

In exocrine cells there is also evidence that members of the canonical transient receptor potential channel (TRPC) family contribute to Ca^{2+} influx following store depletion.[137,153,154] This channel, while Ca^{2+} permeable, is a nonselective cation channel, exhibiting significant Na^+ permeability. STIM-1 has been reported to also function as the ER calcium sensor for this class of proteins.[155] Further, there is molecular and biochemical evidence to suggest that a tertiary complex of STIM-1/TRPC/Orai-1 may contribute to the native SOCE in acinar cells.[137,156] TRPC 3 and 6 are expressed in pancreatic acinar cells.[153] In TRPC 3 *null* transgenic animals, the magnitude of secretagogue-stimulated Ca^{2+} influx is reduced and this is associated with reduced agonist-stimulated amylase secretion.[153] In addition, the frequency of agonist-stimulated Ca^{2+} oscillations was reduced in TRPC 3 *null* animals supporting the view that Ca^{2+} influx through this channel may also serve to maintain Ca^{2+} oscillations.[153] Interestingly, the absence of TRPC 3 is also protective in animal models of pancreatitis indicating that SOCE may be deleterious under these conditions.[153]

While the spatial properties of the Ca^{2+} signal can be largely attributed to the localization of the Ca^{2+} release

machinery, recent data have also suggested that the distribution of the components of the SOCE Ca^{2+} influx pathway may also be spatially heterogeneous. The majority of Orai-1 and TRPC3 is reported to be localized to the apical and lateral plasma membrane.[140,141,153,157] One study has shown that following store depletion some STIM-1 translocates to the apical domain and is colocalized with Orai-1. Further, this study demonstrated that Ca^{2+} influx following store depletion could be detected earlier in the apical region when compared to more basal aspects of the cell.[137] Of note, a majority of STIM-1 does not colocalize with Orai-1 following store depletion.[137,140] Indeed, a study monitoring the localization of adenovirally expressed fluorescently tagged STIM-1 reported little colocalization with Orai-1 and that the vast majority of STIM-1 translocated to the basolateral aspects of cells following stimulation.[140] Further work will be necessary to unequivocally confirm the cellular localization of this influx pathway in acinar cells.

Contribution of ARC channels

As stated earlier, during secretagogue stimulation only relatively minor depletion of the ER pool as a whole occurs.[22] This observation raises the question whether SOCE operates under physiological conditions. If global ER Ca^{2+} is not markedly reduced during physiological stimulation, it follows that for SOCE to be activated under these conditions, local depletion of the strands of ER that infiltrate the apical domain,[158] which are replete with IP_3R,[53] must occur allowing the activation of STIM1/2. A contrary view is that a different mechanism largely controls Ca^{2+} influx during physiological stimulation. In support of this idea, an electrophysiological study of mouse pancreatic acinar cells failed to detect SOCE currents following stimulation with physiologically relevant concentrations of agonists. Under these conditions, it was reported that a channel activated by arachidonic acid was predominately responsible for Ca^{2+} influx.[159] This conductance has been termed I_{ARC} and is present in many nonexcitable cell types.[160,161] Notably, this channel also requires STIM-1 for activation and the pore is formed from multimers of Orai1 and Orai3.[162]

Cellular mechanisms underlying Ca^{2+} clearance from the cytosol

Following [Ca^{2+}]$_i$ elevation, mechanisms must be present to reduce [Ca^{2+}]$_i$ rapidly and efficiently during the falling phase of Ca^{2+} oscillations and ultimately to terminate Ca^{2+} signals following secretagogue removal. Moreover, the resting [Ca^{2+}]$_i$ of 50–200 nM must be maintained in the face of a large concentration gradient from the cell exterior, leak from the ER, and the negative intracellular potential, all of which would tend to drive Ca^{2+} into the cytoplasm. Homeostasis is accomplished by a variety of pumps and transporters that have specific distribution on both the plasma membrane and on the membranes of various organelles.

Ca^{2+} pumping across the PM

Tepikin and colleagues employing a technique where the extracellular $[Ca^{2+}]$ is monitored by an indicator in a small volume of extracellular fluid demonstrated that Ca^{2+} is extruded across the plasma membrane following agonist stimulation.[163,164] This in all probability occurs by Ca^{2+} pumps of the plasma membrane Ca^{2+}-ATPase gene family (PMCA). Ca^{2+}-ATPase activity is present in plasma membrane preparations isolated from pancreatic acinar cells and immunoblotting demonstrates the expression of PMCA family members.[21,165,166] Upon supermaximal stimulation with CCK or Ach, the amount of Ca^{2+} lost from the cell approximates the entire agonist-releasable pool.[167] Ca^{2+} extrusion also occurs during more physiological stimulation with CCK and the activity follows the $[Ca^{2+}]_i$.[163] PMCA is also important for maintaining resting $[Ca^{2+}]_i$ since its activity can be demonstrated at basal $[Ca^{2+}]_i$. Indeed, the rate of pumping has been shown to have a steep dependence on $[Ca^{2+}]_i$ (Hill coefficient of ~3) and is effectively saturated at $[Ca^{2+}]$ above 400 nM.[168] The activity of the PMCA can also be increased by agonist stimulation in a manner independent of Ca^{2+} and this may reflect modulation of the pumps activity by phosphorylation or association with regulatory proteins.[169] PMCA is not homogeneously distributed over the entire plasma membrane and appears to be most abundant in the luminal and lateral plasma membrane.[21] The localization of PMCA correlates with the site of most apparent Ca^{2+} pumping activity since Ca^{2+} extrusion occurs preferentially from the apical aspects of mouse pancreatic acinar cells when monitored using an indicator with limited diffusional mobility.[170]

Ca^{2+} uptake into the ER

Ca^{2+} pumps are also expressed on ER membranes.[171,172] Indeed, the ER in pancreatic acinar cells has been shown to effectively function as a single continuous Ca^{2+} store.[22,173] This has been best illustrated by demonstrating that ER Ca^{2+} pump activity, exclusively at the basal pole, can recharge the ER following maximal agonist stimulation such that Ca^{2+} signals can be initiated at the apical pole.[173] ER Ca^{2+}-ATPases belong to the sarcoplasmic and endoplasmic reticulum Ca^{2+} ATPase gene family (SERCA). Both SERCA 2A and SERCA 2B have been reported to be expressed in pancreatic acinar cells and specific subcellular distributions of specific SERCA pumps have been reported.[21] SERCA 2A appears to be distributed very similar to InsP$_3$R exclusively in the apical pole while SERCA

2B predominately resides in the basal aspects of the cell. All SERCA pumps are inhibited by thapsigargin and treatment results in Ca^{2+} leak from the ER store. In pancreatic acinar cells, thapsigargin treatment results in a uniform rise in $[Ca^{2+}]_i$ and abolishes the characteristic secretagogue-stimulated apically initiated Ca^{2+} wave even before the ER is fully depleted.[21] The later observation indicates that the microenvironment created by SERCA pumps is crucial for the initiation of Ca^{2+} signals in the apical pole during Ca^{2+} oscillations. This could occur as a function of the SERCA pump controlling either the cytosolic $[Ca^{2+}]$ or local luminal ER $[Ca^{2+}]$ in the apical pole as both SERCA pumps and IP$_3$R are markedly influenced by the $[Ca^{2+}]$ on both faces of the ER membrane.[23,174]

Ca^{2+} uptake by mitochondria

A further Ca^{2+} uptake mechanism that plays a significant role in spatially shaping Ca^{2+} signals in pancreatic acinar cells is through Ca^{2+} sequestration into mitochondria. Mitochondrial Ca^{2+} uptake occurs as a function of the large electrical potential across the inner mitochondrial membrane via a ruthenium red-sensitive Ca^{2+} uniporter.[175] Because the Ca^{2+} uniporter is a relatively low-affinity, high-capacity transporter, it was generally believed that mitochondrial Ca^{2+} uptake was only relevant under pathological conditions when the $[Ca^{2+}]$ was significantly elevated for prolonged periods of time. Recently, however, with the advent of mitochondrially targeted indicators, this idea has been reevaluated given that it has become clear that mitochondria function during normal physiological Ca^{2+} signaling in many cell types.[176] Moreover, it is now thought that mitochondrial Ca^{2+} uptake is important not only for shaping cytosolic Ca^{2+} signals but also for stimulating the production of ATP as key enzymes in the TCA cycle are Ca^{2+}-dependent.[177] It appears that the privileged localization of mitochondria very close to Ca^{2+}-release sites, where Ca^{2+} is presumably very high, allows the Ca^{2+} uniporter to function effectively under these conditions.

The long-sought-after molecular components of the mitochondrial uniporter have recently been elucidated. Initially, extensive comparative proteomics for inner mitochondrial membrane proteins present in vertebrates, kinetoplastids (protists containing flagella) but not yeast (which do not have a uniporter) identified MICU1 as essential for mitochondrial uptake.[178] Second, with the knowledge that the uniporter complex contained a highly Ca^{2+}-selective channel,[179] two groups identified an integral membrane protein, MCU, which had coevolved with MICU1.[180,181] Together, these proteins form the core of a complex, which reconstitutes ruthenium red-sensitive mitochondrial Ca^{2+} uptake. A recent study utilizing cells prepared from MCU knockout animals has clearly demonstrated that mitochondrial Ca^{2+} uptake is disrupted, thus providing strong evidence for its role in acini.[182] Interestingly, disruption of mitochondrial Ca^{2+} signaling did not alter the severity of experimental pancreatitis despite compromising mitochondrial bioenergetics.

In pancreatic acinar cells, mitochondria are indeed in close proximity to Ca^{2+}-release sites since stimulation with physiological concentrations of both CCK and Ach, which raise global $[Ca^{2+}]_i$ only to submicromolar levels, leads to mitochondrial Ca^{2+} uptake.[183–185] Furthermore, mitochondrial Ca^{2+} uptake in pancreatic acinar cells is coupled to the conversion of NAD to NADH.[184] Further, stimulation of pancreatic acinar cells with Ca^{2+}-mobilizing secretagogues

has been shown to result in the net generation of ATP as reported by adenovirally expressed mitochondrially targeted luciferase constructs.[186] In common with other Ca^{2+} clearance mechanisms in acini, energized mitochondria also have specific subcellular localization (**Figure 6A**).[187] The majority of studies have reported that mitochondria are concentrated in a perigranular "belt,"[36,188] together with additional further subpopulations surrounding the nucleus and immediately below the basolateral plasma membrane.[183] Strikingly, It appears that these perigranular mitochondria play a role in limiting the spread of Ca^{2+} signals upon stimulation with threshold concentrations of

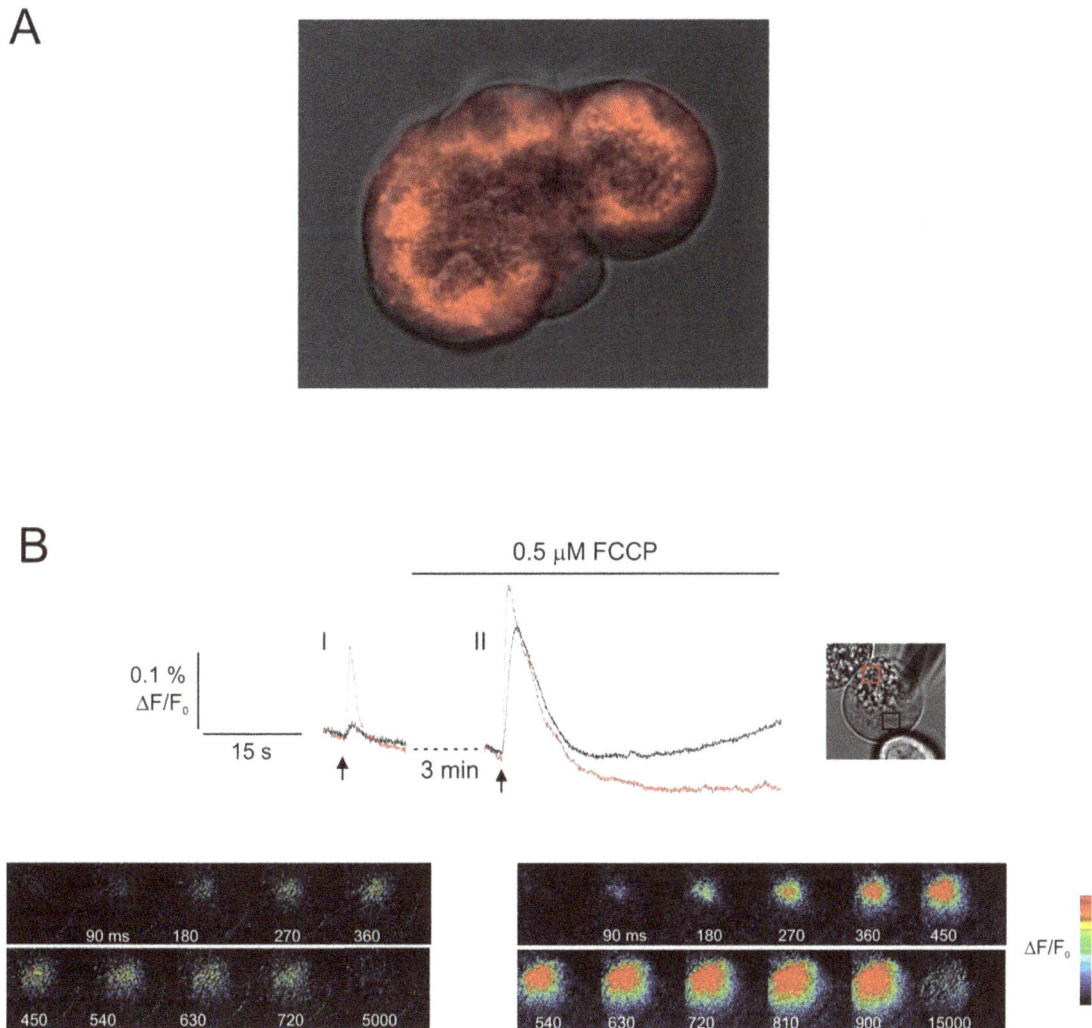

Figure 6. Functional consequences of mitochondrial distribution in pancreatic acini. In (**A**), the localization of active mitochondria was visualized by confocal microscopy in living mouse pancreatic acini loaded with mitotracker red. (**A**) shows mitotracker red fluorescence merged with a phase image of a small mouse acini. The predominant localization of mitochondria is to a perigranular belt surrounding the zymogen granules. (**B**) Photolysis of threshold concentrations of 1,4,5-IP$_3$ results in apically limited Ca^{2+} signals as shown in the images and kinetic traces in (**B**) **I** (red trace apical ROI; black trace basal ROI). In the same cell, following disruption of the mitochondrial membrane potential, and thus mitochondrial Ca^{2+} uptake, an identical exposure to 1,4,5 IP$_3$ results in a global Ca^{2+} signal. These data suggest that functional mitochondria are required to constrain Ca^{2+} signals in the apical portion of the acinar cell (from reference [36]).

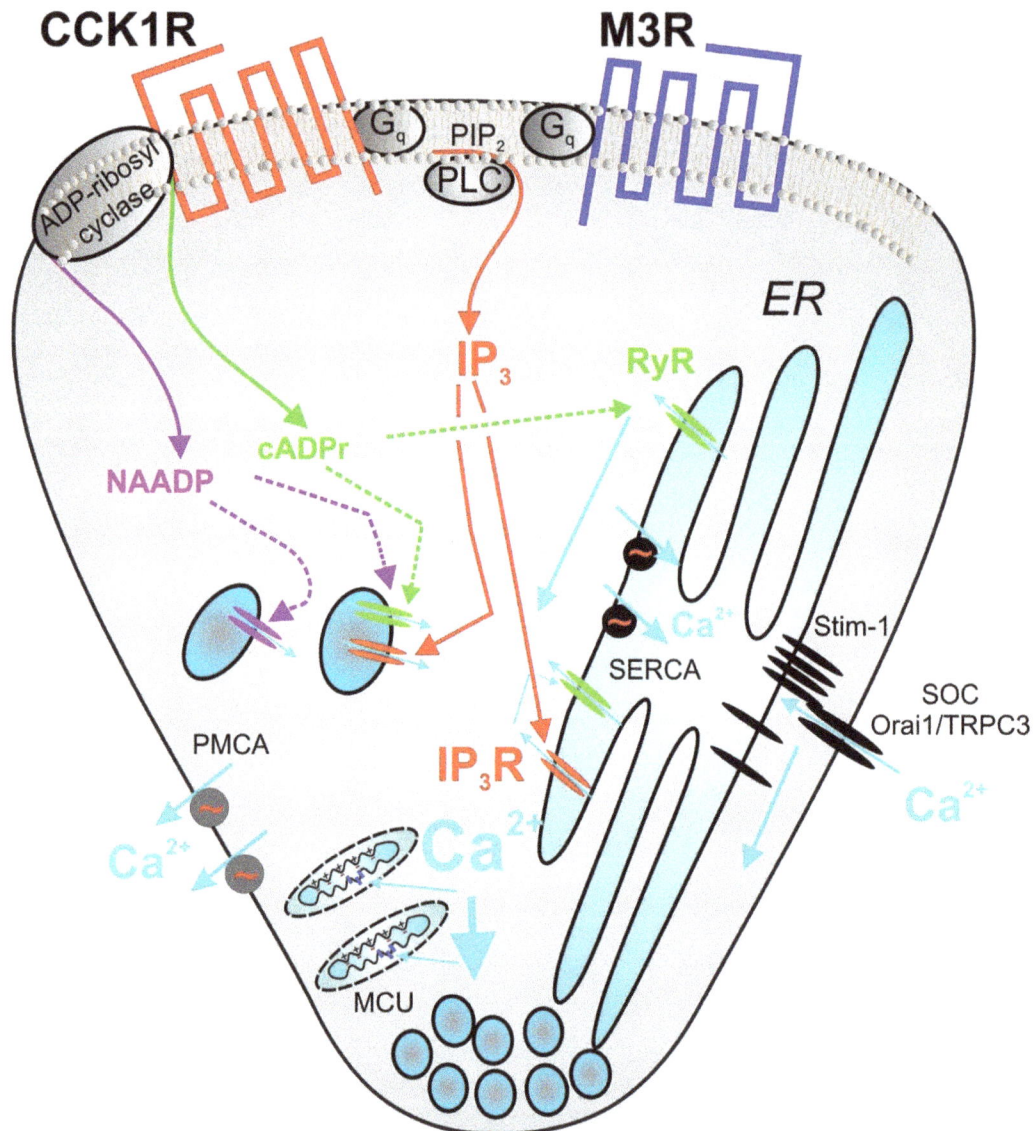

Figure 7. **Intracellular Ca^{2+} signaling in pancreatic acinar cells is initiated by the binding of acetylcholine to muscarinic M3 receptors (M3R) and by cholecystokinin (CCK) to CCK receptors, predominately the CCK1R in mice and rats.** Both receptors are classical seven-transmembrane domain receptors coupled to guanine nucleotide-binding proteins (G proteins). Activation of both receptors leads to stimulation of Gαq and increased activity of phospholipase C-β (PLC), which cleaves the membrane phospholipid phosphatidylinositol,4,5,-bisphosphate (PIP2) into diacylglycerol and inositol 1,4,5-trisphosphate (IP$_3$). IP$_3$ diffuses through the cytoplasm and interacts with InsP3 receptors (IP$_3$Rs), largely type 2 and 3 present predominantly on the apical endoplasmic reticulum (ER) resulting in Ca^{2+} release into the cytoplasm. Ca^{2+} release from IP$_3$R acts a coagonist to increase further IP$_3$R activity and also acts on ryanodine receptors (RyRs) to induce Ca^{2+} release. Depletion of Ca^{2+} within the ER results in Ca^{2+} influx from the extracellular space. The ER Ca^{2+} sensor has been identified as stromal interaction molecule-1 (stim-1). Following ER depletion, Ca^{2+} is released from an EF hand present in a domain of stim-1 within the ER lumen and this results in aggregation of several stim-1 molecules. Aggregation of stim-1 is sufficient to gate plasma membrane Ca^{2+} channels and leads to Ca^{2+} influx. Good candidates for the channel pore are the proteins Orai-1 and TRPC3. CCK receptor stimulation also stimulates ADP-ribosyl cyclase activity resulting in the formation of two additional Ca^{2+}-mobilizing second messengers: nicotinic acid adenine dinucleotide phosphate (NAADP) and cyclic-ADP ribose (cADPr). The particular cyclase responsible has not been defined in pancreas but good candidates include CD38 and CD157. cADPr is generally thought to act on RyR, while the target of NAADP is currently the subject of intense research. Candidates include the RyR and two-pore channel (TPC). In addition to the ER, Ca^{2+} can also be released from a store, which accumulates Ca^{2+} in a manner dependent on a proton gradient—known as the "acidic store." This pool likely represents the endolysosomal compartment. This pool has been reported to be responsive to IP$_3$, cADPr, and NAADP and may represent the store initially released following receptor stimulation. Ca^{2+} is removed from the cytoplasm by the concerted action of SERCA (ER Ca^{2+}-ATPase), PMCA (PM Ca^{2+}-ATPase), and the action of mitochondrial uniporter (MCU).

agonist as disruption of mitochondrial Ca^{2+} uptake facilitates the spread of normally spatially confined Ca^{2+} transients (**Figure 6B**).[36,188]

Summary

A cartoon summarizing the molecular components currently believed to be involved in the processes of Ca^{2+} release, Ca^{2+} influx, and Ca^{2+} clearance from the cytosol of pancreatic acinar cells following secretagogue stimulation is shown in Figure 7.

References

1. Gerasimenko JV, Gerasimenko OV, Petersen OH. The role of Ca^{2+} in the pathophysiology of pancreatitis. J Physiol 2014; 592:269–280.
2. Streb H, Irvine RF, Berridge MJ, Schulz I. Release of Ca^{2+} from a nonmitochondrial intracellular store in pancreatic acinar cells by inositol-1,4,5-trisphosphate. Nature 1983; 306:67–69.
3. Streb H, Bayerdorffer E, Haase W, Irvine RF, Schulz I. Effect of inositol-1,4,5-trisphosphate on isolated subcellular fractions of rat pancreas. J Membr Biol 1984; 81:241–253.
4. Dormer RL, Williams JA. Secretagogue-induced changes in subcellular Ca^{2+} distribution in isolated pancreatic acini. Am J Physiol Gastrointest Liver Physiol 1981; 240:G130–G140.
5. Streb H, Heslop JP, Irvine RF, Schulz I, Berridge MJ. Relationship between secretagogue-induced Ca^{2+} release and inositol polyphosphate production in permeabilized pancreatic acinar cells. J Biol Chem 1985; 260:7309–7315.
6. Matozaki T, Sakamoto C, Nishisaki H, Suzuki T, Wada K, Matsuda K, Nakano O, Konda Y, Nagao M, Kasuga M. Cholecystokinin inhibits phosphatidylcholine synthesis via a Ca^{2+}-calmodulin-dependent pathway in isolated rat pancreatic acini. A possible mechanism for diacylglycerol accumulation. J Biol Chem 1991; 266:22246–22253.
7. Pralong WF, Wollheim CB, Bruzzone R. Measurement of cytosolic free Ca^{2+} in individual pancreatic acini. FEBS Lett 1988; 242:79–84.
8. Stuenkel EL, Tsunoda Y, Williams JA. Secretagogue induced calcium mobilization in single pancreatic acinar cells. Biochem Biophys Res Commun 1989; 158:863–869.
9. Yule DI, Gallacher DV. Oscillations of cytosolic calcium in single pancreatic acinar cells stimulated by acetylcholine. FEBS Lett 1988; 239:358–362.
10. Pandol SJ, Schoeffield MS, Fimmel CJ, Muallem S. The agonist-sensitive calcium pool in the pancreatic acinar cell. Activation of plasma membrane Ca^{2+} influx mechanism. J Biol Chem 1987; 262:16963–16968.
11. Tsunoda Y, Stuenkel EL, Williams JA. Characterization of sustained [Ca^{2+}]$_i$ increase in pancreatic acinar cells and its relation to amylase secretion. Am J Physiol Gastrointest Liver Physiol 1990; 259:G792–G801.
12. Lugea A, Waldron RT, Mareninova OA, Shalbueva N, Deng N, Su HY, Thomas DD, Jones EK, Messenger SW, Yang J, Hu C, Gukovsky I, et al. Human pancreatic acinar cells: proteomic characterization, physiologic responses, and organellar disorders in ex vivo pancreatitis. Am J Pathol 2017; 187:2726–2743.
13. Tsunoda Y, Stuenkel EL, Williams JA. Oscillatory mode of calcium signaling in rat pancreatic acinar cells. Am J Physiol Cell Physiol 1990; 258:C147–C155.
14. Sjodin L, Dahlen HG, Gylfe E. Calcium oscillations in guinea-pig pancreatic acinar cells exposed to carbachol, cholecystokinin and substance P. J Physiol 1991; 444:763–776.
15. Yule DI, Lawrie AM, Gallacher DV. Acetylcholine and cholecystokinin induce different patterns of oscillating calcium signals in pancreatic acinar cells. Cell Calcium 1991; 12:145–151.
16. Osipchuk YV, Wakui M, Yule DI, Gallacher DV, Petersen OH. Cytoplasmic Ca^{2+} oscillations evoked by receptor stimulation, G-protein activation, internal application of inositol trisphosphate or Ca^{2+}: simultaneous microfluorimetry and Ca^{2+} dependent Cl$^-$ current recording in single pancreatic acinar cells. EMBO J 1990; 9:697–704.
17. Giovannucci DR, Bruce JI, Straub SV, Arreola J, Sneyd J, Shuttleworth TJ, Yule DI. Cytosolic Ca^{2+} and Ca^{2+}-activated Cl$^-$ current dynamics: insights from two functionally distinct mouse exocrine cells. J Physiol 2002; 540:469–484.
18. LeBeau AP, Yule DI, Groblewski GE, Sneyd J. Agonist-dependent phosphorylation of the inositol 1,4,5-trisphosphate receptor: a possible mechanism for agonist-specific calcium oscillations in pancreatic acinar cells. J Gen Physiol 1999; 113:851–872.
19. Straub SV, Giovannucci DR, Bruce JI, Yule DI. A role for phosphorylation of inositol 1,4,5-trisphosphate receptors in defining calcium signals induced by Peptide agonists in pancreatic acinar cells. J Biol Chem 2002; 277:31949–31956.
20. Wakui M, Potter BV, Petersen OH. Pulsatile intracellular calcium release does not depend on fluctuations in inositol trisphosphate concentration. Nature 1989; 339:317–320.
21. Lee MG, Xu X, Zeng W, Diaz J, Kuo TH, Wuytack F, Racymaekers L, Muallem S. Polarized expression of Ca^{2+} pumps in pancreatic and salivary gland cells. Role in initiation and propagation of [Ca^{2+}]$_i$ waves. J Biol Chem 1997; 272:15771–15776.
22. Park MK, Petersen OH, Tepikin AV. The endoplasmic reticulum as one continuous Ca^{2+} pool: visualization of rapid Ca^{2+} movements and equilibration. EMBO J 2000; 19:5729–5739.
23. Mogami H, Tepikin AV, Petersen OH. Termination of cytosolic Ca^{2+} signals: Ca^{2+} reuptake into intracellular stores is regulated by the free Ca^{2+} concentration in the store lumen. EMBO J 1998; 17:435–442.
24. Sneyd J, Tsaneva-Atanasova K, Reznikov V, Bai Y, Sanderson MJ, Yule DI. A method for determining the dependence of calcium oscillations on inositol trisphosphate oscillations. Proc Natl Acad Sci U S A 2006; 103:1675–1680.
25. Zeng W, Xu X, Muallem S. Gbetagamma transduces [Ca^{2+}]$_i$ oscillations and Galphaq a sustained response during stimulation of pancreatic acinar cells with [Ca^{2+}]$_i$-mobilizing agonists. J Biol Chem 1996; 271:18520–18526.
26. Zeng W, Xu X, Popov S, Mukhopadhyay S, Chidiac P, Swistok J, Danho W, Yagaloff KA, Fisher SL, Ross EM, Muallem S, Wilkie TM. The N-terminal domain of RGS4 confers receptor-selective inhibition of G protein signaling. J Biol Chem 1998; 273:34687–34690.

27. Xu X, Zeng W, Popov S, Berman DM, Davignon I, Yu K, Yowe D, Offermanns S, Muallem S, Wilkie TM. RGS proteins determine signaling specificity of G_q-coupled receptors. J Biol Chem 1999; 274:3549–3556.

28. Luo X, Popov S, Bera AK, Wilkie TM, Muallem S. RGS proteins provide biochemical control of agonist-evoked $[Ca^{2+}]_i$ oscillations. Mol Cell 2001; 7:651–660.

29. Chen TW, Wardill TJ, Sun Y, Pulver SR, Renninger SL, Baohan A, Schreiter ER, Kerr RA, Orger MB, Jayaraman V, Looger LL, Svoboda K, et al. Ultrasensitive fluorescent proteins for imaging neuronal activity. Nature 2013; 499:295–300.

30. Zariwala HA, Borghuis BG, Hoogland TM, Madisen L, Tian L, De Zeeuw CI, Zeng H, Looger LL, Svoboda K, Chen TW. A Cre-dependent GCaMP3 reporter mouse for neuronal imaging in vivo. J Neurosci 2012; 32:3131–3141.

31. Oshima Y, Imamura T, Shintani A, Kajiura-Kobayashi H, Hibi T, Nagai T, Nonaka S, Nemoto T. Ultrasensitive imaging of Ca^{2+} dynamics in pancreatic acinar cells of yellow cameleon-nano transgenic mice. Int J Mol Sci 2014; 15:19971–19986.

32. Jin K, Imada T, Nakamura S, Izuta Y, Oonishi E, Shibuya M, Sakaguchi H, Adachi T, Tsubota K. Intravital two-photon imaging of Ca^{2+} signaling in secretory organs of yellow cameleon transgenic mice. Sci Rep 2018; 8:15880.

33. Kasai H, Augustine GJ. Cytosolic Ca^{2+} gradients triggering unidirectional fluid secretion from exocrine pancreas. Nature 1990; 348:735–738.

34. Nathanson MH, Padfield PJ, O'Sullivan AJ, Burgstahler AD, Jamieson JD. Mechanism of Ca^{2+} wave propagation in pancreatic acinar cells. J Biol Chem 1992; 267:18118–18121.

35. Toescu EC, Lawrie AM, Petersen OH, Gallacher DV. Spatial and temporal distribution of agonist-evoked cytoplasmic Ca^{2+} signals in exocrine acinar cells analysed by digital image microscopy. EMBO J 1992; 11:1623–1629.

36. Straub SV, Giovannucci DR, Yule DI. Calcium wave propagation in pancreatic acinar cells: functional interaction of inositol 1,4,5-trisphosphate receptors, ryanodine receptors, and mitochondria. J Gen Physiol 2000; 116:547–560.

37. Gonzalez A, Schmid A, Sternfeld L, Krause E, Salido GM, Schulz I. Cholecystokinin-evoked Ca^{2+} waves in isolated mouse pancreatic acinar cells are modulated by activation of cytosolic phospholipase A_2, phospholipase D, and protein kinase C. Biochem Biophys Res Commun 1999; 261:726–733.

38. Gonzalez A, Pfeiffer F, Schmid A, Schulz I. Effect of intracellular pH on acetylcholine-induced Ca^{2+} waves in mouse pancreatic acinar cells. Am J Physiol Cell Physiol 1998; 275:C810–C817.

39. Kasai H, Li YX, Miyashita Y. Subcellular distribution of Ca^{2+} release channels underlying Ca^{2+} waves and oscillations in exocrine pancreas. Cell 1993; 74:669–677.

40. Rosenzweig SA, Miller LJ, Jamieson JD. Identification and localization of cholecystokinin-binding sites on rat pancreatic plasma membranes and acinar cells: a biochemical and autoradiographic study. J Cell Biol 1983; 96:1288–1297.

41. Thorn P, Lawrie AM, Smith PM, Gallacher DV, Petersen OH. Local and global cytosolic Ca^{2+} oscillations in exocrine cells evoked by agonists and inositol trisphosphate. Cell 1993; 74:661–668.

42. Ashby MC, Camello-Almaraz C, Gerasimenko OV, Petersen OH, Tepikin AV. Long distance communication between muscarinic receptors and Ca^{2+} release channels revealed by carbachol uncaging in cell-attached patch pipette. J Biol Chem 2003; 278:20860–20864.

43. Li Q, Luo X, Muallem S. Functional mapping of Ca^{2+} signaling complexes in plasma membrane microdomains of polarized cells. J Biol Chem 2004; 279:27837–27840.

44. Shin DM, Luo X, Wilkie TM, Miller LJ, Peck AB, Humphreys-Beher MG, Muallem S. Polarized expression of G protein-coupled receptors and an all-or-none discharge of Ca^{2+} pools at initiation sites of $[Ca^{2+}]_i$ waves in polarized exocrine cells. J Biol Chem 2001; 276:44146–44156.

45. Ito K, Miyashita Y, Kasai H. Micromolar and submicromolar Ca^{2+} spikes regulating distinct cellular functions in pancreatic acinar cells. EMBO J 1997; 16:242–251.

46. Petersen CC, Toescu EC, Petersen OH. Different patterns of receptor-activated cytoplasmic Ca^{2+} oscillations in single pancreatic acinar cells: dependence on receptor type, agonist concentration and intracellular Ca^{2+} buffering. EMBO J 1991; 10:527–533.

47. Park MK, Lomax RB, Tepikin AV, Petersen OH. Local uncaging of caged Ca^{2+} reveals distribution of Ca^{2+}-activated Cl^- channels in pancreatic acinar cells. Proc Natl Acad Sci U S A 2001; 98:10948–10953.

48. Fogarty KE, Kidd JF, Tuft DA, Thorn P. Mechanisms underlying $InsP_3$-evoked global Ca^{2+} signals in mouse pancreatic acinar cells. J Physiol 2000; 526:515–526.

49. Meda P, Findlay I, Kolod E, Orci L, Petersen OH. Short and reversible uncoupling evokes little change in the gap junctions of pancreatic acinar cells. J Ultrastruct Res 1983; 83:69–84.

50. Iwatsuki N, Petersen OH. Electrical coupling and uncoupling of exocrine acinar cells. J Cell Biol 1978; 79:533–545.

51. Yule DI, Stuenkel E, Williams JA. Intercellular calcium waves in rat pancreatic acini: mechanism of transmission. Am J Physiol Cell Physiol 1996; 271:C1285–C1294.

52. Stauffer PL, Zhao H, Luby-Phelps K, Moss RL, Star RA, Muallem S. Gap junction communication modulates $[Ca^{2+}]_i$ oscillations and enzyme secretion in pancreatic acini. J Biol Chem 1993; 268:19769–19775.

53. Yule DI, Ernst SA, Ohnishi H, Wojcikiewicz RJ. Evidence that zymogen granules are not a physiologically relevant calcium pool. Defining the distribution of inositol 1,4,5-trisphosphate receptors in pancreatic acinar cells. J Biol Chem 1997; 272:9093–9098.

54. Nathanson MH, Fallon MB, Padfield PJ, Maranto AR. Localization of the type 3 inositol 1,4,5-trisphosphate receptor in the Ca^{2+} wave trigger zone of pancreatic acinar cells. J Biol Chem 1994; 269:4693–4696.

55. Lee MG, Xu X, Zeng W, Diaz J, Wojcikiewicz RJ, Kuo TH, Wuytack F, Racymaekers L, Muallem S. Polarized expression of Ca^{2+} channels in pancreatic and salivary gland cells. Correlation with initiation and propagation of $[Ca^{2+}]_i$ waves. J Biol Chem 1997; 272:15765–15770.

56. Furuichi T, Yoshikawa S, Miyawaki A, Wada K, Maeda N, Mikoshiba K. Primary structure and functional expression of the inositol 1,4,5-trisphosphate-binding protein P400. Nature 1989; 342:32–38.

57. Maranto AR. Primary structure, ligand binding, and localization of the human type 3 inositol 1,4,5-trisphosphate receptor expressed in intestinal epithelium. J Biol Chem 1994; 269:1222–1230.

58. Mignery GA, Newton CL, Archer BT, 3rd, Sudhof TC. Structure and expression of the rat inositol 1,4,5-trisphosphate receptor. J Biol Chem 1990; 265:12679–12685.

59. Nagata J, Guerra MT, Shugrue CA, Gomes DA, Nagata N, Nathanson MH. Lipid rafts establish calcium waves in hepatocytes. Gastroenterology 2007; 133:256–267.

60. Gerasimenko JV, Maruyama Y, Yano K, Dolman NJ, Tepikin AV, Petersen OH, Gerasimenko OV. NAADP mobilizes Ca^{2+} from a thapsigargin-sensitive store in the nuclear envelope by activating ryanodine receptors. J Cell Biol 2003; 163:271–282.

61. Krause E, Gobel A, Schulz I. Cell side-specific sensitivities of intracellular Ca^{2+} stores for inositol 1,4,5-trisphosphate, cyclic ADP-ribose, and nicotinic acid adenine dinucleotide phosphate in permeabilized pancreatic acinar cells from mouse. J Biol Chem 2002; 277:11696–11702.

62. Wojcikiewicz RJ. Type I, II, and III inositol 1,4,5-trisphosphate receptors are unequally susceptible to down-regulation and are expressed in markedly different proportions in different cell types. J Biol Chem 1995; 270:11678–11683.

63. De Smedt H, Missiaen L, Parys JB, Henning RH, Sienaert I, Vanlingen S, Gijsens A, Himpens B, Casteels R. Isoform diversity of the inositol trisphosphate receptor in cell types of mouse origin. Biochem J 1997; 322:575–583.

64. Futatsugi A, Nakamura T, Yamada MK, Ebisui E, Nakamura K, Uchida K, Kitaguchi T, Takahashi-Iwanaga H, Noda T, Aruga J, Mikoshiba K. IP_3 receptor types 2 and 3 mediate exocrine secretion underlying energy metabolism. Science 2005; 309:2232–2234.

65. Orabi AI, Luo Y, Ahmad MU, Shah AU, Mannan Z, Wang D, Sarwar S, Muili KA, Shugrue C, Kolodecik TR, Singh VP, Lowe ME, et al. IP_3 receptor type 2 deficiency is associated with a secretory defect in the pancreatic acinar cell and an accumulation of zymogen granules. PLoS One 2012; 7:e48465.

66. Patel S, Joseph SK, Thomas AP. Molecular properties of inositol 1,4,5-trisphosphate receptors. Cell Calcium 1999; 25:247–264.

67. Taylor CW, Genazzani AA, Morris SA. Expression of inositol trisphosphate receptors. Cell Calcium 1999; 26:237–251.

68. Wojcikiewicz RJ, Ernst SA, Yule DI. Secretagogues cause ubiquitination and down-regulation of inositol 1, 4,5-trisphosphate receptors in rat pancreatic acinar cells. Gastroenterology 1999; 116:1194–1201.

69. Alzayady KJ, Wagner LE, 2nd, Chandrasekhar R, Monteagudo A, Godiska R, Tall GG, Joseph SK, Yule DI. Functional inositol 1,4,5-trisphosphate receptors assembled from concatenated homo- and heteromeric subunits. J Biol Chem 2013; 288:29772–29784.

70. Chandrasekhar R, Alzayady KJ, Yule DI. Using concatenated subunits to investigate the functional consequences of heterotetrameric inositol 1,4,5-trisphosphate receptors. Biochem Soc Trans 2015; 43:364–370.

71. Bosanac I, Michikawa T, Mikoshiba K, Ikura M. Structural insights into the regulatory mechanism of IP_3 receptor. Biochim Biophys Acta 2004; 1742:89–102.

72. Seo MD, Velamakanni S, Ishiyama N, Stathopulos PB, Rossi AM, Khan SA, Dale P, Li C, Ames JB, Ikura M, Taylor CW. Structural and functional conservation of key domains in $InsP_3$ and ryanodine receptors. Nature 2012; 483:108–112.

73. Fohr KJ, Wahl Y, Engling R, Kemmer TP, Gratzl M. Decavanadate displaces inositol 1,4,5-trisphosphate (IP_3) from its receptor and inhibits IP_3 induced Ca^{2+} release in permeabilized pancreatic acinar cells. Cell Calcium 1991; 12:735–742.

74. Ferris CD, Huganir RL, Supattapone S, Snyder SH. Purified inositol 1,4,5-trisphosphate receptor mediates calcium flux in reconstituted lipid vesicles. Nature 1989; 342:87–89.

75. Mak DO, Foskett JK. Single-channel inositol 1,4,5-trisphosphate receptor currents revealed by patch clamp of isolated Xenopus oocyte nuclei. J Biol Chem 1994; 269:29375–29378.

76. Boehning D, Mak DO, Foskett JK, Joseph SK. Molecular determinants of ion permeation and selectivity in inositol 1,4,5-trisphosphate receptor Ca^{2+} channels. J Biol Chem 2001; 276:13509–13512.

77. Wagner LE, 2nd, Yule DI. Differential regulation of the $InsP_3$ receptor type-1 and -2 single channel properties by $InsP_3$, Ca^{2+} and ATP. J Physiol 2012; 590:3245–3259.

78. Finch EA, Turner TJ, Goldin SM. Calcium as a coagonist of inositol 1,4,5-trisphosphate-induced calcium release. Science 1991; 252:443–446.

79. Bezprozvanny I, Watras J, Ehrlich BE. Bell-shaped calcium-response curves of $Ins(1,4,5)P_3$- and calcium-gated channels from endoplasmic reticulum of cerebellum. Nature 1991; 351:751–754.

80. Taylor CW, Laude AJ. IP_3 receptors and their regulation by calmodulin and cytosolic Ca^{2+}. Cell Calcium 2002; 32:321–334.

81. Zhang BX, Zhao H, Loessberg PA, Muallem S. Regulation of agonist-evoked $[Ca^{2+}]_i$ oscillation by intracellular Ca^{2+} and Ba^{2+} in AR42J cells. Am J Physiol Cell Physiol 1992; 262:C1125–C1233.

82. Engling R, Fohr KJ, Kemmer TP, Gratzl M. Effect of GTP and Ca^{2+} on inositol 1,4,5-trisphosphate induced Ca^{2+} release from permeabilized rat exocrine pancreatic acinar cells. Cell Calcium 1991; 12:1–9.

83. Kidd JF, Fogarty KE, Tuft RA, Thorn P. The role of Ca^{2+} feedback in shaping $InsP_3$-evoked Ca^{2+} signals in mouse pancreatic acinar cells. J Physiol 1999; 520:187–201.

84. Patterson RL, Boehning D, Snyder SH. Inositol 1,4,5-trisphosphate receptors as signal integrators. Annu Rev Biochem 2004; 73:437–465.

85. Pieper AA, Brat DJ, O'Hearn E, Krug DK, Kaplin AI, Takahashi K, Greenberg JH, Ginty D, Molliver ME, Snyder SH. Differential neuronal localizations and dynamics of phosphorylated and unphosphorylated type 1 inositol 1,4,5-trisphosphate receptors. Neuroscience 2001; 102:433–444.

86. Giovannucci DR, Groblewski GE, Sneyd J, Yule DI. Targeted phosphorylation of inositol 1,4,5-trisphosphate receptors selectively inhibits localized Ca^{2+} release and shapes oscillatory Ca^{2+} signals. J Biol Chem 2000; 275:33704–33711.

87. Camello PJ, Petersen OH, Toescu EC. Simultaneous presence of cAMP and cGMP exert a co-ordinated inhibitory

effect on the agonist-evoked Ca^{2+} signal in pancreatic acinar cells. Pflugers Arch 1996; 432:775–781.

88. Matozaki T, Zhu WY, Tsunoda Y, Goke B, Williams JA. Intracellular mediators of bombesin action on rat pancreatic acinar cells. Am J Physiol Gastrointest Liver Physiol 1991; 260:G858–G864.

89. Yule DI, Betzenhauser MJ, Joseph SK. Linking structure to function: Recent lessons from inositol 1,4,5-trisphosphate receptor mutagenesis. Cell Calcium 2010; 47:469–479.

90. Yule DI, Williams JA. Mastoparan induces oscillations of cytosolic Ca^{2+} in rat pancreatic acinar cells. Biochem Biophys Res Commun 1991; 177:159–165.

91. Willems PH, van Nooij IG, Haenen HE, de Pont JJ. Phorbol ester inhibits cholecystokinin octapeptide-induced amylase secretion and calcium mobilization, but is without effect on secretagogue-induced hydrolysis of phosphatidylinositol 4,5-bisphosphate in rabbit pancreatic acini. Biochim Biophys Acta 1987; 930:230–236.

92. Park HS, Betzenhauser MJ, Won JH, Chen J, Yule DI. The type 2 inositol (1,4,5)-trisphosphate (InsP$_3$) receptor determines the sensitivity of InsP$_3$-induced Ca^{2+} release to ATP in pancreatic acinar cells. J Biol Chem 2008; 283:26081–26088.

93. Chandrasekhar R, Alzayady KJ, Wagner LE, 2nd, Yule DI. Unique regulatory properties of heterotetrameric Inositol 1,4,5-trisphosphate receptors revealed by studying concatenated receptor constructs. J Biol Chem 2016; 291:4846–4860.

94. Xu X, Zeng W, Muallem S. Regulation of the inositol 1,4,5-trisphosphate-activated Ca^{2+} channel by activation of G proteins. J Biol Chem 1996; 271:11737–11744.

95. Zeng W, Mak DO, Li Q, Shin DM, Foskett JK, Muallem S. A new mode of Ca^{2+} signaling by G protein-coupled receptors: gating of IP$_3$ receptor Ca^{2+} release channels by Gbetagamma. Curr Biol 2003; 13:872–876.

96. Ashby MC, Petersen OH, Tepikin AV. Spatial characterisation of ryanodine-induced calcium release in mouse pancreatic acinar cells. Biochem J 2003; 369:441–445.

97. Toescu EC, O'Neill SC, Petersen OH, Eisner DA. Caffeine inhibits the agonist-evoked cytosolic Ca^{2+} signal in mouse pancreatic acinar cells by blocking inositol trisphosphate production. J Biol Chem 1992; 267:23467–23470.

98. Leite MF, Dranoff JA, Gao L, Nathanson MH. Expression and subcellular localization of the ryanodine receptor in rat pancreatic acinar cells. Biochem J 1999; 337:305–309.

99. Fitzsimmons TJ, Gukovsky I, McRoberts JA, Rodriguez E, Lai FA, Pandol SJ. Multiple isoforms of the ryanodine receptor are expressed in rat pancreatic acinar cells. Biochem J 2000; 351:265–271.

100. Lewarchik CM, Orabi AI, Jin S, Wang D, Muili KA, Shah AU, Eisses JF, Malik A, Bottino R, Jayaraman T, Husain SZ. The ryanodine receptor is expressed in human pancreatic acinar cells and contributes to acinar cell injury. Am J Physiol Gastrointest Liver Physiol 2014; 307:G574–G581.

101. Husain SZ, Prasad P, Grant WM, Kolodecik TR, Nathanson MH, Gorelick FS. The ryanodine receptor mediates early zymogen activation in pancreatitis. Proc Natl Acad Sci U S A 2005; 102:14386–14391.

102. Clapper DL, Walseth TF, Dargie PJ, Lee HC. Pyridine nucleotide metabolites stimulate calcium release from sea urchin egg microsomes desensitized to inositol trisphosphate. J Biol Chem 1987; 262:9561–9568.

103. Guse AH, da Silva CP, Berg I, Skapenko AL, Weber K, Heyer P, Hohenegger M, Ashamu GA, Schulze-Koops H, Potter BV, Mayr GW. Regulation of calcium signalling in T lymphocytes by the second messenger cyclic ADP-ribose. Nature 1999; 398:70–73.

104. Cui Y, Galione A, Terrar DA. Effects of photoreleased cADP-ribose on calcium transients and calcium sparks in myocytes isolated from guinea-pig and rat ventricle. Biochem J 1999; 342:269–273.

105. Takasawa S, Nata K, Yonekura H, Okamoto H. Cyclic ADP-ribose in insulin secretion from pancreatic beta cells. Science 1993; 259:370–373.

106. Lee HC, Aarhus R, Graeff RM. Sensitization of calcium-induced calcium release by cyclic ADP-ribose and calmodulin. J Biol Chem 1995; 270:9060–9066.

107. Berridge MJ, Bootman MD, Roderick HL. Calcium signalling: dynamics, homeostasis and remodelling. Nat Rev Mol Cell Biol 2003; 4:517–529.

108. Sitsapesan R, McGarry SJ, Williams AJ. Cyclic ADP-ribose, the ryanodine receptor and Ca^{2+} release. Trends Pharmacol Sci 1995; 16:386–391.

109. Thorn P, Gerasimenko O, Petersen OH. Cyclic ADP-ribose regulation of ryanodine receptors involved in agonist evoked cytosolic Ca^{2+} oscillations in pancreatic acinar cells. EMBO J 1994; 13:2038–2043.

110. Leite MF, Burgstahler AD, Nathanson MH. Ca^{2+} waves require sequential activation of inositol trisphosphate receptors and ryanodine receptors in pancreatic acini. Gastroenterology 2002; 122:415–427.

111. Cancela JM, Gerasimenko OV, Gerasimenko JV, Tepikin AV, Petersen OH. Two different but converging messenger pathways to intracellular Ca^{2+} release: the roles of nicotinic acid adenine dinucleotide phosphate, cyclic ADP-ribose and inositol trisphosphate. EMBO J 2000; 19:2549–2557.

112. Cancela JM, Van Coppenolle F, Galione A, Tepikin AV, Petersen OH. Transformation of local Ca^{2+} spikes to global Ca^{2+} transients: the combinatorial roles of multiple Ca^{2+} releasing messengers. EMBO J 2002; 21:909–919.

113. Ozawa T. Elucidation of the ryanodine-sensitive Ca^{2+} release mechanism of rat pancreatic acinar cells: modulation by cyclic ADP-ribose and FK506. Biochim Biophys Acta 2004; 1693:159–166.

114. Sternfeld L, Krause E, Guse AH, Schulz I. Hormonal control of ADP-ribosyl cyclase activity in pancreatic acinar cells from rats. J Biol Chem 2003; 278:33629–33636.

115. Fukushi Y, Kato I, Takasawa S, Sasaki T, Ong BH, Sato M, Ohsaga A, Sato K, Shirato K, Okamoto H, Maruyama Y. Identification of cyclic ADP-ribose-dependent mechanisms in pancreatic muscarinic Ca2+ signaling using CD38 knockout mice. J Biol Chem 2001; 276:649–655.

116. Cancela JM, Petersen OH. The cyclic ADP ribose antagonist 8-NH$_2$-cADP-ribose blocks cholecystokinin-evoked cytosolic Ca^{2+} spiking in pancreatic acinar cells. Pflugers Arch 1998; 435:746–748.

117. Burdakov D, Cancela JM, Petersen OH. Bombesin-induced cytosolic Ca^{2+} spiking in pancreatic acinar cells depends on

cyclic ADP-ribose and ryanodine receptors. Cell Calcium 2001; 29:211–216.

118. Galione A, Patel S, Churchill GC. NAADP-induced calcium release in sea urchin eggs. Biol Cell 2000; 92:197–204.

119. Yamasaki M, Masgrau R, Morgan AJ, Churchill GC, Patel S, Ashcroft SJ, Galione A. Organelle selection determines agonist-specific Ca²⁺ signals in pancreatic acinar and beta cells. J Biol Chem 2004; 279:7234–7240.

120. Cancela JM, Churchill GC, Galione A. Coordination of agonist-induced Ca²⁺-signalling patterns by NAADP in pancreatic acinar cells. Nature 1999; 398:74–76.

121. Gerasimenko JV, Charlesworth RM, Sherwood MW, Ferdek PE, Mikoshiba K, Parrington J, Petersen OH, Gerasimenko OV. Both RyRs and TPCs are required for NAADP-induced intracellular Ca²⁺ release. Cell Calcium 2015; 58:237–245.

122. Yamasaki M, Thomas JM, Churchill GC, Garnham C, Lewis AM, Cancela JM, Patel S, Galione A. Role of NAADP and cADPR in the induction and maintenance of agonist-evoked Ca²⁺ spiking in mouse pancreatic acinar cells. Curr Biol 2005; 15:874–878.

123. Calcraft PJ, Ruas M, Pan Z, Cheng X, Arredouani A, Hao X, Tang J, Rietdorf K, Teboul L, Chuang KT, Lin P, Xiao R, et al. NAADP mobilizes calcium from acidic organelles through two-pore channels. Nature 2009; 459:596–600.

124. Ruas M, Davis LC, Chen CC, Morgan AJ, Chuang KT, Walseth TF, Grimm C, Garnham C, Powell T, Platt N, Platt FM, Biel M, et al. Expression of Ca²⁺-permeable two-pore channels rescues NAADP signalling in TPC-deficient cells. EMBO J 2015; 34:1743–1758.

125. Lin-Moshier Y, Walseth TF, Churamani D, Davidson SM, Slama JT, Hooper R, Brailoiu E, Patel S, Marchant JS. Photoaffinity labeling of nicotinic acid adenine dinucleotide phosphate (NAADP) targets in mammalian cells. J Biol Chem 2012; 287:2296–2307.

126. Walseth TF, Lin-Moshier Y, Jain P, Ruas M, Parrington J, Galione A, Marchant JS, Slama JT. Photoaffinity labeling of high affinity nicotinic acid adenine dinucleotide phosphate (NAADP)-binding proteins in sea urchin egg. J Biol Chem 2012; 287:2308–2315.

127. Diercks BP, Werner R, Weidemuller P, Czarniak F, Hernandez L, Lehmann C, Rosche A, Kruger A, Kaufmann U, Vaeth M, Failla AV, Zobiak B, et al. ORAI1, STIM1/2, and RYR1 shape subsecond Ca²⁺ microdomains upon T cell activation. Sci Signal 2018; 11:eaat0358.

128. Putney JW, Jr. A model for receptor-regulated calcium entry. Cell Calcium 1986; 7:1–12.

129. Yule DI, Kim ET, Williams JA. Tyrosine kinase inhibitors attenuate "capacitative" Ca²⁺ influx in rat pancreatic acinar cells. Biochem Biophys Res Commun 1994; 202:1697–1704.

130. Metz DC, Patto RJ, Mrozinski JE, Jr., Jensen RT, Turner RJ, Gardner JD. Thapsigargin defines the roles of cellular calcium in secretagogue-stimulated enzyme secretion from pancreatic acini. J Biol Chem 1992; 267:20620–20629.

131. Liou J, Kim ML, Heo WD, Jones JT, Myers JW, Ferrell JE, Jr., Meyer T. STIM is a Ca²⁺ sensor essential for Ca²⁺-store-depletion-triggered Ca²⁺ influx. Curr Biol 2005; 15:1235–1241.

132. Roos J, DiGregorio PJ, Yeromin AV, Ohlsen K, Lioudyno M, Zhang S, Safrina O, Kozak JA, Wagner SL, Cahalan MD,

Velicelebi G, Stauderman KA. STIM1, an essential and conserved component of store-operated Ca²⁺ channel function. J Cell Biol 2005; 169:435–445.

133. Zhang SL, Yeromin AV, Zhang XH, Yu Y, Safrina O, Penna A, Roos J, Stauderman KA, Cahalan MD. Genome-wide RNAi screen of Ca²⁺ influx identifies genes that regulate Ca²⁺ release-activated Ca²⁺ channel activity. Proc Natl Acad Sci U S A 2006; 103:9357–9362.

134. Luik RM, Wu MM, Buchanan J, Lewis RS. The elementary unit of store-operated Ca²⁺ entry: local activation of CRAC channels by STIM1 at ER-plasma membrane junctions. J Cell Biol 2006; 174:815–825.

135. Wu MM, Buchanan J, Luik RM, Lewis RS. Ca²⁺ store depletion causes STIM1 to accumulate in ER regions closely associated with the plasma membrane. J Cell Biol 2006; 174:803–813.

136. Feske S, Gwack Y, Prakriya M, Srikanth S, Puppel SH, Tanasa B, Hogan PG, Lewis RS, Daly M, Rao A. A mutation in Orai1 causes immune deficiency by abrogating CRAC channel function. Nature 2006; 441:179–185.

137. Hong JH, Li Q, Kim MS, Shin DM, Feske S, Birnbaumer L, Cheng KT, Ambudkar IS, Muallem S. Polarized but differential localization and recruitment of STIM1, Orai1 and TRPC channels in secretory cells. Traffic 2011; 12:232–245.

138. Vig M, Peinelt C, Beck A, Koomoa DL, Rabah D, Koblan-Huberson M, Kraft S, Turner H, Fleig A, Penner R, Kinet JP. CRACM1 is a plasma membrane protein essential for store-operated Ca²⁺ entry. Science 2006; 312:1220–1223.

139. Hogan PG, Rao A. Store-operated calcium entry: mechanisms and modulation. Biochem Biophys Res Commun 2015; 460:40–49.

140. Lur G, Haynes LP, Prior IA, Gerasimenko OV, Feske S, Petersen OH, Burgoyne RD, Tepikin AV. Ribosome-free terminals of rough ER allow formation of STIM1 puncta and segregation of STIM1 from IP₃ receptors. Curr Biol 2009; 19:1648–1653.

141. Gwack Y, Srikanth S, Feske S, Cruz-Guilloty F, Oh-hora M, Neems DS, Hogan PG, Rao A. Biochemical and functional characterization of Orai proteins. J Biol Chem 2007; 282:16232–16243.

142. Brandman O, Liou J, Park WS, Meyer T. STIM2 is a feedback regulator that stabilizes basal cytosolic and endoplasmic reticulum Ca²⁺ levels. Cell 2007; 131:1327–1339.

143. Gerasimenko JV, Gryshchenko O, Ferdek PE, Stapleton E, Hebert TO, Bychkova S, Peng S, Begg M, Gerasimenko OV, Petersen OH. Ca²⁺ release-activated Ca²⁺ channel blockade as a potential tool in antipancreatitis therapy. Proc Natl Acad Sci U S A 2013; 110:13186–13191.

144. Kruger B, Albrecht E, Lerch MM. The role of intracellular calcium signaling in premature protease activation and the onset of pancreatitis. Am J Pathol 2000; 157:43–50.

145. Raraty M, Ward J, Erdemli G, Vaillant C, Neoptolemos JP, Sutton R, Petersen OH. Calcium-dependent enzyme activation and vacuole formation in the apical granular region of pancreatic acinar cells. Proc Natl Acad Sci U S A 2000; 97:13126–13131.

146. Muili KA, Wang D, Orabi AI, Sarwar S, Luo Y, Javed TA, Eisses JF, Mahmood SM, Jin S, Singh VP, Ananthanaravanan M, Perides G, et al. Bile acids induce pancreatic acinar cell

148 *David I. Yule*

injury and pancreatitis by activating calcineurin. J Biol Chem 2013; 288:570–580.

147. Muili KA, Ahmad M, Orabi AI, Mahmood SM, Shah AU, Molkentin JD, Husain SZ. Pharmacological and genetic inhibition of calcineurin protects against carbachol-induced pathological zymogen activation and acinar cell injury. Am J Physiol Gastrointest Liver Physiol 2012; 302:G898–G905.

148. Wen L, Voronina S, Javed MA, Awais M, Szatmary P, Latawiec D, Chvanov M, Collier D, Huang W, Barrett J, Begg M, Stauderman K, et al. Inhibitors of ORAI1 prevent cytosolic calcium-associated injury of human pancreatic acinar cells and acute pancreatitis in 3 mouse models. Gastroenterology 2015; 149:481–492 e487.

149. Waldron RT, Chen Y, Pham H, Go A, Su HY, Hu C, Wen L, Husain SZ, Sugar CA, Roos J, Ramos S, Lugea A, et al. The Orai Ca²⁺ channel inhibitor CM4620 targets both parenchymal and immune cells to reduce inflammation in experimental acute pancreatitis. J Physiol 2019; 597:3085–3105.

150. Gerasimenko JV, Peng S, Tsugorka T, Gerasimenko OV. Ca²⁺ signalling underlying pancreatitis. Cell Calcium 2018; 70:95–101.

151. Voronina S, Collier D, Chvanov M, Middlehurst B, Beckett AJ, Prior IA, Criddle DN, Begg M, Mikoshiba K, Sutton R, Tepikin AV. The role of Ca²⁺ influx in endocytic vacuole formation in pancreatic acinar cells. Biochem J 2015; 465:405–412.

152. Huang W, Cane MC, Mukherjee R, Szatmary P, Zhang X, Elliott V, Ouyang Y, Chvanov M, Latawiec D, Wen L, Booth DM, Haynes AC, et al. Caffeine protects against experimental acute pancreatitis by inhibition of inositol 1,4,5-trisphosphate receptor-mediated Ca²⁺ release. Gut 2017; 66:301–313.

153. Kim MS, Hong JH, Li Q, Shin DM, Abramowitz J, Birnbaumer L, Muallem S. Deletion of TRPC3 in mice reduces store-operated Ca²⁺ influx and the severity of acute pancreatitis. Gastroenterology 2009; 137:1509–1517.

154. Liu X, Cheng KT, Bandyopadhyay BC, Pani B, Dietrich A, Paria BC, Swaim WD, Beech D, Yildrim E, Singh BB, Birnbaumer L, Ambudkar IS. Attenuation of store-operated Ca²⁺ current impairs salivary gland fluid secretion in TRPC1⁻/⁻ mice. Proc Natl Acad Sci U S A 2007; 104:17542–17547.

155. Worley PF, Zeng W, Huang GN, Yuan JP, Kim JY, Lee MG, Muallem S. TRPC channels as STIM1-regulated store-operated channels. Cell Calcium 2007; 42:205–211.

156. Ong HL, Cheng KT, Liu X, Bandyopadhyay BC, Paria BC, Soboloff J, Pani B, Gwack Y, Srikanth S, Singh BB, Gill DL, Ambudkar IS. Dynamic assembly of TRPC1-STIM1-Orai1 ternary complex is involved in store-operated calcium influx. Evidence for similarities in store-operated and calcium release-activated calcium channel components. J Biol Chem 2007; 282:9105–9116.

157. Hollinger S, Hepler JR. Cellular regulation of RGS proteins: modulators and integrators of G protein signaling. Pharmacol Rev 2002; 54:527–559.

158. Gerasimenko OV, Gerasimenko JV, Rizzuto RR, Treiman M, Tepikin AV, Petersen OH. The distribution of the endoplasmic reticulum in living pancreatic acinar cells. Cell Calcium 2002; 32:261–268.

159. Mignen O, Thompson JL, Yule DI, Shuttleworth TJ. Agonist activation of arachidonate-regulated Ca²⁺-selective (ARC) channels in murine parotid and pancreatic acinar cells. J Physiol 2005; 564:791–801.

160. Shuttleworth TJ. Orai channels—new insights, new ideas. J Physiol 2012; 590:4155–4156.

161. Thompson JL, Shuttleworth TJ. Exploring the unique features of the ARC channel, a store-independent Orai channel. Channels (Austin) 2013; 7:364–373.

162. Thompson JL, Shuttleworth TJ. Molecular basis of activation of the arachidonate-regulated Ca²⁺ (ARC) channel, a store-independent Orai channel, by plasma membrane STIM1. J Physiol 2013; 591:3507–3523.

163. Tepikin AV, Voronina SG, Gallacher DV, Petersen OH. Pulsatile Ca²⁺ extrusion from single pancreatic acinar cells during receptor-activated cytosolic Ca²⁺ spiking. J Biol Chem 1992; 267:14073–14076.

164. Tepikin AV, Llopis J, Snitsarev VA, Gallacher DV, Petersen OH. The droplet technique: measurement of calcium extrusion from single isolated mammalian cells. Pflugers Arch 1994; 428:664–670.

165. Ansah TA, Molla A, Katz S. Ca²⁺-ATPase activity in pancreatic acinar plasma membranes. Regulation by calmodulin and acidic phospholipids. J Biol Chem 1984; 259:13442–13450.

166. Bayerdorffer E, Eckhardt L, Haase W, Schulz I. Electrogenic calcium transport in plasma membrane of rat pancreatic acinar cells. J Membr Biol 1985; 84:45–60.

167. Tepikin AV, Voronina SG, Gallacher DV, Petersen OH. Acetylcholine-evoked increase in the cytoplasmic Ca²⁺ concentration and Ca²⁺ extrusion measured simultaneously in single mouse pancreatic acinar cells. J Biol Chem 1992; 267:3569–3572.

168. Camello P, Gardner J, Petersen OH, Tepikin AV. Calcium dependence of calcium extrusion and calcium uptake in mouse pancreatic acinar cells. J Physiol 1996; 490:585–593.

169. Zhang BX, Zhao H, Loessberg P, Muallem S. Activation of the plasma membrane Ca²⁺ pump during agonist stimulation of pancreatic acini. J Biol Chem 1992; 267:15419–15425.

170. Belan PV, Gerasimenko OV, Tepikin AV, Petersen OH. Localization of Ca²⁺ extrusion sites in pancreatic acinar cells. J Biol Chem 1996; 271:7615–7619.

171. Ponnappa BC, Dormer RL, Williams JA. Characterization of an ATP-dependent Ca²⁺ uptake system in mouse pancreatic microsomes. Am J Physiol Gastrointest Liver Physiol 1981; 240:G122–G129.

172. Wakasugi H, Kimura T, Haase W, Kribben A, Kaufmann R, Schulz I. Calcium uptake into acini from rat pancreas: evidence for intracellular ATP-dependent calcium sequestration. J Membr Biol 1982; 65:205–220.

173. Mogami H, Nakano K, Tepikin AV, Petersen OH. Ca²⁺ flow via tunnels in polarized cells: recharging of apical Ca²⁺ stores by focal Ca²⁺ entry through basal membrane patch. Cell 1997; 88:49–55.

174. Yano K, Petersen OH, Tepikin AV. Dual sensitivity of sarcoplasmic/endoplasmic Ca²⁺-ATPase to cytosolic and endoplasmic reticulum Ca²⁺ as a mechanism of modulating cytosolic Ca²⁺ oscillations. Biochem J 2004; 383:353–360.

175. Carafoli E. Intracellular calcium homeostasis. Annu Rev Biochem 1987; 56:395–433.
176. Rizzuto R, Pinton P, Carrington W, Fay FS, Fogarty KE, Lifshitz LM, Tuft RA, Pozzan T. Close contacts with the endoplasmic reticulum as determinants of mitochondrial Ca^{2+} responses. Science 1998; 280:1763–1766.
177. Denton RM, McCormack JG. Ca^{2+} as a second messenger within mitochondria of the heart and other tissues. Annu Rev Physiol 1990; 52:451–466.
178. Perocchi F, Gohil VM, Girgis HS, Bao XR, McCombs JE, Palmer AE, Mootha VK. MICU1 encodes a mitochondrial EF hand protein required for Ca^{2+} uptake. Nature 2010; 467:291–296.
179. Kirichok Y, Krapivinsky G, Clapham DE. The mitochondrial calcium uniporter is a highly selective ion channel. Nature 2004; 427:360–364.
180. Baughman JM, Perocchi F, Girgis HS, Plovanich M, Belcher-Timme CA, Sancak Y, Bao XR, Strittmatter L, Goldberger O, Bogorad RL, Koteliansky V, Mootha VK. Integrative genomics identifies MCU as an essential component of the mitochondrial calcium uniporter. Nature 2011; 476:341–345.
181. De Stefani D, Raffaello A, Teardo E, Szabo I, Rizzuto R. A forty-kilodalton protein of the inner membrane is the mitochondrial calcium uniporter. Nature 2011; 476:336–340.
182. Chvanov M, Voronina S, Zhang X, Telnova S, Chard R, Ouyang Y, Armstrong J, Tanton H, Awais M, Latawiec D, Sutton R, Criddle DN, et al. Knockout of the mitochondrial calcium uniporter strongly suppresses stimulus-metabolism coupling in pancreatic acinar cells but does not reduce severity of experimental acute pancreatitis. Cells 2020; 9:1407.
183. Park MK, Ashby MC, Erdemli G, Petersen OH, Tepikin AV. Perinuclear, perigranular and sub-plasmalemmal mitochondria have distinct functions in the regulation of cellular calcium transport. EMBO J 2001; 20:1863–1874.
184. Voronina S, Sukhomlin T, Johnson PR, Erdemli G, Petersen OH, Tepikin A. Correlation of NADH and Ca^{2+} signals in mouse pancreatic acinar cells. J Physiol 2002; 539:41–52.
185. Gonzalez A, Schulz I, Schmid A. Agonist-evoked mitochondrial Ca^{2+} signals in mouse pancreatic acinar cells. J Biol Chem 2000; 275:38680–38686.
186. Voronina SG, Barrow SL, Simpson AW, Gerasimenko OV, da Silva Xavier G, Rutter GA, Petersen OH, Tepikin AV. Dynamic changes in cytosolic and mitochondrial ATP levels in pancreatic acinar cells. Gastroenterology 2010; 138:1976–1987.
187. Voronina S, Tepikin A. Mitochondrial calcium in the life and death of exocrine secretory cells. Cell Calcium 2012; 52:86–92.
188. Tinel H, Cancela JM, Mogami H, Gerasimenko JV, Gerasimenko OV, Tepikin AV, Petersen OH. Active mitochondria surrounding the pancreatic acinar granule region prevent spreading of inositol trisphosphate-evoked local cytosolic Ca^{2+} signals. EMBO J 1999; 18:4999–5008.

Chapter 9

Role of the actin cytoskeleton in acinar cell protein secretion

Robert C. De Lisle

Anatomy and Cell Biology, University of Kansas School of Medicine, Kansas City, KS 66160 USA

Introduction

The actin cytoskeleton that is ubiquitous in nonmuscle cells is largely made up of β-actin and γ-actin, the products of the *ACTB* and *ACTG1* genes, respectively.[1] These two actins differ by only four amino acids but they appear to have different functions, as mutations in the two genes have very different effects.[1] In the mouse acinar cell, *Actb* mRNA is expressed more than twofold greater than *Actg1* (**Table 1**); presumably their protein levels match this expression, although this has not been measured in the pancreas. In gastric parietal cells, β-actin is more localized to the apical cortical cytoskeleton whereas γ-actin is more localized along the basolateral plasma membrane (BLPM).[2] If the same is true in the acinar cell, then β-actin is more likely to be important in protein secretion, which is also localized to the apical surface of the cell. In the acinar cell, γ-actin may be involved in the BLPM blebs that form under supraphysiologic stimulation. From here on, actin will be referred to as if it were a single gene/protein.

Results from studies of the actin cytoskeleton in the exocrine pancreas are the primary source for this review. However, there are many gaps in our knowledge regarding exocrine pancreas function. When data from pancreas are lacking, results from other cell types that can reasonably be assumed to be true of the acinar cell are briefly discussed. It should be noted that even though there are commonalities shared by all cells with a regulated secretory pathway (e.g., regulation by small GTPases and use of SNAREs to mediate granule/vesicle—plasma membrane (PM) recognition and fusion[3]), there can be important differences.

For example, the role of actin coating of granules/vesicles during the content release process may serve different functions related to the speed of product release required (see: *Dynamics of Acinar Actin in the Stimulated State*; below).

Actin Basics

Actin is a 42-kDa protein that exists in two distinct forms in the cell: monomeric (globular or G-actin) and polymeric (filamentous or F-actin). Actin monomers have the inherent behavior of self-assembly into filaments and prevention of inappropriate polymerization is controlled by several G-actin-binding proteins that stoichiometrically outnumber G-actin by more than threefold.[5] However, in general, the majority of a cell's actin is in the form of F-actin, estimated to comprise ~75 percent of the total.[5] The structure of F-actin microfilaments (MFs) is a polarized double helix such that the barbed (+) end is where new monomers are primarily added and the pointed (-) end is where F-actin primarily releases actin monomers. Actin is an ATPase, and for G-actin the ATP-bound form predominates and is thus primed for addition to an actin filament. Once added to the filament, the ATPase is activated and ATP is hydrolyzed to ADP.

F-actin is highly dynamic and there are numerous proteins that regulate the assembly and disassembly of MF. Interrogation of a published microarray study done in the author's lab[4] for actin cytoskeleton elements was performed and is presented in **Table 1**. This yielded a number of genes, the most highly expressed of which not surprisingly were β- and γ-actins as well as several proteins known to bind to or regulate dynamics of actin. Many of these proteins directly interact with actin and are thus called actin-binding proteins (ABPs).[6] The majority of genes on this list and their functions were discovered using other cell systems and have not been investigated in the pancreatic acinar cell. Microfilament (MF) structures in cells may appear to be highly stable structures, but they are constantly undergoing growth at the (+) end and disassembly at the (-) end, a process called treadmilling.[7] Thus, the apparent stability of MFs is actually due to a dynamic equilibrium. It has been estimated that the ~1-μm-long MFs of the intestinal microvillus turns over every 20 minutes by treadmilling, and a similar rate is also believed to be true of the terminal web (TW) that underlies the apical plasma membrane (APM).[8] In addition to proteins that induce disassembly of MFs, there are numerous actin-binding proteins that regulate nucleation of new fibers and growth of existing fibers. Actin can exist as higher-order bundles of multiple MFs in linear arrays as well as branched MF structures, with accessory proteins providing lateral cross-links.[6]

e-mail: rdelisle@kumc.edu

Table 1. **Microarray analysis of mouse pancreas for actin and associated proteins.** From microarray analysis of mouse pancreas (http://www.ncbi.nlm.nih.gov/geo/, accession GDS567). Data are means from three wild-type young adult mice (see reference[4] for details).* Value: "normalized signal count data, reflecting the relative measure of abundance of each transcript" as defined on the GEO website. This list is not comprehensive. Many actin cytoskeletal genes were expressed below detection limits or were not represented on the microarrays.

Gene	Protein	Value*	Function / Notes
Actb	Actin, beta, nonmuscle	20407	Actin
Actg1	Actin, gamma 1, nonmuscle	8659	Actin
Actg2	Actin, gamma 2, smooth muscle	1357	Actin
Actn1	Actinin, alpha 1	466	MF cross-linker
Actn4	Actinin, alpha 4	608	MF cross-linker
Actr2	ARP2 actin-related protein 2	2529	Branched MF nucleation on existing MF
Actr3	ARP3 actin-related protein 3	2952	Branched MF nucleation on existing MF
Arpc2	Actin related protein 2/3 complex, subunit 2	1745	Branched MF nucleation on existing MF
Arpc3	Actin related protein 2/3 complex, subunit 3	1178	Branched MF nucleation on existing MF
Brk1	Brick1	2966	SCAR/WAVE actin-nucleating complex
Capzb	Capping protein (actin filament) muscle Z-line, beta	1517	Caps barbed end of MF, inhibits elongation
Cfl1	Cofilin 1	2089	Depolymerizes MF
Diap1	Diaphanous homolog 1	1500	Formin family; Rho effector; MF nucleation and elongation
Dstn	Destrin	3654	Depolymerizes MF
Ezr	Ezrin	552	MV protein; links PM to MF
Flii	Flightless 1 homolog	2998	Gelsolin-like, MF severing
Flnb	Filamin, beta	650	Actin cross-linker; links PM to cytoskeleton
Fscn1	Fascin	4438	Actin bundling
Gsn	Gelsolin	2960	MF severing
Mprip	Myosin phosphatase Rho interacting protein	2893	Interacts with actin and myosin; function not claer
Mylk	Myosin light chain kinase	861	Activates myosin
Rac1	Rac1	4918	Small GTPase; WASP/WAVE, APR2/3 are effectors
RhoA	RhoA	1372	Small GTPase; Diap1 and ROCK1 are effectors
Rock1	Rho-associated coiled-coil containing protein kinase 1	378	RhoA effector; regulates actin cytoskeleton
Shroom3	Shroom3	1204	ROCK1 activation; MF assembly
Sptbn1	Spectrin beta, nonerythrocytic 1	2808	Actin cross-linker; links PM to cytoskeleton
Svil1	Supervillin	3161	Links PM to cytoskeleton
Tln1	Talin 1	1273	Links integrins to actin cytoskeleton
Tmod3	Tropomodulin 3	2187	Caps pointed end of MF, inhibits disassembly
Vcl	Vinculin	312	Links integrins to actin cytoskeleton

Actin Localization and Dynamics in the Acinar Cell

Methods to visualize F-actin

Most studies of actin in the acinar cell have been of MFs and usually after chemical fixation of the cells in either the unstimulated basal state, with a secretory stimulus, and/ or after a specific experimental manipulation of the cells. The most commonly used technique is fluorescently tagged phalloidin, a fungal toxin with a high affinity for F-actin (see references from Williams et al.,[9] Jang et al.,[10] and Hou et al.[11] on the Pancreapedia website). An example of FITC-phalloidin labeling of mouse acini from the author's lab is

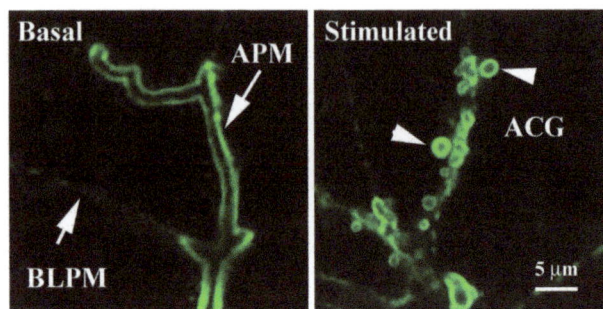

Figure 1. FITC-phalloidin staining of F-actin in mouse acinar cells under basal and stimulated conditions. Mouse acini were incubated without (basal) or with 1 µM carbachol (stimulated). The cells were fixed in suspension with 2 percent formaldehyde and 0.1 percent saponin to permeabilize cell membranes and 1 mM $MgCl_2$ and 5 mM EGTA to stabilize F-actin. The cells were then stained with 0.6 µM FITC-phalloidin and imaged on a confocal microscope, collecting 0.5-µm optical sections. These images are single optical slices. The acinar lumen is bounded by the APM (left panel), under which is the strongly labeled TW. The BLPM (left panel) is much weaker stained than the APM. After stimulus, there are actin-coated granules (ACGs) along the APM (right panel, arrowheads) and the TW is more weakly stained (these images were preliminary data for the work presented in reference [12]).

shown in **Figure 1**, which shows the MFs of acinar cells under basal and stimulated conditions. Phalloidin is a stabilizer of F-actin MFs, so when used in living cells it perturbs the actin cytoskeleton. This property limits use of phalloidin to static imaging of F-actin after fixation and precludes its use to visualize dynamics of actin in live cells where other techniques are used (see below).

F-actin can also be recognized in standard transmission electron microscopy as electron-dense ~6-nm-diameter filaments (smaller than microtubules, ~24 nm; and intermediate filaments, ~10 nm).

Until recently, MFs have been difficult to visualize in living cells without perturbing the structure and function of the cytoskeleton. Adding tags such as green fluorescent protein (GFP) to β-actin perturbs the actin cytoskeleton into which it is incorporated (discussed in reference [13]). A newer approach is to make chimeric fluorescent proteins from ABPs. Using such an approach the Lifeact reporter was made, which consists of a 17-amino-acid actin-binding domain from the ABP140 protein, fused to GFP.[14] This chimeric protein has a 30-fold greater affinity for F-actin than G-actin, was found to not interfere with known activities of the actin cytoskeleton, and it faithfully reproduces known F-actin staining patterns in a variety of cell types even under dynamic situations where the actin cytoskeleton is being remodeled.[13] This construct was later used to make a transgenic mouse expressing the fluorescent F-actin reporter in all tissues, called the Lifeact mouse.[15]

G-actin can also be visualized using fluorescently tagged DNase I, which binds to both G- and F-actin, with higher affinity to the former,[16] although an analysis of DNAse I inhibition by highly purified G- and F-actin concluded that both forms have similar affinity for DNase I.[17] An example of fluorescent DNase I staining of G-actin in the acinar cell shows a diffuse cytoplasmic distribution which is not very informative.[18] Visualization of G-actin has not been used to study the actin cytoskeleton of the acinar cell to any extent. Nor has there been systematic investigation of G-actin ABPs in the acinar cell.

Acinar cell actin in the resting state

The single largest pool of F-actin in the acinar cell is in the TW,[19] also called subapical actin or cortical actin (**Figure 1**; left panel).

The TW is a dense arrangement of MF that forms a continuous sheet-like network ~200 nm thick[21] under the APM. Among ABPs, members of the red blood cell spectrin family cross-link TW MFs and also link the TW to perpendicular MF that extend into the microvilli (MV), as shown in intestinal epithelium.[22] As compared to the villus epithelium of the small intestine where MV are best characterized, the acinar cell MV are relatively sparse. A second actin-based structure is a ring of MFs that are attached to the adherens junctions near the apical surface and encircle each acinar cell. Because this is a ring-shaped structure, it is less easily visualized than the sheet-like TW. There are also recently described MF bundles that project perpendicularly from the TW into the cytoplasm that are called actin cables.[23] Lesser amounts of MFs are found along the BLPM of the acinar cell (**Figure 2**, left panel). Because of the large amount of F-actin in the TW, it can be difficult to visualize the other sites of MFs when using fluorescent labeling approaches. For example, the recently described actin cables that run perpendicular to the TW in acinar cells are not observed using phalloidin staining but are visualized using a chimeric ABP domain fused to GFP (Lifeact; see below).[23]

Effects on secretion of pharmacological agents and molecular approaches that perturb the actin cytoskeleton

There is a large data set about the acinar cell actin cytoskeleton obtained using a variety of approaches. Because the actin cytoskeleton is so important to eukaryotic cells, there are numerous naturally occurring substances produced by a variety of organisms that interfere with the organization of actin, and many of these are available as research tools. Also, investigators have produced constitutively active (CA) and dominant negative (DN) constructs to stimulate or inhibit upstream pathways that control the actin cytoskeleton.

Table 2. Drugs and reagents that affect the actin cytoskeleton and their effects on pancreatic acinar cell cytoskeleton and amylase release.

Empty fields indicate data were not reported for that parameter.

Abbreviations: ACGs: actin-coated granules; BDM: 2,3-butadione monoxime; CA: constitutively active; Cyto: cytochalasin; DN: dominant negative; Lat: latrunculin; TW: terminal web

Agent	Mechanism	Cytoskeleton Effects	Amylase Release / Other	Ref
LatA	Binds G-actin monomers, inhibiting addition to MF	no effect on ACG; loss of actin cables		23
		TW resistant; BPLM MFs lost	inhibits / 28% decrease in MF in 10min; 44% decrease at 2hr	25
			inhibits / prevents BLPM blebs by supramaximal CCK	29
		loss of ACG	/ done at room temperature	18
LatB	Binds G-actin monomers, inhibiting addition to MF		no effect /	32
		loss of ACG; reduces TW	/ early pore closure	33
		loss of ACG	no effect /	37
		loss of TW	/ perturbs Ca^{2+}-signaling	24
Cytochalasin	Binds to barbed end of MF; disassembly of existing MFs		no effect /	32
		reduces lifetime of ACG		18
		loss of TW	reduces / dilation of lumen	43
		loss of TW	reduces; rebounds after washout / dilates acinar lumen with stimulation; compound exocytosis increases	44
Jasplakinolide	Stabilizes existing MFs, enhances MF nucleation		reduces; prevents high-dose CCK inhibition / causes BLPM blebs in absence of stimulation	29
BDM	Inhibits myosin ATPase		totally blocks / prevents loss of MF by latrunculin or cytochalasin	31
BDM & LatB		ACG accumulate		31
BDM & Cyto		no effect on ACG		31
C3 (Rho-toxin)	ADP-ribosylates & inactivates Rho		inhibits /	25
		loss of ACG		18
			/ partially blocks supramaximal CCK inhibition of amylase release	29
CA-Rho	Stimulates downstream effectors	increases total F-actin	enhances / prevents supramaximal CCK inhibition of amylase release; causes BLPM blebs	29
CA-Rac	Stimulates downstream effectors	increases total F-actin	enhances / prevents supramaximal CCK inhibition of amylase release; causes BLPM blebs	29
DN-Rho	Inhibits downstream effectors		reduces, both phases affected /	29
DN-Rac	Inhibits downstream effectors		reduces, primarily late phase / partially blocks supramaximal CCK inhibition of amylase releas	29
DN-Rho & DN-Rac	Inhibits downstream effectors		reduces, slightly greater than sum of individual DN effects /	29
CA-mDia	Recruits G-actin for MF growth	no effect on ACG		23
DN-mDia	Blocks mDia-induced MF growth		no effect /	23
β -thymosin	Binds G-actin monomers, inhibiting addition to MF	no effect on TW		26
Gelsolin fragment	Binds G-actin monomers, inhibiting addition to MF	no effect on TW		26
Phalloidin	Stabilizes existing MF		inhibits /	26

A summary of these toxins and reagents and their effects on the actin cytoskeleton and acinar cell amylase release is given in **Table 2**. Some of the reported effects of a particular agent are contradictory but these are likely attributable to differences in concentration of the agent (e.g., latrunculins[24] have been used over the wide range of 0.3 μM to 100 μM), time of exposure, and temperature at which the experiment was performed (discussed in reference [25]). The

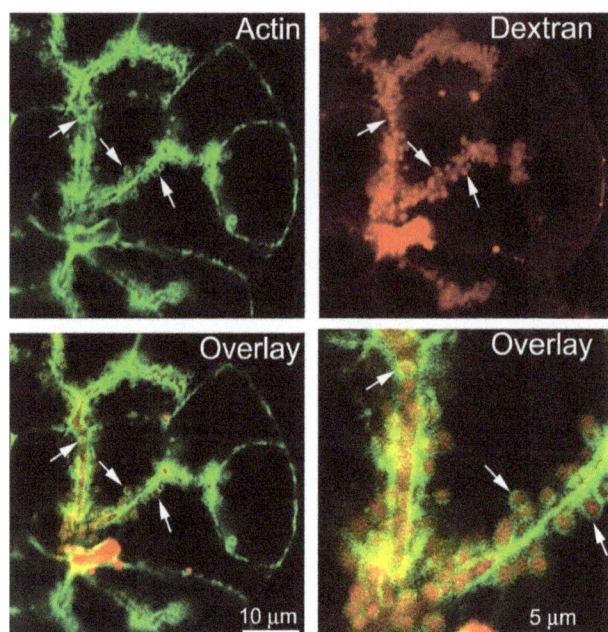

Figure 2. ACGs are open to the lumen and accumulate fluorescent dextran from the medium. Acini were stimulated in the presence of fluorescent lysine-fixable dextran[20] and then fixed and costained with FITC-phalloidin and imaged on a confocal microscope. The enlarged image (lower right) clearly shows that ACGs also have fluorescent dextran in their lumina, demonstrating access of the ZG lumen to the medium during secretory stimulation (courtesy of John A. Williams; unpublished).

importance of a specific agent's concentration is illustrated by the finding discussed below that a small perturbation of the actin cytoskeleton increases amylase release, whereas stronger perturbation blocks amylase release.[26] The data covered here attempt to give the consensus of the available data.

Signals that affect the actin cytoskeleton

The principal mechanism of acinar cell stimulation begins with ligand binding to membrane receptors coupled to G proteins that feed into several pathways. With respect to the acinar cell actin cytoskeleton, the known pathways are activation of the small GTPases RhoA (via $G\alpha_{13}$) and Rac1 (via $G\alpha13$ and $G\alpha q$).[27] These activated small GTPases in turn activate their downstream effectors, such as Diap1 and ROCK1 and WASP/WAVE and ARP2/3, respectively, that lead to actin cytoskeleton remodeling.

As discussed below, a major downstream effector of Rho activation is the formin family member Dia. In drosophila, the formin diaphanous is activated by the small GTPase Rho, resulting in nucleation and formation of linear F-actin filaments that comprise the TW of epithelial cells.[28] Although not yet experimentally verified, Rho activation in the mammalian pancreas is expected

to activate mDia1 and it has been shown in mouse acinar cells that constitutively active (CA)-mDia1 leads to F-actin cables perpendicular to the apical PM.[23] It was suggested these F-actin cables serve to recruit zymogen granules (ZGs) through the TW to the apical membrane during secretory stimulation.[23] Consistent with the requirement for apical MF in secretion, protein release into the tracheal lumen of *Drosophila* lacking Dia, which also lack TW F-actin, was absent. A notable difference between *Drosophila* and mammalian acinar cells is that Dia in drosophila is required for formation of the TW of epithelia,[28] whereas in mouse acini, the only observed role of mDia was in formation of the F-actin cables perpendicular to the APM, and the TW persisted in the presence of a DN-mDia construct.[23] Also in contrast to *Drosophila*, mouse acini expressing dominant negative (DN)-mDia1 have unchanged basal and stimulated amylase release.[23] Clearly, the roles of Dia/mDia1 differ in *Drosophila* and mouse pancreatic acini.

Dynamics of acinar actin in the stimulated state

Actin undergoes a variety of changes during secretory stimulation. The TW in the resting acinar cell is a barrier to ZG exocytosis, and the F-actin here must be rearranged to allow ZG access to the APM when the cell is stimulated.[26] Using streptolysin O-permeabilized acinar cells, it was demonstrated that mild perturbation of the actin cytoskeleton by adding low concentrations of G-actin sequestering proteins (β-thymosin or a gelsolin fragment; **Table 2**) resulted in amylase release, especially the early phase of release.[26] Interestingly, even when Ca^{2+} was chelated at low levels using EGTA, mild perturbation of the actin cytoskeleton resulted in amylase release. If high concentrations of the G-actin ABPs were used that resulted in extensive loss of F-actin, there was reduced amylase release. Similarly, if the F-actin-stabilizing drug phalloidin was introduced to permeabilized acini, amylase release was inhibited. Together, these data indicate that the TW is a barrier to ZG exocytosis and that, while some remodeling of the actin cytoskeleton is required for release, extensive disruption or stabilization of F-actin is detrimental to protein secretion.

The dynamics of G-actin per se have not been investigated in relation to acinar cell stimulation. However, there have been measurements of the relative changes in the total amount of F-actin in acinar cells, which show that an increase in polymerized actin is associated with amylase release.[29] Furthermore, pharmacological reduction of F-actin (1.0 μM latrunculin, a G-actin sequestering agent; **Table 2**) inhibits CCK-stimulated amylase release from mouse acini without affecting Ca^{2+} dynamics.[25] Since an acute increase in F-actin means less G-actin, these data add to the concept that dynamic changes in actin are involved in protein secretion in the acinar cell.

A similar conclusion regarding actin remodeling during secretion can be drawn from experiments using supramaximal CCK which induces large alterations in the actin cytoskeleton and shows inhibition of amylase release.[30,31] When DN-Rho and DN-Rac (**Table 2**) constructs were used, they prevented supramaximal CCK inhibition of amylase release and at the same time lessened reorganization of the actin cytoskeleton.[29] In that study, a low dose of jasplakinolide (1 μM) also prevented supramaximal CCK inhibition of amylase release, but a higher dose (3 μM) inhibited amylase release at all concentrations of CCK, consistent with the effect of jasplakinolide reported by Valentijn et al.[32] (**Table 2**). So again, these results show there is a "sweet spot" for amylase release with respect to organization of the acinar cell actin cytoskeleton.

Remarkably, during exocytosis the ZG membrane (ZGM) becomes "coated" with F-actin (actin-coated granules, ACGs)[32] (**Figure 1**; right panel). In the acinar cell, the role of the F-actin coat has been proposed to stabilize the fused ZG at the PM.[33,34] This may be related to the duration of acinar protein secretion (seconds, which is long relative to neuronal exocytosis which occurs on a millisecond timescale), which may be required for the highly concentrated digestive enzymes to solubilize and diffuse out of the granule lumen into the beginning of the pancreatic ductal system. In other cells with the regulated pathway whose contents are more rapidly solubilized, the F-actin coat appears to flatten the fused vesicle, thus quickly forcing out the secretory product. The potential differences in the roles of F-actin granule/vesicle coats in various secretory cells are discussed in some excellent reviews.[35,36] Interestingly, the G-actin sequestering drug latrunculin blocks coating of ZG with actin but does not prevent ZGM-PM fusion.[37]

Using the C3 Rho toxin from *Clostridium botulinum*, it was shown that ACG formation is Rho-dependent.[18] When acini were pretreated with the C3 toxin and then stimulated to secrete, ACGs were not observed but large vacuoles formed that were filled with the fluorescent extracellular tracer, showing that granule fusion had still occurred but that the individual granules in the absence of actin coating appear to fuse into a large vacuole. Thus, the rapid formation of ACG in stimulated cells appears to be Rho-dependent, and ACG formation may prevent vacuole formation. Amylase release was not measured in these experiments.

The timing of F-actin coating of ZGs relative to ZGM-PM fusion has evolved since its discovery. Early descriptions of ACGs suggested that the coat developed before ZGM-PM fusion, based on failure of entry of a fluorescent lectin (WGA) from the medium into ACGs, indicating their lumens were not yet accessible to the medium.[32] Later experiments concluded that the F-actin coat forms after membrane fusion, while the exocytic pore is still open[18,34] (see also the review from Jang and Thorn[10]).

This conclusion was based on the observation that only ZGs accessible to the acinar lumen (able to take up lysine-fixable fluorescent dextran from the medium) were actin-coated and no ZGs without fluorescent dextran were actin-coated. An example of acini stimulated in the presence of a lysine-fixable dextran followed by fixation and fluorescent phalloidin staining is shown in **Figure 2**. This clearly shows dextran in the lumens of ACGs. Subsequent work using a soluble tracer (sulforhodamine B) with living cells convincingly showed that access to the ZG lumen occurred on average 6.7 seconds before appearance of the actin coat, as visualized in acini from a transgenic mouse expressing Lifeact-GFP[37] (**Figure 3**). This study also showed that latrunculin blocked formation of ACG while ZG fusion was not inhibited (**Figure 4**). Hence, ZGM-PM fusion occurs rapidly followed by actin coating of the fused ZG while it is still in the "Ω" shape, and that fusion per se is independent of ACG formation.

A different interpretation of the timing of ZG fusion and ACG formation has been made recently by Geron et al., using Lifeact-GFP, expressed in acini by adenoviral infection.[23] This study was of the role of the formin family member diaphanous (DIAP1, called mDia in mouse), which is a Rho effector whose activation leads to linear MFs in the pancreatic acinar cell that they called actin cables. They state in the introduction to their paper that "secretory vesicles undergo actin coating ***before*** fusing to the apical membrane . . ." (emphasis added) and they cite these other references.[18,32,35]

One can observe in their videos accompanying the paper ACGs moving apparently along F-actin cables that run perpendicular to the apical PM. Since these ACGs appear deeper inside the cells than the PM, it was concluded that coating preceded ZGM-PM fusion. Even though they had included FM4–64, a fluorescent membrane tag that should enter fused ZGs and label their membranes, in none of the data shown is FM4–64 fluorescence ever observed in ZGs. Therefore, it is not clear that the experiments actually could reveal when ZGM-PM fusion occurred. A more consistent interpretation of the role of mDia-induced actin cables is that fused ZGs are moving along these cables near the apical PM.

Another issue with their interpretation is that it does not concur with what was observed when the cables were interfered with. They showed that expression of a CA- mDia1 resulted in an increase in actin cables, and that a DN-mDia1 mutant caused a strong decrease in the number of cables. Neither of these affected basal or stimulated amylase release. If the cables are an important mechanism to move ZGs in the cytoplasm to the PM, so they can then fuse, one would expect a decrease in stimulated amylase release when the cables are decreased, which was not observed. The DN-mDia1, which reduced cable formation, caused an increase in ACG fused to one

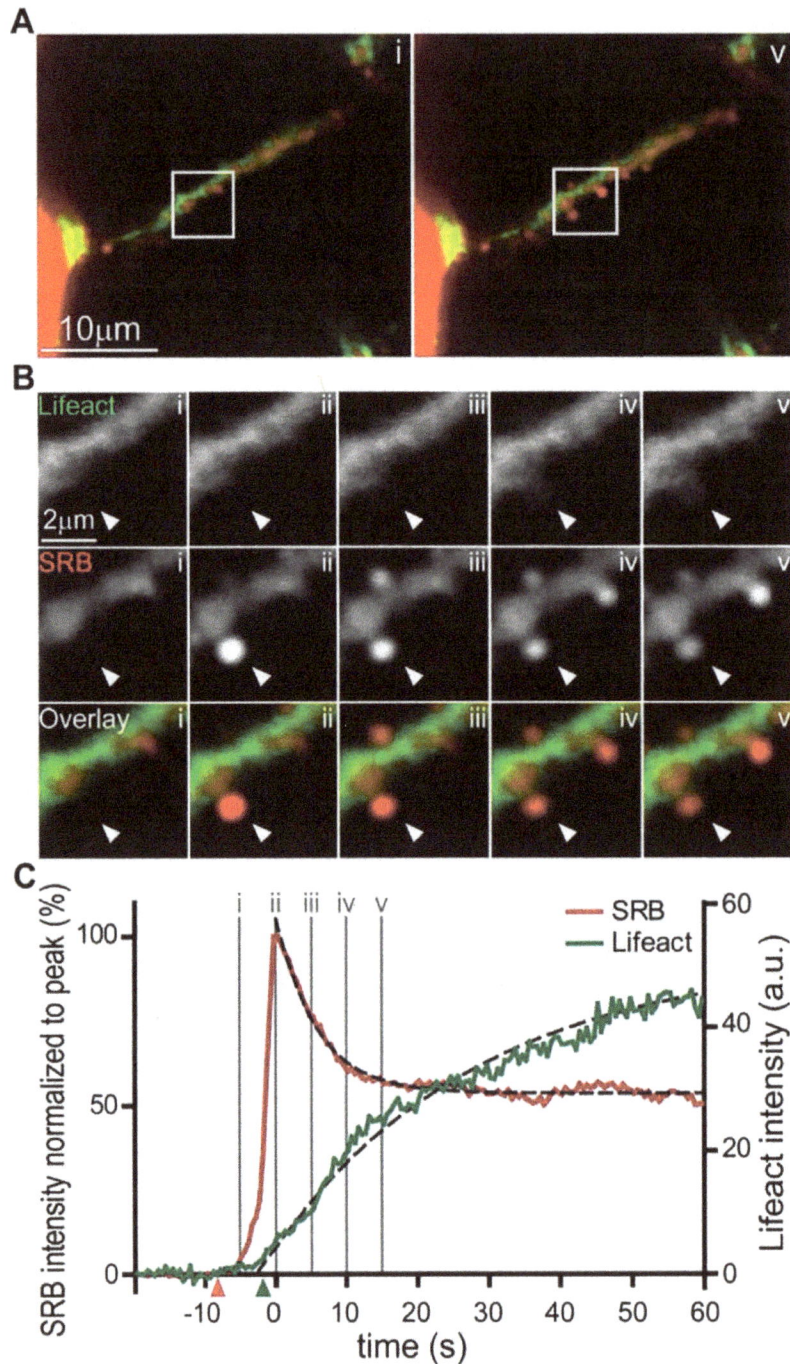

Figure 3. Actin coating of ZGs occurs shortly after ZGM-PM fusion. (**A**) Low-magnification images show a lumen lying diagonally between two acinar cells identified with SRB[20] and Lifeact-EGFP (green) in the subapical region. (**Ai**) is an image taken before the appearance of exocytic events at the time point indicated "i" on the graph of fluorescence intensity over time in panel (**C**). (**Av**) is an image at a time point after induction of a number of exocytic events which can be seen as bright spots of SRB fluorescence along the lumen; the time point v" is indicated on the lower graph (**C**). (**B**) Shows an image sequence from an enlarged region (box shown in **A**) of Lifeact-EGFP and SRB and the overlay, for two exocytic events. The images were taken at the time points i, ii, iii, iv, and v as indicated on the graph in panel (**C**). (**C**) is a graph of fluorescence changes over time taken from a region of interest placed over the lower exocytic event (indicated by an arrowhead). SRB fluorescence is plotted normalized to the peak and rises rapidly to a peak and then decays to a plateau. The simultaneously recorded Lifeact-EGFP signal, plotted in arbitrary fluorescence units, rises slowly and nearly reaches a maximum by the end of the record. The starting points of the SRB and Lifeact-EGFP signals, as determined by a positive deflection of the signal by more than five times the standard deviation of the signal noise, are shown by the color-coded triangles on the X axis. The black dotted lines were mono-exponential fits to the data with τ values of 6.9 s (SRB) and 29.4 s (Lifeact-EGFP) (from reference [37], Figure 4).

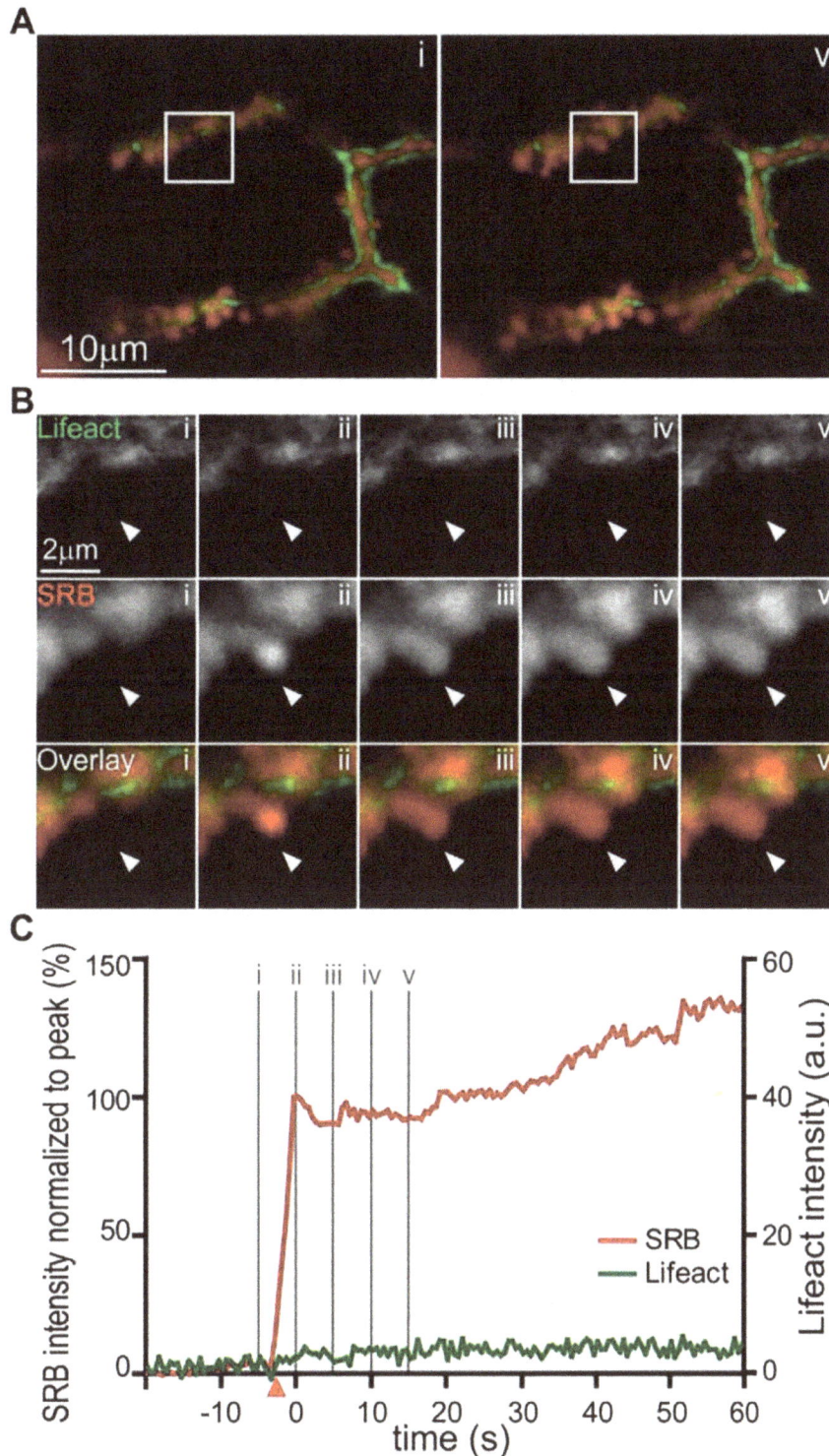

Figure 4. Latrunculin A abolishes actin coating of ZGs but does not inhibit ZGM-PM fusion. (**A**) Low-magnification images show complex lumens, identified by SRB[20] and Lifeact-EGFP fluorescence (green), lying between the cells within a pancreatic fragment. (**Ai**) is an image taken before the appearance of exocytic events, (**Av**) is taken after, at the time points "i" and "v" as indicated on the graph in panel (**C**). (**B**) shows an image sequence from an enlarged region (box shown in **A**) of Lifeact-EGFP and SRB and the overlay, for a single exocytic event. The images were taken at the time points i, ii, iii, iv, and v as indicated on the graph in panel (**C**). (**C**) is a graph of fluorescence changes over time taken from a region of interest placed over the exocytic event (indicated by an arrow). SRB fluorescence is plotted normalized to the first, rapid peak and shows a rapid rise followed by a slower increase. The simultaneously recorded Lifeact-EGFP signal, plotted as arbitrary fluorescence units, shows only very small changes over time (from reference [37], Figure 5).

another, which was interpreted as compound exocytosis. So, it would appear that the role of these mDia1-induced actin cables is to keep ZGs separated such that individual granule fusion predominates and minimizes compound exocytosis.

It is likely an issue in these disparate results regarding the timing of ACG formation whether the acinar preparation was optimal for allowing access from the medium to the acinar lumen. This is discussed in references [38] and [33], where it is shown that the vigor of the tissue dissociation technique can yield groups of cells ranging from multiacinar clusters to single acini or even smaller than acinar-sized groups that retain normal acinar secretory function. To allow access of the bathing solution to the acinar lumen, groups consisting of as few as three to six cells with a short unhindered connection between the acinar lumen and the medium are optimal. (In my lab, we could not get lysine-fixable dextran into fused ZGs[12] as achieved by other labs, likely because my "acini" were actually larger multiacini clusters.)

Although mDia is a downstream effector of activated Rho, introduction of DN-mDia[23] and DN-Rho A[29] in acinar cells does not produce the same effects (**Table 2**). DN-mDia does not affect amylase release whereas DN-Rho inhibits amylase release. These different effects on amylase release indicate that mDia is not the sole target of activated Rho in the acinar cell, and perhaps another of the Rho effectors is involved.[39]

The conclusion that the F-actin coat develops rapidly after ZGM-PM fusion contributes to the idea that this coat functions to stabilize the pore in the open state, providing sufficient time for the stored zymogens to become soluble and diffuse into the acinar lumen. An actin coat attached along the cytosolic surface of the fused ZGs would stiffen this membrane, preventing closure of the energetically unfavorable pore. The presence of the nonclassical myosin Vc (Myo5c) on ZGs, possibly tethered to the membrane by the small GTPase Rab27B,[40,41] is believed to provide a motive force to move ZGs along the TW. Another myosin, Myo2A, is localized to the apical pole of acinar cells, is phosphorylated when the cell is stimulated, and when its activity is inhibited access of the ZG lumen to the soluble markers in the medium is reduced.[42] This was interpreted as more rapid fusion pore closure when this myosin activity is inhibited. Also, when the TW was disassembled using cytochalasin B treatment,[43] the acinar lumen enlarged under stimulation but amylase was not released; when the cytochalasin and secretory stimulus was washed out, the TW recovered and amylase was released from the already fused ZGs.[44] It was suggested that contractile forces on the fused ZGM by actin were needed to force the ZG content out. These results can be reinterpreted in light of the newer data indicating that MFs are required to maintain an open pore for protein release.

In addition to providing sufficient time for zymogens to dissociate and diffuse into the lumen, the actin coat of fused ZGs has been proposed to help counteract the expected hydrostatic pressure of these dissolving contents from the fused ZGs in salivary glands.[45] Also, there is compelling evidence that the fused ZGM transports electrolytes and fluid from the acinar cytosol into the lumen during protein secretion (see review by Frank Thévenod[46]), which will add to hydrostatic pressure within the fused ZG lumen.

While it has been shown that mDia1 is involved in the formation of actin cables in the acinar cell, and is regulated by the small GTPase Rho, it remains to be determined what controls localized disassembly of the TW required to allow passage of ZG to the PM for exocytosis. The F-actin structure that coats granules is likely to consist of a network of short interconnected linear filaments or, alternatively, of branching actin filaments. This is because linear F-actin filaments <10 μm in length are rigid rods[47] and such long linear filaments could not be accommodated on the highly curved surface of granules which are ~1 μm in diameter.[48] Consistent with this idea, a subunit of the ARP2/3 complex and N-WASP, factors involved in nucleation of branches on existing MFs, were immunolocalized to the region of ACGs in stimulated acini.[23] Based on this, Rac1 may be involved, because Rac1 has been shown to regulate WASP/WAVE ABPs that control nucleation/elongation of branched MF in other systems.[49] However, WASP nucleates a new branch on existing F-actin, and proteomic analysis of highly purified ZGMs revealed the presence of small GTPases and molecular motors (dynein, myosins) but not actin or other cytoskeletal elements.[40,41,50] On the other hand, ultrastructural immunogold localization showed actin in the region of ZGs.[20,51] Together these results suggest that actin in the ZG region of the cell is either loosely bound or not bound at all to the ZGM. In any case, it appears F-actin is not present on ZG before fusion, so there needs to be a process to nucleate new MF on the fused ZGM and at this time, how this occurs remains to be identified.

After the stimulus ends

Once the fused ZG has released its content and exocytosis can be considered complete, the actin cytoskeleton has the further role of helping drive compensatory endocytosis. In order to maintain the balance of membranes between apical and basolateral domains, the ZGM added to the apical surface during protein secretion must be taken back into the cell. The endocytic process is believed to retrieve the ZGM in a clathrin-mediated fashion, and this also requires the actin cytoskeleton.[44] It was observed in cytochalasin D-treated acinar cells that the APM became dilated and pits on this surface were immunoreactive for

the clathrin adaptor AP-2, clathrin, dynamin, and cave-olin, all of which are components of various endocytic mechanisms. When cytochalasin was washed out, these proteins rapidly disappeared from the APM and the luminal membrane recovered to its prestimulus size. From this it was suggested that the actin cytoskeleton is also involved in the compensatory endocytosis process that follows protein secretion. The roles of actin in endocytosis are well known.[52]

Since this chapter was written in 2015, a PubMed search was carried out in October 2020 and showed that recent work concerning the actin cytoskeleton in acinar cells was primarily focused on endocytosis and autophagy and not digestive enzyme secretion. By contrast, recent work exists on actin in islet beta cells and its relationship to insulin secretion. This addendum will review this recent work on the acinar cell actin cytoskeleton in endocytosis linked to autophagy during models of pancreatitis and in insulin release from islet β-cells.

Role of Actin in Acinar cell autophagy and pancreatitis

Autophagy is a process whereby cells breakdown and recycle their own components.[53] Autophagy can be induced by starvation, thus providing the cell with energy and substrates to survive this stress. Additionally, and relevant to this discussion, cytotoxic events in the cell can also induce autophagy in an attempt to clear malformed, cytotoxic organelles and materials, again to save the cell and/or protect the wider organism. In the canonical autophagy pathway, this occurs by formation of an autophagosome, a double-membrane structure, to engulf the damaged organelle/material which then fuses with a lysosome to degrade it. In the noncanonical form of autophagy, the machinery that drives phagosome formation (a marker of which is LC3, microtubule-associated protein 1A/1B-light chain 3), is recruited to a single-membrane organelle which promotes their fusion with lysosomes for degradation of the organelle contents.

During pancreatitis, there is widespread disruption of ongoing protein synthesis and the packaging of zymogens into functional secretory granules (for review see reference [54]). For example, during alcohol-induced pancreatitis, zymogen synthesis is increased at the transcriptional level and, at the same time, exocytosis is inhibited resulting in an overabundance of ZGs in the cell. Because of the high synthetic rate, the capacity of the protein folding machinery of the endoplasmic reticulum (ER) is exceeded leading to an unfolded protein response by the cell. These conditions promote activation of zymogens, especially important of which is activation of trypsin from trypsinogen. Active trypsin in turn can lead to cell/organ damage and is accompanied by inflammation, which together enhance the severity of pancreatitis.

One of the cell responses to stress, as mentioned above, is to activate autophagy, in this case to get rid of the excess of ZGs. In pancreatitis, there is a strong body of evidence that lysosomal biogenesis fails to keep up and that there is insufficient flux into the autophagy pathway such that trypsin activation occurs in the cell with its damaging sequelae.

As with all membrane trafficking, formation and movement in the autophagy pathway involves cytoskeletal elements, including actin (see reference [55] for review). Evidence for this includes the observation that actin-depolymerizing agents (e.g., cytochalasin D, latrunculin B) block autophagosome formation. The actin-regulating Rho family GTPases are also involved. RhoA is needed during starvation-induced autophagy, while Rac is inhibitory to induction and itself is inactivated during autophagy. The GTPase Rab1 is also involved, stimulating actin assembly and tubulation of the ER membrane to produce autophagosomes.

Recent studies investigated the role of endocytosis, which is well known to involve actin, in experimental pancreatitis. Chvanov et al. used LifeAct to follow actin dynamics during pancreatitis along with extracellular fluorescent tracers to monitor fluid-phase endocytosis.[56] They found that endocytic structures that form during pancreatitis are much larger than during physiological stimulation, forming endocytic vacuoles (EVs) which are positive for active trypsin. These EVs can be as large as 12 μm in diameter and appear to form from the enlarged apical plasma membrane that result from compound exocytosis during acinar protein release. These EVs are actin-coated, often not continuously. The small "fenestrae" in the actin coat that were sometimes observed appear to be weak spots where rupture of EVs can occur (as revealed by movement of the fluid-phase fluorescent tracer from the EV to the cytosol; see **Figure 5**, reproduced from this paper) and deliver active trypsin to the cytosol. Once rupture occurs, it is quickly followed by plasma membrane blebbing in most of their observations and cell death about 30 minutes later. Importantly, the presence of the F-actin-stabilizing drug jasplakinolide decreased the rupture of EVs from about 30 percent of observed cells down to about 15 percent. This led the authors to suggest that the actin coat on the EVs is a stabilizing force against rupture and is cell protective.

Another interesting observation was that EVs can sometimes fuse with the plasma membrane, as revealed by mixing of preloaded fluorescent tracer of one color with a differently colored tracer in the medium.[56] This exocytosis of EVs occurred at both the apical and basolateral surfaces of the cells (see **Figure 6**; reproduced from reference [56]). Because of the technical setup for these high-resolution confocal studies, not many individual events could be recorded, so it is not known how frequently this occurs or whether the propensity is for apical or basolateral

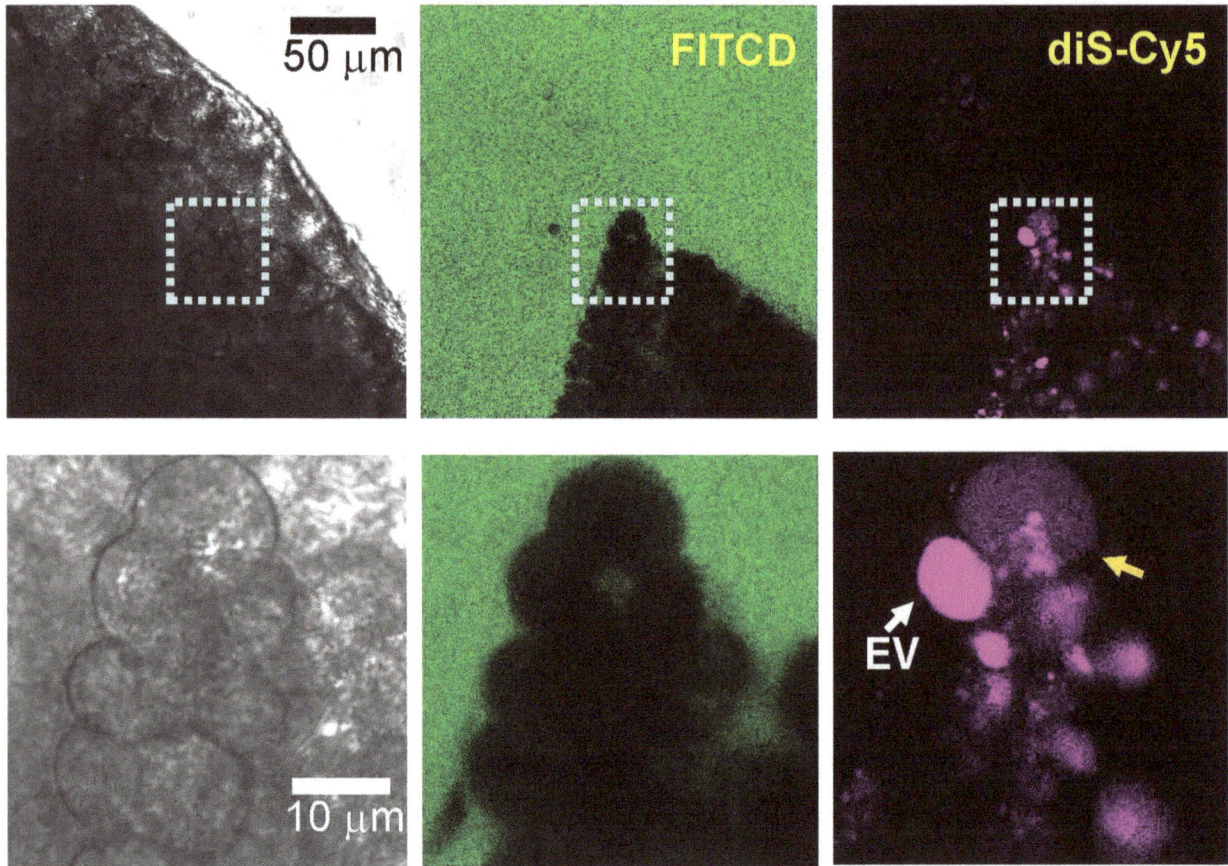

Figure 5. Cytosolic presence of membrane-impermeant fluorescence probe in the cell located in undissociated pancreatic fragment. Small (~1-mm) section of mouse pancreas was stimulated by 100 nM CCK for 2 hours at 35°C in the presence of diS-Cy5 (shown in magenta), washed, and imaged in the presence of FITCD (shown in green). The lower gallery of images depicts the fragment containing two cells within the section: one with a large intact EV (white arrow) and the adjacent cell with increased cytosolic fluorescence of diS-Cy5. The FITCD image indicates that the plasma membrane of this cell is intact, suggesting that the increase of the cytosolic fluorescence occurred as a result of EV rupture. Representative of six similar experiments (from reference [56], with permission).

exocytosis. The basolateral exocytosis is consistent with that investigated by Gaisano and colleagues,[57] which is proposed as mechanism to deliver harmful active trypsin to the interstitial space during pancreatitis. The process of exocytosis from the autophagic pathway has been termed secretory autophagy (see reference [58] for review).

Subsequent work from this group looked in greater detail at the formation of EVs and presented evidence that they form by a noncanonical pathway, LC3-associated phagocytosis (LAP).[59] This process involves an EV with a single membrane which is initially actin-coated and as it loses actin it becomes LC3-positive and is converted into an autophagic organelle. Importantly, most LC3-positive EVs also exhibited activation of trypsinogen.

A different study implicating the actin cytoskeleton in autophagy and pancreatitis studied β1 syntrophin, which binds and regulates dystrophin and is in turn an important linker between the actin cytoskeleton and transmembrane glycoproteins.[60] In this case, actin remodeling driven by β1 syntrophin is involved in curvature of the membrane at the ER to form tubules which become autophagosomes. Using a β1 syntrophin knockout mouse, they observed dilation of the acinar lumen under resting conditions, and the knockout showed an unaltered cerulein dose-response amylase release. However, during cerulein supramaximal stimulation to induce pancreatitis, the β1 syntrophin mice had less F-actin and reduced autophagy: a marked reduction in ER nucleations, autophagosomes, and autolysosomes. The β1 syntrophin knockouts exhibited greater severity of pancreatitis (higher serum amylase and trypsin; more necrosis; more neutrophil infiltration; and greater edema), showing the protective role of autophagy during pancreatitis. At this time, there does not appear to be enough information to know the relative importance of canonical (ER-derived) versus noncanonical (endocytic organelle) autophagic pathways in pancreatic function and in pancreatitis.

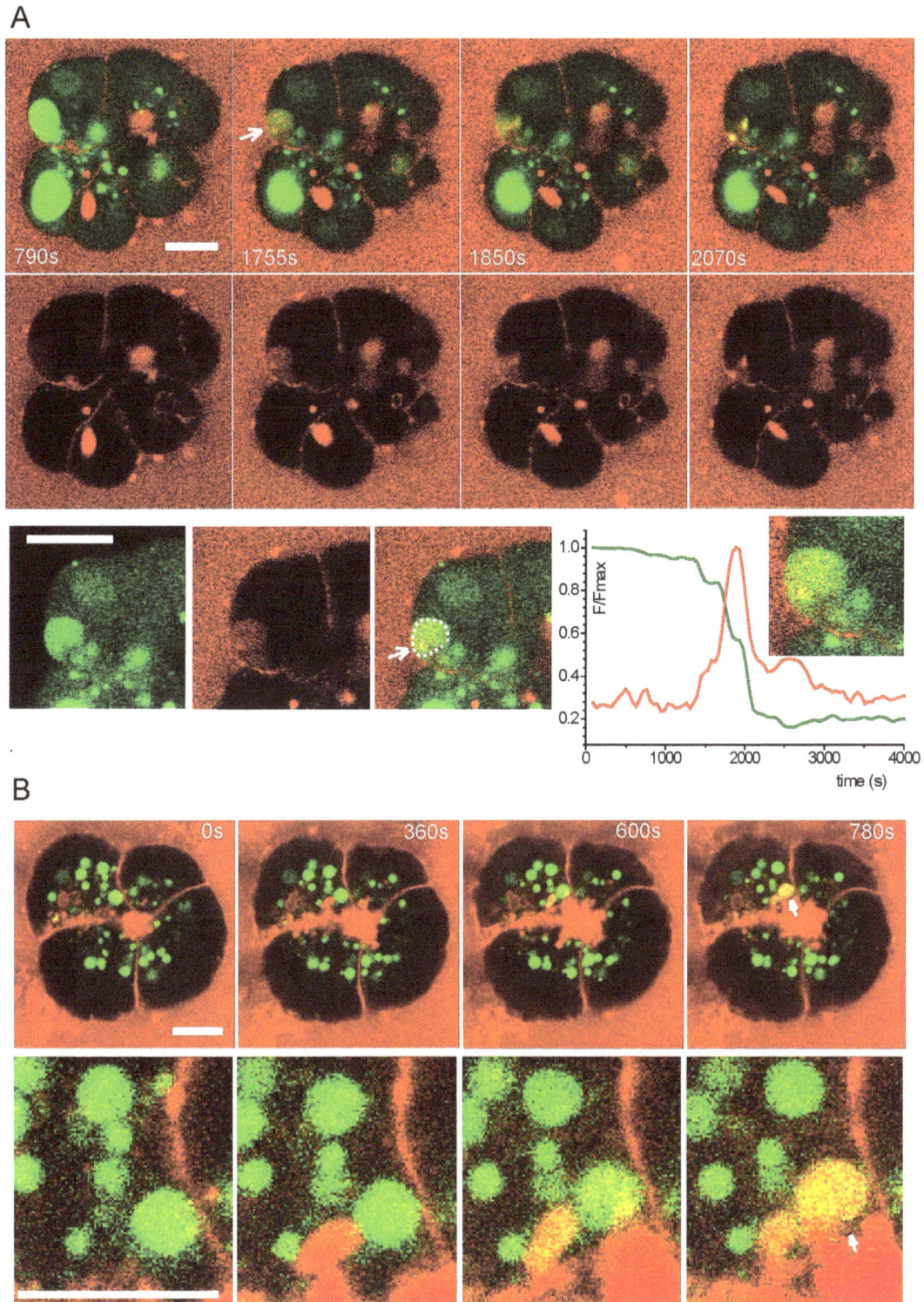

Figure 6. Exocytosis of endocytic vacuoles (EVs). (**A**) Fusion of LY-filled EV (green; the EV undergoing fusion is highlighted by an arrow) with the basal plasma membrane of a CCK-stimulated acinar cell. The extracellular medium contains TRD[20] and a fusion event is characterized by mixing of the probes (yellow). Following fusion, the EV loses LY and gains TRD. The region containing the fusing EV is shown in the bottom row of images. The graph shows the time course of fluorescence of the two dyes recorded in the EV (highlighted by the dashed line). (**B**) LY-filled EV (green) fuses with TRD-filled[20] post-exocytic structure in the apical region of the CCK-stimulated cell. The region containing the fusing EV (white arrow) is shown in the bottom row of images. Following fusion, the composite structure contains both probes (yellow) (representative of n = 7 observations). Scale bars = 10 μm (from reference [56], with permission).

The role of actin in β-cell exocytosis

Although there are known differences in the cell biology of endocrine and exocrine pancreatic cells, there are more recent investigations of the actin cytoskeleton in the β-cell than in the acinar cell. The Pancreapedia: Exocrine Pancreas Knowledge Base includes a review of the physiology and cell biology of insulin release and the intracellular pathways controlling it.[61] The final signaling step in β-cell exocytosis of insulin granules, similar to most secretory cells, is an influx of Ca^{2+} resulting in activation of the SNARE complex (soluble N-ethylmaleimide sensitive factor attachment protein receptor). The complex consists of SNAREs on the granule and plasma membranes whose conformational changes provide the force to fuse these two membranes and release the granule content to the extracellular space.[62] The major physiological stimulus for β-cell exocytosis is elevated concentration of blood nutrients, especially glucose. Glucose-stimulated insulin release proceeds in two phases.[62,63] The initial phase results in a large spike of insulin release and lasts about 10 minutes. If elevated glucose is maintained, the second phase occurs, and this is a lower rate of release but is capable of being sustained for hours, ensuring the circulating insulin remains sufficient throughout nutrient uptake from the intestines to assimilate those nutrients by target organs throughout the body.

As in the acinar cell, glucose-stimulated insulin release from the β-cell involves remodeling of the actin cytoskeleton, which in the case of the β-cell has roles in both phases of exocytosis (see Kalwat and Thurmond for review[63]). As in other secretory cells, the β-cell has a cortical F-actin cytoskeleton which presents a barrier to granule exocytosis, as revealed by actin depolymerizing agents that increase insulin secretion. Actin remodeling is also involved in recruitment of new granules to the plasma membrane during the second, sustained phase of insulin release.

If we look at upstream events, the influx of glucose has two pathways that lead to insulin release.[62,64] Glucose is taken up from the circulation via GLUT glucose channels in the plasma membrane into the cytosol. Metabolism of glucose in the cell increases the ATP/ADP ratio, which inactivates the ATP-sensitive K^+ channel resulting in membrane depolarization and an influx of Ca^{2+}, stimulating SNARE-mediated membrane fusion. The other pathway is that elevated glucose leads to glucosylation of Cdc42, a Rho GTPase, and its glucosylation decreases the level of Cdc42-GTP, which is the active form. Activated Cdc42 interacts with two proteins that affect actin remodeling. One is Wiskott-Aldrich syndrome protein (WASP), which through actin-related protein complex 2/3 (ARP2/3) stimulates branching of F-actin. The other is mDia, which promotes actin polymerization. Thus, inhibition of Cdc42 results in actin remodeling and increases insulin release in a similar fashion as do actin depolymerizing drugs. Further

evidence of the importance of Cdc42 is that a constitutively active form of the protein (Q61L) inhibits glucose-stimulated insulin release; and a constitutively inactive dominant negative form (T17N) is without effect.[64] The authors also used jasplakinolide (nucleates and stabilizes F-actin) which unexpectedly potentiated glucose-stimulated insulin release, without effect on unstimulated (basal) release. Jasplakinolide did not enhance release with KCl depolarization even though the abundance of granules near the plasma membrane was increased. Importantly, glucose stimulation dramatically diminished cortical F-actin whereas KCl depolarization did not. These results indicate that cortical actin is decreased at the same time deeper F-actin is increased during glucose stimulation, thus promoting the initial phase of release and recruiting new granules for the sustained second phase.

They next investigated whether the SNARE syntaxin 1 (plasma membrane localized), which is known to interact with actin, is involved. They immunoprecipitated (IP) syntaxin 1 and measured co-IP actin. Glucose stimulation decreased actin associated with syntaxin 1 approximately twofold. The additional presence of jasplakinolide during glucose stimulation did not affect this decrease in co-IPed actin. Depolarization with KCl had no effect on co-IP of actin, alone or in the presence of jasplakinolide. Therefore, glucose-stimulated actin remodeling occurs even in the presence of the actin nucleating and polymerization effects of this drug. The authors proposed that glucose initiates the transient reduction of filamentous cortical actin to coordinate granule mobilization and pool refilling and thereby promotes granule exocytosis.

Another recent study has investigated F-actin coating of insulin granules with a focus on ARP2/3, a nucleator that leads to branched F-actin filaments.[65] Using endocrine pancreas cells from LifeAct transgenic mice and a fluorescent fluid-phase endocytic marker, with multiphoton microscopy, they provide high-resolution data showing that insulin granules become actin-coated following membrane fusion (labeling with the fluid-phase marker preceded actin coating) and that this is ARP2/3 dependent. Using drugs that block ARP2/3 activity and short hairpin RNA (shRNA) to deplete the cells of the protein, they observed loss of F-actin coating. Interestingly, with this high-resolution technique they found that there was not a global change in the cortical F-actin cytoskeleton but only localized changes to individual fusing granules. When ARP2/3 was inhibited or absent, fused granules accumulated at the plasma membrane and insulin release was decreased even though fusion had occurred. They then examined whether linear F-actin has a role by interfering with formin and find that while actin coating was disrupted, insulin release was unaffected. The authors concluded that ARP2/3-mediated nucleation of branched F-actin plays an active role post-fusion in events involved in insulin release.

Because of the complexity, especially delineating the two phases of insulin release and their regulatory mechanisms, a recent study presented the initial development of a statistically based computer model along the lines of Cellular Autamoton, to integrate what is known and provide a platform for evaluating new data and hypotheses regarding the role of the actin cytoskeleton in insulin granule transport in the β-cell.[66] Since the model they develop needs to incorporate what is known about β-cell insulin release, this paper also serves as an up-to-date review of granule movement and the various roles of the actin cytoskeleton in the β-cell.

The authors interpret recent total internal-reflection fluorescence (TIRF: a very narrowly focused signal with high signal to noise, in close proximity to the plasma membrane that reveals only granules docked/primed and can reveal exocytic events) and confocal data that insulin granules can be docked at the plasma membrane (responsible for first rapid phase of insulin release); can be freely diffusible toward the plasma membrane and back again deeper into the cytosol; and can be anchored via various accessory proteins to F-actin filaments such that these are being transported and will be released during the prolonged second phase.

They set up their initial model such that "*the following processes and hypotheses are considered:*

– *Granules are created inside the cell.*
– *Granules move either diffusely or are directed, with stimuli promoting a directed movement.*
– *Granules can move along actin fibers, and are prevented from moving perpendicular to the strand.*
– *With a certain probability, granules can pass through the actin network, in cases of actin remodeling.*
– *Granules must exist for a certain time before they can fuse at the membrane.*
– *Under constant stimuli, a steady state exists for insulin secretion.*"

In the simulation, they reduce a β-cell to a cube 10 μm on each edge with 343,000 cubic elements approximating the size of a single insulin granule. Focusing on just granules and actin, they define 11 possible states for each element denoting granules in various associations with actin, its movement, and the granule's age (new or in the immediately releasable pool). With this model they can investigate the actin cytoskeleton and how it relates to granule movements and insulin release. To set up initial conditions, they used fluorescence microscopy data from LifeAct expressing cultured mouse islets and MIN6 cells. The simulation is populated with actin filaments and granules, randomly assigned to fit the known data. For changes in granule behavior (movement, actin association, fusion, etc.), the stimulus (glucose, KCl depolarization) and intracellular

Ca^{2+} levels are used to define the probabilities of such changes. They start the simulation by randomly placing 13,000 granules within the actin mesh and assigning the majority to diffusing and a small fraction to different motions, to match data in the literature regarding these parameters.

The simulations were run and compared visually to see how well the simulations approximate actual cells and then quantitatively to see how stimuli affect insulin release. They also used latrunculin to disassemble F-actin and compared results from real cells with the simulations.

In summary, the authors conclude that "*three major results were achieved:*

1. *The simulated pattern of insulin secretion in response to high glucose shows the typical biphasic pattern of cultured islets and cell lines*
2. *The simulated effect of the modifier of the actin cytoskeleton, Latrunculin B, namely the increase of both phases of glucose-induced secretion, concurs with experimental observations, provided the simulated network fulfils a certain set of criteria.*
3. *This network is also compatible with a simulated interaction between glucose- and potassium-stimulated secretion, which mirrors earlier experimental observations.*"

"*Thus, on the one hand, extensions can be implemented informatically in an uncomplicated way, and on the other hand, computing capacities are not yet a problem. Rather, the addition of further findings and hypotheses is an important component in the further development of the model, so that quantitative predictions should also be possible in the future.*"

In conclusion, although there has been little recent progress on investigating the roles of the actin cytoskeleton in acinar cell protein secretion under normal physiological conditions, there are new data in the context of pancreatitis as well as in other systems such as the β-cell. The phasic release of insulin from the β-cell shows several roles for actin and its remodeling. It should be emphasized that much of this progress relied on use of high-resolution live cell imaging. The TIRF signal is limited to ~100 nm from the coverslip surface into the attached cell. Similarly, multiphoton microscopy is able to focus into approximately 600-nm-thick focal plane of interest. These techniques exclude signals outside of these thin focal planes, thus considerably reducing background that would obscure the region of interest. Interestingly, there are also kinetic differences for zymogen secretion from granules that are older or newly synthesized (e.g., see reference [67]). The issue of release of old versus new ZGs has not been explored with respect to actin, but it would seem likely to be important to this process.

References

1. Belyantseva IA, Perrin BJ, Sonnemann KJ, Zhu M, Stepanyan R, McGee J, Frolenkov GI, Walsh EJ, Friderici KH, Friedman TB, Ervasti JM. Gamma-actin is required for cytoskeletal maintenance but not development. Proc Natl Acad Sci U S A 2009; 106:9703–9708.

2. Yao X, Chaponnier C, Gabbiani G, Forte JG. Polarized distribution of actin isoforms in gastric parietal cells. Mol Biol Cell 1995; 6:541–557.

3. Kloepper TH, Kienle CN, Fasshauer D. An elaborate classification of SNARE proteins sheds light on the conservation of the eukaryotic endomembrane system. Mol Biol Cell 2007; 18:3463–3471.

4. Kaur S, Norkina O, Ziemer D, Samuelson LC, De Lisle RC. Acidic duodenal pH alters gene expression in the cystic fibrosis mouse pancreas. Am J Physiol Gastrointest Liver Physiol 2004; 287:G480–G490.

5. Xue B, Robinson RC. Guardians of the actin monomer. Eur J Cell Biol 2013; 92:316–332.

6. Uribe R, Jay D. A review of actin binding proteins: new perspectives. Mol Biol Rep 2009; 36:121–125.

7. Theriot JA. Accelerating on a treadmill: ADF/cofilin promotes rapid actin filament turnover in the dynamic cytoskeleton. J Cell Biol 1997; 136:1165–1168.

8. Brown JW, McKnight CJ. Molecular model of the microvillar cytoskeleton and organization of the brush border. PLoS One 2010; 5:e9406.

9. Williams JA, Ernst SA, Chen X. Rab3. Pancreapedia: Exocrine Pancreas Knowledge Base, 2013. DOI: 10.3998/panc.2013.01.

10. Jang Y, Thorn P. Visualization of exocytosis in pancreatic acinar cells by fluorescence microscopy. Pancreapedia: Exocrine Pancreas Knowledge Base, 2011. DOI: 10.3998/panc.2011.15.

11. Hou Y, Chen X, Ernst SA, Williams JA. Rab27. Pancreapedia: Exocrine Pancreas Knowledge Base, 2015. DOI: 10.3998/panc.2015.12.

12. Tandon C, De Lisle RC. Apactin is involved in remodeling of the actin cytoskeleton during regulated exocytosis. Eur J Cell Biol 2004; 83:79–89.

13. Deibler M, Spatz JP, Kemkemer R. Actin fusion proteins alter the dynamics of mechanically induced cytoskeleton rearrangement. PLoS One 2011; 6:e22941.

14. Riedl J, Crevenna AH, Kessenbrock K, Yu JH, Neukirchen D, Bista M, Bradke F, Jenne D, Holak TA, Werb Z, Sixt M, Wedlich-Soldner R. Lifeact: a versatile marker to visualize F-actin. Nat Methods 2008; 5:605–607.

15. Riedl J, Flynn KC, Raducanu A, Gartner F, Beck G, Bosl M, Bradke F, Massberg S, Aszodi A, Sixt M, Wedlich-Soldner R. Lifeact mice for studying F-actin dynamics. Nat Methods 2010; 7:168–169.

16. Mannherz HG, Goody RS, Konrad M, Nowak E. The interaction of bovine pancreatic deoxyribonuclease I and skeletal muscle actin. Eur J Biochem 1980; 104:367–379.

17. Morrison SS, Dawson JF. A high-throughput assay shows that DNase-I binds actin monomers and polymers with similar affinity. Anal Biochem 2007; 364:159–164.

18. Nemoto T, Kojima T, Oshima A, Bito H, Kasai H. Stabilization of exocytosis by dynamic F-actin coating of zymogen granules in pancreatic acini. J Biol Chem 2004; 279:37544–37550.

19. Bendayan M, Marceau N, Beaudoin AR, Trifaro JM. Immunocytochemical localization of actin in the pancreatic exocrine cell. J Histochem Cytochem 1982; 30:1075–1078.

20. Becich MJ, Bendayan M, Reddy JK. Intracellular transport and storage of secretory proteins in relation to cytodifferentiation in neoplastic pancreatic acinar cells. J Cell Biol 1983; 96:949–960.

21. Clark AG, Dierkes K, Paluch EK. Monitoring actin cortex thickness in live cells. Biophys J 2013; 105:570–580.

22. Glenney JR, Jr., Glenney P, Weber K. The spectrin-related molecule, TW-260/240, cross-links the actin bundles of the microvillus rootlets in the brush borders of intestinal epithelial cells. J Cell Biol 1983; 96:1491–1496.

23. Geron E, Schejter ED, Shilo BZ. Directing exocrine secretory vesicles to the apical membrane by actin cables generated by the formin mDia1. Proc Natl Acad Sci U S A 2013; 110:10652–10657.

24. Shin DM, Zhao XS, Zeng W, Mozhayeva M, Muallem S. The mammalian Sec6/8 complex interacts with Ca^{2+} signaling complexes and regulates their activity. J Cell Biol 2000; 150:1101–1112.

25. Bi Y, Williams JA. A role for Rho and Rac in secretagogue-induced amylase release by pancreatic acini. Am J Physiol Cell Physiol 2005; 289:C22–C32.

26. Muallem S, Kwiatkowska K, Xu X, Yin HL. Actin filament disassembly is a sufficient final trigger for exocytosis in non-excitable cells. J Cell Biol 1995; 128:589–598.

27. Sabbatini ME, Bi Y, Ji B, Ernst SA, Williams JA. CCK activates RhoA and Rac1 differentially through $G\alpha_{13}$ and $G\alpha_q$ in mouse pancreatic acini. Am J Physiol Cell Physiol 2010; 298:C592–C601.

28. Massarwa R, Schejter ED, Shilo BZ. Apical secretion in epithelial tubes of the Drosophila embryo is directed by the Formin-family protein Diaphanous. Dev Cell 2009; 16:877–888.

29. Bi Y, Page SL, Williams JA. Rho and Rac promote acinar morphological changes, actin reorganization, and amylase secretion. Am J Physiol Gastrointest Liver Physiol 2005; 289:G561–G570.

30. Burnham DB, Williams JA. Effects of high concentrations of secretagogues on the morphology and secretory activity of the pancreas: a role for microfilaments. Cell Tissue Res 1982; 222:201–212.

31. Torgerson RR, McNiven MA. The actin-myosin cytoskeleton mediates reversible agonist-induced membrane blebbing. J Cell Sci 1998; 111:2911–2922.

32. Valentijn JA, Valentijn K, Pastore LM, Jamieson JD. Actin coating of secretory granules during regulated exocytosis correlates with the release of rab3D. Proc Natl Acad Sci U S A 2000; 97:1091–1095.

33. Larina O, Bhat P, Pickett JA, Launikonis BS, Shah A, Kruger WA, Edwardson JM, Thorn P. Dynamic regulation of the large exocytotic fusion pore in pancreatic acinar cells. Mol Biol Cell 2007; 18:3502–3511.

34. Turvey MR, Thorn P. Lysine-fixable dye tracing of exocytosis shows F-actin coating is a step that follows granule fusion in pancreatic acinar cells. Pflugers Arch 2004; 448:552–555.

35. Nightingale TD, Cutler DF, Cramer LP. Actin coats and rings promote regulated exocytosis. Trends Cell Biol 2012; 22:329–337.

36. Porat-Shliom N, Milberg O, Masedunskas A, Weigert R. Multiple roles for the actin cytoskeleton during regulated exocytosis. Cell Mol Life Sci 2013; 70:2099–2121.

37. Jang Y, Soekmadji C, Mitchell JM, Thomas WG, Thorn P. Real-time measurement of F-actin remodelling during exocytosis using Lifeact-EGFP transgenic animals. PLoS One 2012; 7:e39815.

38. Williams JA. Isolation of rodent pancreatic acinar cells and acini by collagenase digestion. Pancreapedia: Exocrine Pancreas Knowledge Base, 2010. DOI: 10.3998/panc.2010.18.

39. Thumkeo D, Watanabe S, Narumiya S. Physiological roles of Rho and Rho effectors in mammals. Eur J Cell Biol 2013; 92:303–315.

40. Chen X, Li C, Izumi T, Ernst SA, Andrews PC, Williams JA. Rab27b localizes to zymogen granules and regulates pancreatic acinar exocytosis. Biochem Biophys Res Commun 2004; 323:1157–1162.

41. Chen X, Walker AK, Strahler JR, Simon ES, Tomanicek-Volk SL, Nelson BB, Hurley MC, Ernst SA, Williams JA, Andrews PC. Organellar proteomics: analysis of pancreatic zymogen granule membranes. Mol Cell Proteomics 2006; 5:306–312.

42. Bhat P, Thorn P. Myosin 2 maintains an open exocytic fusion pore in secretory epithelial cells. Mol Biol Cell 2009; 20:1795–1803.

43. Williams JA. Effects of cytochalasin B on pancreatic acinar cell structure and secretion. Cell Tissue Res 1977; 179:453–466.

44. Valentijn KM, Gumkowski FD, Jamieson JD. The subapical actin cytoskeleton regulates secretion and membrane retrieval in pancreatic acinar cells. J Cell Sci 1999; 112 (Pt 1):81–96.

45. Masedunskas A, Sramkova M, Weigert R. Homeostasis of the apical plasma membrane during regulated exocytosis in the salivary glands of live rodents. Bioarchitecture 2011; 1:225–229.

46. Thévenod F. Channels and transporters in zymogen granule membranes and their role in granule function: a critical assessment. Pancreapedia: Exocrine Pancreas Knowledge Base, 2020. DOI: 10.3998/panc.2020.09.

47. Blanchoin L, Boujemaa-Paterski R, Sykes C, Plastino J. Actin dynamics, architecture, and mechanics in cell motility. Physiol Rev 2014; 94:235–263.

48. Risca VI, Wang EB, Chaudhuri O, Chia JJ, Geissler PL, Fletcher DA. Actin filament curvature biases branching direction. Proc Natl Acad Sci U S A 2012; 109:2913–2918.

49. Ridley AJ. Rho GTPases and actin dynamics in membrane protrusions and vesicle trafficking. Trends Cell Biol 2006; 16:522–529.

50. Rindler MJ, Xu CF, Gumper I, Smith NN, Neubert TA. Proteomic analysis of pancreatic zymogen granules: identification of new granule proteins. J Proteome Res 2007; 6:2978–2992.

51. Bendayan M. Ultrastructural localization of cytoskeletal proteins in pancreatic secretory cells. Can J Biochem Cell Biol 1985; 63:680–690.

52. Mooren OL, Galletta BJ, Cooper JA. Roles for actin assembly in endocytosis. Annu Rev Biochem 2012; 81:661–686.

53. Kawabata T, Yoshimori T. Autophagosome biogenesis and human health. Cell Discov 2020; 6:33.

54. Rasineni K, Srinivasan MP, Balamurugan AN, Kaphalia BS, Wang S, Ding WX, Pandol SJ, Lugea A, Simon L, Molina PE, Gao P, Casey CA, et al. Recent advances in understanding the complexity of alcohol-induced pancreatic dysfunction and pancreatitis development. Biomolecules 2020; 10:1–15.

55. Kast DJ, Dominguez R. The cytoskeleton-autophagy connection. Curr Biol 2017; 27:R318–R326.

56. Chvanov M, De Faveri F, Moore D, Sherwood MW, Awais M, Voronina S, Sutton R, Criddle DN, Haynes L, Tepikin AV. Intracellular rupture, exocytosis and actin interaction of endocytic vacuoles in pancreatic acinar cells: initiating events in acute pancreatitis. J Physiol 2018; 596:2547–2564.

57. Dolai S, Liang T, Orabi AI, Holmyard D, Xie L, Greitzer-Antes D, Kang Y, Xie H, Javed TA, Lam PP, Rubin DC, Thorn P, et al. Pancreatitis-induced depletion of syntaxin 2 promotes autophagy and increases basolateral exocytosis. Gastroenterology 2018; 154:1805–1821.

58. Gonzalez CD, Resnik R, Vaccaro MI. Secretory autophagy and its relevance in metabolic and degenerative disease. Front Endocrinol 2020; 11:266.

59. De Faveri F, Chvanov M, Voronina S, Moore D, Pollock L, Haynes L, Awais M, Beckett AJ, Mayer U, Sutton R, Criddle DN, Prior IA, et al. LAP-like non-canonical autophagy and evolution of endocytic vacuoles in pancreatic acinar cells. Autophagy 2020; 16:1314–1331.

60. Ye R, Onodera T, Blanchard PG, Kusminski CM, Esser V, Brekken RA, Scherer PE. beta1 syntrophin supports autophagy initiation and protects against cerulein-induced acute pancreatitis. Am J Pathol 2019; 189:813–825.

61. Mann E, Sunni M, Bellin MD. Secretion of insulin in response to diet and hormones. Pancreapedia: Exocrine Pancreas Knowledge Base, 2020. DOI: 10.3998/panc.2020.16.

62. Thurmond DC, Gaisano HY. Recent insights into beta-cell exocytosis in type 2 diabetes. J Mol Biol 2020; 432:1310–1325.

63. Kalwat MA, Thurmond DC. Signaling mechanisms of glucose-induced F-actin remodeling in pancreatic islet beta cells. Exp Mol Med 2013; 45:e37.

64. Nevins AK, Thurmond DC. Glucose regulates the cortical actin network through modulation of Cdc42 cycling to stimulate insulin secretion. Am J Physiol Cell Physiol 2003; 285:C698–710.

65. Ma W, Chang J, Tong J, Ho U, Yau B, Kebede MA, Thorn P. Arp2/3 nucleates F-actin coating of fusing insulin granules in pancreatic beta cells to control insulin secretion. J Cell Sci 2020; 133:1–11.

66. Muller M, Glombek M, Powitz J, Bruning D, Rustenbeck I. A cellular automaton model as a first model-based assessment of interacting mechanisms for insulin granule transport in beta cells. Cells 2020; 9:1-32

67. Singh M. Nonparallel transport of exportable proteins in rat pancreas in vitro. Can J Physiol Pharmacol 1982; 60:597–603.

Chapter 10

Physiologic exocytosis in pancreatic acinar cells and pathologic fusion underlying pancreatitis

Herbert Y. Gaisano, Subhankar Dolai, and Toshimasa Takahashi

Departments of Medicine and Physiology, University of Toronto, Toronto, Canada, M5S1A8

Pancreatic Acinar Cell, an Important Model for Secretion

The pancreatic acinar cell possesses a robust protein synthetic machinery that synthesizes a large spectrum of enzymes that digest all components of ingested food. Proteinases and some lipases are synthesized as inactive proenzymes. All are sorted through the *trans*-Golgi and then vectorially transferred via a still incompletely defined maturation process to the zymogen granules (ZGs).[1] These mature ZGs, among the largest secretory vesicles (~1 micron), accumulate at the apical pole of the highly polarized acinar cell and occupy ~30 percent of total cell volume. Ca^{2+}-mobilizing neural (acetylcholine) and endocrine (cholecystokinin: CCK) stimulation, and to lesser extent cAMP-acting agonists (vasoactive intestinal polypeptide: VIP, secretin), acts on their respective plasma membrane G-protein-coupled receptors to trigger two signaling cascades.[2] The first is the inositol triphosphate (IP3)-mediated intracellular Ca^{2+} release partnered with diacylglycerol (DAG) formation and protein kinase C (PKC) pathways; the second is elevation of cAMP levels partnered with protein kinase A (PKA)-activated pathways; extensive cross talk exists between these two cascades. Other chapters in this book deal with these topics. These cell signals act on the complex exocytotic fusion machinery discussed below that is comprised of a large number of molecules on the ZGs and the plasma membrane (PM) to culminate in fusion of ZGs with apical PM, releasing the ZG cargo of nutrient hydrolases (e.g., proteases, lipase, amylase) into the ductal lumen, which then transit to the duodenum for a timely meeting with a food bolus whose entry into the duodenum from the stomach is metered by the pylorus.

The pancreatic acinar cell, historically from the work of George Palade (Nobel Prize for Medicine or Physiology in 1999) and coworkers, had long served as the basic model to elucidate the complex processes of secretion.[1] This was however largely overtaken by the neuron and endocrine fields that have exponentially accelerated as a result of the rapid identification of the exocytotic machinery (originally known as the SNARE hypothesis)[3,4] and membrane ion channel machinery that orchestrates plasma membrane Ca^{2+} channel opening with the ensuing Ca^{2+} influx as the major source of Ca^{2+} for secretory granule fusion. The rapid progress in the neural and endocrine fields was further supported by the rapid development of new technology, including genetic models, electrophysiology, and microscopy imaging. The pancreatic acinar cell eventually followed in hot pursuit to initially show that it too has the fundamental components that support the SNARE hypothesis.[5] To the delight of neuroscientists, probably paralleled by the undeclared dismay of earlier workers in pancreatic acinar biology, was that the levels of the SNARE and associated proteins were at much higher levels in neurons than in acinar cells; this made the neuron a better model to isolate and characterize the large number of proteins of the exocytotic machinery.[6] The acinar cell exocytotic machinery however does not simply mimic the neuron and still lends itself to be an outstanding model for nonexcitable cell secretory biology. For instance, the pancreatic acinar apical PM constitutes less than 10 percent of the total cell surface area even when maximally stimulated, therefore exocytosis of the few ZGs abutting the apical PM would be insufficient in exporting enough proteases for delivery into the duodenal lumen to efficiently digest the huge amount of food continually being emptied from the stomach during a meal. Instead, the pancreatic acinar cell has a different secretory granule architecture from the neuronal synapse, which enables an orderly and rapid (within 1 minute) fusion of majority (>70%) of the ZGs within the apical pole, termed sequential exocytosis,[7] whereby fusion pores between homotypically fused ZGs (also called compound exocytosis) remained open for very long periods (>20 minutes)[7,8] to ensure efficient emptying of zymogen cargo from the deepest ZG layers within the apical pole. This provides an exquisitely regulated and metered machinery that can match hydrolase outputs to the varying amounts of food throughout the entire duration of the meal. Similar

e-mail: herbert.gaisano@utoronto.ca

exocytotic machineries can be greatly accelerated in mast cells and neutrophils during an allergic reaction or to mount an effective kill against invading bacteria.[9,10] Furthermore, the exocytotic machinery of the acinar cell can be misdirected (i.e., ectopic exocytosis at the basolateral PM) by disease-causing stimulation such as alcohol, as a major contributing mechanism to alcoholic pancreatitis.[11,12]

SNARE Hypothesis for Membrane Fusion and Exocytosis

The 2013 Nobel Prize in Physiology or Medicine was awarded to James Rothman, Randy Schekman, and Thomas Südhof for their discoveries of the machinery regulating vesicle traffic. This was actually the convergence of four fields (reviewed in reference [13]). James Rothman and coworkers were reconstituting membrane traffic events to show fusion of donor vesicles with target membranes using cell-free assays. Randy Schekman and coworkers were studying the secretory pathway in yeast, which led to the identifying SEC genes and their mutants that accounted for accumulation of various intermediated vesicles along the ER to the Golgi. Thomas Südhof cloned a large number of synaptic vesicle proteins found to mediate neuroexocytosis, the latter actually was pivotally shown by a fourth field of coworkers who found that the synaptic proteins Südhof cloned were actually the substrates for proteolytic cleavage by tetanus and botulinum neurotoxins known to block neurotransmission.

The SNARE hypothesis as originally proposed by abovementioned major contributors stated that cytoplasmic N-ethylmaleimide-sensitive fusion (NSF) proteins and soluble NSF attachment proteins (SNAPs) bind SNAP receptors (SNAREs) on the donor vesicle (v-SNARE) and target membrane (t-SNARE) to form a multimolecular complex capable of mediating fusion between the two membranes.[3,4] The v-SNAREs are vesicle-associated membrane proteins (VAMPs) and t-SNAREs are syntaxins (STX) and SNAP25 (synaptosomal associated protein of 25kDa), which form *trans*-SNARE complexes to constitute the minimal machinery for membrane fusion. The original neuronal SNARE proteins[6] were found to belong to large families that included numerous nonneuronal isoforms which have very broad tissue distribution[14] and with each cell type containing multiple isoforms. Compartmental specificity of membrane fusion was initially thought to be encoded by the specific pairing of these SNARE proteins.[15] This thinking turned out to be inaccurate as promiscuity was increasingly found (i.e., VAMP8 discussed below); and that compartmental specificity actually required many accessory proteins that serve to spatially target and bind distinct SNARE complexes, including Sec1/Munc18 (SM), Munc13, and RIM proteins, which further act to prime the SNARE complex to become fusion-ready.[4,6] For more

exquisite regulation of fusion that achieves a higher level of precision and speed, these primed SNARE complexes would pull the cognate-fusing membranes together only up a hemi-fused state, whereby complete fusion is prevented by fusion clamps (i.e., complexins, synaptotagmins) that also act as Ca^{2+} sensors that are ready to be activated by Ca^{2+} to trigger fusion.[4,6] Many of these proteins turned out to be the precise substrates for specific cell signals, including DAG (Munc13), PKC (SM proteins), Ca^{2+} (SNARE complex, synaptotagmins), and even cAMP/PKA (RIMs and EPACs).[6,16] These fusion clamps also act to superprime the SNARE complex, which is required to ensure speed and precision for neuroexocytosis,[6,16] but probably the full complement of these proteins is not required for the less efficient and less precise secretory cells, such as acinar cells.[16] Nevertheless, the pancreatic acinar cell has been an ideal epithelial cell model to examine these tenets of the SNARE hypothesis, wherein many isoforms of these proteins (i.e., SNARE and SM proteins, complexin, synaptotagmins, EPAC) were indeed reported (discussed below), whereas others that were found in neurons (i.e., Munc13, RIM) have not been yet found in the acinar cell.

Physiologic Apical Exocytosis in the Pancreatic Acinar Cell

SNARE proteins

The pancreatic acinar cell was one of the first nonneuronal cells to show the multiple isoforms of each SNARE protein, along with the accessory proteins described above for the neuron.[5] In fact, the investigation of SNARE-mediated exocytosis in the pancreatic acinar cell somewhat followed the developments reported in neurons as depicted below and shown in **Figure 1**. First, we showed the cellular locations of v-SNARE (VAMPs) and t-SNAREs (syntaxins (STX) and SNAP-25) that mapped out to different exocytotic membrane compartments. The three syntaxin isoforms for each of the three exocytotic compartments include syntaxin 2 (STX-2) for the apical PM, syntaxin 3 (STX-3) for the ZG membrane, and syntaxin 4 (STX-4) for the basolateral PM.[17] SNAP-23 was the ubiquitous nonneuronal SNAP-25 isoform that is present in ZGs, apical, and basolateral PM.[18] Of the first two VAMPs reported, VAMP-2, was first thought to be the main one for regulated exocytosis, whereas VAMP-3 might have a role in constitutive secretion.[19]

To assess the function of these SNARE proteins, we first used the tetanus and botulinum neurotoxins. However, unlike neurons, pancreatic acinar cells do not have PM receptors to internalize these neurotoxins, prompting us and others to employ cell permeabilization strategies (i.e., streptolysin O) to enable internalization of the neurotoxins, showing that tetanus toxin and botulinum neurotoxin C1

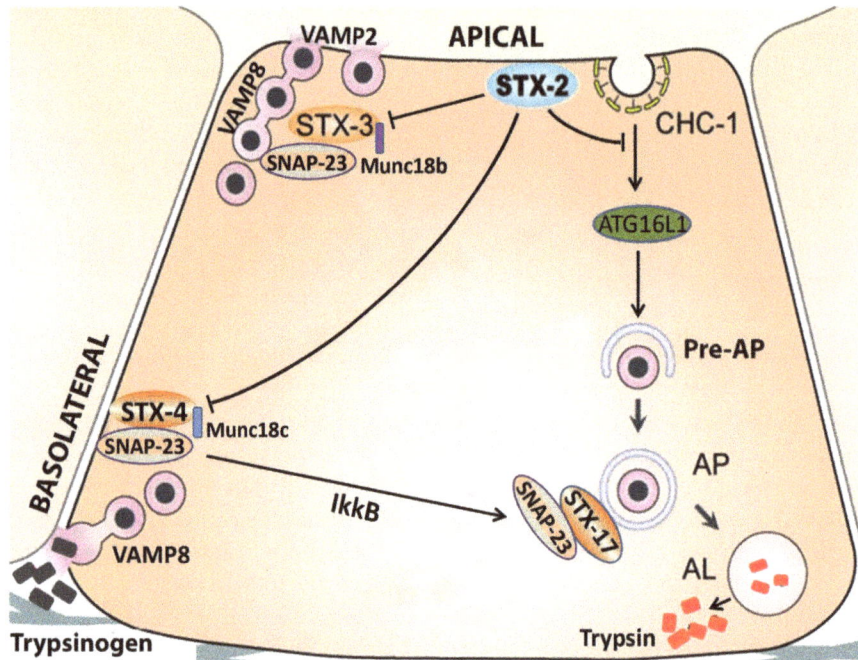

Figure 1. Diagram of pancreatic acinar cell showing the exocytotic machinery for apical exocytosis (Munc18b/STX-3/SNAP23/VAMP2 and VAMP8) and basolateral exocytosis (Munc18c/STX-4/SNAP-23/VAMP8) that have been well characterized and the more recent findings of the roles of STX-2 and SNAP-23 in distorting autophagy in pancreatitis. Apical and pathologic basolateral exocytosis release trypsinogen and other ZG contents into the ductal lumen and interstitial space, respectively. STX-2 acts as an inhibitory SNARE to block apical and basolateral exocytosis and binds ATG16L1 to block the latter from binding clathrin heavy chain (CHC)-1 that would have contributed to pre-autophagosome (pre-AP) formation that drives autophagosome[27] biogenesis. APs fuses with lysosomes to from autolysosomes (ALs), wherein pancreatitis stimuli distort lysosome and/or AL function to cause premature trypsinogen activation into trypsin, which when released into the cytosol would promote acinar cell injury. How trypsinogen is activated into trypsin in the interstitial space to contribute to injury is not yet known. SNAP-23 is a substrate for IKKβ (a major inflammatory signal in pancreatitis)-mediated phosphorylation which enables SNAP-23 translocation from the basolateral plasma membrane to the AP where SNAP-23 would bind and modify syntaxin-17 (STX-17) actions and its SNARE complex in promoting an excess of AP-lysosome fusion.[56]

selectively cleaved pancreatic acinar VAMP-2 and syntaxins (STX-2 and STX-3 only), respectively, which blocked Ca^{2+}-evoked enzyme release.[19,20] However, tetanus toxin, while completely cleaving VAMP-2, resulted in only ~30 percent inhibition of acinar secretion.[19]

It was much later that we discovered a third VAMP, VAMP-8, which when genetically deleted blocked most of the apical exocytosis.[21] This finding was unexpected as VAMP8 is prominent in endosomal vesicles in most cells and was originally known as endobrevin.[22] Here, VAMP-8 accounted for the multiple layers of compound ZG-ZG fusions within the apical pole, whereas exocytosis of the few ZGs with the apical PM is mediated by VAMP-2.[21,23,24] Using a perifusion assay to measure amylase release from acini of VAMP8-KO mice and adenovirus-tetanus toxin-infected acini that acutely cleaved VAMP2, VAMP2 was shown to mediate an early immediate phase of secretion, whereas VAMP8 regulates a later prolonged phase of secretion.[25] These findings matched the exocytosis imaging results wherein VAMP-2-mediated first-phase secretion

corresponds to the fusion of the fewer ZGs abutting the apical PM, and VAMP8-mediated later-phase secretion corresponds to the continued emptying of the much larger number of sequentially fused ZGs that lie behind then fusing with the apical PM-fuse ZGs that would provide a more sustained second-phase secretion.[21,23,24] STX-2 was initially thought to mediate apical exocytosis,[26] which was misled by its location on the apical PM.[17,26] It turned out that STX-2 actually acts as an inhibitory SNARE capable of competing and blocking STX-3 and STX-4 from forming their SNARE complexes, thus blocking apical and basolateral exocytosis (see below), respectively.[27] We had reasoned that this inhibitory SNARE protein STX-2 may be a deliberate mechanism to reduce the efficiency of membrane fusion to a much slower secretory rate (than neurons) of digestive enzyme release over a period of hours during a meal to ensure optimal nutrient digestion.[27] Some SNARE proteins were distinguished by their resistance to botulinum neurotoxin cleavage, of which VAMP8 and also SNAP-23 are two examples. We therefore initially assessed SNAP-23 function with a dominant-negative carboxyl-terminal deleted SNAP-23, which would form nonfunctional SNARE complexes, and indeed this SNAP-23 mutant inhibited acinar secretion.[28] A recent report unequivocally confirmed that SNAP-23 deletion in mice indeed blocked apical exocytosis.[29]

Accessory proteins involved in the apical exocytotic machinery

Although neuronal Munc18a is not present in acinar cells, Munc18b and Munc18c are [21] and were found to bind and prime the predicted SNARE complexes shown to mediate apical exocytosis, ZG-ZG fusion, and basolateral exocytosis. Specifically, Munc18b binds STX-3/SNAP-23 with VAMP2 to mediate exocytosis of ZGs with apical PM, and with VAMP8 to mediate sequential ZG-ZG fusion[21]; whereas Munc18c binds STX-4/VAMP8/SNAP-23 to mediate basolateral exocytosis (discussed below). Of the several isoforms of synaptotagmins (Syt 1, 3, 6, and 7) found in pancreatic acinar cells, Syt-1 found on ZGs and apical PM was the only one involved in enzyme secretion.[31] Pancreatic acinar cells also expressed complexin 2 in its apical pole,[32] and apical VAMP2 could bind both Syt-1 and complexin 2, and their truncated mutants introduced into permeabilized acini inhibited enzyme secretion.[31,32] It is therefore plausible that Syt-1 and complexin 2 may be the fusion clamps and calcium sensors for physiologic exocytoses.[33] However, gene deletion or depletion of Syt-1 or complex 2 in pancreatic acinar cells will be required to definitively show this. Cab45 (the 45-kDa calcium-binding protein) was found to specifically bind the Munc18b/STX-3 complex and also acts as a Ca^{2+} sensor for exocytosis as shown by Cab45 antibody blockade of enzyme secretion by a remarkable 90 percent.[34] cAMP-binding protein, EPAC (exchange protein directly activated by cyclic AMP) partially reversed supramaximal carbachol and CCK-induced inhibition of secretion.[35] While this suggests a role for EPAC in either enhancing apical exocytosis or reversing the apical blockade, further work will be needed to elucidate how EPAC interacts with the apical exocytotic machinery to accomplish these effects.

Other proteins involved in other steps of apical exocytosis

Coordinated interactions of the Rho and Rab family of small G proteins (i.e., Rab3D, Rab 27A/B)[36–40] and synaptotagmin-like proteins,[41] and their subsequent concerted actions on primary exocytotic fusion proteins, including SNARE and accessory proteins (i.e., Munc18b),[42] appear to assist in ZG exocytosis perhaps by their potential ZG-tethering actions. Of these, the stronger evidence seems to building up for Rab27, localized to ZGs, wherein dominant-negative

Rab27 constructs inhibited secretion in pancreatic and parotid acini.[37,39] Whereas Rab27A was shown to be involved only on a minor secretory (lysosomal/endosomal) pathway,[38] Rab27B was found to be more important for ZG exocytosis in pancreatic and parotid acini.[37,39,43] Synaptotagmin-like proteins (SLPs) include various isoforms (Slp1 to Slp5, and rabphilin), of which Slp1 and Slp4 were shown to regulate amylase release in pancreatic and parotid acini in part by their binding to Rab27B.[41,42]

The actin cytoskeleton also plays multiple roles in apical exocytosis, including actin coating of ZGs during acinar stimulation[44] that serves in part to support ZG-ZG fusion[45] and also fusion pore opening[46]—and the actin-myosin complex that facilitates the expulsion of ZG contents.[47] Interestingly, small G-proteins RhoA and Rac1 were shown to also regulate actin rearrangement in pancreatic acinar cells to support secretion.[48]

Pathologic Basolateral Exocytosis Contributing to Pancreatitis

The basolateral PM accounts for 90 percent of the pancreatic acinar cell surface area where little or no exocytosis normally occurs. Nevertheless, the acinar basolateral PM possesses an intact SNARE fusion machinery consisting of t-SNARE proteins (STX-4, SNAP-23) and SM protein Munc18c, which would indicate that exocytosis can potentially occur,[30] as shown in **Figure 1**. In fact, early studies had shown that supramaximal CCK or cholinergic stimulation, capable of inducing pancreatitis *in vivo*, caused apical exocytotic blockade and redirected exocytosis to the lateral PM,[49] releasing enzymes into the interstitial space to cause interstitial pancreatitis.[50] This contribution to pancreatitis that was largely ignored was shown to be of importance when we elucidated the exocytotic machinery mediating this pathologic fusion event (reviewed in reference [11]). Using various fluorescence imaging tools (FM143, syncollin-pHluorin), we were able to observe basolateral exocytosis in rat pancreatic acini in response to supramaximal CCK or carbachol stimulation—and after clinical alcohol sensitization by physiologic CCK or carbachol stimulation[51–53] that mimicked alcoholic pancreatitis. We further showed that not only clinical concentrations of ethanol but also clinical concentrations of ethanol metabolites, acetaldehyde, ethyl palmitate, and ethyl oleate reduced the formation of apical exocytotic complexes with consequent blockade of CCK-8-stimulated apical exocytosis in rat pancreatic acini; and furthermore, acetaldehyde and ethyl oleate redirected VAMP8-containing ZGs toward and fusing with the basolateral PM, along with an increased formation of basolateral PM fusion complexes (Munc18c/STX-4, SNAP-23, VAMP8).[54] It is remarkable that despite only a few ZGs normally close to the basolateral PM, supramaximal stimulation could induce VAMP-8-labeled ZGs to move from the apical pole toward the lateral and basal PM.[21] Mechanistically, these pancreatitis stimuli caused Munc18c on the basolateral PM to be phosphorylated by PKC-α, which induced its binding and activation of the STX-4/VAMP-8/SNAP-23 complex[51] to become receptive to binding VAMP-8 of the approaching ZGs, whereby this complete SM/SNARE complex would then mediate basolateral exocytosis.[21] Eventually, when Munc18c-depleted mice (Munc18c[+/−]) became available, and along with Munc18c knockdown in human pancreas by lenti-Munc18c-shRNA treatment, we confirmed that depletion of Munc18c in pancreatic acini still exhibited normal apical exocytosis during physiologic CCK-8 stimulation, but upon supraphysiologic CCK-8 stimulation, there was reduction in basolateral exocytosis resulting from a decrease in STX-4 SNARE complexes, with consequent alleviation of caerulein pancreatitis.[55] Very recently, we reported that SNAP-23 depletion in rat and human pancreatic acini blocked not only apical exocytosis but also basolateral exocytosis; the latter also led to reduction in the severity of pancreatitis.[56] The Munc18c phosphorylation also caused Munc18c displacement from the basolateral PM into the cytosol where it is rapidly depleted by the cytosolic proteolytic cleavage.[30] This Munc18c displacement from the PM may be of clinical relevance, as in a human case of quiescent chronic alcoholic pancreatitis, Munc18c was similarly displaced into the cytosol of residual intact acinar cells, suggesting that this might be a contributing mechanism predisposing to the clinically observed recurrent pancreatitis.[57] Finally, we showed that VAMP-8 deletion in mice completely blocked basolateral exocytosis, which resulted in reducing not only caerulein pancreatitis but also alcoholic pancreatitis, the latter shown in alcohol diet-fed mice stimulated with postprandial carbachol stimulation that mimicked clinical alcoholic pancreatitis.[21] Whereas the mechanism for pathologic basolateral exocytosis of ZGs is now clear, how the zymogens particularly trypsinogen that are emptied into the interstitial space are activated into damaging trypsin is not and requires further investigation. In a number of our more recent reports, we have performed these studies directly on the human exocrine pancreas using a slice technique that allowed preservation of secretory function over at least four days,[58] enabling us to not only overexpress candidate proteins but also knockdown protein expression with shRNA adenoviruses, thereby providing direct translational value over the conventional use of genetic mouse models.

Dysregulated SNARE Protein-Mediated Fusion Underlying Abnormal Autophagy in Pancreatitis

Autophagy SNARE proteins

Autophagy is the major cellular process that removes damaged organelles and other cellular debris.[59,60] Autophagy is controlled by a series of autophagy-related proteins (ATGs)[59–61] but also involves membrane fusion events which are intrinsically regulated by SNARE proteins[62,63] that also interact with ATGs.[63,64] Autophagy comprises four key stages. The first is phagophore nucleation induced by AMP kinase and PI3K-III complex-I.[59,60] Second is phagophore expansion to engulf damaged organelles (i.e., mitochondria by mitophagy, ER by ER-phagy, ZG by zymophagy). Phagophore expansion is by an aggressive recruitment of membranes from various sources (PM, ER, endosomes), which requires ATG16L1 complexes[59,60] and fusion events mediated by a distinct SNARE complex (STX7/STX8/VAMP7)[62]—and eventual closure of the phagophore ends involving LC3 to form the autophagosome. The third stage is autophagosome fusion with lysosome to form the autolysosome, which is mediated by two other SNARE complexes (STX17/SNAP29/VAMP8/ATG14 and STX7/SNAP29/YKT6).[65,66] STX17 translocates from the cytosol to the autophagosome[65] to initiate fusion with lysosome by first recruiting SNAP29 and then to form a complete complex with lysosomal VAMP8; this SNARE complex is stabilized by ATG14.[62,63] Lysosomal STX7 can also bind SNAP29, which then forms the complete SNARE complex with autophagosome v-SNARE YKT6.[67] The fourth and last stage is autolysosome maturation whereby its cargo is degraded for reuse. Defects in autophagy, particularly in autolysosome formation and maturation, the latter resulting in inadvertent trypsinogen activation, are now well known to occur and has become the current dogma for the cellular mechanism of pancreatitis.[68–70] However, there is no work that directly addresses the role of the STX17 SNARE complex (STX17/SNAP29/VAMP8/ATG14) in autophagy, whose dysregulation could possibly contribute to the defective autolysosome formation or maturation in pancreatitis; and possibly some defects might also arise in phagophore formation and expansion that could be contributed by dysregulation of the STX7/STX8/VAMP7 SNARE complex.

Exocytotic SNARE proteins affecting autophagy

While we were investigating the role of the exocytotic SNARE proteins, we inadvertently ventured into their unexpected roles in autophagy that contributed to pancreatic acinar injury and pancreatitis. Likely as a result of the basolateral exocytotic blockade, Munc18c-depleted acini unexpectedly activated a component of the ER stress response which contributed to autophagy induction, resulting in accumulation of autophagic vacuoles and autolysosomes.[55] We found that STX-2 deletion also led to an accumulation of autophagic vacuoles as a result of increased autophagic induction,[27] but by a different mechanism. ATG16L1 translocates to the PM during intense autophagy to recruit more membranes to expand the phagophore, which is mediated by PM-bound clathrin heavy chain-1 (CHC-1) that forms a complex with pre-autophagosome-linked ATG16L1.[71] As depicted in **Figure 1**, we discovered that STX-2 also binds ATG16L1, which blocked ATG16L1 binding to clathrin in the PM; thus STX-2 deletion resulted in the increased binding of ATG16L1 to clathrin, which leads to increased formation of pre-autophagosomes that drives autophagosome biogenesis.[27] This increased autophagy induction at least in part accounted for the CCK-8-induced increase in autolysosome formation and activation of trypsinogen.[27] As depicted in **Figure 1**, we also discovered a novel link between SNAP-23, autolysosome formation, and their connection to IKK (inhibitor of NF-κB kinase),[56] the latter well-known to mediate NF-κB-mediated production of proinflammatory chemokines and cytokines that initiates and perpetuates the inflammatory response of pancreatitis.[72] Pancreatitis stimuli of acini resulted in the translocation of SNAP-23 from its native location at the PM to the autophagosomes where SNAP-23 is able to perturb the STX-17 SNARE complex-mediated autophagosome-lysosome fusion.[56] This SNAP-23 translocation was mediated by IKKβ-induced phosphorylation of SNAP-23, which was blocked by IKKβ inhibitors and IKKβ-insensitive mutants of SNAP-23.[56] Interestingly, these pathologic SNAP-23-mediated events would not occur during physiologic starvation,[56] which in part explains why this was not previously discovered. SNAP-23 depletion in rats therefore resulted in reduced autolysosome formation with consequent reduction in trypsinogen activation that takes place in autolysosomes, which together with SNAP-23 depletion-induced blockade of basolateral exocytosis contributed to a major reduction in pancreatitis severity.[56] Of note, this is the first report that a single protein could serve as a mechanistic link of the three major cellular pathways for pancreatitis, including inflammatory signaling and the pathologic fusion events of dysregulated autolysosome formation and basolateral exocytosis. This therefore has major therapeutic implication in the targeting of SNAP23 for the treatment of pancreatitis.

Future Directions in the Investigation of Pathologic Fusion Underlying Pancreatitis

While of interest, the above work on exocytotic SNAREs affecting autophagy are nevertheless not the primary SNARE proteins that mediate autophagy. Of note, STX-17 expression was recently reported to be reduced after cerulein pancreatitis.[73] This reduction of STX-17 levels can conceivably perturb autolysosome formation or maturation, which could contribute to the acinar injury. However, there is no reported work on autophagy SNAREs or their role in pancreatitis. Another major pathologic fusion contributing to pancreatitis is the very large (>2.5 micron) trypsin-containing endocytic vacuoles enveloped by F-actin whose

rupture inside the pancreatic acinar cell releases damaging trypsin.[74,75] The membrane fusion machinery that mediates this pathologic endocytosis has not been elucidated. VAMP8, originally discovered as an endosomal VAMP,[22] was shown to play an additional and important role in endosomal trafficking by its interactions with early endosomal proteins D52 and Rab5, whereby VAMP8 depletion resulted in depletion of these endosomal proteins with consequent accumulation of activated trypsin in the autolysosomes.[25,76,77] Remarkably, replenishment of D52 or Rab5 prevented the accumulation of activated trypsin.[25] Finally, there is pathologic endocytosis of zymogens by the macrophages invading the acini during pancreatitis, which led to activation of macrophage NF-κB that in turn induced the production of inflammatory cytokines to perpetuate and accentuate pancreatitis injury.[78] Much more work is needed to pursue these lines of investigation in elucidating the fusion machinery of each of these pathologic fusions that contribute to pancreatic acinar injury in pancreatitis; and from those insights, novel therapeutic strategies could then be devised to treat pancreatitis.

Acknowledgments

This work was supported by a grant from the Canadian Institute of Health Research, CIHR PJT-159542. SD is funded by a postdoctoral fellowship from the American Pancreatic Association/National Pancreas Foundation. We thank Fred Gorelick and Guy Groblewski for help with editing this manuscript.

References

1. Palade G. Intracellular aspects of the process of protein synthesis. Science 1975; 189:347–358.
2. Williams JA. Receptor-mediated signal transduction pathways and the regulation of pancreatic acinar cell function. Curr Opin Gastroenterol 2008; 24:573–579.
3. Rothman JE. Lasker Basic Medical Research Award. The machinery and principles of vesicle transport in the cell. Nat Med 2002; 8:1059–1062.
4. Sudhof TC, Rothman JE. Membrane fusion: grappling with SNARE and SM proteins. Science 2009; 323:474–477.
5. Gaisano HY. A hypothesis: SNARE-ing the mechanisms of regulated exocytosis and pathologic membrane fusions in the pancreatic acinar cell. Pancreas 2000; 20:217–226.
6. Sudhof TC, Rizo J. Synaptic vesicle exocytosis. Cold Spring Harb Perspect Biol 2011; 3:1–14. doi: 10.1101/cshperspect.a005637.
7. Nemoto T, Kimura R, Ito K, Tachikawa A, Miyashita Y, Iino M, Kasai H. Sequential-replenishment mechanism of exocytosis in pancreatic acini. Nat Cell Biol 2001; 3:253–258.
8. Thorn P, Fogarty KE, Parker I. Zymogen granule exocytosis is characterized by long fusion pore openings and preservation of vesicle lipid identity. Proc Natl Acad Sci U S A 2004; 101:6774–6779.

9. Woska JR, Jr., Gillespie ME. SNARE complex-mediated degranulation in mast cells. J Cell Mol Med 2012; 16:649–656.

10. Catz SD, McLeish KR. Therapeutic targeting of neutrophil exocytosis. J Leukoc Biol 2020; 107:393–408.

11. Gaisano HY, Gorelick FS. New insights into the mechanisms of pancreatitis. Gastroenterology 2009; 136:2040–2044.

12. Takahashi T, Miao Y, Kang F, Dolai S, Gaisano HY. Susceptibility factors and cellular mechanisms underlying alcoholic pancreatitis. Alcohol Clin Exp Res 2020; 44:777–789.

13. Pfeffer SR. A prize for membrane magic. Cell 2013; 155:1203–1206.

14. Linial M. SNARE proteins-why so many, why so few? J Neurochem 1997; 69:1781–1792.

15. McNew JA, Parlati F, Fukuda R, Johnston RJ, Paz K, Paumet F, Sollner TH, Rothman JE. Compartmental specificity of cellular membrane fusion encoded in SNARE proteins. Nature 2000; 407:153–159.

16. Kasai H, Takahashi N, Tokumaru H. Distinct initial SNARE configurations underlying the diversity of exocytosis. Physiol Rev 2012; 92:1915–1964.

17. Gaisano HY, Ghai M, Malkus PN, Sheu L, Bouquillon A, Bennett MK, Trimble WS. Distinct cellular locations of the syntaxin family of proteins in rat pancreatic acinar cells. Mol Biol Cell 1996; 7:2019–2027.

18. Gaisano HY, Sheu L, Wong PP, Klip A, Trimble WS. SNAP-23 is located in the basolateral plasma membrane of rat pancreatic acinar cells. FEBS Lett 1997; 414:298–302.

19. Gaisano HY, Sheu L, Foskett JK, Trimble WS. Tetanus toxin light chain cleaves a vesicle-associated membrane protein (VAMP) isoform 2 in rat pancreatic zymogen granules and inhibits enzyme secretion. J Biol Chem 1994; 269:17062–17066.

20. Hansen NJ, Antonin W, Edwardson JM. Identification of SNAREs involved in regulated exocytosis in the pancreatic acinar cell. J Biol Chem 1999; 274:22871–22876.

21. Cosen-Binker LI, Binker MG, Wang CC, Hong W, Gaisano HY. VAMP8 is the v-SNARE that mediates basolateral exocytosis in a mouse model of alcoholic pancreatitis. J Clin Invest 2008; 118:2535–2551.

22. Wong SH, Zhang T, Xu Y, Subramaniam VN, Griffiths G, Hong W. Endobrevin, a novel synaptobrevin/VAMP-like protein preferentially associated with the early endosome. Mol Biol Cell 1998; 9:1549–1563.

23. Behrendorff N, Dolai S, Hong W, Gaisano HY, Thorn P. Vesicle-associated membrane protein 8 (VAMP8) is a SNARE (soluble N-ethylmaleimide-sensitive factor attachment protein receptor) selectively required for sequential granule-to-granule fusion. J Biol Chem 2011; 286:29627–29634.

24. Weng N, Thomas DD, Groblewski GE. Pancreatic acinar cells express vesicle-associated membrane protein 2- and 8-specific populations of zymogen granules with distinct and overlapping roles in secretion. J Biol Chem 2007; 282:9635–9645.

25. Messenger SW, Jones EK, Holthaus CL, Thomas DDH, Cooley MM, Byrne JA, Mareninova OA, Gukovskaya AS, Groblewski GE. Acute acinar pancreatitis blocks vesicle-associated membrane protein 8 (VAMP8)-dependent secretion, resulting in intracellular trypsin accumulation. J Biol Chem 2017; 292:7828–7839.

26. Pickett JA, Thorn P, Edwardson JM. The plasma membrane Q-SNARE syntaxin 2 enters the zymogen granule membrane during exocytosis in the pancreatic acinar cell. J Biol Chem 2005; 280:1506–1511.

27. Dolai S, Liang T, Orabi AI, Holmyard D, Xie L, Greitzer-Antes D, Kang Y, Xie H, Javed TA, Lam PP, Rubin DC, Thorn P, et al. Pancreatitis-induced depletion of syntaxin 2 promotes autophagy and increases basolateral exocytosis. Gastroenterology 2018; 154:1805–1821.

28. Huang X, Sheu L, Tamori Y, Trimble WS, Gaisano HY. Cholecystokinin-regulated exocytosis in rat pancreatic acinar cells is inhibited by a C-terminus truncated mutant of SNAP-23. Pancreas 2001; 23:125–133.

29. Kunii M, Ohara-Imaizumi M, Takahashi N, Kobayashi M, Kawakami R, Kondoh Y, Shimizu T, Simizu S, Lin B, Nunomura K, Aoyagi K, Ohno M, et al. Opposing roles for SNAP23 in secretion in exocrine and endocrine pancreatic cells. J Cell Biol 2016; 215:121–138.

30. Gaisano HY, Lutz MP, Leser J, Sheu L, Lynch G, Tang L, Tamori Y, Trimble WS, Salapatek AM. Supramaximal cholecystokinin displaces Munc18c from the pancreatic acinar basal surface, redirecting apical exocytosis to the basal membrane. J Clin Invest 2001; 108:1597–1611.

31. Falkowski MA, Thomas DD, Messenger SW, Martin TF, Groblewski GE. Expression, localization, and functional role for synaptotagmins in pancreatic acinar cells. Am J Physiol Gastrointest Liver Physiol 2011; 301:G306–G316.

32. Falkowski MA, Thomas DD, Groblewski GE. Complexin 2 modulates vesicle-associated membrane protein (VAMP) 2-regulated zymogen granule exocytosis in pancreatic acini. J Biol Chem 2010; 285:35558–35566.

33. Messenger SW, Falkowski MA, Groblewski GE. Ca^{2+}-regulated secretory granule exocytosis in pancreatic and parotid acinar cells. Cell Calcium 2014; 55:369–375.

34. Lam PP, Hyvarinen K, Kauppi M, Cosen-Binker L, Laitinen S, Keranen S, Gaisano HY, Olkkonen VM. A cytosolic splice variant of Cab45 interacts with Munc18b and impacts on amylase secretion by pancreatic acini. Mol Biol Cell 2007; 18:2473–2480.

35. Chaudhuri A, Husain SZ, Kolodecik TR, Grant WM, Gorelick FS. Cyclic AMP-dependent protein kinase and Epac mediate cyclic AMP responses in pancreatic acini. Am J Physiol Gastrointest Liver Physiol 2007; 292.G1403–G1410.

36. Williams JA, Chen X, Sabbatini ME. Small G proteins as key regulators of pancreatic digestive enzyme secretion. Am J Physiol Endocrinol Metab 2009; 296:E405–E414.

37. Chen X, Li C, Izumi T, Ernst SA, Andrews PC, Williams JA. Rab27b localizes to zymogen granules and regulates pancreatic acinar exocytosis. Biochem Biophys Res Commun 2004; 323:1157–1162.

38. Hou Y, Ernst SA, Stuenkel EL, Lentz SI, Williams JA. Rab27A is present in mouse pancreatic acinar cells and is required for digestive enzyme secretion. PLoS One 2015; 10:e0125596.

39. Imai A, Yoshie S, Nashida T, Shimomura H, Fukuda M. The small GTPase Rab27B regulates amylase release from rat parotid acinar cells. J Cell Sci 2004; 117:1945–1953.

40. Sabbatini ME, Bi Y, Ji B, Ernst SA, Williams JA. CCK activates RhoA and Rac1 differentially through Galpha13 and Galphaq in mouse pancreatic acini. Am J Physiol Cell Physiol 2010; 298:C592–C601.

41. Saegusa C, Kanno E, Itohara S, Fukuda M. Expression of Rab27B-binding protein Slp1 in pancreatic acinar cells and its involvement in amylase secretion. Arch Biochem Biophys 2008; 475:87–92.

42. Fukuda M, Imai A, Nashida T, Shimomura H. Slp4-a/granuphilin-a interacts with syntaxin-2/3 in a Munc18–2-dependent manner. J Biol Chem 2005; 280:39175–39184.

43. Hou Y, Chen X, Tolmachova T, Ernst SA, Williams JA. EPI64B acts as a GTPase-activating protein for Rab27B in pancreatic acinar cells. J Biol Chem 2013; 288:19548–19557.

44. Valentijn JA, Valentijn K, Pastore LM, Jamieson JD. Actin coating of secretory granules during regulated exocytosis correlates with the release of rab3D. Proc Natl Acad Sci U S A 2000; 97:1091–1095.

45. Nemoto T, Kojima T, Oshima A, Bito H, Kasai H. Stabilization of exocytosis by dynamic F-actin coating of zymogen granules in pancreatic acini. J Biol Chem 2004; 279:37544–37550.

46. Jang Y, Soekmadji C, Mitchell JM, Thomas WG, Thorn P. Real-time measurement of F-actin remodelling during exocytosis using Lifeact-EGFP transgenic animals. PLoS One 2012; 7:e39815.

47. Masedunskas A, Sramkova M, Parente L, Sales KU, Amornphimoltham P, Bugge TH, Weigert R. Role for the actomyosin complex in regulated exocytosis revealed by intravital microscopy. Proc Natl Acad Sci U S A 2011; 108:13552–13557.

48. Bi Y, Page SL, Williams JA. Rho and Rac promote acinar morphological changes, actin reorganization, and amylase secretion. Am J Physiol Gastrointest Liver Physiol 2005; 289:G561–G570.

49. Scheele G, Adler G, Kern H. Exocytosis occurs at the lateral plasma membrane of the pancreatic acinar cell during supramaximal secretagogue stimulation. Gastroenterology 1987; 92:345–353.

50. Hartwig W, Jimenez RE, Werner J, Lewandrowski KB, Warshaw AL, Fernandez-del Castillo C. Interstitial trypsinogen release and its relevance to the transformation of mild into necrotizing pancreatitis in rats. Gastroenterology 1999; 117:717–725.

51. Cosen-Binker LI, Lam PP, Binker MG, Gaisano HY. Alcohol-induced protein kinase Cα phosphorylation of Munc18c in carbachol-stimulated acini causes basolateral exocytosis. Gastroenterology 2007; 132:1527–1545.

52. Fernandez NA, Liang T, Gaisano HY. Live pancreatic acinar imaging of exocytosis using syncollin-pHluorin. Am J Physiol Cell Physiol 2011; 300:C1513–1523.

53. Lam PP, Cosen Binker LI, Lugea A, Pandol SJ, Gaisano HY. Alcohol redirects CCK-mediated apical exocytosis to the acinar basolateral membrane in alcoholic pancreatitis. Traffic 2007; 8:605–617.

54. Dolai S, Liang T, Lam PPL, Fernandez NA, Chidambaram S, Gaisano HY. Effects of ethanol metabolites on exocytosis of pancreatic acinar cells in rats. Gastroenterology 2012; 143:832–843.

55. Dolai S, Liang T, Orabi AI, Xie L, Holmyard D, Javed TA, Fernandez NA, Xie H, Cattral MS, Thurmond DC, Thorn P, Gaisano HY. Depletion of the membrane-fusion regulator Munc18c attenuates caerulein hyperstimulation-induced pancreatitis. J Biol Chem 2018; 293:2510–2522.

56. Dolai S, Takahashi T, Qin T, Liang T, Xie L, Kang F, Mio Y, Xie H, Kang Y, J. M, E. W, P R, et al. Pancreas-specific SNAP-23 depletion prevents pancreatitis by attenuating pathologic basolateral exocytosis and formation of trypsin-activating autolysosomes. Autophagy 2020; 7:1–14.

57. Gaisano HY, Sheu L, Whitcomb D. Alcoholic chronic pancreatitis involves displacement of Munc18c from the pancreatic acinar basal membrane surface. Pancreas 2004; 28:395–400.

58. Liang T, Dolai S, Xie L, Winter E, Orabi AI, Karimian N, Cosen-Binker LI, Huang YC, Thorn P, Cattral MS, Gaisano HY. Ex vivo human pancreatic slice preparations offer a valuable model for studying pancreatic exocrine biology. J Biol Chem 2017; 292:5957–5969.

59. Mizushima N. Autophagy: process and function. Genes Dev 2007; 21:2861–2873.

60. Lamb CA, Yoshimori T, Tooze SA. The autophagosome: origins unknown, biogenesis complex. Nat Rev Mol Cell Biol 2013; 14:759–774.

61. Mizushima N, Levine B, Cuervo AM, Klionsky DJ. Autophagy fights disease through cellular self-digestion. Nature 2008; 451:1069–1075.

62. Moreau K, Renna M, Rubinsztein DC. Connections between SNAREs and autophagy. Trends Biochem Sci 2013; 38:57–63.

63. Liu R, Zhi X, Zhong Q. ATG14 controls SNARE-mediated autophagosome fusion with a lysosome. Autophagy 2015; 11:847–849.

64. Diao J, Liu R, Rong Y, Zhao M, Zhang J, Lai Y, Zhou Q, Wilz LM, Li J, Vivona S, Pfuetzner RA, Brunger AT, et al. ATG14 promotes membrane tethering and fusion of autophagosomes to endolysosomes. Nature 2015; 520:563–566.

65. Itakura E, Kishi-Itakura C, Mizushima N. The hairpin-type tail-anchored SNARE syntaxin 17 targets to autophagosomes for fusion with endosomes/lysosomes. Cell 2012; 151:1256–1269.

66. Yu S, Melia TJ. The coordination of membrane fission and fusion at the end of autophagosome maturation. Curr Opin Cell Biol 2017; 47:92–98.

67. Matsui T, Jiang P, Nakano S, Sakamaki Y, Yamamoto H, Mizushima N. Autophagosomal YKT6 is required for fusion with lysosomes independently of syntaxin 17. J Cell Biol 2018; 217:2633–2645.

68. Mareninova OA, Hermann K, French SW, O'Konski MS, Pandol SJ, Webster P, Erickson AH, Katunuma N, Gorelick FS, Gukovsky I, Gukovskaya AS. Impaired autophagic flux mediates acinar cell vacuole formation and trypsinogen activation in rodent models of acute pancreatitis. J Clin Invest 2009; 119:3340–3355.

69. Gukovsky I, Gukovskaya AS. Impaired autophagy underlies key pathological responses of acute pancreatitis. Autophagy 2010; 6:428–429.

70. Habtezion A, Gukovskaya AS, Pandol SJ. Acute pancreatitis: a multi-faceted set of organellar and cellular interactions. Gastroenterology 2019; 156:1941–1950.

71. Ravikumar B, Moreau K, Jahreiss L, Puri C, Rubinsztein DC. Plasma membrane contributes to the formation of pre-autophagosomal structures. Nat Cell Biol 2010; 12:747–757.

72. Huang H, Liu Y, Daniluk J, Gaiser S, Chu J, Wang H, Li ZS, Logsdon CD, Ji B. Activation of nuclear factor-kappaB in acinar cells increases the severity of pancreatitis in mice. Gastroenterology 2013; 144:202–210.

73. Piplani H, Marek-Iannucci S, Sin J, Hou J, Takahashi T, Sharma A, de Freitas Germano J, Waldron RT, Saadaeijahromi H, Song Y, Gulla A, Wu B, et al. Simvastatin induces autophagic flux to restore cerulein-impaired phagosome-lysosome fusion in acute pancreatitis. Biochim Biophys Acta Mol Basis Dis 2019; 1865:165530.

74. Chvanov M, De Faveri F, Moore D, Sherwood MW, Awais M, Voronina S, Sutton R, Criddle DN, Haynes L, Tepikin AV. Intracellular rupture, exocytosis and actin interaction of endocytic vacuoles in pancreatic acinar cells: initiating events in acute pancreatitis. J Physiol 2018; 596:2547–2564.

75. De Faveri F, Chvanov M, Voronina S, Moore D, Pollock L, Haynes L, Awais M, Beckett AJ, Mayer U, Sutton R, Criddle DN, Prior IA, et al. LAP-like non-canonical autophagy and evolution of endocytic vacuoles in pancreatic acinar cells. Autophagy 2020; 16:1314–1331.

76. Messenger SW, Falkowski MA, Thomas DD, Jones EK, Hong W, Gaisano HY, Boulis NM, Groblewski GE. Vesicle associated membrane protein 8 (VAMP8)-mediated zymogen granule exocytosis is dependent on endosomal trafficking via the constitutive-like secretory pathway. J Biol Chem 2014; 289:28040–28053.

77. Messenger SW, Thomas DD, Falkowski MA, Byrne JA, Gorelick FS, Groblewski GE. Tumor protein D52 controls trafficking of an apical endolysosomal secretory pathway in pancreatic acinar cells. Am J Physiol Gastrointest Liver Physiol 2013; 305:G439–G452.

78. Sendler M, Weiss FU, Golchert J, Homuth G, van den Brandt C, Mahajan UM, Partecke LI, Doring P, Gukovsky I, Gukovskaya AS, Wagh PR, Lerch MM, et al. Cathepsin B-mediated activation of trypsinogen in endocytosing macrophages increases severity of pancreatitis in mice. Gastroenterology 2018; 154:704–718.

Chapter 11

Protein composition and biogenesis of pancreatic zymogen granules

Xuequn Chen

Department of Physiology, Wayne State University, Detroit, MI

Introduction

The exocrine pancreas produces and secretes multiple digestive enzymes and has been the model in which the structure and functional organization of the mammalian secretory pathway was originally discovered and intensively studied.[1,2] The pancreatic acinar cells exhibit one of the highest protein synthesis rates among mammalian cells. More than 90 percent of the newly synthesized proteins are targeted to the secretory pathway[3] and packaged into large secretory granules, called zymogen granules (ZGs). In contrast to the smaller neuroendocrine and endocrine granules, ZGs have an averaged diameter of around 1 μm. They are responsible for the transport, storage, and secretion of digestive enzymes and have long been a model for studying the general mechanisms of secretory granule biogenesis and regulated exocytosis. Stimulation of the acinar cells by secretagogues such as acetylcholine and cholecystokinin triggers fusion of ZG membrane with the apical plasma membrane and the release of digestive enzymes into the pancreatic ductal system. In the duodenum, trypsinogen is converted to trypsin by proteolytic cleavage via enterokinase and activated trypsin then proteolytically activates the other zymogen enzymes.[4,5]

Physiological stimulation of acinar cells by secretagogues triggers local apical Ca^{2+} spiking, fusion of ZG membrane with the apical membrane, and exocytosis.[6–9] In contrast to the physiological condition, supramaximal CCK stimulation elicits sustained elevation of cytosolic $[Ca^{2+}]$ and leads to mistrafficking of digestive and lysosomal enzymes, inhibition of apical secretion, and abnormal exocytosis redirected to basolateral plasma membrane, all of which are believed to contribute to the pathogenesis of acute pancreatitis.[10,11] The ZG content contains the digestive enzymes and associated proteins, which are the major protein components of the pancreatic juice secreted into the duodenum. The ZG membrane carries at least part of the molecular machinery responsible for digestive enzyme sorting, granule trafficking, and exocytosis. For example, digestive enzyme sorting and packaging will, at least to some extent, depend on interactions between ZG content and ZG membrane components exposed to the lumen of ZGs. The cytoplasmic surface of the ZG membrane must contain vesicular trafficking proteins including Rabs, SNARE proteins, as well as molecular motors to interact with the cytoskeleton. Defective ZG biogenesis and trafficking can result in various pancreatic diseases such as acute and chronic pancreatitis.[10–12] A comprehensive understanding of the ZG content and membrane protein composition is necessary to provide critical insights in the biogenesis and regulated secretion of pancreatic ZGs.

Protein Composition of Zymogen Granules

Zymogens and digestive enzymes

The major secretory products of the acinar cells, namely the content of ZGs, are digestive enzymes which belong to five functional groups of hydrolytic enzymes including endo and exoproteases, lipases, glycosidases, and nucleases. In contrast to endocrine cells, which often produce a predominant peptide or protein product such as insulin, acinar cells synthesize, package, and secrete a mixture of nearly 20 different enzymes and isoenzymes including amylase, trypsinogens, chymotrypsinogens, carboxypeptidases, esterases, lipases, and ribonucleases. Most pancreatic proteases are synthesized as inactive precursors or zymogens, which only become activated by a cascade of limited proteolysis within the intestinal lumen. Because of the importance of ZGs to digestive enzyme storage and regulated secretion and as a general model for secretory vesicles, the identification and characterization of both the soluble and membrane proteins of ZGs have been of great interest in the field. In an early pioneering study, the secreted ZG contents from the guinea pig exocrine pancreas were analyzed by two-dimensional gel electrophoresis, which resolved 19 distinct high-molecular-weight proteins. Thirteen of the nineteen proteins were identified by actual or potential enzymatic activity.[13] In more recent studies using mass spectrometry-based proteomics analyses, the identities of these enzymes have been confirmed and additional isoforms were found.[14,15] The ZG contents

e-mail: xchen@med.wayne.edu

make up the major protein components of the pancreatic juice secreted into the duodenum. Therefore, the identification of ZG content proteins also significantly impacts biomarker studies in the pancreatic juice.[16]

Components and topology of the ZG membrane proteins

It is believed that the integral and peripheral ZG membrane proteins serve critical functions for zymogen sorting/packaging, vesicular trafficking, and regulated exocytosis. Therefore, comprehensive identification of ZG membrane proteins is expected to shed new light on our understanding of ZG biogenesis and secretion. In early studies, characterization of rat ZG membranes by SDS-PAGE indicated a relatively simple protein composition of about 10 components with GP2 (glycoprotein 2) accounting for 40 percent of the proteins.[17–19] In the past decade, studies have been carried out to characterize ZG membrane proteins using two-dimensional gel electrophoresis. These efforts led to the identifications of two additional major ZG membrane components, GP3 (glycoprotein 3)[20] and membrane dipeptidase,[21] by N-terminal amino acid sequence analysis. However, due to the lack of sensitive tools for protein identification, the identities of many spots resolved on the 2D gels remained unknown. In another study, 14 spots corresponding to small GTP-binding proteins were resolved on a 2D gel of ZG membrane proteins by [^{35}S]GTPγS blotting analysis.[22] However, the identities of the spots were not determined. Different from the above abundant ZG membrane proteins, a number of low-abundance regulatory proteins have been identified on the ZG membrane by immunoblotting and immunocytochemistry. Examples of these proteins included the small GTPase, Rab3D[23,24] and the SNARE proteins, vesicle-associated membrane protein 2 (VAMP2),[25] and syntaxin 3.[26,27] More recently, VAMP 8 was found on ZG membrane and to play a major physiological role in regulated exocytosis.[28] Despite the significant amount of knowledge of ZG membrane proteins accumulated in the past decades on an individual basis, a comprehensive characterization of the membrane protein components of this organelle was not achieved until the application of modern mass spectrometry revealed a much more complex makeup of the ZG membrane.[14,15,29]

Organellar proteomics represents an analytical strategy that combines biochemical fractionation and comprehensive protein identification. Initial purification of organelles leads to reduced sample complexity and links proteomics data to functional analysis.[30–32] In the past decade, organellar proteomic analysis has been carried out for virtually every subcellular compartment in the mammalian secretory pathway, including Golgi, ER, and secretory granules.[30] The first comprehensive analysis of rat ZG membrane was carried out by combining mass spectrometry-based proteomics technologies and a well-established protocol of ZG purification.[14] Using this protocol (outlined in **Figure 1**, *left*), a crude granule pellet (P2) was prepared by two consecutive low-speed centrifugations and then further purified by ultracentrifugation in a self-forming Percoll gradient. A heavy white band, containing highly purified ZGs, was observed and collected just above the bottom of the tube. The ZG membrane and content proteins were then separated by osmotic lysis of ZGs with the ionophore, nigericin, followed by ultracentrifugation. The membrane pellet was washed first with 0.25M KBr and then with 0.1M Na_2CO_3 (pH 11.0) to remove soluble content proteins and loosely associated proteins. The known ZG membrane marker, such as Rab3D was highly enriched in the purified membrane fractions (**Figure 1**, *right*).

These results indicated a much more complex protein composition of the ZG membrane. By combining multiple separation strategies including one-, two-dimensional gel electrophoresis and two-dimensional HPLC with tandem mass spectrometry, over 100 proteins were identified from purified ZG membrane.[14] Most of the known ZG membrane proteins were identified, including high-abundance matrix proteins such as GP2, GP3, ZG16, and syncollin, which are likely involved in ZG sorting and packaging low-abundance proteins such as dynactin2[33] and VAMP2, which are involved in ZG trafficking and exocytosis. Many novel ZG membrane proteins were also identified, including the SNARE protein, SNAP 29, the small G proteins Rab27B, Rab11A, and Rap1, and the molecular motor protein, myosin Vc. Indicative of the interest in understanding the ZG membrane proteome, a later study[15] using 1D SDS-PAGE coupled with 1D LC-MS/MS identified, in a more redundant database, 371 proteins from both ZG membrane and content. A large degree of overlap between proteins was found in these two independent studies. The overlap is higher (nearly 100%) for high-score protein identifications and major new observations, including all the new vesicular trafficking proteins, whereas the disagreement increases in low-score identifications, some of which may be contaminants. Representative ZG proteins found in multiple proteomic analyses[14,15,29] and in some individual studies are summarized in **Table 1**.

As a second step toward a comprehensive architectural model of ZG membrane, a systematic topology analysis of ZG membrane proteins was performed by combining a global protease protection assay with iTRAQ-based quantitative proteomic technology.[29] In this study, isolated ZGs were incubated with or without proteinase K. The control and proteinase K-treated ZG membrane proteins were digested with trypsin and the resulting peptides were labeled with iTRAQ reagents to compare the relative abundance of each peptide in the two samples. Because proteinase K treatment removed cytoplasm-sided ZGM proteins or domains, the iTRAQ ratios (proteinase K treated vs. control) from all peptides fell into two

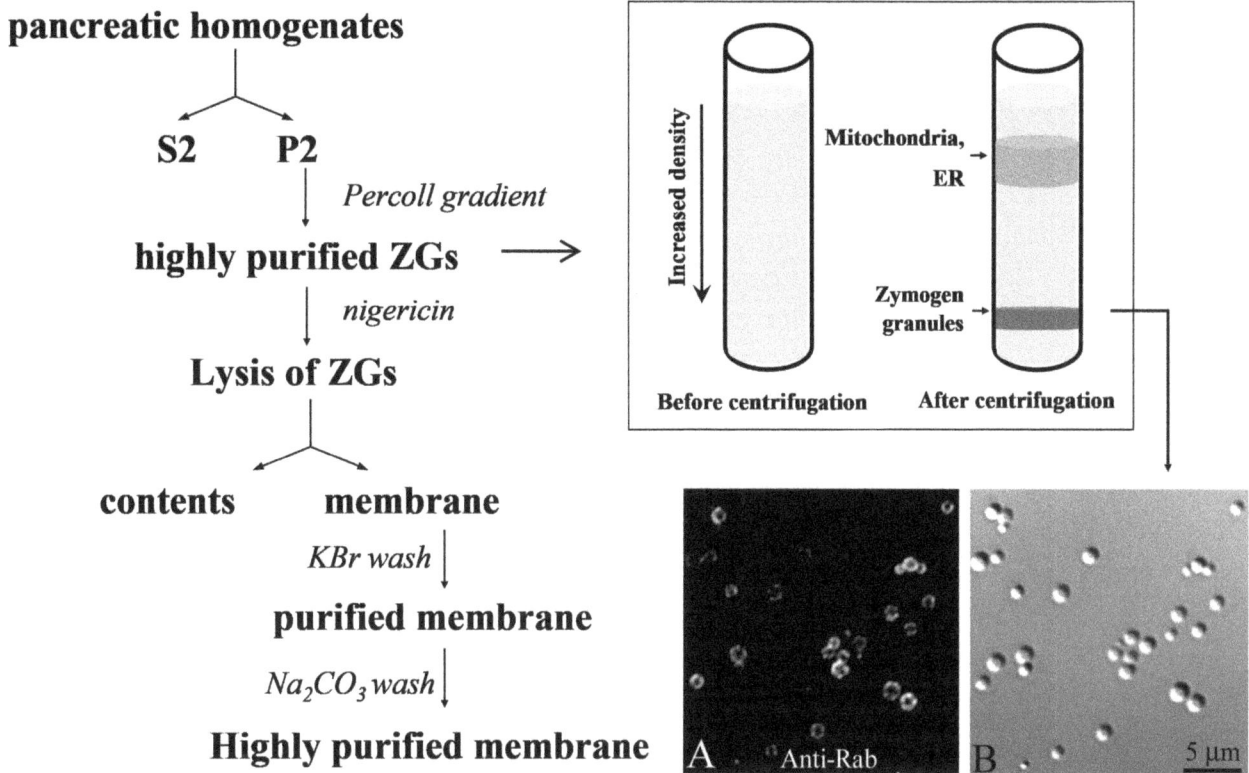

Figure 1. Outline of ZG membrane purification. <u>Left</u>: Rat pancreata were homogenized and then centrifuged in two consecutive low-speed steps to generate a crude particulate fraction (P2) enriched in ZGs. The particulate was resuspended, mixed with equal volume of Percoll, and ultracentrifuged. The dense white ZG band was then collected and washed. To purify ZG membrane, the isolated ZGs were lysed with nigericin and ultracentrifuged to separate contents and membranes. The membrane pellet was then washed sequentially with 250 mM KBr and 0.1 M Na_2CO_3 (pH 11.0). <u>Right</u>: Top shows a cartoon to illustrate the Percoll gradient ultracentrifugation; at the bottom are Nomarski and fluorescent images of purified ZGs to demonstrate the purity of ZGs and the positive staining of a ZG marker, Rab3D (reproduced from reference [14]).

clusters. The peptides from cytoplasm-sided proteins or domains showed reduced iTRAQ ratios (<<1.0), whereas those luminal ZG proteins showed little change (ratios around 1.0) (**Figure 2**). A threshold was determined using a training set of ZG membrane proteins with well-characterized membrane topology. The category with iTRAQ ratios below the threshold included the cytoplasm-sided peripheral membrane proteins, such as synaptotagmin-like protein 1 and Myosin Vc, and membrane proteins with single transmembrane structure or posttranslational lipid modification, such as VAMP 2, 8, Syntaxin 7, and all the Rab proteins. The second category with the ratios above the threshold included the lumen-sided peripheral membrane proteins, such as GP2, GP3, syncollin, and all of the digestive enzymes. Altogether, this analysis was able to successfully assign the membrane topology for 199 identified ZG proteins.[29] The advantage of this technique was to analyze a large number of endogenous proteins simultaneously without the necessity to exogenously express fusion

proteins. In addition, this technique was also able to map both cytoplasm- and lumen-sided domains from the same transmembrane proteins.

The overall goal of these recent proteomics studies is to build a quantitative, architectural model of the pancreatic ZG, which will lead to new hypotheses for subsequent functional analysis of this prototypic secretory granule. The ZG localization of a number of novel proteins, including Rap1, Rab6, Rab11A, Rab27B, SNAP29, and myosin Vc, were confirmed by immunocytochemistry (**Figure 3**). Several such novel observations have already led to hypothesis-driven functional studies. One example was Rab27B, subsequently demonstrated as an essential regulator of acinar exocytosis.[36] Other examples included Rap1[38] and SNAP 29.[52] The fact that Myosin Vc was also identified on ZG membrane in this study led us to hypothesize that Myosin Vc and Rab27B form a complex to tether ZGs at the apical actin web in an analogy to the Rab27A/melanophilin/Myosin Va complex on melanosomes.[53–57] This

Table 1. Representative pancreatic ZG proteins identified from proteomics analysis and their major functional categories. Representative ZG proteins from major functional categories are listed. These proteins were identified on purified rat pancreatic ZGs from multiple proteomics studies.[14,15,29] Proteins potentially copurified from other subcellular organelles are not included. References for their original discovery or with independent functional characterization or immunostaining are also included. Note: (a) SCAMP2, 3, 4, and (b) syntaxin 3 were also identified on ZG membrane.

Protein name	NCBI #	MW	pI	Ref
Digestive enzymes				
Alpha-amylase	62644218	51020	8.42	
Anionic trypsin precursor	67548	28363	4.69	
Chymotrypsin C	1705913	30919	5.64	
Carboxypeptidase A1 precursor	8393183	50282	5.38	
Carboxypeptidase A2 precursor	61556903	50269	5.17	
Cationic trypsinogen	27465583	28821	7.45	
Chymotrypsin B	6978717	25934	4.90	
Colipase	203503	13597	8.04	
Elastase 2	6978803	27274	8.81	
Pancreatic lipase	1865644	54494	6.6	
Pancreatic lipase related protein 1	14091772	57122	5.79	
Similar to elastase 3B	62649890	30806	5.47	
Sterol esterase	1083805	72537	5.37	
Small GTPases				
Rab11A	2463536	24509	5.98	34, 35
Rab14	420272	24078	5.85	
Rab1A	56605816	25670	5.95	
RAB27B	16758202	27382	5.38	36
RAB3D	18034781	26332	4.75	23, 24
Rab8A	77748034	23668	9.15	37
Rap1	52138628	21201	5.37	38
ZG matrix proteins				
Clusterin	46048420	56070	5.53	39
GP2	121538	62355	4.9	18, 40
GP3	171105374	58695	6.03	20, 41
Syncollin	208806121	17780	8.61	42, 43
ZG16	19705541	17316	9.79	44
Transporters, pumps, and ion channels				
Cation–chloride cotransporter 6	13516403	95862	8.09	
Cation–chloride cotransporter 9	23495276	77073	6.2	
Chloride channel protein 3	4762023	90855	5.88	45
L-type amino acid transporter 1	12643400	55903	8.18	
Vacuolar-type H⁺-ATPase 115 kDa subunit, a1 isoform	13928826	102385	6.04	46

Vesicular trafficking proteins

Protein	Accession	MW	pI	Ref
Cysteine string protein	1095322	24892	4.93	47
Dynactin 2	50926127	44148	5.14	33
Myosin Vc	62653910	228341	8.17	
SCAMP1[a]	158749626	37999	7.61	
SNAP29	7769720	29000	5.40	
Synaptotagmin-like protein 1	71043698	59471	5.53	
Synaptotagmin-like protein 4	17939356	75900	9.08	
Syntaxin 7[b]	55741787	29851	5.32	
Syntaxin 12	77695930	31187	5.23	
VAMP 2	51704188	12691	7.84	48
VAMP 8	13929182	12512	8.93	

Other proteins

Protein	Accession	MW	pI	Ref
CD47 antigen	55250722	32995	8.91	
CD59 antigen	6978635	13790	8.9	
CD63 antigen	38648866	29617	7.37	
Dipeptidase 1	16758372	48023	5.68	49
Ectonucleoside triphosphate diphosphohydrolase 1 (CD39)	12018242	57337	7.47	
Gamma-glutamyl transpeptidase	16758696	66667	8.46	50
Itmap1	5916203	72874	6.07	51
Polymeric immunoglobulin receptor	27151742	84798	5.07	

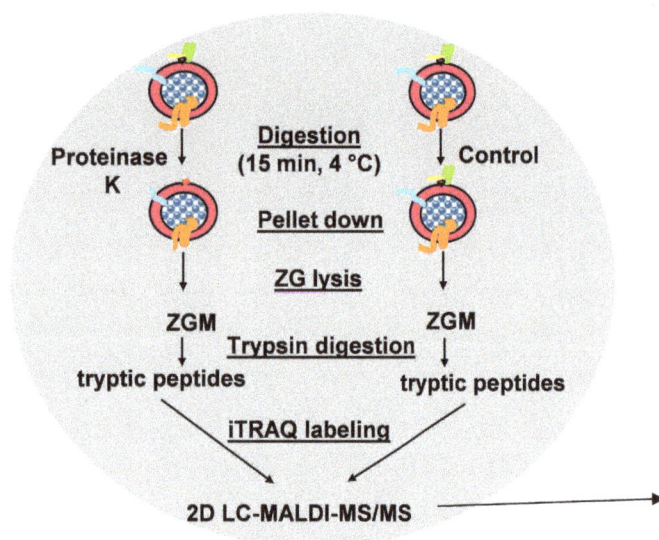
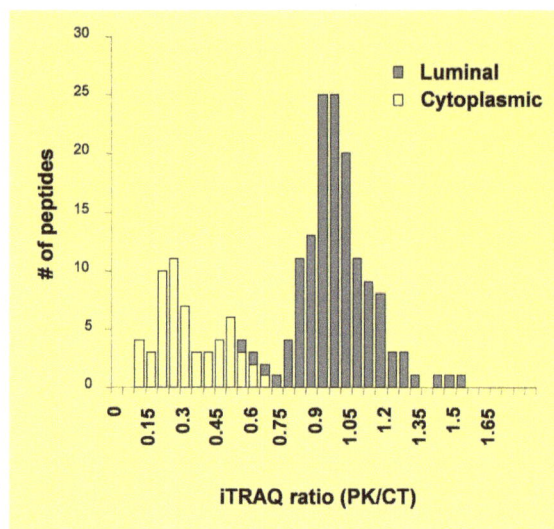

Figure 2. Workflow of iTRAQ-based topology analysis of pancreatic ZG membrane proteins. *Left*: Isolated ZGs were treated with or without proteinase K and then lysed. ZG membrane proteins were digested with trypsin, and resulting peptides were labeled with iTRAQ reagents. The peptides were mixed and analyzed by 2D LC-MALDI-MS/MS. *Right*: iTRAQ ratio distributions of tryptic peptides from ZGM proteins with known topology. A histogram of iTRAQ ratios (proteinase K *versus* control) from identified peptides illustrates the presence of two distinct clusters of tryptic peptides from cytoplasm-orientated and lumen-orientated ZG proteins, respectively (modified from reference [29]).

hypothesis is further supported by the more recent findings of two potential Rab27 effector proteins, synaptotagmin-like protein (Slp) 1 and 4.[15,29]

A comprehensive model of the ZG requires the absolute quantity of each individual ZG protein as well as the stoichiometry among different ZG proteins. However, to date, this information has not yet been determined for any ZG protein. Recently, an absolute quantification (AQUA) proteomics strategy[58,59] using LC-SRM and isotope-labeled synthetic peptides was used to obtain absolute molar abundances for selected mouse ZG proteins, Rab3D and VAMP8.[60] The absolute quantities of mouse Rab3D and VAMP8 were determined as 1242 ± 218 and 2039 ± 151 (mean $\pm$ SEM) copies per ZG. The size distribution and the averaged diameter of ZGs (~750 nm) were determined by atomic force microscopy.[60] For an average-sized ZG, the densities of these two proteins, if evenly distributed on the ZG membrane, were 702 molecules/μm^2 for Rab3D and 1152 molecules/μm^2 for VAMP8, respectively. As a comparison, it was estimated that Rab3A had 10 copies and VAMP2 had 70 copies on a synaptic vesicle (an average diameter of 45.18 nm) with their corresponding membrane densities being 1,572 and 11,003 molecules/μm^2, respectively.[61] Notably, some ZG proteins can be present on a subpopulation of ZGs or concentrated on specific domains of the ZG membrane. The average copy numbers and membrane densities determined here can serve as a starting point

to further examine the relative distributions of specific ZG proteins. In addition to mouse and rat ZGs, ZGs were purified from human acini obtained from pancreatic islet transplantation center, and the constituents of human ZGs were comprehensively characterized for the first time.[60] A total of 180 human ZG proteins were identified, including membrane and the content proteins. The identification of human ZG-specific content and membrane proteins could significantly impact translational studies of biomarkers in pancreatic juice from cancer and pancreatitis patients. From these proteomics analyses in past decades, a comprehensive molecular model of ZG has emerged to include the protein components, membrane topologies, copy numbers per ZG, and protein-protein interactions. A diagram of such a ZG molecular model with selective ZG membrane proteins is shown in **Figure 4**.

Functional categories of ZG proteins

The secretory granules in neuroendocrine, endocrine, and exocrine cells share fundamental molecular mechanisms in granule formation, intracellular trafficking, and regulated exocytosis.[62–64] For ZGs, this multistep process includes budding of immature granules (IGs) off the trans-Golgi network, granule maturation, and granule transport toward the apical pole in the vicinity of plasma membrane, tethering/docking at the plasma membrane, and regulated

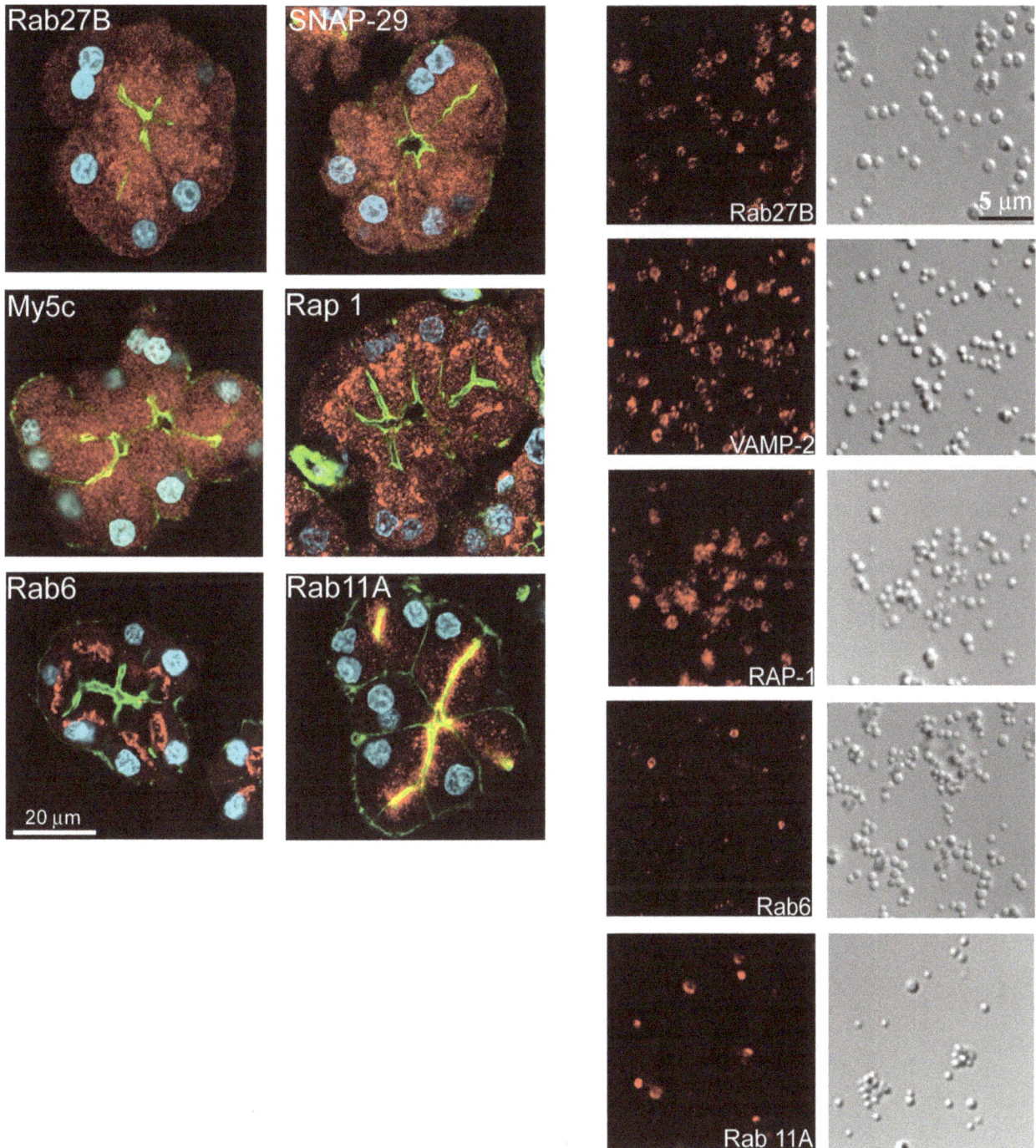

Figure 3. Immunolocalization of novel small GTPases and SNARE proteins to isolated ZGs. The ZG localizations of some novel small GTPases and SNARE proteins were confirmed at isolated ZGs level by immunocytochemistry and confocal microscopy. The immunofluorescent images (red) together with corresponding DIC images for VAMP2, Rab11a, Rap1, and Rab27b are shown. Although VAMP2, Rap1, and Rab27b stained every ZGs as indicated by circles outlining individual ZGs, only a portion of ZGs showed positive Rab11a staining (reproduced from reference [14]).

exocytosis of ZG contents triggered by the rise of local Ca^{2+} concentration upon hormonal and neuronal stimulations (**Figure 5**). Despite the general model outlined above, the detailed molecular mechanisms are not yet completely understood. The ZG membrane is believed to carry at least part of the molecular machinery responsible for each step. Therefore, the recent comprehensive identification of the ZG protein components has shed new light

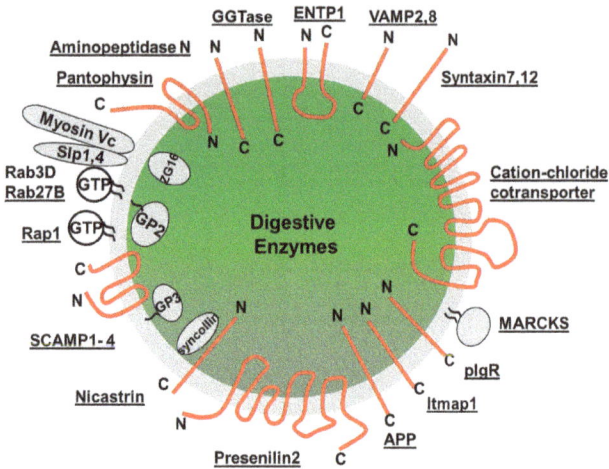

Figure 4. Molecular architecture of pancreatic zymogen granule proteins. A number of identified ZGM proteins and their topology assignments are shown on a single ZG.

matrix and glycoproteins, and small GTP-binding proteins (**Table 1**).

The ZG matrix protein group includes the highly abundant ZG proteins such as GP2, GP3, ZG16, and syncollin, previously known to be present on the inner surface of ZG membrane and likely involved in zymogen sorting and packaging (more detailed discussion in the next section).[44,64,67,68] In addition to the high-abundance structural proteins and enzymes, many of the known functional and regulatory proteins were also identified on the outer surface of the ZGs. These included vesicular trafficking proteins such as SNARE proteins VAMP 2 and VAMP 8, molecular motors myosin Vc, and dynactin 2, a subunit of dynactin adaptor complex for the minus end-driven microtubular transport motor, dynein. A significant number of small GTPases were identified, including multiple Rabs and Rap1. Only Rab3D was reported on ZGs.[23,24,69,70] These newly identified ZG-localized small GTPases represented one of the major novel findings from the extensive proteomics analyses of ZG membrane. Ion channels and transporters are usually low-abundant proteins and multi-transmembrane domains, making them hard to detect by proteomics approaches. Within this category, various subunits of the vacuolar H⁺-ATPase (V-ATPase) have been consistently identified in multiple proteomics studies.[14,15,29]

on ZG biogenesis, trafficking, and exocytosis.[14,15,29,65] The identified ZG proteins fall into several broad functional categories which link their identities to potential functional importance. These categories include proton pumps and ion channels, enzymes, vesicular trafficking proteins,

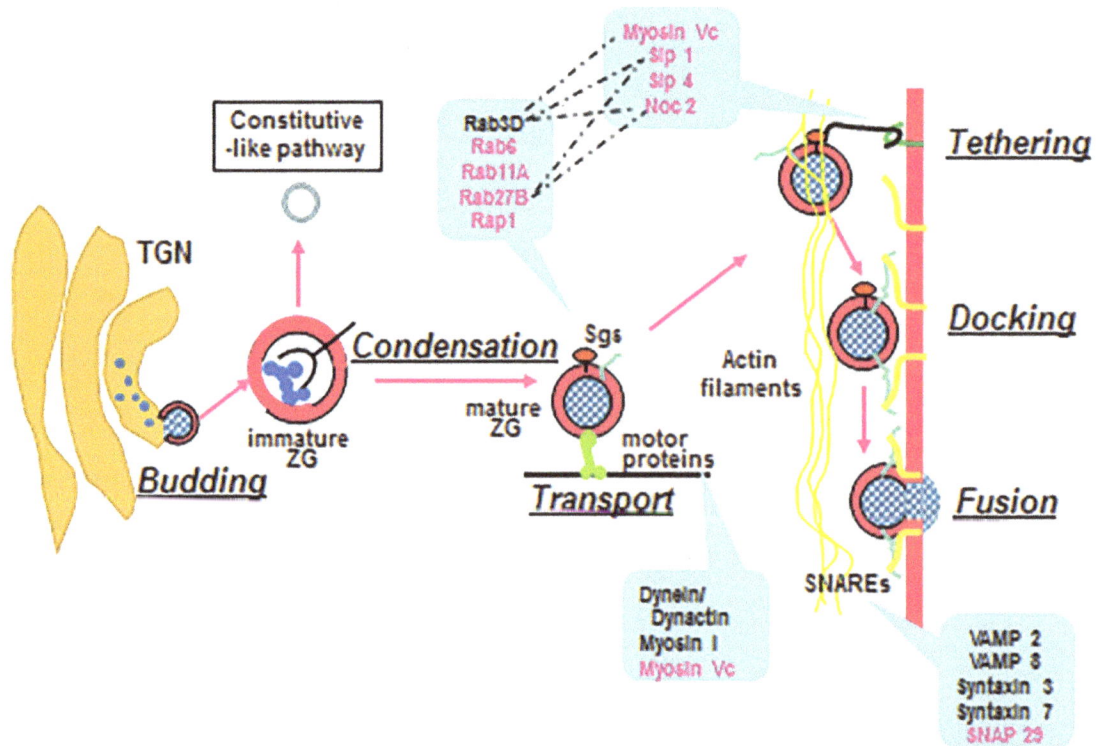

Figure 5. A working model of ZG biogenesis and regulated exocytosis. For ZGs, this multistep process includes the budding of immature granules from the trans-Golgi network (TGN), granule maturation through condensation and transport toward the apical pole in the vicinity of plasma membrane, tethering/docking at the plasma membrane, and regulated exocytosis of ZG contents through membrane fusion (modified from reference [66]).

A few other ion channels and transporters have also been identified in individual proteomics studies (**Table 1**). In addition, functional evidence indicates the presence of several ion channel and transporter proteins in ZG membranes (aquaporins, vacuolar-type H^+-ATPase, zinc influx transporter SLC30A2). The evidence for the K^+ channels, Kv7.1 and Kir6.1, for ClC Cl^- channels and the vesicular nucleotide transporter SLC17A9 in ZG is less strong. A detailed discussion can be found in reference [71].

Zymogen Granules Biogenesis—Sorting and Maturation

Protein sorting at the trans-Golgi network (TGN) is essential for professional secretory cells, such as pancreatic acinar cells. In contrast to most eukaryotic cells that secrete proteins constitutively using TGN-derived vesicles, acinar cells store their secretory product in granules that undergo regulated exocytosis in a stimulus-dependent manner. In acinar cells, a mixture of different zymogens are packaged within the TGN; some zymogens form protein complexes and progressively aggregate in a Ca^{2+}- and pH-dependent manner. Parts of the Golgi cisternae become dilated and these condensing vacuoles (CVs) pinch off to become IGs[72] and then mature into ZGs. The selective aggregation of pancreatic secretory proteins has been documented,[73–75] but the underlying molecular mechanism by which the secretory proteins are sorted into the regulated secretory pathway is still not well understood. Two hypotheses, not necessarily mutually exclusive, have been developed to explain the selection of content proteins for storage in the secretory granules.[62,63]

The sorting-for-entry hypothesis proposes that the TGN acts as the primary protein sorting station in the biosynthetic transport pathway. This is based on the paradigm for newly synthesized lysosomal hydrolases sorted into clathrin-coated vesicles by mannose-6-phosphate receptor (MPR) and then targeted to endosomal membranes. For the regulated secretory proteins, the sorting-for-entry model postulated the presence of TGN membrane-associated sorting receptors to facilitate cargo entry into IG. Through this mechanism, only selected secretory proteins can enter IGs, whereas other proteins, such as those targeting the constitutive secretory pathway, are efficiently excluded. Immunoelectron microscopy observations have demonstrated the segregation of regulated and constitutive cargo at the level of the TGN,[76] supporting an active sorting mechanism.

In the sorting-by-retention hypothesis, the IGs serve as an important post-TGN sorting station. In this model, protein entrance into IGs is largely unselective, and high-order intermolecular associations allow regulated secretory proteins for efficient retention within maturing granules. Concurrently, a subset of protein components is removed

by receptor-mediated sorting or bulk flow. The driving force underlying this subtractive retention involves assembly of granule core proteins within IGs by aggregation/condensation, the progressive protein insolubility within the luminal environment of maturing granules in a Ca^{2+}- and pH-dependent manner. Condensation of regulated secretory proteins allows them to remain in the maturing granules while the lysosomal proteins are removed by constitutive-like vesicle budding.

In pancreatic acinar cells, membrane proteins involved in the sorting and packaging of zymogens are predicted to have their functional domains exposed on the luminal side of the ZG membrane. Studies have been focused on several abundant luminal ZG membrane proteins for their potential roles in zymogen sorting and ZG formation.[64] GP2 represents up to 40 percent of the total ZG membrane proteins in rat ZG.[77] The membrane association of GP2 is via a glycosylphosphatidylinositol (GPI) anchor.[78,79] Since GP2 can form stable complexes with zymogens at mildly acidic pH but not at alkaline pH,[74,80] it was suggested that GP2 may act as a sortase for aggregated secretory proteins.[81] However, the findings that ZGs in mouse pancreas can form in the absence of GP2 indicated it is not required for ZG biogenesis.[82] Another abundant luminal ZG membrane protein, syncollin, is a component of lipid rafts.[67] While the rates of synthesis and intracellular transport of secretory proteins were reduced in syncollin-deficient mice, these mice are viable and showed no detectable changes in pancreatic morphology, regulated exocytosis, or zymogen content.[83] Thus, syncollin does not seem to be required for ZG biogenesis either. Though several other luminal membrane-associated ZG proteins have also been studied in ZG biogenesis, it is still unclear if any of these proteins alone is indispensable for ZG formation. However, it has been shown that the assembly of a proteoglycan scaffold at the ZG membrane promotes the efficient packaging of zymogens and the proper formation of stimulus-competent storage granules in pancreatic acinar cells.[84] Other proteins identified as associated with the inner surface of the ZGM include chymase and peptidyl-prolyl isomerase B.[85]

Zymogen Granules Trafficking and Exocytosis

Upon nervous and/or hormonal stimulation,[86] ZGs move toward the apical plasma membrane in a microtubule- and actin-dependent manner.[33,87] Secretagogue stimulation of the cells causes an elevation of the intracellular Ca^{2+} concentration, which in turn triggers granule fusion. Detailed discussion of ZG exocytosis can be found in Chapter 6 by Cooley and Gaisano et al.[88]

Major new findings from recent studies came from the vesicular trafficking and small GTP-binding proteins groups for ZG membrane proteins.[14,15,29,65,66] Among the small GTPase identified on ZG membrane, only Rab3D was previously reported on ZGs. In addition, punctate subapical labeling of Rab11 just below the apical plasma membrane was observed in pancreatic acinar cells.[34] The ZG localization of the majority of newly identified small G proteins including Rab27B, Rab11A, Rap1, and Rab6 were confirmed by immunocytochemistry at the isolated acini and ZG level (**Figure 3**). In subsequent functional studies, it was shown that Rab27B localizes on ZGs and plays an important role in regulating acinar exocytosis.[36,89] Furthermore, Rap1 was localized on pancreatic ZGs, although it had previously been localized on parotid secretory granules and it was found that Rap1 activation regulates pancreatic amylase secretion.[38] Though Rab3D and Rab27B were present on all the ZGs, Rab6 and Rab11A localized to only a fraction of ZGs (**Figure 3**). This could indicate the existence of different subpopulations of ZGs. However, because corresponding GDIs can extract rabs from the membranes and cycles between membrane and cytosol, an alternative interpretation is that this represents ZGs at different stages in the secretory pathway. Notably, Rab27A was later shown to be present in mouse acinar cells and partially co-localizatized with ZGs and was required for digestive enzyme secretion.[90] In addition to the small G proteins, we also found a novel SNARE protein, SNAP29, on ZGs. The ZG membrane localization of SNAP29 was confirmed by immunocytochemistry and it was also found that SNAP29 and VAMP2 formed a complex on ZG membrane.[52]

Another important category of molecules critical to vesicular trafficking and exocytosis is molecular motors and corresponding adaptors. It was reported previously that myosin Vc localized to the exocrine pancreas and largely overlapped with apical F-actin.[72] Proteomics analysis identified the presence of myosin Vc on ZG membrane. Rab27a forms a complex with myosin Va on melanosome through a synaptotagmin-like linker protein, melanophilin; the complex is required to tether melanosomes to the actin cytoskeleton. Rab27B, myosin Vc, and at least two synaptotagmin-like proteins, slp1 and slp4, are present on the ZG membrane in acinar cells. By analogy, Rab27B, slps, and myosin Vc could form a complex on ZGs to regulate the tethering of ZGs at the apical membrane. In exocrine pancreas, targeting of ZGs to the apical cell surface first requires an intact microtubule system and is then transferred to actin.[91] The minus end-driven microtubular transport is thought mediated by molecular motor, dynein rather than kinesin. Purified ZGs are associated with cytoplasmic dynein intermediate and heavy chain and contained the major components of the dynein activator complex, dynactin. Consistent with this report, proteomics analysis identified a component of the dynactin complex, dynactin2/dynamitin, from highly purified ZG membrane. Interestingly, Rab6 functioned as a tethering factor controlling the recruitment of dynactin to membranes.

Conclusions and Future Directions

The pancreatic ZG has been a prototypic model for all secretory granules in the regulated secretory cells. Since its discovery and initial morphological characterization by electron microscopy, its protein compositions and the related molecular mechanisms that govern its biogenesis, intracellular transport, and regulated exocytosis have been described. Through modern proteomics technologies, a very detailed molecular model of ZG content and membrane is being established. A comprehensive molecular architecture of ZG will be developed in the foreseeable future, including its protein components and protein complexes, their membrane topologies, and protein copy numbers per ZG. In addition, the lipid composition of the ZG membrane will be described using state-of-the-art lipidomic techniques. This comprehensive molecular view will facilitate a thorough mechanistic understanding of how the IGs are formed from TGN, condense into mature ZGs and subsequently released with physiological stimulation.

References

1. Palade G. Intracellular aspects of the process of protein synthesis. Science 1975; 189:347–358.
2. Williams JA. Receptor-mediated signal transduction pathways and the regulation of pancreatic acinar cell function. Curr Opin Gastroenterol 2008; 24:573–579.
3. Scheele GA, Palade GE, Tartakoff AM. Cell fractionation studies on the guinea pig pancreas. Redistribution of exocrine proteins during tissue homogenization. J Cell Biol 1978; 78:110–130.
4. Case RM. Synthesis, intracellular transport and discharge of exportable proteins in the pancreatic acinar cell and other cells. Biol Rev Camb Philos Soc 1978; 53:211–354.
5. Owyang C, Williams JA. Pancreatic secretion. In: Yamada T, Alpers DH, Kaplowitz N, Laine L, Owyang C, Powell DW, eds. Textbook of gastroenterology. Volume 1. 4th ed. Philadelphia: Lippincott Williams & Wilkins, 2003:340–366.
6. Williams JA, Groblewski GE, Ohnishi H, Yule DI. Stimulus-secretion coupling of pancreatic digestive enzyme secretion. Digestion 1997; 1:42–45.
7. Petersen OH. Localization and regulation of Ca^{2+} entry and exit pathways in exocrine gland cells. Cell Calcium 2003; 33:337–344.
8. Petersen OH, Tepikin AV. Polarized calcium signaling in exocrine gland cells. Annu Rev Physiol 2008; 70:273–299.
9. Messenger SW, Falkowski MA, Groblewski GE. Ca^{2+}-regulated secretory granule exocytosis in pancreatic and parotid acinar cells. Cell Calcium 2014; 55:369–375.
10. Gaisano HY, Gorelick FS. New insights into the mechanisms of pancreatitis. Gastroenterology 2009; 136:2040–2044.
11. Sah RP, Saluja A. Molecular mechanisms of pancreatic injury. Curr Opin Gastroenterol 2011; 27:444–451.
12. Gukovsky I, Pandol SJ, Mareninova OA, Shalbueva N, Jia W, Gukovskaya AS. Impaired autophagy and organellar

dysfunction in pancreatitis. J Gastroenterol Hepatol 2012; 27 Suppl 2:27–32.

13. Scheele GA. Two-dimensional gel analysis of soluble proteins. Characterization of guinea pig exocrine pancreatic proteins. J Biol Chem 1975; 250:5375–5385.

14. Chen X, Walker AK, Strahler JR, Simon ES, Tomanicek-Volk SL, Nelson BB, Hurley MC, Ernst SA, Williams JA, Andrews PC. Organellar proteomics: analysis of pancreatic zymogen granule membranes. Mol Cell Proteomics 2006; 5:306–312.

15. Rindler MJ, Xu CF, Gumper I, Smith NN, Neubert TA. Proteomic analysis of pancreatic zymogen granules: identification of new granule proteins. J Proteome Res 2007; 6:2978–2992.

16. Doyle CJ, Yancey K, Pitt HA, Wang M, Bemis K, Yip-Schneider MT, Sherman ST, Lillemoe KD, Goggins MD, Schmidt CM. The proteome of normal pancreatic juice. Pancreas 2012; 41:186–194.

17. MacDonald RJ, Ronzio RA. Comparative analysis of zymogen granule membrane polypeptides. Biochem Biophys Res Commun 1972; 49:377–382.

18. Fukuoka S, Freedman SD, Scheele GA. A single gene encodes membrane-bound and free forms of GP-2, the major glycoprotein in pancreatic secretory (zymogen) granule membranes. Proc Natl Acad Sci U S A 1991; 88:2898–2902.

19. Hoops TC, Rindler MJ. Isolation of the cDNA encoding glycoprotein-2 (GP-2), the major zymogen granule membrane protein. Homology to uromodulin/Tamm-Horsfall protein. J Biol Chem 1991; 266:4257–4263.

20. Wagner AC, Wishart MJ, Mulders SM, Blevins PM, Andrews PC, Lowe AW, Williams JA. GP-3, a newly characterized glycoprotein on the inner surface of the zymogen granule membrane, undergoes regulated secretion. J Biol Chem 1994; 269:9099–9104.

21. Hofken T, Linder D, Kleene R, Goke B, Wagner AC. Membrane dipeptidase and glutathione are major components of pig pancreatic zymogen granules. Exp Cell Res 1998; 244:481–490.

22. Goke B, Williams JA, Wishart MJ, De Lisle RC. Low molecular mass GTP-binding proteins in subcellular fractions of the pancreas: regulated phosphoryl G proteins. Am J Physiol Cell Physiol 1992; 262:C493–C500.

23. Ohnishi H, Ernst SA, Wys N, McNiven M, Williams JA. Rab3D localizes to zymogen granules in rat pancreatic acini and other exocrine glands. Am J Physiol Gastrointest Liver Physiol 1996; 271:G531–G538.

24. Valentijn JA, Sengupta D, Gumkowski FD, Tang LH, Konieczko EM, Jamieson JD. Rab3D localizes to secretory granules in rat pancreatic acinar cells. Eur J Cell Biol 1996; 70:33–41.

25. Gaisano HY, Sheu L, Grondin G, Ghai M, Bouquillon A, Lowe A, Beaudoin A, Trimble WS. The vesicle-associated membrane protein family of proteins in rat pancreatic and parotid acinar cells. Gastroenterology 1996; 111:1661–1669.

26. Gaisano HY, Ghai M, Malkus PN, Sheu L, Bouquillon A, Bennett MK, Trimble WS. Distinct cellular locations of the syntaxin family of proteins in rat pancreatic acinar cells. Mol Biol Cell 1996; 7:2019–2027.

27. Hansen NJ, Antonin W, Edwardson JM. Identification of SNAREs involved in regulated exocytosis in the pancreatic acinar cell. J Biol Chem 1999; 274:22871–22876.

28. Wang CC, Ng CP, Lu L, Atlashkin V, Zhang W, Seet LF, Hong W. A role of VAMP8/endobrevin in regulated exocytosis of pancreatic acinar cells. Dev Cell 2004; 7:359–371.

29. Chen X, Ulintz PJ, Simon ES, Williams JA, Andrews PC. Global topology analysis of pancreatic zymogen granule membrane proteins. Mol Cell Proteomics 2008; 7:2323–2336.

30. Walther TC, Mann M. Mass spectrometry-based proteomics in cell biology. J Cell Biol 2010; 190:491–500.

31. Yan W, Aebersold R, Raines EW. Evolution of organelle-associated protein profiling. J Proteomics 2009; 72:4–11.

32. Au CE, Bell AW, Gilchrist A, Hiding J, Nilsson T, Bergeron JJ. Organellar proteomics to create the cell map. Curr Opin Cell Biol 2007; 19:376–385.

33. Kraemer J, Schmitz F, Drenckhahn D. Cytoplasmic dynein and dynactin as likely candidates for microtubule-dependent apical targeting of pancreatic zymogen granules. Eur J Cell Biol 1999; 78:265–277.

34. Goldenring JR, Smith J, Vaughan HD, Cameron P, Hawkins W, Navarre J. Rab11 is an apically located small GTP-binding protein in epithelial tissues. Am J Physiol Gastrointest Liver Physiol 1996; 270:G515–G525.

35. Hori Y, Takeyama Y, Hiroyoshi M, Ueda T, Maeda A, Ohyanagi H, Saitoh Y, Kaibuchi K, Takai Y. Possible involvement of Rab11 p24, a Ras-like small GTP-binding protein, in intracellular vesicular transport of isolated pancreatic acini. Dig Dis Sci 1996; 41:133–138.

36. Chen X, Li C, Izumi T, Ernst SA, Andrews PC, Williams JA. Rab27b localizes to zymogen granules and regulates pancreatic acinar exocytosis. Biochem Biophys Res Commun 2004; 323:1157–1162.

37. Faust F, Gomez-Lazaro M, Borta H, Agricola B, Schrader M. Rab8 is involved in zymogen granule formation in pancreatic acinar AR42J cells. Traffic 2008; 9:964–979.

38. Sabbatini ME, Chen X, Ernst SA, Williams JA. Rap1 activation plays a regulatory role in pancreatic amylase secretion. J Biol Chem 2008; 283:23884–23894.

39. Min BH, Jeong SY, Kang SW, Crabo BG, Foster DN, Chun BG, Bendayan M, Park IS. Transient expression of clusterin (sulfated glycoprotein-2) during development of rat pancreas. J Endocrinol 1998; 158:43–52.

40. Rindler MJ, Hoops TC. The pancreatic membrane protein GP-2 localizes specifically to secretory granules and is shed into the pancreatic juice as a protein aggregate. Eur J Cell Biol 1990; 53:154–163.

41. Wishart MJ, Andrews PC, Nichols R, Blevins GT, Jr., Logsdon CD, Williams JA. Identification and cloning of GP-3 from rat pancreatic acinar zymogen granules as a glycosylated membrane-associated lipase. J Biol Chem 1993; 268:10303–10311.

42. An SJ, Hansen NJ, Hodel A, Jahn R, Edwardson JM. Analysis of the association of syncollin with the membrane of the pancreatic zymogen granule. J Biol Chem 2000; 275:11306–11311.

43. Edwardson JM, An S, Jahn R. The secretory granule protein syncollin binds to syntaxin in a Ca^{2+}-sensitive manner. Cell 1997; 90:325–333.

44. Kleene R, Dartsch H, Kern HF. The secretory lectin ZG16p mediates sorting of enzyme proteins to the zymogen granule membrane in pancreatic acinar cells. Eur J Cell Biol 1999; 78:79–90.

45. Kelly ML, Abu-Hamzah R, Jeremic A, Cho SJ, Ilie AE, Jena BP. Patch clamped single pancreatic zymogen granules: direct measurements of ion channel activities at the granule membrane. Pancreatology 2005; 5:443–449.

46. Roussa E, Alper SL, Thevenod F. Immunolocalization of anion exchanger AE2, Na$^+$/H$^+$ exchangers NHE1 and NHE4, and vacuolar-type H$^+$-ATPase in rat pancreas. J Histochem Cytochem 2001; 49:463–474.

47. Braun JE, Scheller RH. Cysteine string protein, a DnaJ family member, is present on diverse secretory vesicles. Neuropharmacology 1995; 34:1361–1369.

48. Gaisano HY, Sheu L, Foskett JK, Trimble WS. Tetanus toxin light chain cleaves a vesicle-associated membrane protein (VAMP) isoform 2 in rat pancreatic zymogen granules and inhibits enzyme secretion. J Biol Chem 1994; 269:17062–17066.

49. Hooper NM, Cook S, Laine J, Lebel D. Identification of membrane dipeptidase as a major glycosyl-phosphatidylinositol-anchored protein of the pancreatic zymogen granule membrane, and evidence for its release by phospholipase A. Biochem J 1997; 324:151–157.

50. Beaudoin AR, Grondin G, Laperche Y. Immunocytochemical localization of gamma-glutamyltranspeptidase, GP-2 and amylase in the rat exocrine pancreas: the concept of zymogen granule membrane recycling after exocytosis. J Histochem Cytochem 1993; 41:225–233.

51. Imamura T, Asada M, Vogt SK, Rudnick DA, Lowe ME, Muglia LJ. Protection from pancreatitis by the zymogen granule membrane protein integral membrane-associated protein-1. J Biol Chem 2002; 277:50725–50733.

52. Weng N, Thomas DD, Groblewski GE. Pancreatic acinar cells express vesicle-associated membrane protein 2- and 8-specific populations of zymogen granules with distinct and overlapping roles in secretion. J Biol Chem 2007; 282:9635–9645.

53. Fukuda M, Kuroda TS, Mikoshiba K. Slac2-a/melanophilin, the missing link between Rab27 and myosin Va: implications of a tripartite protein complex for melanosome transport. J Biol Chem 2002; 277:12432–12436.

54. Hammer JA, 3rd, Wu XS. Rabs grab motors: defining the connections between Rab GTPases and motor proteins. Curr Opin Cell Biol 2002; 14:69–75.

55. Strom M, Hume AN, Tarafder AK, Barkagianni E, Seabra MC. A family of Rab27-binding proteins. Melanophilin links Rab27a and myosin Va function in melanosome transport. J Biol Chem 2002; 277:25423–25430.

56. Wu X, Wang F, Rao K, Sellers JR, Hammer JA, 3rd. Rab27a is an essential component of melanosome receptor for myosin Va. Mol Biol Cell 2002; 13:1735–1749.

57. Wu XS, Rao K, Zhang H, Wang F, Sellers JR, Matesic LE, Copeland NG, Jenkins NA, Hammer JA, 3rd. Identification of an organelle receptor for myosin-Va. Nat Cell Biol 2002; 4:271–278.

58. Gerber SA, Rush J, Stemman O, Kirschner MW, Gygi SP. Absolute quantification of proteins and phosphoproteins from cell lysates by tandem MS. Proc Natl Acad Sci U S A 2003; 100:6940–6945.

59. Kirkpatrick DS, Gerber SA, Gygi SP. The absolute quantification strategy: a general procedure for the quantification of proteins and post-translational modifications. Methods 2005; 35:265–273.

60. Lee JS, Caruso JA, Hubbs G, Schnepp P, Woods J, Fang J, Li C, Zhang K, Stemmer PM, Jena BP, Chen X. Molecular architecture of mouse and human pancreatic zymogen granules: protein components and their copy numbers. Biophysics Reports 2018; 4:94–103.

61. Takamori S, Holt M, Stenius K, Lemke EA, Gronborg M, Riedel D, Urlaub H, Schenck S, Brugger B, Ringler P, Muller SA, Rammner B, et al. Molecular anatomy of a trafficking organelle. Cell 2006; 127:831–846.

62. Arvan P, Castle D. Sorting and storage during secretory granule biogenesis: looking backward and looking forward. Biochem J 1998; 332:593–610.

63. Guo Y, Sirkis DW, Schekman R. Protein sorting at the trans-Golgi network. Annu Rev Cell Dev Biol 2014; 30:169–206.

64. Schrader M. Membrane targeting in secretion. Subcell Biochem 2004; 37:391–421.

65. Gomez-Lazaro M, Rinn C, Aroso M, Amado F, Schrader M. Proteomic analysis of zymogen granules. Expert Rev Proteomics 2010; 7:735–747.

66. Williams JA, Chen X, Sabbatini ME. Small G proteins as key regulators of pancreatic digestive enzyme secretion. Am J Physiol Endocrinol Metab 2009; 296:E405–414.

67. Kalus I, Hodel A, Koch A, Kleene R, Edwardson JM, Schrader M. Interaction of syncollin with GP-2, the major membrane protein of pancreatic zymogen granules, and association with lipid microdomains. Biochem J 2002; 362:433–442.

68. Kumazawa-Inoue K, Mimura T, Hosokawa-Tamiya S, Nakano Y, Dohmae N, Kinoshita-Toyoda A, Toyoda H, Kojima-Aikawa K. ZG16p, an animal homolog of beta-prism fold plant lectins, interacts with heparan sulfate proteoglycans in pancreatic zymogen granules. Glycobiology 2012; 22:258–266.

69. Valentijn JA, Gumkowski FD, Jamieson JD. The expression pattern of rab3D in the developing rat exocrine pancreas coincides with the acquisition of regulated exocytosis. Eur J Cell Biol 1996; 71:129–136.

70. Ohnishi H, Samuelson LC, Yule DI, Ernst SA, Williams JA. Overexpression of Rab3D enhances regulated amylase secretion from pancreatic acini of transgenic mice. J Clin Invest 1997; 100:3044–3052.

71. Thévenod F. Channels and transporters in zymogen granule membranes and their role in granule function: a critical assessment. The Pancreapedia: Exocrine Pancreas Knowledge Base, 2020. DOI: 10.3998/panc.2020.09.

72. Rodriguez OC, Cheney RE. Human myosin-Vc is a novel class V myosin expressed in epithelial cells. J Cell Sci 2002; 115:991–1004.

73. Freedman SD, Scheele GA. Regulated secretory proteins in the exocrine pancreas aggregate under conditions that mimic the trans-Golgi network. Biochem Biophys Res Commun 1993; 197:992–999.

74. Leblond FA, Viau G, Laine J, Lebel D. Reconstitution in vitro of the pH-dependent aggregation of pancreatic zymogens en

route to the secretory granule: implication of GP-2. Biochem J 1993; 291:289–296.

75. Dartsch H, Kleene R, Kern HF. In vitro condensation-sorting of enzyme proteins isolated from rat pancreatic acinar cells. Eur J Cell Biol 1998; 75:211–222.

76. Orci L, Ravazzola M, Amherst M, Perrelet A, Powell SK, Quinn DL, Moore HP. The trans-most cisternae of the Golgi complex: a compartment for sorting of secretory and plasma membrane proteins. Cell 1987; 51:1039–1051.

77. Ronzio RA, Kronquist KE, Lewis DS, MacDonald RJ, Mohrlok SH, O'Donnell JJ, Jr. Glycoprotein synthesis in the adult rat pancreas. IV. Subcellular distribution of membrane glycoproteins. Biochim Biophys Acta 1978; 508:65–84.

78. LeBel D, Beattie M. The major protein of pancreatic zymogen granule membranes (GP-2) is anchored via covalent bonds to phosphatidylinositol. Biochem Biophys Res Commun 1988; 154:818–823.

79. Scheele GA, Fukuoka S, Freedman SD. Role of the GP2/THP family of GPI-anchored proteins in membrane trafficking during regulated exocrine secretion. Pancreas 1994; 9:139–149.

80. Colomer V, Kicska GA, Rindler MJ. Secretory granule content proteins and the luminal domains of granule membrane proteins aggregate in vitro at mildly acidic pH. J Biol Chem 1996; 271:48–55.

81. Jacob M, Laine J, LeBel D. Specific interactions of pancreatic amylase at acidic pH. Amylase and the major protein of the zymogen granule membrane (GP-2) bind to immobilized or polymerized amylase. Biochem Cell Biol 1992; 70:1105–1114.

82. Yu S, Michie SA, Lowe AW. Absence of the major zymogen granule membrane protein, GP2, does not affect pancreatic morphology or secretion. J Biol Chem 2004; 279:50274–50279.

83. Antonin W, Wagner M, Riedel D, Brose N, Jahn R. Loss of the zymogen granule protein syncollin affects pancreatic protein synthesis and transport but not secretion. Mol Cell Biol 2002; 22:1545–1554.

84. Aroso M, Agricola B, Hacker C, Schrader M. Proteoglycans support proper granule formation in pancreatic acinar cells. Histochem Cell Biol 2015; 144:331–346.

85. Borta H, Aroso M, Rinn C, Gomez-Lazaro M, Vitorino R, Zeuschner D, Grabenbauer M, Amado F, Schrader M. Analysis of low abundance membrane-associated proteins from rat pancreatic zymogen granules. J Proteome Res 2010; 9:4927–4939.

86. Williams JA, Yule DI. Stimulus-secretion coupling in pancreatic acinar cells. In: Johnson LR, Kaunitz JD, Said HM, Ghishan FK, Merchant JL, Wood JD, eds. Physiology of the gastrointestinal tract. Volume 1. 5th ed. Amsterdam: Elsevier Inc., 2012:1361–1398.

87. Valentijn JA, Valentijn K, Pastore LM, Jamieson JD. Actin coating of secretory granules during regulated exocytosis correlates with the release of rab3D. Proc Natl Acad Sci U S A 2000; 97:1091–1095.

88. Gaisano HY, Dalai S, Takahashi T. Physiologic exocytosis in pancreatic acinar cells and pathologic fusion underlying pancreatitis. The Pancreapedia: Exocrine Pancreas Knowledge Base, 2020. DOI: 10.3998/panc.2020.04.

89. Hou Y, Ernst SA, Lentz SI, Williams JA. Genetic deletion of Rab27B in pancreatic acinar cells affects granules size and has inhibitory effects on amylase secretion. Biochem Biophys Res Commun 2016; 471:610–615.

90. Hou Y, Ernst SA, Stuenkel EL, Lentz SI, Williams JA. Rab27A is present in mouse pancreatic acinar cells and is required for digestive enzyme secretion. PLoS One 2015; 10:e0125596.

91. Geron E, Schejter ED, Shilo BZ. Directing exocrine secretory vesicles to the apical membrane by actin cables generated by the formin mDia1. Proc Natl Acad Sci U S A 2013; 110:10652–10657.

Chapter 12

Channels and transporters in zymogen granule membranes and their role in granule function: A critical assessment

Frank Thévenod

Institute of Physiology, Pathophysiology & Toxicology, Centre for Biomedical Training and Research (ZBAF), Faculty of Health, University of Witten/Herdecke, D-58453 Witten, Germany

Introduction

An update of this review on pancreatic zymogen granule (ZG) channels and transporters and their function is timely since its first publication in 2015[1]. The advent of the post-genomic era has led – similar to other areas of cell biology – to the discovery of a number of ZG transport proteins and the characterization of their function in pancreatic acinar secretion. Recent studies have combined functional and molecular approaches to identify ZG channels and transporters and their contribution to pancreas physiology and pathology. A better understanding of the physiology of pancreatic acinar cell secretion and the role played by ZG in this process is prerequisite to comprehend the pathogenesis of major pancreatic disorders, such as pancreatitis, cystic fibrosis or cancer[2]. Yet the fact that only a very limited number of studies have been published in this area of research is surprising, despite tremendous progress of knowledge and methodologies to investigate the molecular and cellular biology and physiology of the pancreas[3].

In the following paragraphs, I review significant advances in the characterization of ZG channels and transporters in the last two decades and discuss their putative role in pancreatic acinar secretion (**Figure 1**). The review ends with an outlook on unsolved issues that require clarification to move the field forward.

Proteomics

In the past, ion pathways in ZG were either characterized in permeabilized acini in which the effect of the ionic composition of the "extended cytosol" on stimulated secretion was investigated, and/or by recording macroscopic ion fluxes using an osmotic swelling assay (and lysis as an end-point) in suspensions of isolated ZG, and has been previously exhaustively reviewed[4]. That piece of work has provided important 'background' knowledge for the interpretation of the post-genomic work (see below).

A key advance in the last decade has been the widespread availability of proteomics that allows the simultaneous identification of large numbers of expressed proteins. The use of proteomics for analysis of ZG and their membranes has propelled the identification of channels and transporters of pancreatic ZG although various caveats must be contemplated. Several proteomic studies have been published that used highly purified ZG membranes (ZGM) to identify cloned transporters and channel proteins[5–13]. In most of these studies, granules were disrupted by different techniques, and membranes were subjected to carbonate and/or bromide extraction, which is a standardized and reliable procedure to obtain pure membranes without peripheral proteins[14]. A critical step in this isolation procedure represents the initial "purification" of ZG by either differential or continuous Percoll gradient centrifugation. These methods yield fractions that are enriched 4–8-times in α-amylase compared to homogenate, which is close to the theoretical limit of purity[15]. Yet, it is practically impossible to avoid contaminations by lysosomes[6], plasma membranes[16] or membranes of other (disrupted) organelles[6, 13] (also see electron micrographs in references[14, 16, 17]). These contaminants are, however, a major drawback for proteomic analyses because even minimal contaminations by membranes originating from other organelles or plasma membranes may be prominent in a proteomic analysis of ZGM because of proteins that are highly expressed in these contaminating membranes. This issue needs to be considered, particularly for "mitochondrial" proteins such as voltage-dependent anion channels (VDACs) and subunits of ATP-synthase that have been detected in ZGM in several independent studies[5, 7, 10, 11], or for "plasma membrane" proteins, such as the α-subunit of Na^+/K^+-ATPase[6]. Hence, these particular observations need to be taken with caution. In addition, detection of a particular channel or transporter in ZGM may not infer any physiological/functional significance of that identified protein in mature ZG (see the paragraph on *Vacuolar-type H^+-ATPase*). Finally, it should be born in

e-mail: frank.thevenod@uni-wh.de

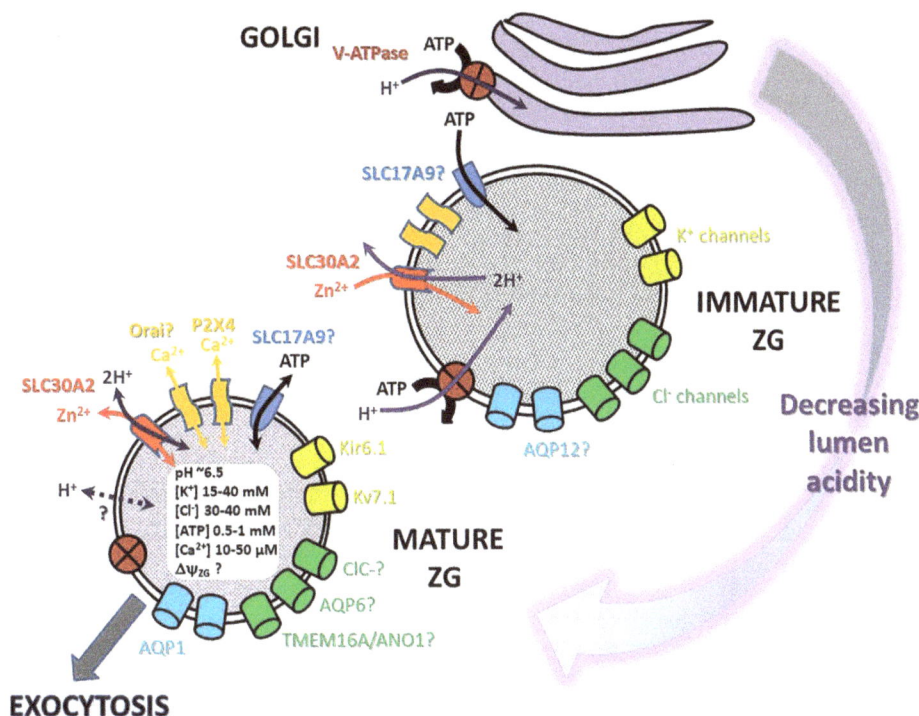

Figure 1. The model summarizes current evidence (strong or weak) for the presence of ion channels and transporters in zymogen granules (ZG) of pancreatic acinar cells. The Figure also describes relevant ion concentrations in mature ZG and emphasizes the decreasing lumen acidity during maturation along the compartments of the secretory pathway that participate in ZG maturation (for further details, see text).

mind that several cloned channel proteins and transporters that have been identified by functional or immunological approaches have not been detected by proteomic analyses. Some studies even claim that no transmembrane proteins or classical membrane-anchored proteins are expressed in ZGM[12]. But this conclusion may be due to the relative insensitivity of the proteomic approaches used because of low expression levels of transport proteins in ZGM.

Interestingly, among a variety of putative novel transporters and channels identified in ZGM by Rindler and coworkers[6], the ATP-gated receptor P2X4 deserves special attention. Indeed, P2X4 has been recently identified in lysosomes and lamellar bodies where it operates as a nonselective cation channel that is inhibited by luminal acidity, while increasing luminal pH in the presence of ATP elicits channel activation[18, 19]. These functional properties of a P2X4 ATP receptor located in ZGM - if not due to lysosomal contamination - are attractive for ZG exocytosis, considering the fact that maturing ZG lose their acidity (see the paragraph on *Vacuolar-type H⁺-ATPase*). Moreover, the P2X4 cation channel is activated by ATP concentrations that have been measured in ZG (see the paragraph on *Vesicular nucleotide transporter SLC17A9* for a discussion of this concept). Nonetheless, the detection of the P2X4 ATP receptor offers an example of a potentially

novel aspect of ZG physiology that is derived from proteomic approaches. Hence, P2X4 is a promising candidate for future studies on ZG exocytosis.

Channels

H₂O Channels

Digestive macromolecules are packaged and condensed together with various osmotically active ions and small organic molecules in the lumen of ZG during maturation, where granular Cl^- and K^+ concentrations display lower concentrations than in the cytosol[20, 21]. It is a long-standing observation that isolated ZG and other secretory granules suspended in isotonic sugar or salt solutions, such as KCl, remain stable for hours[22, 23]. Consequently, the stability of secretory granules in isotonic solutions indicates that the intragranular osmolarity does not exceed cytosolic osmolarity and/or that granules exhibit a low basal permeability for H_2O, K^+ and Cl^- *in situ* and *in vitro*. Reduction of extragranular osmolarity below 245 mOsmol/l[23], however, or addition of cation or anion ionophores to isolated ZG suspended in iso-osmotic KCl buffers[22] elicits granular swelling and lysis. In the latter case, ionophores generate an electrochemical membrane potential that activates endogenous conductive pathways for counterions, thus allowing entry of osmotically active KCl into the granular lumen, which attract H_2O and induce swelling (and lysis) of secretory granules[22, 24]. Thus, these observations imply that H_2O permeable pathways, e.g. aquaporins, are present in ZGM.

The expression and function of aquaporins in pancreas has been recently reviewed[25, 26]. Aquaporin 1 (AQP1) has been identified in rat pancreatic ZGM by a full range of immunodetection methods, namely immunoblotting, immunofluorescence confocal microscopy and immunogold labeling[27]. Furthermore, swelling of ZG suspended in hypo-osmotic KCl solution was detected by atomic force microscopy (AFM) when 40 µM GTP was added and blocked by Hg^{2+}, an inhibitor of AQP1, as well as by a functional AQP1 antibody that was raised against the carboxyl-terminal domain of AQP1. These changes of ZG swelling were paralleled by corresponding effects on fluxes of ^3H_2O[27]. The effect of GTP on AQP1 activation is assumed to be mediated by interaction of GTP with a complex of several proteins that were co-immunoprecipitated with an antibody against AQP1 and identified by immunoblotting. They included the GTP-binding protein subunit $G_{\alpha i3}$, the inwardly rectifying K^+ channel IRK8 (Kir6.1/KCNJ8), the voltage-gated Cl^- channel ClC-2 and phospholipase A_2 (PLA_2)[28]. Furthermore, exposure of ZG to either the K^+ channel blocker glyburide, or the PLA_2 inhibitor ONO-RS-082, blocked GTP-induced ZG swelling measured by AFM, whereas the non-specific Cl^- channel inhibitor 4,4-diisothiocyanatostilbene-2,2-disulfonate (DIDS) had no effect.

Interestingly, proteomic analyses have identified AQP8, but not other AQPs, in rat ZGM[6], thus confirming previous immunohistochemical studies in rat pancreas tissues[29]. AQP8 may be a genuine ZG membrane protein, however a contamination by mitochondria cannot be ruled out, which express AQP8 as well[30, 31].

An intriguing observation has been made by Ohta et al.[32] who found AQP12 expressed in the rough endoplasmic reticulum (rER) of rat pancreatic acinar cells and extends a previous study demonstrating expression of AQP12 in rat pancreatic acinar cells by in situ hybridization and reverse transcription-polymerase chain reaction (RT-PCR)[33]. In rats treated with the secretagogue cholecystokinin octapeptide (CCK-8), Otah et al.[32] detected AQP12 in the rER and also on the membranes of ZG near the rER. Furthermore, AQP12 knockout mice were more prone to caerulein-induced pancreatitis[32]. The authors suggested that AQP12 may be involved in the mechanisms underlying the proper generation, maturation, and trafficking of ZG in the secretory pathway and that adequate H_2O flux may be necessary for this process to function adequately.

Altogether, it seems highly likely that AQP H_2O channels are expressed in pancreatic ZGM because ZG swell and lyse in hypotonic solutions, although the AQP isoform(s) involved remain(s) uncertain. Moreover, the physiological role of AQPs in the complex process of secretory granule maturation, trafficking and exocytosis required for secretagogue-induced pancreatic acinar enzyme and fluid secretion are poorly understood.

K^+ Channels

In pancreatic acinar cells, ZG are about 4–8-fold lower in K^+ compared to the cytosol, as measured by X-ray microanalysis[20] or using a potassium-sensitive fluorescence dye[34]. Hence, any K^+ conductive pathway should allow flux of K^+ into the ZG lumen and increase its osmotic load. Based on the work by Hopfer and coworkers[35], I proposed in 1992 that (ATP- and glibenclamide-sensitive) K^+ channels are expressed in ZGM where they contribute to osmotic swelling of isolated pancreatic ZG[24]. This observation was soon confirmed by others[36]. Subsequent studies have suggested that glibenclamide may modulate ZG K^+ permeability indirectly, possibly by binding to a 65-kDa multidrug resistance P-glycoprotein (ABCB1)-like regulatory protein[37], and/or the ZG membrane-associated protein ZG-16p[38]. Additional support for the presence of K^+ channels in secretory granules came from studies with mucin granules from goblet cells and suggested a Ca^{2+}/K^+ ion-exchange mechanism via parallel operation of a Ca^{2+} channel and a Ca^{2+}-activated K^+ channel[39]. Ca^{2+}/K^+ ion-exchange has been subsequently proposed to play a critical role in the premature activation of trypsin in ZG and development of acute pancreatitis[34].

Jena and coworkers[28, 40] have identified an inwardly rectifying K^+ channel IRK8 (Kir6.1/KCNJ8) in ZGM by immunoblotting and measured whole vesicle currents of patched single ZG with KCl as permeating ions that were reduced by ~15% by either 40 µM quinidine or 20 µM glibenclamide. Proteomic analyses of purified rat ZGM have detected the K^+ channel TWIK-2/KCNK6[6], a member of a two-pore-domain K^+ channel family that produces constitutive inward rectifying K^+ currents of weak amplitude and that is highly expressed in the pancreas[41, 42]. However, their functional properties and their wide distribution suggest that these channels are mainly involved in the control of background K^+ conductive pathways in the plasma membranes of many cell types. Hence, it cannot be excluded that plasma membranes may have contaminated ZGM in the proteomic study reporting TWIK-2/KCNK6 expression in ZGM[6] (see also the paragraph on *Proteomics*).

Kv7.1 (KvLQT1/KCNQ1) is a very low-conductance, voltage-gated six-membrane-spanning K^+ channel distributed widely in epithelial and non-epithelial tissues, including pancreatic acinar cells[43, 44]. KCNQ1 channels associate with all five members of the KCNE β-subunit family, resulting in a β-subunit-specific change of the current characteristics[43, 44]. Kv7.1 K^+ channels are blocked by the chromanol 293B[45] and by the more potent and selective derivative HMR1556[46]. Indeed, we showed that 293B selectively blocks osmotic swelling of ZG induced by activation of the K^+ permeability (IC_{50} ~10 μM)[47]. Upon incorporation of ZGM into planar bilayer membranes, K^+ selective channels were detected with linear current-voltage relationships. Single channel analysis identified several K^+ channel groups with distinct channel behaviors. K^+ channels were inhibited by 100μM 293B or HMR1556, but not by the $_{maxi}K_{Ca}$ channel inhibitor, charybdotoxin (5nM). Kv7.1 protein was demonstrated by immunoperoxidase labelling of rat pancreatic tissue, immunogold labelling of ZG and immunoblotting of ZGM. Moreover, 293B also inhibited cholecystokinin (CCK)-induced amylase secretion of permeabilized acini (IC_{50} ~10μM) when applied together with the ZG nonselective cation conductance blocker, flufenamate. Thus these data indicated that Kv7.1 accounts at least partially for ZG K^+ conductance and contributes to pancreatic secretagogue-stimulated enzyme and fluid secretion[47].

A question of some significance for the discussion of ZG K^+ channels is whether the presence of several K^+ channels with distinct biophysical properties in our bilayer experiments[47] argues against the purity of the ZGM used for single channel studies with the planar bilayer technique. However, the observation by others that single patched ZG also display a mixture of channel types[40] supports our model that ZG carry several different cation and anion conductive pathways (reviewed in[4]). To further exemplify the concept that several conductive pathways are localized in ZGM, a study should be cited that investigated the effect of 293B and HMR-1556 on fluid and enzyme secretion induced by acetylcholine in the perfused rat pancreas[48]. Interestingly, the authors were unable to observe any inhibitory effect of Kv7.1 channel blockers on secretion and concluded that Kv7.1 is not essential for secretagogue-mediated secretion of pancreatic acini. These results appear to question a role of Kv7.1 in enzyme secretion of rat pancreatic acini, as described in our study[47]. However, we showed that CCK-stimulated enzyme secretion in permeabilized rat pancreatic acini was only abolished if flufenamate was applied together with 293B (see above), which is a strong indication that several ZG cation channels with functional redundancy are involved in enzyme secretion[47].

In a more recent study, we followed up the role of the ZG K^+ conductance in CCK-induced enzyme secretion and showed that the immunosuppressive drug cyclosporin A (CsA) (but not FK506, another commonly used immunosuppressant) activates ZG K^+ conductance and selectively increases the open probability of ZG K^+ channels incorporated into planar bilayers[49]. The electrophysiological (and other) data let us conclude that CsA has a direct effect on the underlying K^+ channel[49]. CsA also increased basal enzyme release of permeabilized rat pancreatic acini, but

did not enhance CCK-induced enzyme secretion[49]. These results are concordant with previous studies that had investigated the role of CsA on exocrine pancreas function[50, 51], although these authors suggested a different mode of action of CsA. We speculated that selective activation of ZG K^+ channels by the immunosuppressant CsA may cause an increased intracellular release of digestive enzymes due to osmotic swelling and lysis of ZG *in situ*, and that this pathological process may account for the increased incidence of post-transplant allograft pancreatitis following hypoxia-reperfusion injury in patients treated with CsA (but not with FK506)[49].

In summary, although contamination by other organelles and membranes cannot be completely excluded, it seems likely that several K^+ channels with different biophysical properties and pharmacological profiles are expressed in ZGM. This conclusion is supported by different electrophysiological techniques (incorporation of ZGM in planar bilayers and single ZG patch techniques), as well as by the detection of several different K^+ channel proteins by immunological or proteomic approaches. However, as for AQPs, the significance of ZG K^+ channels for the physiology of secretion by pancreatic acinar cells remains unclear.

Cl^- Channels

ClC Cl^- channels and transporters

X-ray microanalysis studies of pancreatic acinar cells have demonstrated that the ZG Cl^- concentration is about half the cytosolic Cl^- concentration[20]. When measured with a Cl^- sensitive fluorescent dye, cytosolic Cl^- concentrations vary between 63 and 83 mM in resting cells,[48, 52], suggesting that ZG Cl^- concentrations are well below those values. Early osmotic swelling studies with isolated ZG postulated the presence of regulated Cl^- channels (reviewed in reference[4]). Subsequently, several candidates for ClC ion channels[53] have been identified in ZGM that have been linked to ZG Cl^- conductive pathways. The earliest report suggested the presence of the Cl^- channel ClC-2 in ZGM[54]. Using the whole-cell patch-clamp technique, the authors identified a hyperpolarization-activated Cl^- current in isolated pig pancreatic acinar cells. This current had the characteristic biophysical properties of ClC-2[55] and was activated by extracellular hypotonicity, similar to ClC-2[56], suggesting a role in volume regulation. An antiserum raised against the C-terminus of ClC-2 localized the channel to secretory granules containing amylase by immunofluorescence microscopy of acinar cells, and the authors suggested that the channel protein incorporates into the apical plasma membrane following granule exocytosis[54]. By measuring whole vesicle currents of patched single ZG and using KCl as permeating ions, Jena and coworkers[40] detected a conductance that could be reduced by the non-specific anion transport blocker DIDS (40 μM) by ~50% and concluded that Cl⁻ channels/transporters were present in ZGM. Immunoblots of ZG with ClC-2 and ClC-3 antibodies showed several bands, including immunoreactive bands of ~100 kDa, as expected for ClC-2 and ClC-3[57, 58]. However, ClC-2 (as well as ClC-1) is a plasma membrane Cl⁻ channel, whereas ClC-3 through ClC-7 2Cl⁻/H⁺ exchangers are predominantly localized in intracellular vesicles of the endosomal/lysosomal system[53], further arguing against a significant role of ClC proteins in the secretory pathway. Yet proteomic analyses of rat ZGM identified ClC-3 and ClC-5, thus supporting the notion that ClC channel proteins are expressed in ZGM[6, 59]. Although the lack of pancreas or salivary phenotype of ClC-2[58] and ClC-3 knockout mice[60] does not exclude that ZG express these channel proteins - also because of the assumed functional redundancy of ZG ion channel proteins[4] - the evidence provided for ClC-2 and ClC-3 expression in secretory granules largely depends on immunological studies. However, these studies have to be taken with caution as long as negative control experiments with ZG from knockout tissues have not been performed, thus proving specificity of the antibodies (see[61] for a critical discussion). Hence, the decisive experiments proving expression of ClC proteins in ZG are still missing.

Other candidate Cl⁻ channels

Another possible candidate for ZG anion channels may be AQP6 that is unique among the AQPs because it is activated by Hg²⁺ rather than being inhibited as for other AQPs; and it does not operate as a water channel, but as an anion permeable channel with the halide permeability sequence $NO_3^- > I^- \gg Br^- > Cl^- \gg F^-$ [62] (reviewed in[63]). Analysis of sequence alignments led to the identification of a critical amino acid residue responsible for anion permeability: An asparagine residue (Asn-60) at the position corresponding to Gly-57 in other AQP H₂O channels is responsible for anion permeation[62, 64, 65]. Interestingly, AQP6 is almost exclusively localized in acidic intracellular organelles where it co-localizes with vacuolar-type H⁺-ATPases (V-ATPases)[66] and is activated by acidic pH[62, 64] (reviewed in[63]). Although no evidence is available for AQP6 localization in pancreatic ZGM, AQP6 has been detected in secretory granules of parotid acinar cells by immunological methods[67]. Furthermore, osmotic swelling and lysis of secretory granules suspended in iso-osmotic KCl buffer displayed the same conductive properties as the native AQP6 channel when induced by the AQP6 activator Hg²⁺ [68]. Interestingly, osmotic swelling showed a Hg²⁺-independent component (about 33%) and Hg²⁺-dependent swelling was about 50% DIDS-sensitive, again suggesting at least three different Cl⁻ conductive pathways (however without functional redundancy) in these granules.

A family of putative anion channel proteins predominantly expressed in epithelial tissues that are gated by Ca²⁺ were named CLCA (chloride channel, Ca²⁺-activated), but rather represent chloride channel accessory modulatory proteins (see below and[69]). Mouse CLCA1/2 (mCLCA1/2) was detected in ZGM with a variety of immunological techniques using several antibodies that were thoroughly characterized[70, 71]. Nevertheless, the gold standard of antibody characterization (namely to test the specificity of antibodies in knockout tissues used as negative controls) was not performed, because mCLCA1/2 knockout mice were not available at the time of those studies. In one study, an anion permeability was demonstrated using a ZG osmotic swelling assay whose permeability sequence, Ca²⁺ dependence and inhibitor sensitivity was reminiscent of CLCA associated anion currents[70]. It is now established that hCLCA1, the human homolog of mCLCA3 that is expressed in mucous cells of the airway epithelium is a secreted protein rather than being a Cl⁻ channel[72] although it seems to modulate the conductance of endogenous Ca²⁺-activated Cl⁻ channels[73]. It is likely that all the CLCA proteins cannot act in solo as ion channels but instead interact with other proteins in a possible signaling or regulatory capacity (reviewed in[74]). Their role in ZG function is unclear and awaits further elucidation.

Whether a member of the TMEM16/anoctamin (ANO) protein family of Ca²⁺-activated Cl⁻ channels and phospholipid scramblases[75, 76] contributes to ZG anion permeability remains an interesting question[77], especially because the Ca²⁺-activated Cl⁻ channel TMEM16A/ANO1 is expressed in the area of pancreatic acinar ZG[78]. Interestingly, TMEM16A/ANO1 is permeated by HCO₃⁻, a selectivity process which is thought to be regulated by the Ca²⁺/calmodulin complex[79]. Hence, TMEM16A/ANO1-dependent HCO₃⁻ secretion through ZGM fused with the apical plasma membrane may also contribute to solubilization of exocytosed secretory products[80] during secretion of pancreatic acinar cells stimulated by Ca²⁺-dependent secretagogues[81].

In summary, although there is some suggestive evidence for the expression of intracellular ClC Cl⁻ channel proteins, AQP6 and TMEM16A/ANO1 in ZGM, no strong candidate for a ZG Cl⁻ channel has been identified.

Ca²⁺ Channels

Pancreatic ZG are a significant Ca²⁺ store[20, 21, 82], although the majority of ZG Ca²⁺ appears to be bound to proteins[34, 82, 83]. Depending on the Ca²⁺ sensor used, the free Ca²⁺ concentration in the ZG lumen varies between 9 μM[34] and 55 μM[83]. How Ca²⁺ accumulates in ZG is unclear, but Ca²⁺ uptake could occur in compartments of the secretory pathway upstream of secretory granules, e.g. the rER or the Golgi apparatus, and involve active transport mediated by different Ca^{2+}-ATPases, such as SERCA (Sarcoplasmic/endoplasmic Reticulum Calcium ATPase) and SPCA (Secretory Pathway Calcium ATPase) pumps[84]. Ca^{2+} released from ZG into the pancreatic juice may have a paracrine function on pancreatic duct cells that express Ca^{2+}-sensing receptors on their apical surface, as proposed by Bruce *et al.*[85].

Secretagogues have been shown to evoke local increases of cytosolic Ca^{2+} in the secretory pole region of pancreatic acinar cells[86]. Following studies demonstrating IP$_3$-induced Ca^{2+} release from acidic neuroendocrine chromaffin granules induced by the second messenger inositol 1,4,5-trisphosphate (IP$_3$)[87], experimental evidence has suggested that secretory granules from both endocrine and exocrine cells, including pancreatic acinar and airway goblet cells, are IP$_3$-sensitive Ca^{2+} stores that play a significant role in Ca^{2+}-dependent secretion (reviewed in[88, 89]. In particular, using ZG isolated by differential centrifugation or intact pancreatic acinar cells, Petersen and colleagues have demonstrated release of Ca^{2+} from an acidic store in the apical pole of acinar cells induced by IP$_3$, cyclic-ADP ribose (cADPR) or nicotinic acid adenine dinucleotide phosphate (NAADP) (i.e. suggestive of functional IP$_3^-$ and ryanodine receptors) that they associated with ZG[83, 90, 91]. Moreover, Gerasimenko and coworkers[92, 93] (reviewed in[88]) hinted at the importance of this acidic ZG store for the pathogenesis of alcohol-induced pancreatitis. However, two independent studies have challenged these observations in pancreatic[17] and parotid acinar cells[94]. These authors provided functional and biochemical evidence indicating that ZGM do not express IP$_3$- or ryanodine receptors, provided that ZG are not contaminated by other organellar structures, especially ER or mitochondria. Although Gerasimenko *et al.*[83] had shown a three-fold enrichment of the ZG marker α-amylase in the ZG preparation, the assay used to exclude ER contamination relied on the fluorescent dye DiOC6(3) that is generally applied to assess $\Delta\Psi$ of energized mitochondria [*sic*!][95]. Using the ZG isolation procedure of Gerasimenko *et al.*[83], Yule *et al.*[17] showed disrupted mitochondria and ER membranes by electron microscopy that contaminated the ZG preparation. Interestingly, Gerasimenko *et al.*[96] revisited this issue and proposed that the main apical Ca^{2+} pools that are responsible for release of Ca^{2+} by all three intracellular Ca^{2+} messengers IP$_3$, cADPR and NAADP are the ER and ill-defined "acidic stores", where these second messengers elicit Ca^{2+} release by activation of the Ca^{2+} channels IP$_3$- and ryanodine receptors. The two stores interact with each other via Ca^{2+}-induced Ca^{2+} release and can be separated using pharmacological tools. The ER relies on SERCA pumps that can be blocked by the specific inhibitor, thapsigargin; Ca^{2+} uptake in the acidic store relies on a bafilomycin-sensitive V-ATPase and probably a Ca^{2+}/H^+ exchanger, as originally proposed by Thévenod & coworkers for an

IP$_3$-sensitive ER Ca^{2+} pool[97, 98]). Gerasimenko *et al.* suggested that a large part of the acidic store is "co-localized" with ZG, however, other acidic organelles like endosomes and lysosomes may also participate in intracellular Ca^{2+} signaling[96]. This conclusion is more cautious and accurate than the earlier proposal of ZG as acidic Ca^{2+} stores but is still not satisfying. Three lines of evidence argue against ZG being an acidic Ca^{2+} store involved in secretagogue-induced apical Ca^{2+} release. Firstly, the majority of ZG are not acidic[99] (see also the paragraph on *Vacuolar-type H$^+$-ATPase*); secondly, ER also co-localizes with ZG[100, 101]; thirdly, an acidic IP$_3$-sensitive ER Ca^{2+} pool has also been described that takes up Ca^{2+} via parallel activity of a V-type H^+ pump and a Ca^{2+}/H^+ exchanger, both in pancreatic[97, 98] and parotid acinar cells[102]. It also appears unlikely that ZG are an IP$_3$ sensitive Ca^{2+} pool because all proteomic studies of ZGM so far did not detect any IP$_3$ receptors[5-13]. Hence, it remains an open question whether other mechanisms of Ca^{2+} release are operative in ZG.

The ATP-gated cation channel P2X4 is Ca^{2+} permeable[103] and has been identified in ZGM[6] (see the paragraphs on *Proteomics* and *Vesicular nucleotide transporter SLC17A9*). This receptor could play a role in Ca^{2+} release from weakly acidic Ca^{2+} stores[104], such as ZG, because P2X4 receptors are normally inhibited at acidic pH and are activated when intraluminal pH increases[18, 19, 103].

Finally, Hille and coworkers[105] have shown a release of Ca^{2+} via "store-operated" Orai channels in the membrane of secretory granules from neuroendocrine PC12 cells (interestingly, they did not detect IP$_3$ receptors in these secretory granules, in contrast to previous studies[89]). The granule Orai channels are activated by regulator stromal interaction molecule 1 (STIM1) on the ER and may raise local cytoplasmic Ca^{2+} concentrations for refilling of Ca^{2+} stores of the ER and promotion of exocytosis[105]. A subsequent study in RBL-2H3 mast cells identified Orai-2 on secretory granules[106]. Ca^{2+} release from a Ca^{2+} store induced by antigen stimulation was attenuated by Orai-2 silencing, while that induced by thapsigargin was not affected. Furthermore, exocytotic release induced by antigen stimulation was inhibited in Orai-2 silenced cells[106].

Transporters

Vacuolar-type H$^+$-ATPase (V-ATPase)

One of the hallmarks of many vesicular compartments is the expression of a V-ATPase that operates to acidify the vesicular lumen[107]. Subunits of the V-ATPase have been clearly identified in the membrane of rat pancreas ZG by immunological methods[108] and by proteomic analyses[5, 6, 10, 11]. Nevertheless, conflicting results were obtained in several early studies when measuring luminal pH of secretory granules in freshly isolated acini or purified granules.

Using fluorescence microscopy and the weak base acridine orange (AO) that accumulates in acidic organelles and changes its fluorescence from green to orange when concentrated, ZG were found to be acidic whereas parotid secretory granules showed a neutral pH[23, 109–111]. To resolve this issue, in a landmark study Orci et al.[99] used the probe 3-(2,4-dinitroanilino)-3-amino-N-methyldipro-pylamine (DAMP), a weak base that accumulates in the lumen of acidic organelles, where it can be fixed[112]. Quantitative immunostaining of DAMP in conjunction with electron microscopy was used to measure pH in the secretory pathway of fixed isolated pancreatic acini. The data revealed that condensing vacuoles are acidic but lose their acidity during maturation[99]. To account for these results, the authors speculated that the loss of acidity of maturing ZG may be the consequence of either removal of H^+-pump subunits, inactivation of the V-ATPase, or closing of an anion conductance. It is noteworthy that it took more than 20 years to confirm the observation that the acidity of ZG decreases during maturation, using freshly isolated pancreas acini with AO and live-cell two-photon imaging[113], thus indicating that earlier AO measurements with fluorescence microscopic techniques were not accurate. Of note, the study by Thorn and coworkers[113] also showed that the H^+ concentration of mature granules is not affected by pre-incubation with a blocker of the V-ATPase. However, in contrast to the data by Orci et al.[99], labeling of the cells with the ratiometric pH indicator, Lysosensor Yellow/Blue, suggested that mature secretory granules have a pH of less than 6.0 (yet without proof that the labelled acidic structures are indeed mature ZG!)[113]. Furthermore, upon secretagogue-stimulated secretion, the pH in the acinar lumen never dropped below 7.0, which suggests that ZG pH is not or weakly acidic. To account for this discrepancy, Thorn and coworkers have subsequently described a luminal secretagogue-stimulated HCO_3^- conductance, mediated by TMEM16A/ANO1 that is activated coincidently with ZG fusion and exocytosis[81].

An increase of pH in the secretory pathway during maturation is anomalous because along the secretory pathway luminal pH generally drops from a near-cytosolic value of 7.2 within the ER to 6.7–6.0 along the Golgi complex to approximately 5.2–5.7 in secretory granules and is often associated with a reciprocal decrease of a H^+ permeability (see[114, 115] for review). This is particularly the case in neuroendocrine cells where the increasing electrochemical gradient H^+ gradient generated by the V-ATPase is also used to store cargo (e.g. neurotransmitters or metal ions) above equilibrium with the cytosol[107, 116]. In contrast, mature pancreatic ZG do not store ATP[117] or Zn^{2+}[20] above cytosolic concentrations, which is in line with a reduced acidity of mature ZG. How ZG (as well as parotid secretory granules (see[23, 109]) lose their acidity during maturation, remains to be investigated. This could be the consequence

of a differential activity or expression of V-ATPases and/or differences in H^+ permeability, as suggested by Orci et al.[99]. Moreover, activity of exchange pathways for counterions, e.g. Na^+/H^+-, Ca^{2+}/H^+- or Cl^-/H^+-exchangers[118–121] along the secretory pathway of pancreatic (and parotid) acinar cells could also result in a dissipation of the pH gradient during maturation.

Vesicular nucleotide transporter SLC17A9

SLC17A9 (also named VNUT for Vesicular Nucleotide Transporter) is a vesicular ATP transporter widely expressed in various organs, but predominantly in the adrenal gland, brain, and thyroid gland (the pancreas was not tested)[122] (reviewed in[123, 124]). SLC17A9 mediates concentrative accumulation of ATP in secretory vesicles, such as chromaffin or synaptic vesicles, to a luminal concentration of 100 mM or more[125] (reviewed in references[126, 127]). SLC17A9 protein shows a relative mobility varying between 61 and 68 kDa in SDS-PAGE, depending on the origin of the protein (native or recombinant). When reconstituted into proteoliposomes, SLC17A9 transports ATP (as well as ADP, GTP and UTP as shown by *cis* inhibition experiments) and uses a $\Delta\Psi$ (positive inside) but not a ΔpH (the vesicle lumen being acidic) as driving force for uptake. ATP transport depends on external Cl^- and saturates at ~4 mM external Cl^-. ATP uptake is inhibited by DIDS) with an IC_{50} of ~1.5 µM. Evans blue is even more potent (IC_{50} ~40 nM) whereas 200 µM atractyloside inhibits SLC17A9 only in the presence of Mg^{2+}. All these properties are similar to those that have been reported for ATP uptake in chromaffin granules and granule membrane ghosts[127]. Although ATP transport rate appeared to be independent of Ca^{2+} or Mg^{2+}, subsequent studies showed that SLC17A9 also transports both divalent cations in an ATP- and Cl^-- dependent manner, which is driven by an inside-positive $\Delta\Psi$, but not by an inside-positive ΔpH, and with a similar inhibitor sensitivity as ATP transport[128]. Hence, divalent cation transport probably occurs by complexation with ATP and may contribute to vesicular accumulation of divalent cations in secretory vesicles.

It has been suggested that during pancreas secretion evoked by secretagogues, ATP is released from ZG into the primary fluid to mediate paracrine signaling between acinar and duct cells and thereby coordinates ductal and acinar fluid secretion[129]. This may occur via purinergic receptors expressed on the apical membrane of duct cells[130]. Indeed, pancreatic ZG do contain ATP. From measurements with freshly isolated ZG, it has been estimated that the *in vivo* concentration of ATP in ZG is 0.5–1 mM ATP[117], a more than 100-times lower concentration than in neuroendocrine secretory vesicles expressing SLC17A9 and a concentration that is also below the average cytosolic concentration of 2–5 mM

ATP[126], thus questioning concentrative ATP transport in ZG. Nevertheless, Novak and coworkers have suggested that SLC17A9 is expressed in the membrane of ZG where it mediates ATP uptake[117, 131]. Their experimental evidence supports expression of SLC17A9 in ZGM, e.g. by detection of a ca. 65-kDa protein with a SLC17A9-specific antiserum, co-localization of immunolabeling for SLC17A9 with the ZG membrane marker Rab3D, or inhibition of ATP uptake into ZG by 100 μM DIDS and Evans blue[117, 131]. Furthermore, induction of rat AR42J acinar cells into a secretory phenotype by dexamethasone administration increases carbachol-induced ATP release and SLC17A9 expression. In Haanes' and Novak's work on isolated ZG, ATP uptake depends on external Mg^{2+} and high Cl^- [117, 131]. Moreover, ATP uptake is driven by an inside-positive ΔpH (either generated by an artificial pH gradient or as demonstrated by bafilomycin A1-sensitive uptake) and by an inside-positive K^+ diffusion potential. Furthermore, ATP uptake shows no *cis* inhibition by ADP or UTP[117]. As a matter of fact, most of these properties differ from the transport properties of SLC17A9[122] (reviewed in[123, 124]). It may be argued that results obtained in intact ZG may differ from transport of a protein reconstituted into proteoliposomes[117]. However, transport of reconstituted SLC17A9 matches transport in intact chromaffin granules[122, 127]. Hence, ATP transport in ZG may occur independently from SLC17A9 expression.

On the other hand, these observations in ZG have gained recent support from studies in lysosomes[18, 132] and lamellar bodies (LB)[19]. SLC17A9 is expressed in these acidic intracellular organelles, which contain ATP concentrations between <0.5 mM (in lysosomes) and ca. 2 mM (in LB). Interestingly, SLC17A9 expression was functionally linked to the ATP-gated cation channel P2X4[19, 133] (reviewed in[134]), which is also expressed in ZGM[6] (see the paragraphs on *Proteomics* and *Ca^{2+} channels*). The study by Fois *et al.*[19] demonstrates a physiological role of lysosomal P2X4 receptors during the secretion of surfactant from alveolar type II epithelial cells. Apparently, LB fusion triggers P2X4 receptors within the LB membrane to generate a highly localized, cytosolic Ca^{2+} signal in the vicinity of the fused LB, which drives expansion of the fusion pore and facilitates surfactant release (reviewed in[134]). LB are likely lysosome-derived but less acidic than juxtanuclear lysosomes by virtue of reduced V-ATPase activity and an increase in proton leak[135], once more emphasizing the importance of electrogenic conductive pathways for ions, in particular H^+, for concentrative ATP uptake in intracellular vesicles.

In conclusion, experiments providing evidence for a causal relationship between ZG ATP accumulation and SLC17A9 are needed to confirm the postulated role of SLC17A9 in (concentrative?) ATP uptake into ZG. This requires, for instance, silencing of the transporter and

measurements of ZG ATP concentration and/or secretagogue-induced ATP release from acinar cells, as well as measurements of the driving force for ATP accumulation, namely $\Delta\Psi$, in ZG. Moreover, it remains to be tested whether SLC17A9 - if it is indeed expressed in ZGM - may also contribute to the accumulation of Ca^{2+} as nucleotide complex in ZG that, apart from the endoplasmic reticulum represent a significant intracellular Ca^{2+} pool[21, 82, 136] (see also the paragraph on *Ca^{2+} channels*). In summary, the available evidence that SLC17A9 is responsible for concentrative ATP uptake into pancreatic ZG awaits further experimental confirmation.

Zinc transporter SLC30A2

The SLC30A2 gene encodes a zinc (Zn^{2+}) transporter (also named ZnT2 for Zinc Transporter 2) that was cloned from a rat kidney cDNA expression library by complementation of a Zn^{2+}-sensitive Baby Hamster Kidney cell line[137] and is highly expressed in pancreas (reviewed in[138–140]). Using ZnT2-GFP fusion proteins and immunofluorescence microscopy, the transporter was localized in Zn^{2+} accumulating acidic intracellular vesicles (as opposed to the Zn^{2+} efflux transporter SLC30A1/ZnT1 that is expressed in the plasma membrane[141]). The authors suggested that ZnT2 contributes to protection against Zn^{2+} toxicity by facilitating Zn^{2+} influx into an endosomal/lysosomal compartment thereby lowering cytosolic Zn^{2+} jointly with plasma membrane ZnT1[137]. Subsequent studies have revealed that, in addition to the pancreas, SLC30A2 is also expressed in mammary gland, prostate, retina, small intestine and kidney. In these organs, it is localized on the membrane of intracellular organelles (e.g. endosomal/lysosomal and secretory vesicles) and sequesters cytoplasmic Zn^{2+} for secretion, storage, or for use in proteins that require Zn^{2+} for their activities (reviewed in[138–140]). Interestingly, a loss-of-function mutation or silencing of SLC30A2 results in reduced Zn^{2+} secretion from mammary gland epithelial cells *in vivo* and *in vitro*[142]. Subsequently, transient neonatal Zn^{2+} deficiency (TNZD) in infants was found to be caused by a variety of loss-of-function mutations in the *SLC30A2* gene in women, resulting in poor secretion of Zn^{2+} into the breast milk. Consequently, infants exclusively breastfed with Zn^{2+}-deficient breast milk develop severe Zn^{2+} deficiency (reviewed in[143]). The transport mechanism of SLC30A2 has been recently elucidated[144]: Similarly to vesicular Zn^{2+} transport mediated by other mammalian SLC30 transporters (SLC30A1 and SLC30A5)[145, 146], SLC30A2 is an electroneutral proton-coupled vesicular antiporter displaying an apparent stoichiometry of two H^+ per Zn^{2+} ion and is driven by the vesicular H^+ gradient generated by V-ATPases[144].

In pancreatic endocrine β-cells, secretory granules accumulate Zn^{2+} almost 40-fold above cytosolic levels, whereas no difference between cytosolic and acinar ZG Zn^{2+} concentration has been found, as measured by X-ray microanalysis[20]. In β-cell granules (and other islet secretory granules), SLC30A8 is the Zn^{2+} transporter that provides Zn^{2+} for maturation of stored macromolecules, such as insulin, and their crystallization before secretion[147], and its possible importance in the etiology of diabetes has been highlighted[148]. Because insulin containing granules are acidic[149–151], SLC30A8-mediated Zn^{2+} accumulation against a concentration gradient is likely driven by the vesicular H^+ gradient generated by V-ATPases and could occur via H^+/Zn^{2+} exchange, as demonstrated for SLC30A1, SLC30A2 and SLC30A5[144–146].

The situation in ZG is much less clear. In rats, Zn^{2+} deficiency caused Zn^{2+} depletion in acinar but not in β-cells[152], which is in line with the ability of β-cells to store Zn^{2+} in insulin granules in a slowly-exchanging intra-granule compartment and also with the apparent inability of pancreatic ZG to accumulate Zn^{2+} [20]. Liuzzi *et al.*[153] showed that Zn^{2+} depletion in mice is associated with a reduction of pancreatic acinar SLC30A1 and SLC30A2, and the authors speculated that SLC30A1/2 may mediate lowering of cytosolic Zn^{2+} in pancreatic acinar cells of Zn^{2+} replete animals. Furthermore, De Lisle and coworkers showed secretion of the cytosolic Zn^{2+}-binding protein metallothionein into pancreatic juice[154]. All these studies led to the hypothesis by Cousins and colleagues that exocytotic secretion of Zn^{2+} by pancreatic acinar cells may be modulated by Zn^{2+} transporter expression, which would thus affect entero-pancreatic Zn^{2+} trafficking and also regulate physiological Zn^{2+} body homeostasis by controlling Zn^{2+} losses into the pancreatic fluid and intestinal excretion of Zn^{2+} [153, 155]. More recently, Cousins and coworkers demonstrated SLC30A2 expression in ZG, and reduction of the Zn^{2+} diet of mice was paralleled by decreases of Zn^{2+} concentrations and SLC30A2 expression in ZG[156]. Rat AR42J acinar cells, when induced into a secretory phenotype with dexamethasone, exhibited increased SLC30A2 expression that was associated with a reduction of cytosolic[65]Zn^{2+} content and an increase of the[65]Zn^{2+} content of ZG. SLC30A2 silencing of AR42J cells by RNAi (~70% mRNA reduction) increased cytoplasmic[65]Zn^{2+} by 36% and decreased ZG[65]Zn^{2+} by 15%, suggesting that SLC30A2 mediates the sequestration of Zn^{2+} into ZG[156]. The relatively low impact of SLC30A2 silencing on ZG Zn^{2+} content may be explained by low ZG Zn^{2+} below cytosolic concentrations, which is likely caused by the loss of acidity of ZG during maturation (see the paragraph on *Vacuolar-type H^+-ATPase*). Alternatively, other Zn^{2+} transporters may also contribute to Zn^{2+} fluxes across the ZG membrane.

What could be the role of Zn^{2+} in ZG? For instance, it is known that ZG contain Zn^{2+}-metalloenzymes, such as carboxypeptidases that require Zn^{2+} for their function[157, 158], but also for inhibition of their function at higher (micromolar) Zn^{2+} concentrations[159]. Although SLC30A2 is expressed in ZG, Zn^{2+} uptake via SLC30A2 (or other members of the SLC30 protein family) is more likely to be operative in precursor acidic organelles of the secretory pathway, e.g. the Golgi apparatus or immature granules after protein synthesis[160]. Whether secretagogue-evoked exocytotic secretion of Zn^{2+} from ZG[156] and/or secretion of Zn^{2+}-metallothionein[154] are effective means to regulate Zn^{2+} body homeostasis remains to be shown. In fact, regulation of the expression of transporters for Zn^{2+} uptake (from dietary sources and/or the pancreatic juice) in the intestinal mucosa may be a more effective and sensitive means to control Zn^{2+} body homeostasis[161–163]. However, a divergent view, based on zebrafish studies, concluded that the responses of epithelial tissues to Zn^{2+} deficiency and excess are best explained by local epithelial homeostasis, with no evidence of systemic control[164].

Conclusion and Perspectives

From a scientific perspective, the major unresolved issue is not so much the presence/absence of specific pancreatic ZG channels and transporters - there is no doubt that channels and transporters are expressed in the membrane of ZG - but rather a clarification of their physiological role in pancreatic acinar cells during the process of maturation, fusion and exocytosis of ZG. Here the development of organ-specific channel/transporter knockout animal models is mandatory to move the field forward. In this context the anomalous property of the pancreatic acinar cell, which displays a decreasing ZG acidity during maturation (see **Figure 1** and the paragraph on *Vacuolar-type H^+-ATPase*) remains an enigma because other secretory cells, in particular neuroendocrine cells, rather exhibit increasing acidity along the secretory pathway[114, 115]. This is particularly puzzling because studies in neuroendocrine cells have highlighted the important role of the V0 domain of the V-ATPase as a sensor of intragranular pH that identifies and regulates the ability of granules to undergo exocytosis[165, 166]. Hence, the functional implication of a decreasing acidity of maturing ZG for exocytosis needs to be addressed. Moreover, more studies investigating differences in H^+ permeability[114] and/or activity of exchange pathways for counter-ions, e.g. Na^+/H^+-, Ca^{2+}/H^+- or Cl^-/H^+-exchangers[118–121] along the secretory pathway of pancreatic acinar cells and their contribution to dissipation of the pH gradient during maturation are required, to better understand their significance in the process of regulated exocytotic secretion.

As this review has sought to demonstrate, many unsolved – and neglected – problems remain in ZG physiology. They need to be addressed and require clarification before we can frame accurate concepts of efficient preventive and therapeutic strategies for diseases of the pancreas, such as acute and chronic pancreatitis.

Acknowledgements

I thank Prof. John A. Williams for contacting and encouraging me to write this review. I am also grateful to my collaborator, Prof. Wing-Kee Lee, for continuous backing. Financial support for my studies on zymogen granules was obtained from the Deutsche Forschungsgemeinschaft, Deutsche Mukoviszidose e.V. and the University of Witten/Herdecke (ZBAF).

References

1. Thévenod F. Channels and transporters in zymogen granule membranes and their role in granule function: recent progress and a critical assessment. The Pancreapedia: Exocrine Pancreas Knowledge Base; 2015. DOI: 10.3998/panc.2015.01.
2. Lerch MM, Gorelick FS. Models of acute and chronic pancreatitis. Gastroenterology 2013;144:1180–1193.
3. Williams JA. The nobel pancreas: a historical perspective. Gastroenterology 2013;144:1166–1169.
4. Thévenod F. Ion channels in secretory granules of the pancreas and their role in exocytosis and release of secretory proteins. Am J Physiol Cell Physiol 2002;283:C651-C672.
5. Chen X, Walker AK, Strahler JR, Simon ES, Tomanicek-Volk SL, Nelson BB, Hurley MC, Ernst SA, Williams JA, Andrews PC. Organellar proteomics: analysis of pancreatic zymogen granule membranes. Mol Cell Proteomics 2006;5:306–312.
6. Rindler MJ, Xu CF, Gumper I, Smith NN, Neubert TA. Proteomic analysis of pancreatic zymogen granules: identification of new granule proteins. J Proteome Res 2007;6:2978–2992.
7. Berkane AA, Nguyen HT, Tranchida F, Waheed AA, Deyris V, Tchiakpe L, Fasano C, Nicoletti C, Desseaux V, Ajandouz el H, Comeau D, Comeau L, Hiol A. Proteomic of lipid rafts in the exocrine pancreas from diet-induced obese rats. Biochem Biophys Res Commun 2007;355:813–819.
8. Chen X, Ulintz PJ, Simon ES, Williams JA, Andrews PC. Global topology analysis of pancreatic zymogen granule membrane proteins. Mol Cell Proteomics 2008;7:2323–2336.
9. Chen X, Li C, Izumi T, Ernst SA, Andrews PC, Williams JA. Rab27b localizes to zymogen granules and regulates pancreatic acinar exocytosis. Biochem Biophys Res Commun 2004;323:1157–1162.
10. Chen X, Andrews PC. Quantitative proteomics analysis of pancreatic zymogen granule membrane proteins. Methods Mol Biol 2009;528:327–338.
11. Chen X, Andrews PC. Purification and proteomics analysis of pancreatic zymogen granule membranes. Methods Mol Biol 2008;432:275–287.
12. Borta H, Aroso M, Rinn C, Gomez-Lazaro M, Vitorino R, Zeuschner D, Grabenbauer M, Amado F, Schrader M. Analysis of low abundance membrane-associated proteins from rat pancreatic zymogen granules. J Proteome Res 2010;9:4927–4939.
13. Sun X, Jiang X. Combination of FASP and fully automated 2D-LC-MS/MS allows in-depth proteomic characterization of mouse zymogen granules. Biomed Chromatogr 2013;27:407–408.
14. Burnham DB, Munowitz P, Thorn N, Williams JA. Protein kinase activity associated with pancreatic zymogen granules. Biochem J 1985;227:743–751.
15. Meldolesi J. Membranes of pancreatic zymogen granules. Methods Enzymol 1983;98:67–75.
16. Thévenod F, Haase W, Hopfer U. Large-scale purification of calf pancreatic zymogen granule membranes. Anal Biochem 1992;202:54–60.
17. Yule DI, Ernst SA, Ohnishi H, Wojcikiewicz RJ. Evidence that zymogen granules are not a physiologically relevant calcium pool. Defining the distribution of inositol 1,4,5-trisphosphate receptors in pancreatic acinar cells. J Biol Chem 1997;272:9093–9098.
18. Huang P, Zou Y, Zhong XZ, Cao Q, Zhao K, Zhu MX, Murrell-Lagnado R, Dong XP. P2X4 forms functional ATP-activated cation channels on lysosomal membranes regulated by luminal pH. J Biol Chem 2014;289:17658–17667.
19. Fois G, Winkelmann VE, Bareis L, Staudenmaier L, Hecht E, Ziller C, Ehinger K, Schymeinsky J, Kranz C, Frick M. ATP is stored in lamellar bodies to activate vesicular P2X4 in an autocrine fashion upon exocytosis. J Gen Physiol 2018;150:277–291.
20. Norlund R, Roos N, Taljedal IB. Quantitative energy dispersive X-ray microanalysis of eight elements in pancreatic endocrine and exocrine cells after cryo-fixation. Bioscience Rep 1987;7:859–869.
21. Roomans GM, Wei X. X-ray microanalysis of resting and stimulated rat pancreas. Acta Physiol Scand 1985;124:353–359.
22. De Lisle RC, Hopfer U. Electrolyte permeabilities of pancreatic zymogen granules: implications for pancreatic secretion. Am J Physiol Gastrointest Liver Physiol 1986;250:G489-G496.
23. Arvan P, Rudnick G, Castle JD. Osmotic properties and internal pH of isolated rat parotid secretory granules. J Biol Chem 1984;259:13567–13572.
24. Thévenod F, Chathadi KV, Jiang B, Hopfer U. ATP-sensitive K^+ conductance in pancreatic zymogen granules: block by glyburide and activation by diazoxide. J Membr Biol 1992;129:253–266.
25. Delporte C. Aquaporins in salivary glands and pancreas. Biochim Biophys Acta 2014;1840:1524–1532.
26. Mendez-Gimenez L, Ezquerro S, da Silva IV, Soveral G, Fruhbeck G, Rodriguez A. Pancreatic Aquaporin-7: A Novel Target for Anti-diabetic Drugs? Front Chem 2018;6:99.
27. Cho SJ, Sattar AK, Jeong EH, Satchi M, Cho JA, Dash S, Mayes MS, Stromer MH, Jena BP. Aquaporin 1 regulates GTP-induced rapid gating of water in secretory vesicles. Proc Natl Acad Sci U S A 2002;99:4720–4724.
28. Abu-Hamdah R, Cho WJ, Cho SJ, Jeremic A, Kelly M, Ilie AE, Jena BP. Regulation of the water channel aquaporin-1: isolation and reconstitution of the regulatory complex. Cell Biol Int 2004;28:7–17.

29. Tani T, Koyama Y, Nihei K, Hatakeyama S, Ohshiro K, Yoshida Y, Yaoita E, Sakai Y, Hatakeyama K, Yamamoto T. Immunolocalization of aquaporin-8 in rat digestive organs and testis. Arch Histol Cytol 2001;64:159–168.

30. Lee WK, Bork U, Gholamrezaei F, Thévenod F. Cd2+-induced cytochrome c release in apoptotic proximal tubule cells: role of mitochondrial permeability transition pore and Ca^{2+} uniporter. Am J Physiol Renal Physiol 2005;288:F27-F39.

31. Calamita G, Ferri D, Gena P, Liquori GE, Cavalier A, Thomas D, Svelto M. The inner mitochondrial membrane has aquaporin-8 water channels and is highly permeable to water. J Biol Chem 2005;280:17149–17153.

32. Ohta E, Itoh T, Nemoto T, Kumagai J, Ko SB, Ishibashi K, Ohno M, Uchida K, Ohta A, Sohara E, Uchida S, Sasaki S, Rai T. Pancreas-specific aquaporin 12 null mice showed increased susceptibility to caerulein-induced acute pancreatitis. Am J Physiol Cell Physiol 2009;297:C1368-C1378.

33. Itoh T, Rai T, Kuwahara M, Ko SB, Uchida S, Sasaki S, Ishibashi K. Identification of a novel aquaporin, AQP12, expressed in pancreatic acinar cells. Biochem Biophys Res Commun 2005;330:832–838.

34. Yang K, Ding YX, Chin WC. K+-induced ion-exchanges trigger trypsin activation in pancreas acinar zymogen granules. Arch Biochem Biophys 2007;459:256–263.

35. Gasser KW, DiDomenico J, Hopfer U. Potassium transport by pancreatic and parotid zymogen granule membranes. Am J Physiol Cell Physiol 1988;255:C705-C711.

36. Gasser KW, Holda JR. ATP-sensitive potassium transport by pancreatic secretory granule membrane. Am J Physiol Gastrointest Liver Physiol 1993;264:G137-G142.

37. Braun M, Anderie I, Thévenod F. Evidence for a 65 kDa sulfonylurea receptor in rat pancreatic zymogen granule membranes. FEBS Lett 1997;411:255–259.

38. Braun M, Thévenod F. Photoaffinity labeling and purification of ZG-16p, a high-affinity dihydropyridine binding protein of rat pancreatic zymogen granule membranes that regulates a K$^+$-selective conductance. Mol Pharmacol 2000;57:308–316.

39. Nguyen T, Chin WC, Verdugo P. Role of Ca^{2+}/K$^+$ ion exchange in intracellular storage and release of Ca^{2+}. Nature 1998;395:908–912.

40. Kelly ML, Abu-Hamdah R, Jeremic A, Cho SJ, Ilie AE, Jena BP. Patch clamped single pancreatic zymogen granules: direct measurements of ion channel activities at the granule membrane. Pancreatology 2005;5:443–449.

41. Lesage F, Lazdunski M. Molecular and functional properties of two-pore-domain potassium channels. Am J Physiol Renal Physiol 2000;279:F793-F801.

42. Medhurst AD, Rennie G, Chapman CG, Meadows H, Duckworth MD, Kelsell RE, Gloger, II, Pangalos MN. Distribution analysis of human two pore domain potassium channels in tissues of the central nervous system and periphery. Brain Res Mol Brain Res 2001;86:101–114.

43. Jespersen T, Grunnet M, Olesen SP. The KCNQ1 potassium channel: from gene to physiological function. Physiology (Bethesda) 2005;20:408–416.

44. Thompson E, Eldstrom J, Fedida D. Hormonal Signaling Actions on Kv7.1 (KCNQ1) Channels. Annu Rev Pharmacol Toxicol 2020;61:381–400.

45. Lerche C, Bruhova I, Lerche H, Steinmeyer K, Wei AD, Strutz-Seebohm N, Lang F, Busch AE, Zhorov BS, Seebohm G. Chromanol 293B binding in KCNQ1 (Kv7.1) channels involves electrostatic interactions with a potassium ion in the selectivity filter. Mol Pharmacol 2007;71:1503–1511.

46. Thomas GP, Gerlach U, Antzelevitch C. HMR 1556, a potent and selective blocker of slowly activating delayed rectifier potassium current. J Cardiovasc Pharmacol 2003;41:140–147.

47. Lee WK, Torchalski B, Roussa E, Thévenod F. Evidence for KCNQ1 K$^+$ channel expression in rat zymogen granule membranes and involvement in cholecystokinin-induced pancreatic acinar secretion. Am J Physiol Cell Physiol 2008;294:C879-C892.

48. Lee JE, Park HS, Uhm DY, Kim SJ. Effects of KCNQ1 channel blocker, 293B, on the acetylcholine-induced Cl$^-$ secretion of rat pancreatic acini. Pancreas 2004;28:435–442.

49. Lee WK, Braun M, Langeluddecke C, Thévenod F. Cyclosporin a, but not FK506, induces osmotic lysis of pancreas zymogen granules, intra-acinar enzyme release, and lysosome instability by activating K$^+$ channel. Pancreas 2012;41:596–604.

50. Hirano Y, Hisatomi A, Ohara K, Noguchi H. The effects of FK506 and cyclosporine on the exocrine function of the rat pancreas. Transplantation 1992;54:883–887.

51. Groblewski GE, Wagner AC, Williams JA. Cyclosporin A inhibits Ca^{2+}/calmodulin-dependent protein phosphatase and secretion in pancreatic acinar cells. J Biol Chem 1994;269:15111–15117.

52. Zhao H, Muallem S. Na$^+$, K$^+$, and Cl$^-$ transport in resting pancreatic acinar cells. J Gen Physiol 1995;106:1225–1242.

53. Jentsch TJ, Pusch M. CLC Chloride Channels and Transporters: Structure, Function, Physiology, and Disease. Physiol Rev 2018;98:1493–1590.

54. Carew MA, Thorn P. Identification of ClC-2-like chloride currents in pig pancreatic acinar cells. Pflugers Arch 1996;433:84–90.

55. Thiemann A, Grunder S, Pusch M, Jentsch TJ. A chloride channel widely expressed in epithelial and non-epithelial cells. Nature 1992;356:57–60.

56. Grunder S, Thiemann A, Pusch M, Jentsch TJ. Regions involved in the opening of ClC-2 chloride channel by voltage and cell volume. Nature 1992;360:759–762.

57. Stobrawa SM, Breiderhoff T, Takamori S, Engel D, Schweizer M, Zdebik AA, Bosl MR, Ruether K, Jahn H, Draguhn A, Jahn R, Jentsch TJ. Disruption of ClC-3, a chloride channel expressed on synaptic vesicles, leads to a loss of the hippocampus. Neuron 2001;29:185–196.

58. Bosl MR, Stein V, Hubner C, Zdebik AA, Jordt SE, Mukhopadhyay AK, Davidoff MS, Holstein AF, Jentsch TJ. Male germ cells and photoreceptors, both dependent on close cell-cell interactions, degenerate upon ClC-2 Cl$^-$ channel disruption. EMBO J 2001;20:1289–1299.

59. Gomez-Lazaro M, Rinn C, Aroso M, Amado F, Schrader M. Proteomic analysis of zymogen granules. Expert Rev Proteomics 2010;7:735–747.

60. Arreola J, Begenisich T, Nehrke K, Nguyen HV, Park K, Richardson L, Yang B, Schutte BC, Lamb FS, Melvin JE. Secretion and cell volume regulation by salivary acinar cells from mice lacking expression of the Clcn3 Cl⁻ channel gene. J Physiol 2002;545:207–216.

61. Jentsch TJ, Maritzen T, Keating DJ, Zdebik AA, Thévenod F. ClC-3--a granular anion transporter involved in insulin secretion? Cell Metab 2010;12:307–308; author reply 309–310.

62. Ikeda M, Beitz E, Kozono D, Guggino WB, Agre P, Yasui M. Characterization of aquaporin-6 as a nitrate channel in mammalian cells. Requirement of pore-lining residue threonine 63. J Biol Chem 2002;277:39873–39879.

63. Yasui M. pH regulated anion permeability of aquaporin-6. Handb Exp Pharmacol 2009;190:299–308.

64. Yasui M, Hazama A, Kwon TH, Nielsen S, Guggino WB, Agre P. Rapid gating and anion permeability of an intracellular aquaporin. Nature 1999;402:184–187.

65. Liu K, Kozono D, Kato Y, Agre P, Hazama A, Yasui M. Conversion of aquaporin 6 from an anion channel to a water-selective channel by a single amino acid substitution. Proc Natl Acad Sci U S A 2005;102:2192–2197.

66. Yasui M, Kwon TH, Knepper MA, Nielsen S, Agre P. Aquaporin-6: An intracellular vesicle water channel protein in renal epithelia. Proc Natl Acad Sci U S A 1999;96:5808–5813.

67. Matsuki-Fukushima M, Hashimoto S, Shimono M, Satoh K, Fujita-Yoshigaki J, Sugiya H. Presence and localization of aquaporin-6 in rat parotid acinar cells. Cell Tissue Res 2008;332:73–80.

68. Matsuki-Fukushima M, Fujita-Yoshigaki J, Murakami M, Katsumata-Kato O, Yokoyama M, Sugiya H. Involvement of AQP6 in the Mercury-sensitive osmotic lysis of rat parotid secretory granules. J Membr Biol 2013;246:209–214.

69. Loewen ME, Forsyth GW. Structure and function of CLCA proteins. Physiol Rev 2005;85:1061–1092.

70. Thévenod F, Roussa E, Benos DJ, Fuller CM. Relationship between a HCO₃⁻ -permeable conductance and a CLCA protein from rat pancreatic zymogen granules. Biochem Biophys Res Commun 2003;300:546–554.

71. Roussa E, Wittschen P, Wolff NA, Torchalski B, Gruber AD, Thévenod F. Cellular distribution and subcellular localization of mCLCA1/2 in murine gastrointestinal epithelia. J Histochem Cytochem 2010;58:653–668.

72. Gibson A, Lewis AP, Affleck K, Aitken AJ, Meldrum E, Thompson N. hCLCA1 and mCLCA3 are secreted non-integral membrane proteins and therefore are not ion channels. J Biol Chem 2005;280:27205–27212.

73. Hamann M, Gibson A, Davies N, Jowett A, Walhin JP, Partington L, Affleck K, Trezise D, Main M. Human ClCa1 modulates anionic conduction of calcium-dependent chloride currents. J Physiol 2009;587:2255–2274.

74. Patel AC, Brett TJ, Holtzman MJ. The role of CLCA proteins in inflammatory airway disease. Annu Rev Physiol 2009;71:425–449.

75. Pedemonte N, Galietta LJ. Structure and function of TMEM16 proteins (anoctamins). Physiol Rev 2014;94:419–459.

76. Whitlock JM, Hartzell HC. Anoctamins/TMEM16 Proteins: Chloride Channels Flirting with Lipids and Extracellular Vesicles. Annu Rev Physiol 2017;79:119–143.

77. Kunzelmann K, Tian Y, Martins JR, Faria D, Kongsuphol P, Ousingsawat J, Thevenod F, Roussa E, Rock J, Schreiber R. Anoctamins. Pflugers Arch 2011;462:195–208.

78. Yang YD, Cho H, Koo JY, Tak MH, Cho Y, Shim WS, Park SP, Lee J, Lee B, Kim BM, Raouf R, Shin YK, Oh U. TMEM16A confers receptor-activated calcium-dependent chloride conductance. Nature 2008;455:1210–1215.

79. Jung J, Nam JH, Park HW, Oh U, Yoon JH, Lee MG. Dynamic modulation of ANO1/TMEM16A HCO3(-) permeability by Ca²⁺/calmodulin. Proc Natl Acad Sci U S A 2013;110:360–365.

80. Quinton PM. Cystic fibrosis: impaired bicarbonate secretion and mucoviscidosis. Lancet 2008;372:415–417.

81. Han Y, Shewan AM, Thorn P. HCO₃⁻ Transport through Anoctamin/Transmembrane Protein ANO1/TMEM16A in Pancreatic Acinar Cells Regulates Luminal pH. J Biol Chem 2016;291:20345–20352.

82. Clemente F, Meldolesi J. Calcium and pancreatic secretion. I. Subcellular distribution of calcium and magnesium in the exocrine pancreas of the guinea pig. J Cell Biol 1975;65:88–102.

83. Gerasimenko OV, Gerasimenko JV, Belan PV, Petersen OH. Inositol trisphosphate and cyclic ADP-ribose-mediated release of Ca²⁺ from single isolated pancreatic zymogen granules. Cell 1996;84:473–480.

84. Vandecaetsbeek I, Vangheluwe P, Raeymaekers L, Wuytack F, Vanoevelen J. The Ca²⁺ pumps of the endoplasmic reticulum and Golgi apparatus. Cold Spring Harb Perspect Biol 2011;3:a004184.

85. Bruce JI, Yang X, Ferguson CJ, Elliott AC, Steward MC, Case RM, Riccardi D. Molecular and functional identification of a Ca²⁺ (polyvalent cation)-sensing receptor in rat pancreas. J Biol Chem 1999;274:20561–20568.

86. Thorn P, Lawrie AM, Smith PM, Gallacher DV, Petersen OH. Ca²⁺ oscillations in pancreatic acinar cells: spatiotemporal relationships and functional implications. Cell Calcium 1993;14:746–757.

87. Yoo SH, Albanesi JP. Inositol 1,4,5-trisphosphate-triggered Ca²⁺ release from bovine adrenal medullary secretory vesicles. J Biol Chem 1990;265:13446–13448.

88. Petersen OH, Gerasimenko OV, Tepikin AV, Gerasimenko JV. Aberrant Ca²⁺ signalling through acidic calcium stores in pancreatic acinar cells. Cell Calcium 2011;50:193–199.

89. Yoo SH. Role of secretory granules in inositol 1,4,5-trisphosphate-dependent Ca²⁺ signaling: from phytoplankton to mammals. Cell Calcium 2011;50:175–183.

90. Gerasimenko JV, Sherwood M, Tepikin AV, Petersen OH, Gerasimenko OV. NAADP, cADPR and IP₃ all release Ca²⁺ from the endoplasmic reticulum and an acidic store in the secretory granule area. J Cell Sci 2006;119:226–238.

91. Titievsky AV, Takeo T, Tepikin AV, Petersen OH. Decrease of acidity inside zymogen granules inhibits acetylcholine- or inositol trisphosphate-evoked cytosolic Ca²⁺ spiking in pancreatic acinar cells. Pflugers Arch 1996;432:938–940.

92. Gerasimenko JV, Lur G, Ferdek P, Sherwood MW, Ebisui E, Tepikin AV, Mikoshiba K, Petersen OH, Gerasimenko OV. Calmodulin protects against alcohol-induced pancreatic

trypsinogen activation elicited via Ca^{2+} release through IP3 receptors. Proc Natl Acad Sci U S A 2011;108:5873–5878.

93. Gerasimenko JV, Lur G, Sherwood MW, Ebisui E, Tepikin AV, Mikoshiba K, Gerasimenko OV, Petersen OH. Pancreatic protease activation by alcohol metabolite depends on Ca^{2+} release via acid store IP_3 receptors. Proc Natl Acad Sci U S A 2009;106:10758–10763.

94. Nezu A, Tanimura A, Morita T, Irie K, Yajima T, Tojyo Y. Evidence that zymogen granules do not function as an intracellular Ca^{2+} store for the generation of the Ca^{2+} signal in rat parotid acinar cells. Biochem J 2002;363:59–66.

95. Kataoka M, Fukura Y, Shinohara Y, Baba Y. Analysis of mitochondrial membrane potential in the cells by microchip flow cytometry. Electrophoresis 2005;26:3025–3031.

96. Gerasimenko J, Peng S, Gerasimenko O. Role of acidic stores in secretory epithelia. Cell Calcium 2014;55:346–354.

97. Thévenod F, Kemmer TP, Christian AL, Schulz I. Characterization of MgATP-driven H^+ uptake into a microsomal vesicle fraction from rat pancreatic acinar cells. J Membr Biol 1989;107:263–275.

98. Thévenod F, Dehlinger-Kremer M, Kemmer TP, Christian AL, Potter BV, Schulz I. Characterization of inositol 1,4,5-trisphosphate-sensitive (IsCaP) and -insensitive (IisCaP) nonmitochondrial Ca^{2+} pools in rat pancreatic acinar cells. J Membr Biol 1989;109:173–186.

99. Orci L, Ravazzola M, Anderson RG. The condensing vacuole of exocrine cells is more acidic than the mature secretory vesicle. Nature 1987;326:77–79.

100. van de Put FH, Elliott AC. The endoplasmic reticulum can act as a functional Ca^{2+} store in all subcellular regions of the pancreatic acinar cell. J Biol Chem 1997;272:27764–27770.

101. Gerasimenko OV, Gerasimenko JV, Rizzuto RR, Treiman M, Tepikin AV, Petersen OH. The distribution of the endoplasmic reticulum in living pancreatic acinar cells. Cell Calcium 2002;32:261–268.

102. Thévenod F, Schulz I. H^+-dependent calcium uptake into an IP_3-sensitive calcium pool from rat parotid gland. Am J Physiol 1988;255:G429-G440.

103. North RA. Molecular physiology of P2X receptors. Physiol Rev 2002;82:1013–1067.

104. Patel S, Cai X. Evolution of acidic Ca^{2+} stores and their resident Ca^{2+}-permeable channels. Cell Calcium 2015;57:222–230.

105. Dickson EJ, Duman JG, Moody MW, Chen L, Hille B. Orai-STIM-mediated Ca^{2+} release from secretory granules revealed by a targeted Ca^{2+} and pH probe. Proc Natl Acad Sci U S A 2012;109:E3539–3548.

106. Ikeya M, Yamanoue K, Mochizuki Y, Konishi H, Tadokoro S, Tanaka M, Suzuki R, Hirashima N. Orai-2 is localized on secretory granules and regulates antigen-evoked Ca^{2+} mobilization and exocytosis in mast cells. Biochem Biophys Res Commun 2014;451:62–67.

107. Forgac M. Vacuolar ATPases: rotary proton pumps in physiology and pathophysiology. Nat Rev Mol Cell Biol 2007;8:917–929.

108. Roussa E, Alper SL, Thévenod F. Immunolocalization of anion exchanger AE2, Na^+/H^+ exchangers NHE1 and NHE4, and vacuolar type H^+-ATPase in rat pancreas. J Histochem Cytochem 2001;49:463–474.

109. Arvan P, Castle JD. Isolated secretion granules from parotid glands of chronically stimulated rats possess an alkaline internal pH and inward-directed H^+ pump activity. J Cell Biol 1986;103:1257–1267.

110. Niederau C, Van Dyke RW, Scharschmidt BF, Grendell JH. Rat pancreatic zymogen granules. An actively acidified compartment. Gastroenterology 1986;91:1433–1442.

111. De Lisle RC, Williams JA. Zymogen granule acidity is not required for stimulated pancreatic protein secretion. Am J Physiol Gastrointest Liver Physiol 1987;253:G711-G719.

112. Anderson RG, Pathak RK. Vesicles and cisternae in the trans Golgi apparatus of human fibroblasts are acidic compartments. Cell 1985;40:635–643.

113. Behrendorff N, Floetenmeyer M, Schwiening C, Thorn P. Protons released during pancreatic acinar cell secretion acidify the lumen and contribute to pancreatitis in mice. Gastroenterology 2010;139, 1720.

114. Paroutis P, Touret N, Grinstein S. The pH of the secretory pathway: measurement, determinants, and regulation. Physiology (Bethesda) 2004;19:207–215.

115. Casey JR, Grinstein S, Orlowski J. Sensors and regulators of intracellular pH. Nat Rev Mol Cell Biol 2010;11:50–61.

116. Johnson RG, Jr. Proton pumps and chemiosmotic coupling as a generalized mechanism for neurotransmitter and hormone transport. Ann N Y Acad Sci 1987;493:162–177.

117. Haanes KA, Novak I. ATP storage and uptake by isolated pancreatic zymogen granules. Biochem J 2010;429:303–311.

118. Anderie I, Blum R, Haase W, Grinstein S, Thévenod F. Expression of NHE1 and NHE4 in rat pancreatic zymogen granule membranes. Biochem Biophys Res Commun 1998;246:330–336.

119. Anderie I, Thévenod F. Evidence for involvement of a zymogen granule Na^+/H^+ exchanger in enzyme secretion from rat pancreatic acinar cells. J Membr Biol 1996;152:195–205.

120. Orlowski J, Grinstein S. Emerging roles of alkali cation/proton exchangers in organellar homeostasis. Curr Opin Cell Biol 2007;19:483–492.

121. Stauber T, Jentsch TJ. Chloride in vesicular trafficking and function. Annu Rev Physiol 2013;75:453–477.

122. Sawada K, Echigo N, Juge N, Miyaji T, Otsuka M, Omote H, Yamamoto A, Moriyama Y. Identification of a vesicular nucleotide transporter. Proc Natl Acad Sci U S A 2008;105:5683–5686.

123. Hasuzawa N, Moriyama S, Moriyama Y, Nomura M. Physiopathological roles of vesicular nucleotide transporter (VNUT), an essential component for vesicular ATP release. Biochim Biophys Acta Biomembr 2020;1862:183408.

124. Moriyama Y, Hiasa M, Sakamoto S, Omote H, Nomura M. Vesicular nucleotide transporter (VNUT): appearance of an actress on the stage of purinergic signaling. Purinergic Signal 2017;13:387–404.

125. Estevez-Herrera J, Dominguez N, Pardo MR, Gonzalez-Santana A, Westhead EW, Borges R, Machado JD. ATP: The crucial component of secretory vesicles. Proc Natl Acad Sci U S A 2016;113:E4098–4106.

126. Burnstock G. Physiology and pathophysiology of purinergic neurotransmission. Physiol Rev 2007;87:659–797.

127. Johnson RG, Jr. Accumulation of biological amines into chromaffin granules: a model for hormone and neurotransmitter transport. Physiol Rev 1988;68:232–307.

128. Miyaji T, Sawada K, Omote H, Moriyama Y. Divalent cation transport by vesicular nucleotide transporter. J Biol Chem 2011;286:42881–42887.

129. Novak I. ATP as a signaling molecule: the exocrine focus. News Physiol Sci 2003;18:12–17.

130. Luo X, Zheng W, Yan M, Lee MG, Muallem S. Multiple functional P2X and P2Y receptors in the luminal and basolateral membranes of pancreatic duct cells. Am J Physiol 1999;277:C205-C215.

131. Haanes KA, Kowal JM, Arpino G, Lange SC, Moriyama Y, Pedersen PA, Novak I. Role of vesicular nucleotide transporter VNUT (SLC17A9) in release of ATP from AR42J cells and mouse pancreatic acinar cells. Purinergic Signal 2014;10:431–440.

132. Cao Q, Zhao K, Zhong XZ, Zou Y, Yu H, Huang P, Xu TL, Dong XP. SLC17A9 protein functions as a lysosomal ATP transporter and regulates cell viability. J Biol Chem 2014;289:23189–23199.

133. Zhong XZ, Cao Q, Sun X, Dong XP. Activation of lysosomal P2X4 by ATP transported into lysosomes via VNUT/SLC17A9 using V-ATPase generated voltage gradient as the driving force. J Physiol 2016;594:4253–4266.

134. Murrell-Lagnado RD. A role for P2X4 receptors in lysosome function. J Gen Physiol 2018;150:185–187.

135. Johnson DE, Ostrowski P, Jaumouille V, Grinstein S. The position of lysosomes within the cell determines their luminal pH. J Cell Biol 2016;212:677–692.

136. Roos N. A possible site of calcium regulation in rat exocrine pancreas cells: an X-ray microanalytical study. Scanning Microsc 1988;2:323–329.

137. Palmiter RD, Cole TB, Findley SD. ZnT-2, a mammalian protein that confers resistance to zinc by facilitating vesicular sequestration. EMBO J 1996;15:1784–1791.

138. Huang L, Tepaamorndech S. The SLC30 family of zinc transporters - a review of current understanding of their biological and pathophysiological roles. Mol Aspects Med 2013;34:548–560.

139. Seve M, Chimienti F, Devergnas S, Favier A. In silico identification and expression of SLC30 family genes: an expressed sequence tag data mining strategy for the characterization of zinc transporters' tissue expression. BMC genomics 2004;5:32.

140. Liuzzi JP, Cousins RJ. Mammalian zinc transporters. Annu Rev Nutr 2004;24:151–172.

141. Palmiter RD, Findley SD. Cloning and functional characterization of a mammalian zinc transporter that confers resistance to zinc. EMBO J 1995;14:639–649.

142. Chowanadisai W, Lonnerdal B, Kelleher SL. Identification of a mutation in SLC30A2 (ZnT-2) in women with low milk zinc concentration that results in transient neonatal zinc deficiency. J Biol Chem 2006;281:39699–39707.

143. Golan Y, Kambe T, Assaraf YG. The role of the zinc transporter SLC30A2/ZnT2 in transient neonatal zinc deficiency. Metallomics 2017;9:1352–1366.

144. Golan Y, Alhadeff R, Warshel A, Assaraf YG. ZnT2 is an electroneutral proton-coupled vesicular antiporter displaying an apparent stoichiometry of two protons per zinc ion. PLoS Comput Biol 2019;15:e1006882.

145. Ohana E, Hoch E, Keasar C, Kambe T, Yifrach O, Hershfinkel M, Sekler I. Identification of the Zn^{2+} binding site and mode of operation of a mammalian Zn^{2+} transporter. J Biol Chem 2009;284:17677–17686.

146. Shusterman E, Beharier O, Shiri L, Zarivach R, Etzion Y, Campbell CR, Lee IH, Okabayashi K, Dinudom A, Cook DI, Katz A, Moran A. ZnT-1 extrudes zinc from mammalian cells functioning as a Zn^{2+}/H^+ exchanger. Metallomics 2014;6:1656–1663.

147. Chimienti F, Devergnas S, Favier A, Seve M. Identification and cloning of a beta-cell-specific zinc transporter, ZnT-8, localized into insulin secretory granules. Diabetes 2004;53:2330–2337.

148. Lemaire K, Chimienti F, Schuit F. Zinc transporters and their role in the pancreatic beta-cell. J Diabetes Investig 2012;3:202–211.

149. Abrahamsson H, Gylfe E. Demonstration of a proton gradient across the insulin granule membrane. Acta Physiol Scand 1980;109:113–114.

150. Pace CS, Sachs G. Glucose-induced proton uptake in secretory granules of beta-cells in monolayer culture. Am J Physiol Cell Physiol 1982;242:C382–387.

151. Hutton JC. The internal pH and membrane potential of the insulin-secretory granule. Biochem J 1982;204:171–178.

152. Sondergaard LG, Stoltenberg M, Doering P, Flyvbjerg A, Rungby J. Zinc ions in the endocrine and exocrine pancreas of zinc deficient rats. Histol Histopathol 2006;21:619–625.

153. Liuzzi JP, Bobo JA, Lichten LA, Samuelson DA, Cousins RJ. Responsive transporter genes within the murine intestinal-pancreatic axis form a basis of zinc homeostasis. Proc Natl Acad Sci U S A 2004;101:14355–14360.

154. De Lisle RC, Sarras MP, Jr., Hidalgo J, Andrews GK. Metallothionein is a component of exocrine pancreas secretion: implications for zinc homeostasis. Am J Physiol 1996;271:C1103–1110.

155. Cousins RJ. Gastrointestinal factors influencing zinc absorption and homeostasis. Int J Vit Nutr Res 2010;80:243–248.

156. Guo L, Lichten LA, Ryu MS, Liuzzi JP, Wang F, Cousins RJ. STAT5-glucocorticoid receptor interaction and MTF-1 regulate the expression of ZnT2 (Slc30a2) in pancreatic acinar cells. Proc Natl Acad Sci U S A 2010;107:2818–2823.

157. Vallee BL, Neurath H. Carboxypeptidase, a zinc metalloenzyme. J Biol Chem 1955;217:253–261.

158. Hsu JM, Anilane JK, Scanlan DE. Pancreatic carboxypeptidases: activities in zinc-deficient rats. Science 1966;153:882–883.

159. Larsen KS, Auld DS. Characterization of an inhibitory metal binding site in carboxypeptidase A. Biochemistry 1991;30:2613–2618.

160. Pekas JC. Pancreatic incorporation of 65Zn and histidine-14C into secreted proteins of the pig. Am J Physiol 1971;220:799–803.

161. Jou MY, Hall AG, Philipps AF, Kelleher SL, Lonnerdal B. Tissue-specific alterations in zinc transporter expression in intestine and liver reflect a threshold for homeostatic compensation during dietary zinc deficiency in weanling rats. J Nutr 2009;139:835–841.

162. Martin L, Lodemann U, Bondzio A, Gefeller EM, Vahjen W, Aschenbach JR, Zentek J, Pieper R. A high amount of dietary zinc changes the expression of zinc transporters and metallothionein in jejunal epithelial cells in vitro and in vivo but does not prevent zinc accumulation in jejunal tissue of piglets. J Nutr 2013;143:1205–1210.

163. Cragg RA, Phillips SR, Piper JM, Varma JS, Campbell FC, Mathers JC, Ford D. Homeostatic regulation of zinc transporters in the human small intestine by dietary zinc supplementation. Gut 2005;54:469–478.

164. Zheng D, Feeney GP, Handy RD, Hogstrand C, Kille P. Uptake epithelia behave in a cell-centric and not systems homeostatic manner in response to zinc depletion and supplementation. Metallomics 2014;6:154–165.

165. Poea-Guyon S, Ammar MR, Erard M, Amar M, Moreau AW, Fossier P, Gleize V, Vitale N, Morel N. The V-ATPase membrane domain is a sensor of granular pH that controls the exocytotic machinery. J Cell Biol 2013;203:283–298.

166. Wang D, Epstein D, Khalaf O, Srinivasan S, Williamson WR, Fayyazuddin A, Quiocho FA, Hiesinger PR. Ca^{2+}-Calmodulin regulates SNARE assembly and spontaneous neurotransmitter release via v-ATPase subunit V0a1. J Cell Biol 2014;205:21–31.

Chapter 13

Secretory portal—porosome in pancreatic acinar cells

Bhanu P. Jena

Department of Physiology, Wayne State University School of Medicine, Detroit, MI 48201, USA

Introduction

Secretion is a fundamental process through which cells communicate with their environment and exchange information in a multicellular context to reach homeostasis and sustain life. For decades, the prevailing view was that secretion operates as an all-or-none "complete fusion" event, where vesicles are trafficked to the cell surface to fuse and completely incorporate into the plasma membrane. The vesicle contents then diffuse out of the cell. This hypothesis, although attractive at first glance, had several key setbacks. First, it predicted a quantization of secretory products packaged into each secretory vesicle, when in fact, secretory vesicle size greatly varies even within the same cell, sometimes as much as sixfold. Second, the level of additional regulation necessary to rapidly and precisely internalize and sequester vesicle-associated lipids and proteins following incorporation into the cell plasma membrane seems extraordinarily complex, given the tens of thousands of different membrane lipids and their differential distribution even between the same bilayer leaflets. Third, following a secretory episode, partially empty secretory vesicles accumulate within cells as observed in electron micrographs (**Figure 1**), demonstrating that secretory vesicles are capable of transient fusion at the cell plasma membrane and incomplete or partial content release.

We therefore hypothesized some 25 years ago the presence of a tunable dial at the cell plasma membrane for secretory vesicle docking and fusion without full collapse and the generation of hydrostatic pressure within vesicles to drive vesicular contents to the cell exterior. Through the use of atomic force microscope, a force spectroscope, we identified nanoscale transmembrane cup-shaped lipoprotein structures at the cell plasma membrane first in acinar cells of the exocrine pancreas and later in all cells examined including neurons. We named the structures "porosomes" that have since been implicated in a wide range of secretory events. The family of proteins that make up the porosome has been biochemically identified and the mesoscale structure of the complex has been well characterized through

electron microscopy and solution X-ray methods. Defects in one or more porosome components have measurable, often highly potent effects on the regulation of secretion, establishing links between point mutations and secretion-defective disease states such as neurological disorders and cystic fibrosis that were previously correlative and are now causative. The discovery of the porosome solved the conundrum of fractional discharge of intravesicular contents from cells by providing an explanation for regulated graded secretion. Porosomes are cup-shaped unidirectional (inside cells to the outside) transport machineries, which are supramolecular lipoprotein structures at the cell plasma membrane ranging in size from 15 nm in neurons to 100–180 nm in endocrine and exocrine cells and composed of around 30–40 proteins. In comparison, the 120-nm bidirectional transport machinery, the nuclear pore complex, is composed of nearly 1,000 protein molecules.

Porosomes in the Exocrine Pancreas

In 1973, the "transient" or "kiss-and-run" mechanism of secretory vesicle fusion at the cell plasma membrane enabling fractional release of intravesicular contents during secretion was proposed.[2] Nearly two decades later in 1990, it was hypothesized that the fusion pore, a continuity established between the vesicle membrane and the cell plasma membrane, results from a "preassembled ion channel-like structure that could open and close."[3] A later 1992 review[4] opined that the principal difficulty in observing these hypothetical pore structures was the lack of ultrahigh resolution imaging tools to directly monitor their presence and study their activity in live cells.

In the mid-1990s, employing the then newly developed technique of atomic force microscopy (AFM), nanometer-scale cup-shaped pore structures and their dynamics were discovered at the apical plasma membrane in live pancreatic acinar cells.[5] Circular pit-like structures containing 100–180 nm cup-shaped depressions or pores were observed at the apical plasma membrane of pancreatic acinar cells

e-mail: bjena@med.wayne.edu

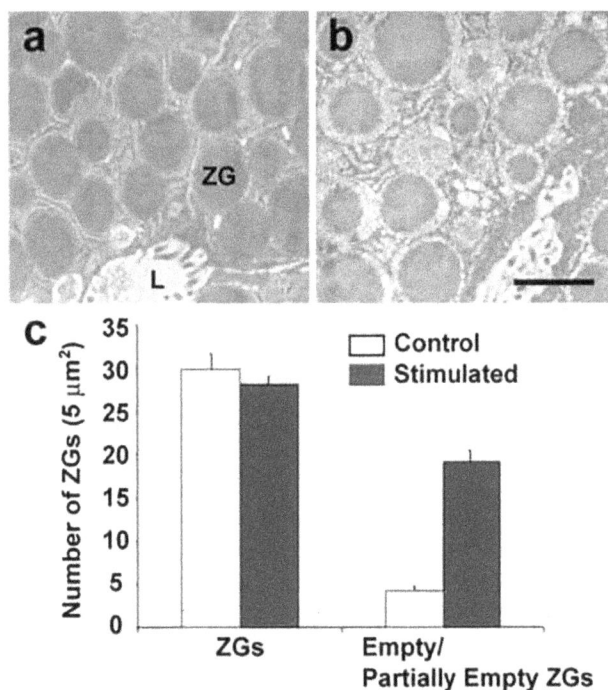

Figure 1. Partial emptying of zymogen granules in acinar cells of the rat exocrine pancreas following secretion. Representative electron micrographs of resting (**A**) and cholecystokinin-stimulated for 15 minutes (**B**) rat pancreatic acinar cells, demonstrating partial loss of zymogen granule (ZG) contents following secretion. The apical lumen (L) of acini demonstrating the presence of microvilli and secreted products is observed. (**C**) These studies using electron microscopy further demonstrate that while the number of ZGs remain unchanged following secretion, an increase in the number of empty and partially empty vesicles is observed. Scale bar = 1 µm (from reference [1]).

where secretion is known to occur.[5] During secretion, the depression or pore openings grew larger, returning to their resting size following completion of cell secretion.

Studies next established the observed depressions to be the secretory portals at the cell plasma membrane.[6,7] Following stimulation of secretion, gold-conjugated amylase antibodies (amylase being one of the major intravesicular enzymes secreted by pancreatic acinar cells) accumulate at depressions. These results established depressions to be the long sought-after secretory portals in cells. The study further reported[7] the presence of t-SNAREs at the porosome base facing the cytosol, firmly establishing depression structures to be secretory portals where zymogen granules (ZGs) transiently dock and fuse for intravesicular content release during secretion.[7] Subsequently, porosomes and their dynamics at the cell plasma membrane in growth hormone (GH)-secreting cells of the pituitary gland and in rat chromaffin cells[8] were reported. In 2003, following immunoisolation of the porosome structures from acinar cells of the exocrine pancreas, its composition was determined,

and isolated porosomes were functionally reconstituted into artificial lipid membranes.[9] Morphological details of porosome associated with docked secretory vesicles were revealed using high-resolution electron microscopy.[9] In the past decade, employing a combination of approaches such as AFM, biochemistry, electrophysiology, conventional EM, mass spectrometry, and small-angle X-Ray solution scattering analysis, this specialized secretory portal has been extensively characterized and documented and found to be present in all cells examined, including neurons.[10–12] Studies[5–29] establish porosomes to be secretory portals that perform the specialized task of fractional discharge of intravesicular contents from cells during secretion. The porosome structure and the associated transient fusion mechanism[30–32] accompanied by fractional discharge of vesicular contents from cells has greatly progressed in the past two decades. It has been demonstrated in exocrine, endocrine, and neuronal cells that "secretory granules are recaptured largely intact following stimulated exocytosis in cultured endocrine cells";[31] that "single synaptic vesicles fuse transiently and successively without loss of identity";[30] and that "zymogen granule exocytosis is characterized by long fusion pore openings and preservation of vesicle lipid identity."[12,32] The past two decades have additionally witnessed much progress in our understanding of Ca^{2+} and SNARE-mediated membrane fusion[31,33–59] and on secretory vesicle volume regulation required for precise fractional release of intravesicular contents from cells[22,60–65] during secretion. These findings have greatly contributed to the progress in our understanding of porosome-mediated fractional release of intravesicular contents from cells during secretion.

Porosome Discovery: AFM Trained on Pancreatic Acini

As noted earlier, the porosome was first discovered in acinar cells of the rat exocrine pancreas nearly 25 years ago.[5] This discovery was made possible by the use of a then new microscope, the AFM.[66,67] In AFM, a probe tip microfabricated from silicon or silicon nitride and mounted on a cantilever spring is used to scan the surface of the sample at a constant force. Either the probe or the sample can be precisely moved in a raster pattern using a xyz piezo to scan the surface of the sample. The deflection of the cantilever measured optically is used to generate an isoforce relief of the sample.[66,67] Force is thus used by the AFM to image surface profiles of objects at nanometer resolution and in real time, objects such as live cells, subcellular organelles, and even biomolecules submerged in physiological buffer solutions.

Compared to neurons and neuroendocrine cells, exocrine pancreas are slow secretory cells, secreting digestive

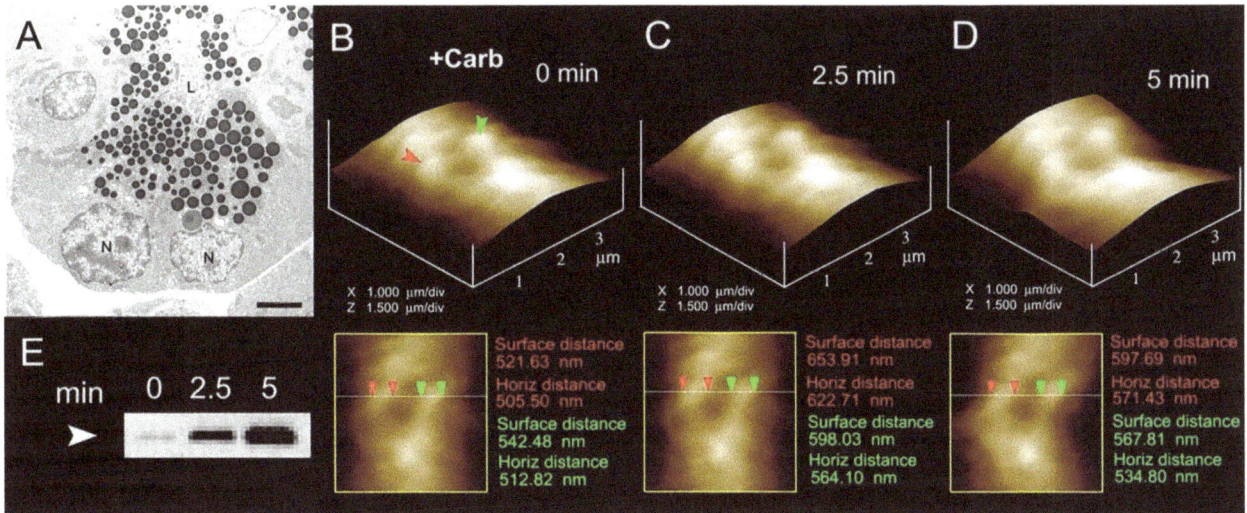

Figure 2. The volume dynamics of zymogen granules (ZGs) in live pancreatic acinar cells demonstrating fractional release of ZG contents during secretion. (A) Electron micrograph of pancreatic acinar cells showing the basolaterally located nucleus (N) and the apically located electron-dense vesicles, the ZGs. The apical end of the cell faces the acinar lumen (L). Bar = 2.5 μm. **(B-D)** Apical ends of live pancreatic acinar cells in physiological buffer imaged by AFM, showing ZGs (red and green arrowheads) lying just below the apical plasma membrane. Exposure of the cell to a secretory stimulus (1 μM carbamylcholine) results in ZG swelling within 2.5 minutes, followed by a decrease in ZG size after 5 minutes. The decrease in size of ZGs after 5 minutes is due to the release of secretory products such as α-amylase, as demonstrated by the immunoblot assay in **(E)**. If ZGs had fused at the plasma membrane and fully merged, it would not be visible, demonstrating transient fusion and fractional discharge on intravesicular contents during secretion in pancreatic acinar cells (from reference [63]).

enzymes over a period of minutes following a meal, as opposed to milliseconds or seconds in a fast secretor such as neurons and endocrine cells. Hence, pancreatic acinar cells were ideal for the discovery of possible plasma membrane structures that could be involved in the fractional release of intravesicular contents from cells during secretion. Similar to neurons, exocrine pancreatic acinar cells are polarized secretory cells possessing an apical and a basolateral end. Digestive enzymes in pancreatic acinar cells are stored within 0.2–1.2 μm in diameter apically located membrane-bound secretory vesicles, called zymogen granules (ZGs). It had been well established that following a secretory stimulus, ZGs dock and fuse at the apical plasma membrane to release their contents. The AFM was therefore trained at the apical plasma membrane of live acinar cells in buffer, before and following stimulation of secretion. These studies demonstrate the presence of pore structures only at the apical plasma membrane of the cell where secretion is known to occur.[5] A group of circular "pits" measuring 0.4–1.2 μm in diameter, contain smaller 100–180 nm in diameter "depressions" structures are identified (**Figure 3**). Typically, three to four depressions are found within each pit structure. Basolateral membrane of acinar cells is devoid of such pit and depression structures.[5] High-resolution AFM images of depressions in live acinar cells further reveal a cup-shaped basket-like morphology,

each cup measuring 35 nm in depth and 100–180 nm in diameter. View of the porosome structure at the cytosolic compartment of the plasma membrane in pancreatic acinar cells has also been determined at near nm resolution[9] (**Figure 4**).

When isolated plasma membrane preparations of pancreatic acinar cells in near-physiological buffered solution are imaged at ultrahigh resolution using the AFM, inverted cup-shaped structures, some with docked ZGs, are demonstrated[9] (**Figure 4**). AFM surface topology maps of inside-out acinar cell plasma membrane preparations reveal scattered circular disks (pits) measuring 0.5–1 μm in diameter, with inverted cup-shaped structures (depressions or porosomes) within.[9] On a number of occasions, ZGs ranging in size from 0.4–1 μm in diameter are observed to be docked at the base of one or more porosomes[62] (**Figure 4**). These results further demonstrated that a single secretory vesicle may dock at more than one porosome at the cell plasma membrane. Porosomes in acinar cells of the exocrine pancreas have also been examined using high-resolution transmission electron microscopy (TEM) (**Figures 5** and **6**), both in isolated cells and tissues (**Figure 5**), and in association with ZGs prepared from stimulated acinar cells (**Figure 6**).[7,9,14,28] In **Figure 5**, the electron micrograph of a porosome sectioned at a certain angle reveals its distinct and separate bilayer and the bilayer membrane

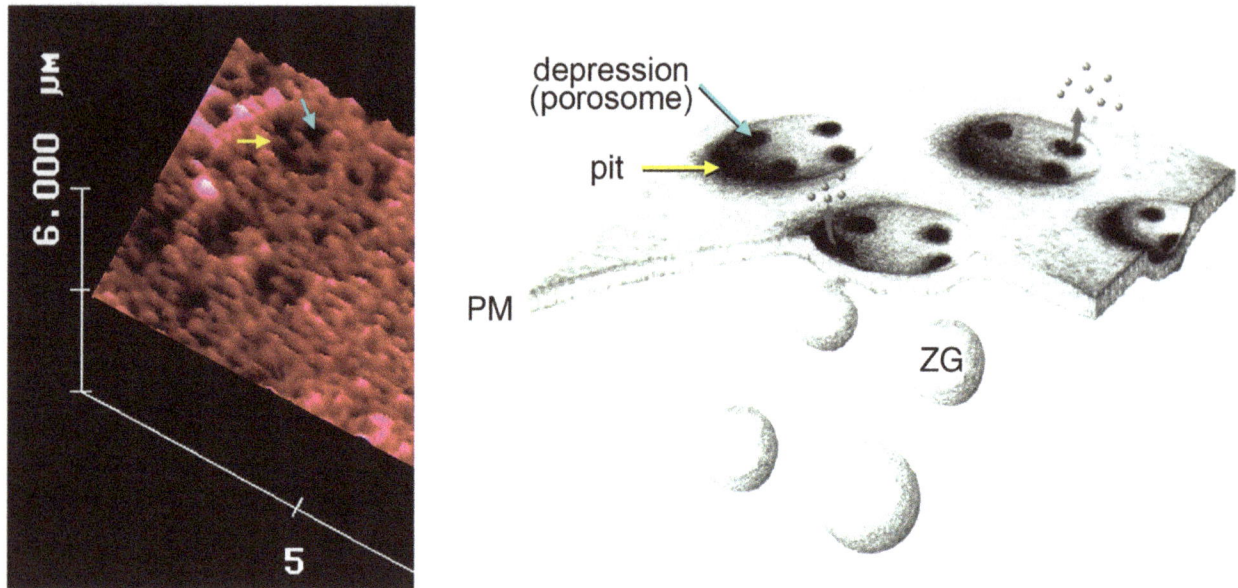

Figure 3. To the left is an AFM micrograph of the apical plasma membrane of a live pancreatic acinar cell demonstrating the presence of a pit (yellow arrow) with depressions or porosomes within (blue arrow). To the right is a schematic drawing demonstrating pits and cup-shaped porosomes where zymogen granules (ZGs), the secretory vesicles in exocrine pancreas, dock and transiently fuse to release intravesicular digestive enzymes from the cell (from reference [5]).

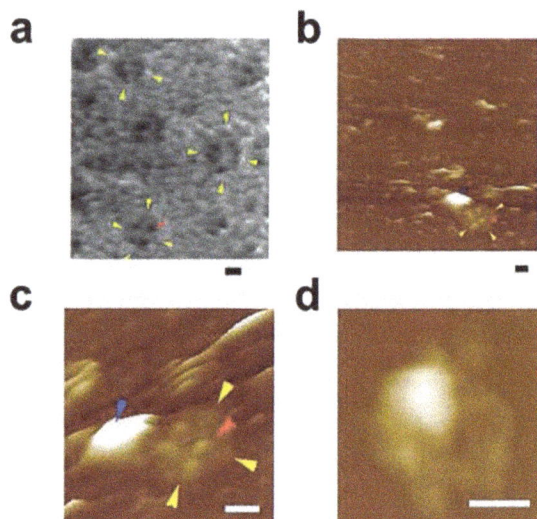

Figure 4. AFM micrograph showing cup-shaped depression or porosome structures at the pancreatic acinar cell plasma membrane. (A) Several circular "pits" (yellow arrowheads) with depressions or porosomes (red arrowheads) are seen in this AFM micrograph of the apical plasma membrane in live pancreatic acinar cell. **(B)** AFM micrograph of the cytosolic compartment of isolated pancreatic plasma membrane preparation depicting a "pit" (yellow arrowheads) containing several inverted cup-shaped porosomes (red arrowhead) within. ZG (blue arrowhead) is found associated with porosomes in figures c and d. **(C)** The "pit" and inverted fusion pores in b is shown at higher magnification. **(D)** AFM micrograph of another "pit" with inverted fusion pores within, and associated with a ZG, is shown. Bar = 200 nm (from reference [9]).

attachment of the associated ZG. A cross-section through three lateral knob-like structures that circle around the porosome cup is clearly delineated. The apical knob-like density at the lip of the porosome appears most prominent and is hypothesized to be composed of actin and myosin, involved in the dilation of the porosome opening during cell secretion. The TEM micrograph further suggests that the lower knob, likely representing the t-/v-SNARE ring or rosette complexes, formed as a consequence of ZG docking and fusion at the base of the porosome complex. Elegant TEM studies in the exocrine pancreas by Craciun and Barbu-Tudoran report docked and post-docked or dissociated ZGs.[14] Similarly, secretory vesicles at the porosome base have been reported at the nerve terminal[12,44,63] and the arrangement of the SNARE complex in a ring or rosette conformation at the porosome base demonstrated.[37] As previously outlined, in vitro studies further reveal the Ca^{2+} and SNARE-mediated membrane fusion to establish the fusion pore[31,33–38,40–59,68,69] and the molecular process of secretory vesicle volume regulation required for the precise fractional release of intravesicular contents from cells[22,60–65] during secretion.

Exposure of pancreatic acinar cells to a secretory stimulus results in a time-dependent increase (20–45%) in both the diameter and relative depth of porosomes (**Figure 7**). Porosomes return to their resting size on completion of cell secretion.[5,6] No demonstrable change in pit size is detected following stimulation of secretion.[5] Enlargement

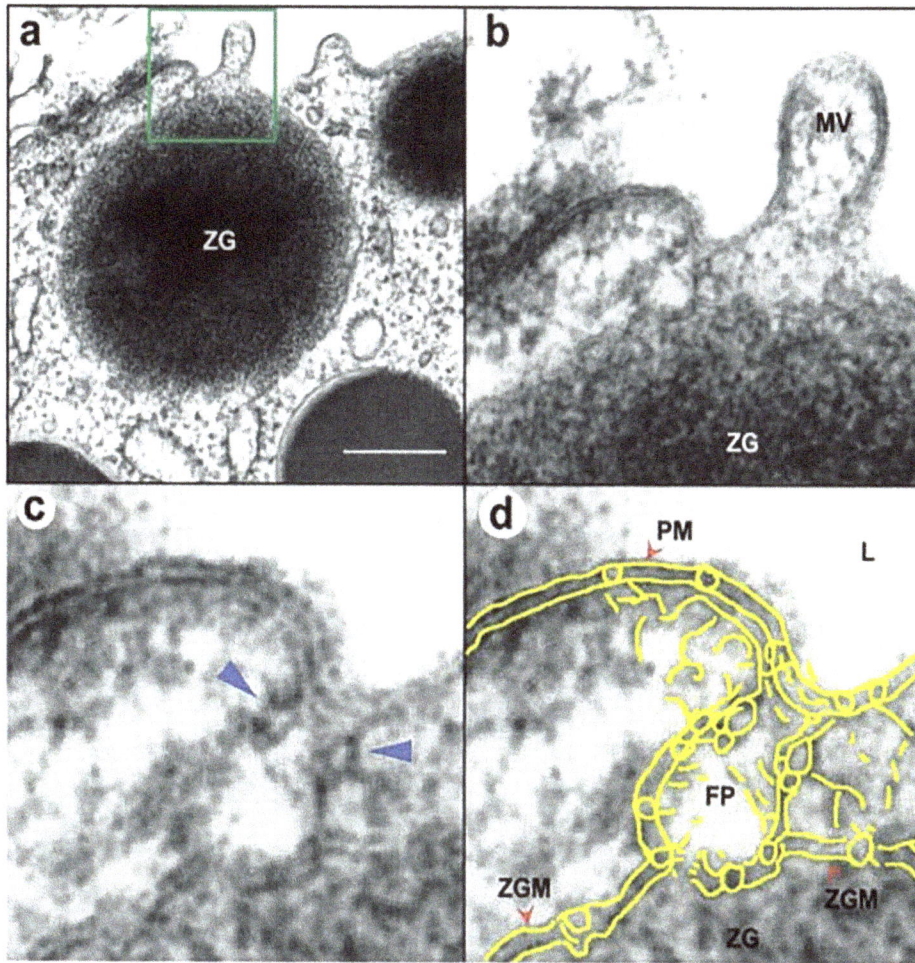

Figure 5. Transmission electron micrograph of a porosome associated with a docked secretory vesicle at the apical end of a pancreatic acinar cell. (**A**) Part of the apical end of a pancreatic acinar cell demonstrating within the green square the presence of a porosome and an associated zymogen granule (ZG) fused at its base (Bar = 400 nm only in figure a). (**B**) The area within the green square in figure a has been enlarged to show the apical microvilli (MV) and a section through the porosome and the ZG. Note the ZG membrane (ZGM) bilayer is fused at the base of the porosome cup. (**C**) A higher magnification the porosome-associated ZG shows in greater detail the porosome bilayer and cross-section through the three protein rings (which appear as knobs in either side of the cup-shaped porosome), with the thicker ring (blue arrowhead) present close to the opening of the porosome to the outside, which may regulate the closing and opening of the structure. The third and the lowest ring away from the porosome opening is attached to the ZGM and may represent the t-/v-SNARE rosette or ring complex. (**D**) Yellow outline of the ZG-fused porosome complex (FP) demonstrating the continuity with the apical plasma membrane (PM) at the apical end of the pancreatic acinar cell facing the lumen (L). The exact points of contact and fusion of the ZGM with the membrane at the porosome base are clearly seen in the micrograph (from reference [9]).

of porosome diameter and an increase in its relative depth, following exposure to a secretory stimulus, correlates with secretion (**Figures 7** and **8**). Additionally, exposure to cytochalasin B, a fungal toxin that inhibits actin polymerization and secretion, results in a 15–20 percent decrease in porosome size and a consequent 50–60 percent loss in cell secretion.[5] These results suggested depressions or porosomes to be the secretory portals in pancreatic acinar cells. Results from these studies further demonstrate actin to compose the porosome in addition to implicating actin in regulation of both the structure and function of porosomes.

To further demine if porosomes serve as cellular secretory portals, the direct observation of the release of secretory products through the structure was required. This was accomplished using immuno-AFM studies in acinar cells of the exocrine pancreas[6] (**Figures 9** and **10**) where gold-conjugated amylase-specific antibody was demonstrated to selectively decorate the porosome opening following a secretory stimulus.[6,7]

To further confirm that the cup-shaped structures are porosomes, where secretory vesicles dock and fuse, additional immuno-AFM studies have been performed.[7] Since

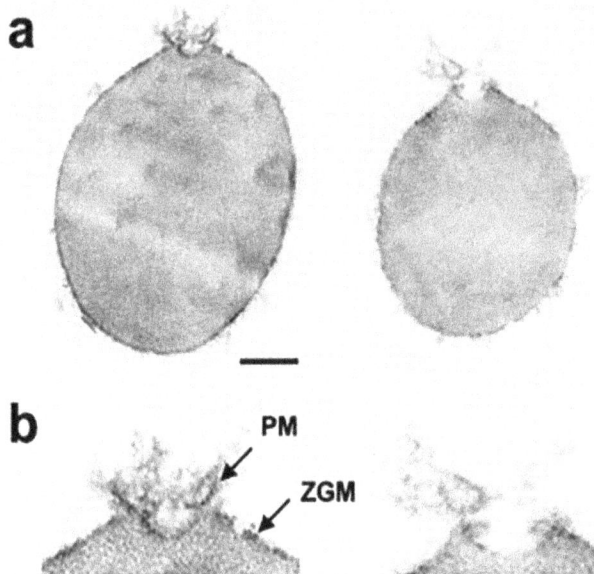

Figure 6. Transmission electron micrographs of zymogen granules (ZGs) co-isolated with porosomes. (a) Porosome associated at the surface of ZGs are shown. (Bar = 120 nm; top panel only). **(b)** At higher magnification details of the porosome complex demonstrating the presence of separate plasma membrane (PM) and the ZG membrane (ZGM). Note the apically arranged ring complex of the porosome, similar to what is observed in electron micrographs of the structure in intact cells as presented in Figure 5. These ZG-associated porosomes are torn off the cell plasma membrane and hence have very little membrane. Therefore, the porosome proteins lining the porosome cup appear as frills in the electron micrograph. Note the ZG size compared to the porosome structure (from reference [9]).

ZGs dock and fuse at the plasma membrane to release vesicular contents, it was hypothesized that if the inverted cups or porosomes are indeed the secretory portals, then plasma membrane-associated t-SNAREs should be present at the base of the porosome cup facing the cytosol. Since the t-SNARE protein SNAP-23 is present in pancreatic acinar cells,[7] a polyclonal monospecific SNAP-23-specific antibody recognizing a single 23-kDa protein in Western blots of pancreatic acinar cell plasma membrane fraction was used in the study. Immuno-AFM studies demonstrate the selective binding of the SNAP-23 antibody to the base of the porosome complex (**Figure 11**).[7] These results establish that the inverted cup-shaped structures in the inside-out pancreatic plasma membrane preparations are indeed secretory portals or porosomes, where secretory vesicles transiently dock and fuse to release their contents during cell secretion.

Immunoisolation studies using SNAP-23-specific antibody on solubilized pancreatic plasma membrane fractions demonstrate the isolation of the porosome complex as assessed both structurally (**Figures 12** and **13**) and functionally (**Figure 14**).[9] Furthermore, immunochemical

characterization of the pancreatic porosome complex demonstrates the presence of SNAP-23, syntaxin 2, actin, fodrin, vimentin, chloride channels CLC2 and CLC3, calcium channels β3 and α1c, and the SNARE regulatory protein NSF, among other proteins.[7,9] Transmission electron micrographs of pancreatic porosomes reconstituted into liposomes exhibit a 150–200-nm cup-shaped basket-like morphology (**Figure 13**), similar to its native structure observed in cells and when coisolated with ZG preparation.[9] To test the functionality of the immunoisolated porosome complex, purified porosomes obtained from exocrine pancreas have been reconstituted in the lipid membrane of the EPC9 electrophysiological bilayer apparatus and exposed to isolated ZGs (**Figure 14**). Electrical activity of the porosome-reconstituted membrane as well as the transport of vesicular contents from the *cis* to the *trans* compartments of the bilayer chambers demonstrate that the lipid membrane-reconstituted porosomes are indeed functional. Thus, in the presence of calcium, isolated secretory vesicles dock and fuse to transfer intravesicular contents from the *cis* to the *trans* compartment of the bilayer chamber (**Figure 14**). ZG fusion and content release through the reconstituted porosome is demonstrated by the increase in capacitance and conductance and a time-dependent transport of the ZG enzyme amylase from *cis* to the *trans* compartment of the bilayer chamber measured using Western blot assays. Amylase is detected using immunoblot analysis of the buffer in the *cis* and *trans* compartments of the bilayer chambers. Chloride channel activity present in the porosome complex has been demonstrated to be critical to porosome function, since the chloride channel blocker DIDS is inhibitory to the function of the reconstituted porosome (**Figure 14**). Interestingly, recent studies[19] demonstrate the interaction between the cystic fibrosis transmembrane conductance regulator (CFTR) and the porosome complex in human airways epithelia. This study implies that the CFTR chloride selective ion channel affects mucus composition and secretion through the porosome complex at the cell plasma membrane of the airways epithelia. Similarly, in cystic fibrosis, CFTR dysfunction is known to reduce secretory activity of pancreatic duct cells,[70] which leads to blockage of the ductal system and eventual fibrosis of the entire gland. This is just one examples of diseases resulting from altered porosome-associated protein.

Porosome Proteins in Secretory Defects and Disease

Porosomes are composed of 30–40 proteins. The composition of proteins in porosomes from different tissue has been demonstrated to be similar, if not identical. Among porosome proteins of the exocrine pancreas are the t-SNAREs (SNAP-23 and syntaxin), actin, vimentin, α-fodrin, Ca^{2+} α1c, and Ca^{2+} β1c.[7] Similarly, in the well-characterized neuronal porosomes, actin, SNAREs, tubulin, and calcium

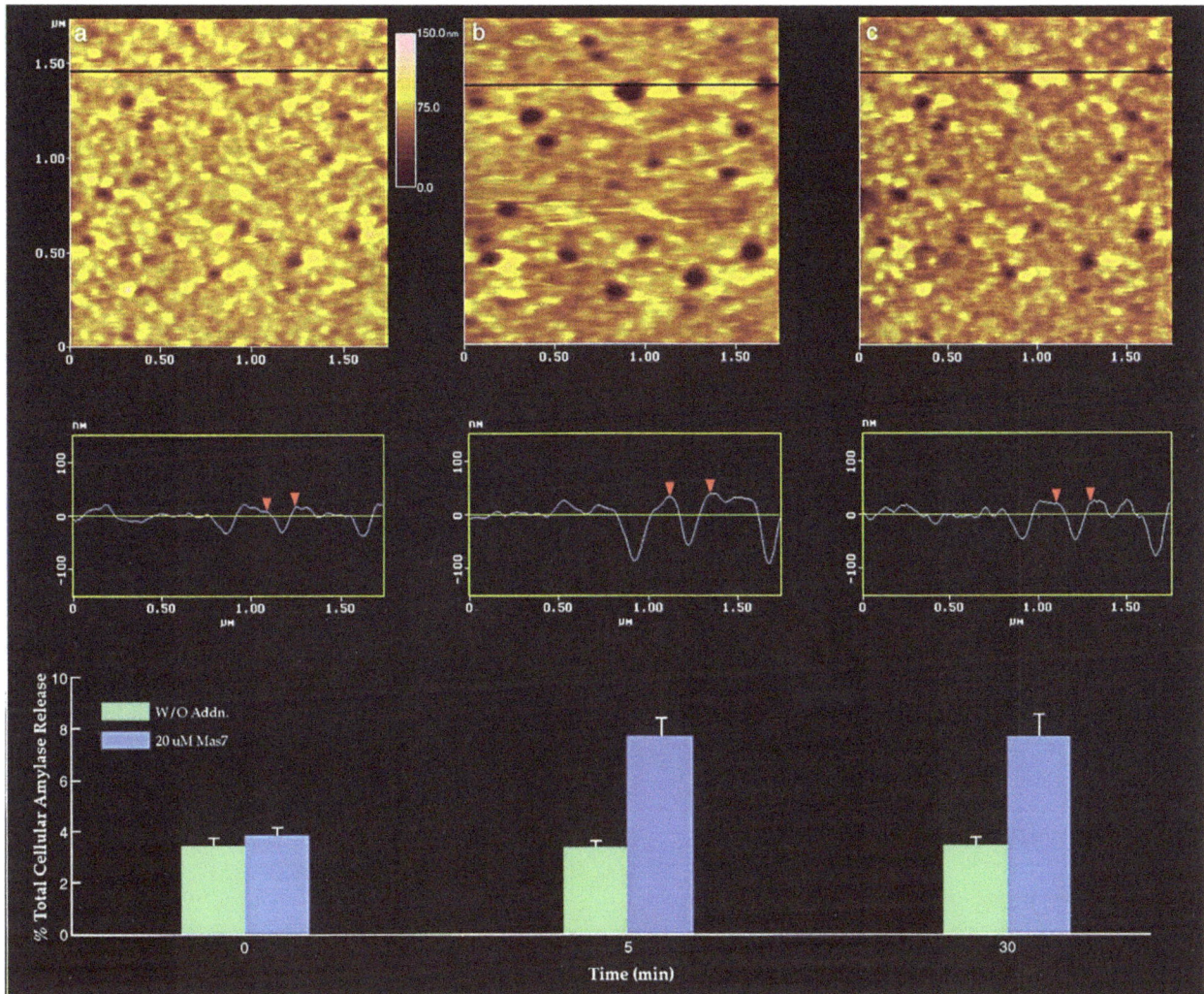

Figure 7. Dynamics of porosomes following stimulation of secretion in live pancreatic acinar cell examined using AFM. (A) Several porosomes within a pit are shown in the AFM micrographs. The scan line across three porosomes in the top panel is represented graphically in the middle panel and defines the diameter and relative depth of three porosomes. The porosome at the center is labeled using red arrowheads. The bottom panel shows the percent total cellular amylase release in the presence (blue bars) and absence (green bars) of the secretion stimulatory tetradecapeptide mastoparan Mas7. **(B)** Note the increase in porosome diameter and relative depth, correlating with an increase in total cellular amylase release at 5 minutes following stimulation of secretion. **(C)** At the 30-minute time point following stimulation, there is a decrease in diameter and depth of porosomes, with no further increase in amylase release following what is observed following the 5-minute time point. No significant changes in amylase secretion or porosome size were observed in control acini, in either the presence or the absence of the non-stimulatory mastoparan analogue Mas17, throughout the time points examined. High-resolution images of both pits and their porosomes were obtained before and after stimulation with Mas7, for up to 30 minutes (from reference [5]).

and potassium transporters have also been reported.[7,10,24] It has long been established that actin is involved in neurotransmission,[71] and our own studies demonstrate its role in secretion of amylase from the exocrine pancreas.[5] Similar to the exocrine pancreas demonstrating the collapse of the porosome opening and loss of amylase secretion following exposure to the actin depolymerizing cytocholasin D,[35] studies performed on neurons in presence of latrinculin A, also an actin depolymerizing agent, partially blocked

neurotransmitter release at the presynaptic terminal of motor neurons.[11] Actin b protein, a posttranslational product of actb mRNA, is important in formation of excitatory synapses, which is promoted by interaction of actb mRNA with Src associated in mitosis (Sam68 protein). Loss in Sam68 diminishes its interaction with actb mRNA leading to lower levels of synaptic actin b, which in turn leads to neurological disorders involved with abnormal synaptic transmissions.[72] Although tubulin involvement in

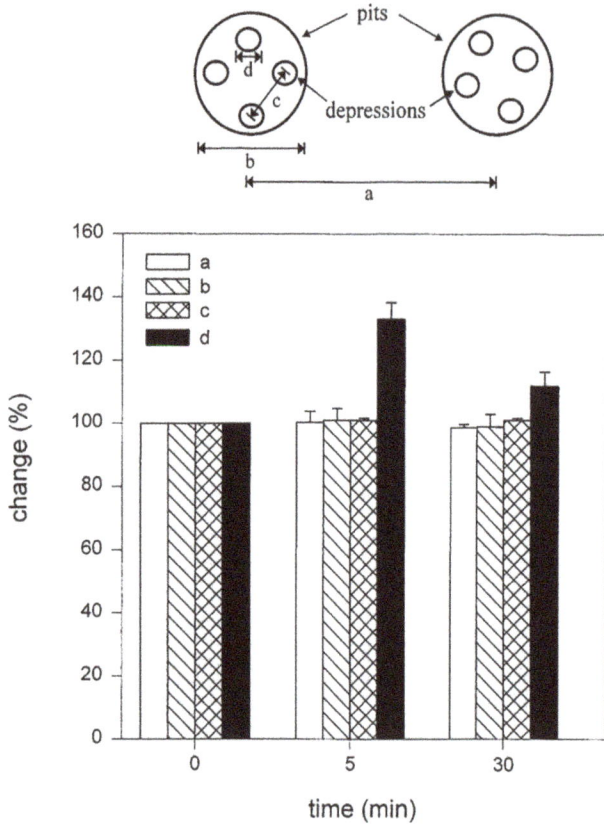

Figure 8. Changes are observed only in depressions or porosomes following stimulation of secretion. Analysis of the dimensions *a–d*, schematically represented at the top and graphically presented below, demonstrates a significant increase in the porosome diameter at 5 minutes and a return toward prestimulatory levels after 30 minutes. No changes (100%) in *a–c* are seen throughout the times examined. Pit and porosome diameters were estimated using section analysis software from Digital Instruments. Each single pit and porosome was measured twice, once in the scan direction and once at 90° to the first (from reference [5]).

neurotransmission is not fully understood, its association with several synaptosomal proteins at the presynaptic membrane[63,73] suggests an important role in neurotransmission. Besides actin and tubulin, other important classes of proteins identified in the neuronal porosome are the membrane-integrated ion-channel proteins that are important in maintaining cellular homeostasis. A good example is the alpha subunit 3 of Na^+/K^+ATPase present in the porosome and plays a critical role in neurotransmission. Na^+/K^+ATPase activity is known to be blocked by dihydrooubain (DHO).[23] Activation of Na^+/K^+ATPase inhibits synaptic transmission, which results from the secretion of a presynaptic protein "follistatin-like 1" which activates the alpha 1 subunit of the ATPase. Transient blocking of Na^+/K^+ATPase pump using DHO results in an increase in both the amplitude and number of action potentials at the nerve terminal. Interestingly, Na^+/K^+ATPase inhibition is

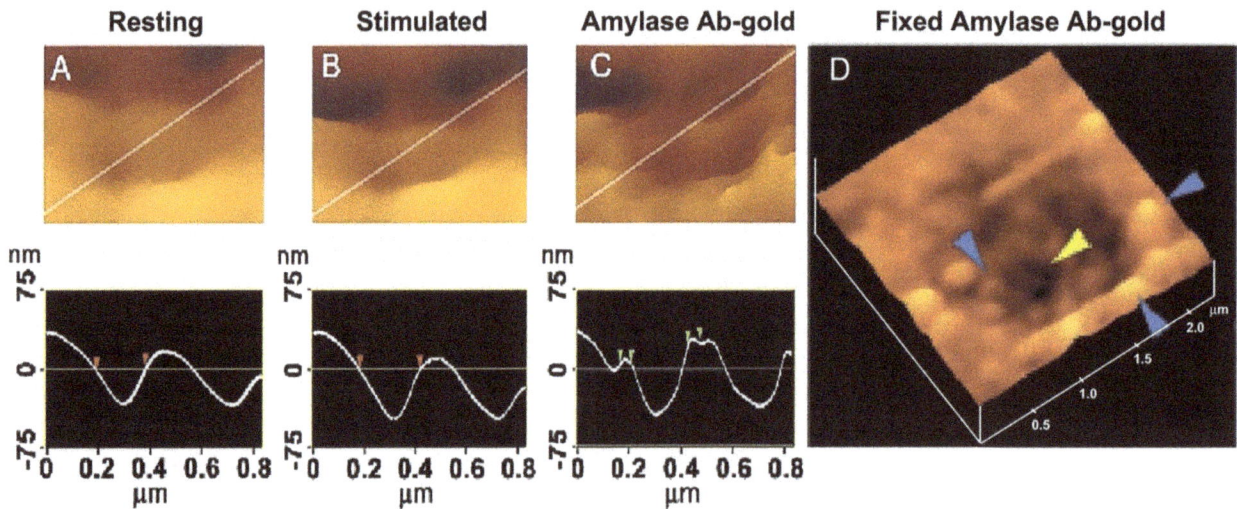

Figure 9. Intravesicular contents are expelled from within cells to the outside through the porosome during cell secretion. (**A**) and (**B**) AFM micrograph and section analysis of a pit and two of the four porosomes, demonstrating enlargement of porosomes following stimulation of cell secretion in the acinar cell of the exocrine pancreas. (**C**) Exposure of live cells to gold-conjugated amylase antibody (Ab) results in selective localization of gold particles to these secretory sites. Note the localization of amylase-specific 30-nm immunogold particles at the edge of porosomes. (**D**) AFM micrograph of pits and porosomes with immunogold localization demonstrated in cells immunolabeled and then fixed. Blue arrowheads point to immunogold clusters and the yellow arrowhead points to a porosome opening (from reference [6]).

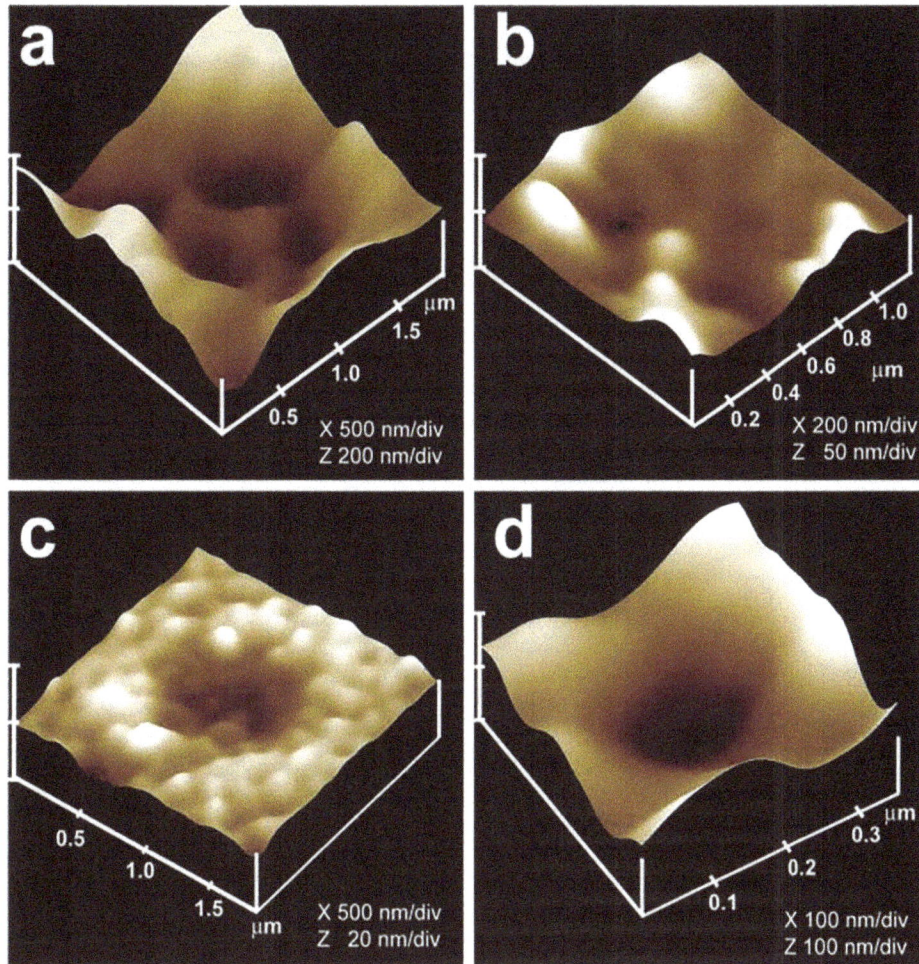

Figure 10. **AFM and immune-AFM micrographs of the pancreatic acinar cell porosome demonstrating pore morphology and the release of secretory products at the site.** (**A**) A pit with four porosomes within, found at the apical surface in a live pancreatic acinar cell. (**B**) After stimulation of secretion, amylase-specific immunogold localize at the pit and porosome openings, demonstrating porosomes to represent the secretory release sites in cells. (**C**) Some porosomes demonstrate greater immunogold localization, suggesting greater secretory product release through them. (**D**) AFM micrograph of a single porosome in a live acinar cell (from reference [7]).

calcium dependent, and it has been demonstrated that increased intracellular calcium levels inhibit Na$^+$/K$^+$ATPase, which increases excitability of neurons.[74] Similarly, the porosome protein plasma membrane calcium ATPases (PMCA) is an important class of proteins known to be involved in maintaining calcium homeostasis within cells including neurons.[24] PMCA2 has been shown to colocalize in synaptosomes with synaptohysin.[23] At the presynaptic membrane, syntaxin-1, also a porosome protein, has been demonstrated to colocalize with PMCA2 and the glycine transporter 2 (GlyT2) that is found coupled to the Na$^+$/K$^+$ pump, suggesting the presence of a protein complex involved in neurotransmission.[24,48,75] Similarly, deletion of PMCA2 generates an ataxic phenotype in mice, where the neurons possess prolonged hyperpolarized membrane, resulting from an increase in the basal levels of calcium within these neurons. Mutation in the PMCA2 encoding

gene is known to result in homozygous deafwaddler mice (dfw/dfw) and they show high levels of calcium accumulation within their synaptic terminals. NAP-22, also known as BASP-1, is a protein found in the neuronal porosome complex. Due to its localization at the pre- and postsynaptic membranes and also in synaptic vesicles, it has been long speculated to be involved in synaptic transmission.[23] NAP-22 is known to bind to the inner leaflet of lipid rafts suggesting interaction with cholesterol. Adenylyl cyclase associated protein-1 (CAP-1) is known to regulate actin polymerization and both actin and CAP-1 are present in the porosomal complex. In cells depleted of CAP-1, using RNAi results in lamellipodia growth and F-actin accumulation along with other cytoskeletal abnormalities. In Alzheimer's, the levels of CNPase (2,3-cyclic nucleotide phosphodiesterase) and the heat shock protein 70 (HSP70) are found to increase while the levels of

212 Bhanu P. Jena

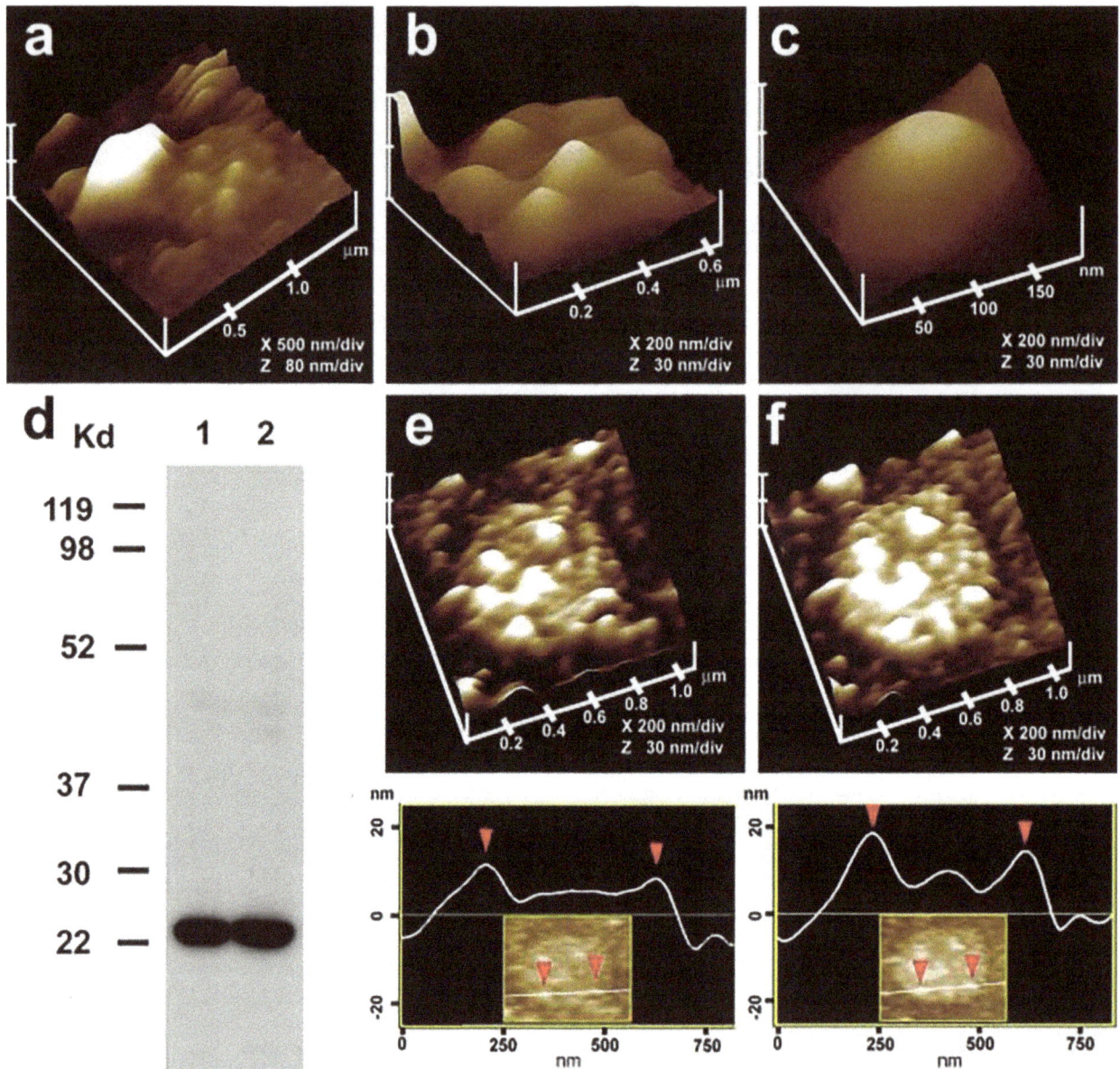

Figure 11. Immuno-AFM of the cytosolic compartment of the porosome complex demonstrate the presence of the t-SNARE protein SNAP-23 at the porosome base in acinar cell of the exocrine pancreas. (A) AFM micrograph of isolated plasma membrane preparation reveals the cytosolic compartment of a pit with inverted cup-shaped porosomes. Note the 600 nm in diameter docked ZG to the left. (B) Higher magnification of the same pit demonstrates the presence of four to five porosomes within. (C) A single porosome is depicted in this AFM micrograph. (D) Western blot analysis of 10 μg and 20 μg of pancreatic plasma membrane preparation resolved using SDS-PAGE demonstrates a single 23-kDa immunoreactive band when probed with SNAP-23-specific antibody. The overexposure of the film is to demonstrate the absence of no other bands, hence specificity of the antibody. (E) and (F) The cytosolic compartment of the plasma membrane with a pit having a number of porosomes within (E) demonstrates binding of the SNAP-23-specific antibody (F), establishing them to represent the docking sites of secretory vesicles at the cell plasma membrane. Note the increase in height of the porosome base revealed by section analysis (bottom panel), demonstrating localization of SNAP-23 antibody to the porosome base (from reference [7]).

dihydropyrimidinase-related protein-2 (DRP-2) decrease. Decreased levels of CNPase have been detected in the frontal and temporal cortex of patients with Alzheimer's disease and/or Down's syndrome. Low CNPase levels have also been detected in postmortem anterior frontal cortex in schizophrenic patients.[23] Additionally, an allele that is associated with low levels of CNPase is also reported to be linked to schizophrenia.[56] CNPase is also detected as a

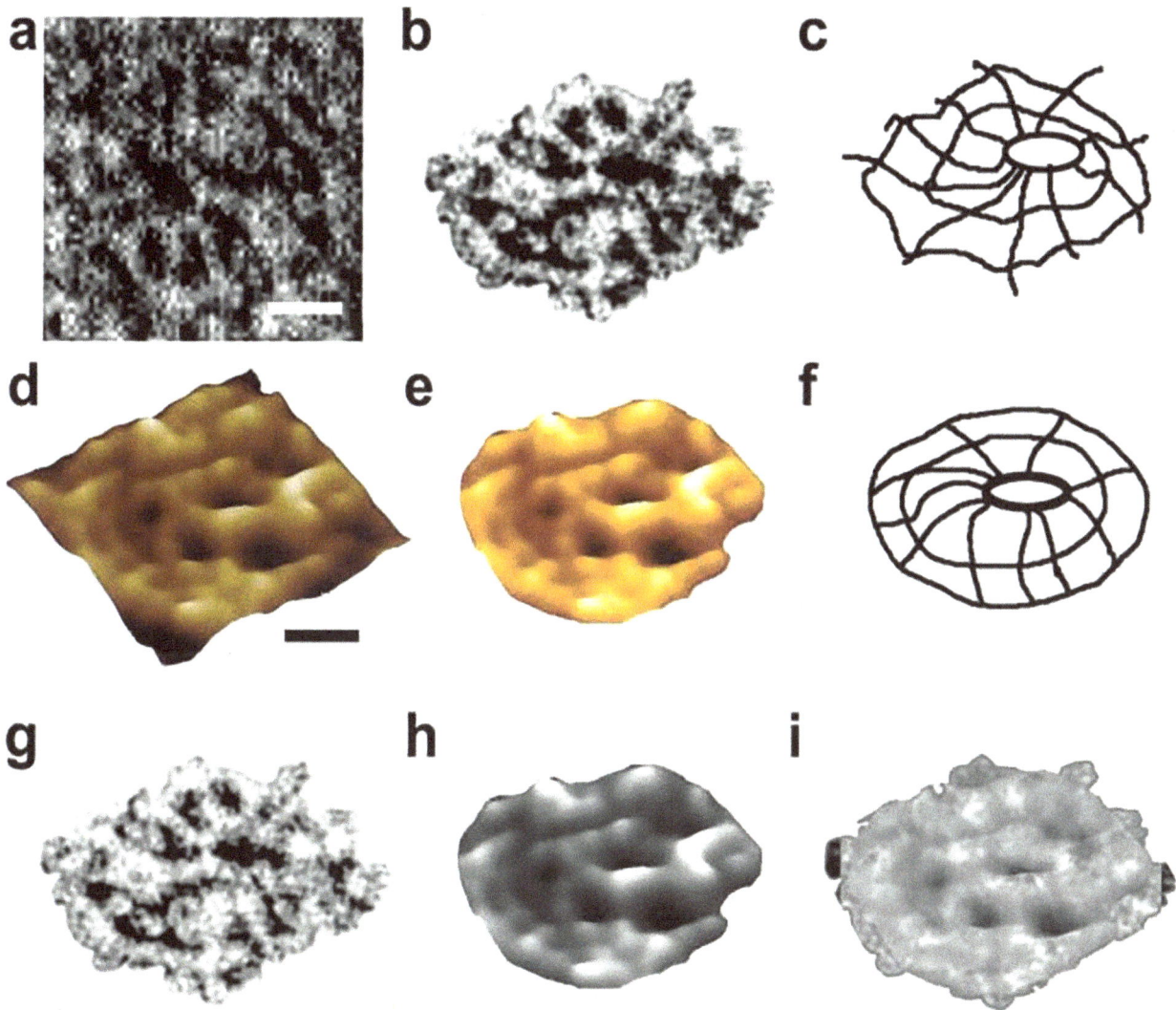

Figure 12. **Negatively stained EM and AFM of the immunoisolated porosome complex from exocrine pancreas.** (**A**) Negatively stained EM of an immunoisolated porosome complex from solubilized pancreatic plasma membrane preparation using a SNAP-23-specific antibody. Note the three rings and the 10 spokes that originate from the inner ring. This structure represents the protein backbone of the pancreatic porosome complex. Bar = 30 nm. (**B**) EM of the isolated porosome complex extracted from figure **A** and (**C**) an outline of the structure presented for clarity. (**D–F**) AFM micrograph of an isolated porosome complex in physiological buffer solution. Bar = 30 nm. Note the structural similarity of the complex when imaged either by EM (**G**) or AFM (**H**). The EM and AFM micrographs are superimposable (**i**) (from reference [9]).

marker for oligodendroglia and myelin and several diseases associated with low levels of CNPase indicating low myelin on neurons.[55] CNPase-positive cells have been shown to increase in corpus callosum of rats exposed to an enriched environment, meaning given to perform a task. Similarly, alterations in the levels of several of the SNARE proteins are associated with various neurological disorders. For instance, SNAP-25 and synaptophysin are significantly reduced in neurons of patients with Alzheimer's disease. Mice that are SNAP-25 (+/-) show disabled learning and memory and exhibit epileptic-like seizures.[23]

Overexpression of SNAP-25 also results in defects in cognitive function. Loss of SNAP-25 is also associated with Huntington's disease and a reduction in rabphilin 3a, another protein involved in vesicle docking and fusion at the presynaptic membrane. SNARE knockout mice are neonatally lethal and mice with a dominant mutation in SNAREs are known to develop ataxia and show impairment in vesicle recycling capability. Increase in synaptophysin levels along with SNAP-25 are also observed in Broddmann's area in the postmortem brain of patients with bipolar disorder I. Protein levels of synaptotagmin and

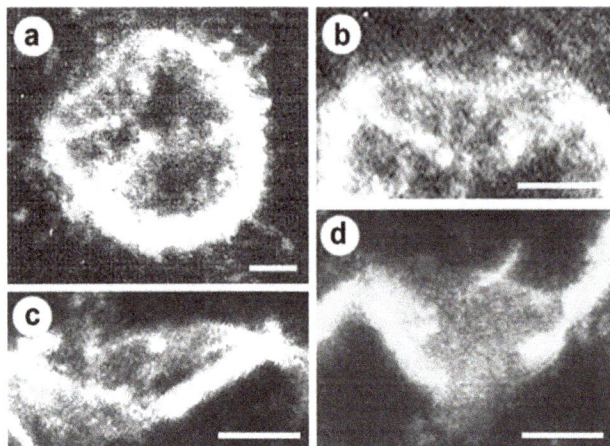

Figure 13. **Electron micrographs of reconstituted porosome complex into liposomes demonstrate a cup-shaped basket-like morphology.** (**A**) A 500-nm vesicle with an incorporated porosome complex is shown. Note the spokes in the complex. The reconstituted complex at higher magnification is shown in **B–D**. Bar = 100 nm (from reference [9]).

syntaxin 1 are highly upregulated in areas of cerebral ischemia, which are known to exhibit highly active levels of neurotransmission. Mutations in certain regions of the protein syntaxin 1A, such as Ca^{+2} channel-binding region, are known to increase neurotransmitter release, which suggests that syntaxin 1A is involved in regulating Ca^{+2} channel function. A point mutation in syntaxin 1A is known to result in augmented release of neurotransmitters in models of the fruit fly *Drosophila melanogaster*. Synaptotagmin similar to syntaxin has Ca^{+2}-binding domains and is known to form dimers and interact with syntaxin to form complexes in the presence of calcium.[23] Binding of synaptotagmin to SNAP 25 is also calcium-dependent.[58,59] Reticulons are proteins which contribute to lipid membrane curvature and are found in the neuronal porosome complex. The involvement of reticulons with the porosome and diseases associated with their deregulation lend credence to the role of membrane curvature in regulation of synaptic vesicle fusion at the porosome complex. These studies reflect on the critical role of various neuronal porosome proteins in protein–protein and protein–lipid interactions within the porosome complex and in their participation in normal porosome-mediated neurotransmission at the nerve terminal and in disease. In addition to the above studies discussed on the involvement of porosome proteins in neurotransmission-related disease, recent morphological studies on the neuronal porosome complex have also shed light on both the health and disease status of the neuronal porosome complex as examined using high-resolution EM.[23] Ultrastructure of the neuronal porosome complex in rats subjected to continuous white noise that is relevant to the increasing random noise encountered by humans in

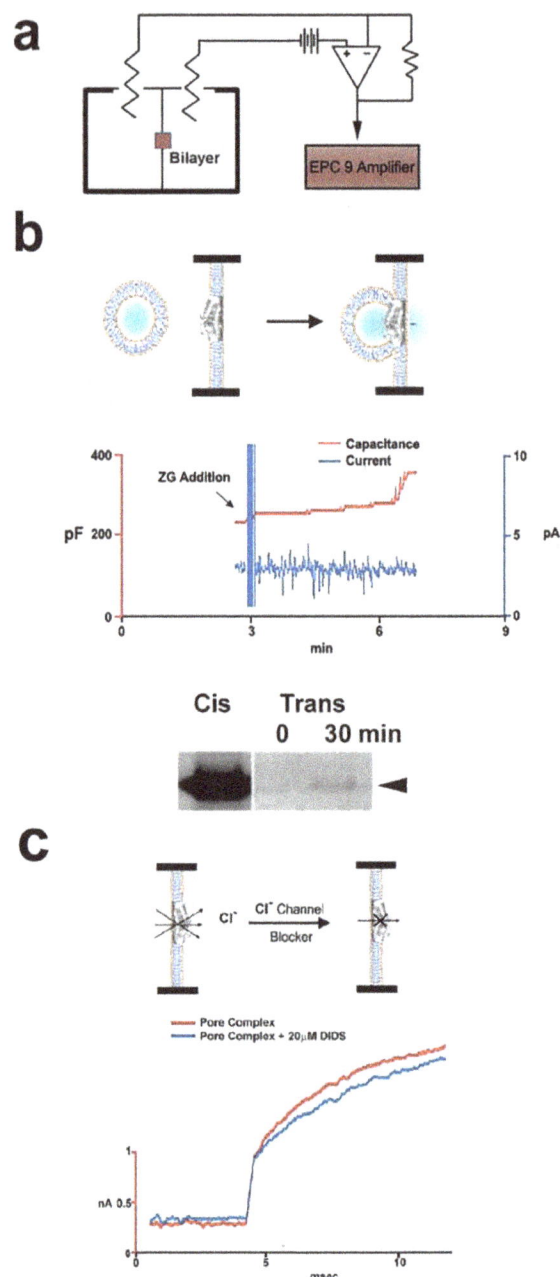

Figure 14. **Functional reconstitution of the pancreatic porosome complex.** (**A**) Schematic drawing of the EPC9 bilayer setup for electrophysiological measurements. (**B**) Zymogen granules (ZGs) added to the cis compartment (left) of the bilayer chamber fuse with the reconstituted porosomes, as demonstrated by the increase in capacitance (red trace) and current (blue trace) activities and a concomitant time-dependent release of amylase from the cis to the trans compartment of the bilayer chamber. The movement of amylase from the cis to the trans compartment in the EPC9 setup was determined by Western blot analysis of the contents in the cis and the trans chamber over time. (**C**) Electrical measurements in the presence and absence of chloride ion channel blocker DIDS demonstrate the presence of chloride channels in association with the complex and its requirement in porosome function (from reference [9]).

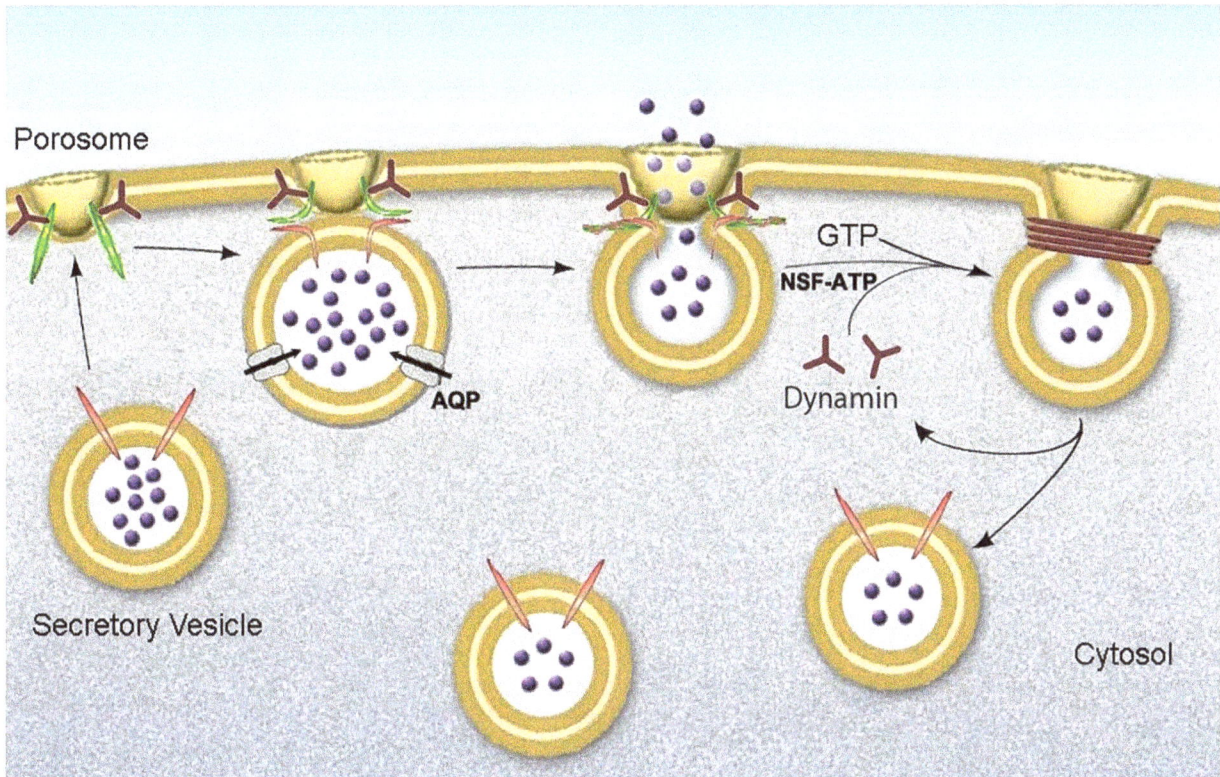

Figure 15. Schematic diagram depicting the transient docking and fusion of secretory vesicles at the porosome base in the cell plasma membrane, increase in turgor pressure of secretory vesicles via the entry of water and ions through water channels called aquaporins (AQP) and ion channels, expulsion of vesicular contents via the porosome to the outside, and the dissociation of the partially empty secretory vesicles from the porosome complex at the cell plasma membrane (from reference [24]).

today's environment is known to provoke diverse effects on different brain regions such as the documented alteration in the length of the porosome complex. Constant exposure to such noise is further known to sabotage the development and normal function of audition, impair hearing, language acquisition, memory performance, and other cognitive functions.

Summary

In summary, these studies demonstrate how the power and scope of the AFM[67] enabled the discovery of the porosome, first in acinar cells of the exocrine pancreas[5,7,9] and later in other cell types including neurons.[10,24] Porosomes are permanent supramolecular lipoprotein structures at the cell plasma membrane in pancreatic acinar cells, where membrane-bound secretory vesicles called ZGs transiently dock and fuse to release intravesicular contents from within cells to the outside. A schematic drawing of porosome-mediated fractional release of intravesicular contents during cell secretion is presented in **Figure 15**. As opposed to the complete merger of secretory vesicles at the cell plasma membrane (via an all or none mechanism), the porosome complex prevents secretory vesicle collapse at the cell plasma membrane during the secretory process and provides precise regulation of intravesicular content release during cell secretion. Whether secretion involves the docking and fusion of a single vesicle or compound exocytosis where a docked vesicle may involve a number of vesicles fused, the porosome would provide specificity and regulation of content release without compromising the integrity of the secretory vesicle membrane and the cell plasma membrane. In this porosome-mediated secretion, following fractional discharge of intravesicular contents, the partially empty vesicle may undergo several docking-fusion-release cycles until empty, prior to being recycled. However, in neurons, synaptic vesicles could be rapidly refilled via neurotransmitter transporters present at the vesicle membrane. A number of secretory defects and disease states are now recognized as a consequence of disfunction of one or more porosome proteins.[23] Further understanding of the molecular structure and regulation of the porosome complex will provide new and novel therapeutic applications. The recent functional reconstitution of the insulin-secreting porosome complex in live beta cells of the endocrine pancreas opens a window in the treatment of diabetes. To determine the distribution of constituent proteins within the complex will require further studies using small-angle X-ray solution scattering,[12] chemical cross-linking followed my mass spectrometry, immuno-EM, immuno-AFM, and single-particle cryo-electron tomography.

Acknowledgments

The work described in this manuscript was supported in part by grants from the NIH R01 DK56212 and NS39918 and NSF EB00303, CBET1066661 to BPJ.

References

1. Cho SJ, Cho J, Jena BP. The number of secretory vesicles remains unchanged following exocytosis. Cell Biol Int 2002; 26:29–33.

2. Ceccarelli B, Hurlbut WP, Mauro A. Turnover of transmitter and synaptic vesicles at the frog neuromuscular junction. J Cell Biol 1973; 57:499–524.

3. Almers W, Tse FW. Transmitter release from synapses: does a preassembled fusion pore initiate exocytosis? Neuron 1990; 4:813–818.

4. Monck JR, Fernandez JM. The exocytotic fusion pore. J Cell Biol 1992; 119:1395–1404.

5. Schneider SW, Sritharan KC, Geibel JP, Oberleithner H, Jena BP. Surface dynamics in living acinar cells imaged by atomic force microscopy: identification of plasma membrane structures involved in exocytosis. Proc Natl Acad Sci U S A 1997; 94:316–321.

6. Cho SJ, Quinn AS, Stromer MH, Dash S, Cho J, Taatjes DJ, Jena BP. Structure and dynamics of the fusion pore in live cells. Cell Biol Int 2002; 26:35–42.

7. Jena BP, Cho SJ, Jeremic A, Stromer MH, Abu-Hamdah R. Structure and composition of the fusion pore. Biophys J 2003; 84:1337–1343.

8. Cho SJ, Wakade A, Pappas GD, Jena BP. New structure involved in transient membrane fusion and exocytosis. Ann N Y Acad Sci 2002; 971:254–256.

9. Jeremic A, Kelly M, Cho SJ, Stromer MH, Jena BP. Reconstituted fusion pore. Biophys J 2003; 85:2035–2043.

10. Cho WJ, Jeremic A, Rognlien KT, Zhvania MG, Lazrishvili I, Tamar B, Jena BP. Structure, isolation, composition and reconstitution of the neuronal fusion pore. Cell Biol Int 2004; 28:699–708.

11. Jena BP. Nano cell biology of secretion: imaging its cellular and molecular underpinnings. SpringerBriefs in Biological Imaging 2012; 1:1–70.

12. Kovari LC, Brunzelle JS, Lewis KT, Cho WJ, Lee JS, Taatjes DJ, Jena BP. X-ray solution structure of the native neuronal porosome-synaptic vesicle complex: implication in neurotransmitter release. Micron 2014; 56:37–43.

13. Cho SJ, Jeftinija K, Glavaski A, Jeftinija S, Jena BP, Anderson LL. Structure and dynamics of the fusion pores in live GH-secreting cells revealed using atomic force microscopy. Endocrinology 2002; 143:1144–1148.

14. Craciun C, Barbu-Tudoran L. Identification of new structural elements within "porosomes" of the exocrine pancreas: a detailed study using high-resolution electron microscopy. Micron 2013; 44:137–142.

15. Drescher DG, Cho WJ, Drescher MJ. Identification of the porosome complex in the hair cell. Cell Biol Int Rep 2011; 18:31–34.

16. Elshennawy WW. Image processing and numerical analysis approaches of porosome in mammalian pancreatic acinar cell. J American Sci 2011; 6:835–843.

17. Hammel I, Meilijson I. Function suggests nano-structure: electrophysiology supports that granule membranes play dice. J R Soc Interface 2012; 9:2516–2526.

18. Horber JK, Miles MJ. Scanning probe evolution in biology. Science 2003; 302:1002–1005.

19. Hou X, Lewis KT, Wu Q, Wang S, Chen X, Flack A, Mao G, Taatjes DJ, Sun F, Jena BP. Proteome of the porosome complex in human airway epithelia: interaction with the cystic fibrosis transmembrane conductance regulator (CFTR). J Proteomics 2014; 96:82–91.

20. Japaridze NJ, Okuneva VG, Qsovreli MG, Surmava AG, Lordkipanidze TG, Kiladze MT, Zhvania MG. Hypokinetic stress and neuronal porosome complex in the rat brain: the electron microscopic study. Micron 2012; 43:948–953.

21. Jena BP. Porosome: the secretory portal. Exp Biol Med (Maywood) 2012; 237:748–757.

22. Jena BP, Schneider SW, Geibel JP, Webster P, Oberleithner H, Sritharan KC. G$_i$ regulation of secretory vesicle swelling examined by atomic force microscopy. Proc Natl Acad Sci U S A 1997; 94:13317–13322.

23. Jena BP. Cellular nanomachines. From discovery to structure-function and therapeutic applications. Springer Nature, 2020;1–105, ISBN 978-3-030-44495-2.

24. Lee JS, Jeremic A, Shin L, Cho WJ, Chen X, Jena BP. Neuronal porosome proteome: molecular dynamics and architecture. J Proteomics 2012; 75:3952–3962.

25. Paredes-Santos TC, de Souza W, Attias M. Dynamics and 3D organization of secretory organelles of Toxoplasma gondii. J Struct Biol 2012; 177:420–430.

26. Savigny P, Evans J, McGrath KM. Cell membrane structures during exocytosis. Endocrinology 2007; 148:3863–3874.

27. Siksou L, Rostaing P, Lechaire JP, Boudier T, Ohtsuka T, Fejtova A, Kao HT, Greengard P, Gundelfinger ED, Triller A, Marty S. Three-dimensional architecture of presynaptic terminal cytomatrix. J Neurosci 2007; 27:6868–6877.

28. Taatjes DJ, Quinn AS, Rand JH, Jena BP. Atomic force microscopy: high resolution dynamic imaging of cellular and molecular structure in health and disease. J Cell Physiol 2013; 228:1949–1955.

29. Deng Z, Lulevich V, Liu FT, Liu GY. Applications of atomic force microscopy in biophysical chemistry of cells. J Phys Chem B 2010; 114:5971–5982.

30. Aravanis AM, Pyle JL, Tsien RW. Single synaptic vesicles fusing transiently and successively without loss of identity. Nature 2003; 423:643–647.

31. Sutton RB, Fasshauer D, Jahn R, Brunger AT. Crystal structure of a SNARE complex involved in synaptic exocytosis at 2.4 A resolution. Nature 1998; 395:347–353.

32. Taraska JW, Perrais D, Ohara-Imaizumi M, Nagamatsu S, Almers W. Secretory granules are recaptured largely intact

after stimulated exocytosis in cultured endocrine cells. Proc Natl Acad Sci U S A 2003; 100:2070–2075.

33. Bennett MK, Calakos N, Scheller RH. Syntaxin: a synaptic protein implicated in docking of synaptic vesicles at presynaptic active zones. Science 1992; 257:255–259.

34. Brunger AT, Weninger K, Bowen M, Chu S. Single-molecule studies of the neuronal SNARE fusion machinery. Annu Rev Biochem 2009; 78:903–928.

35. Chapman ER. How does synaptotagmin trigger neurotransmitter release? Annu Rev Biochem 2008; 77:615–641.

36. Cho SJ, Kelly M, Rognlien KT, Cho JA, Horber JK, Jena BP. SNAREs in opposing bilayers interact in a circular array to form conducting pores. Biophys J 2002; 83:2522–2527.

37. Cho WJ, Lee JS, Zhang L, Ren G, Shin L, Manke CW, Potoff J, Kotaria N, Zhvania MG, Jena BP. Membrane-directed molecular assembly of the neuronal SNARE complex. J Cell Mol Med 2011; 15:31–37.

38. Cho WJ, Jeremic A, Jena BP. Size of supramolecular SNARE complex: membrane-directed self-assembly. J Am Chem Soc 2005; 127:10156–10157.

39. Cook JD, Cho WJ, Stemmler TL, Jena BP. Circular dichroism (CD) spectroscopy of the assembly and disassembly of SNAREs: the proteins involved in membrane fusion in cells. Chem Phys Lett 2008; 462:6–9.

40. Diao J, Ishitsuka Y, Bae WR. Single-molecule FRET study of SNARE-mediated membrane fusion. Biosci Rep 2011; 31:457–463.

41. Fasshauer D, Margittai M. A transient N-terminal interaction of SNAP-25 and syntaxin nucleates SNARE assembly. J Biol Chem 2004; 279:7613–7621.

42. Hui E, Johnson CP, Yao J, Dunning FM, Chapman ER. Synaptotagmin-mediated bending of the target membrane is a critical step in Ca^{2+}-regulated fusion. Cell 2009; 138:709–721.

43. Issa ZK, Manke CW, Jena BP, Potoff JJ. Ca^{2+} bridging of apposed phospholipid bilayers. J Phys Chem B 2010; 114:13249–13254.

44. Jahn R, Scheller RH. SNAREs-engines for membrane fusion. Nat Rev Mol Cell Biol 2006; 7:631–643.

45. Jeremic A, Cho W-J. JBP. Membrane fusion: what may transpire at the atomic level. J Biol Phys Chem 2004; 4:139–142.

46. Jeremic A, Kelly M, Cho JA, Cho SJ, Horber JK, Jena BP. Calcium drives fusion of SNARE-apposed bilayers. Cell Biol Int 2004; 28:19–31.

47. Jeremic A, Quinn AS, Cho WJ, Taatjes DJ, Jena BP. Energy-dependent disassembly of self-assembled SNARE complex: observation at nanometer resolution using atomic force microscopy. J Am Chem Soc 2006; 128:26–27.

48. Martens S, Kozlov MM, McMahon HT. How synaptotagmin promotes membrane fusion. Science 2007; 316:1205–1208.

49. Misura KM, Scheller RH, Weis WI. Three-dimensional structure of the neuronal-Sec1-syntaxin 1a complex. Nature 2000; 404:355–362.

50. Oyler GA, Higgins GA, Hart RA, Battenberg E, Billingsley M, Bloom FE, Wilson MC. The identification of a novel synaptosomal-associated protein, SNAP-25, differentially expressed by neuronal subpopulations. J Cell Biol 1989; 109:3039–3052.

51. Pobbati AV, Stein A, Fasshauer D. N- to C-terminal SNARE complex assembly promotes rapid membrane fusion. Science 2006; 313:673–676.

52. Shen J, Tareste DC, Paumet F, Rothman JE, Melia TJ. Selective activation of cognate SNAREpins by Sec1/Munc18 proteins. Cell 2007; 128:183–195.

53. Shin L, Cho WJ, Cook JD, Stemmler TL, Jena BP. Membrane lipids influence protein complex assembly-disassembly. J Am Chem Soc 2010; 132:5596–5597.

54. Stein A, Weber G, Wahl MC, Jahn R. Helical extension of the neuronal SNARE complex into the membrane. Nature 2009; 460:525–528.

55. Sudhof TC. The synaptic vesicle cycle. Annu Rev Neurosci 2004; 27:509–547.

56. Sudhof TC, Rothman JE. Membrane fusion: grappling with SNARE and SM proteins. Science 2009; 323:474–477.

57. Weber T, Zemelman BV, McNew JA, Westermann B, Gmachl M, Parlati F, Sollner TH, Rothman JE. SNAREpins: minimal machinery for membrane fusion. Cell 1998; 92:759–772.

58. Wickner W, Schekman R. Membrane fusion. Nat Struct Mol Biol 2008; 15:658–664.

59. Yao J, Gaffaney JD, Kwon SE, Chapman ER. Doc2 is a Ca^{2+} sensor required for asynchronous neurotransmitter release. Cell 2011; 147:666–677.

60. Chen ZH, Lee JS, Shin L, Cho WJ, Jena BP. Involvement of beta-adrenergic receptor in synaptic vesicle swelling and implication in neurotransmitter release. J Cell Mol Med 2011; 15:572–576.

61. Cho SJ, Sattar AK, Jeong EH, Satchi M, Cho JA, Dash S, Mayes MS, Stromer MH, Jena BP. Aquaporin 1 regulates GTP-induced rapid gating of water in secretory vesicles. Proc Natl Acad Sci U S A 2002; 99:4720–4724.

62. Jeremic A, Cho WJ, Jena BP. Involvement of water channels in synaptic vesicle swelling. Exp Biol Med (Maywood) 2005; 230:674–680.

63. Kelly ML, Cho WJ, Jeremic A, Abu-Hamdah R, Jena BP. Vesicle swelling regulates content expulsion during secretion. Cell Biol Int 2004; 28:709–716.

64. Lee JS, Cho WJ, Shin L, Jena BP. Involvement of cholesterol in synaptic vesicle swelling. Exp Biol Med (Maywood) 2010; 235:470–477.

65. Shin L, Basi N, Jeremic A, Lee JS, Cho WJ, Chen Z, Abu-Hamdah R, Oupicky D, Jena BP. Involvement of vH+-ATPase in synaptic vesicle swelling. J Neurosci Res 2010; 88:95–101.

66. Alexander S, Hellemans L MO, Schneir J, Elings V, Hansma PK. An atomic resolution atomic force microscope implemented using an optical lever. J Appl Physics 1989; 65:164–167.

67. Binnig G, Quate CF, Gerber C. Atomic force microscope. Phys Rev Lett 1986; 56:930–933.

68. Shin W, Ge L, Arpino G, Villarreal SA, Hamid E, Liu H, Zhao WD, Wen PJ, Chiang HC, Wu LG. Visualization of membrane pore in live cells reveals a dynamic-pore theory governing fusion and endocytosis. Cell 2018; 173:934–945 e912.

69. Sudhof TC, Rizo J. Synaptic vesicle exocytosis. Cold Spring Harb Perspect Biol 2011; 3:1–14 a005637.

70. Gray MA, Winpenny JP, Verdon B, McAlroy H, Argent BE. Chloride channels and cystic fibrosis of the pancreas. Biosci Rep 1995; 15:531–541.

71. Cole JC, Villa BR, Wilkinson RS. Disruption of actin impedes transmitter release in snake motor terminals. J Physiol 2000; 525:579–586.

72. Klein ME, Younts TJ, Castillo PE, Jordan BA. RNA-binding protein Sam68 controls synapse number and local beta-actin mRNA metabolism in dendrites. Proc Natl Acad Sci U S A 2013; 110:3125–3130.

73. Khanna R, Zougman A, Stanley EF. A proteomic screen for presynaptic terminal N-type calcium channel (CaV2.2) binding partners. J Biochem Mol Biol 2007; 40:302–314.

74. Scuri R, Lombardo P, Cataldo E, Ristori C, Brunelli M. Inhibition of Na^+/K^+ ATPase potentiates synaptic transmission in tactile sensory neurons of the leech. Eur J Neurosci 2007; 25:159–167.

75. Lee JS, Hou X, Bishop N, Wang S, Flack A, Cho WJ, Chen X, Mao G, Taatjes DJ, Sun F, Zhang K, Jena BP. Aquaporin-assisted and ER-mediated mitochondrial fission: a hypothesis. Micron 2013; 47:50–58.

Exocrine Pancreas Integrated Responses

Chapter 14

Regulation of pancreatic secretion

Rashmi Chandra and Rodger A. Liddle

Department of Medicine, Division of Gastroenterology, Duke University and VA Medical Centers, Durham North Carolina 27710, USA

Introduction

The exocrine pancreas secretes digestive enzymes, fluid, and bicarbonate in response to food ingestion. This is a critical digestive process that is regulated by neural reflexes, gastrointestinal hormones, and absorbed nutrients. Secretion is highly regulated by both stimulatory and inhibitory influences that coordinate the delivery of digestive enzymes with food emptying into the intestine to assure adequate digestion of a meal. In the absence of proper pancreatic secretion, maldigestion and malabsorption of nutrients may cause malnutrition and associated complications. This review will describe the physiological processes that regulate pancreatic exocrine secretion.

Phases of Meal Response

Pancreatic secretion in response to a meal occurs in four distinct but overlapping phases which are named based on the location of ingested food. The four phases of pancreatic secretion are cephalic, gastric, intestinal, and absorbed nutrient. Considerable crosstalk and inter-regulation is associated within the phases, thereby ensuring adequate, but not excessive, enzyme and bicarbonate secretion. Each phase is regulated by a complex network of neural, humoral, and paracrine feedback mechanisms which help to maintain an optimal environment for food digestion and absorption.

Cephalic phase

Sensory inputs such as sight, smell, taste, and mastication (prior to swallowing) lead to the anticipation of food. These sensations initiate the first phase of pancreatic secretion known as the cephalic phase. In addition to sensory input, interaction of certain food molecules such as long-chain fatty acids (but not triglycerides or medium-chain fatty acids) with receptors in the oral cavity also induce the cephalic phase.[1] Furthermore, studies in animals have implicated a gustatory vagopancreatic reflex in mediating the cephalic phase of pancreatic secretion.[2,3] Approximately 20–25 percent of the total pancreatic exocrine secretion occurs during the cephalic phase.[4–6] This estimate is based on data obtained by sham feeding, a process by which food is anticipated by sight, smell, and taste, but not ingested. Sham feeding in animals, such as dogs, has been evaluated by inserting a surgically prepared gastric fistula that diverts food from the esophagus, allowing swallowing but not entry of food into the stomach. In humans, sham feeding involves chewing but not swallowing. The pancreatic response to sham feeding in humans lasts approximately 60 minutes while in dogs it can last for more than 4 hours.[7,8] Sham feeding stimulates pancreatic secretion which is low in bicarbonate but rich in enzymes, suggesting that pancreatic acinar, rather than ductal, cells are stimulated in this phase.[5]

The cephalic phase of exocrine secretion is under the control of the vagus nerve. Sensory inputs arising from anticipation of food are integrated in the dorsal vagal complex (located in the brainstem) and transmitted to the exocrine pancreas via the vagus nerve.[9,10] Cholinergic agonists produce secretory responses similar to cephalic stimulation while vagotomy blocks the cephalic responses, suggesting that acetylcholine released by vagal efferents is the primary mechanism by which sensory inputs lead to exocrine secretion.[11,12] Secretion of the islet hormone, pancreatic polypeptide (PP), increases with sham feeding and serves as an indicator of vagal innervation of the pancreas, as its secretion is inhibited by cholinergic blockers.[13,14] When sham feeding is accompanied by swallowing, the pancreatic secretory and PP responses are much greater, implying that chewing and swallowing stimulate PP secretion by cholinergic mechanisms.[15]

e-mail: rodger.liddle@duke.edu

The exocrine pancreas also contains peptidergic nerve terminals and there is some evidence to suggest that neuropeptides such as vasoactive intestinal peptide (VIP) and gastrin-releasing peptide (GRP) may influence the cephalic phase. In addition, thyrotropin-releasing hormone stimulates pancreatic exocrine secretion of protein and bicarbonate through vagal efferents and this process involves both muscarinic and VIP receptors.[5,16] In contrast, the effects of inhibitory cerebral calcitonin gene-related peptide (CGRP) are mediated by sympathetic noradrenergic efferents acting upon α-adrenergic receptors.[17] Sham feeding and electrical vagus nerve stimulation in dogs triggers the release of cholecystokinin (CCK), although this response may be absent in humans.[5,18,19] Endogenous CCK was shown to enhance PP release in humans during sham feeding.[14] Therefore, although peptidergic neurotransmitters are released during vagal stimulation, acetylcholine is believed to be the main neurotransmitter which regulates the cephalic phase.

Gastric phase

Entry of food into the stomach initiates the gastric phase of pancreatic secretion. This phase has been difficult to study in anaesthetized animals because presence of food in the stomach initiates neural reflexes and release of hormones. Therefore, physiological data regarding this phase have been collected by gastric distention induced either by balloon dilation or instillation of inert substances in the antrum. Experiments in which gastric contents were prevented from emptying into the duodenum demonstrated that the gastric phase accounted for approximately 10 percent of pancreatic secretion. Secretions induced during this phase consist mainly of enzymes with minimal release of bicarbonate suggesting that acinar cells are primarily involved in the induction of this phase.[5,20–22]

The role of gastrin in this phase of pancreatic secretion remains unclear. Stepwise alkaline distension of the antrum induced graded release of gastrin and pancreatic enzymes.[23] However, when exogenous gastrin was administered to dogs the amount required to stimulate exocrine secretion was much greater than normal postprandial gastrin levels, suggesting that gastrin did not have a physiological role.[24] These findings are supported by other studies demonstrating that gastrin release is not required for pancreatic enzyme secretion during this phase.

The vagus nerve plays an important role in the gastric phase of pancreatic secretion. Early experiments in anesthetized cats demonstrated that stimulation of the antrum resulted in vagal stimulation of pancreatic amylase release.[25] Antral distension in dogs also increased pancreatic secretion by long-route vagal pathways.[23] An antropancreatic short reflex pathway that is blocked by hexamethonium and atropine also mediates this phase.[26]

In addition, atropine and vagotomy block the gastric phase providing further evidence that gastric contributions to pancreatic secretion are mediated by vagovagal cholinergic reflexes that originate in the stomach and terminate in the pancreas.[21,27,28] CCK release plays an important role in antral motility and gastrin release in humans as suggested by sham feeding experiments.[14]

In the stomach, pepsin and gastric lipases catabolize proteins and fats into peptides and triglycerides plus fatty acids, respectively, while salivary amylase contributes to the continued digestion of carbohydrates. Peptic digests of proteins are effective in stimulating the intestinal phase.[29] Thus when gastric chyme enters the duodenum, it stimulates the intestinal phase of pancreatic secretion. In a clinical setting, surgical procedures that slow the rate of gastric emptying reduce pancreatic secretion.[30,31] Therefore, the rate of gastric emptying regulates the discharge of nutrients into the intestine and consequently the activation of the intestinal phase through neural and hormonal pathways.

Intestinal phase

Digestion of food in the stomach is followed by release of acidic chyme into the duodenum, which initiates the intestinal phase of pancreatic secretion. The pancreas has already been primed by cephalic and gastric influences, which enhance blood flow and initiate exocrine secretion. A majority of the pancreatic secretory response (50–80%) occurs during the intestinal phase and is regulated by hormonal and neural mechanisms.

The intestinal phase is more easily studied than the gastric phase as food can be instilled directly into the intestinal lumen without concern for gastric emptying. Stimulation of both acinar and ductal cells results in the production of enzyme and bicarbonate secretion. Pancreatic digestive enzyme secretion is stimulated by food molecules such as sodium oleate, monoglycerides, peptides, and amino acids (particularly tryptophan and phenylalanine).[32–37] In the duodenum the high volume of bicarbonate released neutralizes the acidity of gastric chyme, while pancreatic enzymes catabolize partially digested food into molecules that are easily absorbed by intestinal enterocytes.

In the intestinal phase, pancreatic response is regulated primarily by the hormones secretin and CCK and by neural influences including the enteropancreatic reflex which is mediated by the enteric nervous system and amplifies the pancreatic secretory response. Entry of low pH gastric chyme into the intestine stimulates release of secretin from S cells into the blood.[38] The main action of secretin is to stimulate bicarbonate release from pancreatic duct cells, but it also has a direct effect on acinar cells and potentiates enzyme secretion. CCK is released by proteins and fats and their partial digestion products: peptides and fatty acids.

Experiments in dogs with chronic pancreatic fistulae have shown that CCK antagonism diminishes pancreatic protein response to a meal and duodenal perfusion suggesting that CCK plays an important role in this phase.[39] Similar results were also obtained in humans, where CCK receptor antagonism reduced pancreatic enzyme secretion during the intestinal phase.[40,41]

Cholinergic regulation plays a critical role during this phase of pancreatic secretion. In the absence of secretin, atropine partially inhibits pancreatic bicarbonate secretion stimulated by low pH due to acidic chyme in the duodenum.[42,43] In addition, the amount of bicarbonate produced by infusion of secretin is lower than that released by entry of food into the duodenum suggesting that other factors contribute to meal-stimulated pancreatic bicarbonate secretion.[44] Atropine inhibited pancreatic enzyme secretion from 30 to 120 minutes following meal ingestion, implicating cholinergic mechanisms.[44] Vagovagal enteropancreatic reflexes mediated by M1 and M3 muscarinic receptors and CCK receptors play an important role in the intestinal phase of secretion.[7,45] These vagovagal enteropancreatic reflexes are modulated by input from the dorsal motor nucleus of the vagus projecting into the pancreas. Vagal stimulation activates pancreatic bicarbonate secretion through both cholinergic muscarinic and noncholinergic transmission.

Role of gastric acid

The physiological effects of acid on pancreatic secretion were evaluated by various methods such as diversion of gastric and pancreatic contents with fistulae and instillation of acidic solutions into the duodenum. Both gastric acid and exogenous HCl are powerful regulators of postprandial pancreatic bicarbonate secretion and their effects are potentiated by intrapancreatic and vagovagal neural pathways as well as by CCK.[46]

Intraduodenal infusion of hydrochloric acid elicited a concentration-dependent increase in both the amount of bicarbonate and volume of pancreatic secretion. Secretion was similar to that attained with intravenous infusion of exogenous secretin suggesting that pH changes resulting from entry of acidic contents into the duodenum are important in inducing pancreatic secretion. Administration of a peptone meal of varying pH (pH 1–5) produced a maximal secretory response at pH 3.0, which was comparable in magnitude to that obtained with exogenous secretin.[47] Acid infusion in both the duodenum and upper jejunum elicited pancreatic secretion suggesting that the proximal small intestine responds to this stimulus.[38]

Entry of gastric contents into the duodenum creates an acidic environment with a pH of 2.0–3.0 in the initial segment of the duodenum, while the pH of the distal segment remains alkaline.[48,49] This difference in pH is largely due to pancreatic bicarbonate release, which is augmented in large part by gastric acid-induced secretin release from the intestinal mucosa. In conscious rats with gastric and pancreatic fistulae, diversion through a gastric fistula produced a small increase in pancreatic secretion. However, instilling hydrochloric acid into the duodenum with an open gastric fistula augmented pancreatic secretion.[50,51] In addition, pancreatic bicarbonate secretion was much greater when pancreatic juice was diverted from the intestine signifying a correlation between intestinal pH and quantity of pancreatic bicarbonate release.[52,53]

The pancreatic bicarbonate response is dependent on the concentration of free unbuffered hydrogen ions and not on the total load of buffered acid entering the duodenum. Inhibition of gastric acid production by cimetidine (an histamine H2 receptor blocker) or omeprazole (an H^+/K^+ ATPase inhibitor) substantially reduced the pancreatic bicarbonate response to a meal.[50,54] The pH of a liquid gastric meal also plays a significant role in pancreatic bicarbonate secretion; in cats and dogs, pH > 4.5 resulted in little pancreatic bicarbonate secretion, while at pH < 4.0 secretion increased substantially suggesting that a pH threshold of <4.5 is critical for stimulation of pancreatic secretion.[47,55] This evidence implies that gastric acid is an important regulator of pancreatic bicarbonate secretion, which neutralizes the acid to create an alkaline environment optimal for the action of pancreatic enzymes and continued digestion of food.

Role of dietary fat in pancreatic secretion

Dietary fats stimulate pancreatic enzyme and bicarbonate secretion. Intestinal perfusion of monoolein stimulated pancreatic enzyme secretion in humans and this effect was similar in potency to that observed with intravenous CCK injection.[56] In contrast, triglycerides administered directly into the duodenum (in the absence of endogenous lipase) were unable to induce pancreatic secretion. However, following lipase digestion of fatty acids, monoglycerides stimulated pancreatic secretion but glycerol was ineffective indicating that fatty acids are the major component of ingested fats that stimulate pancreatic secretion.[56,57]

There is some evidence to suggest that both free and saponified fatty acids induce pancreatic secretion, while other experiments suggest effectiveness only in a micellar form. Secretion has been shown to be dependent on fatty acid chain length, with C4 being least effective and C18 being most effective.[58] Although the reason for this difference in potency is not entirely clear, it is not believed to be related to the efficiency of absorption.[59] Other studies have demonstrated that intraduodenal administration of propionate (C3) was more effective than oleate (C18) in stimulating acinar cell secretion.[60] The reason for the differences between the two studies is not entirely clear but could be species related as these experiments have been performed in humans, rats, and rabbits. Both oleate and neutral fats stimulate bicarbonate and fluid secretion, whereas only neutral fats stimulate pancreatic enzyme secretion. In dogs, oleic acid was shown to potentiate acidified protein-mediated pancreatic enzyme and bicarbonate secretion.[61] Fat emulsions given to conscious rats produced a threefold increase in pancreatic protein secretion. The route of fat administration also has an impact on pancreatic secretion. Intravenous administration of fat did not produce pancreatic secretion, whereas intraduodenal administration led to elevated protein, bicarbonate, and fluid secretion.[62,63]

Administration of fat emulsions increases plasma CCK and secretin levels. Fat-mediated pancreatic secretion was blocked by proglumide, a CCK receptor antagonist, implicating the importance of CCK in stimulating pancreatic secretion.[64] Both C12 and C18 fatty acids augment the effects of secretin-induced bicarbonate secretion.[65] In humans, introduction of different concentrations of oleic acid into the duodenum induces pancreatic secretion, although the threshold for CCK stimulation is much lower than for secretin.[66] Secretin release is physiologically important since injection of anti-secretin antibodies in conscious rats greatly reduces fat-mediated protein and bicarbonate secretion.[67]

A critical fatty acid chain length of C12 was required for CCK release from STC-1 cells, a neuroendocrine tumor cell line. Fatty acids with less than 10 carbon atoms did not augment secretion. This dependence on fatty acid chain length is similar to that observed previously for *in vivo* CCK release in humans. In addition to the fatty acid carbon chain length, a free carboxyl terminus is also important as esterification of the carboxylic terminus abolished CCK secretion, while modification of the methyl terminus had no effect.[68–70] Two cell surface receptors have been identified and demonstrated to promote fat-mediated CCK release. Mice with global deletion of GPR40 show partial reduction in CCK secretion following fatty acid administration.[71] The recently discovered immunoglobulin-like domain-containing receptor (ILDR) is expressed in the duodenum in a subtype of EECs known as I cells that secrete CCK. ILDR appears to play an essential role in fat-stimulated CCK release as deletion of ILDR in mice completely eliminates fatty acid-stimulated CCK secretion.[72]

Thus, fats and fatty acids are important regulators of pancreatic secretion. Experimental evidence suggests that the degree and extent of acinar and ductal cell activation may vary depending on the animal species and the route of fat administration.

Contributions of proteins, peptides, and amino acids to pancreatic secretion

Studies performed in dogs, rats, and humans have shown that proteins, peptides, and amino acids stimulate pancreatic secretion.[73] In rats, proteins such as casein stimulate pancreatic secretion in proportion to their trypsin inhibitor capacity.[74] However, in dogs, intact, undigested proteins such as casein, albumin, and gelatin did not stimulate pancreatic secretion, whereas protease digests of these proteins were very effective.[34] In contrast, studies in rats suggested that intestinal administration of hydrolyzed casein produced a smaller response than some of the other proteins which potently stimulated pancreatic enzyme secretion, suggesting that the amino acid composition of a protein is relevant in determining the extent of stimulation.[74,75]

Although intravenous infusion of amino acids in humans stimulated pancreatic enzyme and bicarbonate secretion, a mixture of L-amino acids when infused intravenously in dogs was not effective. In contrast to intravenous infusion, intraduodenal delivery of amino acids in dogs induced pancreatic fluid, bicarbonate, and protein secretion which was comparable to an elemental diet suggesting the importance of the route of administration on pancreatic secretion.[62,76] Only L-amino acids stimulate pancreatic secretion which is consistent with the overall physiological importance of these stereoisomers. Of all the amino acids, aromatic amino acids such as phenylalanine and tryptophan have the greatest potency.[36,77,78]

Acidification of amino acid[77,79] and peptide[78] preparations with hydrochloric acid potentiates the bicarbonate response but pancreatic enzyme secretion is not influenced beyond that observed in the absence of acid. Aromatic amino acids are capable of inducing maximal secretory response as potentiation of pancreatic enzyme secretion is not observed when amino acids or peptides are administered concomitantly with lipid molecules such as oleate or monoolein.[56,61]

The pancreatic secretory response to intraduodenal administration of amino acids appears to be concentration dependent. A minimal concentration of 8 mM is necessary for stimulation by most amino acids,[36] although the more potent aromatic amino acids such as tryptophan stimulate secretion at concentrations as low as 3 mM.[80] The length of the intestine exposed to amino acids also plays a critical role in pancreatic secretion. In dogs, exposure of the first 10 cm was least effective, while perfusion of the whole intestine produced significant enzyme output,[36] suggesting that the pancreatic response was dependent upon the entire load of nutrients, not just their concentration. The majority of stimuli responsible for pancreatic stimulation originate in the proximal small intestine. In humans, amino acids stimulated pancreatic secretion only when perfused into the duodenum and no response was observed upon perfusion in the ileum.[81] Therefore, similar to fats, the primary mechanisms that stimulate pancreatic secretion are limited to the proximal regions of the small intestine.

The amount of bicarbonate released by intraluminal administration of tryptophan is similar to that produced by maximal doses of exogenously infused CCK indicating that release of CCK by tryptophan leads to pancreatic secretion.[34,56,82] Similarly, intraduodenal administration of liver extracts in dogs mediated CCK release along with pancreatic enzyme and bicarbonate secretion, both of which were blocked by CCK receptor antagonists.[83] Bile acids released from the gallbladder can significantly inhibit pancreatic stimulation induced by intraluminal amino acids. This inhibition of pancreatic secretion by bile acids appears to be due to inhibition of CCK release and serves as a feedback mechanism in regulating pancreatic and gallbladder function.[56] By using a sensitive bioassay for CCK measurement, it was shown that one of the pathways by which proteins stimulate CCK release is by their ability to inhibit intraluminal trypsin activity.[75] Another mechanism by which aromatic amino acids mediate CCK release is by activation of the calcium-sensing receptor (CaSR), a known nutrient sensor.[84–87] In addition to stimulating the release of hormones such as CCK and secretin, amino acids also activate cholinergic neural mechanisms which regulate pancreatic bicarbonate secretion.[80]

Role of bile and bile acids in pancreatic secretion

Bile is produced by hepatocytes as a complex mixture of bile acids, cholesterol, and organic molecules. It is stored and concentrated in the gall bladder and released into the duodenum upon entry of chyme. Bile acids such as cholate, deoxycholate, and chenodeoxycholate are conjugated with glycine or taurine amino acids, which increase their solubility. In the intestine, bile acids assist in the emulsification and absorption of fatty acids, monoacylglycerols, and lipids and stimulate lipolysis by facilitating binding of pancreatic lipase with its colipase.

Under basal conditions, intraduodenal administration of physiological concentrations of bile or the bile salt sodium taurocholate elevated plasma secretin and stimulated pancreatic fluid secretion in cats.[88,89] Secretin was released only in response to perfusion of sodium taurocholate in the duodenum. Perfusion in the upper jejunum produced a significantly diminished pancreatic response, while no response was observed upon ileal perfusion.[90] Pancreatic fluid secretion was stimulated by the free ionized form of taurocholate and was not dependent on its detergent properties.[91] In humans, infusion of bovine bile augmented secretin release along with pancreatic exocrine secretions of fluid, bicarbonate, and enzymes.[92,93]

In addition to secretin, infusion of bovine bile and bile acids in humans and dogs was shown to stimulate the release of several hormones and neuropeptides such as CCK, neurotensin, VIP, gastric inhibitory peptide (GIP), PP, and somatostatin.[94–97] Fluid and bicarbonate release was enhanced when elevated levels of VIP were present in the plasma, suggesting that bile activates peptidergic nerves

resulting in pancreatic secretion. Additionally, cholinergic mechanisms are also important as atropine blocked bile- and taurocholate-stimulated exocrine pancreatic secretion.[94] The composition of bile is important in mechanisms regulating this secretory response as some differences in hydrokinetic and ecbolic responses were observed with administration of bile versus various bile acids.[98]

However, a stimulatory effect of bile acids on pancreatic fluid secretion was not observed in the presence of digestive intraluminal contents.[99] In some studies, where bile acids were administered concomitantly with amino acids or fat, an inhibition of pancreatic enzyme secretion was observed. The mechanism underlying this observation is not completely understood, although it is possible that bile acids inhibit CCK release by a negative feedback mechanism, which helps to relax and refill the gallbladder.[56,100–102] Chemical sequestration of bile acids in dogs augmented the release of CCK and pancreatic enzyme secretion in response to amino acids and addition of taurocholate reversed this effect.[103] Long-term diversion of bile in dogs also augmented basal and oleate-stimulated pancreatic fluid, bicarbonate, and enzyme secretion along with plasma CCK levels, further supporting the role of bile acids in inhibiting CCK release.[104]

Other studies have shown that the bile salt chenodeoxycholate when infused in humans inhibited bombesin- and CCK-stimulated gallbladder emptying along with elevation of plasma CCK levels. These results led the authors to hypothesize that chenodeoxycholate, by a yet unknown mechanism, reduced the sensitivity of the gallbladder to stimulation by bombesin and CCK.[105]

In contrast to many species including mice and humans, rats do not possess a gallbladder and multiple pancreatic ducts join the lower end of the common bile duct. In rats, diversion of bile and pancreatic juice stimulates the release of pancreatic enzymes. This augmentation of enzyme secretion has been suggested to compensate for the increased degradation of proteolytic enzymes in the absence of bile. Thus, exocrine secretion in rats is regulated by a luminal feedback mechanism.[106,107] Additional experiments have shown that certain bile salts stimulate bicarbonate secretion via CCK release, whereas other bile salts inhibit exocrine secretion.[108–110] Two inhibitory mechanisms have been proposed—one dependent on the stabilization of luminal proteases and the other independent of protease activity.[111] Stimulation of pancreatic fluid secretion in anesthetized rats has been demonstrated to be mediated by taurocholate-induced transcription of $Na^+/K^+/2Cl^-$ cotransporter, which plays a key role in regulating the entry of Cl^- from the basolateral surface of acinar cells.

The physiological role of bile and bile salts in regulating pancreatic secretions is not completely understood and appears to be dependent on multiple factors, including the chemical properties of bile salts, the animal model being evaluated, and prandial status of the animal being studied.[112]

Absorbed nutrient phase

Once nutrients are absorbed from the intestinal lumen, they can directly stimulate pancreatic secretion as part of the absorbed nutrient phase. Nutrients can either directly stimulate pancreatic acinar cells or they may indirectly activate hormonal and neural pathways to further regulate exocrine secretion. In contrast to intraduodenal administration of amino acids, which produces large increases in pancreatic secretion,[113–115] intravenous administration of amino acids stimulates trypsin and chymotrypsin secretion but not lipase or amylase. Less is known about lipids and glucose, although they appear to have little if any direct effect on pancreatic secretion.[63,116]

Feedback regulation of pancreatic secretion

The concept of feedback regulation of pancreatic secretion emanated from a series of studies demonstrating that[117] instillation of trypsin inhibitor into the upper small intestine or[118] surgical diversion of the bile-pancreatic duct removing bile and pancreatic juice from the duodenum of rats stimulated pancreatic enzyme secretion.[119] Conversely, infusion of trypsin into the duodenum during bile-pancreatic juice diversion suppressed pancreatic enzyme release. Thus, the protease concentration in the upper small intestine appears to be intimately linked to pancreatic secretion through a negative feedback system in which active proteases within the duodenum limit pancreatic secretion but reduced protease activity stimulates pancreatic secretion. When assays for CCK became available, it was shown that CCK mediated the effects of proteases on pancreatic secretion[120] through protease-sensitive CCK-releasing factors[121,122] (see **Figure 1**). In the absence of proteases, CCK-releasing factor can stimulate CCK cells, but in the presence of proteases, the releasing factors are inactivated and CCK secretion is low. Negative feedback regulation of pancreatic secretion has been shown to exist in many species, although other proteases such as elastase may be more important in regulating pancreatic secretion in humans.

Pancreatic exocrine secretion is also influenced through a positive feedback mechanism. Monitor peptide is a 61-amino acid peptide produced by pancreatic acinar cells and possessing CCK-releasing activity. Although monitor peptide has modest trypsin inhibitor capability, its ability to stimulate CCK is independent of this action because monitor peptide can directly stimulate CCK secretion from isolated CCK cells in vitro.[123,124] Monitor peptide is secreted in pancreatic juice, therefore, it does not stimulate CCK secretion unless pancreatic secretion is underway. Thus, monitor peptide cannot account for the increase in CCK during bile-pancreatic juice diversion, but it may serve to reinforce pancreatic secretion once the process has been initiated.

Pancreatic Exocrine Secretion

The exocrine pancreas delivers its secretions of digestive enzymes, fluid, and bicarbonate ions to the duodenum

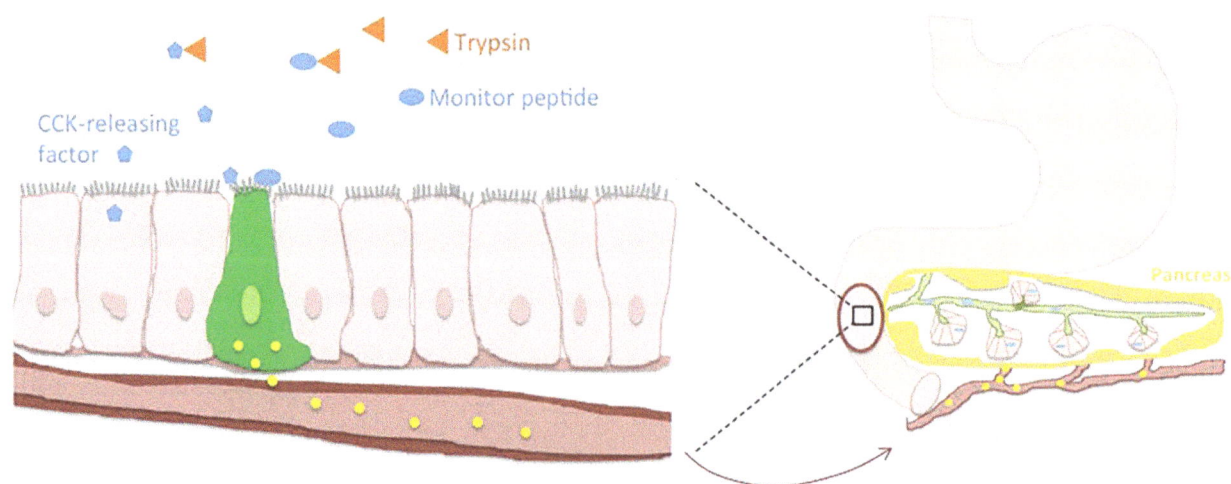

Figure 1. Feedback regulation of pancreatic exocrine secretion is mediated by positive and negative mechanisms. Positive feedback: Monitor peptide is secreted by acinar cells and directly stimulates CCK cells in the small intestine and amplifies pancreatic secretion once it has been initiated. Negative feedback: Trypsin-sensitive CCK-releasing factors are produced by the intestine and stimulate CCK secretion when trypsin is temporarily consumed by ingested protein or other trypsin "inhibitors."

following ingestion of food. The pancreas is composed of both endocrine and exocrine components. The endocrine pancreas is comprised of α, β, δ, ε, and PP (F) cells, which are located in the islets of Langerhans. These specialized cells secrete the hormones glucagon, insulin, somatostatin, ghrelin, amylin, and pancreatic polypeptide into the blood, which exert endocrine and paracrine actions within the pancreas. Ninety percent of the pancreas is composed of acinar cells which secrete digestive enzymes such as trypsin, chymotrypsin, and amylase for digestion of food in the small intestine. The acinar cells are triangular in shape and arranged in clusters with the apex of the cell opening into a centrally located terminal duct. The terminal or intercalated ducts merge to form interlobular ducts, which in turn congregate to form the main pancreatic duct. The pancreatic duct delivers exocrine secretions into the duodenum. The ductal cells secrete fluid and bicarbonate ions, which neutralize acinar cell secretions, as well as the acidic gastric contents entering the duodenum.[125] The pancreas is heavily innervated by sympathetic and parasympathetic peripheral nerves and contains a dense network of blood vessels which regulate blood flow and modulate pancreatic secretion.

Pancreatic exocrine secretion is a highly integrated process mediated by neural and hormonal signals arising from the gut as well as by factors secreted by other tissues and hormones released from pancreatic islets. The secretory pathways can be stimulatory or inhibitory in nature and represent a highly regulated system that responds to ingestive signals. The agents that modulate pancreatic exocrine secretion are discussed in **Table 1**.

Neural mechanisms

Neural Innervation

The pancreas is innervated by parasympathetic nerve fibers, postganglionic sympathetic neurons, as well as a network of intrapancreatic nerves. Together these nerves regulate pancreatic exocrine function by releasing neurotransmitters such as acetylcholine, serotonin, and neuropeptides such as VIP, GRP, and neuropeptide Y (NPY). The pancreatic ganglia receive input from pre- and postganglionic nerve fibers and regulate exocrine and endocrine secretion.

Intrapancreatic postganglionic neurons are activated by central input during the cephalic phase and by vagovagal responses initiated during the gastric and intestinal phases of stimulation. They stimulate enzyme and bicarbonate secretion primarily by releasing acetylcholine, which activates muscarinic receptors located on acinar and duct cells.

Vagal innervation

The dorsal vagal complex in the brainstem is comprised of the nucleus of the solitary tract (NTS) and the dorsal motor nucleus of the vagus (DMV) and exerts parasympathetic control on pancreatic secretion. Information relayed by sensory vagal afferent nerves innervating the pancreas is first processed in the NTS, which then projects onto the preganglionic motor neurons of the DMV. The DMV receives inputs from other regions of the brain such as the hypothalamus and from numerous hormones and neuropeptides through the afferent limb of the vagus nerve.[126]

Parasympathetic preganglionic efferent vagal nerves innervating the pancreas originate primarily from the DMV and terminate in the pancreatic ganglion. Electrical and chemical stimulation of the DMV induces rapid pancreatic secretion, and this response is inhibited by vagotomy or blockade of muscarinic receptors by atropine.[127] It has been suggested that vagal cholinergic neurons mediate pancreatic secretion during low loads of intestinal stimulants whereas hormones mediate the response during high loads of intestinal stimuli.[45,128]

CCK affects pancreatic secretion through both a direct effect on pancreatic acinar cells and an indirect effect on the vagus nerve (**Figure 2**). However, the effects on the vagus nerve are complex and the firing response of neurons in the DMV appears to be dictated by their spatial location. In one study, neurons in the caudal region were activated; those in the rostral region were unaffected, while neurons in the intermediate region were inhibited by a direct action of CCK.[129] Although not fully understood, it appears that CCK's effects on the vagus nerve influence the overall pancreatic secretory response.

The exocrine pancreas is regulated directly by the vagus. Studies with muscarinic receptor knockout mice demonstrated that both M1 and M3 receptors mediate amylase release from dispersed acini. It is likely that M3 receptors are more relevant physiologically since the level of M3 receptor expression was significantly higher in acinar cells[130] and M1 receptors were found to have only a minor effect on bicarbonate secretion in conscious dogs.[131] The vagus nerve also possesses group II metabotropic glutamate receptors that couple primarily to $G_{i/o}$. These receptors are located on excitatory and inhibitory presynaptic terminals of pancreas-projecting DMV neurons[132] that are also activated by CCK and pancreatic polypeptide. Thus, in addition to γ-amino butyric acid, glutamate also modulates pancreatic exocrine secretion through distinct vagal neurons.

Vasoactive intestinal peptide

Vasoactive intestinal peptide (VIP) is a 28-amino acid neuropeptide that is found throughout the body. Immunocytochemical evidence suggests that VIP is localized in pancreatic nerve fibers and functions as a vagal neurotransmitter. In the chick, VIP immunoreactive nerve endings are found in close proximity to acinar cells and

Table 1. Regulators of pancreatic exocrine secretion

AGENT	TYPE OF SECRETION	SITE OF ACTION
STIMULATE		
Neurotransmitter		
Acetylcholine	Protein	Acinar cell
Vasoactive intestinal peptide	Bicarbonate	Acinar cell
Gastrin-releasing peptide	Protein	Acinar cells
Neuromedin C	Protein	Acinar cell
PACAP	Bicarbonate, protein	Acinar cell
Neurotensin	Bicarbonate, protein	Direct effect on acinar cells; indirect effect via dopamine and bile acids
Substance P	Protein, fluid, bicarbonate	Acinar cells, duct cells
Calcitonin gene-related peptide	Protein	Perivascular nerves
Cholecystokinin	Protein	Acinar cell
Peptide histidine isoleucine	Bicarbonate, fluid	Pancreatic nerves, ganglia, blood vessels, acinar cells
Catecholamine	Protein	Acinar cell
Dopamine	Protein, fluid, bicarbonate	Acinar cell
Serotonin	Protein	Paracrine effect via vagal afferent fibers in duodenal mucosa
Nitric oxide	Fluid, protein	Acinar cell, nerve terminals
Hormone		
Cholecystokinin	Protein	Acinar cell
Secretin	Bicarbonate. fluid, protein	Acinar cell, duct cell
A-natriuretic peptide	Fluid, protein	Acinar cell
C-natriuretic peptide	Protein	Acinar cell
Insulin	Protein	Acinar cell
Bombesin	Protein or Bicarbonate	Acinar cell
Melatonin	Protein	Acinar cell
Amylin	Amylase release from AR42J cells	?
Histamine	Fluid, protein	Presynaptic parasympathetic nerve terminals; acinar cell
INHIBIT		
Neurotransmitter		
Calcitonin gene-related peptide	Protein	Neurons
Neuropeptide Y	Protein	Neurons
Hormone		
Peptide YY	Fluid, protein	Smooth muscle cells of blood vessels
Pancreatic polypeptide	Protein	?
Somatostatin	Protein	Acinar cells, ganglia, cholinergic neurons
Galanin	Fluid, protein	Acinar cell
Pancreastatin	Fluid, protein	Blood vessels?
Glucagon	Bicarbonate, protein	Acinar cell
Ghrelin	Protein	Intrapancreatic neurons
Leptin	Protein	Neurons
Adrenomedullin	Protein	Acinar cell

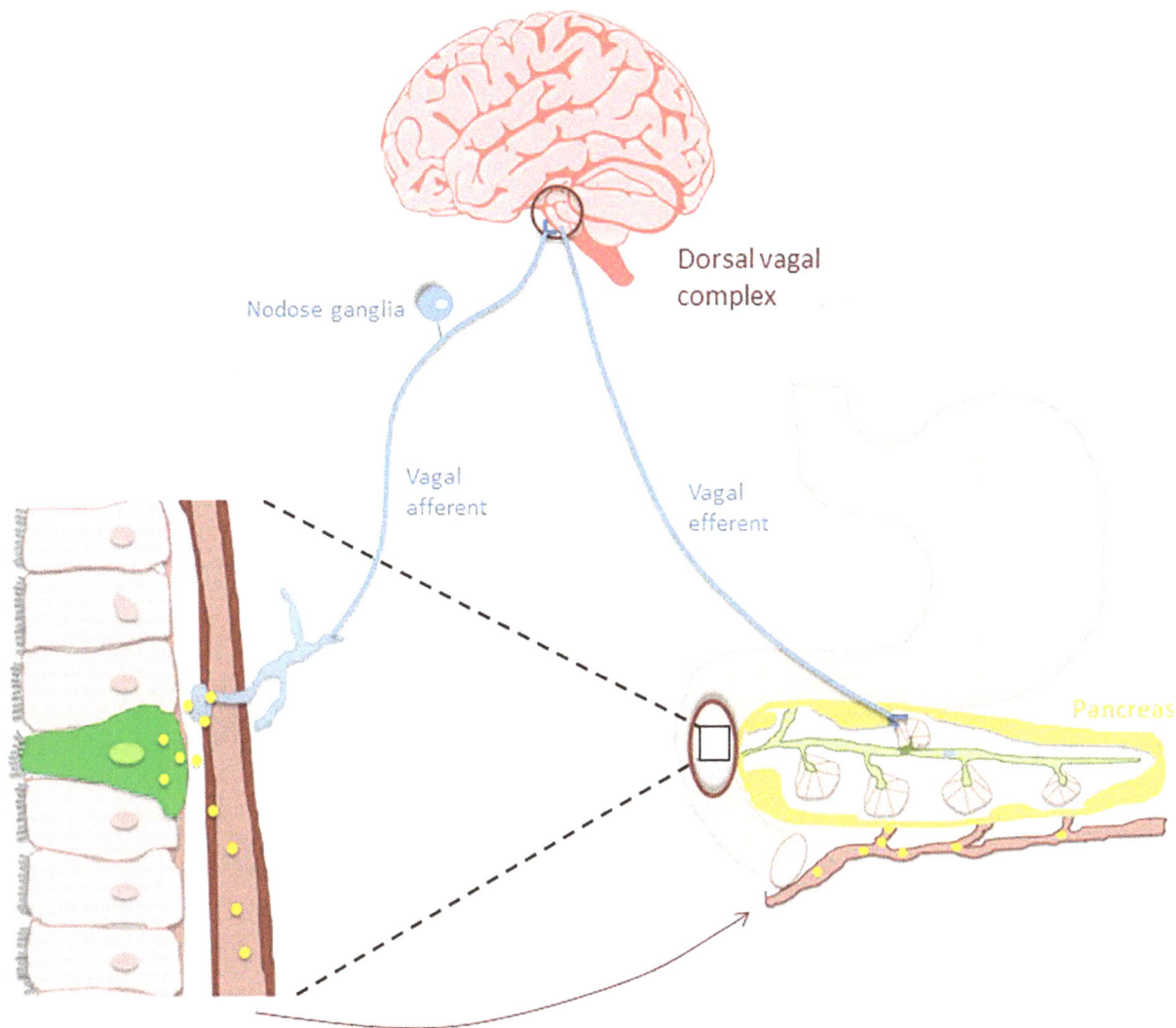

Figure 2. CCK stimulates pancreatic secretion through hormonal and neuronal pathways. CCK is released from I cells of the small intestine and diffuses into the bloodstream where it is carried to the pancreas. CCK binds to receptors on acinar cells to stimulate pancreatic enzyme secretion. Secreted CCK also diffuses through the paracellular space and binds to CCK1-bearing nerves in the submucosa. Vagal afferent signals are integrated in the dorsal vagal complex which also receives signals from other regions of the brain (e.g., hypothalamus). Vagal efferent fibers transmit cholinergic signals to the pancreas to stimulate pancreatic secretion.

epithelial cells of arterioles. Small clear vesicles were present in VIP-positive nerves indicating that these neurons are were cholinergic in nature.[133] In normal human pancreas, autonomic ganglia receive an abundant supply of VIP-positive fiber plexi and VIP-positive nerves and appear to innervate acinar cells, ducts, and blood vessels.[134] After atropine treatment, electrical stimulation of the vagus still increased bicarbonate secretion concurrent with detection of VIP in pancreatic venous effluent suggesting that VIP release is coupled with bicarbonate secretion.[135] The effects of VIP are especially prominent in the pig as perfusion of the pancreas with VIP antibodies inhibited fluid

and bicarbonate secretion, and treatment of rats with a VIP antagonist reduced bicarbonate secretion concomitant with vasodepression further supporting a direct relationship.[136,137] High- and low-affinity VIP receptors have been identified on pancreatic acinar membranes. The high-affinity receptors are coupled to cAMP-mediated amylase release, while activation of low-affinity receptors did not cause cAMP elevation or amylase release, suggesting that only high-affinity receptors are important in protein secretion.[138] These effects of VIP were attenuated by somatostatin and galanin, which reduced VIP-mediated fluid and protein output.[139] One of the main functions of VIP appears to be increasing blood flow by vasodilation, and as a result its effects on pancreatic secretion independent of blood flow in the pancreas are difficult to evaluate.[140,141]

Gastrin-releasing peptide

Gastrin-releasing peptide (GRP) is a 27-amino acid neuropeptide that is present in postganglionic vagal afferents and has been detected in neurons innervating the feline, porcine, rodent, and human pancreas.[142] Receptors responsive to GRP have been identified in rat pancreatic membranes and cancer cells where they mediate enzyme secretion.[143,144] In the cat, GRP is present in intrapancreatic ganglia, acinar and stromal regions, and occasionally on the vasculature and ducts.[145] In humans, the pattern of GRP expression was similar to that of VIP; GRP was localized on nerve fibers in proximity to pancreatic acini, capillaries, ductules, and arterial walls.[146] Several studies in different species have demonstrated that GRP modulates exocrine secretion. Vagal stimulation of porcine pancreas resulted in GRP release, which enhanced pancreatic exocrine secretion.[147] In isolated perfused rat pancreatic preparations, electrical field-stimulated GRP release potentiated secretin-mediated fluid and amylase secretion through a noncholinergic pathway.[148] The effects of GRP on rat pancreatic exocrine secretion were enhanced by γ-amino butyric acid.[149] Neuromedin C, a decapeptide of GRP, also enhanced pancreatic secretion by direct action on canine acinar cells as well as indirectly by stimulating CCK release.[150,151] However, since bombesin (GRP analog) does not stimulate pancreatic secretion in dogs, it appears that its effects may be dependent on the species being evaluated.[152]

Other peptide neurotransmitters

Immunohistochemical staining has revealed that the neuropeptides listed below are present in pancreatic nerves and their functional significance and ability to regulate pancreatic secretion has been demonstrated by *in vitro* and/or *in vivo* studies.

PACAP: Pituitary adenylate cyclase-activating polypeptide (PACAP) has been identified in nerve fibers and intrapancreatic ganglion in rodents.[153,154] PACAP has been shown to evoke bicarbonate and enzyme secretion from the pancreas albeit with a slower time course than VIP.[137,155] In the acinar cell line AR42J, PACAP activated phospholipase C, which led to elevation of intracellular Ca^{2+} and amylase release.[156]

Neurotensin: Neurotensin is a 13-amino acid neuropeptide that is widely expressed in the central nervous system and is also present in pancreatic nerves. It stimulates amylase secretion and its effects are potentiated by carbachol (a cholinergic agonist), secretin, and caerulein (a CCK analog).[157,158] Other studies demonstrated that neurotensin

stimulates bicarbonate but not protein secretion in dogs and may act indirectly by stimulating dopamine release.[159]

Substance P: Substance P is expressed in periductal nerves in the guinea pig pancreas and inhibits ductal bicarbonate secretion by modulating neurokinin 2 and 3 receptors.[160–162] It enhanced caerulein-stimulated enzyme secretion in isolated perfused pancreas as well as in anesthetized rodents.[163]

CGRP: CGRP is a 37-amino acid peptide that is present in central and peripheral neurons. The effect of CGRP on exocrine secretion is not clear and may be species-specific. Interaction of CGRP with receptors on guinea pig acinar cells led to amylase release, although its effect was not as potent as VIP.[164] In rat acinar cell preparations, CGRP inhibited amylase release by a mechanism involving cholinergic (muscarinic) neural pathways.[165]

NPY: NPY, a 36-residue peptide, is expressed in intrapancreatic ganglia and nerve fibers that surround exocrine pancreatic tissue.[166] NPY inhibited CCK- and vagally mediated amylase secretion from intact pancreas and lobules, but not from dispersed acini, suggesting that its actions were mediated by neurons innervating the exocrine pancreas.[167] Other evidence suggests that NPY plays at best only a modest role in pancreatic exocrine secretion.[168]

CCK: The presence of CCK in intrapancreatic nerves has led to the suggestion that it may serve a dual role as neurotransmitter and hormone in pancreatic secretion. However, CCK could not be detected in the venous effluent of isolated perfused porcine pancreas after vagal stimulation following a meal[136] suggesting that synaptic release of CCK does not occur within the pancreas. Therefore, the role of CCK as a neurotransmitter in pancreatic secretion merits further investigation.

Peptide histidine isoleucine: Peptide histidine isoleucine (PHI) is a 27-amino acid peptide with an N-terminal histidine and C-terminal isoleucine and is derived from the same precursor molecule as VIP. It has been reported in pancreatic nerves and ganglia and stimulates fluid and bicarbonate secretion in a cAMP-dependent fashion.[166,169]

Adrenergic nerves

Compared to cholinergic stimulation, adrenergic nerves play a relatively minor role in pancreatic exocrine secretion. Catecholamine-containing nerves are found in the celiac ganglion and extend to intrapancreatic ganglia, blood vessels, ducts, and islets.[170] Epinephrine and norepinephrine (NE) evoked amylase release from superfused rat pancreatic preparations, similar to that induced by electrical stimulation in the presence of cholinergic blockade.[171] This process is dependent on elevated intracellular Ca^{2+} and inhibited by propranolol, suggesting that β-adrenergic receptors are involved.[172] Catecholamines also interact with α-adrenergic receptors expressed in pancreatic acini and inhibit amylase secretion.[173] NPY is coexpressed with NE in some nerve fibers, and stimulation of splanchnic nerves leads to the release of NPY and NE.[166,174] Infusion of PACAP into the pancreatoduodenal artery enhanced release of NE after electrical stimulation of nerves. However, the physiological relevance of this observation is not clear.[175]

Celiac denervation reduces pancreatic secretion by ~70 percent while increasing blood flow by 350 percent. This dissonance presumably occurs by the disruption of stimulatory fibers and sympathetic fibers that maintain tonic constriction of pancreatic vessels.[176] The effect of adrenergic transmitters on pancreatic secretion has been difficult to discern due to the wide-ranging effects of norepinephrine on multiple processes including blood pressure, blood flow, neural reflexes, and release of hormones. Even though high concentrations of norepinephrine have been found in rabbit pancreatic ganglia, ducts, and blood vessels, its effects are controversial.[177] Norepinephrine has been reported to stimulate, inhibit, or have no effect on pancreatic secretion.[170,178–183] Unfortunately α- and β- adrenergic receptor agonists and antagonists have not provided information that could be used to delineate mechanisms important in adrenergic regulation of pancreatic secretion.[184,185]

Dopamine

Dopamine was detected in pancreatic ducts and ampullae and dopamine β-hydroxylase (DBH)positive fibers were identified along the vasculature, ducts, and ganglia suggesting it may play some role in pancreatic secretion.[177,186] There is conflicting evidence regarding the role of dopamine in pancreatic secretion. Dopamine stimulates pancreatic secretion in anesthetized dogs and rats, although the effect is negligible in conscious animals.[180,182,187–190] Other data suggest that the secretory response to dopamine differs between dogs, cats, rabbits, and rats, and species-specific effects must be taken into consideration when evaluating its effects.[191]

Serotonin

Like dopamine, serotonin is present in pancreatic ducts and ampullae. Autoradiography of tissue sections after tritiated serotonin uptake demonstrated the presence of serotonergic innervation of blood vessels, but not exocrine parenchyma in rats, suggesting a limited role in pancreatic secretion.[177,192] However, phenylbiguanide, a 5-HT3 receptor agonist, activated preganglionic neurons located in the caudal DMV, inhibited those in the intermediate DMV, and had no effect on rostral DMV neurons, suggesting complex spatial regulation of pancreatic vagal neurons by serotonin.[129] Intraduodenal infusion of melatonin (a serotonin derivative) increased pancreatic amylase secretion, while pretreatment with 5-HT2 serotonin antagonist ketanserin

or the 5-HT3 antagonist MDL72222 decreased amylase release. Serotonin-induced amylase release was blocked by bilateral vagotomy supporting a role for serotonergic mechanisms on exocrine secretion.[193]

Nitric oxide

Nitric oxide (NO) is a gaseous signaling molecule that is synthesized by NO synthase from L-arginine in the presence of nicotinamide adenine dinucleotide hydrogen phosphate (NADPH). It is a potent vasodilator and modulates secretory activity as well as pancreatic blood flow in the pancreas.[194] Because it is not practical to directly measure NO in biological tissues, the presence of NO has been identified by expression of NOS or histochemistry of NADPH diaphorase (NADPH-d), since NOS and NADPH-d colocalize in neurons of the peripheral and central nervous systems. The actions of NO in tissues have been identified by the use of NO donors, NOS inhibitors, and agents that inactivate (e.g., superoxide-generating compounds) or stabilize NO (e.g., superoxide dismutase). Unlike ligands that signal through cell surface receptors, NO penetrates cells and activates guanylate cyclase to generate the second messenger cGMP.[195]

Immunostaining of pancreas from a wide range of mammals (mouse, rat, hamster, guinea pig, cat, and man) indicates that NOS is expressed in the cell bodies of intrapancreatic ganglia, interlobular nerve fibers, and along blood vessels. VIP is sometimes coexpressed with NOS in ganglia and nerve fibers. These studies suggest that NO is important in pancreatic exocrine secretion.[196] In newborn guinea pig, nitrergic neurons were present primarily in the head and body of the pancreas, along blood vessels, the main pancreatic duct, and in association with pancreatic acini.[197] These nerves also immunostained with antibodies against NPY, VIP, and DBH indicating complex coregulation of pancreatic secretion by various neurotransmitters.

In rat pancreas, the NO donor sodium nitroprusside and cGMP analog 8-bromo cGMP inhibited basal and vagal amylase secretion through a Ca^{2+}-dependent mechanism.[198,199] The G protein-coupled receptor, protease-activated receptor-2 (PAR-2), modulates NO-mediated amylase release in mice, and inhibition of NOS abolished PAR-2-mediated amylase release suggesting that the effects of NO may be mediated by neuronal release of a PAR-2 agonist. Ablation of sensory nerves by capsaicin did not affect PAR-2-mediated amylase release, suggesting that TRPV1-expressing sensory vagal fibers are not involved in this pathway.[200] Analysis of the effects of NO on pancreatic secretions in pigs supports the findings that NO is essential for pancreatic fluid and amylase secretion mediated by the vagus nerve.[201] Thus, in addition to maintaining the vascular tone, nitrergic nerves play an important role in pancreatic fluid and amylase release.

Stimulatory hormones

Cholecystokinin

Cholecystokinin (CCK) is released from specialized enteroendocrine cells (I cells) located mainly in the upper small intestine. Using confocal microscopy to examine the intestine of transgenic CCK-enhanced green fluorescent protein mice (*Cck-eGFP*), it was noted that enteroendocrine cells (EECs) possessed basal processes that extended to other cells[202] (**Figure 3**). In addition to hormones like CCK, it was found that these cells possessed many neuron-like properties including neurotransmitters and synaptic proteins. Their basal extensions, called neuropods, were filled with neurofilaments[203] and connected to nerves.[204] Thus, synaptic connections between EECs and nerves, including the vagus nerve,[205] provide a way for EECs to sense gut contents and directly regulate neural function.

The major stimulants of CCK release are dietary fats and proteins. In the rat, intraluminal proteases via an active feedback system participate in the release of a putative intestinal CCK-releasing factor (e.g., LCRF), which in turn causes CCK secretion.[121,122] Various molecular forms of CCK, ranging in size from CCK-8 [CCK-(26–33)-NH$_2$] to CCK-58, have been described in dogs, rats, and humans.[206–208] CCK-58 was determined to be the predominant peptide in dogs and humans and the only form detected in rats after employing CCK isolation techniques that prevented degradation of CCK in blood.[209] The actions

of CCK-8 and CCK-58 appear to be functionally identical suggesting that CCK-8 retains the biological activity ascribed to this hormone.[210] CCK is posttranslationally modified and has an amidated C-terminus. A sulfated tyrosine residue in CCK-8 is important for its biological actions including exocrine secretion.[211,212] C-terminal amidation is critical for binding of CCK to its receptors and removal of the amide group decreases CCK activity. Other studies have reported that deamidation and desulfation do not significantly impair the ability of CCK to stimulate amylase release and these discrepancies may arise from differences between species. Shorter forms of CCK such as the tetrapeptide CCK-4 are generally much less effective in mediating exocrine secretion than CCK-8, while longer forms of CCK, such as CCK-33, are equally effective.[213–216] CCK mediates its hormonal effects through two G-coupled protein receptors, CCK-1 and CCK-2, previously known as CCK-A and CCK-B, respectively. The contribution of these receptors to pancreatic secretion has been evaluated in order to delineate the molecular mechanisms of CCK action. CCK receptors have been proposed to exist in two states, a high-affinity (picomolar) but low-capacity state and a low-affinity (nanomolar) but high-capacity state.[217,218] Autoradiography of pancreatic membranes incubated with radioiodinated CCK-8 demonstrated that CCK-1 receptors are highly expressed in rat pancreas, while CCK-2 receptors are less abundant. CCK-1 receptors appear to modulate pancreatic secretion

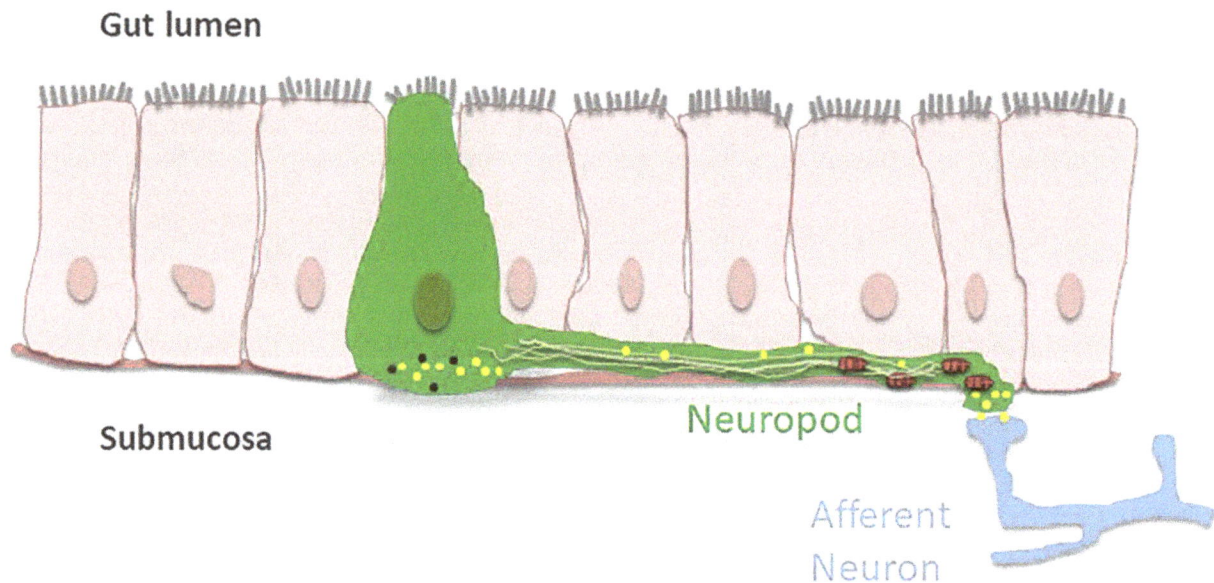

Figure 3. Model of an enteroendocrine cell (EEC) with a neuropod. EECs (green) are sensory cells that reside in the intestinal mucosa. Some EECs possess neuropods that contain neurofilaments (light green), synaptic proteins, mitochondria (red), and neurotransmitters (yellow) and synapse with neurons in the submucosa. In this manner, EECs may sense intraluminal contents such as nutrients or bacteria and signal directly to nerves, including the vagus nerve.

as oral administration of loxiglumide, a potent CCK-1 receptor antagonist, reduces protein and fluid output in rats.[219] Similarly, caerulein-induced pancreatic amylase release was blocked by CCK-1 receptor antagonists.[220] CCK-8 did not induce amylase release in CCK-1 receptor knockout mice confirming that CCK-1 receptors are critical for CCK-mediated protein secretion.[221] Since bicarbonate secretion was not observed from dispersed acinar cells, this effect is not believed to be regulated by CCK receptors.[222] Pancreatic responses in CCK-2 receptor knockout mice were similar to wild-type mice suggesting that CCK-2 is not important for amylase release, although it may be involved in augmenting vascular flow.[220,223] In pigs, where a majority of receptors are of the CCK-2 subtype, acinar cells demonstrated a low responsivity to CCK and did not secrete amylase in response to caerulein or a CCK-1 agonist.[224] In humans, the actions of CCK on pancreatic secretion have been variously reported. Infusion of CCK, caerulein, and secretin in the presence of amino acids substantially increased output of fluid, bicarbonate, and enzyme.[225,226] Similar to porcine pancreas, CCK-2 is the major CCK receptor subtype expressed in human pancreas, although interestingly CCK-1 receptor antagonists are able to inhibit amylase secretion.[227–229] In dispersed human acini, which responded to carbamylcholine and neuromedin C, CCK did not stimulate amylase release presumably because of a paucity of cellular membrane receptors. It has been proposed that the effects of CCK on human pancreatic secretion are mediated through CCK-1 receptors on nerves which innervate the pancreas.[230,231] However, recent data demonstrated that application of physiologic concentrations of CCK-8 and CCK-58 to human acinar cells produced intracellular Ca^{2+} oscillations and normal exocytosis of pancreatic enzyme, suggesting that functional CCK receptors are expressed on human pancreas.[232] By visualizing Ca^{2+} oscillations in a unique slice preparation of human pancreas obtained from cancer resections, it has been unequivocally demonstrated that CCK-8 stimulates pancreatic secretions through direct action on CCK-1 receptors expressed on acinar cells and does not require a neuronal component.[233] Thus, it appears that CCK receptors are expressed on acinar cells of both human and rodent pancreas, and the differences between the two may not be as great as previously predicted.

Recently, it was shown that rat and human pancreatic stellate cells express CCK receptors and secrete acetylcholine in response to CCK stimulation. This source of acetylcholine was sufficient to stimulate amylase release from acinar cells. Pancreatic stellate cells may represent a previously unrecognized intrapancreatic pathway regulating CCK-induced pancreatic exocrine secretion.[234]

Several hormones and neuropeptides regulate CCK-mediated exocrine secretion. Locally, insulin has been shown to influence exocrine secretion. Intra-arterial infusion of canine pancreas with anti-insulin antibodies prevented CCK release as well as secretin-mediated protein and fluid secretion from canine pancreas.[235] Secretin potentiated, as well as attenuated, CCK-mediated amylase secretion by the inositol signaling pathway while VIP enhanced CCK-mediated enzyme secretion.[236,237] Peptide YY (PYY), PP, and somatostatin also inhibited CCK-mediated protein secretion and their effects are discussed later in this review.

The mechanism of CCK-induced amylase secretion involves transient elevation in intracellular Ca^{2+}.[238] It also requires phospholipase C activation and generation of second messengers inositol trisphosphate and diacylglycerol.[239] In some instances, CCK activates secretion by elevation of cAMP, as 8-bromo-cAMP and a phosphodiesterase inhibitor augmented CCK-mediated amylase release.[236] Heterotrimeric G proteins $G\alpha_{13}$ and $G\alpha_q$ through downstream interactions with small GTP-binding proteins RhoA and Rac1 regulate actin cytoskeleton reorganization which is required for exocytosis.[240,241]

Secretin

Secretin is a 27-amino acid hormone released by S cells of the small intestine.[242] Secretin release is stimulated during the intestinal phase upon entry of gastric acid and ingested fatty acids into the duodenum.[243] It augments fluid and bicarbonate secretion and is one of the most potent stimulators of pancreatic secretion.[244] Examination of pancreatic ultrastructure shortly after secretin injection revealed that fluid is secreted by duct as well as acinar cells.[245]

Large increases in pancreatic fluid and bicarbonate secretion have been demonstrated with secretin infusions as low as 1–2.8 pmol/kg·hr.[246–249] In humans, bolus injections of secretin as low as 0.125 pmol/kg stimulated fluid and bicarbonate secretion.[250,251] Although secretin is believed to be the single most powerful stimulator of pancreatic bicarbonate secretion, infusion of exogenous secretin equivalent to postprandial blood levels only produced 10 percent of the maximal pancreatic bicarbonate secretory response suggesting that other hormones and neurotransmitters play important roles in postprandial pancreatic bicarbonate secretion in humans.[247,250] Secretin receptors have been localized in acinar and duct cells in the rat pancreas.[118]

Secretin stimulates the release of fluid and bicarbonate and, to a lesser extent, protein from acinar cells by a cholinergic mechanism. Perfusion of acetic and lactic acids in the duodenum of anesthetized rats increased fluid and protein output from the pancreas concomitant with elevation of plasma secretin levels. In addition, treatment of rats with atropine decreased plasma secretin levels and inhibited fluid (but not protein) secretion, indicating that only fluid secretion is dependent on cholinergic input.[252] Electrical field stimulation of isolated perfused rat pancreas demonstrated that secretin-mediated exocrine secretion was sensitive to tetrodotoxin and atropine blockade, further suggesting cholinergic regulation. Nicotinic acetylcholine receptors are not involved in this mechanism as hexamethonium did not exert an inhibitory effect on pancreatic secretion.[253]

Both cAMP-dependent and independent pathways contribute to secretin-mediated exocrine secretion. Interaction of secretin with its receptor induced a three- to fourfold increase in adenylate cyclase activity, which was abolished in the presence of secretin antagonists.[254] Secretin did not stimulate pancreatic fluid release or elevate acinar cell cAMP levels in secretin receptor knockout mice.[255] Exocrine protein secretion by secretin was associated with phospholipase C activation in one report.[256] At secretin concentrations $>10^{-8}$ M, inositol trisphosphate and diacylglycerol were generated in acinar cells, which caused release of Ca^{2+} from intracellular stores and activated protein kinase C. However, not all investigators have observed this effect.

Several hormones and peptides modulate the effects of secretin on pancreatic secretion. CCK augmented secretin-induced pancreatic fluid and protein output by stimulating acetylcholine release and this effect was blocked by atropine or by dispersion of acini.[257,258] Venous drainage from pancreatic islets bathes the exocrine pancreas with high concentrations of islet hormones. Several of these hormones have potent effects on pancreatic secretion. Insulin enhances secretin-mediated fluid and protein secretion through an ouabain-sensitive Na^+,K^+-ATPase while glucagon inhibits secretin-stimulated release of fluid and protein.[259] Addition of anti-somatostatin antibodies increased secretion from perfused rat pancreas implying that somatostatin inhibits secretin-induced fluid and enzyme secretion.[253,259]

Glucagon-like peptide 1

Glucagon-like peptide 1 (GLP-1) is a 29-amino acid peptide that is produced by enteroendocrine cells in the distal small intestine.[260] Until recently the role of GLP-1 in pancreatic exocrine secretion was not clear.[261] Using real-time PCR it has now been shown that GLP-1 receptor (GLP-1R) is expressed in both pancreatic acinar cells and islets. GLP-1 stimulated amylase release from wild-type but not GLP-1R knockout mice through a cAMP-dependent mechanism.[262] These results have been supported by Satoh et al.,[263] who showed that when isolated rat pancreatic acini were stimulated by GLP, amylase was released in conjunction with phosphorylation of myristoylated alanine-rich C kinase substrate (MARCKS) protein. Inhibition of MARCKS phosphorylation prevented GLP-1-mediated amylase

secretion, suggesting it was a critical step in this signaling pathway.

Atrial natriuretic factor and C-natriuretic peptide

Atrial natriuretic factor (ANF) is a peptide hormone that is secreted by atrial stretch and regulates blood pressure and volume by inhibiting reabsorption of sodium by the kidney.[264] Immunochemical studies showed that ANF is present in acinar cells. Early studies suggested that ANF did not influence protein or fluid secretion. However, incubation of rat acini with ANF caused a dose-dependent elevation of cellular cGMP[265] showing that guanylate cyclase receptors transduce ANF signaling.[266] Injection of human ANF in dogs induced bicarbonate but not sodium or protein secretion.[267] ANF also stimulates pancreatic natriuretic peptide receptor-C (NPR-C)-mediated phosphoinositide-dependent pathway in rats, causing the release of fluid and protein.[268] NPR-C is a nonguanylyl cyclase receptor and is coupled to adenylyl cyclase inhibition or phospholipase C activation through G_i proteins. ANF attenuated secretin- and VIP-induced elevation of intracellular cAMP and this effect was blocked by inhibitors of protein kinase C and phospholipase C.[269] Along with elevating intracellular cAMP, secretin mediates the efflux of cAMP from intact pancreas and acinar cells. ANF augmented secretin-induced cAMP efflux and caused the rapid elimination of cAMP from cells. The multidrug resistance protein 4 (MRP4) has been reported to play a role in the extrusion of cAMP in many cellular systems. MRP4 is also expressed in the pancreas and genetic knockdown of MRP4 expression reduced intracellular cAMP levels in acinar cells by ANF and an NPR-C dependent mechanism.[270]

C-natriuretic peptide (CNP) is structurally similar to ANF and is expressed in the CNS and gastrointestinal tract. CNP increases pancreatic protein, chloride, and fluid secretion (without influencing bicarbonate output) suggesting that it acts on acinar rather than duct cells. Truncal vagotomy or perivagal application of capsaicin or hexamethonium attenuated chloride secretion, demonstrating that the effect of CNP is modulated by the parasympathetic nervous system.[271] At low concentrations, CNP induced protein secretion by activation of NPR C. Similar to ANF, CNP-induced amylase release was inhibited by PLC and PKC inhibitors. CNP also elevated intracellular cGMP and reduced cAMP concentrations suggesting that CNP can interact directly with receptors located on pancreatic acini.[272]

Insulin

Insulin modulates pancreatic exocrine function and insulin receptors are present in high density on the basolateral surfaces of acinar cells.[273] Insulin increases pancreatic enzyme synthesis and secretion and its effects are enhanced by CCK and secretin.[274–278] Since CCK induces insulin release in the presence of glucose and amino acids, it is possible that these two hormones act in conjunction on exocrine stimulation following food intake.[212,279]

Limited data suggest that insulin regulates exocrine secretion by potentiation of ouabain-sensitive Na^+,K^+-ATPase and by vagal cholinergic input.[259,276,280] The action of insulin on exocrine secretion is modulated by PP that exerts an inhibitory effect on pancreatic secretion.[281] Since galanin, pancreastatin, and somatostatin are known to inhibit insulin secretion, it is possible that these peptides also regulate insulin-mediated amylase release.[282–284]

Bombesin

Bombesin is a 14-amino acid peptide homolog of GRP and neuromedin B and has the ability to suppress appetite.[285] The effects of bombesin on pancreatic exocrine secretion appear to vary based on the species. In pigs, administration of bombesin alone or in combination with secretin induced protein but not fluid secretion.[286] In guinea pigs, bombesin was very effective in inducing bicarbonate release from interlobular ducts, and this effect was blocked by a GRP receptor antagonist.[226] Administration of bombesin to rats resulted in pancreatic hypertrophy with increased pancreatic weight, protein, RNA, and enzyme content and this effect was not regulated by CCK or secretin.[287,288]

Melatonin

Melatonin is a lipophilic hormone produced by the pineal gland as well as by certain neuroendocrine cells located in the gastrointestinal tract. Melatonin receptors are present on acinar cells and melatonin protects the pancreas against caerulein-induced acute pancreatitis.[289] Initial studies showed that melatonin induced pancreatic amylase release, which was mediated by CCK, vagal sensory nerves, and melatonin type 2 receptors. However, melatonin did not appear to have a direct effect on pancreatic acinar cells.[290–293] The extent and importance of melatonin-induced pancreatic secretion is not well understood and has recently been proposed to involve activation of the enteropancreatic reflex and CCK release from the duodenum.[294]

Amylin

Amylin is a 37-amino acid hormone that is cosecreted along with insulin from pancreatic β-cells in response to nutrients. Amylin stimulates CCK-independent pancreatic secretion in rats, and this effect is blocked by proton pump inhibitors and atropine, perhaps due to inhibition of somatostatin release.[295] Amylin was also shown to stimulate protein secretion from pancreatic AR42J cells by a mechanism involving activation of GPCRs and release of Ca^{2+} from intracellular stores.[296] Others investigations have suggested that amylin has no effect on pancreatic exocrine secretion from isolated perfused pancreas, acinar preparations, or AR42J cells.[297–299] Hence, the effects of amylin on exocrine secretion remain unresolved.

Fibroblast growth factor 21

Fibroblast growth factor 21 (FGF21) was initially discovered in the liver and thymus of mouse embryos[300] and was later found to be expressed at high levels in pancreatic acinar cells. FGF21 overexpression protected mice against caerulein and alcohol-induced pancreatitis.[301–303] By evaluating the function of FGF21 in transgenic and knockout mice, it was discovered that FGF21 regulates pancreatic enzyme secretion from acinar cells in a paracrine/autocrine manner.[304] FGF21 binds its heteromeric receptor (comprised of FGFR1c and β-klotho) to elevate intracellular calcium levels by signaling through phospholipase C and inositol trisphosphate. Zymogen granule accumulation was observed in both FGF21 knockout and β-klotho knockout mice. FGF21 is secreted in acinar cells in response to ER stress. Unlike CCK, FGF21's role in acinar cells is limited to augmentation of amylase secretion, as it does not increase synthesis of pancreatic enzymes.

Histamine

The amino acid histamine is a potential mediator of pancreatic exocrine secretion, although it may have a gender-dependent role.[305] Activation of H1 receptors and inhibition of H2 receptors in the rabbit pancreas led to an increase in fluid and protein secretion suggesting differential action based on regulation and coupling of the two receptors.[306] The effect of histamine on pancreatic secretion is considered to be minor at best under normal physiological conditions.

Inhibitory hormones

Peptide YY and pancreatic peptide

The NPY family of peptides consists of three hormones: NPY, PYY, and PP.[307,308] All three peptides contain 36 residues several of which are tyrosines and share a tertiary structural motif known as the PP fold. The N-terminal amino acids of PYY and NPY can be cleaved by peptidases to generate truncated forms, PYY_{3-36} and NPY_{3-36}, which are biologically active. NPY has been localized to sympathetic pancreatic nerves and its role has been discussed previously in this review.[174] In islets, PYY is coexpressed with glucagon in α-cells, whereas PP is secreted postprandially by F cells of the islets of Langerhans. In certain species, PP immunopositive cells are also present in the exocrine pancreas and some of these PP cells also express PYY.[309] These three hormones exert their effects through a family of five GPCRs denoted Y1–5. NPY and PYY possess similar affinity for Y1, Y2, and Y5, PYY_{3-36} interacts preferentially with Y2, whereas PP is the preferred ligand for Y4.[310].

PYY levels in blood are elevated postprandially and following instillation of fatty acids into the distal small intestine.[311] Physiologically relevant concentrations of PYY in the circulation inhibit both meal- and hormone-stimulated pancreatic secretion.[312–314] Intravenous administration of PYY significantly diminishes secretin- and secretin plus CCK-mediated pancreatic protein and fluid secretion concomitant with a reduction in pancreatic blood flow.[313,315–317] However, PYY does not inhibit 2-diacylglycerol-stimulated pancreatic secretion, suggesting that suppression of CCK-stimulated exocrine secretion occurs prior to activation by 2-diacylglycerol or does not involve protein kinase C-activated signaling.[318] In denervated pancreas, PYY_{1-36} but not PYY_{3-36} reduced CCK-stimulated amylase release suggesting that hormonal effects of PYY are mediated by Y1 receptors.[319,320] Autoradiographic analysis of rat pancreas with radio-iodinated PYY ligand demonstrated that Y1 receptors are present primarily on smooth muscle cells of blood vessels. Specific staining was not observed in acinar cells indicating that decreased protein and fluid secretion could be due to reduced blood flow.[321]

PP secretion is also stimulated by ingesting a meal and can be reproduced by intraduodenal infusion of acid, aromatic amino acids, or fatty acids.[322–326] Like PYY, PP attenuated secretin- and CCK-mediated exocrine secretion in dogs independent of cholinergic blockade.[320,327,328] In one study, PP decreased secretin- and secretin plus CCK-mediated amylase release in dispersed acini, suggesting that PP can act directly on acinar cells.[329] However, although bovine and rat PP inhibited CCK-stimulated protein secretion in vivo, both peptides were ineffective in vitro, and binding of bovine PP to rat acinar cells or lobules was not observed.[330] Other studies have also not shown direct effects of PP.[308] In humans, unlike dogs, infusion of PP decreased pancreatic protein output but did not influence bicarbonate secretion suggesting species-specific differences in PP action.[331] However, unlike PYY, PP does not affect pancreatic blood flow and therefore inhibits exocrine secretion at the cellular level.[320,327,328]

Somatostatin

Somatostatin is composed of 14 or 28 amino acids and is produced by δ-cells of pancreatic islets. It is also secreted by certain intestinal cells and by neurons in the hypothalamus. It is released into the blood after a meal but functions primarily through a paracrine mechanism. It has broad inhibitory actions on the release of several hormones and their target organs.

Somatostatin and its analogs inhibited secretin- and CCK-induced protein secretion in a dose-dependent fashion. Low doses of somatostatin exerted a greater inhibitory effect on CCK-stimulated pancreatic secretion compared to secretin-stimulated secretion.[284,332–334] Secretin-mediated activation of slowly activating voltage-dependent K^+ channels (present on the basolateral surface of pancreatic acini) resulted in cAMP generation and secretion of chloride ions. Addition of somatostatin to acini decreased intracellular cAMP production as well as secretin-mediated enhancement of K^+ current suggesting that somatostatin regulates exocrine secretion through this pathway.[335] Additionally, somatostatin inhibited Ca^{2+}-dependent and cAMP-stimulated amylase release by inhibiting exocytosis through a G_i protein-dependent mechanism.[336]

Somatostatin also inhibits exocrine secretion via a neural mechanism. Based on atropine, hexamethonium, and tetrodotoxin sensitivities, it appears that peptidergic but not cholinergic and nicotinic acetylcholine receptors present on sympathetic and parasympathetic ganglia mediate somatostatin action.[337] Somatostatin-mediated inhibition of secretin-stimulated fluid and protein secretion was not influenced by denervation, suggesting that extrapancreatic nerves are not involved. Bethanechol, a muscarinic receptor agonist, reversed the inhibitory effects of somatostatin, indicating that its actions are mediated primarily by intrapancreatic cholinergic neurons.[338] The mechanisms by which somatostatin inhibits pancreatic secretion are not completely understood. However, it is believed that somatostatin has an inhibitory effect on the release of hormones and neurotransmitters that normally stimulate pancreatic secretion.

Galanin

Galanin is a 29-amino acid peptide that plays diverse roles including inhibition of insulin, somatostatin, and PP secretion from the pancreas.[339] It is found in the secretory granules of central and peripheral neurons suggesting that it functions as a neurotransmitter. Galanin immunoreactivity was present in nerve fibers surrounding acini, ductules, and blood vessels, with 73 percent of fibers being dual positive for galanin and VIP.[340] Galanin receptor 3 mRNA is present in acinar cells indicating that galanin may act directly on acini.[341] Consistent with this finding, galanin inhibited CCK- and carbachol-stimulated amylase release from acinar cells.[342] Galanin inhibited the sustained phase of amylase release stimulated by carbachol, suggesting that it attenuates cholinergic action possibly by a mechanism that involves pertussis toxin-sensitive G_i proteins.[282,343–345] Extrapancreatic nerves are not involved in its action since galanin inhibited food-, secretin- and CCK-mediated fluid release, as well as food- and CCK-mediated protein release in both innervated and denervated dogs.[346]

Pancreastatin

Pancreastatin is derived from the cleavage of chromogranin A and is expressed in many neuroendocrine tissues. It has been localized to duct cells of the exocrine pancreas and its numerous roles include inhibition of pancreatic exocrine secretion.[117] Initial studies showed that pancreastatin inhibited postprandial fluid and protein secretion in rats with bile-pancreatic juice diversion. No effect was observed on basal secretion, secretin-stimulated secretion in conscious rats, or CCK-stimulated secretion from dispersed acini. However, pancreastatin inhibited CCK-stimulated pancreatic secretion in conscious rats, although it did not influence plasma CCK levels. These results suggest that pancreastatin does not have a direct effect on acinar cells but may regulate the intestinal phase of pancreatic secretion.[283,347–349] Pancreastatin inhibited caerulein-induced blood flow in the exocrine pancreas raising the possibility that its inhibitory effects are derived from its role in regulating pancreatic blood flow.[350]

Glucagon

Glucagon is released from the endocrine pancreas after ingestion of a meal.[351] Early investigations suggested that glucagon inhibited secretin- or secretin- and CCK-stimulated pancreatic protein but not bicarbonate secretion.[352,353] However other studies have demonstrated that glucagon inhibits postprandial protein and bicarbonate secretion.[354–356] The effect of glucagon on isolated pancreatic lobules and acini appears to be direct and stimulatory, instead of inhibitory, suggesting complex action at the cellular versus physiological levels.[357–360] The experimentally observed effects of glucagon on exocrine secretion are inconclusive and merit further investigation. Another proglucagon-derived peptide known as oxyntomodulin is a dual agonist of GLP-1R and glucagon receptors. Although this peptide did not influence plasma amylase concentration on its own, it attenuated CCK-mediated plasma amylase release. The mechanism of action of oxyntomodulin has not been identified.[361]

Ghrelin

Ghrelin is a 28-amino acid orexigenic hormone released by gastric endocrine cells under fasting conditions. Ghrelin stimulates acid secretion by oxyntic cells in the stomach, and plasma levels of ghrelin rise immediately before a meal suggesting a role in modulating ingestive behavior. In the pancreas, both ghrelin and its receptor have been identified in acini by evaluation of protein and mRNA expression.[362]

Ghrelin expression was not altered by gastric acid inhibition, acute pancreatitis, or food deprivation, although its receptor was upregulated by gastric acid inhibition and downregulated during acute pancreatitis.[362] Experimentally, ghrelin did not affect basal or CCK-stimulated amylase release from dispersed acini. However, ghrelin inhibited CCK-stimulated protein secretion in normal and vagotomized rats and amylase secretion from lobules exposed to depolarizing potassium concentrations, suggesting that it modulates intrapancreatic neurons.[363]

Leptin

Leptin is a 16-kDa orexigenic peptide that is secreted by adipocytes and regulates energy homeostasis by reducing food intake while increasing energy expenditure. In the pancreas, intravenous or intraperitoneal administration of leptin reduced basal and CCK-stimulated protein output *in vivo*. This effect was attenuated by CCK-1 receptor blockade, vagotomy, and capsaicin pretreatment suggesting that it inhibited pancreatic exocrine secretion through a CCK-dependent vagal pathway. Leptin had no effect on dispersed acini *in vitro*, further supporting a neural mechanism of action.[364,365] In contrast, intraduodenal infusion of leptin in fasted rats augmented pancreatic protein output possibly by elevating plasma CCK levels leading to activation of sensory neurons.[366]

Adrenomedullin

Adrenomedullin is colocalized with PP in F cells of pancreatic islets and inhibits insulin.[367] In one study, it interacted directly with acinar cells and inhibited CCK-stimulated pancreatic amylase release.[368] The mechanism of adrenomedullin action is not well understood.

Conclusion

Pancreatic secretion is a complex process that is initiated by the sight and smell of food and progresses until food enters the duodenum. At each level of food digestion, this process is regulated by various stimuli which affect neuronal and hormonal pathways. These pathways are both stimulatory and inhibitory and optimize the release of enzymes, bicarbonate, and fluid.

References

1. Hiraoka T, Fukuwatari T, Imaizumi M, Fushiki T. Effects of oral stimulation with fats on the cephalic phase of pancreatic enzyme secretion in esophagostomized rats. Physiol Behav 2003; 79:713–717.
2. Niijima A. Effect of umami taste stimulations on vagal efferent activity in the rat. Brain Res Bull 1991; 27:393–396.

3. Powers MA, Schiffman SS, Lawson DC, Pappas TN, Taylor IL. The effect of taste on gastric and pancreatic responses in dogs. Physiol Behav 1990; 47:1295–1297.

4. Defilippi C, Solomon TE, Valenzuela JE. Pancreatic secretory response to sham feeding in humans. Digestion 1982; 23:217–223.

5. Anagnostides A, Chadwick VS, Selden AC, Maton PN. Sham feeding and pancreatic secretion. Evidence for direct vagal stimulation of enzyme output. Gastroenterology 1984; 87:109–114.

6. Konturek SJ, Bielanski W, Solomon TE. Effects of an antral mucosectomy, L-364,718 and atropine on cephalic phase of gastric and pancreatic secretion in dogs. Gastroenterology 1990; 98:47–55.

7. Singer M. Neurohormonal control of pancreatic enzyme secretion in Animals. In: Go VLW, ed. The pancreas: biology, pathobiology, and disease. 2nd ed. New York: Raven Press, 1993.

8. Sarles H, Dani R, Prezelin G, Souville C, Figarella C. Cephalic phase of pancreatic secretion in man. Gut 1968; 9:214–221.

9. Furukawa N, Okada H. Effects of stimulation of the hypothalamic area on pancreatic exocrine secretion in dogs. Gastroenterology 1989; 97:1534–1543.

10. Pandol SJ. The exocrine pancreas. San Rafael, CA: Morgan & Claypool Life Sciences, 2010.

11. Becker S, Niebel W, Singer MV. Nervous control of gastric and pancreatic secretory response to 2-deoxy-D-glucose in the dog. Digestion 1988; 39:187–196.

12. Holtmann G, Singer MV, Kriebel R, Stacker KH, Goebell H. Differential effects of acute mental stress on interdigestive secretion of gastric acid, pancreatic enzymes, and gastroduodenal motility. Dig Dis Sci 1989; 34:1701–1707.

13. Schwartz TW. Pancreatic polypeptide: a hormone under vagal control. Gastroenterology 1983; 85:1411–1425.

14. Katschinski M, Dahmen G, Reinshagen M, Beglinger C, Koop H, Nustede R, Adler G. Cephalic stimulation of gastrointestinal secretory and motor responses in humans. Gastroenterology 1992; 103:383–391.

15. Schwartz TW, Stenquist B, Olbe L. Cephalic phase of pancreatic-polypeptide secretion studied by sham feeding in man. Scand J Gastroenterol 1979; 14:313–320.

16. Holst JJ, Knuhtsen S, Jensen SL, Fahrenkrug J, Larsson LI, Nielsen OV. Interrelation of nerves and hormones in stomach and pancreas. Scand J Gastroenterol 1983; 82:85–99.

17. Messmer B, Zimmerman FG, Lenz HJ. Regulation of exocrine pancreatic secretion by cerebral TRH and CGRP: role of VIP, muscarinic, and adrenergic pathways. Am J Physiol Gastrointest Liver Physiol 1993; 264:G237–G242.

18. Schafmayer A, Nustede R, Pompino A, Kohler H. Vagal influence on cholecystokinin and neurotensin release in conscious dogs. Scand J Gastroenterol 1988; 23:315–320.

19. Kim CK, Lee KY, Wang T, Sun G, Chang TM, Chey WY. Role of endogenous cholecystokinin on vagally stimulated pancreatic secretion in dogs. Am J Physiol Gastrointest Liver Physiol 1989; 257:G944–G949.

20. Cargill JM, Wormsley KG. Effect of gastric distension on human pancreatic secretion. Acta Hepatogastroenterol (Stuttg) 1979; 26:235–238.

21. White TT, Lundh G, Magee DF. Evidence for the existence of a gastropancreatic reflex. Am J Physiol 1960; 198:725–728.

22. Vagne M, Grossman MI. Gastric and pancreatic secretion in response to gastric distention in dogs. Gastroenterology 1969; 57:300–310.

23. Debas HT, Yamagishi T. Evidence for pyloropancreatic reflux for pancreatic exocrine secretion. Am J Physiol Endocrinol Metab 1978; 234:E468–E471.

24. Kohler E, Beglinger C, Eysselein V, Grotzinger U, Gyr K. Gastrin is not a physiological regulator of pancreatic exocrine secretion in the dog. Am J Physiol Gastrointest Liver Physiol 1987; 252:G40–G44.

25. Blair EL, Brown JC, Harper AA, Scratcherd T. A gastric phase of pancreatic secretion. J Physiol 1966; 184:812–824.

26. Furukawa N, Okada H. Effects of antral distension on pancreatic exocrine secretion in dogs: evidence for a short reflex. Jpn J Physiol 1987; 37:671–685.

27. White TT, Mc AR, Magee DF. Gastropancreatic reflex after various gastric operations. Surg Forum 1962; 13:286–288.

28. Kreiss C, Schwizer W, Erlacher U, Borovicka J, Lochner-Kuery C, Muller R, Jansen JB, Fried M. Role of antrum in regulation of pancreaticobiliary secretion in humans. Am J Physiol Gastrointest Liver Physiol 1996; 270:G844–G851.

29. Guan D, Green GM. Significance of peptic digestion in rat pancreatic secretory response to dietary protein. Am J Physiol Gastrointest Liver Physiol 1996; 271:G42–G47.

30. MacGregor I, Parent J, Meyer JH. Gastric emptying of liquid meals and pancreatic and biliary secretion after subtotal gastrectomy or truncal vagotomy and pyloroplasty in man. Gastroenterology 1977; 72:195–205.

31. Mayer EA, Thompson JB, Jehn D, Reedy T, Elashoff J, Meyer JH. Gastric emptying and sieving of solid food and pancreatic and biliary secretion after solid meals in patients with truncal vagotomy and antrectomy. Gastroenterology 1982; 83:184–192.

32. Go VL, Hofmann AF, Summerskill WH. Pancreozymin bioassay in man based on pancreatic enzyme secretion: potency of specific amino acids and other digestive products. J Clin Invest 1970; 49:1558–1564.

33. Liddle RA, Goldfine ID, Rosen MS, Taplitz RA, Williams JA. Cholecystokinin bioactivity in human plasma. Molecular forms, responses to feeding, and relationship to gallbladder contraction. J Clin Invest 1985; 75:1144–1152.

34. Meyer JH, Kelly GA. Canine pancreatic responses to intestinally perfused proteins and protein digests. Am J Physiol 1976; 231:682–691.

35. Meyer JH, Kelly GA, Jones RS. Canine pancreatic response to intestinally perfused oligopeptides. Am J Physiol 1976; 231:678–681.

36. Meyer JH, Kelly GA, Spingola LJ, Jones RS. Canine gut receptors mediating pancreatic responses to luminal L-amino acids. Am J Physiol 1976; 231:669–677.

37. Dale WE, Turkelson CM, Solomon TE. Role of cholecystokinin in intestinal phase and meal-induced pancreatic secretion. Am J Physiol Gastrointest Liver Physiol 1989; 257:G782–G790.

38. Konturek SJ, Dubiel J, Gabrys B. Effect of acid infusion into various levels of the intestine on gastric and pancreatic secretion in the cat. Gut 1969; 10:749–753.

39. Konturek SJ, Tasler J, Cieszkowski M, Szewczyk K, Hladij M. Effect of cholecystokinin receptor antagonist on pancreatic

responses to exogenous gastrin and cholecystokinin and to meal stimuli. Gastroenterology 1988; 94:1014–1023.

40. Hildebrand P, Beglinger C, Gyr K, Jansen JB, Rovati LC, Zuercher M, Lamers CB, Setnikar I, Stalder GA. Effects of a cholecystokinin receptor antagonist on intestinal phase of pancreatic and biliary responses in man. J Clin Invest 1990; 85:640–646.

41. Gabryelewicz A, Kulesza E, Konturek SJ. Comparison of loxiglumide, a cholecystokinin receptor antagonist, and atropine on hormonal and meal-stimulated pancreatic secretion in man. Scand J Gastroenterol 1990; 25:731–738.

42. You CH, Rominger JM, Chey WY. Effects of atropine on the action and release of secretin in humans. Am J Physiol Gastrointest Liver Physiol 1982; 242:G608–G611.

43. Singer MV, Solomon TE, Rammert H, Caspary F, Niebel W, Goebell H, Grossman MI. Effect of atropine on pancreatic response to HCl and secretin. Am J Physiol Gastrointest Liver Physiol 1981; 240:G376–G380.

44. Bozkurt T, Adler G, Koop I, Arnold R. Effect of atropine on intestinal phase of pancreatic secretion in man. Digestion 1988; 41:108–115.

45. Singer MV, Niebergall-Roth E. Secretion from acinar cells of the exocrine pancreas: role of enteropancreatic reflexes and cholecystokinin. Cell Biol Int 2009; 33:1–9.

46. Singer MV. Pancreatic secretory response to intestinal stimulants: a review. Scand J Gastroenterol Suppl 1987; 139:1–13.

47. Dembinski A, Konturek SJ, Thor P. Gastric and pancreatic responses to meals varying in pH. J Physiol 1974; 243:115–128.

48. Brooks AM, Grossman MI. Postprandial pH and neutralizing capacity of the proximal duodenum in dogs. Gastroenterology 1970; 59:85–89.

49. Rune SJ. pH in the human duodenum. Its physiological and pathophysiological significance. Digestion 1973; 8:261–268.

50. Bilski J, Konturek PK, Krzyzek E, Konturek SJ. Feedback control of pancreatic secretion in rats. Role of gastric acid secretion. J Physiol Pharmacol 1992; 43:237–257.

51. Green GM. Role of gastric juice in feedback regulation of rat pancreatic secretion by luminal proteases. Pancreas 1990; 5:445–451.

52. Cooke AR, Nahrwold DL, Grossman MI. Diversion of pancreatic juice on gastric and pancreatic response to a meal stimulus. Am J Physiol 1967; 213:637–639.

53. Henriksen FW, Worning H. External pancreatic response to food and its relation to the maximal secretory capacity in dogs. Gut 1969; 10:209–214.

54. Moore EW, Verine HJ, Grossman MI. Pancreatic bicarbonate response to a meal. Acta Hepatogastroenterol (Stuttg) 1979; 26:30–36.

55. Meyer JH, Way LW, Grossman MI. Pancreatic bicarbonate response to various acids in duodenum of the dog. Am J Physiol 1970; 219:964–970.

56. Malagelada JR, Go VL, DiMagno EP, Summerskill WH. Interactions between intraluminal bile acids and digestive products on pancreatic and gallbladder function. J Clin Invest 1973; 52:2160–2165.

57. Meyer JH, Jones RS. Canine pancreatic responses to intestinally perfused fat and products of fat digestion. Am J Physiol 1974; 226:1178–1187.

58. Demol P, Sarles H. Action of fatty acids on the exocrine pancreatic secretion of the conscious rat: further evidence for a protein pancreatic inhibitory factor. J Physiol 1978; 275:27–37.

59. Malagelada JR, DiMagno EP, Summerskill WH, Go VL. Regulation of pancreatic and gallbladder functions by intraluminal fatty acids and bile acids in man. J Clin Invest 1976; 58:493–499.

60. Navas JM, Calvo JJ, Lopez MA, De Dios I. Exocrine pancreatic response to intraduodenal fatty acids and fats in rabbits. Comp Biochem Physiol Comp Physiol 1993; 105:141–145.

61. Fink AS, Meyer JH. Intraduodenal emulsions of oleic acid augment acid-induced canine pancreatic secretion. Am J Physiol Gastrointest Liver Physiol 1983; 245:G85–G91.

62. Stabile BE, Borzatta M, Stubbs RS, Debas HT. Intravenous mixed amino acids and fats do not stimulate exocrine pancreatic secretion. Am J Physiol Gastrointest Liver Physiol 1984; 246:G274–G280.

63. Niederau C, Sonnenberg A, Erckenbrecht J. Effects of intravenous infusion of amino acids, fat, or glucose on unstimulated pancreatic secretion in healthy humans. Dig Dis Sci 1985; 30:445–455.

64. Green GM, Taguchi S, Friestman J, Chey WY, Liddle RA. Plasma secretin, CCK, and pancreatic secretion in response to dietary fat in the rat. Am J Physiol Gastrointest Liver Physiol 1989; 256:G1016–G1021.

65. Fink AS, Luxenburg M, Meyer JH. Regionally perfused fatty acids augment acid-induced canine pancreatic secretion. Am J Physiol Gastrointest Liver Physiol 1983; 245:G78–G84.

66. Olsen O, Schaffalitzky de Muckadell OB, Cantor P. Fat and pancreatic secretion. Scand J Gastroenterol 1989; 24:74–80.

67. Guan D, Spannagel A, Ohta H, Nakano I, Chey WY, Green GM. Role of secretin in basal and fat-stimulated pancreatic secretion in conscious rats. Endocrinology 1991; 128:979–982.

68. McLaughlin J, Grazia Luca M, Jones MN, D'Amato M, Dockray GJ, Thompson DG. Fatty acid chain length determines cholecystokinin secretion and effect on human gastric motility. Gastroenterology 1999; 116:46–53.

69. McLaughlin JT, Lomax RB, Hall L, Dockray GJ, Thompson DG, Warhurst G. Fatty acids stimulate cholecystokinin secretion via an acyl chain length-specific, Ca^{2+}-dependent mechanism in the enteroendocrine cell line STC-1. J Physiol 1998; 513:11–18.

70. McLaughlin J. Long-chain fatty acid sensing in the gastrointestinal tract. Biochem Soc Trans 2007; 35:1199–1202.

71. Liou AP, Lu X, Sei Y, Zhao X, Pechhold S, Carrero RJ, Raybould HE, Wank S. The G-protein-coupled receptor GPR40 directly mediates long-chain fatty acid-induced secretion of cholecystokinin. Gastroenterology 2011; 140:903–912.

72. Chandra R, Wang Y, Shahid RA, Vigna SR, Freedman NJ, Liddle RA. Immunoglobulin-like domain containing receptor 1 mediates fat-stimulated cholecystokinin secretion. J Clin Invest 2013; 123:3343–3352.

73. Solomon TE. Regulation of pancreatic secretion. Clin Gastroenterol 1984; 13:657–678.

74. Green GM, Nasset ES. Role of dietary protein in rat pancreatic enzyme secretory response to a meal. J Nutr 1983; 113:2245–2252.

75. Liddle RA, Green GM, Conrad CK, Williams JA. Proteins but not amino acids, carbohydrates, or fats stimulate

cholecystokinin secretion in the rat. Am J Physiol Gastrointest Liver Physiol 1986; 251:G243–G248.

76. Wolfe BM, Keltner RM, Kaminski DL. The effect of an intraduodenal elemental diet on pancreatic secretion. Surg Gynecol Obstet 1975; 140:241–245.

77. Meyer JH, Grossman MI. Comparison of D- and L-phenylalanine as pancreatic stimulants. Am J Physiol 1972; 222:1058–1063.

78. Fink AS, Miller JC, Jehn DW, Meyer JH. Digests of protein augment acid-induced canine pancreatic secretion. Am J Physiol Gastrointest Liver Physiol 1982; 242:G634–G641.

79. Konturek SJ, Radecki T, Mikos E, Thor J. The effect of exogenous and endogenous secretin and cholecystokinin on pancreatic secretion in cats. Scand J Gastroenterol 1971; 6:423–428.

80. Singer MV, Solomon TE, Grossman MI. Effect of atropine on secretion from intact and transplanted pancreas in dog. Am J Physiol Gastrointest Liver Physiol 1980; 238:G18–G22.

81. DiMagno EP, Go VL, Summerskill HJ. Intraluminal and postabsorptive effects of amino acids on pancreatic enzyme secretion. J Lab Clin Med 1973; 82:241–248.

82. Debas HT, Grossman MI. Pure cholecystokinin: pancreatic protein and bicarbonate response. Digestion 1973; 9:469–481.

83. Moriyasu M, Lee YL, Lee KY, Chang TM, Chey WY. Effect of digested protein on pancreatic exocrine secretion and gut hormone release in the dog. Pancreas 1994; 9:129–133.

84. Hira T, Nakajima S, Eto Y, Hara H. Calcium-sensing receptor mediates phenylalanine-induced cholecystokinin secretion in enteroendocrine STC-1 cells. FEBS J 2008; 275:4620–4626.

85. Nakajima S, Hira T, Hara H. Calcium-sensing receptor mediates dietary peptide-induced CCK secretion in enteroendocrine STC-1 cells. Mol Nutr Food Res 2012; 56:753–760.

86. Wang Y, Chandra R, Samsa LA, Gooch B, Fee BE, Cook JM, Vigna SR, Grant AO, Liddle RA. Amino acids stimulate cholecystokinin release through the Ca^{2+}-sensing receptor. Am J Physiol Gastrointest Liver Physiol 2011; 300:G528–G537.

87. Liou AP, Sei Y, Zhao X, Feng J, Lu X, Thomas C, Pechhold S, Raybould HE, Wank SA. The extracellular calcium-sensing receptor is required for cholecystokinin secretion in response to L-phenylalanine in acutely isolated intestinal I cells. Am J Physiol Gastrointest Liver Physiol 2011; 300:G538–G546.

88. Hanssen LE, Hotz J, Hartmann W, Nehls W, Goebell H. Immunoreactive secretin release following taurocholate perfusions of the cat duodenum. Scand J Gastroenterol 1980; 15:89–95.

89. Hanssen LE, Hotz J, Layer P, Goebell H. Bile-stimulated secretin release in cats. Scand J Gastroenterol 1986; 21:886–890.

90. Hartmann W, Hotz J, Ormai S, Aufgebauer J, Schneider F, Goebell H. Stimulation of bile and pancreatic secretion by duodenal perfusion with Na-taurocholate in the cat compared with jejunal and ileal perfusion. Scand J Gastroenterol 1980; 15:433–442.

91. Gries E, Hotz J, Goebell H. Pancreatic exocrine secretion in response to intraduodenal infusion of different detergent agents in anesthetized cats. Digestion 1986; 34:61–67.

92. Osnes M, Hanssen LE, Flaten O, Myren J. Exocrine pancreatic secretion and immunoreactive secretin (IRS) release after intraduodenal instillation of bile in man. Gut 1978; 19:180–184.

93. Osnes M, Hanssen LE, Lehnert P, Flaten O, Larsen S, Londong W, Otte M. Exocrine pancreatic secretion and immunoreactive secretin release after repeated intraduodenal infusions of bile in man. Scand J Gastroenterol 1980; 15:1033–1039.

94. Riepl RL, Reichardt B, Rauscher J, Tzavella K, Teufel J, Lehnert P. Mediators of exocrine pancreatic secretion induced by intraduodenal application of bile and taurodeoxycholate in man. Eur J Med Res 1997; 2:23–29.

95. Riepl RL, Fiedler F, Teufel J, Lehnert P. Effect of intraduodenal bile and taurodeoxycholate on exocrine pancreatic secretion and on plasma levels of vasoactive intestinal polypeptide and somatostatin in man. Pancreas 1994; 9:109–116.

96. Burhol PG, Lygren I, Waldum HL, Jorde R. The effect of duodenal infusion of bile on plasma VIP, GIP, and secretin and on duodenal bicarbonate secretion. Scand J Gastroenterol 1980; 15:1007–1011.

97. Chayvialle JA, Miyata M, Rayford PL, Thompson JC. Effects of test meal, intragastric nutrients, and intraduodenal bile on plasma concentrations of immunoreactive somatostatin and vasoactive intestinal peptide in dogs. Gastroenterology 1980; 79:844–852.

98. Riepl RL, Fiedler F, Kowalski C, Teufel J, Lehnert P. Exocrine pancreatic secretion and plasma levels of cholecystokinin, pancreatic polypeptide, and somatostatin after single and combined intraduodenal application of different bile salts in man. Ital J Gastroenterol 1996; 28:421–429.

99. Bondesen S, Christensen H, Lindorff-Larsen K, Schaffalitzky de Muckadell OB. Plasma secretin in response to pure bile salts and endogenous bile in man. Dig Dis Sci 1985; 30:440–444.

100. Bjornsson OG, Maton PN, Fletcher DR, Chadwick VS. Effects of duodenal perfusion with sodium taurocholate on biliary and pancreatic secretion in man. Eur J Clin Invest 1982; 12:97–105.

101. Koop I, Schindler M, Bosshammer A, Scheibner J, Stange E, Koop H. Physiological control of cholecystokinin release and pancreatic enzyme secretion by intraduodenal bile acids. Gut 1996; 39:661–667.

102. Nustede R, Schmidt WE, Kohler H, Folsch UR, Schafmayer A. The influence of bile acids on the regulation of exocrine pancreatic secretion and on the plasma concentrations of neurotensin and CCK in dogs. Int J Pancreatol 1993; 13:23–30.

103. Gomez G, Upp JR, Jr., Lluis F, Alexander RW, Poston GJ, Greeley GH, Jr., Thompson JC. Regulation of the release of cholecystokinin by bile salts in dogs and humans. Gastroenterology 1988; 94:1036–1046.

104. Takahashi M, Naito H, Sasaki I, Funayama Y, Shibata C, Matsuno S. Long-term bile diversion enhances basal and duodenal oleate-stimulated pancreatic exocrine secretion in dogs. Tohoku J Exp Med 2004; 203:87–95.

105. Thimister PW, Hopman WP, Tangerman A, Rosenbusch G, Willems HL, Jansen JB. Effect of intraduodenal bile salt on

pancreaticobiliary responses to bombesin and to cholecystokinin in humans. Hepatology 1998; 28:1454–1460.

106. Green G, Nasset ES. Effect of bile duct obstruction on pancreatic enzyme secretion and intestinal proteolytic enzyme activity in the rat. Am J Dig Dis 1977; 22:437–444.

107. Miyasaka K, Green GM. Effect of partial exclusion of pancreatic juice on rat basal pancreatic secretion. Gastroenterology 1984; 86:114–119.

108. Miyasaka K, Kitani K. A difference in stimulatory effects on pancreatic exocrine secretion between ursodeoxycholate and trypsin inhibitor in the rat. Dig Dis Sci 1986; 31:978–986.

109. Miyasaka K, Funakoshi A, Shikado F, Kitani K. Stimulatory and inhibitory effects of bile salts on rat pancreatic secretion. Gastroenterology 1992; 102:598–604.

110. Miyasaka K, Kitani K. Effects of bile salts on pancreatic secretion in rabbits: ursodeoxycholate infused into the duodenum stimulates pancreas. Pancreas 1986; 1:264–269.

111. Miyasaka K, Sazaki N, Funakoshi A, Matsumoto M, Kitani K. Two mechanisms of inhibition by bile on luminal feedback regulation of rat pancreas. Gastroenterology 1993; 104:1780–1785.

112. Riepl RL, Lehnert P. The role of bile in the regulation of exocrine pancreatic secretion. Scand J Gastroenterol 1992; 27:625–631.

113. Rothman SS. Molecular regulation of digestion: short term and bond specific. Am J Physiol 1974; 226:77–83.

114. Konturek SJ, Tasler J, Cieszkowski M, Jaworek J, Konturek J. Intravenous amino acids and fat stimulate pancreatic secretion. Am J Physiol Endocrinol Metab 1979; 236:E678–E684.

115. Meyer JH, Spingola J, Grossman MI. Endogenous cholecystokinin potentiates exogenous secretin on pancreas of dog. Am J Physiol 1971; 221:742–747.

116. Liddle RA. Regulation of pancreatic secretion. In: Leonard R. Johnson FKG, Jonathan DK, Juanita LM, Hamid MS, and Jackie DW, eds. Physiology of the gastrointestinal tract. Volume 2. 5th ed. London: Academic Press, 2012:1425–1460.

117. Adeghate E, Ember Z, Donath T, Pallot DJ, Singh J. Immunohistochemical identification and effects of atrial natriuretic peptide, pancreastatin, leucine-enkephalin, and galanin in the porcine pancreas. Peptides 1996; 17:503–509.

118. Ulrich CD, 2nd, Wood P, Hadac EM, Kopras E, Whitcomb DC, Miller LJ. Cellular distribution of secretin receptor expression in rat pancreas. Am J Physiol Gastrointest Liver Physiol 1998; 275:G1437–G1444.

119. Green GM, Lyman RL. Feedback regulation of pancreatic enzyme secretion as a mechanism for trypsin inhibitor-induced hypersecretion in rats. Proc Soc Exp Biol Med 1972; 140:6–12.

120. Liddle RA. Regulation of cholecystokinin secretion by intraluminal releasing factors. Am J Physiol Gastrointest Liver Physiol 1995; 269:G319–G327.

121. Spannagel AW, Green GM, Guan D, Liddle RA, Faull K, Reeve JR, Jr. Purification and characterization of a luminal cholecystokinin-releasing factor from rat intestinal secretion. Proc Natl Acad Sci U S A 1996; 93:4415–4420.

122. Herzig KH, Schon I, Tatemoto K, Ohe Y, Li Y, Folsch UR, Owyang C. Diazepam binding inhibitor is a potent cholecystokinin-releasing peptide in the intestine. Proc Natl Acad Sci U S A 1996; 93:7927–7932.

123. Bouras EP, Misukonis MA, Liddle RA. Role of calcium in monitor peptide-stimulated cholecystokinin release from perifused intestinal cells. Am J Physiol Gastrointest Liver Physiol 1992; 262:G791–G796.

124. Liddle RA, Misukonis MA, Pacy L, Balber AE. Cholecystokinin cells purified by fluorescence-activated cell sorting respond to monitor peptide with an increase in intracellular calcium. Proc Natl Acad Sci U S A 1992; 89:5147–5151.

125. Hegyi P, Maleth J, Venglovecz V, Rakonczay Z, Jr. Pancreatic ductal bicarbonate secretion: challenge of the acinar Acid load. Front Physiol 2011; 2:36.

126. Babic T, Travagli RA. Neural control of the pancreas. Pancreapedia: Exocrine Pancreas Knowledge Base, 2016. DOI: 10.3998/panc.2016.27.

127. Mussa BM, Verberne AJ. Activation of the dorsal vagal nucleus increases pancreatic exocrine secretion in the rat. Neurosci Lett 2008; 433:71–76.

128. Niebergall-Roth E, Singer MV. Central and peripheral neural control of pancreatic exocrine secretion. J Physiol Pharmacol 2001; 52:523–538.

129. Mussa BM, Sartor DM, Verberne AJM. Dorsal vagal preganglionic neurons: differential responses to CCK_1 and $5-HT_3$ receptor stimulation. Auton Neurosci 2010; 156:36–43.

130. Gautam D, Han SJ, Heard TS, Cui Y, Miller G, Bloodworth L, Wess J. Cholinergic stimulation of amylase secretion from pancreatic acinar cells studied with muscarinic acetylcholine receptor mutant mice. J Pharmacol Exp Ther 2005; 313:995–1002.

131. Teyssen S, Niebergall E, Chari ST, Singer MV. Comparison of two dose-response techniques to study the pancreatic secretory response to intraduodenal tryptophan in the absence and presence of the M1-receptor antagonist telenzepine. Pancreas 1995; 10:368–373.

132. Babic T, Browning KN, Kawaguchi Y, Tang X, Travagli RA. Pancreatic insulin and exocrine secretion are under the modulatory control of distinct subpopulations of vagal motoneurones in the rat. J Physiol 2012; 590:3611–3622.

133. Hiramatsu K, Oshima K. Immunocytochemical study on the innervation of the chicken pancreas by vasoactive intestinal polypeptide (VIP)-containing nerves. Histol Histopathol 1997; 12:961–965.

134. Marongiu L, Perra MT, Pinna AD, Sirigu F, Sirigu P. Peptidergic (VIP) nerves in normal human pancreas and in pancreatitis: an immunohistochemical study. Histol Histopathol 1993; 8:127–132.

135. Fahrenkrug J, Schaffalitzky de Muckadell OB, Holst JJ, Jensen SL. Vasoactive intestinal polypeptide in vagally mediated pancreatic secretion of fluid and HCO_3^-. Am J Physiol Endocrinol Metab 1979; 237:E535–E540.

136. Holst JJ, Fahrenkrug J, Knuhtsen S, Jensen SL, Poulsen SS, Nielsen OV. Vasoactive intestinal polypeptide (VIP) in the pig pancreas: role of VIPergic nerves in control of fluid and bicarbonate secretion. Regul Pept 1984; 8:245–259.

137. Wheeler S, Eardley JE, McNulty KF, Sutcliffe CP, Morrison JD. An investigation into the relative merits of pituitary adenylate cyclase-activating polypeptide (PACAP-27) and vasoactive intestinal polypeptide as vagal neuro-transmitters in exocrine pancreas of rats. Exp Physiol 1997; 82:729–747.

138. Bissonnette BM, Collen MJ, Adachi H, Jensen RT, Gardner JD. Receptors for vasoactive intestinal peptide and secretin on rat pancreatic acini. Am J Physiol Gastrointest Liver Physiol 1984; 246:G710–G717.

139. Holst JJ, Rasmussen TN, Harling H, Schmidt P. Effect of intestinal inhibitory peptides on vagally induced secretion from isolated perfused porcine pancreas. Pancreas 1993; 8:80–87.

140. Ashton N, Argent BE, Green R. Effect of vasoactive intestinal peptide, bombesin and substance P on fluid secretion by isolated rat pancreatic ducts. J Physiol 1990; 427:471–482.

141. Konturek SJ, Yanaihara N, Pawlik W, Jaworek J, Szewczyk K. Comparison of helodermin, VIP and PHI in pancreatic secretion and blood flow in dogs. Regul Pept 1989; 24:155–166.

142. Moghimzadeh E, Ekman R, Hakanson R, Yanaihara N, Sundler F. Neuronal gastrin-releasing peptide in the mammalian gut and pancreas. Neuroscience 1983; 10:553–563.

143. Kane MA, Kelley K, Ross SE, Portanova LB. Isolation of a gastrin releasing peptide receptor from normal rat pancreas. Peptides 1991; 12:207–213.

144. Hajri A, Koenig M, Balboni G, Damge C. Expression and characterization of gastrin-releasing peptide receptor in normal and cancerous pancreas. Pancreas 1996; 12:25–35.

145. De Giorgio R, Sternini C, Brecha NC, Widdison AL, Karanjia ND, Reber HA, Go VL. Patterns of innervation of vasoactive intestinal polypeptide, neuropeptide Y, and gastrin-releasing peptide immunoreactive nerves in the feline pancreas. Pancreas 1992; 7:376–384.

146. Shimosegawa T, Asakura T, Kashimura J, Yoshida K, Meguro T, Koizumi M, Mochizuki T, Yanaihara N, Toyota T. Neurons containing gastrin releasing peptide-like immunoreactivity in the human pancreas. Pancreas 1993; 8:403–412.

147. Knuhtsen S, Holst JJ, Jensen SL, Knigge U, Nielsen OV. Gastrin-releasing peptide: effect on exocrine secretion and release from isolated perfused porcine pancreas. Am J Physiol Gastrointest Liver Physiol 1985; 248:G281–G286.

148. Park HS, Kwon HY, Lee YL, Chey WY, Park HJ. Role of GRPergic neurons in secretin-evoked exocrine secretion in isolated rat pancreas. Am J Physiol Gastrointest Liver Physiol 2000; 278:G557–G562.

149. Park YD, Cui ZY, Park HJ. Effects of gamma-aminobutyric acid on action of gastrin-releasing peptidergic neurons in exocrine secretion of isolated, perfused rat pancreas. Pancreas 2002; 25:308–313.

150. Hosotani R, Inoue K, Fujii N, Yajima H, Tobe T. Effect of synthetic neuromedin C, a decapeptide of gastrin-releasing peptide (GRP [18–27]), on blood flow and exocrine secretion of the pancreas in dogs. Life Sci 1985; 36:2429–2434.

151. Hosotani R, Inoue K, Kogire M, Suzuki T, Otsuki M, Rayford PL, Yajima H, Tobe T. Synthetic neuromedin C stimulates exocrine pancreatic secretion in dogs and rats. Pancreas 1987; 2:414–421.

152. Konturek SJ, Tasler J, Bilski J, Cieszkowski M, Cai RZ, Schally AV. Antagonism of receptors for gastrin, cholecystokinin and GRP/bombesin in postprandial stimulation of exocrine pancreas in dogs. J Physiol Pharmacol 1993; 44:43–53.

153. Fridolf T, Sundler F, Ahren B. Pituitary adenylate cyclase-activating polypeptide (PACAP): occurrence in rodent pancreas and effects on insulin and glucagon secretion in the mouse. Cell Tissue Res 1992; 269:275–279.

154. Goyal D, Pisegna JR. Pituitary adenylate cyclase activating polypeptide (PACAP). Pancreapedia: Exocrine Pancreas Knowledge Base, 2014. DOI: 10.3998/panc.2014.12.

155. Alonso RM, Rodriguez AM, Garcia LJ, Lopez MA, Calvo JJ. Comparison between the effects of VIP and the novel peptide PACAP on the exocrine pancreatic secretion of the rat. Pancreas 1994; 9:123–128.

156. Barnhart DC, Sarosi GA, Jr., Mulholland MW. PACAP-38 causes phospholipase C-dependent calcium signaling in rat acinar cell line. Surgery 1997; 122:465–474.

157. Feurle GE, Reinecke M. Neurotensin interacts with carbachol, secretin, and caerulein in the stimulation of the exocrine pancreas of the rat in vitro. Regul Pept 1983; 7:137–143.

158. Baca I, Feurle GE, Haas M, Mernitz T. Interaction of neurotensin, cholecystokinin, and secretin in the stimulation of the exocrine pancreas in the dog. Gastroenterology 1983; 84:556–561.

159. Iwatsuki K, Horiuchi A, Ren LM, Chiba S. Direct and indirect stimulation of pancreatic exocrine secretion by neurotensin in anaesthetized dogs. Clin Exp Pharmacol Physiol 1991; 18:475–481.

160. Hegyi P, Rakonczay Z, Jr., Tiszlavicz L, Varro A, Toth A, Racz G, Varga G, Gray MA, Argent BE. Protein kinase C mediates the inhibitory effect of substance P on HCO_3^- secretion from guinea pig pancreatic ducts. Am J Physiol Cell Physiol 2005; 288:C1030–C1041.

161. Kemeny LV, Hegyi P, Rakonczay Z, Jr., Borka K, Korompay A, Gray MA, Argent BE, Venglovecz V. Substance P inhibits pancreatic ductal bicarbonate secretion via neurokinin receptors 2 and 3 in the guinea pig exocrine pancreas. Pancreas 2011; 40:793–795.

162. Koh YH, Bhatia M. Substance P. Pancreapedia: Exocrine Pancreas Knowledge Base, 2011. DOI: 10.3998/panc.2011.23.

163. Katoh K, Murai K, Nonoyama T. Effects of substance P on fluid and amylase secretion in exocrine pancreas of rat and mouse. Res Vet Sci 1984; 36:147–152.

164. Seifert H, Sawchenko P, Chesnut J, Rivier J, Vale W, Pandol SJ. Receptor for calcitonin gene-related peptide: binding to exocrine pancreas mediates biological actions. Am J Physiol Gastrointest Liver Physiol 1985; 249:G147–G151.

165. Bunnett NW, Mulvihill SJ, Debas HT. Calcitonin gene-related peptide inhibits exocrine secretion from the rat pancreas by a neurally mediated mechanism. Exp Physiol 1991; 76:115–123.

166. Sheikh SP, Holst JJ, Skak-Nielsen T, Knigge U, Warberg J, Theodorsson-Norheim E, Hokfelt T, Lundberg JM, Schwartz TW. Release of NPY in pig pancreas: dual parasympathetic and sympathetic regulation. Am J Physiol Gastrointest Liver Physiol 1988; 255:G46–G54.

167. Mulholland MW, Lally K, Taborsky GJ, Jr. Inhibition of rat pancreatic exocrine secretion by neuropeptide Y: studies in vivo and in vitro. Pancreas 1991; 6:433–440.

168. Holst JJ, Orskov C, Knuhtsen S, Sheikh S, Nielsen OV. On the regulatory functions of neuropeptide Y (NPY) with respect to vascular resistance and exocrine and endocrine secretion in the pig pancreas. Acta Physiol Scand 1989; 136:519–526.

169. Iwatsuki K, Ren LM, Chiba S. Effects of peptide histidine isoleucine on pancreatic exocrine secretion in anaesthetized dogs. Clin Exp Pharmacol Physiol 1993; 20:501–507.

170. Larsson LI, Rehfeld JF. Peptidergic and adrenergic innervation of pancreatic ganglia. Scand J Gastroenterol 1979; 14:433–437.

171. Singh J, Pearson GT. Effects of nerve stimulation on enzyme secretion from the in vitro rat pancreas and 3H-release after preincubation with catecholamines. Naunyn Schmiedebergs Arch Pharmacol 1984; 327:228–233.

172. Pearson GT, Singh J, Petersen OH. Adrenergic nervous control of cAMP-mediated amylase secretion in the rat pancreas. Am J Physiol Gastrointest Liver Physiol 1984; 246:G563–G573.

173. Varga G, Papp M, Vizi ES. Cholinergic and adrenergic control of enzyme secretion in isolated rat pancreas. Dig Dis Sci 1990; 35:501–507.

174. Carlei F, Allen JM, Bishop AE, Bloom SR, Polak JM. Occurrence, distribution and nature of neuropeptide Y in the rat pancreas. Experientia 1985; 41:1554–1557.

175. Yamaguchi N, Fukushima Y. PACAP enhances stimulation-induced norepinephrine release in canine pancreas in vivo. Can J Physiol Pharmacol 1998; 76:788–797.

176. Klein E, Salinas A, Shemesh E, Dreiling DA. Effects of autonomic denervation on canine exocrine pancreatic secretion and blood flow. Int J Pancreatol 1988; 3:165–170.

177. Yi E, Smith TG, Baker RC, Love JA. Catecholamines and 5-hydroxytryptamine in tissues of the rabbit exocrine pancreas. Pancreas 2004; 29:218–224.

178. Barlow TE, Greenwell JR, Harper AA, Scratcherd T. The effect of adrenaline and noradrenaline on the blood flow, electrical conductance and external secretion of the pancreas. J Physiol 1971; 217:665–678.

179. Solomon TE, Solomon N, Shanbour LL, Jacobson ED. Direct and indirect effects of nicotine on rabbit pancreatic secretion. Gastroenterology 1974; 67:276–283.

180. Furuta Y, Hashimoto K, Washizaki M. beta-Adrenoceptor stimulation of exocrine secretion from the rat pancreas. Br J Pharmacol 1978; 62:25–29.

181. Chariot J, Roze C, de la Tour J, Souchard M, Vaille C. Modulation of stimulated pancreatic secretion by sympathomimetic amines in the rat. Pharmacology 1983; 26:313–323.

182. Demol P, Sarles H. Action of catecholamines on exocrine pancreatic secretion of conscious rats. Arch Int Pharmacodyn Ther 1980; 243:149–163.

183. Lenz HJ, Messmer B, Zimmerman FG. Noradrenergic inhibition of canine gallbladder contraction and murine pancreatic secretion during stress by corticotropin-releasing factor. J Clin Invest 1992; 89:437–443.

184. Chariot J, Appia F, del Tacca M, Tsocas A, Roze C. Central and peripheral inhibition of exocrine pancreatic secretion by alpha-2 adrenergic agonists in the rat. Pharmacol Res Commun 1988; 20:707–717.

185. Demol P, Sarles H. Inhibition of exocrine pancreatic secretion by alpha-adrenergic blocking agents in conscious rats. Arch Int Pharmacodyn Ther 1980; 243:164–176.

186. Love JA, Szebeni K. Morphology and histochemistry of the rabbit pancreatic innervation. Pancreas 1999; 18:53–64.

187. Iwatsuki K, Horiuchi A, Ren LM, Chiba S. D-1 dopamine receptors mediate dopamine-induced pancreatic exocrine secretion in anesthetized dogs. Hypertens Res 1995; 18 Suppl 1:S173–174.

188. Iijima F, Iwatsuki K, Chiba S. Effects of dopamine on exocrine secretion and cyclic nucleotide concentration in the dog pancreas. Eur J Pharmacol 1983; 92:191–197.

189. Bastie MJ, Vaysse N, Brenac B, Pascal JP, Ribet A. Effects of catecholamines and their inhibitors on the isolated canine pancreas. II. Dopamine. Gastroenterology 1977; 72:719–723.

190. Delcenserie R, Devaux MA, Sarles H. Action of dopamine on the exocrine pancreatic secretion of the intact dog. Br J Pharmacol 1986; 88:189–195.

191. Hashimoto K, Oguro K, Furuta Y. Species difference in the secretory response to dopamine in the pancreas of dogs, cats, rabbits and rats. Arch Histol Jpn 1977; 40 Suppl:129–132.

192. Koevary SB, McEvoy RC, Azmitia EC. Specific uptake of tritiated serotonin in the adult rat pancreas: evidence for the presence of serotonergic fibers. Am J Anat 1980; 159:361–368.

193. Nawrot-Porabka K, Jaworek J, Leja-Szpak A, Szklarczyk J, Konturek SJ, Reiter RJ. Luminal melatonin stimulates pancreatic enzyme secretion via activation of serotonin-dependent nerves. Pharmacol Rep 2013; 65:494–504.

194. Chey WY, Chang T. Neural hormonal regulation of exocrine pancreatic secretion. Pancreatology 2001; 1:320–335.

195. Yago MD, Manas M, Ember Z, Singh J. Nitric oxide and the pancreas: morphological base and role in the control of the exocrine pancreatic secretion. Mol Cell Biochem 2001; 219:107–120.

196. Ekblad E, Alm P, Sundler F. Distribution, origin and projections of nitric oxide synthase-containing neurons in gut and pancreas. Neuroscience 1994; 63:233–248.

197. Liu HP, Tay SS, Leong SK. Nitrergic neurons in the pancreas of newborn guinea pig: their distribution and colocalization with various neuropeptides and dopamine-beta-hydroxylase. J Auton Nerv Syst 1996; 61:248–256.

198. Ember Z, Yago MD, Singh J. Distribution of nitric oxide synthase and secretory role of exogenous nitric oxide in the isolated rat pancreas. Int J Pancreatol 2001; 29:77–84.

199. Yago MD, Tapia JA, Salido GM, Adeghate E, Juma LM, Martinez-Victoria E, Manas M, Singh J. Effect of sodium nitroprusside and 8-bromo cyclic GMP on nerve-mediated and acetylcholine-evoked secretory responses in the rat pancreas. Br J Pharmacol 2002; 136:49–56.

200. Kawabata A, Kuroda R, Nishida M, Nagata N, Sakaguchi Y, Kawao N, Nishikawa H, Arizono N, Kawai K. Protease-activated receptor-2 (PAR-2) in the pancreas and parotid gland: immunolocalization and involvement of nitric oxide in the evoked amylase secretion. Life Sci 2002; 71:2435–2446.

201. Holst JJ, Rasmussen TN, Schmidt P. Role of nitric oxide in neurally induced pancreatic exocrine secretion in pigs. Am J Physiol Gastrointest Liver Physiol 1994; 266:G206–G213.

202. Chandra R, Samsa L, Vigna S, Liddle R. Pseudopod-like basal cell processes in intestinal cholecystokinin cells. Cell and Tissue Research 2010; 341:289–297.

203. Bohorquez DV, Samsa LA, Roholt A, Medicetty S, Chandra R, Liddle RA. An enteroendocrine cell-enteric glia connection revealed by 3D electron microscopy. PLoS One 2014; 9:e89881.

204. Bohorquez DV, Shahid RA, Erdmann A, Kreger AM, Wang Y, Calakos N, Wang F, Liddle RA. Neuroepithelial circuit formed by innervation of sensory enteroendocrine cells. J Clin Invest 2015; 125:782–786.

205. Kaelberer MM, Buchanan KL, Klein ME, Barth BB, Montoya MM, Shen X, Bohorquez DV. A gut-brain neural circuit for nutrient sensory transduction. Science 2018; 361(6408):eaat5236. doi: 10.1126/science.aat5236.

206. Eysselein VE, Eberlein GA, Hesse WH, Singer MV, Goebell H, Reeve JR, Jr. Cholecystokinin-58 is the major circulating form of cholecystokinin in canine blood. J Biol Chem 1987; 262:214–217.

207. Turkelson CM, Solomon TE, Bussjaeger L, Turkelson J, Ronk M, Shively JE, Ho FJ, Reeve JR, Jr. Chemical characterization of rat cholecystokinin-58. Peptides 1988; 9:1255–1260.

208. Eberlein GA, Eysselein VE, Hesse WH, Goebell H, Schaefer M, Reeve JR, Jr. Detection of cholecystokinin-58 in human blood by inhibition of degradation. Am J Physiol Gastrointest Liver Physiol 1987; 253:G477–G482.

209. Reeve JR, Jr., Green GM, Chew P, Eysselein VE, Keire DA. CCK-58 is the only detectable endocrine form of cholecystokinin in rat. Am J Physiol Gastrointest Liver Physiol 2003; 285:G255–G265.

210. Criddle DN, Booth DM, Mukherjee R, McLaughlin E, Green GM, Sutton R, Petersen OH, Reeve JR. Cholecystokinin-58 and cholecystokinin-8 exhibit similar actions on calcium signaling, zymogen secretion, and cell fate in murine pancreatic acinar cells. Am J Physiol Gastrointest Liver Physiol 2009; 297:G1085–G1092.

211. Jensen SL, Holst JJ, Nielsen OV, Rehfeld JF. Effect of sulfation of CCK-8 on its stimulation of the endocrine and exocrine secretion from the isolated perfused porcine pancreas. Digestion 1981; 22:305–309.

212. Sakamoto C, Otsuki M, Ohki A, Yuu H, Maeda M, Yamasaki T, Baba S. Glucose-dependent insulinotropic action of cholecystokinin and caerulein in the isolated perfused rat pancreas. Endocrinology 1982; 110:398–402.

213. Gardner JD, Knight M, Sutliff VE, Tamminga CA, Jensen RT. Derivatives of CCK-(26–32) as cholecystokinin receptor antagonists in guinea pig pancreatic acini. Am J Physiol Gastrointest Liver Physiol 1984; 246:G292–G295.

214. Otsuki M, Okabayashi Y, Ohki A, Oka T, Fujii M, Nakamura T, Sugiura N, Yanaihara N, Baba S. Action of cholecystokinin analogues on exocrine and endocrine rat pancreas. Am J Physiol Gastrointest Liver Physiol 1986; 250:G405–G411.

215. Okabayashi Y, Otsuki M, Ohki A, Sakamoto C, Baba S. Effects of C-terminal fragments of cholecystokinin on exocrine and endocrine secretion from isolated perfused rat pancreas. Endocrinology 1983; 113:2210–2215.

216. Doi R, Inoue K, Kogire M, Sumi S, Yun M, Futaki S, Fujii N, Yajima H, Tobe T. Effect of synthetic human cholecystokinin-33 on exocrine pancreas. Biochem Biophys Res Commun 1988; 150:1251–1255.

217. Van Dijk A, Richards JG, Trzeciak A, Gillessen D, Mohler H. Cholecystokinin receptors: biochemical demonstration and autoradiographical localization in rat brain and pancreas using [3H] cholecystokinin8 as radioligand. J Neurosci 1984; 4:1021–1033.

218. Tang C, Biemond I, Lamers CB. Visualization and characterization of CCK receptors in exocrine pancreas of rat with storage phosphor autoradiography. Pancreas 1996; 13:311–315.

219. Imoto I, Yamamoto M, Jia DM, Otsuki M. Effect of chronic oral administration of the CCK receptor antagonist loxiglumide on exocrine and endocrine pancreas in normal rats. Int J Pancreatol 1997; 22:177–185.

220. Griesbacher T, Rainer I, Heinemann A, Groisman D. Haemodynamic and exocrine effects of caerulein at submaximal and supramaximal levels on the rat pancreas: role of cholecystokinin receptor subtypes. Pancreatology 2006; 6:65–75.

221. Kopin AS, Mathes WF, McBride EW, Nguyen M, Al-Haider W, Schmitz F, Bonner-Weir S, Kanarek R, Beinborn M. The cholecystokinin-A receptor mediates inhibition of food intake yet is not essential for the maintenance of body weight. J Clin Invest 1999; 103:383–391.

222. Takiguchi S, Suzuki S, Sato Y, Kanai S, Miyasaka K, Jimi A, Shinozaki H, Takata Y, Funakoshi A, Kono A, Minowa O, Kobayashi T, Noda T. Role of CCK-A receptor for pancreatic function in mice: a study in CCK-A receptor knockout mice. Pancreas 2002; 24:276–283.

223. Miyasaka K, Shinozaki H, Suzuki S, Sato Y, Kanai S, Masuda M, Jimi A, Nagata A, Matsui T, Noda T, Kono A, Funakoshi A. Disruption of cholecystokinin (CCK)-B receptor gene did not modify bile or pancreatic secretion or pancreatic growth: a study in CCK-B receptor gene knockout mice. Pancreas 1999; 19:114–118.

224. Morisset J, Levenez F, Corring T, Benrezzak O, Pelletier G, Calvo E. Pig pancreatic acinar cells possess predominantly the CCK-B receptor subtype. Am J Physiol Endocrinol Metab 1996; 271:E397–E402.

225. Konturek JW, Gabryelewicz A, Kulesza E, Konturek SJ, Domschke W. Cholecystokinin (CCK) in the amino acid uptake and enzyme protein secretion by the pancreas in humans. Int J Pancreatol 1995; 17:55–61.

226. Szalmay G, Varga G, Kajiyama F, Yang XS, Lang TF, Case RM, Steward MC. Bicarbonate and fluid secretion evoked by cholecystokinin, bombesin and acetylcholine in isolated guinea-pig pancreatic ducts. J Physiol 2001; 535:795–807.

227. Tang C, Biemond I, Lamers CB. Cholecystokinin receptors in human pancreas and gallbladder muscle: a comparative study. Gastroenterology 1996; 111:1621–1626.

228. Pisegna JR, de Weerth A, Huppi K, Wank SA. Molecular cloning of the human brain and gastric cholecystokinin receptor: structure, functional expression and chromosomal localization. Biochem Biophys Res Commun 1992; 189:296–303.

229. Adler G, Beglinger C, Braun U, Reinshagen M, Koop I, Schafmayer A, Rovati L, Arnold R. Interaction of the cholinergic system and cholecystokinin in the regulation of endogenous and exogenous stimulation of pancreatic secretion in humans. Gastroenterology 1991; 100:537–543.

230. Ji B, Bi Y, Simeone D, Mortensen RM, Logsdon CD. Human pancreatic acinar cells lack functional responses to cholecystokinin and gastrin. Gastroenterology 2001; 121:1380–1390.

231. Miyasaka K, Shinozaki H, Jimi A, Funakoshi A. Amylase secretion from dispersed human pancreatic acini: neither

cholecystokinin A nor cholecystokinin B receptors mediate amylase secretion in vitro. Pancreas 2002; 25:161–165.

232. Murphy JA, Criddle DN, Sherwood M, Chvanov M, Mukherjee R, McLaughlin E, Booth D, Gerasimenko JV, Raraty MG, Ghaneh P, Neoptolemos JP, Gerasimenko OV, et al. Direct activation of cytosolic Ca²⁺ signaling and enzyme secretion by cholecystokinin in human pancreatic acinar cells. Gastroenterology 2008; 135:632–641.

233. Liang T, Dolai S, Xie L, Winter E, Orabi AI, Karimian N, Cosen-Binker LI, Huang YC, Thorn P, Cattral MS, Gaisano HY. Ex vivo human pancreatic slice preparations offer a valuable model for studying pancreatic exocrine biology. J Biol Chem 2017; 292:5957–5969.

234. Phillips PA, Yang L, Shulkes A, Vonlaufen A, Poljak A, Bustamante S, Warren A, Xu Z, Guilhaus M, Pirola R, Apte MV, Wilson JS. Pancreatic stellate cells produce acetylcholine and may play a role in pancreatic exocrine secretion. Proc Natl Acad Sci U S A 2010; 107:17397–17402.

235. Lee KY, Krusch D, Zhou L, Song Y, Chang TM, Chey WY. Effect of endogenous insulin on pancreatic exocrine secretion in perfused dog pancreas. Pancreas 1995; 11:190–195.

236. Burnham DB, McChesney DJ, Thurston KC, Williams JA. Interaction of cholecystokinin and vasoactive intestinal polypeptide on function of mouse pancreatic acini in vitro. J Physiol 1984; 349:475–482.

237. Camello PJ, Salido GM. Inhibitory interactions between stimulus-secretion pathways in the exocrine rat pancreas. Biochem Pharmacol 1993; 46:1005–1009.

238. Pandol SJ, Schoeffield MS, Sachs G, Muallem S. Role of free cytosolic calcium in secretagogue-stimulated amylase release from dispersed acini from guinea pig pancreas. J Biol Chem 1985; 260:10081–10086.

239. Pandol SJ, Thomas MW, Schoeffield MS, Sachs G, Muallem S. Role of calcium in cholecystokinin-stimulated phosphoinositide breakdown in exocrine pancreas. Am J Physiol Gastrointest Liver Physiol 1985; 248:G551–G560.

240. Williams JA, Chen X, Sabbatini ME. Small G proteins as key regulators of pancreatic digestive enzyme secretion. Am J Physiol Endocrinol Metab 2009; 296:E405–E414.

241. Sabbatini ME, Bi Y, Ji B, Ernst SA, Williams JA. CCK activates RhoA and Rac1 differentially through Gα₁₃ and Gα_q in mouse pancreatic acini. Am J Physiol Cell Physiol 2010; 298:C592-C601.

242. Williams JA. Secretin. Pancreapedia: Exocrine Pancreas Knowledge Base, 2013. DOI: 10.3998/panc.2013.15.

243. Faichney A, Chey WY, Kim YC, Lee KY, Kim MS, Chang TM. Effect of sodium oleate on plasma secretin concentration and pancreatic secretion in dog. Gastroenterology 1981; 81:458–462.

244. Chey WY, Chang T-M. Secretin, 100 years later. J Gastroenterol 2003; 38:1025–1035.

245. Blomfield J, Settree PJ. Ultrastructural responses of rat exocrine pancreas to cholecystokinin octapeptide and secretin. Exp Mol Pathol 1983; 38:389–397.

246. Beglinger C, Fried M, Whitehouse I, Jansen JB, Lamers CB, Gyr K. Pancreatic enzyme response to a liquid meal and to hormonal stimulation. Correlation with plasma secretin and cholecystokinin levels. J Clin Invest 1985; 75:1471–1476.

247. Schaffalitzky de Muckadell OB, Fahrenkrug J, Watt-Boolsen S, Worning H. Pancreatic response and plasma secretin concentration during infusion of low dose secretin in man. Scand J Gastroenterol 1978; 13:305–311.

248. Hacki WH, Bloom SR, Mitznegg P, Domschke W, Domschke S, Belohlavek D, Demling L, Wunsch E. Plasma secretin and pancreatic bicarbonate response to exogenous secretin in man. Gut 1977; 18:191–195.

249. You CH, Rominger JM, Chey WY. Potentiation effect of cholecystokinin-octapeptide on pancreatic bicarbonate secretion stimulated by a physiologic dose of secretin in humans. Gastroenterology 1983; 85:40–45.

250. Schaffalitzky de Muckadell OB, Fahrenkrug J, Matzen P, Rune SJ, Worning H. Physiological significance of secretin in the pancreatic bicarbonate secretion. II. Pancreatic bicarbonate response to a physiological increase in plasma secretin concentration. Scand J Gastroenterol 1979; 14:85–90.

251. Jowell PS, Robuck-Mangum G, Mergener K, Branch MS, Purich ED, Fein SH. A double-blind, randomized, dose response study testing the pharmacological efficacy of synthetic porcine secretin. Aliment Pharmacol Ther 2000; 14:1679–1684.

252. Sanchez-Vicente C, Rodriguez-Nodal F, Minguela A, Garcia LJ, San Roman JI, Calvo JJ, Lopez MA. Cholinergic pathways are involved in secretin and VIP release and the exocrine pancreatic response after intraduodenally perfused acetic and lactic acids in the rat. Pancreas 1995; 10:93–99.

253. Park HS, Lee YL, Kwon HY, Chey WY, Park HJ. Significant cholinergic role in secretin-stimulated exocrine secretion in isolated rat pancreas. Am J Physiol Gastrointest Liver Physiol 1998; 274:G413–G418.

254. Meuth-Metzinger VL, Philouze-Rome V, Metzinger L, Gespach C, Guilloteau P. Differential activation of adenylate cyclase by secretin and VIP receptors in the calf pancreas. Pancreas 2005; 31:174–181.

255. Sans MD, Sabbatini ME, Ernst SA, D'Alecy LG, Nishijima I, Williams JA. Secretin is not necessary for exocrine pancreatic development and growth in mice. Am J Physiol Gastrointest Liver Physiol 2011; 301:G791–G798.

256. Trimble ER, Bruzzone R, Biden TJ, Farese RV. Secretin induces rapid increases in inositol trisphosphate, cytosolic Ca2+ and diacylglycerol as well as cyclic AMP in rat pancreatic acini. Biochem J 1986; 239:257–261.

257. Sommer H, Kasper H. The action of synthetic secretin, cholecystokinin-octapeptide and combinations of these hormones on the secretion of the isolated perfused rat pancreas. Hepatogastroenterology 1981; 28:311–315.

258. Alcon S, Rosado JA, Garcia LJ, Pariente JA, Salido GM, Pozo MJ. Secretin potentiates guinea pig pancreatic response to cholecystokinin by a cholinergic mechanism. Can J Physiol Pharmacol 1996; 74:1342–1350.

259. Hasegawa H, Okabayashi Y, Koide M, Kido Y, Okutani T, Matsushita K, Otsuki M, Kasuga M. Effect of islet hormones on secretin-stimulated exocrine secretion in isolated perfused rat pancreas. Dig Dis Sci 1993; 38:1278–1283.

260. Müller TD, Finan B, Bloom SR, D'Alessio D, Drucker DJ, Flatt PR, Fritsche A, Gribble F, Grill HJ, Habener JF, Holst JJ, Langhans W, et al. Glucagon-like peptide 1 (GLP-1). Mol Metab 2019; 30:72–130.

261. Williams JA. GLP-1. Pancreapedia: Exocrine Pancreas Knowledge Base, 2014. DOI: 10.3998/panc.2014.07.

262. Hou Y, Ernst SA, Heidenreich K, Williams JA. Glucagon-like peptide-1 receptor is present in pancreatic acinar cells and regulates amylase secretion through cAMP. Am J Physiol Gastrointest Liver Physiol 2016; 310:G26–G33.

263. Satoh K, Ouchi M, Morita A, Kashimata M. MARCKS phosphorylation and amylase release in GLP-1-stimulated acini isolated from rat pancreas. J Physiol Sci 2019; 69:143–149.

264. Ventimiglia MS, Najenson AC, Rodríguez MR, Davio CA, Vatta MS, Bianciotti LG. Natriuretic peptides and their receptors. Pancreapedia: Exocrine Pancreas Knowledge Base, 2011. DOI: 10.3998/panc.2011.36.

265. Heisler S, Kopelman H, Chabot JG, Morel G. Atrial natriuretic factor and exocrine pancreas: effects on the secretory process. Pancreas 1987; 2:243–251.

266. Maack T. Role of atrial natriuretic factor in volume control. Kidney Int 1996; 49:1732–1737.

267. Oguchi H, Iwatsuki K, Horiuchi A, Furuta S, Chiba S. Effects of human atrial natriuretic polypeptide on pancreatic exocrine secretion in the dog. Biochem Biophys Res Commun 1987; 146:757–763.

268. Sabbatini ME, Villagra A, Davio CA, Vatta MS, Fernandez BE, Bianciotti LG. Atrial natriuretic factor stimulates exocrine pancreatic secretion in the rat through NPR-C receptors. Am J Physiol Gastrointest Liver Physiol 2003; 285:G929–G937.

269. Sabbatini ME, Vatta MS, Davio CA, Bianciotti LG. Atrial natriuretic factor negatively modulates secretin intracellular signaling in the exocrine pancreas. Am J Physiol Gastrointest Liver Physiol 2007; 292:G349–G357.

270. Rodríguez MR, Diez F, Ventimiglia MS, Morales V, Copsel S, Vatta MS, Davio CA, Bianciotti LG. Atrial natriuretic factor stimulates efflux of cAMP in rat exocrine pancreas via multidrug resistance-associated proteins. Gastroenterology 2011; 140:1292–1302.

271. Sabbatini ME, Rodriguez MR, Dabas P, Vatta MS, Bianciotti LG. C-type natriuretic peptide stimulates pancreatic exocrine secretion in the rat: role of vagal afferent and efferent pathways. Eur J Pharmacol 2007; 577:192–202.

272. Sabbatini ME, Rodriguez M, di Carlo MB, Davio CA, Vatta MS, Bianciotti LG. C-type natriuretic peptide enhances amylase release through NPR-C receptors in the exocrine pancreas. Am J Physiol Gastrointest Liver Physiol 2007; 293:G987–G994.

273. Bergeron JJ, Rachubinski R, Searle N, Sikstrom R, Borts D, Bastian P, Posner BI. Radioautographic visualization of in vivo insulin binding to the exocrine pancreas. Endocrinology 1980; 107:1069–1080.

274. Adler G, Kern HF. Regulation of exocrine pancreatic secretory process by insulin in vivo. Horm Metab Res 1975; 7:290–296.

275. Lee YL, Kwon HY, Park HS, Lee TH, Park HJ. The role of insulin in the interaction of secretin and cholecystokinin in exocrine secretion of the isolated perfused rat pancreas. Pancreas 1996; 12:58–63.

276. Matsushita K, Okabayashi Y, Koide M, Hasegawa H, Otsuki M, Kasuga M. Potentiating effect of insulin on exocrine secretory function in isolated rat pancreatic acini. Gastroenterology 1994; 106:200–206.

277. Lee KY, Lee YL, Kim CD, Chang TM, Chey WY. Mechanism of action of insulin on pancreatic exocrine secretion in perfused rat pancreas. Am J Physiol Gastrointest Liver Physiol 1994; 267:G207–G212.

278. Kim C, Kim K, Lee H, Song C, Ryu H, Hyun J. Potentiation of cholecystokinin and secretin-induced pancreatic exocrine secretion by endogenous insulin in humans. Pancreas 1999; 18:410–414.

279. Liddle RA, Rushakoff RJ, Morita ET, Beccaria L, Carter JD, Goldfine ID. Physiological role for cholecystokinin in reducing postprandial hyperglycemia in humans. J Clin Invest 1988; 81:1675–1681.

280. Patel R, Singh J, Yago MD, Vilchez JR, Martinez-Victoria E, Manas M. Effect of insulin on exocrine pancreatic secretion in healthy and diabetic anaesthetised rats. Mol Cell Biochem 2004; 261:105–110.

281. Park HJ, Lee YL, Kwon HY. Effects of pancreatic polypeptide on insulin action in exocrine secretion of isolated rat pancreas. J Physiol 1993; 463:421–429.

282. Barreto SG, Woods CM, Carati CJ, Schloithe AC, Jaya SR, Toouli J, Saccone GT. Galanin inhibits caerulein-stimulated pancreatic amylase secretion via cholinergic nerves and insulin. Am J Physiol Gastrointest Liver Physiol 2009; 297:G333–G339.

283. Miyasaka K, Funakoshi A, Yasunami Y, Nakamura R, Kitani K, Tamamura H, Funakoshi S, Fujii N. Rat pancreastatin inhibits both pancreatic exocrine and endocrine secretions in rats. Regul Pept 1990; 28:189–198.

284. Lee W, Miyazaki K, Funakoshi A. Effects of somatostatin and pancreatic polypeptide on exocrine and endocrine pancreas in the rats. Gastroenterol Jpn 1988; 23:49–55.

285. Williams JA. Bombesin. Pancreapedia: Exocrine Pancreas Knowledge Base, 2015. DOI: 10.3998/panc.2015.10.

286. Lhoste EF, Levenez F, Chabanet C, Fiszlewicz M, Corring T. Effect of bombesin at low doses on the secretion of the exocrine pancreas and on plasma gastrin concentration in the conscious pig. Regul Pept 1998; 74:41–45.

287. Lhoste E, Aprahamian M, Pousse A, Hoeltzel A, Stock-Damge C. Combined effect of chronic bombesin and secretin or cholecystokinin on the rat pancreas. Peptides 1985; 6 Suppl 3:83–87.

288. Stock-Damge C, Lhoste E, Aprahamian M, Loza E. Influence of repeated administration of bombesin on rat pancreatic secretion. Pancreas 1987; 2:658–663.

289. Jaworek J, Szklarczyk J, Jaworek AK, Nawrot-Porabka K, Leja-Szpak A, Bonior J, Kot M. Protective effect of melatonin on acute pancreatitis. Int J Inflam 2012; 2012:173675.

290. Jaworek J, Nawrot K, Konturek SJ, Leja-Szpak A, Thor P, Pawlik WW. Melatonin and its precursor, L-tryptophan: influence on pancreatic amylase secretion in vivo and in vitro. J Pineal Res 2004; 36:155–164.

291. Leja-Szpak A, Jaworek J, Nawrot-Porabka K, Palonek M, Mitis-Musiol M, Dembinski A, Konturek SJ, Pawlik WW. Modulation of pancreatic enzyme secretion by melatonin and its precursor; L-tryptophan. Role of CCK and afferent nerves. J Physiol Pharmacol 2004; 55 Suppl 2:33–46.

292. Jaworek J, Nawrot-Porabka K, Leja-Szpak A, Bonior J, Szklarczyk J, Kot M, Konturek SJ, Pawlik WW. Melatonin as modulator of pancreatic enzyme secretion and pancreatoprotector. J Physiol Pharmacol 2007; 58 Suppl 6:65–80.

293. Nawrot-Porabka K, Jaworek J, Leja-Szpak A, Szklarczyk J, Kot M, Mitis-Musiol M, Konturek SJ, Pawlik WW. Involvement of vagal nerves in the pancreatostimulatory effects of luminal melatonin, or its precursor L-tryptophan. Study in the rats. J Physiol Pharmacol 2007; 58 Suppl 6:81–95.

294. Jaworek J, Leja-Szpak A, Nawrot-Porabka K, Szklarczyk J, Kot M, Pierzchalski P, Goralska M, Ceranowicz P, Warzecha Z, Dembinski A, Bonior J. Effects of melatonin and its analogues on pancreatic inflammation, enzyme secretion, and tumorigenesis. Int J Mol Sci 2017; 18:1014–1026.

295. Funakoshi A, Miyasaka K, Kitani K, Nakamura J, Funakoshi S, Fukuda H, Fujii N. Stimulatory effects of islet amyloid polypeptide (amylin) on exocrine pancreas and gastrin release in conscious rats. Regul Pept 1992; 38:135–143.

296. Huang Y, Fischer JE, Balasubramaniam A. Amylin mobilizes [Ca2+]i and stimulates the release of pancreatic digestive enzymes from rat acinar AR42J cells: evidence for an exclusive receptor system of amylin. Peptides 1996; 17:497–502.

297. Fehmann H-C, Weber V, Göke R, Göke B, Eissele R, Arnold R. Islet amyloid polypeptide (IAPP;Amylin) influences the endocrine but not the exocrine rat pancreas. Biochem Biophys Res Commun 1990; 167:1102–1108.

298. Kikuchi Y, Koizumi M, Shimosegawa T, Kashimura J, Suzuki S, Toyota T. Islet amyloid polypeptide has no effect on amylase release from rat pancreatic acini stimulated by CCK-8, secretin, carbachol and VIP. Tohoku J Exp Med 1991; 165:41–48.

299. Young AA, Jodka C, Pittner R, Parkes D, Gedulin BR. Dose-response for inhibition by amylin of cholecystokinin-stimulated secretion of amylase and lipase in rats. Regul Pept 2005; 130:19–26.

300. Nishimura T, Nakatake Y, Konishi M, Itoh N. Identification of a novel FGF, FGF-21, preferentially expressed in the liver. Biochim Biophys Acta (BBA) - Gene Structure and Expression 2000; 1492:203–206.

301. Johnson CL, Weston JY, Chadi SA, Fazio EN, Huff MW, Kharitonenkov A, Köester A, Pin CL. Fibroblast growth factor 21 reduces the severity of cerulein-induced pancreatitis in mice. Gastroenterology 2009; 137:1795–1804.

302. Tu HJ, Zhao CF, Chen ZW, Lin W, Jiang YC. Fibroblast growth factor (FGF) signaling protects against acute pancreatitis-induced damage by modulating inflammatory responses. Med Sci Monit 2020; 26:e920684.

303. Hernandez G, Luo T, Javed TA, Wen L, Kalwat MA, Vale K, Ammouri F, Husain SZ, Kliewer SA, Mangelsdorf DJ. Pancreatitis is an FGF21-deficient state that is corrected by replacement therapy. Sci Transl Med 2020; 12:eaay5186.

304. Coate KC, Hernandez G, Thorne CA, Sun S, Le TDV, Vale K, Kliewer SA, Mangelsdorf DJ. FGF21 is an exocrine pancreas secretagogue. Cell Metab 2017; 25:472–480.

305. Singh J, Pariente JA, Salido GM. The physiological role of histamine in the exocrine pancreas. Inflamm Res 1997; 46:159–165.

306. Pariente JA, Madrid JA, Salido GM. Role of histamine receptors in rabbit pancreatic exocrine secretion stimulated by cholecystokinin and secretin. Exp Physiol 1990; 75:657–667.

307. Gomez GA, Greeley GHJ. Peptide YY. Pancreapedia: Exocrine Pancreas Knowledge Base, 2013. DOI: 10.3998/panc.2013.06.

308. Williams JA. Pancreatic polypeptide. Pancreapedia: Exocrine Pancreas Knowledge Base, 2014. DOI: 10.3998/panc.2014.04.

309. Ekblad E, Sundler F. Distribution of pancreatic polypeptide and peptide YY. Peptides 2002; 23:251–261.

310. Holzer P, Reichmann F, Farzi A. Neuropeptide Y, peptide YY and pancreatic polypeptide in the gut-brain axis. Neuropeptides 2012; 46:261–274.

311. Mannon P, Taylor IL. The pancreatic polypeptide family. In: Walsh JH, Dockray GJ, eds. Gut peptides: biochemistry and Physiology. New York: Raven Press, 1994:341–370.

312. Tatemoto K. Isolation and characterization of peptide YY (PYY), a candidate gut hormone that inhibits pancreatic exocrine secretion. Proc Natl Acad Sci U S A 1982; 79:2514–2518.

313. Pappas TN, Debas HT, Taylor IL. Peptide YY: metabolism and effect on pancreatic secretion in dogs. Gastroenterology 1985; 89:1387–1392.

314. Pappas TN, Debas HT, Goto Y, Taylor IL. Peptide YY inhibits meal-stimulated pancreatic and gastric secretion. Am J Physiol Gastrointest Liver Physiol 1985; 248:G118–G123.

315. Inoue K, Hosotani R, Tatemoto K, Yajima H, Tobe T. Effect of natural peptide YY on blood flow and exocrine secretion of pancreas in dogs. Dig Dis Sci 1988; 33:828–832.

316. Sumi S, Inoue K, Tobe T. Experimental studies on the interrelationship between organs mediated by peptide YY: effect on splanchnic circulation and exocrine pancreas in dogs. Nihon Geka Hokan 1990; 59:224–233.

317. Jin H, Cai L, Lee K, Chang TM, Li P, Wagner D, Chey WY. A physiological role of peptide YY on exocrine pancreatic secretion in rats. Gastroenterology 1993; 105:208–215.

318. Deng X, Guarita DR, Wood PG, Kriess C, Whitcomb DC. PYY potently inhibits pancreatic exocrine secretion mediated through CCK-secretin-stimulated pathways but not 2-DG-stimulated pathways in awake rats. Dig Dis Sci 2001; 46:156–165.

319. Grandt D, Siewert J, Sieburg B, al Tai O, Schimiczek M, Goebell H, Layer P, Eysselein VE, Reeve JR, Jr., Muller MK. Peptide YY inhibits exocrine pancreatic secretion in isolated perfused rat pancreas by Y1 receptors. Pancreas 1995; 10:180–186.

320. DeMar AR, Taylor IL, Fink AS. Pancreatic polypeptide and peptide YY inhibit the denervated canine pancreas. Pancreas 1991; 6:419–426.

321. Sheikh SP, Roach E, Fuhlendorff J, Williams JA. Localization of Y1 receptors for NPY and PYY on vascular smooth muscle cells in rat pancreas. Am J Physiol Gastrointest Liver Physiol 1991; 260:G250–G257.

322. Kayasseh L, Haecki WH, Gyr K, Stalder GA, Rittmann WW, Halter F, Girard J. The endogenous release of pancreatic polypeptide by acid and meal in dogs. Effect of somatostatin. Scand J Gastroenterol 1978; 13:385–391.

323. Taylor IL, Byrne WJ, Christie DL, Ament ME, Walsh JH. Effect of individual l-amino acids on gastric acid secretion and serum gastrin and pancreatic polypeptide release in humans. Gastroenterology 1982; 83:273–278.

324. Schmid R, Schulte-Frohlinde E, Schusdziarra V, Neubauer J, Stegmann M, Maier V, Classen M. Contribution of postprandial amino acid levels to stimulation of insulin, glucagon, and pancreatic polypeptide in humans. Pancreas 1992; 7:698–704.

325. Choi BR, Palmquist DL. High fat diets increase plasma cholecystokinin and pancreatic polypeptide, and decrease plasma insulin and feed intake in lactating cows. J Nutr 1996; 126:2913–2919.

326. Beglinger C, Taylor IL, Grossman MI, Solomon TE. Pancreatic polypeptide release: role of stimulants of exocrine pancreatic secretion in dogs. Gastroenterology 1984; 87:530–536.

327. Köhler H, Nustede, R., Barthel, M., Müller, C., Schafmayer, A. Total denervation of the pancreas does not alter the pancreatic polypeptide-induced inhibition of pancreatic exocrine secretion in dogs. Res Exp Med (Berl) 1991; 191:359–369.

328. DeMar AR, Lake R, Fink AS. The effect of pancreatic polypeptide and peptide YY on pancreatic blood flow and pancreatic exocrine secretion in the anesthetized dog. Pancreas 1991; 6:9–14.

329. Joehl RJ, DeJoseph MR. Pancreatic polypeptide inhibits amylase release by rat pancreatic acini. J Surg Res 1986; 40:310–314.

330. Louie DS, Williams JA, Owyang C. Action of pancreatic polypeptide on rat pancreatic secretion: in vivo and in vitro. Am J Physiol Gastrointest Liver Physiol 1985; 249:G489–G495.

331. Konturek SJ, Meyers CA, Kwiecien N, Obtulowicz W, Tasler J, Oleksy J, Kopp B, Coy DH, Schally AV. Effect of human pancreatic polypeptide and its C-terminal hexapeptide on pancreatic secretion in man and in the dog. Scand J Gastroenterol 1982; 17:395–399.

332. Shiratori K, Watanabe S, Takeuchi T. Somatostatin analog, SMS 201–995, inhibits pancreatic exocrine secretion and release of secretin and cholecystokinin in rats. Pancreas 1991; 6:23–30.

333. Shiratori K, Watanabe S, Takeuchi T. Inhibitory effect of intraduodenal administration of somatostatin analogue SDZ CO 611 on rat pancreatic exocrine secretion. Pancreas 1993; 8:471–475.

334. Lin TM, Evans DC, Shaar CJ, Root MA. Action of somatostatin on stomach, pancreas, gastric mucosal blood flow, and hormones. Am J Physiol Gastrointest Liver Physiol 1983; 244:G40–G45.

335. Lee E, Gerlach U, Uhm DY, Kim J. Inhibitory effect of somatostatin on secretin-induced augmentation of the slowly activating K$^+$ current (IKs) in the rat pancreatic acinar cell. Pflugers Arch 2002; 443:405–410.

336. Ohnishi H, Mine T, Kojima I. Inhibition by somatostatin of amylase secretion induced by calcium and cyclic AMP in rat pancreatic acini. Biochem J 1994; 304 (Pt 2):531–536.

337. Mulvihill SJ, Bunnett NW, Goto Y, Debas HT. Somatostatin inhibits pancreatic exocrine secretion via a neural mechanism. Metabolism 1990; 39:143–148.

338. Kuvshinoff BW, Brodish RJ, James L, McFadden DW, Fink AS. Somatostatin inhibits secretin-induced canine pancreatic

339. response via a cholinergic mechanism. Gastroenterology 1993; 105:539–547.

339. Barreto SG. Galanin. Pancreapedia: Exocrine Pancreas Knowledge Base, 2015. DOI: 10.3998/panc.2015.21.

340. Shimosegawa T, Moriizumi S, Koizumi M, Kashimura J, Yanaihara N, Toyota T. Immunohistochemical demonstration of galaninlike immunoreactive nerves in the human pancreas. Gastroenterology 1992; 102:263–271.

341. Barreto SG, Bazargan M, Zotti M, Hussey DJ, Sukocheva OA, Peiris H, Leong M, Keating DJ, Schloithe AC, Carati CJ, Smith C, Toouli J, et al. Galanin receptor 3—a potential target for acute pancreatitis therapy. Neurogastroenterol Motil 2011; 23:e141–151.

342. Ahren B, Andren-Sandberg A, Nilsson A. Galanin inhibits amylase secretion from isolated rat pancreatic acini. Pancreas 1988; 3:559–562.

343. Flowe KM, Lally KM, Mulholland MW. Galanin inhibits rat pancreatic amylase release via cholinergic suppression. Peptides 1992; 13:487–492.

344. Kashimura J, Shimosegawa T, Kikuchi Y, Yoshida K, Koizumi M, Mochizuki T, Yanaihara N, Toyota T. Effects of galanin on amylase secretion from dispersed rat pancreatic acini. Pancreas 1994; 9:258–262.

345. Herzig KH, Brunke G, Schon I, Schaffer M, Folsch UR. Mechanism of galanin's inhibitory action on pancreatic enzyme secretion: modulation of cholinergic transmission-studies in vivo and in vitro. Gut 1993; 34:1616–1621.

346. Brodish RJ, Kuvshinoff BW, Fink AS, McFadden DW. Inhibition of pancreatic exocrine secretion by galanin. Pancreas 1994; 9:297–303.

347. Miyasaka K, Funakoshi A, Nakamura R, Kitani K, Shimizu F, Tatemoto K. Effects of porcine pancreastatin on postprandial pancreatic exocrine secretion and endocrine functions in the conscious rat. Digestion 1989; 43:204–211.

348. Funakoshi A, Miyasaka K, Nakamura R, Kitani K, Tatemoto K. Inhibitory effect of pancreastatin on pancreatic exocrine secretion in the conscious rat. Regul Pept 1989; 25:157–166.

349. von Schonfeld J, Muller MK, Runzi M, Geling M, Neisius I, Kleimann J, Goebell H. Pancreastatin—a mediator in the islet-acinar axis? Metabolism 1993; 42:552–555.

350. Migita Y, Nakano I, Goto M, Ito T, Nawata H. Effect of pancreastatin on cerulein-stimulated pancreatic blood flow and exocrine secretion in anaesthetized rats. J Gastroenterol Hepatol 1999; 14:583–587.

351. Bozadjieva N, Williams JA, Bernal-Mizrachi E. Glucagon. Pancreapedia: Exocrine Pancreas Knowledge Base, 2013. DOI: 10.3998/panc.2013.23.

352. Clain JE, Barbezat GO, Waterworth MM, Bank S. Glucagon inhibition of secretin and combined secretin and cholecystokinin stimulated pancreatic exocrine secretion in health and disease. Digestion 1978; 17:11–17.

353. Miller TA, Watson LC, Rayford PL, Thompson JC. The effect of glucagon on pancreatic secretion and plasma secretin in dogs. World J Surg 1977; 1:93–97.

354. Singer MV, Tiscornia OM, Mendes de Oliveiro JP, Demol P, Levesque D, Sarles H. Effect of glucagon on canine exocrine pancreatic secretion stimulated by a test meal. Can J Physiol Pharmacol 1978; 56:1–6.

355. Harada H, Kochi F, Hanafusa E, Kobayashi T, Oka H, Kimura I. Studies on the effect of glucagon on human pancreatic secretion by analysis of endoscopically obtained pure pancreatic juice. Gastroenterol Jpn 1985; 20:28–36.

356. Fontana G, Costa PL, Tessari R, Labo G. Effect of glucagon on pure human exocrine pancreatic secretion. Am J Gastroenterol 1975; 63:490–494.

357. Adler G. Effect of glucagon on the secretory process in the rat exocrine pancreas. Cell Tissue Res 1977; 182:193–204.

358. Singh M. Effect of glucagon on digestive enzyme synthesis, transport and secretion in mouse pancreatic acinar cells. J Physiol 1980; 306:307–322.

359. Pandol SJ, Sutliff VE, Jones SW, Charlton CG, O'Donohue TL, Gardner JD, Jensen RT. Action of natural glucagon on pancreatic acini: due to contamination by previously undescribed secretagogues. Am J Physiol Gastrointest Liver Physiol 1983; 245:G703–G710.

360. Horiuchi A, Iwatsuki K, Ren LM, Kuroda T, Chiba S. Dual actions of glucagon: direct stimulation and indirect inhibition of dog pancreatic secretion. Eur J Pharmacol 1993; 237:23–30.

361. Akalestou E, Christakis I, Solomou AM, Minnion JS, Rutter GA, Bloom SR. Proglucagon-derived peptides do not significantly affect acute exocrine pancreas in rats. Pancreas 2016; 45:967–973.

362. Lai KC, Cheng CH, Leung PS. The ghrelin system in acinar cells: localization, expression, and regulation in the exocrine pancreas. Pancreas 2007; 35:e1–8.

363. Zhang W, Chen M, Chen X, Segura BJ, Mulholland MW. Inhibition of pancreatic protein secretion by ghrelin in the rat. J Physiol 2001; 537:231–236.

364. Matyjek R, Herzig KH, Kato S, Zabielski R. Exogenous leptin inhibits the secretion of pancreatic juice via a duodenal CCK1-vagal-dependent mechanism in anaesthetized rats. Regul Pept 2003; 114:15–20.

365. Jaworek J, Bonior J, Konturek SJ, Bilski J, Szlachcic A, Pawlik WW. Role of leptin in the control of postprandial pancreatic enzyme secretion. J Physiol Pharmacol 2003; 54:591–602.

366. Nawrot-Porabka K, Jaworek J, Leja-Szpak A, Palonek M, Szklarczyk J, Konturek SJ, Pawlik WW. Leptin is able to stimulate pancreatic enzyme secretion via activation of duodeno-pancreatic reflex and CCK release. J Physiol Pharmacol 2004; 55 Suppl 2:47–57.

367. Lopez J, Cuesta N. Adrenomedullin as a pancreatic hormone. Microsc Res Tech 2002; 57:61–75.

368. Tsuchida T, Ohnishi H, Tanaka Y, Mine T, Fujita T. Inhibition of stimulated amylase secretion by adrenomedullin in rat pancreatic acini. Endocrinology 1999; 140:865–870.

Chapter 15

Secretion of the human exocrine pancreas in health and disease

Phil A. Hart and Darwin L. Conwell

Section of Pancreatic Disorders, Division of Gastroenterology, Hepatology, and Nutrition, The Ohio State University Wexner Medical Center, Columbus, OH

Introduction

Secretion from the human exocrine pancreas is highly regulated and essential for nutrient digestion. Pancreatic secretions are coordinated with responses occurring in both digestive and interdigestive periods and vary with meal composition and patient-related factors. In diseases affecting the exocrine pancreas, the amount and/or composition of pancreatic secretion can change and lead to maldigestion. Although our understanding of pancreatic secretion in the healthy state has been well-characterized, knowledge of the changes that occur during disease states remains limited. Management of exocrine pancreatic insufficiency is primarily focused on replacement of digestive enzymes, particularly lipase, with exogenous pancreatic enzymes. There is renewed interest in the study of pancreatic secretion for studies of both pancreatic physiology and disease due to advances in its collection and analysis and the growing recognition that pancreatic insufficiency often follows acute pancreatitis, particularly when it's severe. In this chapter, we review various aspects of human pancreatic secretion in health and disease.

Anatomy of the Exocrine Pancreas

The anatomy of the exocrine pancreas has been comprehensively reviewed.[1] In brief, the functional exocrine unit of the pancreas consists of groups of acinar cells and a network of pancreatic ductules and ducts. The production of the vast majority of pancreatic juice proteins occurs in the acinar cells. Of the pancreatic enzymes, proteases (e.g., trypsin, carboxypeptidases, chymotrypsin, and elastase) are the most abundant (according to mass) and comprise around 90 percent of the enzymes found in human pancreas fluid; amylase, lipase, and nucleases are relatively less abundant.[2] Following production and storage of inactive proenzymes (termed zymogens) and active enzymes in acinar cell zymogen granules, they are secreted into a network of pancreatic ductules and ducts in response to a meal. These conduits are lined with cuboidal ductal epithelial cells, which secrete a bicarbonate-rich fluid needed to solubilize proteins secreted from the acinar cell and carry them to the duodenum. Secretion from acinar and ductal cells is highly regulated, primarily in response to nutrient ingestion and neural stimulation.[3] Diseases that interfere with either the function of the ductal cells or the continuity of the ductal network will lead to exocrine pancreatic insufficiency, as discussed below.

Normal Secretion of the Exocrine Pancreas

During the interdigestive (i.e., fasting) state, exocrine secretion is somewhat cyclic and associated with the three phases of gastrointestinal motility.[4] Pancreatic secretion in the digestive period is much greater than the interdigestive period. This secretion is regulated by neurohormonal responses and occurs in three primary phases—cephalic, gastric, and intestinal.[3,5] The initial cephalic phase (stimulated by the sight, smell, or taste of food) is primarily controlled by the vagal nerve and results in levels of enzyme secretion that account for 20–25 percent of the total elicited by a meal.[5] Next, the gastric phase is activated by gastric distention, which only produces a small increase in pancreatic enzyme secretion. Finally, the intestinal phase is activated by chyme in the duodenum, resulting in the majority (50–80%) of the pancreatic stimulus associated with a meal.[5] Together these responses cause a rapid increase in enzyme secretion to a maximal output about 1 hour after meal ingestion.[6,7] The enzyme output generally remains elevated for approximately 3–4 hours but can vary depending on multiple factors.[6–8] The cessation of pancreatic enzyme secretion is believed to be primarily the result of nutrient exposure to the distal small intestine based on experiments demonstrating the inhibition of endogenously stimulated secretion following ileal perfusion of carbohydrates or lipids.[9] Hormones believed to be involved in this feedback inhibition include peptide YY and glucagon-like peptide-1.[5]

Several meal-related factors influence exocrine pancreatic secretion, including caloric content, nutrient

e-mails: philip.hart@osumc.edu, darwin.conwell@osumc.edu

composition, and physical properties of the meal. In regards to caloric content, there appears to be a minimal and maximal threshold of pancreas stimulation, with the maximal enzyme response occurring after consumption of approximately 500 kcal.[10] The meal composition of fats, carbohydrates, and proteins can also influence the relative and absolute levels of enzymes in pancreatic secretion. For example, healthy subjects exposed to a high-fat diet for 2 weeks had a total enzyme output that was two to four times greater than those on a high-carbohydrate diet in the immediate postprandial and interdigestive periods, respectively.[11] Although temporary exposure to a high-fat diet for 24 hours also increased enzyme output in this study, the observation was not replicated with intraduodenal infusion of different nutrient compositions.[12] Another key observation is that ingestion of a solid meal results in a more sustained enzyme response than an identical homogenized meal.[8,11]

Pancreatic secretion of a bicarbonate-rich fluid has a distinct mechanism and has important consequences. The transport of bicarbonate by duct cells into the pancreatic duct lumen is mediated by the cystic fibrosis transmembrane conductance regulator (CFTR). This transporter can either secrete chloride to drive a chloride-bicarbonate exchanger on duct cells or directly transport bicarbonate into the duct lumen. The relative levels of ion secretion depend on the amount of stimulation; at low rates, chloride predominates but at high secretory rates, bicarbonate is dominant. Reduced bicarbonate and fluid secretion are seen in chronic pancreatitis, cystic fibrosis, and in some individuals with alcohol abuse possibly due to inhibition of CFTR function (discussed below). Duodenal pH is the primary driving force for ductal fluid and electrolyte secretion though nutrient digestive products can also enhance bicarbonate secretion. After ingestion of a meal in a healthy subject, the duodenal pH changes from a baseline near 7.0 to a nadir of 5.0–4.5 in the early postprandial phase.[6,8] When a threshold of about pH 4.5 is reached in the duodenum, secretin is released and secretion of bicarbonate-rich fluids into the duodenum follows. The alkaline fluid in the duodenum has several functions including inactivating pepsin, increasing the solubility of fatty acids and bile acids, maintaining an optimal pH (>4.0) for pancreatic and brush border enzymes, and preventing intestinal mucosal damage.

Alteration of Exocrine Secretion in Disease

Pancreatic enzyme secretion is a highly controlled process in health, so it is not surprising that its dysfunction is often observed with disorders of the exocrine pancreas. Although enzyme response to different nutrients has been well characterized in a healthy state, much less is known regarding the extent of change in disease. Rather, most studies have

focused on the clinical ramifications of inadequate enzyme production and/or secretion.

Mechanisms of abnormal exocrine function

The multiple mechanisms whereby pancreatic disorders can lead to decreased pancreatic enzyme activity can be categorized as:

1. Decreased enzyme production,
2. Impaired enzyme secretion, and/or
3. Impaired enzyme mixing with food.

Reduced acinar cell mass leads to decreased enzyme production and arises from disorders of primary and secondary exocrine pancreatic insufficiency (EPI). Primary disorders are illustrated by Shwachman-Diamond syndrome and Johanson-Blizzard syndrome; secondary causes include chronic pancreatitis, severe acute pancreatitis, or pancreatic surgical resection. Patients with impaired enzyme secretion can have impairment at the cellular (e.g., impaired duct cell function in cystic fibrosis) or macroscopic level (e.g., obstruction of the main pancreatic duct secondary to a stricture, intraductal stone, or mass). Impaired mixing of pancreatic enzymes with food often occurs in the postoperative setting, such as following a gastrojejunostomy. Additionally, impaired bicarbonate secretion by ductal cells can lead to impaired mixing, a suboptimal pH for enzyme activities, and reduced formation of micelles.[5,13]

Chronic pancreatitis

Chronic pancreatitis is one of the most common causes of exocrine pancreatic insufficiency (EPI), with insufficiency observed in 30 to 90 percent depending on the clinical severity of disease.[14] In a study examining the natural history of chronic pancreatitis, the onset of EPI was often delayed 10–15 years after symptom onset.[14] This delay is attributable to the dramatic functional reserve of the exocrine pancreas, requiring loss of more than 90 percent of exocrine pancreatic function before developing malabsorption. Whether or not the reduction of the various pancreatic enzymes occurs in parallel remains unclear. However, the clinical abnormalities are predominantly related to fat maldigestion. This is primarily a consequence of limited levels of lipase in the normal pancreas and minimal redundancy of lipase production from nonpancreatic sources. Further, other luminal conditions for fat digestion and absorption, especially the formation of micelles needed for both optimal lipid hydrolysis and lipid delivery to mucosal cells, are disordered in chronic pancreatitis. In contrast, only 5 percent of normal amylase content is needed to maintain starch digestion; this amount can typically be reached by a combination of salivary gland and brush border oligosaccharidase production of amylase.[15] Similar levels of pancreatic proteases are likely needed for efficient protein digestion. In a landmark study, DiMagno and colleagues demonstrated that steatorrhea (>7 gram fat/24 hours) develops when the lipase output is <10 percent of normal.[16] The reported lipase output in chronic pancreatitis may range from 1 to 60 percent of levels found in healthy controls.[5] The wide variability is in part due to methodologic differences in the studies (e.g., method of pancreas stimulation and collection methods) but also likely reflects subjects studied at various stages of disease.

Chronic pancreatitis can alter all aspects of pancreatic enzyme synthesis and secretion, so EPI is typically multifactorial. First, parenchymal fibrosis can lead to acinar cell injury and loss of exocrine mass with a resultant loss of enzyme production. Second, duct cell function is impaired in chronic pancreatitis leading to decreased bicarbonate secretion.[17] As a result, postprandial duodenal pH is lower in chronic pancreatitis than in healthy subjects and can exceed the pH threshold for irreversible lipase inactivation (pH 4.5).[6] Obstruction of the main pancreatic duct can also occur due to fibrotic strictures and intraductal calculi and likely contributes to reduced duct pancreatic secretion. However, neither endoscopic nor surgical correction of main pancreatic duct obstruction has been convincingly demonstrated to restore pancreatic enzyme secretion. Since the functional and tissue loss found in chronic pancreatitis is currently not reversible with the possible exception of cystic fibrosis, clinical treatment of exocrine insufficiency is limited to oral supplementation with pancreatic enzyme replacement therapy.

Cystic fibrosis

The CFTR gene codes for a chloride channel, which is central to the production of fluid and electrolyte secretion in various exocrine glands. In the pancreas, the primary pathophysiologic result of CFTR mutations is decreased electrolyte and fluid secretion into the pancreatic duct lumen. The consequence of these changes is pancreas fluid that is more viscous and acidic than normal secretion. Although acinar cells may not be directly affected in early or mild disease, both pancreatic enzyme secretion and function are impaired. In addition to impaired flow, the acidic pancreatic duct lumen can result in precipitation of mucins, secretory protein aggregation, and premature activation of enzymes.[18] Furthermore, the decreased bicarbonate secretion leads to a more acidic intraluminal pH in the duodenum. It has been observed that the pH is lower by 1–2 units in both the interdigestive and digestive periods in cystic fibrosis (CF) compared to controls.[19] Collectively, EPI is prevalent in classic CF with estimates that 80 percent of patients with CF will develop EPI by 2 years of life.[20]

Importantly, EPI appears to be mostly limited to those subjects with two severe deleterious mutations, though pancreatic disease can rarely occur in those with less severe mutations.[20] Recent studies reported that correction of EPI can occur with treatment using CFTR modulators.[21,22] How often this response is seen in CF patients and whether this class of drugs might impact EPI in those having other pancreatic diseases remains to be studied. There is also evidence that prominent levels of alcohol ingestion can rapidly reduce ductal CFTR function.[23]

As in other diseases, the EPI in CF and impaired lipase secretion may at least partially be compensated by increased activity of gastric lipase.[24] Studies of pancreatic enzyme secretion in CF are limited. This is likely due to the multiple methodologic challenges to completing these studies, the greatest of which is that EPI manifests at a young age. This poses technical and ethical issues for research study design. In these patients with EPI, lifelong pancreatic enzyme replacement therapy is recommended and improvement in nutritional status has been associated with improvements in pulmonary function and overall survival.[25]

Acute pancreatitis

In contrast to chronic pancreatitis and cystic fibrosis, which are chronic, progressive diseases, acute pancreatitis is a multiphasic disease. Thus, impairment of pancreas enzyme secretion is variable and depends on the clinical severity of acute pancreatitis and chronologic proximity to the inciting insult. Knowledge of pancreatic enzyme secretion during the early phase of acute pancreatitis is limited in part because of the belief that pancreatic stimulation in this setting may be harmful. Epidemiological studies have characterized EPI prevalence following acute pancreatitis. It is estimated that 62 percent of these patients have EPI during admission with persistent EPI in 35 percent of patients during follow-up.[26] This prevalence trend suggests that transient exocrine pancreatic dysfunction is common after acute pancreatitis and that a smaller number will develop long-term EPI. Persistent EPI appears to be more common in patients with pancreatic necrosis or an alcohol etiology.[26] This is a commonly overlooked sequela of acute pancreatitis, and additional understanding of the pathophysiology is needed.

Pancreatic cancer

The changes in pancreas enzyme secretion are not well characterized in pancreatic cancer. Although the prevalence of EPI in pancreatic cancer has not been characterized in a large clinical cohort, the pooled prevalence in a recent meta-analysis was over 70 percent.[27] In addition to tumor-mediated parenchymal destruction, abnormal secretion can develop as a consequence of pancreatic duct obstruction and following surgical resection.[28] The location of the tumor is an important factor in determining whether pancreatic secretion remains adequate. In one study involving direct PFTs, a shorter length of unaffected pancreatic duct (i.e., a tumor located closer to the pancreatic head) was associated with decreased bicarbonate and enzyme output in response to CCK stimulation.[29]

Pancreatic surgery

Alterations in exocrine secretion following pancreatic surgery are influenced by the type of surgery, extent of resection, and health of the remnant pancreatic parenchyma. The exocrine function remains largely unchanged, and occasionally improves, in those undergoing a drainage procedure with or without minor resection of the parenchyma.[30] Conversely, clinically diagnosed EPI is observed in approximately 25–50 percent of those following a pancreaticoduodenectomy.[30–34] In patients undergoing surgery for a diffuse disease, such as chronic pancreatitis, the risk of developing EPI following surgery is increased. For example, in a clinical series limited to subjects with chronic pancreatitis, over 50 percent developed clinically overt steatorrhea following pancreaticoduodenectomy and up to 40 percent following distal pancreatectomy.[33,35,36]

Nonpancreatic etiologies of impaired exocrine pancreatic function

In addition to the preceding pancreatic etiologies, there are other nonpancreatic diseases that can indirectly impair exocrine pancreatic function. For example, in those undergoing a partial or total gastrectomy there are multiple factors that decrease intraluminal enzyme activity. The greatest contributing factor is postcibal asynchrony, which describes disordered mixing of meal contents and pancreatic enzymes in the small intestine.[5] The asynchrony is further compounded by pancreatic denervation during surgical dissection and the loss of compensatory gastric lipase production. The dilution of pancreatic enzymes in the duodenal lumen that results from rapid delivery of an osmotic load also reduces nutrient digestion. Asynchrony is also a contributing factor to the EPI observed following pancreaticoduodenectomy. Although steatorrhea is not universally present in patients with Zollinger-Ellison syndrome (ZES), it is important to consider the effect on exocrine pancreatic function due to its unique pathophysiologic mechanism.[37] In ZES there is excessive gastrin production, which leads to inappropriate gastric acid secretion. This lowers the duodenal pH beyond the threshold at which normally secreted pancreatic enzymes are active and leads to inactivation of

some, most importantly lipase.[38] The relatively acidic duodenal pH in ZES also leads to bile salt precipitation. This reduces the effectiveness of lipase-dependent triglyceride hydrolysis as well as formation of micelles that are needed to carry lipids to enterocytes. Last, impaired exocrine function has been shown in patients without clinically evident pancreatic disease. For example, patients with type 2 diabetes mellitus can develop fibrosis that is histologically similar to chronic pancreatitis.[39] This condition, currently referred to as diabetic exocrine pancreatopathy, differs from chronic pancreatitis in that chronic inflammatory cells are absent and overt EPI is rare. Exocrine pancreatic insufficiency may also occur in celiac disease (when measured using fecal elastase levels) despite the absence of structural changes in the pancreas.[40,41] The EPI in celiac disease is generally attributed with mucosal injury, which may disrupt an enteric-mediated cholecystokinin stimulation of the pancreas and gallbladder emptying. In children, short-term support with oral pancreatic enzymes is often considered. Pancreatic secretion usually resolves with mucosal healing on a gluten-free diet.[40] EPI has also been reported to occur in patients with inflammatory bowel disease and following gastrointestinal surgeries, but additional studies are needed to further characterize the potential mechanisms of exocrine dysfunction in these scenarios.[42]

Measurement of Human Pancreatic Secretion (Direct Pancreatic Function Testing)

Our knowledge of normal human pancreatic secretion primarily comes from human studies using luminal tube(s) to collect intestinal fluids following pancreatic stimulation with either a standardized meal or an intravenous secretagogue such as secretin or cholecystokinin. The use of exogenous stimulation to measure pancreatic function is referred to as direct pancreatic function testing (PFT) (**Table 1**). Although different types of tubes have been used, one of the most recognized is the Dreiling tube, which permitted simultaneous aspiration from the gastric and duodenal lumens. One drawback of this approach was that the enzyme output could not be determined because the flow rate could not be calculated. To resolve this issue, a method was developed using a second tube placed with ports in both proximal and distal duodenum. A nonabsorbable marker (typically polyethylene glycol) was infused through the proximal port and aspirated with pancreatic secretions through the distal port.[43] Using this marker-perfusion method, permitted determination of correction coefficient for the amount of fluid not collected distally and accurate calculation of pancreatic enzyme output. A variety of endogenous and exogenous stimuli were administered, and the changes in pancreatic secretions could be characterized. Significant drawbacks to these techniques included

Table 1. **Comparison of direct and indirect pancreatic function tests.** Estimates of sensitivity and specificity are assigned semiquantitatively due to extreme heterogeneity in study designs, which precludes accurate pooling.[44, 45]

Name	Sensitivity	Specificity	Advantages	Disadvantages	Cost
Direct Pancreatic Function Tests:					
CCK-stimulated pancreatic function test	High	High	• Provides the most direct measure of pancreatic enzyme output	• Tubes are no longer available to measure enzyme output • Invasive	$$$$
Secretin-stimulated pancreatic function test	High	Moderate	• Highly sensitive in early stages of disease • High negative predictive value to rule out chronic pancreatitis	• Invasive • False positives: CFTR mutations and cigarette smoking • Does not directly assess acinar cell function	$$$$
Indirect Pancreatic Function Tests:					
Coefficient of fat absorption (CFA)	High	Moderate	• Highly accurate for fat malabsorption • Can be used to monitor PERT • Noninvasive	• Requires 3 day stool collection • False positives: any cause of fat malabsorption (which must be excluded before diagnosing EPI)	$
Fecal elastase (FE-1)	Mild - Moderate	Moderate-High	• Convenient collection • Noninvasive	• Lower accuracy in those with mild EPI or history of pancreatic resection • Cannot be used to monitor PERT • False positives: bacterial overgrowth, watery stool	$
^{13}C-mixed triglyceride breath test (MTBT)	Moderate	Moderate	• Can be used to monitor PERT • Noninvasive	• Limited availability • Time consuming test	$$

EPI, exocrine pancreatic insufficiency; PERT, pancreatic enzyme replacement therapy.

patient discomfort and prolonged fluoroscopy time to maintain proper tube location. Despite these challenges, the majority of the data related to human pancreatic secretory physiology were acquired using this technique.

Further studies regarding the mechanisms of human pancreatic secretion became possible following adaptation of early work demonstrating the ability to isolate functioning pancreatic acini from rodents.[46] The ability to isolate acini from human pancreases and study them in a controlled environment has led to further understanding of the cellular mechanisms of pancreatic secretion in health and disease, which are described elsewhere.[47]

More than a decade ago, an endoscopic-based pancreatic function test (ePFT) was developed that greatly improved patient tolerance and eliminated the need for fluoroscopy.[17] Although the stimuli for pancreatic secretion remained similar (i.e., cholecystokinin (CCK) and/or secretin), the method of fluid collection was through the suction channel of the endoscope rather than an enteric tube. Since there is only one "port" for collecting samples through the endoscope, this technique permits determination of enzyme and analyte concentration but not enzyme

output (or flow rate). The most common use of the ePFT is for evaluation of suspected chronic pancreatitis. However, there is emerging use of this methodology for translational science, including the study of disease biomarkers.[48]

Currently, the most commonly used measurement made using ePFT is the peak bicarbonate concentration following secretin stimulation. Although this measurement is more directly an assay of pancreatic duct-cell function than acinar-cell function, previous studies have demonstrated that peak bicarbonate corresponds to peak lipase concentrations in the pancreatic fluid of healthy subjects as well as chronic pancreatitis.[49] For the evaluation of chronic pancreatitis, measurement of peak bicarbonate concentration (following secretin stimulation) has improved discrimination compared to peak lipase or amylase concentrations (following CCK stimulation).[50] Importantly, the negative predictive value (97%) of a normal bicarbonate response to secretin stimulation to exclude chronic pancreatitis is very high; however, the positive predictive value (45%) is much lower.[51] Thus, the primary clinical use for endoscopic function tests is to "rule out" a diagnosis of chronic pancreatitis.

In addition to the previously mentioned electrolytes and pancreatic enzymes, there is a large number of other potential analytes in pancreatic fluid. During an ePFT, there is typically an abundant volume of fluid collected, which provides the opportunity for other studies. Investigators have begun to explore various molecular targets in pancreas fluid, including protein expression, cytokines, DNA methylation markers, microRNAs, and genetic mutations.[48] Various markers are being examined for the purposes of identifying diagnostic or disease biomarkers, as well as observations that may lead to novel therapeutic approaches. Although early studies were primarily limited to proteomics, there has been a recent resurgence in other areas with exciting preliminary findings.[52,53] For example, Abu Dayyeh et al. demonstrated that prostaglandin E2 (PGE2) is a promising disease biomarker for the various stages of chronic pancreatitis.[54] Levels are different in early and advanced chronic pancreatitis compared to healthy controls, with areas under the curve (AUCs) of 0.62 and 0.9, respectively. When used in combination with the pancreatic fluid bicarbonate, the AUCs for diagnosis of early and advanced chronic pancreatitis were 0.94 and 1.0, respectively. Another example of biomarker discovery includes the identification of DNA hypermethylation markers that identify patients with pancreatic cancer compared to controls with AUCs ranging from 0.62 to 0.92.[55]

Indirect Pancreatic Function Testing

Tests measuring pancreatic function without the use of hormonal stimulation are referred to as indirect PFTs. Since the indirect tests are typically less accurate for detecting early stages of exocrine dysfunction, they are more helpful for quantifying the degree of insufficiency in those with known pancreatic disease, rather than diagnosis. Indirect PFTs are noninvasive and typically less expensive, so the selection of the pancreatic function test to be employed in the clinical setting requires considering the tradeoff between diagnostic test performance, invasiveness, and costs (**Table 1**). For example, indirect PFTs are generally adequate to identify EPI in patients with overt morphological changes (e.g., calcifications and/or main pancreatic duct dilation). In contrast, a direct PFT would be preferred to identify EPI in a patient with clinical suspicion of chronic pancreatitis and normal or equivocal imaging tests.[44] Poorly studied indirect tests (including serum trypsin, fecal chymotrypsin, and qualitative fecal fat analysis) are not discussed here.

Coefficient of fat absorption

Among indirect testing methods, the coefficient of fat absorption (CFA) is considered the gold standard to diagnose fat malabsorption from any cause and to document and quantify EPI in those with pancreatic disease.[56–58] This test involves consumption of a high-fat diet (100 grams/day) for at least 5 days with stool collection during the final 3 days of the diet. The daily dietary fat intake is recorded and factored into the final CFA calculation, using the following equation:

$$\text{CFA (\%)} = 100 \times [(\text{mean daily fat intake} - \text{mean daily stool fat})/\text{mean daily fat intake}]$$

A CFA of <0.93 (which corresponds to >7 grams of stool fat per 24-hour period while on a 100 g fat diet) is considered abnormal.[45] Although this test is accurate when carefully performed, the three-day collection period can be inconvenient for patients and errors may occur when performed in a noncontrolled environment with either collection or processing. Since fat excretion is directly related to fat ingestion over a broad range of values (excretion ~7% of intake over a broad range of intakes), subjects must have an exact record of their intakes. Though this test is also useful to assess adequacy of PERT dosing, follow-up testing is rarely performed outside of the research setting.

Fecal elastase-1

In contrast to the CFA, a fecal elastase-1 (FE-1) level can be determined from a single formed stool sample. The test's convenience makes it the most widely used indirect PFT in clinical practice. Pancreatic elastase is resistant to degradation as it passes through the gut, so it can be measured in the stool.[59] Although early studies demonstrated strong correlation with pancreatic enzyme output during direct PFT, more recent studies have demonstrated only fair accuracy in mild EPI.[45] False positives occur in about 10 percent of patients with low pretest probability for EPI.[60] Similarly, levels are falsely low any time a liquid stool specimen is analyzed. Also, low levels are commonly observed in diabetes mellitus (both type 1 and type 2 diabetes), but it is uncertain if this truly represents EPI.[61,62] Last, since FE-1 levels detected by monoclonal (but not polyclonal) assays are unaffected by PERT, this test is not useful for monitoring the adequacy of therapy.

^{13}C-mixed triglyceride breath test

The ^{13}C-mixed triglyceride breath test (MTBT) is another indirect test that measures the intraluminal lipolytic activity as an estimate of exocrine pancreatic function. The test involves ingesting a standardized meal (including triglycerides with radiolabeled carbon tracers).[45] The triglycerides are hydrolyzed by pancreatic lipase releasing ^{13}C-fatty acid, which is absorbed, then transported to the liver. Lipolysis and beta-oxidation occur in the liver leading to the formation of $^{13}CO_2$. These molecules are exhaled by the lungs and measured in serial breath samples. A decrease in the recovery of $^{13}CO_2$ is associated with decreased pancreatic lipase secretion and fat malabsorption as measured by CFA.[63-65] Although this is a noninvasive test, it lasts approximately 6 hours and the radiotracers are not universally available, so the clinical utility of this test worldwide is limited.

Summary

Pancreas exocrine secretion represents a complex response to a meal which involves the coordinated release and transport of enzymes from acinar cells and fluid and electrolytes from duct cells into the pancreatic ductal system and then into the duodenal lumen where they are required for normal digestion. Essentially all pancreatic disorders may alter this process to different degrees and do so through a variety of mechanisms. There are various methods for determining enzyme output and bicarbonate secretion in response to endogenous and exogenous pancreatic stimulation. These tools have shaped our current understanding of pancreas physiology and hold significant potential for biomarker discovery and identification of novel therapeutic targets.

Acknowledgments

This publication was supported by National Cancer Institute and National Institute of Diabetes and Digestive and Kidney Diseases of the National Institutes of Health under award numbers U01DK108327 and U01DK127388 (DC, PH).

References

1. Longnecker DS. Anatomy and histology of the pancreas. Pancreapedia: Exocrine Pancreas Knowledge Base, 2021. DOI: 10.3998/panc.2021.01.
2. Scheele G, Bartelt D, Bieger W. Characterization of human exocrine pancreatic proteins by two-dimensional isoelectric focusing/sodium dodecyl sulfate gel electrophoresis. Gastroenterology 1981; 80:461–473.
3. Chandra R, Liddle RA. Regulation of pancreatic secretion. Pancreapedia: Exocrine Pancreas Knowledge Base, 2020. DOI: 10.3998/panc.2020.14.
4. Keller J, Groger G, Cherian L, Gunther B, Layer P. Circadian coupling between pancreatic secretion and intestinal motility in humans. Am J Physiol Gastrointest Liver Physiol 2001; 280:G273–G278.
5. Keller J, Layer P. Human pancreatic exocrine response to nutrients in health and disease. Gut 2005; 54 Suppl 6:vi1–28.
6. DiMagno EP, Malagelada JR, Go VL, Moertel CG. Fate of orally ingested enzymes in pancreatic insufficiency. Comparison of two dosage schedules. N Engl J Med 1977; 296:1318–1322.
7. Fried M, Mayer EA, Jansen JB, Lamers CB, Taylor IL, Bloom SR, Meyer JH. Temporal relationships of cholecystokinin

8. Malagelada JR, Go VL, Summerskill WH. Different gastric, pancreatic, and biliary responses to solid-liquid or homogenized meals. Dig Dis Sci 1979; 24:101–110.
9. Layer P, Peschel S, Schlesinger T, Goebell H. Human pancreatic secretion and intestinal motility: effects of ileal nutrient perfusion. Am J Physiol Gastrointest Liver Physiol 1990; 258:G196–G201.
10. Brunner H, Northfield TC, Hofmann AF, Go VL, Summerskill WH. Gastric emptying and secretion of bile acids, cholesterol, and pancreatic enzymes during digestion. Duodenal perfusion studies in healthy subjects. Mayo Clin Proc 1974; 49:851–860.
11. Boivin M, Lanspa SJ, Zinsmeister AR, Go VL, DiMagno EP. Are diets associated with different rates of human interdigestive and postprandial pancreatic enzyme secretion? Gastroenterology 1990; 99:1763–1771.
12. Holtmann G, Kelly DG, DiMagno EP. Nutrients and cyclical interdigestive pancreatic enzyme secretion in humans. Gut 1996; 38:920–924.
13. Andersen JR, Bendtsen F, Ovesen L, Pedersen NT, Rune SJ, Tage-Jensen U. Pancreatic insufficiency. Duodenal and jejunal pH, bile acid activity, and micellar lipid solubilization. Int J Pancreatol 1990; 6:263–270.
14. Layer P, Yamamoto H, Kalthoff L, Clain JE, Bakken LJ, DiMagno EP. The different courses of early- and late-onset idiopathic and alcoholic chronic pancreatitis. Gastroenterology 1994; 107:1481–1487.
15. Layer P, Zinsmeister AR, DiMagno EP. Effects of decreasing intraluminal amylase activity on starch digestion and postprandial gastrointestinal function in humans. Gastroenterology 1986; 91:41–48.
16. DiMagno EP, Go VL, Summerskill WH. Relations between pancreatic enzyme outputs and malabsorption in severe pancreatic insufficiency. N Engl J Med 1973; 288:813–815.
17. Conwell DL, Zuccaro G, Jr., Vargo JJ, Trolli PA, Vanlente F, Obuchowski N, Dumot JA, O'Laughlin C. An endoscopic pancreatic function test with synthetic porcine secretin for the evaluation of chronic abdominal pain and suspected chronic pancreatitis. Gastrointest Endosc 2003; 57:37–40.
18. Pallagi P, Hegyi P, Rakonczay Z, Jr. The physiology and pathophysiology of pancreatic ductal secretion: the background for clinicians. Pancreas 2015; 44:1211–1233.
19. Gregory PC. Gastrointestinal pH, motility/transit and permeability in cystic fibrosis. J Pediatr Gastroenterol Nutr 1996; 23:513–523.
20. Durno C, Corey M, Zielenski J, Tullis E, Tsui LC, Durie P. Genotype and phenotype correlations in patients with cystic fibrosis and pancreatitis. Gastroenterology 2002; 123:1857–1864.
21. Munce D, Lim M, Akong K. Persistent recovery of pancreatic function in patients with cystic fibrosis after ivacaftor. Pediatr Pulmonol 2020; 55:3381–3383.
22. Nichols AL, Davies JC, Jones D, Carr SB. Restoration of exocrine pancreatic function in older children with cystic fibrosis on ivacaftor. Paediatr Respir Rev 2020; 35:99–102.
23. Venglovecz V, Hegyi P, Gray MA. Ion channels in pancreatic duct epithelial cells in health and disease. Pancreapedia:

Exocrine Pancreas Knowledge Base, 2021. DOI: 10.3998/panc.2021.08.

24. Balasubramanian K, Zentler-Munro PL, Batten JC, Northfield TC. Increased intragastric acid-resistant lipase activity and lipolysis in pancreatic steatorrhoea due to cystic fibrosis. Pancreas 1992; 7:305–310.

25. Sankararaman S, Schindler T, Sferra TJ. Management of exocrine pancreatic insufficiency in children. Nutr Clin Pract 2019; 34 Suppl 1:S27–S42.

26. Huang W, de la Iglesia-Garcia D, Baston-Rey I, Calvino-Suarez C, Larino-Noia J, Iglesias-Garcia J, Shi N, Zhang X, Cai W, Deng L, Moore D, Singh VK, et al. Exocrine pancreatic insufficiency following acute pancreatitis: systematic review and meta-analysis. Dig Dis Sci 2019; 64:1985–2005.

27. Iglesia D, Avci B, Kiriukova M, Panic N, Bozhychko M, Sandru V, de-Madaria E, Capurso G. Pancreatic exocrine insufficiency and pancreatic enzyme replacement therapy in patients with advanced pancreatic cancer: a systematic review and meta-analysis. United European Gastroenterol J 2020;8:1115–1125.

28. Vujasinovic M, Valente R, Del Chiaro M, Permert J, Lohr JM. Pancreatic exocrine insufficiency in pancreatic cancer. Nutrients 2017; 9:183.

29. DiMagno EP, Malagelada JR, Go VL. The relationships between pancreatic ductal obstruction and pancreatic secretion in man. Mayo Clin Proc 1979; 54:157–162.

30. Jalleh RP, Williamson RC. Pancreatic exocrine and endocrine function after operations for chronic pancreatitis. Ann Surg 1992; 216:656–662.

31. Kachare SD, Fitzgerald TL, Schuth O, Vohra NA, Zervos EE. The impact of pancreatic resection on exocrine homeostasis. Am Surg 2014; 80:704–709.

32. Matsumoto J, Traverso LW. Exocrine function following the whipple operation as assessed by stool elastase. J Gastrointest Surg 2006; 10:1225–1229.

33. Tseng DS, Molenaar IQ, Besselink MG, van Eijck CH, Borel Rinkes IH, van Santvoort HC. Pancreatic exocrine insufficiency in patients with pancreatic or periampullary cancer: a systematic review. Pancreas 2016; 45:325–330.

34. Wollaeger EE, Comfort MW, et al. Efficiency of gastrointestinal tract after resection of head of pancreas. J Am Med Assoc 1948; 137:838–848.

35. Frey CF, Child CG, Fry W. Pancreatectomy for chronic pancreatitis. Ann Surg 1976; 184:403–413.

36. Kalser MH, Leite CA, Warren WD. Fat assimilation after massive distal pancreatectomy. N Engl J Med 1968; 279:570–576.

37. Roy PK, Venzon DJ, Shojamanesh H, Abou-Saif A, Peghini P, Doppman JL, Gibril F, Jensen RT. Zollinger-Ellison syndrome. Clinical presentation in 261 patients. Medicine (Baltimore) 2000; 79:379–411.

38. Go VL, Poley JR, Hofmann AF, Summerskill WH. Disturbances in fat digestion induced by acidic jejunal pH due to gastric hypersecretion in man. Gastroenterology 1970; 58:638–646.

39. Mohapatra S, Majumder S, Smyrk TC, Zhang L, Matveyenko A, Kudva YC, Chari ST. Diabetes Mellitus is associated with an exocrine pancreatopathy: conclusions from a review of literature. Pancreas 2016; 45:1104–1110.

40. Rana SS, Dambalkar A, Chhabra P, Sharma R, Nada R, Sharma V, Rana S, Bhasin DK. Is pancreatic exocrine insufficiency in celiac disease related to structural alterations in pancreatic parenchyma? Ann Gastroenterol 2016; 29:363–366.

41. Vujasinovic M, Tepes B, Volfand J, Rudolf S. Exocrine pancreatic insufficiency, MRI of the pancreas and serum nutritional markers in patients with coeliac disease. Postgrad Med J 2015; 91:497–500.

42. Singh VK, Haupt ME, Geller DE, Hall JA, Quintana Diez PM. Less common etiologies of exocrine pancreatic insufficiency. World J Gastroenterol 2017; 23:7059–7076.

43. Go VL, Hofmann AF, Summerskill WH. Simultaneous measurements of total pancreatic, biliary, and gastric outputs in man using a perfusion technique. Gastroenterology 1970; 58:321–328.

44. Conwell DL, Lee LS, Yadav D, Longnecker DS, Miller FH, Mortele KJ, Levy MJ, Kwon R, Lieb JG, Stevens T, Toskes PP, Gardner TB, et al. American Pancreatic Association practice guidelines in chronic pancreatitis: evidence-based report on diagnostic guidelines. Pancreas 2014; 43:1143–1162.

45. Hart PA, Conwell DL. Diagnosis of exocrine pancreatic insufficiency. Curr Treat Options Gastroenterol 2015; 13:347–353.

46. Williams JA, Korc M, Dormer RL. Action of secretagogues on a new preparation of functionally intact, isolated pancreatic acini. Am J Physiol 1978; 235:E517–E524.

47. Lerch MM, Gorelick FS. Models of acute and chronic pancreatitis. Gastroenterology 2013; 144:1180–1193.

48. Hart PA, Topazian M, Raimondo M, Cruz-Monserrate Z, Fisher WE, Lesinski GB, Steen H, Conwell DL. Endoscopic pancreas fluid collection: methods and relevance for clinical care and translational science. Am J Gastroenterol 2016; 111:1258–1266.

49. Stevens T, Dumot JA, Zuccaro G, Jr., Vargo JJ, Parsi MA, Lopez R, Kirchner HL, Purich E, Conwell DL. Evaluation of duct-cell and acinar-cell function and endosonographic abnormalities in patients with suspected chronic pancreatitis. Clin Gastroenterol Hepatol 2009; 7:114–119.

50. Law R, Lopez R, Costanzo A, Parsi MA, Stevens T. Endoscopic pancreatic function test using combined secretin and cholecystokinin stimulation for the evaluation of chronic pancreatitis. Gastrointest Endosc 2012; 75:764–768.

51. Ketwaroo G, Brown A, Young B, Kheraj R, Sawhney M, Mortele KJ, Najarian R, Tewani S, Dasilva D, Freedman S, Sheth S. Defining the accuracy of secretin pancreatic function testing in patients with suspected early chronic pancreatitis. Am J Gastroenterol 2013; 108:1360–1366.

52. Paulo JA, Kadiyala V, Banks PA, Steen H, Conwell DL. Mass spectrometry-based (GeLC-MS/MS) comparative proteomic analysis of endoscopically (ePFT) collected pancreatic and gastroduodenal fluids. Clin Transl Gastroenterol 2012; 3:e14.

53. Paulo JA, Kadiyala V, Lee LS, Banks PA, Conwell DL, Steen H. Proteomic analysis (GeLC-MS/MS) of ePFT-collected pancreatic fluid in chronic pancreatitis. J Proteome Res 2012; 11:1897–1912.

54. Abu Dayyeh BK, Conwell D, Buttar NS, Kadilaya V, Hart PA, Baumann NA, Bick BL, Chari ST, Chowdhary S, Clain

JE, Gleeson FC, Lee LS, et al. Pancreatic juice prostaglandin E2 concentrations are elevated in chronic pancreatitis and improve detection of early disease. Clin Transl Gastroenterol 2015; 6:e72.

55. Kisiel JB, Raimondo M, Taylor WR, Yab TC, Mahoney DW, Sun Z, Middha S, Baheti S, Zou H, Smyrk TC, Boardman LA, Petersen GM, et al. New DNA methylation markers for pancreatic cancer: discovery, tissue validation, and pilot testing in pancreatic juice. Clin Cancer Res 2015; 21:4473–4481.

56. Forsmark CE. Management of chronic pancreatitis. Gastroenterology 2013; 144:1282–1291.

57. Pezzilli R, Andriulli A, Bassi C, Balzano G, Cantore M, Delle Fave G, Falconi M, Exocrine Pancreatic Insufficiency collaborative G. Exocrine pancreatic insufficiency in adults: a shared position statement of the Italian Association for the Study of the Pancreas. World J Gastroenterol 2013; 19:7930–7946.

58. Wollaeger EE, Comfort MW, Osterberg AE. Total solids, fat and nitrogen in the feces; a study of normal persons taking a test diet containing a moderate amount of fat: comparison with results obtained with normal persons taking a test diet containing a large amount of fat. Gastroenterology 1947; 9:272–283.

59. Sziegoleit A, Krause E, Klor HU, Kanacher L, Linder D. Elastase 1 and chymotrypsin B in pancreatic juice and feces. Clin Biochem 1989; 22:85–89.

60. Vanga RR, Tansel A, Sidiq S, El-Serag HB, Othman MO. Diagnostic performance of measurement of fecal elastase-1 in detection of exocrine pancreatic insufficiency: Systematic review and meta-analysis. Clin Gastroenterol Hepatol 2018; 16:1220–1228 e1224.

61. Hahn JU, Kerner W, Maisonneuve P, Lowenfels AB, Lankisch PG. Low fecal elastase 1 levels do not indicate exocrine pancreatic insufficiency in type-1 diabetes mellitus. Pancreas 2008; 36:274–278.

62. Leeds JS, Oppong K, Sanders DS. The role of fecal elastase-1 in detecting exocrine pancreatic disease. Nat Rev Gastroenterol Hepatol 2011; 8:405–415.

63. Dominguez-Munoz JE, Iglesias-Garcia J, Castineira Alvarino M, Luaces Regueira M, Larino-Noia J. EUS elastography to predict pancreatic exocrine insufficiency in patients with chronic pancreatitis. Gastrointest Endosc 2015; 81:136–142.

64. Dominguez-Munoz JE, Iglesias-Garcia J, Vilarino-Insua M, Iglesias-Rey M. 13C-mixed triglyceride breath test to assess oral enzyme substitution therapy in patients with chronic pancreatitis. Clin Gastroenterol Hepatol 2007; 5:484–488.

65. Keller J, Meier V, Wolfram KU, Rosien U, Layer P. Sensitivity and specificity of an abbreviated (13)C-mixed triglyceride breath test for measurement of pancreatic exocrine function. United European Gastroenterol J 2014; 2:288–294.

Chapter 16

Neural control of the pancreas

Tanja Babic and R. Alberto Travagli

Department of Neural and Behavioral Sciences, Penn State College of Medicine
500 University Drive, mail code H109. Hershey, PA 17033

Introduction

The pancreas plays a critical role in the control of nutritional homeostasis. It consists of two major parts: the exocrine pancreas, which secretes digestive enzymes, and the endocrine pancreas, which releases hormones such as insulin, glucagon, pancreatic polypeptide, and somatostatin and maintains glucose homeostasis. Cells in the endocrine pancreas are organized in pancreatic clusters of cells, the islets of Langerhans. Within the islets, the β-cells, which secrete insulin, are the predominant cell type and comprise approximately 70 percent of the cells within the islets. The remaining cells consist of α-cells that secrete glucagon, δ-cells that secrete somatostatin, and cells that secrete pancreatic polypeptide. The main function of the exocrine pancreas is to aid in digestion by secreting digestive enzymes and bicarbonate into the duodenum. The exocrine pancreas consists of only two major cell types, namely acinar cells that synthesize, store, and secrete digestive enzymes and ductal cells that secrete chloride and bicarbonate.

Both parts of the pancreas are innervated by the sympathetic and parasympathetic nervous system, with separate pathways regulating the exocrine and the endocrine pancreas. In this chapter, we provide an overview of the central neural pathways that control the pancreas and the main neurotransmitters expressed in these pathways.

Sensory Innervation of the Pancreas

Sensory information from the pancreas is transmitted to the central nervous system (CNS) *via* both vagal and spinal pathways. Cell bodies of the spinal afferent pancreatic neurons are located in the T6-L2 dorsal root ganglia (DRG) and their axons traverse the splanchnic nerves and celiac plexus, before they enter the pancreas. These fibers comprise small myelinated (Aδ) and unmyelinated (C) fibers that transmit both mechanoreceptive and nociceptive information to the preganglionic sympathetic neurons in the intermediolateral cell column (IML) *via* interneurons in the spinal cord laminae I and IV.[1] Most DRG neurons are capsaicin-sensitive and contain substance P,[2] calcitonin gene-related peptide (CGRP), or both.[3–8] Mechanosensitive fibers are primarily associated with blood vessels and although their axons are located within the pancreatic parenchyma, they do not appear to innervate the ductal system.[9] SP and CGRP may be involved in pain associated with chronic pancreatitis, as intrathecal administration of their antagonists attenuated behavioral pain responses in a rat model of chronic pancreatitis.[10]

Pancreatic vagal afferent neurons originate in the nodose ganglia and are relatively sparse compared to spinal afferents. Most of these neurons are capsaicin-sensitive and contain substance P and calcitonin gene-related peptide or both.[3,5,6] Anterograde tracing studies have shown that axons originating in the nodose ganglion supply large blood vessels, pancreatic ducts, acini, and islets and are only sparsely distributed in the pancreatic ganglia.[1] Interestingly, injections of an anterograde tracer into the right nodose ganglion resulted in labeling primarily in the duodenal pancreatic lobe, whereas injections into the left ganglion predominantly labeled the splenic lobe, indicating that sensory innervation of the pancreas is distributed in a regionally specific manner.[11]

The role of sensory nerves on pancreatic functions in control conditions is not completely understood, however. Chemical ablation of pancreatic sensory nerves has been shown to increase[12] or have no effect[13] on glucose-stimulated insulin secretion, suggesting that sensory nerves may exert tonic inhibition of insulin secretion. Similarly, substance P has been shown to either stimulate[14,15] or inhibit[16] insulin secretion. Effects on glucagon secretion are equally contradictory. Calcitonin gene-related peptide has been reported to either stimulate or inhibit glucagon release.[17,18] Furthermore, ablation of the sensory nerves with capsaicin has been reported to either reduce[13] or have no effect[18] on stimulated glucagon secretion. Although the role of sensory afferents in the regulation of exocrine secretion has not been fully established, it has been shown that calcitonin gene-related peptide and substance P inhibit pancreatic exocrine secretion indirectly *via* actions on ganglionic transmission.[19]

Sympathetic Nervous System Control of the Pancreas

Anatomy of the sympathetic pathways regulating pancreatic functions

Sympathetic innervation of the pancreas originates from the sympathetic preganglionic neurons in the lower thoracic and upper lumbar segments of the spinal cord. Axons from these neurons exit the spinal cord through the ventral roots and supply either the paravertebral ganglia of the sympathetic chain via communicating rami of the thoracic and lumbar nerves or the celiac and mesenteric ganglia via the splanchnic nerves. The catecholaminergic neurons of these ganglia innervate the intrapancreatic ganglia, islets, and blood vessels and, to a lesser extent, the ducts and acini. These differences in the innervation of various portions of the pancreas are evident following sympathetic nerve activation, as sympatho-activation decreases insulin secretion and results in vasoconstriction, while it has little or no effect on ductal and acinar cell secretions. The principal neurotransmitters released by the postganglionic sympathetic neurons that innervate the pancreas are noradrenaline, galanin, and neuropeptide Y (NPY).

Retrograde tracing studies using transsynaptic tracers such as the Bartha strain of pseudorabies virus have revealed the distribution of neurons that supply the sympathetic innervation to the pancreas. Unlike traditional retrograde tracers, transsynaptic tracers can cross synapses and therefore enable identification of higher-order neurons in the neurocircuits that innervate the locus of injection.[20] Injections of the virus into the pancreas of vagotomized rats have demonstrated that second-order neurons in the sympathetic circuits to the pancreas are located in the brainstem, specifically in the A5 cell group, locus coeruleus, ventrolateral medulla, and the caudal raphe, as well as in the paraventricular, lateral, and retrochiasmatic nuclei of the hypothalamus and the prefrontal cortex. Third-order neurons are located in the bed nucleus of the stria terminalis, medial preoptic area, and subfornical organ, in the dorsomedial, ventromedial, and arcuate nuclei of the hypothalamus and the central nucleus of the amygdala.[21] A schematic representation of the sympathetic innervation of the pancreas is shown in **Figure 1**.

Effects of sympathetic nervous system activation on pancreatic functions

The role of the sympathetic nervous system in the regulation of pancreatic functions still remains somewhat controversial. Stimulation of the sympathetic nerves elicits diverse effects, including effects on blood pressure, blood flow, and hormone release and therefore direct effects of sympathetic nervous system stimulation are difficult to discern from effects secondary to changes in blood flow or hormone release. Nonetheless, the sympathetic nervous system has been shown to affect the function of the endocrine and, to a lesser extent, exocrine pancreas.

Stimulation of the splanchnic nerve, which supplies the sympathetic innervation to the pancreas has been shown to decrease plasma insulin levels, possibly *via* direct actions of noradrenaline on pancreatic β-cells.[22–26] Splanchnic nerve stimulation also increases catecholamine levels, which have been shown to decrease insulin secretion via α2 adrenoreceptors on pancreatic β-cells.[25,27] Furthermore, both splanchnic nerve stimulation and adrenaline administration/release increase glucagon secretion.[23,26,28] In contrast, disruption of the splanchnic nerve increases insulin levels, suggesting that the sympathetic nervous system exerts a tonic inhibition of the endocrine pancreas. Taken together, these findings indicate that the overall effect of sympathetic nervous system stimulation is to maintain glycemic levels during stressful conditions by decreasing insulin and increasing glucagon secretion.

The effects of sympathetic nerve stimulation on pancreatic exocrine secretions (PES) are not as clear. Although the sparse innervation of acinar and ductal by the sympathetic nervous system would suggest that the sympathetic nervous system does not play a major role in the regulation of the exocrine pancreas, some studies have reported that the sympathetic nervous system may exert profound effects on exocrine secretions.[1] Electrical stimulation of the splanchnic nerves inhibits, whereas cutting the splanchnic nerves in pigs increases PES, suggesting a tonic inhibition of pancreatic exocrine secretion by the sympathetic nervous system.[29] However, studies using more selective stimulation of the sympathetic nervous system have reported conflicting results. Noradrenaline as well as selective α- and β-adrenoreceptor agonists or antagonists have been shown to decrease, increase, or have no effect on pancreatic exocrine secretion.[1] These conflicting findings may be due to the fact that these agents influence blood flow, which exerts secondary effects on PES. For example, vasoconstriction induced by activation of α-adrenoreceptors would result in reduced blood flow to the exocrine pancreas, thus causing a decrease in the amount of fluid secreted by the exocrine pancreas. In support of this suggestion, noradrenergic vasoconstriction has been shown to decrease pancreatic exocrine secretion.[30] In addition, denervation of the celiac ganglion in the dog reduced pancreatic secretions by approximately 70 percent but increased blood flow by approximately 350 percent, suggesting that the sympathetic nervous system exerts a tonic effect on both pancreatic exocrine secretion and vasoconstriction.[31] Considering

Sympathetic pathways

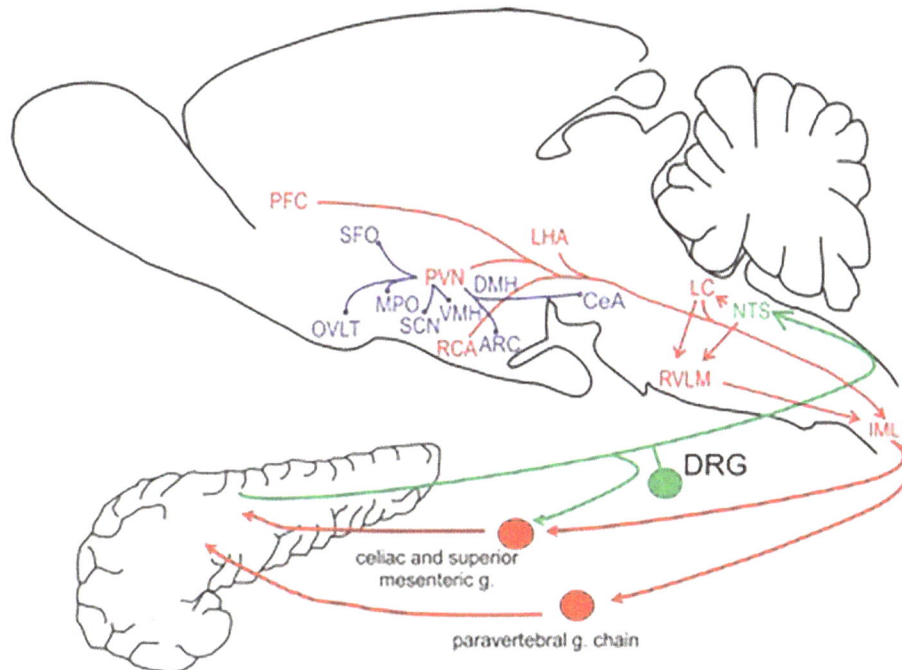

Figure 1. Sympathetic innervation of the pancreas. Abbreviations: arcuate nucleus (ARC), dorsal root ganglion (DRG), dorsomedial hypothalamic nucleus (DMH), lateral hypothalamic area (LHA), medial preoptic area (MPO), nucleus of the tractus solitaries (NTS), organum vasculosum of the lamina terminalis (OVLT), prefrontal cortex (PFC), retrociasmatic area (RCA), suprachiasmatic nucleus (SCN), subfornical organ (SFO), ventromedial hypothalamus (VMH).

these constraints of studying pancreatic exocrine secretion independently of vasoconstriction, it is not clear how much influence the sympathetic nervous system has in the regulation of PES.

Parasympathetic Innervation of the Pancreas

Anatomy of parasympathetic pathways innervating the pancreas

The parasympathetic nervous system provides the major excitatory input to the pancreas. Preganglionic parasympathetic neurons that innervate the pancreas originate in the dorsal motor nucleus of the vagus (DMV) and activate parasympathetic postganglionic neurons in the pancreatic ganglia, primarily via activation of nicotinic acetylcholine receptors. Vagal motor output from DMV neurons is conveyed to the GI tract via two pathways, which can be distinguished based on their postganglionic neurotransmitters. The excitatory cholinergic pathway releases acetylcholine, which acts on muscarinic M3 and M1 receptors and provides a tonic input to the gastrointestinal viscera. The inhibitory nonadrenergic, noncholinergic pathway uses nitric oxide, vasointestinal peptide, gastrin-releasing peptide,

or pituitary adenylate cyclase-activating polypeptide.[1,32] Nicotinic transmission between pre- and postganglionic neurons can be modulated by various neurotransmitters and neuromodulators.[1,33] It should also be kept in mind that species differences in the parasympathetic innervation of the pancreas have been reported. In the mouse, parasympathetic axons provide input to both alpha and beta cells, while parasympathetic axons are rare in the human islets.[2]

The DMV, which contains preganglionic parasympathetic neurons that supply various regions of the GI tract, shows viscerotopic organization, with neurons that project to different parts of the GI tract distributed in anatomically distinct mediolateral columns. Neurons in the medial part of the DMV project to the proximal GI tract, whereas neurons in the lateral DMV project to the more distal parts of the GI tract.[32] Vagal preganglionic DMV neurons that innervate the pancreas are usually located in the left DMV in the area that comprises the hepatic and anterior gastric branches of the vagus, although a few scattered neurons innervating the splenic end of the pancreas are located in the areas corresponding to the celiac branches. Pancreas-projecting DMV neurons can be distinguished from gastric- and intestinal-projecting DMV neurons based on their morphological and electrophysiological properties, further reinforcing the observation that DMV neurons display a highly specialized organization with respect to regulation of various GI functions.[34] Some pancreas-projecting DMV neurons display a slowly developing apamin-insensitive afterhyperpolarization, which is not present in other DMV neurons.[35] Compared to gastric-projecting neurons, pancreas-projecting neurons have a longer action potential duration and longer afterhyperpolarization decay time. Pancreas-projecting neurons also have higher input resistance, smaller afterhyperpolarization amplitude, and a higher firing rate in response to current injections compared to intestinal-projecting neurons. Furthermore, pancreas-projecting neurons have a smaller soma area and a larger diameter than gastric-projecting neurons and fewer segments than gastric- or intestine-projecting DMV neurons.[35]

The major input to DMV neurons originates in the adjacent nucleus tractus solitarius (NTS) (**Figure 2**). Although NTS neurons express a wide variety of neurotransmitters and neuromodulators, NTS projects to the DMV primarily via glutamatergic, GABAergic, and catecholaminergic inputs.[32] Despite this relatively simple neurochemistry, NTS-DMV synapses display a great deal of plasticity and can be modulated by numerous

neurotransmitters, neuromodulators, hormones, and physiological conditions.[36]

Studies using injections of transsynaptic retrograde tracers into the pancreas of sympathectomized rats have demonstrated the distribution of higher-order neurons that innervate the pancreas.[21,37,38] These studies have revealed that neurons that comprise the parasympathetic circuitry to the pancreas show a wider distribution compared to the neurons involved in the sympathetic innervation to the pancreas, with some regions overlapping those that comprise sympathetic inputs to the pancreas.[1] In addition to the NTS, second-order neurons that innervate the pancreas are located in the area postrema, accessory nucleus of the spinal trigeminal nerve, raphe pallidus, raphe obscurus, substantia reticulata, ventrolateral medulla, and the A5 area (**Figure 2**). Parasympathetic second-order neurons are also located in the hypothalamic areas, namely the paraventricular, lateral, dorsomedial, and arcuate nuclei; medial preoptic area, retrochiasmatic area, subfornical organ, bed nucleus of stria terminalis. Furthermore, higher-order neurons have been detected in the prefrontal, piriform, and gustatory cortices, and these neurons provide anatomical basis for the cephalic phase of exocrine secretion.[21,39]

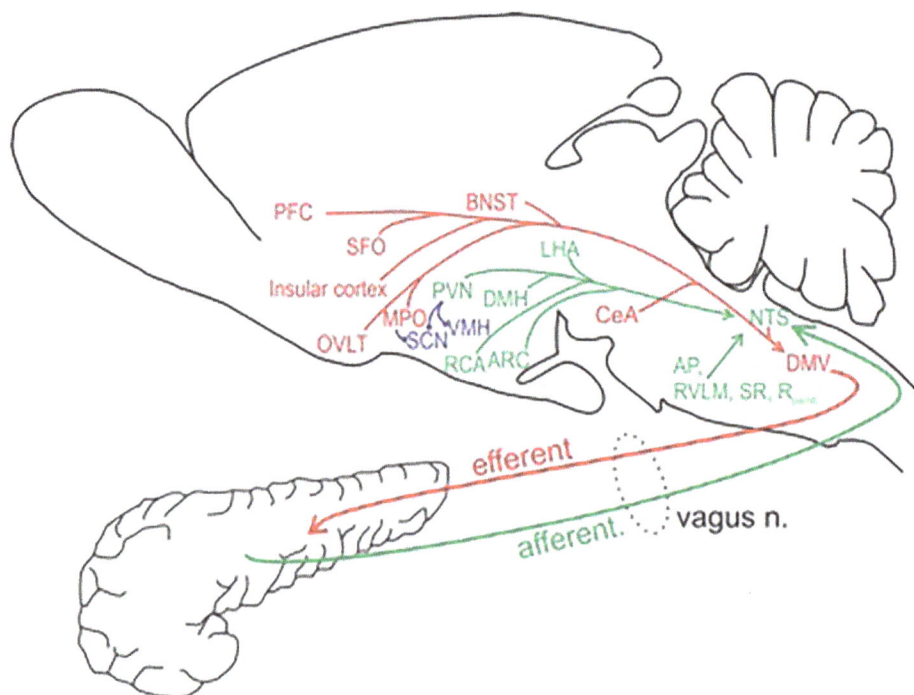

Figure 2. Parasympathetic pathways innervating the pancreas. Abbreviations: area postrema (AP), arcuate nucleus (ARC), bed nucleus of the stria terminalis (BNST), dorsomedial nucleus of the hypothalamus (DMH), dorsal motor nucleus of the vagus (DMV), lateral hypothalamic area (LHA), medial preoptic area (MPO), nucleus of the tractus solitarius (NTS), organum vasculosum of the lamina terminalis (OVLT), prefrontal cortex (PFC), paraventricular nucleus (PVN), retrochiasmatic area (RCA), suprachiasmatic nucleus (SCN), subfornical organ (SFO), ventromedial hypothalamus (VMH).

Effects of parasympathetic stimulation on pancreatic functions

Parasympathetic innervation plays a major role in the regulation of pancreatic functions. Activation of the vagus nerve directly affects pancreatic exocrine and endocrine secretion.[40–44] Electrical stimulation of the DMV or the NTS increases insulin secretion,[45] as do microinjections of the GABA$_A$ receptor antagonist bicuculline.[46,47] In addition, the vagus nerve modulates the intrinsic pacemaker activity of the pancreas, which is responsible for pulsatile insulin secretion, indeed patients with complete resection of the subdiaphragmatic vagus display a longer periodicity of plasma insulin oscillations.[48]

The vagus nerve also plays a crucial role in the regulation of PES. Effects of peptides that modulate pancreatic secretions, such as cholecystokinin (CCK), somatostatin, calcitonin gene-related peptide (CGRP), and pancreatic polypeptide (PP), are vagally mediated.[49] Furthermore, vagotomy has been shown to almost completely abolish pancreatic exocrine secretion induced by feeding or by pharmacological or electrical stimulation,[50–52] whereas disinhibition of the DMV by microinjections of the GABA$_A$ receptor antagonist bicuculline increases pancreatic exocrine secretion.[53]

The cephalic phase response, which refers to the release of gut hormones and digestive enzymes before the ingested nutrients have induced a systemic hormonal response, is also dependent on the vagus nerve and its inputs from the gustatory, piriform, and prefrontal cortices.[33] In fact, vagally mediated exocrine secretion in the cephalic phase accounts for a significant portion of total postprandial enzyme secretion,[54] suggesting that inputs from higher CNS centers to pancreas-projecting DMV neurons play an important role in regulation of PES.

Neurotransmitters in Central Pathways Regulating the Pancreas

The brainstem plays an important role in the regulation of autonomic outflow to the pancreas. The NTS and the DMV have reciprocal connections with higher CNS regions, and these connections contain many neurotransmitters and neuromodulators that influence efferent outflow to the pancreas. In addition to receiving inputs from other CNS regions, the dorsal vagal complex is a circumventricular organ with fenestrated capillaries, which expose it to the influence of circulating hormones.[32] While autonomic output to the pancreas can be regulated by numerous substances, we will focus on the neurotransmitters that have been most extensively studied.

GABA and Glutamate

GABA and glutamate provide major inhibitory and excitatory synaptic inputs to pancreas projecting DMV neurons,

respectively. GABA is the main inhibitory neurotransmitter in the CNS and is the principal neurotransmitter regulating vagal outflow to the pancreas. Microinjections of the GABA$_A$ receptor antagonist bicuculline into the dorsal vagal complex increase pancreatic exocrine secretion[47,55] and glucose-stimulated insulin secretion,[56] suggesting that GABA exerts a tonic inhibition on both pancreatic exocrine secretion and insulin release.[56] In addition to modulating pancreatic functions directly, GABAergic synapses in the DMV are subject to modulation by various other neurotransmitters and hormones. Studies from our laboratory have shown that GABAergic synapses impinging on pancreas-projecting DMV neurons can be modulated by PP, GLP-1, CCK, as well as metabotropic glutamate receptor agonists.[36]

Although glutamate is one of the principal neurotransmitters in synapses impinging onto pancreas-projecting DMV neurons, it does not appear to exert a major role on pancreatic functions under control conditions. In fact, microinjections of ionotropic glutamate receptor antagonist kynurenic acid into the DMV do not affect pancreatic exocrine secretion in control rats.[53] However, glutamatergic synapses impinging on pancreas-projecting DMV neurons are subject to modulation by various neurotransmitters and hormones.[36] Similar to GABAergic synapses, glutamatergic synapses are modulated by PP, GLP-1, and CCK as well as by metabotropic glutamate receptors.[34,46,57,58]

Both GABAergic and glutamatergic synapses impinging on pancreas-projecting DMV neurons express metabotropic glutamate receptors (mGluR), which have also been shown to affect pancreatic functions.[46] Unlike the ionotropic glutamate receptors, which couple to ion channels and mediate fast synaptic transmission, mGluRs are members of G-protein coupled receptor (GPCR) family of receptors and couple to different second messenger systems. There are eight known subtypes of mGluRs, which belong to three different groups (group I, II, and III mGluRs), each of which has unique pharmacological characteristics.[36] Both GABAergic and glutamatergic synapses impinging on pancreas-projecting DMV neurons express group II and group III mGluRs and activation of either receptor type decreases inhibitory and excitatory synaptic transmission.[46] These observations suggest that glutamate released from the synaptic terminals in the DMV not only activates pancreas-projecting neurons postsynaptically but also modulates synaptic transmission onto these neurons. Microinjections of the group II mGluR agonist into the dorsal vagal complex increases pancreatic exocrine secretion and decreases plasma insulin levels, whereas microinjections of the group III mGluR agonist decreases plasma insulin, without affecting pancreatic exocrine secretion.[46] These findings further support the suggestion that mGluRs modulate pancreatic functions via actions on pancreas-projecting DMV neurons. Our laboratory has shown that the responsiveness of DMV neurons to the group II mGluR agonist is altered in a rat model of acute pancreatitis, suggesting that these receptors may play a role in the development of pathological conditions of the exocrine pancreas.[53]

Pancreatic polypeptide

Pancreatic polypeptide (PP) is released by the cells of the pancreatic islets of Langerhans after ingestion of a meal. The release of PP is vagally mediated and involves activation of postganglionic muscarinic acetylcholine receptors.[59] Circulating PP inhibits PES, not via direct actions on the pancreatic acini but rather via actions on the dorsal vagal complex.[60] PP receptors are not expressed by acinar or ductal cells, and isolated acini or ducts are not inhibited by PP.[44,60] Instead, PP receptors are expressed in the dorsal vagal complex, in the area postrema, NTS, and DMV.[43,60,61] Microinjections of PP in the dorsal vagal complex inhibit pancreatic exocrine secretion by modulating vagal cholinergic output but does not affect basal plasma insulin, somatostatin, or glucagon secretion,[41,43] suggesting that PP modulates PES but not endocrine pancreatic secretions. Electrophysiological studies from our laboratory have demonstrated that approximately half of the identified pancreas-projecting DMV neurons respond to PP. In these experiments, PP inhibited both excitatory and inhibitory postsynaptic currents elicited by the stimulation of the NTS and reduced the amplitude of currents stimulated by chemical activation of the area postrema.[34] Interestingly, pancreas-projecting DMV neurons that responded to PP did not respond to GLP-1, suggesting that these two peptides affect separate populations of pancreas-projecting neurons.

Cholecystokinin

Cholecystokinin (CCK) is released from enteroendocrine cells in the small intestine in response to ingestion of a meal and exerts various effects along the GI tract, including increased PES, gastric relaxation, decreased gastric acid secretion, and reduction of food intake.[62,63] CCK exerts its effects both via paracrine actions on vagal sensory neurons and via actions in the dorsal vagal complex.[32] In addition, CCK1 receptors are also present on acinar cells and CCK can therefore directly influence acinar cell function, at least in rodents (reviewed in references[49,64,65]). CCK-1 receptors are expressed on neurons of the dorsal vagal complex and are activated by exogenous administration of CCK. Intraduodenal infusions of casein, a protein known to release endogenous CCK, increased pancreatic exocrine secretion even after vagal afferent fibers were surgically removed, although the response was attenuated. Furthermore, the casein-induced increase in pancreatic exocrine secretion was attenuated after application of CCK-1 receptor blocker

in the dorsal vagal complex, suggesting that CCK increases pancreatic exocrine secretion via centrally mediated mechanisms.[66] Electrophysiological studies from our laboratory have shown that CCK excites pancreas-projecting neurons via direct effects on DMV neurons and via effects on excitatory synapses impinging onto these neurons. Neurons that were excited by CCK were also inhibited by PP, suggesting that these peptides affect the same population of pancreas-projecting neurons.[58]

Glucagon-like peptide-1

Glucagon-like peptide-1 (GLP-1) is released from intestinal cells into the circulation, where it binds to receptors on the pancreatic β-cells to stimulate insulin release. In addition to its actions on pancreatic β-cells, GLP-1 acts via central mechanisms to decrease food intake and increase insulin secretion.[67,68] GLP-1 increases the discharge of fibers from the hepatic branch of the vagus nerve and selective hepatic branch vagotomy attenuated GLP-1-induced increase in insulin secretion.[69,70] GLP-1 administration also increases the expression of c-fos, an intermediate early gene and a marker of neuronal activation, in the NTS,[71] providing further evidence for central effects of GLP-1. Studies from our laboratory have shown that GLP-1 increases the frequency of excitatory and inhibitory synaptic inputs to pancreas-projecting neurons in the DMV[46,57] and that microinjections of exendin-4, a GLP-1 analogue, into the dorsal vagal complex increased plasma insulin levels.[46] Taken together, these findings suggest that GLP-1 increases pancreatic endocrine secretions via actions on DMV neurons as well as pancreatic β-cells.

Serotonin

Serotonin (5-hydroxytryptamine [5-HT]) modulates pancreatic secretions via both direct and indirect actions. Serotonin-containing neurons innervate the pancreas, stomach, and small intestine and it has been suggested that serotonin inhibits pancreatic exocrine secretion via activation of presynaptic receptors on cholinergic neurons, although this mechanism has not been fully investigated.[1]

Serotonin also modulates pancreatic exocrine secretion via excitation of vagal afferent fibers.[72,73] Vagal deafferentation and serotonin-3 receptor antagonists have been shown to block an increase in pancreatic exocrine secretion induced by intraduodenal carbohydrates or mucosal stimulation.[72] It has also been demonstrated that serotonin and CCK have synergistic actions in the regulation of pancreatic secretion. This suggestion is supported by the finding that the CCK-1 receptor antagonists attenuate the ability of serotonergic agonist to excite pancreatic vagal afferent fibers.[74] This interaction between serotonin and CCK may provide a means to finely tune the regulation of the neural control of pancreatic functions.

Thyrotropin-releasing hormone

Thyrotropin-releasing hormone (TRH) receptors, as well as TRH-immunoreactive axons that originate from the medullary raphe, the parapyramidal nuclei, and the hypothalamus, are expressed in the dorsal vagal complex.[75] Intracerebroventricular and intra-DVC injections of TRH increase pancreatic exocrine secretion and this effect is prevented by vagotomy, ganglionic blockade with hexamethonium, blockade of postganglionic transmission with atropine, or by a VIP antagonist,[76–78] suggesting that TRH-induced increase in pancreatic exocrine secretion is vagally mediated.

Acetylcholine

The hypothalamus plays an important role in modulation of pancreatic secretions. Electrical stimulation of the ventromedial anterior hypothalamus increases, whereas stimulation of the posterior hypothalamus decreases pancreatic secretions.[79] It has been suggested that hypothalamic nuclei that modulate pancreatic secretions receive cholinergic inputs from higher centers in the CNS. Microinjections of muscarinic receptor antagonists into the lateral hypothalamus or the paraventricular nucleus of the hypothalamus inhibited basal and stimulated pancreatic exocrine secretion and central depletion of neuronal acetylcholine stores had similar effects.[80] In contrast, microinjection of muscarinic receptor agonists into the hypothalamus increased pancreatic exocrine secretion.[48] Cholinergic inputs to the hypothalamus originate in the lateral septum and the lateral parabrachial nucleus, which provide a major influence on hypothalamic neurons that project to the dorsal vagal complex.[80]

Orexin

Neurons in several regions in the CNS, including the ventromedial hypothalamus, NTS, and the DMV, can directly sense changes in glucose levels. The lateral hypothalamic area contains neurons that are activated by hypoglycemia (glucose-inhibited neurons) and those that are activated by hyperglycemia (glucose-excited neurons) and is known to modulate the efferent outflow to the pancreas.[81,82] Orexin-containing neurons in the lateral hypothalamic area project to the parasympathetic and sympathetic preganglionic neurons that innervate the pancreas[21] and microinjection of orexin-A antagonist into the lateral hypothalamic area decreases pancreatic vagal nerve activity.[82] These

observations suggest that changes in peripheral glucose levels activate glucose-sensitive orexin neurons in the lateral hypothalamic area, which, in turn, activates pancreas-projecting neurons in the DMV.

Evidence for Distinct Regulation of Endocrine and Exocrine Pancreas

Several lines of evidence suggest that vagal circuits that modulate pancreatic exocrine secretion are separate from those that regulate pancreatic endocrine secretions. At the level of the pancreas, vagal innervation shows an anatomical gradient, with innervation being more dense at the head compared to the tail of the pancreas.[40] The influence of vagal stimulation on pancreatic exocrine secretion and endocrine secretions depends on either the frequency of vagal stimulation or the frequency of action potentials in DMV neurons.[36] Furthermore, although vagal celiac branches innervate the splenic end of the pancreas, electrical stimulation of the hepatic and gastric branches of the vagus are solely responsible for insulin and glucagon secretion.[83] This finding suggests that celiac branches innervate targets other than pancreatic α- and β-cells. Taken together, these observations suggest that vagal circuits are organized in a highly specific manner and that separate circuits may regulate different pancreatic functions.

Further evidence for distinct circuits regulating exocrine and endocrine pancreatic secretions came from recent studies in our laboratory. Pancreas-projecting neurons in the DMV that regulate pancreatic exocrine secretion can be distinguished from those regulating insulin secretion based on their neurochemical and pharmacological properties.[46,55,57] Electrophysiological studies from our laboratory have shown that pancreas-projecting neurons that respond to GLP-1 do not respond to PP or CCK,[55,66] whereas the majority of neurons that respond to CCK also respond to PP.[55] This observation suggested that pancreas-projecting neurons in the DMV comprise at least two distinct neuronal populations, one of which responds to GLP-1 and the other to CCK and PP. This finding also raised the possibility that the two populations of neurons may regulate separate pancreatic functions. In support of this suggestion, CCK and PP have been shown to modulate PES, whereas GLP-1 modulates insulin release.[36] We have shown that microinjections of GLP-1 into the DVC increase plasma insulin levels, but have no effect on PES, whereas microinjections of CCK and PP in the DVC increase pancreatic exocrine secretion.[46] Furthermore, we have demonstrated that in rats with copper deficiency, which selectively destroys the exocrine pancreas while leaving the islets of Langerhans unaffected, DMV neurons display a diminished responsiveness to CCK and PP, peptides that selectively regulate pancreatic exocrine secretion.[55] This evidence further supports the idea that separate neuronal populations within the DMV regulate pancreatic exocrine secretion and insulin release.

DMV neurons that regulate pancreatic exocrine secretion can also be distinguished from those that regulate insulin release based on their responses to metabotropic glutamate receptor (mGluR) agonists and antagonists.[46] Using single-cell patch-clamp, we demonstrated that both group II and group III mGluRs are present on excitatory (glutamatergic) and inhibitory (GABAergic) synapses impinging on identified pancreas-projecting neurons in the DMV.[46] Application of a group II mGluR (mGluRII) agonist reduced the frequency of postsynaptic currents in the vast majority of excitatory (89%) and inhibitory (71%) synaptic terminals, whereas application of mGluRIII agonist affected a smaller proportion of excitatory (65%) and inhibitory (58%) synapses. All neurons that responded to the mGluRIII agonist also responded to the mGluRII agonist, whereas another population of neurons responded only to mGluRII agonist. Further analysis revealed that a majority of neurons that responded to the mGluRIII agonist also responded to the GLP-1 analogue exendin-4 but not to CCK or PP. Conversely, neurons that did not respond to the mGluRIII agonist responded to PP and CCK but not to exendin-4. These findings suggested that group III mGluRs modulate the activity of a specific subpopulation of pancreas-projecting neurons in the DMV that has a unique neurochemical phenotype and raised the possibility that this population of neurons modulates a specific pancreatic function, namely insulin secretion.[46]

In order to determine the roles of these neuronal populations in modulating pancreatic functions, we conducted a series of *in vivo* experiments using DVC microinjections while monitoring pancreatic exocrine secretion and insulin secretion. Microinjections of the mGluRII agonist into the DVC dose-dependently increased pancreatic exocrine secretion and decreased plasma insulin levels, whereas microinjections of the mGluRIII agonist decreased insulin levels but had no effect on PES. Taken together with the patch-clamp data described earlier, these findings suggested that DMV synaptic terminals that express mGluRIII modulate insulin release, whereas terminals that express mGluRII modulate both pancreatic exocrine secretion and insulin release.[46]

Further support for the regulation of endocrine and exocrine function by separate pathways came from studies using models of pancreatic disorders. A study from our laboratory has shown that copper deficiency, which selectively destroys the exocrine pancreas while leaving the endocrine pancreas intact, affects DMV neurons that regulate pancreatic exocrine secretion.[55] Intraduodenal infusions of CCK or casein, potent stimulators of pancreatic exocrine secretion in control conditions, failed to increase pancreatic exocrine secretion in copper-deficient rats. This

lack of an effect was accompanied by a reduction in the number of tyrosine hydroxylase-immunoreactive neurons in the DMV, suggesting that there was a reduction in catecholaminergic regulation of pancreatic exocrine secretion.[55] Furthermore, electrophysiological evidence showed that fewer pancreas-projecting DMV neurons responded to CCK and PP in copper-deficient rats compared to controls. Interestingly, while copper deficiency affected postsynaptic responses to these peptides, it did not affect presynaptic responses, suggesting that copper deficiency selectively affects pancreas-projecting neurons in the DMV, while leaving the sensory synaptic inputs onto these neurons intact.[55]

Synaptic inputs to pancreas-projecting DMV neurons are also affected by acute pancreatitis, a severe, and sometimes fatal, disorder of the exocrine pancreas. Acute pancreatitis is characterized by premature activation of zymogens leading to acinar cell injury, release of chemokines and cytokines, and an inflammatory response.[84] Although early events that lead to the development of acute pancreatitis are initiated in the pancreas, it has also been shown that severity of acute pancreatitis is modulated by the CNS. Our laboratory, for example, has demonstrated that acute pancreatitis alters the sensitivity of pancreas-projecting DMV neurons to group II mGluR agonist, which, in turn, changes the balance of glutamatergic and GABAergic synaptic inputs to DMV neurons. Specifically, we demonstrated that acute pancreatitis decreases the response of glutamatergic synaptic terminals in the DMV to group II mGluR agonist. In contrast, group III mGluRs do not appear to be affected by acute pancreatitis.[53] These findings suggest that acute pancreatitis selectively affects DMV neurons involved in the regulation of pancreatic exocrine secretion and further supports the notion that exocrine and endocrine pancreatic secretions are regulated by separate neuronal populations.

Summary

The pancreas plays an important role in the control of nutritional homeostasis. Pancreatic functions are regulated by finely tuned inputs from the sympathetic and parasympathetic branches of the autonomic nervous system, which perform as an integrated neural circuit to adapt exocrine and endocrine secretions to constantly change environmental and physiological conditions.

An increasing amount of experimental evidence indicates that autonomic pathways involved in regulation of pancreatic function are organized in a highly specific manner, with distinct pathways regulating endocrine and exocrine secretions. It is therefore important to understand how specific neural pathways regulate pancreatic secretions and to identify neurotransmitter and receptor phenotypes involved in regulation of specific pancreatic functions. Data from our laboratory have shown that DMV neurons that regulate exocrine and endocrine secretions can be differentiated by their responses to CCK, PP, and GLP-1, as well as their responses to group II and group III mGluRs. Thus, in order to completely understand the role of the central nervous system in the regulation of pancreatic functions, future studies should be aimed at further characterizing neuropeptides and receptors involved in regulation of various pancreatic functions. Data from animal models suggests that pathological conditions that affect the pancreas, including diabetes and acute pancreatitis, induce neurochemical changes in DMV neurons. Therefore, understanding of specific pathways that regulate exocrine and endocrine secretions would provide novel targets for the treatment of these disorders. Further studies of neuropeptides, their receptors and receptor pharmacology in pathological conditions are needed to fully understand the contribution of neural regulation in disorders of the pancreas.

References

1. Love JA, Yi E, Smith TG. Autonomic pathways regulating pancreatic exocrine secretion. Auton Neurosci 2007; 133:19–34.

2. Battaglia G, Rustioni A. Coexistence of glutamate and substance P in dorsal root ganglion neurons of the rat and monkey. J Comp Neurol 1985; 277:302–312.

3. Fasanella KE, Christianson JA, Chanthaphavong RS, Davis BM. Distribution and neurochemical identification of pancreatic afferents in the mouse. J Comp Neurol 2008; 509:42–52.

4. Rinaman L, Miselis RR. The organization of vagal innervation of rat pancreas using cholera toxin-horseradish peroxidase conjugate. J Auton Nerv Syst 1987; 21:109–125.

5. Sharkey KA, Williams RG. Extrinsic innervation of the rat pancreas: demonstration of vagal sensory neurones in the rat by retrograde tracing. Neurosci Lett 1983; 42:131–135.

6. Sharkey KA, Williams RG, Dockray GJ. Sensory substance P innervation of the stomach and pancreas. Demonstration of capsaicin-sensitive sensory neurons in the rat by combined immunohistochemistry and retrograde tracing. Gastroenterology 1984; 87:914–921.

7. Su HC, Bishop AE, Power RF, Hamada Y, Polak JM. Dual intrinsic and extrinsic origins of CGRP- and NPY-immunoreactive nerves of rat gut and pancreas. J Neurosci 1987; 7:2674–2687.

8. Won MH, Park HS, Jeong YG, Park HJ. Afferent innervation of the rat pancreas: retrograde tracing and immunohistochemistry in the dorsal root ganglia. Pancreas 1998; 16:80–87.

9. Schloithe AC, Sutherland K, Woods CM, Blackshaw LA, Davison JS, Toouli J, Saccone GT. A novel preparation to study rat pancreatic spinal and vagal mechanosensitive afferents in vitro. Neurogastroenterol Motil 2008; 20:1060–1069.

10. Liu L, Shenoy M, Pasricha PJ. Substance P and calcitonin gene related peptide mediate pain in chronic pancreatitis and their expression is driven by nerve growth factor. JOP 2011; 12:389–394.

11. Neuhuber WL. Vagal afferent fibers almost exclusively innervate islets in the rat pancreas as demonstrated by anterograde tracing. J Auton Nerv Syst 1989; 29:13–18.

12. Karlsson S, Scheurink AJ, Steffens AB, Ahren B. Involvement of capsaicin-sensitive nerves in regulation of insulin secretion and glucose tolerance in conscious mice. Am J Physiol Regul Integr Comp Physiol 1994; 267:R1071–R1077.

13. Karlsson S, Sundler F, Ahren B. Neonatal capsaicin-treatment in mice: effects on pancreatic peptidergic nerves and 2-deoxy-D-glucose-induced insulin and glucagon secretion. J Auton Nerv Syst 1992; 39:51–59.

14. Hermansen K. Effects of substance P and other peptides on the release of somatostatin, insulin, and glucagon in vitro. Endocrinology 1980; 107:256–261.

15. Schmidt PT, Tornoe K, Poulsen SS, Rasmussen TN, Holst JJ. Tachykinins in the porcine pancreas: potent exocrine and endocrine effects via NK-1 receptors. Pancreas 2000; 20:241–247.

16. Brown M, Vale W. Effects of neurotensin and substance P on plasma insulin, glucagon and glucose levels. Endocrinology 1976; 98:819–822.

17. Ahren B, Holst JJ. The cephalic insulin response to meal ingestion in humans is dependent on both cholinergic and noncholinergic mechanisms and is important for postprandial glycemia. Diabetes 2001; 50:1030–1038.

18. Jaworek J, Konturek SJ, Szlachcic A. The role of CGRP and afferent nerves in the modulation of pancreatic enzyme secretion in the rat. Int J Pancreatol 1997; 22:137–146.

19. Kirkwood KS, Kim EH, He XD, Calaustro EQ, Domush C, Yoshimi SK, Grady EF, Maa J, Bunnett NW, Debas HT. Substance P inhibits pancreatic exocrine secretion via a neural mechanism. Am J Physiol Gastrointest Liver Physiol 1999; 277:G314–G320.

20. Card JP, Rinaman L, Lynn RB, Lee BH, Meade RP, Miselis RR, Enquist LW. Pseudorabies virus infection of the rat central nervous system: ultrastructural characterization of viral replication, transport, and pathogenesis. J Neurosci 1993; 13:2515–2539.

21. Buijs RM, Chun SJ, Niijima A, Romijn HJ, Nagai K. Parasympathetic and sympathetic control of the pancreas: a role for the suprachiasmatic nucleus and other hypothalamic centers that are involved in the regulation of food intake. J Comp Neurol 2001; 431:405–423.

22. Ahren B, Martensson H, Nobin A. Effects of calcitonin gene-related peptide (CGRP) on islet hormone secretion in the pig. Diabetologia 1987; 30:354–359.

23. Ahren B, Veith RC, Paquette TL, Taborsky GJ, Jr. Sympathetic nerve stimulation versus pancreatic norepinephrine infusion in the dog: 2). Effects on basal release of somatostatin and pancreatic polypeptide. Endocrinology 1987; 121:332–339.

24. Andersson PO, Holst J, Jarhult J. Effects of adrenergic blockade on the release of insulin, glucagon and somatostatin from the pancreas in response to splanchnic nerve stimulation in cats. Acta Physiol Scand 1982; 116:403–409.

25. Dunning BE, Ahren B, Veith RC, Taborsky GJ, Jr. Nonadrenergic sympathetic neural influences on basal pancreatic hormone secretion. Am J Physiol Endocrinol Metab 1988; 255:E785–E792.

26. Holst JJ, Schwartz TW, Knuhtsen S, Jensen SL, Nielsen OV. Autonomic nervous control of the endocrine secretion from the isolated, perfused pig pancreas. J Auton Nerv Syst 1986; 17:71–84.

27. Holst JJ, Jensen SL, Knuhtsen S, Nielsen OV. Autonomic nervous control of pancreatic somatostatin secretion. Am J Physiol Endocrinol Metab 1983; 245:E542–E548.

28. Ahren B, Veith RC, Taborsky GJ, Jr. Sympathetic nerve stimulation versus pancreatic norepinephrine infusion in the dog: 1). Effects on basal release of insulin and glucagon. Endocrinology 1987; 121:323–331.

29. Holst JJ, Schaffalitzky de Muckadell OB, Fahrenkrug J. Nervous control of pancreatic exocrine secretion in pigs. Acta Physiol Scand 1979; 105:33–51.

30. Barlow TE, Greenwell JR, Harper AA, Scratcherd T. The influence of the splanchnic nerves on the external secretion, blood flow and electrical conductance of the cat pancreas. J Physiol 1974; 236:421–433.

31. Klein E, Salinas A, Shemesh E, Dreiling DA. Effects of autonomic denervation on canine exocrine pancreatic secretion and blood flow. Int J Pancreatol 1988; 3:165–170.

32. Travagli RA, Anselmi L. Vagal neurocircuitry and its influence on gastric motility. Nat Rev Gastroenterol Hepatol 2016; 13:389–401.

33. Browning KN, Travagli RA. Plasticity of vagal brainstem circuits in the control of gastrointestinal function. Auton Neurosci 2011; 161:6–13.

34. Browning KN, Coleman FH, Travagli RA. Effects of pancreatic polypeptide on pancreas-projecting rat dorsal motor nucleus of the vagus neurons. Am J Physiol Gastrointest Liver Physiol 2005; 289:G209–G219.

35. Browning KN, Coleman FH, Travagli RA. Characterization of pancreas-projecting rat dorsal motor nucleus of vagus neurons. Am J Physiol Gastrointest Liver Physiol 2005; 288:G950–G955.

36. Babic T, Travagli RA. Role of metabotropic glutamate receptors in the regulation of pancreatic functions. Biochem Pharmacol 2014; 87:535–542.

37. Loewy AD, Franklin MF, Haxhiu MA. CNS monoamine cell groups projecting to pancreatic vagal motor neurons: a transneuronal labeling study using pseudorabies virus. Brain Res 1994; 638:248–260.

38. Streefland C, Maes FW, Bohus B. Autonomic brainstem projections to the pancreas: a retrograde transneuronal viral tracing study in the rat. J Auton Nerv Syst 1998; 74:71–81.

39. Loewy AD, Haxhiu MA. CNS cell groups projecting to pancreatic parasympathetic preganglionic neurons. Brain Res 1993; 620:323–330.

40. Berthoud HR, Fox EA, Powley TL. Localization of vagal preganglionics that stimulate insulin and glucagon secretion. Am J Physiol Regul Integr Comp Physiol 1990; 258:R160–R168.

41. Krowicki ZK, Hornby PJ. Pancreatic polypeptide, microinjected into the dorsal vagal complex, potentiates glucose-stimulated insulin secretion in the rat. Regul Pept 1995; 60:185–192.

42. Li Y, Hao Y, Owyang C. High-affinity CCK-A receptors on the vagus nerve mediate CCK-stimulated pancreatic secretion in rats. Am J Physiol Gastrointest Liver Physiol 1997; 273:G679–G685.

43. Okumura T, Pappas TN, Taylor IL. Pancreatic polypeptide microinjection into the dorsal motor nucleus inhibits pancreatic secretion in rats. Gastroenterology 1995; 108:1517–1525.

44. Putnam WS, Liddle RA, Williams JA. Inhibitory regulation of rat exocrine pancreas by peptide YY and pancreatic polypeptide. Am J Physiol Gastrointest Liver Physiol 1989; 256:G698–G703.

45. Ionescu E, Rohner-Jeanrenaud F, Berthoud HR, Jeanrenaud B. Increases in plasma insulin levels in response to electrical stimulation of the dorsal motor nucleus of the vagus nerve. Endocrinology 1983; 112:904–910.

46. Babic T, Browning KN, Kawaguchi Y, Tang X, Travagli RA. Pancreatic insulin and exocrine secretion are under the modulatory control of distinct subpopulations of vagal motoneurones in the rat. J Physiol 2012; 590:3611–3622.

47. Mussa BM, Verberne AJ. Activation of the dorsal vagal nucleus increases pancreatic exocrine secretion in the rat. Neurosci Lett 2008; 433:71–76.

48. Travagli RA, Browning KN. Central autonomic control of the pancreas In: Llewellyn-Smith IJ, Verberne AJM, eds. Central regulation of autonomic functions. Oxford: Oxford University Press, 2011:274–291.

49. Chandra R, Liddle RA. Regulation of pancreatic secretion. The Pancreapedia: Exocrine Pancreas Knowledge Base, 2020. DOI: 10.3998/panc.2020.14.

50. DiMagno EP. Regulation of interdigestive gastrointestinal motility and secretion. Digestion 1997; 58 Suppl 1:53–55.

51. Li Y, Owyang C. Somatostatin inhibits pancreatic enzyme secretion at a central vagal site. Am J Physiol Gastrointest Liver Physiol 1993; 265:G251–G257.

52. Li Y, Jiang YC, Owyang C. Central CGRP inhibits pancreatic enzyme secretion by modulation of vagal parasympathetic outflow. Am J Physiol Gastrointest Liver Physiol 1998; 275:G957–G963.

53. Babic T, Travagli RA. Acute pancreatitis decreases the sensitivity of pancreas-projecting dorsal motor nucleus of the vagus neurones to group II metabotropic glutamate receptor agonists in rats. J Physiol 2014; 592:1411–1421.

54. Anagnostides A, Chadwick VS, Selden AC, Maton PN. Sham feeding and pancreatic secretion. Evidence for direct vagal stimulation of enzyme output. Gastroenterology 1984; 87:109–114.

55. Babic T, Bhagat R, Wan S, Browning KN, Snyder M, Fortna SR, Travagli RA. Role of the vagus in the reduced pancreatic exocrine function in copper-deficient rats. Am J Physiol Gastrointest Liver Physiol 2013; 304:G437–G448.

56. Mussa BM, Sartor DM, Rantzau C, Verberne AJ. Effects of nitric oxide synthase blockade on dorsal vagal stimulation-induced pancreatic insulin secretion. Brain Res 2011; 1394:62–70.

57. Wan S, Browning KN, Travagli RA. Glucagon-like peptide-1 modulates synaptic transmission to identified pancreas-projecting vagal motoneurons. Peptides 2007; 28:2184–2191.

58. Wan S, Coleman FH, Travagli RA. Cholecystokinin-8s excites identified rat pancreatic-projecting vagal motoneurons. Am J Physiol Gastrointest Liver Physiol 2007; 293:G484–G492.

59. Jung G, Louie DS, Owyang C. Pancreatic polypeptide inhibits pancreatic enzyme secretion via a cholinergic pathway. Am J Physiol Gastrointest Liver Physiol 1987; 253:G706–G710.

60. Whitcomb DC, Taylor IL. A new twist in the brain-gut axis. Am J Med Sci 1992; 304:334–338.

61. Deng X, Wood PG, Sved AF, Whitcomb DC. The area postrema lesions alter the inhibitory effects of peripherally infused pancreatic polypeptide on pancreatic secretion. Brain Res 2001; 902:18–29.

62. Dufresne M, Seva C, Fourmy D. Cholecystokinin and gastrin receptors. Physiol Rev 2006; 86:805–847.

63. Dockray GJ. Cholecystokinin and gut-brain signalling. Regul Pept 2009; 155:6–10.

64. Wang BJ, Cui ZJ. How does cholecystokinin stimulate exocrine pancreatic secretion? From birds, rodents, to humans. Am J Physiol Regul Integr Comp Physiol 2007; 292:R666–R678.

65. Williams JA. Regulation of acinar cell function in the pancreas. Curr Opin Gastroenterol 2010; 26:478–483.

66. Viard E, Zheng Z, Wan S, Travagli RA. Vagally mediated, nonparacrine effects of cholecystokinin-8s on rat pancreatic exocrine secretion. Am J Physiol Gastrointest Liver Physiol 2007; 293:G493–G500.

67. Drucker DJ. The biology of incretin hormones. Cell Metab 2006; 3:153–165.

68. Holst JJ. The physiology of glucagon-like peptide 1. Physiol Rev 2007; 87:1409–1439.

69. Nishizawa M, Nakabayashi H, Kawai K, Ito T, Kawakami S, Nakagawa A, Niijima A, Uchida K. The hepatic vagal reception of intraportal GLP-1 is via receptor different from the pancreatic GLP-1 receptor. J Auton Nerv Syst 2000; 80:14–21.

70. Nishizawa M, Nakabayashi H, Uehara K, Nakagawa A, Uchida K, Koya D. Intraportal GLP-1 stimulates insulin secretion predominantly through the hepatoportal-pancreatic vagal reflex pathways. Am J Physiol Endocrinol Metab 2013; 305:E376–E387.

71. Van Dijk G, Thiele TE, Donahey JC, Campfield LA, Smith FJ, Burn P, Bernstein IL, Woods SC, Seeley RJ. Central infusions of leptin and GLP-1-(7–36) amide differentially stimulate c-FLI in the rat brain. Am J Physiol Regul Integr Comp Physiol 1996; 271:R1096–R1100.

72. Li Y, Hao Y, Zhu J, Owyang C. Serotonin released from intestinal enterochromaffin cells mediates luminal non-cholecystokinin-stimulated pancreatic secretion in rats. Gastroenterology 2000; 118:1197–1207.

73. Zhu JX, Zhu XY, Owyang C, Li Y. Intestinal serotonin acts as a paracrine substance to mediate vagal signal transmission evoked by luminal factors in the rat. J Physiol 2001; 530:431–442.

74. Mussa BM, Sartor DM, Verberne AJ. Activation of cholecystokinin (CCK 1) and serotonin (5-HT 3) receptors increases the discharge of pancreatic vagal afferents. Eur J Pharmacol 2008; 601:198–206.

75. Tache Y, Yang H, Miampamba M, Martinez V, Yuan PQ. Role of brainstem TRH/TRH-R1 receptors in the vagal gastric cholinergic response to various stimuli including sham-feeding. Auton Neurosci 2006; 125:42–52.

76. Kato Y, Kanno T. Thyrotropin-releasing hormone injected intracerebroventricularly in the rat stimulates exocrine pancreatic secretion via the vagus nerve. Regul Pept 1983; 7:347–356.

77. Messmer B, Zimmerman FG, Lenz HJ. Regulation of exocrine pancreatic secretion by cerebral TRH and CGRP: role of VIP, muscarinic, and adrenergic pathways. Am J Physiol Gastrointest Liver Physiol 1993; 264:G237–G242.

78. Okumura T, Taylor IL, Pappas TN. Microinjection of TRH analogue into the dorsal vagal complex stimulates pancreatic secretion in rats. Am J Physiol Gastrointest Liver Physiol 1995; 269:G328–G334.

79. Furukawa N, Okada H. Effects of stimulation of the hypothalamic area on pancreatic exocrine secretion in dogs. Gastroenterology 1989; 97:1534–1543.

80. Li Y, Wu X, Zhu J, Yan J, Owyang C. Hypothalamic regulation of pancreatic secretion is mediated by central cholinergic pathways in the rat. J Physiol 2003; 552:571–587.

81. Miyasaka K, Masuda M, Kanai S, Sato N, Kurosawa M, Funakoshi A. Central Orexin-A stimulates pancreatic exocrine secretion via the vagus. Pancreas 2002; 25:400–404.

82. Wu X, Gao J, Yan J, Owyang C, Li Y. Hypothalamus-brain stem circuitry responsible for vagal efferent signaling to the pancreas evoked by hypoglycemia in rat. J Neurophysiol 2004; 91:1734–1747.

83. Berthoud HR, Powley TL. Identification of vagal preganglionics that mediate cephalic phase insulin response. Am J Physiol Regul Integr Comp Physiol 1990; 258:R523–R530.

84. Szatmary P, Gukovsky I. The role of cytokines and inflammation in the genesis of experimental pancreatitis. The Pancreapedia: Exocrine Pancreas Knowledge Base, 2016. DOI: 10.3998/panc.2016.29.

Chapter 17

Regulation of pancreatic exocrine function by islet hormones

Maria Dolors Sans,[1] Jason I. E. Bruce,[2] and John A. Williams[1]

[1]*Department of Molecular and Integrative Physiology, University of Michigan, Ann Arbor, MI* [2]*Faculty of Biology, Medicine, and Health, School of Medical Sciences, University of Manchester, Manchester, UK*

Introduction

The exocrine pancreas is regulated by nerves, mainly the vagus, and gastrointestinal hormones, especially cholecystokinin and secretin. A third specific source of regulation is by peptide hormones from the islets of Langerhans. Islet hormones are relevant because of structural relations within the pancreas. Except for some fishes that have a single larger collection of endocrine cells, islets are dispersed within the pancreas and have the ability to regulate exocrine cells both by local diffusion that manifests as peri-insular halos of acinar cells and by a specific vascular relationship, the islet-acinar portal system. The latter results in acinar cells being exposed to high concentrations of islet hormones and other regulatory molecules. Of these the best defined relationship is with insulin produced in islet beta cells, which serves as a trophic factor for the exocrine pancreas, as a short-term signal promoting digestive enzyme synthesis at the translational level, and as part of a long-term control system where dietary carbohydrate leads to the synthesis of the carbohydrate-targeted digestive enzyme, pancreatic amylase. Insulin also facilitates digestive enzyme secretion and regulates the membrane transport of glucose and Ca^{2+}. Earlier reviews of this topic have been published.[1-3] Evidence for a direct effect of other islet peptides is less clear and some may exert an effect through regulating insulin secretion.

Structure and Vascular Supply of the Pancreas

Structure and distribution of islets of the pancreas

Pancreatic islets are collections of endocrine cells scattered through the exocrine pancreas making up 2–3 percent of the gland. Islets are made up of five types of endocrine cells known as α-, β-, δ-, ε-, and PP/F cells that synthesize and secrete, respectively, glucagon, insulin, somatostatin, ghrelin, and pancreatic polypeptide, all of which are considered peptide hormones. Beta cells are the most numerous cell type (60–70%) and form the core in rodent islets but are scattered throughout the human islet. Alpha cells

make up about 20 percent and along with δ-cells (1–2%) form a mantle on the outside of rodent islets. Some of the cell types are geographically distributed with PP/F cells primarily in the ventral lobe of the pancreas. By contrast, ε-cells are developmentally regulated, making up about 10 percent of the fetal islet but then differentiate into α- and β-cells so that in the adult they are rare. Some of the endocrine cells also express a second biologically active peptide.[2] Amylin and galanin are synthesized in β-cells and are cosecreted with insulin. Pancreastatin is present in α-, β-, and δ-cells where it is derived from chromogranin, and adrenomedullin is produced in PP/F cells. In understanding the effects of these islet peptides on the exocrine pancreas, it is important to realize that some (insulin, glucagon, and PP) are essentially unique to the islets but others like somatostatin are more broadly distributed in the GI tract, which could affect the exocrine pancreas locally or from systemic sources. Because islet peptides can regulate the other cells within the islet, effects of a particular peptide could be mediated directly on exocrine cells or indirectly by affecting the secretion of insulin or another islet peptide. Furthermore, some of these peptides can affect the CNS and vagal regulation of the exocrine pancreas. Receptors on exocrine pancreatic cells along with direct in vitro effects of islet peptides have been identified for insulin and somatostatin but not for glucagon or PP. Further details with references will be given in the sections on individual peptides.

Islet—acinar portal system

Blood flow to the pancreas has been reported to be 0.4–1.1 ml/g/min with most data from rats, rabbits, mice, and dogs.[4] Early studies injecting dye or ink into arteries feeding the pancreas showed rapid and intense perfusion of glomerular structures that were identified as islets of Langerhans.[5-7] Islet blood flow has been most often quantitated using microspheres. It shows that 5–20 percent of blood flow goes to the islets which have a relative blood flow 5–10 times higher than the exocrine pancreas.[8,9] The regulation

e-mail: jawillms@umich.edu

of blood flow to the two portions of the pancreas is distinct with exocrine secretagogues such as CCK and VIP increasing blood flow to the exocrine pancreas[10,11] and glucose increasing islet blood flow.[12] In a study in rabbits, Lifson and colleagues concluded that essentially all blood flow to the islets goes subsequently to the acinar capillaries.[13] Considerable attention has also been paid to the pathway of capillaries within the islets as it pertains to which islet cells are upstream of the others.[14,15]

Early dye injection vascular perfusion studies also showed that small capillary vessels exited the islets and carried blood to the exocrine pancreas and these have been named the insulo-acinar portal system. This directional flow is facilitated by the fact that islet capillaries are slightly larger in diameter, which should lead to higher capillary pressure.[16,17] In some species, there are also direct arterioles feeding the acinar tissue and venules draining the islets so not all islet effluent passes to the exocrine tissue. Similar results have been seen for rabbits, rats, mice, guinea pigs, cats, dogs, and baboons.[18] Evidence for the portal system also comes from scanning electron microscopy of vascular casts and from functional studies. Vascular casts are made by injecting resin or acrylate into the arteries and after polymerization the residual tissue is digested away leaving a three-dimensional vascular cast which is coated with gold and viewed in a scanning electron microscope. Each islet has one or sometimes two afferent vessels which break down into a glomus-like capillary network. Efferent vessels similar in size to capillaries leave the islet and connect to the acinar capillaries.[19–23] Similar findings have been made in the rat, rabbit, monkey, horse, baboon, and human. Some controversy exists over the extent of this phenomenon especially in rats where Bonner-Weir reported that a large extent of islet blood was drained directly into venules[24] but this was subsequently disputed.[25] Lifson and Lassa also described an acinar-ductal portal system in the rabbit pancreas.[26] In any case, all blood from exocrine pancreas exits in venules which combine to form the pancreatic veins.

A different type of evidence for the islet-acinar portal system comes from physiological studies particularly using the perfused rat pancreas. Insulin added to the vascular perfusate enhances the exocrine response to CCK and secretin.[27–30] Glucose, which by itself does not affect exocrine secretion, when added to the vascular perfusate, enhances the release of insulin and potentiates exocrine secretion.[30,31] Other sugars that do not stimulate insulin secretion had no effect, and epinephrine, which blocks insulin secretion, blocked the potentiating effect of glucose. Since insulin can only come from the islets and would be carried away in the venous drainage in the single-pass perfusion system, an effect on exocrine secretion can only come from endogenous insulin reaching exocrine cells by the portal system. Similar studies depleting endogenous somatostatin with cysteamine have also provided evidence for endogenous somatostatin acting through the portal system.[32]

Based on the presence of the insulo-acinar portal system, an important question is what is the concentration of islet peptides in the exocrine interstitial compartment? Two estimates indicate that the concentration of insulin and somatostatin is in the nanomolar range. The first comes from a study evaluating saturable binding of iodinated tracer for insulin and somatostatin in the rat pancreas perfused anterograde or retrograde.[33] Binding was higher for retrograde perfusion, which was attributed to endogenous hormone released during anterograde perfusion. Displacement curves during retrograde perfusion indicated an interstitial concentration of insulin during anterograde perfusion with glucose of 7.5 nM and for somatostatin of 1.1 nM. A second study used a microdialysis technique which showed a unstimulated concentration of insulin of 0.4 nM during both retrograde and anterograde perfusion of dog pancreas.[34] The response to stimulation with glucose plus arginine was markedly less during retrograde perfusion. Although the stimulated concentrations did not reach steady state, if insulin release went up 10-fold, the interstitial exocrine concentration during glucose stimulation would be about 4 nM. Unfortunately, reported plasma insulin levels in fasting rats and dogs are quite variable with most ranging from 50 to 400 pM. Much better data are available for humans where most fasting levels are reported to be 25–150 pM.

In summary, although the islet-acinar portal system does not account for all blood flow to exocrine pancreas cells, it does allow a significant fraction of exocrine acinar and duct cells to be exposed to higher concentrations of islet hormones than is the case for other organs. In addition, islet hormones can act directly on acinar cells via specific receptors or indirectly by affecting the release of other islet hormones.

Action of Insulin on the Exocrine Pancreas

Insulin appears to be the most important of the classical islet hormones as a regulator of the exocrine pancreas and especially acinar cells. Evidence exists from both human studies and animal models and includes both physiology and disease. Experimental studies have been carried out at different levels of integration from the intact organism to cellular and molecular studies of acinar cells.

Clinical studies of the effects of diabetes on the exocrine pancreas

For over 100 years, the pancreas of diabetic patients has been known to be smaller with increased fibrosis consistent with pathology of the entire pancreas and not just the islets.[35] Diabetes is now divided into type 1 or insulin-dependent diabetes mellitus (IDDM) where destruction of beta cells leads to lack of insulin, type 2 whose hallmark is insulin resistance and often does not require additional insulin, and most recently type 3 which is due to exocrine pathology such as chronic pancreatitis. It has long been recognized that type 1 and 2 patients may also suffer exocrine abnormalities ranging from subclinical to pancreatic exocrine insufficiency (PEI).

Loss of pancreatic mass in diabetes

Autopsy studies of type 1 diabetes have shown decreased pancreatic weight and volume.[36,37] Methods have been developed to use CT and MRI imaging to determine pancreatic volume and distinguish fat within the pancreas. Ultrasound has also been used but is not as quantitative and is more operator dependent. Studies of diabetic patients with CT and MRI have consistently shown a decrease in pancreatic volume of 30–55 percent in type 1 and 15–30 percent in type 2 diabetes.[38–42] Some studies have shown a dependence on the duration of diabetes but others have not. Several studies in which type 3 diabetes or a history of alcoholism have been omitted have shown a smaller loss in pancreatic volume. These changes have been confirmed in a recent meta-analysis of 17 studies.[43] The authors noted, however, that many of the individual studies were small and classified as low- to moderate-quality data. Only type 2 diabetes has shown increased pancreatic fat which reduces the volume of parenchymal tissue from the total volume.[40,42] In some studies, with functional as well as imaging data, the decreased pancreatic volume correlated with decreased function as measured by fecal elastase.[41] To date, there are no longitudinal studies, although this could be done with MRI since there is no ionizing radiation.

Effects of diabetes on pancreatic function

Direct measurement of pancreatic juice bicarbonate ion and digestive enzymes in response to secretin and CCK using multilumen tube collection has been applied to diabetes. Ewald and Hardt list nine such studies from 1943 to 1996.[44] These studies all show decreased exocrine pancreatic secretion with a bigger effect seen in type 1 diabetes. An effect of the duration of diabetes was seen in some[45] but not in other studies.[46] Most studies show a bigger effect on amylase than on other digestive enzymes and the smallest effect is on bicarbonate.

Indirect pancreatic function testing on stool samples has allowed study of greater number of patients. The most commonly used test is fecal elastase measured by an ELISA assay where <200 µg/g is considered evidence for pancreatic exocrine insufficiency (PEI) and <100 µg/g severe deficiency. Elastase is used because it is more resistant to protease digestion, although some studies have measured fecal trypsin or chymotrypsin activity, which yields similar results relative to controls. In the first large study of more than 1,000 diabetics, 40.7 percent showed fecal elastase levels of less than 200 µg/g with 22.9 percent being <100 µg/g.[47] Other studies have reported similar results,[48–50] although the frequency of PEI is lower when patients with type 3 diabetes were excluded.[51,52] Type 1 diabetics with reduced fecal elastase show a higher frequency of steatorrhea up to 60 percent, although the amount of stool fat is modest[53,54] and clinical PEI requiring enzyme supplementation is rare.[55]

Overall, diabetes is accompanied by decreased pancreatic mass and reduced secretion of digestive enzymes along with a diffuse pancreatic fibrosis characterized by intra-acinar fibrosis whose pattern is distinct from that of chronic pancreatitis. These changes may be the result of reduced number of acinar cells, loss of insulin as a trophic factor, or alteration in neural control.[56] These authors proposed that these changes be recognized as distinct from chronic pancreatitis and be termed "diabetic exocrine pancreatopathy."

Animal studies of the effects of insulin and diabetes on the exocrine pancreas

Insulin receptors in the exocrine pancreas

Insulin receptors (IRs) on pancreatic acinar cells were first demonstrated by the binding of ^{125}I-insulin in a saturable manner to isolated acini from mouse, rat, and guinea pig.[57–59] Scatchard-plot analysis of binding were biphasic and showed high-affinity sites with a Kd of about 1.5 nM and capacity of about 10,000 receptors per cell; low-affinity sites were much more numerous with a Kd of 88 nM for mouse and 998 nM for rat. Addition of unlabeled insulin accelerated dissociation from rat acini consistent with negative cooperativity rather than two distinct sites.[58] The receptor bound insulin analogs with varying potency (insulin > desdipeptide insulin > proinsulin > desoctapeptide insulin) that was similar to the ability of the analogs to increase protein synthesis in acini from diabetic rats.[60] Guinea pig insulin is significantly different from other species and guinea pig acinar receptors bound bovine insulin with a higher affinity than guinea pig insulin or porcine proinsulin.[59] Mouse acini were also shown to possess distinct receptors for IGF-I and IGF-II.[61]

The localization of insulin binding to acinar and other cells in the pancreas has been demonstrated by light and EM autoradiography. When ^{125}I-Insulin was injected intravenously, saturable uptake was observed by the pancreas.[62–64] EM autoradiographs showed silver grains localized both over the plasma membrane and intracellularly in acinar cells. Saturable binding was also observed over duct cells.[64] In an *in vitro* study of ^{125}I-Insulin binding to rat acini, silver grains from ^{125}I were primarily over the plasma membrane at 3 minutes but by 30 minutes were primarily intracellular.[65] Studies in other cell types have shown that both insulin and its receptor are internalized, although the function of internalized insulin is unclear.

Insulin receptors on acinar cells are also subject to regulation. Insulin downregulates its own receptor in a variety of cells, including acinar cells, when they are exposed to high concentrations of insulin through the islet-acinar portal blood system.[2,66] Acini prepared from diabetic mice showed an increased number of IRs compared to normal mice and addition of insulin to 24-hour cultured acini decreased insulin binding.[67] Using pancreatic AR42J cells, insulin decreased the biosynthesis of insulin receptors in part by decreasing IR mRNA levels.[67,68] The binding of insulin was also affected by CCK and other secretagogues.[69] CCK, carbamylcholine, active phorbol ester, and calcium ionophore (A23187) all decreased high-affinity binding without affecting insulin internalization or degradation.

The occupancy of insulin receptors has also been related to the biological effects of insulin on acinar cells. Insulin increased glucose transport (uptake) by acini from normal and diabetic mice[57,70] with a greater effect on acini from diabetic animals that was seen at lower concentrations of insulin. This greater effect is due both to lower-contaminating insulin levels and IR upregulation. In acini from diabetic rats and mice, insulin increased protein synthesis as shown by incorporation of radioactive amino acids.[60] Of likely relevance to protein synthesis, insulin rapidly increased the phosphorylation of ribosomal protein S6, which is associated with increased protein synthesis.[71,72] The concentration dependence and analog specificity of these actions of insulin is similar to the data for receptor occupancy.[73]

More recently, IRs have been demonstrated in pancreatic acini by immunoblotting of the beta subunit of the receptor.[74] The subunit had an apparent size of about 95 kDa and was essentially absent after deletion of the IR gene in acinar cells. The biological response to insulin in the pancreas of these mice was greatly reduced.

Insulin regulation of exocrine pancreas biosynthetic and metabolic effects

GENE EXPRESSION

Insulin is known to regulate both the amount and activity of a number of anabolic processes and specific metabolic enzymes. In the exocrine pancreas, the best-studied tissue-specific regulation is that of the digestive enzyme amylase which acts on dietary starch and glycogen.[75] The pancreatic amylase content is known to decrease by over 90 percent in experimental diabetes in rats and the fall can be reversed by administration of insulin. This has been studied primarily in rats given alloxan or streptozotocin that induces beta cell death and diabetes.[76–82] At the same time, other digestive enzymes including trypsinogen, chymotrypsinogen, and lipase increase slightly and ribonuclease decreases slightly. Amylase also decreases in mouse pancreas in experimental diabetes but by only about 50 percent. Pancreatic amylase also decreases in the Zucker fatty rat and the ob/ob mouse, well-characterized models of insulin resistance.[83]

The decrease in pancreatic amylase in experimental diabetes is accompanied by a decrease in amylase synthesis.[80] Moreover, insulin increased amylase biosynthesis in rat pancreatic-derived AR42J cells.[84] The major cause of the decreased amylase synthesis and tissue content is a decrease in amylase mRNA, which was first reported by Korc et al.[85,86] The decrease in amylase mRNA could be reversed by insulin. Chymotrypsinogen mRNA showed a small change in the opposite direction and there was essentially no change in salivary amylase or its mRNA. Subsequently, Brannon and colleagues showed that dietary glucose and insulin together regulated pancreatic amylase (Amy2) gene expression.[87] More detailed information on the mechanism of amylase gene regulation comes from studies in mice. A 30-bp region in the proximal amylase promoter overlapping with the PTF1-binding site present in most pancreatic digestive enzymes contains the insulin-dependent element and can transfer insulin sensitivity to the elastase promoter.[88,89] This regulatory region is not present in all mouse amylase alleles and this can explain why some mouse strains are not as sensitive to diabetes and insulin.[90] Unfortunately, there has been little subsequent work defining how insulin receptor signaling regulates the insulin response element.

MEMBRANE TRANSPORT

Insulin is known to stimulate membrane transport of many substances usually by effects on the amount of or properties of specific transport proteins. The best studied and probably the most important physiologically is the uptake of glucose into fat and muscle cells that is mediated by the glucose transporter, GLUT4. GLUT4 is present at rest in intracellular vesicles that translocate to the plasma membrane in response to insulin, increasing glucose uptake 5–20-fold. Most other cell types contain GLUT1 that is regulated by another mechanism but to a lesser extent. In mouse pancreatic acini, insulin stimulated the uptake of the nonmetabolizable sugar 2-deoxy-glucose (2DG) about twofold.[57] A more robust effect of lower concentrations of insulin was seen when acini were prepared from diabetic mice and uptake of both 2DG and 3-O-methyl glucose, a nonmetabolizable glucose analog, was increased.[70] Glucose uptake was not affected by inhibitors of protein synthesis but was reduced by the actin inhibitor, cytochalasin B.

In some tissues, insulin also stimulates the uptake of amino acids through effects on specific amino acid transporters. Uptake of amino acids have been studied in many pancreatic preparations. One of the earliest studies using microdisected pieces of mouse pancreas showed that amino acids were taken up and oxidized in preference to glucose.[91] Neither CCK or insulin affected uptake and transport did not appear to be separated from protein synthesis. Latter studies used perfused pancreas or isolated acini and showed that the basolateral membranes of pancreatic acinar cells possess four to five distinct transporters as measured by physiological studies or gene expression.[92,93] Insulin stimulation studies in the perfused rat pancreas show effects of exogenous insulin and diabetes on two Na^+-independent transporters, the Asc transporter used by serine[94,95] and a basic amino acid transporter termed y^+ for lysine/arginine[96] while no effect was seen on the Na^+-dependent transporters such as the system A transporter for AIB and glutamine.[97] However, increasing glucose in the perfusate that released endogenous insulin had no effect.[95] Studies in isolated rat pancreatic acini also showed no effect of insulin on AIB and cycloleucine uptake, which are not used for protein synthesis, although it stimulated incorporation of leucine into protein.[60] In a more recent study of the mouse pancreas, Rooman et al. analyzed the expression of 37 genes encoding transporters and found expression of a number with the highest expression of slc7a8 and slc3a2 and confirmed the expression of five by western blot.[93]

PROTEIN SYNTHESIS

Insulin is known to stimulate protein synthesis by translational effects in many tissues.[98] Early studies in the pancreas involved in vivo studies of the incorporation of radioactive amino acids into total protein or specific enzymes of normal and diabetic rodents.[99–101] While most of the studies showed a positive effect of insulin, the results are complicated by changes in the precursor pool or changes in the levels of mRNAs. When protein synthesis in mice was quantitated by autoradiography, there was more incorporation into peri-insular acinar cells then into tele-insular acinar cells.[102] This difference was lost following treatment with streptozotocin and the authors ascribed it to insulin. One study overcame some of these issues by changing plasma insulin and glucose in vivo but measuring protein synthesis in vitro using pancreatic lobules incubated in a constant medium.[76] It showed that either glucose infusion in vivo or treatment with a sulfonylurea drug both of which stimulate insulin release increased protein synthesis by 25–40 percent. When Zucker fatty rats were studied as a model of insulin resistance, overall pancreatic protein synthesis was reduced by nearly 50 percent.[103] There was considerable difference in synthesis between different digestive enzymes separated by 2-D gel electrophoresis. A morphological study of prolonged diabetes in rats showed gross abnormality in the secretory pathway 28 days after STZ

treatment that was partially reversed by insulin administration.[104] However, shorter studies have not shown significant structural alterations 1 week after STZ other than the appearance of cytoplasmic lipid droplets.[60]

More recent studies have been able to overcome some of the methodological issues by measuring the effects of insulin over short times to prevent changes in mRNA and under conditions where the precursor pool is large and constant.[105] In vitro studies have used isolated pancreatic acini under dilute incubation to keep the precursor pool constant. Insulin stimulates the incorporation of multiple different amino acids (leucine, methionine, phenylalanine) in isolated acini from diabetic rats.[60,106–108] Similarly, insulin increased methionine incorporation into total protein and immunoprecipitated amylase in rat pancreatic-derived AR42J cells.[84] The effect of insulin on both cell types occurred over concentrations from 30 pM to 100 nM and was mediated by the insulin receptor as insulin analogs stimulated in parallel to their ability to bind to the insulin receptor.[60,107] Insulin was also shown to have nonparallel effects on different digestive enzymes and structural proteins such as myosin and LDH.[107]

Although the mechanism of insulin signaling is well studied in a number of target tissues especially liver, fat, and muscle, only a few studies have been carried out in the exocrine pancreas. The main signaling pathway regulating protein synthesis is the mTOR pathway and this pathway has been documented in pancreatic acinar cells primarily in mediating the actions of CCK and acetylcholine to stimulate digestive enzyme synthesis.[109,110] Insulin has been shown to activate S6 kinase and S6 phosphorylation downstream of TORC1 in rat and mouse acini.[71,72] Akt is upstream of mTOR and insulin activates the phosphorylation of Akt on S473 and T308 in rat acini.[111] This action appeared mediated by PI3K. That these actions are important physiologically is shown by over a 50 percent reduction in mTOR pathway activation after feeding in mice without acinar cell insulin receptors.[74]

SECRETION

The effects of insulin administration and diabetes on pancreatic exocrine secretion have been studied in a number of species and systems ranging from intact awake animals down to isolated acinar cells. The more isolated systems have better-defined parameters but lack neural control and other integrative aspects found in the intact organism.

In vivo studies have largely been carried out in rats with some studies in dogs and other species and often have compared streptozotocin (STZ) treated and intact animals. STZ and alloxan treatment damage islets, thus lowering insulin levels and leading to hyperglycemia. In STZ-diabetic rats, pancreatic juice flow has been reported to be moderately increased.[112–114] In two studies, this effect was reversed by transplanting syngeneic islets into the liver.[115,116] However, in mice STZ treatment or genetic models of diabetes lead to reduced juice flow.[74] and in sheep, alloxan treatment led to a substantial decrease that could be restored by insulin treatment.[117] In all of these studies the effects of the diabetic state on plasma glucose and lipids are not clear. The effects of CCK on digestive enzyme secretion in diabetes are complicated by changes in the pancreatic content of specific digestive enzymes. The concentration of secreted amylase is decreased and trypsin increased, although both respond to CCK.[112,114] The most valid results may be in animals with chronic pancreatic fistulas where juice can be collected in the unanesthetized state but such studies have not been carried out in diabetes.

Other studies have focused on the effects of insulin in normal animals with intact islets. When a high concentration of insulin (4 U per 100 g) was injected into the femoral vein, subsequent pancreatic stimulation by CCK was potentiated.[27] In another study, insulin (1U as a bolus) induced hypoglycemia, increased fluid flow, and increased amylase and total protein output; this effect was blocked by atropine and ascribed to hypoglycemia-induced vagal cholinergic activation.[113] In a third study, an intravenous infusion of insulin also increased pancreatic secretion.[118] Not all studies however have reported this effect.[119] In a different approach using unanesthetized rats, IV injection of rabbit anti-insulin serum but not control serum blocked the pancreatic response to a liquid meal and to exogenous CCK or secretin stimulation.[120] In a study in conscious dogs which secrete a high-bicarbonate pancreatic juice, most of which comes from the ducts, insulin administration diminished secretin-stimulated bicarbonate secretion and this effect was blocked by prior pancreatic denervation.[121] This study illustrates the complexity of administering insulin in vivo and shows that some of the effects of high-dose insulin in vivo are mediated by neural pathways. However, overall they suggest a positive effect of insulin on pancreatic exocrine secretion.

Studying an in vitro pancreas preparation removes the effects of plasma glucose and the brain-initiated neural control. While some studies have been carried out with isolated pancreas segments from animals with thin pancreas, the perfused pancreas provides better oxygenation and when single-pass perfusion is used the secretion of both endocrine and exocrine secretions can be separately measured. Because the islets are present in their normal position, they can be stimulated independent of the exocrine pancreas and the islet hormones will reach the exocrine cells by local diffusion and even more importantly by the islet-acinar portal system. Most studies of the perfused normal rat pancreas or isolated pancreatic segments incubated with normal glucose have shown that exogenous insulin has little or no effect on basal fluid and protein secretion but potentiates the secretory effect of CCK, ACh, or secretin.[27–31,122–125] In STZ-induced diabetic animals with high plasma glucose, the *in*

vitro response to secretin and CCK was reduced compared to normal rats.[126,127] This change could be partially reversed by treatment with insulin. To study the effect of endogenous insulin, investigators have used a low-plasma glucose background and then a pulse of high glucose to stimulate insulin release; this endogenous insulin which must reach the acinar cells by the islet-acinar portal system stimulated secretion of fluid, protein, and amylase in response to CCK or Ach.[29–31] This effect was induced only by sugars that stimulate insulin release and could be blocked with epinephrine or somatostatin which blocks glucose-induced insulin release. In an alternative approach, Chey and colleagues added anti-insulin antibody to the perfusate in the rat- or dog-perfused pancreas and showed that this resulted in reduced juice, bicarbonate, and protein secretion.[120,128,129] All of these data are consistent with insulin having an action on acinar and duct cells to facilitate secretion. However, other studies have reported that both high glucose and exogenous insulin inhibit CCK-stimulated amylase secretion in the mouse[130] or caerulein-stimulated perfused rat pancreas.[131] In the perfused cat pancreas, insulin had no effect on the basal fluid or enzyme secretion.[132] While a stimulatory effect of insulin on acinar cells is consistent with the presence of insulin receptors on acinar and duct cells and the ability of insulin to act on isolated pancreatic acini (see below), other studies have suggested effects within the islet can explain much of the actions of insulin. In the studies of immunoneutralization of insulin in perfused rat and dog pancreas, Chey's group showed that insulin antibody increased the concentration of somatostatin and pancreatic polypeptide in the venous effluent. In their rat study, addition of a antisomatostatin antibody partially reversed the effect of the anti-insulin and the addition of somatostatin partially reversed the effect of combined stimulation by CCK and secretin.[123,129] Using a perfused canine pancreas, insulin antiserum blocked secretion but when SS and PP antibodies were added together, the effect on fluid and bicarbonate was fully reversed and the inhibition of protein secretion was partially reversed.[128] These investigators concluded that the inhibitory effect of insulin antisera was mediated by the local release of and action on the exocrine pancreas of somatostatin and PP.[128] In a related study, Nakagawa et al. concluded that the net effect of islet peptides on the exocrine pancreas was negative because exocrine secretion was greater with retrograde than anterograde perfusion.[33] Thus, the action of insulin could be mediated at least in part by suppressing the secretion of inhibitory islet peptides. One study of duct function showed that insulin increased pancreatic juice secretion and that this effect was blocked by ouabain.[28]

Isolated pancreatic acini and ducts can be used to study cellular function in the absence of neural, hormonal, and islet control. Isolated acini are exquisitely sensitive to CCK, acetylcholine, bombesin, and respond through an increase in intracellular Ca^{2+} to increase digestive enzyme secretion. Acini also respond through an increase in cyclic AMP to secretin and VIP, which can increase enzyme secretion or potentiate the actions of agonists that mobilize Ca^{2+}. Isolated acini from STZ-induced diabetic rats show reduced secretion of amylase and ribonuclease in response to CCK and cholinergic agonists.[81,133-135] This decreased secretion is also seen with alloxan-induced diabetes and can be reversed by treatment of the animals with insulin. One cause of the decreased secretion is a decrease in the synthesis and content of digestive enzymes. When secretion is normalized to content and expressed as percent of content released per 30 minutes, maximal secretion is similar but the dose response to CCK is shifted to the right while the response to carbachol is similar. Thus, there is also an effect on CCK receptors, which was confirmed by radioactive binding of CCK showing changes in affinity and capacity of the two affinity states.[133] Interestingly, RT-PCR of the CCK1 receptors showed no change in pancreatic receptor mRNA in rat STZ-induced diabetes.[136] Acinar cell postreceptor signaling is also altered in diabetes with changes in CCK-induced IP_3 formation.[136-138] This is accompanied by changes in the amount and phosphorylation of IP_3 receptors. The increase in intracellular free Ca^{2+} measured in suspension or single cells is also reduced in diabetes.[135,136]

An effect of insulin in vitro on both diabetic and normal rat acini has also been demonstrated. An effect of insulin to increase amylase release was seen as early as 30 minutes and increased to a maximum at 2 hours.[134] Over this time, the number and affinity of CCK receptors was also changed. In another study in acini from normal rats, insulin after 90 minutes potentiated the secretion in response to CCK plus secretin but had no effect on either alone; this potentiating effect of insulin was inhibited by the Na^+-K^+ ATPase inhibitor, ouabain.[139] Insulin also has been shown to directly activate the plasma membrane Ca^{2+} ATPase (PMCA) in rat.[140]

These studies along with other metabolic effects of insulin on isolated acinar cells indicate that insulin can directly affect acinar cell secretion. However, the mechanisms of this effect are not fully established. Moreover, all studies have been carried out in rats and rat tissue and the results need to be extended to other species. The cellular sites where insulin acts on pancreatic acinar cells are shown in **Figure 1**. Finally, data are needed from isolated ducts to establish whether there is a direct effect of insulin on ductal bicarbonate secretion.

Effects of insulin on pancreatitis and other pancreatic diseases

Clinical and animal studies linking diabetes with pancreatitis

There is emerging evidence from clinical studies[141-146] and animal studies[147-149] that preexisting diabetes may predispose, increase the risk, or make pancreatitis more severe. This implies that endogenous insulin may have a protective role during pancreatitis. Type-2 diabetics have an approximately threefold increased risk of developing acute pancreatitis (AP),[141,145,146,150] which could be explained by a loss of direct protection of insulin on pancreatic acinar cells, due to insulin resistance. Preexisting diabetes increases the severity of acute pancreatitis[144] and diabetes increases the mortality in patients with chronic pancreatitis.[151,152] Around half of type-1 diabetic patients exhibit lesions within the exocrine pancreas that are reminiscent of chronic pancreatitis.[143]

Moreover, acute pancreatic patients with hyperglycemia are at higher risk of multiple organ failure.[153] Hyperglycemia frequently accompanies severe acute pancreatitis and is used in the Ranson score as a predictor of disease severity.[154] Furthermore, the incidence of AP is higher among type 2 diabetics compared to the normal population and the risk of AP was reduced among diabetic patients treated with insulin or metformin compared to those treated with other drugs such as sulphonylurea-based antidiabetic drugs.[142]

For many years, there had been numerous anecdotal histological observations of the pancreas in animals that the acinar cells surrounding the islets (peri-islet acinar cells) were morphologically distinct from acinar cells distant from the islets. This was further investigated in L-arginine-induced experimental animal models of pancreatitis in which peri-islet acinar cells remained relatively intact compared to distal injured acinar cells.[147,148,155,156] This peri-islet acinar cell protection was abolished in streptozotocin (STZ)-induced diabetic rats, with impaired insulin secretion.[147,148,157] These studies suggest that insulin release has a paracrine protective role in the pancreas. Moreover, in addition to the acute injury, the regeneration of exocrine pancreatic tissue was abolished in diabetic rats and restored following the administration of exogenous insulin.[147,148,155,156] This was further investigated in the seminal study by Zechener et al. (2012),[149] in which caerulein-induced pancreatitis was aggravated in streptozotocin (STZ)-induced type 1 diabetic mice.[149] Specifically, they showed that numerous markers of the acute phase of pancreatitis injury (within 24 hours) were markedly potentiated—including plasma amylase, lipase, and trypsinogen, pancreatic edema (wet/dry weight ratio), pancreatic histological tissue injury score (H&E), inflammation, and cell death. In addition, regeneration of the pancreas (7 days later) was delayed.[149] Moreover, this diabetic-induced potentiation of pancreatitis phenotype was partially corrected by exogenous administration of insulin (slow-release pellet implant). They also suggested that the mechanism for this more severe pancreatitis phenotype in STZ-induced diabetic mice was due at least in part to depleted pancreatic regenerating islet-derived 3β (REG3β), which has important anti-inflammatory and antimicrobial properties. REG3β

Sites of Insulin Action on Pancreatic Acinar Cells

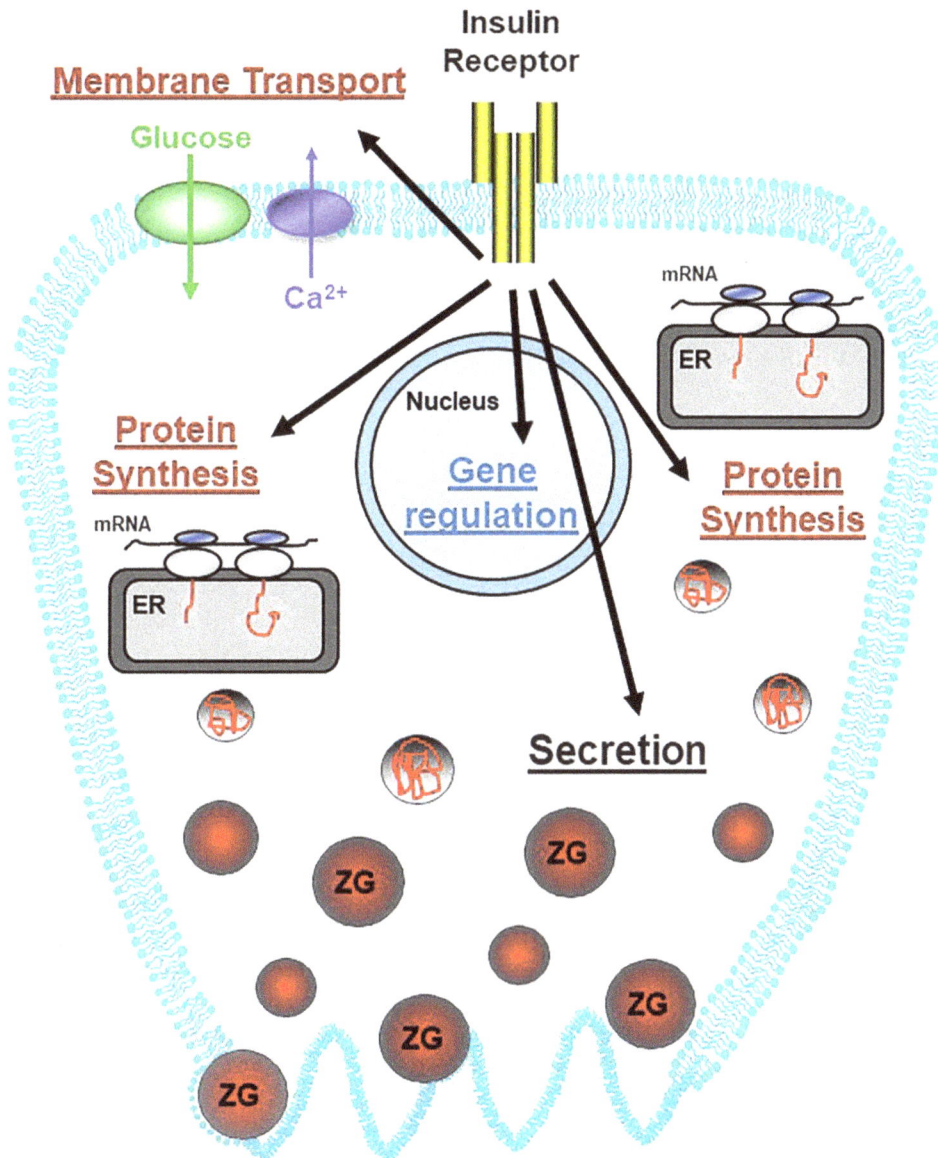

Figure 1. Sites of insulin action on pancreatic acinar cells. Insulin has receptor-dependent action to regulate protein synthesis at the translational level, gene expression at the transcriptional level, membrane transport of glucose and calcium, and potentiation of digestive enzyme secretion. Not all of the steps between receptor occupancy and the end biological effect are known and most of the data are from rodent pancreas (see text for details and references).

is upregulated during pancreatic injury and is suggested to act as an acute emergency program to circumvent the inflammatory response during pancreatitis.[158–161] However, REG3β expression was found to be severely blunted in the pancreas of STZ-induced diabetic mice, suggesting that this may contribute to the more severe pancreatitis phenotype.[149]

Insulin protection during cellular models of pancreatitis

From a clinical perspective, and to some extent the diabetic animal studies, it is very difficult to separate the confounding effects of hyperglycemia or reduced systemic effects of insulin from a loss of direct insulin protection of acinar cells. This is important because hyperglycemia is associated with a more severe pancreatitis, by providing a pro-inflammatory environment and by facilitating sepsis.[162] Moreover, insulin itself exhibits anti-inflammatory properties,[163] therefore the systemic loss of insulin secretion or insulin effectiveness would be predicted to promote systemic inflammation. However, studies using acutely isolated pancreatic acinar cells provide compelling evidence that insulin directly protects acinar cells from cellular injury induced by pancreatitis-inducing agents, such as oxidative stress (H_2O_2) or the alcohol/fatty acid metabolite, palmitoleic acid (POA) (cellular models of acute pancreatitis).[140,164] Specifically, insulin markedly attenuated ATP depletion, the inhibition of plasma membrane calcium ATPase (PMCA) that transports Ca^{2+} out of cells, cytotoxic Ca^{2+} overload, and necrotic cell death. This protection was partially PI3K/Akt-dependent and due to an acute metabolic switch from mitochondrial to glycolytic metabolism.[140,164] Consistent with this, insulin enhanced PMCA inhibition by glycolytic inhibitors and abolished PMCA inhibition by mitochondrial inhibitors.[140] Therefore, this switch to glycolysis appears to maintain cytosolic ATP concentration sufficiently to fuel the PMCA and thus prevent cytotoxic Ca^{2+} overload, even in the face of impaired mitochondrial function.

The most likely downstream signaling pathways responsible for insulin's protective effects during pancreatitis are likely to be mediated by tyrosine kinase or PI3K/Akt. It's also interesting to note that several related growth factors (GFs), that also couple to tyrosine kinase and PI3K/Akt, similar to insulin, are also reported to be protective in animal models of pancreatitis.[165–167] These include insulin-like growth factor (IGF-1), fibroblast growth factor (FGF), hepatocyte growth factor (HGF), and epidermal growth factor (EGF). Although, altered expression of signaling proteins and metabolic enzymes may contribute to insulins protection *in vivo*, most of the protective effects of insulin observed in acinar cells occurred over a relatively short term (15–30 minutes), suggesting a more rapid effect of posttranslational effects, such as tyrosine kinase or PI3K/Akt phosphorylation.

Arguably, the most important mechanism for insulin's protection is the regulation of glycolytic ATP. Glycolytic flux is primarily regulated by the activity of phosphofructokinase-1 (PFK1), which catalyzes the conversion of fructose-6-phosphate (F6P) to fructose-1,6-bisphosphate (F1,6BP) and represents the first rate-limiting irreversible step in glycolysis. Insulin has been shown to lead to the Akt-mediated phosphorylation and direct activation of phosphofructokinase-2 (PFK2)—otherwise known as phosphofructokinase-fructose bisphosphatase (PFKFB). PFKFB consists of four separate bifunctional glycolytic enzymes (PFKFB1–4), with varying catalytic and functional activities.[168] PFKFB catalyzes the conversion of F6P to fructose-2,6-bisphosphate (F2,6BP), via their

kinase activity, but also the reverse reaction, by converting F2,6BP back to F6P, via their bisphosphatase activity.[168] F2,6BP is a potent positive allosteric activator of PFK-1, which maintains high glycolytic flux.[169] Therefore, Akt-mediated phosphorylation of PFKFB2 may represent the major molecular mechanism by which insulin increases glycolytic ATP supply in the face of impaired mitochondrial metabolism during acute pancreatitis.

The effect of insulin on the gut microbiome-pancreatitis link

Severe acute pancreatitis is frequently accompanied by infected pancreatic necrosis and systemic bacteremia or sepsis.[170] The source of the bacteria is thought to be from "leaky" gut—particularly the colon which houses a large reservoir of pathogenic and commensal bacteria.[171] This caused by gut dysbiosis, inflammation, altered mucus secretion, and the consequent loss of gut barrier function and bacterial translocation into the blood.[170] The gut microbiome consists of a delicate balance of trillions of bacteria from thousands of different bacterial species that can be broadly categorized into "good" (anti-inflammatory) or "bad" (pro-inflammatory). Moreover, bacteria can aid in the breakdown and absorption of undigested carbohydrates and produce numerous beneficial anti-inflammatory metabolites, such as short-chain fatty acids (SCFAs), such as butyrate, acetate, and propionate that promote epithelial integrity and gut barrier function.[172]

This delicate balance of the gut microbiome and bacterial diversity is maintained by numerous anti-microbial peptides (AMPs) secreted mainly from the Paneth cells within the crypts of Lieberkuhn of the small intestine.[173] However, several studies show that AMPs are secreted from pancreatic acinar cells, which may contribute to maintaining a healthy gut microbiome. These include the related C-type lectin family members: lithostathine (REG1α/β),[174] REG3 family members (REG3α/β/γ)—sometimes referred to as pancreatitis-associated peptide (PAP)/hepatocarcinoma-intestine-pancreas (HIP),[158–161] cathelicidin-related AMP, (CRAMP[175]) and defensins.[176]

PAP was originally discovered in pancreas homogenates, secretory granules, and in the pancreatic juice during experimentally induced acute pancreatitis in rats but was absent in control rats.[177] PAP expression was also increased in pancreatic acinar cells in response to both acute and chronic pancreatitis.[178] Moreover, PAP/REG3β expression is regulated by dietary carbohydrates[179] and is blunted in diabetic mice,[149] suggesting that insulin regulates the expression of PAP/REG3β. Therefore, during pancreatitis the loss of PAP/REG3β secretion into the gut may accentuate the gut dysbiosis, inflammation, and loss of barrier function and bacterial translocation, thereby further contributing to the severity of acute pancreatitis.

It is also possible that gut dysbiosis in diabetics may be caused by reduced pancreatic secretion of amylase. Amylase is a highly abundant digestive enzyme that breaks down starch in the gut and its expression and secretion is controlled by insulin and high-carbohydrate diet[180] and is reduced in diabetic animals.[60] Such loss of amylase secretion may also alter the gut microbiome, due to huge amounts of undigested starch entering the colon leading to overgrowth of bacteria that break down undigested starch (*Bacteroides spp*). However, this has never been fully and specifically investigated and it remains unclear whether reduced amylase per se in the gut would be beneficial or detrimental. Even though pancreatic enzyme insufficiency (PEI) leads to numerous comorbidities that might be explained by an altered gut microbiome, disentangling the specific effects of reduced amylase secretion from the loss of other digestive enzymes would be difficult.

It is also interesting that the gut microbiome is able to signal to and thus contribute to the maintenance of a healthy inflammatory and immune landscape within the pancreas and when disrupted may ultimately leading to autoimmune type 1 diabetes.[181] Specifically, SCFAs, such as butyrate, released from gut bacteria are able to leak into the circulation and reach sufficient concentration to promote the secretion of CRAMP from pancreatic β-cells. CRAMP induced a positive immunoregulatory phenotype in pancreatic resident macrophages by promoting the production of anti-inflammatory cytokine, TGFβ, to maintain immune homeostasis via Treg induction. However, the loss of SCFAs following gut dysbiosis disrupts this process, leading to reduced CRAMP secretion which promotes a pro-inflammatory environment in which activated macrophages secrete TNFα and the ultimate autoimmune destruction of pancreatic β-cells and the consequent type 1 diabetes.[181] This therefore suggests a complex reciprocal regulation between the gut microbiome, the exocrine, and the endocrine pancreas.

Insulin as a therapy to treat acute pancreatitis

Insulin therapy has been used to specifically treat hypertriglyceridemia (HTG)-induced pancreatitis with promising outcomes.[182,183] HTG occurs when plasma lipids exceeds 1,000 mg/dL (normal range, 101–150 mg/dL).[184] HTG-induced pancreatitis is relatively rare, accounting for 2.3 percent to 10 percent of all cases of acute pancreatitis but is a well-documented etiological risk factor leading to severe disease.[185] The rationale for using insulin therapy to treat HTG-induced pancreatitis is that it lowers plasma triglycerides, by activating lipoprotein lipase (convert triglycerides into free fatty acids) and inhibiting the hormone-sensitive lipase (liberates adipocyte triglyceride), thereby limiting inflammation.[186] However, given the evidence presented above, it's entirely possible that the beneficial

effects of insulin therapy in HTG-induced pancreatitis patients could be due to a direct protection of acinar cells.

Insulin is also used as the standard of care for all critical care patients, including those with severe acute pancreatitis, with the aim of targeting hyperglycemia associated with the acute phase of injury, which facilitates inflammation and sepsis.[187] There have been numerous clinical trials and meta-analyses testing the effectiveness of intensive insulin therapy in critical care patients and some studies question whether there is any overall patient benefit.[188–193] However, this is likely because in patient groups receiving insulin, fewer patients die from septicemia, but more die from the complications of inadvertent hypoglycemia, an independent risk factor of mortality.[194–197]

However, given the direct protective effect of insulin on acinar cells described above there is a strong rationale to test high-dose insulin infusion with very tight moment-to-moment glucose control to specifically target the acinar injury. This could be achieved using the hyperinsulinemic euglycemic clamp with automated insulin minipumps combined with continuous closed-loop subcutaneous glucose-monitoring devices.[198] It could be argued that there is a greater requirement for high-dose insulin because pancreatic acinar cells receive a portal blood flow[13] with ~10 times higher insulin concentration than the systemic circulation.[33, 34] Furthermore, stress hormones, such as adrenaline, cortisol, and glucagon, and inflammatory cytokines (TNFα and IL-1β) reduce tissue sensitivity of insulin, so a higher dose of insulin may be necessary to overcome this. Moreover, high-dose insulin infusion (8 mU/kg/min; adjusted for surface area to weight ratio) with tight physiological glucose control has been tested in healthy human volunteers during endurance exercise studies with no reported adverse effects.[199]

Action of Other Islet Hormones/Peptides on the Exocrine Pancreas

Glucagon

Glucagon was originally identified in pancreatic extracts as a hyperglycemic-glycogenolytic factor that came from the α-cells of the pancreatic islets.[200,201] Foa and colleagues showed by cross-circulation experiments in dogs that hypoglycemia induced by insulin triggered glucagon release which causes hyperglycemia in the recipient animal.[202] The subsequent development of a radioimmunoassay for glucagon by Unger made more detailed studies of the physiology and pathophysiology of glucagon possible.[203]

Glucagon is a 29-amino acid peptide derived from proglucagon in islet α-cells through the tissue-specific processing by prohormone convertase 2 (PCSK2).[204] By contrast, in intestinal enteroendocrine cells proglucagon is processed to GLP-1, GLP-2, oxyntomodulin, and glicentin.[205] The major role of glucagon is to antagonize the effects of insulin and maintain plasma glucose homeostasis by promoting hepatic gluconeogenesis and glycogenolysis and inhibiting glycogen synthesis.[206] The major regulators of glucagon secretion by islet alpha cells are glucose which inhibits and amino acids as well as parasympathetic and sympathetic nerves which stimulate glucagon release. The suppression by high glucose, however, is not direct but mediated by insulin/GABA secreted by islet beta cells. Somatostatin also suppresses glucagon secretion.

The effect of exogenous glucagon on exocrine pancreatic secretion has been studied in vivo with a consistent effect of high concentrations of purified crystalline glucagon to inhibit exocrine secretion of fluid, bicarbonate, and protein stimulated by food, CCK, or secretin in dogs,[207–210] cats,[211] rats,[212–214] and humans.[215–217] The mechanism of this inhibition is unclear in these studies. In vitro studies have been carried out in the perfused pancreas, pancreatic segments, and lobules with mixed results.[206] With the development of isolated pancreatic acini and acinar cells, natural purified glucagon was shown to stimulate amylase secretion and increase cyclic AMP in rat, mouse, and guinea pig acini.[218–220] However, the stimulatory principle did not elute with synthetic glucagon, so its nature is unknown. Most importantly, synthetic glucagon has no effect on amylase secretion by isolated acini.[220,221] Glucagon receptor mRNA has been identified in the pancreas but not in isolated acini or ducts.[222,223] Thus, these studies do not support a direct effect of glucagon on acinar cells. Whether the islets or the nervous system is involved is not yet established.

Somatostatin

Somatostatin (SS) was originally identified in hypothalamic extracts as a factor able to inhibit growth hormone release and was subsequently purified as a cyclic peptide containing 14 amino acids (SS-14).[224] A second form extended at the amino terminal and containing 28 amino acids (SS-28) was later characterized.[225] Subsequently, somatostatin was found to be widely distributed in the body including other regions of the brain, the small intestine, pancreatic islets, and the stomach. For information on the mRNA sequence and the somatostatin precursor, see reference [226]. Outside the nervous system, including islets, somatostatin is produced in D cells and is now considered to be both a hormone and a paracrine regulator and generally acts as an inhibitor of specific physiological processes such as gastrin secretion in the stomach and insulin secretion in the islet. The effect of somatostatin is mediated by specific receptors which most often act to inhibit adenylyl cyclase but can also regulate ion channels and activate tyrosine phosphatase.[227]

In all species studied, plasma somatostatin goes up after a meal; in humans, the basal concentration is about 10 pM and this doubles after eating.[228,229] The majority of released

SS comes from the gut and not the islets.[230] However, somatostatin released from islet D cells can have direct effects on other islet cells and pancreatic exocrine cells. Somatostatin release is controlled by cholinergic and adrenergic nerves as well as gastrointestinal hormones including CCK, secretin, and VIP.[226] The half-life of SS-14 in plasma is 1–2 minutes while that of SS-28 is 3–4 minutes.[231]

Pancreatic exocrine secretion of fluid and enzymes is inhibited by in vivo administration of SS-14 or SS-28 in conjunction with stimulation by meal feeding or administration of CCK and/or secretin in dogs,[232–237] rats,[238,239] cats,[240] rabbits,[241] and humans.[242–244] Additional studies using long-acting SS analogs such as Octreotide (SMS201–995) and RC160 have shown similar results.[245–248] That an effect of endogenous SS release on the pancreas may have physiological importance is suggested by the fact that exogenous SS mimicking postprandial levels is able to inhibit insulin and amylase secretion.[229] Moreover, immunoneutralization of circulating SS leads to an enhancement of CCK-stimulated pancreatic amylase secretion.[249]

The site at which SS inhibits exocrine pancreatic secretion and its physiological importance is unclear. Suggested sites include the central or peripheral nervous system, within the islet through effects on insulin secretion and at the level of pancreatic exocrine cells. It is of course possible that SS acts at more than one site. Because this review focuses on regulation of exocrine pancreas by islet peptides and because somatostatin receptors have been identified in the exocrine pancreas, that possibility will be considered first.

Using iodinated analogs of SS-14 and SS-28, high-affinity binding sites with the characteristics of receptors have been identified in isolated pancreatic acini, pancreatic membranes, and purified plasma membranes.[250–253] Covalent cross-linking studies showed that the binding protein was a 90-kDa glycoprotein.[252] Saturable binding sites were also demonstrated in the perfused rat pancreas with autoradiography showing uptake by islets and acinar cells with the acini showing the highest density.[31] Molecular cloning has identified six different SS receptors, SSRT1–SSRT5 with SSRT2 having two subtypes.[227] All of these are seven transmembrane proteins that are G protein-coupled whose action is mediated by G_i/G_o.[254] SSR2 and SSR5 appear to be the main SS receptor isoforms in acinar and islet cells, although further work is needed.[255,256]

Studies directed at actions of somatostatin on pancreatic exocrine cells have primarily been carried out using isolated pancreatic acini that have been separated from islets. SS-14, SS-28, and synthetic analogs have consistently been shown to inhibit cAMP formation through a pertussis toxin-sensitive mechanism and in most cases to inhibit amylase secretion stimulated by VIP or secretin that acts through cAMP.[257–261] However, most studies reported no inhibition by somatostatin of amylase release stimulated by CCK or cholinergic analogues. [258,259,262,263] Rather SS inhibited the potentiating effect of secretin or VIP on CCK stimulation.[261,264] This effect whose size is species-dependent may explain a portion of SS inhibition in vivo but it does not appear to be a major effect. Moreover, it is not clear if it involves islet SS or systemic SS.

The major in vitro technique that allows study of islet SS on the exocrine pancreas is the isolated perfused pancreas where secretion of both islet hormones and exocrine secretion can be measured. Insulin in the perfused pancreas potentiates CCK-stimulated exocrine secretion and somatostatin has been established to inhibit both exocrine secretion and insulin secretion.[31,265] In the perfused rat pancreas, high glucose is known to induce secretion of both insulin and SS, and a SS antagonist enhanced amylase secretion induced by glucose plus CCK.[32] Depleting SS with cysteamine pretreatment also enhanced CCK-stimulated secretion that could be inhibited with exogenous SS. Interestingly, endogenous SS did not modify exocrine secretion stimulated by CCK alone.[32,266] In the perfused pancreas, SS also appears to inhibit duct function as fluid secretion stimulated by secretin was potentiated by insulin and inhibited by somatostatin in a manner dependent on Na^+- K^+ ATPase function.[28]

The other proposed locus of somatostatin action to inhibit the exocrine pancreas is at the level of efferent neural control, which is initiated in the dorsal motor nucleus of the vagus (DMV) and travels through the vagus nerve to reach pancreatic parenchymal cells. Studies suggesting inhibition at a central vagal site include the finding in rats that somatostatin inhibits the action of 2-deoxyglucose, a known central vagal stimulator[267] and that microinjection of somatostatin into the DMV inhibits pancreatic secretion evoked by CCK or 2DG.[268] However, in dogs and rats, other investigators have reported that somatostatin still blocked CCK-stimulated secretion after complete pancreatic denervation.[269,270] These and other authors concluded that the inhibitory action of somatostatin is not dependent on the extrinsic nervous system but that the intrinsic nervous system could be involved.[263]

Pancreatic polypeptide

Pancreatic polypeptide (PP) was discovered as a contaminant in the purification of insulin in 1975.[271] It contains 36 amino acids and is part of a family of peptides that includes peptide YY (PYY) and neuropeptide Y (NPY) and has about 50 percent homology to these other peptides.[272] The biologically active part of the molecule is in the carboxyl terminal and this part of the molecule extends out from the main globular portion.[273] PP is immunogenic and immunohistochemistry has localized it in the islets to a specific cell type originally known as F cells, and now referred to as PP cells.[274] These cells are most abundant in the original

ventral lobe of the pancreas and some are present outside the islet in the duct epithelium. Small amounts of PP are found in other tissues including the brain.[275]

In humans, fasting pancreatic polypeptide plasma levels are 10–30 pM; they increase rapidly after feeding and remain elevated for 4–5 hours. The vagal nerve is the main stimulator of PP secretion and secretion can be blocked with atropine. Electrical stimulation of the vagus, sham feeding, 2-deoxyglucose, and insulin-induced hypoglycemia all stimulate PP secretion in a vagal-dependent manner.[276,277] The half-life of PP in plasma is about 6 minutes. Reported actions of PP include effects on the GI tract, metabolism, most notably reversal of hepatic insulin resistance, and as a satiety factor.[277] The latter is based in part on Prader-Willi syndrome where loss of PP secretion is associated with obesity.[278] The actions of PP are mediated by specific receptors that belong to a family of receptors that bind NPY, PYY, and PP.[279] The various receptors are denoted by a capital Y with a numerical subscript. The Y_4 receptor has specificity for PP with a high affinity and a hundred-fold lower affinity for PYY.

Purified bovine PP was shown by Lin et al. in 1977 to reduce exocrine pancreatic secretion in dogs with pancreatic fistulas.[280] This was confirmed by Taylor et al. who showed that inhibition of exocrine secretion occurred at doses of porcine PP that raised plasma levels less than seen after a meat meal.[281] Similar actions in dogs have been reported by others.[282–284] Shiratori et al. showed a similar effect of human synthetic PP to inhibit canine pancreatic secretion and also showed that immunoneutralization of endogenous PP enhanced both interdigestive and postprandial secretion.[285] Similar effects of exogenous PP have been reported in humans[286,287] and in rats.[288,289]

Although PP acts to inhibit pancreatic secretion in vivo, this effect appears to be indirect as exogenous PP had no effect on amylase release from isolated rat or mouse pancreatic acini,[288,290] the perfused cat pancreas,[291] or incubated uncinate pancreas of young rats or pancreatic fragments.[288,291] Binding studies with ^{125}I-PP also failed to reveal high-affinity binding sites on rat pancreatic acini. Although this lack of in vitro effects is generally accepted,[272] there are several differing reports all using isolated rat pancreatic acini showing a small amount of inhibition,[292] stimulation by high concentrations of human PP,[293] and inhibition of carbachol but not CCK stimulation by bovine PP.[294] Some of these effects could possibly have been due to contaminants in purified PP.

More recent studies have focused on a neural locus for the action of PP to inhibit pancreatic exocrine secretion. Most studies indicate a site of action in the brain stem. Receptors for PP are present in the area postrema, the nucleus tractus solitarius (NTS), and the dorsal motor nucleus of the vagus,[295,296] and intravenous PP inhibits pancreatic enzyme secretion in the rat stimulated with

2-deoxyglucose which acts centrally.[289] More definitively, PP microinjected into the DMV in a site-specific manner inhibited pancreatic secretion.[297] PP directly spritzed on individual DMV neurons revealed a subset where PP reduced postsynaptic currents.[298] This finding suggest that PP in the circulation gains access to the brain through the area postrema and reaches the adjacent DMV where it inhibits vagal excitatory output to the pancreas.[299] As an alternative neural site for PP inhibition, Jung et al. presented data that rat PP inhibited potassium-stimulated amylase release and the presynaptic release of acetylcholine in rat pancreatic slices.[300] They suggested that PP acts on postganglionic cholinergic neurons to prevent acetylcholine release. The function of inhibition of pancreatic secretion by PP is unclear, although several authors have suggested it is to prevent overstimulation of the pancreas.

Adrenomedullin

Adrenomedullin (AM) is a 52-amino acid peptide that was isolated from a human pheochromocytoma and has slight primary homology but stronger tertiary resemblance to calcitonin gene-related peptide (CGRP). When injected intravenously, AM causes a potent and long-lasting hypotensive effect in anesthetized rats.[301,302] AM is present in various organs but most abundantly in the adrenal medulla. It is present diffusely in endocrine cells of the pancreatic islet but most abundantly colocalizes with PP cells where it is present in secretory granules.[303] These authors also showed that it is a strong inhibitor of insulin secretion both in vivo and with isolated islets. The AM receptor is made up of a previously orphan receptor, the calcitonin receptor-like receptor (CTL), combined with a receptor-activity-modifying protein (RAMP).[304] There are two AM receptors: AM1 which includes CTLR and RAMP2 and AM2 which is made up of CTLR and RAMP3; the CGRP receptor is CTLR plus RAMP1.[305] In most cells that respond to AM, there is an increase in cAMP.[306] In aortic endothelial cells, AM stimulates two signal transduction pathways that increase cAMP and mobilize Ca^{2+}.[307]

Because AM released from PP cells might affect the exocrine pancreas via local blood flow, Tsuchida et al. studied the effect of AM on pancreatic acini.[308] Radioiodinated AM was shown to bind to pancreatic acini in a manner inhibited by picomolar concentrations of AM but not by CGRP. AM at concentrations of 1 pM to 1 nM inhibited CCK-stimulated amylase secretion. AM did not affect CCK binding to acini. In contrast to some other cell types, AM did not affect intracellular cAMP or the rise in Ca^{2+} induced by CCK. Rather, the effect of AM was blocked by pertussis toxin (PTX) suggesting the participation of a PTX-sensitive G protein in the action of AM. Additional work on this effect of AM would seem warranted. More recent work on AM in the pancreas has focused on AM

produced by pancreatic cancer cells as a mediator of diabetes that often associates with pancreatic cancer, possibly by inhibiting insulin secretion.[309]

Amylin

Amylin or islet amyloid polypeptide (IAP) was isolated from amyloid-rich pancreatic extracts of type 2 diabetic pancreas and shown to be a 37-amino acid peptide that can form oligomers, fibrils, and amyloid deposits.[310] The amyloid deposits are injurious to the pancreas and are also seen in the brain where they resemble the plaques of Alzheimer's disease.[311] Amyloid is a term for protein aggregation state in which the proteins form a β-sheet. The amyloid deposits are injurious to the pancreas and are also seen in the brain of type 2 diabetics.[312] Although amylin is not homologous to the Alzheimer Aβ protein, both can interact with the Amylin receptor which is widely distributed in the brain.

Amylin is produced by islet beta cells, has been localized to the secretory granules,[313] and is secreted along with insulin at a ratio of 1:10 to 1:100.[312] Amylin in vivo can inhibit glucagon secretion that occurs at low glucose concentrations and inhibits glucagon secretion from isolated mouse islets.[314] The amylin receptor was identified as a heterodimer of a calcitonin seven transmembrane receptor and a single transmembrane receptor activity-modifying protein or RAMP.[315] These receptors are widely distributed in the brain. Amylin receptor antagonist increased glucagon secretion.[316] The acute effects on insulin secretion are more complex as amylin at low concentrations enhanced insulin release while high concentrations inhibited insulin release from mouse islets.[314]

A few reports have evaluated the effect of amylin on the exocrine pancreas. In one report, amylin had a small effect to increase basal pancreatic secretion in conscious rats.[317] In a study in anesthetized rats, Young et al. found no effect on basal secretion but a dose-dependent inhibition of CCK-stimulated fluid, amylase, and lipase secretion up to 67 percent.[318] This effect was seen at low concentrations of amylin (ED50 was 0.11 μg) comparable to effects reported on gastric emptying. However, multiple studies have shown no effect of amylin on secretion by isolated pancreatic acini or AR42J cells.[318–320] Thus, there does not appear to be a direct effect of amylin on pancreatic exocrine cells and the in vivo effects may be on the CNS. There is also no information available on whether amylin receptors exist in the exocrine pancreas.

Pancreastatin

Pancreastatin (PST) was first isolated from pig pancreas as a result of screening for peptides with a C-terminal amide structure and shown to inhibit stimulated insulin release.[321]

While porcine PST is a 49-amino acid peptide, the biological activity resides in the carboxyl half and requires the C-terminal glycine-amide. PST is believed to be derived from chromogranin A as the result of further processing by prohormone convertase 1. The 52-amino acid sequence predicted by the gene structure of human chromogranin A is homologous to porcine PST and when synthesized the full-length peptide and its 29-amino acid carboxy terminal had similar activity to porcine PST to inhibit insulin release.[322] Subsequent studies confirmed the inhibition of insulin secretion in vivo, in the perfused pancreas and in isolated pancreatic islets. For reviews, see references [323, 324]. By immunocytochemistry, pancreastatin was shown to be present in the pituitary, adrenal glands, pancreas, and throughout the gastrointestinal tract in cells known to contain chromogranin. In the pancreas, PST colocalizes with all four major endocrine cell types and elsewhere is found in neuroendocrine cells.[324,325] PST can be measured in plasma by RIA and may serve as a predictive marker for neuroendocrine tumors where PST is elevated.[326,327]

Along with the inhibition of insulin secretion, multiple natural and synthetic forms of PST (human, porcine, bovine, rat) have been shown to inhibit pancreatic digestive enzyme secretion stimulated by CCK in vivo in rats[322,328,329] and dogs.[330] PST also inhibited pancreatic exocrine secretion stimulated by 2-deoxy-D-Glucose, a central vagal activator.[331] By contrast to CCK, pancreatic exocrine secretion stimulated by the cholinergic analog bethanechol in vivo was not blocked by PST. Studies of isolated rat pancreatic acini showed no inhibitory action of PST on CCK stimulation,[328–332] although there is one report of inhibition of CCK stimulation in guinea pig acini.[333] Because PST inhibited exocrine pancreatic secretion in vivo but not on isolated cells, the effect of PST is presumed to be indirect.

One possible site of action is on pancreatic blood flow as shown in anesthetized rats using the hydrogen clearance method.[334] These investigators found that caerulein enhanced pancreatic blood flow and this increase was dose-dependently inhibited by PST at 100–500 pmol/kg/h. However, another study in anesthetized dogs showed no effect of PST at similar concentrations using the laser Doppler flowmeter method, although protein and amylase secretion were inhibited.[330] The other possible site is on the neural stimulatory pathway. Herzig et al. showed that when pancreatic lobules which contained neural elements were incubated in vitro, high K$^+$ concentrations stimulated the release of acetylcholine and amylase secretion—both of which were partially inhibited by PST.[329] Further studies are necessary to establish pancreastatin as a regulatory peptide and to follow up these potential mechanisms of action. It is also not clear whether PST from a pancreatic or systemic source plays a physiological role in the regulation of exocrine pancreatic function. Absent at present is any information about a PST receptor molecule or PST-sensitive ion

channel. Moreover, it is not clear what is a physiological concentration of PST. Finally, there is little new information carried out in the last twenty years beyond using PST levels as a tumor marker.

Galanin

Galanin is a 29-amino acid peptide isolated from porcine intestine in 1983 and named for its amino terminal glycine and carboxyl terminal alanine.[335] It is considered to be a neuropeptide and is found throughout the body with the highest concentrations in the brain, spinal cord, and enteric nervous system.[336] Galanin is often coexpressed with other peptides or neurotransmitters, including norepinephrine, serotonin, GABA, and acetylcholine. Galanin acts via three major G protein-coupled receptors, GalR1, GalR2, and GalR3.[337] In the pancreas, galanin has been localized mainly to nerves present in both the exocrine and endocrine components but also in some studies to islet cells where colocalization with insulin has been observed.[338–340] In dogs, rodents, and humans, galanin suppresses insulin release both in vivo and in vitro.[337]

A number of studies have evaluated the effect of galanin on isolated rodent acini and the results are presented in the review by Barreto.[336] Most show no effect or inhibition. In a study of the isolated, perfused rat pancreas, galanin at low concentrations enhanced both insulin secretion and CCK-stimulated amylase secretion.[341] On the other hand, using isolated mouse pancreas lobules, galanin had no effect on carbachol stimulation but inhibited CCK-stimulated release suggesting an effect on cholinergic transmitter release.[342] Galanin also inhibited acetylcholine release stimulated by veratridine in rat pancreatic lobules.[343] Overall, these studies suggest galanin is more likely to affect the neural control of exocrine secretion rather than directly affecting acinar cells. Galanin could also affect amylase secretion by blocking somatostatin release.[336]

References

1. Williams JA, Goldfine ID. The insulin-pancreatic acinar axis. Diabetes 1985; 34:980–986.
2. Barreto SG, Carati CJ, Toouli J, Saccone GT. The islet-acinar axis of the pancreas: more than just insulin. Am J Physiol Gastrointest Liver Physiol 2010; 299:G10–G22.
3. Shiratori K, Shimizu K. Insulo-acinar relationship. In: Beger HG, Warshaw AL, Hruban RH, Buchler MW, Lerch MM, Neoptolemos JP, Shimosegawa T, Whitcomb DC, eds. The pancreas: an integrated textbook of basic science, medicine, and Surgery. 3rd ed. Hoboken, NJ: John Wiley & Sons Ltd, 2018:123–131.
4. Studley JG, Mathie RT, Blumgart LH. Blood flow measurement in the canine pancreas. J Surg Res 1987; 42:101–115.
5. McCuskey RS, Chapman TM. Microscopy of the living pancreas in situ. Am J Anat 1969; 126:395–407.

6. Henderson JR, Daniel PM, Fraser PA. The pancreas as a single organ: the influence of the endocrine upon the exocrine part of the gland. Gut 1981; 22:158–167.

7. El-Gohary Y, Gittes G. Structure of islets and vascular relationship to the exocrine pancreas. Pancreapedia: Exocrine Pancreas Knowledge Base, 2017. DOI: 10.3998/panc.2017.10.

8. Jansson L, Barbu A, Bodin B, Drott CJ, Espes D, Gao X, Grapensparr L, Kallskog O, Lau J, Liljeback H, Palm F, Quach M, et al. Pancreatic islet blood flow and its measurement. Ups J Med Sci 2016; 121:81–95.

9. Lifson N, Lassa CV, Dixit PK. Relation between blood flow and morphology in islet organ of rat pancreas. Am J Physiol Endocrinol Metab 1985; 249:E43–E48.

10. Doi R, Hosotani R, Inoue K, Kogire M, Sumi S, Fujii N, Yajima H, Rayford PL, Tobe T. Effect of synthetic human cholecystokinin-33 on pancreatic blood flow in dogs. Pancreas 1990; 5:615–620.

11. Jansson L. Vasoactive intestinal polypeptide increases whole pancreatic blood flow but does not affect islet blood flow in the rat. Acta Diabetol 1994; 31:103–106.

12. Jansson L, Hellerstrom C. Stimulation by glucose of the blood flow to the pancreatic islets of the rat. Diabetologia 1983; 25:45–50.

13. Lifson N, Kramlinger KG, Mayrand RR, Lender EJ. Blood flow to the rabbit pancreas with special reference to the islets of Langerhans. Gastroenterology 1980; 79:466–473.

14. Liu YM, Guth PH, Kaneko K, Livingston EH, Brunicardi FC. Dynamic in vivo observation of rat islet microcirculation. Pancreas 1993; 8:15–21.

15. Brunicardi FC, Stagner J, Bonner-Weir S, Wayland H, Kleinman R, Livingston E, Guth P, Menger M, McCuskey R, Intaglietta M, Charles A, Ashley S, et al. Microcirculation of the islets of Langerhans. Long Beach Veterans Administration Regional Medical Education Center Symposium. Diabetes 1996; 45:385–392.

16. Henderson JR, Moss MC. A morphometric study of the endocrine and exocrine capillaries of the pancreas. Q J Exp Physiol 1985; 70:347–356.

17. Vetterlein F, Petho A, Schmidt G. Morphometric investigation of the microvascular system of pancreatic exocrine and endocrine tissue in the rat. Microvasc Res 1987; 34:231–238.

18. Henderson JR, Daniel PM. A comparative study of the portal vessels connecting the endocrine and exocrine pancreas, with a discussion of some functional implications. Q J Exp Physiol Cogn Med Sci 1979; 64:267–275.

19. Fujita T. Insulo-acinar portal system in the horse pancreas. Arch Histol Jpn 1973; 35:161–171.

20. Fujita T, Murakami T. Microcirculation of monkey pancreas with special reference to the insulo-acinar portal system. A scanning electron microscope study of vascular casts. Arch Histol Jpn 1973; 35:255–263.

21. Ohtani O, Ushiki T, Kanazawa H, Fujita T. Microcirculation of the pancreas in the rat and rabbit with special reference to the insulo-acinar portal system and emissary vein of the islet. Arch Histol Jpn 1986; 49:45–60.

22. Murakami T, Fujita T, Taguchi T, Nonaka Y, Orita K. The blood vascular bed of the human pancreas, with special reference to the insulo-acinar portal system. Scanning electron microscopy of corrosion casts. Arch Histol Cytol 1992; 55:381–395.

23. Zhou ZG, Gao XH, Wayand WU, Xiao LJ, Du Y. Pancreatic microcirculation in the monkey with special reference to the blood drainage system of Langerhans islets: light and scanning electron microscopic study. Clin Anat 1996; 9:1–9.

24. Bonner-Weir S, Orci L. New perspectives on the microvasculature of the islets of Langerhans in the rat. Diabetes 1982; 31:883–889.

25. Miyake T, Murakami T, Ohtsuka A. Incomplete vascular casting for a scanning electron microscope study of the microcirculatory patterns in the rat pancreas. Arch Histol Cytol 1992; 55:397–406.

26. Lifson N, Lassa CV. Note on the blood supply of the ducts of the rabbit pancreas. Microvasc Res 1981; 22:171–176.

27. Kanno T, Saito A. The potentiating influences of insulin on pancreozymin-induced hyperpolarization and amylase release in the pancreatic acinar cell. J Physiol 1976; 261:505–521.

28. Hasegawa H, Okabayashi Y, Koide M, Kido Y, Okutani T, Matsushita K, Otsuki M, Kasuga M. Effect of islet hormones on secretin-stimulated exocrine secretion in isolated perfused rat pancreas. Dig Dis Sci 1993; 38:1278–1283.

29. Saito A, Williams JA, Kanno T. Potentiation by insulin of the acetylcholine-induced secretory response of the perfused rat pancreas. Biomedical Research 1980; 1:101–103.

30. Saito A, Williams JA, Kanno T. Potentiation of cholecystokinin-induced exocrine secretion by both exogenous and endogenous insulin in isolated and perfused rat pancreata. J Clin Invest 1980; 65:777–782.

31. Garry DJ, Garry MG, Williams JA, Mahoney WC, Sorenson RL. Effects of islet hormones on amylase secretion and localization of somatostatin binding sites. Am J Physiol Gastrointest Liver Physiol 1989; 256:G897–G904.

32. Park HS, Yoon HS, Park YD, Cui ZY, Lee YL, Park HJ. Endogenous somatostatin inhibits interaction of insulin and cholecystokinin on exocrine secretion of isolated, perfused rat pancreas. Pancreas 2002; 24:373–379.

33. Nakagawa A, Stagner JI, Samols E. In situ binding of islet hormones in the isolated perfused rat pancreas: evidence for local high concentrations of islet hormones via the islet-acinar axis. Diabetologia 1995; 38:262–268.

34. Nakagawa A, Samols E, Stagner JI. Exocrine interstitial insulin and somatostatin in the perfused dog pancreas. Am J Physiol Gastrointest Liver Physiol 1993; 264:G728–G734.

35. Cecil RL. A study of the pathological anatomy of the pancreas in ninety cases of diabetes mellitus. J Exp Med 1909; 11:266–290.

36. Gepts W. Pathologic anatomy of the pancreas in juvenile diabetes mellitus. Diabetes 1965; 14:619–633.

37. Lohr M, Kloppel G. Residual insulin positivity and pancreatic atrophy in relation to duration of chronic type 1 (insulin-dependent) diabetes mellitus and microangiopathy. Diabetologia 1987; 30:757–762.

38. Goda K, Sasaki E, Nagata K, Fukai M, Ohsawa N, Hahafusa T. Pancreatic volume in type 1 and type 2 diabetes mellitus. Acta Diabetol 2001; 38:145–149.

39. Williams AJ, Chau W, Callaway MP, Dayan CM. Magnetic resonance imaging: a reliable method for measuring pancreatic volume in type 1 diabetes. Diabet Med 2007; 24:35–40.

40. Lim S, Bae JH, Chun EJ, Kim H, Kim SY, Kim KM, Choi SH, Park KS, Florez JC, Jang HC. Differences in pancreatic volume, fat content, and fat density measured by multidetector-row computed tomography according to the duration of diabetes. Acta Diabetol 2014; 51:739–748.

41. Philippe MF, Benabadji S, Barbot-Trystram L, Vadrot D, Boitard C, Larger E. Pancreatic volume and endocrine and exocrine functions in patients with diabetes. Pancreas 2011; 40:359–363.

42. Macauley M, Percival K, Thelwall PE, Hollingsworth KG, Taylor R. Altered volume, morphology and composition of the pancreas in type 2 diabetes. PLoS One 2015; 10:e0126825.

43. Garcia TS, Rech TH, Leitao CB. Pancreatic size and fat content in diabetes: a systematic review and meta-analysis of imaging studies. PLoS One 2017; 12:e0180911.

44. Ewald N, Hardt PD. Alterations in exocrine pancreatic function in diabetes mellitus. Pancreapedia: Exocrine Pancreas Knowledge Base, 2015. DOI: 10.3998/panc.2015.7.

45. Lankisch PG, Manthey G, Otto J, Koop H, Talaulicar M, Willms B, Creutzfeldt W. Exocrine pancreatic function in insulin-dependent diabetes mellitus. Digestion 1982; 25:211–216.

46. Frier BM, Saunders JH, Wormsley KG, Bouchier IA. Exocrine pancreatic function in juvenile-onset diabetes mellitus. Gut 1976; 17:685–691.

47. Hardt PD, Hauenschild A, Nalop J, Marzeion AM, Jaeger C, Teichmann J, Bretzel RG, Hollenhorst M, Kloer HU, Group SSS. High prevalence of exocrine pancreatic insufficiency in diabetes mellitus. A multicenter study screening fecal elastase 1 concentrations in 1,021 diabetic patients. Pancreatology 2003; 3:395–402.

48. Larger E, Philippe MF, Barbot-Trystram L, Radu A, Rotariu M, Nobecourt E, Boitard C. Pancreatic exocrine function in patients with diabetes. Diabet Med 2012; 29:1047–1054.

49. Ewald N, Raspe A, Kaufmann C, Bretzel RG, Kloer HU, Hardt PD. Determinants of exocrine pancreatic function as measured by fecal elastase-1 concentrations (FEC) in patients with diabetes mellitus. Eur J Med Res 2009; 14:118–122.

50. Nunes AC, Pontes JM, Rosa A, Gomes L, Carvalheiro M, Freitas D. Screening for pancreatic exocrine insufficiency in patients with diabetes mellitus. Am J Gastroenterol 2003; 98:2672–2675.

51. Terzin V, Varkonyi T, Szabolcs A, Lengyel C, Takacs T, Zsori G, Stajer A, Palko A, Wittmann T, Palinkas A, Czako L. Prevalence of exocrine pancreatic insufficiency in type 2 diabetes mellitus with poor glycemic control. Pancreatology 2014; 14:356–360.

52. Vujasinovic M, Zaletel J, Tepes B, Popic B, Makuc J, Epsek Lenart M, Predikaka M, Rudolf S. Low prevalence of exocrine pancreatic insufficiency in patients with diabetes mellitus. Pancreatology 2013; 13:343–346.

53. Hardt PD, Hauenschild A, Jaeger C, Teichmann J, Bretzel RG, Kloer HU, Group SSS. High prevalence of steatorrhea in 101 diabetic patients likely to suffer from exocrine

pancreatic insufficiency according to low fecal elastase 1 concentrations: a prospective multicenter study. Dig Dis Sci 2003; 48:1688–1692.

54. Cavalot F, Bonomo K, Fiora E, Bacillo E, Salacone P, Chirio M, Gaia E, Trovati M. Does pancreatic elastase-1 in stools predict steatorrhea in type 1 diabetes? Diabetes Care 2006; 29:719–721.

55. Chakraborty PP, Chowdhury S. A look inside the pancreas: The "endocrine-exocrine cross-talk." Endocrinol Metab Synd 2015; 4:1–4.

56. Mohapatra S, Majumder S, Smyrk TC, Zhang L, Matveyenko A, Kudva YC, Chari ST. Diabetes mellitus is associated with an exocrine pancreatopathy: conclusions from a review of literature. Pancreas 2016; 45:1104–1110.

57. Korc M, Sankaran H, Wong KY, Williams JA, Goldfine ID. Insulin receptors in isolated mouse pancreatic acini. Biochem Biophys Res Commun 1978; 84:293–299.

58. Sankaran H, Iwamoto Y, Korc M, Williams JA, Goldfine ID. Insulin action in pancreatic acini from streptozotocin-treated rats. II. Binding of 125I-insulin to receptors. Am J Physiol Gastrointest Liver Physiol 1981; 240:G63–G68.

59. Sjodin L, Holmberg K, Lyden A. Insulin receptors on pancreatic acinar cells in guinea pigs. Endocrinology 1984; 115:1102–1109.

60. Korc M, Iwamoto Y, Sankaran H, Williams JA, Goldfine ID. Insulin action in pancreatic acini from streptozotocin-treated rats. I. Stimulation of protein synthesis. Am J Physiol Gastrointest Liver Physiol 1981; 240:G56–G62.

61. Williams JA, Bailey A, Humbel R, Goldfine ID. Insulinlike growth factors bind to specific receptors in isolated pancreatic acini. Am J Physiol Gastrointest Liver Physiol 1984; 246:G96–G99.

62. Bergeron JJ, Rachubinski R, Searle N, Sikstrom R, Borts D, Bastian P, Posner BI. Radioautographic visualization of in vivo insulin binding to the exocrine pancreas. Endocrinology 1980; 107:1069–1080.

63. Cruz J, Posner BI, Bergeron JJ. Receptor-mediated endocytosis of [125I]insulin into pancreatic acinar cells in vivo. Endocrinology 1984; 115:1996–2008.

64. Sakamoto C, Williams JA, Roach E, Goldfine ID. In vivo localization of insulin binding to cells of the rat pancreas. Proc Soc Exp Biol Med 1984; 175:497–502.

65. Goldfine ID, Kriz BM, Wong KY, Hradek G, Jones AL, Williams JA. Insulin action in pancreatic acini from streptozotocin-treated rats. III. Electron microscope autoradiography of 125I-insulin. Am J Physiol Gastrointest Liver Physiol 1981; 240:G69–G75.

66. Williams JA, Goldfine ID. The insulin-acinar relationship. In: Go VLW, ed. The exocrine pancreas: biology, pathobiology, and diseases. New York: Raven Press, 1986:347–360.

67. Mossner J, Logsdon CD, Goldfine ID, Williams JA. Regulation of pancreatic acinar cell insulin receptors by insulin. Am J Physiol Gastrointest Liver Physiol 1984; 247:G155–G160.

68. Okabayashi Y, Maddux BA, McDonald AR, Logsdon CD, Williams JA, Goldfine ID. Mechanisms of insulin-induced insulin-receptor downregulation. Decrease of receptor biosynthesis and mRNA levels. Diabetes 1989; 38:182–187.

69. Okabayashi Y, Otsuki M, Nakamura T, Koide M, Hasegawa H, Okutani T, Kido Y. Regulatory effect of cholecystokinin on subsequent insulin binding to pancreatic acini. Am J Physiol Endocrinol Metab 1990; 258:E562–E568.

70. Williams JA, Bailey AC, Preissler M, Goldfine ID. Insulin regulation of sugar transport in isolated pancreatic acini from diabetic mice. Diabetes 1982; 31:674–682.

71. Burnham DB, Williams JA. Effects of carbachol, cholecystokinin, and insulin on protein phosphorylation in isolated pancreatic acini. J Biol Chem 1982; 257:10523–10528.

72. Sung CK, Williams JA. Insulin and ribosomal protein S6 kinase in rat pancreatic acini. Diabetes 1989; 38:544–549.

73. Goldfine ID, Williams JA. Receptors for insulin and CCK in the acinar pancreas: relationship to hormone action. Int Rev Cytol 1983; 85:1–38.

74. Sans MD, Amin RK, Vogel NL, D'Alecy LG, Kahn RC, Williams JA. Specific deletion of insulin receptors on pancreatic acinar cells defines the insulin-acinar axis: implications for pancreatic insufficiency in diabetes. Gastroenterology 2011; 140 Suppl 1:A-233.

75. Williams JA. Amylase. Pancreapedia: Exocrine Pancreas Knowledge Base, 2019. DOI: 10.3998/panc.2019.02.

76. Adler G, Kern HF. Regulation of exocrine pancreatic secretory process by insulin in vivo. Horm Metab Res 1975; 7:290–296.

77. Bazin R, Lavau M. Diet composition and insulin effect on amylase to lipase ratio in pancreas of diabetic rats. Digestion 1979; 19:386–391.

78. Ben Abdeljlil A, Palla JC, Desnuelle P. Effect of insulin on pancreatic amylase and chymotrypsinogen. Biochem Biophys Res Commun 1965; 18:71–75.

79. Snook JT. Effect of diet, adrenalectomy, diabetes, and actinomycin D on exocrine pancreas. Am J Physiol 1968; 215:1329–1333.

80. Soling HD, Unger KO. The role of insulin in the regulation of α-amylase synthesis in the rat pancreas. Eur J Clin Invest 1972; 2:199–212.

81. Otsuki M, Williams JA. Effect of diabetes mellitus on the regulation of enzyme secretion by isolated rat pancreatic acini. J Clin Invest 1982; 70:148–156.

82. Duan RD, Erlanson-Albertsson C. Pancreatic lipase and colipase activity increase in pancreatic acinar tissue of diabetic rats. Pancreas 1989; 4:329–334.

83. Trimble ER, Bruzzone R, Belin D. Insulin resistance is accompanied by impairment of amylase-gene expression in the exocrine pancreas of the obese Zucker rat. Biochem J 1986; 237:807–812.

84. Mossner J, Logsdon CD, Williams JA, Goldfine ID. Insulin, via its own receptor, regulates growth and amylase synthesis in pancreatic acinar AR42J cells. Diabetes 1985; 34:891–897.

85. Korc M, Owerbach D, Quinto C, Rutter WJ. Pancreatic islet-acinar cell interaction: amylase messenger RNA levels are determined by insulin. Science 1981; 213:351–353.

86. Duan RD, Poensgen J, Wicker C, Westrom B, Erlanson-Albertsson C. Increase in pancreatic lipase and trypsin activity and their mRNA levels in streptozotocin-induced diabetic rats. Dig Dis Sci 1989; 34:1243–1248.

87. Tsai A, Cowan MR, Johnson DG, Brannon PM. Regulation of pancreatic amylase and lipase gene expression by diet and insulin in diabetic rats. Am J Physiol Gastrointest Liver Physiol 1994; 267:G575–G583.

88. Keller SA, Rosenberg MP, Johnson TM, Howard G, Meisler MH. Regulation of amylase gene expression in diabetic mice is mediated by a cis-acting upstream element close to the pancreas-specific enhancer. Genes Dev 1990; 4:1316–1321.

89. Johnson TM, Rosenberg MP, Meisler MH. An insulin-responsive element in the pancreatic enhancer of the amylase gene. J Biol Chem 1993; 268:464–468.

90. Dranginis A, Morley M, Nesbitt M, Rosenblum BB, Meisler MH. Independent regulation of nonallelic pancreatic amylase genes in diabetic mice. J Biol Chem 1984; 259:12216–12219.

91. Danielsson A, Sehlin J. Transport and oxidation of amino acids and glucose in the isolated exocrine mouse pancreas: effects of insulin and pancreozymin. Acta Physiol Scand 1974; 91:557–565.

92. Mailliard ME, Stevens BR, Mann GE. Amino acid transport by small intestinal, hepatic, and pancreatic epithelia. Gastroenterology 1995; 108:888–910.

93. Rooman I, Lutz C, Pinho AV, Huggel K, Reding T, Lahoutte T, Verrey F, Graf R, Camargo SM. Amino acid transporters expression in acinar cells is changed during acute pancreatitis. Pancreatology 2013; 13:475–485.

94. Mann GE, Norman PS. Regulatory effects of insulin and experimental diabetes on neutral amino acid transport in the perfused rat exocrine pancreas. Kinetics of unidirectional L-serine influx and efflux at the basolateral plasma membrane. Biochim Biophys Acta 1984; 778:618–622.

95. Norman PS, Habara Y, Mann GE. Paradoxical effects of endogenous and exogenous insulin on amino acid transport activity in the isolated rat pancreas: somatostatin-14 inhibits insulin action. Diabetologia 1989; 32:177–184.

96. Munoz M, Sweiry JH, Mann GE. Insulin stimulates cationic amino acid transport activity in the isolated perfused rat pancreas. Exp Physiol 1995; 80:745–753.

97. Mann GE, Habara Y, Peran S. Characteristics of L-glutamine transport in the perfused rat exocrine pancreas: lack of sensitivity to insulin and streptozotocin-induced experimental diabetes. Pancreas 1986; 1:239–245.

98. Jefferson LS. Lilly Lecture 1979: role of insulin in the regulation of protein synthesis. Diabetes 1980; 29:487–496.

99. Couture Y, Dunnigan J, Morisset. Stimulation of pancreatic amylase secretion and protein synthesis by insulin. Scand J Gastroenterol 1972; 7:257–263.

100. Kramer MF, Poort C. The effect of insulin on amino acid incorporation into exocrine pancreatic cells of the rat. Horm Metab Res 1975; 7:389–393.

101. Duan RD, Wicker C, Erlanson-Albertsson C. Effect of insulin administration on contents, secretion, and synthesis of pancreatic lipase and colipase in rats. Pancreas 1991; 6:595–602.

102. Aughsteen AA, Kataoka K. Quantitative radioautographic study on 3H-leucine uptake of peri- and tele-insular acinar cells of the pancreas in normal and streptozotocin-diabetic mice. Acta Histochem. Cytochem. 1994; 27:67–74.

103. Trimble ER, Rausch U, Kern HF. Changes in individual rates of pancreatic enzyme and isoenzyme biosynthesis in the obese Zucker rat. Biochem J 1987; 248:771–777.

104. Yagihashi S. Exocrine pancreas of streptozotocin-diabetes rats treated with insulin. Tohoku J Exp Med 1976; 120:31–42.

105. Sans MD. Measurement of pancreatic protein synthesis. Pancreapedia: Exocrine Pancreas Knowledge base, 2010. DOI: 10.3998/panc.2010.16.

106. Korc M, Bailcy AC, Williams JA. Regulation of protein synthesis in normal and diabetic rat pancreas by cholecystokinin. Am J Physiol Gastrointest Liver Physiol 1981; 241:G116–G121.

107. Okabayashi Y, Moessner J, Logsdon CD, Goldfine ID, Williams JA. Insulin and other stimulants have nonparallel translational effects on protein synthesis. Diabetes 1987; 36:1054–1060.

108. Lahaie RG. Translational control of protein synthesis in isolated pancreatic acini: role of CCK8, carbachol, and insulin. Pancreas 1986; 1:403–410.

109. Sans MD, Williams JA. The mTOR signaling pathway and regulation of pancreatic function. Pancreapedia: Exocrine Pancreas Knowledge Base, 2017. DOI: 10.3998/panc.2017.08.

110. Williams JA. Cholecystokinin (CCK) regulation of pancreatic acinar cells: physiological actions and signal transduction mechanisms. Compr Physiol 2019; 9:535–564.

111. Berna MJ, Tapia JA, Sancho V, Thill M, Pace A, Hoffmann KM, Gonzalez-Fernandez L, Jensen RT. Gastrointestinal growth factors and hormones have divergent effects on Akt activation. Cell Signal 2009; 21:622–638.

112. Sofrankova A, Dockray GJ. Cholecystokinin- and secretin-induced pancreatic secretion in normal and diabetic rats. Am J Physiol Gastrointest Liver Physiol 1983; 244:G370–G374.

113. Patel R, Singh J, Yago MD, Vilchez JR, Martinez-Victoria E, Manas M. Effect of insulin on exocrine pancreatic secretion in healthy and diabetic anaesthetised rats. Mol Cell Biochem 2004; 261:105–110.

114. Berg T, Johansen L, Brekke IB. Insulin potentiates cholecystokinin (CCK)-induced secretion of pancreatic kallikrein. Acta Physiol Scand 1985; 123:89–95.

115. Dodi G, Militello C, Pedrazzoli S, Zannini G, Lise M. Exocrine pancreatic function in diabetic rats treated with intraportal islet transplantation. Eur Surg Res 1984; 16:9–14.

116. Lee PC, Jordan M, Pieper GM, Roza AM. Normalization of pancreatic exocrine enzymes by islet transplantation in diabetic rats. Biochem Cell Biol 1995; 73:269–273.

117. Pierzynowski S, Barej W. The dependence of exocrine pancreatic secretion on insulin in sheep. Q J Exp Physiol 1984; 69:35–39.

118. Ferrer R, Medrano J, Diego M, Calpena R, Graells L, Molto M, Perez T, Perez F, Salido G. Effect of exogenous insulin and glucagon on exocrine pancreatic secretion in rats in vivo. Int J Pancreatol 2000; 28:67–75.

119. Sofrankova A. Effect of exogenous and endogenous insulin on the secretory response of the pancreas to the octapeptide of cholecystokinin (CCK8) in normal rats. Physiol Bohemoslov 1984; 33:391–398.

120. Lee KY, Zhou L, Ren XS, Chang TM, Chey WY. An important role of endogenous insulin on exocrine pancreatic

secretion in rats. Am J Physiol Gastrointest Liver Physiol 1990; 258:G268–G274.

121. Berry SM, Fink AS. Insulin inhibits secretin-stimulated pancreatic bicarbonate output by a dose-dependent neurally mediated mechanism. Am J Physiol Gastrointest Liver Physiol 1996; 270:G163–G170.

122. Singh J. Mechanism of action of insulin on acetylcholine-evoked amylase secretion in the mouse pancreas. J Physiol 1985; 358:469–482.

123. Lee YL, Kwon HY, Park HS, Lee TH, Park HJ. The role of insulin in the interaction of secretin and cholecystokinin in exocrine secretion of the isolated perfused rat pancreas. Pancreas 1996; 12:58–63.

124. Juma LM, Singh J, Pallot DJ, Salido GM, Adeghate E. Interactions of islet hormones with acetylcholine in the isolated rat pancreas. Peptides 1997; 18:1415–1422.

125. Patel R, Shervington A, Pariente JA, Martinez-Burgos MA, Salido GM, Adeghate E, Singh J. Mechanism of exocrine pancreatic insufficiency in streptozotocin-induced type 1 diabetes mellitus. Ann N Y Acad Sci 2006; 1084:71–88.

126. Okabayashi Y, Otsuki M, Ohki A, Nakamura T, Tani S, Baba S. Secretin-induced exocrine secretion in perfused pancreas isolated from diabetic rats. Diabetes 1988; 37:1173–1180.

127. Okabayashi Y, Otsuki M, Ohki A, Suehiro I, Baba S. Effect of diabetes mellitus on pancreatic exocrine secretion from isolated perfused pancreas in rats. Dig Dis Sci 1988; 33:711–717.

128. Lee KY, Krusch D, Zhou L, Song Y, Chang TM, Chey WY. Effect of endogenous insulin on pancreatic exocrine secretion in perfused dog pancreas. Pancreas 1995; 11:190–195.

129. Lee KY, Lee YL, Kim CD, Chang TM, Chey WY. Mechanism of action of insulin on pancreatic exocrine secretion in perfused rat pancreas. Am J Physiol Gastrointest Liver Physiol 1994; 267:G207–G212.

130. Danielsson A. Effects of glucose, insulin and glucagon on amylase secretion from incubated mouse pancreas. Pflugers Arch 1974; 348:333–342.

131. Bruzzone R, Trimble ER, Gjinovci A, Renold AE. Glucose-insulin interactions on exocrine secretion from the perfused rat pancreas. Gastroenterology 1984; 87:1305–1312.

132. Wizemann V, Weppler P, Mahrt R. Effect of glucagon and insulin on the isolated exocrine pancreas. Digestion 1974; 11:432–435.

133. Otsuki M, Goldfine ID, Williams JA. Diabetes in the rat is associated with a reversible postreceptor defect in cholecystokinin action. Gastroenterology 1984; 87:882–887.

134. Otsuki M, Williams JA. Direct modulation of pancreatic CCK receptors and enzyme secretion by insulin in isolated pancreatic acini from diabetic rats. Diabetes 1983; 32:241–246.

135. Patel R, Pariente JA, Martinez MA, Salido GM, Singh J. Effect of insulin on acetylcholine-evoked amylase release and calcium mobilization in streptozotocin-induced diabetic rat pancreatic acinar cells. Ann N Y Acad Sci 2006; 1084:58–70.

136. Ryu GR, Sung CH, Kim MJ, Sung JH, Lee KH, Park DW, Sim SS, Min DS, Rhie DJ, Yoon SH, Hahn SJ, Kim MS, et al. Changes in IP3 receptor are associated with altered

calcium response to cholecystokinin in diabetic rat pancreatic acini. Pancreas 2004; 29: e106–112.

137. Chandrasekar B, Korc M. Alteration of cholecystokinin-mediated phosphatidylinositol hydrolysis in pancreatic acini from insulin-deficient rats. Evidence for defective G protein activation. Diabetes 1991; 40:1282–1291.

138. Komabayashi T, Sawada H, Izawa T, Kogo H. Altered intracellular Ca^{2+} regulation in pancreatic acinar cells from acute streptozotocin-induced diabetic rats. Eur J Pharmacol 1996; 298:299–306.

139. Matsushita K, Okabayashi Y, Koide M, Hasegawa H, Otsuki M, Kasuga M. Potentiating effect of insulin on exocrine secretory function in isolated rat pancreatic acini. Gastroenterology 1994; 106:200–206.

140. Mankad P, James A, Siriwardena AK, Elliott AC, Bruce JI. Insulin protects pancreatic acinar cells from cytosolic calcium overload and inhibition of the plasma membrane calcium pump. J Biol Chem 2012; 287:1823–1836.

141. Girman CJ, Kou TD, Cai B, Alexander CM, O'Neill EA, Williams-Herman DE, Katz L. Patients with type 2 diabetes mellitus have higher risk for acute pancreatitis compared with those without diabetes. Diabetes Obes Metab 2010; 12:766–771.

142. Gonzalez-Perez A, Schlienger RG, Rodriguez LA. Acute pancreatitis in association with type 2 diabetes and antidiabetic drugs: a population-based cohort study. Diabetes Care 2010; 33:2580–2585.

143. Hardt PD, Brendel MD, Kloer HU, Bretzel RG. Is pancreatic diabetes (type 3c diabetes) underdiagnosed and misdiagnosed? Diabetes Care 2008; 31 Suppl 2:S165–169.

144. Miko A, Farkas N, Garami A, Szabo I, Vincze A, Veres G, Bajor J, Alizadeh H, Rakonczay Z, Jr, Vigh E, Marta K, Kiss Z, et al. Preexisting diabetes elevates risk of local and systemic complications in acute pancreatitis: systematic review and meta-analysis. Pancreas 2018; 47:917–923.

145. Renner IG, Savage WT, 3rd, Pantoja JL, Renner VJ. Death due to acute pancreatitis. A retrospective analysis of 405 autopsy cases. Dig Dis Sci 1985; 30:1005–1018.

146. Seicean A, Tantau M, Grigorescu M, Mocan T, Seicean R, Pop T. Mortality risk factors in chronic pancreatitis. J Gastrointestin Liver Dis 2006; 15:21–26.

147. Hegyi P, Rakonczay-Jr Z, Sari R, Czako L, Farkas N, Gog C, Nemeth J, Lonovics J, Takacs T. Insulin is necessary for the hypertrophic effect of cholecystokinin-octapeptide following acute necrotizing experimental pancreatitis. World J Gastroenterol 2004; 10:2275–2277.

148. Hegyi P, Takacs T, Tiszlavicz L, Czako L, Lonovics J. Recovery of exocrine pancreas six months following pancreatitis induction with L-arginine in streptozotocin-diabetic rats. J Physiol Paris 2000; 94:51–55.

149. Zechner D, Spitzner M, Bobrowski A, Knapp N, Kuhla A, Vollmar B. Diabetes aggravates acute pancreatitis and inhibits pancreas regeneration in mice. Diabetologia 2012; 55:1526–1534.

150. Noel RA, Braun DK, Patterson RE, Bloomgren GL. Increased risk of acute pancreatitis and biliary disease observed in patients with type 2 diabetes: a retrospective cohort study. Diabetes Care 2009; 32:834–838.

151. Koizumi M, Yoshida Y, Abe N, Shimosegawa T, Toyota T. Pancreatic diabetes in Japan. Pancreas 1998; 16:385–391.

152. Levy P, Milan C, Pignon JP, Baetz A, Bernades P. Mortality factors associated with chronic pancreatitis. Unidimensional and multidimensional analysis of a medical-surgical series of 240 patients. Gastroenterology 1989; 96:1165–1172.

153. Mentula P, Kylanpaa ML, Kemppainen E, Puolakkainen P. Obesity correlates with early hyperglycemia in patients with acute pancreatitis who developed organ failure. Pancreas 2008; 36:e21–25.

154. Rajaratnam SG, Martin IG. Admission serum glucose level: an accurate predictor of outcome in gallstone pancreatitis. Pancreas 2006; 33:27–30.

155. Takacs T, Hegyi P, Jarmay K, Czako L, Gog C, Rakonczay Z, Jr., Nemeth J, Lonovics J. Cholecystokinin fails to promote pancreatic regeneration in diabetic rats following the induction of experimental pancreatitis. Pharmacol Res 2001; 44:363–372.

156. Hegyi P, Takacs T, Jarmay K, Nagy I, Czako L, Lonovics J. Spontaneous and cholecystokinin-octapeptide-promoted regeneration of the pancreas following L-arginine-induced pancreatitis in rat. Int J Pancreatol 1997; 22:193–200.

157. Hegyi P, Rakonczay Z, Jr., Sari R, Gog C, Lonovics J, Takacs T, Czako L. L-arginine-induced experimental pancreatitis. World J Gastroenterol 2004; 10:2003–2009.

158. Vasseur S, Folch-Puy E, Hlouschek V, Garcia S, Fiedler F, Lerch MM, Dagorn JC, Closa D, Iovanna JL. p8 improves pancreatic response to acute pancreatitis by enhancing the expression of the anti-inflammatory protein pancreatitis-associated protein I. J Biol Chem 2004; 279:7199–7207.

159. Closa D, Motoo Y, Iovanna JL. Pancreatitis-associated protein: from a lectin to an anti-inflammatory cytokine. World J Gastroenterol 2007; 13:170–174.

160. Gironella M, Folch-Puy E, LeGoffic A, Garcia S, Christa L, Smith A, Tebar L, Hunt SP, Bayne R, Smith AJ, Dagorn JC, Closa D, et al. Experimental acute pancreatitis in PAP/HIP knock-out mice. Gut 2007; 56:1091–1097.

161. Zhang H, Kandil E, Lin YY, Levi G, Zenilman ME. Targeted inhibition of gene expression of pancreatitis-associated proteins exacerbates the severity of acute pancreatitis in rats. Scand J Gastroenterol 2004; 39:870–881.

162. Roberts SR, Hamedani B. Benefits and methods of achieving strict glycemic control in the ICU. Crit Care Nurs Clin North Am 2004; 16:537–545.

163. Krogh-Madsen R, Moller K, Dela F, Kronborg G, Jauffred S, Pedersen BK. Effect of hyperglycemia and hyperinsulinemia on the response of IL-6, TNF-alpha, and FFAs to low-dose endotoxemia in humans. Am J Physiol Endocrinol Metab 2004; 286:E766–E772.

164. Samad A, James A, Wong J, Mankad P, Whitehouse J, Patel W, Alves-Simoes M, Siriwardena AK, Bruce JI. Insulin protects pancreatic acinar cells from palmitoleic acid-induced cellular injury. J Biol Chem 2014; 289:23582–23595.

165. Dembinski A, Warzecha Z, Ceranowicz P, Cieszkowski J, Pawlik WW, Tomaszewska R, Kusnierz-Cabala B, Naskalski JW, Kuwahara A, Kato I. Role of growth hormone and insulin-like growth factor-1 in the protective effect of ghrelin

in ischemia/reperfusion-induced acute pancreatitis. Growth Horm IGF Res 2006; 16:348–356.

166. Warzecha Z, Dembinski A, Ceranowicz P, Konturek SJ, Tomaszewska R, Stachura J, Konturek PC. IGF-1 stimulates production of interleukin-10 and inhibits development of caerulein-induced pancreatitis. J Physiol Pharmacol 2003; 54:575–590.

167. Warzecha Z, Dembinski A, Konturek PC, Ceranowicz P, Konturek SJ, Tomaszewska R, Schuppan D, Stachura J, Nakamura T. Hepatocyte growth factor attenuates pancreatic damage in caerulein-induced pancreatitis in rats. Eur J Pharmacol 2001; 430:113–121.

168. Deprez J, Vertommen D, Alessi DR, Hue L, Rider MH. Phosphorylation and activation of heart 6-phosphofructo-2-kinase by protein kinase B and other protein kinases of the insulin signaling cascades. J Biol Chem 1997; 272:17269–17275.

169. Bartrons R, Simon-Molas H, Rodriguez-Garcia A, Castano E, Navarro-Sabate A, Manzano A, Martinez-Outschoorn UE. Fructose 2,6-bisphosphate in cancer cell metabolism. Front Oncol 2018; 8:331.

170. Cen ME, Wang F, Su Y, Zhang WJ, Sun B, Wang G. Gastrointestinal microecology: a crucial and potential target in acute pancreatitis. Apoptosis 2018; 23:377–387.

171. Walker AW, Ince J, Duncan SH, Webster LM, Holtrop G, Ze X, Brown D, Stares MD, Scott P, Bergerat A, Louis P, McIntosh F, et al. Dominant and diet-responsive groups of bacteria within the human colonic microbiota. ISME J 2011; 5:220–230.

172. Tan J, McKenzie C, Potamitis M, Thorburn AN, Mackay CR, Macia L. The role of short-chain fatty acids in health and disease. Adv Immunol 2014; 121:91–119.

173. Ayabe T, Satchell DP, Wilson CL, Parks WC, Selsted ME, Ouellette AJ. Secretion of microbicidal alpha-defensins by intestinal Paneth cells in response to bacteria. Nat Immunol 2000; 1:113–118.

174. De Reggi M, Gharib B. Protein-X, pancreatic Stone-, pancreatic thread-, reg-protein, P19, lithostathine, and now what? Characterization, structural analysis and putative function(s) of the major non-enzymatic protein of pancreatic secretions. Curr Protein Pept Sci 2001; 2:19–42.

175. Ahuja M, Schwartz DM, Tandon M, Son A, Zeng M, Swaim W, Eckhaus M, Hoffman V, Cui Y, Xiao B, Worley PF, Muallem S. Orai1-mediated antimicrobial secretion from pancreatic acini shapes the gut microbiome and regulates gut innate immunity. Cell Metab 2017; 25:635–646.

176. Schnapp D, Reid CJ, Harris A. Localization of expression of human beta defensin-1 in the pancreas and kidney. J Pathol 1998; 186:99–103.

177. Keim V, Rohr G, Stockert HG, Haberich FJ. An additional secretory protein in the rat pancreas. Digestion 1984; 29:242–249.

178. Orelle B, Keim V, Masciotra L, Dagorn JC, Iovanna JL. Human pancreatitis-associated protein. Messenger RNA cloning and expression in pancreatic diseases. J Clin Invest 1992; 90:2284–2291.

179. Dusetti NJ, Frigerio JM, Keim V, Dagorn JC, Iovanna JL. Structural organization of the gene encoding the rat pancreatitis-associated protein. Analysis of its evolutionary history reveals an ancient divergence from the other carbohydrate-recognition domain-containing genes. J Biol Chem 1993; 268:14470–14475.

180. Reboud JP, Marchis-Mouren G, Cozzone A, Desnuelle P. Variations in the biosynthesis rate of pancreatic amylase and chymotrypsinogen in response to a starch-rich or a protein-rich diet. Biochem Biophys Res Commun 1966; 22:94–99.

181. Sun J, Furio L, Mecheri R, van der Does AM, Lundeberg E, Saveanu L, Chen Y, van Endert P, Agerberth B, Diana J. Pancreatic beta-cells limit autoimmune diabetes via an immunoregulatory antimicrobial peptide expressed under the influence of the gut microbiota. Immunity 2015; 43:304–317.

182. Inayat F, Zafar F, Riaz I, Younus F, Baig AS, Imran Z. Hypertriglyceridemic pancreatitis: Is insulin monotherapy a feasible therapeutic option? Cureus 2018; 10:e3461.

183. Rawla P, Sunkara T, Thandra KC, Gaduputi V. Hypertriglyceridemia-induced pancreatitis: updated review of current treatment and preventive strategies. Clin J Gastroenterol 2018; 11:441–448.

184. Lithell H, Vessby B, Walldius G, Carlson LA. Hypertriglyceridemia-acute pancreatitis-ischemic heart disease. A case study in a pair of monozygotic twins. Acta Med Scand 1987; 221:311–316.

185. Tsuang W, Navaneethan U, Ruiz L, Palascak JB, Gelrud A. Hypertriglyceridemic pancreatitis: presentation and management. Am J Gastroenterol 2009; 104:984–991.

186. Eckel RH. Lipoprotein lipase. A multifunctional enzyme relevant to common metabolic diseases. N Engl J Med 1989; 320:1060–1068.

187. Song F, Zhong LJ, Han L, Xie GH, Xiao C, Zhao B, Hu YQ, Wang SY, Qin CJ, Zhang Y, Lai DM, Cui P, et al. Intensive insulin therapy for septic patients: a meta-analysis of randomized controlled trials. Biomed Res Int 2014; 2014:698265.

188. Griesdale DE, de Souza RJ, van Dam RM, Heyland DK, Cook DJ, Malhotra A, Dhaliwal R, Henderson WR, Chittock DR, Finfer S, Talmor D. Intensive insulin therapy and mortality among critically ill patients: a meta-analysis including NICE-SUGAR study data. CMAJ 2009; 180:821–827.

189. Ling Y, Li X, Gao X. Intensive versus conventional glucose control in critically ill patients: a meta-analysis of randomized controlled trials. Eur J Intern Med 2012; 23:564–574.

190. Wiener RS, Wiener DC, Larson RJ. Benefits and risks of tight glucose control in critically ill adults: a meta-analysis. JAMA 2008; 300:933–944.

191. Yamada T, Shojima N, Hara K, Noma H, Yamauchi T, Kadowaki T. Glycemic control, mortality, secondary infection, and hypoglycemia in critically ill pediatric patients: a systematic review and network meta-analysis of randomized controlled trials. Intensive Care Med 2017; 43:1427–1429.

192. Yamada T, Shojima N, Noma H, Yamauchi T, Kadowaki T. Glycemic control, mortality, and hypoglycemia in critically ill patients: a systematic review and network meta-analysis of randomized controlled trials. Intensive Care Med 2017; 43:1–15.

193. Yatabe T, Inoue S, Sakaguchi M, Egi M. The optimal target for acute glycemic control in critically ill patients: a network meta-analysis. Intensive Care Med 2017; 43:16–28.

194. Egi M, Bellomo R, Stachowski E, French CJ, Hart GK, Taori G, Hegarty C, Bailey M. Hypoglycemia and outcome in critically ill patients. Mayo Clin Proc 2010; 85:217–224.

195. Investigators CS, Annane D, Cariou A, Maxime V, Azoulay E, D'Honneur G, Timsit JF, Cohen Y, Wolf M, Fartoukh M, Adrie C, Santre C, et al. Corticosteroid treatment and intensive insulin therapy for septic shock in adults: a randomized controlled trial. JAMA 2010; 303:341–348.

196. Krinsley JS, Egi M, Kiss A, Devendra AN, Schuetz P, Maurer PM, Schultz MJ, van Hooijdonk RT, Kiyoshi M, Mackenzie IM, Annane D, Stow P, et al. Diabetic status and the relation of the three domains of glycemic control to mortality in critically ill patients: an international multicenter cohort study. Crit Care 2013; 17:R37.

197. van den Berghe G, Wouters P, Weekers F, Verwaest C, Bruynninckx F, Schetz M, Vlasselaers D, Ferdinande P, Lauwers P, Bouillon R. Intensive insulin therapy in critically ill patients. N Engl J Med 2001; 345:1359–1367.

198. Leelarathna L, English SW, Thabit H, Caldwell K, Allen JM, Kumareswaran K, Wilinska ME, Nodale M, Mangat J, Evans ML, Burnstein R, Hovorka R. Feasibility of fully automated closed-loop glucose control using continuous subcutaneous glucose measurements in critical illness: a randomized controlled trial. Crit Care 2013; 17:R159.

199. Maclaren DP, Mohebbi H, Nirmalan M, Keegan MA, Best CT, Perera D, Harvie MN, Campbell IT. Effect of a 2-h hyperglycemic-hyperinsulinemic glucose clamp to promote glucose storage on endurance exercise performance. Eur J Appl Physiol 2011; 111:2105–2114.

200. Sutherland EW, De Duve C. Origin and distribution of the hyperglycemic-glycogenolytic factor of the pancreas. J Biol Chem 1948; 175:663–674.

201. Gromada J, Franklin I, Wollheim CB. Alpha-cells of the endocrine pancreas: 35 years of research but the enigma remains. Endocr Rev 2007; 28:84–116.

202. Foa PP, Santamaria L, Weinstein HR, Berger S, Smith JA. Secretion of the hyperglycemic-glycogenolytic factor in normal dogs. Am J Physiol 1952; 171:32–36.

203. Unger RH, Eisentraut AM, McCall M, Madison LL. Glucagon antibodies and an immunoassay for glucagon. J Clin Invest 1961; 40:1280–1289.

204. Furuta M, Zhou A, Webb G, Carroll R, Ravazzola M, Orci L, Steiner DF. Severe defect in proglucagon processing in islet A-cells of prohormone convertase 2 null mice. J Biol Chem 2001; 276:27197–27202.

205. Lefebvre PJ. Glucagon and its family revisited. Diabetes Care 1995; 18:715–730.

206. Bozadjieva N, Williams JA, Bernal-Mizrachi E. Glucagon. Pancreapedia: Exocrine Pancreas Knowledge Base, 2013. DOI: 10.3998/panc.2013.23.

207. Konturek SJ, Tasler J, Obtulowicz W. Characteristics of inhibition of pancreatic secretion by glucagon. Digestion 1974; 10:138–149.

208. Horiguchi Y. Interaction of secretin and glucagon on exocrine pancreatic secretion. Gastroenterol Jpn 1979; 14:63–73.

209. Dyck WP, Rudick J, Hoexter B, Janowitz HD. Influence of glucagon on pancreatic exocrine secretion. Gastroenterology 1969; 56:531–537.

210. Singer MV, Tiscornia OM, Mendes de Oliveiro JP, Demol P, Levesque D, Sarles H. Effect of glucagon on canine exocrine pancreatic secretion stimulated by a test meal. Can J Physiol Pharmacol 1978; 56:1–6.

211. Konturek SJ, Demitrescu T, Radecki T, Thor P, Pucher A. Effect of glucagon on gastric and pancreatic secretion and peptic ulcer formation in cats. Am J Dig Dis 1974; 19:557–564.

212. Shaw HM, Heath TJ. The effect of glucagon on the formation of pancreatic juice and bile in the rat. Can J Physiol Pharmacol 1973; 51:1–5.

213. Adler G. Effect of glucagon on the secretory process in the rat exocrine pancreas. Cell Tissue Res 1977; 182:193–204.

214. Biedzinski TM, Bataille D, Devaux MA, Sarles H. The effect of oxyntomodulin (glucagon-37) and glucagon on exocrine pancreatic secretion in the conscious rat. Peptides 1987; 8:967–972.

215. Clain JE, Barbezat GO, Waterworth MM, Bank S. Glucagon inhibition of secretin and combined secretin and cholecystokinin stimulated pancreatic exocrine secretion in health and disease. Digestion 1978; 17:11–17.

216. DiMagno EP, Go VL, Summerskill HJ. Intraluminal and postabsorptive effects of amino acids on pancreatic enzyme secretion. J Lab Clin Med 1973; 82:241–248.

217. Harada H, Kochi F, Hanafusa E, Kobayashi T, Oka H, Kimura I. Studies on the effect of glucagon on human pancreatic secretion by analysis of endoscopically obtained pure pancreatic juice. Gastroenterol Jpn 1985; 20:28–36.

218. Singh M. Effect of glucagon on digestive enzyme synthesis, transport and secretion in mouse pancreatic acinar cells. J Physiol 1980; 306:307–322.

219. Singh M. Amylase release from dissociated mouse pancreatic acinar cells stimulated by glucagon: effect of membrane stabilizers. J Physiol 1980; 309:81–91.

220. Pandol SJ, Sutliff VE, Jones SW, Charlton CG, O'Donohue TL, Gardner JD, Jensen RT. Action of natural glucagon on pancreatic acini: due to contamination by previously undescribed secretagogues. Am J Physiol Gastrointest Liver Physiol 1983; 245:G703–G710.

221. Bandisode MS, Singh M. Amylase secretion from isolated pure acinar cells. Biochem Biophys Res Commun 1985; 129:63–69.

222. Dunphy JL, Taylor RG, Fuller PJ. Tissue distribution of rat glucagon receptor and GLP-1 receptor gene expression. Mol Cell Endocrinol 1998; 141:179–186.

223. Hansen LH, Abrahamsen N, Nishimura E. Glucagon receptor mRNA distribution in rat tissues. Peptides 1995; 16:1163–1166.

224. Brazeau P, Vale W, Burgus R, Ling N, Butcher M, Rivier J, Guillemin R. Hypothalamic polypeptide that inhibits the secretion of immunoreactive pituitary growth hormone. Science 1973; 179:77–79.

225. Pradayrol L, Jornvall H, Mutt V, Ribet A. N-terminally extended somatostatin: the primary structure of somatostatin-28. FEBS Lett 1980; 109:55–58.

226. Morisset J. Somatostatin. Pancreapedia: Exocrine Pancreas Knowledge Base, 2015. DOI: 10.3998/panc.2015.43.

227. Reisine T, Bell GI. Molecular properties of somatostatin receptors. Neuroscience 1995; 67:777–790.

228. Zyznar ES, Pietri AO, Harris V, Unger RH. Evidence for the hormonal status of somatostatin in man. Diabetes 1981; 30:883–886.

229. Gyr K, Beglinger C, Kohler E, Trautzl U, Keller U, Bloom SR. Circulating somatostatin. Physiological regulator of pancreatic function? J Clin Invest 1987; 79:1595–1600.

230. Taborsky GJ, Jr, Ensinck JW. Contribution of the pancreas to circulating somatostatin-like immunoreactivity in the normal dog. J Clin Invest 1984; 73:216–223.

231. Vaysse N, Chayvialle JA, Pradayrol L, Esteve JP, Susini C, Lapuelle J, Descos F, Ribet A. Somatostatin 28: comparison with somatostatin 14 for plasma kinetics and low-dose effects on the exocrine pancreas in dogs. Gastroenterology 1981; 81:700–706.

232. Boden G, Sivitz MC, Owen OE, Essa-Koumar N, Landor JH. Somatostatin suppresses secretin and pancreatic exocrine secretion. Science 1975; 190:163–165.

233. Susini C, Esteve JP, Vaysse N, Pradayrol L, Ribet A. Somatostatin 28: effects on exocrine pancreatic secretion in conscious dogs. Gastroenterology 1980; 79:720–724.

234. Susini C, Esteve JP, Bommelaer G, Vaysse N, Ribet A. Inhibition of exocrine pancreatic secretion by somatostatin in dogs. Digestion 1978; 18:384–393.

235. Wilson RM, Boden G, Shore LS, Essa-Koumar N. Effect of somatostatin on meal-stimulated pancreatic exocrine secretions in dogs. Diabetes 1977; 26:7–10.

236. Kayasseh L, Gyr K, Stalder GA, Rittmann WW, Girard J. Effect of somatostatin on exocrine pancreatic secretion stimulated by pancreozymin-secretin or by a test meal in the dog. Horm Res 1978; 9:176–184.

237. Kohler E, Beglinger C, Ribes G, Grotzinger U, Loubatieres-Mariani MM, Gyr K. Effect of circulating somatostatin on exocrine pancreatic secretion in conscious dogs. Pancreas 1986; 1:455–459.

238. Folsch UR, Lankisch PG, Creutzfeldt W. Effect of somatostatin on basal and stimulated pancreatic secretion in the rat. Digestion 1978; 17:194–203.

239. Chariot J, Roze C, Vaille C, Debray C. Effects of somatostatin on the external secretion of the pancreas of the rat. Gastroenterology 1978; 75:832–837.

240. Albinus M, Blair EL, Case RM, Coy DH, Gomez-Pan A, Hirst BH, Reed JD, Schally AV, Shaw B, Smith PA, Smy JR. Comparison of the effect of somatostatin on gastrointestinal function in the conscious and anaesthetized cat and on the isolated cat pancreas. J Physiol 1977; 269:77–91.

241. Miller TA, Tepperman FS, Fang WF, Jacobson ED. Effect of somatostatin on pancreatic protein secretion induced by cholecystokinin. J Surg Res 1979; 26:488–493.

242. Dollinger HC, Raptis S, Pfeiffer EF. Effects of somatostatin on exocrine and endocrine pancreatic function stimulated by intestinal hormones in man. Horm Metab Res 1976; 8:74–78.

243. Hildebrand P, Ensinck JW, Gyr K, Mossi S, Leuppi J, Eggenberger C, Beglinger C. Evidence for hormonal inhibition of exocrine pancreatic function by somatostatin 28 in humans. Gastroenterology 1992; 103:240–247.

244. Emoto T, Miyata M, Izukura M, Yumiba T, Mizutani S, Sakamoto T, Matsuda H. Simultaneous observation of endocrine and exocrine functions of the pancreas responding to somatostatin in man. Regul Pept 1997; 68:1–8.

245. Maouyo D, Morisset J. Modulation of pancreatic secretion of individual digestive enzymes in octreotide (SMS 201–995)-infused rats. Pancreas 1997; 14:47–57.

246. Suzuki T, Naruse S, Yanaihara N. The inhibitory effect of octreotide on exocrine pancreatic secretion in conscious dogs. Pancreas 1993; 8:226–232.

247. Konturck SJ, Bilski J, Jaworek J, Tasler J, Schally AV. Comparison of somatostatin and its highly potent hexa- and octapeptide analogs on exocrine and endocrine pancreatic secretion. Proc Soc Exp Biol Med 1988; 187:241–249.

248. Kemmer TP, Malfertheiner P, Buchler M, Friess H, Meschenmoser L, Ditschuneit H. Inhibition of human exocrine pancreatic secretion by the long-acting somatostatin analogue octreotide (SMS 201–995). Aliment Pharmacol Ther 1992; 6:41–50.

249. Varga G, Kisfalvi I, Jr., Kordas K, Wong H, Walsh JH, Solomon TE. Effect of somatostatin immunoneutralization on gastric acid and pancreatic enzyme secretion in anesthetized rats. J Physiol Paris 1997; 91:223–227.

250. Taparel D, Esteve JP, Susini C, Vaysse N, Balas D, Berthon G, Wunsch E, Ribet A. Binding of somatostatin to guinea-pig pancreatic membranes: regulation by ions. Biochem Biophys Res Commun 1983; 115:827–833.

251. Esteve JP, Susini C, Vaysse N, Antoniotti H, Wunsch E, Berthon G, Ribet A. Binding of somatostatin to pancreatic acinar cells. Am J Physiol Gastrointest Liver Physiol 1984; 247:G62–G69.

252. Sakamoto C, Goldfine ID, Williams JA. The somatostatin receptor on isolated pancreatic acinar cell plasma membranes. Identification of subunit structure and direct regulation by cholecystokinin. J Biol Chem 1984; 259:9623–9627.

253. Zeggari M, Viguerie N, Susini C, Esteve JP, Vaysse N, Rivier J, Wunsch E, Ribet A. Characterization of pancreatic somatostatin binding sites with a 125I-somatostatin 28 analog. Peptides 1986; 7:953–959.

254. Law SF, Manning D, Reisine T. Identification of the subunits of GTP-binding proteins coupled to somatostatin receptors. J Biol Chem 1991; 266:17885–17897.

255. Braun M. The somatostatin receptor in human pancreatic beta-cells. Vitam Horm 2014; 95:165–193.

256. Hunyady B, Hipkin RW, Schonbrunn A, Mezey E. Immunohistochemical localization of somatostatin receptor SST2A in the rat pancreas. Endocrinology 1997; 138:2632–2635.

257. Sakamoto C, Matozaki T, Nagao M, Baba S. Coupling of guanine nucleotide inhibitory protein to somatostatin receptors on pancreatic acinar membranes. Am J Physiol Gastrointest Liver Physiol 1987; 253:G308–G314.

258. Esteve JP, Vaysse N, Susini C, Kunsch JM, Fourmy D, Pradayrol L, Wunsch E, Moroder L, Ribet A. Bimodal regulation of pancreatic exocrine function in vitro by somatostatin-28. Am J Physiol Gastrointest Liver Physiol 1983; 245:G208–G216.

259. Ishiguro H, Hayakawa T, Kondo T, Shibata T, Kitagawa M, Sakai Y, Sobajima H, Nakae Y, Tanikawa M. The effect of

somatostatin analogue octreotide on amylase secretion from mouse pancreatic acini. Digestion 1993; 54:207–212.

260. Singh P, Asada I, Owlia A, Collins TJ, Thompson JC. Somatostatin inhibits VIP-stimulated amylase release from perifused guinea pig pancreatic acini. Am J Physiol Gastrointest Liver Physiol 1988; 254:G217–G223.

261. Matsushita K, Okabayashi Y, Hasegawa H, Koide M, Kido Y, Okutani T, Sugimoto Y, Kasuga M. In vitro inhibitory effect of somatostatin on secretin action in exocrine pancreas of rats. Gastroenterology 1993; 104:1146–1152.

262. Singh M. Effect of somatostatin on amylase secretion from in vivo and in vitro rat pancreas. Dig Dis Sci 1986; 31:506–512.

263. Mulvihill SJ, Bunnett NW, Goto Y, Debas HT. Somatostatin inhibits pancreatic exocrine secretion via a neural mechanism. Metabolism 1990; 39:143–148.

264. Ohnishi H, Mine T, Kojima I. Inhibition by somatostatin of amylase secretion induced by calcium and cyclic AMP in rat pancreatic acini. Biochem J 1994; 304 (Pt 2):531–536.

265. Ensinck JW, Laschansky EC, Vogel RE, D'Alessio DA. Effect of somatostatin-28 on dynamics of insulin secretion in perfused rat pancreas. Diabetes 1991; 40:1163–1169.

266. Muller MK, Kessel B, Hutt T, Kath R, Layer P, Goebell H. Effects of somatostatin-14 on gastric and pancreatic responses to hormonal and neural stimulation using an isolated perfused rat stomach and pancreas preparation. Pancreas 1988; 3:303–310.

267. Li Y, Owyang C. Somatostatin inhibits pancreatic enzyme secretion at a central vagal site. Am J Physiol Gastrointest Liver Physiol 1993; 265:G251–G257.

268. Liao Z, Li ZS, Lu Y, Wang WZ. Microinjection of exogenous somatostatin in the dorsal vagal complex inhibits pancreatic secretion via somatostatin receptor-2 in rats. Am J Physiol Gastrointest Liver Physiol 2007; 292:G746–G752.

269. Schonfeld JV, Muller MK, Demirtas B, Soukup J, Runzi M, Goebell H. Effect of neural blockade on somatostatin-induced inhibition of exocrine pancreatic secretion. Digestion 1989; 43:81–86.

270. Brodish RJ, Kuvshinoff BW, McFadden DW, Fink AS. Somatostatin inhibits cholecystokinin-induced pancreatic protein secretion via cholinergic pathways. Pancreas 1995; 10:401–406.

271. Kimmel JR, Hayden LJ, Pollock HG. Isolation and characterization of a new pancreatic polypeptide hormone. J Biol Chem 1975; 250:9369–9376.

272. Taylor IL. Pancreatic polypeptide family: pancreatic polypeptide, neuropeptide Y, and peptide YY. In: Makhlouf GE, ed. Handbook of Physiology: the Gastrointestinal System: Section VI: Volume 2. Oxford, UK: Oxford University Press, 1989.

273. Glover ID, Barlow DJ, Pitts JE, Wood SP, Tickle IJ, Blundell TL, Tatemoto K, Kimmel JR, Wollmer A, Strassburger W, Zhang Y-S. Conformational studies on the pancreatic polypeptide hormone family. Eur J Biochem 1984; 142:379–385.

274. Fiocca R, Sessa F, Tenti P, Usellini L, Capella C, O'Hare MM, Solcia E. Pancreatic polypeptide (PP) cells in the PP-rich lobe of the human pancreas are identified ultrastructurally and immunocytochemically as F cells. Histochemistry 1983; 77:511–523.

275. Holzer P, Reichmann F, Farzi A. Neuropeptide Y, peptide YY and pancreatic polypeptide in the gut-brain axis. Neuropeptides 2012; 46:261–274.

276. Schwartz TW. Pancreatic polypeptide: a hormone under vagal control. Gastroenterology 1983; 85:1411–1425.

277. Williams JA. Pancreatic Polypeptide. Pancreapedia: Exocrine Pancreas Knowledge Base, 2014. DOI: 10.3998/panc.2014.04.

278. Zipf WB, O'Dorisio TM, Cataland S, Sotos J. Blunted pancreatic polypeptide responses in children with obesity of Prader-Willi syndrome. J Clin Endocrinol Metab 1981; 52:1264–1266.

279. Michel MC, Beck-Sickinger A, Cox H, Doods HN, Herzog H, Larhammar D, Quirion R, Schwartz T, Westfall T. XVI. International Union of Pharmacology recommendations for the nomenclature of neuropeptide Y, peptide YY, and pancreatic polypeptide receptors. Pharmacol Rev 1998; 50:143–150.

280. Lin TM, Evans DC, Chance RE, Spray GF. Bovine pancreatic peptide: action on gastric and pancreatic secretion in dogs. Am J Physiol Endocrinol Metab 1977; 232:E311–E315.

281. Taylor IL, Solomon TE, Walsh JH, Grossman MI. Pancreatic polypeptide. Metabolism and effect on pancreatic secretion in dogs. Gastroenterology 1979; 76:524–528.

282. Beglinger C, Taylor IL, Grossman MI, Solomon TE. Pancreatic polypeptide inhibits exocrine pancreatic responses to six stimulants. Am J Physiol Gastrointest Liver Physiol 1984; 246:G286–G291.

283. Chance RE, Cieszkowski M, Jaworek J, Konturek SJ, Swierczek J, Tasler J. Effect of pancreatic polypeptide and its C-terminal hexapeptide on meal and secretin induced pancreatic secretion in dogs. J Physiol 1981; 314:1–9.

284. Lonovics J, Guzman S, Devitt PG, Hejtmancik KE, Suddith RL, Rayford PL, Thompson JC. Action of pancreatic polypeptide on exocrine pancreas and on release of cholecystokinin and secretin. Endocrinology 1981; 108:1925–1930.

285. Shiratori K, Lee KY, Chang TM, Jo YH, Coy DH, Chey WY. Role of pancreatic polypeptide in the regulation of pancreatic exocrine secretion in dogs. Am J Physiol Gastrointest Liver Physiol 1988; 255:G535–G541.

286. Adrian TE, Besterman HS, Mallinson CN, Greenberg GR, Bloom SR. Inhibition of secretin stimulated pancreatic secretion by pancreatic polypeptide. Gut 1979; 20:37–40.

287. Greenberg GR, McCloy RF, Adrian TE, Chadwick VS, Baron JH, Bloom SR. Inhibition of pancreas and gallbladder by pancreatic polypeptide. Lancet 1978; 2:1280–1282.

288. Louie DS, Williams JA, Owyang C. Action of pancreatic polypeptide on rat pancreatic secretion: in vivo and in vitro. Am J Physiol Gastrointest Liver Physiol 1985; 249:G489–G495.

289. Putnam WS, Liddle RA, Williams JA. Inhibitory regulation of rat exocrine pancreas by peptide YY and pancreatic polypeptide. Am J Physiol Gastrointest Liver Physiol 1989; 256:G698–G703.

290. Gettys TW, Tanaka I, Taylor IL. Modulation of pancreatic exocrine function in rodents by treatment with pancreatic polypeptide. Pancreas 1992; 7:705–711.

291. Kim KH, Case RM. Effects of pancreatic polypeptide on the secretion of enzymes and electrolytes by in vitro preparations of rat and cat pancreas. Yonsei Med J 1980; 21:99–105.

292. Joehl RJ, DeJoseph MR. Pancreatic polypeptide inhibits amylase release by rat pancreatic acini. J Surg Res 1986; 40:310–314.

293. Duan RD, Erlanson-Albertsson C. Stimulatory effects of human pancreatic polypeptide on rat pancreatic acini. Regul Pept 1985; 12:215–222.

294. Pan GZ, Lu L, Qian JM, Xue BG. Bovine pancreatic polypeptide as an antagonist of muscarinic cholinergic receptors. Am J Physiol Gastrointest Liver Physiol 1987; 252:G384–G391.

295. Whitcomb DC, Taylor IL, Vigna SR. Characterization of saturable binding sites for circulating pancreatic polypeptide in rat brain. Am J Physiol Gastrointest Liver Physiol 1990; 259:G687–G691.

296. Whitcomb DC, Puccio AM, Vigna SR, Taylor IL, Hoffman GE. Distribution of pancreatic polypeptide receptors in the rat brain. Brain Res 1997; 760:137–149.

297. Okumura T, Pappas TN, Taylor IL. Pancreatic polypeptide microinjection into the dorsal motor nucleus inhibits pancreatic secretion in rats. Gastroenterology 1995; 108:1517–1525.

298. Browning KN, Coleman FH, Travagli RA. Effects of pancreatic polypeptide on pancreas-projecting rat dorsal motor nucleus of the vagus neurons. Am J Physiol Gastrointest Liver Physiol 2005; 289:G209–G219.

299. Mussa BM, Verberne AJ. The dorsal motor nucleus of the vagus and regulation of pancreatic secretory function. Exp Physiol 2013; 98:25–37.

300. Jung G, Louie DS, Owyang C. Pancreatic polypeptide inhibits pancreatic enzyme secretion via a cholinergic pathway. Am J Physiol Gastrointest Liver Physiol 1987; 253:G706–G710.

301. Kitamura K, Kangawa K, Kawamoto M, Ichiki Y, Nakamura S, Matsuo H, Eto T. Adrenomedullin: a novel hypotensive peptide isolated from human pheochromocytoma. Biochem Biophys Res Commun 1993; 192:553–560.

302. Kitamura K, Sakata J, Kangawa K, Kojima M, Matsuo H, Eto T. Cloning and characterization of cDNA encoding a precursor for human adrenomedullin. Biochem Biophys Res Commun 1993; 194:720–725.

303. Martinez A, Weaver C, Lopez J, Bhathena SJ, Elsasser TH, Miller MJ, Moody TW, Unsworth EJ, Cuttitta F. Regulation of insulin secretion and blood glucose metabolism by adrenomedullin. Endocrinology 1996; 137:2626–2632.

304. Muff R, Born W, Fischer JA. Adrenomedullin and related peptides: receptors and accessory proteins. Peptides 2001; 22:1765–1772.

305. Hay DL, Garelja ML, Poyner DR, Walker CS. Update on the pharmacology of calcitonin/CGRP family of peptides: IUPHAR Review 25. Br J Pharmacol 2018; 175:3–17.

306. Hinson JP, Kapas S, Smith DM. Adrenomedullin, a multifunctional regulatory peptide. Endocr Rev 2000; 21:138–167.

307. Shimekake Y, Nagata K, Ohta S, Kambayashi Y, Teraoka H, Kitamura K, Eto T, Kangawa K, Matsuo H. Adrenomedullin stimulates two signal transduction pathways, cAMP accumulation and Ca^{2+} mobilization, in bovine aortic endothelial cells. J Biol Chem 1995; 270:4412–4417.

308. Tsuchida T, Ohnishi H, Tanaka Y, Mine T, Fujita T. Inhibition of stimulated amylase secretion by adrenomedullin in rat pancreatic acini. Endocrinology 1999; 140:865–870.

309. Sah RP, Nagpal SJ, Mukhopadhyay D, Chari ST. New insights into pancreatic cancer-induced paraneoplastic diabetes. Nat Rev Gastroenterol Hepatol 2013; 10:423–433.

310. Cooper GJ, Willis AC, Clark A, Turner RC, Sim RB, Reid KB. Purification and characterization of a peptide from amyloid-rich pancreases of type 2 diabetic patients. Proc Natl Acad Sci U S A 1987; 84:8628–8632.

311. Fu W, Patel A, Jhamandas JH. Amylin receptor: a common pathophysiological target in Alzheimer's disease and diabetes mellitus. Front Aging Neurosci 2013; 5:42.

312. Caillon L, Hoffmann AR, Botz A, Khemtemourian L. Molecular structure, membrane interactions, and toxicity of the islet amyloid polypeptide in type 2 diabetes mellitus. J Diabetes Res 2016; 2016:5639875.

313. Lukinius A, Wilander E, Westermark GT, Engstrom U, Westermark P. Co-localization of islet amyloid polypeptide and insulin in the B cell secretory granules of the human pancreatic islets. Diabetologia 1989; 32:240–244.

314. Akesson B, Panagiotidis G, Westermark P, Lundquist I. Islet amyloid polypeptide inhibits glucagon release and exerts a dual action on insulin release from isolated islets. Regul Pept 2003; 111:55–60.

315. Muff R, Buhlmann N, Fischer JA, Born W. An amylin receptor is revealed following co-transfection of a calcitonin receptor with receptor activity modifying proteins-1 or -3. Endocrinology 1999; 140:2924–2927.

316. Gedulin BR, Jodka CM, Herrmann K, Young AA. Role of endogenous amylin in glucagon secretion and gastric emptying in rats demonstrated with the selective antagonist, AC187. Regul Pept 2006; 137:121–127.

317. Funakoshi A, Miyasaka K, Kitani K, Nakamura J, Funakoshi S, Fukuda H, Fujii N. Stimulatory effects of islet amyloid polypeptide (amylin) on exocrine pancreas and gastrin release in conscious rats. Regul Pept 1992; 38:135–143.

318. Young AA, Jodka C, Pittner R, Parkes D, Gedulin BR. Dose-response for inhibition by amylin of cholecystokinin-stimulated secretion of amylase and lipase in rats. Regul Pept 2005; 130:19–26.

319. Fehmann HC, Weber V, Goke R, Goke B, Eissele R, Arnold R. Islet amyloid polypeptide (IAPP;amylin) influences the endocrine but not the exocrine rat pancreas. Biochem Biophys Res Commun 1990; 167:1102–1108.

320. Kikuchi Y, Koizumi M, Shimosegawa T, Kashimura J, Suzuki S, Toyota T. Islet amyloid polypeptide has no effect on amylase release from rat pancreatic acini stimulated by CCK-8, secretin, carbachol and VIP. Tohoku J Exp Med 1991; 165:41–48.

321. Tatemoto K, Efendic S, Mutt V, Makk G, Feistner GJ, Barchas JD. Pancreastatin, a novel pancreatic peptide that inhibits insulin secretion. Nature 1986; 324:476–478.

322. Funakoshi A, Miyasaka K, Nakamura R, Kitani K, Funakoshi S, Tamamura H, Fujii N, Yajima H. Bioactivity of synthetic human pancreastatin on exocrine pancreas. Biochem Biophys Res Commun 1988; 156:1237–1242.

323. von Schonfeld J, Muller MK. The effect of pancreastatin on endocrine and exocrine pancreas. Scand J Gastroenterol 1991; 26:993–999.

324. Schmidt WE, Creutzfeldt W. Pancreastatin—a novel regulatory peptide? Acta Oncol 1991; 30:441–449.

325. Bretherton-Watt D, Ghatei MA, Bishop AE, Facer P, Fahey M, Hedges M, Williams G, Valentino KL, Tatemoto K, Roth K, Polak JM, Bloom SR. Pancreastatin distribution and plasma levels in the pig. Peptides 1988; 9:1005–1014.

326. Rustagi S, Warner RR, Divino CM. Serum pancreastatin: the next predictive neuroendocrine tumor marker. J Surg Oncol 2013; 108:126–128.

327. O'Dorisio TM, Krutzik SR, Woltering EA, Lindholm E, Joseph S, Gandolfi AE, Wang YZ, Boudreaux JP, Vinik AI, Go VL, Howe JR, Halfdanarson T, et al. Development of a highly sensitive and specific carboxy-terminal human pancreastatin assay to monitor neuroendocrine tumor behavior. Pancreas 2010; 39:611–616.

328. Funakoshi A, Miyasaka K, Nakamura R, Kitani K, Tatemoto K. Inhibitory effect of pancreastatin on pancreatic exocrine secretion in the conscious rat. Regul Pept 1989; 25:157–166.

329. Herzig KH, Louie DS, Tatemoto K, Owyang C. Pancreastatin inhibits pancreatic enzyme secretion by presynaptic modulation of acetylcholine release. Am J Physiol Gastrointest Liver Physiol 1992; 262:G113–G117.

330. Doi R, Inoue K, Hosotani R, Higashide S, Takori K, Funakoshi S, Yajima H, Rayford PL, Tobe T. Effects of synthetic human pancreastatin on pancreatic secretion and blood flow in rats and dogs. Peptides 1991; 12:499–502.

331. Miyasaka K, Funakoshi A, Kitani K, Tamamura H, Funakoshi S, Fujii N. Inhibitory effect of pancreastatin on pancreatic exocrine secretions. Pancreastatin inhibits central vagal nerve stimulation. Gastroenterology 1990; 99:1751–1756.

332. Miyasaka K, Funakoshi A, Yasunami Y, Nakamura R, Kitani K, Tamamura H, Funakoshi S, Fujii N. Rat pancreastatin inhibits both pancreatic exocrine and endocrine secretions in rats. Regul Pept 1990; 28:189–198.

333. Ishizuka J, Asada I, Poston GJ, Lluis F, Tatemoto K, Greeley GH, Jr, Thompson JC. Effect of pancreastatin on pancreatic endocrine and exocrine secretion. Pancreas 1989; 4:277–281.

334. Migita Y, Nakano I, Goto M, Ito T, Nawata H. Effect of pancreastatin on cerulein-stimulated pancreatic blood flow and exocrine secretion in anaesthetized rats. J Gastroenterol Hepatol 1999; 14:583–587.

335. Tatemoto K, Rokaeus A, Jornvall H, McDonald TJ, Mutt V. Galanin—a novel biologically active peptide from porcine intestine. FEBS Lett 1983; 164:124–128.

336. Barreto SG. Galanin. Pancreapedia: Exocrine Pancreas Knowledge Base, 2015. DOI: 10.3998/panc.2015.21.

337. Lang R, Gundlach AL, Kofler B. The galanin peptide family: receptor pharmacology, pleiotropic biological actions, and implications in health and disease. Pharmacol Ther 2007; 115:177–207.

338. Baltazar ET, Kitamura N, Hondo E, Narreto EC, Yamada J. Galanin-like immunoreactive endocrine cells in bovine pancreas. J Anat 2000; 196 (Pt 2):285–291.

339. Dunning BE, Ahren B, Veith RC, Bottcher G, Sundler F, Taborsky GJ, Jr. Galanin: a novel pancreatic neuropeptide. Am J Physiol Endocrinol Metab 1986; 251:E127–E133.

340. Lindskog S, Ahren B, Dunning BE, Sundler F. Galanin-immunoreactive nerves in the mouse and rat pancreas. Cell Tissue Res 1991; 264:363–368.

341. Runzi M, Muller MK, Schmid P, von Schonfeld J, Goebell H. Stimulatory and inhibitory effects of galanin on exocrine and endocrine rat pancreas. Pancreas 1992; 7:619–623.

342. Barreto SG, Woods CM, Carati CJ, Schloithe AC, Jaya SR, Toouli J, Saccone GT. Galanin inhibits caerulein-stimulated pancreatic amylase secretion via cholinergic nerves and insulin. Am J Physiol Gastrointest Liver Physiol 2009; 297:G333–G339.

343. Herzig KH, Brunke G, Schon I, Schaffer M, Folsch UR. Mechanism of galanin's inhibitory action on pancreatic enzyme secretion: modulation of cholinergic transmission-studies in vivo and in vitro. Gut 1993; 34:1616–1621.

Chapter 18

Regulation of pancreatic fluid and electrolyte secretion

Hiroshi Ishiguro, Makoto Yamaguchi, and Akiko Yamamoto

Department of Human Nutrition, Nagoya University Graduate School of Medicine, Nagoya, Japan

Introduction

Pancreatic juice is the product of two distinct secretory processes. Enzymes are secreted by exocytosis from acinar cells. Fluid and electrolyte secretion is achieved primarily by the vectorial transport of ions across the ductal epithelium accompanied by water in isotonic proportions (**Figure 1A**). Each day, the human pancreas delivers 6–20 g of digestive enzymes to the duodenum in approximately 2.5 liters of HCO_3^--rich fluid. The HCO_3^--rich fluid acts as a vehicle for transporting enzymes to the duodenum where the HCO_3^- neutralizes gastric acid. Moreover, pancreatic HCO_3^- secretion is thought to aid disaggregation of secreted enzymes following their exocytosis in the lumen of pancreatic duct.

Species-Dependent Regulation of Pancreatic Fluid and Electrolyte Secretion

The regulation of pancreatic fluid and electrolyte secretion and the volume and composition of the secreted fluid differ considerably from species to species (**Table 1**).[1] In all species, secretin evokes the secretion of HCO_3^--rich pancreatic juice. However, the amount of fluid is small in the rat and mouse (approximately fivefold less than in the cat, per gram of tissue). Maximal HCO_3^- concentration ($[HCO_3^-]$) reaches 130 mM or more in all species except the rat and mouse, in which 70 and 40 mM are the values. The secreted fluid is nearly isotonic ($[HCO_3^-] + [Cl^-] = 150–160$ mM) because of the high transepithelial water permeability and the Na^+ conductance of the paracellular pathway.

$[HCO_3^-]$ in human pancreatic juice reaches ~140 mM under stimulation. It is generally thought that most of the HCO_3^- secretion originates from epithelial cells lining the proximal pancreatic ducts (centroacinar cells, intralobular ducts, and small interlobular ducts) (**Figure 1A**). Pancreatic HCO_3^- secretion depends on the activity of cystic fibrosis transmembrane conductance regulator (CFTR) anion channel localized in the apical membrane of pancreatic duct cells (**Figure 1B**). While acinar cells add a small volume of Cl^- and protein-rich secretion to the lumen, luminal

$[HCO_3^-]$ quickly increases with time and distance along the duct as a result of HCO_3^- secretion by ductal epithelium. Thus, pancreatic duct cells can transport HCO_3^- against five to sixfold concentration gradients.

Vectorial Transport in Isolated Pancreatic Duct

Studies using isolated interlobular pancreatic ducts from guinea pig pancreas (**Figure 1C**) demonstrated that pancreatic duct cell is capable to secrete HCO_3^- into already HCO_3^--rich fluid under cAMP stimulation (**Figure 2A**).[2,3] The isolated pancreatic ducts respond to physiological concentrations (1–10 pM) of secretin (**Figure 2B**).[2,4]

Measurement of intracellular pH in luminally microperfused pancreatic duct demonstrates a polarity that is necessary to achieve vectorial transport of HCO_3^-.[5] While basolaterally applied HCO_3^- easily gains access to the cell, intraluminally applied HCO_3^- does not. HCO_3^- accumulation across the basolateral membrane is largely mediated by an electrogenic Na^+-coupled HCO_3^- transporter (NBCe1-B, SLC4A4) and the contribution of Na^+-H^+ exchanger 1 (NHE1, SLC9A1) is small (**Figure 3**).[1] Although the CFTR-dependent HCO_3^- permeability of the apical membrane is large enough to mediate the observed HCO_3^- secretion,[6] the cell is relatively hyperpolarized[7] and the electrochemical gradient for HCO_3^- is outwardly directed (does not allow entry of luminal HCO_3^-) (**Figure 3**).

Cholinergic stimulation also stimulates HCO_3^- secretion in guinea pig pancreatic duct.[8] Ca^{2+}-activated Cl^- channels (CaCCs) in the apical membrane[9] instead of CFTR are thought to be involved in HCO_3^- secretion. The molecular identity of CaCCs in pancreatic duct cell is not known at present.[10]

Regulation of Pancreatic Ductal Secretion

This section focuses on the regulatory mechanisms of pancreatic juice secretion, which directly act on pancreatic duct cells. *In vitro* studies using isolated pancreatic ducts or

e-mails: ishiguro@htc.nagoya-u.ac.jp; yamaguchi.makoto@e.mbox.nagoya-u.ac.jp; akikoy@htc.nagoya-u.ac.jp

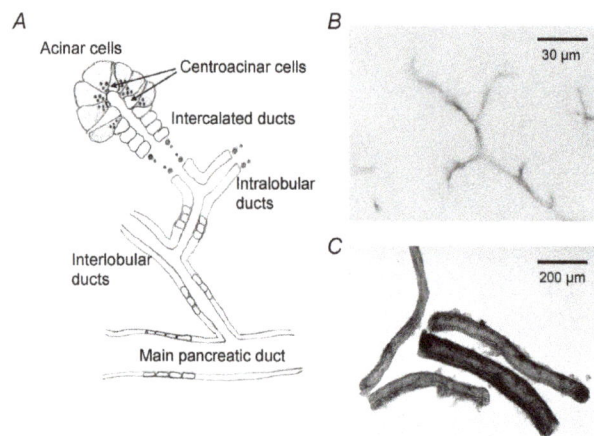

Figure 1. Pancreatic duct system. (**A**) Schematic architecture of pancreatic ductal tree. Centroacinar cells belong to duct cells and represent the terminal cells of the ductal tree. (**B**) Immunostaining of CFTR in guinea pig pancreas indicates apical membrane of pancreatic ductal epithelium. (**C**) Interlobular duct segments isolated from guinea pig pancreas (adapted from reference [1]).

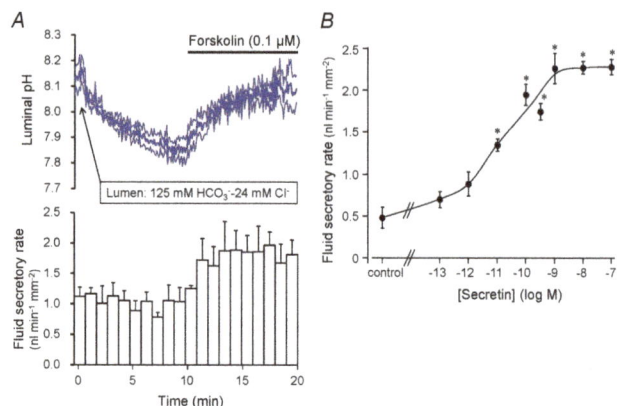

Figure 2. Fluid and HCO_3^- secretion by interlobular duct segments isolated from guinea pig pancreas. (**A**) When isolated segments of interlobular pancreatic ducts are cultured overnight, both ends of the ducts seal spontaneously. The lumen of sealed ducts is micropunctured and membrane-impermeable dextran conjugate of the pH sensor BCECF is injected. Time course changes in luminal pH (upper panel) and fluid secretory rates, determined from expansion speed of the lumen (lower panel), in ducts filled with high-HCO_3^- solution (125 mM HCO_3^--24 mM Cl^-, pH ~8.2), before and after cAMP stimulation (0.1 μM forskolin), are shown (means ± SE of four experiments) (adapted from reference [3]). Before cAMP stimulation, there is a slight decrease of luminal pH accompanied by fluid secretion (likely Cl-rich). Upon stimulation, luminal pH starts to rise, which is accompanied by increase of fluid secretion (HCO_3^--rich). (**B**) Effects of secretin concentration on the rate of fluid secretion into the lumen of sealed ducts (means ± SE of five experiments) (adapted from reference [4]).

Table 1. Species-dependent patterns of pancreatic HCO_3^- and fluid secretion. This table shows the response to stimuli given alone. Potentiation often occurs when stimuli are given together. Most data were obtained from studies on anesthetized animals. Quantitative differences may occur in conscious animals, especially in the rat, in which secretion is increased fivefold in conscious animals (adapted from reference [1]).

Species	Stimulus	Volume	[HCO₃] (mM)
Dog, cat, human	Spontaneous	0 (+)	–
	+ Secretin	+++++ +++++	145
	+ CCK	+	60
	+ Vagus	+	?
Rat, mouse	Spontaneous	+	25–30
	+ Secretin	++	40–70
	+ CCK	+++	30
	+ Vagus	++	?
Guinea pig	Spontaneous	+	95
	+ Secretin	+++++	150
	+ CCK	+++	140
	+ Vagus	+++	120

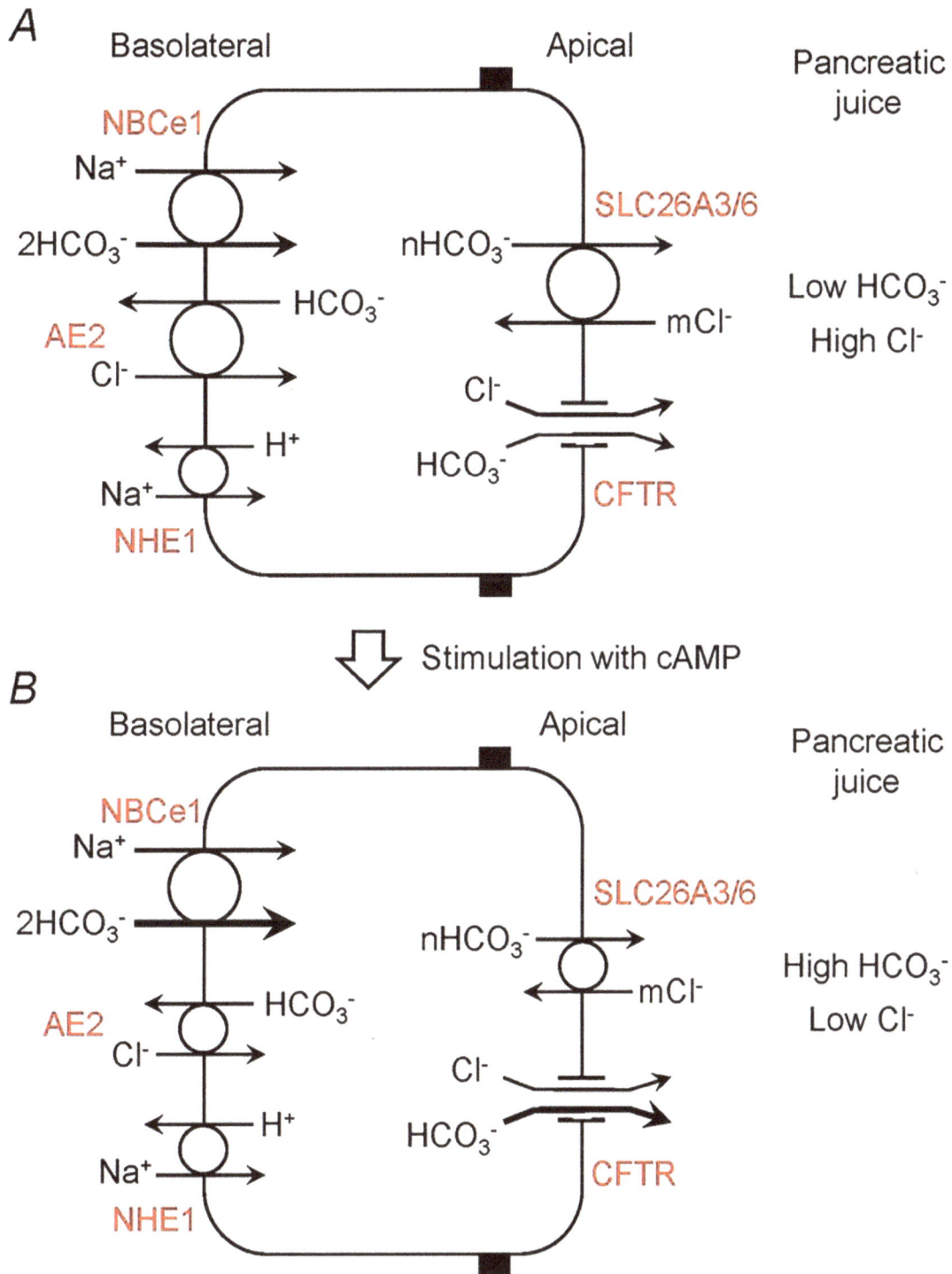

Figure 3. A model of HCO$_3^-$ secretion by guinea pig pancreatic duct cell. (**A**) Unstimulated duct cell whose apical membrane faces low-HCO$_3^-$, high-Cl$^-$ pancreatic juice. Accumulation of HCO$_3^-$ in the cell is achieved mainly by basolateral NBCe1, while the contribution of NHE1 is small. Some HCO$_3^-$ is lost via basolateral AE2. HCO$_3^-$ secretion across the apical membrane is mediated by SLC26A3/6 and CFTR. CFTR mediates both Cl$^-$ and HCO$_3^-$ secretion. (**B**) Cyclic AMP-stimulated duct cell of which the apical membrane faces high-HCO$_3^-$, low-Cl$^-$ pancreatic juice. NBCe1, SLC26A3/6, and CFTR are activated, while AE2 is inhibited. At the basolateral membrane, HCO$_3^-$ accumulation via NBCe1 is increased, while Cl$^-$ uptake (and HCO$_3^-$ loss) via AE2 is reduced. As luminal [HCO$_3^-$] increases, the contribution of SLC26A3/6 becomes smaller and CFTR mediates most of HCO$_3^-$ secretion. The Na$^+$:HCO$_3^-$ stoichiometry of NBCe1 is 1:2. SLC26A3 and SLC26A6 are reported to be electrogenic with opposite polarities: SLC26A3 mediates 2Cl-1HCO$_3^-$ exchange, while SLC26A6 mediates 1Cl-2HCO$_3^-$ exchange (adapted from reference [1]).

cell lines derived from human pancreatic ductal adenocarcinoma (Capan-1, PANC-1, CFPAC-1 from a patient with cystic fibrosis homozygous for F508del) demonstrated the presence of receptors of various hormones and neurotransmitters and their effects on intracellular signaling, ion transport, and HCO_3^-/fluid secretion.[9] Enhancement of HCO_3^-/fluid secretion is usually associated with accumulation of intracellular cAMP or elevation of intracellular Ca^{2+} ($[Ca^{2+}]_i$). However, we should note that elevation of $[Ca^{2+}]_i$ is not necessarily associated with the increase in HCO_3^-/fluid secretion and that pancreatic ductal regulation is also regulated by various intraluminal substances (**Table 2**).

The major regulators are secretin, vasoactive intestinal peptide (VIP), and acetylcholine, which increase intracellular Ca^{2+}/cAMP and stimulate HCO_3^- and fluid secretion. Cholecystokinin (CCK) and bombesin stimulate not only enzyme secretion by acinar cells but also HCO_3^- and fluid secretion by duct cells.[11] In isolated vascularly perfused guinea pig pancreas, arginine vasopressin (AVP) increases the vascular resistance and inhibited fluid secretion.[12] But AVP also directly acts on pancreatic duct cell and inhibits fluid secretion which is associated with $[Ca^{2+}]_i$ elevation. Reduction of pancreatic juice secretion by AVP may be physiologically important for body fluid conservation.

The long-term regulation of HCO_3^- and fluid secretion by pancreatic duct cells is thought to be under neurohormonal control and mediated by receptors localized in the basolateral membranes. However, in some occasions, the ion composition and pH of the luminal fluid as well as intraductal pressure needs to be controlled to avoid mechanical injury to ductal epithelium, autoactivation of trypsin, or precipitation of pancreatic stone protein and $CaCO_3$ in the lumen. **Table 2** summarizes the candidate list of local (stimulatory and inhibitory) regulators which are locally produced in the pancreas and play autocrine or paracrine roles.

Pancreatic duct epithelium expresses purinergic (P2Y and P2X) and adenosine receptors.[13–15] Thus, pancreatic ductal secretion is likely influenced by luminal ATP and ATPase (ectonucleotidase) secreted by acinar cells together with the exocytotic release of digestive enzymes[16,17] as well as basolateral ATP released as neurotransmitter. In isolated pancreatic ducts, ATP causes $[Ca^{2+}]_i$ increase when applied to either the apical or basolateral membrane.[18] However, apical and basolateral purinoceptors have opposite effects on secretion: luminally applied ATP enhances HCO_3^- and fluid secretion while basolaterally applied ATP inhibits secretion. Luminal adenosine activates CFTR in Capan-1 cells.[15]

Table 2. Local regulation of pancreatic ductal secretion. This table shows the list of substances that are thought to be locally released in the pancreas and directly act on pancreatic duct cells.

Substances	Receptor	Membrane localization	Effects	Physiological/pathological role	References
ATP	$P2Y_2$	Basolateral	Inhibitory	ATP released from nerve endings	13, 14, 18
	P2X and $P2Y_2$	Apical	Stimulatory	ATP secreted from acinar cells	13, 14, 18
Adenosine	A_{2B}	Apical	Stimulatory	Adenosine produced by the hydrolysis of ATP	15
Serotonin (5-HT)	$5-HT_3$	Basolateral	Inhibitory	Released from EC cells in ductal epithelium upon elevation of intraluminal pressure	19
Substance P	NK2, NK3	Basolateral	Inhibitory	Released from sensory nerve endings in inflammatory condition	22
Angiotensin II	AT1	Basolateral and apical	Stimulatory	Local renin-angiotensin system is activated in response to hypoxia, pancreatitis, etc.	26
Trypsin	PAR-2	Basolateral	Stimulatory	Transiently evoke HCO_3^- secretion Induce exocytosis of mucins to the lumen	27, 28
		Apical	Inhibitory	Reduce HCO_3^- secretion and enhance trypsinogen activation	29
Ca^{2+}	CaSR	Apical	Stimulatory	Stimulate HCO_3^- and fluid secretion and prevent the formation of pancreatic stone	31
Guanylin	GC-C	Apical	Stimulatory	?	33

5-hydroxytryptamine (5-HT, serotonin)-immuno-reactive cells with the morphological characteristics of enterochromaffin cells (EC cells) are found scattered in the pancreatic duct epithelium (**Figure 4D**).[19] These cells may sense the elevation of intraductal pressure (as EC cells do in the intestine), and the released 5-HT may inhibit fluid secretion by pancreatic duct (**Figure 4C**). In fact, basolateral application of 5-HT at relatively low concentrations (IC50: ~30 nM) inhibited fluid secretion (~75%) by isolated pancreatic ducts (**Figure 4A**). A small elevation (+ 3 cm H_2O) of intraluminal pressure reversibly reduced pancreatic fluid secretion *in vivo* (anesthetized guinea pig), and the effect was attenuated by intravenous granisetron, a 5-HT$_3$ receptor antagonist (**Figure 4B**). The intraductal pressure of the human pancreas increases after feeding by as much as ~20 cm of H_2O.[20] The pressure is elevated in some patients with chronic pancreatitis. Thus, 5-HT may regulate pancreatic fluid secretion under physiological and pathological conditions.

During the course of acute and chronic pancreatitis, substance P and calcitonin gene-related peptide (CGRP) are released from intrapancreatic endings of sensory nerves, which cause vasodilatation and mediate hyperalgesia.[21] Substance P strongly inhibits HCO_3^- and fluid secretion by isolated pancreatic duct.[22] Pancreatic fluid hypersecretion was observed during the early stage of caerulein-induced

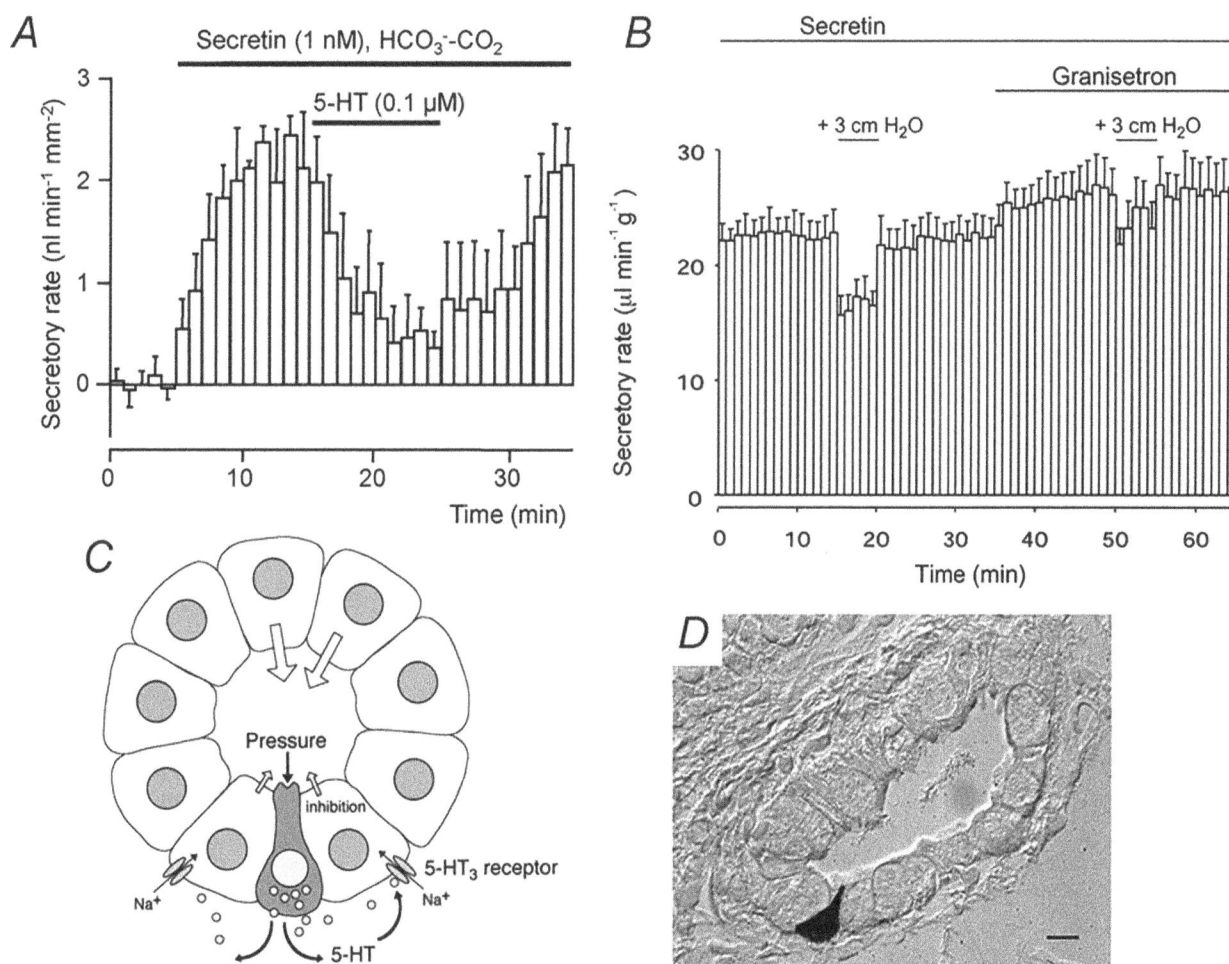

Figure 4. Serotonin (5-HT) inhibits pancreatic fluid secretion in an autocrine manner. (**A**) Effects of serotonin (5-HT 0.1 μM) on secretin (1 nM)-stimulated fluid secretion by isolated guinea pig pancreatic ducts (means ± SE of four experiments). (**B**) Effects of small elevations of intraductal pressure and 5-HT$_3$ receptor antagonist (granisetron 40 μg/kg/h) on *in vivo* secretin (1 μg/kg/h)-stimulated fluid secretion from guinea pig pancreas (means ± SE of four experiments). (**C**) A hypothetical mechanism. EC cells in the duct epithelium sense the elevation of intraductal pressure and release 5-HT into the interstitium. The released 5-HT binds to 5-HT$_3$ receptor in the basolateral membrane of neighboring duct cells, which facilitates Na^+ entry into the cell. That reduces the inward gradient of Na^+, which is necessary for intracellular HCO_3^- accumulation. (**D**) A 5-HT-immunoreactive cell embedded in pancreatic duct epithelium (adapted from reference [19]).

experimental acute pancreatitis in rats.[23] While it is not known whether such fluid hypersecretion occurs in human acute pancreatitis, substance P may protect the pancreas by inhibiting ductal fluid secretion

A local intrapancreatic renin-angiotensin system (RAS) is activated in response to hypoxia, pancreatitis, and so on, and involved in apoptosis and fibrosis.[24] Inhibition of RAS in stellate cells attenuates pancreatic fibrosis in a rat model of chronic pancreatitis.[25] Angiotensin II (either luminal or basolateral) induces anion secretion in CFPAC-1 cells, which is associated with $[Ca^{2+}]_i$ elevation.[26] Thus, the activation of intrapancreatic RAS may increase the volume of pancreatic juice.

The membrane localization of proteinase-activated receptor-2 (PAR-2) in pancreatic duct cells and the effect of trypsin on HCO_3^- secretion are controversial. Basolateral trypsin causes $[Ca^{2+}]_i$ increase and induces HCO_3^- secretion via activation of CFTR-dependent apical $Cl-HCO_3^-$ exchange in Capan-1 cells[27] and stimulates exocytosis of mucins to the lumen in canine pancreatic duct cells.[28] On the contrary, apical trypsin causes $[Ca^{2+}]_i$ increase but reduces HCO_3^- secretion by inhibition of CFTR and apical $Cl-HCO_3^-$ exchange in guinea pig pancreatic duct cells.[29] Consequent acidification of luminal pH may promote the autoactivation of trypsin. Those controversial data may be related to the conflicting effects of genetic PAR-2 deletion on experimental pancreatitis.[30] While caerulein-induced pancreatitis is worse in PAR-2$^{-/-}$ mice, taurocholate-induced pancreatitis is milder in PAR-2$^{-/-}$ mice.

Ca^{2+}-sensing receptor (CaSR) is localized to the apical membrane of human and rat pancreatic duct cells.[31] Luminal application of Gd^{3+}, a CaSR agonist, induces $[Ca^{2+}]_i$ elevation and stimulates HCO_3^- secretion in rat pancreatic duct. $[HCO_3^-]$ in human pancreatic juice reaches ~140 mM and $[CO_3^{2-}]$ is estimated to be 0.03~1 mM. The pancreatic juice also contains millimolar quantities of Ca^{2+}, which are released from acinar cells along with digestive enzymes. CaSR-mediated fluid secretion by pancreatic duct may be crucial to prevent precipitation of $CaCO_3$ by dilution. Some studies suggest that *CaSR* variants increase the risk of chronic pancreatitis.[32]

In human pancreas, guanylin is specifically localized to centroacinar cells and proximal duct cells.[33] Guanylin activates CFTR via cGMP in Capan-1 cells. Thus, guanylin may regulate ductal cell secretion in a luminocrine mode, although the signal to induce guanylin secretion is not known.

Pancreatic Ductal Secretion and Lifestyle Disease

Effects of ethanol

Ethanol abuse is the leading cause of acute and chronic pancreatitis. The more the daily amount of alcohol drinking, the larger the risk of developing chronic pancreatitis.[34] However, there is no known threshold dose of ethanol that induces chronic pancreatitis and patients drinking even small quantities of ethanol (1–20 g/day) run a higher risk. Studies using isolated pancreatic ducts from guinea pig show that low and high concentrations of ethanol have opposite effects on fluid secretion. While low concentrations of ethanol (0.1–30 mM relevant to normal drinking conditions) enhance fluid secretion,[4,35] high concentrations of ethanol (100 mM relevant to lethal level) reduce fluid secretion.[36]

The objective effects of ethanol on the central nervous system (CNS) appear at a blood ethanol level of ~5 mM, and they are attributed to the modulation of ion channels.[37] Ethanol reversibly and strongly augments secretin-stimulated fluid secretion in guinea pig pancreatic duct (**Figure 5A**) where the maximal augmentative effect is achieved with 1 mM ethanol (**Figure 5B**).[4] The augmentation of fluid secretion involves a transient increase of $[Ca^{2+}]_i$ (**Figure 5D**) likely via the activation of plasma membrane Ca^{2+} channels. The effects of *n*-alcohols on pancreatic duct cells are dependent on the length of alkyl chain. While methanol (C1) and ethanol (C2) augment fluid secretion and induce a transient $[Ca^{2+}]_i$ increase, propanol (C3) and butanol (C4) inhibit fluid secretion and reduce $[Ca^{2+}]_i$ (**Figure 5C and D**).[35] The observation is similar to the so-called "cutoff" effects of *n*-alcohols on neurotransmitter-gated ion channels in CNS neurons that increase in potency, with an increasing alkyl chain length of *n*-alcohol up to a point but then disappears with a further increase in chain length.[37] Concerning the pathogenesis of alcoholic chronic pancreatitis, the ductal fluid hypersecretion by ethanol may raise the intraductal pressure when the flow of pancreatic juice is blocked by the presence of highly viscous juice or protein plugs.

Ethanol (100 mM) inhibits cAMP-stimulated CFTR currents, apical $Cl-HCO_3^-$ exchange, and fluid secretion in guinea-pig pancreatic ductal epithelial cells.[36] Moreover, prolonged treatment with 100 mM ethanol decreases membrane expression of CFTR. Those effects of ethanol involve sustained elevation of $[Ca^{2+}]_i$ and depletion of ATP, which is probably associated with the pathogenesis of alcoholic pancreatitis.

Effects of glucose

Pancreatic exocrine dysfunction is frequently found in patients with type 1 and type 2 diabetes mellitus. Both enzyme secretion from acinar cells and fluid/HCO_3^- secretion by duct cells are impaired. It was reported that exposure to high glucose (for 72 hours) activates polyol metabolism and decreases the activity of Na^+,K^+-ATPase in Capan-1 cells.[38] The resultant elevation of $[Na^+]_i$ would attenuate the inward Na^+ gradient across the basolateral membrane,

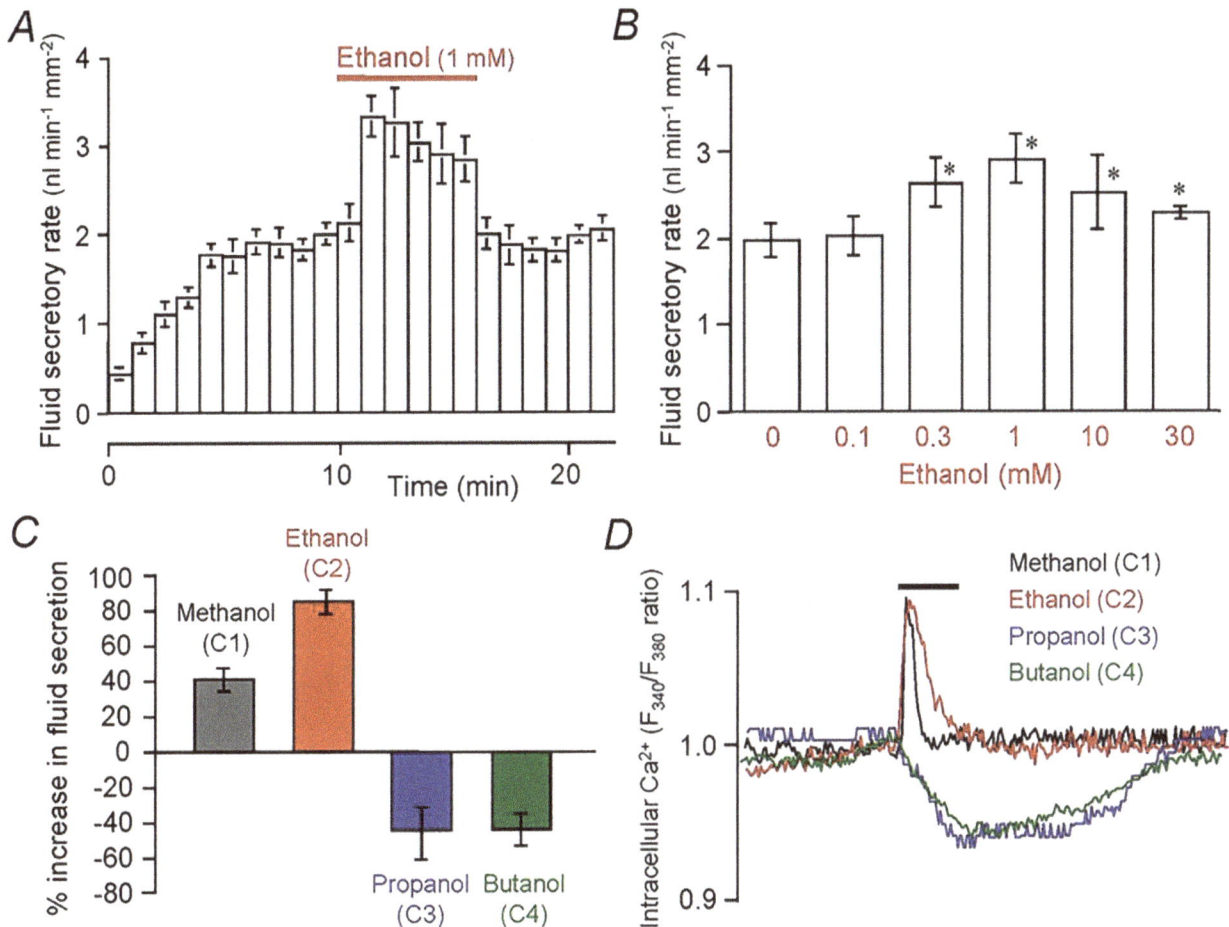

Figure 5. Effects of low concentrations of alcohol on fluid secretion and intracellular Ca²⁺ response in isolated pancreatic ducts.
(**A**) Rapid and reversible effects of ethanol on secretin (0.3 nM)-stimulated fluid secretion in isolated guinea pig pancreatic ducts (means ± SE of six experiments). (**B**) Effects of various concentrations of ethanol (0.1–30 mM) on secretin (1 nM)-stimulated fluid secretion (means ± SE of four to six experiments). (**C**) Effects of n-alcohols (C1 methanol, C2 ethanol, C3 propanol, C4 butanol at 1 mM) on secretin-stimulated fluid secretion (means ± SE of four experiments). (**D**) Effects of n-alcohols (C1-C4) on intracellular Ca²⁺ concentration under secretin stimulation. Each trace is representative of four experiments (adapted from references [4, 35]).

which reduces HCO_3^- uptake via Na^+-HCO_3^- cotransport and Na^+-H^+ exchange (**Figure 3**). This leads to a reduction of fluid secretion.

Even acute exposure to high glucose concentrations (15 minutes) inhibits fluid secretion and reduces basolateral HCO_3^- uptake in secretin-stimulated rat pancreatic ducts.[39] This inhibition is most likely attributed to transepithelial absorption of glucose. Rat interlobular pancreatic ducts express Na^+-glucose cotransporter (SGLT1) and glucose transporters (GLUT1 and GLUT2) (**Figure 6A**) and absorb luminal glucose iso-osmotically (**Figure 6B**). The absorption of luminal glucose is abolished by phlorizin, an inhibitor of SGLT1 in the lumen. Absorption of luminal Na^+ and glucose via SGLT1 would increase $[Na^+]_i$ and depolarize the apical membrane. $[Na^+]_i$ elevation reduces HCO_3^- uptake and apical depolarization reduces the driving force

for HCO_3^- secretion via CFTR, both of which leads to a reduction of fluid secretion.

The glucose concentration in human pancreatic juice (0.5~1 mM) is much lower than in plasma. Under physiological conditions, pancreatic duct epithelium probably absorbs luminal glucose via apical SGLT1 to maintain the glucose concentration at a low level in the pancreatic juice.

Computer Simulation of Pancreatic Ductal Secretion

On the basis of previous modelling studies,[40,41] a new computational model has been developed as a set of simultaneous ordinary differential equations in MATLAB (MathWorks, Natick, MA, USA) using the Simulink interface to provide a modular structure and facilitate the simulation of time-course experiments.[42] The activities and

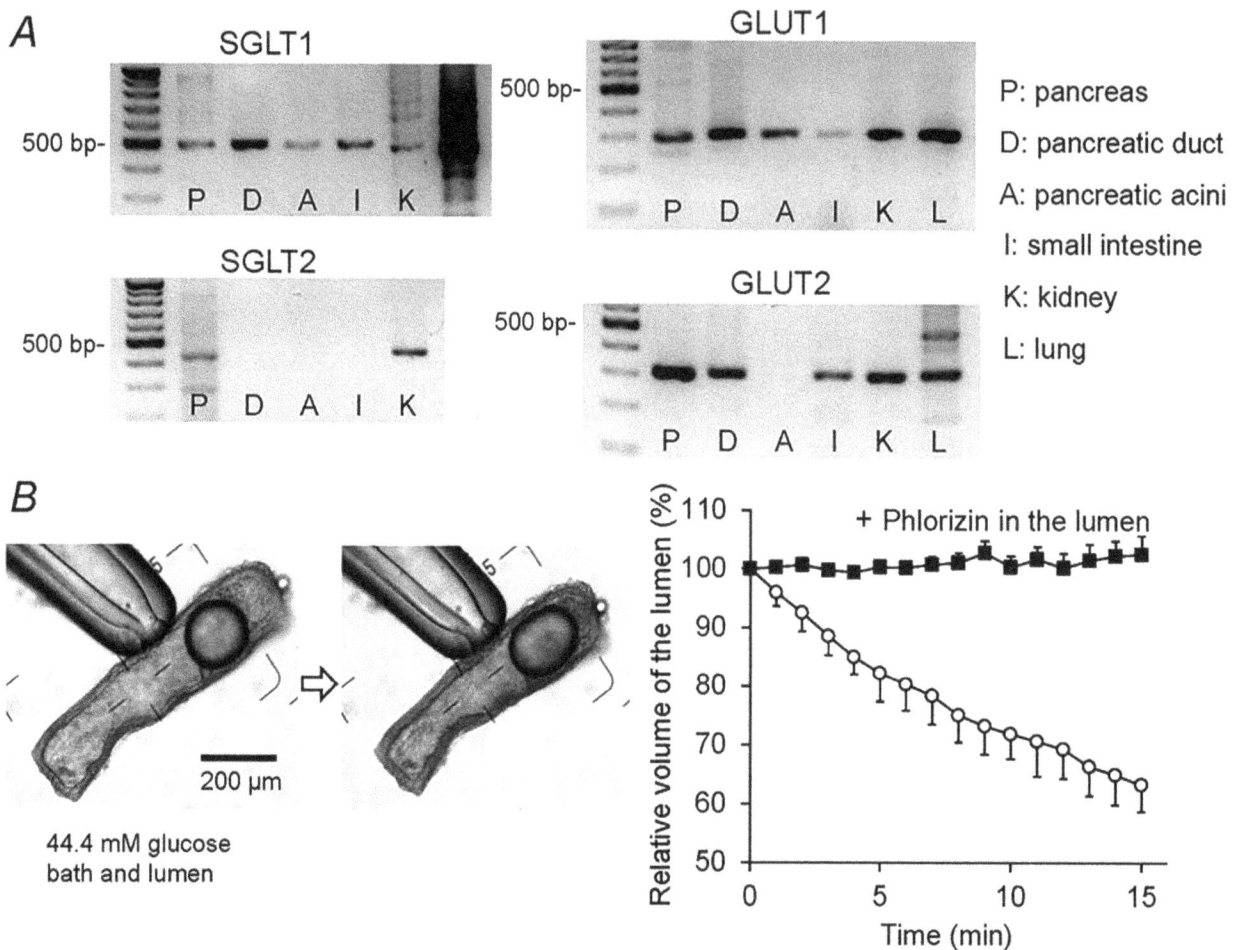

Figure 6. Expression of glucose transporters and transepithelial glucose absorption by isolated pancreatic ducts. (A) mRNA expression of glucose transporters (SGLT1, SGLT2, GLUT1, GLUT2) in the pancreas (P), pancreatic duct (D), pancreatic acini (A), small intestine (I), kidney (K), and lung (L) of the rat. (B) Time course shrinkage of the lumen indicates transepithelial absorption of luminal glucose by isolated rat pancreatic ducts. The lumen is filled with the HCO_3-free HEPES-buffered solution containing high glucose (44.4 mM). The bath is also perfused with the HCO_3-free HEPES-buffered solution containing 44.4 mM glucose. The left panel shows images of a duct at the beginning and end of a representative experiment (15 minutes) showing 33 percent shrinkage of the lumen. Time-course changes of the luminal volume (right panel) demonstrate that phlorizin (0.5 mM) in the lumen completely inhibits glucose absorption (means ± SE of four experiments) (adapted from reference [39]).

permeabilities of individual ion channels and transporters are estimated by least-squares fitting of the model predictions to the experimental data of isolated guinea pig pancreatic duct (intracellular pH, Cl^-, membrane potential, luminal pH, and volume).

The pancreatic duct epithelium is represented as a four-compartment system comprising the bath, the lateral intercellular space, the cytoplasm, and the lumen. In the perfused duct model, the composition of the luminal fluid is set to predefined values. Alternatively, in the secreting duct model (**Figure 7A**), the composition of the luminal fluid is allowed to evolve with time and defined by the fluid secreted by the epithelial cells. Stimulation

is replicated by increasing the activities of the basolateral Na^+-HCO_3^- cotransporter (NBCe1-B, SLC4A4) and apical Cl^--HCO_3^- exchanger (SLC26A6), increasing the basolateral K^+ permeability and apical Cl^- and HCO_3^- permeabilities (CFTR) and reducing the activity of the basolateral Cl^--HCO_3^- exchanger (AE2, SLC4A2). Under these conditions, the model secretes ~140 mM HCO_3^- at a rate of ~3.5 nl min^{-1} mm^{-2}, which is consistent with experimental observations.[2,4]

Figure 7A shows steady-state fluxes of HCO_3^-, H^+, and Cl^- at rest and following cAMP stimulation in the model. Intracellular/luminal/bath Cl^-/HCO_3^- concentrations, osmolarities, and basolateral potential differences

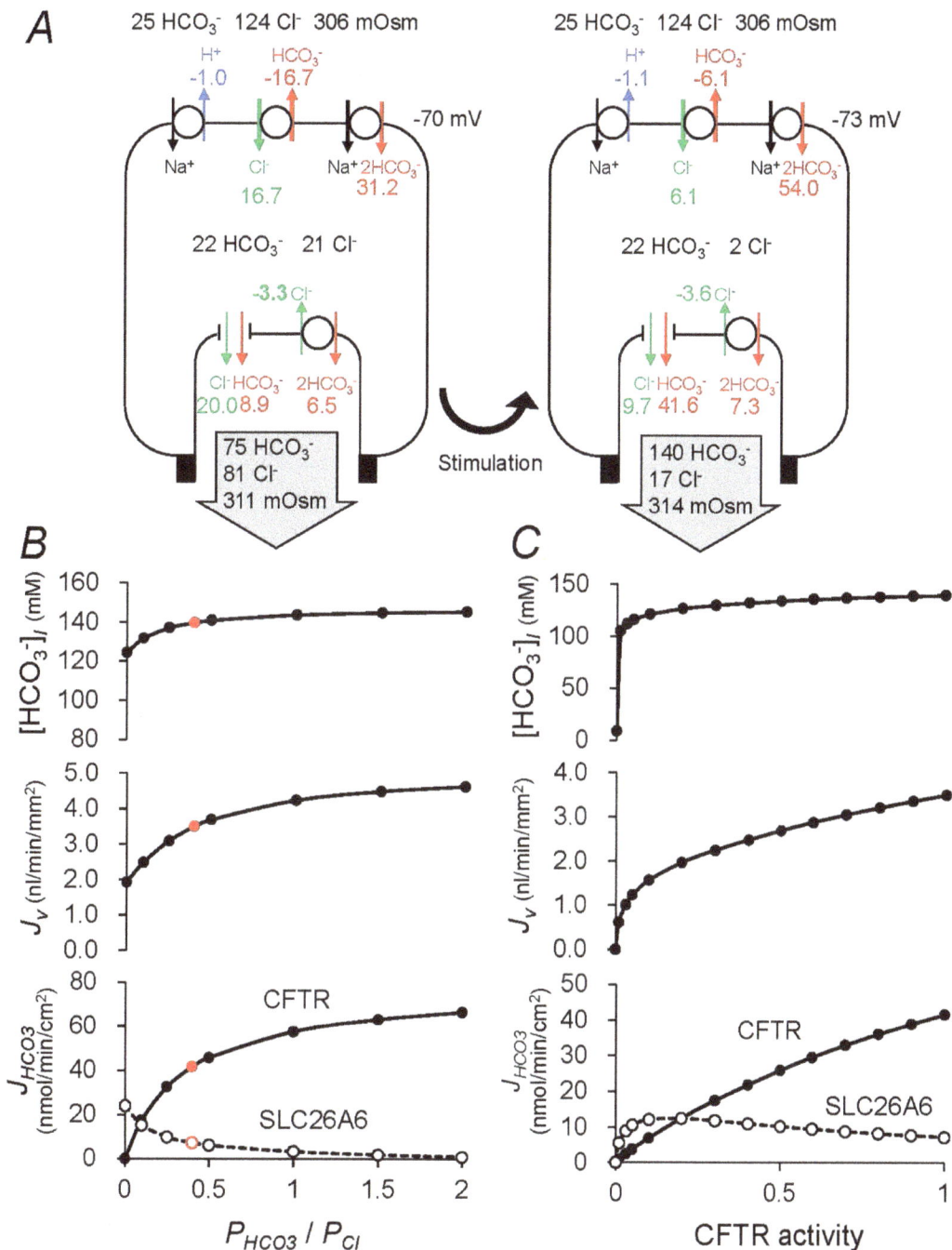

Figure 7. Computational model of HCO$_3^-$ secretion by pancreatic duct cell. (**A**) A secreting model of guinea pig pancreatic duct epithelium shows steady-state fluxes of HCO$_3^-$ (red font), H$^+$ (blue font), and Cl$^-$ (green font) mediated by (clockwise from top left) NHE1, AE2, NBC1, SLC26A6, and CFTR. Also shown are the steady-state intracellular/luminal/bath Cl$^-$/HCO$_3^-$ concentrations, osmolarities, and basolateral potential difference (black font). The left panel represents the unstimulated duct and the right panel shows the effect of altering the transporter/channel activities to represent cAMP stimulation. The Cl$^-$:HCO$_3^-$ stoichiometry of the SLC26A6 exchanger is assumed to be 1:2. Fluxes are given in nmol min^{-1} cm^{-2} (normalized to the luminal surface area of the epithelium). (**B**) Effects of varying the HCO$_3^-$/Cl$^-$ permeability ratio of CFTR on HCO$_3^-$ and fluid secretion in the secreting duct model. Steady-state values in the cAMP-stimulated condition of secreted HCO$_3^-$ concentration ([HCO$_3^-$]$_l$), secretory volume flow (Jv), and net apical HCO$_3^-$ fluxes via CFTR and SLC26A6 (J_{HCO3}) as a function of the HCO$_3^-$/Cl$^-$ permeability ratio (P_{HCO3}/P_{Cl}) of CFTR. The sum of the permeabilities ($P_{HCO3} + P_{Cl}$) is maintained at a constant value. (**C**) Effects of CFTR activity on HCO$_3^-$ and fluid secretion in the secreting duct model. Steady-state values in the cAMP-stimulated condition of [HCO$_3^-$]$_l$, Jv, and J_{HCO3} via CFTR and SLC26A6 as a function of the CFTR activity (adapted from reference [42]).

are shown as well. In the steady state, ion compositions of the effluent (pancreatic juice), the luminal fluid, and the fluid secreted by the epithelial cells are identical. While the apical HCO_3^- flux at rest is divided between CFTR and SLC26A6 (assumed to be $1Cl^-$:$2HCO_3^-$ stoichiometry in the standard model), the apical HCO_3^- flux under cAMP stimulation is predominantly (~90%) via CFTR rather than SLC26A6. The luminal fluid is ~3 percent hypertonic compared to the bath under stimulation.

The HCO_3^-/Cl^- permeability ratio (P_{HCO3}/P_{Cl}) of CFTR has been estimated to be 0.2~0.5 and reported to be regulated by Cl^--sensitive kinases.[43] P_{HCO3}/P_{Cl} is set at 0.4 in the standard model (**Figure 7A**). Increasing the ratio to 1.0 has a little effect on the volume (J_v) and HCO_3^- concentration ($[HCO_3^-]_l$) of the secreted fluid (**Figure 7B**). When the ratio is decreased to zero, SLC26A6 compensates for the defective HCO_3^- transport of by CFTR and the model still secretes ~120 mM HCO_3^- at a rate of ~2 nl min^{-1} mm^{-2}.

CFTR Cl^- permeability is set at 60×10^{-6} cm s^{-1} in the standard model. **Figure 7C** shows the effects of varying CFTR activity on $[HCO_3^-]_l$, J_v, and apical HCO_3^- fluxes (J_{HCO3}) via CFTR and SLC26A6. When CFTR activity is completely lost, SLC26A6 by itself cannot support HCO_3^- and Cl^- secretion and there is no fluid secretion. Residual CFTR activity as low as 10 percent in combination with SLC26A6 accomplishes fluid secretion containing ~120 mM HCO_3^- at a rate of ~1.5 nl min^{-1} mm^{-2}. Effective treatment of cystic fibrosis requires restoration of CFTR function and different organs have different requirements for CFTR function. Ion transport in airway epithelium is normalized when 6–10 percent of cells have corrected CFTR function.[44] Recent studies in young children with cystic fibrosis suggest improvement of pancreatic exocrine function by CFTR modulators.[45]

Although the computational model is simple and lacks molecular interactions between CFTR and other transporters and regulation by intracellular Cl^-,[10] it well reproduces the experimental data including the redundancy of acid/base transporters: CFTR and SLC26A6 in the apical membrane and NBC and NHE in the basolateral membrane.[42]

Acknowledgments

This work was supported by grants from the Japanese Society for the Promotion of Science and the Japanese study group for pediatric rare and intractable hepato-biliary-pancreatic diseases provided by the Ministry of Health, Labor, and Welfare of Japan.

References

1. Ishiguro H, Yamamoto A, Steward, M.C. Pancreatic bicarbonate secretion. In: Kuipers EJ, ed. Encyclopedia of gastroenterology. Volume 4. 2nd ed. Oxford: Academic Press, 2021:24–29.

2. Ishiguro H, Naruse S, Steward MC, Kitagawa M, Ko SB, Hayakawa T, Case RM. Fluid secretion in interlobular ducts isolated from guinea-pig pancreas. J Physiol 1998; 511:407–422.

3. Ko SB, Yamamoto A, Azuma S, Song H, Kamimura K, Nakakuki M, Gray MA, Becq F, Ishiguro H, Goto H. Effects of CFTR gene silencing by siRNA or the luminal application of a CFTR activator on fluid secretion from guinea-pig pancreatic duct cells. Biochem Biophys Res Commun 2011; 410:904–909.

4. Yamamoto A, Ishiguro H, Ko SB, Suzuki A, Wang Y, Hamada H, Mizuno N, Kitagawa M, Hayakawa T, Naruse S. Ethanol induces fluid hypersecretion from guinea-pig pancreatic duct cells. J Physiol 2003; 551:917–926.

5. Ishiguro H, Naruse S, Kitagawa M, Suzuki A, Yamamoto A, Hayakawa T, Case RM, Steward MC. CO$_2$ permeability and bicarbonate transport in microperfused interlobular ducts isolated from guinea-pig pancreas. J Physiol 2000; 528:305–315.

6. Ishiguro H, Steward MC, Naruse S, Ko SB, Goto H, Case RM, Kondo T, Yamamoto A. CFTR functions as a bicarbonate channel in pancreatic duct cells. J Gen Physiol 2009; 133:315–326.

7. Ishiguro H, Steward MC, Sohma Y, Kubota T, Kitagawa M, Kondo T, Case RM, Hayakawa T, Naruse S. Membrane potential and bicarbonate secretion in isolated interlobular ducts from guinea-pig pancreas. J Gen Physiol 2002; 120:617–628.

8. Ishiguro H, Steward MC, Wilson RW, Case RM. Bicarbonate secretion in interlobular ducts from guinea-pig pancreas. J Physiol 1996; 495:179–91.

9. Argent BE, Gray MA, Steward MC, Case RM. Cell physiology of pancreatic ducts. In: Johnson LR, ed. Physiology of the gastrointestinal tract. 4th ed. San Diego: Elsevier, 2006:1371–1396.

10. Lee MG, Kim Y, Jun I, Aoun J, Muallem S. Molecular mechanisms of pancreatic bicarbonate secretion. Pancreapedia: Exocrine Pancreas Knowledge Base, 2020. DOI: 10.3998/panc.2020.06.

11. Szalmay G, Varga G, Kajiyama F, Yang XS, Lang TF, Case RM, Steward MC. Bicarbonate and fluid secretion evoked by cholecystokinin, bombesin and acetylcholine in isolated guinea-pig pancreatic ducts. J Physiol 2001; 535:795–807.

12. Ko SB, Naruse S, Kitagawa M, Ishiguro H, Murakami M, Hayakawa T. Arginine vasopressin inhibits fluid secretion in guinea pig pancreatic duct cells. Am J Physiol Gastrointest Liver Physiol 1999; 277:G48–G54.

13. Luo X, Zheng W, Yan M, Lee MG, Muallem S. Multiple functional P2X and P2Y receptors in the luminal and basolateral membranes of pancreatic duct cells. Am J Physiol Cell Physiol 1999; 277:C205–C215.

14. Novak I. Purinergic receptors in the endocrine and exocrine pancreas. Purinergic Signal 2008; 4:237–253.

15. Hayashi M, Inagaki A, Novak I, Matsuda H. The adenosine A2B receptor is involved in anion secretion in human pancreatic duct Capan-1 epithelial cells. Pflugers Arch 2016; 468:1171–1181.

16. Sorensen CE, Novak I. Visualization of ATP release in pancreatic acini in response to cholinergic stimulus. Use of fluorescent probes and confocal microscopy. J Biol Chem 2001; 276:32925–32932.

17. Kordas KS, Sperlagh B, Tihanyi T, Topa L, Steward MC, Varga G, Kittel A. ATP and ATPase secretion by exocrine pancreas in rat, guinea pig, and human. Pancreas 2004; 29:53–60.

18. Ishiguro H, Naruse S, Kitagawa M, Hayakawa T, Case RM, Steward MC. Luminal ATP stimulates fluid and HCO_3^- secretion in guinea-pig pancreatic duct. J Physiol 1999; 519:551–558.

19. Suzuki A, Naruse S, Kitagawa M, Ishiguro H, Yoshikawa T, Ko SB, Yamamoto A, Hamada H, Hayakawa T. 5-hydroxytryptamine strongly inhibits fluid secretion in guinea pig pancreatic duct cells. J Clin Invest 2001; 108:749–756.

20. Hallenbeck GA. Biliary and pancreatic intraductal pressures. In: Code CF eds. Handbook of physiology. Alimentary Canal Section 6, Vol 4. Washington, DC: American Physiological Society, 1967:1007–1025.

21. Ikeura T, Kataoka Y, Wakabayashi T, Mori T, Takamori Y, Takamido S, Okazaki K, Yamada H. Effects of sensory denervation by neonatal capsaicin administration on experimental pancreatitis induced by dibutyltin dichloride. Med Mol Morphol 2007; 40:141–149.

22. Hegyi P, Rakonczay Z, Jr, Tiszlavicz L, Varro A, Toth A, Racz G, Varga G, Gray MA, Argent BE. Protein kinase C mediates the inhibitory effect of substance P on HCO_3^- secretion from guinea pig pancreatic ducts. Am J Physiol Cell Physiol 2005; 288:C1030–C1041.

23. Czako L, Yamamoto M, Otsuki M. Pancreatic fluid hypersecretion in rats after acute pancreatitis. Dig Dis Sci 1997; 42:265–272.

24. Leung PS. The physiology of a local renin-angiotensin system in the pancreas. J Physiol 2007; 580:31–37.

25. Kuno A, Yamada T, Masuda K, Ogawa K, Sogawa M, Nakamura S, Nakazawa T, Ohara H, Nomura T, Joh T, Shirai T, Itoh M. Angiotensin-converting enzyme inhibitor attenuates pancreatic inflammation and fibrosis in male Wistar Bonn/Kobori rats. Gastroenterology 2003; 124:1010–1019.

26. Chan HC, Law SH, Leung PS, Fu LX, Wong PY. Angiotensin II receptor type I-regulated anion secretion in cystic fibrosis pancreatic duct cells. J Membr Biol 1997; 156:241–249.

27. Namkung W, Lee JA, Ahn W, Han W, Kwon SW, Ahn DS, Kim KH, Lee MG. Ca^{2+} activates cystic fibrosis transmembrane conductance regulator- and Cl^--dependent HCO_3^- transport in pancreatic duct cells. J Biol Chem 2003; 278:200–207.

28. Kim MH, Choi BH, Jung SR, Sernka TJ, Kim S, Kim KT, Hille B, Nguyen TD, Koh DS. Protease-activated receptor-2 increases exocytosis via multiple signal transduction pathways in pancreatic duct epithelial cells. J Biol Chem 2008; 283:18711–18720.

29. Pallagi P, Venglovecz V, Rakonczay Z, Jr, Borka K, Korompay A, Ozsvari B, Judak L, Sahin-Toth M, Geisz A, Schnur A, Maleth J, Takacs T, et al. Trypsin reduces pancreatic ductal bicarbonate secretion by inhibiting CFTR Cl^- channels and luminal anion exchangers. Gastroenterology 2011; 141:2228–2239.

30. Laukkarinen JM, Weiss ER, van Acker GJ, Steer ML, Perides G. Protease-activated receptor-2 exerts contrasting model-specific effects on acute experimental pancreatitis. J Biol Chem 2008; 283:20703–20712.

31. Bruce JI, Yang X, Ferguson CJ, Elliott AC, Steward MC, Case RM, Riccardi D. Molecular and functional identification of a Ca^{2+} (polyvalent cation)-sensing receptor in rat pancreas. J Biol Chem 1999; 274:20561–20568.

32. Larusch J, Whitcomb DC. Genetics of pancreatitis with a focus on the pancreatic ducts. Minerva Gastroenterol Dietol 2012; 58:299–308.

33. Kulaksiz H, Schmid A, Honscheid M, Eissele R, Klempnauer J, Cetin Y. Guanylin in the human pancreas: a novel luminocrine regulatory pathway of electrolyte secretion via cGMP and CFTR in the ductal system. Histochem Cell Biol 2001; 115:131–145.

34. Lin Y, Tamakoshi A, Hayakawa T, Ogawa M, Ohno Y, Research Committee on Intractable Pancreatic Disease. Associations of alcohol drinking and nutrient intake with chronic pancreatitis: findings from a case-control study in Japan. Am J Gastroenterol 2001; 96:2622–2627.

35. Hamada H, Ishiguro H, Yamamoto A, Shimano-Futakuchi S, Ko SB, Yoshikawa T, Goto H, Kitagawa M, Hayakawa T, Seo Y, Naruse S. Dual effects of n-alcohols on fluid secretion from guinea pig pancreatic ducts. Am J Physiol Cell Physiol 2005; 288:C1431–C1439.

36. Maleth J, Balazs A, Pallagi P, Balla Z, Kui B, Katona M, Judak L, Nemeth I, Kemeny LV, Rakonczay Z, Jr, Venglovecz V, Foldesi I, et al. Alcohol disrupts levels and function of the cystic fibrosis transmembrane conductance regulator to promote development of pancreatitis. Gastroenterology 2015; 148:427–439.

37. Korpi ER, Makela R, Uusi-Oukari M. Ethanol: novel actions on nerve cell physiology explain impaired functions. News Physiol Sci 1998; 13:164–170.

38. Busik JV, Hootman SR, Greenidge CA, Henry DN. Glucose-specific regulation of aldose reductase in capan-1 human pancreatic duct cells in vitro. J Clin Invest 1997; 100:1685–1692.

39. Futakuchi S, Ishiguro H, Naruse S, Ko SB, Fujiki K, Yamamoto A, Nakakuki M, Song Y, Steward MC, Kondo T, Goto H. High glucose inhibits HCO_3^- and fluid secretion in rat pancreatic ducts. Pflugers Arch 2009; 459:215–226.

40. Sohma Y, Gray MA, Imai Y, Argent BE. HCO_3^- transport in a mathematical model of the pancreatic ductal epithelium. J Membr Biol 2000; 176:77–100.

41. Whitcomb DC, Ermentrout GB. A mathematical model of the pancreatic duct cell generating high bicarbonate concentrations in pancreatic juice. Pancreas 2004; 29:e30–e40.

42. Yamaguchi M, Steward MC, Smallbone K, Sohma Y, Yamamoto A, Ko SB, Kondo T, Ishiguro H. Bicarbonate-rich fluid secretion predicted by a computational model of guinea-pig pancreatic duct epithelium. J Physiol 2017; 595:1947–1972.

43. Park HW, Nam JH, Kim JY, Namkung W, Yoon JS, Lee JS, Kim KS, Venglovecz V, Gray MA, Kim KH, Lee MG. Dynamic regulation of CFTR bicarbonate permeability by $[Cl^-]i$ and its role in pancreatic bicarbonate secretion. Gastroenterology 2010; 139:620–631.

44. Cutting GR. Cystic fibrosis genetics: from molecular understanding to clinical application. Nat Rev Genet 2015; 16:45–56.

45. Megalaa R, Gopalareddy V, Champion E, Goralski JL. Time for a gut check: pancreatic sufficiency resulting from CFTR modulator use. Pediatr Pulmonol 2019; 54:E16–E18.

Chapter 19

Molecular mechanisms of pancreatic bicarbonate secretion

Min Goo Lee,[1] Yonjung Kim,[1] Ikhyun Jun,[1]* Joydeep Aoun,[2] and Shmuel Muallem[2]

[1]*Department of Pharmacology, Brain Korea 21 PLUS Project for Medical Sciences, Severance Biomedical Science Institute, Yonsei University College of Medicine, Seoul 03722, Korea.*

[2]*The Epithelial Signaling and Transport Section, National Institute of Dental and Craniofacial Research, National Institutes of Health, Bethesda, Maryland, 20892*

* Present Address: The Institute of Vision Research, Department of Ophthalmology, Yonsei University College of Medicine, Seoul 03722, Korea

Introduction

The human exocrine pancreas secretes 1–2 liters of pancreatic juice each day. When stimulated, the pancreas secretes alkaline pancreatic juice containing copious amounts of bicarbonate (HCO_3^-).[1,2] HCO_3^- plays essential roles in the digestive system. HCO_3^- determines the pH of bodily fluids as the major buffer system that guards against toxic pH fluctuations.[3] HCO_3^- in pancreatic juice neutralizes gastric acid and provides an optimal pH environment for digestive enzymes to function in the duodenum.[2] In addition, HCO_3^- acts as a moderate chaotropic ion that facilitates the solubilization of macromolecules, such as digestive enzymes and mucins.[4] The importance of pancreatic HCO_3^- secretion is highlighted in the abnormal HCO_3^- secretion in several forms of pancreatitis[5,6] and in cystic fibrosis (CF), which causes poor mucin hydration and solubilization leading to obstruction of ductal structures of the pancreas, intestine, vas deferens, and the lung.[7,8]

The exocrine pancreas is composed of three major cell types, acinar, duct, and stellate cells. Acinar cells secrete a small volume of isotonic, plasma-like, NaCl-rich fluid and digestive enzymes. Duct cells modify the ionic composition of the fluid and secrete the bulk of the fluid and HCO_3^- of the pancreatic juice. Stellate cells may aid the pancreas recovery from injury.[9] As the main HCO_3^- secretor, the duct has key roles in the development of acute and chronic pancreatitis. At pH 7.4 and 5 percent CO_2, the HCO_3^- concentration in plasma is approximately 25 mM. In human, dog, cat, and guinea pig, HCO_3^- concentration in postprandial pancreatic juice is higher than 140 mM.[1,2] This remarkable transport performance has attracted much attention from pancreatologists and physiologists. Current understanding of the molecular mechanism of pancreatic HCO_3^- secretion was improved by the recent identification of ion transporters and channels, including the cystic fibrosis transmembrane conductance regulator (CFTR),[10] the electrogenic Na^+-HCO_3^- cotransporter NBCe1-B (also known an pNBC1),[11] and the solute-linked carrier 26 (SLC26) transporters,[12,13] together with regulatory proteins, such as with-no-lysine kinase 1 (WNK1),[14] STE20/SPS1-related proline/alanine-rich kinase (SPAK),[15] and the inositol-1,4,5-triphosphate (IP_3) receptor-binding protein released with IP_3 (IRBIT)[16] and their role in pancreatic HCO_3^- secretion.

Control of Pancreatic HCO_3^- Secretion

Pancreatic HCO_3^- secretion increases in response to ingestion of a meal and is regulated by multiple neurohumoral inputs. Fluid and enzyme secretion by acinar cells are controlled predominantly by an increase in cytoplasmic free Ca^{2+} concentration ($[Ca^{2+}]_i$).[17–19] Fluid and HCO_3^- secretion by duct cells are regulated by the second messengers cAMP[2,20] that synergizes with Ca^{2+} to generate the physiological response.[21–23] Pancreatic ductal cells express receptors for a battery of hormones and neurotransmitters. The two major hormones controlling pancreatic fluid and HCO_3^- secretion are the G_s-coupled, cAMP-generating hormone secretin and the G_q-coupled, Ca^{2+}-mobilizing hormone cholecystokinin (CCK), which are released from gastrointestinal endocrine cells in the upper duodenum. Cholinergic vagal output via an enteropancreatic vagovagal reflex also has an important role in controlling ductal fluid and HCO_3^- secretion. In addition to these classic stimuli, several other humoral agents are released by the pancreas for fine-tuning its secretion, including insulin, somatostatin, purines, and prostaglandins.[24] Additional information on hormonal control of pancreatic secretion can be found in a previous review[2] and the "Regulation of Pancreatic Secretion" section in Pancreapedia.[25]

e-mail: mlee@yuhs.ac, Shmuel.Muallem@nih.gov

Humoral control

Secretin

The low pH (below 4.5) of gastric chyme stimulates the release of secretin from duodenal S cells into the blood.[26,27] Secretin stimulates ductal fluid and HCO_3^- secretion and synergizes with Ca^{2+}-mobilizing agonists to potentiate enzyme secretion by acinar cells. Plasma secretin levels rise after a meal[27,28] and correlate with HCO_3^- output.[29] Secretin-stimulated fluid and HCO_3^- secretion is modulated directly or indirectly by both peptide hormones, such as CCK and somatostatin, and by vagal stimulation.[30–32]

CCK

CCK is a major stimulator of acinar cell enzyme and fluid secretion, which is mediated by the Ca^{2+}-dependent exocytosis of zymogen granules and activation of apical (luminal) Cl^- channels, respectively. The synaptotagmins are the Ca^{2+} sensors that convey the Ca^{2+} signal for pancreatic exocytosis[33] and Ca^{2+} activates the Ca^{2+}-activated anoctamin 1 (TMEM16A) to initiate acinar cells fluid secretion.[34] CCK also acts on pancreatic duct secretion; however, the effects of CCK on pancreatic duct differ among species. In humans, the effect of CCK alone on ductal fluid secretion is weak; however, CCK greatly potentiates the effects of secretin.[32]

Purines

Pancreatic duct cells express multiple purinergic type 2 receptor (P2R) types, including ionotropic P2X and metabotropic P2Y receptors at the apical and basolateral membranes.[35] P2Rs are stimulated by purinergic ligands released from nerve terminals at the basolateral space, zymogen granules of acinar cells into the luminal space, or efflux by ductal ATP transporters to both the basolateral and luminal compartments.[36] Stimulation of P2Rs increases $[Ca^{2+}]_i$ in duct cells.[37,38] Several studies have examined effects of P2Rs on ion transporters in ductal cell lines, but there are almost no studies on ductal fluid and HCO_3^- secretion. Ishiguro et al. demonstrated that luminal ATP stimulated while basolateral ATP inhibited fluid and HCO_3^- secretion in guinea pig pancreatic duct.[39] More recent studies examined the effect of various stimuli and ion channels of ATP release from ductal cell lines[36] that will be important to verify in native ducts.

Neuronal control

Pancreatic secretion is regulated by the enteric nervous system, which is composed of a gut-brain axis and an intrapancreatic system. The major neurotransmitter acting on pancreatic duct cells is acetylcholine released by vagal parasympathetic fibers. Duct cells express both M1 and M3 muscarinic receptors, which act through changes in $[Ca^{2+}]_i$. The M3 receptors maybe more prominent based on their higher expression level relative to the M1 receptors.[40,41] In humans, cholinergic stimulation enhances ductal secretion stimulated by secretin, likely by synergistic mechanism that is mediated by IRBIT.[21,22] Vasoactive intestinal peptide (VIP) and ATP are also localized in parasympathetic nerve terminals.[42,43] Vagal stimulation causes VIP release that is associated with fluid and HCO_3^- secretion.[42,44,45]

Key Transporters Involved in Pancreatic HCO_3^- Secretion

Pancreatic HCO_3^- secretion is mediated by a coordinated function of transporters expressed in the apical and basolateral membranes of duct cells. Pancreatic HCO_3^- secretion can be divided into two steps. The first step is uptake of HCO_3^- into duct cells from the blood through the basolateral membrane. The second step is efflux of HCO_3^- across the apical membrane of duct cells. Regulatory mechanisms in the cytosol that include ions like Cl^- and several kinases and phosphatases act on the transporters to coordinate and integrate the secretory process. Recent advances in molecular, cellular, and physiological techniques have enhanced our understanding of the molecular identity, localization, function, and regulatory mechanisms of ductal ion transporters.[22,23,46] The major ion transporters expressed in the apical and basolateral membranes of the pancreatic duct cells are summarized in **Table 1** and **Figure 1**.

Na⁺/K⁺ ATPase and K⁺ channels

The main driving force for fluid secretion is achieved by the Na^+/K^+ ATPase pump and K^+ channels which generate the transmembrane Na^+ and K^+ gradients and the negative intracellular membrane potential.[6,46] The Na^+/K^+ ATPase pump is expressed in the basolateral membrane of the ducts[47–50] and generates the Na^+ and K^+ gradients by extruding three Na^+ ions in exchange for uptake of two extracellular K^+ ions using the energy of ATP hydrolysis. K^+ channels in both the basolateral and apical membranes use the K^+ gradient generated by the pump to generate a negative membrane potential. The Na^+ gradient is used to drive several Na^+-coupled solutes, including HCO_3^- absorption by the basolateral Na^+-HCO_3^- cotransporter NBCe1-B and basolateral and luminal Na^+/H^+ exchangers (NHEs). The negative membrane potential aids in controlling HCO_3^- uptake by NBCe1-B and in HCO_3^- efflux through luminal electrogenic transporters. MaxiK channels (KCNMA1) have been identified on the basolateral membrane of rat pancreatic duct cells and are likely candidates for maintaining the membrane potential during agonist-stimulated

Table 1. Transporters of pancreatic duct

Transporters in the luminal membrane of pancreatic duct		
Transporters	**Gene**	**Function**
cAMP-activated Cl^- channel	CFTR (ABCC7)	Fluid and HCO_3^- secretion
Ca^{2+}-activated Cl^- channel	TMEM16/ANO family	Cl^- and HCO_3^-(?) secretion, lipids flipping
Anion exchangers	SLC26A3 (DRA/CLD)	HCO_3^- secretion, electrogenic Cl^-/FICO3$^-$ exchanger (Cl^-:HCO_3^- stoichiometry of 2:1 or higher)
	PAT1 (SLC26A6)	HCO_3^- secretion, electrogenic Cl^-/FICO3$^-$ exchanger (Cl:HCO_3^- stoichiometry of 1:2)
Na^+/H^+ exchangers	NHE3 (SLC9A3)	HCO_3^- reabsorption (HCO_3^- salvage mechanism)
	NHE2 (SLC9A2)	HCO_3^- reabsorption (?)
Na^+-HCO_3^- Cotransporter	NBCn1-A (NBC3, SLC4A7)	HCO_3 reabsorption (HCO_3^- salvage mechanism)
K^+ channels	Maxi-K^+ channels (KCNMA1?)	Maintain membrane potential during stimulated secretion Modulate luminal HCO_3^- secretion
Water channel	Aquaporin 5 (AQP5)	H_2O flow
Transporters in the basolateral membrane of pancreatic duct		
Transporters	**Gene**	**Function**
Na^+/H^+ exchangers	NHE1(SLC9A1)	Na^+-coupled H^+ extrusion, pH_{in} homeostasis contribute to basolateral HCO_3^- influx
	NHE4 (SLC9A4)	Role uncertain
Na^+-HCO_3^- cotrans porters	NBCe1-B (pNBC1, SLC4A4)	Basolateral HCO_3^- entry
Anion exchangers	AE2 (SLC4A2)	pH_{in} homeostasis, Cl^-_{in} supplier (?)
Cation-chloride cotransporters	Na^+-K^+-2Ch cotransporter (NKCC1, SLC12A2) K^+-Cl^- cotransporter (KCC1, SLC12A4)	Basolateral Cl^- uptake (in mouse and rat ducts, but not in guinea pig and human) Basolateral K^+ and Cl^- efflux Cell volume regulation
K^+ channels	Maxi-K^+ channels (KCNMA1)	Maintain membrane potential during stimulated secretion
	Small or intermediate conductance K^+ channels (KCNN4)	Maintain resting membrane potential
Na^+, K^+-ATPase	Na^+, K^+-ATPase (ATP1B1–3)	Maintain inward Na^+ gradient and outward K^+ gradient that determines the membrane potential
Water channels	Aquaporin 1 (AQP1)	Water transport
	Aquaporin 5 (AQP5)	Water transport
Carbonic anhydrases	CAXII	HCO_3^- supply to AE2 and NBCe1-B

HCO_3^- secretion.[51] A Ba^{2+}-sensitive channel of 82 pS conductance (KCNN4) appears to be a basolateral K^+ channel, which is responsible for the resting K^+ permeability.[52] Apical membrane K^+ channels were identified in acinar cells[53] and in pancreatic duct cells, with the later having a role in ductal HCO_3^- secretion.[54]

Na^+-HCO_3^- cotransporters (NBCs)

The main ductal basolateral membrane HCO_3^- accumulation transporter is NBCe1-B.[46] NBCe1-B was cloned from pancreas and was named pNBC1.[11] It was later renamed NBCe1-B as part of classification of the NBC family.[55] NBCe1-B is an electrogenic transporter with a 1 Na^+:2

HCO_3^- stoichiometry in pancreatic duct cells.[56] NBCe1-B can be regulated by cAMP-dependent protein kinase A (PKA) phosphorylation at Ser1026 and Thr49.[57] In principle, Na^+/H^+ exchangers in the basolateral membrane (e.g., NHE1) can also mediate HCO_3^- influx in duct cells. However, the electrogenic NBCe1-B utilizes the Na^+ gradient more efficiently than the electroneutral NHE1 (1 Na^+:1 HCO_3^-). Indeed, NBCe1-B contributes up to ~75 percent of the HCO_3^- influx during secretin-induced ductal fluid and HCO_3^- secretion in guinea pig.[58,59] The activity of NBCe1-B is controlled by multiple inputs, including IRBIT and the WNK/Ste20-related proline/alanine-rich kinase (SPAK) pathway[60,61] and most notably intracellular Cl^-.[62] A more recent analysis revealed an intricate regulation of NBCe1-B

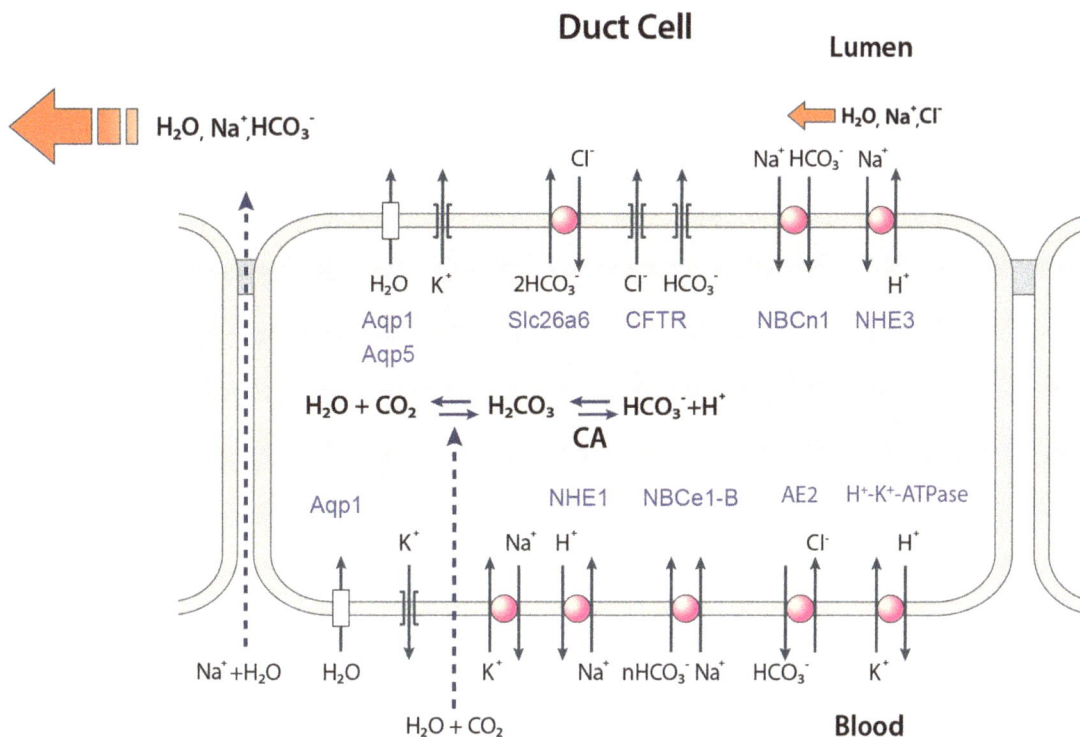

Figure 1. A schematic diagram depicting the transporters and channels in the apical (luminal) and basolateral membranes of pancreatic duct cells. The main driving force for HCO_3^- secretion is achieved by the Na^+ gradient generated by the Na^+/K^+ ATPase pump and K^+ channels at the basolateral membrane, which generate the intracellular negative membrane potential. HCO_3^- is loaded mainly through the electrogenic ($1Na^+$-$2HCO_3^-$) NBCe1-B and partly by NHE1 located in the basolateral membrane. Basolateral AE2 may act to supply Cl^-_{in} to maintain the secretion. Apical HCO_3^- secretion is performed by the interacting and functionally interrelated CFTR and Slc26a6. Transcellular HCO_3^- movement generates a lumen-negative electrical potential that results in paracellular Na^+ secretion through the paracellular pathway. Water follows Na^+ and HCO_3^- osmotically via paracellular and transcellular (aquaporins) pathways. In the resting state, luminal NHE3 and NBCn1-A function to salvage luminal HCO_3^- (modified from reference [46]).

by the WNK and CaMKII (Ca^{2+}- and calmodulin-activated kinase II) kinases and the SPAK along with the calcineurin phosphatase that dephosphorylate the serine residues phosphorylated by the respective kinases.[63] The kinases/phosphatases pairs determine regulation of NBCe1-B by Cl^-_{in},[63] which emerges as a new general form of signaling ion.[23] In addition to NBCe1-B, the duct expresses electroneutral NBCn1-A (NBC3) on the apical (luminal) membrane.[64,65] This transporter may mediate HCO_3^- salvage in the resting state to maintain acidified pancreatic juice.[66,67]

CFTR

The discovery of acidic pancreatic juice in patients with cystic fibrosis (CF) was a milestone in understanding the mechanism of pancreatic HCO_3^- secretion.[68,69] The CF transmembrane conductance regulator (CFTR) was discovered as the protein mutated in patients with CF.[10,70,71] Although CFTR is a member of the ATP-binding cassette[72] transporter superfamily that usually acts as membrane pumps that transport their substrates against the

electrochemical gradient,[73] CFTR functions as an anion (Cl^- and HCO_3^-) channel, through which ions diffuse down the electrochemical gradient. CFTR is located at the apical membrane of pancreatic ducts[74–76] (and all secretory epithelia) and is activated by the cAMP/PKA pathway. At $[Cl^-]_i$ higher than 10 mM, CFTR functions as a Cl^- channel that has limited permeability to HCO_3^-.[72,77,78] However, when $[Cl^-]_i$ drops to below 10 mM, CFTR anionic selectivity changes to increase HCO_3^- permeability and mediate luminal HCO_3^- exit.[14,79] Indeed, as has been shown in patients with CF,[74,75,80] CFTR is critically involved in epithelial HCO_3^- secretion. This leads to revision of the original model of ductal HCO_3^- secretion, in which Cl^-/HCO_3^- exchangers mediate apical HCO_3^- efflux and CFTR facilitates the apical Cl^-/HCO_3^- exchangers by recycling the Cl–.[81] This continues to be the case at high Cl^-_{in}. However, at low $[Cl^-]_i$, HCO_3^- efflux via CFTR driven by the membrane potential has essential role in HCO_3^- efflux and HCO_3-driven fluid secretion in the pancreatic duct.[82,83] The dynamic change in CFTR Cl^-/HCO_3^- permeability is mediated by the protein kinase WNK1.[14,84] WNK1[85] and other members of the WNK kinases are regulated by Cl^-_{in}, with high Cl^-_{in} in the low (WNK4) and high (WNK1) range inhibiting the WNKs. Reduction in Cl^-_{in} activates the WNKs that act directly, or through SPAK, on CFTR and other HCO_3^-, Na^+, Cl^-, and K^+ transporters[23] to regulate their activity and selectivity. If is of interest that the WNKs show differential sensitivity to Cl^-_{in} and effect on the transporters.[16,86] Thus, a modest reduction in Cl^-_{in} is sufficient to activate WNK1 and increase HCO_3^- transport by CFTR.[14] Further reduction in Cl^-_{in} will activate WNK4 that inhibits CFTR activity,[16] perhaps to prevent excess HCO_3^- secretion, that is energetically very expensive involving transport by multiple electrogenic transporters. The significance of CFTR-dependent HCO_3^- secretion in CFTR-expressing epithelia, including the pancreas, has been established in a study correlating CFTR-dependent HCO_3^- transport and severity of the CF disease.[87] The importance of the shift in the WNK1-mediated shift in CFTR HCO_3^- selectivity has been clearly demonstrated in a study that examined pancreatitis-associated CFTR mutations with altered WNK1-mediated increase in HCO_3^- permeability and found clear correlation between reduced HCO_3^- permeability and chronic pancreatitis in humans.[88]

CFTR has a more global role in ductal fluid and HCO_3^- secretion. In addition to functioning as a Cl^- and HCO_3^- channel, CFTR functions as a scaffold forming macromolecular complexes with other transporters and regulatory proteins at the apical membrane.[46] CFTR has a PSD95/Discs-large/ZO-1 (PDZ) ligand at the C-terminus and binds to PDZ domains of adapter proteins, such as Na^+/H^+ exchanger regulatory factors (NHERFs). It also has SH3 and multiple ankyrin repeat domains 2 (Shank2),[89,90] through which CFTR interacts and regulates the activity of slc26a6, slc26a3,[91] NHE3,[92] and NBCn1-A,[64] Other interactions of CFTR are with soluble NSF attachment protein receptor (SNARE) proteins, A-kinase anchor proteins (AKAPs), kinases, and phosphatases[93] that may serve to regulate CFTR activity and the activity of the transporters interacting with CFTR. The interaction with the SLC26 transporters is of particular significance since the two transporters are mutually activated when interacting.[91,94] The mutual regulation is mediated by interaction of the CFTR R domain with the SLC26 transporters STAS domain.[94]

Cl^-/HCO_3^- exchangers

Cl^-/HCO_3^- exchangers mediate the bulk of HCO_3^- exit across the apical membranes of the pancreatic duct cells until the last portion of HCO_3^- exit that is mediated by CFTR once it gains HCO_3^- permeability. In humans, members of the solute-linked carrier 4 (SLC4) and the SLC26 families function as Cl^-/HCO_3^- exchangers. Among the SLC4 transporters, duct cells express AE2 (SLC4A2) at the basolateral

membrane that regulates pH_i and protects against alkaline load.[6] However, our studies revealed an essential role for AE2 in ductal fluid and HCO_3^- secretion.[95] Intuitively, basolateral HCO_3^- efflux mechanism should inhibit rather than stimulate ductal HCO_3^- secretion. It is not clear why AE2 is essential for ductal fluid secretion. Maintaining stable pH_{in} that neutralize acid load by the Na^+/H^+ exchangers and high pH next to the plasma membrane is one potential critical function of AE2. Another possibility is that AE2 may provide the duct with Cl^- that is needed to keep the luminal slc26a6 functioning in a face of limited Cl^- provided by acinar secretion.[95]

Among the SLC26 family transporters, SLC26A3 and SLC26A6 are located on the apical membrane of the pancreatic duct cells and mediate Cl^-/HCO_3^- exchange. Interestingly, SLC26A3 has a $2Cl^-/1HCO_3^-$ stoichiometry,[94,96] while SLC26A6 functions as a $2HCO_3^-/1Cl^-$ exchanger.[96,97] A persistent osmotic gradient is needed to support the copious fluid secretion by the pancreatic duct. This is satisfied by the coupled action of NBCe1-B and SLC26A6 that results in a continuous net HCO_3^- (osmolyte) transcellular transport and thus transcellular H_2O flow.[98–100] In addition, as indicated above, SLC26 transporters interact with CFTR through the sulfate transporter and anti-sigma factor antagonist (STAS) domain and regulate pancreatic secretion by activating CFTR.[94] This form of regulation is critical for pancreatic and other exocrine glands HCO_3^- secretion, including the pancreas, salivary glands, the kidney, and the lung[101].

Other transporters, channels, and carbonic anhydrases

Na^+/H^+ exchangers (NHEs)

The SLC9A NHE family contains electroneutral $1Na^+/1H^+$ exchangers. The ubiquitous NHE1 (SLC9A1) is essential for intracellular pH homeostasis and supplies Na^+ to the Na^+/K^+ ATPase pump on the basolateral membrane of the pancreatic duct.[102] Diffusion of CO_2 from the blood into the duct and CO_2 generated by metabolism is hydrated by the action of carbonic anhydrases to generate HCO_3^- and H^+. Consequently, H^+ efflux by NHE1 may contribute to basolateral HCO_3^- uptake. However, NHE1 does not have a major role in basolateral HCO_3^- influx as revealed by minimal inhibition of fluid and HCO_3^- secretion by inhibition of NHE1 in pancreatic duct of most species.[103,104] The NHE3 isoform is expressed in the apical membrane of pancreatic duct and is thought to mediate HCO_3^- salvage at the resting state.[105] At the resting state, the pancreatic juice is acidic, indicating an active H^+ secretion[66,67] that may be mediated by the combined action of NHE3 and NBCn1-A. Similar to NBCn1-A,[64] NHE3 interacts with CFTR via PDZ domain containing proteins[92] and is regulated by IRBIT.[106,107] However, the physiological significance of these transporters await evaluation in mouse models with targeted pancreatic deletion of ductal NHEs and NBCn1-A.

Ca^{2+}-activated Cl^- channels (CaCCs)

Several members of the anoctamin (TMEM16/ANO) family function as CaCCs.[108–110] TMEM16A/ANO1, TMEM16B/ANO2, TMEM16F/ANO6, TMEM16H/ANO8, and TMEM16K/ANO10 are expressed in pancreas.[46] However, ANO1 is expressed in acinar but not duct cells,[109] ANO6 functions as a flipase and as a Cl^- channel,[50] and ANO8 is a tether at the ER/PM junctions that controls assembly of Ca^{2+}-signaling complexes.[111] The function of ANO2 and ANO10 in the pancreas is not clear at this time. Nevertheless, ample evidence shows that the pancreatic duct (and ducts of other secretory glands) has CaCC activity in the apical membrane.[112–115] The molecular identity of this channel is not known at present, nor its function in HCO_3^- secretion. ANO6 appears to function as a Ca^{2+}-activated Cl^- channel in the intestine that participates in fluid and electrolyte secretion[116] and may have a similar function in the pancreatic duct. Several other CaCCs are known and are candidates for the ductal CaCC. In pancreatic acinar cells and other serous cells, ANO1 may have a role in HCO_3^- transport. At physiological $[Ca^{2+}]_i$ concentrations, ANO1 functions as a Cl^- channel. However, at high $[Ca^{2+}]_i$ and perhaps at high $[Ca^{2+}]_i$ microdomains, ANO1 HCO_3^- permeability is increased by Ca^{2+}/calmodulin,[117, 118] raising the possibility that ANO1 can provide an alternative Cl^- and HCO_3^- conduction in acinar cells.[119]

Aquaporins

Although the paracellular pathway is permeable to H_2O, H_2O flows mostly transcellularly via the water channels aquaporins (AQP) family. This is best illustrated in salivary glands, where knockout of AQP5 markedly reduces salivation.[120] Among the 13 AQPs, AQP1 and AQP5 are the major aquaporins in pancreatic duct.[121–123] AQP1 is expressed in the luminal membrane of human acinar and duct cells and is significantly reduced in chronic pancreatitis.[124] Moreover, deletion of AQP1 in mice prominently inhibits ductal and pancreatic fluid and HCO_3^- secretion, due to both reduction in fluid transport and in CFTR expression and activity and thus HCO_3^- secretion.[124] The role of AQP5 in the duct and pancreatic secretion has not been established yet.

Carbonic anhydrases

A poorly studied topic that deserve more attention is the role of the ductal carbonic anhydrases (CAs) in fluid and electrolyte secretion, in particular with the emerging secretory epithelial diseases due to mutations in CAs. Mutations that affect the action of CA4 cause retinitis pigmentosa[125] and a mutation in CA12 causes salt wasting.[126,127] All transporters involved in fluid and HCO_3^- secretion are affected by HCO_3^- concentration at the cellular compartments and microdomains that determine HCO_3^- availability at plasma membrane inner and outer surfaces. Hydration of CO_2 by CAs determines local HCO_3^- concentration both at the outer and inner plasma membrane surfaces.[128] Several CAs are localized in the cytoplasm (such as CA2 and CA7) and several are anchored at the plasma membrane (such as CA4, CA12, and CA14) with the catalytic site at the extracellular surface and regulate HCO_3^- concentration at the basolateral (CA4 and CA12) or the luminal (CA4) membrane surfaces.[129]

CAs localized in the plasma membrane and cytoplasm interact with H^+ and HCO_3^- transporters that mediate ductal fluid and HCO_3^- secretion and regulate their activity. CA4 interacts with the C terminus of NBCe1-A to increase its activity.[130] The C terminus of NBCe1-A and NBCe1-B are conserved and thus it is likely that CA4 regulates NBCe1-B. NBCn1-A recruits the cytoplasmic CA2 to the plasma membrane, where CA2 increases the activity of NBCn1-A.[131] CA2 is closely associated with NHE3 and increases NHE3 activity.[132] CA2 interacts with a novel site at the C terminus of NHE1 to regulate NHE1 activity.[133] CA2 has been reported to interact with the C terminus of slc26a6 to increase its activity. However, the role of other CAs, in particular the plasma membrane-anchored CAs, on the activity of the slc26a6 and other SLC26 transporters has not been investigated yet. Finally, CA2 also interacts with AQP1 to increase water flux by AQP1 by an unknown mechanism.[134] A particularly interesting CA is the basolateral membrane-anchored CA12 with its catalytic site at the extracellular membrane surface. A human mutation in CA12(E143K) is the cause of an autosomal recessive form of salt wasting, which leads to hyponatremia with hyperkalemia, high sweat Cl^-, dehydration, and failure to thrive.[126,127,135] A recent work to understand the cause of the disease established a prominent role for CA12 in ductal fluid and HCO_3^- secretion. Thus, CA12 increased, while CA12 (E143K) markedly reduced ductal fluid secretion in isolated ducts and *in vivo*. This could be attributed to a potent stimulation of ductal and topically expressed AE2 by CA12.[95] The E143K mutation is a folding mutation that resulted in retention of CA12(E143K) in the ER.[95] How exactly CA12 with an external catalytic site activates AE2 is not obvious. CA12 may clear the extruded HCO_3^- from the membrane surface to prevent its buildup at the mouth of the AE2. If this can be established, it will be a new mode of regulating HCO_3^- transporters by CAs.

Regulation and Mechanism of Pancreatic HCO$_3^-$ Secretion

Intracellular signaling pathways: cAMP and Ca^{2+}

The cAMP/PKA pathway is central in inducing ductal HCO$_3^-$ secretion. Secretin is the major hormone that activates the cAMP pathway. VIP also signals to increase cAMP via VIP receptors (VPAC1).[136,137] At maximal receptor stimulation, the cAMP/PKA pathway can fully activate fluid and HCO$_3^-$ secretion by activation of the apical CFTR and the basolateral Na$^+$-HCO$_3^-$ cotransporter, NBCe1-B.[138] However, at physiological conditions the cAMP/PKA pathway synergizes with the Ca^{2+} signaling pathway to activate the secretory process (see below).

Several agonists that act on the pancreatic duct engage the Ca^{2+} signaling pathway. These include CCK, cholinergic stimuli, P2Rs, and protease-activated receptor 2 (PAR2) receptors.[19,41] When activated, the CCK and muscarinic receptors activate PLCβ to generate IP$_3$ that releases Ca^{2+} from intracellular stores, mainly the endoplasmic reticulum (ER), and activates the membrane Ca^{2+} influx channels, Orai and TRPC. P2Rs[35,139] and PAR2[140–143] also act through activation of the Ca^{2+} signaling pathway. At physiological stimulus intensity, the cAMP and Ca^{2+} signaling pathways synergize to activate ductal secretion.[79] Early studies *in vivo* already noted the synergistic action of ductal stimuli. Application of secretin at a level observed in the postprandial state only produces modest HCO$_3^-$ and fluid output.[144,145] Application of CCK and stimulation of M1 and M3 receptors markedly augmented secretin-stimulated pancreatic fluid secretion, although alone CCK and muscarinic stimulation have minimal effect on ductal secretion.[2,32] The molecular mechanism of synergism was resolved with the discovery of regulation of ductal secretion by IRBIT, which is discussed below. The cAMP and Ca^{2+} signaling pathways crosstalk on several additional levels to modulate the activity of each other.[79,146] cAMP/PKA phosphorylates IP$_3$R2 to augment Ca^{2+} release from the ER.[147] Ca^{2+} influx through the Orai1 channels activates the Ca^{2+}-dependent adenylyl cyclase (AC) AC8.[148] Ca^{2+} can also activate the CFTR-dependent Cl$^-$/HCO$_3^-$ exchange activity in CAPAN-1 human pancreatic duct cells,[149] which may involve activation by IRBIT.

Regulation by IRBIT

Activation of NBCe1-B, Slc26a6, and CFTR

IRBIT was isolated as a protein that interacts with the IP$_3$ binding pocket of the receptors (IP$_3$Rs) and it can be dissociated from the IP$_3$Rs by IP$_3$.[150] IRBIT competes with IP$_3$ for binding to the IP$_3$Rs[151] to inhibit Ca^{2+} release. In fact, the IP$_3$Rs appear to function as IRBIT buffers to prevent IRBIT access to many transporters and targets regulated by IRBIT.[46] The C-terminal region of IRBIT family proteins shows ~50 percent homology with the ubiquitous housekeeping enzyme S-adenosyl-l-homocysteine hydrolase (AHCY), with IRBIT having additional N terminal sequence while it lacks the hydrolase activity.[152] The main known domains of IRBIT are PP1 and calcineurin-binding motif, a PEST domain, a coiled-coil domain, and a PDZ ligand at the end of C terminus.[46,63]

IRBIT plays an important role in pancreatic ductal secretion by regulating multiple transporters and mediating the synergistic action of the cAMP/PKA and Ca^{2+} signaling pathways (**Figure 2**). Knockdown of IRBIT in ducts and knockout in mice modestly inhibit fully stimulated pancreatic duct fluid and HCO$_3^-$ secretion[138] and eliminates the physiological synergistic action of the cAMP/PKA and Ca^{2+} signaling pathways.[21] IRBIT accumulates at the apical pole where IP$_3$Rs are highly expressed, but it can be found all over the cell where IP$_3$Rs are present.[153] A search for IRBIT-binding proteins identified NBCe1-B as a binding partner, where IRBIT binds to the N terminus autoinhibitory domain of NBCe1-B to activate it by removing the autoinhibition.[60] Subsequent detailed studies, in particular with the pancreatic duct, revealed that IRBIT at the apical pole potently activates the apical CFTR,[16,138] SLC26A6,[21] and NHE3.[107] At the basal side, IRBIT regulates NBCe1-B.[16,60,138] IRBIT activates the transporters by multiple mechanisms. First, IRBIT recruits protein phosphatase 1 (PP1) to the transporters to dephosphorylate serine residue 75 in NBCe1-B and yet to be identified residue in CFTR that are phosphorylated by the kinase SPAK. For these phosphorylations, SPAK must be activated by phosphorylated by the two kinases WNK1 and/or WNK4.[62] IRBIT also recruits the phosphatase calcineurin to dephosphorylate serine residue 12 that is phosphorylated by CaMKII.[63] This enhances the plasma membrane relocation of NBCe1-B, CFTR,[16] and slc26a6[21] from intracellular vesicular pools. At the plasma membrane, IRBIT directly interacts with the transporters to further increase their activity. Moreover, phosphorylation by SPAK and CaMKII and dephosphorylation by the respective phosphatases PP1 and calcineurin determine regulation of NBCe1-B, and likely other IRBIT-regulated transporters, by Cl$^-_{in}$.[63] The mechanism by which IRBIT activates the other transporters is not known at this time beyond the need for the PDZ-binding motif of IRBIT for assembling the IRBIT-NBCe1-B and IRBIT-CFTR complex.[138]

IRBIT and synergism

An important action of IRBIT is mediating the synergistic action of the cAMP/PKA and Ca^{2+} signaling pathways[21] (see **Figure 2**). Physiological stimulus intensity must be quite weak to prevent cell toxicity that occurs under strong stimulation of all signaling pathways. Indeed, at

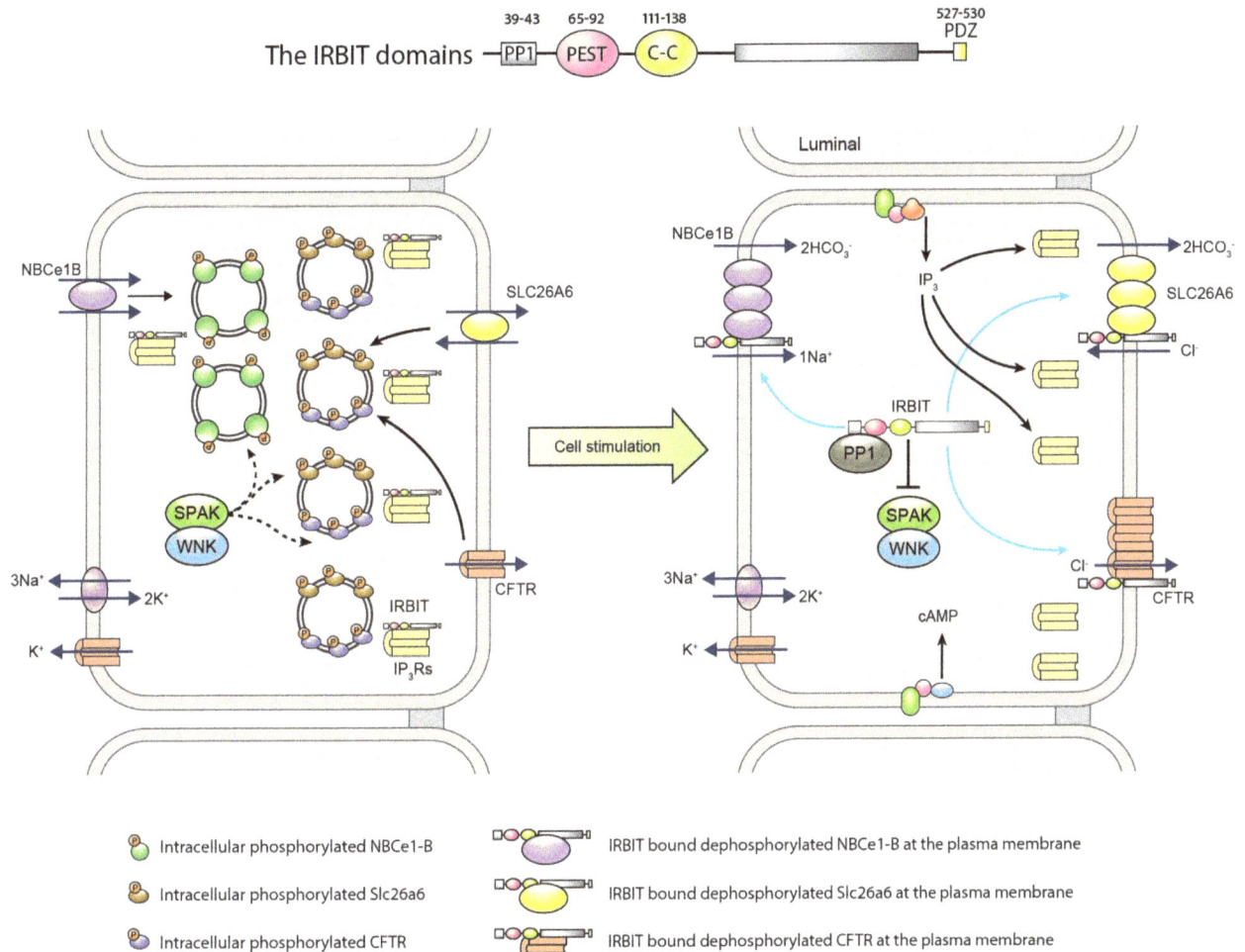

Figure 2. A model for IRBIT-associated pathway of pancreatic ductal fluid and HCO$_3^-$ secretion. Key domains of IRBIT related to HCO$_3^-$ secretion are illustrated at the top of the figure. In the resting state, IRBIT is bound to IP$_3$Rs, and SPAK phosphorylates NBCe1-B, Slc26a6, and CFTR located at intracellular organelle. When the duct cells are stimulated, IP$_3$ is released and bound to IP$_3$Rs, while IRBIT is disengaged from IP$_3$Rs. PP1 recruited to IRBIT dephosphorylates transporters located at the plasma membrane. IRBIT also binds to the autoinhibitory domain of NBCe1-B to activate it. Increased surface expression of the transporters also aids pancreatic ductal HCO$_3^-$ secretion (modified from reference [146]. See text for details).

physiological stimulus intensity the secretory process is activated only by about 5–10 percent or less of maximal stimulation. Synergism between weakly stimulated signaling is used to generate the maximal response while avoiding cell toxicity and increasing fidelity. IRBIT mediates the synergism between the cAMP/PKA and Ca^{2+} signaling pathways by translocation between cellular compartments and transporters. At the resting state, IRBIT is sequestered by the high level of IP$_3$Rs at the ductal ER apical pole and is not available for interaction with the transporters. The affinity of the IP$_3$Rs for IRBIT and IP$_3$ is regulated by PKA-mediated phosphorylation of specific IP$_3$Rs serine residues. Phosphorylation of the serine residues increases the affinity for IP$_3$ and at the same time decreases the affinity

for IRBIT. Now, a small increase in IP$_3$ evoked by weak stimulation of the Ca^{2+} signaling pathway is sufficient to dissociate IRBIT from the IP$_3$Rs.[21] The released IRBIT can bind to CFTR and Slc26a6 first in intracellular vesicles to dephosphorylate them by the IRBIT-recruited PP1 and calcineurin and promote their translocation to the luminal membrane. At the luminal membrane, IRBIT activates the transporters and reduce their inhibition by Cl$^-_{in}$ to initiate ductal fluid and HCO$_3^-$ secretion.[21,63] Of note, the synergistic action of the cAMP/PKA and Ca^{2+} signaling pathways is eliminated by the knockout of IRBIT,[21] highlighting the key role of IRBIT in the synergistic action of the cAMP/PKA and Ca^{2+} signaling pathways, which is the physiological way that ductal fluid and HCO$_3^-$ secretion take place.

Regulation by [Cl⁻]ᵢ

WNK1 and dynamic regulation of CFTR HCO₃⁻ permeability

The WNK proteins consist of four members (WNK1–WNK4) with a conserved kinase domain that is noted for the unique position of the catalytic lysine residue.[154] The discovery that mutations in WNK1 and WNK4 cause hypertension in humans has attracted much attention to these kinases function and regulation.[155] The main function of the WNKs is the regulation of Na^+, K^+, Cl^-, HCO_3^-, and Ca^{2+} transporters in epithelia and brain.[146,156–158] The WNKs act either by regulating surface expression of membrane transporters through modulation of their endocytosis or by phosphorylating the transporters and other target proteins directly or indirectly through affecting the effect of other kinases.[157] Several functions of WNKs are mediated by phosphorylating and activating the downstream oxidative stress-responsive kinase 1 (OSR1) and SPAK.[159] WNK1, WNK3, WNK4, SPAK, and OSR1 are expressed in the pancreatic duct[14,16] and participate in the regulation of HCO_3^- transporters and channels.[46] Accordingly, knockdown of WNK4 alone or a combined knockdown of WNK1, WNK3, and WNK4 increase pancreatic duct fluid secretion by removing a tonic negative effect on ductal HCO_3^- transporters.[16] However, the role of the WNKs, in particular WNK1, changes at the terminal portion on the duct when [Cl⁻]ᵢ is reduced to below 10 mM. WNK1 and the other WNKs bind [Cl⁻]ᵢ and their activity is regulated by [Cl⁻]ᵢ.[85,86,160]

The role of WNK1 in pancreatic HCO_3^- secretion is illustrated in the left portion of **Figure 3**. Osmotic stress or

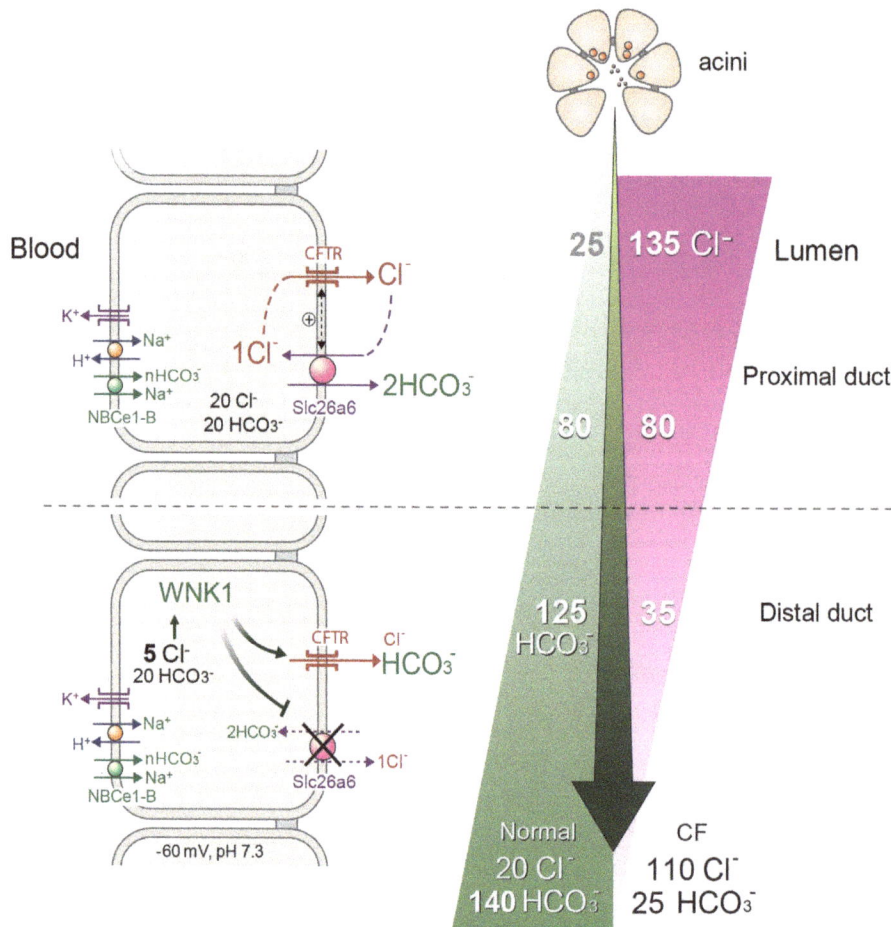

Figure 3. A model depicting WNK1-mediated regulation of CFTR in pancreatic ductal function. During active pancreatic HCO_3^- secretion, Cl^- concentration in the pancreatic juice is progressively reduced due to Cl^-/HCO_3^- exchange activities at the apical membrane of duct cells. Because the basolateral membrane of duct cells has poor Cl^- permeability but the apical Cl^- permeability is very high due to activation of CFTR, [Cl⁻]ᵢ rapidly decreases in response to the reduction in luminal Cl^- concentration. Activation of WNK1 by low [Cl⁻]ᵢ increases the P_{HCO3}/P_{Cl} of CFTR to over 1.0, which greatly augments HCO_3^- flux through the CFTR pore. Simultaneously, WNK1/SPAK pathway inhibits Slc26a6 to prevent HCO_3^- reabsorption. This mechanism enables an increase to as much as 140 mM HCO_3^- in pancreatic juice. See text for details (modified from reference [46]).

low $[Cl^-]_i$ activates WNK1.[85] Notably, activation of WNK1 by low $[Cl^-]_i$ greatly increases the HCO_3^- permeability of CFTR.[14,117] During active pancreatic HCO_3^- secretion, lower Cl^- concentration in the pancreatic juice progressively reduces Cl^- absorption. Because of the low basolateral and high luminal Cl^- permeability,[14,161] $[Cl^-]_i$ rapidly decreases in response to the reduction in luminal duct Cl^- concentration. At a membrane potential of -60 mV, $[Cl^-]_i$ is less than 1/10 of luminal Cl^- concentration. Indeed, ductal $[Cl^-]_i$ was estimated to be about 5 mM during cAMP-induced active secretion.[14,161]

WNK1 modulates the anion selectivity of CFTR by changing its pore size.[117] Stimulation by WNK1 increases CFTR pore size from 4.8 Å to 5.3 Å, which facilitates the passage of HCO_3^- (4.3 Å, diameter) more than the smaller anion, Cl^- (3.7 Å, diameter). Changes in pore size affect the energy barrier of ion dehydration by altering the electric permittivity of the water-filled cavity in the pore. Dielectric constant (relative permittivity, ε) is a unit of electric permittivity, and the dielectric constant of water (ε_w) is approximately 80 at room temperature. Water molecules in confined geometry like ion channels exhibit a space-dependent reduction in the pore water ε_w down to 20, due to the restriction of the translational and rotational mobility of water molecules.[162] Pore dilation relieves this restriction of water molecule movement and increases ε_w, which eventually leads to an increase in the overall ε of the anion channel pore. Indeed, the pore dilation induced by WNK1 activation increased the ε of the CFTR pore from 16 to 43.[117] In general, ions pass through the channel after dehydration (at least partial dehydration). Asymmetrically charged ions, such as HCO_3^-, show lower permeability than the symmetrically charged ions, such as Cl^-, due to the high hydration/dehydration energy barrier. The increase in anion channel pore ε greatly alleviates the dehydration penalty of the asymmetrically charged HCO_3^- and increases P_{HCO3}/P_{Cl} (**Figure 4**). In an initial study, WNK-OSR1 or WNK1-SPAK complex was suggested to increase the CFTR P_{HCO3}/P_{Cl}.[14] However, subsequent study showed that WNK1 alone was sufficient to increase CFTR P_{HCO3}/P_{Cl}.[84] Molecular dissection of the WNK1 domains revealed that the WNK1 kinase domain is responsible for CFTR P_{HCO3}/P_{Cl} selectivity by direct association with CFTR, while the surrounding N-terminal regions mediate the $[Cl^-]_i$ sensitivity of WNK1.[84]

Although the precise WNK1-binding sites on the CFTR are not fully defined, examining pancreatitis-causing CFTR mutations revealed that R74 and R75 located in the first

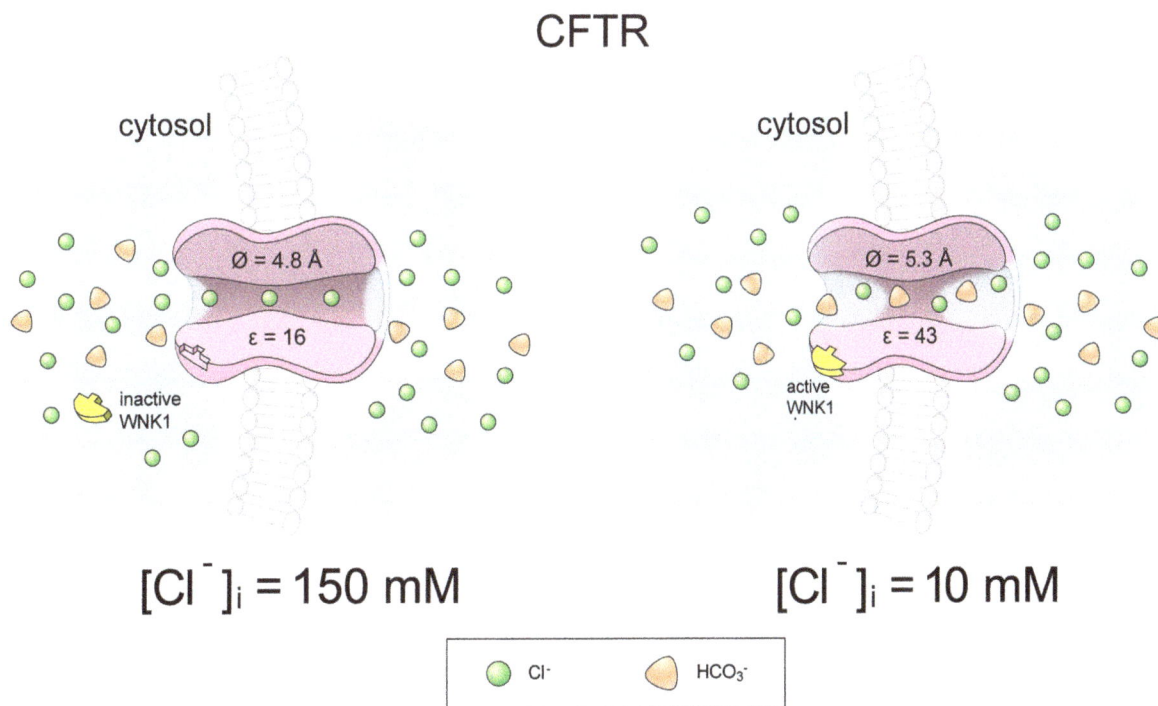

Figure 4. WNK1 modulates the anion selectivity of CFTR by changing the pore size. Stimulation by WNK1 increases the pore size of CFTR from 4.8 Å to 5.3 Å and the pore dilation increases the dielectric constant (ε) of the CFTR pore from 16 to 43. The increase in pore size facilitates the passage of the larger anion, HCO_3^- (4.3 Å, diameter), more than the smaller anion, Cl^- (3.7 Å, diameter) by reducing the energy barriers of size exclusion. More importantly, the dielectric constant increase enhances the HCO_3^- permeability of CFTR by reducing energy barriers required for ion dehydration of HCO_3^{-}[117] (see text for details).

elbow helix region of the CFTR are involved in the WNK1-CFTR association [39]. A computational protein-protein docking analysis using the Protein Data Bank-deposited structures of human CFTR and the WNK1 kinase domain showed that WNK1 S231–T234 and I384–E388 can potentially bind to an intracellular CFTR region near R74–R75 in the elbow helix 1 (**Figure 5**). The handle-like elbow helix 1 is located immediately ahead of transmembrane domain 1 and contacts a proximally located lasso motif that has been suggested to play a role in CFTR gating and regulation of the R domain.[163] Therefore, it appears that the binding of WNK1 kinase domain to the elbow helix 1 region of CFTR affects the CFTR open structure to facilitate HCO_3^- permeation.[119] Notably, WNK1 not only increases P_{HCO3}/P_{Cl} of CFTR HCO_3^- channel but also the HCO_3^- conductance (G_{HCO3}) and P_o in single-channel recording.[84] Increase in CFTR G_{HCO3} may also significantly contribute

to augmenting HCO_3^- flux across the apical membrane of pancreatic duct cells.

Interestingly, activated WNK1 while increasing CFTR P_{HCO3}/P_{Cl} and G_{HCO3} does not lose the inhibitory effect on SLC26A6 and SLC26A3.[14] When the luminal HCO_3^- concentration is greater than 140 mM, continuous activation of apical Cl^-/HCO_3^- exchange would reverse to absorb HCO_3^- from the lumen. This is more of a problem for the $2Cl^-/1HCO_3^-$ exchange by Slc26a3 and less, if at all, for the $1Cl^-/2HCO_3^-$ Slc26a6, especially at membrane potential of -60 mV across the luminal membrane. However, inhibition of the apical Cl^-/HCO_3^- exchangers is required to prevent the reverse mode of Cl^-/HCO_3^- exchange activity if Slc26a3 dominates the exchange when ductal $[Cl^-]_i$ is below 10 mM and ultimately achieves HCO_3^- concentration above 140 mM in pancreatic juice.[75,164]

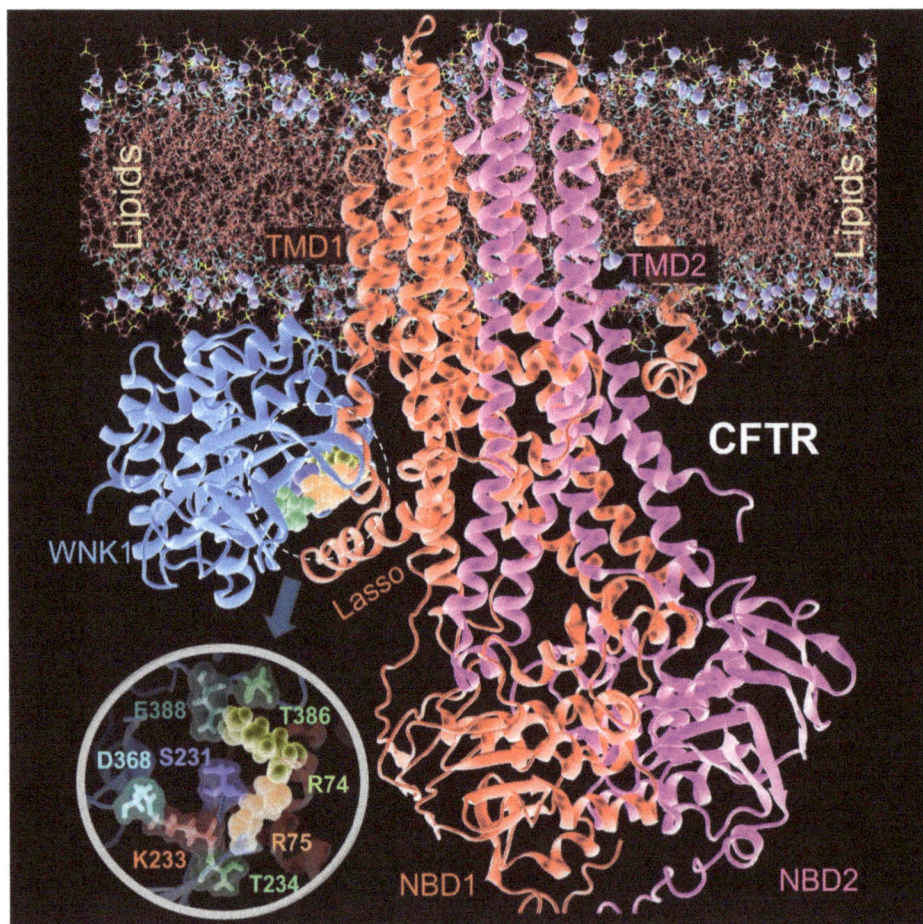

Figure 5. Structural model for the molecular complex between hCFTR and the WNK1 kinase domain in the presence of a lipid bilayer. The R74 (*yellow balls*) and R75 (*orange balls*) residues from hCFTR participate in the binding interface. The figure shows the hCFTR-WNK1 complex predicted by ClusPro, after equilibration in the MD simulation system where it is embedded into the membrane lipids (lines with their phosphorus atoms shown as *purple spheres*) and solvated by 0.1 M NaCl solution. The snapshot was taken after 100 ns MD simulations. The inset figure shows a close-up view of interfacial interactions. WNK1 residues at the interface include S231, K233, T234, T386, and E388 (modified from reference [84])

Cl⁻ₗₙ as a signaling ion, the case for NBCe1-B

[Cl⁻]ᵢ emerged as a signaling ion by regulating several ion transporters and channels. A comprehensive review of Cl⁻ as a *bona fide* signaling ion can be found in reference [23]. Here we discuss the signaling function of Cl⁻ᵢₙ with respect to ductal function. By virtue of regulating the function of the WNK kinases [Cl⁻]ᵢ may affect other transporters regulated by these kinases. A significant discovery is that [Cl⁻]ᵢ profoundly regulates the activity of several Na⁺-HCO₃⁻ cotransporters (NBCs) at the [Cl⁻]ᵢ physiological range.[62,63] [Cl⁻]ᵢ regulates the activity of all NBCs tested NBCe1-B, NBCe1-C, and NBCe1-A. The IRBIT-independent activity of NBCe1-B is inhibited by [Cl⁻]ᵢ between 60 and 140 mM that is outside the physiological range and may function to inhibit NBCe1-B activity under pathological conditions. Most notably, when activated by IRBIT, NBCe1-B activity is reduced by [Cl⁻]ᵢ in the range of 5–20 mM, where at 20 mM [Cl⁻]ᵢ, NBCe1-B activity is reduced to the basal, IRBIT-independent level. Molecular analysis identified two Cl⁻ interacting motifs at the N terminus of NBCe1-B that mediate high- and low-affinity inhibition by [Cl⁻]ᵢ. Regulation of NBCe1-B is mediated by sites that contain the GXXXP motif. The first site mediates the high [Cl⁻]ᵢ affinity (5–20 mM) regulation of NBCe1-B and the second site mediates the low [Cl⁻]ᵢ affinity (60–140 mM) regulation of NBCe1-B.[62] NBCe2-C activity is not regulated by IRBIT and in this case regulation of NBCe1-C is mediated by a single site containing the GXXXP motif and takes place at [Cl⁻]ᵢ between 10 and 30 mM. Regulation of NBCe1-A by [Cl⁻]ᵢ is mediated by a cryptic Cl⁻ interacting site containing the GXXXP motif. The cryptic NBCe1-A [Cl⁻]ᵢ interacting sites were unmasked by deletion of residues 29–41. Further analysis showed that interaction of Cl⁻ᵢₙ with the GXXXP sites is regulated by phosphorylation/dephosphorylation events with SPAK and PP1 acting on serine 65 to affect Cl⁻ sensing by the ³²GXXXP³⁶ site, while CaMKII and calcineurin acting on serine 12 to affect Cl⁻ sensing by the ¹⁹⁴GXXXP¹⁹⁸ site.[63] Other phosphorylation sites affecting NBCe1-B activity are Ser232, Ser233, and Ser235 with the phosphorylation status of Ser232, Ser233, and Ser235 is regulated by IRBIT to determine whether NBCe1 transporters are in active or inactive conformations.[63]

Hence, cells have a [Cl⁻]ᵢ sensing mechanism that plays an important role in the regulation of Na⁺ and HCO₃⁻ transporters that mediate the critical step of HCO₃⁻ influx in the process of ductal fluid and HCO₃⁻ secretion. At [Cl⁻]ᵢ of up to 20 mM, CFTR functions mostly as a Cl⁻ channel and slc26a6 mediates most ductal HCO₃⁻ secretion. As [Cl⁻]ᵢ is reduced below 20 mM and additional HCO₃⁻ secretion takes place in the face of unfavorable Cl⁻ and HCO₃⁻ gradients across the apical membrane, there is an increased demand for HCO₃⁻ entry across the basolateral membrane. Pancreatic duct cells achieve this by [Cl⁻]ᵢ-mediated regulation of NBCe1-B and CFTR, at which NBCe1-B activity and CFTR HCO₃⁻ permeability gradually increase as [Cl⁻]ᵢ is reduced toward 5 mM.

A model for pancreatic HCO₃⁻ secretion

Electrogenic HCO₃⁻ transporters can generate higher gradients of HCO₃⁻ than electroneutral transporters when the electrorepulsive force generated by the negative membrane potential is coupled to the basolateral uptake of HCO₃⁻ by NBCe1-B and the luminal efflux of HCO₃⁻ through SLC26A6 and CFTR. The electrogenic SLC26A6 exchanger with the stoichiometry of 1Cl⁻:2HCO₃⁻ can achieve luminal HCO₃⁻ concentration of up to about 120 mM at apical membrane potential of −60 mV.[75] To drive luminal HCO₃⁻ concentration to 140 mM, the physiologic HCO₃⁻ concentrations in pancreatic juice, another mechanism is needed.[75] Such a mechanism should be Cl⁻ independent, since significant fraction of pancreatic HCO₃⁻ secretion is retained in the absence of luminal Cl⁻.[82,165] The WNK1-activated CFTRs satisfy these requirements. At HCO₃⁻ᵢₙ in stimulated duct cells above 25 mM and membrane potential of −60 mV, CFTR mediates HCO₃⁻ efflux even at luminal HCO₃⁻ concentrations of above 140 mM. Transport by NBCe1-B and Slc26a6 results in osmotic transport of HCO₃⁻ that is essential for the transcellular water transport by the duct. The transcellular basal to luminal electrogenic HCO₃⁻ transport by both Slc26a6 and CFTR generates a lumen-negative electrical potential that results in paracellular Na⁺ secretion. Water flows down the osmotic gradient generated by the Na⁺ and HCO₃⁻ fluxes via paracellular and transcellular (aquaporins) pathways (**Figure 3**). Overall, these processes generate an efficient mechanism for HCO₃⁻-driven ductal fluid secretion to generate the volume and HCO₃⁻ content of the pancreatic juice.

Conclusions

The mechanism by which the human pancreatic duct secretes nearly isotonic HCO₃⁻ solution has long been an enigmatic question for both physiologists and clinicians.[2,75] When Bayliss and Starling first noticed that the exocrine pancreas secretes alkaline fluid, they assumed that carbonate is responsible for the strong alkalinity of the pancreatic juice.[166] Later, with better understanding of the carbonate/HCO₃⁻/CO₂ buffer system,[167] it became clear that the exocrine pancreas secretes fluid in which the dominant anion is HCO₃⁻, and HCO₃⁻ secretion is coupled to fluid secretion.[144,168,169] Current understanding indicates that activation of three key transporters, the basolateral NBCe1-B (and likely AE2), and the luminal SLC26A6 and CFTR, and their synergistic regulation by the cAMP and Ca²⁺ signaling pathways through IRBIT and WNK1 perform vectorial pancreatic HCO_3^- secretion that drives fluid secretion. NBCe1-B, with a 1 Na^+/2 HCO_3^- stoichiometry, is the main HCO_3^--concentrating transporter in the basolateral membrane and can achieve the necessary HCO_3^- influx.[11,58,170] Basolateral AE2 activity is also required to support ductal HCO_3^- fluid and HCO_3^- secretion probably by controlling cytoplasmic and near membrane pH_{in}, although the exact role of AE2 is not known at present. The electrogenic SLC26A6, with a 1 Cl^-/2 HCO_3^- stoichiometry, is the major apical Cl^-/HCO_3^- exchanger, which mediates most HCO_3^- efflux in the early step of pancreatic HCO_3^- secretion.[94,171] Activated WNK1 increases HCO_3^- permeability and conductance of CFTR, allowing further apical HCO_3^- efflux and setting the pancreatic juice HCO_3^- concentrations above 140 mM.[14] Ductal fluid and HCO_3^- secretion is essential for the function of the pancreas and is severely altered in all forms of pancreatitis.[5,172] Our understanding of the mechanism of pancreatic fluid and HCO_3^- secretion will continue to improve as our knowledge of existing pathways increases and new mechanisms are identified and delineated, to provide a better scientific basis for therapeutic approaches to treat diseases like cystic fibrosis and acute and chronic pancreatitis.

Acknowledgments

We thank Dong-Su Jang for assistance in illustrations. This work was supported by grant, 2013R1A3A2042197, from the National Research Foundation, the Ministry of Science, ICT & Future Planning, Republic of Korea, and by Intramural NIH/NIDCR grant DE000735–010.

References

1. Domschke S, Domschke W, Rosch W, Konturek SJ, Sprugel W, Mitznegg P, Wunsch E, Demling L. Inhibition by somatostatin of secretin-stimulated pancreatic secretion in man: a study with pure pancreatic juice. Scand J Gastroenterol 1977; 12:59–63.
2. Lee MG, Muallem S. Physiology of duct cell secretion. In: Beger HG, Buchler MW, Kozarek RA, Lerch MM, Neoptolemos JP, Warshaw AL, Whitcomb DC, Shiratori K, eds. The pancreas: an integrated textbook of basic science, medicine, and surgery. 2nd Edition, Oxford: Blackwell Publishing, 2008:78–90.
3. Roos A, Boron WF. Intracellular pH. Physiol Rev 1981; 61:296–434.
4. Hatefi Y, Hanstein WG. Solubilization of particulate proteins and nonelectrolytes by chaotropic agents. Proc Natl Acad Sci U S A 1969; 62:1129–1136.
5. Zeng M, Szymczak M, Ahuja M, Zheng C, Yin H, Swaim W, Chiorini JA, Bridges RJ, Muallem S. Restoration of CFTR activity in ducts rescues acinar cell function and reduces inflammation in pancreatic and salivary glands of mice. Gastroenterology 2017; 153:1148–1159.
6. Pallagi P, Hegyi P, Rakonczay Z, Jr. The physiology and pathophysiology of pancreatic ductal secretion: the background for clinicians. Pancreas 2015; 44:1211–1233.
7. Quinton PM. The neglected ion: HCO_3^-. Nat Med 2001; 7:292–293.
8. Quinton PM. Cystic fibrosis: impaired bicarbonate secretion and mucoviscidosis. Lancet 2008; 372:415–417.
9. Kusiak AA, Szopa MD, Jakubowska MA, Ferdek PE. Signaling in the physiology and pathophysiology of pancreatic stellate cells—a brief review of recent advances. Front Physiol 2020; 11:78.
10. Kerem B, Rommens JM, Buchanan JA, Markiewicz D, Cox TK, Chakravarti A, Buchwald M, Tsui LC. Identification of the cystic fibrosis gene: genetic analysis. Science 1989; 245:1073–1080.
11. Abuladze N, Lee I, Newman D, Hwang J, Boorer K, Pushkin A, Kurtz I. Molecular cloning, chromosomal localization, tissue distribution, and functional expression of the human pancreatic sodium bicarbonate cotransporter. J Biol Chem 1998; 273:17689–17695.
12. Dorwart MR, Shcheynikov N, Yang D, Muallem S. The solute carrier 26 family of proteins in epithelial ion transport. Physiology (Bethesda) 2008; 23:104–114.
13. Ohana E, Yang D, Shcheynikov N, Muallem S. Diverse transport modes by the solute carrier 26 family of anion transporters. J Physiol 2009; 587:2179–2185.
14. Park HW, Nam JH, Kim JY, Namkung W, Yoon JS, Lee JS, Kim KS, Venglovecz V, Gray MA, Kim KH, Lee MG. Dynamic regulation of CFTR bicarbonate permeability by [Cl^-]i and its role in pancreatic bicarbonate secretion. Gastroenterology 2010; 139:620–631.
15. Gagnon KB, Delpire E. Molecular physiology of SPAK and OSR1: two Ste20-related protein kinases regulating ion transport. Physiol Rev 2012; 92:1577–1617.
16. Yang D, Li Q, So I, Huang CL, Ando H, Mizutani A, Seki G, Mikoshiba K, Thomas PJ, Muallem S. IRBIT governs epithelial secretion in mice by antagonizing the WNK/SPAK kinase pathway. J Clin Invest 2011; 121:956–965.
17. Melvin JE, Yule D, Shuttleworth T, Begenisich T. Regulation of fluid and electrolyte secretion in salivary gland acinar cells. Annu Rev Physiol 2005; 67:445–469.
18. Petersen OH. Stimulus-excitation coupling in plasma membranes of pancreatic acinar cells. Biochim Biophys Acta 1982; 694:163–184.
19. Petersen OH, Tepikin AV. Polarized calcium signaling in exocrine gland cells. Annu Rev Physiol 2008; 70:273–299.
20. Martinez JR. Ion transport and water movement. J Dent Res 1987; 66 Spec No:638–647.
21. Park S, Shcheynikov N, Hong JH, Zheng C, Suh SH, Kawaai K, Ando H, Mizutani A, Abe T, Kiyonari H, Seki G, Yule D, et al. Irbit mediates synergy between Ca^{2+} and cAMP signaling pathways during epithelial transport in mice. Gastroenterology 2013; 145:232–241.
22. Ahuja M, Chung WY, Lin WY, McNally BA, Muallem S. Ca^{2+} signaling in exocrine cells. Cold Spring Harb Perspect Biol 2020; 12:a035279.

23. Luscher BP, Vachel L, Ohana E, Muallem S. Cl⁻ as a bona fide signaling ion. Am J Physiol Cell Physiol 2020; 318:C125–C136.

24. Lifson N, Kramlinger KG, Mayrand RR, Lender EJ. Blood flow to the rabbit pancreas with special reference to the islets of Langerhans. Gastroenterology 1980; 79:466–473.

25. Chandra R, Liddle RA. Regulation of pancreatic secretion. The Pancreapedia: Exocrine Pancreas Knowledge Base, 2020. DOI: 10.3998/panc.2020.14.

26. Brooks AM, Grossman MI. Postprandial pH and neutralizing capacity of the proximal duodenum in dogs. Gastroenterology 1970; 59:85–89.

27. Chey WY, Lee YH, Hendricks JG, Rhodes RA, Tai HH. Plasma secretin concentrations in fasting and postprandial state in man. Am J Dig Dis 1978; 23:981–988.

28. Preshaw RM, Cooke AR, Grossman MI. Quantitative aspects of response of canine pancreas to duodenal acidification. Am J Physiol 1966; 210:629–634.

29. Schaffalitzky de Muckadell OB, Fahrenkrug J, Watt-Boolsen S, Worning H. Pancreatic response and plasma secretin concentration during infusion of low dose secretin in man. Scand J Gastroenterol 1978; 13:305–311.

30. Grundy D, Hutson D, Scratcherd T. The response of the pancreas of the anaesthetized cat to secretin before, during and after reversible vagal blockade. J Physiol 1983; 342:517–526.

31. Kohler H, Nustede R, Barthel M, Schafmayer A. Exocrine pancreatic function in dogs with denervated pancreas. Pancreas 1987; 2:676–683.

32. You CH, Rominger JM, Chey WY. Potentiation effect of cholecystokinin-octapeptide on pancreatic bicarbonate secretion stimulated by a physiologic dose of secretin in humans. Gastroenterology 1983; 85:40–45.

33. Messenger SW, Falkowski MA, Groblewski GE. Ca²⁺-regulated secretory granule exocytosis in pancreatic and parotid acinar cells. Cell Calcium 2014; 55:369–375.

34. Ousingsawat J, Martins JR, Schreiber R, Rock JR, Harfe BD, Kunzelmann K. Loss of TMEM16A causes a defect in epithelial Ca²⁺-dependent chloride transport. J Biol Chem 2009; 284:28698–28703.

35. Luo X, Zheng W, Yan M, Lee MG, Muallem S. Multiple functional P2X and P2Y receptors in the luminal and basolateral membranes of pancreatic duct cells. Am J Physiol Cell Physiol 1999; 277:C205–C215.

36. Kowal JM, Yegutkin GG, Novak I. ATP release, generation and hydrolysis in exocrine pancreatic duct cells. Purinergic Signal 2015; 11:533–550.

37. North RA. Molecular physiology of P2X receptors. Physiol Rev 2002; 82:1013–1067.

38. Novak I. Purinergic receptors in the endocrine and exocrine pancreas. Purinergic Signal 2008; 4:237–253.

39. Ishiguro H, Naruse S, Kitagawa M, Hayakawa T, Case RM, Steward MC. Luminal ATP stimulates fluid and HCO₃⁻ secretion in guinea-pig pancreatic duct. J Physiol 1999; 519:551–558.

40. Gautam D, Han SJ, Heard TS, Cui Y, Miller G, Bloodworth L, Wess J. Cholinergic stimulation of amylase secretion from pancreatic acinar cells studied with muscarinic acetylcholine receptor mutant mice. J Pharmacol Exp Ther 2005; 313:995–1002.

41. Kiselyov K, Wang X, Shin DM, Zang W, Muallem S. Calcium signaling complexes in microdomains of polarized secretory cells. Cell Calcium 2006; 40:451–459.

42. Konturek SJ, Zabielski R, Konturek JW, Czarnecki J. Neuroendocrinology of the pancreas; role of brain-gut axis in pancreatic secretion. Eur J Pharmacol 2003; 481:1–14.

43. Novak I. ATP as a signaling molecule: the exocrine focus. News Physiol Sci 2003; 18:12–17.

44. Holst JJ, Fahrenkrug J, Knuhtsen S, Jensen SL, Poulsen SS, Nielsen OV. Vasoactive intestinal polypeptide (VIP) in the pig pancreas: role of VIPergic nerves in control of fluid and bicarbonate secretion. Regul Pept 1984; 8:245–259.

45. Jin C, Naruse S, Kitagawa M, Ishiguro H, Nakajima M, Mizuno N, Ko SB, Hayakawa T. The effect of calcitonin gene-related peptide on pancreatic blood flow and secretion in conscious dogs. Regul Pept 2001; 99:9–15.

46. Lee MG, Ohana E, Park HW, Yang D, Muallem S. Molecular mechanism of pancreatic and salivary gland fluid and HCO₃⁻ secretion. Physiol Rev 2012; 92:39–74.

47. Madden ME, Sarras MP, Jr. Distribution of Na⁺,K⁺-ATPase in rat exocrine pancreas as monitored by K⁺-NPPase cytochemistry and [³H]-ouabain binding: a plasma membrane protein found primarily to be ductal cell associated. J Histochem Cytochem 1987; 35:1365–1374.

48. Roussa E. Channels and transporters in salivary glands. Cell Tissue Res 2011; 343:263–287.

49. Smith ZD, Caplan MJ, Forbush B, 3rd, Jamieson JD. Monoclonal antibody localization of Na⁺-K⁺-ATPase in the exocrine pancreas and parotid of the dog. Am J Physiol Gastrointest Liver Physiol 1987; 253:G99–G109.

50. Suzuki J, Umeda M, Sims PJ, Nagata S. Calcium-dependent phospholipid scrambling by TMEM16F. Nature 2010; 468:834–838.

51. Gray MA, Greenwell JR, Garton AJ, Argent BE. Regulation of maxi-K⁺ channels on pancreatic duct cells by cyclic AMP-dependent phosphorylation. J Membr Biol 1990; 115:203–215.

52. Novak I, Greger R. Electrophysiological study of transport systems in isolated perfused pancreatic ducts: properties of the basolateral membrane. Pflugers Arch 1988; 411:58–68.

53. Almassy J, Diszhazi G, Skaliczki M, Marton I, Magyar ZE, Nanasi PP, Yule DI. Expression of BK channels and Na⁺-K⁺ pumps in the apical membrane of lacrimal acinar cells suggests a new molecular mechanism for primary tear-secretion. Ocul Surf 2019; 17:272–277.

54. Venglovecz V, Hegyi P, Rakonczay Z, Jr, Tiszlavicz L, Nardi A, Grunnet M, Gray MA. Pathophysiological relevance of apical large-conductance Ca²⁺-activated potassium channels in pancreatic duct epithelial cells. Gut 2011; 60:361–369.

55. Boron WF, Chen L, Parker MD. Modular structure of sodium-coupled bicarbonate transporters. J Exp Biol 2009; 212:1697–1706.

56. Gross E, Hawkins K, Abuladze N, Pushkin A, Cotton CU, Hopfer U, Kurtz I. The stoichiometry of the electrogenic sodium bicarbonate cotransporter NBC1 is cell-type dependent. J Physiol 2001; 531:597–603.

57. Gross E, Fedotoff O, Pushkin A, Abuladze N, Newman D, Kurtz I. Phosphorylation-induced modulation of pNBC1 function: distinct roles for the amino- and carboxy-termini. J Physiol 2003; 549:673–682.

58. Ishiguro H, Steward MC, Lindsay AR, Case RM. Accumulation of intracellular HCO_3^- by Na^+-HCO_3^- cotransport in interlobular ducts from guinea-pig pancreas. J Physiol 1996; 495:169–178.

59. Ishiguro H, Steward MC, Wilson RW, Case RM. Bicarbonate secretion in interlobular ducts from guinea-pig pancreas. J Physiol 1996; 495:179–191.

60. Shirakabe K, Priori G, Yamada H, Ando H, Horita S, Fujita T, Fujimoto I, Mizutani A, Seki G, Mikoshiba K. IRBIT, an inositol 1,4,5-trisphosphate receptor-binding protein, specifically binds to and activates pancreas-type Na^+/HCO_3^- cotransporter 1 (pNBC1). Proc Natl Acad Sci U S A 2006; 103:9542–9547.

61. Yang D, Shcheynikov N, Muallem S. IRBIT: it is everywhere. Neurochem Res 2011; 36:1166–1174.

62. Shcheynikov N, Son A, Hong JH, Yamazaki O, Ohana E, Kurtz I, Shin DM, Muallem S. Intracellular Cl^- as a signaling ion that potently regulates Na^+/HCO_3^- transporters. Proc Natl Acad Sci U S A 2015; 112:E329–E337.

63. Vachel L, Shcheynikov N, Yamazaki O, Fremder M, Ohana E, Son A, Shin DM, Yamazaki-Nakazawa A, Yang CR, Knepper MA, Muallem S. Modulation of Cl^- signaling and ion transport by recruitment of kinases and phosphatases mediated by the regulatory protein IRBIT. Sci Signal 2018; 11:eaat5018.

64. Park M, Ko SB, Choi JY, Muallem G, Thomas PJ, Pushkin A, Lee MS, Kim JY, Lee MG, Muallem S, Kurtz I. The cystic fibrosis transmembrane conductance regulator interacts with and regulates the activity of the HCO_3^- salvage transporter human Na^+-HCO_3^- cotransport isoform 3. J Biol Chem 2002; 277:50503–50509.

65. Pushkin A, Abuladze N, Lee I, Newman D, Hwang J, Kurtz I. Cloning, tissue distribution, genomic organization, and functional characterization of NBC3, a new member of the sodium bicarbonate cotransporter family. J Biol Chem 1999; 274:16569–16575.

66. Gerolami A, Marteau C, Matteo A, Sahel J, Portugal H, Pauli AM, Pastor J, Sarles H. Calcium carbonate saturation in human pancreatic juice: possible role of ductal H^+ secretion. Gastroenterology 1989; 96:881–884.

67. Marteau C, Blanc G, Devaux MA, Portugal H, Gerolami A. Influence of pancreatic ducts on saturation of juice with calcium carbonate in dogs. Dig Dis Sci 1993; 38:2090–2097.

68. Johansen PG, Anderson CM, Hadorn B. Cystic fibrosis of the pancreas. A generalised disturbance of water and electrolyte movement in exocrine tissues. Lancet 1968; 1:455–460.

69. Kunzelmann K, Schreiber R, Hadorn HB. Bicarbonate in cystic fibrosis. J Cyst Fibros 2017; 16:653–662.

70. Riordan JR, Rommens JM, Kerem B, Alon N, Rozmahel R, Grzelczak Z, Zielenski J, Lok S, Plavsic N, Chou JL, Drumm ML, Iannuzzi MC, et al. Identification of the cystic fibrosis gene: cloning and characterization of complementary DNA. Science 1989; 245:1066–1073.

71. Rommens JM, Iannuzzi MC, Kerem B, Drumm ML, Melmer G, Dean M, Rozmahel R, Cole JL, Kennedy D, Hidaka N,

72. Zsiga M, Buchwald M, et al. Identification of the cystic fibrosis gene: chromosome walking and jumping. Science 1989; 245:1059–1065.

72. Linsdell P, Tabcharani JA, Rommens JM, Hou YX, Chang XB, Tsui LC, Riordan JR, Hanrahan JW. Permeability of wild-type and mutant cystic fibrosis transmembrane conductance regulator chloride channels to polyatomic anions. J Gen Physiol 1997; 110:355–364.

73. Deeley RG, Westlake C, Cole SP. Transmembrane transport of endo- and xenobiotics by mammalian ATP-binding cassette multidrug resistance proteins. Physiol Rev 2006; 86:849–899.

74. Catalan MA, Nakamoto T, Gonzalez-Begne M, Camden JM, Wall SM, Clarke LL, Melvin JE. Cftr and ENaC ion channels mediate NaCl absorption in the mouse submandibular gland. J Physiol 2010; 588:713–724.

75. Steward MC, Ishiguro H, Case RM. Mechanisms of bicarbonate secretion in the pancreatic duct. Annu Rev Physiol 2005; 67:377–409.

76. Zeng W, Lee MG, Yan M, Diaz J, Benjamin I, Marino CR, Kopito R, Freedman S, Cotton C, Muallem S, Thomas P. Immuno and functional characterization of CFTR in submandibular and pancreatic acinar and duct cells. Am J Physiol Cell Physiol 1997; 273:C442–C455.

77. Poulsen JH, Fischer H, Illek B, Machen TE. Bicarbonate conductance and pH regulatory capability of cystic fibrosis transmembrane conductance regulator. Proc Natl Acad Sci U S A 1994; 91:5340–5344.

78. Shcheynikov N, Kim KH, Kim KM, Dorwart MR, Ko SB, Goto H, Naruse S, Thomas PJ, Muallem S. Dynamic control of cystic fibrosis transmembrane conductance regulator Cl^-/HCO_3^- selectivity by external Cl^-. J Biol Chem 2004; 279:21857–21865.

79. Jung J, Lee MG. Role of calcium signaling in epithelial bicarbonate secretion. Cell Calcium 2014; 55:376–384.

80. Hug MJ, Tamada T, Bridges RJ. CFTR and bicarbonate secretion by [correction of to] epithelial cells. News Physiol Sci 2003; 18:38–42.

81. Argent BE, Case RM. Pancreatic ducts. Cellular mechanism and control of bicarbonate secretion. In: Johnson LR, Alpers DH, Christensen J, Jacobs ED, Walsh JH, eds. Physiology of the gastrointestinal tract. Volume 1. 3rd ed. New York: Raven Press, 1994:1473–1497.

82. Ishiguro H, Steward MC, Naruse S, Ko SB, Goto H, Case RM, Kondo T, Yamamoto A. CFTR functions as a bicarbonate channel in pancreatic duct cells. J Gen Physiol 2009; 133:315–326.

83. Sohma Y, Gray MA, Imai Y, Argent BE. A mathematical model of the pancreatic ductal epithelium. J Membr Biol 1996; 154:53–67.

84. Kim Y, Jun I, Shin DH, Yoon JG, Piao H, Jung J, Park HW, Cheng MH, Bahar I, Whitcomb DC, Lee MG. Regulation of CFTR bicarbonate channel activity by WNK1: implications for pancreatitis and CFTR-related disorders. Cell Mol Gastroenterol Hepatol 2020; 9:79–103.

85. Piala AT, Moon TM, Akella R, He H, Cobb MH, Goldsmith EJ. Chloride sensing by WNK1 involves inhibition of autophosphorylation. Sci Signal 2014; 7:ra41.

86. Terker AS, Zhang C, Erspamer KJ, Gamba G, Yang CL, Ellison DH. Unique chloride-sensing properties of WNK4

permit the distal nephron to modulate potassium homeostasis. Kidney Int 2016; 89:127–134.

87. Choi JY, Muallem D, Kiselyov K, Lee MG, Thomas PJ, Muallem S. Aberrant CFTR-dependent HCO3⁻ transport in mutations associated with cystic fibrosis. Nature 2001; 410:94–97.

88. LaRusch J, Jung J, General IJ, Lewis MD, Park HW, Brand RE, Gelrud A, Anderson MA, Banks PA, Conwell D, Lawrence C, Romagnuolo J, et al. Mechanisms of CFTR functional variants that impair regulated bicarbonate permeation and increase risk for pancreatitis but not for cystic fibrosis. PLoS Genet 2014; 10:e1004376.

89. Lee JH, Richter W, Namkung W, Kim KH, Kim E, Conti M, Lee MG. Dynamic regulation of cystic fibrosis transmembrane conductance regulator by competitive interactions of molecular adaptors. J Biol Chem 2007; 282:10414–10422.

90. Short DB, Trotter KW, Reczek D, Kreda SM, Bretscher A, Boucher RC, Stutts MJ, Milgram SL. An apical PDZ protein anchors the cystic fibrosis transmembrane conductance regulator to the cytoskeleton. J Biol Chem 1998; 273:19797–19801.

91. Ko SB, Shcheynikov N, Choi JY, Luo X, Ishibashi K, Thomas PJ, Kim JY, Kim KH, Lee MG, Naruse S, Muallem S. A molecular mechanism for aberrant CFTR-dependent HCO3⁻ transport in cystic fibrosis. EMBO J 2002; 21:5662–5672.

92. Ahn W, Kim KH, Lee JA, Kim JY, Choi JY, Moe OW, Milgram SL, Muallem S, Lee MG. Regulatory interaction between the cystic fibrosis transmembrane conductance regulator and HCO3⁻ salvage mechanisms in model systems and the mouse pancreatic duct. J Biol Chem 2001; 276:17236–17243.

93. Guggino WB. The cystic fibrosis transmembrane regulator forms macromolecular complexes with PDZ domain scaffold proteins. Proc Am Thorac Soc 2004; 1:28–32.

94. Ko SB, Zeng W, Dorwart MR, Luo X, Kim KH, Millen L, Goto H, Naruse S, Soyombo A, Thomas PJ, Muallem S. Gating of CFTR by the STAS domain of SLC26 transporters. Nat Cell Biol 2004; 6:343–350.

95. Hong JH, Muhammad E, Zheng C, Hershkovitz E, Alkrinawi S, Loewenthal N, Parvari R, Muallem S. Essential role of carbonic anhydrase XII in secretory gland fluid and HCO3⁻ secretion revealed by disease causing human mutation. J Physiol 2015; 593:5299–5312.

96. Shcheynikov N, Wang Y, Park M, Ko SB, Dorwart M, Naruse S, Thomas PJ, Muallem S. Coupling modes and stoichiometry of Cl⁻/HCO3⁻ exchange by slc26a3 and slc26a6. J Gen Physiol 2006; 127:511–524.

97. Knauf F, Yang CL, Thomson RB, Mentone SA, Giebisch G, Aronson PS. Identification of a chloride-formate exchanger expressed on the brush border membrane of renal proximal tubule cells. Proc Natl Acad Sci U S A 2001; 98:9425–9430.

98. Shcheynikov N, Yang D, Wang Y, Zeng W, Karniski LP, So I, Wall SM, Muallem S. The Slc26a4 transporter functions as an electroneutral Cl⁻/I⁻/HCO3⁻ exchanger: role of Slc26a4 and Slc26a6 in I⁻ and HCO3⁻ secretion and in regulation of CFTR in the parotid duct. J Physiol 2008; 586:3813–3824.

99. Stewart AK, Yamamoto A, Nakakuki M, Kondo T, Alper SL, Ishiguro H. Functional coupling of apical Cl⁻/

HCO3⁻ exchange with CFTR in stimulated HCO3⁻ secretion by guinea pig interlobular pancreatic duct. Am J Physiol Gastrointest Liver Physiol 2009; 296:G1307–G1317.

100. Wang Y, Soyombo AA, Shcheynikov N, Zeng W, Dorwart M, Marino CR, Thomas PJ, Muallem S. Slc26a6 regulates CFTR activity in vivo to determine pancreatic duct HCO3⁻ secretion: relevance to cystic fibrosis. EMBO J 2006; 25:5049–5057.

101. Lin WY, Muallem S. No zoom required: meeting at the beta-intercalated cells. J Am Soc Nephrol 2020; 31:1655–1657.

102. Zhao H, Muallem S. Na⁺, K⁺, and Cl⁻ transport in resting pancreatic acinar cells. J Gen Physiol 1995; 106:1225–1242.

103. Veel T, Villanger O, Holthe MR, Cragoe EJ, Jr., Raeder MG. Na⁺-H⁺ exchange is not important for pancreatic HCO3⁻ secretion in the pig. Acta Physiol Scand 1992; 144:239–246.

104. Wizemann V, Schulz I. Influence of amphotericin, amiloride, ionophores, and 2,4-dinitrophenol on the secretion of the isolated cat's pancreas. Pflugers Arch 1973; 339:317–338.

105. Lee MG, Ahn W, Choi JY, Luo X, Seo JT, Schultheis PJ, Shull GE, Kim KH, Muallem S. Na⁺-dependent transporters mediate HCO3⁻ salvage across the luminal membrane of the main pancreatic duct. J Clin Invest 2000; 105:1651–1658.

106. He P, Klein J, Yun CC. Activation of Na⁺/H⁺ exchanger NHE3 by angiotensin II is mediated by inositol 1,4,5-triphosphate (IP3) receptor-binding protein released with IP3 (IRBIT) and Ca²⁺/calmodulin-dependent protein kinase II. J Biol Chem 2010; 285:27869–27878.

107. He P, Zhang H, Yun CC. IRBIT, inositol 1,4,5-triphosphate (IP3) receptor-binding protein released with IP3, binds Na⁺/H⁺ exchanger NHE3 and activates NHE3 activity in response to calcium. J Biol Chem 2008; 283:33544–33553.

108. Caputo A, Caci E, Ferrera L, Pedemonte N, Barsanti C, Sondo E, Pfeffer U, Ravazzolo R, Zegarra-Moran O, Galietta LJ. TMEM16A, a membrane protein associated with calcium-dependent chloride channel activity. Science 2008; 322:590–594.

109. Schroeder BC, Cheng T, Jan YN, Jan LY. Expression cloning of TMEM16A as a calcium-activated chloride channel subunit. Cell 2008; 134:1019–1029.

110. Yang YD, Cho H, Koo JY, Tak MH, Cho Y, Shim WS, Park SP, Lee J, Lee B, Kim BM, Raouf R, Shin YK, et al. TMEM16A confers receptor-activated calcium-dependent chloride conductance. Nature 2008; 455:1210–1215.

111. Jha A, Chung WY, Vachel L, Maleth J, Lake S, Zhang G, Ahuja M, Muallem S. Anoctamin 8 tethers endoplasmic reticulum and plasma membrane for assembly of Ca²⁺ signaling complexes at the ER/PM compartment. EMBO J 2019; 38:e101452.

112. Gray MA, Harris A, Coleman L, Greenwell JR, Argent BE. Two types of chloride channel on duct cells cultured from human fetal pancreas. Am J Physiol Cell Physiol 1989; 257:C240–C251.

113. Gray MA, Winpenny JP, Porteous DJ, Dorin JR, Argent BE. CFTR and calcium-activated chloride currents in pancreatic duct cells of a transgenic CF mouse. Am J Physiol Cell Physiol 1994; 266:C213–C221.

114. Venglovecz V, Rakonczay Z, Jr., Ozsvari B, Takacs T, Lonovics J, Varro A, Gray MA, Argent BE, Hegyi P. Effects of bile acids on pancreatic ductal bicarbonate secretion in guinea pig. Gut 2008; 57:1102–1112.

115. Zeng W, Lee MG, Muallem S. Membrane-specific regulation of Cl⁻ channels by purinergic receptors in rat submandibular gland acinar and duct cells. J Biol Chem 1997; 272:32956–32965.

116. Aoun J, Hayashi M, Sheikh IA, Sarkar P, Saha T, Ghosh P, Bhowmick R, Ghosh D, Chatterjee T, Chakrabarti P, Chakrabarti MK, Hoque KM. Anoctamin 6 contributes to Cl⁻ secretion in Accessory cholera enterotoxin (Ace)-stimulated diarrhea: An essential role for phosphatidylinositol 4,5-bisphosphate (PIP_2) signaling in cholera. J Biol Chem 2016; 291:26816–26836.

117. Jun I, Cheng MH, Sim E, Jung J, Suh BL, Kim Y, Son H, Park K, Kim CH, Yoon JH, Whitcomb DC, Bahar I, et al. Pore dilatation increases the bicarbonate permeability of CFTR, ANO1 and glycine receptor anion channels. J Physiol 2016; 594:2929–2955.

118. Jung J, Nam JH, Park HW, Oh U, Yoon JH, Lee MG. Dynamic modulation of ANO1/TMEM16A HCO_3^- permeability by Ca^{2+}/calmodulin. Proc Natl Acad Sci U S A 2013; 110:360–365.

119. Shin DH, Kim M, Kim Y, Jun I, Jung J, Nam JH, Cheng MH, Lee MG. Bicarbonate permeation through anion channels: its role in health and disease. Pflugers Arch 2020; 472:1003–1018.

120. Ma T, Song Y, Gillespie A, Carlson EJ, Epstein CJ, Verkman AS. Defective secretion of saliva in transgenic mice lacking aquaporin-5 water channels. J Biol Chem 1999; 274:20071–20074.

121. Burghardt B, Elkaer ML, Kwon TH, Racz GZ, Varga G, Steward MC, Nielsen S. Distribution of aquaporin water channels AQP1 and AQP5 in the ductal system of the human pancreas. Gut 2003; 52:1008–1016.

122. Ko SB, Mizuno N, Yatabe Y, Yoshikawa T, Ishiguro H, Yamamoto A, Azuma S, Naruse S, Yamao K, Muallem S, Goto H. Corticosteroids correct aberrant CFTR localization in the duct and regenerate acinar cells in autoimmune pancreatitis. Gastroenterology 2010; 138:1988–1996.

123. Ko SB, Naruse S, Kitagawa M, Ishiguro H, Furuya S, Mizuno N, Wang Y, Yoshikawa T, Suzuki A, Shimano S, Hayakawa T. Aquaporins in rat pancreatic interlobular ducts. Am J Physiol Gastrointest Liver Physiol 2002; 282:G324–G331.

124. Venglovecz V, Pallagi P, Kemeny LV, Balazs A, Balla Z, Becskehazi E, Gal E, Toth E, Zvara A, Puskas LG, Borka K, Sendler M, et al. The importance of aquaporin 1 in pancreatitis and its relation to the CFTR Cl⁻ Channel. Front Physiol 2018; 9:854.

125. Alvarez BV, Vithana EN, Yang Z, Koh AH, Yeung K, Yong V, Shandro HJ, Chen Y, Kolatkar P, Palasingam P, Zhang K, Aung T, et al. Identification and characterization of a novel mutation in the carbonic anhydrase IV gene that causes retinitis pigmentosa. Invest Ophthalmol Vis Sci 2007; 48:3459–3468.

126. Feldshtein M, Elkrinawi S, Yerushalmi B, Marcus B, Vullo D, Romi H, Ofir R, Landau D, Sivan S, Supuran CT, Birk OS. Hyperchlorhidrosis caused by homozygous mutation in CA12, encoding carbonic anhydrase XII. Am J Hum Genet 2010; 87:713–720.

127. Muhammad E, Leventhal N, Parvari G, Hanukoglu A, Hanukoglu I, Chalifa-Caspi V, Feinstein Y, Weinbrand J, Jacoby H, Manor E, Nagar T, Beck JC, et al. Autosomal recessive hyponatremia due to isolated salt wasting in sweat associated with a mutation in the active site of carbonic anhydrase 12. Hum Genet 2011; 129:397–405.

128. McKenna R, Frost SC. Overview of the carbonic anhydrase family. Subcell Biochem 2014; 75:3–5.

129. Frost SC. Physiological functions of the alpha class of carbonic anhydrases. Subcell Biochem 2014; 75:9–30.

130. Alvarez BV, Loiselle FB, Supuran CT, Schwartz GJ, Casey JR. Direct extracellular interaction between carbonic anhydrase IV and the human NBC1 sodium/bicarbonate co-transporter. Biochemistry 2003; 42:12321–12329.

131. Loiselle FB, Morgan PE, Alvarez BV, Casey JR. Regulation of the human NBC3 Na^+/HCO_3^- cotransporter by carbonic anhydrase II and PKA. Am J Physiol Cell Physiol 2004; 286:C1423–C1433.

132. Krishnan D, Liu L, Wiebe SA, Casey JR, Cordat E, Alexander RT. Carbonic anhydrase II binds to and increases the activity of the epithelial sodium-proton exchanger, NHE3. Am J Physiol Renal Physiol 2015; 309:F383–F392.

133. Li X, Liu Y, Alvarez BV, Casey JR, Fliegel L. A novel carbonic anhydrase II binding site regulates NHE1 activity. Biochemistry 2006; 45:2414–2424.

134. Vilas G, Krishnan D, Loganathan SK, Malhotra D, Liu L, Beggs MR, Gena P, Calamita G, Jung M, Zimmermann R, Tamma G, Casey JR, et al. Increased water flux induced by an aquaporin-1/carbonic anhydrase II interaction. Mol Biol Cell 2015; 26:1106–1118.

135. Feinstein Y, Yerushalmi B, Loewenthal N, Alkrinawi S, Birk OS, Parvari R, Hershkovitz E. Natural history and clinical manifestations of hyponatremia and hyperchlorhidrosis due to carbonic anhydrase XII deficiency. Horm Res Paediatr 2014; 81:336–342.

136. Evans RL, Perrott MN, Lau KR, Case RM. Elevation of intracellular cAMP by noradrenaline and vasoactive intestinal peptide in striated ducts isolated from the rabbit mandibular salivary gland. Arch Oral Biol 1996; 41:689–694.

137. Ulrich CD, 2nd, Holtmann M, Miller LJ. Secretin and vasoactive intestinal peptide receptors: members of a unique family of G protein-coupled receptors. Gastroenterology 1998; 114:382–397.

138. Yang D, Shcheynikov N, Zeng W, Ohana E, So I, Ando H, Mizutani A, Mikoshiba K, Muallem S. IRBIT coordinates epithelial fluid and HCO_3^- secretion by stimulating the transporters pNBC1 and CFTR in the murine pancreatic duct. J Clin Invest 2009; 119:193–202.

139. Nguyen TD, Meichle S, Kim US, Wong T, Moody MW. $P2Y_{11}$, a purinergic receptor acting via cAMP, mediates secretion by pancreatic duct epithelial cells. Am J Physiol Gastrointest Liver Physiol 2001; 280:G795–G804.

140. Alvarez C, Regan JP, Merianos D, Bass BL. Protease-activated receptor-2 regulates bicarbonate secretion by pancreatic duct cells in vitro. Surgery 2004; 136:669–676.

141. Namkung W, Han W, Luo X, Muallem S, Cho KH, Kim KH, Lee MG. Protease-activated receptor 2 exerts local protection and mediates some systemic complications in acute pancreatitis. Gastroenterology 2004; 126:1844–1859.

142. Namkung W, Yoon JS, Kim KH, Lee MG. PAR2 exerts local protection against acute pancreatitis via modulation of MAP

kinase and MAP kinase phosphatase signaling. Am J Physiol Gastrointest Liver Physiol 2008; 295:G886–894.

143. Nguyen TD, Moody MW, Steinhoff M, Okolo C, Koh DS, Bunnett NW. Trypsin activates pancreatic duct epithelial cell ion channels through proteinase-activated receptor-2. J Clin Invest 1999; 103:261–269.

144. Domschke S, Domschke W, Rosch W, Konturek SJ, Wunsch E, Demling L. Bicarbonate and cyclic AMP content of pure human pancreatic juice in response to graded doses of synthetic secretin. Gastroenterology 1976; 70:533–536.

145. Gyr K, Beglinger C, Fried M, Grotzinger U, Kayasseh L, Stalder GA, Girard J. Plasma secretin and pancreatic response to various stimulants including a meal. Am J Physiol Gastrointest Liver Physiol 1984; 246:G535–G542.

146. Park S, Hong JH, Ohana E, Muallem S. The WNK/SPAK and IRBIT/PP1 pathways in epithelial fluid and electrolyte transport. Physiology (Bethesda) 2012; 27:291–299.

147. Bruce JI, Shuttleworth TJ, Giovannucci DR, Yule DI. Phosphorylation of inositol 1,4,5-trisphosphate receptors in parotid acinar cells. A mechanism for the synergistic effects of cAMP on Ca^{2+} signaling. J Biol Chem 2002; 277:1340–1348.

148. Willoughby D, Cooper DM. Organization and Ca^{2+} regulation of adenylyl cyclases in cAMP microdomains. Physiol Rev 2007; 87:965–1010.

149. Namkung W, Lee JA, Ahn W, Han W, Kwon SW, Ahn DS, Kim KH, Lee MG. Ca^{2+} activates cystic fibrosis transmembrane conductance regulator- and Cl^--dependent HCO_3 transport in pancreatic duct cells. J Biol Chem 2003; 278:200–207.

150. Dekker JW, Budhia S, Angel NZ, Cooper BJ, Clark GJ, Hart DN, Kato M. Identification of an S-adenosylhomocysteine hydrolase-like transcript induced during dendritic cell differentiation. Immunogenetics 2002; 53:993–1001.

151. Ando H, Mizutani A, Matsu-ura T, Mikoshiba K. IRBIT, a novel inositol 1,4,5-trisphosphate (IP_3) receptor-binding protein, is released from the IP_3 receptor upon IP_3 binding to the receptor. J Biol Chem 2003; 278:10602–10612.

152. Ando H, Kawaai K, Mikoshiba K. IRBIT: a regulator of ion channels and ion transporters. Biochim Biophys Acta 2014; 1843:2195–2204.

153. Lee MG, Xu X, Zeng W, Diaz J, Wojcikiewicz RJ, Kuo TH, Wuytack F, Racymaekers L, Muallem S. Polarized expression of Ca^{2+} channels in pancreatic and salivary gland cells. Correlation with initiation and propagation of $[Ca^{2+}]_i$ waves. J Biol Chem 1997; 272:15765–15770.

154. Min X, Lee BH, Cobb MH, Goldsmith EJ. Crystal structure of the kinase domain of WNK1, a kinase that causes a hereditary form of hypertension. Structure 2004; 12:1303–1311.

155. Wilson FH, Disse-Nicodeme S, Choate KA, Ishikawa K, Nelson-Williams C, Desitter I, Gunel M, Milford DV, Lipkin GW, Achard JM, Feely MP, Dussol B, et al. Human hypertension caused by mutations in WNK kinases. Science 2001; 293:1107–1112.

156. Huang CL, Cha SK, Wang HR, Xie J, Cobb MH. WNKs: protein kinases with a unique kinase domain. Exp Mol Med 2007; 39:565–573.

157. Huang CL, Yang SS, Lin SH. Mechanism of regulation of renal ion transport by WNK kinases. Curr Opin Nephrol Hypertens 2008; 17:519–525.

158. Furusho T, Uchida S, Sohara E. The WNK signaling pathway and salt-sensitive hypertension. Hypertens Res 2020; 43:733–743.

159. Delpire E, Gagnon KB. SPAK and OSR1: STE20 kinases involved in the regulation of ion homoeostasis and volume control in mammalian cells. Biochem J 2008; 409:321–331.

160. Shekarabi M, Zhang J, Khanna AR, Ellison DH, Delpire E, Kahle KT. WNK kinase signaling in ion homeostasis and human disease. Cell Metab 2017; 25:285–299.

161. Ishiguro H, Naruse S, Kitagawa M, Mabuchi T, Kondo T, Hayakawa T, Case RM, Steward MC. Chloride transport in microperfused interlobular ducts isolated from guinea-pig pancreas. J Physiol 2002; 539:175–189.

162. Aguilella-Arzo M, Andrio A, Aguilella VM, Alcaraz A. Dielectric saturation of water in a membrane protein channel. Phys Chem Phys 2009; 11:358–365.

163. Liu F, Zhang Z, Csanady L, Gadsby DC, Chen J. Molecular structure of the human CFTR ion channel. Cell 2017; 169:85–95 e88.

164. Sohma Y, Gray MA, Imai Y, Argent BE. HCO_3^- transport in a mathematical model of the pancreatic ductal epithelium. J Membr Biol 2000; 176:77–100.

165. Ishiguro H, Naruse S, Steward MC, Kitagawa M, Ko SB, Hayakawa T, Case RM. Fluid secretion in interlobular ducts isolated from guinea-pig pancreas. J Physiol 1998; 511:407–422.

166. Bayliss WM, Starling EH. The mechanism of pancreatic secretion. J Physiol 1902; 28:325–353.

167. Henderson LJ. The regulation of neutrality in the animal body. Science 1913; 37:389–395.

168. Case RM, Harper AA, Scratcherd T. The secretion of electrolytes and enzymes by the pancreas of the anaesthetized cat. J Physiol 1969; 201:335–348.

169. Hart WM, Thomas JE. Bicarbonate and chloride of pancreatic juice secreted in response to various stimuli. Gastroenterology 1945; 4:409–420.

170. Zhao H, Star RA, Muallem S. Membrane localization of H^+ and HCO_3^- transporters in the rat pancreatic duct. J Gen Physiol 1994; 104:57–85.

171. Lohi H, Kujala M, Kerkela E, Saarialho-Kere U, Kestila M, Kere J. Mapping of five new putative anion transporter genes in human and characterization of SLC26A6, a candidate gene for pancreatic anion exchanger. Genomics 2000; 70:102–112.

172. Hegyi P, Wilschanski M, Muallem S, Lukacs GL, Sahin-Toth M, Uc A, Gray MA, Rakonczay Z, Jr, Maleth J. CFTR: a new horizon in the pathomechanism and treatment of pancreatitis. Rev Physiol Biochem Pharmacol 2016; 170:37–66.

Chapter 20

Ion channels in pancreatic duct physiology

Viktoria Venglovecz,[1] Péter Hegyi,[2] and Michael A. Gray[3]

[1]Department of Pharmacology and Pharmacotherapy, University of Szeged, Szeged, Hungary
[2]Medical School, Institute for Translational Medicine, University of Pécs, Pécs, Hungary
[3]Biosciences Institute, Faculty of Medical Sciences, Newcastle University, Newcastle upon Tyne, UK

Physiology of Pancreatic Ductal Cells

The main function of pancreatic ductal cells is to secrete a HCO_3^--rich, isotonic fluid that washes out the inactive form of digestive enzymes from the ductal system, as well as provides pH conditions that are essential for normal pancreatic function. The rate of HCO_3^- secretion is influenced by several factors (such as secretory rate, species, or the location of the cell in the ductal tree), and with stimulation, HCO_3^- can reach up to 140 mM. This means a significant concentration difference exists between the outside and the inside of the ductal cell, which poses a physiologic challenge to duct cell homeostasis. The high level of HCO_3^- secretion is achieved through the coordinated action of ion transporters and channels, in which the Cl^-/HCO_3^- exchanger and the cystic fibrosis transmembrane conductance regulator (CFTR) Cl^- channel play a central role.

Ion transporters and channels on ductal cells are differentially expressed on the luminal and basolateral membranes, resulting in functional polarization of the ductal cells (**Figure 1**).[1] The major ion transporters on the duct cell basolateral membrane include the Na^+/H^+ exchanger (NHE), the Na^+/HCO_3^- cotransporter (NBC), the Na^+/K^+ ATPase, and various types of K^+ channels.[2] The electroneutral, NHE1 isoform can be found on the basolateral membrane of ductal cells and acts as a proton extruder with a $1 Na^+ : 1 H^+$ stoichiometry.[3] In addition to playing an important role in the regulation of intracellular pH, NHE1 promotes the formation of HCO_3^- from carbonic acid by removing excess H^+ from the cell. The NHE3 isoform is located on the luminal membrane of ductal cells and, unlike NHE1, is involved in HCO_3^- salvage.[4,5] HCO_3^- also enters the cell directly through NBC. The electrogenic NBC isoform NBCe1B has been identified on the basolateral membrane of ductal cells, with $1 Na^+ / 2 HCO_3^-$ stoichiometry.[6,7] Through this transporter, more HCO_3^- enters the cell than is formed during the dissociation of carbonic acid. The electroneutral form of NBC, the NBCn1 or NBC3, has

been shown to be active on the luminal membrane of ductal cells, where it plays a role in HCO_3^- salvage.[8] In addition, Na^+/K^+ ATPase, together with basolateral K^+ channels, maintains a negative membrane potential using the energy from the hydrolysis of ATP, which provides a driving force for anion secretion across the luminal membrane.

There are two major transporters on the luminal side of the ductal cells: the CFTR chloride channel and the Cl^-/HCO_3^- exchanger. Several mechanisms have been proposed to explain how ductal cells can secrete up to five times as much HCO_3^- as is present in the cytosol.[9,10] The most generally accepted view is that HCO_3^- secretion occurs through both the Cl^-/HCO_3^- exchanger and the CFTR channel.[11,12] Among the Cl^-/HCO_3^- exchangers, the Slc26a6 (PAT1) and the Slc26a3 (DRA) isoforms are present on pancreatic ductal cells.[13–15] PAT1 and DRA have different $Cl^-:HCO_3^-$ stoichiometry (2:1 for DRA and 1:2 for PAT1) and are expressed in different parts of the ductal tree. In the initial stage of HCO_3^- secretion, HCO_3^- is secreted into the lumen through the Cl^-/HCO_3^- exchanger in exchange for Cl^- that is returned to the lumen through the CFTR Cl^- channel. Luminal concentration of HCO_3^- and Cl^- follows a reciprocal pattern that is stimulation dependent. With increased secretory volume the concentration of HCO_3^- increases in the lumen and Cl^- decreases. At the same time, the WNK1-OSR1/SPAK signaling pathway is activated, which increases the permeability of the CFTR channel to HCO_3^-, which allows HCO_3^- concentration to reach 140 mM HCO_3 in the lumen.[12]

Pathophysiology of Pancreatic Duct Cells

Normal electrolyte secretion is required to maintain pancreatic exocrine cell homeostasis and solubilization of secretory protein; impairment of pancreatic fluid and electrolyte secretion contributes to tissue destruction in diseases such as cystic fibrosis (CF).[16–19] An important consequence of impaired

e-mails: vviki3@gmail.com; hegyi.peter@pte.hu; m.a.gray@newcastle.ac.uk

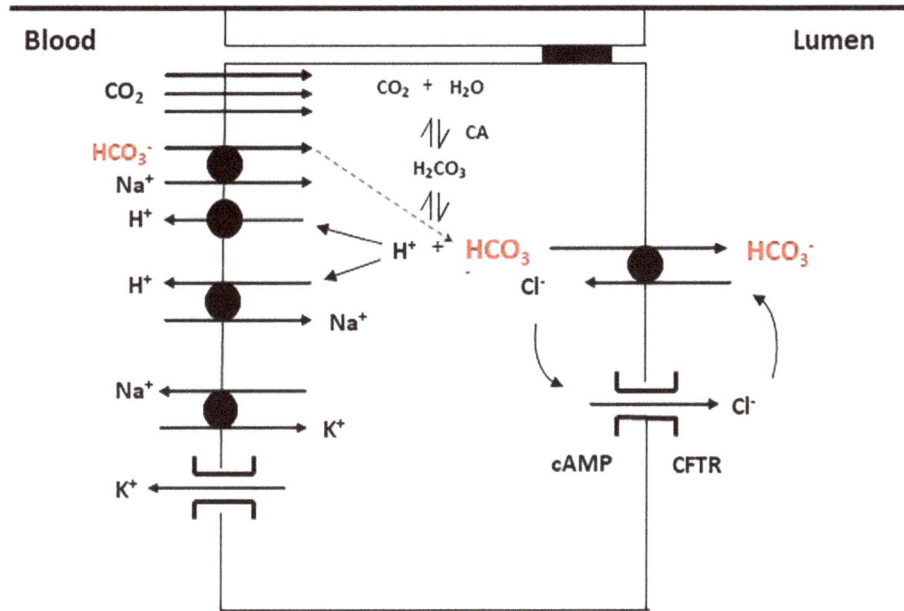

Figure 1. Major ion transporters of pancreatic ductal cells. HCO_3^- enters the ductal cell directly via the basolateral NBC. In addition, carbonic anhydrase (CA) is involved in the intracellular accumulation of HCO_3^- by catalyzing the formation of HCO_3^- and H^+ from carbonic acid. The resulting H^+ leaves the cell via the H^+ pump or NHE. HCO_3^- is secreted into the lumen through the Cl^-/HCO_3^- exchangers and the CFTR Cl^- channel, respectively.

HCO_3^- secretion is an acidic pancreatic juice (less than 6.5). This increases mucus viscosity and decreases the solubility of secreted digestive enzymes, which predisposes to the formation of mucin/protein plugs and eventually cysts within the ductal tree. A more acidic pH may also induce premature activation of digestive enzymes within the ductal tree, leading to the development of pancreatitis. Therefore, intensive research is underway to develop drug molecules capable of restoring the function of transporters, especially in CF.

Cystic fibrosis

CF is the most common, life-limiting, inherited disease in Caucasian populations (1 in 3,500 newborns in Europe).[16,20] Over 2,000 CF-causing mutations have been identified in the *cftr* gene (http://www.genet.sickkids.on.ca/), although to date only ~360 of these variants have been fully annotated (https://cftr2.org/). However, 70–90 percent of CF individuals harbor the F508del mutation on at least one allele,[21] which results in misfolding and incorrect processing of CFTR to the apical membrane. For those mutations that have been studied in detail, the genetic alteration leads to a variety of functional defects in the CFTR protein. These functional defects have been grouped into six classes.[22] Classes 1–3 cause severe CFTR dysfunction, while Classes 4–6 produce less severe effects on CFTR, and, in general, the mutated protein retains some level of channel activity.

In relation to pancreatic pathology, ~85 percent of people with CF are born pancreatic insufficient (PI), which equates to a reduction in pancreatic function of more than 95 percent. In these people, there is a very good correlation between pancreatic disease severity and the class of mutation[23–25] with "severe" CF mutations, such as the most common Class 2 CF mutation, F508del, and the Class 3 gating mutant, G511D, strongly correlate with PI. For those with "milder" mutations (some residual channel activity such as Class 4, R117H), pancreatic function is preserved (pancreatic sufficient (PS)), albeit to differing levels. However, in general, these PS individuals require less pancreatic enzyme replacement supplements but can become PI with increasing age. Mutations that cause moderate deficiencies in CFTR activity (10–50% of normal function) can increase the risk of developing pancreatitis, alone or when combined with other risk factors such as alcohol.

As described in the introduction, CFTR plays a fundamental role in pancreatic ductal $NaHCO_3$ and fluid secretion, where it regulates HCO_3^- secretion in two fundamentally different ways: first, as a direct exit pathway for HCO_3^- secretion and, second, as a regulator of SLC26A-mediated Cl^-/HCO_3^- exchange.[1] For the latter process, this involves physical interaction between CFTR and the SLC26A6 anion exchanger,[14,26–28] and loss of functional CFTR leads to loss of anion exchange activity. The exact mechanism underlying this complete loss of anion exchange activity is not fully understood, but studies from polarized cultures of CFPAC cells, a human CF pancreatic ductal cell line homozygous for F508del, showed that apical SLC26A Cl^-/HCO_3^- exchange activity was absent, despite evidence for mRNA expression. Importantly, viral-mediated CFTR transduction of the CFPAC cells restored anion exchange activity, suggesting that CFTR may be required for the trafficking of SLC26A6 to the apical membrane. Furthermore, it is interesting that the anion exchanger is activated by a number of CFTR mutants that lack Cl^- channel activity[27] and that this correlates with a good retention of pancreatic function in patients carrying those mutations.[29] Taken together, these results strongly suggest that a functional CFTR at the apical plasma membrane is a prerequisite for SLC26A6-mediated anion exchange and that mild CFTR mutations are likely to preserve Cl^-/HCO_3^- exchange activity, although this needs formal demonstration.

Strategies for improving HCO_3^- secretion in the CF pancreas are limited because of the marked tissue destruction at birth in the majority of people with CF. However, the last decade has seen a major improvement in the treatment of the basic defect in CF, through the development of small molecule CFTR modulators.[30] Clinically approved drugs include the CFTR potentiator, Ivacaftor, which enhances CFTR open probability for a number of gating mutants, as well as CFTR correctors, which help restore processing CFTR trafficking mutants (including F508del-CFTR) to the apical membrane. These include the first-generation corrector drug Lumacaftor as well as more efficacious second-generation correctors, such as tezacaftor and elexacaftor. These drugs are given either alone or in combination (double and triple combinations), depending on the functional defects(s), and have produced substantial improvements in lung function, number of hospitalizations and exacerbations, as well as BMI.[31,32] Since lung dysfunction is the major cause of morbidity and mortality in people with CF, median survival rates are predicted to significantly improve. However, there are only a few studies that have directly assessed if these CFTR modulators also improve pancreatic function.[33] For example, pancreatic function measurements in young children with CF taking Ivacaftor over 24 weeks showed a significant restoration of enzyme-secreting capacity (increased fecal elastase-1 levels) and, by inference, pancreatic tissue regeneration, which is an extremely exciting finding[34] that warrants further research. A more recent but limited study also provided evidence that Ivacaftor restored some pancreatic function in an adult with CF (http://dx.doi.org/10.3390/).

Acute pancreatitis

Acute pancreatitis (AP) is a sudden inflammation of the pancreas, which in most cases is mild, but in approximately 20 percent of patients, a life-threatening, severe form can develop in which the mortality rate can reach up to 20–40 percent.[35] Large individual differences can be observed in the development and course of the disease, in which the disturbed balance of protective and damaging factors presumably plays a significant role.[36] Pancreatic ductal cells are considered as protective mechanism in the pancreas by the secretion of a HCO_3^- rich, isotonic fluid. The two main etiological factors in the development of AP are gallstone obstruction and excessive alcohol consumption. Both etiologies associated with marked changes in HCO_3^- and fluid secretion based on *in vitro* intracellular pH and fluid transport studies from isolated microdissected ducts.[37–42] At low concentrations, both agents increased HCO_3^- secretion, a response that required CFTR and Cl^-/HCO_3^- exchange activity.[41–43] However, higher levels of these agents led to a severe inhibition of CFTR-dependent HCO_3^- secretion, which was due to profound mitochondrial damage and a consequent reduction in intracellular ATP levels[38,40] (**Table 1**). These studies were the first to suggest that ductal HCO_3^- secretion could play a protective role against these noxious agents.

Bile acids

Under normal conditions, bile cannot enter the pancreatic ductal system; however, in the case of gallstone obstruction it may regurgitate into the pancreas through the common bile duct. Chenodeoxycholate (CDCA) is the most abundant hydrophobic, unconjugated bile acid in human bile. Using isolated guinea pig pancreatic ducts, it has been shown that luminal administration of CDCA at low concentrations (0.1 mM) significantly increases ductal HCO_3^- secretion, (**Table 1**) in which CDCA-induced intracellular Ca^{2+} oscillation plays an important role.[42] The major source of CDCA-induced Ca^{2+} is the endoplasmic reticulum (ER), from which Ca^{2+} is released via an IP_3-mediated pathway. The released Ca^{2+} activates large-conductance Ca^{2+}-activated K^+ channels (BK_{Ca}) on the luminal membrane of ductal cells, and opening of these channels hyperpolarizes the cell membrane, thereby enhancing the electrochemical driving force for anion secretion through the luminal membrane.[41] This stimulatory effect of CDCA has also been demonstrated in the CFPAC-1 cell line and has been shown to be highly dependent on CFTR expression.[43] In contrast, high concentration of this bile acid (1 mM) causes a toxic increase in Ca^{2+} and strongly inhibits HCO_3^- secretion.[42] The inhibitory effect of high concentrations of bile acids presumably results from the fact that at this concentration CDCA damages mitochondria, resulting in ATP depletion and ultimately apoptosis.[40,42,44,45] This dual effect of bile acid is thought to be important in the pathogenesis of AP. In the early stages of the disease, when bile acids reach ductal cells at low concentrations, the increased fluid secretion tries to wash out the toxic bile acids from the ductal

Table 1. The effects of bile, ethanol, and trypsin on pancreatic ductal HCO_3^- secretion.

Agent	concentration	Ductal HCO_3 secretion	Mechanism of action	Reference
Bile acids	0.1 mM	Increase	Induction of physiological Ca^{2+} signaling and activation of BK_{Ca} channels at the luminal membrane of the ductal cells	41, 42, 43
	1 mM	Decrease	Induction of a huge, sustained $(Ca^{2+})^i$ signal, damage of mitochondria and consequently ATP depletion	40, 42, 46
Ethanol	10 mM	Increase	Induction of physiological Ca^{2+} signaling	38, 39, 48
	100 mM	Decrease	Induction of a huge, sustained $(Ca^{2+})^i$ signal, damage of mitochondria and consequently ATP depletion. Inhibition of CFTR activity and expression	38, 39, 48
Trypsin	10 μM	Decrease	Elevation of $(Ca^{2+})i$ and inhibition of the Cl-/HCO3-exchanger and CFTR, which leads to the acidification of the luminal pH and autoactivation of trypsinogen	49

tree to avoid pancreatic damage. If the concentration of bile acids further increases, it energetically destabilizes the cell, inhibits the function of ion transporters and induces apoptosis. Bile acids reach the acini where they induce inflammatory processes. Interestingly, the toxic effect of bile acids highly depends on their hydrophobicity. The hydrophilic bile acid, ursodeoxycholic acid, the 7-alpha enantiomer of CDCA, is able to counteract the cell-damaging effects of a high dose of CDCA through stabilization of the mitochondrial membrane that raises the possibility of therapeutic use of this bile acid.[46,47]

Ethanol

Ethanol (EtOH) also dose-dependently affects ductal HCO_3^- secretion (**Table 1**). Low concentration of EtOH (0.3–30 mM) enhances both basal and secretin-stimulated ductal fluid secretion from intra-interlobular ducts isolated from guinea pigs, in which cAMP activation and Ca^{2+} release play a key role.[48] The stimulatory effect of low concentrations of EtOH has also been demonstrated in the Capan-1 cell line and has been shown to be dependent on Ca^{2+} release from the ER via an IP_3-mediated pathway.[39] In contrast, high concentrations of EtOH (100 mM) strongly inhibited the rate of HCO_3^- secretion and the activity of CFTR.[39,48] The inhibitory effect of EtOH is presumably mediated by fatty acids (FAs) and fatty acid ethyl esters (FAEEs) formed during the non-oxidative metabolism of EtOH. Similar to the effect of bile acids, EtOH and its non-oxidative metabolites induce toxic Ca^{2+} signaling in ductal cells by completely depleting ER stores and promoting extracellular Ca^{2+} uptake into cells. Persistently elevated Ca^{2+} causes mitochondrial Ca^{2+} overload, resulting in decreased mitochondrial membrane potential and ATP production. Chelation of Ca^{2+} abolished the inhibitory effect of EtOH and FA on HCO_3^- secretion, suggesting that the inhibitory effect of high dose of these agents is mediated by toxic intracellular Ca^{2+}.[39] Moreover,

EtOH and its metabolites profoundly inhibit CFTR function on the ductal cells which can be prevented by the supplementation of intracellular ATP (ATP_i), indicating that the inhibitory effect of EtOH on CFTR is mediated by ATP_i depletion. Since CFTR works in close coordination with the Cl^-/HCO_3^- exchanger, improper CFTR function may contribute to decreased fluid and HCO_3^- secretion and thus to the pathogenesis of AP. Consequently, maintenance of ATP_i may represent a therapeutic option in the treatment of the disease.[38,39] Decreased expression of CFTR has also been observed on the luminal membrane of human pancreatic ductal cells in alcohol-induced acute and chronic pancreatitis, and in response to FAs and FAEEs in which the accelerated turnover and decreased biosynthesis of the channel play a role. The importance of CFTR in the alcohol-induced pancreatic damage has been further confirmed in CFTR knockout mice, where the absence of CFTR caused much more severe pancreatitis.[39]

Trypsin

One of the most important roles of ductal HCO_3^- secretion is to prevent intraductal activation of trypsinogen. Although there is no significant trypsin in the lumen under physiological conditions, it may leak from the acinar cells in the early stages of pancreatitis. By binding to PAR-2 receptors on the luminal membrane of ductal cells, trypsin or trypsin-activating peptide (PAR-2-AP) increases $[Ca^{2+}]_i$ levels and inhibits Cl^-/HCO_3^- exchange and CFTR function, resulting in lower luminal pH.[49] Low luminal pH favors premature activation of trypsinogen, which will activate additional trypsinogen molecules. This process leads to a vicious cycle in which the more trypsin is formed, the more trypsinogen will be activated, resulting in even more inhibition of the activity of the luminal transporters.[49] The importance of PAR-2 activation in the pathobiology of pancreatitis has been also demonstrated using PAR-2 knockout mice, in which luminal administration of either trypsin or PAR-2-AP had significantly lower effect on both $[Ca^{2+}]_i$ and pH_i.

Therapeutic Perspectives and Clinical Significance

The CFTR chloride channel is clearly the most investigated and the most utilized ductal channel that has been targeted for therapy.[50,51] In the first decades after the discovery of the CFTR gene, only symptomatic therapy was available. Difficulties over the years have been caused by the heterogeneity of CFTR gene mutations. Therefore, it is almost needless to say, that CF is typically a disease where personalized therapy needs to be considered. However, one approach that would be potentially suitable for all people with CF is gene therapy. The first randomized clinical trial, of a non-viral-based gene therapy for CF, was performed by E. Alton et al. They found that a 12-month treatment by pGM169/G67A gene therapy formulation improved lung function among the CF patients.[52] Although the results were very promising, no further trials have taken place since the original observation, although preclinical development of a viral-mediated gene therapy treatment for CF is underway (https://www.cff.org/Trials/Pipeline/details/10160/Spirovant-Sciences). However, in the last decade the orally bioavailable correctors, potentiators, and suppressors of CFTR gene mutations have become available for treatment.[53]

CFTR-directed therapies may also be useful for the treatment of pancreatitis, since recent animal studies have suggested that strategies that help maintain levels of HCO_3^- secretion would limit the extent of pathology induced by bile and alcohol.[39,40,54] Furthermore, the effects of ethanol and ethanol metabolites on CFTR are consistent with reduced biogenesis, accelerated plasma membrane turnover, as well as channel inhibition.[39] Thus, restoring cell surface expression and activity of CFTR could partially alleviate the ethanol-induced damage. This potentially could be through the use of CFTR correctors (Lumacaftor, Tezacaftor) and potentiators (Ivacaftor) to improve channel activity. We have recently found that Ivacaftor (VX-770) and Lumacaftor (VX-809) restore CFTR expression defect associated with alcohol in pancreatic ductal cells, suggesting that their combination sold as Orkambi® may serve as a therapeutic option in acute, recurrent, or chronic pancreatitis.[55] Akshintala et al. recently showed that CFTR modulators, alone or in combination, reduced the risk of recurrent acute pancreatitis within a three-year follow-up therapy in adult CF patients.[56] A 24-year-old CF patient with R117H/7 T/F508del mutations with recurrent acute pancreatitis were reported pancreatitis free during Ivacaftor therapy.[57] Carrion et al. also found a reduced frequency and recurrence rate of pancreatitis in patients with CF during Ivacaftor therapy.[58] Both the basic research results and the pancreatitis-free periods achieved in CF patients suggest that the drugs used for treating CF patients should be tested in randomized clinical trials in non-CF patients with recurrent pancreatitis as well.

Another possible way to compensate for defective CFTR would be to target alternative ion channels, such as the calcium-activated Cl channel, TMEM16A,[59,60] or transporters such as the SLC26 family members (A3, A6, or A9), or short-circuiting their regulation by CFTR and rebalancing exocrine homeostasis. It has been shown that variants (SNPs) in the SLC26A9 anion transporter influence disease severity in the CF lungs and intestinal tract and therefore act as gene modifiers. Importantly, a recent study has suggested that SNPs in SLC26A9 also influence the degree of pancreatic insufficiency.[61] Furthermore, variants of SLC26A9 also influence the extent of CF-related diabetes, which may be due to beneficial effects of restoring ductal bicarbonate secretion on endocrine (islet) function in CF.[62] This opens up the possibility of targeting this anion transporter as a potential therapeutic target to slow the progression of exocrine dysfunction in CF. One important advantage of this "alternate non-CFTR approach" is that it would benefit all CF patients regardless of genotype.

Summary

One of the most important functions of ductal cells is their ability to neutralize acidic pH within the pancreas and duodenum. This ability of ductal cells is due to the secretion of a HCO_3^--rich fluid, which results from the coordinated action of ductal ion channels and transporters. Impairment of transport processes can result in a decrease in both the volume and the pH of pancreatic fluid, which can predispose to inflammatory processes and consequently the development of various diseases. In the case of CF, it is well known that the inadequate function of the CFTR channel underlies the disease; however, it is only recently that research has shed light on the pathological role of ductal cells in pancreatitis. Various etiological factors such as bile acids and EtOH are now known to impair ductal HCO_3^- and fluid secretion, which are likely to play an important role in initiating pancreatitis by creating an unfavorable pH environment. Consequently, drugs that enhance the function of ion transporters may be of great importance not only in CF therapy but also in the treatment of pancreatitis.

References

1. Lee MG, Kim Y, Jun I, Aoun J, Muallem S. Molecular mechanisms of pancreatic bicarbonate secretion. Pancreapedia: Exocrine Pancreas Knowledge Base 2020. DOI: 10.3998/panc.2020.06.
2. Venglovecz V, Rakonczay Z, Jr., Gray MA, Hegyi P. Potassium channels in pancreatic duct epithelial cells: their role, function and pathophysiological relevance. Pflugers Arch 2015; 467:625–640.

3. Roussa E, Alper SL, Thevenod F. Immunolocalization of anion exchanger AE2, Na⁺/H⁺ exchangers NHE1 and NHE4, and vacuolar type H⁺-ATPase in rat pancreas. J Histochem Cytochem 2001; 49:463–474.

4. Lee MG, Ahn W, Choi JY, Luo X, Seo JT, Schultheis PJ, Shull GE, Kim KH, Muallem S. Na⁺-dependent transporters mediate HCO₃⁻ salvage across the luminal membrane of the main pancreatic duct. J Clin Invest 2000; 105:1651–1658.

5. Marteau C, Silviani V, Ducroc R, Crotte C, Gerolami A. Evidence for apical Na⁺/H⁺ exchanger in bovine main pancreatic duct. Dig Dis Sci 1995; 40:2336–2340.

6. Satoh H, Moriyama N, Hara C, Yamada H, Horita S, Kunimi M, Tsukamoto K, Iso ON, Inatomi J, Kawakami H, Kudo A, Endou H et al. Localization of Na⁺HCO₃⁻ cotransporter (NBC-1) variants in rat and human pancreas. Am J Physiol Cell Physiol 2003; 284:C729–C737.

7. Thevenod F, Roussa E, Schmitt BM, Romero MF. Cloning and immunolocalization of a rat pancreatic Na⁺ bicarbonate cotransporter. Biochem Biophys Res Commun 1999; 264:291–298.

8. Park M, Ko SB, Choi JY, Muallem G, Thomas PJ, Pushkin A, Lee MS, Kim JY, Lee MG, Muallem S, Kurtz I. The cystic fibrosis transmembrane conductance regulator interacts with and regulates the activity of the HCO₃⁻ salvage transporter human Na⁺HCO₃⁻ cotransport isoform 3. J Biol Chem 2002; 277:50503–50509.

9. Sohma Y, Gray MA, Imai Y, Argent BE. HCO₃⁻ transport in a mathematical model of the pancreatic ductal epithelium. J Membr Biol 2000; 176:77–100.

10. Sohma Y, Gray MA, Imai Y, Argent BE. 150 mM HCO₃⁻—how does the pancreas do it? Clues from computer modelling of the duct cell. JOP 2001; 2:198–202.

11. Lee MG, Ohana E, Park HW, Yang D, Muallem S. Molecular mechanism of pancreatic and salivary gland fluid and HCO₃⁻ secretion. Physiol Rev 2012; 92:39–74.

12. Park HW, Nam JH, Kim JY, Namkung W, Yoon JS, Lee JS, Kim KS, Venglovecz V, Gray MA, Kim KH, Lee MG. Dynamic regulation of CFTR bicarbonate permeability by [Cl⁻]i and its role in pancreatic bicarbonate secretion. Gastroenterology 2010; 139:620–631.

13. Greeley T, Shumaker H, Wang Z, Schweinfest CW, Soleimani M. Downregulated in adenoma and putative anion transporter are regulated by CFTR in cultured pancreatic duct cells. Am J Physiol Gastrointest Liver Physiol 2001; 281:G1301–G1308.

14. Ko SB, Shcheynikov N, Choi JY, Luo X, Ishibashi K, Thomas PJ, Kim JY, Kim KH, Lee MG, Naruse S, Muallem S. A molecular mechanism for aberrant CFTR-dependent HCO₃⁻ transport in cystic fibrosis. EMBO J 2002; 21:5662–5672.

15. Stewart AK, Yamamoto A, Nakakuki M, Kondo T, Alper SL, Ishiguro H. Functional coupling of apical Cl-/HCO3-exchange with CFTR in stimulated HCO₃⁻ secretion by guinea pig interlobular pancreatic duct. Am J Physiol Gastrointest Liver Physiol 2009; 296:G1307–G1317.

16. Southern KW, Munck A, Pollitt R, Travert G, Zanolla L, Dankert-Roelse J, Castellani C, Group ECNSW. A survey of newborn screening for cystic fibrosis in Europe. J Cyst Fibros 2007; 6:57–65.

17. Durie PR. The pathophysiology of the pancreatic defect in cystic fibrosis. Acta Paediatr Scand Suppl 1989; 363:41–44.

18. Durie PR. Pancreatitis and mutations of the cystic fibrosis gene. N Engl J Med 1998; 339:687–688.

19. Scheele GA, Fukuoka SI, Kern HF, Freedman SD. Pancreatic dysfunction in cystic fibrosis occurs as a result of impairments in luminal pH, apical trafficking of zymogen granule membranes, and solubilization of secretory enzymes. Pancreas 1996; 12:1–9.

20. Salvatore D, Buzzetti R, Baldo E, Forneris MP, Lucidi V, Manunza D, Marinelli I, Messore B, Neri AS, Raia V, Furnari ML, Mastella G. An overview of international literature from cystic fibrosis registries. Part 3. Disease incidence, genotype/phenotype correlation, microbiology, pregnancy, clinical complications, lung transplantation, and miscellanea. J Cyst Fibros 2011; 10:71–85.

21. De Boeck K, Zolin A, Cuppens H, Olesen HV, Viviani L. The relative frequency of CFTR mutation classes in European patients with cystic fibrosis. J Cyst Fibros 2014; 13:403–409.

22. Veit G, Avramescu RG, Chiang AN, Houck SA, Cai Z, Peters KW, Hong JS, Pollard HB, Guggino WB, Balch WE, Skach WR, Cutting GR et al. From CFTR biology toward combinatorial pharmacotherapy: expanded classification of cystic fibrosis mutations. Mol Biol Cell 2016; 27:424–433.

23. De Boeck K, Weren M, Proesmans M, Kerem E. Pancreatitis among patients with cystic fibrosis: correlation with pancreatic status and genotype. Pediatrics 2005; 115:e463–e469.

24. Durno C, Corey M, Zielenski J, Tullis E, Tsui LC, Durie P. Genotype and phenotype correlations in patients with cystic fibrosis and pancreatitis. Gastroenterology 2002; 123:1857–1864.

25. Zielenski J. Genotype and phenotype in cystic fibrosis. Respiration 2000; 67:117–133.

26. Lee MG, Choi JY, Luo X, Strickland E, Thomas PJ, Muallem S. Cystic fibrosis transmembrane conductance regulator regulates luminal Cl⁻/HCO₃⁻ exchange in mouse submandibular and pancreatic ducts. J Biol Chem 1999; 274:14670–14677.

27. Lee MG, Wigley WC, Zeng W, Noel LE, Marino CR, Thomas PJ, Muallem S. Regulation of Cl⁻/ HCO₃⁻ exchange by cystic fibrosis transmembrane conductance regulator expressed in NIH 3T3 and HEK 293 cells. J Biol Chem 1999; 274:3414–3421.

28. Ko SB, Zeng W, Dorwart MR, Luo X, Kim KH, Millen L, Goto H, Naruse S, Soyombo A, Thomas PJ, Muallem S. Gating of CFTR by the STAS domain of SLC26 transporters. Nat Cell Biol 2004; 6:343–350.

29. Choi JY, Muallem D, Kiselyov K, Lee MG, Thomas PJ, Muallem S. Aberrant CFTR-dependent HCO₃⁻ transport in mutations associated with cystic fibrosis. Nature 2001; 410:94–97.

30. Patel SD, Bono TR, Rowe SM, Solomon GM. CFTR targeted therapies: recent advances in cystic fibrosis and possibilities in other diseases of the airways. Eur Respir Rev 2020; 29.

31. Heijerman HGM, McKone EF, Downey DG, Van Braeckel E, Rowe SM, Tullis E, Mall MA, Welter JJ, Ramsey BW, McKee CM, Marigowda G, Moskowitz SM et al. Efficacy and safety of the elexacaftor plus tezacaftor plus Ivacaftor combination regimen in people with cystic fibrosis

homozygous for the F508del mutation: a double-blind, randomised, phase 3 trial. Lancet 2019; 394:1940–1948.

32. Middleton PG, Mall MA, Drevinek P, Lands LC, McKone EF, Polineni D, Ramsey BW, Taylor-Cousar JL, Tullis E, Vermeulen F, Marigowda G, McKee CM et al. Elexacaftor-Tezacaftor-Ivacaftor for cystic fibrosis with a single Phe508del allele. N Engl J Med 2019; 381:1809–1819.

33. Megalaa R, Gopalareddy V, Champion E, Goralski JL. Time for a gut check: Pancreatic sufficiency resulting from CFTR modulator use. Pediatr Pulmonol 2019; 54:E16–E18.

34. Davies JC, Cunningham S, Harris WT, Lapey A, Regelmann WE, Sawicki GS, Southern KW, Robertson S, Green Y, Cooke J, Rosenfeld M, Group KS. Safety, pharmacokinetics, and pharmacodynamics of Ivacaftor in patients aged 2–5 years with cystic fibrosis and a CFTR gating mutation (KIWI): an open-label, single-arm study. Lancet Respir Med 2016; 4:107–115.

35. Boxhoorn L, Voermans RP, Bouwense SA, Bruno MJ, Verdonk RC, Boermeester MA, van Santvoort HC, Besselink MG. Acute pancreatitis. Lancet 2020; 396:726–734.

36. Barreto SG, Habtezion A, Gukovskaya A, Lugea A, Jeon C, Yadav D, Hegyi P, Venglovecz V, Sutton R, Pandol SJ. Critical thresholds: key to unlocking the door to the prevention and specific treatments for acute pancreatitis. Gut 2021;70:194–203.

37. Hegyi P, Pandol S, Venglovecz V, Rakonczay Z, Jr. The acinar-ductal tango in the pathogenesis of acute pancreatitis. Gut 2011; 60:544–552.

38. Judak L, Hegyi P, Rakonczay Z, Jr., Maleth J, Gray MA, Venglovecz V. Ethanol and its non-oxidative metabolites profoundly inhibit CFTR function in pancreatic epithelial cells which is prevented by ATP supplementation. Pflugers Arch 2014; 466:549–562.

39. Maleth J, Balazs A, Pallagi P, Balla Z, Kui B, Katona M, Judak L, Nemeth I, Kemeny LV, Rakonczay Z, Jr., Venglovecz V, Foldesi I, et al. Alcohol disrupts levels and function of the cystic fibrosis transmembrane conductance regulator to promote development of pancreatitis. Gastroenterology 2015; 148:427–439.

40. Maleth J, Venglovecz V, Razga Z, Tiszlavicz L, Rakonczay Z, Jr., Hegyi P. Non-conjugated chenodeoxycholate induces severe mitochondrial damage and inhibits bicarbonate transport in pancreatic duct cells. Gut 2011; 60:136–138.

41. Venglovecz V, Hegyi P, Rakonczay Z, Jr., Tiszlavicz L, Nardi A, Grunnet M, Gray MA. Pathophysiological relevance of apical large-conductance Ca^{2+}-activated potassium channels in pancreatic duct epithelial cells. Gut 2011; 60:361–369.

42. Venglovecz V, Rakonczay Z, Jr., Ozsvari B, Takacs T, Lonovics J, Varro A, Gray MA, Argent BE, Hegyi P. Effects of bile acids on pancreatic ductal bicarbonate secretion in guinea pig. Gut 2008; 57:1102–1112.

43. Ignath I, Hegyi P, Venglovecz V, Szekely CA, Carr G, Hasegawa M, Inoue M, Takacs T, Argent BE, Gray MA, Rakonczay Z, Jr. CFTR expression but not Cl⁻ transport is involved in the stimulatory effect of bile acids on apical Cl⁻/HCO_3^- exchange activity in human pancreatic duct cells. Pancreas 2009; 38:921–929.

44. Maleth J, Hegyi P, Rakonczay Z, Jr., Venglovecz V. Breakdown of bioenergetics evoked by mitochondrial damage in acute pancreatitis: mechanisms and consequences. Pancreatology 2015; 15:S18–S22.

45. Maleth J, Rakonczay Z, Jr., Venglovecz V, Dolman NJ, Hegyi P. Central role of mitochondrial injury in the pathogenesis of acute pancreatitis. Acta Physiol (Oxf) 2013; 207:226–235.

46. Katona M, Hegyi P, Kui B, Balla Z, Rakonczay Z, Jr., Razga Z, Tiszlavicz L, Maleth J, Venglovecz V. A novel, protective role of ursodeoxycholate in bile-induced pancreatic ductal injury. Am J Physiol Gastrointest Liver Physiol 2016; 310:G193–G204.

47. Tsubakio K, Kiriyama K, Matsushima N, Taniguchi M, Shizusawa T, Katoh T, Manabe N, Yabu M, Kanayama Y, Himeno S. Autoimmune pancreatitis successfully treated with ursodeoxycholic acid. Intern Med 2002; 41:1142–1146.

48. Yamamoto A, Ishiguro H, Ko SB, Suzuki A, Wang Y, Hamada H, Mizuno N, Kitagawa M, Hayakawa T, Naruse S. Ethanol induces fluid hypersecretion from guinea-pig pancreatic duct cells. J Physiol 2003; 551:917–926.

49. Pallagi P, Venglovecz V, Rakonczay Z, Jr., Borka K, Korompay A, Ozsvari B, Judak L, Sahin-Toth M, Geisz A, Schnur A, Maleth J, Takacs T, et al. Trypsin reduces pancreatic ductal bicarbonate secretion by inhibiting CFTR Cl⁻ channels and luminal anion exchangers. Gastroenterology 2011; 141:2228–2239.

50. Kleizen B, Hunt JF, Callebaut I, Hwang TC, Sermet-Gaudelus I, Hafkemeyer S, Sheppard DN. CFTR: New insights into structure and function and implications for modulation by small molecules. J Cyst Fibros 2020; 19 (suppl 1):S19–S24.

51. Schneider-Futschik EK. Beyond cystic fibrosis transmembrane conductance regulator therapy: a perspective on gene therapy and small molecule treatment for cystic fibrosis. Gene Ther 2019; 26:354–362.

52. Alton E, Armstrong DK, Ashby D, Bayfield KJ, Bilton D, Bloomfield EV, Boyd AC, Brand J, Buchan R, Calcedo R, Carvelli P, Chan M, et al. Repeated nebulisation of non-viral CFTR gene therapy in patients with cystic fibrosis: a randomised, double-blind, placebo-controlled, phase 2b trial. Lancet Respir Med 2015; 3:684–691.

53. Nichols DP, Donaldson SH, Frederick CA, Freedman SD, Gelfond D, Hoffman LR, Kelly A, Narkewicz MR, Pittman JE, Ratjen F, Sagel SD, Rosenfeld M, et al. PROMISE: Working with the CF community to understand emerging clinical and research needs for those treated with highly effective CFTR modulator therapy. J Cyst Fibros 2021; 20:205–212.

54. Pallagi P, Balla Z, Singh AK, Dosa S, Ivanyi B, Kukor Z, Toth A, Riederer B, Liu Y, Engelhardt R, Jarmay K, Szabo A, et al. The role of pancreatic ductal secretion in protection against acute pancreatitis in mice. Crit Care Med 2014; 42:e177–e88.

55. Grassalkovich A MJ, Madácsy T, Venglovecz V, Hegyi P. VX-770 and VX-809 restore the alcohol-induced CFTR expression defect in pancreatic ductal cells. European Pancreatic Club. Volume 19. Bergen, Norway: Elsevier, 2019:S28.

56. Akshintala VS, Kamal A, Faghih M, Cutting GR, Cebotaru L, West NE, Jennings MT, Dezube R, Whitcomb DC, Lechtzin N, Merlo CA, Singh VK. Cystic fibrosis transmembrane conductance regulator modulators reduce the risk of recurrent

acute pancreatitis among adult patients with pancreas sufficient cystic fibrosis. Pancreatology 2019; 19:1023–1026.

57. Johns JD, Rowe SM. The effect of CFTR modulators on a cystic fibrosis patient presenting with recurrent pancreatitis in the absence of respiratory symptoms: a case report. BMC Gastroenterol 2019; 19:123.

58. Carrion A, Borowitz DS, Freedman SD, Siracusa CM, Goralski JL, Hadjiliadis D, Srinivasan S, Stokes DC. Reduction of recurrence risk of pancreatitis in cystic fibrosis with Ivacaftor: case series. J Pediatr Gastroenterol Nutr 2018; 66:451–454.

59. Danahay H, Gosling M. TMEM16A: an alternative approach to restoring airway anion secretion in cystic fibrosis? Int J Mol Sci 2020; 21:2386.

60. Quesada R, Dutzler R. Alternative chloride transport pathways as pharmacological targets for the treatment of cystic fibrosis. J Cyst Fibros 2020; 19 (suppl 1):S37–S41.

61. Miller MR, Soave D, Li W, Gong J, Pace RG, Boelle PY, Cutting GR, Drumm ML, Knowles MR, Sun L, Rommens JM, Accurso F, et al. Variants in solute carrier SLC26A9 modify prenatal exocrine pancreatic damage in cystic fibrosis. J Pediatr 2015; 166:1152–1157.

62. Lam AN, Aksit MA, Vecchio-Pagan B, Shelton CA, Osorio DL, Anzmann AF, Goff LA, Whitcomb DC, Blackman SM, Cutting GR. Increased expression of anion transporter SLC26A9 delays diabetes onset in cystic fibrosis. J Clin Invest 2020; 130:272–286.

Chapter 21

Structure-function relationships of CFTR in health and disease: The pancreas story

Mohamad Bouhamdan,[1] Xie Youming,[2] and Fei Sun[1]

[1]*Department of Physiology*
[2]*Karmanos Cancer Institute and Department of Oncology, Wayne State University School of Medicine, Detroit, MI*

Introduction

In 1938, Dr. Dorothy Anderson published a paper describing the characteristics of cystic fibrosis (CF) in the pancreas; the term "cystic fibrosis" refers to the autopsy findings of fibrosis with cysts in the pancreas of children who died early in life with this disease. In 1949, Anderson also discovered that CF is a genetic condition. More recently, this was understood to be caused by mutations in the CF transmembrane conductance regulator (CFTR).[1] Other than the pancreas, CF also affects the lungs, liver, kidneys, male reproductive system, and gastrointestinal tract.[2,3] The disease leads to shortened life expectancy, most often due to respiratory failure resulting from airway obstruction, bacterial infections, and inflammation.[4,5]

CFTR, a member of the ATP-binding cassette (ABC) transporter protein family, is the cAMP-dependent Cl channel at the apical membranes of most epithelial cells.[6] Mutations of CFTR gene cause CF, which is the most common fatal autosomal recessive disorder with a disease frequency of 1 in 2,500 live births and a carrier rate of approximately 5 percent in Caucasian population.[7,8] The disease is characterized by a malfunction of exocrine tissues due to dysregulation of an epithelial chloride (Cl) channel. The major clinical features include chronic pulmonary disease, pancreatic exocrine insufficiency, intestinal disease (especially constipation), and an increase in the concentration of sweat chloride.[9] In the lungs, airways become colonized with bacteria, and repeated pulmonary infections ensue. The recurrent infections and inflammation result in submucosal gland (SMG) hypertrophy and excessive mucus secretion. The impaired mucociliary clearance and plugging of small airways cause progressive bronchiectasis and ultimately lead to respiratory failure.[10] Following the lung, the pancreas is the most affected organ in CF. It has been documented that most of the CF patients have pancreatic exocrine insufficiency, which leads to maldigestion and potentially malnutrition.[11] In this context, malabsorption of fat and fat-soluble vitamins are the most common nutritional deficient seen in this disease.

In this chapter, we will focus on how decreased CFTR function leads to protein plugging of the ducts and pancreatic atrophy. We will also shed light on the latest animal models to better understand the CF pancreatic disease and its relationship to chronic pulmonary disease and intestinal disease.

CF Animal Models

Mice, rats, pigs, and ferrets for CF research

Since the discovery of the CF gene, many animal models have been generated to mimic the CF symptoms in human patients. The earliest models were mice with ΔF508 mutation—*cftr* mutation.[12–15] CF mouse models have made significant contributions toward our understanding of the disease and the development of therapies.[14] Different CF mouse models have been developed, such as the exon 10 knockout (KO) model, the ΔF508 model,[15,16] and the G551D model.[17] However, significant limitations have been acknowledged in translating the information gained from CF mice to the humans. For example, unlike human CF patients, CF mice show neither pulmonary pathophysiology nor obvious pancreatic pathology or liver problems.[18] Recently established CFTR KO rats recapitulate several features of human CF disease; however, they do not develop spontaneous lung infections.[14] CFTR KO ferrets and CFTR KO and ΔF508 pigs generated by nuclear transfer have shown a similar pathology to that observed in human CF patients,[19] including lung, pancreatic, and liver phenotypes that were not often found in CF mice. However, neither pigs nor ferrets are convenient laboratory species. Both CF ferrets and CF pigs suffer from meconium ileus (MI), which causes these animals to die within a few days after birth; therefore, they are associated with high maintenance cost and require special animal-handling skills.[20] These factors have limited the applicability of CF pigs and ferrets almost exclusively to the labs originally producing these animals and a few groups closely associated with them (**Table 1**). An ideal animal CF model would

e-mail: mbouhamdan@gmail.com; fsun@med.wayne.edu

Table 1. **Characteristics of CF and key clinical consequences noted in animal models.** With emphasis on the rabbit model the table summarizes the differences of phenotypes of CF in all known animal models. In addition, we showed the type of laxatives that will help overcome the gut impaction. The pancreas defect was also shown in the different models. * represents the cost of the animal model on a 1–5 scale.

	Knock-out	DF508 KI	Pancreas	Lung, liver Phenotypes	Meconium ileus	Laxatives	Cost	Special Care
CF Mice	Yes	Yes	Yes/mild	?	Yes	Golytely	*	Yes
CF Pigs	Yes	Yes	Yes/severe	Yes	Yes, 100% death <48hrs	Surgery, or gut-corrected	*****	Yes
CF Ferrets	Yes	NO	Yes/severe	Yes	Yes, 75% death <72hrs	gut-corrected	****	No
CF Rats	Yes	NO?	No	?	Yes	Golytely	•	No
CF Zebrafish	Yes	?	Yes/Severe	?	?	?	*	No
CF Rabbits	Yes	Yes	Yes/Severe	Yes	Yes	Golytely	**	Yes

mimic the characteristics of human CF patients, including the pancreatic insufficiency, but not require exceptional expertise or resources.

Rabbits for CF research

There are anatomical, genetic, and biochemical similarities between rabbits and humans,[21] making rabbits potentially more relevant models for biochemical, molecular, and physiological characterization of CF pathology and for the development of CF therapies than mice. As shown in **Table 2**, the amino acid sequence of rabbit CFTR is about 93 percent identical to that in humans (**Table 2**). Rabbits also have a chromosome arrangement that is similar to humans: 44 chromosomes in rabbit versus 46 in human; both rabbit and human CFTR genes are present on chromosome 7. Compared to other large animals, such as the pig and ferret, a rabbit is a standard lab animal species that can be easily housed in most research institutes and is relatively economically affordable. Though rabbit airways have some anatomic and physiologic features similar to that found in humans, the main concern associated with using rabbits for CF research is the absence of airway SMGs in rabbits.[22] Since CFTR is abundantly expressed in SMGs of human airways, it has been hypothesized that dysfunction of SMGs initiates CF-like lung disease, leading to mucus accumulation observed in CF patients, as well as in CF pigs and ferrets, both of which contain SMGs in their airways. On the other hand, the absence of these glands from mouse airways has been cited as one of the explanations for the lack of lung disease in CFTR KO mice. Therefore, it has been predicted that CFTR-deficient/defective rabbits are unlikely to display mucociliary defects and spontaneous lung infections associated with CF. However, our preliminary data reveal that CF rabbits have a similar airway pathology to that in human CF patients (**Tables 1 and 2**).

Although CFTR[-/-] KO rabbits eventually develop distal intestinal obstruction, MI is rarely observed in the animal within the first month after birth because rabbit has a large functional cecum. To some extent, our work challenges the traditional view on the importance of SMGs in CF pathology, suggesting that SMGs may not be a critical player for the development of CF lung disease. In support of this view, overexpression of βENaC in mice can produce mucus obstruction in the small, non-glandular airways,[23] which are thought to be the site of disease initiation in CF neonates.[24] In CF pigs, the mucus appears to arise from goblet cells in the surface of the epithelium of the airways.[25] Mucus accumulation in CF ferret airways is associated with variable levels of goblet and mucus cell hyperplasia in the surface airway epithelium and SMGs.[19] More recent data indicated that defective goblet cell exocytosis in CFTR KO mice contributes to CF-associated disease in the intestine.[26] In fact, though SMGs in airways might be a primary site for CF pathogenesis, a critical role for mucus-producing goblet cells in CF airway pathology has not been excluded. Indeed, our preliminary data revealed that the goblet cells, which exist in rabbit airways, maybe the primary contributor to mucus accumulation in the airways of CFTR KO rabbits (**Tables 1 and 2**).

ΔF508 mutation

Mutations of the gene encoding CFTR lead to CF, and more than 2,023 CF mutations (disease related or not) have been identified (http://www.genletsickkids.on.ca/cftr/) in the CFTR gene. The most common mutation in CF is the deletion of the phenylalanine residue at position 508 (ΔF508),[27] which is in the first nucleotide-binding domain of CFTR. The ΔF508 mutation is present in more than 90 percent of CF patients. A critical issue in CF disease is the inability of ΔF508CFTR to achieve the native, folded state required

Table 2. CFTR-related characteristics among species.

	No. of aminoacid	Identity to human (%)	CFTR locus (Ch*)	Days of gestation	Avg. litter size	Sexual maturation	Avg. life expectancy (years)	no of Chromosome
Human	1480	100	7	280	1	10–16 yrs	~78	46
Mouse	1476	78	6	21	6	6–8 wks	2	40
Rat	1476	78	4	22	6	8 wks	2–3	42
Pig	1482	92	18	114	10	6–8 months	10–15	38
Ferret	1484	91	?	42	8	4–6 months	8–10	40
Zebrafish	1485	55	18	3	185	3 months	3–5	25
Rabbit	1481	93	7	30	8	4–6 months	8–10	44

for its export from the endoplasmic reticulum (ER) and traffic to the cell surface. Instead, ΔF508 protein is exclusively retained in the ER and degraded by the ubiquitin-proteasome system.[27–29] A therapeutic strategy aimed at facilitating ΔF508 folding and trafficking is highly desirable for treatment of the disease because ΔF508 mutant has a substantial CFTR Cl current if it reaches the cell surface.[30] A new drug, Orkambi, that combines a CF corrector (Lumacaftor, which acts as a chaperone for a correct protein folding which increases the number of CFTR proteins to the cell surface) and potentiator (Ivacaftor, increases the activity of the CFTR (conductance) at the cell surface) recently has received breakthrough therapy designation to treat CF patients with ΔF508CFTR. However, this drug only improves lung function assayed by forced expiratory volume in 1 second (FEV1) by 2.6–4 percent. Therefore, further research on ΔF508 mutation is needed to develop a better drug to treat CF patients.

CFTR Deficiency in CF

The *Cftr* gene encodes the CFTR protein, a member of the ABC transporter superfamily, which is the cAMP-dependent Cl channel at the apical membranes of most epithelial cells, making it unique among members of this protein family.[6] The CFTR protein migrates to the surface of cells that line the pancreatic duct, airways, gastrointestinal tract, biliary tract, part of the male reproductive tract, and cells that are part of sweat glands.[6,31,32] CFTR forms a pore or channel that allows ions, including chloride and bicarbonate, to move from one side of the cell membrane to the other (**Figure 1**).[1,33] Channel activation is mediated by cycles of regulatory (R) domain phosphorylation by PKA/PKC, ATP binding to the nucleotide-binding domains, and ATP hydrolysis (**Figure 1**). Demonstration that CFTR functions as a chloride channel regulated by cyclic AMP[34]-dependent phosphorylation is consistent with the ion transport disturbances documented in CF tissues (for review, see[35]). These

disturbances in ions change the concentration of molecules in the fluid within the ducts or organs.[11]

CFTR Protein Structure

The CFTR protein is comprised of 1,480 amino acids organized into five functional domains.[7,9] As other ABC transporters, CFTR has two membrane-spanning domains (TMD1 and TMD2), two nucleotide-binding domains (NBD1 and NBD2), and one regulatory domain (R) (**Figure 1**). For more insights regarding the structure of CFTR, see the review by Patrick and Thomas.[36] The two TMDs, each composed of six transmembrane segments, form the CFTR channel pore, and the two NBDs interact with nucleotides to regulate channel activity opening and closing of the TMDs (**Figure 1**).[7,36] The R domain, through interactions with the N-terminal cytosolic region of TMD1, also controls the channel activity.[7,37,38] NBDs are responsible for the binding and hydrolysis of the ATP, which causes a conformational change in the TMDs leading to the transport of substrates across cell membranes.[39] CFTR mutations can occur in any of the five protein domains. However, many mutations occur in NBD1, including the ΔF508 mutation. The location of the CFTR mutations can affect the formation or function of the CFTR protein (**Figure 2, Table 3**).[7] In **Table 3** we summarized the role of not only the major domains described for CFTR but also the connecting sequences, which include N-terminal, intracellular loops (ICLs), extracellular loops (ECLs), transmembrane (TM) helices 1 through 12, C-terminal domain. The N-terminal domain was shown to be involved in the folding and the trafficking of the CFTR protein through protein–protein interactions (e.g. syntaxin 1A), and mutations in this domain reduced the function of the channel.[40,41] Some TM helices have been studied more than the others. For instance, the results from many studies suggested that transmembrane segments TM1, TM2, TM6, TM11, and TM12 form the pore lining and regulate the pore function

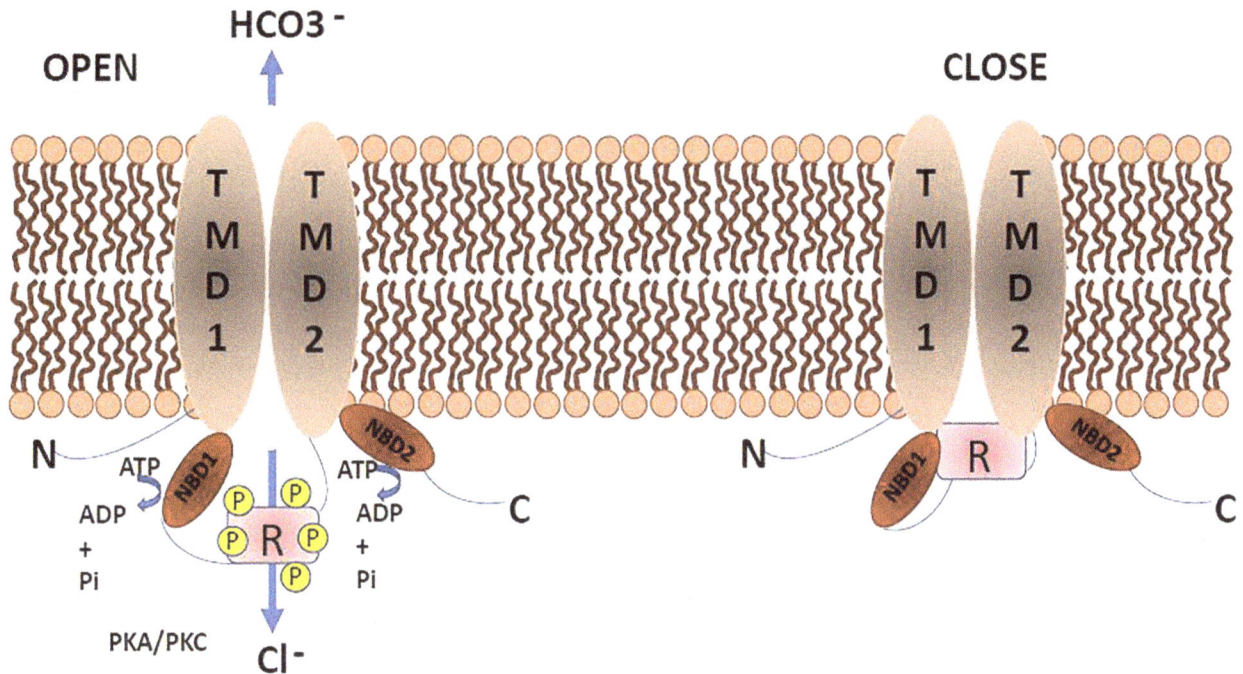

Figure 1. Activation of the CFTR channel. The CF transmembrane conductance regulator (CFTR) protein channel is a member of the ABC transporter superfamily. It acts in apical part of the epithelial cells as a plasma membrane, cyclic AMP-activated chloride, and bicarbonate anion channel. CFTR has two membrane-spanning domains (TMD1 and TMD2), two nucleotide-binding domains (NBD1 and NBD2), and one regulatory domain (R). CFTR is a key regulator for cell surface water-salt homeostasis of the apical membranes of epithelial cells in multiple organs, including the pancreas that produces alkaline fluid in pancreatic ducts. The open status of the CFTR is initiated by ATP binding at the NBD domains. The activation of the channel is dependent on phosphorylation by cyclic AMP-dependent protein kinase (PKA) at multiple sites in the R domain. The magnitude of response to PKA is amplified by phosphorylation of CFTR by protein kinase C (PKC).

by selecting the anions,[42–44] whereas TM5 plays a role in the anion binding.[43] Also, V232D mutation in TM4 leads to a loss of function because it forms a bond with Q207 in TM3, which does not occur in the WT form of CFTR.[45,46] The mutation V232D is not the only mutation that involves a change from a neutral/hydrophobic residue to a polar or charged residue, causing CF. Therien et al. studied more than 31 mutations in TMD1 including all six TM helices and concluded that CFTR mutations in the TMs lead to a loss of function through the formation of membrane-buried interhelical hydrogen bonds.[45] ECLs 1 through 6 represent about 4 percent of the CFTR protein, whereas 77 percent is in the cytoplasm and the rest, 19 percent, includes TM1 to 12. CF-associated mutations in the ECLs have been shown to affect channel gating[47] and interactions with extracellular anions.[48,49] For example, mutations D110H/E and R117C/H/L/P in ECL1 are associated with CF disease. These mutations affect the stability of the CFTR ion pore, resulting in reduced conductance of CFTR.[47,50] Q890X and K892C mutations in ECL4 have been reported to affect channel gating and extracellular anion interaction.[50–52] ECL4 is the only ECL that contains N-linked glycosylation sites (N894

and N900).[51] Many studies have been performed on the function of CFTR's intracytoplasmic loops (ICLs), including their roles in regulating inter-molecular interactions as well as CFTR interactions with other proteins. E193K and I148T mutations in ICL1 have been reported to affect the pore opening.[53,54] Recent studies showed that ICL2/3-NBD2 interface and ICL1/4-NBD1 interface have a role in protein folding and processing in the ER.[55,56] In addition, S945L, H949Y, and G970R mutations in ICL3; L1065P, R1070Q, and Q1071P mutations in ICL4; and D1152H, D1154G, and W1204X mutations in ICL5 have been shown to affect the conductance of CFTR.[52,57,58] As for the rest of the domains, ATP-binding events occurring in both NBDs allow the hydrolysis of intracellular ATP to ADP.[59] This event allows the conformational changes, and that change in structure allows the CFTR channel to transition from an opened to closed state, thus controlling the gating kinetics of the channel.[59] NBD1 was studied more than NBD2 due to the presence of frequent ΔF508 deletion of NBD1 in the CF patients. The ΔF508 is present in more than 70 percent of the CF patients, resulting in destabilization of the CFTR protein.[59–61] Also, mutations ΔF508 and G551D modify

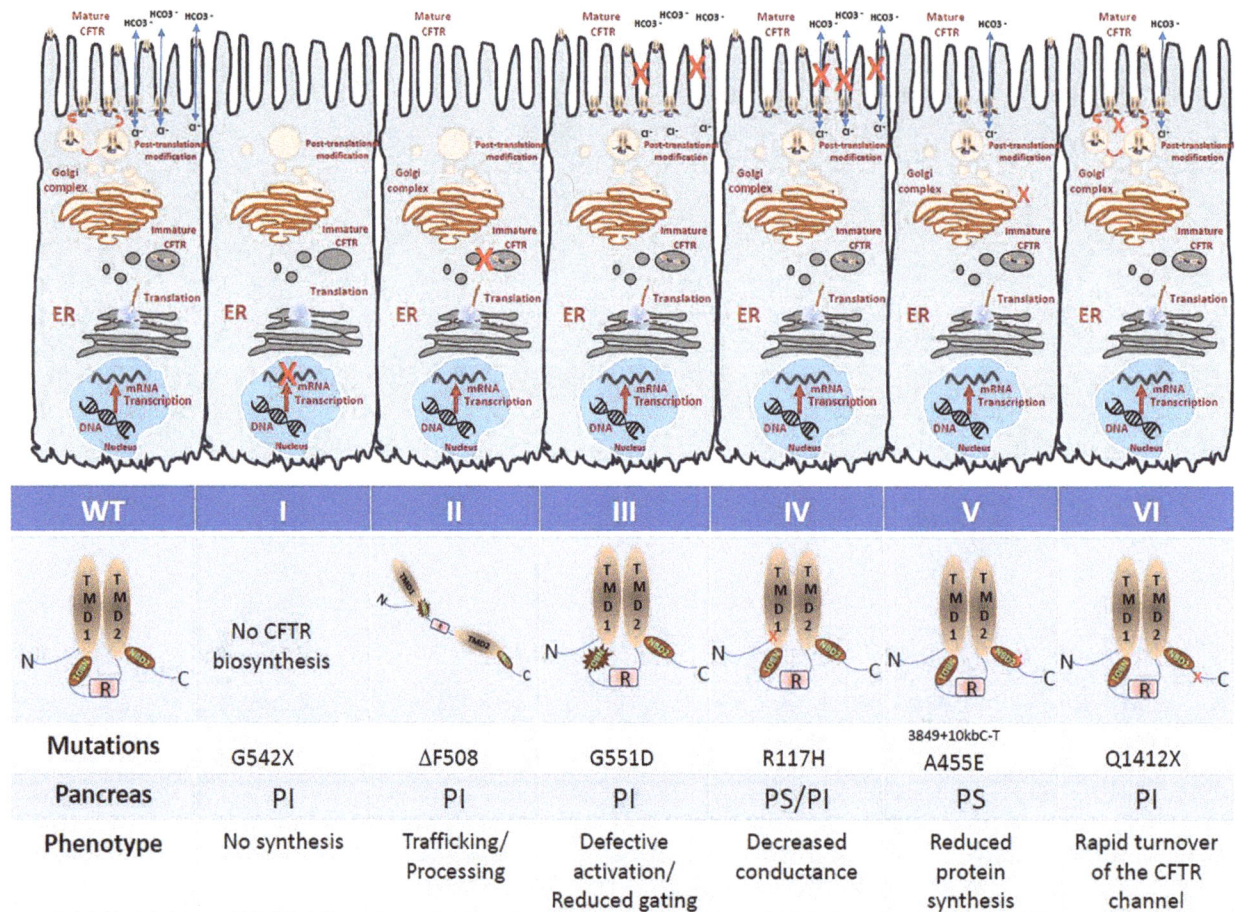

Figure 2. Classes of CFTR mutations. Class I mutations lead to no protein synthesis, which includes mutations that involve premature stop codons and nonsense mutations. **Class II** mutations include the most frequent mutation of CF disease, ΔF508, which lead to trafficking, improper folding, and processing defects of the CFTR protein. This class is the primary target in the CF research and the main target by the pharmaceuticals companies. **Class III** mutations affect the ATP binding at one of the two binding sites in the NBDs. The CFTR protein reaches the cell surface, but the mutations render the CFTR channel nonfunctional, which impairs the opening of the channel. **Class IV** mutations also involve CFTR protein reaching the cell surface but with reduced ion passage through the channel because of the structural defect caused by the mutations in the CFTR channel. **Class V** mutations affect the amount of CFTR protein that reaches the cell surface because of splicing problems or inefficient trafficking. **Class VI** mutations lead to a rapid turnover of the CFTR channel at the cell surface. Examples of CFTR mutations regarding their pancreatic defects, PI or PS. The mutations mentioned in the table are representative of each class, but we have to keep in mind that according to the new classification some mutations are classified in more than one class, which will result in having more than one defect.

the interactions between NBDs and NBDs–ICLs.[59] Other well-studied NBD1 mutations such as G542X and G551D result in channel-gating problems.[59–62] Similarly, NBD2 mutations like N1303K, p.Ile1234_Arg1239del, G1244E, S1251N, S1255P, and G1349D are CF disease related and are considered to be gating mutations.[62] The R domain, along with NBDs, controls the channel activity of CFTR. The activation of the channel is dependent on the phosphorylation by cyclic AMP-dependent protein kinase (PKA).[63] To date, more than 15 phosphorylation sites in the R domain have been attributed to PKA phosphorylation, which contributes in varying proportions to the response to activation of the CFTR channel.[63] It has been reported that CFTR also can be phosphorylated by several other protein kinases, including protein kinase C (PKC), casein kinase II, cyclic GMP-activated protein kinase, and Src kinase.[63] The R domain has multiple phosphorylation sites for PKC, which modulate PKA-induced domain–domain interactions.[64] Many CF-related mutations like D648V, E664X, E656X, and 2108delA in the R domain disrupt the normal function of the R domain, for example, the transport of HCO_3^- in secretory epithelia and in CF.[65,66] The same authors showed that mutants reported to be associated with CF with pancreatic insufficiency do not support HCO_3^- transport and

Table 3. **Domains structure of CFTR.** A detailed description of the CFTR domains consists of an N-terminal, 6 ICLs, 6 ECLs, 12 transmembrane (TM) arranged into TMD1 and TMD2 consisting of 6 TM for each domain, NBD1 and NBD2, and the c-terminal domain. The specified aa for each domain, their known function, positive charge residues, some mutations causing CF, and their effects on the pancreas are as well described.

Domains	Amino acids	Functions	Positive charge residues	Mutations causing CF	References	Pancreas
N-terminal	1 to 81	Folding and/or trafficking Protein Interaction			40, 41	PI
TM1	82 to 103	Regulation of pore function; Pore lining	K95		42–44, 67	PS/PI
ECL1	104 to 117	Stability of the CFTR ion pore		R117C/H/L/P	42,47, 50, 68	PS/PI
TM2	118 to 138	CFTR pore lining	R134		42–44, 69	PS/PI
ICL1	139 to 194	Pore opening		E193K, I148T	53–55	PS/PI
TM3	195 to 215	ND		Q207 form a bond with the mutation V232D	45	
ECL2	216 to 220				47	
TM4	221 to 241	Loss of function of pore		V232D	45, 46, 70	PS/PI
ICL2	242 to 307	Protein folding, Processing in the ER			55,56	
TM5	308 to 328	Anion binding			43	
ECL3	329	ND				
TM6	330 to 350	Pore lining; anion selectivity	R334, K335, R347		43, 44	PS/PI
ICl2.5	351 to 432	ND				
NBD1	433 to 586	Hydrolyzation of ATP, Channel opening, regulation of the sodium ion channel		Delta(F508); G551D; G542X	60, 61	PI
R	587 to 859	Phosphorylation sites for PKA/PKC		D648V, E664X, E656X and 2108delA	66	PI
TM7	860 to 870	ND				
ECL4	881 to 911	Glycosylation at N894 and N900		Q890X, K892C	50–52	
TM8	912 to 932	ND		S912L	52	
ICL3	933 to 990	Conductance		S945L, H949Y, G970R	52, 57	PS/PI
TM9	991 to 1011	ND				
ECL5	1012	ND				
TM10	1013 to 1034	Processing	R1030			PI
ICl4	1035 to 1102			L1065P, R1070Q, Q1071P	52	
TM11	1103 to 1123	Pore lining; anion selectivity			42–44, 67	PS/PI
ECL6	1124 to 1128	ND				
TM12	1129 to 1150	Pore lining; anion selectivity		M1137V, M1137R, I11139V and deltaM1140	43, 44, 58	PS/PI
ICL5	1151 to 1218	Conductance		D1152H, D1154G, W1204X	52, 58	PS/PI
NBD2	1219 to 1386	Maturation, gating		N1303K, G1349D, G1244E, S1251N, S1255P, and G1349D	58	PI
C-terminal	1387 to 1480	ND				

PI: pancreas insufficient, PS: pancreas sufficient.

those associated with pancreatic sufficiency show reduced HCO$_3^-$ transport.[65]

Classification of CF Patients

Traditional classification

As mentioned earlier, more than 2023 CF mutations have been identified (http://www.genletsickkids.on.ca/cftr/) in the CFTR gene. These mutations are categorized based on the dysfunctions of CFTR at different levels of the maturation and function of the CFTR channels (**Figure 2**). Traditionally, these dysfunctions were divided into 6 groups based on function (Classes III, IV, and VI) and processing (Classes I, II, and V) of the CFTR (**Figure 2**).[33,71,72] The class I mutations include a nonsense, frame-shift, or splicing mutations which prevent CFTR biosynthesis by introducing a premature termination codon (PTC). The most common mutation in this class is the G542X (**Figure 2**).

The class II mutations include a missense mutation which causes misfolding of the CFTR to lead to its degradation in the ER by quality-control machinery, resulting in the absence of functional protein at the cell surface. The most important mutation in this class is the ΔF508, present in more than 90 percent of CF patients (**Figure 2**).

The class III mutations include a missense mutation which leads to a non-functioning CFTR at the cell surface, resulting in unstable and reduced channel gating characterized by a lower open probability. The most common mutation representing this class is the G551D (**Figure 2**).

The class IV mutations include missense mutation which leads to a reduced CFTR channel conductance. The decrease in conductance is caused by an abnormal conformation of the pore resulting in disruption of the ion flow. The most common mutation in this class is the R117H (**Figure 2**).

The class V mutations introduce splicing or promoter defects in the CFTR gene, resulting in a reduced amount of CFTR protein at the cell membrane caused by reduced protein synthesis. Those mutations affect the gene expression but do not change the conformation of the channel. The most representative mutations in this class are the 3849 + 10kbC-T and A455E (**Figure 2**).

The class VI mutations include missense mutation which leads to a decrease in the CFTR stability. These mutations result in an accelerated turnover of CFTR protein at the cell membrane and reduced apical cell surface expression.[72] The most representative mutation in this class is Q1412X (**Figure 2**).

Though more than 2,023 mutations/variants have reported for CFTR, whether each of these can cause channel dysfunction and disease is largely unknown. However, studies to predict the functional consequences and clinical outcome of individual patients carrying these mutations are

being conducted.[73,74] These interpretations of such studies have been challenged by the general lack of correlation between the genotype and the clinical severity[73] (**Table 1**).

New classification

The lack of correlation between the genotype and the phenotype of the CF patients led to a new classification based on the severity and the clinical symptoms of the CF patients. Recently, Dupuis et al. studied meconium ileus (MI) in CF patients and reported that only a subset of patients with CF develop MI.[73] Furthermore, MI demonstrates notable heritability. Although studies have shown that non-CFTR genes contribute to susceptibility, the CFTR genotype itself affects the occurrence of this complication; only patients with the more severe CFTR variants are at risk for MI.[73] It was hypothesized that the susceptibility to MI is influenced by specific CFTR genotypes, and that the prevalence of MI can be used to discriminate among severe CFTR mutations.[73] The pleiotropic molecular defects of a single mutation in the CFTR have limited the drug therapy effects for some mutants which have been categorized as class I, II, or II/IV.[72] The authors proposed a modification of the traditional class I–VI CF mutations classification. This expanded classification of the major mechanistic categories[7,9,75] accommodates the unusually complex, combinatorial molecular/cellular phenotypes of CF alleles. The new classification consists of 31 possible classes of mutations, including the original classes I, II, III/IV, V, and VI, as well as their 26 combinations.[72] For example, according to the expanded classification, G551D will be designated as a class III mutation as before,[7] while ΔF508 will be classified as class II–III–VI, W1282X as class I–II–III–VI, P67L as class II–III, Q1412X as class III–VI, and R117H as class II–III/IV, reflecting the composite defects in mutant CFTR biology.[72] More evidence supporting the new classification came from a study by Vertex Pharmaceuticals where they tested 54 missense mutations and found that 24 of them have both processing and gating defects.[76]

CFTR Function and Its Role in Pancreas

CF and exocrine pancreas

As mentioned earlier, CFTR is predominantly expressed on the apical membrane of epithelial cells in the small pancreatic ducts. CFTR acts as a selective ion channel involved in chloride and bicarbonate (HCO$_3^-$) transport across the apical membranes of epithelial cells in multiple organs, including the pancreas, which produces alkaline fluid in pancreatic ducts.[11,77,78] HCO$_3^-$ is an important ion in the pancreatic juice. It also facilitates solubilization of the digestive enzymes and mucins.[69] Indeed, aberrant HCO$_3^-$ transport has a crucial role in human diseases.[31,79] In his review, Quinton proposed that in CF patients, the HCO_3^- is required to form normal mucus. His explanation is that once a granule is released, HCO_3^- sequesters Ca^{2+} and H^+ ions away from the mucin anions to form a complex with them. Therefore, the lack of secreted HCO_3^- in CF patients impairs Ca^{2+} removal, prevents normal mucin expansion, and promotes stasis of mucus in the ducts or on the luminal surfaces of affected organs.[31] In addition, reduced secretion of HCO_3^- and chloride (Cl^-) leads to a more acidic and viscous luminal content.[67] MUC6 is one of the pancreatic mucins expressed by 13 weeks of gestation and shows a very similar distribution to that of CFTR. In addition, MUC6 mucin is the main constituent of the complexes that form in small ducts and cause obstruction.[80] Therefore, CF patients carrying mutations in the *CFTR* gene showed a lower pH, low flow of secretions, and high protein concentration in the pancreas duct secretions, which lead to precipitates, obstruction, and injury in the duct lumina.[11,78] Meyerholz and his colleagues showed in the CF pig model that the changes (obstruction) could be detected in gestations as early as week 17.[81] They showed that the site of obstruction ranged from the distal jejunum to the proximal spiral colon, similar to that reported in humans with MI.[81,82] The obstruction in the acini and ducts lead to dilatation which causes epithelial injury and destruction, inflammation, fibrosis, and fatty infiltration.[78,81,83] Tucker and colleagues reported that acinar plugs developed before mucous metaplasia and found that early acinar plugs are composed of zymogen granules and were distinct from mucus in pancreatic tissue of CF patients.[84] These findings then indicate that zymogen material from the acinar cell, not mucus, may become inspissated in the acinus in early CF and that subsequent mucous metaplasia occurs as the obstruction and exocrine atrophy progress.[84]

CFTR and Hyperinflammation

As mentioned earlier, airways of the lungs become colonized with bacteria, and repeated pulmonary infections ensue. The recurrent infections and inflammation result in SMG hypertrophy and excessive mucus secretion. The impaired mucociliary clearance and plugging of small airways cause progressive bronchiectasis and ultimately lead to respiratory failure.[10] Many studies have been done to explain the cause of the hyperinflammation in the CF patients. Many have suggested that the balance between Th1, Th2, and Th 17 could play a role in the CF disease.[85] It has been established that Th17 is known to be a key player in autoimmune diseases.[86] In addition, the authors showed that Th17 is regulated by miR-183C via inhibition of Foxo1.[86] For a long time, it has been thought that CFTR mutations in epithelial cells have an indirect effect on the immune system which causes the inflammation in the lungs, because of the colonization of bacteria like

Pseudomonas aeruginosa, Burkholderia cenocepacia, and Mycobacterium abscessus which causes the infections in the CF patients.[85] Recently, emerging evidence points the problem to be directly affecting the immune cells. Since it has been shown that CFTR is expressed in lymphocyte T cells (like Th2, Th17, and Tregs) and macrophages, CFTR mutations causing CF may have a direct impact on these cells. Grumlli et al. summarized in their review that CFTR mutations have a direct effect on the T cells function which results in an enhanced Th2 response, a reduced Treg population, and elevated Th17 response which translate by an increase of neutrophils and recruitments by IL-17 to the lung which leads to the destruction of alveoli in the lungs of CF patients.[85] In addition to the lymphocytes and neutrophils, macrophages also have been shown to participate in the lung decay found in CF patients through activation of MMp12.[85] Understanding better the mechanisms behind the infections and the immune response could lead to a better drug therapy targeted to each patient depending on the severity of the CF disease.

CF and endocrine pancreas

CF is also recognized to affect the endocrine pancreas. There is a correlation between glucose abnormalities, morbidity, and mortality in CF patients.[87,88] Glucose abnormalities include cystic fibrosis–related diabetes (CFRD) and impaired glucose tolerance (IGT). CFRD is one complication in the CF patients occurring in more than 40 percent of adults and 25 percent of adolescents, which is preceded by episodes of IGT.[88–91] Both reduced insulin secretion and insulin resistance are observed in CFRD.[34,92–94] CFRD has characteristics of both type I and II diabetes and does not belong to either one of the diabetes classes. It is characterized by the loss of functional β-cell mass and varying degrees of insulin resistance (**Figure 3**).[34,78,92] Mutations in CFTR that lead to both a decrease in islet cell mass and dysfunction in the β-cell are the cause of CFRD.[88,95,96] It is believed that cross-talk between pancreatic exocrine and endocrine components can also contribute to the CFDR (**Figure 3**). Destruction of pancreatic exocrine tissue caused by a decrease in the islet cell mass evolves to dysfunctional endocrine β-cells.[97] β-cell dysfunction may be caused by increased oxidative and ER stress which is associated with CFRD.[98–101] It is very well documented that glucose deprivation leads to ER stress.[101] Some CFTR mutations (like ΔF508) can cause the accumulation of unfolded proteins in the ER, triggering an evolutionarily conserved response, termed the "unfolded protein response" (UPR).[102] In addition, aberrant Ca^{2+} regulation in the ER lumen causes protein unfolding, and rapid degradation of mutated CFTR proteins may contribute to the complex multi-organ CF pathology.[103] It is known that ER stress triggers β-cell death typically by apoptosis

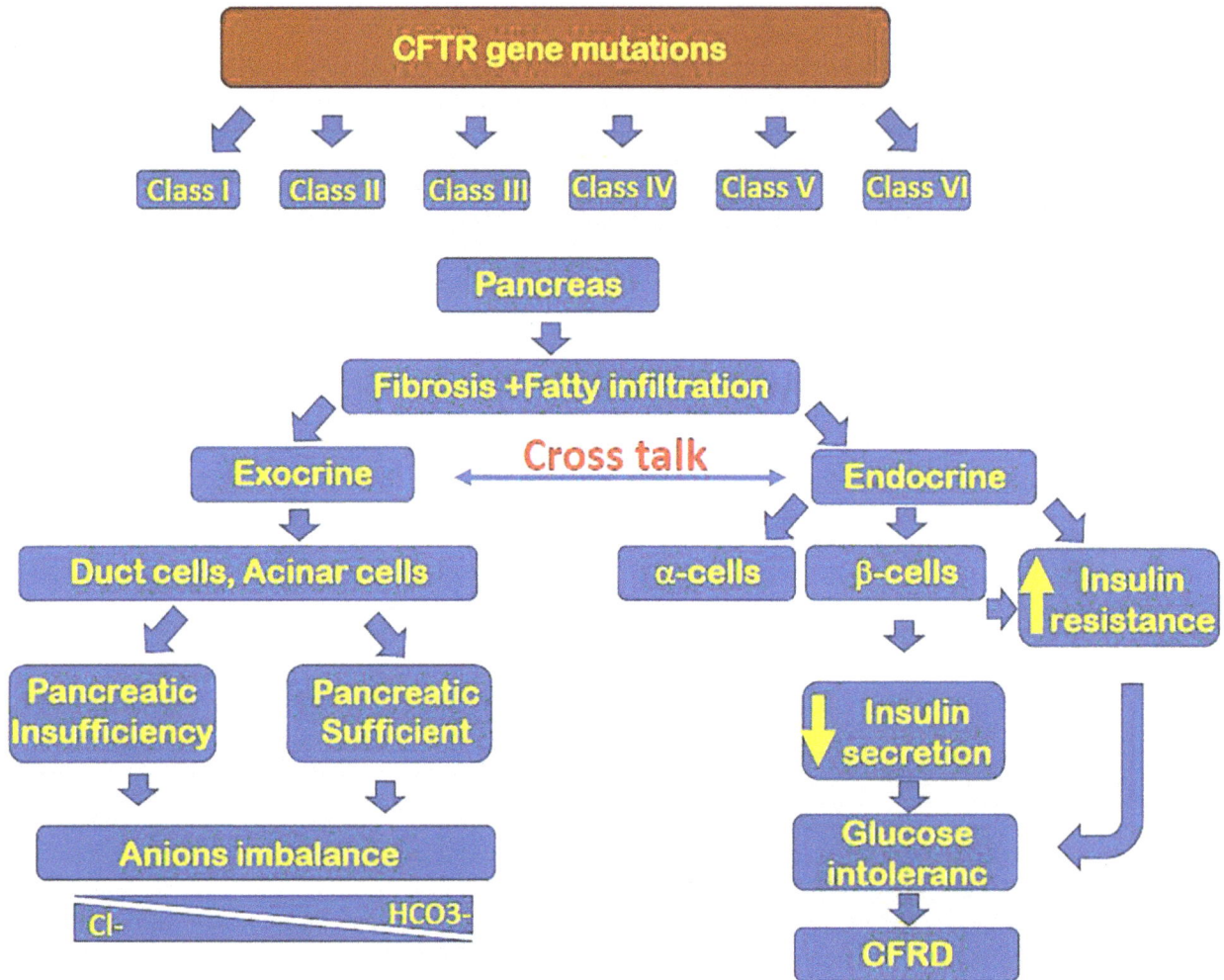

Figure 3. CF and pancreas defects. Schematic representation of the crosstalk between exocrine and endocrine of the pancreas. Defects in the exocrine pancreas lead to PI and PS, causing the anions imbalance. Defects in the endocrine pancreas lead to CFRD. As summarized in this diagram, many factors, defects, and the crosstalk between endocrine and exocrine leads to the phenotypes described in the CF patients, including PI, PS, and CFRD.

when protein misfolding is persistent or excessive.[104] As Harding et al. stated, the special sensitivity of insulin-producing cells to a mutation (like CFTR mutations) that affects a signaling protein responsive to ER stress may also be relevant to the development of more common forms of human diabetes mellitus. The major abnormality in most patients with CFRD is resistance to the action of insulin. However, glucose intolerance develops only after β-cell decompensation renders the endocrine pancreas unable to keep up with the demand imposed by IR.[105] Because of their high rates of protein synthesis, β-cells are particularly susceptible to ER stress, which may trigger CFRD.[88] The expression of CFTR was reported to be required in β-cells for glucose-induced secretion. Therefore, CFTR plays a significant role in the normal function of pancreatic β-cells.[106] Ntimbane and his colleagues have summarized

the factors leading to an abnormal glucose homeostasis in CF patients: (a) impairment of β-cell function with progressive fibrosis of islets of Langerhans with resultant distortion, ischemia, cell death, and a decrease in islet numbers; (b) impairment of other islet cell functions; (c) impairment of the insulinotropic gut hormone secretin; (d) changes in insulin sensitivity; and (e) altered insulin clearance rate.[88] The exact cause and mechanism of CFRD are still largely unknown. The most probable cause of CFRD is a combination of many events unlikely to be attributable to one defect. The clinical effects and disease states associated with CF patients with CFRD include chronic pancreatic inflammation, dysfunction of the immune system, oxidative stress, impaired insulin production and secretion, variable state of IR, and altered entero–insular axis hormones.[78,88]

Insulin secretion and CFTR

All the proposed causes mentioned earlier are supported by evidence to explain the pathogenesis of impaired insulin secretion in CFRD. An additional cause not mentioned here is the expression and direct effect of CFTR on insulin secretion in the β-cells.[70,107–109] The expression of CFTR has been reported in cultured β-cells derived from mice (MIN6) or rat (RINm5F).[70,107,110] But the earliest report came in 2007 from studies by Boom and his colleagues showing the expression of CFTR protein in rat islet cells and the significantly higher level in non-β than in β-cell populations.[111] They also showed by immunohistochemistry studies that CFTR expression also occurs in glucagon-secreting alpha cells.[111] Guo and his colleagues demonstrated that glucose-induced whole-cell currents, membrane depolarization, electrical bursts or action potentials, Ca^{2+} oscillations, and insulin secretion in β-cells are dependent on CFTR, indicating an essential role of CFTR in the regulation of insulin secretion.[107] Their studies showed that specific inhibitors of CFTR (GlyH-101 and CFTRinh-172) blocked a CFTR Cl⁻ gating needed for insulin secretion in primary β-cells and ΔF508-CFTR mutant mouse islets.[107] In addition, Edlund and her colleagues detected small CFTR conductance in both human and mouse β-cells. The augmentation of insulin secretion by activation of CFTR by cAMP (forskolin or GLP-1) in the presence of glucose was significantly inhibited by the specific CFTR inhibitors. They also demonstrated reduced cAMP-dependent exocytosis upon CFTR inhibition, concomitant with fewer docked insulin granules.[70] These reports and others from the patients with CFTR mutations showed insufficiency of secreted insulin. However, these studies did not describe the molecular mechanism that causes a decrease in insulin secretion. In addition to the role of CFTR in regulating insulin secretion and exocytosis after glucose-induced membrane depolarization leading to insulin secretion, the study also demonstrated that CFTR molecules act upstream of the chloride channel Anoctamin 1 (ANO1; TMEM16A) in the regulation of cAMP- and glucose-stimulated insulin secretion.[70] Thus, the impaired insulin secretion seen in patients with CF would be caused by the lack of glucose-induced Cl⁻ efflux through both CFTR Cl⁻ channels and ANO1 due to a decreased membrane depolarization.[70,107] In summary, these studies showed that CFTR is an important regulator of pancreatic β-cell insulin secretion, exocytosis, and membrane depolarization and is induced by glucose via elevation of cytosolic Ca^{2+} concentration.[33,107,108]

Marunaka in his recent review summarized new studies connecting the role of CFTR and ANO1 in insulin secretion.[33] Briefly, it has been established that intracellular Cl⁻ concentration ([Cl⁻]) is a very useful marker of channel activity. In these recent studies, the authors showed that [Cl⁻] measured using N-(ethoxycarbonylmethyl)-6-methoxyquinolinium bromide (MQAE) is about 100 mM under the basal condition in RINm5F β-cell line, and application of CFTRinh-172 (an inhibitor of the CFTR channel) increased [Cl⁻] about 26 mM.[107] This means that CFTR indeed mediates Cl⁻ efflux under basal condition, which may maintain a relatively depolarized membrane potential in the β-cells at rest, and the electrochemical potential of Cl⁻ in the intracellular space is larger than that in the extracellular space.[33,107] This was confirmed by showing that membrane potentials of pancreatic β-cells expressing wild-type CFTR Cl⁻ channels are −61~−67 mV [70] but hyperpolarized to −75 mV when using CFTRinh-172 or ΔF508.[107] Thus, CFTR has an important role in determining the resting membrane potential of the β-cells by acting as a Cl⁻ channel to maintain the membrane depolarization.[107] The same study also showed that ΔF508 CFTR Cl⁻ channel decreases membrane depolarization induced by glucose and increases [Ca^{2+}] due to the activation of voltage-dependent Ca^{2+} channels.[33,107] In his review, Marunaka concludes that the loss of CFTR function leads to insulin insufficiency which is caused by the higher intracellular Cl⁻ electrochemical potential in pancreatic β-cells. In general, Cl⁻ uptake into the intracellular space is mediated via active Cl⁻ transporting systems, such as Na^+-Cl⁻ cotransporter (NCC) and/or NKCC, driven by the Na^+, K^+-ATPase-generated Na^+ chemical potential difference between the intracellular and extracellular spaces: the intracellular Na^+ chemical potential < the extracellular Na^+ chemical potential. Therefore, if we could increase the [Cl⁻] by elevating the NCC- and/or NKCC-mediated Cl⁻ uptake, the insufficiency of insulin secretion would be improved via membrane depolarization due to an elevation of Cl⁻ efflux from pancreatic β-cells of ΔF508 CFTR-expressing CF patients.

Recent studies contradict the findings by Guo et al. by presenting some evidence using the ferret's pancreas that β-cells do not express CFTR.[109] The authors showed that CFTR RNA is expressed in exocrine and not in endocrine cell types of islets and pancreas. They used a different approach, smFISH, to show the expression of CFTR. WT and CFTR-KO neonatal ferret pancreas were used to perform *CFTR* and *INS* dual smFISH. As expected, the *INS* was present in both genotypes, but *CFTR* was present in the WT pancreas.[109] More importantly, *CFTR* RNA was not co-expressed in *INS* (β-cell), GCG (α-cell), PPY (PP cell), or SST (δ-cell) expressing cells but was expressed in KRT7-expressing ductal cells in the WT pancreas. The same findings were shown in dissociated cells from isolated adult ferret and human islets.[109] These findings contradict the previous findings by Guo et al. and demonstrate that exocrine-derived duct cells associated with isolated islets express the highest levels of CFTR and support a mechanism by which CFTR-dependent duct/islet crosstalk might influence β-cell insulin secretion.[109]

Although there is evidence that supported each of these contradicting studies, a clear resolution to the question of whether CFTR directly or indirectly functions within the β-cells or other islet cell types to support insulin secretion needs further clarifications.

Role of glucose transporters (GLUTs) in CF and their impact in the pancreas

As mentioned earlier, glucose abnormalities in CF include CFRD and IGT. The relationship between CFTR and the causes of CFRD is still not very well established. Studies are also emerging regarding the implication of GLUT transporters in developing the diabetic state in CF patients.[112] In studies of obesity and diabetes in mice, it was reported that CFTR was significantly decreased, while GLUT5 and Villin were increased in the jejunum.[112] It is known that CFTR Cl– channel provides the major route for Cl– exit across the apical membrane in normal murine intestine, and disturbance in the anion exchange and recycling of K^+ is thought to be one of the causes of diabetes.[112]

Recently, it was reported that GLUT-4 is expressed in normal human airway epithelial cells.[113] The authors also showed that cells expressing F508del-CFTR have impaired glucose uptake, elevating action on the transepithelial resistance and diminishing action on paracellular flux of small molecules after insulin stimulation.[113] In a different study, GLUT-4 subfractionation demonstrated that, despite insulin stimulation, the GLUT-4 content of intracellular-associated subfraction was significantly higher in CF subjects compared with controls, corresponding to significantly lower GLUT-4 content in cell surface–associated subfraction.[114] These findings are consistent with the abnormal subcellular localization of GLUT-4. Impairment of GLUT-4 translocation in CF correlated with higher TNF-α levels in all CF subjects than in controls.[114] CF patients that have CFRD with a decreased insulin secretion could be explained by the elevation of TNF-α and impaired translocation of GLUT-4. In addition, the results indicate that the function of CFTR Cl⁻ channels is required for insulin to stimulate glucose uptake, elevate the transepithelial resistance, and diminish the paracellular flux of small molecules in airway epithelial cells.[33]

Pancreatic Defect and CFTR

In general, the exocrine pancreatic disease and its progression correlates well with the genetic factors of the CF patients.[68,115] CF patients have been divided into two classes, pancreatic insufficient (PI) and pancreatic sufficient (PS).[68,115,116] Approximately 85 percent of patients with CF have PI which is categorized as the "severe" CF phenotype, the rest 15 percent are PS patients and thus "mild" CF phenotype. The exocrine pancreas in PI patients no longer secretes the required digestive enzymes. Therefore, CF patients often require an oral pancreatic enzyme supplement each meal.[11] Pancreatic damage from CF can be detected in utero in subjects with PI and show obstruction of the small ducts and acini which will lead to the destruction of the pancreas with only a few islets or ducts left in a sea of adipose tissue.[78] In contrast, PS-CF patients do have pancreatic damage, as measured by the high levels of serum immunoreactive trypsinogen (IRT), but retain normal digestion due to a sufficient endogenous function of exocrine pancreatic ducts.[11,78] As we have mentioned earlier, CF patients are classified into six traditional classes from I to VI based on their *CFTR* mutations. The exocrine pancreatic phenotypes PI and PS are directly linked to genotype.[7,68,115,116]

Wilschanski and his colleagues describe that CF patients homozygous or compound heterozygous for severe alleles belonging to classes I, II, III, or VI confer PI. Whereas a mild class IV or V allele sustains pancreatic function in a dominant fashion even if the second mutation is severe and falls into PS.[11] They go further and explained that this observation appears plausible because all known mild alleles belong to class IV or V, all of which are (or predicted to be) associated with some residual chloride channel activity at the epithelial apical membranes.[11] On the other hand, in a recent review by Gibson-Corley and his colleagues, they described the PI CF patients to be in the classes I, II, III, IV, and VI because they have mutations that render CFTR to be absent or nonfunctional.[78] The remainder of the patients belonging to class V or mild class IV considered PS, due to less severe *CFTR* mutations.[78]

In both reviews, it was mentioned that this classification of the PI and PS does not entirely fit into the six classes of CFTR mutations. Those who are considered PS still show pancreatic destruction as the serum level of IRT is elevated but will not require enzyme replacement for normal digestion.[78] Some class I mutations with the stop codon at the end of the gene are CF patients with PS.[11] In addition, a small portion (~3%) of CF patients with a severe mutation on both alleles are considered PS at diagnosis but eventually transition from PS to PI.[11,117] To further exemplify G85E a missense mutation and few other mutations do show variable pancreatic phenotypes.[11] PS CF patients are more susceptible to developing pancreatitis than the PI patients.[117] It is known that pancreatic ducts have an essential part in CF and chronic pancreatitis, and only PS patients develop pancreatitis, suggesting that partially impaired the function of pancreatic ducts is retained in PS patients.[118] Druno and colleagues showed that there is a strong correlation between genotype and phenotype in patients with CF and pancreatitis.[117] They showed from a CF cohort study of about 1,000 patients followed over a period of 30 years that PS CF patients carry at least one mild mutant allele and are at a significant risk of developing pancreatitis. Symptoms of pancreatitis may precede the diagnosis of CF. Pancreatitis is associated with an otherwise mild CF phenotype.[117] A larger CF cohort study of about 10,071 patients had reported that out of 331 patients with PS, 34 cases had pancreatitis, whereas the occurrence of pancreatitis among patients with PI was 15 cases out of 2,971 patients.[119] More evidence goes toward the correlation of PS CF patients and the development of pancreatitis, with a novel pancreatic insufficiency prevalence score where they divided the patients into three groups: severe, moderate-severe, and mild, with the mild mutations more susceptible to the risk of developing pancreatitis.[11,120] The new classification of CFTR mutations into 31 new classes reflects the composite defects in mutant CFTR biology.[72] We will need to take into consideration the complexity of the disease and the severity of the CFTR genotype and their relationship with risk of pancreatitis. In a very recent study, for example, three sibling patients with a novel missense mutation, the R248G in exon 6 of the *CFTR* gene, present a recurrent acute pancreatitis.[121] A similar missense mutation, R248T, has been previously reported as a mild CFTR-RD mutation that is not associated with pancreatitis. The R248G mutation may alter the normal function of CFTR more than the R248T mutation based on the clinical phenotypes of the three patients.[121] As the authors conclude, future structure-functional studies on the CFTR protein can provide further insight into the impact of the R248G mutation at the molecular level.

CFTR Mutations and Pancreatitis

It is very well established by now that CFTR is a key protein in the pancreatic duct, which regulates the exchange of anions between the luminal surface and the cytoplasm of the duct cells. As mentioned, in CF patients the pancreas is one of the first organs to fail because mutations in CFTR play a critical role in pancreatic pathophysiology. The large number of mutations known to date in CFTR led scientists to tackle this complex disease with so many symptoms from a different angle, that is, to make a specific correlation between CFTR mutations with certain symptoms such as pancreatitis in its both forms, chronic and acute, to determine the severity of the disease. In an Austrian cohort study of 133 pancreatitis patients, the frequency of CFTR mutations was 11.2 percent.[122] In patients classified as "idiopathic definitive chronic pancreatitis," the frequency of mutations was 12.7 percent, whereas patients with "acute pancreatitis" or "possible chronic pancreatitis" had a frequency of CFTR mutations of 10 percent and 9.1 percent, respectively.[122] The authors concluded that the frequency of CFTR mutations is highest in patients with definitive chronic pancreatitis and may, therefore, be regarded as a risk factor for the development of chronic pancreatitis.[122]

Another large Canadian CF cohort study of 2,481 subjects with PS-CF (with and without pancreatitis) showed some correlation between the severity of CFTR genotype and the risk of pancreatitis.[120] They showed that patients carrying mild mutations are more likely to develop pancreatitis than those who had moderate-severe mutations.[120] Therefore, patients with mild mutations had 71 percent increase in the risk of developing pancreatitis at any given time than those with moderate-severe mutations.[120] Thus, approximately 20 percent of PS-CF patients develop pancreatitis.[120] Coffey et al. in their review summarized the complexity of the correlation between the CFTR mutations and the development of pancreatitis by categorizing the mutations into four groups based on the clinical status of the patients: (i) CF-causing mutations, (ii) mutations associated with CFTR-related disease, (iii) mutations with no known clinical consequence, and (iv) mutations with unknown clinical relevance.[123] Also, Ooi and colleagues found that certain diseases that resemble CF at an organ-specific level (e.g., pancreatitis) are also strongly associated with mutations in the CFTR gene.[123,124] In conclusion, pancreatitis in the CF patients and the relationship with multiple mutations of CFTR are very complex due to the multiple levels of the disease symptoms, which are different from mutation to another due to the extended classification of those different mutations into the 27 different classes according to the new classification.

Conclusion

In CF patients the pancreas is one of the first organs to fail because mutations in CFTR have a critical role in pancreatic pathophysiology. CFTR is the key regulator of the pancreatic duct that regulates the anion exchange between the luminal surface and the cytoplasm of the duct cells. The large number of CFTR mutations is leading scientists to approach an understanding of their functional impact from a different angle with the goal of making a specific correlation between CFTR mutations and certain symptoms such as PI and PS. The lack of correlation between the genotype and the phenotype of the CF patients has led to a new classification (31 possible classes of mutations) based on the severity and the clinical symptoms of the CF patients. The pleiotropic molecular defects of a single mutation in CFTR have limited the effects of drug therapy for some mutants which have been categorized as class I, II, or II/IV.[72] The expanded classification of the major mechanistic categories will accommodate the unusually complex, combinatorial molecular/cellular phenotypes of CF alleles. In addition to the new proposed classification, one more level of complexity is to categorize the mutations into four groups based on the clinical status of the patients: (i) CF-causing mutations, (ii) mutations associated with CFTR-related disease,

(iii) mutations with no known clinical consequence, and (iv) mutations with unknown clinical relevance.[123] All the recent discoveries and the new hypothesis will help shed light on the complex CF disease from a new perspective, which will help develop a new combined therapy to rectify the mutation or mutations at different levels of CFTR defects.

Acknowledgments

We thank Dr. Xuequn Chen, Jawad Bouhamdan, and Noor Charara for assistance in reviewing the manuscript and English editing. This work was supported by grants to Fei Sun, from the National Heart, Lung, & Blood Institute:1RO1HL133162, and the Cystic Fibrosis Foundation: SUN15XXO.

References

1. Riordan JR, Rommens JM, Kerem BS, Alon NOA, Rozmahel R, Grzelczak Z, Zielenski J, Lok SI, Plavsic N, Chou JL, Drumm ML, Iannuzzi MC, et al. Identification of the cystic fibrosis gene: Cloning and characterization of complementary DNA. Science 1989; 245:1066–1073.

2. Ballard S, Spadafora D. Fluid secretion by submucosal glands of the tracheobronchial airways. Respir Physiol Neurobiol. 2008; 159:271–277.

3. Riordan JR. CFTR function and prospects for therapy. Annual Review of Biochemistry 2008; 77:701–726.

4. Brennan AL, Beynon J. Clinical updates in cystic fibrosis–related diabetes. Semin Respir Crit Care Med 2015; 36:236–250.

5. Stecenko AA, Moran A. Update on cystic fibrosis-related diabetes. Current Opinion in Pulmonary Medicine 2010; 16:611–615.

6. Pilewski JM, Frizzell RA. Role of CFTR in airway disease. Physiol Rev 1999; 79:S215–S255.

7. Welsh MJ, Smith AE. Molecular mechanisms of CFTR chloride channel dysfunction in cystic fibrosis. Cell 1993; 73:1251–1254.

8. Hameed S, Jaffe A, Verge CF. Advances in the detection and management of cystic fibrosis related diabetes. Curr Opin Pediatr 2015; 27:525–533.

9. Rowe SM, Miller S, Sorscher EJ. Cystic fibrosis. N Engl J Med 2005; 352:1992–2001.

10. Stoltz DA, Meyerholz DK, Welsh MJ. Origins of cystic fibrosis lung disease. N Engl J Med 2015; 372:351–362.

11. Wilschanski M, Novak I. The cystic fibrosis of exocrine pancreas. Cold Spring Harb Perspect Med 2013; 3:a009746.

12. Scholte BJ, Davidson DJ, Wilke M, De Jonge HR. Animal models of cystic fibrosis. J Cyst Fibros 2004; 3 (suppl 2):183–190.

13. Wilke M, Buijs-Offerman RM, Aarbiou J, Colledge WH, Sheppard DN, Touqui L, Bot A, Jorna H, de Jonge HR, Scholte BJ. Mouse models of cystic fibrosis: Phenotypic analysis and research applications. J Cyst Fibros 2011; 10 (suppl 2):S152–S171.

14. Snouwaert JN, Brigman KK, Latour AM, Malouf NN, Boucher RC, Smithies O, Koller BH. An animal model for cystic fibrosis made by gene targeting. Science 1992; 257:1083–1088.

15. Colledge WH, Abella BS, Southern KW, Ratcliff R, Jiang C, Cheng SH, MacVinish LJ, Anderson JR, Cuthbert AW, Evans MJ. Generation and characterization of a delta F508 cystic fibrosis mouse model. Nat Genet 1995; 10:445–452.

16. van Doorninck JH, French PJ, Verbeek E, Peters RH, Morreau H, Bijman J, Scholte BJ. A mouse model for the cystic fibrosis delta F508 mutation. EMBO J 1995; 14:4403–4411.

17. Delaney SJ, Alton EW, Smith SN, Lunn DP, Farley R, Lovelock PK, Thomson SA, Hume DA, Lamb D, Porteous DJ, Dorin JR, Wainwright BJ. Cystic fibrosis mice carrying the missense mutation G551D replicate human genotype-phenotype correlations. EMBO J 1996; 15:955–963.

18. Rozmahel R, Wilschanski M, Matin A, Plyte S, Oliver M, Auerbach W, Moore A, Forstner J, Durie P, Nadeau J, Bear C, Tsui LC. Modulation of disease severity in cystic fibrosis transmembrane conductance regulator deficient mice by a secondary genetic factor. Nat Genet 1996; 12:280–287.

19. Sun X, Olivier AK, Liang B, Yi Y, Sui H, Evans TI, Zhang Y, Zhou W, Tyler SR, Fisher JT, Keiser NW, Liu X, et al. Lung phenotype of juvenile and adult cystic fibrosis transmembrane conductance regulator-knockout ferrets. Am J Respir Cell Mol Biol 2014; 50:502–512.

20. Keiser NW, Engelhardt JF. New animal models of cystic fibrosis: what are they teaching us? Curr Opin Pulm Med 2011; 17:478–483.

21. Kamaruzaman NA, Kardia E, Kamaldin N, Latahir AZ, Yahaya BH. The rabbit as a model for studying lung disease and stem cell therapy. Biomed Res Int 2013; 2013:Art ID 691830.

22. Widdicombe JH, Chen LL, Sporer H, Choi HK, Pecson IS, Bastacky SJ. Distribution of tracheal and laryngeal mucous glands in some rodents and the rabbit. J Anat 2001; 198:207–221.

23. Mall M, Grubb BR, Harkema JR, O'Neal WK, Boucher RC. Increased airway epithelial Na+ absorption produces cystic fibrosis-like lung disease in mice. Nat Med 2004; 10:487–493.

24. Zuelzer WW, Newton WA, Jr. The pathogenesis of fibrocystic disease of the pancreas; a study of 36 cases with special reference to the pulmonary lesions. Pediatrics 1949; 4:53–69.

25. Stoltz DA, Meyerholz DK, Pezzulo AA, Ramachandran S, Rogan MP, Davis GJ, Hanfland RA, Wohlford-Lenane C, Dohrn CL, Bartlett JA, Nelson GAt, Chang EH, et al. Cystic fibrosis pigs develop lung disease and exhibit defective bacterial eradication at birth. Sci Transl Med 2010; 2:29ra31.

26. Liu J, Walker NM, Ootani A, Strubberg AM, Clarke LL. Defective goblet cell exocytosis contributes to murine cystic fibrosis-associated intestinal disease. J Clin Invest 2015; 125:1056–1068.

27. Cheng SH, Gregory RJ, Marshall J, Paul S, Souza DW, White GA, O'Riordan CR, Smith AE. Defective intracellular transport and processing of CFTR is the molecular basis of most cystic fibrosis. Cell 1990; 63:827–834.

28. Jensen TJ, Loo MA, Pind S, Williams DB, Goldberg AL, Riordan JR. Multiple proteolytic systems, including the

proteasome, contribute to CFTR processing. Cell 1995; 83:129–135.

29. Ward CL, Omura S, Kopito RR. Degradation of CFTR by the ubiquitin-proteasome pathway. Cell 1995; 83:121–127.

30. Denning GM, Anderson MP, Amara JF, Marshall J, Smith AE, Welsh MJ. Processing of mutant cystic fibrosis transmembrane conductance regulator is temperature-sensitive. Nature 1992; 358:761–764.

31. Quinton PM. Cystic fibrosis: impaired bicarbonate secretion and mucoviscidosis. Lancet 2008; 372:415–417.

32. Kunzelmann K, Schreiber R, Hadorn HB. Bicarbonate in cystic fibrosis. J Cyst Fibros 2017; 66:653–662.

33. Marunaka Y. The mechanistic links between insulin and cystic fibrosis transmembrane conductance regulator (CFTR) Cl⁻ channel. Int J Mol Sci 2017; 18:E1767.

34. Moran A, Hardin D, Rodman D, Allen HF, Beall RJ, Borowitz D, Brunzell C, Campbell PW, 3rd, Chesrown SE, Duchow C, Fink RJ, Fitzsimmons SC, et al. Diagnosis, screening and management of cystic fibrosis related diabetes mellitus: a consensus conference report. Diabetes Res Clin Pract 1999; 45:61–73.

35. Cutting GR. Cystic fibrosis genetics: from molecular understanding to clinical application. Nat Rev Genet 2015; 16:45–56.

36. Patrick AE, Thomas PJ. Development of CFTR structure. Front Pharmacol 2012; 3:162.

37. Naren AP, Cormet-Boyaka E, Fu J, Villain M, Blalock JE, Quick MW, Kirk KL. CFTR chloride channel regulation by an interdomain interaction. Science 1999; 286:544–548.

38. Chappe V, Irvine T, Liao J, Evagelidis A, Hanrahan JW. Phosphorylation of CFTR by PKA promotes binding of the regulatory domain. EMBO J 2005; 24:2730.

39. Sheppard DN, Welsh MJ. Structure and function of the CFTR chloride channel. Physiol Rev 1999; 79:S23–45.

40. Naren AP, Quick MW, Collawn JF, Nelson DJ, Kirk KL. Syntaxin 1A inhibits CFTR chloride channels by means of domain-specific protein-protein interactions. Proc Natl Acad Sci U S A 1998; 95:10972–10977.

41. Gene GG, Llobet A, Larriba S, de Semir D, Martinez I, Escalada A, Solsona C, Casals T, Aran JM. N-terminal CFTR missense variants severely affect the behavior of the CFTR chloride channel. Hum Mutat 2008; 29:738–749.

42. Wang G, Linsley R, Norimatsu Y. External Zn²⁺ binding to cysteine-substituted cystic fibrosis transmembrane conductance regulator constructs regulates channel gating and curcumin potentiation. FEBS J 2016; 283:2458–2475.

43. Wang W, El Hiani Y, Rubaiy HN, Linsdell P. Relative contribution of different transmembrane segments to the CFTR chloride channel pore. Pflugers Arch 2014; 466:477–490.

44. Zang X, Perez JJ, Jones CM, Monge ME, McCarty NA, Stecenko AA, Fernandez FM. Comparison of ambient and atmospheric pressure ion sources for cystic fibrosis exhaled breath condensate ion mobility-mass spectrometry metabolomics. J Am Soc Mass Spectrom 2017; 28:1489–1496.

45. Therien AG, Grant FE, Deber CM. Interhelical hydrogen bonds in the CFTR membrane domain. Nat Struct Biol 2001; 8:597–601.

46. Partridge AW, Melnyk RA, Deber CM. Polar residues in membrane domains of proteins: molecular basis for helix-helix association in a mutant CFTR transmembrane segment. Biochemistry 2002; 41:3647–3653.

47. Hammerle MM, Aleksandrov AA, Riordan JR. Disease-associated mutations in the extracytoplasmic loops of cystic fibrosis transmembrane conductance regulator do not impede biosynthetic processing but impair chloride channel stability. J Biol Chem 2001; 276:14848–14854.

48. Li MS, Cowley EA, Linsdell P. Pseudohalide anions reveal a novel extracellular site for potentiators to increase CFTR function. Br J Pharmacol 2012; 167:1062–1075.

49. Zhou JJ, Linsdell P. Evidence that extracellular anions interact with a site outside the CFTR chloride channel pore to modify channel properties. Can J Physiol Pharmacol 2009; 87:387–395.

50. Infield DT, Cui G, Kuang C, McCarty NA. Positioning of extracellular loop 1 affects pore gating of the cystic fibrosis transmembrane conductance regulator. Am J Physiol Lung Cell Mol Physiol 2016; 310:L403-L414.

51. Broadbent SD, Wang W, Linsdell P. Interaction between 2 extracellular loops influences the activity of the cystic fibrosis transmembrane conductance regulator chloride channel. Biochem Cell Biol 2014; 92:390–396.

52. Ghanem N, Costes B, Girodon E, Martin J, Fanen P, Goossens M. Identification of eight mutations and three sequence variations in the cystic fibrosis transmembrane conductance regulator (CFTR) gene. Genomics 1994; 21:434–436.

53. Caputo A, Hinzpeter A, Caci E, Pedemonte N, Arous N, Di Duca M, Zegarra-Moran O, Fanen P, Galietta LJ. Mutation-specific potency and efficacy of cystic fibrosis transmembrane conductance regulator chloride channel potentiators. J Pharmacol Exp Ther 2009; 330:783–791.

54. Cremonesi L, Ferrari M, Belloni E, Magnani C, Seia M, Ronchetto P, Rady M, Russo MP, Romeo G, Devoto M. Four new mutations of the CFTR gene (541delC, R347H, R352Q, E585X) detected by DGGE analysis in Italian CF patients, associated with different clinical phenotypes. Hum Mutat 1992; 1:314–319.

55. Loo TW, Clarke DM. The transmission interfaces contribute asymmetrically to the assembly and activity of human P-glycoprotein. J Biol Chem 2015; 290:16954–16963.

56. Billet A, Mornon JP, Jollivet M, Lehn P, Callebaut I, Becq F. CFTR: effect of ICL2 and ICL4 amino acids in close spatial proximity on the current properties of the channel. J Cyst Fibros 2013; 12:737–745.

57. Seibert FS, Linsdell P, Loo TW, Hanrahan JW, Riordan JR, Clarke DM. Cytoplasmic loop three of cystic fibrosis transmembrane conductance regulator contributes to regulation of chloride channel activity. J Biol Chem 1996; 271:27493–27499.

58. Vankeerberghen A, Wei L, Teng H, Jaspers M, Cassiman JJ, Nilius B, Cuppens H. Characterization of mutations located in exon 18 of the CFTR gene. FEBS Lett 1998; 437:1–4.

59. Belmonte L, Moran O. On the interactions between nucleotide binding domains and membrane spanning domains in cystic fibrosis transmembrane regulator: A molecular dynamic study. Biochimie 2015; 111:19–29.

60. Schreiber R, Hopf A, Mall M, Greger R, Kunzelmann K. The first-nucleotide binding domain of the cystic-fibrosis transmembrane conductance regulator is important for inhibition

of the epithelial Na+ channel. Proc Natl Acad Sci U S A 1999; 96:5310–5315.

61. Yuan YR, Blecker S, Martsinkevich O, Millen L, Thomas PJ, Hunt JF. The crystal structure of the MJ0796 ATP-binding cassette. Implications for the structural consequences of ATP hydrolysis in the active site of an ABC transporter. J Biol Chem 2001; 276:32313–32321.

62. Vernon RM, Chong PA, Lin H, Yang Z, Zhou Q, Aleksandrov AA, Dawson JE, Riordan JR, Brouillette CG, Thibodeau PH, Forman-Kay JD. Stabilization of a nucleotide-binding domain of the cystic fibrosis transmembrane conductance regulator yields insight into disease-causing mutations. J Biol Chem 2017; 292:14147–14164.

63. Seibert FS, Chang XB, Aleksandrov AA, Clarke DM, Hanrahan JW, Riordan JR. Influence of phosphorylation by protein kinase A on CFTR at the cell surface and endoplasmic reticulum. Biochimica et Biophysica Acta (BBA)—Biomembranes 1999; 1461:275–283.

64. Seavilleklein G, Amer N, Evagelidis A, Chappe F, Irvine T, Hanrahan JW, Chappe V. PKC phosphorylation modulates PKA-dependent binding of the R domain to other domains of CFTR. Am J Physiol Cell Physiol 2008; 295:C1366-C1375.

65. Choi JY, Muallem D, Kiselyov K, Lee MG, Thomas PJ, Muallem S. Aberrant CFTR-dependent HCO₃⁻ transport in mutations associated with cystic fibrosis. Nature 2001; 410:94–97.

66. Aznarez I, Chan EM, Zielenski J, Blencowe BJ, Tsui LC. Characterization of disease-associated mutations affecting an exonic splicing enhancer and two cryptic splice sites in exon 13 of the cystic fibrosis transmembrane conductance regulator gene. Hum Mol Genet 2003; 12:2031–2040.

67. Hohwieler M, Perkhofer L, Liebau S, Seufferlein T, Muller M, Illing A, Kleger A. Stem cell-derived organoids to model gastrointestinal facets of cystic fibrosis. United European Gastroenterol J 2017; 5:609–624.

68. Kristidis P, Bozon D, Corey M, Markiewicz D, Rommens J, Tsui LC, Durie P. Genetic determination of exocrine pancreatic function in cystic fibrosis. Am J Hum Genet 1992; 50:1178–1184.

69. Lee MG, Ohana E, Park HW, Yang D, Muallem S. Molecular mechanism of pancreatic and salivary gland fluid and HCO3 secretion. Physiol Rev 2012; 92:39–74.

70. Edlund A, Esguerra JL, Wendt A, Flodstrom-Tullberg M, Eliasson L. CFTR and Anoctamin 1 (ANO1) contribute to cAMP amplified exocytosis and insulin secretion in human and murine pancreatic beta-cells. BMC Med 2014; 12:87.

71. Lavelle GM, White MM, Browne N, McElvaney NG, Reeves EP. Animal Models of Cystic Fibrosis Pathology: Phenotypic Parallels and Divergences. BioMed Res Int 2016; 2016:5258727.

72. Veit G, Avramescu RG, Chiang AN, Houck SA, Cai Z, Peters KW, Hong JS, Pollard HB, Guggino WB, Balch WE, Skach WR, Cutting GR, et al. From CFTR biology toward combinatorial pharmacotherapy: expanded classification of cystic fibrosis mutations. Mol Bio Cell 2016; 27:424–433.

73. Dupuis A, Keenan K, Ooi CY, Dorfman R, Sontag MK, Naehrlich L, Castellani C, Strug LJ, Rommens JM, Gonska T. Prevalence of meconium ileus marks the severity of

mutations of the cystic fibrosis transmembrane conductance regulator (CFTR) gene. Genet Med 2016; 18:333–340.

74. Sosnay PR, Siklosi KR, Van Goor F, Kaniecki K, Yu H, Sharma N, Ramalho AS, Amaral MD, Dorfman R, Zielenski J, Masica DL, Karchin R, et al. Defining the disease liability of variants in the cystic fibrosis transmembrane conductance regulator gene. Nat Genet 2013; 45:1160–1167.

75. Zielenski J. Genotype and phenotype in cystic fibrosis. Respiration 2000; 67:117–133.

76. Van Goor F, Yu H, Burton B, Hoffman BJ. Effect of Ivacaftor on CFTR forms with missense mutations associated with defects in protein processing or function. J Cyst Fibros 2014; 13:29–36.

77. Olivier AK, Gibson-Corley KN, Meyerholz DK. Animal models of gastrointestinal and liver diseases. Animal models of cystic fibrosis: gastrointestinal, pancreatic, and hepatobiliary disease and pathophysiology. Am J Physiol Gastrointest Liver Physiol 2015; 308:G459-G471.

78. Gibson-Corley KN, Meyerholz DK, Engelhardt JF. Pancreatic pathophysiology in cystic fibrosis. J Pathol 2016; 238:311–320.

79. Quinton PM. The neglected ion: HCO₃⁻. Nat Med 2001; 7:292–293.

80. Reid CJ, Hyde K, Ho SB, Harris A. Cystic fibrosis of the pancreas: involvement of MUC6 mucin in obstruction of pancreatic ducts. Mol Med 1997; 3:403–411.

81. Meyerholz DK, Stoltz DA, Pezzulo AA, Welsh MJ. Pathology of gastrointestinal organs in a porcine model of cystic fibrosis. Am J Pathol 2010; 176:1377–1389.

82. Mushtaq I, Wright VM, Drake DP, Mearns MB, Wood CB. Meconium ileus secondary to cystic fibrosis. The East London experience. Pediatr Surg Int 1998; 13:365–369.

83. Imrie JR, Fagan DG, Sturgess JM. Quantitative evaluation of the development of the exocrine pancreas in cystic fibrosis and control infants. Am J Path 1979; 95:697–707.

84. Tucker JA, Spock A, Spicer SS, Shelburne JD, Bradford W. Inspissation of pancreatic zymogen material in cystic fibrosis. Ultrastruct Pathol 2003; 27:323–335.

85. Grumelli S IG, Castro GR Consequences of cystic fibrosis transmembrane regulator mutations on inflammatory cells. Pulm Crit Care Med Volume 1, 2016.

86. Ichiyama K, Gonzalez-Martin A, Kim BS, Jin HY, Jin W, Xu W, Sabouri-Ghomi M, Xu S, Zheng P, Xiao C, Dong C. The MicroRNA-183-96-182 Cluster Promotes T Helper 17 Cell Pathogenicity by Negatively Regulating Transcription Factor Foxo1 Expression. Immunity 2016; 44:1284–1298.

87. Ode KL, Moran A. New insights into cystic fibrosis-related diabetes in children. Lancet Diabetes Endocrinol 2013; 1:52–58.

88. Ntimbane T, Comte B, Mailhot G, Berthiaume Y, Poitout V, Prentki M, Rabasa-Lhoret R, Levy E. Cystic fibrosis-related diabetes: from CFTR dysfunction to oxidative stress. Clin Biochem Rev 2009; 30:153–177.

89. Moran A, Doherty L, Wang X, Thomas W. Abnormal glucose metabolism in cystic fibrosis. J Pediatr 1998; 133:10–17.

90. Schwarzenberg SJ, Thomas W, Olsen TW, Grover T, Walk D, Milla C, Moran A. Microvascular complications in cystic fibrosis-related diabetes. Diabetes Care 2007; 30:1056–1061.

91. Cucinotta D, De Luca F, Scoglio R, Lombardo F, Sferlazzas C, Di Benedetto A, Magazzu G, Raimondo G, Arrigo T. Factors affecting diabetes mellitus onset in cystic fibrosis: evidence from a 10-year follow-up study. Acta Paediatr 1999; 88:389–393.

92. Nathan BM, Laguna T, Moran A. Recent trends in cystic fibrosis-related diabetes. Curr Opin Endocrinol Diabetes Obes 2010; 17:335–341.

93. Preumont V, Hermans MP, Lebecque P, Buysschaert M. Glucose homeostasis and genotype-phenotype interplay in cystic fibrosis patients with CFTR gene deltaF508 mutation. Diabetes Care 2007; 30:1187–1192.

94. Hardin DS, Moran A. Diabetes mellitus in cystic fibrosis. Endocrinol Metab Clin North Am 1999; 28:787–800, ix.

95. Consortium CFG-P. Correlation between genotype and phenotype in patients with cystic fibrosis. N Engl J Med 1993; 329:1308–1313.

96. Rosenecker J, Eichler I, Kuhn L, Harms HK, von der Hardt H. Genetic determination of diabetes mellitus in patients with cystic fibrosis. Multicenter Cystic Fibrosis Study Group. J Pediatr 1995; 127:441–443.

97. Barrio R. Management of endocrine disease: Cystic fibrosis-related diabetes: novel pathogenic insights opening new therapeutic avenues. Eur J Endocrinol 2015; 172:R131–141.

98. Poitout V, Tanaka Y, Reach G, Robertson RP. [Oxidative stress, insulin secretion, and insulin resistance]. J Annu Diabetol Hotel Dieu 2001:75–86.

99. Robertson RP. Chronic oxidative stress as a central mechanism for glucose toxicity in pancreatic islet beta cells in diabetes. J Biol Chem 2004; 279:42351–42354.

100. Galli F, Battistoni A, Gambari R, Pompella A, Bragonzi A, Pilolli F, Iuliano L, Piroddi M, Dechecchi MC, Cabrini G. Oxidative stress and antioxidant therapy in cystic fibrosis. Biochim Biophys Acta 2012; 1822:690–713.

101. Xu C, Bailly-Maitre B, Reed JC. Endoplasmic reticulum stress: cell life and death decisions. J Clin Invest 2005; 115:2656–2664.

102. Araki E, Oyadomari S, Mori M. Impact of endoplasmic reticulum stress pathway on pancreatic beta-cells and diabetes mellitus. Exp Biol Med (Maywood) 2003; 228:1213–1217.

103. Kopito RR. Aggresomes, inclusion bodies and protein aggregation. Trends Cell Biol 2000; 10:524–530.

104. Pirot P, Eizirik DL, Cardozo AK. Interferon-gamma potentiates endoplasmic reticulum stress-induced death by reducing pancreatic beta cell defence mechanisms. Diabetologia 2006; 49:1229–1236.

105. Harding HP, Zeng H, Zhang Y, Jungries R, Chung P, Plesken H, Sabatini DD, Ron D. Diabetes mellitus and exocrine pancreatic dysfunction in Perk$^{-/-}$ mice reveals a role for translational control in secretory cell survival. Molecular Cell 2001; 7:1153–1163.

106. Robinson J, Yates R, Harper A, Kelly C. [beta]-cells require CFTR for glucose-induced insulin secretion. Endocrine Abstracts, 2014:34 OC32.35.

107. Guo JH, Chen H, Ruan YC, Zhang XL, Zhang XH, Fok KL, Tsang LL, Yu MK, Huang WQ, Sun X, Chung YW, Jiang X, et al. Glucose-induced electrical activities and insulin secretion in pancreatic islet beta-cells are modulated by CFTR. Nat Commun 2014; 5:4420.

108. Koivula FN, McClenaghan NH, Harper AG, Kelly C. Islet-intrinsic effects of CFTR mutation. Diabetologia 2016; 59:1350–1355.

109. Sun X, Yi Y, Xie W, Liang B, Winter MC, He N, Liu X, Luo M, Yang Y, Ode KL, Uc A, Norris AW, Engelhardt JF. CFTR Influences beta cell function and insulin secretion through non-cell autonomous exocrine-derived factors. Endocrinology 2017; 158:3325–3338.

110. Ntimbane T, Mailhot G, Spahis S, Rabasa-Lhoret R, Kleme ML, Melloul D, Brochiero E, Berthiaume Y, Levy E. CFTR silencing in pancreatic beta-cells reveals a functional impact on glucose-stimulated insulin secretion and oxidative stress response. Am J Physiol Endocrinol Metab 2016; 310:E200-E212.

111. Boom A, Lybaert P, Pollet JF, Jacobs P, Jijakli H, Golstein PE, Sener A, Malaisse WJ, Beauwens R. Expression and localization of cystic fibrosis transmembrane conductance regulator in the rat endocrine pancreas. Endocrine 2007; 32:197–205.

112. Leung L, Kang J, Rayyan E, Bhakta A, Barrett B, Larsen D, Jelinek R, Willey J, Cochran S, Broderick TL, Al-Nakkash L. Decreased basal chloride secretion and altered cystic fibrosis transmembrane conductance regulatory protein, Villin, GLUT5 protein expression in jejunum from leptin-deficient mice. Diabetes Metab Syndr Obes 2014; 7:321–330.

113. Molina SA, Moriarty HK, Infield DT, Imhoff BR, Vance RJ, Kim AH, Hansen JM, Hunt WR, Koval M, McCarty NA. Insulin signaling via the PI3-kinase/Akt pathway regulates airway glucose uptake and barrier function in a CFTR-dependent manner. Am J Physiol Lung Cell Mol Physiol 2017; 312:L688-L702.

114. Hardin DS, Leblanc A, Marshall G, Seilheimer DK. Mechanisms of insulin resistance in cystic fibrosis. Am J Physiol Endocrinol Metab 2001; 281:E1022-E1028.

115. Ahmed N, Corey M, Forstner G, Zielenski J, Tsui LC, Ellis L, Tullis E, Durie P. Molecular consequences of cystic fibrosis transmembrane regulator (CFTR) gene mutations in the exocrine pancreas. Gut 2003; 52:1159–1164.

116. Kerem E, Corey M, Kerem BS, Rommens J, Markiewicz D, Levison H, Tsui LC, Durie P. The relation between genotype and phenotype in cystic fibrosis-analysis of the most common mutation (delta F508). N Engl J Med 1990; 323:1517–1522.

117. Durno C, Corey M, Zielenski J, Tullis E, Tsui L-C, Durie P. Genotype and phenotype correlations in patients with cystic fibrosis and pancreatitis. Gastroenterology 2002; 123:1857–1864.

118. Hegyi P, Rakonczay Z, Jr. The role of pancreatic ducts in the pathogenesis of acute pancreatitis. Pancreatology 2015; 15:S13–17.

119. De Boeck K, Weren M, Proesmans M, Kerem E. Pancreatitis among patients with cystic fibrosis: correlation with pancreatic status and genotype. Pediatrics 2005; 115:e463-e469.

120. Ooi CY, Dorfman R, Cipolli M, Gonska T, Castellani C, Keenan K, Freedman SD, Zielenski J, Berthiaume Y, Corey

M, Schibli S, Tullis E, et al. Type of CFTR mutation determines risk of pancreatitis in patients with cystic fibrosis. Gastroenterology 2011; 140:153–161.

121. Villalona S, Glover-Lopez G, Ortega-Garcia JA, Moya-Quiles R, Mondejar-Lopez P, Martinez-Romero MC, Rigabert-Montiel M, Pastor-Vivero MD, Sanchez-Solis M. R248G cystic fibrosis transmembrane conductance regulator mutation in three siblings presenting with recurrent acute pancreatitis and reproductive issues: a case series. J Med Case Rep 2017; 11:42.

122. Zoller H, Egg M, Graziadei I, Creus M, Janecke AR, Löffler-Ragg J, Vogel W. CFTR gene mutations in pancreatitis: Frequency and clinical manifestations in an Austrian patient cohort. Wiener Klinische Wochenschrift 2007; 119:527–533.

123. Coffey MJ, Ooi CY. Pancreatitis in cystic fibrosis and CFTR-related disorder. In: Rodrigo L, ed. Acute Pancreatitis. Rijeka: InTech, 2012:67–90.

124. Ooi CYT, E. & Durie, P. Diagnostic Approach to CFTR-related Disorders, In: Cystic Fibrosis. Lung Biology in Health and Disease Series. Informa Healthcare, 2010:103–122.

Chapter 22

Regulation of normal and adaptive pancreatic growth

John A. Williams

Department of Molecular and Integrative Physiology, The University of Michigan, Ann Arbor, Michigan 48109–5622

Introduction

It is now established that almost all vertebrate organs show dynamic replacement or turnover at the level of individual proteins, organelles, and cells. This allows replacement of damaged components and also regulation of functional mass.[1] Regulatory loops where the expression of a mitogen is modulated by a feedback mechanism are common to many tissues. This is consistent with the observation that many tissues show both autonomous and regulated growth.[2] Furthermore, in vertebrates, cell division and cell size are independent processes.

Regulation of pancreatic size is similar to other organs and occurs through genetic programing and environmental influences (especially food intake), with both directed at assuring that the organ can carry out its function in an energy-efficient manner.[3,4] In most vertebrates, the islets of Langerhans are scattered through the gland but make up only 2–3 percent of the volume. Acinar cells make up about 85 percent of volume and ducts 5 percent, so overall pancreatic size reflects the exocrine component. However, acinar cells make up a smaller fraction of total cells and nuclei (approximately 50%) as individual acinar cells are larger than myofibroblasts and duct cells. In the adult, islet mass and exocrine mass appear to be regulated independently. Assessment of islet volume generally requires histological analysis and will not be further considered here.

Measurement of Pancreatic Size

Overall pancreatic size is expressed as cm^3 if volume is measured or g if weighed. Because the pancreas increases in size with age, the weight or volume is usually normalized to body weight. Until recently, the most common technique was dissecting and weighing the pancreas. This yields values for young adults, usually males, fed a control diet, of 8–12 mg pancreatic weight/g body weight for mice,[5–9] 22.5 mg PW/gBW for Syrian hamster,[10] 2.5–4 mg PW/g BW for rats,[11–16] and 0.5–1.1 g PW/kg BW for adult humans.[17,18]

Note that the relative pancreas size gets smaller as the animal gets larger. Smaller animals have a larger body surface area and higher metabolic rate and food consumption relative to body size. It seems reasonable that the higher food consumption requires more digestive enzymes and hence a larger pancreas.

Recently, imaging studies using CT or MRI have been used and give similar values for pancreatic size in humans as earlier autopsy studies.[19–22] A study in pigs validated the method by comparing the pancreatic volume obtained by MRI to water displacement after pancreas removal.[23] A recent study adopted MRI to determine the volume of mouse pancreas, and the results were similar to water displacement.[24] The main advantage of the imaging techniques is that they are noninvasive and can be carried out repetitively. Also, the volume for pancreatic parenchyma and fat can be determined separately. Such studies have shown that both total pancreatic mass and pancreatic fat increase with increasing BMI in humans[21] and decrease in both type 1 and type 2 diabetes.[25–28] The technique has been adapted to mice by use of a micro-MRI with a 7 Tesla magnet and worked best with the Rapid Acquisition with Relaxation (RARE) protocol.[24] In this study, mice were perfusion-fixed initially, so further work is necessary to establish the technique in living mice. In another study the pancreas architecture including the islet component was carried out by micro-magnetic resonance microscopy.[29]

Prenatal Pancreatic Growth

We can divide the lifespan of the pancreas into the prenatal, postnatal to adolescence and adult phases. The prenatal phase is dominated by organogenesis in which outgrowth from the foregut produces undifferentiated branching tubules expressing the transcription factor Pdx1 (pancreatic and duodenal homeobox 1). These cells, particularly the buds at the end of the tubules, are considered to be transient progenitor cells. They will develop into acini, islets,

Email: jawillms@umich.edu

and mature ducts under the influence of mesenchyme and a number of other transcriptional regulators.[30] However, these progenitor cells are not totipotent and probably should not be considered stem cells. During development, cell division and differentiation result in an increase in organ size, and at birth, or within a few days, the pancreatic cells are able to assume their physiological function in response to environmental stimuli which activate neurohormonal control mechanisms. Some aspects of the pancreas change in mammals with weaning when a shift from a high fat to a high carbohydrate or high-protein diet occurs, and these changes can be considered as either completion of development or an environmental (dietary) adaptation. The size of the pancreas at birth depends on the original number of progenitor cells as ablation of part of these cells during embryogenesis by targeting diphtheria toxin to the Pdx1 expressing cells reduces the size of the pancreas at birth and through postnatal life.[31] This implies that each progenitor cell can give rise to a fixed amount of pancreas. At the time of birth the rat pancreas weighs about 5–10 mg and the relative size is 3–3.8 mg/g body weight.[14,32]

Prenatal pancreatic growth has recently been shown to be regulated by the Hippo pathway. Originally discovered in Drosophilla, the mammalian pathway consists of mammalian sterile 20-like kinase 1/2 (MST1/2), which phosphorylates large tumor suppressor 1/2 (LATS1/2) which then phosphorylates Yes associated protein (YAP) and its homolog TAZ leading to inactivation and cytosolic retention.[33] When the pathway is activated by cell–cell interaction or certain chemical factors, hypophosphorylated YAP translocates to the nucleus and induces gene expression. YAP1 increases growth and inhibits differentiation in most cell types.[34,35] In pancreatic cells the pathway is primarily active before birth, especially in the secondary transition to differentiated cells.[36] Activation in the adult liver leads to a fourfold increase in mass, but a similar increase in pancreas has little or no effect.[37] Functional deletion of this pathway by gene deletion of the upstream kinase MST1/2 leads to acinar cell dedifferentiation.[38] YAP expression is lost in adult endocrine pancreas but remains in the exocrine compartment, especially ducts, and may play a role in pathology and recovery from acute pancreatitis.[39] The role of the Hippo pathway in islet beta cells has been recently reviewed.[40]

Postnatal pancreatic growth

In the period from birth to adulthood the pancreas grows roughly in proportion to body growth. This has been quantitated both by dissecting and weighing the pancreas and by imaging using CT or MRI followed by volume calculation. In a detailed study in Wistar rats, Iovanna et al. evaluated pancreatic weight, protein, and RNA from birth to 90 days.[14] Pancreas weight increased faster than body

mass until reaching a peak at 35 days and then declined (**Figure 1A**). A similar pattern of pancreas weight increase can be derived from the data of Snook carried out using CD strain rats which showed a peak PW/BW at 28 days.[32,41] Using micro-MRI, Paredes et al. (120) showed a linear relation in mice between pancreas volume and body weight between 20 days and 9 months of age (**Figure 1B**). By contrast, Stanger et al. demonstrated a hyperbolic relationship with pancreatic growth faster than body growth in the first three weeks of life with gradual slowing thereafter.[31]

These differences are not major and may reflect differences in species, strain, or diet. In mice postnatal growth is not dependent on two of the major pancreas regulators, cholecystokinin (CCK) and secretin, as the pancreas in young adult mice is similar when these peptides or their receptors undergo gene deletion.[6-8] Similarly, in rats, pancreatic growth with age is not altered by a CCK antagonist.[42] By contrast, the growth of the neonatal pancreas can be increased by glucocorticoids and thyroxine which probably reflects a role of these hormones on development.[43] The human pancreas also shows a linear relationship between size and age from birth to 25 years, after which there is little change in size in adults until it begins declining around age 60.[21] In humans, due to the larger size, imaging can distinguish fat from parenchymal tissue, and from midlife on the amount of fat in the pancreas

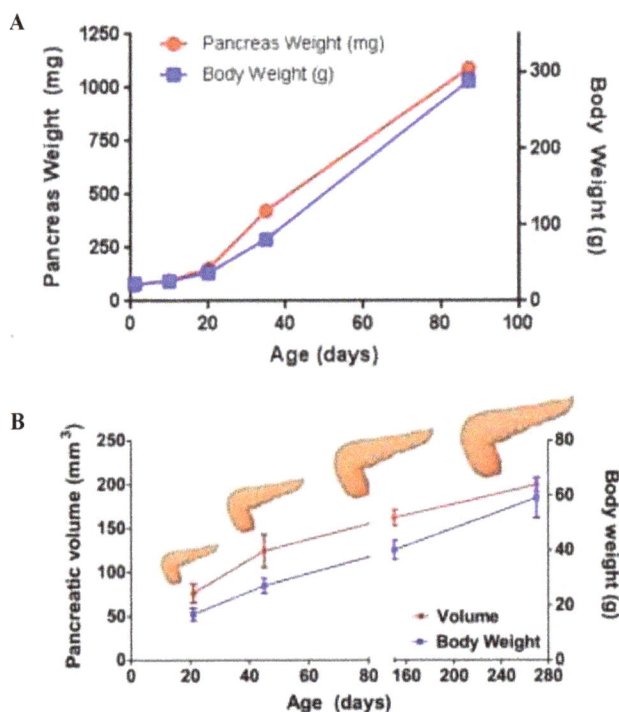

Figure 1. Pancreatic and body weight as a function of age. (**A**) Data for rats from 1 day postnatal to 87 days (replotted from Iovanna et al.[14]). (**B**) Data for mice showing pancreatic volume and body weight from 20 to 280 days (from Paredes et al.[24])

increases with a concomitant decrease in the mass of exocrine tissue. During the growth phase, the increase in pancreatic mass primarily reflects cell division, and the size of individual cells changes little after the first few days of life.[44]

During the postnatal growth phase, pancreatic DNA increases in parallel to pancreatic weight. Acinar cell size shows a transient increase around birth due to the cells being packed with zymogen granules before secretion matures, but it is otherwise relatively constant for a particular species, with cell volume inversely related to the size and longevity of the animal—for example, cell volume varying from 4,288 um^3 for the tree shrew to only 457 um^3 for a rhinoceros.[45] In mice, cell size increases at the time of weaning.[45] Humans, however, maintain a constant acinar cell size over the lifespan. In support of the concept that pancreatic growth is primarily driven by cell proliferation, a number of studies have evaluated ^{3}H-thymidine incorporation by autoradiography or the incorporation of the thymidine analog BrdU visualized by immunohistochemistry. Other studies evaluated specific proteins expressed by cells undergoing mitogenesis including proliferating cell nuclear antigen (PCNA) and Ki67. Finally, some early studies counted cells undergoing mitosis to obtain the mitotic index. These approaches generally parallel each other, although showing different absolute numbers where the percentage of cells expressing PCNA is higher than those synthesizing DNA. Almost all of the earlier studies up to the year 2000 or later were carried out in rats. All reported measures showed a high rate of acinar cell division in the last week of gestation, with the rate falling around birth and then increasing postnatally to a peak index of 5–20 percent from 5 to 20 days, and then a decline starting around weaning to adult rates at 30–60 days.[44,46–49] In adult rats and mice on a nutritionally sound diet, the indices of proliferation are under 1 percent.[44,46,48,50] Lineage tracing has shown that acinar cell growth is due to self-duplication.[51] Similar to acinar cells, new adult islet beta cells also arise by self-duplication from existing beta cells.[52] In a study of thymidine incorporation by autoradiography in ICR mice, the authors estimated a half-life for adult acinar cells of 70 days, with 40 days for duct cells.[53]

One additional complication is that starting around day 20 in rats the number of bi-nucleate cells increases to near 50 percent at 50 days, and these cells synthesize little more DNA.[54] In a study in mice, polyploidy first began at 20 days and increased to 9 percent at 90 days.[55] Recently, most studies of pancreatic growth and regeneration are being carried out in mice, and immunostaining with antibody to Ki67 has become the most common way to show which cell types are proliferating. However, there is much less quantitative data over the life span in mice, compared to rats, and only a little in other species.

Adaptive Pancreatic Growth

Need for regulation of pancreatic size

As discussed earlier, the size of the pancreas and the spectrum of digestive enzymes that it produces are generally related to the animal's size and diet. The pancreas must provide sufficient enzymes for digestion, and for trypsin, a specific feedback system mediated by CCK is present to maintain the activity of trypsin in the intestinal lumen.[56] Thus, the conditions that induce adaptive pancreatic growth are generally related to changes in diet. An increase in food intake such as occurs in pregnancy, lactation, and cold exposure leads to an increase in pancreatic size in rodents.[57–60] Whether heavy exercise, which can double caloric intake, would affect pancreatic size would also be worthy of study. Also, there is little information as to whether pancreatic size adjusts to increased caloric consumption in humans.

The converse of increased food intake is starvation. Starvation is accompanied by loss of body and organ weights in all species examined, from fish to mammals.[61] While the pancreas has not always been evaluated, all organs in the GI tract decrease their weight during starvation, especially the liver and intestine. In rats starved for 4 or 7 days, pancreatic weight decreased slightly more than body weight.[62–64] Tissue weight and protein decreased more than DNA[65] consistent with cellular atrophy. Smaller acini with fewer ZG have also been observed by electron microscopy with starvation.[66–68] This was most prominent in phase 3 of starvation, when fat reserves have been utilized. Similar results have been reported in catfish[69] and sparrows.[70] A study in obese humans who were fasted for 20 days showed a decrease in pancreatic juice volume and enzyme content upon pancreatic function testing consistent with pancreatic atrophy, although pancreatic size was not measured.[71] An extreme case is seen in snakes such as pythons that feed on a large meal and then carry out an extended fast.[72] In this case, the involution of the GI tract occurs, while the animal has adequate nutrients but food is not entering the gut. That the latter is important is shown by the results of parenteral feeding in rats where the intestine and pancreas atrophy even though normal nutrition is maintained.[73,74] Total pancreatic weight and protein were decreased in total parental nutrition (TPN), while results for DNA were not clear. This effect of TPN has been ascribed to the loss of GI hormones that normally would be released in response to feeding as well as the absence of nutrients in the lumen. Both, exogenous CCK and bombesin have been shown to reduce the pancreatic atrophy induced by TPN in rats,[75,76] and the atrophy was shown to be due to the loss of luminal protein, as the combination of luminal fat, carbohydrate, and amino acids could not reverse the pancreatic atrophy induced by TPN.[77]

Another case of altered pancreatic function as a result of altered gut function is celiac disease. In celiac disease, ingestion of gluten, present in wheat and some other grains, triggers an immune attack on the duodenal mucosa which loses its villi, resulting in a decreased surface area leading to malabsorption. Twenty to thirty percent of patients showed reduced exocrine pancreatic function, and a portion of these show pancreatic insufficiency.[78–80] While the mechanism is not clear, the most favored explanation is a reduction of GI hormone secretion, especially CCK.[81] Other postulated mechanisms include reduced response to CCK, amino acid deficiency, and protein malnutrition. There is no evidence at present, however, that the pancreas is smaller in celiac disease.

Models for studying adaptive growth

Exogenous secretagogue stimulation

A large number of studies have shown that exogenous CCK can stimulate pancreatic growth. Because of the short half-life of CCK *in vivo*, CCK or its analogue caerulein have been given as two or three subcutaneous injections per day or by continuous infusion. Studies in rats, mice, and hamsters all show an increase in pancreatic size of 30–50 percent.[10,12,82–85] Secretin administration potentiates the increase in response to CCK in some but not all studies,[63,85–87] while somatostatin inhibits growth stimulated by CCK.[15] This secretagogue driven increase in pancreatic mass was accompanied by increased protein content, thymidine incorporation into acinar cells peaking at 2–3 days accompanied by an increase in pancreatic DNA content.[88] However, higher doses of CCK can induce apoptosis and a reduction in pancreatic weight.[89,90] The trophic action of CCK is mediated through CCK1 (CCK-A) receptors.[91]

The effect of exogenous administration of other secretagogues has also been studied. Bombesin, an analog of gastrin-releasing peptide,[92] when administered to rat, mouse, and rabbit induced an increase in pancreatic wet weight and protein with or without an increase in DNA.[93–96] Because bombesin can stimulate the release of CCK, several studies showed that the growth effect of bombesin was not blocked by a CCK antagonist which blocked the effect of exogenous CCK.[96,97] The effects of gastrin on pancreatic growth has been studied, but no consistent effect has emerged.[43,98,99] Secretin by itself has no effect but may potentiate the action of CCK. The effect of administration of cholinergic analogs such as bethanechol and carbachol has been studied and shown to induce modest pancreatic growth which is less than the response to CCK.[60,100] However, it is not clear whether this effect is direct on acinar cells or mediated by release of some other regulator.

Although having considerable differences from *in vivo* pancreatic growth, primary cultures of acinar and duct cells have shown that CCK can directly stimulate acinar cell growth. Caerulein, CCK, and gastrin stimulated

^{3}H-thymidine incorporation into monolayer cultures of mouse acinar cells on a collagen matrix (**Figure 2**), while carbachol, bombesin, and substance P had no effect.[101,102] The effect involved both high- and low-affinity CCK receptors in rat acini and could be inhibited by TGF-β.[103,104] CCK has also been shown to stimulate the growth of pancreatic cancer cell lines.[105] Growth stimulatory effects of IGF-1 on mouse cells cultured on a laminin-coated membrane have also been reported.[106] However, the acinar cells in these cultures appear to have dedifferentiated,[107] so they may be more of a model for recovery from damage than adaptive growth. Similar cultures of isolated duct cells have been studied, but only growth factors and not secretagogues stimulated their growth.[108]

Trypsin inhibitor feeding

It has been known for many years that chicks and rats fed raw soybean flour developed large pancreases with increased capacity to secrete digestive enzymes.[109–112] This was found due to the presence of active trypsin inhibitor (TI), and the effect could be reproduced by a variety of purified TIs[113,114] and in a variety of animal species, but with most subsequent work in rats.[111,115,116] Green and Lyman[117] showed that soybean trypsin inhibitor (SBTI) stimulated pancreatic secretion and presented evidence for a feedback mechanism in which active trypsin in the gut lumen inhibited pancreatic secretion. Thus, the feedback loop works to maintain an adequate level of trypsin and other proteases in the intestinal lumen. With the development of sensitive and specific CCK assays, it was shown that release of CCK is part of the feedback loop.[118,119] Diversion of bile pancreatic juice from the duodenum to the lower intestine also stimulates CCK release and pancreatic growth by removing active trypsin from the duodenum.[120–124] Further information and references on this feedback loop can be found in Liddle, RA (1995) and Logsdon, C (1999).[56,125]

The feeding of either natural or orally active chemical trypsin inhibitors such as camostat (also known as FOY-305) can be used as a model to stimulate CCK-driven pancreatic adaptive growth. The effects of TI are bigger than the response to exogenous CCK and will stimulate pancreatic growth when given two or three times a day[83,85,126] probably because of desensitization to large exogenous boluses of CCK. Rodent and hamster pancreases will double in size in about a week in response to feeding TI (**Figure 3**), and the growth consists of hypertrophy and hyperplasia with the relative amount dependent on species. Current evidence suggests islet morphology and insulin content of the pancreas are unaffected.[127] Co-administration of CCK antagonists blocks the effect of TI.[11,128] In rats the action of CCK was shown to be directly on acinar cells.[16] In mice the effect of TI is not seen following the genetic deletion of either

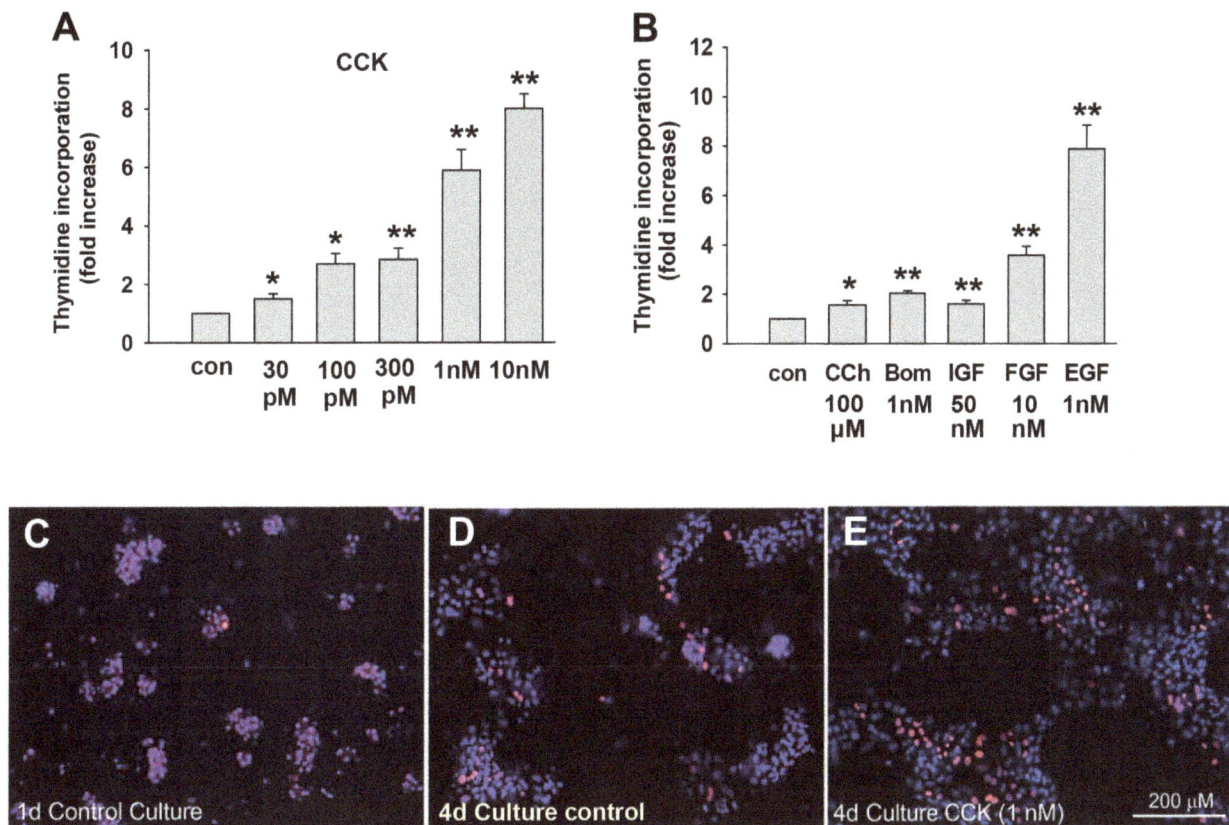

Figure 2. Stimulation of mitogenesis in pancreatic acinar cells in monolayer culture on a collagen substrate. Top panels show thymidine incorporation into DNA when cells are stimulated by different secretagogues and growth factors. Bottom panels (C–E) show BrdU incorporation visualized by immunofluorescence (from Guo et al.[107]).

CCK or the CCK receptor.[6,8,9] In both rats and mice the adaptive growth is reversible, and when the TI is removed the pancreas returns to its original size.[129–131] No dependence on secretin was seen in mice as TI had similar effects in mice where secretin signaling was genetically deleted.[7] The evidence as to whether such a feedback loop involving CCK occurs in humans is mixed and may involve proteases other than trypsin.[19,132,133]

High-protein diet

Consumption of a high-protein diet is well established to cause pancreatic growth, although most of the information is from rodents. Switching from a 5–10 percent casein diet to one containing 70–75 percent casein caused an increase in pancreatic weight, protein, and DNA content reaching a plateau in 7–14 days.[13,134,135] This was accompanied by a transient increase in plasma CCK, and the growth response could be partially blocked with a CCK antagonist.[134,135] Protein is a primary stimulus for CCK release in the rat, with casein being the most potent protein tested after trypsin inhibitor.[136] These experiments led to the concept

that protein stimulated pancreatic growth through CCK and that this growth involved both hypertrophy and hyperplasia of acinar cells and was accompanied by an increase in pancreatic protease content.

However, it has been demonstrated that feeding a diet high in amino acids to rats stimulated pancreatic growth without stimulating CCK release.[136,137] This growth was also not blocked by a CCK antagonist. More definitively, a CCK-deficient mouse fed a high-protein diet showed an increase in pancreatic weight and protein content and a smaller increase in DNA (**Figure 4**). That the growth in CCK-deficient mice was primarily due to cellular hypertrophy was shown by the increase in protein-to-DNA ratio, an increased acinar cell size, and no increase in DNA synthesis as assessed by BrdU incorporation.[5]

The converse of feeding a high-protein diet is feeding a protein-free or low-protein diet in the presence of adequate calories. Normal rodent chow contains 18–22 percent protein, most often casein, but what should be the normal amount is not always clear. Rat studies of high protein feeding have most often used 5 or 10 percent protein as the control or low protein to maximize the effect of the

Plasma CCK (pM)

Pancreatic Weight (mg/g)

Days on Camostat Diet

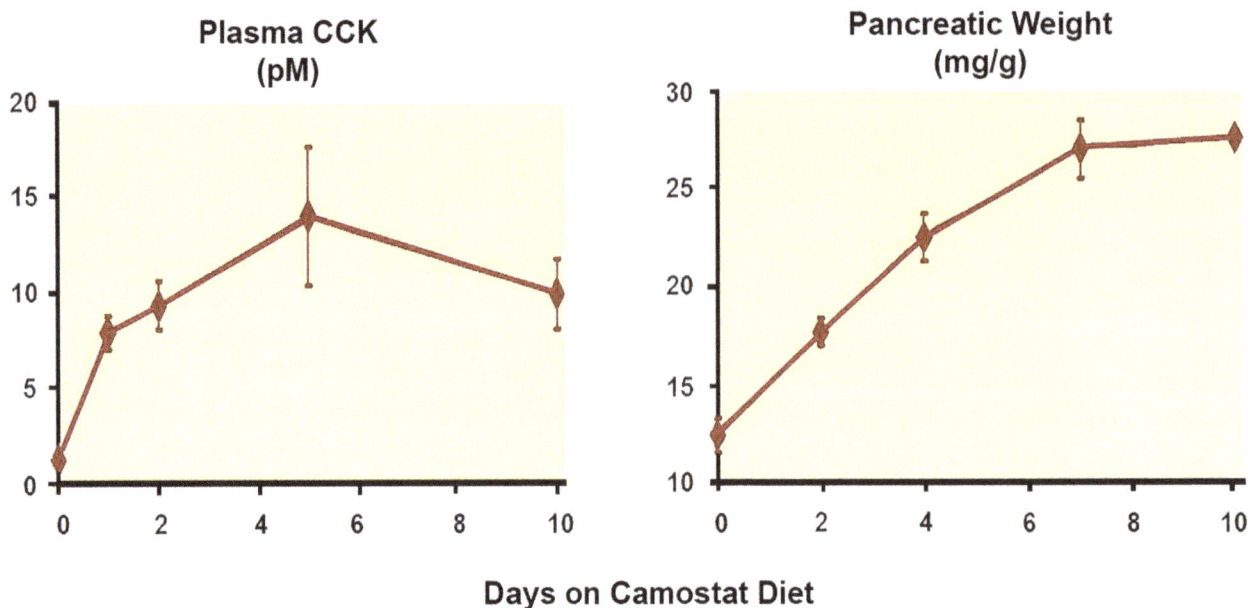

Figure 3. Adaptive growth of mouse pancreas and change in plasma CCK in response to feeding a chemical trypsin inhibitor, camostat. Data from Tashiro et al.[9]

high protein. In rats on 5 percent protein the body weight failed to increase over the 14-day experimental period and the pancreas weight decreased slightly.[135] In rats fed trypsin inhibitor or injected with caerulein to stimulate pancreatic growth, there was no pancreatic growth in animals on the 5 percent protein diet.[135,138] It has also been shown that gavage feeding of 9 percent casein increased plasma CCK but that 2 percent did not.[136] Thus, rodents on diets with less than 5 percent protein may be compromised due to shortage of amino acids and lack of secretion of the pancreatic trophic hormone CCK. Rats fed a protein-free diet lost pancreatic weight in excess of body weight, and there was an increased apoptosis of acinar cells.[139] A more detailed study has been carried out in mice where feeding protein-free diet was restricted to four days at which time there was a 11 percent decrease in body weight, no change in heart weight, but pancreas weight decreased by 24 percent.[140] Most strikingly, the protein content of the pancreas decreased almost 60 percent, while there was little fall in DNA (**Figure 5**).

These changes were reversible when protein was returned to the diet. Electron microscopy has shown decreased acinar size and a large decrease in the number of zymogen granules in mice, rats, rabbits, and monkeys subjected to protein deficiency.[140–143] The changes were reversible and interpreted as cellular atrophy. Small-sized acinar cells and failure of digestion have also been reported for children with Kwashiorkor.[144–146] This maldigestion was interpreted as a failure of the pancreas to secrete digestive enzymes.

Signaling Pathways Mediating Adaptive Growth

Pathways regulating cell size

The primary pathway regulating adaptive cell size growth is centered on the mechanistic target of rapamycin complex 1 (mTORC1).[147–149] This pathway was discovered in yeast where TOR1 and TOR2 were identified as the targets for the macrolide rapamycin which was isolated from a soil bacterium discovered on Rapa Nui (Easter Island) and found to inhibit growth. Mammals have one form of TOR named mTOR (m originally stood for mammalian but now for mechanistic), which forms two distinct complexes: mTORC1 being primarily related to growth and is sensitive to rapamycin, while mTORC2 regulates the actin cytoskeleton and is relatively insensitive to rapamycin. Each complex has a number of additional regulatory proteins. TORC1 contains six known proteins with the scaffolding protein Raptor, and the inhibitory protein proline-rich Akt substrate of 40 kDa (Pras40) being unique to mTORC1. Other subunits including DEP-domain-containing mTOR interacting protein (Deptor, a negative regulator of mTOR) and mammalian lethal with sec13 protein 8 (mLST8) and are shared with mTORC2 which also contains its own unique protein, Rictor. mTOR the catalytic subunit is a 280 kDa protein that belongs to the phosphoinositide 3-kinase-related kinase family and contains multiple domains, including one binding FKBP-12, the receptor for both FK506 and rapamycin. The mTORC1 pathway integrates input from growth factors, energy status, oxygen, and amino acids. When these are present, mTORC1 stimulates

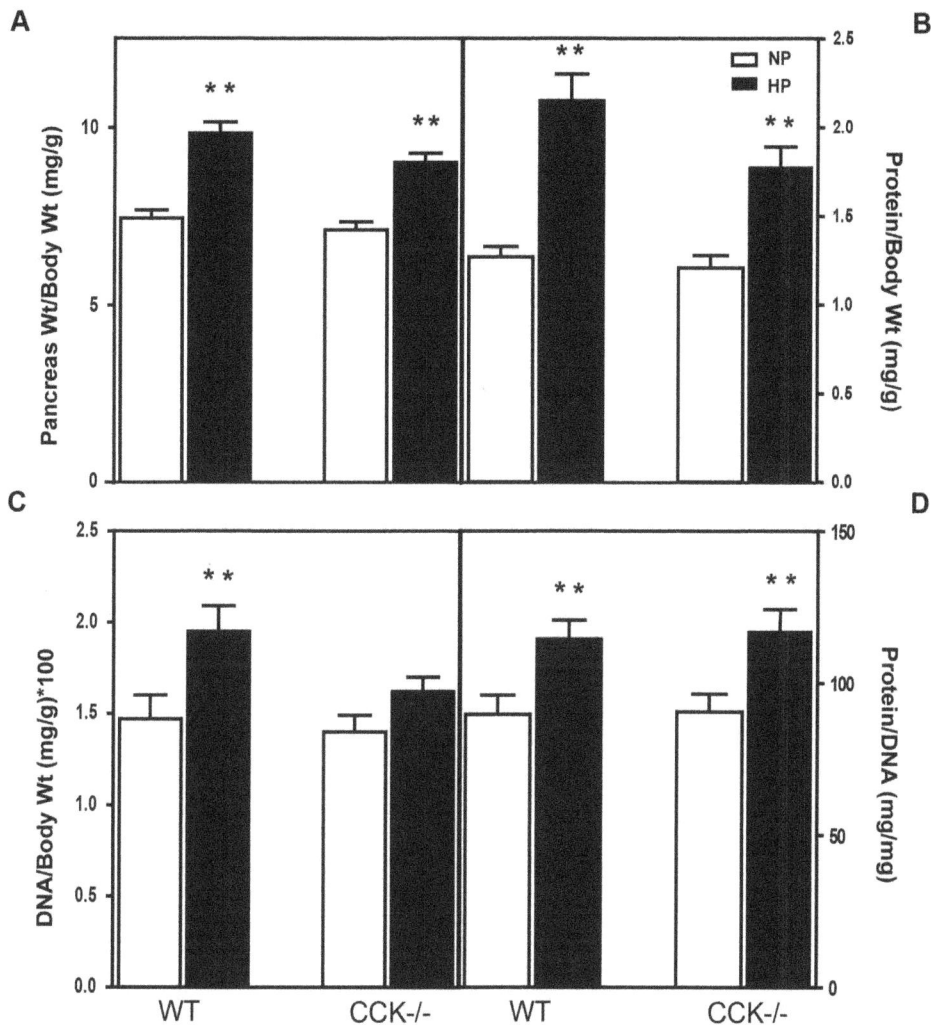

Figure 4. Effect of a high-protein diet on pancreatic size and composition in normal and CCK-deficient mice. Changes are shown for pancreas weight (**A**), protein (**B**), DNA (**C**), and protein-to-DNA ratio (**D**) after seven days on normal or high-protein diet (from Crozier et al.[5]).

protein synthesis, lipid synthesis, nucleotide synthesis, and cell cycle progression and inhibits autophagy.[147,150]

A simplified diagram of the pathway as known for pancreatic acinar cells is shown in **Figure 6**.

The CCK1 receptor, M3 muscarinic receptors and insulin, and other growth factor receptors all activate phosphatidyl inositol 3-kinase (PI3-K). PI3-K phosphorylates and activates Akt, also known as PKB, which phosphorylates and inactivates the tuberous sclerosis complex (TSC). In other cells hypoxia and low cellular energy status are known to activate TSC through REDD and AMPK, respectively. The TSC complex inhibits activation of the G protein Rheb, which when liganded by GTP activates TORC1.[151] This activation process in cells that have been studied involves migration to the surface of lysosomes, which is the site where amino acids, especially leucine and

arginine, activate TORC1 by a process involving a number of other proteins.[152,153] TORC1 in pancreatic acinar cells is known to be activated by branched chain amino acids.[154] Moreover, in acini incubated in amino acid–free media, basal TORC1 activity declines to near zero and cannot be activated by CCK (Crozier, S, and Williams, JA; unpublished data).

The best characterized substrates of TORC1 are ribosomal protein S6 kinase (S6K) 1 and initiation factor 4E binding protein 1 (4E-BP1). S6K1 phosphorylates a number of substrates including ribosomal small subunit 6 (S6), while phosphorylation of 4E-BP1 releases IF4E which binds the 5' cap on mRNA to initiate cap-dependent translation which accounts for about 90 percent of mRNA species. Both of these events are readily demonstrable in intact pancreas and pancreatic acini in response to CCK, growth

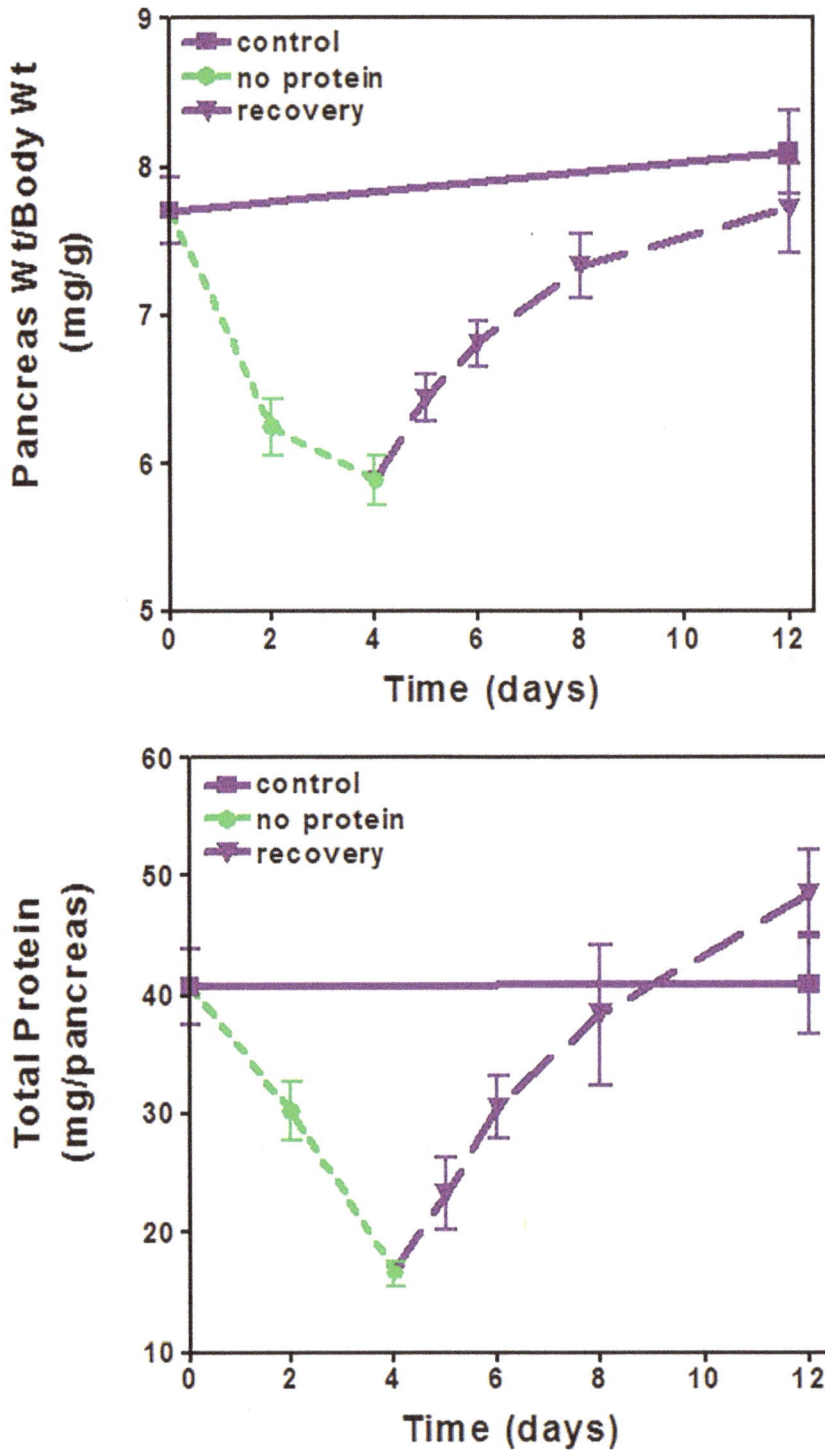

Figure 5. Feeding a protein-free diet induces pancreatic atrophy as shown by a decrease in pancreatic weight (**A**) and protein content (**B**) (from Crozier et al.[140]).

factors, and amino acids.[154–157] A third action in acini contributing to protein synthesis is the activation of elongation factor 2 (eEF2).[158]

The importance of TORC1 for adaptive pancreatic growth is shown by the fact that pancreatic growth stimulated by feeding TI is blocked by rapamycin[157] or by genetic deletion of raptor in acinar cells (Crozier, S, and Williams, JA; unpublished data), both of which selectively block the TORC1 pathway without affecting other pathways such as the activation of ERK and JNK leading to

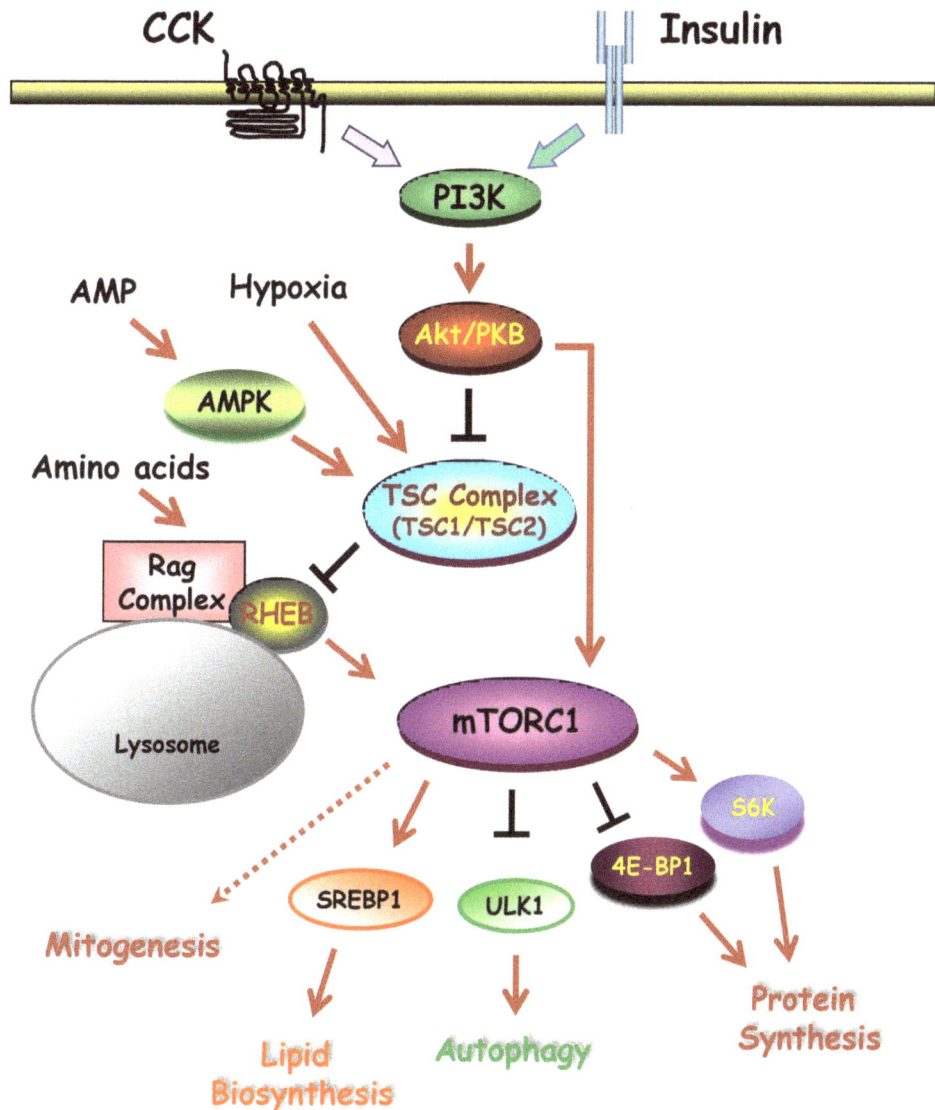

Figure 6. Schematic drawing of mTORC1 signaling pathway in pancreatic acinar cells. The major hormonal stimulants are CCK and insulin.

early response genes.[157] Interestingly, rapamycin blocks adaptive growth but has little effect on the baseline size of the pancreas. By contrast, feeding a protein-free diet blocks the TORC1 pathway and causes a reduction in pancreatic size due to cellular atrophy.[140] This difference may be due to the fact that some actions of TORC1 are known to be insensitive to rapamycin.[159] When raptor is deleted in acinar cells leading to a loss of TORC1 components, there is both an increase in apoptosis and a dedifferentiation of acinar cells. Thus, there is clearly more to learn of the importance of the TORC1 pathway in acinar cells. Although not presented in detail here, islet beta and alpha cells also are dependent on TORC1 to maintain their mass and demonstrate adaptive growth.[160]

Pathways regulating pancreatic mitogenesis

Because of the prominent role of CCK in inducing adaptive pancreatic growth, attention has focused on how CCK receptors activate acinar cell mitogenesis. Mature acinar cells appear to exist in a Go state, and mitogenic stimuli cause them to enter the cell cycle. Early response genes are a class of genes known to be rapidly upregulated following cellular stimulation and function as mediators coupling short-term signals to longer-term cellular responses such as growth by regulating other genes.[161] Initial work showed that CCK induced various early response genes in rat pancreas or isolated acini including c-Fos, c-Jun, Egr-1, and c-myc.[162,163] This led to enhanced activity of AP-1 transcription factor, which in its most common form is a Fos-Jun heterodimer.

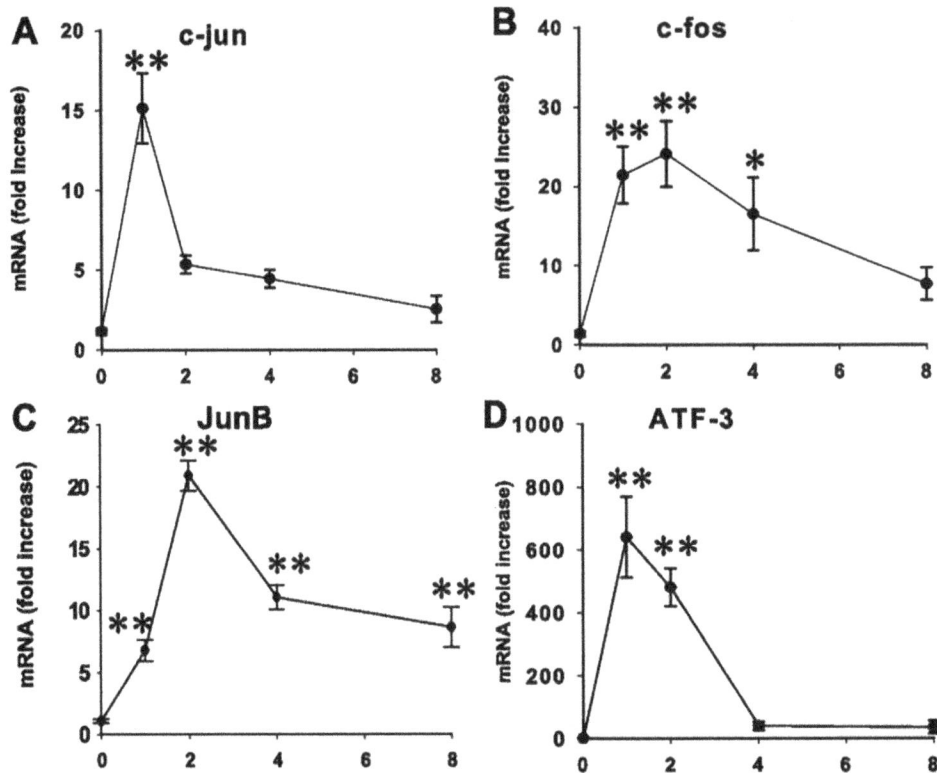

Figure 7. Expression of four early response genes peaking at one or two hours after refeeding chow containing trypsin inhibitor to mice after an overnight fast (from Guo et al.[164]).

A more complete study of the response of the mouse pancreas to feeding TI in chow evaluated mRNA for 18 early response genes, with 17 showing a rapid increase with a maximum effect seen at 1 to 4 hours followed by a decline (**Figure 7**).[164] Protein expression for c-Jun, c-Fos, ATF-3, Egr-1, and JunB peaked at two hours. Immunohistochemistry showed nuclear localization, and AP-1 activity was increased in nuclei. The effect of feeding TI was largely blocked by a CCK antagonist, and there was no effect of refeeding regular chow. That this is a direct effect of CCK was supported by similar findings in monolayer cultures of acinar cells.[107] In the *in vitro* study CCK also induced c-Jun, AP-1 activity, and cyclin D1 expression. Dominant negative JNK or shRNA blocked the c-Jun increase and prevented enhanced mitogenesis. In this culture system acinar cells dedifferentiate before mitogenesis, so it is not a perfect model but shows that CCK can directly stimulate mitogenesis as shown earlier by Logsdon.[101,102] For further information on signaling pathways activated by CCK in pancreatic acinar cells, see the review by Williams.[165]

Calcineurin-NFAT pathway

CCK is known to activate a number of signaling pathways, but a central role has been assigned to the activation of intracellular Ca^{2+} signaling for the stimulation of digestive enzyme secretion. A sustained increase in Ca^{2+} is known to activate the Ca^{2+}-regulated phosphatase calcineurin $(CN)^{94}$ in a variety of cells. Utilizing the pharmacological inhibitors cyclosporine A (CsA) and FK-506, Tashiro et al. showed that active CN was necessary for TI-stimulated pancreatic growth that occurs by activation of mitogenesis.[9] A major target of CN are NFATs (nuclear factor of activated T cells). In the basal state NFATs are phosphorylated and remain in the cytoplasm. Upon stimulation, which increases intracellular Ca^{2+}, activated CN dephosphorylates NFAT which move into the nucleus and either alone or in partnership with AP-1 activates specific gene expression (**Figure 8**). In isolated pancreatic acini, CCK stimulates NFAT translocation into nuclei with the effect dependent on CN.[166] An NFAT reporter was also used to show increased NFAT activity. Similar activation of NFATs with nuclear translocation was seen *in vivo* in response to feeding TI.

Further evidence for the importance of CN came from the discovery that Rcan1 (Regulator of calcineurin 1), also known as DSCR1, was highly induced by feeding TI, with the induction almost totally inhibitable by FK-506. Rcan1 blocked CN action in cell models, and when induced in pancreatic acinar cells *in vivo* using a LSL expression

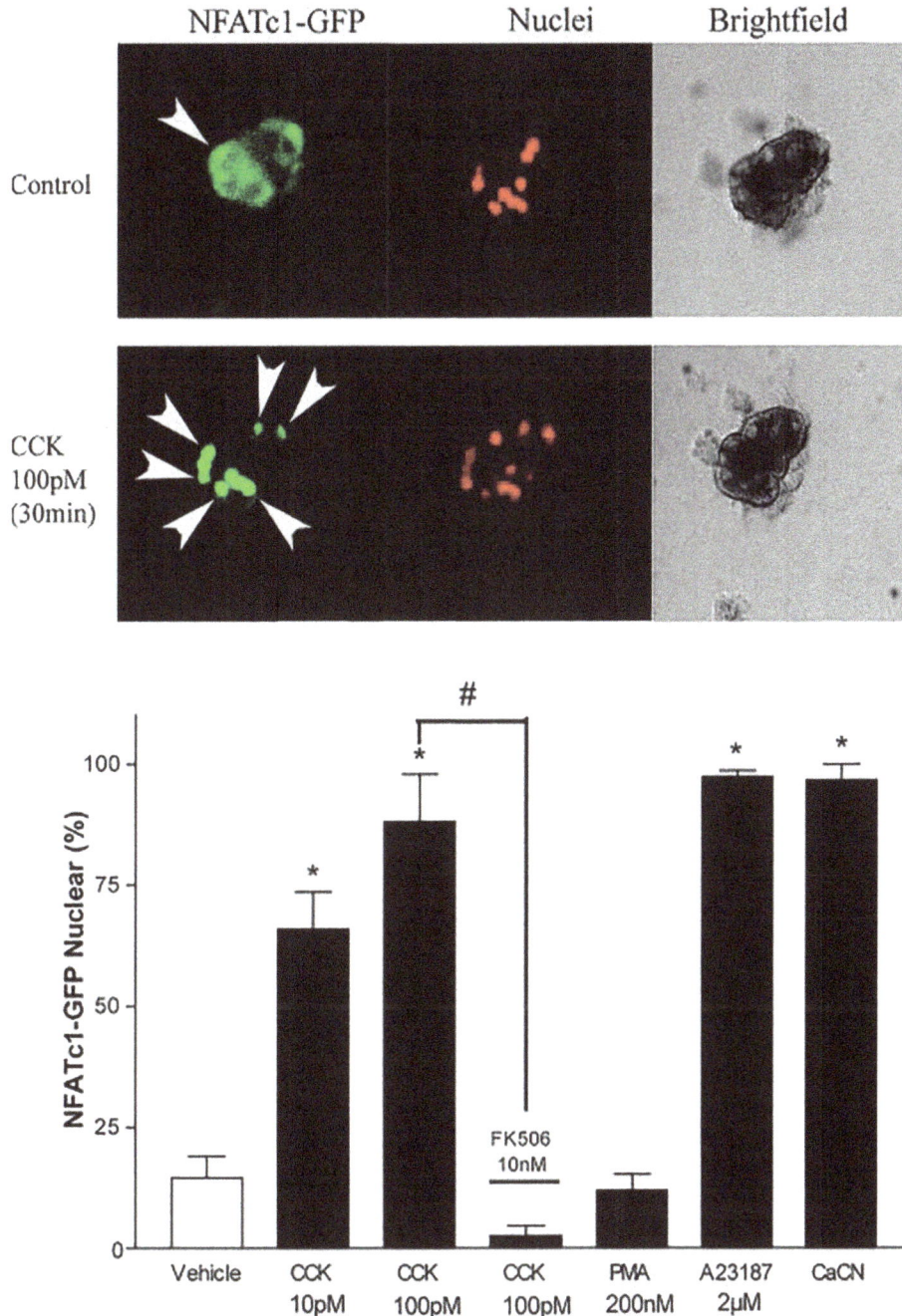

Figure 8. CCK stimulates nuclear translocation of NFATc1-GFP in mouse pancreatic acini (from Gurda et al.[166]).

of and Rcan1 construct activated by Ela-Cre, it completely blocked adaptive growth in response to feeding TI (**Figure 9**).[167] Thus, both pharmacological and genetic inhibition of CN blocked adaptive growth. Rcan1 functions as an endogenous feedback inhibitor of CN action. Though it is reasonable to presume that adaptive growth is mediated by NFATs, other CN substrates could also participate.

RNA microarray analysis of pancreatic genes revealed 38 genes induced by TI two hours after feeding that were at least 70 percent inhibited by FK-506.[167] These changes were not observed in CCK-deficient mice. The 38 genes included feedback inhibitors (Rcan1, Socs3, Rgs2, Socs 2, Lif), growth factors (FGF21, HBEGF), and transcriptional regulators. About 60 percent had predicted NFAT

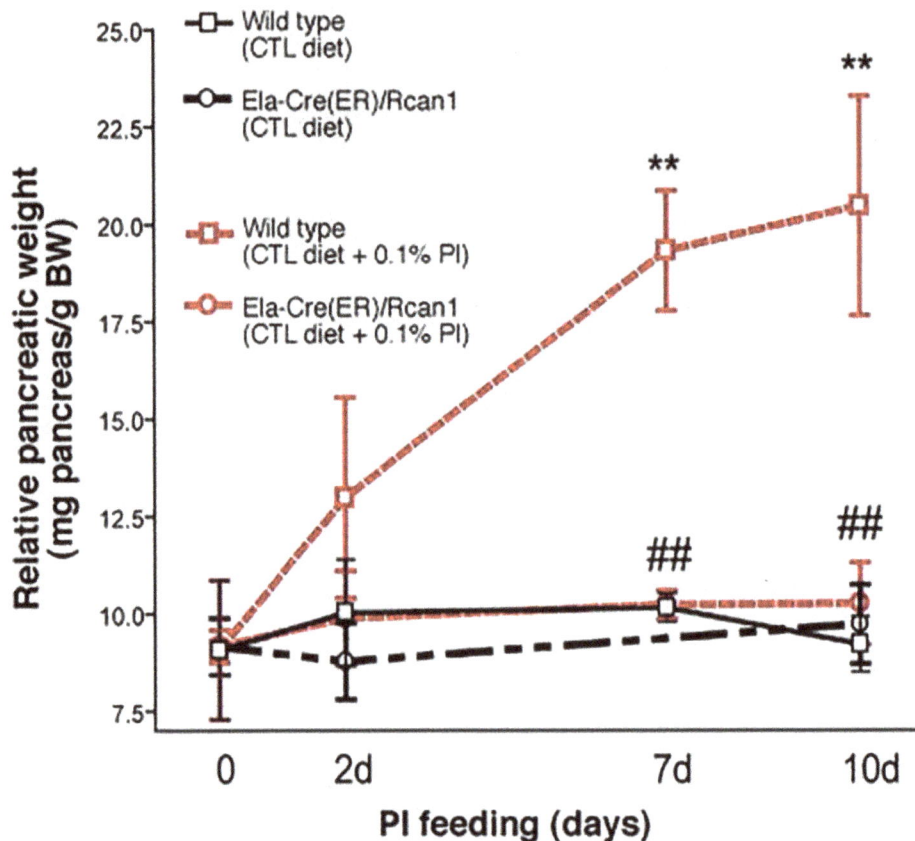

Figure 9. Overexpression of Rcan1 in acinar cells blocks TI-stimulated pancreatic growth (from Gurda et al.[167]).

regulation, and CHIP analysis confirmed this for five genes in pancreatic 266–6 acinar cells stimulated with calcium ionophore.[167]

TORC1 Pathway

The TORC1 pathway, discussed earlier as playing a role in the increased cell size seen in response to high-protein feeding, is also necessary for adaptive growth involving cell proliferation. This has been shown through the use of rapamycin, which blocks both mitogenesis and the increase in pancreatic weight, protein, and DNA in response to TI feeding.[157] Genetically, deletion of raptor, an essential component of TORC1, also blocks adaptive growth (Crozier, S, and Williams, JA; unpublished data). Similarly, mTORC1 is required for islet beta cell growth.[168] In yeast, rapamycin blocks cell division because mitogenesis cannot occur without the increase in cell protein required to convert one cell into two. This does not appear to be the case in mammalian cells, and attention has focused instead on actions of TORC1 on cell cycle regulators such as Cyclin D1 and CDK4 which are sensitive to rapamycin.

MAPK Pathways

CCK and other pancreatic secretagogues activate multiple MAPK pathways including ERK, JNK, and p38 MAPK.[169] Of these, both the ERK and JNK pathways are also activated by feeding TI.[170] Interestingly, the activation of JNK was blocked by FK-506 pretreatment, while activation of ERK, which occurred rapidly with peak activation at 30 min, was unaffected. A later study also showed a more prolonged activation of ERK.[171] JNK activation is known to be important for mitogenesis in cultured acinar cells[107] but has not been studied in normal animals. A requirement for ERK for adaptive growth *in vivo* in mice has been shown using new ERK inhibitors developed for *in vivo* administration as potential cancer therapeutic agents (**Figure 10**).

PD-0325901 and GSK-11202012 (Trametenib) can be mixed in chow or gavage fed and inhibited by binding to the catalytic site of MEK, thereby blocking ERK with low toxicity. The MEK inhibitors blocked pancreatic ERK signaling, adaptive growth, and the increased DNA synthesis stimulated by feeding TI to mice.[171] Interestingly, the MEK inhibitor blocked induction of c-Fos but had no effect on

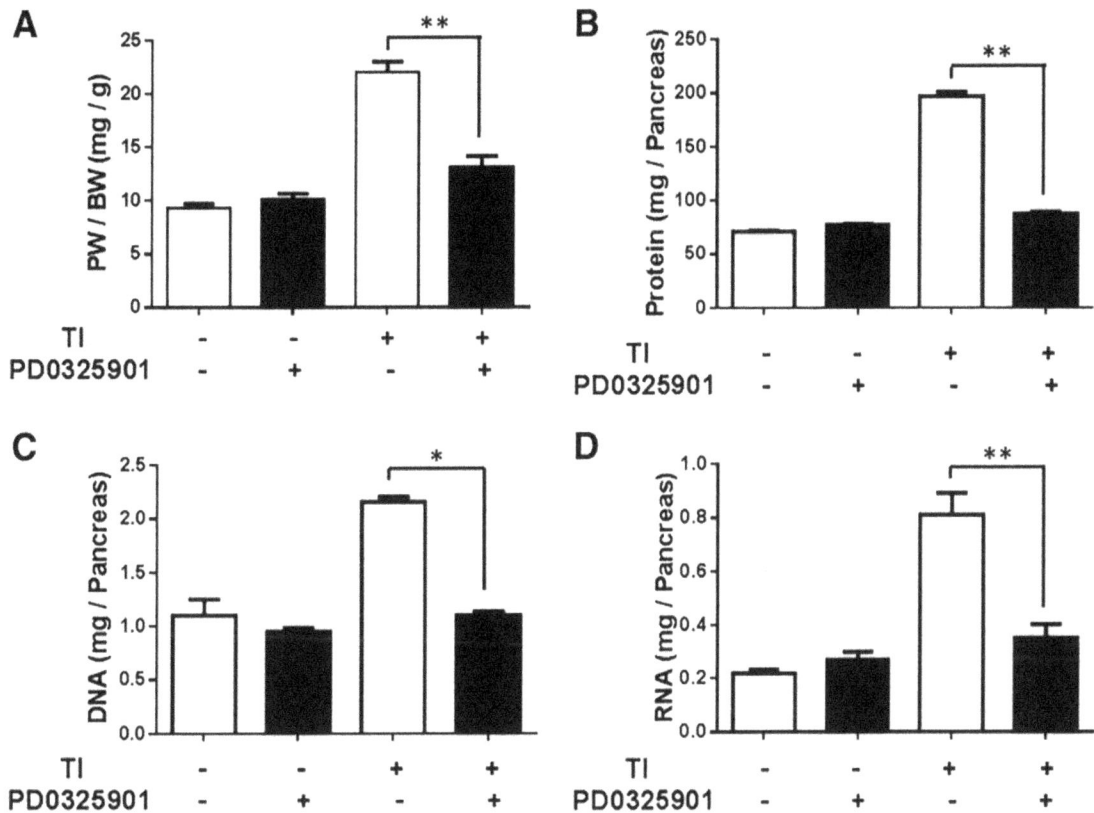

Figure 10. Inhibition of ERK blocks pancreatic growth. Mice were fed TI for five days after gavaging with the ERK inhibitor PD0325901 (from Holtz et al.[171]).

c-Jun induction. It also blocked the expression of cyclins D1, D3, E, and A along with PCNA.[171] The ERK pathway has previously been shown to play a proliferative role in multiple other cells and tissues,[172] although its mechanistic target is not clearly established.

Stat3

Bioinformatics analysis of 318 genes which were upregulated in pancreas from 1 to 8 hours after feeding TI suggested involvement of Jak-STAT signaling.[173] An increase in phospho STAT3 showed a strong nuclear occupancy peaking at 75 percent of acinar cell nuclei at 2 h. Western blotting showed a large increase in p-STAT3 without a change in total STAT3. Socs2, which acts as a negative feedback regulator of STAT3, is a transcriptional target of STAT3, and its mRNA was found to also be enhanced. Interestingly, in isolated acini CCK fails to activate Jak-STAT signaling while growth factors can. Since growth factors such as HB-EGF are induced by TI feeding, it is possible that CCK does not work directly but rather through a growth factor whose receptor can activate Jak-Stat signaling. At present, there is no direct evidence that STAT3 is involved in adaptive pancreatic growth. However, it

seems likely that it will be involved as STAT3 is involved in the malignant transformation of acinar cells.[174,175] Also, STAT5 and Socs3 have been shown to regulate growth of pancreatic islet beta cells.[176]

Overall model of mitogenesis

It seems clear that acinar cell mitogenesis is under different and more complex control than is secretion of digestive enzymes. Blockage of three different pathways—the calcineurin-NFAT, the TORC1, and the ERK pathways—blocks adaptive growth, indicating that all must be activated for mitogenesis to occur (**Figure 11**). It is not clear whether all three converge on a single rate-limiting step or that multiple key steps exist and each is activated by a different pathway. It appears that the induction of multiple cyclins is blocked by inhibiting each pathway, so the regulated step is most likely connected with moving cells out of Go and initiating the cell cycle. In addition to completing the understanding of how acinar cell mitogenesis is initiated, work is needed on whether and how there is a parallel increase in ducts and blood vessels as the pancreas grows. Similar studies have also yielded an incomplete map for regulation of islet beta cell growth.[177,178]

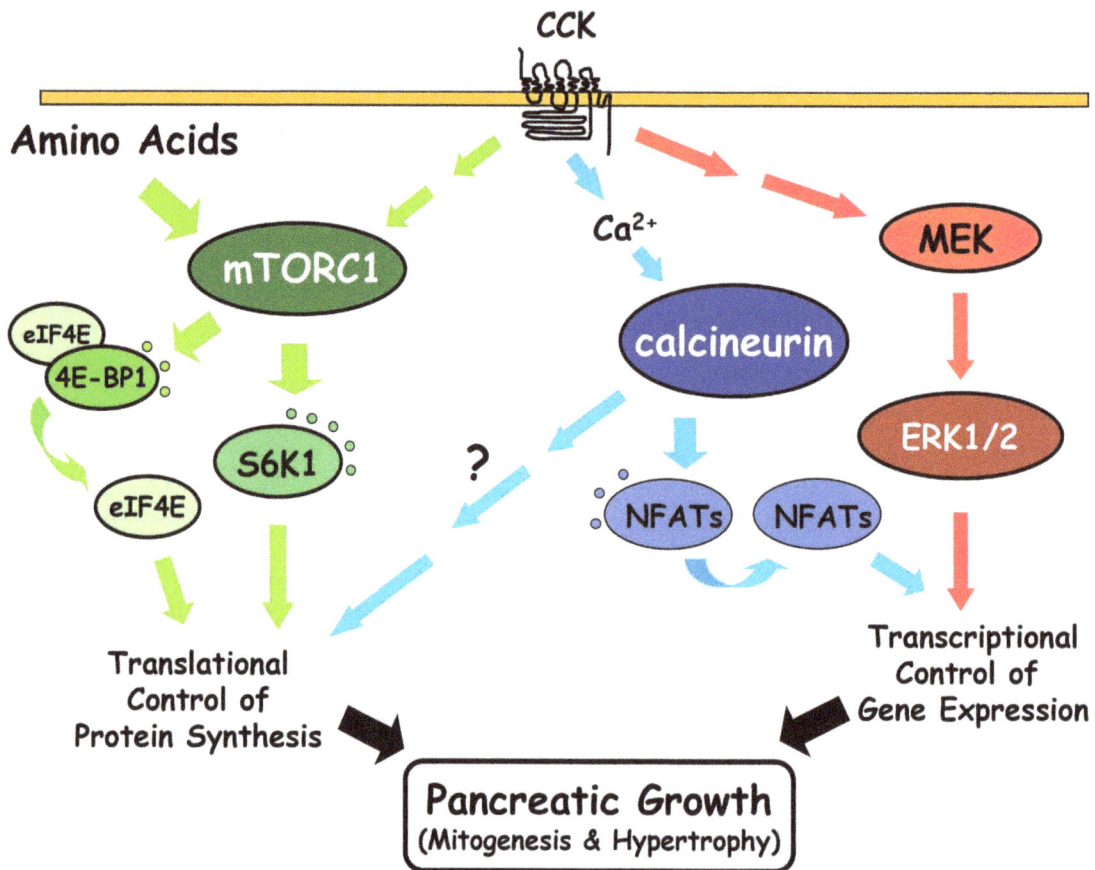

Figure 11. Three CCK-activated signaling pathways (mTORC1, calcineurin-NFAT, and the ERK pathways) required for pancreatic acinar cell growth by proliferation of acinar cells.

References

1. Stanger BZ. Organ size determination and the limits of regulation. Cell Cycle 2008; 7:318–324.
2. Penzo-Mendez AI, Stanger BZ. Organ-Size Regulation in Mammals. Cold Spring Harb Perspect Biol 2015; 7:a019240.
3. Conlon I, Raff M. Size control in animal development. Cell 1999; 96:235–244.
4. Lloyd AC. The regulation of cell size. Cell 2013; 154:1194–1205.
5. Crozier SJ, Sans MD, Wang JY, Lentz SI, Ernst SA, Williams JA. CCK-independent mTORC1 activation during dietary protein-induced exocrine pancreas growth. Am J Physiol Gastrointest Liver Physiol 2010; 299:G1154–G1163.
6. Lacourse KA, Swanberg LJ, Gillespie PJ, Rehfeld JF, Saunders TL, Samuelson LC. Pancreatic function in CCK-deficient mice: adaptation to dietary protein does not require CCK. Am J Physiol Gastrointest Liver Physiol 1999; 276:G1302–G1309.
7. Sans MD, Sabbatini ME, Ernst SA, D'Alecy LG, Nishijima I, Williams JA. Secretin is not necessary for exocrine pancreatic development and growth in mice. Am J Physiol Gastrointest Liver Physiol 2011; 301:G791–G798.
8. Sato N, Suzuki S, Kanai S, Ohta M, Jimi A, Noda T, Takiguchi S, Funakoshi A, Miyasaka K. Different effects of oral administration of synthetic trypsin inhibitor on the pancreas between cholecystokinin-A receptor gene knockout mice and wild type mice. Jpn J Pharmacol 2002; 89:290–295.
9. Tashiro M, Samuelson LC, Liddle RA, Williams JA. Calcineurin mediates pancreatic growth in protease inhibitor-treated mice. Am J Physiol Gastrointest Liver Physiol 2004; 286:G784–G790.
10. Satake K, Mukai R, Kato Y, Umeyama K. Effects of cerulein on the normal pancreas and on experimental pancreatic carcinoma in the Syrian golden hamster. Pancreas 1986; 1:246–253.
11. Douglas BR, Woutersen RA, Jansen JB, Rovati LC, Lamers CB. Comparison of the effect of lorglumide on pancreatic growth stimulated by camostate in rat and hamster. Life Sci 1990; 46:281–286.
12. Folsch UR, Winckler K, Wormsley KG. Influence of repeated administration of cholecystokinin and secretin on the pancreas of the rat. Scand J Gastroenterol 1978; 13:663–671.
13. Green GM, Jurkowska G, Berube FL, Rivard N, Guan D, Morisset J. Role of cholecystokinin in induction and maintenance of dietary protein-stimulated pancreatic growth. Am J Physiol Gastrointest Liver Physiol 1992; 262:G740–G746.

14. Iovanna JL, Dusetti N, Cadenas B, Calvo EL. Changes in growth and pancreatic mRNA concentrations during postnatal development of rat pancreas. Pancreas 1990; 5:421–426.

15. Sarfati PD, Genik P, Morisset J. Caerulein and secretin induced pancreatic growth: a possible control by endogenous pancreatic somatostatin. Regul Pept 1985; 11:261–273.

16. Yamamoto M, Otani M, Jia DM, Fukumitsu K, Yoshikawa H, Akiyama T, Otsuki M. Differential mechanism and site of action of CCK on the pancreatic secretion and growth in rats. Am J Physiol Gastrointest Liver Physiol 2003; 285:G681–G687.

17. Caglar V K B, Uygur R, Alkoc OA, Ozen OA and Demirel H. Study of Volume, Weight and Size of Normal Pancreas, Spleen and Kidney in Adults Autopsies.. Forensic Medicine and Anatomy Research 2014; 2:63–69.

18. Schaefer J. The normal weight of the pancreas in the adult human being: A biometric study. The Anatomical Record 32: 119–132 1926.

19. Liddle RA, Toskes PP, Horrow J, Ghali J, Dachman A, Stong D. Lack of trophic pancreatic effects in humans with long-term administration of ximelagatran. Pancreas 2006; 32:205–210.

20. Nakamura Y, Higuchi S, Maruyama K. Pancreatic volume associated with endocrine and exocrine function of the pancreas among Japanese alcoholics. Pancreatology 2005; 5:422–431.

21. Saisho Y, Butler AE, Meier JJ, Monchamp T, Allen-Auerbach M, Rizza RA, Butler PC. Pancreas volumes in humans from birth to age one hundred taking into account sex, obesity, and presence of type-2 diabetes. Clin Anat 2007; 20:933–942.

22. Sato T, Ito K, Tamada T, Sone T, Noda Y, Higaki A, Kanki A, Tanimoto D, Higashi H. Age-related changes in normal adult pancreas: MR imaging evaluation. Eur J Radiol 2012; 81:2093–2098.

23. Szczepaniak EW, Malliaras K, Nelson MD, Szczepaniak LS. Measurement of pancreatic volume by abdominal MRI: a validation study. PLoS One 2013; 8:e55991.

24. Paredes JL, Orabi AI, Ahmad T, Benbourenane I, Tobita K, Tadros S, Bae KT, Husain SZ. A non-invasive method of quantifying pancreatic volume in mice using micro-MRI. PLoS One 2014; 9:e92263.

25. Gaglia JL, Guimaraes AR, Harisinghani M, Turvey SE, Jackson R, Benoist C, Mathis D, Weissleder R. Noninvasive imaging of pancreatic islet inflammation in type 1A diabetes patients. J Clin Invest 2011; 121:442–445.

26. Lim S, Bae JH, Chun EJ, Kim H, Kim SY, Kim KM, Choi SH, Park KS, Florez JC, Jang HC. Differences in pancreatic volume, fat content, and fat density measured by multidetector-row computed tomography according to the duration of diabetes. Acta Diabetol 2014; 51:739–748.

27. Macauley M, Percival K, Thelwall PE, Hollingsworth KG, Taylor R. Altered volume, morphology and composition of the pancreas in type 2 diabetes. PLoS One 2015; 10:e0126825.

28. Williams AJ, Chau W, Callaway MP, Dayan CM. Magnetic resonance imaging: a reliable method for measuring pancreatic volume in Type 1 diabetes. Diabet Med 2007; 24:35–40.

29. Grippo PJ, Venkatasubramanian PN, Knop RH, Heiferman DM, Iordanescu G, Melstrom LG, Adrian K, Barron MR, Bentrem DJ, Wyrwicz AM. Visualization of mouse pancreas architecture using MR microscopy. Am J Pathol 2011; 179:610–618.

30. Kim SK, Hebrok M. Intercellular signals regulating pancreas development and function. Genes Dev 2001; 15:111–127.

31. Stanger BZ, Tanaka AJ, Melton DA. Organ size is limited by the number of embryonic progenitor cells in the pancreas but not the liver. Nature 2007; 445:886–891.

32. Snook JT. Effect of diet on development of exocrine pancreas of the neonatal rat. Am J Physiol 1971; 221:1388–1391.

33. Yu FX, Zhao B, Guan KL. Hippo pathway in organ size control, tissue homeostasis, and cancer. Cell 2015; 163:811–828.

34. Camargo FD, Gokhale S, Johnnidis JB, Fu D, Bell GW, Jaenisch R, Brummelkamp TR. YAP1 increases organ size and expands undifferentiated progenitor cells. Curr Biol 2007; 17:2054–2060.

35. Yu FX, Meng Z, Plouffe SW, Guan KL. Hippo pathway regulation of gastrointestinal tissues. Annu Rev Physiol 2015; 77:201–227.

36. George NM, Day CE, Boerner BP, Johnson RL, Sarvetnick NE. Hippo signaling regulates pancreas development through inactivation of Yap. Mol Cell Biol 2012; 32:5116–5128.

37. Patel SH, Camargo FD, Yimlamai D. Hippo Signaling in the Liver Regulates Organ Size, Cell Fate, and Carcinogenesis. Gastroenterology 2017; 152:533–545.

38. Gao T, Zhou D, Yang C, Singh T, Penzo-Mendez A, Maddipati R, Tzatsos A, Bardeesy N, Avruch J, Stanger BZ. Hippo signaling regulates differentiation and maintenance in the exocrine pancreas. Gastroenterology 2013; 144:1543–1553.

39. Gu L, Liu J, Xu D, Lu Y. Reciprocal feedback loop of the MALAT1-MicroRNA-194-YAP1 pathway regulates progression of acute pancreatitis. Med Sci Monit 2019; 25:6894–6904.

40. Ardestani A, Maedler K. The Hippo signaling pathway in pancreatic beta-cells: functions and regulations. Endocr Rev 2018; 39:21–35.

41. Snook JT. Effect of age and long-term diet on exocrine pancreas of the rat. Am J Physiol 1975; 228:262–268.

42. Zucker KA, Adrian TE, Bilchik AJ, Modlin IM. Effects of the CCK receptor antagonist L364,718 on pancreatic growth in adult and developing animals. Am J Physiol Gastrointest Liver Physiol 1989; 257:G511–G516.

43. Morisset J. Hormonal control of pancreatic growth during fetal, neonatal and adult life. Adv Med Sci 2008; 53:99–118.

44. Elsasser HP, Biederbick A, Kern HF. Growth of rat pancreatic acinar cells quantitated with a monoclonal antibody against the proliferating cell nuclear antigen. Cell Tissue Res 1994; 276:603–609.

45. Anzi S, Stolovich-Rain M, Klochendler A, Fridlich O, Helman A, Paz-Sonnenfeld A, Avni-Magen N, Kaufman E, Ginzberg MB, Snider D, Ray S, Brecht M, et al. Postnatal exocrine pancreas growth by cellular hypertrophy correlates with a shorter lifespan in mammals. Dev Cell 2018; 45:726–737.

46. Elsässer HP, Adler G, Kern HF. Replication and Regeneration of the Pancreas. In: Go VLW, DiMagno EP, Gardner JD, Lebenthal E, Reber HA, Scheele GA, eds. The Pancreas: Biology, Pathobiology, and Disease. 2nd ed. New York: Raven Press Ltd., 1993:75–86.

47. Muller R, Laucke R, Trimper B, Cossel L. Pancreatic cell proliferation in normal rats studied by in vivo autoradiography with 3H-thymidine. Virchows Arch B Cell Pathol Incl Mol Pathol 1990; 59:133–136.

48. Sesso A, Abrahamsohn PA, Tsanaclis A. Acinar cell proliferation in the rat pancreas during early postnatal growth. Acta Physiol Lat Am 1973; 23:37–50.

49. Wenzel G, Stocker E, Heine WD. [Cell proliferation in pancreatic acinar epithelia of rats. Autoradiographic studies with 3 H-thymidine]. Virchows Arch B Cell Pathol 1972; 10:118–126.

50. Messier B, Leblond CP. Cell proliferation and migration as revealed by radioautography after injection of thymidine-H3 into male rats and mice. Am J Anat 1960; 106:247–285.

51. Houbracken I, Bouwens L. Acinar cells in the neonatal pancreas grow by self-duplication and not by neogenesis from duct cells. Sci Rep 2017; 7:12643.

52. Dor Y, Brown J, Martinez OI, Melton DA. Adult pancreatic beta-cells are formed by self-duplication rather than stem-cell differentiation. Nature 2004; 429:41–46.

53. Magami Y, Azuma T, Inokuchi H, Moriyasu F, Kawai K, Hattori T. Heterogeneous cell renewal of pancreas in mice: [(3)H]-thymidine autoradiographic investigation. Pancreas 2002; 24:153–160.

54. Oates PS, Morgan RG. Cell proliferation in the exocrine pancreas during development. J Anat 1989; 167:235–241.

55. Webb SR, Dore BA, Grogan WM. Cell cycle analysis of the postnatal mouse pancreas. Biol Neonate 1982; 42:73–78.

56. Liddle RA. Regulation of cholecystokinin secretion by intraluminal releasing factors. Am J Physiol Gastrointest Liver Physiol 1995; 269:G319–G327.

57. Harada E, Kanno T. Progressive enhancement in the secretory functions of the digestive system of the rat in the course of cold acclimation. J Physiol 1976; 260:629–645.

58. Kim YS, Kim WJ, Kim HK, Hong SS. Effect of cold and hot environments on the exocrine pancreas of rats. Yonsei Med J 1970; 11:1–9.

59. McLaughlin CL, Baile CA, Peikin SR. Hyperphagia during lactation: satiety response to CCK and growth of the pancreas. Am J Physiol Endocrinol Metab 1983; 244:E61–E65.

60. Morisset J, Jolicoeur L, Caussignac Y, Solomon TE. Trophic effects of chronic bethanechol on pancreas, stomach, and duodenum in rats. Can J Physiol Pharmacol 1982; 60:871–876.

61. McCue MD. Starvation physiology: reviewing the different strategies animals use to survive a common challenge. Comp Biochem Physiol A Mol Integr Physiol 2010; 156:1–18.

62. Mainz DL, Parks NM, Webster PD, 3rd. Effect of fasting and refeeding on pancreatic DNA synthesis and content. Proc Soc Exp Biol Med 1977; 156:340–344.

63. Nagy I, Pap A, Varro V. Time-course of changes in pancreatic size and enzyme composition in rats during starvation. Int J Pancreatol 1989; 5:35–45.

64. Webster PD, Singh M, Tucker PC, Black O. Effects of fasting and feeding on the pancreas. Gastroenterology 1972; 62:600–605.

65. Morisset JA, Webster PD. Effects of fasting and feeding on protein synthesis by the rat pancreas. J Clin Invest 1972; 51:1–8.

66. Kitagawa T, Ono K. Ultrastructure of pancreatic exocrine cells of the rat during starvation. Histol Histopathol 1986; 1:49–57.

67. Nevalainen TJ, Janigan DT. Degeneration of mouse pancreatic acinar cells during fasting. Virchows Arch B Cell Pathol 1974; 15:107–118.

68. Weiss JM. The ergastoplasm; its fine structure and relation to protein synthesis as studied with the electron microscope in the pancreas of the Swiss albino mouse. J Exp Med 1953; 98:607–618.

69. Zeng LQ, Li FJ, Li XM, Cao ZD, Fu SJ, Zhang YG. The effects of starvation on digestive tract function and structure in juvenile southern catfish (Silurus meridionalis Chen). Comp Biochem Physiol A Mol Integr Physiol 2012; 162:200–211.

70. Chediack JG, Funes SC, Cid FD, Filippa V, Caviedes-Vidal E. Effect of fasting on the structure and function of the gastrointestinal tract of house sparrows (Passer domesticus). Comp Biochem Physiol A Mol Integr Physiol 2012; 163:103–110.

71. Folsch UR, Dreessen UW, Talaulicar M, Willms B, Creutzfeldt W. Effect of long-term fasting of obese patients on pancreatic exocrine function, gastrointestinal hormones and bicarbonate concentration in plasma. Z Gastroenterol 1984; 22:357–364.

72. Secor SM, Diamond J. A vertebrate model of extreme physiological regulation. Nature 1998; 395:659–662.

73. Baumler MD, Nelson DW, Ney DM, Groblewski GE. Loss of exocrine pancreatic stimulation during parenteral feeding suppresses digestive enzyme expression and induces Hsp70 expression. Am J Physiol Gastrointest Liver Physiol 2007; 292:G857–G866.

74. Fan BG, Salehi A, Sternby B, Axelson J, Lundquist I, Andren-Sandberg A, Ekelund M. Total parenteral nutrition influences both endocrine and exocrine function of rat pancreas. Pancreas 1997; 15:147–153.

75. Fan BG, Axelson J, Sternby B, Rehfeld J, Ihse I, Ekelund M. Total parenteral nutrition affects the tropic effect of cholecystokinin on the exocrine pancreas. Scand J Gastroenterol 1997; 32:380–386.

76. Pierre JF, Neuman JC, Brill AL, Brar HK, Thompson MF, Cadena MT, Connors KM, Busch RA, Heneghan AF, Cham CM, Jones EK, Kibbe CR, et al. The gastrin-releasing peptide analog bombesin preserves exocrine and endocrine pancreas morphology and function during parenteral nutrition. Am J Physiol Gastrointest Liver Physiol 2015; 309:G431–G442.

77. Baumler MD, Koopmann MC, Thomas DD, Ney DM, Groblewski GE. Intravenous or luminal amino acids are insufficient to maintain pancreatic growth and digestive enzyme expression in the absence of intact dietary protein. Am J Physiol Gastrointest Liver Physiol 2010; 299:G338–G347.

78. Evans KE, Leeds JS, Morley S, Sanders DS. Pancreatic insufficiency in adult celiac disease: do patients require long-term enzyme supplementation? Dig Dis Sci 2010; 55:2999–3004.

79. Freeman HJ. Pancreatic endocrine and exocrine changes in celiac disease. World J Gastroenterol 2007; 13:6344–6346.

80. Regan PT, DiMagno EP. Exocrine pancreatic insufficiency in celiac sprue: a cause of treatment failure. Gastroenterology 1980; 78:484–487.

81. Low-Beer TS, Harvey RF, Davies ER, Read AF. Abnormalities of serum cholecystokinin and gallbladder emptying in celiac disease. N Engl J Med 1975; 292:961–963.

82. Arnesjo, II, Lundquist I. Effects on exocrine and endocrine rat pancreas of long-term administration of CCK-PZ (cholecystokinin-pancreozymin) or synthetic octapeptide-CCK-PZ. Scand J Gastroenterol 1976; 11:529–535.

83. Niederau C, Liddle RA, Williams JA, Grendell JH. Pancreatic growth: interaction of exogenous cholecystokinin, a protease inhibitor, and a cholecystokinin receptor antagonist in mice. Gut 1987; 28 Suppl:63–69.

84. Pfeiffer CJ, Chernenko GA, Kohli Y, Barrowman JA. Trophic effects of cholecystokinin octapeptide on the pancreas of the Syrian hamster. Can J Physiol Pharmacol 1982; 60:358–362.

85. Solomon TE, Petersen H, Elashoff J, Grossman MI. Interaction of caerulein and secretin on pancreatic size and composition in rat. Am J Physiol Endocrinol Metab 1978; 235:E714–E719.

86. Haarstad H, Petersen H. The effects of graded doses of a cholecystokinin-like peptide with and without secretin on pancreatic growth and synthesis of RNA and polyamines in rats. Scand J Gastroenterol 1989; 24:907–915.

87. Nagy I, Takacs T, Mohacsi G. One-year treatment with cholecystokinin-octapeptide and secretin: effects on pancreatic trophism in the rat. Pharmacol Res 1997; 36:77–85.

88. Solomon TE, Vanier M, Morisset J. Cell site and time course of DNA synthesis in pancreas after caerulein and secretin. Am J Physiol Gastrointest Liver Physiol 1983; 245:G99–G105.

89. Ohlsson B, Borg K, Rehfeld JF, Axelson J, Sundler F. The method of administration of cholecystokinin determines the effects evoked in the pancreas. Pancreas 2001; 23:94–101.

90. Trulsson LM, Svanvik J, Permert J, Gasslander T. Cholecystokinin octapeptide induces both proliferation and apoptosis in the rat pancreas. Regul Pept 2001; 98:41–48.

91. Povoski SP, Zhou W, Longnecker DS, Jensen RT, Mantey SA, Bell RH, Jr. Stimulation of in vivo pancreatic growth in the rat is mediated specifically by way of cholecystokinin-A receptors. Gastroenterology 1994; 107:1135–1146.

92. Williams JA. Bombesin. Pancreapedia: Exocrine Pancreas Knowledge Base, 2015. DOI: 10.3998/panc.2015.18.

93. Damge C, Hajri A, Lhoste E, Aprahamian M. Comparative effect of chronic bombesin, gastrin-releasing peptide and caerulein on the rat pancreas. Regul Pept 1988; 20:141–150.

94. Karkashan EM, MacNaughton WK, Gall DG. Evidence of a physiological role for bombesin in the postnatal development of the rabbit pancreas. Biol Neonate 1992; 62:395–401.

95. Lhoste EF, Aprahamian M, Balboni G, Damge C. Evidence for a direct trophic effect of bombesin on the mouse pancreas: in vivo and cell culture studies. Regul Pept 1989; 24:45–54.

96. Schmidt WE, Choudhury AR, Siegel EG, Loser C, Conlon JM, Folsch UR, Creutzfeldt W. CCK-antagonist L-364,718: influence on rat pancreatic growth induced by caerulein and bombesin-like peptides. Regul Pept 1989; 24:67–79.

97. Borysewicz R, Ren KJ, Mokotoff M, Lee PC. Direct effect of bombesin on pancreatic and gastric growth in suckling rats. Regul Pept 1992; 41:157–169.

98. Niederau C, Niederau M, Klonowski H, Luthen R, Ferrell LD. Effect of hypergastrinemia and blockade of gastrin-receptors

99. on pancreatic growth in the mouse. Hepatogastroenterology 1995; 42:423–431.

99. Solomon TE, Morisset J, Wood JG, Bussjaeger LJ. Additive interaction of pentagastrin and secretin on pancreatic growth in rats. Gastroenterology 1987; 92:429–435.

100. Carling RC, Templeton D. The effect of carbachol and isoprenaline on cell division in the exocrine pancreas of the rat. Q J Exp Physiol 1982; 67:577–585.

101. Logsdon CD. Stimulation of pancreatic acinar cell growth by CCK, epidermal growth factor, and insulin in vitro. Am J Physiol Gastrointest Liver Physiol 1986; 251:G487–G494.

102. Logsdon CD, Williams JA. Pancreatic acinar cells in monolayer culture: direct trophic effects of caerulein in vitro. Am J Physiol Gastrointest Liver Physiol 1986; 250:G440–G447.

103. Hoshi H, Logsdon CD. Both low- and high-affinity CCK receptor states mediate trophic effects on rat pancreatic acinar cells. Am J Physiol Gastrointest Liver Physiol 1993; 265:G1177–G1181.

104. Logsdon CD, Keyes L, Beauchamp RD. Transforming growth factor-beta (TGF-beta 1) inhibits pancreatic acinar cell growth. Am J Physiol Gastrointest Liver Physiol 1992; 262:G364–G368.

105. Smith JP, Kramer ST, Solomon TE. CCK stimulates growth of six human pancreatic cancer cell lines in serum-free medium. Regul Pept 1991; 32:341–349.

106. Watanabe H, Saito H, Rychahou PG, Uchida T, Evers BM. Aging is associated with decreased pancreatic acinar cell regeneration and phosphatidylinositol 3-kinase/Akt activation. Gastroenterology 2005; 128:1391–1404.

107. Guo L, Sans MD, Hou Y, Ernst SA, Williams JA. c-Jun/AP-1 is required for CCK-induced pancreatic acinar cell dedifferentiation and DNA synthesis in vitro. Am J Physiol Gastrointest Liver Physiol 2012; 302:G1381–G1396.

108. Bhattacharyya E, Panchal A, Wilkins TJ, de Ondarza J, Hootman SR. Insulin, transforming growth factors, and substrates modulate growth of guinea pig pancreatic duct cells in vitro. Gastroenterology 1995; 109:944–952.

109. Lyman RL, Lepkovsky S. The effect of raw soybean meal and trypsin inhibitor diets on pancreatic enzyme secretion in the rat. J Nutr 1957; 62:269–284.

110. Chernick SS, Lepkovsky S, Chaikoff IL. A dietary factor regulating the enzyme content of the pancreas; changes induced in size and proteolytic activity of the chick pancreas by the ingestion of raw soy-bean meal. Am J Physiol 1948; 155:33–41.

111. Oates PS, Morgan RG. Pancreatic growth and cell turnover in the rat fed raw soya flour. Am J Pathol 1982; 108:217–224.

112. Booth AN, Robbins DJ, Ribelin WE, Deeds F, Smith AK, Rackis JJ. Prolonged Pancreatic Hypertrophy and Reversibility in Rats Fed Raw Soybean Meal. Proc Soc Exp Biol Med 1964; 116:1067–1069.

113. Melmed RN, Bouchier IA. A further physiological role for naturally occurring trypsin inhibitors: the evidence for a trophic stimulant of the pancreatic acinar cell. Gut 1969; 10:973–979.

114. Richter BD, Schneeman BO. Pancreatic response to long-term feeding of soy protein isolate, casein or egg white in rats. J Nutr 1987; 117:247–252.

115. Yen JT, Jensen AH, Simon J. Effect of dietary raw soybean and soybean trypsin inhibitor on trypsin and chymotrypsin

activities in the pancreas and in small intestinal juice of growing swine. J Nutr 1977; 107:156–165.

116. Hasdai A, Liener IE. Growth, digestibility and enzymatic activities in the pancreas and intestines of hamsters fed raw and heated soy flour. J Nutr 1983; 113:662–668.

117. Green GM, Lyman RL. Feedback regulation of pancreatic enzyme secretion as a mechanism for trypsin inhibitor-induced hypersecretion in rats. Proc Soc Exp Biol Med 1972; 140:6–12.

118. Goke B, Printz H, Koop I, Rausch U, Richter G, Arnold R, Adler G. Endogenous CCK release and pancreatic growth in rats after feeding a proteinase inhibitor (camostate). Pancreas 1986; 1:509–515.

119. Liddle RA, Goldfine ID, Williams JA. Bioassay of plasma cholecystokinin in rats: effects of food, trypsin inhibitor, and alcohol. Gastroenterology 1984; 87:542–549.

120. Rivard N, Guan D, Maouyo D, Grondin G, Berube FL, Morisset J. Endogenous cholecystokinin release responsible for pancreatic growth observed after pancreatic juice diversion. Endocrinology 1991; 129:2867–2874.

121. Chu M, Borch K, Lilja I, Blomqvist L, Rehfeld JF, Ihse I. Endogenous hypercholecystokininemia model in the hamster: trophic effect on the exocrine pancreas. Pancreas 1992; 7:220–225.

122. Bamba T, Ishizuka Y, Hosoda S. Effect of intrinsic CCK and CCK antagonist on pancreatic growth and pancreatic enzyme secretion in pancreaticobiliary diversion rats. Dig Dis Sci 1993; 38:653–659.

123. Gasslander T, Axelson J, Hakanson R, Ihse I, Lilja I, Rehfeld JF. Cholecystokinin is responsible for growth of the pancreas after pancreaticobiliary diversion in rats. Scand J Gastroenterol 1990; 25:1060–1065.

124. Watanapa P, Efa EF, Beardshall K, Calam J, Sarraf CE, Alison MR, Williamson RC. Inhibitory effect of a cholecystokinin antagonist on the proliferative response of the pancreas to pancreatobiliary diversion. Gut 1991; 32:1049–1054.

125. Logsdon C. Role of Cholecystokinin in Physiologic and Pathophysiologic Growth of the Pancreas. In: Greeley G, ed. Gastrointestinal endocrinology. Totowa, NJ: Humana Press, 1999:393–422.

126. Dembinski AB, Johnson LR. Stimulation of pancreatic growth by secretin, caerulein, and pentagastrin. Endocrinology 1980; 106:323–328.

127. Muller MK, Goebell H, Alfen R, Ehlers J, Jager M, Plumpe H. Effects of camostat, a synthetic protease inhibitor, on endocrine and exocrine pancreas of the rat. J Nutr 1988; 118:645–650.

128. Wisner JR, Jr., McLaughlin RE, Rich KA, Ozawa S, Renner IG. Effects of L-364,718, a new cholecystokinin receptor antagonist, on camostate-induced growth of the rat pancreas. Gastroenterology 1988; 94:109–113.

129. Oates PS, Morgan RG, Light AM. Cell death (apoptosis) during pancreatic involution after raw soya flour feeding in the rat. Am J Physiol Gastrointest Liver Physiol 1986; 250:G9–G14.

130. Crozier SJ, Sans MD, Lang CH, D'Alecy LG, Ernst SA, Williams JA. CCK-induced pancreatic growth is not limited by mitogenic capacity in mice. Am J Physiol Gastrointest Liver Physiol 2008; 294:G1148–G1157.

131. Oates PS, Morgan RG. Random or selective cell death during pancreatic involution following withdrawal of raw soya flour feeding in the rat. Pathology 1986; 18:234–236.

132. Liener IE, Goodale RL, Deshmukh A, Satterberg TL, Ward G, DiPietro CM, Bankey PE, Borner JW. Effect of a trypsin inhibitor from soybeans (Bowman-Birk) on the secretory activity of the human pancreas. Gastroenterology 1988; 94:419–427.

133. Friess H, Kleeff J, Isenmann R, Malfertheiner P, Buchler MW. Adaptation of the human pancreas to inhibition of luminal proteolytic activity. Gastroenterology 1998; 115:388–96.

134. Morisset J, Guan D, Jurkowska G, Rivard N, Green GM. Endogenous cholecystokinin, the major factor responsible for dietary protein-induced pancreatic growth. Pancreas 1992; 7:522–529.

135. Green GM, Levan VH, Liddle RA. Plasma cholecystokinin and pancreatic growth during adaptation to dietary protein. Am J Physiol Gastrointest Liver Physiol 1986; 251:G70–G74.

136. Liddle RA, Green GM, Conrad CK, Williams JA. Proteins but not amino acids, carbohydrates, or fats stimulate cholecystokinin secretion in the rat. Am J Physiol Gastrointest Liver Physiol 1986; 251:G243–G248.

137. Hara H, Narakino H, Kiriyama S, Kasai T. Induction of pancreatic growth and proteases by feeding a high amino acid diet does not depend on cholecystokinin in rats. J Nutr 1995; 125:1143–1149.

138. Green GM, Sarfati PD, Morisset J. Lack of effect of cerulein on pancreatic growth of rats fed a low-protein diet. Pancreas 1991; 6:182–189.

139. Fitzgerald PJ, Herman L, Carol B,Roque A, Marsh WH, Rosenstock L, et al. Pancreatic acinar cell regeneration part 1 cytologic, cytochemical, and pancreatic weight changes. Am J Pathol 52: 983–1012.

140. Crozier SJ, D'Alecy LG, Ernst SA, Ginsburg LE, Williams JA. Molecular mechanisms of pancreatic dysfunction induced by protein malnutrition. Gastroenterology 2009; 137:1093–1101.

141. Lazarus SS, Volk BW. Electron microscopy and histochemistry of rabbit pancreas in protein malnutrition (experimental kwashiorkor). Am J Pathol 1964; 44:95–111.

142. Racela AS, Jr., Grady HJ, Higginson J, Svoboda DJ. Protein deficiency in rhesus monkeys. Am J Pathol 1966; 49:419–443.

143. Weisblum B, Herman L, Fitzgerald PJ. Changes in pancreatic acinar cells during protein deprivation. J Cell Biol 1962; 12:313–327.

144. Williams CD. A nutritional disease of childhood associated with a maize diet. Arch Dis Child 1933; 8:423–433.

145. Blackburn WR, Vinijchaikul K. The pancreas in kwashiorkor. An electron microscopic study. Lab Invest 1969; 20:305–318.

146. Pitchumoni CS. Pancreas in primary malnutrition disorders. Am J Clin Nutr 1973; 26:374–379.

147. Laplante M, Sabatini DM. mTOR signaling in growth control and disease. Cell 2012; 149:274–293.

148. Fingar DC, Blenis J. Target of rapamycin (TOR): an integrator of nutrient and growth factor signals and coordinator of cell growth and cell cycle progression. Oncogene 2004; 23:3151–3171.

149. Wullschleger S, Loewith R, Hall MN. TOR signaling in growth and metabolism. Cell 2006; 124:471–484.

150. Sengupta S, Peterson TR, Sabatini DM. Regulation of the mTOR complex 1 pathway by nutrients, growth factors, and stress. Mol Cell 2010; 40:310–322.

151. Li Y, Corradetti MN, Inoki K, Guan KL. TSC2: filling the GAP in the mTOR signaling pathway. Trends Biochem Sci 2004; 29:32–38.

152. Bar-Peled L, Sabatini DM. Regulation of mTORC1 by amino acids. Trends Cell Biol 2014; 24:400–406.

153. Chantranupong L, Scaria SM, Saxton RA, Gygi MP, Shen K, Wyant GA, Wang T, Harper JW, Gygi SP, Sabatini DM. The CASTOR proteins are arginine sensors for the mTORC1 pathway. Cell 2016; 165:153–164.

154. Sans MD, Tashiro M, Vogel NL, Kimball SR, D'Alecy LG, Williams JA. Leucine activates pancreatic translational machinery in rats and mice through mTOR independently of CCK and insulin. J Nutr 2006; 136:1792–1799.

155. Bragado MJ, Groblewski GE, Williams JA. Regulation of protein synthesis by cholecystokinin in rat pancreatic acini involves PHAS-I and the p70 S6 kinase pathway. Gastroenterology 1998; 115:733–742.

156. Bragado MJ, Tashiro M, Williams JA. Regulation of the initiation of pancreatic digestive enzyme protein synthesis by cholecystokinin in rat pancreas in vivo. Gastroenterology 2000; 119:1731–1739.

157. Crozier SJ, Sans MD, Guo L, D'Alecy LG, Williams JA. Activation of the mTOR signalling pathway is required for pancreatic growth in protease-inhibitor-fed mice. J Physiol 2006; 573:775–786.

158. Sans MD, Xie Q, Williams JA. Regulation of translation elongation and phosphorylation of eEF2 in rat pancreatic acini. Biochem Biophys Res Commun 2004; 319:144–151.

159. Kang SA, Pacold ME, Cervantes CL, Lim D, Lou HJ, Ottina K, Gray NS, Turk BE, Yaffe MB, Sabatini DM. mTORC1 phosphorylation sites encode their sensitivity to starvation and rapamycin. Science 2013; 341:1236566.

160. Xie J, El Sayed NM, Qi C, Zhao X, Moore CE, Herbert TP. Exendin-4 stimulates islet cell replication via the IGF1 receptor activation of mTORC1/S6K1. J Mol Endocrinol 2014; 53:105–115.

161. Rozengurt E. Early signals in the mitogenic response. Science 1986; 234:161–166.

162. Lu L, Logsdon CD. CCK, bombesin, and carbachol stimulate c-fos, c-jun, and c-myc oncogene expression in rat pancreatic acini. Am J Physiol Gastrointest Liver Physiol 1992; 263:G327–G332.

163. Ji B, Chen XQ, Misek DE, Kuick R, Hanash S, Ernst S, Najarian R, Logsdon CD. Pancreatic gene expression during the initiation of acute pancreatitis: identification of EGR-1 as a key regulator. Physiol Genomics 2003; 14:59–72.

164. Guo L, Sans MD, Gurda GT, Lee SH, Ernst SA, Williams JA. Induction of early response genes in trypsin inhibitor-induced pancreatic growth. Am J Physiol Gastrointest Liver Physiol 2007; 292:G667–G677.

165. Williams JA. Cholecystokinin (CCK) regulation of pancreatic acinar cells: physiological actions and signal transduction mechanisms. Compr Physiol 2019; 9:535–564.

166. Gurda GT, Guo L, Lee SH, Molkentin JD, Williams JA. Cholecystokinin activates pancreatic calcineurin-NFAT signaling in vitro and in vivo. Mol Biol Cell 2008; 19:198–206.

167. Gurda GT, Crozier SJ, Ji B, Ernst SA, Logsdon CD, Rothermel BA, Williams JA. Regulator of calcineurin 1 controls growth plasticity of adult pancreas. Gastroenterology 2010; 139:609–619.

168. Blandino-Rosano M, Chen AY, Scheys JO, Alejandro EU, Gould AP, Taranukha T, Elghazi L, Cras-Meneur C, Bernal-Mizrachi E. mTORC1 signaling and regulation of pancreatic beta-cell mass. Cell Cycle 2012; 11:1892–1902.

169. Williams JA. Intracellular signaling mechanisms activated by cholecystokinin-regulating synthesis and secretion of digestive enzymes in pancreatic acinar cells. Annu Rev Physiol 2001; 63:77–97.

170. Tashiro M, Dabrowski A, Guo L, Sans MD, Williams JA. Calcineurin-dependent and calcineurin-independent signal transduction pathways activated as part of pancreatic growth. Pancreas 2006; 32:314–320.

171. Holtz BJ, Lodewyk KB, Sebolt-Leopold JS, Ernst SA, Williams JA. ERK activation is required for CCK-mediated pancreatic adaptive growth in mice. Am J Physiol Gastrointest Liver Physiol 2014; 307:G700–G710.

172. Meloche S, Pouyssegur J. The ERK1/2 mitogen-activated protein kinase pathway as a master regulator of the G1- to S-phase transition. Oncogene 2007; 26:3227–3239.

173. Gurda GT, Wang JY, Guo L, Ernst SA, Williams JA. Profiling CCK-mediated pancreatic growth: the dynamic genetic program and the role of STATs as potential regulators. Physiol Genomics 2012; 44:14–24.

174. Nagathihalli NS, Castellanos JA, Shi C, Beesetty Y, Reyzer ML, Caprioli R, Chen X, Walsh AJ, Skala MC, Moses HL, Merchant NB. Signal Transducer and Activator of Transcription 3, Mediated Remodeling of the Tumor Microenvironment Results in Enhanced Tumor Drug Delivery in a Mouse Model of Pancreatic Cancer. Gastroenterology 2015; 149:1932–1943.

175. Gruber R, Panayiotou R, Nye E, Spencer-Dene B, Stamp G, Behrens A. YAP1 and TAZ Control Pancreatic Cancer Initiation in Mice by Direct Up-regulation of JAK-STAT3 Signaling. Gastroenterology 2016; 151:526–539.

176. Lindberg K, Ronn SG, Tornehave D, Richter H, Hansen JA, Romer J, Jackerott M, Billestrup N. Regulation of pancreatic beta-cell mass and proliferation by SOCS-3. J Mol Endocrinol 2005; 35:231–243.

177. Kulkarni RN, Mizrachi EB, Ocana AG, Stewart AF. Human beta-cell proliferation and intracellular signaling: driving in the dark without a road map. Diabetes 2012; 61:2205–2213.

178. Stewart AF, Hussain MA, Garcia-Ocana A, Vasavada RC, Bhushan A, Bernal-Mizrachi E, Kulkarni RN. Human beta-cell proliferation and intracellular signaling: part 3. Diabetes 2015; 64:1872–1885.

Chapter 23

Pancreatic regeneration: Models, mechanisms, and inconsistencies

Mairobys Socorro[1] and Farzad Esni[1,2,3,4]

[1]*Department of Surgery, Division of Pediatric General and Thoracic Surgery, Children's Hospital of Pittsburgh, University of Pittsburgh Medical Center, Pittsburgh, PA, USA*

[2]*Department of Developmental Biology, University of Pittsburgh, Pittsburgh, PA, USA*

[3]*Department of Microbiology & Molecular Genetics, University of Pittsburgh, Pittsburgh, PA, USA*

[4]*University of Pittsburgh Cancer Institute, Pittsburgh, PA, USA*

Introduction

While skin, liver, and gut have the ability to regenerate and heal, other organs such as heart and brain do not display similar regenerative capacities. The adult pancreas displays a limited capacity to regenerate, although this regenerative capacity declines with age.[1–5] Thus, with respect to the pancreas, the disagreement is not so much about the overall ability of the adult pancreas to regenerate but rather about which cells may act as cell(s) of origin in this process. For example, it is widely accepted that under physiologic conditions β-cell regeneration in the adult mouse pancreas originates from β-cell self-duplication.[6,7] However, depending on the type of injury model, it appears that new β-cells can arise from cells either residing within the ducts,[8–12] in proximity to the ductal network,[13] or from other pancreatic endocrine cells.[14–17] This uncertainty regarding the types of cells that may potentially give rise to new β-cells comes in part from the fact that in each experimental model of regeneration, the exact target cells and the severity of the injury are different. Here, we will first review some of the injury models that have been used to study the mechanisms leading to replacement of acinar- or β-cells, followed by a discussion of the discrepancies in these reports.

Injury Models to Study Pancreatic Regeneration

Over the years multiple models of pancreatic injury have been used by different investigators to explore the regenerative capacity of exocrine or endocrine compartments of the adult pancreas.[18] Among these models, some of the commonly used are pancreatic duct ligation, partial pancreatectomy, caerulein-induced pancreatitis, alloxan- or streptozotocin-induced diabetes, and diphtheria toxin-mediated cell ablation. These models are of varying specificity, and they entail surgical, chemical, or genetic methods.

Pancreatectomy

Pancreatectomy (Px) is the oldest model with which to examine the regenerative capacity of the pancreas.[19] The first documented removal of the pancreas was performed on dogs by Johann C. Brunner in 1683.[20] However, it was first in 1890s that Px was reported to result in diabetes, and hence a link between the pancreas and glucose homeostasis was established.[19,20] Px can be used to study acinar and β-cell recovery in both rats and mice; however, because of the increased islet mass, this injury model has been extensively used to study β-cell regeneration.[3,5,21–27] Partial pancreatectomy (PPx) involves resection of less than 90 percent (often 50–75%) of the adult mouse or rat pancreas.[5,6,23,28–33] Here, the remnant of the pancreas displays normal gross morphology. A more severe form of Px that has been performed on rats is subtotal pancreatectomy (SPx) and entails removal of 90–95 percent of the gland through tissue abrasion.[25,28,34–36] Thus, in contrast to PPx, here the acinar cells in the remnant of the pancreas undergo rapid atrophy, which results in a desmoplastic reaction.[25,37] Regardless of the extent of resection, Px is associated with an IGF/PI3K-dependent upregulation of Pdx1 expression in duct cells.[5,26,36] Interestingly, the limited organ recovery that follows Px appears to be proportional to the size of the excision.[28,34,35,38,39] SPx in rats leads to ductal cell proliferation and induction of an extensive regenerative process that promotes mature duct cells to regress and reexpress embryonic genes such as *Pdx1*, *Ptf1a*, and *Ngn3* before differentiating to the different pancreatic cell types.[25,36] While these data imply that the duct cells might contribute to the observed increased islet mass following Px, other studies would argue against the involvement of duct cells in this process.[6,30,32,33] Using lineage-tracing studies, independent investigators have not been able to find any evidence for β-cell neogenesis following 50–75% PPx.[6,33] Accordingly, 50 percent PPx in Ngn3-GFP transgenic mice failed to

e-mail: farzad.esni@chp.edu

induce Ngn3-expression in cells within islets or ducts.[30] All together, although the potential contribution of other cell types cannot be ruled out, the current literature supports the notion that the main source for acinar or β-cell regeneration that follows Px is preexisting acinar or β-cells, respectively.[6,32,33,40]

Pancreatic duct ligation

As in the case of Px, ductal obstruction and ligation have historically been used in investigating pancreatic regeneration.[19] Pancreatic duct ligation (PDL) involves ligation of one of the main ducts, which leads to acinar cell death and inflammation in the area distal to the ligation. An advantage with PDL is that the unligated portion of the pancreas remains unaffected and thus can be used as an internal control. However, the regenerative process in this model, particularly the acinar regeneration, appears to be species-dependent. PDL in rats is associated with near complete acinar recovery through a process that involves appearance of ductular structures and their differentiation into acinar cells.[34,41] In mice, although PDL results in the formation of similar metaplastic ducts, the acinar compartment does not regenerate.[32,41–43] Lineage-tracing studies in mice show that both surviving acinar cells[32] and Hnf1β-expressing duct cells[42] in the ligated part of the pancreas can contribute to the formation of these ductular structures. In other words, the tubular structures observed in the ligated part consist of acinar-derived metaplastic ducts and preexisting ducts that have changed morphology. Similarly, in rats it is believed that preexisting acinar cells transdifferentiate into these ductal structures,[34,41] whereas the contribution of duct cells in this process has yet to be determined.

PDL has been primarily used to provide insights on islet β-cell generation, as it is reported to stimulate β-cell regeneration in both mouse and rat.[12,13,33,34,41,44–46] Nevertheless, there is a controversy with respect to the mechanism allowing the observed β-cell generation. While some studies favor β-cell proliferation as the main mechanism for β-cell formation after PDL,[33,41–43,47] others support the potential contribution of non β-cells (in particular, cells within or in proximity of ductal network) to β-cell neogenesis in a PDL setting.[9,12,13,48]

Caerulein-induced pancreatitis

Caerulein is a cholecystokinin orthologue, which when administered repeatedly at high concentrations leads to the death and dropout of over half of the acinar compartment which can cause acute or chronic pancreatitis.[18,49,50] In mice, the pancreas regains its normal histology within a week after caerulein treatment. The rapid regenerative process associated with this injury model has been used by

many investigators to follow the course of acinar recovery.[51–55] Lineage-tracing studies have shown that following caerulein-induced pancreatitis the surviving acinar cells contribute to the recovery of the acinar compartment.[32,51,53,55] Acinar regeneration in this model is through acinar-to-ductal metaplasia (ADM), a process that requires a transient reactivation of various developmental genes and signaling pathways, including notch, hedgehog, and wnt.[51–53,56–58] ADM involves dedifferentiation of acinar cells into duct-like cells, proliferation of metaplastic ducts, and finally re-differentiation of duct-like cells into acinar cells.[53] A first step toward transdifferentiation of one cell type to another cell type is that cells have to lose their original identity in order to acquire a new one. Accordingly, the dedifferentiation of acinar cells is associated with expression of ductal markers such as Hnf6, Sox9, and cytokeratin-19 and concomitant repression of acinar markers Ptf1a, Mist1, amylase, and Cpa.[58] Wnt/β-catenin signaling is one of the embryonic pathways, which is reactivated in ADM following caerulein-induced pancreatitis.[53,56] However, for the ADM to re-differentiate into acinar cells wnt signaling has to be eventually downregulated, as persistent wnt/β-catenin activity leads to impaired acinar recovery.[53,59] The precise mechanism for the dynamic activity of wnt/β-catenin in ADM is not clear, but a recent report implicates HDACs as an important epigenetic switch required for controlling nuclear β-catenin transcriptional activity.[59] The transcriptional factor PDX1 has primarily been associated with the embryonic pancreas and mature β-cells in the adult pancreas. However, a more recent study highlights the importance of PDX1 in maintaining acinar cell identity.[60] PDX1 displays similar dynamic expression as wnt/β-catenin during ADM, and accordingly its downregulation is necessary for re-differentiation of ADM into acinar cells.[60]

Other models of pancreatitis

In addition to the aforementioned caerulein-induced pancreatitis, other rodent models commonly used to study acute pancreatitis entail bile salt infusion, duct obstruction, the choline-deficient ethionine supplemented diet (CDE), or administration of basic amino acids such as L-arginine. These injury models have been extensively described and reviewed by Lerch and Gorelick elsewhere.[18]

Alloxan- or streptozotocin-induced diabetes

Alloxan and streptozotocin (STZ) are used to induce diabetes by chemical ablation of pancreatic β-cells. Alloxan was first described in the early 1800s, but its diabetogenic property was reported in 1943, and since then alloxan treatment has been used as an experimental model for diabetes.[61] STZ was initially used as a chemotherapeutic agent in pancreatic islet cell tumors and other malignancies,[62] but since its discovery as a diabetogenic agent in 1963, it has been widely used in diabetes research.[63] Alloxan and STZ are both toxic glucose analogues that preferentially accumulate in insulin producing β-cells via the Glut2 glucose transporter.[62] Diabetes as the result of alloxan or STZ-treatment is not associated with β-cell regeneration.[33,61] Because of the absence of spontaneous β-cell recovery, these models have been useful tools to study a given treatment on β-cell regeneration. In addition, alloxan- or STZ-treatment can be combined with PDL to study the effect of hyperglycemia on the regenerative process in the ligated portion of the pancreas.[15,41,48] While the combination of alloxan- or STZ-treatment and PDL in mice led to transformation of glucagon-producing α-cells or acinar cells into β-cells,[15,48] no such α-to β-cell conversion could be found when rats were subjected to a combined PDL and STZ-treatment.[41]

Diphtheria toxin-mediated cell ablation

A relatively new method which enables cell-specific ablation is transgenic activation of the diphtheria toxin cell death pathway using a cell-specific promoter.[64,65] Mature diphtheria toxin (DT) is composed of subunits A and B (DTA and DTB).[66,67] DT binds a toxin receptor on the cell surface of toxin-sensitive cells and is endocytosed.[68–70] Upon entry into the cytoplasm, the DTA subunit is released and it catalyzes the inactivation of elongation factor 2, resulting in termination of all protein synthesis, with rapid apoptotic death of the target cell.[71,72] The toxicity of DTA is sufficiently high that only one molecule of DTA in the cytosol may be enough to kill the cell.[73] The DT receptor (DTR) is a membrane-anchored form of the heparin binding EGF-like growth factor (HB-EGF precursor).[74] The human and simian HB-EGF precursors bind DT and function as toxin receptors, whereas HB-EGF from mice and rats do not bind the toxin and therefore remain insensitive to DT.[75] Thus, transgenic expression of the simian or human DTR in mice can render naturally DT-resistant mouse cells DT-sensitive.[69,76,77] Recently, a mouse strain was generated (R26^{DTR}), in which a loxP-flanked STOP cassette and the open reading frame of simian DTR had been introduced into the *ROSA26* locus.[72] In the R26^{DTR} strain, the gene encoding DTR is under the control of the potent Rosa promoter, but DTR expression is dependent first on Cre-recombinase removal of the STOP cassette.[72] Following Cre-recombinase activity, the DTR-expressing cells, that is, cells expressing Cre, and all of their progeny, are viable and function normally. However, these cells are rapidly killed upon DT administration. Noteworthy, the HB-EGF is no longer active as an EGFR ligand, as transgenic lines expressing DTR in different pancreatic lineages do not display any abnormal phenotype.[1,10] In the adult pancreas, DTR/DTA-mediated β-cell ablation has been used to study regeneration following α- or β-cell specific losses,[1,14,17,78,79] acinar,[1,10] or acinar and endocrine cell ablation.[1,10]

The Inconsistencies

A brief look at **Table 1** highlights the inconsistencies that currently exist in the literature regarding the regenerative capacity of the adult mouse pancreas. For example, there is a complete recovery of the acinar compartment within a week after caerulein-induced pancreatitis, whereas there is principally no acinar regeneration following PDL. Additionally, α-cells can differentiate into β-cells; however, this plasticity has been observed only upon total β-cell ablation but not following partial loss of β-cell mass. Variations between different studies are likely due to the disparities in the nature, extent, and, perhaps more importantly, the severity of injury used. In this review, we will argue that the combined effects of these parameters not only may determine whether or not regeneration would occur but also dictate which cell type(s) should contribute to this process.

The nature of injury

Here, the question is not so much about whether the injury is chemically, mechanically, or genetically induced but rather about what kind of cell death it triggers. Apoptosis is programmed cell death generally associated with retention of plasma membrane integrity, condensation, and cleavage of nuclear and cytoplasmic proteins and cell shrinkage or the formation of apoptotic bodies.[80] It is a highly coordinated process which requires significant amount of energy and therefore relies on mitochondrial respiration and ATP production.[81] Necrosis, on the other hand, is invoked in response to external stimuli and ATP deficiency.[81] Pancreatic injury and ensuing regeneration invariably depend on proper clearance of the dead cells.[1] Because of its nature (loss of cytoplasmic membrane integrity, cellular fragmentation, and release of lysosomal and granular contents into surrounding extracellular space), necrosis does not allow for proper removal of cell organelles, and as a result it is followed by reactive inflammation.[81–86] In contrast, apoptosis involves debriding the tissue without generating massive inflammation that is usually induced by the degeneration of dead cells.[81] The effect of apoptosis or necrosis on regeneration is perhaps best manifested when one compares β-cell regeneration following STZ (or alloxan) treatment with DT-mediated β-cell ablation. In these two models, the target cells (β-cells) and the degree of β-cell loss (75–80% sub-optimal condition for STZ, or 75% β-cell ablation using PdxCreERT) are similar.[1,6] Interestingly, STZ-induced necrosis is accompanied with a massive inflammatory response and the absence of β-cell regeneration, whereas DT-induced apoptosis leads to almost complete β-cell mass recovery.[78]

Table 1. Response of mouse pancreas to different damage regimens.

Injury	Nature	Extent	Severity	Regeneration	References
Px	Surgical resection	Exocrine and Endocrine	50–90% resection	Limited organ recovery	6, 9, 11, 20, 24, 32, 43, 50–52, 54, 59, 63, 66, 68, 70, 71, 76, 80, 92
PDL	Apoptosis	Acinar cells	Area distal to the ligation	β-cells	18, 20, 24, 29, 30, 38, 41, 49, 57, 62, 83, 94
Alloxan	Necrosis	β-cells	Vast majority	No	17, 48, 64, 66
STZ	Necrosis	β-cells	Dose dependent	No	48, 57, 64, 66
Caerulein	Necrosis	Acinar cells	< 50%	Yes	10, 20, 24, 37, 40, 53, 56, 74, 75, 78, 84, 85, 89, 91

Overall, compared to the necrosis, apoptosis provides an environment that would favor β-cell regeneration. Notably, acinar regeneration appears to be less sensitive to the nature of injury, as robust acinar tissue recovery has been reported in both necrotic (caerulein) and apoptotic (DT-mediated) environments.[1,10,51–56]

The extent of injury

Another factor that may influence the regenerative process is the extent of injury. In other words, how many different cell types are affected by the insult? In addition to the ductal, acinar, and five different hormone-producing endocrine cell types, the adult pancreas is home to numerous endothelial, stellate, and neuronal cells. The type of injured cells is important as recent reports have shed light on the role of macrophages in inducing ADM, as well as promoting β-cell, or acinar cell regeneration.[1,45,87–89] Macrophages appear to have differential functions in diverse phases of regeneration: first, to debride the tissue following injury and, second, to convert injury signals into lineage-specific regenerative signals.[1,90,91] One exception to this rule is PDL, which as mentioned earlier is associated with acinar atrophy. Thus, one would expect that this model would lead to acinar regeneration, but instead it stimulates β-cell regeneration. Clearly, the effect of PDL on non-acinar cells residing in the ligated part cannot be ruled out. Therefore, it is possible that the combined regenerative signals released by macrophages (as the result of engulfing damaged acinar- and non-acinar cells) would serendipitously create an environment that would promote β-cell regeneration instead of acinar. In fact, combined PDL and β-cell ablation by STZ has been reported to enhance acinar to β-cell transdifferentiation.[48]

Based on our current understanding of the involvement of macrophages in regeneration, simultaneous ablation of many cell types would make direct interpretation of cellular mechanism of regeneration difficult. Thus, cell-type specific ablation may be a better method for analyzing the *in vivo* function of cells during regeneration, which can be achieved by using streptozotocin, alloxan (for β-cells), or caerulein (for acinar cells). Alternatively, transgenic activation of the diphtheria toxin (DTR/DTA) cell death pathway, which depending on the promoter, can target one specific (e.g., Elastase promoter for acinar cell ablation) or more than one cell type (Pdx1 promoter to target all pancreatic epithelial cells).[1,10]

The severity of injury

Mounting evidence suggests that the severity of injury is perhaps one of the most important elements that dictate whether the mechanism for repair should include replication of preexisting cells or neogenesis from other cell types. DT-mediated cell ablation has been used by a number of investigators to vary the extent and the severity of injury, while keeping other variables (such as the nature of the injury) relatively constant. Collectively, it appears that regardless of cell type, as long as ablation of a specific cell type does not reach near 100 percent, the mechanism for regeneration mainly involves the preexisting cells.[1,6,14,51,53,55] Therefore, following 75 percent ablation of β-cells, surviving β-cells proliferate to generate new β-cells, whereas complete loss of insulin-producing cells promotes conversion of other endocrine cell types into β-cells.[1,6,14,17,92,93] Consistently, acinar cell recovery following caerulein-induced pancreatitis is through preexisting acinar cells. However, near complete loss of both acinar and endocrine cells stimulates cells within the ductal compartment to form new acinar and endocrine cells.[1,10] One could also argue that the extent of the surgical intervention may be important also in the Px setting and could explain discrepancies in some reports describing absence or vigorous pancreatic regeneration after partial or SPx, respectively.[6,25,35] However, as mentioned earlier unlike PPx, SPx is associated with acinar atrophy and a desmoplastic reaction. Of note, this inflammatory reaction has been reported to be important for the robust regeneration that follows SPx, as inhibition of the inflammation prevented regeneration.[34,94] Therefore, it is likely that the inconsistencies between PPx and SPx are due to the presence or the absence of inflammation rather than the extent of injury.

Conclusions

Pancreatic regeneration relies on a complex interaction between cells that provide necessary regenerative signals and cells that are receptive to those signals. As discussed here, the nature, extent, and the severity of injury are three important parameters that determine whether tissue recovery is achieved. β-cell regeneration seems to be more sensitive to the nature of injury than acinar regeneration. Finally, the extent of injury determines which cell types would respond to these regenerative signals.

Acknowledgments

Supported by NIH/NIDDK grants DK101413–01A1, and DK103002–01.

References

1. Criscimanna A, Coudriet GM, Gittes GK, Piganelli JD, Esni F. Activated macrophages create lineage-specific microenvironments for pancreatic acinar- and beta-cell regeneration in mice. Gastroenterology 2014; 147:1106–1118.

2. Takahashi H, Okamura D, Starr ME, Saito H, Evers BM. Age-dependent reduction of the PI3K regulatory subunit p85alpha suppresses pancreatic acinar cell proliferation. Aging Cell 2012; 11:305–314.

3. Tanigawa K, Nakamura S, Kawaguchi M, Xu G, Kin S, Tamura K. Effect of aging on B-cell function and replication in rat pancreas after 90% pancreatectomy. Pancreas 1997; 15:53–59.

4. Teta M, Long SY, Wartschow LM, Rankin MM, Kushner JA. Very slow turnover of beta-cells in aged adult mice. Diabetes 2005; 54:2557–2567.

5. Watanabe H, Saito H, Rychahou PG, Uchida T, Evers BM. Aging is associated with decreased pancreatic acinar cell regeneration and phosphatidylinositol 3-kinase/Akt activation. Gastroenterology 2005; 128:1391–1404.

6. Dor Y, Brown J, Martinez OI, Melton DA. Adult pancreatic beta-cells are formed by self-duplication rather than stem-cell differentiation. Nature 2004; 429:41–46.

7. Teta M, Rankin MM, Long SY, Stein GM, Kushner JA. Growth and regeneration of adult beta cells does not involve specialized progenitors. Dev Cell 2007; 12:817–826.

8. Bonner-Weir S, Inada A, Yatoh S, Li WC, Aye T, Toschi E, Sharma A. Transdifferentiation of pancreatic ductal cells to endocrine beta-cells. Biochem Soc Trans 2008; 36:353–356.

9. Inada A, Nienaber C, Katsuta H, Fujitani Y, Levine J, Morita R, Sharma A, Bonner-Weir S. Carbonic anhydrase II-positive pancreatic cells are progenitors for both endocrine and exocrine pancreas after birth. Proc Natl Acad Sci U S A 2008; 105:19915–19919.

10. Criscimanna A, Speicher JA, Houshmand G, Shiota C, Prasadan K, Ji B, Logsdon CD, Gittes GK, Esni F. Duct cells contribute to regeneration of endocrine and acinar cells following pancreatic damage in adult mice. Gastroenterology 2011; 141:1451–1462.

11. Al-Hasani K, Pfeifer A, Courtney M, Ben-Othman N, Gjernes E, Vieira A, Druelle N, Avolio F, Ravassard P, Leuckx G, Lacas-Gervais S, Ambrosetti D, et al. Adult Duct-Lining Cells Can Reprogram into beta-like Cells Able to Counter Repeated Cycles of Toxin-Induced Diabetes. Dev Cell 2013; 26:86–100.

12. Zhang M, Lin Q, Qi T, Wang T, Chen CC, Riggs AD, Zeng D. Growth factors and medium hyperglycemia induce Sox9+ ductal cell differentiation into beta cells in mice with reversal of diabetes. Proc Natl Acad Sci U S A 2016; 113:650–655.

13. Xu X, D'Hoker J, Stange G, Bonne S, De Leu N, Xiao X, Van de Casteele M, Mellitzer G, Ling Z, Pipeleers D, Bouwens L, Scharfmann R, et al. Beta cells can be generated from endogenous progenitors in injured adult mouse pancreas. Cell 2008; 132:197–207.

14. Thorel F, Nepote V, Avril I, Kohno K, Desgraz R, Chera S, Herrera PL. Conversion of adult pancreatic alpha-cells to beta-cells after extreme beta-cell loss. Nature 2010; 464:1149–1154.

15. Chung CH, Hao E, Piran R, Keinan E, Levine F. Pancreatic beta-cell neogenesis by direct conversion from mature alpha-cells. Stem Cells 2010; 28:1630–1638.

16. Yang YP, Thorel F, Boyer DF, Herrera PL, Wright CV. Context-specific {alpha}-to-{beta}-cell reprogramming by forced Pdx1 expression. Genes Dev 2011; 25:1680–1685.

17. Chera S, Baronnier D, Ghila L, Cigliola V, Jensen JN, Gu G, Furuyama K, Thorel F, Gribble FM, Reimann F, Herrera PL. Diabetes recovery by age-dependent conversion of pancreatic delta-cells into insulin producers. Nature 2014; 514:503–507.

18. Lerch MM, Gorelick FS. Models of acute and chronic pancreatitis. Gastroenterology 2013; 144:1180–1193.

19. Weaver CV, Garry DJ. Regenerative biology: a historical perspective and modern applications. Regen Med 2008; 3:63–82.

20. Granger A, Kushner JA. Cellular origins of beta-cell regeneration: a legacy view of historical controversies. J Intern Med 2009; 266:325–338.

21. Bonner-Weir S, Toschi E, Inada A, Reitz P, Fonseca SY, Aye T, Sharma A. The pancreatic ductal epithelium serves as a potential pool of progenitor cells. Pediatr Diabetes 2004; 5 (suppl 2):16–22.

22. Brockenbrough JS, Weir GC, Bonner-Weir S. Discordance of exocrine and endocrine growth after 90% pancreatectomy in rats. Diabetes 1988; 37:232–236.

23. De Leon DD, Deng S, Madani R, Ahima RS, Drucker DJ, Stoffers DA. Role of endogenous glucagon-like peptide-1 in islet regeneration after partial pancreatectomy. Diabetes 2003; 52:365–371.

24. Hayashi KY, Tamaki H, Handa K, Takahashi T, Kakita A, Yamashina S. Differentiation and proliferation of endocrine cells in the regenerating rat pancreas after 90% pancreatectomy. Arch Histol Cytol 2003; 66:163–174.

25. Li WC, Rukstalis JM, Nishimura W, Tchipashvili V, Habener JF, Sharma A, Bonner-Weir S. Activation of pancreatic-duct-derived progenitor cells during pancreas regeneration in adult rats. J Cell Sci 2010; 123:2792–2802.

26. Watanabe H, Saito H, Nishimura H, Ueda J, Evers BM. Activation of phosphatidylinositol-3 kinase regulates pancreatic duodenal homeobox-1 in duct cells during pancreatic regeneration. Pancreas 2008; 36:153–159.

27. Watanabe H, Saito H, Ueda J, Evers BM. Regulation of pancreatic duct cell differentiation by phosphatidylinositol-3 kinase. Biochem Biophys Res Commun 2008; 370:33–37.

28. Pearson KW, Scott D, Torrance B. Effects of partial surgical pancreatectomy in rats. I. Pancreatic regeneration. Gastroenterology 1977; 72:469–473.

29. Leahy JL, Bonner-Weir S, Weir GC. Minimal chronic hyperglycemia is a critical determinant of impaired insulin secretion after an incomplete pancreatectomy. J Clin Invest 1988; 81:1407–1414.

30. Lee CS, De Leon DD, Kaestner KH, Stoffers DA. Regeneration of pancreatic islets after partial pancreatectomy in mice does not involve the reactivation of neurogenin-3. Diabetes 2006; 55:269–272.

31. Peshavaria M, Larmie BL, Lausier J, Satish B, Habibovic A, Roskens V, Larock K, Everill B, Leahy JL, Jetton TL. Regulation of pancreatic beta-cell regeneration in the normoglycemic 60% partial-pancreatectomy mouse. Diabetes 2006; 55:3289–3298.

32. Desai BM, Oliver-Krasinski J, De Leon DD, Farzad C, Hong N, Leach SD, Stoffers DA. Preexisting pancreatic acinar cells contribute to acinar cell, but not islet beta cell, regeneration. J Clin Invest 2007; 117:971–977.

33. Xiao X, Chen Z, Shiota C, Prasadan K, Guo P, El-Gohary Y, Paredes J, Welsh C, Wiersch J, Gittes GK. No evidence for beta cell neogenesis in murine adult pancreas. J Clin Invest 2013; 123:2207–2217.

34. Bouwens L. Beta cell regeneration. Curr Diabetes Rev 2006; 2:3–9.

35. Bonner-Weir S, Baxter LA, Schuppin GT, Smith FE. A second pathway for regeneration of adult exocrine and endocrine pancreas. A possible recapitulation of embryonic development. Diabetes 1993; 42:1715–1720.

36. Sharma A, Zangen DH, Reitz P, Taneja M, Lissauer ME, Miller CP, Weir GC, Habener JF, Bonner-Weir S. The homeodomain protein IDX-1 increases after an early burst of proliferation during pancreatic regeneration. Diabetes 1999; 48:507–513.

37. Migliorini RH. Two-stage procedure for total pancreatectomy in the rat. Diabetes 1970; 19:694–697.

38. Lehv M, Fitzgerald PJ. Pancreatic acinar cell regeneration. IV. Regeneration after resection. Am J Pathol 1968; 53:513–535.

39. Plachot C, Movassat J, Portha B. Impaired beta-cell regeneration after partial pancreatectomy in the adult Goto-Kakizaki rat, a spontaneous model of type II diabetes. Histochem Cell Biol 2001; 116:131–139.

40. Murtaugh LC, Keefe MD. Regeneration and repair of the exocrine pancreas. Annu Rev Physiol 2015; 77:229–249.

41. Cavelti-Weder C, Shtessel M, Reuss JE, Jermendy A, Yamada T, Caballero F, Bonner-Weir S, Weir GC. Pancreatic duct ligation after almost complete beta-cell loss: exocrine regeneration but no evidence of beta-cell regeneration. Endocrinology 2013; 154:4493–4502.

42. Solar M, Cardalda C, Houbracken I, Martin M, Maestro MA, De Medts N, Xu X, Grau V, Heimberg H, Bouwens L, Ferrer J. Pancreatic exocrine duct cells give rise to insulin-producing beta cells during embryogenesis but not after birth. Dev Cell 2009; 17:849–860.

43. Kopp JL, Dubois CL, Schaffer AE, Hao E, Shih HP, Seymour PA, Ma J, Sander M. Sox9+ ductal cells are multipotent progenitors throughout development but do not produce new endocrine cells in the normal or injured adult pancreas. Development 2011; 138:653–665.

44. Wang RN, Kloppel G, Bouwens L. Duct- to islet-cell differentiation and islet growth in the pancreas of duct-ligated adult rats. Diabetologia 1995; 38:1405–1411.

45. Xiao X, Gaffar I, Guo P, Wiersch J, Fischbach S, Peirish L, Song Z, El-Gohary Y, Prasadan K, Shiota C, Gittes GK. M2 macrophages promote beta-cell proliferation by upregulation of SMAD7. Proc Natl Acad Sci U S A 2014; 111:E1211-E1220.

46. Van de Casteele M, Leuckx G, Cai Y, Yuchi Y, Coppens V, De Groef S, Van Gassen N, Baeyens L, Heremans Y, Wright CV, Heimberg H. Partial duct ligation: beta-cell proliferation and beyond. Diabetes 2014; 63:2567–2577.

47. Rankin MM, Wilbur CJ, Rak K, Shields EJ, Granger A, Kushner JA. beta-Cells are not generated in pancreatic duct ligation-induced injury in adult mice. Diabetes 2013; 62:1634–1645.

48. Pan FC, Bankaitis ED, Boyer D, Xu X, Van de Casteele M, Magnuson MA, Heimberg H, Wright CV. Spatiotemporal patterns of multipotentiality in Ptf1a-expressing cells during pancreas organogenesis and injury-induced facultative restoration. Development 2013; 140:751–764.

49. Lampel M, Kern HF. Acute interstitial pancreatitis in the rat induced by excessive doses of a pancreatic secretagogue. Virchows Arch A Pathol Anat Histol 1977; 373:97–117.

50. Eisses JF, Davis AW, Tosun AB, Dionise ZR, Chen C, Ozolek JA, Rohde GK, Husain SZ. A computer-based automated algorithm for assessing acinar cell loss after experimental pancreatitis. PLoS One 2014; 9:e110220.

51. Fendrich V, Esni F, Garay MV, Feldmann G, Habbe N, Jensen JN, Dor Y, Stoffers D, Jensen J, Leach SD, Maitra A. Hedgehog signaling is required for effective regeneration of exocrine pancreas. Gastroenterology 2008; 135:621–631.

52. Jensen JN, Cameron E, Garay MV, Starkey TW, Gianani R, Jensen J. Recapitulation of elements of embryonic development in adult mouse pancreatic regeneration. Gastroenterology 2005; 128:728–741.

53. Morris JPt, Cano DA, Sekine S, Wang SC, Hebrok M. Beta-catenin blocks Kras-dependent reprogramming of acini into pancreatic cancer precursor lesions in mice. J Clin Invest 2010; 120:508–520.

54. Hess DA, Humphrey SE, Ishibashi J, Damsz B, Lee AH, Glimcher LH, Konieczny SF. Extensive Pancreas Regeneration Following Acinar-Specific Disruption of Xbp1 in Mice. Gastroenterology 2011; 141:1463–1472.

55. Strobel O, Dor Y, Alsina J, Stirman A, Lauwers G, Trainor A, Castillo CF, Warshaw AL, Thayer SP. In vivo lineage tracing

defines the role of acinar-to-ductal transdifferentiation in inflammatory ductal metaplasia. Gastroenterology 2007; 133:1999–2009.

56. Morris JPt, Wang SC, Hebrok M. KRAS, Hedgehog, Wnt and the twisted developmental biology of pancreatic ductal adenocarcinoma. Nat Rev Cancer 2010; 10:683–695.

57. Kopp JL, von Figura G, Mayes E, Liu FF, Dubois CL, Morris JPt, Pan FC, Akiyama H, Wright CV, Jensen K, Hebrok M, Sander M. Identification of Sox9-dependent acinar-to-ductal reprogramming as the principal mechanism for initiation of pancreatic ductal adenocarcinoma. Cancer Cell 2012; 22:737–750.

58. Prevot PP, Simion A, Grimont A, Colletti M, Khalaileh A, Van den Steen G, Sempoux C, Xu X, Roelants V, Hald J, Bertrand L, Heimberg H, et al. Role of the ductal transcription factors HNF6 and Sox9 in pancreatic acinar-to-ductal metaplasia. Gut 2012; 61:1723–1732.

59. Eisses JF, Criscimanna A, Dionise ZR, Orabi AI, Javed TA, Sarwar S, Jin S, Zhou L, Singh S, Poddar M, Davis AW, Tosun AB, et al. Valproic acid limits pancreatic recovery after pancreatitis by inhibiting histone deacetylases and preventing acinar redifferentiation programs. Am J Pathol 2015; 185:3304–3315.

60. Roy N, Takeuchi KK, Ruggeri JM, Bailey P, Chang D, Li J, Leonhardt L, Puri S, Hoffman MT, Gao S, Halbrook CJ, Song Y, et al. PDX1 dynamically regulates pancreatic ductal adenocarcinoma initiation and maintenance. Genes Dev 2016; 30:2669–2683.

61. Szkudelski T. The mechanism of alloxan and streptozotocin action in B cells of the rat pancreas. Physiol Res 2001; 50:537–546.

62. Lenzen S. The mechanisms of alloxan- and streptozotocin-induced diabetes. Diabetologia 2008; 51:216–226.

63. Eleazu CO, Eleazu KC, Chukwuma S, Essien UN. Review of the mechanism of cell death resulting from streptozotocin challenge in experimental animals, its practical use and potential risk to humans. J Diabetes Metab Disord 2013; 12:60.

64. Palmiter RD, Behringer RR, Quaife CJ, Maxwell F, Maxwell IH, Brinster RL. Cell lineage ablation in transgenic mice by cell-specific expression of a toxin gene. Cell 1987; 50:435–443.

65. Breitman ML, Clapoff S, Rossant J, Tsui LC, Glode LM, Maxwell IH, Bernstein A. Genetic ablation: targeted expression of a toxin gene causes microphthalmia in transgenic mice. Science 1987; 238:1563–1565.

66. Greenfield L, Bjorn MJ, Horn G, Fong D, Buck GA, Collier RJ, Kaplan DA. Nucleotide sequence of the structural gene for diphtheria toxin carried by corynebacteriophage beta. Proc Natl Acad Sci U S A 1983; 80:6853–6857.

67. Tsuneoka M, Nakayama K, Hatsuzawa K, Komada M, Kitamura N, Mekada E. Evidence for involvement of furin in cleavage and activation of diphtheria toxin. J Biol Chem 1993; 268:26461–26465.

68. Gill DM, Dinius LL. Observations on the structure of diphtheria toxin. J Biol Chem 1971; 246:1485–1491.

69. Saito M, Iwawaki T, Taya C, Yonekawa H, Noda M, Inui Y, Mekada E, Kimata Y, Tsuru A, Kohno K. Diphtheria toxin receptor-mediated conditional and targeted cell ablation in transgenic mice. Nat Biotechnol 2001; 19:746–750.

70. Drazin R, Kandel J, Collier RJ. Structure and activity of diphtheria toxin. II. Attack by trypsin at a specific site within the intact toxin molecule. J Biol Chem 1971; 246:1504–1510.

71. Honjo T, Nishizuka Y, Kato I, Hayaishi O. Adenosine diphosphate ribosylation of aminoacyl transferase II and inhibition of protein synthesis by diphtheria toxin. J Biol Chem 1971; 246:4251–4260.

72. Buch T, Heppner FL, Tertilt C, Heinen TJ, Kremer M, Wunderlich FT, Jung S, Waisman A. A Cre-inducible diphtheria toxin receptor mediates cell lineage ablation after toxin administration. Nat Methods 2005; 2:419–426.

73. Yamaizumi M, Uchida T, Okada Y, Furusawa M. Neutralization of diphtheria toxin in living cells by microinjection of antifragment A contained within resealed erythrocyte ghosts. Cell 1978; 13:227–232.

74. Naglich JG, Metherall JE, Russell DW, Eidels L. Expression cloning of a diphtheria toxin receptor: identity with a heparin-binding EGF-like growth factor precursor. Cell 1992; 69:1051–1061.

75. Mitamura T, Higashiyama S, Taniguchi N, Klagsbrun M, Mekada E. Diphtheria toxin binds to the epidermal growth factor (EGF)-like domain of human heparin-binding EGF-like growth factor/diphtheria toxin receptor and inhibits specifically its mitogenic activity. J Biol Chem 1995; 270:1015–1019.

76. Jung S, Unutmaz D, Wong P, Sano G, De los Santos K, Sparwasser T, Wu S, Vuthoori S, Ko K, Zavala F, Pamer EG, Littman DR, et al. In vivo depletion of CD11c+ dendritic cells abrogates priming of CD8+ T cells by exogenous cell-associated antigens. Immunity 2002; 17:211–220.

77. Cha JH, Chang MY, Richardson JA, Eidels L. Transgenic mice expressing the diphtheria toxin receptor are sensitive to the toxin. Mol Microbiol 2003; 49:235–240.

78. Nir T, Melton DA, Dor Y. Recovery from diabetes in mice by beta cell regeneration. J Clin Invest 2007; 117:2553–2561.

79. Shiota C, Prasadan K, Guo P, El-Gohary Y, Wiersch J, Xiao X, Esni F, Gittes GK. Alpha-cells are dispensable in postnatal morphogenesis and maturation of mouse pancreatic islets. Am J Physiol Endocrinol Metab 2013; 305:E1030-E1040.

80. Erwig LP, Henson PM. Immunological consequences of apoptotic cell phagocytosis. Am J Pathol 2007; 171:2–8.

81. Dorn GW, 2nd. Molecular mechanisms that differentiate apoptosis from programmed necrosis. Toxicol Pathol 2013; 41:227–234.

82. Bhatia M. Apoptosis versus necrosis in acute pancreatitis. Am J Physiol Gastrointest Liver Physiol 2004; 286:G189-G196.

83. Buja LM, Eigenbrodt ML, Eigenbrodt EH. Apoptosis and necrosis. Basic types and mechanisms of cell death. Arch Pathol Lab Med 1993; 117:1208–1214.

84. Majno G, Joris I. Apoptosis, oncosis, and necrosis. An overview of cell death. Am J Pathol 1995; 146:3–15.

85. Padanilam BJ. Cell death induced by acute renal injury: a perspective on the contributions of apoptosis and necrosis. Am J Physiol Renal Physiol 2003; 284:F608-F627.

86. Scaffidi P, Misteli T, Bianchi ME. Release of chromatin protein HMGB1 by necrotic cells triggers inflammation. Nature 2002; 418:191–195.

87. Liou GY, Doppler H, Necela B, Krishna M, Crawford HC, Raimondo M, Storz P. Macrophage-secreted cytokines drive

pancreatic acinar-to-ductal metaplasia through NF-kappaB and MMPs. J Cell Biol 2013; 202:563–577.

88. Brissova M, Aamodt K, Brahmachary P, Prasad N, Hong JY, Dai C, Mellati M, Shostak A, Poffenberger G, Aramandla R, Levy SE, Powers AC. Islet microenvironment, modulated by vascular endothelial growth factor-A signaling, promotes beta cell regeneration. Cell Metab 2014; 19:498–4511.

89. Cruz AF, Rohban R, Esni F. Macrophages in the pancreas: villains by circumstances, not necessarily by actions. Immun Inflamm Dis 2020; 8:807–824.

90. Boulter L, Govaere O, Bird TG, Radulescu S, Ramachandran P, Pellicoro A, Ridgway RA, Seo SS, Spee B, Van Rooijen N, Sansom OJ, Iredale JP, et al. Macrophage-derived Wnt opposes Notch signaling to specify hepatic progenitor cell fate in chronic liver disease. Nat Med 2012; 18:572–579.

91. Lin SL, Li B, Rao S, Yeo EJ, Hudson TE, Nowlin BT, Pei H, Chen L, Zheng JJ, Carroll TJ, Pollard JW, McMahon AP, et al. Macrophage Wnt7b is critical for kidney repair and regeneration. Proc Natl Acad Sci U S A 2010; 107:4194–4199.

92. Furuyama K, Chera S, van Gurp L, Oropeza D, Ghila L, Damond N, Vethe H, Paulo JA, Joosten AM, Berney T, Bosco D, Dorrell C, et al. Diabetes relief in mice by glucose-sensing insulin-secreting human alpha-cells. Nature 2019; 567:43–48.

93. Thorel F, Damond N, Chera S, Wiederkehr A, Thorens B, Meda P, Wollheim CB, Herrera PL. Normal glucagon signaling and {beta}-cell function after near-total {alpha}-cell ablation in adult mice. Diabetes 2011; 60:2872–2882.

94. Lampeter EF, Gurniak M, Brocker U, Klemens C, Tubes M, Friemann J, Kolb H. Regeneration of beta-cells in response to islet inflammation. Exp Clin Endocrinol Diabetes 1995; 103 (suppl 2):74–78.

Pancreatic Islet and Stellate Cell Structure and Function

Chapter 24

Structure of islets and vascular relationship to the exocrine pancreas

Yousef El-Gohary[1] and George Gittes[2]

[1]Department of Surgery, St. Jude Children's Research Hospital, 262 Danny Thomas Pl, Memphis, TN 38105

[2]Department of Surgery, Division of Pediatric General and Thoracic Surgery, Children's Hospital of Pittsburgh of UPMC, One Children's Hospital Drive, 4401 Penn Ave., Pittsburgh, PA 15224

Introduction

The pancreas is a retroperitoneal organ with its function being dictated by the possession of two morphologically distinct tissues, the exocrine and endocrine pancreas. The endocrine pancreas is organized into islets of Langerhans, consisting of five cell subtypes: alpha, beta, delta, epsilon, and PP (pancreatic polypeptide) cells that secrete glucagon, insulin, somatostatin, ghrelin, and pancreatic polypeptide, respectively. Islet cells make up only 2 percent of the adult pancreatic mass. The exocrine pancreas, which is composed of acinar and ductal epithelial cells, accounts for nearly 98 percent of the adult pancreatic mass.[1] The structure of the exocrine and endocrine pancreas has been extensively studied due to the clinical importance of pancreas-specific diseases such as diabetes, pancreatic cancer, and pancreatitis. However, many of these findings have been extrapolated from mice, which remains the most studied animal model in pancreas research. This focus on mouse is due to difficulty in procuring human pancreas. The mouse pancreas contains approximately 1000–5000 islets, whereas the human pancreas contains approximately 1–15 million islets; their size is similar with an upper diameter of 500–700 μm in both mice and humans.[2] Pancreatic islets are highly vascularized micro-organs with a capillary network that is five to ten times denser than that of the exocrine pancreas (**Figure 1**).[3] Other structures within the pancreas that appear highly vascularized are the pancreatic ducts which are enveloped by a dense network of vessels that are much denser than in the surrounding acinar tissue (**Figure 1**).[3] Blood vessel formation plays an essential role in adult tissue function as well as during embryonic development.[4]

History

Over time, the pancreas has evolved from relative obscurity to possessing the most studied cell in the world, the beta cell. The term pancreas, "all-flesh," was first coined by Aristotle (384–322 B.C.) and later described by the Greek anatomist and surgeon Herophilus (335 B.C.–280 B.C.).[5] However, this organ was then overlooked for many years, partly because of its location as opposed to other more prominent organs such as the liver or kidneys, but mostly because its function was unknown for many centuries. Galen's (Claudius Galenus, 129–216 A.D.) false dogma that the organ functioned as a mechanical protective cushion for the mesenteric vessels was unchallenged until the 17th century.[6] In 1642, the main pancreatic duct was discovered by the German anatomist Johann Georg Wirsüng (1589–1643) during an autopsy of a human subject, engraving his finding on a copper plate as a method of reproducing the anatomical illustration.[6] Although its function was never known to him, it was eventually named after him in honor of his discovery. Though Wirsüng himself and others did make the observation of occasionally seeing the duct to be double, the discovery of the accessory pancreatic duct as a normal variant is credited to Giovanni Domenico Santorini (1681–1737).[6] The pancreatic duct discovery was a milestone in pancreatic history because it made Galen's theory obsolete and led to studies of the pancreas as an exocrine gland. Reignier de DeGraaf (1641–1673) showed that the pancreas was undoubtedly an exocrine gland by cannulating the duct, which abolished the previous theory that the pancreas acted as a cushion of the stomach. Paul Langerhans (1847–1888), a German pathologist, was working on his thesis entitled "Contributions to the microscopic anatomy of the pancreas" when he discovered discrete areas of "clear cells" within the pancreas. Langerhans documented the histology of the pancreas by injecting the ductal system in the rabbit pancreas with Berlin blue dye, identifying and describing the endocrine cells as "small irregular-polygonal structures with a cytoplasm perfectly shiny." However, the first to discover that these "islands of clear cells" were well vascularized and communicated with the capillary bed of the exocrine tissue were Wilhelm F. Kuhne and A. S. Lea in 1882 who compared these distinct

e-mail: yousef.gohary@gmail.com, gittesgk@upmc.edu

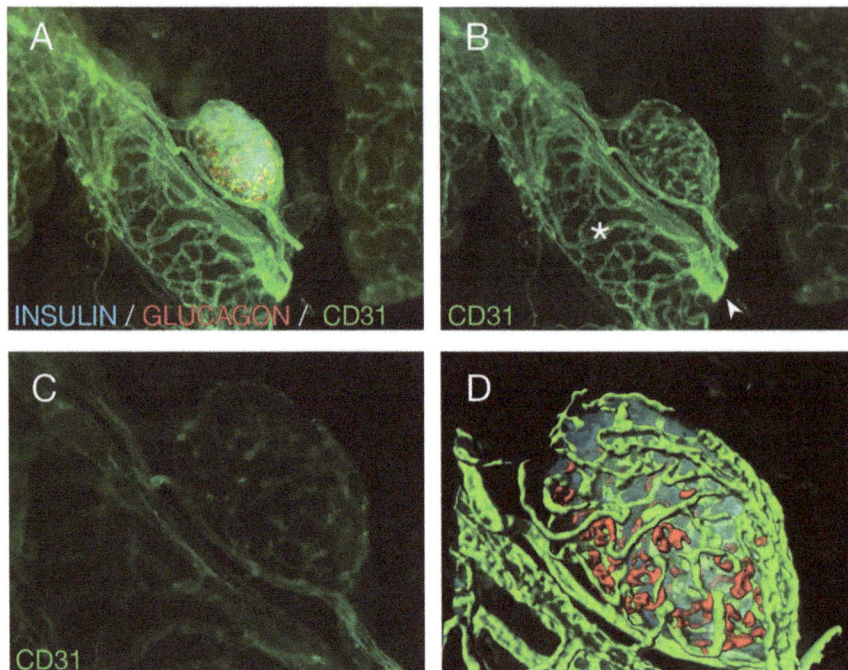

Figure 1. (**A**) Whole-mount image of wild-type mouse pancreatic tissue demonstrating the hypervascular microcapillary network of the islet with glucagon in the periphery of the islet. (**B**) Pancreatic duct (*) running parallel to a vessel (arrow head) and enveloped by a dense network of vessels, almost as dense as the islet, as opposed to the acinar vessels (arrow). (**C**) A higher magnification to demonstrate the hypervascularity of the islet and pancreatic duct. (**D**) A three-dimensional reconstruction of (**A**). CD31: an endothelial marker, also known as anti-platelet/endothelial cell adhesion molecule-1 (PECAM).

vascular regions to the renal glomeruli.[7] In 1893, the French histologist Edouard Laguesse suggested that these vascular regions be named "islets of Langerhans" in memory of their original discoverer.[7]

Since then, a variety of injection techniques have been used by a variety of authors to demonstrate the islet microvascular anatomy including Berlin blue dye, methylene blue, dyed collagenase, and fluorescent dyes such as Lucifer Yellow or tomato lectin.[8–11] Other methods utilized to visualize the microvascular anatomy included corrosion casting coupled with scanning electron microscopy.[12] The latter method allowed a beautiful three-dimensional view of the pancreatic and islet vasculature but was labor intensive. Then came the development of diffusible tracer techniques to estimate blood flow distributions within organs, including the islets.[13] Microspheres measuring 10–15 μm introduced through an intra-arterial injection became the standard marker for islet blood flow measurements. The number of microspheres within each organ is proportional to their blood perfusion. Microspheres are quantified here by either directly counting the spheres or by quantifying a signal (radioactivity, fluorescence, or different colors), that is incorporated into the spheres.[14] However, there was a risk of overestimating blood flow due to shunting, and if the number and size of the spheres exceeded a certain threshold,

there would be a risk for emboli and tissue ischemia.[14] At the turn of the 21st century, novel imaging techniques were developed to visualize the intricate islet microvascular anatomy in three dimensions, such as whole-mount immunohistochemistry (**Figure 1 A-D**) and real-time *in vivo* imaging.[3,15–18] A major challenge now is to quantify the endocrine cell mass *in vivo* to better monitor beta cell loss in diabetics or after islet transplant. New pancreas-specific biomarkers are being sought for this purpose.[19]

Embryonic Pancreatic Vasculature Development

Despite the well-established role that blood vessels play in the development of different organ systems, their role in pancreatic morphogenesis and differentiation remains poorly understood. However, it is clear that blood vessels play different roles during different stages of pancreatic development, with the potent vasculogenic factor, vascular endothelial growth factor (VEGF), having an especially important function.[20] In the developing pancreas, there are abundant VEGFR2-positive endothelial cells throughout the early embryonic mesenchyme.[21,22] Furthermore, VEGF receptor (VEGFR2) inhibition resulted in abnormal epithelial growth and differentiation, whereas Pdx1-driven overexpression of VEGF resulted in pancreatic growth arrest and islet disruption.[20,21,23] The avascular dorsal pancreatic bud first evaginates at embryonic day 9.5 (E9.5) toward the aorta and subsequently undergoes extensive morphogenesis, giving rise to a highly branched and tubular epithelium with small capillaries being observed within the surrounding mesenchyme of the pancreatic bud epithelium. At E10.5, some blood flow is seen in the capillaries surrounding the pancreatic bud. Then at E11.5–E12.0, the pancreatic epithelium further remodels, forming bulging "tips" with blood vessels becoming intercalated in between them. These observations suggest that the pancreas gets vascularized in the early stages through a combination of epithelial growth into a preexisting mesenchymal vascular plexus of patent capillaries, as well as peripheral vasculogenesis (i.e. de novo development of vessels from angioblasts). The latter process is directed by the early expression of VEGF throughout the pancreatic epithelium at E11.5. The developing endocrine islets and ductal trunk exhibit high VEGF expression, whereas the acini at the epithelial tips lack VEGF expression. This explains why both islets and pancreatic ducts appear highly vascular in the postnatal pancreas compared to the surrounding acini (**Figure 1**). A primitive primary honeycomb-like primary plexus of vessels appears in the surrounding pancreatic mesenchyme at E10.5, which later forms the veins, followed by the formation of a large central vessel, which is formed along the center of the pancreatic bud by E12.5, becoming an artery. By E13.5, the central vessel becomes progressively integrated into the expanding pancreas and by E14.5, the main central artery becomes more established, with the fusion of more vessels progressing from proximal to distal pancreas. Thus, the veins localize to the periphery, and the arteries more centrally within the pancreas.[20]

Shah et al.[22] nicely demonstrated in their whole embryo "angiogram," utilizing a novel in utero intracardiac injection, a relative paucity of blood flow in early pancreatic development (E11.5), with the presence of blood flow correlating with the differentiation of pancreatic epithelial cells in the region of the blood flow. Specifically, during E12.5–E14.0 glucagon-positive cells tended to localize specifically near perfused blood vessels, and by early E14 insulin-positive cells tended to localize near the blood flow. Thus, endothelial cells and blood flow appear to help direct pancreatic differentiation.[21,22]

Gross Anatomy of the Pancreas

The pancreas is an endodermally derived retroperitoneal organ that is formed by the fusion of the ventral and dorsal buds of the foregut. In the human it is divided into the head, neck, body, and the tail, with no clear-cut borders between them. However, it is generally accepted that the border between the head/neck and body is marked by the superior mesenteric artery, whereas the border between the body and tail is considered to be the midpoint between the neck and the tip of the tail. The head has an extension known as the uncinate process that hooks beneath the superior mesenteric artery, with the head nestled in the curvature of the duodenum. The body of pancreas sits beneath the stomach, extending horizontally and crossing over the inferior vena cava, abdominal aorta, superior mesenteric artery, and vein. The neck is the thin section of the gland between head and body overlying the superior mesenteric artery. The pancreatic tail abuts the splenic hilum. The total length of the gland is about 15 cm and it weighs 50–100 gm[2].

The mouse pancreas is divided into three major parts: the duodenal, splenic, and gastric lobes, with the splenic lobe making up over half of the total pancreatic volume. The duodenal lobe is homologous with the head of the human pancreas, whereas the splenic lobe is homologous with the body and tail of the human pancreas, which extends horizontally between the duodenum and spleen. The gastric lobe is the third and smallest part of the mouse pancreas.[2]

The distribution of islets in different lobes of the mouse pancreas is heterogeneous and strain dependent. The number of islets per unit volume is highest in the gastric lobe, followed by the duodenal and splenic lobe. In the human pancreas, the density of islets per unit volume is higher in the tail compared to the head and body.[2]

Macroscopic Vascular Anatomy of the Pancreas

The blood supply of the pancreas is primarily derived from the celiac artery, representing approximately 1 percent of the total cardiac output.[24] The head of the pancreas is supplied by an arterial arcade system formed by the anterior and posterior superior pancreaticoduodenal arteries arising from the gastroduodenal artery of the common hepatic artery and anastomosing with the anterior and posterior inferior pancreaticoduodenal artery, which is the first branch of the superior mesenteric artery. The body and pancreatic tail, on the other hand, primarily receive their blood supply from the pancreatic branches of the splenic artery, which is the left most branch of the celiac artery, and by the dorsal pancreatic artery that branches off near the origin of either celiac, hepatic, or splenic artery. Its right branch anastomoses with the anterior superior pancreaticoduodenal artery, whereas its left branch forms the transverse pancreatic artery that runs along the inferior border of the pancreas and usually anastomoses with the pancreatica magna artery, the largest pancreatic branch of the splenic artery in the middle of the gland.[25,26]

The venous drainage of the pancreas mainly follows the arterial system and drains into the portal vein via either the splenic vein, from the body and tail of the pancreas, or via the superior mesenteric vein from the pancreatic head venous arcade system.[2]

Microscopic Vascular Anatomy of the Pancreas

The islets of Langerhans, which compromise nearly 2 percent of the total pancreatic volume, are highly vascularized structures within the pancreas, receiving approximately tenfold the blood supply of the exocrine tissue by volume, with the intra-islet microcapillary network constituting

7–8 percent of the total islet volume[27,28] (**Figure 1**). The islets are irregular in shape, but mostly circular to oval, scattered individual endocrine cells, not contained within an islet, are found throughout the pancreatic acinar and ductal tissue (**Figure 2**). The ratio of beta to alpha cells is higher in mice than in humans, with beta cells making up the majority of endocrine cells within an islet (50–70% and 60–80% in humans and mice, respectively). Alpha cells compose 20–40 percent and 10–20 percent of endocrine cells in human and mouse islets, respectively[2,3] (**Figure 3**). Depending on the size of the islet, 1–5 arterioles enter each

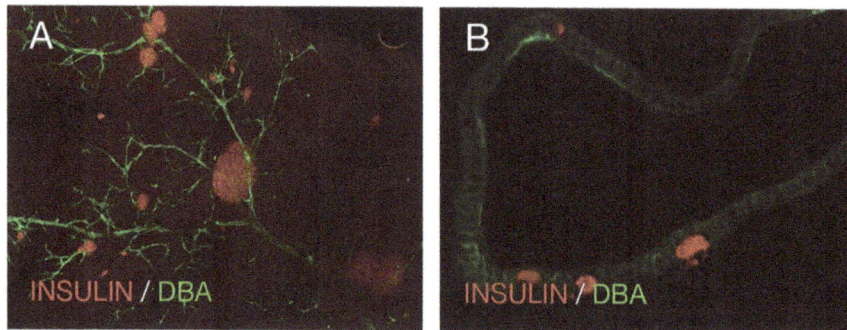

Figure 2. (**A**) Whole-mount image of wild-type mouse pancreatic tissue demonstrating the pancreatic duct (DBA) and different shapes and sizes of the islets (red). (**B**) A pancreatic section from a wild-type mouse showing scattered insulin positive endocrine cells (red) in a pancreatic duct (DBA). DBA: dolichos biflorus agglutinin, a ductal marker.

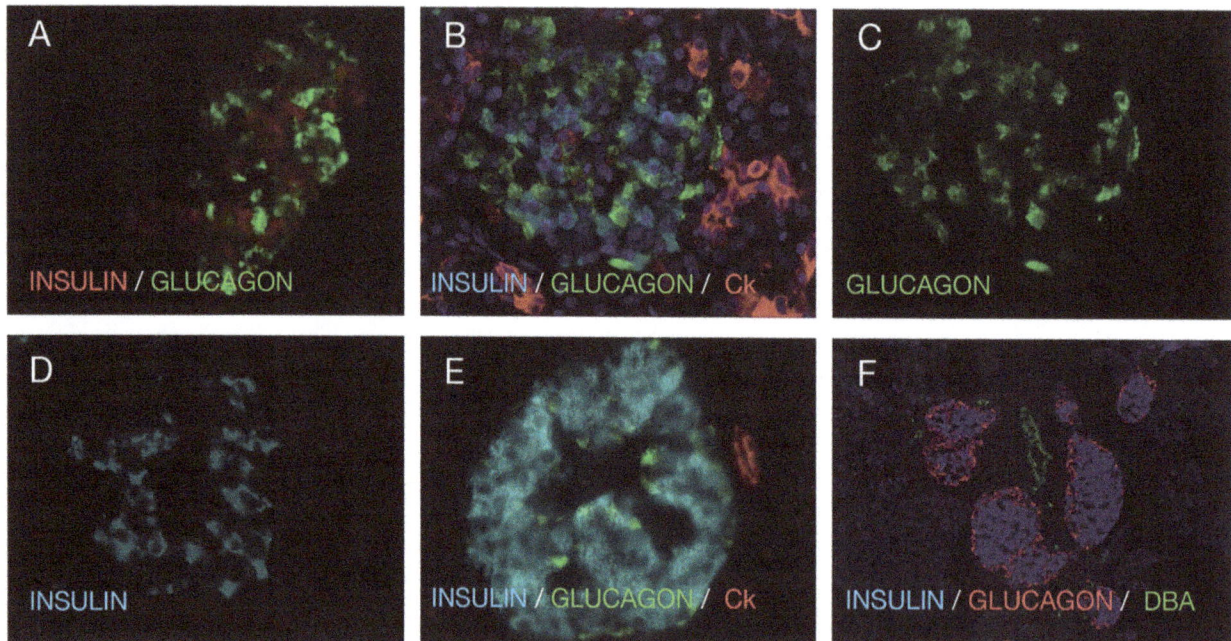

Figure 3. (**A**) Whole-mount image of a two-month-old human pancreatic islet. (**B–D**) Pancreatic section of a 16-year-old human pancreatic islet revealing that up to 40 percent of the islet is composed of alpha cells (**C**). (**E**) Section of a two-month-old small (<60um) human pancreatic islet demonstrating the mantle-core distribution of alpha and beta cells. (**F**) Section of a wild-type mouse pancreas revealing the typical mantle distribution of the alpha cells with 80 percent of the islet composed of insulin-secreting beta cells. Pancreatic duct cells are labeled by DBA.

islet and divide into a dense fenestrated capillary network[15] (**Figure 4**). In most rodents, the distribution of endocrine cells within an islet follows a mantle distribution of non-beta cells, including alpha, delta, and pancreatic polypeptide, with beta cells filling up the core (**Figure 3F**). This unique distribution appears to have functional implications. When beta cells are intermingled with non-beta cells instead of the mantle-core segregation, the beta-cell function was reduced.[29] Bosco et al.[30] demonstrated in their morphological analysis of the human pancreas that the alpha cell distribution was not random. In small human islets (40–60 μm in diameter) the typical mantle-core distribution was seen with alpha cells in the periphery and beta cells occupying the core. In larger human islets (>60 μm in diameter) it was noted that alpha cells were found in a mantle, but they also always lined the inner vascular channels, which are the empty spaces found in the middle of the islet, thus always juxtaposed to endothelial cells. Their three-dimensional analysis revealed that islet cells are organized into a trilaminar plate composed of one layer of beta cells sandwiched between two layers of alpha cells. The islets are folded onto each other with different degrees of complexity and vessels invaginating the "empty spaces" within the islets.[30]

There have been numerous islet vascular perfusion studies conducted and it appears, at least in mice, that there are three microcirculatory flow patterns. Firstly, from center to periphery (i.e. alpha cells are exposed to blood first). Secondly, from center to periphery (i.e. beta cells are exposed to blood first) and lastly, from one pole of the islet to the other. Roughly, 66 percent of islets have the center-to-periphery pattern[16,31] (**Figure 5**).

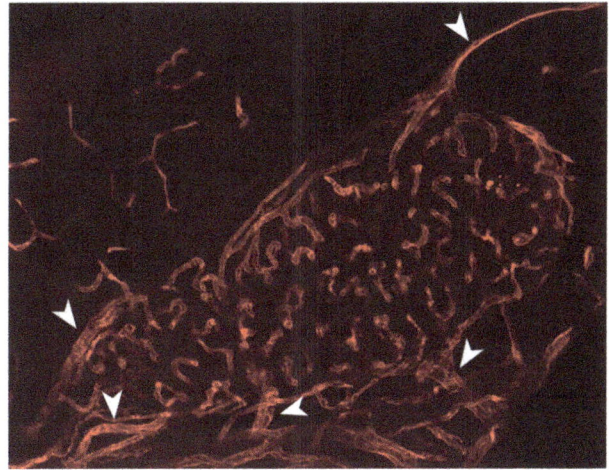

Figure 4. Whole mount confocal microscopy of a mouse islet with arrowheads indicating five vessels, two arterioles (thin caliber vessel), and three venules (larger caliber vessel).

References

1. Gittes GK. Developmental biology of the pancreas: a comprehensive review. Dev Biol 2009; 326:4–35.
2. Dolensek J, Rupnik MS, Stozer A. Structural similarities and differences between the human and the mouse pancreas. Islets 2015; 7:e1024405.
3. El-Gohary Y, Tulachan S, Branca M, Sims-Lucas S, Guo P, Prasadan K, Shiota C, Gittes GK. Whole-mount imaging demonstrates hypervascularity of the pancreatic ducts and other pancreatic structures. Anat Rec (Hoboken) 2012; 295:465–473.
4. Herbert SP, Stainier DY. Molecular control of endothelial cell behaviour during blood vessel morphogenesis. Nat Rev Mol Cell Biol 2011; 12:551–564.

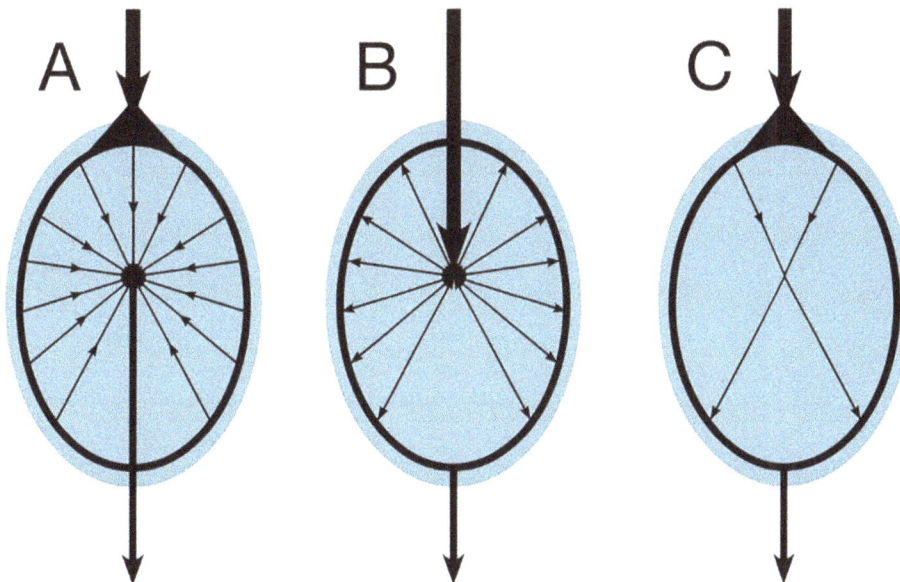

Figure 5. Three blood flow patterns of the islet microcirculation. (**A**) From periphery to center; (**B**) from center to periphery; (**C**) from one pole to the other.

5. Tsuchiya R, Fujisawa N. On the etymology of "pancreas". Int J Pancreatol 1997; 21:269–272.

6. Howard JM, Hess W. History of the pancreas: mysteries of a hidden organ. New York: Kluwer Academic, 2002.

7. Henderson JR, Daniel PM, Fraser PA. The pancreas as a single organ: the influence of the endocrine upon the exocrine part of the gland. Gut 1981; 22:158–167.

8. Henderson JR, Daniel PM. A comparative study of the portal vessels connecting the endocrine and exocrine pancreas, with a discussion of some functional implications. Q J Exp Physiol Cogn Med Sci 1979; 64:267–275.

9. Shimoda M, Itoh T, Sugimoto K, Iwahashi S, Takita M, Chujo D, Sorelle JA, Naziruddin B, Levy MF, Grayburn PA, Matsumoto S. Improvement of collagenase distribution with the ductal preservation for human islet isolation. Islets 2012; 4:130–137.

10. Michaels RL, Sheridan JD. Islets of Langerhans: dye coupling among immunocytochemically distinct cell types. Science 1981; 214:801–803.

11. Bank HL. Assessment of islet cell viability using fluorescent dyes. Diabetologia 1987; 30:812–816.

12. Bonner-Weir S, Orci L. New perspectives on the microvasculature of the islets of Langerhans in the rat. Diabetes 1982; 31:883–889.

13. Prinzen FW, Bassingthwaighte JB. Blood flow distributions by microsphere deposition methods. Cardiovasc Res 2000; 45:13–21.

14. Jansson L, Barbu A, Bodin B, Drott CJ, Espes D, Gao X, Grapensparr L, Kallskog O, Lau J, Liljeback H, Palm F, Quach M, et al. Pancreatic islet blood flow and its measurement. Ups J Med Sci 2016; 121:81–95.

15. El-Gohary Y, Sims-Lucas S, Lath N, Tulachan S, Guo P, Xiao X, Welsh C, Paredes J, Wiersch J, Prasadan K, Shiota C, Gittes GK. Three-dimensional analysis of the islet vasculature. Anat Rec (Hoboken) 2012; 295:1473–1481.

16. Nyman LR, Wells KS, Head WS, McCaughey M, Ford E, Brissova M, Piston DW, Powers AC. Real-time, multidimensional in vivo imaging used to investigate blood flow in mouse pancreatic islets. J Clin Invest 2008; 118:3790–3797.

17. Berclaz C, Pache C, Bouwens A, Szlag D, Lopez A, Joosten L, Ekim S, Brom M, Gotthardt M, Grapin-Botton A, Lasser T. Combined Optical Coherence and Fluorescence Microscopy to assess dynamics and specificity of pancreatic beta-cell tracers. Sci Rep 2015; 5:10385.

18. Li G, Wu B, Ward MG, Chong AC, Mukherjee S, Chen S, Hao M. Multifunctional in vivo imaging of pancreatic islets during diabetes development. J Cell Sci 2016; 129:2865–2875.

19. Balhuizen A, Massa S, Mathijs I, Turatsinze JV, De Vos J, Demine S, Xavier C, Villate O, Millard I, Egrise D, Capito C, Scharfmann R, et al. A nanobody-based tracer targeting DPP6 for non-invasive imaging of human pancreatic endocrine cells. Sci Rep 2017; 7:15130.

20. Azizoglu DB, Chong DC, Villasenor A, Magenheim J, Barry DM, Lee S, Marty-Santos L, Fu S, Dor Y, Cleaver O. Vascular development in the vertebrate pancreas. Dev Biol 2016; 420:67–78.

21. Lammert E, Cleaver O, Melton D. Induction of pancreatic differentiation by signals from blood vessels. Science 2001; 294:564–567.

22. Shah SR, Esni F, Jakub A, Paredes J, Lath N, Malek M, Potoka DA, Prasadan K, Mastroberardino PG, Shiota C, Guo P, Miller KA, et al. Embryonic mouse blood flow and oxygen correlate with early pancreatic differentiation. Dev Biol 2011; 349:342–349.

23. Magenheim J, Ilovich O, Lazarus A, Klochendler A, Ziv O, Werman R, Hija A, Cleaver O, Mishani E, Keshet E, Dor Y. Blood vessels restrain pancreas branching, differentiation and growth. Development 2011; 138:4743–4752.

24. Lewis MP, Reber HA, Ashley SW. Pancreatic blood flow and its role in the pathophysiology of pancreatitis. J Surg Res 1998; 75:81–89.

25. Woodburne RT, Olsen LL. The arteries of the pancreas. Anat Rec 1951; 111:255–270.

26. Bertelli E, Di Gregorio F, Mosca S, Bastianini A. The arterial blood supply of the pancreas: a review. V. The dorsal pancreatic artery. An anatomic review and a radiologic study. Surg Radiol Anat 1998; 20:445–452.

27. Saito K, Yaginuma N, Takahashi T. Differential volumetry of A, B and D cells in the pancreatic islets of diabetic and nondiabetic subjects. Tohoku J Exp Med 1979; 129:273–283.

28. Henderson JR, Moss MC. A morphometric study of the endocrine and exocrine capillaries of the pancreas. Q J Exp Physiol 1985; 70:347–356.

29. Gannon M, Ray MK, Van Zee K, Rausa F, Costa RH, Wright CV. Persistent expression of HNF6 in islet endocrine cells causes disrupted islet architecture and loss of beta cell function. Development 2000; 127:2883–2895.

30. Bosco D, Armanet M, Morel P, Niclauss N, Sgroi A, Muller YD, Giovannoni L, Parnaud G, Berney T. Unique arrangement of alpha- and beta-cells in human islets of Langerhans. Diabetes 2010; 59:1202–1210.

31. Nyman LR, Ford E, Powers AC, Piston DW. Glucose-dependent blood flow dynamics in murine pancreatic islets in vivo. Am J Physiol Endocrinol Metab 2010; 298:E807–E814.

Chapter 25

Secretion of insulin in response to diet and hormones

Elizabeth Mann,[1] Muna Sunni,[2] and Melena D. Bellin[2]

[1]*Division of Pediatric Endocrinology and Diabetes, University of Wisconsin-Madison*
[2]*Division of Pediatric Endocrinology, Department of Pediatrics, University of Minnesota*

The Dual Nature of the Pancreas

The pancreas is a complex gland active in digestion and metabolism through secretion of digestive enzymes from the exocrine portion and hormones from the endocrine portion. The exocrine pancreas, which accounts for more than 95–98 percent of the pancreas mass,[1] is structurally comprised of lobules, with acinar cells surrounding a duct system. The endocrine pancreas makes up only 2 percent of the pancreatic mass and is organized into the islets of Langerhans—small semi-spherical clusters of about 1,500 cells[2] dispersed throughout the pancreatic parenchyme—which produce and secrete hormones critical for glucose homeostasis. The existence of islets was described by Paul Langerhans in 1869, and the functional role of islets in glucose homeostasis was first demonstrated in 1890 when Joseph von Mering and colleagues showed that dogs developed diabetes mellitus following pancreatectomy.[3] Though islet mass may vary between individuals—an example is the increase in the setting of adult obesity[4]—the average adult human pancreas is estimated to contain one to two million islets.[5,6] In humans, the concentration of islets is up to two times higher in the tail compared to the head and neck. However, the cellular composition and architectural organization of cell types within the islets are preserved throughout the pancreas.[7]

Pancreatic islets are composed of α, β, δ, ε, and PP (pancreatic polypeptide) cells; these are primarily endocrine (hormone-secreting) cells, containing numerous secretory granules with stored hormone molecules, ready for release upon receipt of the appropriate stimulus. Insulin-producing β-cells are the most common cell type, making up 50–70 percent of islet mass, with small islets containing a greater percentage of β-cells in contrast to moderate or large islets.[8,9] β-cells were first discovered in 1907 by silver staining[10] and were the second islet cell type discovered, thus designated "β" cells. In addition to insulin, β-cells also produce islet amyloid polypeptide (IAPP), or amylin, which is packaged and released within insulin-containing granules.[11] Amylin reduces post-prandial hyperglycemia by slowing gastric emptying and promoting satiety.

Glucagon-producing α cells were discovered before β-cells, by alcohol fixation, thereby garnering their name "α" –cells.[10] As the second most abundant islet cell type, they make up about 35 percent of islet mass in humans[12] but less in rodents. Glucagon's primary function is to prevent hypoglycemia by stimulating glycogenolysis and hepatic gluconeogenesis.[13] Somatostatin-producing δ-cells comprise less than 10 percent of islet mass and are evenly distributed throughout the pancreas.[14] Somatostatin is an inhibitory peptide hormone, inhibiting both endocrine and gastrointestinal hormones. Pancreatic polypeptide[15] producing cells, also known as PP or F cells,[1,16] comprise less than 5 percent of islet mass and, like α cells, are most prominent in the head of the pancreas. Pancreatic polypeptide (PP) has roles in exocrine and endocrine secretion functions of the pancreas.[17] Ghrelin-producing ε cells are the last discovered islet endocrine cell type. Although present in islets, ghrelin is predominately produced in the stomach; ghrelin suppresses insulin release and plays a role in regulating energy homeostasis.[18]

The close proximity of the acini and the islets of Langerhans mirrors their functional interplay. The anatomic structure of the pancreatic parenchyme allows for a paracrine effect of the islet hormones on adjacent acinar cells, termed the "islet-acinar" axis.[19,20] Notably, the islets are highly vascularized—receiving 15 percent of pancreatic arterial blood flow despite composing only 2 percent of the pancreatic mass.[21] Via the islet-acinar portal system, blood bathing the pancreatic islets flows into a capillary bed within the pancreatic acini, thus exposing the acinar pancreas to the islet hormones.[22] Insulin binds to an insulin receptor on acinar tissue and potentiates amylase secretion.[23] In contrast, somatostatin inhibits pancreatic exocrine secretion;[24] endogenous PP is also largely noted to inhibit pancreatic exocrine secretion.[17,25] Studies have been inconsistent with regard to the effect of glucagon, some suggesting a

e-mail: bell0130@umn.edu

stimulatory effect while many suggesting an inhibitor effect of glucagon on secretion of zymogen granules.[19]

Insulin Structure

The hormone insulin was first isolated in the 1920s by Dr. Frederick Banting and a medical student Charles Best, garnering Banting (jointly with John James Rickard Macleod) the Nobel Prize in Medicine in 1923. This was a critical step forward in diabetes care, as porcine insulin therapy was then made available for human use to treat type 1 diabetes, an otherwise fatal disease. In the 1950s Frederick Sanger determined its primary amino acid structure, consisting of an A and a B chain connected by disulfide bonds.[26,27] Ten years following this discovery, these chains were found to be from the same polypeptide precursor, preproinsulin. In the 1960s Dorothy Hodgkin defined its tertiary structure. During translation of preproinsulin from its mRNA, the N-terminal signal peptide is cleaved to yield proinsulin. The proinsulin molecule is a single-chain polypeptide containing both the A-chain (21 amino acids long) and the B-chain (30 amino acids long). In proinsulin, two chains are connected by C-peptide, which is cleaved to release C-peptide, and the remaining insulin molecule, which contains the A- and B-chains connected via two disulfide bonds.[26] Although insulin and C-peptide are co-released from β-cell secretory vesicles into circulation,[28] only insulin is biologically active in regulating blood glucose. C-peptide, however, can serve as a useful clinical and research measure of endogenous insulin production in patients receiving exogenous insulin injections.

Insulin Gene Transcription

The *insulin* gene on chromosome 11 is primarily expressed in pancreatic β-cells but is expressed in low levels in the brain, thymus, and in the yolk sac during fetal development.[29–31] It has three exons and two introns, and its transcription results in the 446 base pair preproinsulin mRNA (**Figure 1**).

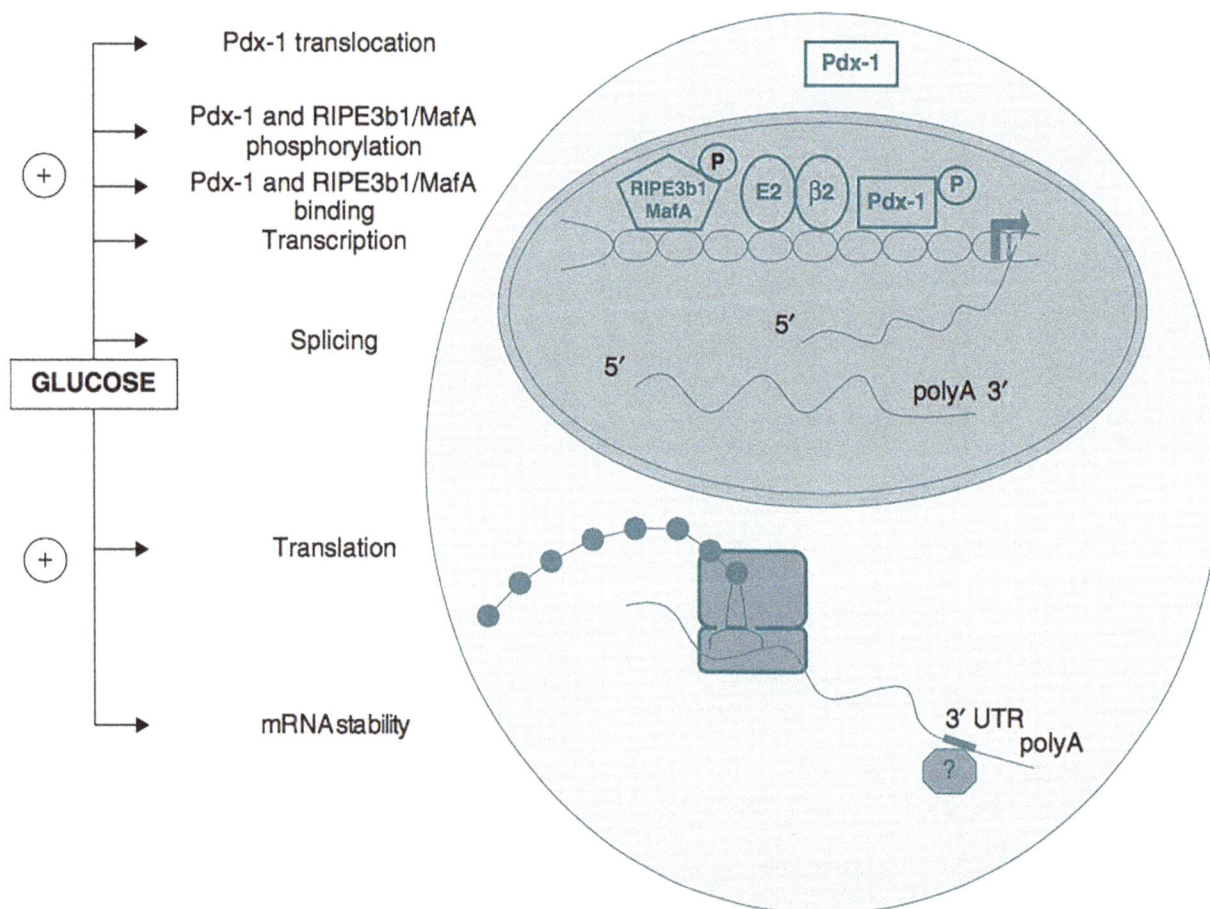

Figure 1. Various levels of glucose regulation of *insulin* gene expression. Glucose stimulates nuclear translocation of Pdx-1; promotes Pdx-1 and MafA phosphorylation and binding to the insulin promoter; and stimulates transcription of the *insulin* gene, pre-mRNA splicing, translation, and mRNA stability (used with permission from Poitout et al.[32]).

Transcription of the *insulin* gene to preproinsulin mRNA is sophisticated and reflects the tight regulation by transcription factors and recruited coactivators. Pdx-1, NeuroD1, and MafA are important transcription cofactors in beta-cell function, and their activity is responsive to elevated glucose levels. Individual β-cells respond to ambient glucose with differential insulin secretion, and these changes are apparent at the level of gene transcription.[33] At the level of the islet, rapid increase in blood glucose results in rapid elevation in preproinsulin mRNA in the endocrine pancreas. A rapid decrease in blood glucose results in a slow decline in preproinsulin mRNA. This is due to the unusual stability of preproinsulin mRNA, further stabilized by increased glucose concentrations.[34] The specific regulation of this molecule's translation is the primary mechanism of insulin production control.[32]

Mature insulin-containing granules are retained from a few hours up to several days within the β-cell, ready for transport to plasma membrane and exocytosis when stimulated. The storage of insulin in mature β granules is far greater than that secreted.[35,36] During a one-hour glucose stimulation only ~1–2 percent of insulin within a primary islet β-cell is released.[37] The insulin content within a given β-cell remains relatively constant in the short term, but in the long term it will adapt in response to physiologic demands.[37]

Insulin Function

In an evolutionary milieu of sporadic access to nutrients, insulin became critical in facilitating survival. As an anabolic hormone, insulin controls metabolism of carbohydrates, lipids, and protein. It mediates the availability of energy sources in both fasting and fed states. Insulin promotes energy storage in the fasting state and energy utilization and uptake in the fed state (**Table 1**). In so doing, it maintains serum glucose levels within a narrow physiologic range despite variation in energy intake and expenditure. Insulin acts at extracellular insulin receptors in multiple organ tissues including the liver, muscle, and adipose tissue,[1] and its effect depends on interstitial insulin concentration, which is influenced by insulin secretion rate from β-cells and clearance from circulation.[38]

The liver serves as the primary storage site for glucose, accounting for 80 percent of glucose production in fasting states, with the kidney only contributing 20 percent.[40,41] To preserve glucose stores, the low insulin concentrations in the portal venous blood—as seen in the fasting state—allows minimal glucose production, only enough to match the needs of essential glucose-dependent tissues including the red blood cells and the central and peripheral nervous systems. The liver also clears insulin more rapidly in the fasting state, thus maintaining low circulating insulin levels. Low insulin concentrations also contribute to lipolysis in adipocytes, releasing free fatty acids to encourage utilization of lipid over glucose to meet resting energy needs. Hepatic glucose release during fasting states through glycogenolysis and gluconeogenesis is stimulated by counterregulatory, or "anti-insulin" hormones. Glucagon plays a major role, with synergistic effects from catecholamines, cortisol, and growth hormone.[38]

Table 1. Endocrine effects of insulin.

Tissue	Effect of insulin

Liver	Catabolic Pathways
	    Inhibits glycogenolysis
	    Inhibits conversion of fatty acids and amino acids to keto acids
	    Inhibits conversion of amino acids to glucose
	Anabolic Pathways
	    Promotes glucose storage as glycogen (induces glucokinase and glycogen synthase, inhibits phosphorylase)
	    Increases triglyceride synthesis and VLDL formation

Muscle	Protein Synthesis
	    Increases amino acid transport
	    Increases ribosomal protein synthesis
	Glycogen Synthesis
	    Increases glucose transport
	    Induces glycogen synthetase
	    Inhibits phosphorylase

Adipose Tissue	Triglyceride Storage
	    Lipoprotein lipase is induced by insulin to hydrolyze triglycerides in circulating lipoproteins for delivery of fatty acids to the adipocytes
	    Glucose transport into cell provides glycerol phosphate to permit esterification of fatty acids supplied by lipoprotein transport
	    Intracellular lipase is inhibited by insulin

Brain	Decreased appetite Increased energy expenditure

Adapted from Masharani and German.[39]

By contrast, in the fed state—in response to digestion and absorption of nutrients—circulating insulin concentration increases in the portal vein secondary to insulin secretion from pancreatic β-cells. The increased insulin and glucose concentrations normally limit hepatic glucose production and stimulate liver glucose uptake through glycogen deposition.[42–44] Insulin causes upregulation of hexokinase, phosphofructokinase, and glycogen synthase within hepatocytes, thus inhibiting glycogenolysis and gluconeogenesis and stimulating glycogen synthesis.[40,45]

The effect of insulin on gluconeogenesis can be direct (via its effect on the liver) or indirect via its effect on islet α cells (by decreasing glucagon secretion), adipose tissue (by suppressing lipolysis), skeletal muscle (by reducing proteolysis), and the brain (pleiotropic effect).[44,46]

In situations when there is poor insulin response such as type 2 diabetes mellitus or insulin resistance, the process of gluconeogenesis continues even in the fed state, thus further compounding hyperglycemia.[44] Liver clearance of insulin is decreased in the fed state, thus further increasing the circulating insulin concentration. In adipocytes, insulin upregulates lipoprotein lipase and downregulates hormone sensitive lipase, which inhibits lipolysis and subsequent free fatty acid release.[47] In hepatocytes, insulin instead stimulates hepatic free fatty acid synthesis from glucose, thereby increasing lipid stores. Proteolysis of skeletal muscle is also inhibited by insulin, which along with lipolysis inhibition, limits delivery of glucose precursors (glycerol and amino acids) to the liver. Systemic circulation of insulin stimulates glucose uptake and utilization in skeletal muscle and adipocytes.

In summary, the release of insulin in the fed state (1) promotes accumulation of energy stores through glycogenesis and lipogenesis, (2) reduces new hepatic glucose output by preventing glycogenolysis and gluconeogenesis (in the non-insulin resistant, non-diabetic individual), and (3) promotes uptake of glucose by skeletal muscle and fat, the net effect of which is to maintain a normal circulating serum glucose levels while storing extra energy for use during later periods of fasting (**Figure 2**).

Glucose movement into cells is made possible by specific protein transporters within the plasma membrane of glucose-responsive cells that reversibly bind glucose and transport it bidirectionally across the cell membrane. There

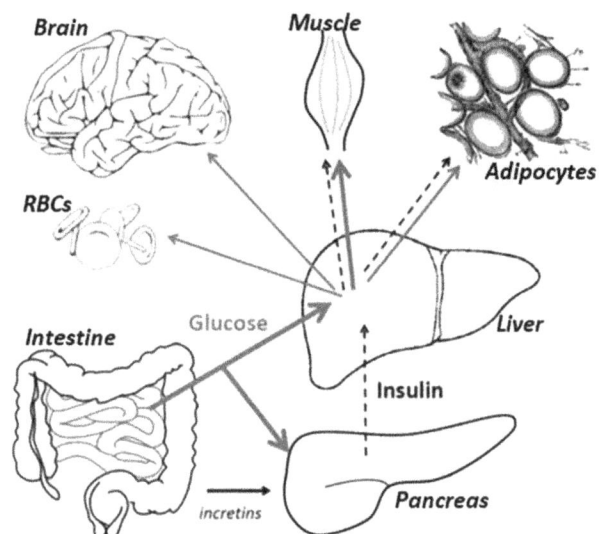

Figure 2. Glucose homeostasis in the fed state. Glucose absorbed from the digestive tract enters the portal blood flow and then systemic circulation. In the fed state, increased glucose stimulates insulin release from the pancreatic β-cells. Insulin acts at the level of the liver to inhibit hepatic gluconeogenesis, at the skeletal muscle to promote storage of glucose as glycogen, and in the adipocytes to stimulate lipogenesis. High insulin levels inhibit the release of non-esterified fatty acids. Incretin hormones released from small intestine in response to a meal augment pancreatic glucose-stimulated insulin secretion. Brain and red blood cells take up glucose independently of insulin in the fasting and fed state. In the fasting state (not shown), in the setting of low circulating insulin, hepatic gluconeogenesis, glycogenolysis, and release of non-esterified fatty acids occur. Solid line stimulation; dashed lines denote inhibition.

are 14 known glucose transporters (GLUTs).[48,49] They are present in different concentrations and in different tissues, with varying sensitivity to insulin (**Table 2**).

Tissues such as muscle and adipocytes carry the insulin-dependent glucose transporter GLUT-4 and uptake of glucose into these tissues occurs only under conditions of adequate circulating insulin. In contrast, vital organs such as red blood cells, brain, placenta, and kidney carry insulin-independent glucose transporters. Thus, these latter essential organs can continue to function even in states of insulin deficiency. β-cells also depend upon on a

Table 2. Most common glucose transporters (GLUT) in human tissues.

Glucose transporter	Insulin sensitivity	Tissue(s)
GLUT-1	Insulin-independent	RBCs; blood brain barrier; more recently identified in human β-cell
GLUT-2	Insulin-independent	β-cell; kidney, liver, intestinal cells
GLUT-3	Insulin-independent	Blood-brain barrier, placenta
GLUT-4	Insulin-dependent	Skeletal muscle, smooth muscle, cardiac muscle, adipocytes

glucose-independent transporter, GLUT2, to allow ambient blood glucose to freely transverse the β-cell membrane in order to stimulate insulin production.

Insulin Secretory Pathway

The pancreatic β-cells act as a self-contained system to secrete insulin in response to changes in ambient blood glucose concentration, in order to maintain glucose homeostasis. Glucose is freely taken up into the β-cell via GLUT transporters, metabolized to produce ATP, which triggers a cascade of signals within the β-cell necessary for glucose-induced insulin secretion. While GLUT2 has been traditionally assumed as the major mediator of glucose uptake into β-cells based on extrapolation from rodent studies and subsequent confirmation of GLUT2 transporters on human β-cells,[50–52] more recent studies in human islets suggest that the other insulin-independent glucose transporters GLUT1 and GLUT3 play a more important role, and are the main glucose transporters in human islet β-cells.[53,54] This redundancy explains why individuals with variants in the gene encoding GLUT2 (SLC2A2 mutations, or Fanconi–Bickel syndrome) do not have significant abnormalities in insulin secretion.[55]

As blood glucose increases (e.g., after a meal), there is a resultant flux of glucose across the GLUT transporters in the β-cell. Subsequently, within the β-cell, glucose is phosphorylated to glucose-6-phosphate by glucokinase. This is the rate-limiting step of insulin secretion, and as such, glucokinase is considered the "glucose sensor" for the β-cell.[50,56] Because of this critical role of glucokinase, individuals with heterozygous mutations in the glucokinase gene have a mild to moderate non-progressive hyperglycemia (maturity onset of diabetes in the young, type 2).[57] Once in the mitochondria, glucose-6-phosphate is metabolized by the Krebs cycle to produce ATP. The resultant ATP binds and closes the ATP-dependent potassium channel, a pore across the cell membrane, which consists of four Kir6.2 subunits and has four regulatory SUR (sulfonylurea receptor) subunits. Channel closure blocks potassium exit from the β-cell, thus depolarizing the cell membrane.

Once the cell is depolarized, the L-type voltage-gated calcium channels are triggered, increasing influx of calcium and resultant cellular calcium concentrations. Increased cytoplasmic calcium concentrations triggers release of insulin and C-peptide from a pool of insulin-containing docked secretory vesicles and stimulates the migration of additional vesicles to the cell membrane (**Figure 3**). Though simple glucose-stimulated insulin secretion (GSIS), as described earlier, is considered the primary pathway for insulin secretion, the full picture is more nuanced. GSIS is augmented by amplifying pathways including (1) metabolic amplification by amino acids, free fatty acids, and glucose itself;

and (2) neurohormonal amplifiers such as GLP-1 and parasympathetic innervation.[59–62] More recent data from mice suggest a role for skeletal muscle in regulating β-cell insulin secretion via production of an anorexic factor typically derived from the hypothalamus in the brain called BDNF (brain-derived neurotrophic factor).[63] This effect is mediated via the BDNF receptor (TrkB.T1), which is expressed on β-cells, and is thought to play a potential role in exercise-induced glucose metabolism.[63] These physiologic, and pharmacologic, triggers for insulin secretion are further described in the following sections. About half of insulin secretion occurs as basal insulin release, while the other half occurs as "bolus" insulin responses to a meal.[64] This basal-bolus dynamic of insulin secretion is important in considering clinical management of the patient with diabetes (**Figure 4**). In those with complete insulin deficiency— for example, type 1 diabetes, late-stage type 2 diabetes, or late-stage chronic pancreatitis diabetes—insulin analogs are administered by multiple daily injections or a continuous subcutaneous insulin infusion (insulin pump) to mimic this basal-bolus pattern of endogenous insulin secretion.

Regulation of Insulin Release

Insulin release from pancreatic β-cells is tightly regulated and allows the sensitive response of insulin levels to calorigenic nutrients in the body. Glucose, free fatty acids, and amino acids serve as fuel stimuli for insulin release, promoting insulin granule exocytosis. Additional hormonal factors influence the regulation pathway. Pharmacological agents can also be used to augment insulin release.

Glucose-stimulated insulin secretion

Glucose-stimulated β-cell insulin release is the primary mechanism of insulin regulation (**Figure 3**).[66,67] In humans, this is illustrated by use of the hyperglycemic clamp technique (**Figure 5**), in which individuals are made rapidly hyperglycemic by injection of intravenous dextrose, and hyperglycemia is maintained by variable rate dextrose infusion at a predefined target glucose.[68] Hyperglycemic clamp studies demonstrate a dose-response of insulin secretion in response to glucose concentration, with greater degrees of hyperglycemia eliciting a more robust insulin secretory response in the non-diabetic individual.[69,70] Using this research technique, two distinct phases of insulin secretion are observed. During the first-phase insulin response (otherwise referred to as the acute insulin response to glucose, AIRglu), there is an immediate and transient rise in insulin secretion, peaking by five minutes and lasting no more than ten minutes. This first phase of insulin secretion is hypothesized to largely result from the immediate release of insulin from insulin secretory vesicles that are already docked and

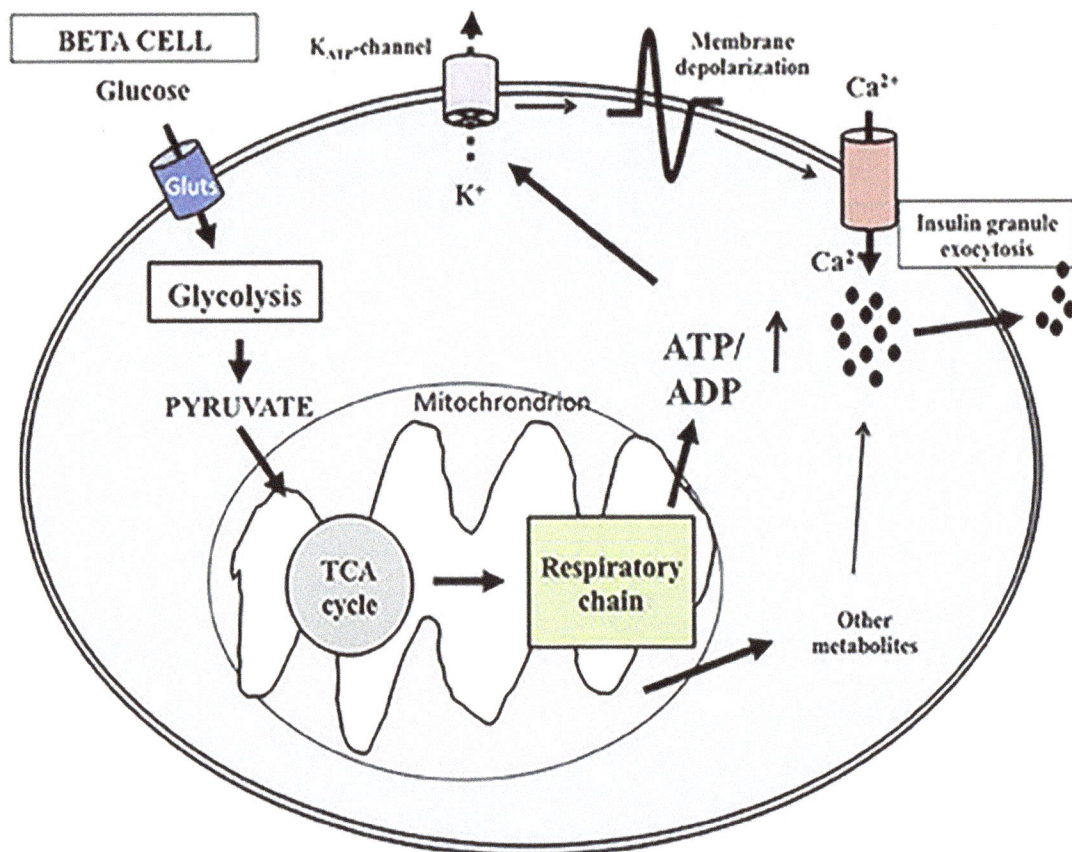

Figure 3. Glucose stimulated insulin-secretion coupling in the β-cell. The drawing shows the main pathway of glucose-stimulated insulin secretion in the beta cell. Glucose enters the beta cell through GLUT transporters. Glucose metabolism results in an enhanced cytoplasmic ATP/ADP ratio which prompts closure of ATP-sensitive K^+ (K_{ATP}) channels in the plasma membrane evoking membrane depolarization and subsequent opening of voltage-gated Ca^{2+} channels. This culminates in an increase in cellular Ca^{2+} influx—a primary driver of insulin exocytosis. Ca^{2+} and vesicle docking and fusion events can also be modulated by agents acting through the phospholipase C (PLC)/protein kinase C (PKC) or adenylate cyclase (AC)/protein kinase A (PKA) pathways, via neuro-hormonal and metabolic amplification (not illustrated) (used with permission from Marchetti et al.[58]).

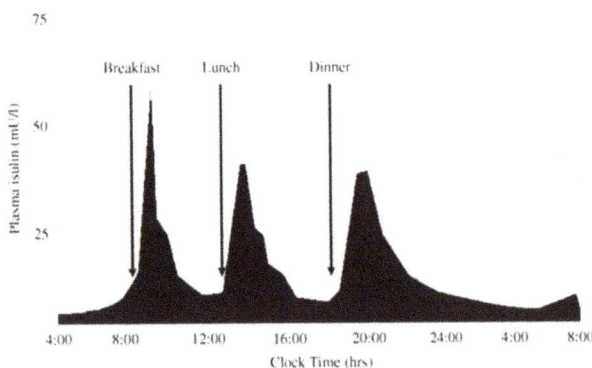

Figure 4. Diagrammatic illustration of insulin secretion. A low background secretion exists upon which is superimposed insulin secretory bursts stimulated by food intake (used with permission from Thompson et al.[65]).

primed at the β-cell membrane. This first-phase response is lost under conditions of diabetes mellitus, when β-cell reserves are exhausted.[71] The second sustained phase begins at this ten-minute time-point and lasts as long as the glucose elevation is elevated. The second phase results from recruitment of insulin secretory vesicles to the β-cell membrane, and is also controlled by intracellular calcium levels.[38]

It is unclear how significant first- and second-phase insulin responses are in the 'real world' setting. In contrast to this scenario of rapid infusion of intravenous glucose, ingestion of a physiologic meal results in a much more gradual rise of serum glucose.[72] However, characterization of first-phase insulin response is critically important in diabetes research. In progression to type 1 and type 2 diabetes mellitus, the earliest abnormality is a loss in the first-phase insulin secretion (measured as the AIRglu). Although chronic pancreatitis

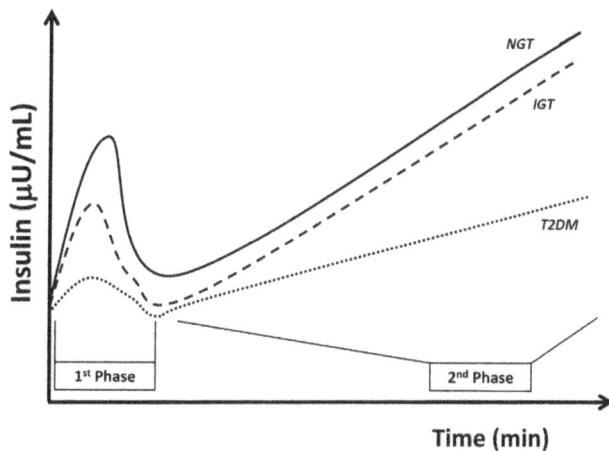

Figure 5. Hyperglycemic clamp illustration. Example of hyperglycemic clamp testing in obese adolescents with normal glucose tolerance (NGT, solid line), impaired glucose tolerance (IGT, dashed line), and type 2 diabetes (T2DM, dotted line). In the hyperglycemic clamp in healthy, non-diabetic individuals, glucose concentration is briskly elevated by administering a suitable intravenous glucose infusion at time 0. This elicits a rapid and short-lived insulin secretion peak (first-phase secretion) due to release of preformed insulin vesicles, followed by a drop toward basal levels and then by a relatively rapid return to a sustained increase in insulin in the second half of the clamp (second-phase secretion) as dextrose infusion is continued. This example illustrates the loss, in first and second-phase insulin secretion, as individual progress from normal to impaired glucose tolerance, to type 2 diabetes. In the latter, the first-phase insulin response is essentially lost and the second-phase insulin response is reduced (with permission from Weiss et al.[71]).

diabetes is much less studied, this appears likely also to be the case in patients with chronic pancreatitis progressing to diabetes based on limited studies, and often in patients with chronic pancreatitis who have diabetes or pre-diabetes.[40] The AIRglu can be elicited experimentally by administering a 0.3 g/kg dextrose bolus and sampling insulin levels at baseline and at +2, 3, 4, 5, 7 and 10 min after the rapid IV administration of dextrose. The AIRglu can be calculated using various methods, including, but not limited to, the area under the curve minus baseline or mean of the 2–5 min values minus baseline.

Interestingly, glucose also appears to be a "metabolic amplifier" for insulin secretion, in addition to the classic pathway of glucose-stimulated insulin secretion. Glucose amplifies insulin secretion, a process called time-dependent potentiation of insulin secretion—when β-cells are exposed to hyperglycemia this augments subsequent insulin secretory responses to glucose.[73]

Proteins and amino acids

Pancreatic β-cells adjust insulin secretion based on other nutrients, including amino acids, fatty acids, and ketone bodies. Oral protein intake and subsequent rise in serum amino acids stimulate insulin release by direct β-cell stimulation.[74–76] The insulinotropic effect varies among amino acids, and there appears to be a synergistic effect of mixed amino acids versus individual administration.[77]

Some amino acids stimulate insulin secretion by acting as substrates in the Krebs cycle, metabolizing glucose-6-phosphate to create ATP. Enzymes active in β-cell mitochondrial amino acid metabolism have been implicated in hyperinsulinemic hypoglycemic syndromes associated with high-protein containing meals.[62] The ATP binds to and closes the potassium channel, leading to cell depolarization and insulin secretion. There seems to be a direct effect of proteins and amino acids on β-cell glucose sensitivity, because ingestion of amino acids with glucose results in the same plasma insulin concentrations as elicited by a lower level of glucose alone.[78]

Lipids and free fatty acids

It is generally accepted that lipids play a role in insulin secretion signaling, but the precise pathways and molecules involved in the process remain less well understood. Lipid breakdown and metabolism to signaling molecules have been linked to glucose metabolism through enhanced membrane phospholipid metabolism turnover and other pathways yet to be firmly established. It is thought that free fatty acids (FFA) modulate β-cell insulin secretion either directly via GPR40 (G-protein coupled receptor on the β-cell) leading to insulin secretion, or indirectly via oxidation of FFA to acyl coA, which enters the Krebs cycle and generates ATP.[1]

Glucose and FFA metabolism have been shown to be tightly linked and likely includes malonyl-CoA/carnitine palmitoyltransferase I/fatty acyl-CoA metabolic signaling network and the glycerolipid/free fatty acid (GL/FFA) cycle.[79,80] The GL/FFA cycle along with the Krebs cycle and pyruvate cycling are the three likely interlined metabolic cycles that play essential roles in insulin secretion promoted by glucose, FFA, and amino acids.[62] In so doing, FFA work synergistically with glucose-stimulated insulin secretion to enhance insulin secretion in nutrient-abundant states.

Chronic elevation of fatty acids may increase basal insulin secretion levels but inhibits glucose-stimulated insulin secretion. Chain length and degree of saturation affect the role free fatty acids play in regard to insulin plasma levels.[81] Adipose tissue responds to insulin resistance with a persistently elevated rate of lipolysis, thus increasing the plasma free fatty acid levels. This is believed to be important in type 2 diabetes development.[82]

Incretin hormones

Though glucose concentrations can account for the majority of changes in insulin concentrations, complex studies evaluating *in vivo* insulin concentrations following meals have identified other factors.[83] Indeed, insulin secretion following an oral glucose tolerance test is directly related to blood glucose levels, but is considerably higher than predicted following intravenous glucose infusion. These findings suggest a role for potentiating effects on insulin release by hormones that specifically respond to oral glucose, the "incretin effect". This terminology is derived from intestinal hormones called *incretins*, which are credited with facilitating this response.

The most active incretins are glucagon-like peptide-1 (GLP-1) and glucose-dependent insulinotropic polypeptide (GIP),[84,85] but gastrin, secretin, and cholecystokinin may also have minor roles. In response to glucose and other nutrients, intestinal L cells secrete GLP-1 and K cells secrete GIP. These hormones then bind their specific receptors on the pancreatic β-cell membrane. GLP-1 binds a G protein-coupled receptor. This results in direct activation of *insulin* gene's cyclic-AMP (cAMP) response element (CRE) of the 5'-proximal control sequence.[86,87] It can also augment Pdx-1 binding in the setting of a glucose-stimulus and stimulate transcription of the PDX-1 gene.[88] Finally, it potentiates glucose-induced *insulin* gene transcription by activating NFAT (nuclear factor of activated T-cell).[89] The incretin effect is also mediated by glucose concentration, stimulating more insulin secretion in more extreme hyperglycemic states. GIP and GLP-1 receptors also exist on neuronal cells (e.g. nodose ganglion of the Vagus nerve), suggesting an additional indirect role in β-cell regulation.[90] GLP-1 and GIP are cleared by dipeptidyl peptidase-4 (DPP-4) which is present on vascular endothelium. As a result, their half-life in the circulation is 2–3 minutes and 4–5 minutes, respectively.[91]

Insulin's Counterregulatory Hormones

The tight control of energy utilization and stores by insulin is balanced by the counterregulatory hormones glucagon, pancreatic polypeptide, somatostatin, cortisol, catecholamines, and growth hormone. There is asymmetry in the glucose regulation hormones, as insulin is the only hormone to prevent hyperglycemia, while at least three other hormones (cortisol, glucagon, and adrenaline) prevent hypoglycemia. Collectively, these counter-regulatory hormones act to promote glucose release from the liver by glycogenolysis and gluconeogenesis, and inhibit glucose storage during times of starvation.

Glucagon is formed within pancreatic α islet cells and has a hyperglycemic effect on the body.[13] Its name is derived from *glucose agonist*.[92] Glucagon carries out its effects via activating its G-protein-coupled receptor that is found in various organs/tissues such as the liver, adipose tissue, kidneys, gastrointestinal (GI) tract, brain, and islet α- and β-cells.[93] It stimulates glucose production from

amino acids and glycerol through gluconeogenesis and from the liver through glycogenolysis. Glucagon also acts at the adipocyte to upregulate hormone-sensitive lipase, thereby enhancing lipolysis and free fatty acid delivery to the liver.[94] In the brain it increases satiety,[95] and in the GI tract it reduces GI motility.[96] Glucagon, via its autocrine role, stimulates further glucagon secretion through its effect on α-cells.[97] Interestingly, glucagon stimulates insulin secretion via glucagon's effect on β-cells. It is not clear if this effect is mediated mainly via glucagon's effect on glucagon receptors or on GLP-1 receptors.[93] This effect on insulin secretion occurs in the fed state.[98]

Mechanisms explaining glucagon secretion are not as well understood as those of insulin secretion, although the direct effect of reduced glucose on cAMP,[99] and the sodium-glucose cotransporters (SGLT) are thought to play a role in α-cell glucose transport.[54] Mice and human data suggest that α-cell inhibition can occur, at least in part, due to the paracrine action of somatostatin from δ-cells as a result of gap junction-dependent activation by adjacent β-cells.[100]

Cortisol antagonizes insulin's function by promoting protein catabolism to provide amino acid substrate for gluconeogenesis and also impairs peripheral insulin-mediated glucose uptake.

Catecholamines directly affect β-cell secretion of insulin, as activation of α-2 receptors inhibits insulin secretion and β stimulation increases it. Catecholamines promote adipocyte lipolysis, hepatic glycogenolysis and peripheral insulin resistance. Epinephrine inhibits insulin secretion through inhibiting the rate of *insulin* gene transcription.[101] Somatostatin also destabilizes the preproinsulin mRNA, resulting in premature degradation.[15]

Somatostatin is released from pancreatic islet δ-cells and exerts inhibitory effect on pancreatic β-cells. Once bound to specific somatostatin receptors, β-cell membrane repolarization is induced, resulting in reduction of calcium influx and thereby inhibiting insulin release.[67,101]

Pancreatic polypeptide[15] is secreted by PP, or F, cells in pancreatic islets.[17] In addition to its effects reducing gastric acid secretion, decreasing gastric emptying, and slowing upper intestinal motility, PP acts within the pancreas to self-regulate pancreatic insulin secretion.

Pharmacologic Modulators of Insulin Response

There is a plethora of pharmacologic agents designed to target various aspects of glucose metabolism. In this chapter, we provide examples of pharmacologic agents that directly or indirectly modulate insulin response.

Incretin mimetics

Diabetes therapeutics have recently utilized the role of incretin hormones for pharmacologic benefit. Due to the desirable effect of GLP-1 on hemoglobin A1c (HbA1c) reduction and weight loss,[102] GLP-1 receptor agonists and inhibitors of its degradation via dipeptidyl peptidase-4 (DPP-4) inhibitors have been used to treat type 2 diabetes since 2005.

Short-acting GLP-1 receptor agonists (such as exenatide and Liraglutide) and long-acting GLP-1 receptor agonists (such as weekly exenatide and Semaglutide) potentiate insulin secretion and reduce gastric motility.[103] Given that GLP-1 receptor agonists potentiate glucose-induced insulin gene transcription, they, alone, do not induce hypoglycemia when used as monotherapy.[16,104]

DPP-4 inhibitors (such as sitagliptin) can significantly increase the peak post-prandial concentration of GLP-1.[92] Additionally, sitagliptin has been found to potentiate GSIS independently of GLP-1 via islet peptide tyrosine tyrosine (PYY).[105]

Sulfonylureas

Through a direct action on pancreatic islet cells, sulfonylureas are pharmacological agents that stimulate insulin secretion, thereby lowering blood glucose levels. This class of medication was discovered by happenstance in 1942 when Marcel Janbon, a clinician at the Clinic of the Montpellier Medical School in France, found his patients treated for typhoid fever with a new sulfonamide (2254 RP) developed hypoglycemia. Shortly after this, his colleague Professor August Loubatieres established the hypoglycemic property of 2254 RP, and its analogues were by direct action on pancreatic islets. This marked the birth of sulfonylureas for treatment of certain forms of diabetes.[106] It was not until 50 years later that the mechanism of action was discovered. Sulfonylurea was found to bind to and block the potassium ATP channel on the β-cell surface, thus depolarizing the membrane and provoking calcium influx, raising intracellular calcium concentration, and triggering insulin secretion.[107,108] Sulfonylurea binding to the sulfonylurea receptor associated with the K-ATP channel stimulates events similar to those in response to glucose stimulation.

Sulfonylureas are also used in the chronic treatment of type 2 diabetes mellitus for their effects on both insulin release and blood glucose reduction. In contrast to acute use of sulfonylureas, chronic use results in improved blood glucose control, but with less rather than more insulin secretion.[109] Assessments of its chronic effects are difficult to interpret, given that the magnitude of sulfonylurea stimulation of insulin secretion are multifactorial.[110]

Insulin sensitizers

Biguanides (such as metformin) and Thiazolidenediones (such as pioglitazone) improve hepatic and peripheral (muscle and fat tissue) insulin sensitivity, respectively. Metformin is by far the most widely used pharmacologic agent as first-line therapy in patients with type 2 diabetes mellitus. Similar to thiazolidenediones, metformin has an effect on improving peripheral insulin sensitivity in addition to reducing hepatic glucose output. Contrary to thiazolidenediones and sufonylureas, metformin does not cause weight gain, and, in fact, it has a modest weight loss effect. When used as monotherapy, metformin does not induce hypoglycemia.[111]

Diazoxide

Diazoxide is a sulfonamide pharmacological agent used in the treatment of hyperinsulinism, insulinoma, and hypoglycemia due to overtreatment with sulfonylureas. It works by opening β-cell membrane potassium ATP channels, hyperpolarizing the β-cells, thus decreasing intracellular calcium concentration and inhibiting insulin secretion.[112]

Conclusion

In conclusion, although the pancreatic islets comprise only a small portion of the pancreas, pancreatic islets play a vital role in our well-being and survival through control of glucose homeostasis. Most critically, loss of insulin production from the pancreatic β-cells, whether due to autoimmune destruction in type 1 diabetes mellitus, exhaustion and genetic predisposition to failure in type 2 diabetes mellitus, or bystander fibrotic destruction in pancreatic exocrine disease, results in diabetes. Insulin secretion is tightly regulated in healthy non-diabetic individuals, with both insulin gene transcription and exocytosis from insulin-containing granules responsive to rises in ambient circulating blood glucose. Other nutrients (protein and lipid) play a smaller role. In contrast, the other pancreatic islet cells, particularly the glucagon-producing α-cells, play a key role in glucose counter-regulation to avoid dangerous hypoglycemia.

References

1. Jouvet N, Estall JL. The pancreas: Bandmaster of glucose homeostasis. Exp Cell Res 2017; 360:19–23.
2. Pisania A, Weir GC, O'Neil JJ, Omer A, Tchipashvili V, Lei J, Colton CK, Bonner-Weir S. Quantitative analysis of cell composition and purity of human pancreatic islet preparations. Lab Invest 2010; 90:1661–1675.
3. Dittrich HM. [History of the discovery of pancreatic diabetes by von Mering and Minkowski 1889. A historical overview on the occasion of the 100th anniversary]. Z Gesamte Inn Med 1989; 44:335–340.
4. Saisho Y, Butler AE, Manesso E, Elashoff D, Rizza RA, Butler PC. beta-cell mass and turnover in humans: effects of obesity and aging. Diabetes Care 2013; 36:111–117.

5. Hellman B. The frequency distribution of the number and volume of the islets Langerhans in man. I. Studies on non-diabetic adults. Acta Soc Med Ups 1959; 64:432–460.

6. Stefan Y, Orci L, Malaisse-Lagae F, Perrelet A, Patel Y, Unger RH. Quantitation of endocrine cell content in the pancreas of nondiabetic and diabetic humans. Diabetes 1982; 31:694–700.

7. Wang X, Misawa R, Zielinski MC, Cowen P, Jo J, Periwal V, Ricordi C, Khan A, Szust J, Shen J, Millis JM, Witkowski P, Hara M. Regional differences in islet distribution in the human pancreas-preferential beta-cell loss in the head region in patients with type 2 diabetes. PLoS One 2013; 8:e67454.

8. Bonner-Weir S, Sullivan BA, Weir GC. Human Islet Morphology Revisited: Human and Rodent Islets Are Not So Different After All. J Histochem Cytochem 2015; 63:604–612.

9. Bosco D, Armanet M, Morel P, Niclauss N, Sgroi A, Muller YD, Giovannoni L, Parnaud G, Berney T. Unique arrangement of alpha- and beta-cells in human islets of Langerhans. Diabetes 2010; 59:1202–1210.

10. Lane MA. The cytological characters of the areas of Langerhans. Am J Anat 1907; 7:409–422.

11. Kahn SE, D'Alessio DA, Schwartz MW, Fujimoto WY, Ensinck JW, Taborsky GJ, Jr., Porte D, Jr. Evidence of cosecretion of islet amyloid polypeptide and insulin by beta-cells. Diabetes 1990; 39:634–638.

12. Brissova M, Fowler MJ, Nicholson WE, Chu A, Hirshberg B, Harlan DM, Powers AC. Assessment of human pancreatic islet architecture and composition by laser scanning confocal microscopy. J Histochem Cytochem 2005; 53:1087–1097.

13. Bozadjieva N, Williams JA, Bernal-Mizrachi E. Glucagon. Pancreapedia: Exocrine Pancreas Knowledge Base 2013, DOI: 10.3998/panc.2013.23

14. Baetens D, Malaisse-Lagae F, Perrelet A, Orci L. Endocrine pancreas: three-dimensional reconstruction shows two types of islets of Langerhans. Science 1979; 206:1323–1325.

15. Philippe J. Somatostatin inhibits insulin-gene expression through a posttranscriptional mechanism in a hamster islet cell line. Diabetes 1993; 42:244–249.

16. Röder PV, Wu B, Liu Y, Han W. Pancreatic regulation of glucose homeostasis. Exp Mol Med 2016; 48:e219.

17. Williams JA. Pancreatic Polypeptide. Pancreapedia: Exocrine Pancreas Knowledge Base 2014, DOI: 10.3998/panc.2014.7

18. Tong J, Prigeon RL, Davis HW, Bidlingmaier M, Kahn SE, Cummings DE, Tschop MH, D'Alessio D. Ghrelin suppresses glucose-stimulated insulin secretion and deteriorates glucose tolerance in healthy humans. Diabetes 2010; 59:2145–2151.

19. Barreto SG, Carati CJ, Toouli J, Saccone GT. The islet-acinar axis of the pancreas: more than just insulin. Am J Physiol Gastrointest Liver Physiol 2010; 299:G10–G22.

20. Williams JA, Goldfine ID. The insulin-pancreatic acinar axis. Diabetes 1985; 34:980–986.

21. Jansson L, Hellerstrom C. Stimulation by glucose of the blood flow to the pancreatic islets of the rat. Diabetologia 1983; 25:45–50.

22. Murakami T, Fujita T, Taguchi T, Nonaka Y, Orita K. The blood vascular bed of the human pancreas, with special reference to the insulo-acinar portal system. Scanning electron microscopy of corrosion casts. Arch Histol Cytol 1992; 55:381–395.

23. Williams JA, Sankaran H, Korc M, Goldfine ID. Receptors for cholecystokinin and insulin in isolated pancreatic acini: hormonal control of secretion and metabolism. Fed Proc 1981; 40:2497–2502.

24. Morisset J. Somatostatin. Pancreapedia: Exocrine Pancreas Knowledge Base 2015, DOI: 10.3998.panc.2015.43

25. Shiratori K, Lee KY, Chang TM, Jo YH, Coy DH, Chey WY. Role of pancreatic polypeptide in the regulation of pancreatic exocrine secretion in dogs. Am J Physiol Gastrointest Liver Physiol 1988; 255:G535–G541.

26. Jameson JL, De Kretser DM, DeGroot LJ. Endocrinology: adult and pediatric. Philadelphia: Saunders/Elsevier, 2010.

27. Sanger F. Chemistry of insulin; determination of the structure of insulin opens the way to greater understanding of life processes. Science 1959; 129:1340–1344.

28. Rubenstein AH, Clark JL, Melani F, Steiner DF. Secretion of Proinsulin C-Peptide by Pancreatic [beta] Cells and its Circulation in Blood. Nature 1969; 224:697–699.

29. Giddings SJ, Carnaghi L. Rat insulin II gene expression by extraplacental membranes. A non-pancreatic source for fetal insulin. J Biol Chem 1989; 264:9462–9469.

30. Le Roith D, Hendricks SA, Lesniak MA, Rishi S, Becker KL, Havrankova J, Rosenzweig JL, Brownstein MJ, Roth J. Insulin in brain and other extrapancreatic tissues of vertebrates and nonvertebrates. Adv Metab Disord 1983; 10:303–340.

31. Smith KM, Olson DC, Hirose R, Hanahan D. Pancreatic gene expression in rare cells of thymic medulla: evidence for functional contribution to T cell tolerance. Int Immunol 1997; 9:1355–1365.

32. Poitout V, Stein R, Rhodes CJ. Insulin gene expression and biosynthesis. In: DeFronzo RA, Ferrannini E, Zimmet P, Alberti G, eds. International Textbook of Diabetes Mellitus. Volume 1.4 ed: John Wiley & Sons, Ltd, 2015:82.

33. de Vargas LM, Sobolewski J, Siegel R, Moss LG. Individual beta cells within the intact islet differentially respond to glucose. J Biol Chem 1997; 272:26573–26577.

34. Fred RG, Welsh N. The importance of RNA binding proteins in preproinsulin mRNA stability. Mol Cell Endocrinol 2009; 297:28–33.

35. MacDonald PE, Rorsman P. The ins and outs of secretion from pancreatic beta-cells: control of single-vesicle exo- and endocytosis. Physiology (Bethesda) 2007; 22:113–121.

36. Rorsman P, Renstrom E. Insulin granule dynamics in pancreatic beta cells. Diabetologia 2003; 46:1029–1045.

37. Uchizono Y, Alarcon C, Wicksteed BL, Marsh BJ, Rhodes CJ. The balance between proinsulin biosynthesis and insulin secretion: where can imbalance lead? Diabetes Obes Metab 2007; 9 (suppl 2):56–66.

38. Natali A, Del Prato S, Mari A. Normal β-cell function. In: DeFronzo RA, Ferrannini E, Zimmet P, Alberti G, eds. International textbook of diabetes mellitus. Volume 1.4 ed: John Wiley & Sons, Ltd, 2015:108.

39. Masharani U, German, M.S. Pancreatic Hormones and Diabetes Melitus. In: Gardner DG and Shobach D ed

Greenspan's Basic & Clinical Endocrinology. 10th ed. New York, NY: The McGraw-Hill Companies, 2017.

40. DeFronzo RA. Banting Lecture. From the triumvirate to the ominous octet: a new paradigm for the treatment of type 2 diabetes mellitus. Diabetes 2009; 58:773–795.

41. Stumvoll M, Meyer C, Mitrakou A, Nadkarni V, Gerich JE. Renal glucose production and utilization: new aspects in humans. Diabetologia 1997; 40:749–757.

42. Edgerton DS, Lautz M, Scott M, Everett CA, Stettler KM, Neal DW, Chu CA, Cherrington AD. Insulin's direct effects on the liver dominate the control of hepatic glucose production. J Clin Invest 2006; 116:521–527.

43. Sindelar DK, Chu CA, Venson P, Donahue EP, Neal DW, Cherrington AD. Basal hepatic glucose production is regulated by the portal vein insulin concentration. Diabetes 1998; 47:523–529.

44. Hatting M, Tavares CDJ, Sharabi K, Rines AK, Puigserver P. Insulin regulation of gluconeogenesis. Ann N Y Acad Sci 2018; 1411:21–35.

45. DeFronzo RA. Lilly lecture 1987. The triumvirate: beta-cell, muscle, liver. A collusion responsible for NIDDM. Diabetes 1988; 37:667–687.

46. Morton GJ, Schwartz MW. Leptin and the central nervous system control of glucose metabolism. Physiol Rev 2011; 91:389–411.

47. Groop LC, Bonadonna RC, DelPrato S, Ratheiser K, Zyck K, Ferrannini E, DeFronzo RA. Glucose and free fatty acid metabolism in non-insulin-dependent diabetes mellitus. Evidence for multiple sites of insulin resistance. J Clin Invest 1989; 84:205–213.

48. Leto D, Saltiel AR. Regulation of glucose transport by insulin: traffic control of GLUT4. Nat Rev Mol Cell Biol 2012; 13:383–396.

49. Thorens B, Mueckler M. Glucose transporters in the 21st Century. Am J Physiol Endocrinol Metab 2010; 298:E141–E145.

50. De Vos A, Heimberg H, Quartier E, Huypens P, Bouwens L, Pipeleers D, Schuit F. Human and rat beta cells differ in glucose transporter but not in glucokinase gene expression. J Clin Invest 1995; 96:2489–2495.

51. Ohtsubo K, Takamatsu S, Minowa MT, Yoshida A, Takeuchi M, Marth JD. Dietary and genetic control of glucose transporter 2 glycosylation promotes insulin secretion in suppressing diabetes. Cell 2005; 123:1307–1321.

52. Thorens B, Sarkar HK, Kaback HR, Lodish HF. Cloning and functional expression in bacteria of a novel glucose transporter present in liver, intestine, kidney, and beta-pancreatic islet cells. Cell 1988; 55:281–290.

53. Thorens B. GLUT2, glucose sensing and glucose homeostasis. Diabetologia 2015; 58:221–232.

54. Berger C, Zdzieblo D. Glucose transporters in pancreatic islets. Pflugers Arch 2020; 472:1249–1272.

55. Seker-Yilmaz B, Kor D, Bulut FD, Yuksel B, Karabay-Bayazit A, Topaloglu AK, Ceylaner G, Onenli-Mungan N. Impaired glucose tolerance in Fanconi-Bickel syndrome: Eight patients with two novel mutations. Turk J Pediatr 2017; 59:434–441.

56. Matschinsky FM, Wilson DF. The Central Role of Glucokinase in Glucose Homeostasis: A Perspective

57. Chakera AJ, Steele AM, Gloyn AL, Shepherd MH, Shields B, Ellard S, Hattersley AT. Recognition and Management of Individuals With Hyperglycemia Because of a Heterozygous Glucokinase Mutation. Diabetes Care 2015; 38:1383–1392.

58. Marchetti P, Bugliani M, De Tata V, Suleiman M, Marselli L. Pancreatic Beta Cell Identity in Humans and the Role of Type 2 Diabetes. Front Cell Dev Biol 2017; 5:55.

59. Curi R, Lagranha CJ, Doi SQ, Sellitti DF, Procopio J, Pithon-Curi TC, Corless M, Newsholme P. Molecular mechanisms of glutamine action. J Cell Physiol 2005; 204:392–401.

60. Henquin JC. The dual control of insulin secretion by glucose involves triggering and amplifying pathways in beta-cells. Diabetes Res Clin Pract 2011; 93 (suppl 1):S27–31.

61. Komatsu M, Takei M, Ishii H, Sato Y. Glucose-stimulated insulin secretion: A newer perspective. J Diabetes Investig 2013; 4:511–516.

62. Prentki M, Matschinsky FM, Madiraju SR. Metabolic signaling in fuel-induced insulin secretion. Cell Metab 2013; 18:162–185.

63. Fulgenzi G, Hong Z, Tomassoni-Ardori F, Barella LF, Becker J, Barrick C, Swing D, Yanpallewar S, Croix BS, Wess J, Gavrilova O, Tessarollo L. Novel metabolic role for BDNF in pancreatic β-cell insulin secretion. Nat Commun 2020; 11:1950.

64. Meglasson MD, Matschinsky FM. Pancreatic islet glucose metabolism and regulation of insulin secretion. Diabetes Metab Rev 1986; 2:163–214.

65. Thompson R, Christie D, Hindmarsh PC. The role for insulin analogues in diabetes care. Paediatrics and Child Health; 16:117–122.

66. Henquin JC. Triggering and amplifying pathways of regulation of insulin secretion by glucose. Diabetes 2000; 49:1751–1760.

67. Seino S, Shibasaki T, Minami K. β-Cell biology of insulin secretion. In: DeFronzo RA, Ferrannini E, Zimmet P, Alberti G, eds. International Textbook of Diabetes Mellitus. Volume 1.4 ed: John Wiley & Sons, Ltd, 2015:96.

68. DeFronzo RA, Tobin JD, Andres R. Glucose clamp technique: a method for quantifying insulin secretion and resistance. Am J Physiol Endocrinol Metab 1979; 237:E214–E223.

69. O'Rahilly SO, Hosker JP, Rudenski AS, Matthews DR, Burnett MA, Turner RC. The glucose stimulus-response curve of the beta-cell in physically trained humans, assessed by hyperglycemic clamps. Metabolism 1988; 37:919–923.

70. Rudenski AS, Hosker JP, Burnett MA, Matthews DR, Turner RC. The beta cell glucose stimulus-response curve in normal humans assessed by insulin and C-peptide secretion rates. Metabolism 1988; 37:526–534.

71. Weiss R, Santoro N, Giannini, G., Galderial, A., Umano, G.R., Caprio, S. Prediabetes in youths: mechanisms and biomarkers. The Lancet Child & Adolescent Health 2017; 1:240–248.

72. Curry DL, Bennett LL, Grodsky GM. Dynamics of insulin secretion by the perfused rat pancreas. Endocrinology 1968; 83:572–584.

73. Zawalich WS, Zawalich KC. Species differences in the induction of time-dependent potentiation of insulin secretion. Endocrinology 1996; 137:1664–1669.

74. Carr RD, Larsen MO, Winzell MS, Jelic K, Lindgren O, Deacon CF, Ahren B. Incretin and islet hormonal responses to fat and protein ingestion in healthy men. Am J Physiol Endocrinol Metab 2008; 295:E779–E784.

75. Karamanlis A, Chaikomin R, Doran S, Bellon M, Bartholomeusz FD, Wishart JM, Jones KL, Horowitz M, Rayner CK. Effects of protein on glycemic and incretin responses and gastric emptying after oral glucose in healthy subjects. Am J Clin Nutr 2007; 86:1364–1368.

76. Nuttall FQ, Gannon MC, Wald JL, Ahmed M. Plasma glucose and insulin profiles in normal subjects ingesting diets of varying carbohydrate, fat, and protein content. J Am Coll Nutr 1985; 4:437–450.

77. Floyd JC, Jr., Fajans SS, Conn JW, Knopf RF, Rull J. Stimulation of insulin secretion by amino acids. J Clin Invest 1966; 45:1487–1502.

78. Gannon MC, Nuttall FQ. Amino acid ingestion and glucose metabolism-a review. IUBMB Life 2010; 62:660–668.

79. Chen S, Ogawa A, Ohneda M, Unger RH, Foster DW, McGarry JD. More direct evidence for a malonyl-CoA-carnitine palmitoyltransferase I interaction as a key event in pancreatic beta-cell signaling. Diabetes 1994; 43:878–883.

80. Prentki M, Madiraju SR. Glycerolipid/free fatty acid cycle and islet beta-cell function in health, obesity and diabetes. Mol Cell Endocrinol 2012; 353:88–100.

81. Stein DT, Stevenson BE, Chester MW, Basit M, Daniels MB, Turley SD, McGarry JD. The insulinotropic potency of fatty acids is influenced profoundly by their chain length and degree of saturation. J Clin Invest 1997; 100:398–403.

82. Kashyap SR, Belfort R, Berria R, Suraamornkul S, Pratipranawatr T, Finlayson J, Barrentine A, Bajaj M, Mandarino L, DeFronzo R, Cusi K. Discordant effects of a chronic physiological increase in plasma FFA on insulin signaling in healthy subjects with or without a family history of type 2 diabetes. Am J Physiol Endocrinol Metab 2004; 287:E537–E546.

83. Muscelli E, Mari A, Natali A, Astiarraga BD, Camastra S, Frascerra S, Holst JJ, Ferrannini E. Impact of incretin hormones on beta-cell function in subjects with normal or impaired glucose tolerance. Am J Physiol Endocrinol Metab 2006; 291:E1144–1150.

84. Williams JA. GLP-1. Pancreapedia: Exocrine Pancreas Knowledge Base 2014, DOI: 10.3998/panc.2014.4

85. Holst JJ, Gromada J. Role of incretin hormones in the regulation of insulin secretion in diabetic and nondiabetic humans. Am J Physiol Endocrinol Metab 2004; 287:E199–206.

86. Lamont BJ, Li Y, Kwan E, Brown TJ, Gaisano H, Drucker DJ. Pancreatic GLP-1 receptor activation is sufficient for incretin control of glucose metabolism in mice. J Clin Invest 2012; 122:388–402.

87. Smith PA, Sakura H, Coles B, Gummerson N, Proks P, Ashcroft FM. Electrogenic arginine transport mediates stimulus-secretion coupling in mouse pancreatic beta-cells. J Physiol 1997; 499 (Pt 3):625–635.

88. Hussain MA, Daniel PB, Habener JF. Glucagon stimulates expression of the inducible cAMP early repressor and suppresses insulin gene expression in pancreatic beta-cells. Diabetes 2000; 49:1681–1690.

89. Lawrence MC, Bhatt HS, Easom RA. NFAT regulates insulin gene promoter activity in response to synergistic pathways induced by glucose and glucagon-like peptide-1. Diabetes 2002; 51:691–698.

90. Holst JJ. The physiology of glucagon-like peptide 1. Physiol Rev 2007; 87:1409–1439.

91. Meier JJ, Nauck MA, Kranz D, Holst JJ, Deacon CF, Gaeckler D, Schmidt WE, Gallwitz B. Secretion, degradation, and elimination of glucagon-like peptide 1 and gastric inhibitory polypeptide in patients with chronic renal insufficiency and healthy control subjects. Diabetes 2004; 53:654–662.

92. Herman GA, Bergman A, Stevens C, Kotey P, Yi B, Zhao P, Dietrich B, Golor G, Schrodter A, Keymeulen B, Lasseter KC, Kipnes MS, et al. Effect of single oral doses of sitagliptin, a dipeptidyl peptidase-4 inhibitor, on incretin and plasma glucose levels after an oral glucose tolerance test in patients with type 2 diabetes. J Clin Endocrinol Metab 2006; 91:4612–4619.

93. Wendt A, Eliasson L. Pancreatic α-cells - The unsung heroes in islet function. Semin Cell Dev Biol 2020; 103:41–50.

94. Lefèbvre PJ. Biosynthesis secretion and action of glucagon. In: DeFronzo RA, Ferrannini E, Zimmet P, Alberti G, eds. International Textbook of Diabetes Mellitus. Volume 1.4 ed: John Wiley & Sons, Ltd, 2015:136.

95. Campbell JE, Drucker DJ. Islet α cells and glucagon-critical regulators of energy homeostasis. Nat Rev Endocrinol 2015; 11:329–338.

96. Kock NG, Darle N, Dotevall G. Inhibition of intestinal motility in man by glucagon given intraportally. Gastroenterology 1967; 53:88–92.

97. Leibiger B, Moede T, Muhandiramlage TP, Kaiser D, Vaca Sanchez P, Leibiger IB, Berggren PO. Glucagon regulates its own synthesis by autocrine signaling. Proc Natl Acad Sci U S A 2012; 109:20925–20930.

98. Capozzi ME, Wait JB, Koech J, Gordon AN, Coch RW, Svendsen B, Finan B, D'Alessio DA, Campbell JE. Glucagon lowers glycemia when β-cells are active. JCI Insight 2019; 5.

99. Yu Q, Shuai H, Ahooghalandari P, Gylfe E, Tengholm A. Glucose controls glucagon secretion by directly modulating cAMP in alpha cells. Diabetologia 2019; 62:1212–1224.

100. Briant LJB, Reinbothe TM, Spiliotis I, Miranda C, Rodriguez B, Rorsman P. Delta-cells and beta-cells are electrically coupled and regulate alpha-cell activity via somatostatin. J Physiol 2018; 596:197–215.

101. Winzell MS, Ahren B. G-protein-coupled receptors and islet function-implications for treatment of type 2 diabetes. Pharmacol Ther 2007; 116:437–448.

102. Jones B, Bloom SR, Buenaventura T, Tomas A, Rutter GA. Control of insulin secretion by GLP-1. Peptides 2018; 100:75–84.

103. Guo XH. The value of short- and long-acting glucagon-like peptide-1 agonists in the management of type 2 diabetes mellitus: experience with exenatide. Curr Med Res Opin 2016; 32:61–76.

104. Dhillon S. Semaglutide: First Global Approval. Drugs 2018; 78:275–284.

105. Guida C, McCulloch LJ, Godazgar M, Stephen SD, Baker C, Basco D, Dong J, Chen D, Clark A, Ramracheya RD. Sitagliptin and Roux-en-Y gastric bypass modulate insulin secretion via regulation of intra-islet PYY. Diabetes Obes Metab 2018; 20:571–581.

106. Loubatieres-Mariani MM. [The discovery of hypoglycemic sulfonamides]. J Soc Biol 2007; 201:121–125.

107. Schmid-Antomarchi H, de Weille J, Fosset M, Lazdunski M. The antidiabetic sulfonylurea glibenclamide is a potent blocker of the ATP-modulated K+ channel in insulin secreting cells. Biochem Biophys Res Commun 1987; 146:21–25.

108. Schmid-Antomarchi H, De Weille J, Fosset M, Lazdunski M. The receptor for antidiabetic sulfonylureas controls the activity of the ATP-modulated K+ channel in insulin-secreting cells. J Biol Chem 1987; 262:15840–15844.

109. Reaven G, Dray J. Effect of chlorpropamide on serum glucose and immunoreactive insulin concentrations in patients with maturity-onset diabetes mellitus. Diabetes 1967; 16:487–492.

110. Lebovitz HE, Melander A. Sulfonylureas and meglitinides: Insights into physiology and translational clinical utility. In: DeFronzo RA, Ferrannini E, Zimmet P, Alberti G, eds. International textbook of diabetes mellitus. Volume 1.4 ed: John Wiley & Sons, Ltd, 2015:615.

111. Schernthaner G, Schernthaner GH. The right place for metformin today. Diabetes Res Clin Pract 2020; 159:107946.

112. Quesada I, Nadal A, Soria B. Different effects of tolbutamide and diazoxide in alpha, beta-, and delta-cells within intact islets of Langerhans. Diabetes 1999; 48:2390–2397.

Chapter 26

Pancreatic stellate cells in health and disease

Alpha R. Mekapogu, Srinivasa P. Pothula, Romano C. Pirola, Jeremy S. Wilson, and Minoti V. Apte

Pancreatic Research Group, South Western Sydney Clinical School, Faculty of Medicine, The University of New South Wales, Ingham Institute for Applied Medical Research, Sydney, Australia

Introduction

Pancreatic fibrosis is a well-recognized histopathological feature of two major diseases of the pancreas—chronic pancreatitis and pancreatic cancer. In health, the process of fibrogenesis is a well-regulated dynamic process which is necessary for regular turnover of extracellular matrix (ECM) that allows remodeling and maintenance of normal pancreatic architecture. However, during injury, the equilibrium between production and degradation of fibrous tissue is disrupted, leading to excessive deposition of ECM proteins resulting in fibrosis.

Pancreatic stellate cells (PSCs) are now considered to be the key contributors of pancreatic fibrosis.[1–3] These cells were first observed by Watari et al.[4] in 1982 and later confirmed by Ikejiri[5] in 1990. The development of isolation and culture methods for PSCs in 1998 helped to unravel the mechanisms involved in the process of pancreatic fibrogenesis[1,2] and also helped researchers to investigate the functions of these cells in both health and disease.

Pancreatic Stellate Cells

PSCs in health

In a healthy pancreas, PSCs make up 4–7 percent[1] of the total parenchyma and are located around the basolateral aspects of the acinar cells (**Figure 1A**). PSCs are also found around small pancreatic ducts,[1,2] pancreatic islets,[6] and blood vessels. PSCs in their native state express abundant vitamin A (retinoids) containing droplets in their cytoplasm which is a characteristic feature of the "stellate cell system" in the body.[7] The buoyancy imparted by the vitamin A-containing vesicles was utilized by Apte et al.[1] to develop a density gradient centrifugation method for the isolation of quiescent PSCs from the pancreas. In early culture, PSCs are polygonal in shape with abundant lipid droplets in the cytoplasm (**Figure 1B**) and express stellate cell selective markers such as desmin, glial fibrillary acidic

protein (GFAP), nestin, neural cell adhesion molecule, nerve growth factor, and synemin.[8,9] Due to the expression of neural markers, PSCs were initially thought to be of neuroectodermal origin. However, lineage-tracing studies with hepatic stellate cells (counterparts of PSCs in the liver) have confirmed a mesenchymal origin for these cells.[10] The fact that a proportion of PSCs are replenished from bone marrow further supports a mesenchymal origin for PSCs.[11]

Islet stellate cells (ISCs) are similar to but differ in certain aspects from PSCs. ISCs have fewer vitamin A-containing droplets and undergo a more rapid activation than PSCs. Upon activation, ISCs express more α-SMA but have reduced rates of proliferation and migration compared with exocrine PSCs.[6]

PSCs are the primary producers of ECM proteins, such as collagen I–IV, fibronectin, and laminin, in the pancreas. At the same time, they also produce ECM-degrading enzymes like matrix metalloproteinases (MMPs) and their inhibitors—tissue inhibitors of matrix metalloproteinases (TIMPs).[12] In the normal pancreas, these enzymes maintain an equilibrium between the deposition and degradation of ECM proteins.

It is now recognized that PSCs have functions beyond ECM remodeling in the normal pancreas. PSCs express receptors for the secretagogue cholecystokinin (CCK1 and CCK2)[13] and have the capacity to synthesize the neurotransmitter acetylcholine (ACh). Upon exposure to physiological levels of CCK, PSCs have been shown to secrete ACh which, in turn, acts on the muscarinic receptors on acinar cells and induces amylase secretion.[14] PSCs also play an important role in innate immunity through the expression of Toll-like receptors (TLRs) 2, 3, 4, and 5[15] which can recognize pathogen-associated molecular patterns (PAMPs), alarmins, and endogenous molecules released by tissue damage (DAMPs). These cells also possess the ability to phagocytose polymorphonuclear neutrophils and cell debris, which is mediated by CD36 via peroxisome proliferator-activated receptor γ.[16] PSCs may

1A

1B

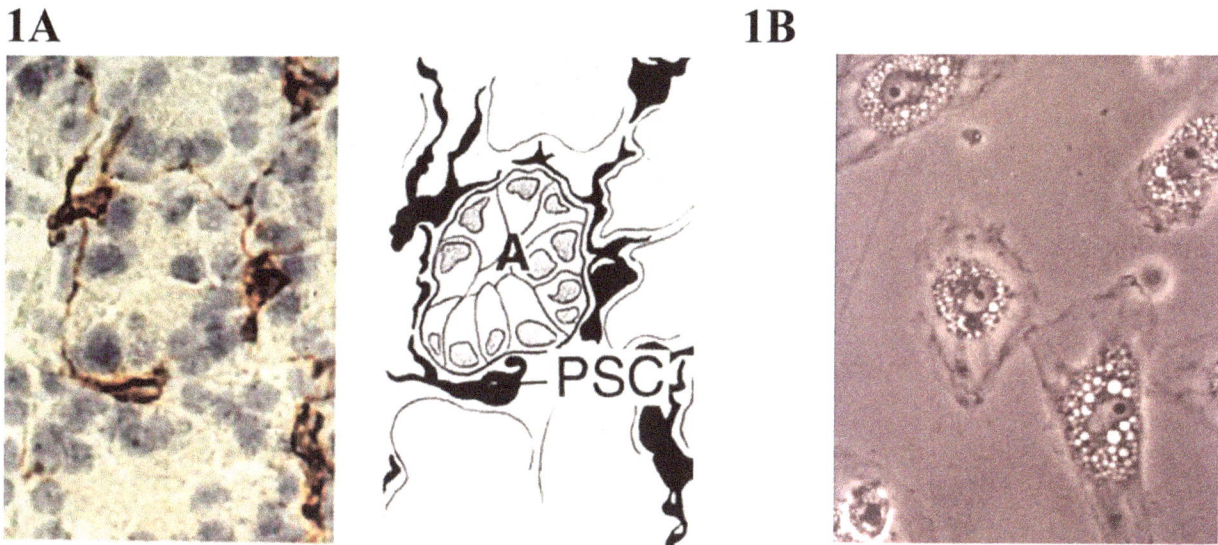

Figure 1. (**A**) Normal rat pancreatic stellate cells stained for the cytoskeletal protein desmin. Photomicrograph showing a normal rat pancreatic section immunostained for the stellate cell selective marker desmin (left) and a corresponding line diagram (right). Desmin-positive (brown) PSCs can be seen surrounding the basolateral aspects of pancreatic acini. (**B**) PSCs in an early culture showing a typical flattened polygonal shape with abundant vitamin A-containing lipid droplets in the cytoplasm (from reference[1]; reproduced with permission from BMJ Publishing Group Ltd.).

also possess progenitor-like capabilities since they express several stem cell markers, including CD133, SOX9, nestin, and GDF3.[17,18] They can differentiate into other cell types including insulin-secreting cells under the influence of relevant growth factors.[13,18–21]

PSCs in disease

In the event of pancreatic injury, PSCs are activated, leading to loss of the vitamin A-containing lipid droplets, and transform to a myofibroblast-like phenotype that expresses alpha smooth muscle actin (αSMA), fibroblast activation protein-α (FAP-α), fibroblast specific protein-1 (FSP-1), and fibrinogen.[8] These cells also exhibit increased proliferation, migration, and ECM synthesis.[22,23] Activated PSCs produce collagenous stroma during pancreatic injury[24–26] which can overwhelm the ECM-degrading capacity of MMPs, leading to fibrosis. A comparison of resting and activated PSCs is given in **Table 1**.

PSCs can be activated by several factors such as proinflammatory cytokines;[27] oxidant stress;[28,29] ethanol and its metabolites, acetaldehyde and fatty acid ethyl esters;[30,31] fatty acids (oleate):[32] and endotoxins[33] (**Table 2**). In addition, in the setting of pancreatic cancer, factors such as acidosis,[34] hypoxia,[35] increased interstitial pressure[36] and hyperglycemia[37] are also involved in the activation of PSCs. Recently, epigenetic modifications (increased acetylation of histones) have also been shown to play

an important role in the activation of PSCs and collagen synthesis.[38]

In addition to being activated by exogenous factors, PSCs are capable of sustaining their own activation (even in the absence of the initial triggers) through the production of cytokines (such as transforming growth factor beta (TGFβ), connective tissue growth factor (CTGF), interleukins 6, 7, 8 and 15, CXCR1, monocyte chemotactic protein 1 (MCP1), and regulated-on-activation-normal T cell expressed-and-secreted (RANTES)).[39–41] These cytokines act via corresponding receptors on the PSCs themselves (autocrine effects) leading to a state of perpetual activation, which further facilitates pathological fibrosis.

Several pathways that mediate PSC activation have now been identified (summarized in **Table 3**). A number of these pathways interact with each other leading to significant redundancy in terms of the modulation of PSC activation.[22,42] Furthermore, studies have also shown that numerous signaling pathways can converge, causing sustained increase in intracellular calcium within the PSCs.[43–45] Recently, microRNAs (small noncoding RNAs), which are implicated in cell functions such as proliferation, differentiation, apoptosis, and protein synthesis, have been shown to be differentially expressed between quiescent and activated PSCs.[46] Other studies have implicated miR15b and 16 (known to regulate PSC apoptosis)[47] and miR21 as possible cofactors in connective tissue growth factor (CCN2)-mediated PSC activation.[48]

Table 1. Characteristics of quiescent and activated PSCs.

Characteristics	Quiescent PSCs	Activated PSCs
Vitamin A lipid containing droplets	Abundant	Absent
Alpha smooth muscle actin	Not expressed	Expressed
Proliferation	Basic	Enhanced
Ability to migrate	No	Yes
Collagen production	Limited	Increased
Activity of MMPs and TIMPs	In equilibrium	TIMP > MMP
Cytokine production	Limited	Increased inflammatory cytokines (PDGF, TGFβ, CTGF, IL-1, IL-6, IL-15)
Ability for phagocytosis	No	Present (CD6 mediated)
Protein expression	Basal expression	Differential expression

Table 2. Factors causing activation of pancreatic stellate cells.

Angiotensin
Cyclooxygenase 2 (COX-2)
Endothelin-1
Endotoxin
Ethanol and its metabolites (acetaldehyde, fatty acid ethyl esters)
Fibrinogen
Galectin-1
Hyperglycemia
Hypoxia
Inflammatory mediators (cytokines, growth factors, complement) Oxidant stress
Parathyroid hormone-related protein (PTHrP)
Pigment epithelium-derived factor Proteases

The past two decades have seen a considerable improvement in our understanding of the role of PSCs in pancreatic inflammation (both acute and chronic pancreatitis) as well as pancreatic cancer. Acute pancreatitis is usually a self-limiting disease where PSCs aid remodeling and repair of tissue architecture, while both chronic pancreatitis and pancreatic cancer are characterized by pathological accumulation of fibrous tissue.

Acute Pancreatitis

Acute pancreatitis is a condition which results from acute inflammation of the pancreatic parenchyma, ranging from mild inflammation to extensive pancreatic necrosis. Most of the cases are mild and self-limiting with a mortality rate <1 percent. The most common causes of acute pancreatitis are gallstones and alcohol excess.[49]

In response to inflammatory cytokines and chemokines secreted by damaged acinar cells and inflammatory cells, PSCs are activated and proliferate early in acute pancreatitis.[50] The majority of the increased numbers of activated PSCs, in the inflamed areas, are a result of local proliferation of resident PSCs. However, a small proportion (7–18%) is derived from circulating bone marrow cells.[11] Activated PSCs secrete increased ECM proteins that act as a scaffold for regenerating ductal and acinar cells during the repair process. These ECM proteins are also important for differentiation and cell growth via integrin-mediated interactions between cell membranes and the surrounding matrix.[51] Indeed, using β1-integrin knockout animals, it has been shown that the lack of integrin-mediated interactions results in decreased ECM production by PSCs and increased apoptosis and decreased proliferation of acinar cells, thereby impeding pancreatic repair.[52]

Table 3. **Signaling pathways involved in the interactions between PSCs and cancer cells.** α-SMA, alpha smooth muscle actin; c-MET, tyrosine-protein kinase of Met; CCL, chemokine ligand; CXCL, chemotactic cytokine ligand; ECM, extracellular matrix; JAK/STAT, Janus kinase/signal transducers and activators of transcription; HGF, hepatocyte growth factor; PSC, pancreatic stellate cells; PCC, pancreatic cancer cells; SDF, stromal derived factor; PDGF, platelet-derived growth factor; SMAD, small worm mothers against decapentaplegic.

Mediator	Signaling pathway	Functional role	Reference
PSC-derived CXCL12 (SDF-1)	CXCL12 (SDF-1) signaling	Immunosuppression	154
PSC-derived SDF- 1	Galectin-1	Proliferation of PSC and chemokine production facilitating PCC metastasis	155
Smo, Gli	Hedgehog pathway	PSC activation, ECM synthesis, migration, desmoplasia, angiogenesis; PCC proliferation, migration and chemoresistance	140
HGF	HGF/c-MET pathway	PSC promote proliferation and metastasis of tumor cells	123
CCL2	Hypoxia inducible factor 1 (HIF-1)	PSC activation, Macrophage recruitment	156
IL-6	IL6/JAK/STAT	PSC activation and proliferation	157
Kindl in-2	Integrin	Cytokines production in PSCs facilitating progression and migration of PCC	158
MAPK	Mitogen-activated protein kinase (MAPK) signaling pathway	PSC proliferation, TIMP-1 production	159
Periostin	Periostin pathway	Periostin secreted by PSCs promoting PCC proliferation, EMT and resistance to nutrient deprivation and hypoxia	160
PPAR-y ligands	Peroxisome proliferator-activated gamma (PPARy)	Inhibition of PSC activation, proliferation, and collagen synthesis Increased phagocytosis	16
PDGF	PI3-Kinase pathway	PSC migration and proliferation	42
Hyperglycemia	Protein kinase C	PSC proliferation, a-SMA, collagen-I production, angiogenesis	161
Suppression of miRNA-21	Reactive oxygen species (ROS)	PSC activation and induction of glycolysis	91
Rho - associated protein kinase	Rho-ROCK pathway	Activation of PSC, collagen I synthesis and fibrosis	162
SMAD-2,3	SMADS	PSC activation, proliferation, ECM deposition, transdifferentiation, TGF-β expression	163
TLR9	Toll-like receptor (TLR) signaling	Immunosuppression; PSC-derived cytokine production	164
Vitamin D	Vitamin D receptor	PSC quiescence; Decreased chemoresistance	96
β-catenin	Wnt/β-catenin signaling	Invasion of PCCs	165

PSCs may also play an important role in remodeling and regeneration after acute necrotizing pancreatitis. Zimmerman et al.[53] observed that pilot ductules that originate from hypercellular regenerative spheres (islands of vascular granulation tissue, ductular cells, stellate cells, and residual lobular elements) grow outwards in close association with activated PSCs. The authors have suggested that these PSCs may be able to support differentiation of cells to mature duct and acinar cells, thereby playing a role in the reconstitution of the cell populations after acute injury. The process of regeneration also requires removal of excess ECM, a task that is aided by the fibrinolytic activity of PSCs via the production of matrix degrading enzymes (MMPs).

Chronic Pancreatitis

Chronic pancreatitis (CP) is a progressive disease characterized by fibrosis, acinar cell loss, distorted ducts, and inflammatory cell infiltration, eventually leading to significant exocrine and endocrine dysfunction[54] (**Figure 2**). The presence of activated PSCs was confirmed in the areas of fibrosis using dual staining techniques (Sirius Red for collagen and immunostaining for αSMA)[23] (**Figure 3**), while dual staining for αSMA and procollagen mRNA has shown that activated PSCs are the predominant source of collagen I in the fibrotic areas[23] (**Figure 5**). Interestingly, the fibrosis of chronic pancreatitis was found not to be limited to the exocrine parenchyma. Activated PSCs have also been demonstrated within and surrounding pancreatic

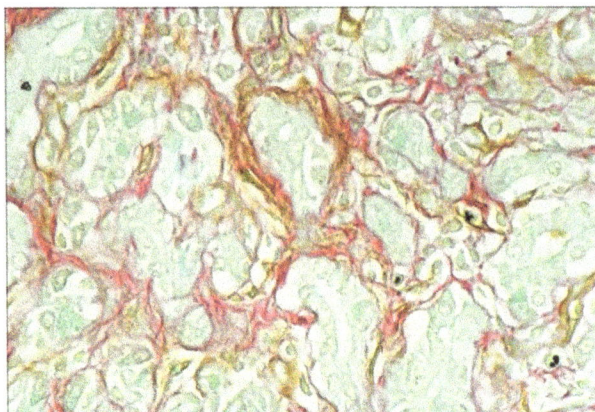

Figure 2. Activated PSCs in chronic pancreatitis. A section from a patient with chronic pancreatitis showing colocalization of staining for the PSC activation marker alpha smooth muscle actin (αSMA, brown) and collagen (using Sirius Red) (red) (dual staining) in fibrotic areas of the pancreas (from reference[23]; reproduced with permission of Elsevier).

islets and are co-localized in fibrotic areas around the islets in diabetic Goto–Kakizaki rats[6] and db/db mice (obesity, diabetes, and dyslipidemia model due to leptin receptor deficiency).[55] Additionally, activated PSCs can inhibit insulin secretion and cause beta cell apoptosis, effects which are further aggravated by hyperglycemia.[56] These findings suggest that PSCs can modulate islet cell function, suggesting a direct role for these cells in the diabetes of chronic pancreatitis.

Many factors are known to activate PSCs during chronic pancreatitis, including the profibrogenic growth factor TGFβ, platelet-derived growth factor (PDGF),[23] nerve growth factor (NGF),[57] and oxidative stress.[58] It is now known that PSC-activating factors such as TGFβ, PDGF, oxidant stress, and other cytokines are upregulated early during pancreatic injury leading to PSC proliferation and ECM synthesis. As noted earlier, PSCs can be maintained in a perpetually activated state through autocrine effects of endogenous cytokines, leading to pathological fibrosis.

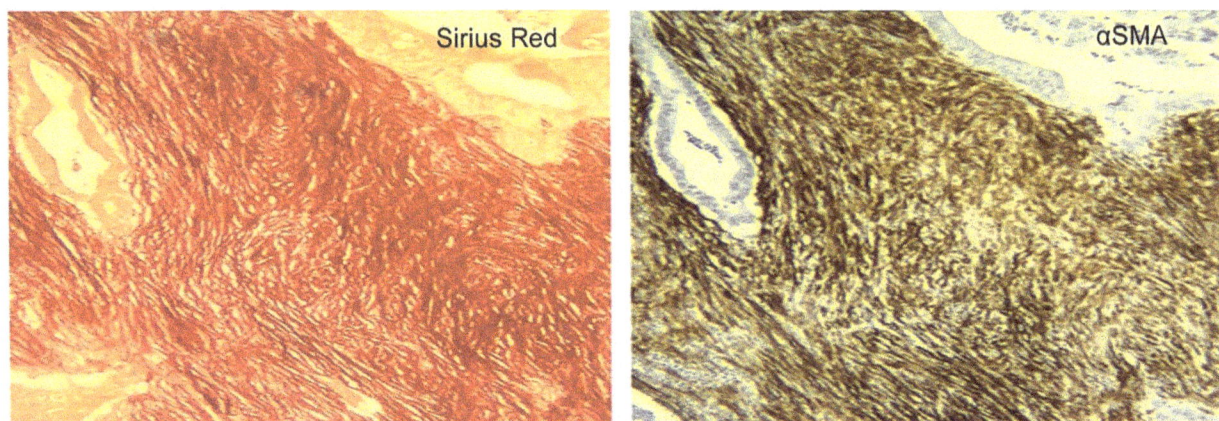

Figure 3. Colocalization of collagen and alpha SMA staining in pancreatic cancer. A representative pair of serial paraffin sections of the pancreas from a patient with pancreatic cancer demonstrating that stromal areas exhibit strong positive staining for collagen as well as for alpha SMA, indicative of the presence of activated PSCs in the desmoplastic reaction in pancreatic cancer. Original magnification × 100 (from reference[59]; reproduced with permission of Wolters Kluwer).

Animal models provide an invaluable tool for understanding the pathogenesis of chronic pancreatitis. Several methods of inducing experimental chronic pancreatitis have been reported in the literature: repeated intravenous injections of caerulein[60] or superoxide dismutase inhibitor;[61] instillation of toxins into the pancreatic duct;[23,48,62] and chronic ethanol administration followed by secondary challenge with caerulein,[63,64] cyclosporin[65] or endotoxin.[33]

Regarding alcoholic chronic pancreatitis, endotoxin is a relevant trigger factor, given that gut permeability is known to be increased by alcohol and elevated serum endotoxin levels have been reported in heavy drinkers.[66] Further, endotoxin (lipopolysaccharide, LPS) challenge was shown

to enhance fibrosis in Sprague-Dawley rats fed Lieber de Carli alcohol liquid diets.[67]

Transgenic models of CP include animals overexpressing cytokines or profibrogenic factors such as IL-1β, TGFβ, and heparin-binding EGF-like growth factor (HB-EGF)[68,69] and WBN–Kob rats[70] and Goto–Kakizaki rats with type 2 diabetes.[71] More recently, the effects of smoking on pancreatic fibrosis have been studied, given the close association of drinking and smoking as lifestyle factors. Lugea and colleagues[72] have shown that smoking significantly worsened tissue injury as well as fibrosis in a rat model of alcoholic pancreatitis. The increased fibrosis could be due to direct activation of PSCs by (i) nicotine and NNK via nicotinic

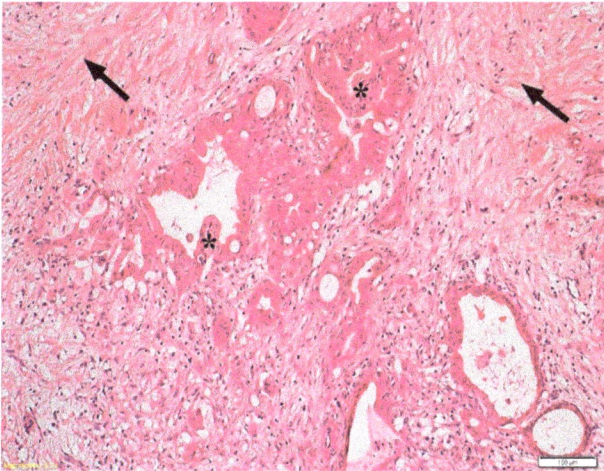

Figure 4. Desmoplasia in a human pancreatic cancer section. A representative photomicrograph of a hematoxylin and eosin-stained human pancreatic cancer section showing malignant elements (duct-like and tubular structures—indicated by asterisks) embedded in highly fibrotic stroma (indicated by arrows) (from reference[24]; reproduced with permission from Elsevier).

acetylcholine receptors (nAChRs) on the cells and/or (ii) IL-22 secreted by infiltrating macrophages in response to smoke compounds (via aryl hydrolase receptors).[73]

Reversal of Pancreatic Fibrosis in Chronic Pancreatitis

The study of mechanisms underlying PSC activation and fibrosis can help in developing novel anti-fibrotic treatments. Several antifibrotic strategies have been successfully evaluated to counter CP in experimental animal models including: (i) inhibition of profibrogenic growth factors TGFβ and tumor necrosis factor alpha (TNFα);[73–76] (ii) antioxidants such as vitamin E,[77] ellagic acid, a plant polyphenol,[78] and salvianolic acid, a herbal medicine;[79] (iii) protease inhibitors;[80] (iv) modulation of signaling molecules (e.g., troglitazone binding to the peroxisome proliferator receptor gamma, PPARγ);[81] retinoic acid-induced PSC quiescence via suppression of the Wnt–catenin pathway;[82] (v) collagen siRNA;[83] (vi) an anthraquinone derivative Rhein and a flavonoid, apigenin; (vii) a prostacyclin analogue ONO-1301, which inhibits proinflammatory and profibrogenic cytokine production;[84] and (viii) alcohol withdrawal in alcohol-induced pancreatitis;[85,71] (ix) amygdalin was shown to inhibit PSC activation and attenuate fibrosis by decreasing production of pro-fibrotic cytokines in a rat CP model induced by injecting dibutyltin dichloride (DBTC);[86] (x) an experimental compound miR-200a inhibited TGF-β1-induced pancreatic stellate cell activation and ECM formation through inhibition of PTEN/Akt/mTOR pathway;[87] (xi) in chronic pancreatitis induced by dibutyltin dichloride in rats, 3-methyladenine (3-MA), a

PI3K inhibitor decreased fibrosis by decreasing autophagy in PSCs.[88]

Several plant compounds have shown promising antifibrotic activity in several studies. Curcumin (a polyphenol found in turmeric) was reported to inhibit activation of PSCs through the inhibition of IL1β and TNFα-induced activation of activator protein-1 (AP-1) and mitogen-activated protein (MAP) kinases (ERK, c-Jun N-terminal kinase (JNK), and p38 MAP kinase).[89] Conophylline, a plant alkaloid, decreased fibrosis through the inhibition of ERK1/2 in PSCs.[90] In an *in vitro* study, resveratrol, a natural polyphenol, decreased oxidative stress-induced activation and glycolysis in PSCs mediated by ROS/miR-21.[91] Genestein, a natural isoflavone, decreased fibrosis in human PSCs transfected with let-7d to express thrombospondin 1, a marker of fibrosis.[92] Saikosponin A, the active component of the Chinese medicine *chaihu*, was shown to decrease PSC activation, viability, proliferation, and migration and to promote apoptosis by inhibiting autophagy and the formation of NLRP3 inflammasome via the AMPK/mTOR pathway.[93] Similarly, date palm fruit extract decreased fibronectin-1 and αSMA, markers of fibrosis in PSCs activated by TNF-α.[94]

Vitamins D and its isoforms D2 and D3 are reported to decrease *in vitro* PSC activation by decreasing IL-6.[95] Furthermore, *in vivo* studies with the vitamin D ligand calcipotriol have shown significant attenuation of the fibrosis of chronic pancreatitis in mice.[96] Since storage of vitamin A is associated with PSC quiescence, administration of vitamin A (retinol), or its metabolites such as retinoic acid has been assessed in models of chronic pancreatitis. In this regard, vitamin A-containing liposomes combined with TLR4-silencing shRNA have been reported to inhibit pancreatic fibrosis in mouse models of chronic pancreatitis.[97]

Other compounds known to induce quiescence of activated PSCs include melatonin, the anthraquinone derivative rhein,[98] bone morphogenic protein,[99] and troglitazone (a ligand for the peroxisome proliferator activated receptor PPARγ).[81] Recently, kinase inhibitors such as sorafenib, sunitinib, trametinib, and dactolisib have been shown to inhibit PSC proliferation and ECM synthesis.[100,101] Interestingly, trametinib also decreases the expression of two autocrine mediators of PSC activation, IL6 and TGFβ.[100] Coenzyme Q10 suppresses PSC activation by inhibiting autophagy through PI3K/ATK/mTOR signaling.[102] Inhibition of cyclo-oxygenase by indomethacin, an anti-inflammatory drug, also results in decreased activation of PSCs.[103]

Pancreatic Cancer

Pancreatic ductal adenocarcinoma (PDAC) is characterized by extensive stroma/desmoplasia (**Figure 4**) which

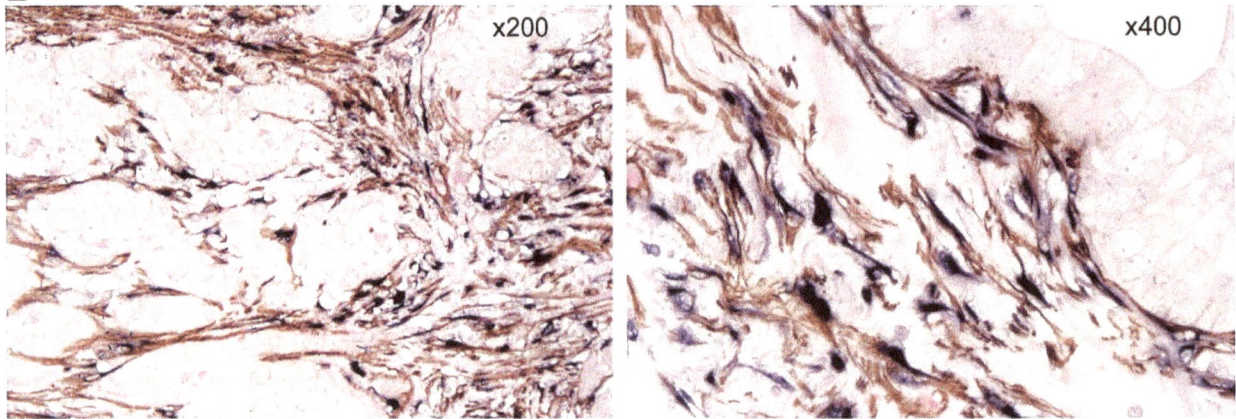

Figure 5. Dual staining for alpha smooth muscle actin (αSMA) and collagen mRNA in a human pancreatic cancer section. Immunostaining for αSMA (brown) combined with in situ hybridization for collagen mRNA (blue) reveals colocalization of the two stains in stromal areas of the section with no staining in tumor cells. This pattern of staining indicates that pancreatic stellate cells are the main source of collagen in pancreatic cancer stroma (from reference[59]; reproduced with permission of Wolters Kluwer).

constitutes 80–90 percent of the tumor volume.[104] This stroma consists of ECM proteins including collagen type I, fibronectin and laminin, non-collagenous factors such as glycoaminoglycans (e.g., hyaluronan), glycoproteins, and proteoglycans, and several cell types, including stellate cells, endothelial cells, neural elements and immune cells.[105] Just as with the fibrosis of chronic pancreatitis, activated PSCs are now also recognized as the primary source of the collagenous stroma in PDAC.[59] The detection of activated PSCs expressing periostin (a cell adhesion protein), galectin-1 (a glycan binding protein) and αSMA (an activation marker) in precursor pancreatic intraepithelial neoplasms (PanINs) (**Figure 6**) and intra-ductal papillary mucinous neoplasms (IPMNs) suggests that activation of PSCs is an early event in PDAC.[106,107] Studies have also shown a positive correlation between extent of PSC activation in the stroma and poor clinical outcome.[108,109]

It is now known that functionally different subsets of PSCs exist within the stroma of pancreatic cancer. Yuzawa et al.[109] observed the presence of fibroblasts with varying expression of αSMA and PDGFRβ, while Ohlund et al.[110] observed that cancer-associated fibroblasts (CAFs)/pancreatic stellate cells in proximity to cancer cells exhibit higher αSMA expression compared to those located at a distance from cancer cells. The authors termed the PSCs close to cancer cells as (myofibroblastic) myPSCs and those at some distance from cancer cells as (inflammatory) iPSCS, based on their cytokine secretion profile.

Studies involving *in vitro* and *in vivo* models have established that PSCs interact bidirectionally with cancer cells to facilitate local tumor growth and distant metastasis.[24] PSCs also interact with other cells in the stroma, including immune cells resulting in a complex cascade of events (**Figure 7**). PSCs promote cancer cell progression,

migration, and cancer cell survival[111,112] while, in turn, cancer cells promote PSC proliferation, migration, and ECM synthesis.[111,113] Several *in vitro* studies have shown that cancer cells increase PSC proliferation and ECM synthesis mediated by PDGF, FGF2, and TGFβ1[114] and also promote the secretion of MMPs by PSCs[115] through ECM metalloproteinase inducer (EMMPRIN) secretion[116] and TGFβ1 signaling.[111,117] Cancer cells can also induce autophagy in PSCs leading to release of alanine which acts as an alternative carbon source for the TCA cycle and lipid synthesis in cancer cells, thus improving cancer cell survival in the

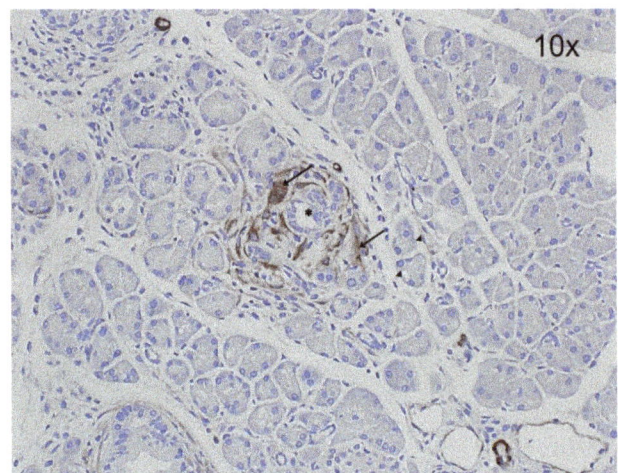

Figure 6. Activated PSCs in early pancreatic intraepithelial neoplasia (PanIN). A section from an eight-week-old transgenic Kras[G12D] P53[R172H] Cre (KPC) mouse showing staining for the PSC-activation marker alpha smooth muscle actin (αSMA, brown) around PanINs in a section of the pancreas. Magnification 20×.

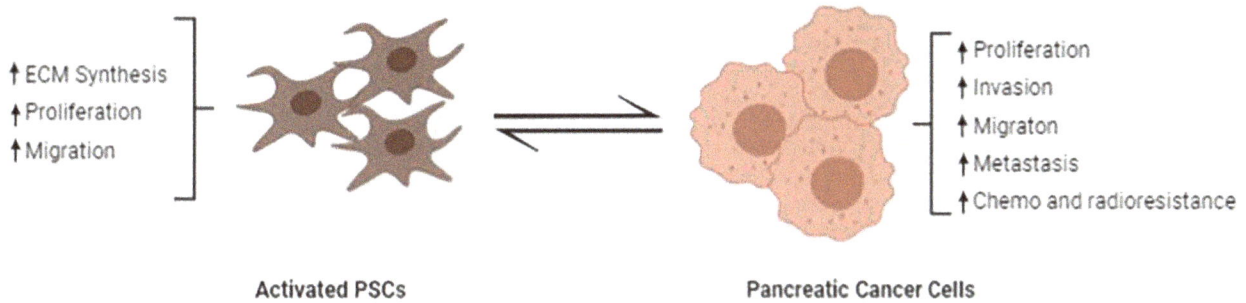

Figure 7. Bidirectional interactions of pancreatic stellate cells with cancer cells. Activated PSCs promote proliferation, invasion, metastasis, survival, and chemoresistance as well as radio-resistance of cancer cells, while cancer cells in turn promote ECM synthesis, proliferation, and migration of PSCs.

nutrient-poor and hypoxic environment of PDAC.[118,119] Epithelial–mesenchymal transition (EMT) and stemness, two processes that play an important role in cancer metastasis and recurrence, are known to be induced by PSCs.[120,121]

Studies have shown that stromal signaling is indispensable for cancer progression. Sherman and colleagues have reported that KRas mutation alone is insufficient for PDAC and that stromal cues—collagen and cytokines derived from cancer associated PSCs—are essential to this process.[122] Furthermore, in both subcutaneous and orthotopic xenograft models, co-injection of cancer cells and stromal cells (PSCs) resulted in larger tumors with significant desmoplasia compared to injection of cancer cells alone.[111,113,123]

An intriguing report published some years ago suggested that PSCs may facilitate seeding of cancer cells at metastatic sites. Using a gender mismatch approach and an orthotopic model of pancreatic cancer, the authors of this study found PSCs from the primary tumor within metastatic nodules at distant sites, indicating that, in addition to cancer cells, PSCs can also disseminate via the circulation.[112] Similar observations were made by Suetsugu et al.[124] who injected a mixture of PSCs and pancreatic cancer cells into the spleen and observed that both cell types comigrated from the spleen to metastatic nodules. In a recent groundbreaking study, Pang et al.[125] have identified the presence of circulating PSCs alongside circulating tumor cells (CTCs) in the blood of orthotopic tumor-bearing mice.

In recent years, there has been considerable focus on the role of exosomes (nano size vesicles secreted by most cell types) in a variety of diseases. These exosomes carry a cargo of proteins, lipids, DNA, mRNA, and microRNA, which can influence the function of cells remote from the source of the exosomes. Regarding pancreatic cancer, Takikawa et al.[126] have demonstrated that exosomes derived from PSCs contained a variety of microRNAs and an abundance of miR-21–5p and miR-451a which mediated PSC-induced proliferation and migration of PSCs.

Similarly, exosomes derived from PSCs were found to stimulate activation, proliferation, and migration of PSCs through upregulation of transforming growth factor β1 (TGFβ1) and tumor necrosis factor (TNF).

Several studies have shown that pancreatic desmoplasia can contribute to chemoresistance by affecting the kinetics of chemotherapeutic agents. Hessmann et al.[127] found that cancer-associated activated fibroblasts are resistant to gemcitabine and tend to accumulate and rapidly convert gemcitabine into an inactive metabolite 2′,2′-difluorodeoxyuridine, thus contributing to decreased efficacy of gemcitabine on tumor cells.

Moreover, there is additional evidence that PSCs may also directly alter the response of cancer cells to chemotherapeutic agents. Autocrine secretion of IL-6 by PSCs in response to stromal-derived factor 1α (SDF-1α) demonstrated a protective effect on cancer cells from the apoptotic effect of gemcitabine.[128] Owing to their chemoresistance, post therapy, PSCs can facilitate proliferation of residual cancer stem-cells leading to recurrence.[129]

Strategies to Counter Stromal–Tumor Interactions in Pancreatic Cancer

As PSC-cancer cell interactions play an important role in the progression of PDAC, several strategies have been developed to counter these interactions. Several chemicals including phytochemicals and hormones have been shown to decrease stromal–tumor interactions in experimental studies. Metformin, an activator of AMP-activated protein kinase, reduced the production of fibrogenic cytokines from cancer cells, inhibited PSC activation in co-culture studies, decreased tumor growth, and enhanced chemosensitivity to gemcitabine in an orthotopic xenograft model.[130] ProAgioQ20, which targets an integrin α(v)β (3) overexpressed in PSCs, caused apoptosis of cancer-associated PSCs, reabsorption of collagen, inhibition of tumor growth, and inhibition of drug penetration in a subcutaneous, orthotopic xenograft and transgenic KPC model.[131] Similarly, inhibition of cyclic AMP-response element binding protein-binding protein (CBP)/β-catenin signaling using ICG-001 resulted in decreased activation of PSCs and PSC-induced migration of PANC-1 cancer cells in a transwell migration assay.[132] Curcumin, the active compound of turmeric, was shown to inhibit IL-6 in PSCs under hypoxic conditions, leading to suppression of EMT and PSC-mediated cancer cell migration *in vitro*.[133] Relaxin-2, an endogenous hormone combined with supermagnetic iron oxide nanoparticles decreased fibrosis and tumor growth besides potentiating the effects of co-administered gemcitabine in an orthotopic xenograft model.[134]

Several pathways are shown to be key regulators of cancer cell–PSC interactions, and inhibition of these pathways in animal studies has been reported to decrease fibrosis, tumor growth, and metastasis. In this regard, the Hedgehog pathway is significant as the ligand Sonic hedgehog (Shh) is expressed in tumor cells while the receptor, Smoothened (Smo), is located mainly on cancer-associated fibroblasts.[135] In animal models, inhibitors of this pathway such as ormeloxifene[136] and IPI-926[137] decreased desmoplasia and tumor growth and improved chemosensitivity to co-administered gemcitabine. Interestingly, in an orthotopic tumor model where cancer and PSCs are co-injected, a triple combination of CXCR4 antagonist (AMD3100), Hedgehog inhibitor (GDC0449), and gemcitabine resulted in complete remission of tumor growth.[138]

Similar to the Hedgehog pathway, the HGF/c-MET pathway is involved in cancer proliferation, invasion, and metastasis. The ligand HGF is produced by PSCs while its receptor, c-MET, is located on cancer cells and endothelial cells.[139] In transgenic KPC mice, dual inhibition of both Sonic Hedgehog and HGF/c-MET pathways were found to sensitize PDAC to gemcitabine.[140] In a subcutaneous xenograft mouse model, combination of the c-Met inhibitor XL184 with gemcitabine significantly decreased tumor growth compared to individual treatments.[141] Xu et al.[142] used a triple therapy approach consisting of HGF and c-Met inhibitors plus gemcitabine in an advanced model of orthotopic pancreatic cancer. Triple therapy not only reduced tumor size but, importantly, eliminated metastasis. Further, the authors reported that the combination therapy was successful in overcoming gemcitabine-stimulated stemness and aggressiveness of cancer cells. These findings corroborated the findings of previous studies that reported that gemcitabine alone increases stemness of cancer cells in PDAC.[143,144]

Extracellular signal-regulated kinases (ERKs) regulate cellular processes such as proliferation, differentiation, transcription, and development, and this pathway has been shown to be strongly expressed in cancer-associated PSCs. Dual inhibition of PSC ERK1/2 and autophagy with S7101

and chloroquine, respectively, resulted in PSC senescence and suppressed liver metastasis in a splenic pancreatic cancer organoid mouse model.[145]

Vitamins inducing PSC quiescence have been successfully demonstrated to improve the outcome of pancreatic cancer in both preclinical and clinical studies. PEGylated gold nanoparticles containing all trans retinoic acid (ATRA) combined with siRNA against heat shock protein (HSP-47, known to activate PSCs) was used to induce PSC quiescence and inhibit ECM hyperplasia, thereby promoting drug delivery to pancreatic tumors in an orthotopic model developed by co-inoculation of cancer cells and PSCs.[146] Similarly, the vitamin D analogue calcipotriol, combined with gemcitabine, induced quiescence of activated PSCs and decreased fibrosis in KPC mice.[96] Based on this study, a vitamin D analogue paricalcitol, in combination with gemcitabine and nab-paclitaxel, is currently being evaluated in a clinical trial (NCT03520790).[147]

As epigenetic modifications also play an important role in PDAC, some studies have targeted epigenetic pathways to reprogram the tumor microenvironment in a bid to improve outcomes.[148,149] Common epigenetic targets relevant to PDAC include DNA methylation, histone modifications, and bromodomain and extra-terminal domain (BET) family proteins. A DNA methyl transferase inhibitor, 5-azacitidine, in combination with abraxane and gemcitabine is currently being tested in a Phase II clinical trial in PDAC patients (NCT01845805) undergoing resection. Inhibiting epigenetic bromodomain and the extraterminal domain (BET) family of proteins using iBET151, and SB 203580, a MAPK inhibitor, decreased YAP-1 levels resulting in decreased PSC activation.[119]

Myeloid-derived suppressor cells (MDSC) help to maintain an immunosuppressive environment in PDAC by eliminating effector T cells.[150] Therefore, therapies that boost effector T cell infiltration into tumor are gaining attention. The desmoplastic reaction in PDAC is known to sequester cytotoxic CD8+ T cells and hinder drug penetration. In transgenic spontaneous pancreatic cancer-bearing KC mice (Kras[G12D], Cre) challenged with caerulein, an inhibitor of rho/MRTF pathway, CCG-222740 not only decreased PSC activation but also decreased infiltration of macrophages, CD4+ T cells, and B cells, suggesting inhibition of PSC activation improves local immunity.[151] Similarly, inhibiting stroma through an acidic-pH-responsive-nanoparticle cluster containing TGF-β inhibitor (LY2157299) and PD-L1 (siRNA) improved local tumor immunity by increasing CD8+ T cell infiltration in a subcutaneous xenograft orthotopic model.[152] Tumor-associated macrophages (TAM), which are derived from MDSCs, are shown to rapidly inactivate gemcitabine, and depletion of TAM using clodronate increased the concentration of active gemcitabine in desmoplastic tumors of KPC mice compared to either PanINs or normal pancreas.[153]

Conclusion

PSCs play manifold roles in both health and disease. In the healthy pancreas they are responsible for maintaining ECM turnover and may also have additional roles in innate immunity and CCK-mediated exocrine secretion. In diseased conditions, PSCs are activated by a variety of factors known to be upregulated in the injured pancreas. The last two decades have seen an explosion of findings regarding the interaction of PSCs with other cells in the pancreas, in health, and in disease. These findings have vastly improved our understanding of the physiology and pathology of pancreatic disease, raising hopes for innovative new treatments. Once activated, PSCs become the primary contributors to the pathological fibrosis of both pancreatic inflammation and PDAC and impact on the function of beta cells of islets. Research efforts are now directed toward understanding the mechanisms underlying PSC-mediated fibrogenesis to develop novel therapies to treat fibrotic diseases of the pancreas.

References

1. Apte MV, Haber PS, Applegate TL, Norton ID, McCaughan GW, Korsten MA, Pirola RC, Wilson JS. Periacinar stellate shaped cells in rat pancreas: identification, isolation, and culture. Gut 1998; 43:128–133.
2. Bachem MG, Schneider E, Gross H, Weidenbach H, Schmid RM, Menke A, Siech M, Beger H, Grunert A, Adler G. Identification, culture, and characterization of pancreatic stellate cells in rats and humans. Gastroenterology 1998; 115:421–432.
3. Vonlaufen A, Phillips PA, Yang L, Xu Z, Fiala-Beer E, Zhang X, Pirola RC, Wilson JS, Apte MV. Isolation of quiescent human pancreatic stellate cells: a promising in vitro tool for studies of human pancreatic stellate cell biology. Pancreatology 2010; 10:434–443.
4. Watari N, Hotta Y, Mabuchi Y. Morphological studies on a vitamin A-storing cell and its complex with macrophage observed in mouse pancreatic tissues following excess vitamin A administration. Okajimas Folia Anat Jpn 1982; 58:837–858.
5. Ikejiri N. The vitamin A-storing cells in the human and rat pancreas. Kurume Med J 1990; 37:67–81.
6. Zha M, Xu W, Jones PM, Sun Z. Isolation and characterization of human islet stellate cells. Exp Cell Res 2016; 341:61–66.
7. Wake K. Perisinusoidal stellate cells (fat-storing cells, interstitial cells, lipocytes), their related structure in and around the liver sinusoids, and vitamin A-storing cells in extrahepatic organs. Int Rev Cytol 1980; 66:303–353.
8. Fu Y, Liu S, Zeng S, Shen H. The critical roles of activated stellate cells-mediated paracrine signaling, metabolism and onco-immunology in pancreatic ductal adenocarcinoma. Mol Cancer 2018; 17:62.
9. Zhao L, Burt AD. The diffuse stellate cell system. J Mol Histol 2007; 38:53–64.
10. Asahina K, Tsai SY, Li P, Ishii M, Maxson RE, Jr., Sucov HM, Tsukamoto H. Mesenchymal origin of hepatic stellate cells, submesothelial cells, and perivascular mesenchymal cells during mouse liver development. Hepatology 2009; 49:998–1011.
11. Sparmann G, Kruse ML, Hofmeister-Mielke N, Koczan D, Jaster R, Liebe S, Wolff D, Emmrich J. Bone marrow-derived pancreatic stellate cells in rats. Cell Res 2010; 20:288–298.
12. Phillips PA, McCarroll JA, Park S, Wu MJ, Pirola R, Korsten M, Wilson JS, Apte MV. Rat pancreatic stellate cells secrete matrix metalloproteinases: implications for extracellular matrix turnover. Gut 2003; 52:275–282.
13. Berna MJ, Seiz O, Nast JF, Benten D, Blaker M, Koch J, Lohse AW, Pace A. CCK1 and CCK2 receptors are expressed on pancreatic stellate cells and induce collagen production. J Biol Chem 2010; 285:38905–38914.
14. Phillips PA, Yang L, Shulkes A, Vonlaufen A, Poljak A, Bustamante S, Warren A, Xu Z, Guilhaus M, Pirola R, Apte MV, Wilson JS. Pancreatic stellate cells produce acetylcholine and may play a role in pancreatic exocrine secretion. Proc Natl Acad Sci U S A 2010; 107:17397–17402.
15. Masamune A, Kikuta K, Watanabe T, Satoh K, Satoh A, Shimosegawa T. Pancreatic stellate cells express Toll-like receptors. J Gastroenterol 2008; 43:352–362.
16. Shimizu K, Kobayashi M, Tahara J, Shiratori K. Cytokines and peroxisome proliferator-activated receptor gamma ligand regulate phagocytosis by pancreatic stellate cells. Gastroenterology 2005; 128:2105–2118.
17. Kordes C, Sawitza I, Haussinger D. Hepatic and pancreatic stellate cells in focus. Biol Chem 2009; 390:1003–1012.
18. Mato E, Lucas M, Petriz J, Gomis R, Novials A. Identification of a pancreatic stellate cell population with properties of progenitor cells: new role for stellate cells in the pancreas. Biochem J 2009; 421:181–191.
19. Sicchieri RD, da Silveira WA, Mandarano LR, de Oliveira TM, Carrara HH, Muglia VF, de Andrade JM, Tiezzi DG. ABCG2 is a potential marker of tumor-initiating cells in breast cancer. Tumour Biol 2015; 36:9233–9243.
20. Morishita K, Shimizu K, Haruta I, Kawamura S, Kobayashi M, Shiratori K. Engulfment of gram-positive bacteria by pancreatic stellate cells in pancreatic fibrosis. Pancreas 2010; 39:1002–1007.
21. Gonzalez-Villasana V, Rodriguez-Aguayo C, Arumugam T, Cruz-Monserrate Z, Fuentes-Mattei E, Deng D, Hwang RF, Wang H, Ivan C, Garza RJ, Cohen E, Gao H, et al. Bisphosphonates inhibit stellate cell activity and enhance antitumor effects of nanoparticle albumin-bound paclitaxel in pancreatic ductal adenocarcinoma. Mol Cancer Ther 2014; 13:2583–2594.
22. Apte M, Pirola RC, Wilson JS. Pancreatic stellate cell: physiologic role, role in fibrosis and cancer. Curr Opin Gastroenterol 2015; 31:416–423.
23. Haber PS, Keogh GW, Apte MV, Moran CS, Stewart NL, Crawford DH, Pirola RC, McCaughan GW, Ramm GA, Wilson JS. Activation of pancreatic stellate cells in human and experimental pancreatic fibrosis. Am J Pathol 1999; 155:1087–1095.
24. Apte MV, Xu Z, Pothula S, Goldstein D, Pirola RC, Wilson JS. Pancreatic cancer: The microenvironment needs attention too! Pancreatology 2015; 15:S32–38.

25. Wilson JS, Pirola RC, Apte MV. Stars and stripes in pancreatic cancer: role of stellate cells and stroma in cancer progression. Front Physiol 2014; 5:52.

26. Xu Z, Pothula SP, Wilson JS, Apte MV. Pancreatic cancer and its stroma: a conspiracy theory. World J Gastroenterol 2014; 20:11216–11229.

27. Mews P, Phillips P, Fahmy R, Korsten M, Pirola R, Wilson J, Apte M. Pancreatic stellate cells respond to inflammatory cytokines: potential role in chronic pancreatitis. Gut 2002; 50:535–541.

28. Uden S, Bilton D, Nathan L, Hunt LP, Main C, Braganza JM. Antioxidant therapy for recurrent pancreatitis: placebo-controlled trial. Aliment Pharmacol Ther 1990; 4:357–371.

29. Casini A, Galli A, Pignalosa P, Frulloni L, Grappone C, Milani S, Pederzoli P, Cavallini G, Surrenti C. Collagen type I synthesized by pancreatic periacinar stellate cells (PSC) co-localizes with lipid peroxidation-derived aldehydes in chronic alcoholic pancreatitis. J Pathol 2000; 192:81–89.

30. Apte M, Pirola R, Wilson J. The fibrosis of chronic pancreatitis: new insights into the role of pancreatic stellate cells. Antioxid Redox Signal 2011; 15:2711–2722.

31. Masamune A, Satoh A, Watanabe T, Kikuta K, Satoh M, Suzuki N, Satoh K, Shimosegawa T. Effects of ethanol and its metabolites on human pancreatic stellate cells. Dig Dis Sci 2010; 55:204–211.

32. Ben-Harosh Y, Anosov M, Salem H, Yatchenko Y, Birk R. Pancreatic stellate cell activation is regulated by fatty acids and ER stress. Exp Cell Res 2017; 359:76–85.

33. Vonlaufen A, Xu Z, Daniel B, Kumar RK, Pirola R, Wilson J, Apte MV. Bacterial endotoxin: a trigger factor for alcoholic pancreatitis? Evidence from a novel, physiologically relevant animal model. Gastroenterology 2007; 133:1293–1303.

34. Wang T, Wang Q-q, Pan G-x, Jia G-r, Li X, Wang C, Zhang L-m, C-j Z. ASIC1a involves acidic microenvironment-induced activation and autophagy of pancreatic stellate cells. RSC Advances 2018; 8:30950–30956.

35. Masamune A, Kikuta K, Watanabe T, Satoh K, Hirota M, Shimosegawa T. Hypoxia stimulates pancreatic stellate cells to induce fibrosis and angiogenesis in pancreatic cancer. Am J Physiol Gastrointest Liver Physiol 2008; 295:G709–G717.

36. Provenzano PP, Hingorani SR. Hyaluronan, fluid pressure, and stromal resistance in pancreas cancer. Br J Cancer 2013; 108:1–8.

37. Nomiyama Y, Tashiro M, Yamaguchi T, Watanabe S, Taguchi M, Asaumi H, Nakamura H, Otsuki M. High glucose activates rat pancreatic stellate cells through protein kinase C and p38 mitogen-activated protein kinase pathway. Pancreas 2007; 34:364–372.

38. Kumar K, DeCant BT, Grippo PJ, Hwang RF, Bentrem DJ, Ebine K, Munshi HG. BET inhibitors block pancreatic stellate cell collagen I production and attenuate fibrosis in vivo. JCI Insight 2017; 2:e88032.

39. Karger A, Fitzner B, Brock P, Sparmann G, Emmrich J, Liebe S, Jaster R. Molecular insights into connective tissue growth factor action in rat pancreatic stellate cells. Cell Signal 2008; 20:1865–1872.

40. Shek FW, Benyon RC, Walker FM, McCrudden PR, Pender SL, Williams EJ, Johnson PA, Johnson CD, Bateman AC, Fine DR, Iredale JP. Expression of transforming growth factor-β 1 by pancreatic stellate cells and its implications for matrix secretion and turnover in chronic pancreatitis. Am J Pathol 2002; 160:1787–1798.

41. Andoh A, Takaya H, Saotome T, Shimada M, Hata K, Araki Y, Nakamura F, Shintani Y, Fujiyama Y, Bamba T. Cytokine regulation of chemokine (IL-8, MCP-1, and RANTES) gene expression in human pancreatic periacinar myofibroblasts. Gastroenterology 2000; 119:211–219.

42. McCarroll JA, Phillips PA, Kumar RK, Park S, Pirola RC, Wilson JS, Apte MV. Pancreatic stellate cell migration: role of the phosphatidylinositol 3-kinase(PI3-kinase) pathway. Biochem Pharmacol 2004; 67:1215–1225.

43. Gryshchenko O, Gerasimenko JV, Gerasimenko OV, Petersen OH. Ca²⁺ signals mediated by bradykinin type 2 receptors in normal pancreatic stellate cells can be inhibited by specific Ca²⁺ channel blockade. J Physiol 2016; 594:281–293.

44. Hennigs JK, Seiz O, Spiro J, Berna MJ, Baumann HJ, Klose H, Pace A. Molecular basis of P2-receptor-mediated calcium signaling in activated pancreatic stellate cells. Pancreas 2011; 40:740–746.

45. Won JH, Zhang Y, Ji B, Logsdon CD, Yule DI. Phenotypic changes in mouse pancreatic stellate cell Ca²⁺ signaling events following activation in culture and in a disease model of pancreatitis. Mol Biol Cell 2011; 22:421–436.

46. Masamune A, Nakano E, Hamada S, Takikawa T, Yoshida N, Shimosegawa T. Alteration of the microRNA expression profile during the activation of pancreatic stellate cells. Scand J Gastroenterol 2014; 49:323–331.

47. Shen J, Wan R, Hu G, Yang L, Xiong J, Wang F, Shen J, He S, Guo X, Ni J, Guo C, Wang X. miR-15b and miR-16 induce the apoptosis of rat activated pancreatic stellate cells by targeting Bcl-2 in vitro. Pancreatology 2012; 12:91–99.

48. Charrier A, Chen R, Chen L, Kemper S, Hattori T, Takigawa M, Brigstock DR. Connective tissue growth factor (CCN2) and microRNA-21 are components of a positive feedback loop in pancreatic stellate cells (PSC) during chronic pancreatitis and are exported in PSC-derived exosomes. J Cell Commun Signal 2014; 8:147–156.

49. Frossard JL, Steer ML, Pastor CM. Acute pancreatitis. Lancet 2008; 371:143–152.

50. Elsasser HP, Adler G, Kern HF. Time course and cellular source of pancreatic regeneration following acute pancreatitis in the rat. Pancreas 1986; 1:421–429.

51. Brizzi MF, Tarone G, Defilippi P. Extracellular matrix, integrins, and growth factors as tailors of the stem cell niche. Curr Opin Cell Biol 2012; 24:645–651.

52. Riopel MM, Li J, Liu S, Leask A, Wang R. β1 integrin-extracellular matrix interactions are essential for maintaining exocrine pancreas architecture and function. Lab Invest 2013; 93:31–40.

53. Zimmermann A, Gloor B, Kappeler A, Uhl W, Friess H, Buchler MW. Pancreatic stellate cells contribute to regeneration early after acute necrotising pancreatitis in humans. Gut 2002; 51:574–578.

54. Kloppel G. Pathology of chronic pancreatitis and pancreatic pain. Acta Chir Scand 1990; 156:261–265.

55. Xu W, Li W, Wang Y, Zha M, Yao H, Jones PM, Sun Z. Regenerating islet-derived protein 1 inhibits the activation

of islet stellate cells isolated from diabetic mice. Oncotarget 2015; 6:37054–37065.

56. Kikuta K, Masamune A, Hamada S, Takikawa T, Nakano E, Shimosegawa T. Pancreatic stellate cells reduce insulin expression and induce apoptosis in pancreatic β-cells. Biochem Biophys Res Commun 2013; 433:292–297.

57. Friess H, Zhu ZW, di Mola FF, Kulli C, Graber HU, Andren-Sandberg A, Zimmermann A, Korc M, Reinshagen M, Buchler MW. Nerve growth factor and its high-affinity receptor in chronic pancreatitis. Ann Surg 1999; 230:615–624.

58. Schneider E, Schmid-Kotsas A, Zhao J, Weidenbach H, Schmid RM, Menke A, Adler G, Waltenberger J, Grunert A, Bachem MG. Identification of mediators stimulating proliferation and matrix synthesis of rat pancreatic stellate cells. Am J Physiol Cell Physiol 2001; 281:C532–C543.

59. Apte MV, Park S, Phillips PA, Santucci N, Goldstein D, Kumar RK, Ramm GA, Buchler M, Friess H, McCarroll JA, Keogh G, Merrett N, et al. Desmoplastic reaction in pancreatic cancer: role of pancreatic stellate cells. Pancreas 2004; 29:179–187.

60. Neuschwander-Tetri BA, Burton FR, Presti ME, Britton RS, Janney CG, Garvin PR, Brunt EM, Galvin NJ, Poulos JE. Repetitive self-limited acute pancreatitis induces pancreatic fibrogenesis in the mouse. Dig Dis Sci 2000; 45:665–674.

61. Matsumura N, Ochi K, Ichimura M, Mizushima T, Harada H, Harada M. Study on free radicals and pancreatic fibrosis-pancreatic fibrosis induced by repeated injections of superoxide dismutase inhibitor. Pancreas 2001; 22:53–57.

62. Emmrich F, Stahl HD, Altrichter S. [Leipzig Center for Therapy Studies-a cooperative structure for realizing clinical studies]. Z Rheumatol 2000; 59:57–60.

63. Tsukamoto H, Towner SJ, Yu GS, French SW. Potentiation of ethanol-induced pancreatic injury by dietary fat. Induction of chronic pancreatitis by alcohol in rats. Am J Pathol 1988; 131:246–257.

64. Uesugi T, Froh M, Gabele E, Isayama F, Bradford BU, Ikai I, Yamaoka Y, Arteel GE. Contribution of angiotensin II to alcohol-induced pancreatic fibrosis in rats. J Pharmacol Exp Ther 2004; 311:921–928.

65. Gukovsky I, Lugea A, Shahsahebi M, Cheng JH, Hong PP, Jung YJ, Deng QG, French BA, Lungo W, French SW, Tsukamoto H, Pandol SJ. A rat model reproducing key pathological responses of alcoholic chronic pancreatitis. Am J Physiol Gastrointest Liver Physiol 2008; 294:G68–G79.

66. Fukui H, Brauner B, Bode JC, Bode C. Plasma endotoxin concentrations in patients with alcoholic and non-alcoholic liver disease: reevaluation with an improved chromogenic assay. J Hepatol 1991; 12:162–169.

67. Sun L, Xiu M, Wang S, Brigstock DR, Li H, Qu L, Gao R. Lipopolysaccharide enhances TGF-β1 signalling pathway and rat pancreatic fibrosis. J Cell Mol Med 2018; 22:2346–2356.

68. Blaine SA, Ray KC, Branch KM, Robinson PS, Whitehead RH, Means AL. Epidermal growth factor receptor regulates pancreatic fibrosis. Am J Physiol Gastrointest Liver Physiol 2009; 297:G434–G441.

69. Marrache F, Tu SP, Bhagat G, Pendyala S, Osterreicher CH, Gordon S, Ramanathan V, Penz-Osterreicher M, Betz KS, Song Z, Wang TC. Overexpression of interleukin-1β

in the murine pancreas results in chronic pancreatitis. Gastroenterology 2008; 135:1277–1287.

70. Ohashi K, Kim JH, Hara H, Aso R, Akimoto T, Nakama K. WBN/Kob rats. A new spontaneously occurring model of chronic pancreatitis. Int J Pancreatol 1990; 6:231–247.

71. Calderari S, Irminger JC, Giroix MH, Ehses JA, Gangnerau MN, Coulaud J, Rickenbach K, Gauguier D, Halban P, Serradas P, Homo-Delarche F. Regenerating 1 and 3b gene expression in the pancreas of type 2 diabetic Goto-Kakizaki (GK) rats. PLoS One 2014; 9:e90045.

72. Lugea A, Gerloff A, Su HY, Xu Z, Go A, Hu C, French SW, Wilson JS, Apte MV, Waldron RT, Pandol SJ. The combination of alcohol and cigarette smoke induces endoplasmic reticulum stress and cell death in pancreatic acinar cells. Gastroenterology 2017; 153:1674–1686.

73. Xue J, Zhao Q, Sharma V, Nguyen LP, Lee YN, Pham KL, Edderkaoui M, Pandol SJ, Park W, Habtezion A. Aryl hydrocarbon receptor ligands in cigarette smoke induce production of interleukin-22 to promote pancreatic fibrosis in models of chronic pancreatitis. Gastroenterology 2016; 151:1206–1217.

74. Pereda J, Sabater L, Cassinello N, Gomez-Cambronero L, Closa D, Folch-Puy E, Aparisi L, Calvete J, Cerda M, Lledo S, Vina J, Sastre J. Effect of simultaneous inhibition of TNF-α production and xanthine oxidase in experimental acute pancreatitis: the role of mitogen activated protein kinases. Ann Surg 2004; 240:108–116.

75. Menke A, Yamaguchi H, Gress TM, Adler G. Extracellular matrix is reduced by inhibition of transforming growth factor beta1 in pancreatitis in the rat. Gastroenterology 1997; 113:295–303.

76. Hughes CB, Gaber LW, Mohey el-Din AB, Grewal HP, Kotb M, Mann L, Gaber AO. Inhibition of TNFα improves survival in an experimental model of acute pancreatitis. Am Surg 1996; 62:8–13.

77. Gomez JA, Molero X, Vaquero E, Alonso A, Salas A, Malagelada JR. Vitamin E attenuates biochemical and morphological features associated with development of chronic pancreatitis. Am J Physiol Gastrointest Liver Physiol 2004; 287:G162–G169.

78. Suzuki N, Masamune A, Kikuta K, Watanabe T, Satoh K, Shimosegawa T. Ellagic acid inhibits pancreatic fibrosis in male Wistar Bonn/Kobori rats. Dig Dis Sci 2009; 54:802–810.

79. Lu XL, Dong XY, Fu YB, Cai JT, Du Q, Si JM, Mao JS. Protective effect of salvianolic acid B on chronic pancreatitis induced by trinitrobenzene sulfonic acid solution in rats. Pancreas 2009; 38:71–77.

80. Gibo J, Ito T, Kawabe K, Hisano T, Inoue M, Fujimori N, Oono T, Arita Y, Nawata H. Camostat mesilate attenuates pancreatic fibrosis via inhibition of monocytes and pancreatic stellate cells activity. Lab Invest 2005; 85:75–89.

81. Shimizu K, Shiratori K, Kobayashi M, Kawamata H. Troglitazone inhibits the progression of chronic pancreatitis and the profibrogenic activity of pancreatic stellate cells via a PPARgamma-independent mechanism. Pancreas 2004; 29:67–74.

82. Xiao W, Jiang W, Shen J, Yin G, Fan Y, Wu D, Qiu L, Yu G, Xing M, Hu G, Wang X, Wan R. Retinoic acid ameliorates

pancreatic fibrosis and inhibits the activation of pancreatic stellate cells in mice with experimental chronic pancreatitis via suppressing the Wnt/β-catenin signaling pathway. PLoS One 2015; 10:e0141462.

83. Ishiwatari H, Sato Y, Murase K, Yoneda A, Fujita R, Nishita H, Birukawa NK, Hayashi T, Sato T, Miyanishi K, Takimoto R, Kobune M, et al. Treatment of pancreatic fibrosis with siRNA against a collagen-specific chaperone in vitamin A-coupled liposomes. Gut 2013; 62:1328–1339.

84. Niina Y, Ito T, Oono T, Nakamura T, Fujimori N, Igarashi H, Sakai Y, Takayanagi R. A sustained prostacyclin analog, ONO-1301, attenuates pancreatic fibrosis in experimental chronic pancreatitis induced by dibutyltin dichloride in rats. Pancreatology 2014; 14:201–210.

85. Vonlaufen A, Phillips PA, Xu Z, Zhang X, Yang L, Pirola RC, Wilson JS, Apte MV. Withdrawal of alcohol promotes regression while continued alcohol intake promotes persistence of LPS-induced pancreatic injury in alcohol-fed rats. Gut 2011; 60:238–246.

86. Zhang X, Hu J, Zhuo Y, Cui L, Li C, Cui N, Zhang S. Amygdalin improves microcirculatory disturbance and attenuates pancreatic fibrosis by regulating the expression of endothelin-1 and calcitonin gene-related peptide in rats. J Chin Med Assoc 2018; 81:437–443.

87. Xu M, Wang G, Zhou H, Cai J, Li P, Zhou M, Lu Y, Jiang X, Huang H, Zhang Y, Gong A. TGF-β1-miR-200a-PTEN induces epithelial-mesenchymal transition and fibrosis of pancreatic stellate cells. Mol Cell Biochem 2017; 431:161–168.

88. Li CX, Cui LH, Zhuo YZ, Hu JG, Cui NQ, Zhang SK. Inhibiting autophagy promotes collagen degradation by regulating matrix metalloproteinases in pancreatic stellate cells. Life Sci 2018; 208:276–283.

89. Masamune A, Suzuki N, Kikuta K, Satoh M, Satoh K, Shimosegawa T. Curcumin blocks activation of pancreatic stellate cells. J Cell Biochem 2006; 97:1080–1093.

90. Tezuka T, Ota A, Karnan S, Matsuura K, Yokoo K, Hosokawa Y, Vigetti D, Passi A, Hatano S, Umezawa K, Watanabe H. The plant alkaloid conophylline inhibits matrix formation of fibroblasts. J Biol Chem 2018; 293:20214–20226.

91. Yan B, Cheng L, Jiang Z, Chen K, Zhou C, Sun L, Cao J, Qian W, Li J, Shan T, Lei J, Ma Q, Ma J. Resveratrol inhibits ROS-promoted activation and glycolysis of pancreatic stellate cells via suppression of miR-21. Oxid Med Cell Longev 2018; 2018:1346958.

92. Asama H, Suzuki R, Hikichi T, Takagi T, Masamune A, Ohira H. MicroRNA let-7d targets thrombospondin-1 and inhibits the activation of human pancreatic stellate cells. Pancreatology 2019; 19:196–203.

93. Cui L, Li C, Zhuo Y, Yang L, Cui N, Li Y, Zhang S. Saikosaponin A inhibits the activation of pancreatic stellate cells by suppressing autophagy and the NLRP3 inflammasome via the AMPK/mTOR pathway. Biomed Pharmacother 2020; 128:110216.

94. Al Alawi R, Alhamdani MSS, Hoheisel JD, Baqi Y. Antifibrotic and tumor microenvironment modulating effect of date palm fruit (Phoenix dactylifera L.) extracts in pancreatic cancer. Biomed Pharmacother 2020; 121:109522.

95. Wallbaum P, Rohde S, Ehlers L, Lange F, Hohn A, Bergner C, Schwarzenbock SM, Krause BJ, Jaster R. Antifibrogenic effects of vitamin D derivatives on mouse pancreatic stellate cells. World J Gastroenterol 2018; 24:170–178.

96. Sherman MH, Yu RT, Engle DD, Ding N, Atkins AR, Tiriac H, Collisson EA, Connor F, Van Dyke T, Kozlov S, Martin P, Tseng TW, et al. Vitamin D receptor-mediated stromal reprogramming suppresses pancreatitis and enhances pancreatic cancer therapy. Cell 2014; 159:80–93.

97. Zhang Y, Yue D, Cheng L, Huang A, Tong N, Cheng P. Vitamin A-coupled liposomes carrying TLR4-silencing shRNA induce apoptosis of pancreatic stellate cells and resolution of pancreatic fibrosis. J Mol Med (Berl) 2018; 96:445–458.

98. Tsang SW, Bian ZX. Anti-fibrotic and anti-tumorigenic effects of rhein, a natural anthraquinone derivative, in mammalian stellate and carcinoma cells. Phytother Res 2015; 29:407–414.

99. Gao X, Cao Y, Staloch DA, Gonzales MA, Aronson JF, Chao C, Hellmich MR, Ko TC. Bone morphogenetic protein signaling protects against cerulein-induced pancreatic fibrosis. PLoS One 2014; 9:e89114.

100. Witteck L, Jaster R. Trametinib and dactolisib but not regorafenib exert antiproliferative effects on rat pancreatic stellate cells. Hepatobiliary Pancreat Dis Int 2015; 14:642–650.

101. Elsner A, Lange F, Fitzner B, Heuschkel M, Krause BJ, Jaster R. Distinct antifibrogenic effects of erlotinib, sunitinib and sorafenib on rat pancreatic stellate cells. World J Gastroenterol 2014; 20:7914–7925.

102. Xue R, Yang J, Wu J, Meng Q, Hao J. Coenzyme Q10 inhibits the activation of pancreatic stellate cells through PI3K/AKT/mTOR signaling pathway. Oncotarget 2017; 8:92300–92311.

103. Sun L, Chen K, Jiang Z, Chen X, Ma J, Ma Q, Duan W. Indomethacin inhibits the proliferation and activation of human pancreatic stellate cells through the downregulation of COX-2. Oncol Rep 2018; 39:2243–2251.

104. Hartel M, Di Mola FF, Gardini A, Zimmermann A, Di Sebastiano P, Guweidhi A, Innocenti P, Giese T, Giese N, Buchler MW, Friess H. Desmoplastic reaction influences pancreatic cancer growth behavior. World J Surg 2004; 28:818–825.

105. Feig C, Gopinathan A, Neesse A, Chan DS, Cook N, Tuveson DA. The pancreas cancer microenvironment. Clin Cancer Res 2012; 18:4266–4276.

106. Apte MV, Wilson JS, Lugea A, Pandol SJ. A starring role for stellate cells in the pancreatic cancer microenvironment. Gastroenterology 2013; 144:1210–1219.

107. Kakizaki Y, Makino N, Tozawa T, Honda T, Matsuda A, Ikeda Y, Ito M, Saito Y, Kimura W, Ueno Y. Stromal fibrosis and expression of matricellular proteins correlate with histological grade of intraductal papillary mucinous neoplasm of the pancreas. Pancreas 2016; 45:1145–1152.

108. Erkan M, Michalski CW, Rieder S, Reiser-Erkan C, Abiatari I, Kolb A, Giese NA, Esposito I, Friess H, Kleeff J. The activated stroma index is a novel and independent prognostic marker in pancreatic ductal adenocarcinoma. Clin Gastroenterol Hepatol 2008; 6:1155–1161.

109. Yuzawa S, Kano MR, Einama T, Nishihara H. PDGFRβ expression in tumor stroma of pancreatic adenocarcinoma as a reliable prognostic marker. Med Oncol 2012; 29:2824–2830.

110. Ohlund D, Handly-Santana A, Biffi G, Elyada E, Almeida AS, Ponz-Sarvise M, Corbo V, Oni TE, Hearn SA, Lee EJ, Chio, II, Hwang CI, et al. Distinct populations of inflammatory fibroblasts and myofibroblasts in pancreatic cancer. J Exp Med 2017; 214:579–596.

111. Vonlaufen A, Joshi S, Qu C, Phillips PA, Xu Z, Parker NR, Toi CS, Pirola RC, Wilson JS, Goldstein D, Apte MV. Pancreatic stellate cells: partners in crime with pancreatic cancer cells. Cancer Res 2008; 68:2085–2093.

112. Xu Z, Vonlaufen A, Phillips PA, Fiala-Beer E, Zhang X, Yang L, Biankin AV, Goldstein D, Pirola RC, Wilson JS, Apte MV. Role of pancreatic stellate cells in pancreatic cancer metastasis. Am J Pathol 2010; 177:2585–2596.

113. Hwang RF, Moore T, Arumugam T, Ramachandran V, Amos KD, Rivera A, Ji B, Evans DB, Logsdon CD. Cancer-associated stromal fibroblasts promote pancreatic tumor progression. Cancer Res 2008; 68:918–926.

114. Bachem MG, Schunemann M, Ramadani M, Siech M, Beger H, Buck A, Zhou S, Schmid-Kotsas A, Adler G. Pancreatic carcinoma cells induce fibrosis by stimulating proliferation and matrix synthesis of stellate cells. Gastroenterology 2005; 128:907–921.

115. Pandol S, Gukovskaya A, Edderkaoui M, Dawson D, Eibl G, Lugea A. Epidemiology, risk factors, and the promotion of pancreatic cancer: role of the stellate cell. J Gastroenterol Hepatol 2012; 27 (suppl 2):127–134.

116. Schneiderhan W, Diaz F, Fundel M, Zhou S, Siech M, Hasel C, Moller P, Gschwend JE, Seufferlein T, Gress T, Adler G, Bachem MG. Pancreatic stellate cells are an important source of MMP-2 in human pancreatic cancer and accelerate tumor progression in a murine xenograft model and CAM assay. J Cell Sci 2007; 120:512–519.

117. Vonlaufen A, Phillips PA, Xu Z, Goldstein D, Pirola RC, Wilson JS, Apte MV. Pancreatic stellate cells and pancreatic cancer cells: an unholy alliance. Cancer Res 2008; 68:7707–7710.

118. Sousa CM, Biancur DE, Wang X, Halbrook CJ, Sherman MH, Zhang L, Kremer D, Hwang RF, Witkiewicz AK, Ying H, Asara JM, Evans RM, et al. Pancreatic stellate cells support tumour metabolism through autophagic alanine secretion. Nature 2016; 536:479–483.

119. Hu C, Yang J, Su HY, Waldron RT, Zhi M, Li L, Xia Q, Pandol SJ, Lugea A. Yes-associated protein 1 plays major roles in pancreatic stellate cell activation and fibroinflammatory responses. Front Physiol 2019; 10:1467.

120. Hamada S, Masamune A, Takikawa T, Suzuki N, Kikuta K, Hirota M, Hamada H, Kobune M, Satoh K, Shimosegawa T. Pancreatic stellate cells enhance stem cell-like phenotypes in pancreatic cancer cells. Biochem Biophys Res Commun 2012; 421:349–354.

121. Kikuta K, Masamune A, Watanabe T, Ariga H, Itoh H, Hamada S, Satoh K, Egawa S, Unno M, Shimosegawa T. Pancreatic stellate cells promote epithelial-mesenchymal transition in pancreatic cancer cells. Biochem Biophys Res Commun 2010; 403:380–384.

122. Sherman MH, Yu RT, Tseng TW, Sousa CM, Liu S, Truitt ML, He N, Ding N, Liddle C, Atkins AR, Leblanc M, Collisson EA, et al. Stromal cues regulate the pancreatic cancer epigenome and metabolome. Proc Natl Acad Sci U S A 2017; 114:1129–1134.

123. Pothula SP, Xu Z, Goldstein D, Merrett N, Pirola RC, Wilson JS, Apte MV. Targeting the HGF/c-MET pathway: stromal remodelling in pancreatic cancer. Oncotarget 2017; 8:76722–76739.

124. Suetsugu A, Snyder CS, Moriwaki H, Saji S, Bouvet M, Hoffman RM. Imaging the interaction of pancreatic cancer and stellate cells in the tumor microenvironment during metastasis. Anticancer Res 2015; 35:2545–2551.

125. Pang T, Xu Z, Pothula S, Yang MH, Becker T, Goldstein D, Heeschen C, Pirola R, Wilson J, MV. A. 517-World-first identification of circulating pancreatic stellate cells in metastatic pancreatic cancer. Gastroenterology 2018; 154:S-114.

126. Takikawa T, Masamune A, Yoshida N, Hamada S, Kogure T, Shimosegawa T. Exosomes derived from pancreatic stellate cells: microRNA signature and effects on pancreatic cancer cells. Pancreas 2017; 46:19–27.

127. Hessmann E, Patzak MS, Klein L, Chen N, Kari V, Ramu I, Bapiro TE, Frese KK, Gopinathan A, Richards FM, Jodrell DI, Verbeke C, et al. Fibroblast drug scavenging increases intratumoural gemcitabine accumulation in murine pancreas cancer. Gut 2018; 67:497–507.

128. Zhang H, Wu H, Guan J, Wang L, Ren X, Shi X, Liang Z, Liu T. Paracrine SDF-1α signaling mediates the effects of PSCs on GEM chemoresistance through an IL-6 autocrine loop in pancreatic cancer cells. Oncotarget 2015; 6:3085–3097.

129. Cabrera MC, Tilahun E, Nakles R, Diaz-Cruz ES, Charabaty A, Suy S, Jackson P, Ley L, Slack R, Jha R, Collins SP, Haddad N, Kallakury BV, et al. Human pancreatic cancer-associated stellate cells remain activated after in vivo chem-oradiation. Front Oncol 2014; 4:102.

130. Duan W, Chen K, Jiang Z, Chen X, Sun L, Li J, Lei J, Xu Q, Ma J, Li X, Han L, Wang Z, et al. Desmoplasia suppression by metformin-mediated AMPK activation inhibits pancreatic cancer progression. Cancer Lett 2017; 385:225–233.

131. Chakra Turaga R, Sharma M, Mishra F, Krasinskas A, Yuan Y, Yang JJ, Wang S, Liu C, Li S, Liu ZR. Modulation of cancer-associated fibrotic stroma by an integrin $\alpha_v\beta_3$ targeting protein for pancreatic cancer treatment. Cell Mol Gastroenterol Hepatol 2021; 11:161–179.

132. Che M, Kweon SM, Teo JL, Yuan YC, Melstrom LG, Waldron RT, Lugea A, Urrutia RA, Pandol SJ, Lai KKY. Targeting the CBP/β-Catenin interaction to suppress activation of cancer-promoting pancreatic stellate cells. Cancers (Basel) 2020; 12:1476.

133. Li W, Sun L, Lei J, Wu Z, Ma Q, Wang Z. Curcumin inhibits pancreatic cancer cell invasion and EMT by interfering with tumorstromal crosstalk under hypoxic conditions via the IL6/ERK/NFkappaB axis. Oncol Rep 2020; 44:382–392.

134. Mardhian DF, Storm G, Bansal R, Prakash J. Nano-targeted relaxin impairs fibrosis and tumor growth in pancreatic cancer and improves the efficacy of gemcitabine in vivo. J Control Release 2018; 290:1–10.

135. Walter K, Omura N, Hong SM, Griffith M, Vincent A, Borges M, Goggins M. Overexpression of smoothened activates the sonic hedgehog signaling pathway in pancreatic cancer-associated fibroblasts. Clin Cancer Res 2010; 16:1781–1789.

136. Khan S, Ebeling MC, Chauhan N, Thompson PA, Gara RK, Ganju A, Yallapu MM, Behrman SW, Zhao H, Zafar N, Singh MM, Jaggi M, et al. Ormeloxifene suppresses desmoplasia and enhances sensitivity of gemcitabine in pancreatic cancer. Cancer Res 2015; 75:2292–2304.

137. Olive KP, Jacobetz MA, Davidson CJ, Gopinathan A, McIntyre D, Honess D, Madhu B, Goldgraben MA, Caldwell ME, Allard D, Frese KK, Denicola G, et al. Inhibition of hedgehog signaling enhances delivery of chemotherapy in a mouse model of pancreatic cancer. Science 2009; 324:1457–1461.

138. Khan MA, Srivastava SK, Zubair H, Patel GK, Arora S, Khushman M, Carter JE, Gorman GS, Singh S, Singh AP. Co-targeting of CXCR4 and hedgehog pathways disrupts tumor-stromal crosstalk and improves chemotherapeutic efficacy in pancreatic cancer. J Biol Chem 2020; 295:8413–8424.

139. Pothula SP, Xu Z, Goldstein D, Biankin AV, Pirola RC, Wilson JS, Apte MV. Hepatocyte growth factor inhibition: a novel therapeutic approach in pancreatic cancer. Br J Cancer 2016; 114:269–280.

140. Rucki AA, Xiao Q, Muth S, Chen J, Che X, Kleponis J, Sharma R, Anders RA, Jaffee EM, Zheng L. Dual inhibition of Hedgehog and c-Met pathways for pancreatic cancer treatment. Mol Cancer Ther 2017; 16:2399–2409.

141. Li C, Wu JJ, Hynes M, Dosch J, Sarkar B, Welling TH, Pasca di Magliano M, Simeone DM. c-Met is a marker of pancreatic cancer stem cells and therapeutic target. Gastroenterology 2011; 141:2218–2227 e2215.

142. Xu Z, Pang TCY, Liu AC, Pothula SP, Mekapogu AR, Perera CJ, Murakami T, Goldstein D, Pirola RC, Wilson JS, Apte MV. Targeting the HGF/c-MET pathway in advanced pancreatic cancer: a key element of treatment that limits primary tumour growth and eliminates metastasis. Br J Cancer 2020; 122:1486–1495.

143. Quint K, Tonigold M, Di Fazio P, Montalbano R, Lingelbach S, Ruckert F, Alinger B, Ocker M, Neureiter D. Pancreatic cancer cells surviving gemcitabine treatment express markers of stem cell differentiation and epithelial-mesenchymal transition. Int J Oncol 2012; 41:2093–2102.

144. Hermann PC, Huber SL, Herrler T, Aicher A, Ellwart JW, Guba M, Bruns CJ, Heeschen C. Distinct populations of cancer stem cells determine tumor growth and metastatic activity in human pancreatic cancer. Cell Stem Cell 2007; 1:313–323.

145. Yan Z, Ohuchida K, Fei S, Zheng B, Guan W, Feng H, Kibe S, Ando Y, Koikawa K, Abe T, Iwamoto C, Shindo K, et al. Inhibition of ERK1/2 in cancer-associated pancreatic stellate cells suppresses cancer-stromal interaction and metastasis. J Exp Clin Cancer Res 2019; 38:221.

146. Han X, Li Y, Xu Y, Zhao X, Zhang Y, Yang X, Wang Y, Zhao R, Anderson GJ, Zhao Y, Nie G. Reversal of pancreatic desmoplasia by re-educating stellate cells with a tumour microenvironment-activated nanosystem. Nat Commun 2018; 9:3390.

147. Perez K. Vitamin D receptor agonist paricalcitol plus gemcitabine and Nab-paclitaxel in patients with metastatic pancreatic cancer. Identifier NCT03520790, edited by https://clinicaltrials.gov/ct2/show/NCT03520790 2018.

148. Nguyen AH, Elliott IA, Wu N, Matsumura C, Vogelauer M, Attar N, Dann A, Ghukasyan R, Toste PA, Patel SG, Williams JL, Li L, et al. Histone deacetylase inhibitors provoke a tumor supportive phenotype in pancreatic cancer associated fibroblasts. Oncotarget 2017; 8:19074–19088.

149. Yamamoto K, Tateishi K, Kudo Y, Hoshikawa M, Tanaka M, Nakatsuka T, Fujiwara H, Miyabayashi K, Takahashi R, Tanaka Y, Ijichi H, Nakai Y, et al. Stromal remodeling by the BET bromodomain inhibitor JQ1 suppresses the progression of human pancreatic cancer. Oncotarget 2016; 7:61469–61484.

150. Gabrilovich DI, Ostrand-Rosenberg S, Bronte V. Coordinated regulation of myeloid cells by tumours. Nat Rev Immunol 2012; 12:253–268.

151. Leal AS, Misek SA, Lisabeth EM, Neubig RR, Liby KT. The Rho/MRTF pathway inhibitor CCG-222740 reduces stellate cell activation and modulates immune cell populations in Kras^{G12D}; Pdx1-Cre (KC) mice. Sci Rep 2019; 9:7072.

152. Wang Y, Gao Z, Du X, Chen S, Zhang W, Wang J, Li H, He X, Cao J, Wang J. Co-inhibition of the TGF-β pathway and the PD-L1 checkpoint by pH-responsive clustered nanoparticles for pancreatic cancer microenvironment regulation and antitumor immunotherapy. Biomater Sci 2020; 8:5121–5132.

153. Buchholz SM, Goetze RG, Singh SK, Ammer-Herrmenau C, Richards FM, Jodrell DI, Buchholz M, Michl P, Ellenrieder V, Hessmann E, Neesse A. Depletion of macrophages improves therapeutic response to gemcitabine in murine pancreas cancer. Cancers (Basel) 2020; 12:1978.

154. Garg B, Giri B, Modi S, Sethi V, Castro I, Umland O, Ban Y, Lavania S, Dawra R, Banerjee S, Vickers S, Merchant NB, et al. NFκB in pancreatic stellate cells reduces infiltration of tumors by cytotoxic T cells and killing of cancer cells, via up-regulation of CXCL$_{12}$. Gastroenterology 2018; 155:880–891.

155. Qian D, Lu Z, Xu Q, Wu P, Tian L, Zhao L, Cai B, Yin J, Wu Y, Staveley-O'Carroll KF, Jiang K, Miao Y, et al. Galectin-1-driven upregulation of SDF-1 in pancreatic stellate cells promotes pancreatic cancer metastasis. Cancer Lett 2017; 397:43–51.

156. Li N, Li Y, Li Z, Huang C, Yang Y, Lang M, Cao J, Jiang W, Xu Y, Dong J, Ren H. Hypoxia Inducible Factor 1 (HIF-1) recruits macrophage to activate pancreatic stellate cells in pancreatic ductal adenocarcinoma. Int J Mol Sci 2016; 17:799.

157. Komar HM, Serpa G, Kerscher C, Schwoegl E, Mace TA, Jin M, Yang MC, Chen CS, Bloomston M, Ostrowski MC, Hart PA, Conwell DL, et al. Inhibition of Jak/STAT signaling reduces the activation of pancreatic stellate cells in vitro and limits caerulein-induced chronic pancreatitis in vivo. Sci Rep 2017; 7:1787.

158. Yoshida N, Masamune A, Hamada S, Kikuta K, Takikawa T, Motoi F, Unno M, Shimosegawa T. Kindlin-2 in pancreatic stellate cells promotes the progression of pancreatic cancer. Cancer Lett 2017; 390:103–114.

159. Yoshida S, Yokota T, Ujiki M, Ding XZ, Pelham C, Adrian TE, Talamonti MS, Bell RH, Jr., Denham W. Pancreatic

cancer stimulates pancreatic stellate cell proliferation and TIMP-1 production through the MAP kinase pathway. Biochem Biophys Res Commun 2004; 323:1241–1245.

160. Liu Y, Li F, Gao F, Xing L, Qin P, Liang X, Zhang J, Qiao X, Lin L, Zhao Q, Du L. Periostin promotes the chemotherapy resistance to gemcitabine in pancreatic cancer. Tumour Biol 2016; 37:15283–15291.

161. Sugimoto R, Enjoji M, Kohjima M, Tsuruta S, Fukushima M, Iwao M, Sonta T, Kotoh K, Inoguchi T, Nakamuta M. High glucose stimulates hepatic stellate cells to proliferate and to produce collagen through free radical production and activation of mitogen-activated protein kinase. Liver Int 2005; 25:1018–1026.

162. Masamune A, Kikuta K, Satoh M, Satoh K, Shimosegawa T. Rho kinase inhibitors block activation of pancreatic stellate cells. Br J Pharmacol 2003; 140:1292–1302.

163. Ohnishi H, Miyata T, Yasuda H, Satoh Y, Hanatsuka K, Kita H, Ohashi A, Tamada K, Makita N, Iiri T, Ueda N, Mashima H, et al. Distinct roles of Smad2-, Smad3-, and ERK-dependent pathways in transforming growth factor-β_1 regulation of pancreatic stellate cellular functions. J Biol Chem 2004; 279:8873–8878.

164. Zambirinis CP, Levie E, Nguy S, Avanzi A, Barilla R, Xu Y, Seifert L, Daley D, Greco SH, Deutsch M, Jonnadula S, Torres-Hernandez A, et al. TLR9 ligation in pancreatic stellate cells promotes tumorigenesis. J Exp Med 2015; 212:2077–2094.

165. Froeling FE, Feig C, Chelala C, Dobson R, Mein CE, Tuveson DA, Clevers H, Hart IR, Kocher HM. Retinoic acid-induced pancreatic stellate cell quiescence reduces paracrine Wnt-β-catenin signaling to slow tumor progression. Gastroenterology 2011; 141:1486–1497.

www.ingramcontent.com/pod-product-compliance
Lightning Source LLC
Chambersburg PA
CBHW040142200326

41458CB00025B/6347

* 9 7 8 1 6 0 7 8 5 7 1 6 7 *